HANDBOOK OF ORGANIC CONDUCTIVE MOLECULES AND POLYMERS

About the Editor

Hari Singh Nalwa is a materials scientist at the Hitachi Research Laboratory, Hitachi Ltd, Japan. Dr. Nalwa is also an honorary visiting professor at the Indian Institute of Technology in New Delhi. Recently cited in 'Who's Who in the World', editor/ co-editor of three books: *Ferroelectric Polymers* (1995), *Nonlinear Optics of Organic Molecules and Polymers* (1997) and *Organic Electroluminescent Materials and Devices* (1997), he has authored over 100 scientific publications in leading refereed journals and books, and has 18 patents either issued or applied for on electronic and photonic materials. He is the founder and Editor-in-Chief of the *Journal of Porphyrins and Phthalocyanines* and serves on the editorial board of *Applied Organometallic Chemistry, Journal of Macromolecular Science-Physics* and *Photonics Science News*. A member of the American Chemical Society and American Association for the Advancement of Science, he has also been awarded a number of prestigious fellowships in India and abroad: National Merit Scholarship, Indian Space Research Organization (ISRO) Fellowship, Council of Scientific and Industrial Research (CSIR) Senior Fellowship, NEC Fellowship and Japanese Government Science and Technology Agency (STA) Fellowship. He was a guest scientist at Hahn-Meitner Institute in Berlin, Germany (1983), Research Associate at the University of Southern California in Los Angeles, USA (1984–1987) and lecturer at Tokyo University of Agriculture and Technology, Japan (1988–90). He has been working with Hitachi Ltd since 1990. Dr. Nalwa received a B.Sc. degree in biosciences from the Meerut University in 1974, and M.Sc. in organic chemistry from the University of Roorkee in 1977, and a Ph.D. in polymer science from the Indian Institute of Technology in New Delhi, India in 1983. His research work encompasses ferroelectric polymers, electrets, electrically conductive polymers, nonlinear optical materials for integrated optics, low and high dielectric constant materials, Langmuir–Blodgett films and nanostructured materials.

HANDBOOK OF ORGANIC CONDUCTIVE MOLECULES AND POLYMERS

WITHDRAWN

Volume 4

Conductive Polymers: Transport, Photophysics and Applications

Edited by

Hari Singh Nalwa, M.Sc., Ph.D.

Hitachi Research Laboratory, Hitachi Ltd.
Ibaraki, Japan

JOHN WILEY & SONS
Chichester • New York • Weinheim • Brisbane • Singapore • Toronto

National   01243 779777
International (+44) 1243 779777
e-mail (for orders and customer service enquiries): cs-book@wiley.co.uk
Visit our Home Page on http://www.wiley.co.uk
                or http://www.wiley.com

Other Wiley Editorial Offices

John Wiley & Sons, Inc., 605 Third Avenue,
New York, NY 10158-0012, USA

VCH Verlagsgesellschaft mbH,
Pappelallee 3, D-69469 Weinheim, Germany

Jacaranda Wiley Ltd, 33 Park Road, Milton,
Queensland 4064, Australia

John Wiley & Sons (Asia) Pte Ltd, 2 Clementi Loop #02-01,
Jin Xing Distripark, Singapore 0512

John Wiley & Sons (Canada) Ltd, 22 Worcester Road,
Rexdale, Ontario M9W IL1, Canada

620·11797 NAL

Library of Congress Cataloguing-in-Publication Data

Handbook of organic conductive molecules and polymers/edited by Hari
    Singh Nalwa.
        p.   cm.
    Includes bibliographical references and index.
    Contents: v. 1. Charge-transfer salts, fullerenes, and
photoconductors—v. 2. Conductive polymers: synthesis and
electrical properties—v. 3. Conductive polymers: spectroscopy and
physical properties—v. 4. Conductive polymers: transport,
photophysics and applications.
    ISBN 0-471-96275-9 (set).—ISBN 0-471-96593-6 (v. 1).—ISBN
0-471-96594-4 (v. 2).—ISBN 0-471-96595-2 (v. 3).—ISBN
0-471-96813-7 (v. 4)
    1. Organic conductors.   2. Conducting polymers.   I. Nalwa, Hari
Singh, 1954–
QD382.C66H38   1997
620.1'1797–dc20   96-30337
                                                    CIP

British Library Cataloguing in Publication Data

A catalogue record for this book is available from the British Library

ISBN 0-471 96593 6 (Vol 1)
ISBN 0 471 96594 4 (Vol 2)
ISBN 0 471 96595 2 (Vol 3)
ISBN 0 471 96813 7 (Vol 4)
ISBN 0 471 96275 9 (Set)

Cover illustration courtesy of H. Naarmann
Data and information in this handbook is published from reliable sources but authors, editor and the publisher cannot assume any
responsibility for the validity of all materials or for the consequences of their use.
Typeset in Great Britain by Techset Composition Ltd, Salisbury, Wiltshire
Printed and bound in Great Britain by Antony Rowe Ltd, Chippenham, Wiltshire
This book is printed on acid-free paper responsibly manufactured from sustainable forestation, for which at least two trees are planted for
each one used for paper production.
Reprinted with corrections August 1997
971101175

To the memory of my younger brother
Jaivir Singh Tomar

Contents

Contributors List

M. S. A. Abdou
Department of Chemistry, Simon Fraser University, Burnaby, B.C., Canada, V5A 1S6

C. Arbizzani
Department of Chemistry 'G. Ciamician', University of Bologna, Via Selmi 2, 40126 Bologna, Italy

Wai Kin Chan
Department of Chemistry and James Frank Institute, University of Chicago, 5735 S. Ellis Avenue, Chicago, IL 60637, USA

Esther M. Conwell
Xerox Webster Research Center, 800 Phillips Road, 0114/22D, Webster, NY 14580, USA

A. R. Gharavi
Department of Chemistry and James Frank Institute, University of Chicago, 5735 S. Ellis Avenue, Chicago, IL 60637, USA

W. Göpel
Institute of Physical and Theoretical Chemistry and Center of Interface Analysis, University of Tübingen, D-72076 Tübingen, Germany

T. F. Gaurr
Gentex Corporation, 600 N. Centennial, Zeeland, MI 59575, USA

S. Holdcroft
Department of Chemistry, Simon Fraser University, Burnaby, B.C., Canada, V5A 1S6

A. Ivaska
Laboratory of Analytical Chemistry, Åbo Akademi University, Biskopsgatan 8, FIN-20500 Åbo-Turku, Finland

M. Kaneko
Faculty of Science, Ibaraki University, Bunkyo, Mito, 310 Japan

M. Kertesz
Department of Chemistry, Georgetown University, Washington DC, 20057-2222, USA

T. Kobayashi
Department of Physics, Faculty of Science, University of Tokyo, Hongo 7-3-1, Bunkyo-ku, Tokyo 113, Japan

C. Kvarnström
Laboratory of Analytical Chemistry, Åbo Akademi University, Biskopsgatan 8, FIN-20500 Åbo-Turku, Finland

Wenjie Li
Department of Chemistry and James Frank Institute, University of Chicago, 5735 S. Ellis Avenue, Chicago, IL 60637, USA

M. Mastragostino
Department of Physical Chemistry, University of Palermo, Via Archirafi 26, 90123 Palermo, Italy

R. Menon
Institute for Polymers and Organic Solids, University of California at Santa Barbara, Santa Barbara, CA 93106, USA

Hari Singh Nalwa
Hitachi Research Laboratory, Hitachi Ltd, 7-1-1 Ohmika-cho, Hitachi City, Ibaraki 319-12, Japan

T. F. Otero
Laboratorio Electroquímica, Facultad de Química de San Sebastián, Universidad del País Vasco (UPV/EHU), PO Box 1072, 20080 San Sebastián, Spain

Z. Peng
Department of Chemistry and James Frank Institute, University of Chicago, 5735 S. Ellis Avenue, Chicago, IL 60637, USA

K.-D. Schierbaum
Institute of Physical and Theoretical Chemistry and Center of Interface Analysis, University of Tübingen, D-72076 Tübingen, Germany

B. Scrosati
Department of Chemistry, 'La Sapienza' University of Rome, P. le Moro 5, 00185 Rome, Italy

Luping Yu
Department of Chemistry and James Frank Institute, University of Chicago, 5735 S. Ellis Avenue, Chicago, IL 60637, USA

Preface

The past decade has witnessed tremendous advances in the development of organic conductive molecular and polymeric materials and this field continues to be of great scientific and commercial interest. This field of science flourished on discovering new π-conjugated materials and by tailoring their electrical conductivity from semiconductive to metallic to a superconductive regime when doped. The term 'Synthetic Metals' therefore originated from this theme. The first category of organic conductive and superconductive molecules are based on charge-transfer salts where BEDT-TTF [bis(ethylenedithio)tetrathiafulvalene], $M(dmit)_2$ (dmit = 1,3-dithio-2-thione-4,5-dithiolate), DCNQI (N,N'-dicyano-p-quinonediimine), perylene, tetrachalcogenafulvalenes and their structurally related analogs are just a few of the most commonly studied in the past decade. Fullerenes constitute another class which also exhibit superconductivity when doped with alkali metals. Research activities are being carried out around the world to produce pure fullerenes (C_{60}, C_{70}, C_{76}, C_{78}, C_{82}, C_{84}, C_{90}, and higher fullerenes) and to prepare their derivatives as well as polymers. Interestingly the 1996 Nobel Prize for Chemistry was awarded to the research team who discovered fullerenes in 1985. Attempts continue in both charge-transfer salts and fullerenes to develop new superconductive materials with a record-high transition temperature. In 1977, a team led by A. J. Heeger, A. G. MacDiarmid and H. Shirakawa demonstrated that the electrical conductivity of polyacetylene can be increased by 13 orders of magnitude upon doping with electron acceptors and electron-donors, which created virtually the new field of conducting polymers. Achieving conductivity as high as that of copper metal in polyacetylene by H. Naarmann and coworkers, was a milestone discovery. Inspired by polyacetylenes, new conductive polymers such as poly-thiophene, polypyrrole, poly-(p-phenylene), poly(p-phenylene sulphide) and polyaniline were also developed since 1978.

From a chemistry point of view, conductive polymers possess a highly delocalized π-electron system, whereas charge-transfer salts are composed of π-molecules. The superiority of organic conductive materials over their counterpart inorganic materials resides in their tremendous architectural flexibility, inexpensiveness and ease of processing and fabrica-tion. When I first thought of editing these volumes, I considered a reference work including every aspect of conductive and superconductive materials dealing with synthesis, processing, electrical properties, spectroscopy, physical properties and applications. Now combining all these aspects these volumes, with the most valuable contributions in this field of science, consist of 64 chapters with the 110 authors coming from 21 different countries.

These volumes, written by leading experts of the international scientific community in industry and academia, give the most comprehensive coverage of the whole field of organic conductive materials and their based applications as part of a four-volume set. Each chapter is self-contained with cross references. Some overlap may inevitably exist in chapters but it was kept at a minimum. This book illustrates in a clear and concise way the structure–property relationship to understand a broader range of organic molecular and polymeric materials with exciting potential for future electronics and photonics industries. Almost all aspects of organic supramolecular structures specifically tailored to optimize electrical responses are discussed with an emphasis on chemical synthesis, molecular engineering, spectroscopic analysis, characterization techniques, physical properties and devices. The state-of-the-art in this field is presented by some of the most renowned scientists in the world in four parts:

Volume 1: Charge-Transfer Salts, Fullerenes and Photoconductors

Volume 2: Conductive Polymers: Synthesis and Electrical Properties

Volume 3: Conductive Polymers: Spectroscopy and Physical Properties

Volume 4: Conductive Polymers: Transport, Photophysics and Applications

Volume 1 contains 15 chapters on the recent developments in charge-transfer salts, fullerenes, photoconductors and molecular conductors. The topics covered include electron acceptor molecules, photoinduced intermolecular electron transfer systems, perylene based conductors, tetrachalcogenafulvalenes, metal 1,2-dichalcogenolenes and their conductive salts, conductive hetero-TCNQs (tetracyano-p-quinodimethane derivatives), molecular metals and superconductors based on transition metal complexes, growth and preparation of single crystals, crystal structures, fabrication of thin films, conductivity and superconductivity in doped fullerenes, electrochemistry of fullerenes, photophysics of conjugated polymer/fullerene composites, photoconductivity in fullerenes, organic photoconductive materials for xerographic photoreceptors, photoconductive polymers, graphite intercalation compounds, electrically conductive metallophthalocyanines, electrically conductive Langmuir–Blodgett films and magnetism of stable organic radical crystals. By careful selection of topics, the contents of this volume will be useful for scientists involved in any aspect of the science and technology of molecular conductors, superconductors and photoconductors.

Volume 2 contains 20 chapters and the subject coverage includes polyacetylene and its copolymers, electrochemical synthesis of polyheterocycles, π-conductive polymers prepared by organometallic polycondensation, poly(p-phenylenes) and poly(p-phenylene vinylene), polythiophenes, oligothiophenes, thienyl polyenylene oligomers and polymers, polypyrroles, polypyrrole films containing transition metal complexes as counteranions, polythiophene and polypyrrole copolymers, polyanilines, polytoluidines, silicon containing thiophene oligomers and polymers, silicon and germanium containing conductive polymers, polyazines, metallophthalocyanine polymers, conductive polymer blends and composites, organometallic conductive polymers and self-doped conductive polymers.

Volume 3 contains 16 chapters dealing with photoelectron spectroscopy, spectroelectrochemistry, crystallography scanning force microscopy, optically detected magnetic resonance studies of conductive polymers and fullerenes, magnetic properties, microwave properties, electrochemistry, electrocatalytic properties, metallic properties of conductive polymers due to dispersion, thin film properties of oligothiophenes, electrochromism, thermochromism, solvatochromism, degradation and stability of conductive polymers.

Conductive polymers have become valuable electronic components of the modern age because they possess a broad range of applications in thin film transistors, batteries, antistatic coatings, protective and camouflage coatings, packaging, electromagnetic shielding, artificial muscles, light-emitting diodes, coaxial cable, condensers, biosensors, solder, fuel cells, solar cells, printed circuit boards, encapsulation, radiation detectors, fillers, Schottky diodes, electroplating, flexible displays and backlight source for displays; therefore they render tremendous future marketing opportunities. Volume 4 contains 13 chapters dealing

with charge transport in conductive polymers, electronic structures, photochemical processes, photorefractive effects, second- and third-order non-linear optical properties of π-conjugated oligomeric and polymeric materials and their ultrafast spectroscopy. A wide variety of applications of conductive polymers are discussed in Volume 4 and partly in Volumes 2 and 3, together with individual conductive polymers.

This book thoroughly covers today's rapidly growing topics on the chemistry of conductive materials and provides the most up-to-date, comprehensive overview of their materials science, solid-state physics, technology and device applications. My intention in assembling these volumes is to provide a useful reference book to scientists interested in recent developments in organic superconductors and conductive polymers, with each chapter having an extensive list of current references. Contributors have spent a great deal of time reviewing the last 18 years of scientific research on organic conductive materials with major emphasis on in-depth structure–property relationships. With over 13 300 bibliographic citations, I hope that this comprehensive text will be stimulating for academic and industrial researchers working in chemistry, polymer science, semiconductor physics, materials science, surface science, spectroscopy, crystallography, electrochemistry, biology, xerography, superconductivity, electronics, photonics and device engineering.

As we have such a huge volume of literature (the contributors used materials from many different sources in preparing their chapters), every effort has been made by all of us to obtain copyright permissions from the owners. I still take this opportunity to offer my sincere apologies to any copyright holder whose rights may have been infringed and whose names are cited here unwittingly. The following publishers kindly provided us with permission to reproduce originally published materials: Academic Press, American Association for the Advancement of Science, American Chemical Society, American Physical Society, American Institute of Physics, Australian Chemical Society, CRC Press, Chapman & Hall, Chemical Society of Japan, Electrochemical Society, Elsevier Science, Gordon & Breach Science Publishers, Huthig & Wepf Verlag, Indian Academy of Science, IOP Publishers, IEEE Industry Applications Association, International Union of Crystallography, Japan Society of Applied Physics, John Wiley & Sons, Kluwer Academic Publishers, Materials Research Society, Marcel Dekker, Macmillan, North-Holland, National Research Council of Canada, Oil & Colour Chemists Association, Pergamon Press, Royal Society of Chemistry, Springer Verlag, Taylor &

Francis, The Society of Imaging Science and Technology, The Society of Polymer Science Japan, Salamander Books Ltd, Trans Tech Publications and VCH Publishers.

I wish to express my sincere gratitude to all contributors who devoted their valuable time and expertise in preparing excellent state-of-the-art reviews which led this book to fruition. I am very grateful to Professor Werner F. Schmidt of Hahn-Meitner Institute, Berlin, Germany, for introducing me to the exciting area of conductive polymers during postdoctoral studies and for his excellent guidance. I would also like to give special thanks to the following: Alan J. Heeger, Seizo Miyata, H. Shirakawa, K. Yoshino, A. B. P. Lever, Cliff Leznoff, Richard T. Keys, Larry R. Dalton, Hans Thomann, Akio Mukoh, Atsushi Kakuta, Toshiyuki Watanabe, Claus Jessen, Robert Lewis, Arnt I. Vistnes, Duane Whitney, Ron Biegel, John Aklonis, Jean Lee, G. K. Surya Prakash, N. S. Dalal, E. T. Kang, Sri Ram Singh, Jiley Singh, Braham Singh, Sardar Singh, the late Iqbal Singh, Jagmer Singh, Dharam Pal Singh, Ranvir Singh Chaudhary, Yash Pal Singh, Braj Pal Singh, Satyendra Singh, Ashish Kumar, Khilari Singh, Vikram Singh, Harbir Singh, Dharam Vir Singh, Karan Singh, Rohtas Singh, Dr. Harbir Singh Malik, Chinta, Ram Dhama, Ram Pal Singh Malik, Kamala Devi, Rishi Pal, Chandaro Devi, Yogendra Malik, Bhanwar Singh Chaudhary, Arvind Kumar, Amar Singh, Satya Vir Arya, Brahm Pal Singh Randhi, Indra Kumari Varma, Prem Vrat, Raj Pal Dahiya, Padma Vasudevan, Dr. Dharam Pal Singh, Deepak Singal, Rakesh Misra, Krishi Pal Raghuvanshi, Yogesh Malik, Butchi Reddy, Subhash Chand Chauhan, Indra Prakash, Deo Pal Singh, Rajbal Singh, Ami Chand, Naresh Gill, Tej Pal Singh, Jagmer Singh, Rishi Pal Singh Tomar, Vinod Kumar. Finally, I would like to acknowledge the warmest cooperation and patience of my wife Dr. Beena Singh Nalwa, the moral support of my parents Sri Kadam Singh and Srimati Sukh Devi and love of my children, Surya and Ravina, in this exciting enterprise.

Hari Singh Nalwa

CHAPTER 1
Transport in Conducting Polymers

Esther M. Conwell
Xerox Webster Research Center, Webster, New York, USA

1 INTRODUCTION

The conductivity of conducting polymers, by which we mean π-bonded or conjugated polymers, spans a very wide range ($<10^{-12}$ to $\sim 10^5$ ohm^{-1} cm^{-1}) depending on doping. It is clear that a wide variety of phenomena is involved in transport in this group of materials. A major source of the variety stems from the quasi-one-dimensional (ID) nature of the materials. The chain relaxation or deformation that results from adding electrons or holes to a chain produces a variety of dressed particles—solitons, polarons and bipolarons —each with its own characteristic transport properties.

In this section, some conducting polymers that have been studied extensively will be first introduced, and then the properties of solitons, polarons and bipolarons will be briefly reviewed. Following these preliminaries, transport will be discussed in three sections: pristine material, moderately doped and heavily doped material.

Handbook of Organic Conductive Molecules and Polymers: Vol. 4. Conductive Polymers: Transport, Photophysics and Applications.
Edited by H. S. Nalwa. © 1997 John Wiley & Sons Ltd

1.1   The conducting polymers

The basic polymers, on which the most work has been done, are shown in Table 1.1. Note also that their acronyms are included. A great deal of work has been done also on derivatives of these polymers, in which the hydrogens have been replaced by side groups of varying complexity. The common feature is the alternation of single and double bonds, i.e. conjugation, at least in the backbone. For the polymers that support excitons, the quantity listed in the third column, the optical absorption edge, is smaller that the band gap by the amount of the exciton binding energy. In polyacetylene the lowest excited state has the same symmetry as the ground state (A_g) so radiative transitions are not permitted and few excitons are formed. The band gap is then the absorption edge, 1.4 eV. There is generally some uncertainty in the absorption edges because they are not sharply defined experimentally, presumably due to variations in the conjugation length and other disorder. For poly(phenylene vinylene, PPV, there is considerable evidence that the exciton binding energy is a few tenths of an electron volt. This is likely to be the case also for the other conducting polymers listed, apart from polyacetylene [1].

Trans-Polyacetylene differs from the other polymers in Table 1.1 in that two bonding configurations can occur in the ground state. The second configuration differs from the one shown in the Table by the interchange of single and double bonds. It is evident that this interchange does not change the energy of the

Table 1.1. The basic polymers. Adapted by permission of E.M Conwell and H.A. Mizes, *Handbook of Semiconductors* (ed. T.S. Moss), Vol 1 (ed. P.T. Landsbert), pp. 583–625, Elsevier Science SA, Switzerland (1992)

Polymer	Structure	Optical absorption edge (eV)
trans-Polyacetylene (*t*-PA)	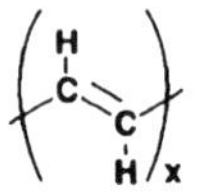	1.4
cis-Polyacetylene (*c*-PA)	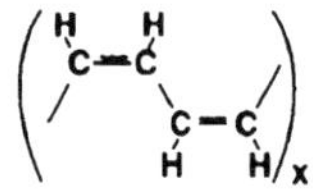	2.0
Polypyrrole (PPy)	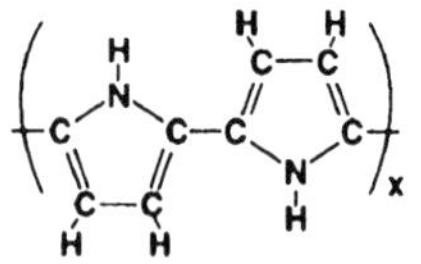	2.5
Polythiophene (PT)	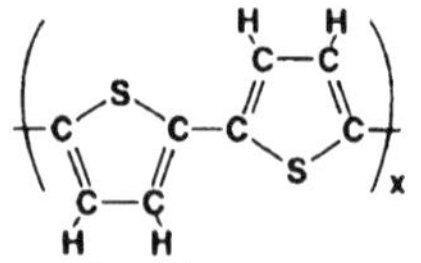	2.0
Poly(*p*-phenylene) (PPP)	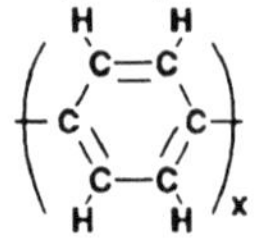	3.0
Poly(phenylenevinylene) (PPV)	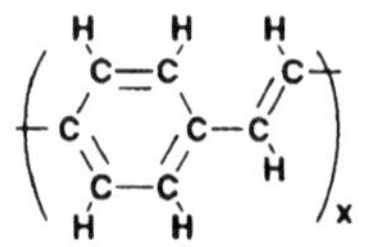	2.4
Polyaniline (PANI) emeraldine form	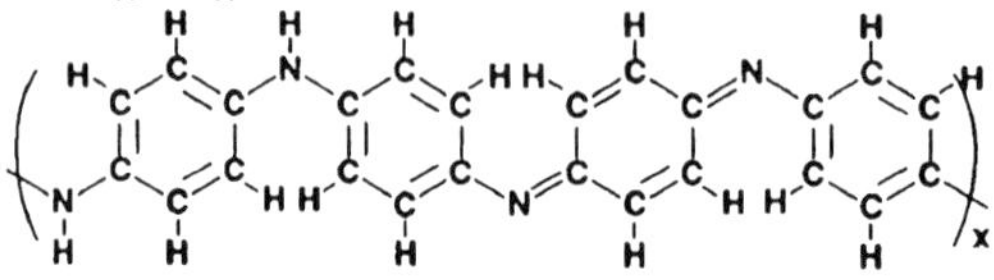	1.6

polymer (we neglect possible end effects), making the ground state degenerate. The degeneracy of the ground state has important consequences for transport, as well as other properties, as will be discussed. In all of the other cases shown in Table 1.1 the ground state is non-degenerate, interchange of single and double bonds resulting in a higher energy state.

1.2 Solitons, polarons and bipolarons

When an electron is added to a perfect chain, within fractions of a picosecond it causes the chain to deform, creating a characteristic pattern of bond deformation about 20 sites long. The electron plus the pattern constitute a polaron. In the case of t-PA the deformation is a weakening of the dimerization pattern, the difference between single and double bonds decreasing gradually, but not vanishing, in the progression from either end of the polaron to its center. For the other polymers shown the deformation is better described as a tendency for the interchange of single and double bonds. Along with the chain deformation there is a change in the energy level structure. A level is pulled out of the valence band into the gap with its two electrons and a level is pulled out of the conduction band. To form an electron polaron, P^-, an electron taken from a donor, or a result of photoexcitation, is added to the upper level. To form a hole polaron, P^+, an electron from the lower polaron level leaves, going into an acceptor or into a hole created by photoexcitation. Having one half-filled level, the polaron has spin $1/2$. The energy interval between a polaron level and the nearest band edge depends, as does the band gap, on the chain length or, more accurately, conjugation length. For typical conjugation lengths, say 40 Å, the polaron level is a few tenths of an electron volt from the band edge.

A bipolaron is formed by the union of two polarons of like sign. The bipolaron also has two levels in the forbidden gap. For a negative bipolaron, BP^{2-}, they are each occupied by two electrons, while for a positive bipolaron, BP^{2+}, they are both empty, i.e. occupied by two holes. The bipolaron has no spin. Because of the repulsion of the two like charges the stability of an isolated bipolaron has been questioned. Bipolarons that result from doping and are therefore close to the charged donors or acceptors that produced them are undoubtedly stable.

The soliton is an excitation that occurs only in degenerate polymers. Let a t-PA chain with double bonds sloping downward in going from left to right, as in Table 1.1, be denoted type A, one with the double bonds sloping the other way, type B. A type A region next to a type B region may occur on the same t-PA chain. Characteristically there is a region between them, approximately 14 sites long, in which the transition takes place. In this region the dimerization weakens progressively, vanishing at the center, i.e. at the center the bonds are of equal length. The domain wall between the different directions of the double bonds, i.e. between regions A and B, is called a soliton. Although as described so far the soliton is neutral, S^0, it is characterized by a bound electron whose wave-function is spread over the region in which the dimerization is varying. In a simple picture, where Coulomb forces between electrons are neglected, this bound electron, being neither bonding or antibonding, occupies a level at the center of the gap. The neutral soliton therefore has spin $1/2$. A second electron may occupy this level, resulting in a negatively charged soliton S^-, which has no spin. Alternatively the electron may leave this level, resulting in a positively charged soliton, S^+, with no spin.

A polaron, bipolaron or soliton can travel along a chain as an entity, the atoms in its path changing their positions so that the deformation travels with the electron or hole. Except for the metallic state, these are the entities through which charge transport is accomplished in conducting polymers. The fact that the atoms must move as these particles drift along a chain results in an increment to the effective mass of the particles. The increments are small, however, because the excitations cover many sites and the atom displacements required to establish the deformation pattern are quite small. For example, the mass of a soliton in t-PA is about six times the mass of the electron, m_0 [2]. The mass of a polaron in t-PA is smaller, $\sim m_0$, because the polaron involves smaller deformations.

Because it is a topological defect, the soliton has some special properties [2] that polarons and bipolarons do not possess. There is a difference between a soliton that connects a type A region on the left and B on the right and a soliton that connects B on the left and A on the right. To distinguish between them it is customary to call one, say the one with A on the left, a soliton, S and the one with B on the left an anti-soliton, $\bar{S}$. It is apparent that on a chain with more than two solitons, S and $\bar{S}$ must alternate. A single soliton cannot hop between chains because it would require all the bonds to one side or the other to change. Solitons cannot go through each other, a feat that is possible for polarons.

So far we have assumed that the properties of the particles are determined by a single chain, i.e. that interchain interactions have no effect. This assumption has been questioned, particularly for the case of polarons. Vogl and Campbell [3], from their local density functional calculations for polyacetylene, concluded that the effect of interchain interactions was sufficient in that case to destabilize the polaron, and thus make the electron into a conduction band electron of the kind usually found in three-dimensional semiconductors. These calculations were done for an infinite lattice of perfect polyacetylene. It was pointed out, however, that the many defects and short conjugation lengths in these materials would tend to stabilize the polaron [4]. Also, the interchain interaction is usually smaller for the other conducting polymers than polyacetylene.

At the other extreme, it was suggested by Emin and Ngai [5] that the effect of defects and disorder would be sufficient to localize the polaron, turning it from a large polaron spread over about 20 atomic sites to a small polaron occupying about one atomic site. The effect of conjugation, which allows the π-electrons to spread out over the length of a conjugated segment, is apparently sufficient to prevent such localization. At any rate there has been no evidence found for small polarons in the conducting polymers. On the other hand, electron nuclear double resonance (ENDOR) experiments on phenylene vinylene oligomers in the solid state, for example, have shown that an electron added to an oligomer spreads over the length of the oligomer to lengths up to seven phenyl rings, the longest investigated [6].

For a more detailed picture of solitons, polarons and bipolarons the reader is referred to review articles by Heeger *et al.* [7], Conwell [8] and Conwell and Mizes [9].

2 TRANSPORT IN PRISTINE POLYMERS

The transport properties that have been measured and/ or calculated theoretically for pristine polymers are mobility, diffusion coefficient and thermoelectric power. These quantities reflect the fact that carriers move in response to an electric field, a concentration gradient or a temperature gradient, respectively. Mobility μ is defined as the drift velocity developed by a carrier in an electric field per unit electric field. It is a measure of the relative ease with which the carrier moves in an electric field. Mobility is related to conductivity σ by $\sigma = ne\mu$, where n is the number of carriers per unit volume and e is the charge on the

electron. Carriers are introduced by doping or by the action of light. Discussion of n is beyond the scope of this article, however. The diffusion coefficient D is defined as the carrier current (number of carriers crossing 1 cm^2 s^{-1}) developed per unit concentration gradient. It is related to the mobility by the Einstein relation

$$D = \mu kT/e \qquad (1.1)$$

where k is Boltzmann's constant and T is the absolute temperature. Thermoelectric power Q is defined as the potential difference developed per unit thermal gradient.

In this section these transport properties will be considered, first for the case of carriers moving by hopping and then for the case of carrier drift or band motion. For ordered material the critical factor determining whether carrier motion is hopping or band-like is the width of the band in which it moves. For temperatures where the bandwidth $W \gg kT$, band motion dominates. When $kT > W$ the band loses its meaning and hopping dominates. The bandwidth is determined by the overlap between the carrier wavefunctions on the initial site or chain and on the final site or chain. A calculation of the bandwidth for polaron motion between chains in t-PA, including the wavefunctions of all of the valence band electrons and the nuclei, will be described briefly in Section 2.2.1. For other polymers or oligomers there is not sufficient information to carry out such a calculation, however. The decision as to whether transport for a particular material in a particular temperature range is hopping or band motion is frequently made on the basis of the size of the mobility. Hopping mobilities are smaller, usually orders of magnitude less, than 1 cm^2 V^{-1} s^{-1}, while band mobilities, at least around room temperature, are greater than 1 cm^2 V^{-1} s^{-1}.

It will be seen that room temperature values of hopping mobility reported for polymers and oligomers range from less than 10^{-7} to ~ 0.5 cm^2 V^{-1} s^{-1}. The values are strongly dependent on both the technique used for measurement, as will be discussed, and on the perfection of the polymer or oligomer sample. At the high end are oligomers with a high degree of structural organization and purity. This is not surprising. It is known from studies of transport in molecular organic crystals, such as naphthalene, that mobilities are highly dependent on both the perfection of the structural organization and purity, reaching values of a few hundred cm^2 V^{-1} s^{-1} at low temperatures for highly perfect and pure crystals [10]. The low mobilities are found in materials with poor structural organization, varying chain lengths, impurities, defects of all sorts. Since these effects have not been studied in detail, we will usually lump them together under the heading 'disorder'. Some of the consequences of disorder for transport will be discussed later in this section.

Hopping mobility is usually characterized by an activation energy and by field dependence. The activation energy is in general made up of intermolecular and intramolecular contributions. The former represents the difference in the energy of the hopping sites, in this case conjugated chain segments, due to differences in length, presence of defects and local polarization energy [11], in addition of course to the potential drop due to the electric field. The intramolecular contribution arises from the change in conformation of the chain segments due to the addition or removal of an electron, i.e. the polaronic effect.

The field dependence of the mobility may be quite strong. The form of the dependence found over a large field range for a wide variety of polymers and molecularly doped polymers is

$$\ln \mu \propto E^{0.5} \qquad (1.2)$$

where E is the electric field strength [12]. This dependence is apparently characteristic of disordered systems but its universality and wide range of applicability are not yet understood.

Hopping mobility may not be activated, in the sense that the dependence of $\log \mu$ on $1/T$ need not be exponential. Pristine t-PA is a case where transport is undoubtedly by hopping. Conductivity decreases rapidly with T but not exponentially with $1/T$. We will discuss mobility of t-PA in Sections 2.1.2 and 2.3.

2.1 Experimental studies of polaron hopping

The most direct way of measuring mobility is the time-of-flight method. In this method the carriers, in this case polarons, are injected into a sample either by a contact, or by illuminating one side of the sample, which usually has a transparent contact, with strongly absorbed light. In the conventional method the mobility is obtained by measuring electrically the time it takes for the pulse so generated to travel to the other contact in a known field. Unfortunately this method will not work when the transport is dispersive, i.e. the carriers straggle into the second contact, the result of their having undergone trapping by deep traps, so that there is no clear time-of-flight. It has been the experience, with a few exceptions, that this method does not work for conducting polymer samples. However, it is still possible to obtain a time-of-flight mobility for a light emitting diode (LED) by measuring the time it takes after application of the voltage for luminescence to appear. μ is then determined from the distance the carrier must travel to meet one of the opposite sign. The next section reviews the results of time-of-flight measurements on PPV, a derivative of PPV and films made of the oligomer sexithiophene, α-6T, which consists of six thiophene rings connected as shown for polythiophene with $x = 3$ in Table 1.1.

Photoconductivity has been studied extensively in these polymers. It is particularly complex in that it may be due to two different types of carrier—solitons and polarons in t-PA, polarons and bipolarons in the non-degenerate ground state polymers. Even if there were only one type of carrier it would not be possible to determine its number n and mobility separately. In cases where n, although unknown, can be expected to stay constant with temperature we can, however, extract some information, notably the temperature dependence of μ. We report some data of this kind for t-PA in Section 2.2.

Another device that has aroused considerable interest is the thin film transistor (TFT) with a conducting polymer or oligomer as the active layer. As will be discussed in Section 2.1.3, the operation of this device is, in many respects, similar to that of a field effect transistor (FET) made of a three-dimensional semiconductor such as silicon. From the measured characteristics of the TFT it is possible to deduce a value of the mobility, usually called μ_{FE}, the field effect mobility. Values of μ_{FE} obtained for number of conducting polymers and oligomers will be presented in Section 2.1.3. It will be seen that the value of μ_{FE} may be much higher than the values of μ obtained from time-of-flight measurements. The reasons for this will be discussed.

The mobility in vacuum evaporated films of α-6T has also been obtained from the observation of space charge limited current at high voltages in this material. This will be discussed in Section 2.1.4.

2.1.1 Time-of-flight mobility measurements

There are only a few polymers on which time-of-flight measurements have been reported, as is seen in Table 1.2. In most cases the measurements were made using the conventional method. Measurements using the time delay of luminescence in an LED have been made on α-6T as well as PPV.

A number of measurements of time-of-flight room temperature hole mobility in PPV are summarized in Table 1.2. Because the transit was found to be more or less dispersive, with the exception of ref. 13, and the electric field not necessarily uniform, the values and field dependence are to some extent uncertain. Nevertheless the small size of μ ensures that it is hopping mobility that is being observed. Also, from the scatter in the values it is clear that there are a variety of PPVs or, stated differently, PPV with different degrees of disorder. As indicated earlier, disorder is a common feature of conducting polymers and has important effects on transport, as will be discussed.

The PPV samples of Table 1.2 were all made by heating, and thus converting, a precursor, non-conjugated polymer. The choice of the precursor polymer and conversion protocol, i.e. heat treatment time and temperature, has considerable influence on the electronic properties of the resulting PPV [21]. The average conjugation length increases with heat treatment time and temperature to a limiting value generally agreed to be about seven monomers in typical samples. It is the short conjugation length that results in transport being primarily by hopping. It is quite possible that there are some chains that are long enough for some of the transport to involve drift along them. The mobility for such drift should be within an order of magnitude of that calculated for polyacetylene, ~ 1 cm^2 V^{-1}s^{-1}, as will be discussed in Section 2. The rate limiting process is clearly hopping and drift will not be considered further in this section.

The field dependence of mobility given by equation (1.2) was found for PPV by Karg et al. [17] using the time delay of luminescence. Fitting their data to the form $\mu \propto \exp(BE^{1/2}/kT)$ they found $B = 1.6 \times$ 10^{-4} eV$^{1/2}$ cm$^{1/2}$ for room temperature. In the other two cases where the field dependence was studied it was not found to fit equation (1.2).

Data on temperature dependence of the hopping mobility were obtained for two samples of PPV, converted at temperatures of 250°C and 200°C, by Takiguchi et al. [13]. Over the temperature range from 285 K to 385 K they found the mobility to vary as $\exp(-E_a/kT)$ for both samples. For the sample converted at 250°C, $E_a = 0.12$ eV. E_a was a little larger for the sample converted at 200°C.

For α-6T the time delay for luminescence gave a hole mobility of 5×10^{-6} cm^2 V^{-1} s^{-1} [20]. As noted earlier, this value is orders of magnitudes smaller than the value of μ_{FE} and the reason for the difference will be discussed in Section 2.1.3.

For a couple of substituted polyphenylacetylenes where trapping was too severe to obtain time-of-flight, Zhou et al.. [22] were able to obtain mobility values with the use of Scher–Montroll theory. They found electron and hole mobilities of 1.8×10^{-5} and 1.7×10^{-5} cm^2 V^{-1} s^{-1}, respectively, in poly(2,6-dimethyl-4-*t*butyl phenylacetylene), 4.8×10^{-7} and 4.0×10^{-7} cm^2 V^{-1}s^{-1} in poly(O-trimethylsilyl phenylacetylene).

2.1.2 Hopping mobility of polarons in t-PA

Picosecond photoconductivity vs. reciprocal temperature measured in oriented *t*-PA is shown in Figure 1.1. As will be discussed in Section 2.4 picosecond photoconductivity parallel to the chains represents drift band motion of the polarons along the chains. Calculations of the drift mobility of polarons parallel to the chains in *t*-PA [25,26] indicate that it does not vary much with temperature in the range 80–300 K. From the lack of temperature variation of the photo-

Table 1.2. Time-of-flight hole mobilities

Material	Mobility μ (cm^2 V^{-1} s^{-1})	Measurement field E; E-Dependence (V cm^{-1})	Reference
PPV	10^{-4} to 10^{-3} $285 < T < 385$ K	Independent of E $10^4 < E < 2 \times 10^5$	13
PPV	$10^{-4} < \mu < 10^{-3}$	10^4	14
PPV	3×10^{-3}	1.5×10^3	15
PPV	$\sim 5 \times 10^{-6} < \mu < 5 \times 10^{-5}$	4.5×10^4	16
PPV	$10^{-7} < \mu < 3 \times 10^{-6}$	$\ln \mu \propto E^{1/2}$ $10^5 < E < 10^6$	17
PPV	$\sim 10^{-6}$	2×10^4	18
PPPV	$10^{-3} < \mu < 10^{-4}$	$\ln \mu \propto E^{-0.85}$ $3 \times 10^4 < E < 3 \times 10^5$	19
α−6T	5×10^{-6}	$3 \times 10^5 < E < 8 \times 10^5$	20

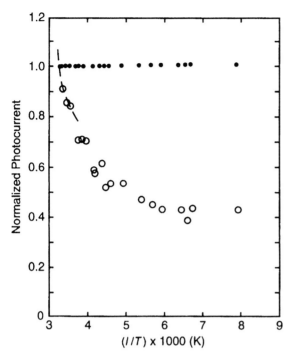

Figure 1.1. Temperature dependence of picosecond photo-current in oriented *t*-PA for fields (●) parallel to the chains and (○) perpendicular to the chains according to the data of Walser *et al.* [23]. The dashed line represents μ perpendicular to the chains calculated as described in ref. 24. (See Section 2.2.1). (Adapted by permission from E.M. Conwell, H.-Y. Choi and S. Jeyadev, *Synth. Met.* 49–50, 359 1992).

conductivity parallel to the chains we infer that carrier concentration has little dependence on T, at least within the picosecond range. From this it follows that the T dependence of the data for photoconductivity perpendicular to the chains represents the T dependence of $\mu_\perp$, the mobility perpendicular to the chains. The facts that interchain coupling is small, and that $\mu_\perp$ increases with T, identify $\mu_\perp$ as a hopping mobility, at least above $(10^3/T) = 6$. A plot of log μ vs. $10^3/T$, using the data of Figure 1.1, gives a straight line, albeit with a lot of scatter, for $(10^3/T) \leqslant 6$. The corresponding activation energy is 0.048 eV, about half of that found for PPV according to Section 2.1.1. For $(10^3/T) > 6$, i.e. below ~ 167 K, the photoconductivity does not appear activated. A possible explanation for this is presented in Section 2.2.1.

2.1.3 Field effect mobility

A TFT may be made by depositing a series of layers. A common design, shown schematically in Figure 1.2, has

a suitable substrate, e.g. glass, on which is deposited the gate electrode. This electrode is frequently a highly conducting doped silicon layer. Above this is deposited an insulating layer, usually SiO_2 if the gate electrode is silicon although it may be an organic insulator [27]. Above this are the source and drain electrodes, of width W and spaced L apart. The organic semiconductor, a polymer or oligomer film, may be deposited by spin coating or evaporation to occupy the space between the source and drain. The semiconductor is generally a hole conductor and is designed to have as few carriers as possible to minimize the ohmic current between the source and drain. Application of voltage V_G between the gate and source contacts causes charging of the condenser whose dielectric consists of the insulating layer and some of the adjoining high resistivity semiconductor layer. If V_G is of the proper polarity it causes hole injection, introducing additional holes into this adjoining layer (accumulation layer) in an amount depending of V_G. It has been shown experimentally that the accumulation layer is quite thin, ~ 50 Å [28].

Application of a voltage V_D between the source and drain results in a drain current I_D. I_D depends on the number of holes in the accumulation layer and their mobility. The drain current saturates when V_D is large enough to prevent further hole injection near the drain. From I_D, the capacitance/area, the dimensions W and L and the voltages, it is possible to determine the mobility in the accumulation layer. This mobility is denoted μ_{FE} because of the similarity of this device to the FET. Table 1.3 gives the values of μ_{FE} for various oligomers and polymers.

It is seen that the highest values are obtained for sexithiophene and its dihexyl derivative. Consistent with the high μ_{FE} values, X-ray studies of evaporated thin films of these two oligomers show that they are

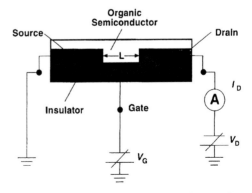

Figure 1.2. Schematic diagram of the thin film transistor.

highly ordered, consisting of layered structures in a monoclinic arrangement with all-*trans* planar molecules standing on the substrate [30]. The higher mobility of the dihexyl derivative is due to longer range order, the result of the self-assembly properties created by adding the alkyl groups [30]. Another example of this is the larger μ_{FE} of diethylquinquethio-diethylquinquethiophene compared to quinquethio-quinquethiophene.

It is noteworthy that μ_{FE} increases with increasing conjugation length, as seen in comparing quater-, quinque- and sexi-thiophene. However μ_{FE} decreases in going to octithiophene. Horowitz *et al.* [34] suggest this is due to an increase in conjugation defects in the octamer. Fuchigami *et al.* [39] attribute the high μ_{FE} values that they obtained for poly-(thienylene vinylene), PTV, to long conjugation length, having found μ_{FE} to increase with increasing degree of conversion from the insulating precursor polymer. Their result is surprising, however, in view of the disorder usually found in polymers, which is manifested in low μ_{FE} values for polythiophene and *trans*-polyacetylene, as seen in Table 1.3.

It is worthwhile to discuss the relation of μ_{FE} to bulk mobility. The first question to consider is the difference for α-6T between μ_{FE}, $\sim 10^{-2}$ cm^{-1} s^{-1}, and its time-of-flight mobility of 5×10^{-6} cm^2 V^{-1} s^{-1}. Delannoy *et al.* [20] suggested that the difference is due to traps that are operative under the conditions of the time-of-flight experiment but not under those that hold during TFT operation. Such traps would be ones that in thermal equilibrium are below the Fermi energy, thus filled with electrons and able to trap holes. That is the situation that holds under the transient conditions characteristic of the time-of-flight experiment, where the current due to background carriers in the sample is negligible. In the case of TFT the mobility is determined in a steady state in which a large number of holes have been injected into the current-carrying region. For α-6T this results in the traps having been emptied of electrons and thus no longer effective for trapping holes. This is consistent with α-6T having a relatively small number of traps, due to its ordered nature, concentrated at an energy close to the Fermi energy. In other experiments, to be described in the next section, Horowitz *et al.* [41] located a trap level in α-6T 0.3 eV above the valence band edge. Another example of the effect of trap filling is probably the finding of Burroughes *et al.* [40] that μ_{FE} in *t*-PA depends on carrier concentration. In this case, however, the value of μ_{FE} is only 6.5×10^{-4} cm^2 V^{-1} s^{-1} at the higher carrier concentration, as would be expected for a greater density of traps, probably spread over a wider energy range, characteristic of a more disordered material.

The presence of traps is not the only reason for the difference between the mobility values obtained by different types of measurements. The mobility depends

Table 1.3. Field effect hole mobilities

Material	μ_{FE} (cm^2 V^{-1} s^{-1})	Reference
Sexithiophene	4.3×10^{-1}	29
Sexithiophene	2×10^{-3}	30
Sexithiophene	3×10^{-2}	28
Dihexylsexithiophene	7×10^{-2}	31
Dimethylsexithiophene	2×10^{-2}	32
Quaterthiophene	2×10^{-7}	33
Quinquethiophene	10^{-5}	34
Octithiophene	2×10^{-4}	34
Diethylquaterthiophene	9×10^{-5}	33
Diethylquinquethiophene	9×10^{-4}	33
Polythiophene	10^{-5}	35
Polythiophene	2×10^{-4}	36
Poly(3-hexylthiophene)	10^{-4} to 10^{-5}	37
Poly(3-hexylthiophene)-based Langmuir–Blodgett film	7×10^{-7}	38
Quinquethiophene-based Langmuir–Blodgett film	1×10^{-5}	38
Polythienylene vinylene	2.2×10^{-1}	39
trans-Polyacetylene	$< 10^{-5}$ to 6.5×10^{-4} depending on carrier concentration	40

strongly on direction in these oriented films, and the direction may be different for the different measurements. The fact that current flows in a very narrow channel in the TFT causes μ_{FE} to be dependent on the properties of the interface between the insulator and the channel [34,28]. Also μ_{FE} is determined by comparison of experimental results with a theory in which various approximations have been made, such as neglecting the effect of parasitic series resistance [28].

The importance of crystalline order presumably explains why μ_{FE} was found to increase considerably on annealing [42]. Dodabalapur et al. [28] note that crystallite size, normally 100 nm, was increased to $> 5 \ \mu m$ by short time annealing near the melting point. A goal of such efforts would be crystallite size greater than channel length. It is estimated that the maximum intrinsic mobility in α-6T is at least $0.08 \ cm^2 \ V^{-1} \ s^{-1}$ [28]. The mobility could also be increased by high field operation, as pointed out by Dodabalapur et al. [28], who found an increase of μ_{FE} with field.

2.1.4 Mobility from space charge limited current

Measurements of current density vs. voltage have been carried out for evaporated films of α-6T sandwiched between gold electrodes. Over a wide range of voltage it was found that the variation of j with V was linear at low voltages and quadratic at high voltages [41]. The quadratic behavior is evidence for space charge limited current. Theory gives the relation between space charge limited current density j_{sc} and voltage as

$$j_{sc} = (9/8)\kappa\theta\mu V^2/L^3$$

where κ is the dielectric constant, θ is the fraction of charges free to move (i.e. untrapped) and L is the length of the sample. Thus from j vs. V in the quadratic region, and a knowledge of κ and L it is possible to deduce the produce $\theta\mu$, the effective mobility. For 300 K Horowitz et al. found $\theta\mu = 7.34 \times 10^{-3} \ cm^2 \ V^{-1} \ s^{-1}$. Since $\theta < 1$ this value is consistent with the higher values of μ_{FE} in Table 1.3. On decreasing the temperature $\theta\mu$ was found to decrease, showing a linear decrease of $\ln \theta\mu$ with $1/T$. From these data the activation energy for the effective mobility was determined as 0.28 eV. If μ is assumed to be constant with T, this can be ascribed to a trap level 0.28 eV above the valence band [41].

2.2 Theory of polaron hopping mobility

Section 2.2.1 will review a calculation carried out for an ordered system in the low field limit, specifically that of interchain hopping of a polaron in t-PA that is assumed ordered. Although that may not be the case, the calculation is of interest if only because it illustrates the difficulties involved in calculating the hopping rate for even the simplest case. Also, it makes clearer the significance of some of the approximations made in the other calculations to be described.

For a hopping theory it is necessary to calculate the overlap between the initial and final wavefunctions or, more accurately, for interchain hopping the transverse bandwidth W. This was first done by Jeyadev and Schrieffer [42] for t-PA at $T = 0$. The calculation involves the changes in the wavefunctions of the nuclei and valence electrons as well as the electron that hops. These calculations can be done for t-PA because there are only two important phonon branches and their dispersion, with or without the polaron on the chain, can be calculated for $T = 0$ from the Su Schrieffer and Heeger (SSH) Hamiltonian. After allowing for the effect of finite temperature it is possible to calculate the hopping probability by first order perturbation theory. This will be described in Section 2.2.1.

The elementary jump rate is much more difficult to evaluate for the cases of greatest interest—more complicated, and disordered, polymers in high fields. We will essentially follow the disorder model pioneered by Bässler [11] and colleagues.

2.2.1 Hopping in ordered t-PA

For their calculation of the transverse bandwidth Jeyadev and Schrieffer [42] used the SSH Hamiltonian. This does not allow extension to temperatures above 0 K. To study transport at higher temperatures it is necessary to convert the SSH Hamiltonian to a normal mode representation, which was carried out in ref. 43.

The dependence of wavefunctions on the number of phonons on the chains makes it necessary to calculate a thermally averaged bandwidth, W_T, given by

$$W_T = 4 \sum_i \exp(-\beta\varepsilon_i)\langle\psi_f|H_\perp|\psi_i\rangle / \sum_i \exp(-\beta\varepsilon_i) \quad (1.3)$$

where $\beta = 1/kT$, i refers to a state of the system, $H_\perp$ is the operator for the interchain electron hop and the matrix element is taken between the state i with the polaron on chain l and the state f with the polaron on chain $l \pm 1$. The factor four is required to obtain a bandwidth from the matrix element or hopping integral.

Operation by $H_\perp$ on the wavefunction of the electron in the polaron was as usual taken to give a factor $-t_\perp$, $4|t_\perp|$ being the transverse bandwidth of the π-band for the hopping direction. Within the Born–Oppenheimer approximation the remainder of ψ_i or ψ_f may be written as the product of the valence electron wavefunctions and a nuclear part X expressed as a product of normal modes. With the notation Φ for the overlap of the valence band wavefunctions with and without the extra electron on the chain the bandwidth may be written

$$W_T = 4t_\perp \Phi^2$$
$$\times \left[\sum_i \exp(-\beta\varepsilon_i)|\langle X_l^d(m_i)|X_l^p(m_i)\rangle|^2 \Big/ \sum_i \exp(\beta\varepsilon_i) \right]$$
$$(1.4)$$

where $X_l^p(m_i)$ and $X_l^d(m_i)$ represent the wavefunctions for the normal modes on chain l with and without (thus simply dimerized) the polaron, respectively, and m_i the number of phonons in the various normal modes for the state with total energy ε_i. The factors involving the values of X and Φ are both squared because chains l and $l \pm 1$ give identical contributions. The quantity in angular brackets is the Franck–Condon factor (to be denoted FC) for the transition. The result of carrying out the integrations within the angular brackets is [43]

$$FC = \exp\left[-\sum_k \Delta_k^2(m_k + \tfrac{1}{2}) \right] \equiv \exp(-S_T) \quad (1.5)$$

where equation (5) defines S_T, Δ_k being given by

$$\Delta_k = (M\omega_k/\hbar)^{1/2}\delta_k \qquad (1.6a)$$

and

$$m_k = [\exp(\beta\hbar\omega_k) - 1]^{-1} \qquad (1.6b)$$

the average number of phonons in the normal mode with frequency ω_k; δ_k is the transformation onto the kth normal coordinate of the set of displacements δ_n of the C–H bonds due to the polaron. To evaluate FC the dynamical matrices were calculated in the presence and absence of the polaron using the method formulated by Terai and Ono [44]. The ω_k values that resulted were in excellent agreement with those obtained by Terai and Ono and others. With δ_k calculated from δ_n and the dynamical matrix in the presence of the polaron it was found that summation over k of Δ_k^2 gave 10, leading to a Franck-Condon factor at $T=0$ of 5.6×10^{-3}. The major contributions to the summation over k of Δ_k^2 are from the optical modes and the high frequency acoustic modes [45].

The overlap Φ of the valence band wavefunctions and without a polaron on the chain depends on the number of valence electrons, and thus on the chain length. For a t-PA chain length of 70 sites Φ has been estimated as 0.2 using the results of ref. 46. However this may be an underestimate. With $4|t_\perp|$ as the maximum transverse bandwidth calculated for t-PA by Vogl and Campbell [3] (0.5 eV) and with the above values for FC and Φ, equation (1.5) gives $W_T = 1 \times 10^{-4}$ eV. Even if $\Phi \simeq 1, W_T$ would only be 30 K and the transverse motion of the polaron should certainly be hopping over the temperature range of the data in Figure 1.1.

The procedure used to calculate the hopping probability is similar to that used by Holstein for the small polaron [47]. By first order perturbation theory the probability $w_\perp$ of a hop per unit time from state i with the polaron on chain l to state f with the polaron on chain $l \pm 1$ is given by

$$w_\perp = \frac{2\pi}{\hbar} \sum_{i,f} \exp(-\beta\varepsilon_i)|\langle\psi_f|H_\perp|\psi_i\rangle|^2$$
$$\times \delta(\varepsilon_i - \varepsilon_f) \Big/ \sum_i \exp(-\beta\varepsilon_i) \qquad (1.7)$$

Evaluation of the matrix element in equation (1.7) differs from that carried out for the bandwidth because the number of phonons in the normal modes is now allowed to vary, energy conservation in the process being ensured by the δ function. Terms in which the number of phonons in each normal mode is the same in i and f correspond to band motion and are subtracted out. The result finally is [43]

$$w_\perp = \frac{W_T^2}{16\hbar^2} \int_{-\infty}^{\infty} dt$$
$$\times \left\{ \exp\left[\sum_k \Delta_k^2 \operatorname{cosech}\left(\frac{\beta\hbar\omega_k}{2}\right) \cos\omega_k t \right] - 1 \right\}$$
$$(1.8)$$

The significance of the terms in the integrand is easily understood. The -1 results from the subtraction of the terms corresponding to band motion. The quantity Δ_k^2 represents the coupling to the mode k; $\operatorname{cosech}(\beta\hbar\omega_k/2)$ is related to the number of phonons in mode k at temperature T; and $\cos \omega_k t$ represents the time variation of energy in mode k. The modes have been chosen as standing waves in this calculation, with phases such that all modes have their maximum at $t=0$. In the case discussed by Holstein, where only optical modes with small dispersion are included, the integrand has a series of peaks corresponding to

$\omega_k t = 0$, 2π, 4π, etc. and is quite small in between. When Δ_k^2 is large, as is the case for our calculations, and β is small (high temperature) the peak centered at $t = 0$ is much larger than the succeeding peaks even when acoustic modes are included, and makes the major contribution to the integral. As a result only small values of $\omega_k t$ contribute at high temperatures and, following Holstein, we expand $\cos \omega_k t$, dropping terms higher than quadratic. Integration of equation (1.8) thus yields [47,43]

$$w_\perp = \frac{(2\pi)^{1/2} W_T^2}{16\hbar^2} \frac{\exp\left[\sum_k \Delta_k^2 \, \text{cosech}(\beta \hbar \omega_k / 2)\right]}{\left[\sum_k \Delta_k^2 \omega_k^2 \, \text{cosech}(\beta \hbar \omega_k / 2)\right]^{1/2}} \quad (1.9)$$

It has been noted by de Wit [48] and others that the integral in equation (1.8) actually has a logarithmic divergence. Of course $w_\perp$ must be finite. As discussed by de Wit, cutting off the integral after the first peak, which is essentially what we have done, yields a good value for $w_\perp$ when the coefficient of $\cos \omega_k t$ in equation (1.8) is large.

Given the hopping rate, the mobility $\mu_\perp$ can be obtained from the diffusivity D and the Einstein relation:

$$\mu_\perp = \frac{D_\perp e}{kT} = \frac{eb^2 w_\perp}{kT} \quad (1.10)$$

where b is the interchain spacing. With b taken as 4 Å, we find using equation (1.10) that from ~ 200 to 400 K, $\mu_\perp$ is in the range ~ 2 to 6×10^{-4} cm^2 V^{-1} s^{-1}. This is a typical value for hopping mobility. Note that the high temperature approximation made above is not valid until temperatures somewhat higher than 200 K are reached.

As noted in Section 2.1.2, the variation with temperature of photoconductivity perpendicular to the chains shown in Figure 1.1 should represent that of $\mu_\perp$ because the carrier concentration appears to be independent of T within the range of the data. The theoretical results for $(10^3/T) \leqslant 4$, where the high temperature approximation should be valid, were therefore plotted in Figure 1.1 as a dashed line by matching the theory with experiment at $(10^3/T) = 3.5$. It is seen that over the small range of overlap there is an excellent fit except for $10^3/T$ close to 4.0, where apparently the high temperature approximation is not yet good. Although a good approximation for the integral in equation (1.8) at lower temperatures has not yet been found, the smaller rate of increase of $\mu_\perp$ with T at lower temperatures is what would be expected.

The lowest frequency modes have very small Δ_k values. For the hopping rate to increase markedly beyond the value dictated by the zero-point vibrations, T must be high enough to excite modes with sizable values of Δ_k, thus higher frequencies. It was concluded that the transport responsible for photoconductivity perpendicular to the chains is phonon-assisted hopping, to quite low temperatures according to these calculations.

Within the accuracy of their data Walser et al. [23] found that the high temperature photocurrent could be fitted with an activation energy of 0.048 eV. Interestingly, Phillips and Heeger [49] found that photoconductivity in t-PA parallel to the chains has an activated temperature dependence at 1.5 ns, with an activation energy of 0.045 eV. They suggest that this may be due to oxygen traps. However, such traps should be much deeper [26]. In the light of the present results it appears likely that the activation energy arises from phonon-assisted interchain hops of polarons, which could be the rate limiting step in $\mu_\perp$ at long times due to finite chain lengths.

2.2.2 The disorder model

In the disorder model the basic quantity, the elementary jump rate v_{ij} between sites i and j, is assumed to be of the form

$$v_{ij} = v_0 W_{ij} R(\Delta_{ij}) \quad (1.11)$$

Here the prefactor v_0 establishes the dimensions and allows for effects not included in the other two factors. W_{ij} represents the overlap of the hopping electron wavefunction between the two sites, a sort of dimensionless bond. For the polymer case the 'site' may be considered a conjugation length. The factor R is a function of the energy difference between sites $\Delta_{ij} = E_i - E_j$, including the potential drop due to the electric field. It may include the changes in overlap of the nuclear wavefunctions due to the distortion caused by the polaron, i.e. polaronic effects. In general $R(\Delta)$ depends on the phonons with which the electron interacts. For PPV, for example, it is known that at least two phonon branches are effectively coupled to the π electrons. The spacing between exciton emission peaks identifies one of these modes as the ring stretching mode, with a frequency of ~ 0.2 eV. It has been suggested that the small variation of this spacing with temperature, and the thermal broadening of the exciton peaks, are due to torsional modes, with a characteristic energy of ~ 5 meV [50]. Strong librations of the phenyl rings about the C−C vinyl linkages have also been seen in X-ray studies of PPV [51] from which it has been concluded that the polymer chains are not planar, even in a solid. Since the polaron wavefunction involves distortions of the vinyl group as well as the rings, it is likely that phonon modes in addition to the two cited are also involved in the hopping. Unfortunately there is little information about coupling to other modes.

$R(\Delta)$ has been calculated for a number of different cases of non-adiabatic polaron hops, usually where there is interaction with only one optical phonon branch of frequency ω for a zero wavevector. One particular case is that of the high temperature regime, $kT > \hbar\omega_0$, where [52]

$$R(\Delta)\alpha(\varepsilon_b k_B T)^{-1/2} \exp\left[-\frac{\varepsilon_b}{2kT}\left(1 + \frac{\Delta}{2\varepsilon_b}\right)^2 \right] \quad (1.12)$$

Here E_b is the polaron binding energy, the energy required for thermal dissociation. It involves the electron–phonon coupling and the phonon frequency. Equation (1.12) is valid for electron–phonon coupling strong enough so that multi-phonon processes dominate. This equation is well known from the theory of charge transfer in chemical reactions due to Marcus [53]. The reorganization energy λ of Marcus' formulation is equal to $2\varepsilon_b$.

It is evident that the information about the phonons involved in the hopping process is not available to write $R(\Delta)$ for the actual many phonon case. For this reason, among others, it has been customary to use an expression for the jump rate introduced by Miller and Abrahams [54].

$$\begin{aligned} R(\Delta) &\propto \exp(-\Delta/kT), & \Delta > 0, \\ R(\Delta) &\propto 1, & \Delta \leqslant 0 \end{aligned} \quad (1.13)$$

A significant difference between the rates given by equations (1.12) and (1.13) is that according to (1.12) a jump downhill in energy, as well as an uphill jump, requires an activation energy. This results from inclusion of the polaronic effect in equation (1.12). As suggested earlier, the importance of including polaronic effects depends on the amount of their contribution to the activation energy compared to that of the site energy variation.

Although information on site energy variations is not directly available, Bässler reasons that they should be similar to the variations found for excitonic absorption and fluorescence bands, which are also affected by conjugation length variations and presence of defects [11]. These bands are of Gaussian shape. Additionally,

the fact that polarisation energy contribution to the site energy is determined by a large number of internal coordinates that vary randomly also suggests a Gaussian distribution of site energies [11]. The distribution of site energies $\rho(\varepsilon)$ is thus taken as [11]

$$\rho(\varepsilon) = (2\pi\sigma^2)^{-1/2} \exp(-\varepsilon^2/2\sigma^2) \quad (1.14)$$

where σ is the variance. For PPV Gailberger and Bässler claim that spectral diffusion of singlet excitations indicates a variance of 650 to 1000 cm^{-1}, on average ~ 0.1 eV [19]. As shown by Bässler [11], this variance leads to an activation energy of ~ 0.4 eV.

I return now to the question of relative contributions of site energy variation and polaronic effects to the activation energy of PPV. In a similar spirit to that used for estimating the site energy contribution the polaronic contribution may be estimated from the calculated value of the relaxation energy of the singlet exciton. In a first approximation the singlet exciton consists of an electron polaron and a hole polaron that do not overlap greatly in PPV due to the large extent of this exciton. The relaxation energy of the singlet exciton in PPV is calculated to be ~ 0.1–0.2 eV [55]. The contribution of the polaronic effect to the activation energy should be roughly half as large according to equation (1.12). To a first approximation, the polaronic contribution to the activation energy compared to that of the site energy variation can be neglected. The jump rate given by equations (1.11) and (1.13) should therefore be sufficiently accurate for further discussion.

In addition to site energy disorder (also called diagonal disorder) there is disorder due to local variations in the overlap integral, or intersite coupling W_{ij}. This results from randomness in chain orientation and in distance between hopping sites. This type of disorder is also called off-diagonal disorder, ODD. In Bässler's work ODD was incorporated by random assignment of an overlap decay rate to each site independently [11]. The rate was described by a Gaussian attached to each site with variance Σ. This procedure results in a correlation between all hops to or from a given site. A physically more reasonable procedure is to assign a random overlap to each bond W_{ij}. It was shown that attaching a decay rate to each site can lead to an overestimate of the mobility at relatively low fields by orders of magnitude [56]. The reason is an overestimate of the contribution of longer, thus less probable, hops [56]. Nevertheless, in what follows Bässler's procedure is used because it leads to a convenient analytic formula for μ and, as will be seen, ODD is less important than site disorder for the data to be analyzed. The disorder model now set up can be used to obtain some results for mobility variation with temperature and field by the use of Monte Carlo simulations.

2.2.3 Temperature and field variation of mobility

The simulations following Bässler's procedure are briefly discussed. The sites are assumed to be on a cubic lattice, variations in distance between them and relative orientation taken into account by a value Γ_i for each site taken from a Gaussian with variance Σ. The site energies E_i are taken from equation (1.14) and the hopping rate from equations (1.11) and (1.13). For further details see refs. 11 or 56.

Consider now the effect of disorder on the field variation of mobility μ. In the low field limit Ohm's law dictates that μ goes to a constant value and drift velocity $v_d = \mu E$ increases linearly with field. When there is no disorder, with the hopping rate given by equation (1.13) v_d must reach a constant value, or saturate, in the high field limit because, according to (1.13), there is no field dependence for downhill hops. In that limit μ must decrease as $1/E$. When there is site disorder only, $\sigma \neq 0$, beyond the low field region μ increases as E increases because the field helps the carrier overcome barriers in the field direction. In the high field limit, however, saturation of the drift velocity must be reached for the same reason as for $\sigma = 0$, and μ must decrease with increasing field. In between the two extremes of field there is a limited field range in which the increase of μ with field can be fitted by $\ln \mu \propto E^{1/2}$.

When there is no site disorder, the effect of ODD at low fields is to increase μ and v_d over the values for the ordered lattice. This is a familiar result from percolation theory [57]; ODD allows for some more favorable jumps. Apart from the increase in μ, the course of μ vs. E is quite similar to that for the case of no disorder; in the high field limit $\mu \propto 1/E$, corresponding to saturation of v_d. With somewhat higher ODD μ and v_d behave initially in a similar fashion to the lower ODD case, but v_d may not saturate. Rather it reaches a maximum and then decreases, a case of negative differential conductivity. With still larger ODD the initial mobility does not increase and the negative differential conductivity is more pronounced [56].

In the presence of both site energy variation and ODD the mobility may either increase or decrease with the field depending on the relative amounts of the two types of disorder, and the bevavior need not be

monotonic [11,56]. Whichever type dominates it may still be possible to find a limited range of fields over which the relation $\ln \mu \propto E^{1/2}$ is satisfied. For this range Bässler has set up, from the Monte Carlo simulations and experimental data, an approximate formula for the temperature and field dependence of μ [11]:

$$\mu(\hat{\sigma}, \Sigma, E) = \mu_0 \exp\left[-\left(\frac{2\hat{\sigma}}{3}\right)^2\right]$$
$$\times \begin{cases} \exp[C(\hat{\sigma}^2 - \Sigma^2)E^{1/2}] & \Sigma \geqslant 1.5 \\ \exp[C(\hat{\sigma}^2 - 2.25)E^{1/2}] & \Sigma < 1.5 \end{cases}$$
$$(1.15)$$

where $\hat{\sigma} = \sigma/KT$ and $C = 2.9 \times 10^{-4}$ (cm V^{-1})$^{1/2}$. The first exponential, which is dependent on temperature and independent of field, represents the effect of activation energy. Note, however, that μ is not in the Arrhenius form, but is super-Arrhenius, i.e.

$$\mu \propto \exp(-T_0/T)^2 \tag{1.16}$$

where $T_0 = 2\sigma/3k$. Thus equation (1.15) predicts that T_0 is a function of σ only and not of Σ. We have shown, however, that the value of T_0 is also affected by ODD, and that diagonal and off-diagonal disorders produce a synergetic effect on T_0 at high fields [58]. Another shortcoming of equation (1.15), pointed out earlier, is the use of a single parameter Γ_i attached to each site to account for the effects of ODD. As noted earlier, it is nevertheless convenient to use equation (1.15) for a first order understanding of the experimental data.

2.2.4 Comparison with experimental results for PPV

Data on temperature dependence of hole mobility in PPV are available only from Takiguchi et al.. [13]. It was noted earlier that over the temperature range from 285 to 385 K they find the mobility to vary as $\exp(-E_a/kT)$, in contrast with the prediction of equation (1.16). As pointed out by Bässler, however, over a small range in temperature the distinction between the exponent being linear in $1/T$ and quadratic in $1/T$ would be difficult to make [11]. Given a plot of $\ln \mu$ vs. $1/T$, a value of σ can be obtained from the slope at temperature T according to

$$E_a = -k\frac{\partial \ln \mu}{\partial(1/T)} = 2kT_0^2/T = \frac{8}{9}\frac{\sigma^2}{kT} \tag{1.17}$$

With E_a taken from Figure 7 of ref. 13 for the sample converted at 250°C, and the temperature taken as that of the middle of the measurement range, 335°C, σ

equals 0.07 eV. At room temperature then $\hat{\sigma} = 2.7$. This is a large enough value that, if there were no ODD, according to equation (1.16) a considerable increase in μ would be seen. At the highest field for which measurements were taken, 2×10^5 V cm^{-1} [13], μ would be double its low field value. To account for the observation that no increase in μ is seen, according to equation (1.15) Σ must be somewhat greater than 1.5 although still less than $\hat{\sigma}$.

Going through the same procedure for the lower conversion temperature, 200°C, results in $\sigma = 0.085$ eV. Consistent with the larger value of σ the mobility is lower. To account for the finding that μ is independent of E to fields of 2×10^5 V cm^{-1} for this conversion temperature would require that Σ also is larger. Thus the lower conversion temperature led to greater disorder for these samples.

It is significant that for PPPV μ was found to decrease sharply with increasing field, being reduced by an order of magnitude at a field of 3×10^5 V cm^{-1} [19]. Thus in this case $\Sigma > \hat{\sigma}$. Gailberger and Bässler suggest that this is to be expected because the presence of the solubilizing phenyl groups should reduce the degree of crystallinity significantly as compared to PPV.

As indicated in Table 1.2, Karg et al. found that ln μ increased linearly with $E^{1/2}$ over the range of their measurements $\sim 10^5$ to 10^6 V cm^{-1}. The low value of their low field mobility (10^{-7} cm^2 V^{-1} s^{-1}) and the considerable increase they saw in μ with E, both indicate much larger values of σ than th Takiguchi samples. Although their field values are somewhat in doubt because the field was non-uniform (their LED had a Schottky barrier) their results are compared with equation (1.15) to obtain some information on σ and Σ. Fitting their results to the form $\mu \propto \exp(BE^{1/2}/kT)$, Karg et al. obtained $B = 1.6 \times 10^{-4}$ eV$^{1/2}$ cm$^{1/2}$ for room temperature. Equating their coefficient of $E^{1/2}$ to that in equation (1.15) leads to $\hat{\sigma}^2 = 22 + \Sigma^2$ with the minimum value of Σ^2 being 2.25. Corresponding to this the minimum value of σ at room temperature is 0.12 eV, almost twice the value obtained for the Takiguchi samples with the higher conversion temperature.

The σ values of ~ 0.1 eV found in the last section are in agreement with the estimate of Gailberger and Bässler [19] from the data on spectral diffusion of singlet excitations in PPV cited earlier. Also in agreement with a σ value of ~ 0.1 eV is the fact that the calculated variations of the polaron energy level with chain length, from the long chain limit down to two monomers, are of that order of magnitude [59].

It is evident that the choice of precursor polymer and the details of the conversion from the precursor to PPV have a very important effect on the hole mobility and in fact whether the transit is sufficiently non-dispersive to make time-of-flight measurements possible. Takiguchi et al. found that increasing conversion by increasing the heating temperature increased μ, up to some saturation level [13], whereas Hsieh et al. found that increased conversion due to increased heating time decreased μ and also made the sample transit more dispersive [16]. To account for their findings Hsieh et al. suggest that at low conversion mostly di- and tri-PPV units are formed. As the degree of conversion increases, higher PPV oligomers are formed but the distribution of oligomer lengths grows wider. This results in increasing σ and therefore decreasing μ with increasing conversion time. Since polaron energy is lower on the longer oligomers they may also act as traps, causing dispersive transport. Hsieh et al. suggest that the PPV samples used by Takiguchi et al. contained a large number of non-conjugated defects because of the particular precursor used and thus were incompletely converted even at the higher heating temperature [16]. Consistent with this there have been reports in the literature of light emission efficiency increasing with decreasing conversion [60] or with insertion of non-conjugated segments [61]. It is plausible that, if all conjugation segments were the same length, samples with shorter segments would have higher mobility than ones with longer segments. One reason for this, other things being equal, is the overlap of the valence band wavefunctions, thus W_{ij}, is larger for a pair of shorter segments than for a pair of longer segments [45]. It must be remembered also that the Takiguchi samples were oriented, which in itself would enhance the mobility.

It has been reported by a number of investigators that the LEDs still operate at very low temperatures, although voltages higher by an order of magnitude are required [62]. At low temperatures $\hat{\sigma}$ is very large, about several hundred at 4 K, for example. The quantity T_0/T of equation (1.16), or the effective activation energy, is then very large. Nevertheless at high enough fields the potential drop due to the field can overcome the energy difference. The field required can be estimated from equation (1.15) if it is assumed to be still valid at low temperatures. For the large values of $\hat{\sigma}$ involved, the term in Σ can be neglected even at 77 K. The condition that the field be large enough to ensure conduction is then $CE^{1/2} > (2/3)^2$, which leads to $E > 2 \times 10^6$ V cm^{-1}.

In conclusion, the temperature and field dependence of hole mobility in PPV appear consistent with theory based on disorder. The variance of site energy disorder, from the data for two different sets of samples, is found to be ~ 0.1 eV. This value is in agreement with expectations based on the behavior of singlet excitons, notably spectral diffusion. Significantly, the variation of the polaron energy level with conjugation length, one of the contributions to site energy variation, is of this order of magnitude (0.1 to 0.2 eV). Choice of the precursor polymer and the heat treatment for conversion strongly affect the distribution of chain lengths and therefore the mobility. Narrowing the distribution of chain lengths should increase μ, and thus decrease the voltage required for LED operation. One possibility for doing this might be by assembling the film from oligomers of the same length. This is one of the features contributing to the high mobility (close to 0.5 cm^2 V^{-1} s^{-1}) of α-sexithiophene.

Electron, or more accurately negative polaron, mobility in PPV has been estimated to be two orders of magnitude smaller than positive polaron mobility as a result of trapping. The traps have been identified as carbonyl groups introduced photochemically [63] or thermally [64]. These traps can be eliminated by carrying out device processing in the absence of a carbonyl forming environment.

2.3   Transport in *t*-PA—intersoliton hopping?

As discussed in the last section, upon injection at a contact an electron or hole becomes a polaron, which may then carry current by hopping. In *t*-PA the existence of solitons provides a couple of other options for conduction. For reasons that are not clear, many workers in this area have assumed that transport in pristine *t*-PA is by means of solitons.

An isolated soliton cannot hop between chains because it would require that all the bonds to one side or the other be altered. However if a sample had both positive and negative solitons, conduction could occur by electrons jumping from S$^-$ levels to S$^+$ levels. As pointed out by Kivelson, the existence of neutral solitons provides another conduction mechanism [65]. To illustrate this, consider the case of a negatively charged soliton S$^-$ bound to a donor ion. If a neutral soliton, S^0, came close to a nearby donor ion, the electron on the negatively charged soliton could jump to the S^0, causing it to become bound to the nearby donor, and leaving behind the original S$^-$ as an unbound S^0. The conductivity and thermoelectric power arising from this mechanism were calculated by

Kivelson [65] under various approximations and assumptions. Among these was the assumption of isoenergetic hopping, i.e. that the energy level of the electron is the same on every S$^-$ bound to an impurity. (For further discussion of the assumption see Conwell [8] and Ngai and Rendell [66]). It was also assumed, essentially as in equation (1.11), that the average hopping rate v_{ij} could be written as a product of a factor $\rho^2(R)$ depending on R, the distance between the sites, as does W_{ij} of equation (1.11), and another factor $\gamma(T)$, i.e.

$$\bar{v}_{ij} = \rho^2(R)\gamma(T)/N \qquad (1.18)$$

where N is the number of carbons on a chain. $\gamma(T)$, similar to $R(\Delta_{ij})$ of equation (1.11), involves a complicated thermal average of the electron–phonon coupling and the soliton states. With this, Kivelson finds [65]

$$\sigma_{DC} = Ae^2[\gamma(T)/kTN]$$
$$\times (\xi/R_0^2)y_n y_{ch}(y_n + y_{ch})^{-2} \exp(-2BR_0/\xi) \qquad (1.19)$$

where $A = 0.45$, $B = 1.39$, y_n and y_{ch} are the concentrations of neutral and charged solitons, respectively, per C atom; $R_0 \equiv [4(4\pi/3)C_{im}]^{-1/3}$ is the typical separation between impurities; C_{im} is the impurity concentration and ξ is the dimensionally averaged decay length of the soliton wavefunction, $(\xi_\parallel \xi_\perp^2)^{1/3}$.

As discussed in the earlier sections on the theory of hopping, the evaluation of $\gamma(T)$ requires a knowledge of all the phonon wavefunctions. Kivelson circumvented this by making an approximation to the effective electron–phonon coupling, chosen so that it would lead to a value of σ in agreement with experimental values. This led to $\gamma(T) \propto T^{11}$. The theory was compared to σ_{dc} measured in samples of Shirakawa transpolyacetylene, to be abbreviated S-*t*-PA, by Epstein *et al.* [67]. They found that in the range from 300 to ~ 120 K the conductivity decreased sharply with decreasing T, but was not activated. The decrease was fitted to $T^{13.7}$, which was considered to be in reasonable agreement with T^{10} predicted by Kivelson. A quantitative fit to equation (1.19) was not possible because C_{im} and y_{ch} were not known but the values of the parameters required for the theory to agree with experiment appeared reasonable [67].

A better test for the theory is provided by measurement of the complex conductivity σ^*, which was carried out for S-*t*-PA by Epstein *et al.* [67,68]. They performed the analysis of their results on σ_{ac}, defined as $Re(\sigma^*) - \sigma_{dc}$, as is frequently done. Comparing the results with a calculation of σ_{ac} due to intersoliton hopping by Kivelson [65] they concluded that their results for σ_{ac} were also in good agreement with Kivelson's theory [67,68].

Emin and Ngai [5] have reanalyzed the ac conductivity data of Epstein et al.. Using a hopping theory developed for small polarons they were able to choose the parameters in the theory so as to obtain a good fit to the data. They concluded from this that the charge transport involves extremely localized states, i.e. small polarons, strongly coupled to the electronic motion. As discussed in Section 1, there is no reason, however, to believe that small polarons exist in conducting polymers.

There were also experiments that raised questions about, or directly contradicted, Kivelson's theory. Summerfield and Chroboczek [69] found different T dependence and order of magnitude of σ_{dc} than the results cited above. They were able to account for their results with a more orthodox hopping theory based on the EPA (extended pair approximation) and not involving solitons. They also showed experimentally that Kivelson's model is not in agreement with the frequency and T dependence of σ^*. Epstein, citing behavior similar to that seen by Summerfield and Chroboczek in samples with remanent cis-PA inclusions, suggested that their samples might have contained sufficient cis-PA to confine neutral solitons, making Kivelson's theory inapplicable. A similar reason, short conjugation lengths with resultant neutral soliton confinement, was presented [68] to account for the fact that t-PA made by the Durham procedure [70] had much lower σ than predicted by Kivelson's model [71]. It should be mentioned also that, from studies of nuclear spin-lattice relaxation rates for ^1H and ^{13}C in polyacetylene, Scott and Clarke [72] concluded that only a small fraction of the t-PA chains contain solitons.

Additional measurements of σ^* of S-t-PA have been carried out more recently by Ito et al. [73]. Pointing out that the analysis of σ_{ac} [obtained by the subtraction of σ_{dc} from Re(σ^*)] only yields valid results when the hopping processes responsible for σ_{dc} have a single hopping relaxation time [66] (completely unlikely for these materials) Ito et al. performed their analysis on Re(σ^*) directly. They concluded that the measured T dependence could not be the result of isoenergetic hopping as assumed in Kivelson's model, but rather required a model with energy differences between hopping sites.

Ito et al. [73] also carried out measurements of σ_{dc} and σ^* on t-PA made by varying the Shirakawa process (specifically by using aged catalyst and polymerizing by

a non-solvent method) to give much higher σ_{dc}, bulk density and Young's modulus. This material is also highly stretchable, for which reason it is called hs-t-PA. Rather than the fibrillar structure observed for S-t-PA it has a densely packed granular structure. Undoubtedly the hs-t-PA is better ordered than S-t-PA.

Plots of log σ_{dc} vs. $10^3/T$ for a pristine hs-t-PA sample and several lightly doped samples are shown in Figure 1.3. The doped samples contain iodine, the concentration y being defined by their formula, $(CHI_y)_n$. It is seen that σ_{dc} decreases sharply with decreasing T but in no case is simply activated, in agreement with earlier data on S-t-PA. Plots of Re(σ^*) and Re(ε^*) (=Im $\sigma^*/\omega\varepsilon_0$) vs. frequency for one of the samples of Figure 1.3 are shown in Figure 1.4. It is seen that Re(σ^*) is proportional to f^s, with the slope s varying with frequency. Interestingly, when the data of Figure 1.4 are replotted with normalized scales, as log [Re(σ^*)/ σ_{dc}] vs. log [$f\varepsilon_0/\sigma_{dc}$], the points for all temperatures fall on a single curve. Ito et al. [73] stated that this behaviour was found for all the hs-t-PA samples but not for any of their S-t-PA samples.

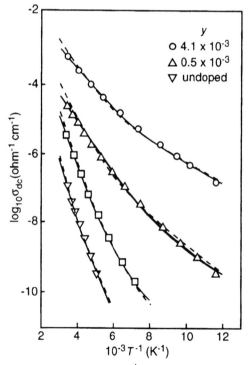

Figure 1.3. Plot of σ_{dc} vs. T^{-1} for a set of hs-PA samples with doping y as indicated. The solid curves represent theoretical fits. (Adapted by permission from K. Ito, Y. Tanabe, K. Akagi and H. Shirakawa, *Phys. Rev.* **B45**, 1246, 1992).

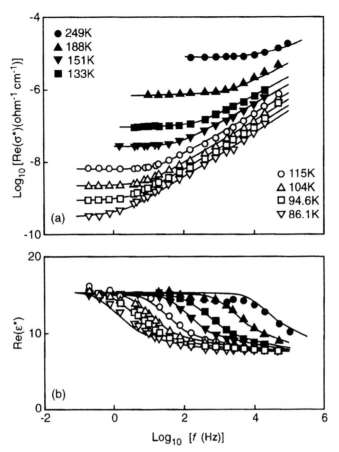

Figure 1.4. The frequency dependence of (a) Re(σ^*) and (b) Re(ε^*) at different temperatures for a lightly-doped *hs*-PA sample ($y = 5.0 \times 10^{-4}$). The solid curves are theoretical fits. (Adapted by permision from K. Ito, Y. Tanaka, K. Akagi and H. Shirakawa, *Phys. Rev.* **B45**, 1246, 1992.)

From the fact that in the normalized plots the data for Re(σ^*) vs. *f* for all of the temperatures fell on the same curve, Ito *et al.* reached the conclusion that the hopping in hs-t-PA, although not fitted by the Kivelson mechanism, is isoenergetic. This conclusion is not uniquely called for by the data. They then showed, much as Emin and Ngai had done earlier for the Epstein data, that the temperature variation of σ_{dc} could be fitted with a hopping rate calculated by Emin [52] for small polaron isoenergetic hopping between nearest neighbor sites on a lattice by means of optical phonons. Using a generalization of the formula to acoustic mode-assisted hopping, because the frequencies of the optical modes are too high for them to be excited at room temperature and below, they concluded that the conductivity is due to acoustic phonon-assisted soliton hopping between randomly located solitons bound to impurities. It is not clear that this conclusion is

required by the data. It was seen in earlier sections that variations in the energy levels of the hopping entity with variance as small as 0.1 eV have major effects on the hopping mobility. It is implausible that the energy levels of solitons bound to impurity levels, or solitons on conjugation segments (which undoubtedly vary in length in hs-t-PA as in any other conducting polymer sample) or polarons in these situations would not have variations of this order of magnitude. It was noted also by Ito *et al.* that the hs-t-PA samples show strongly non-ohmic behavior. The logical source of this behavior in these samples is disorder, as discussed in Section 2.2.3. In terms of the discussion in that section, the smaller non-ohmic effects found in S-t-PA [73] could be due to a closer balance between energetic (diagonal) disorder of the sites and orientational (off-diagonal) disorder in that case, due to the fibrillar nature perhaps.

The only clear conclusion that emerges from the discussion of this section is that transport in *t*-PA is sample dependent, which means disorder is important. Neither theory nor experiment gives us grounds to decide to what extent solitons contribute to transport in *t*-PA, and of course their role may be different in different samples. It is quite possible that, except perhaps for photoconductivity, the contribution of solitons is small and that hopping transport is dominated by polarons in pristine and weakly doped *t*-PA. As noted earlier, based on the fact that theories of small polaron hopping could lead to good agreement with experimental σ_{dc}, Emin and Ngai suggested that the hopping entity is the small polaron. It is noted, however, that it is quite possible that calculations for the large polarons found in conducting polymers give results similar in form to those obtained for small polarons. This is exemplified by the fact that equation (1.8), for the hopping rate of large polarons, is of precisely the form obtained from Holstein's small polaron hopping theory [47]. (This theory formed the basis also for the small polaron hopping theory of Emin [52] used by Emin and Ngai [5].) The only difference between equation (1.8) and the small polaron hopping rate obtained by Holstein is in the definition of Δ_k, the coupling term, as might be expected.

2.4 Drift mobility

In addition to hopping between sites, solitons, polarons and bipolarons may, as discussed in Section 1.2, drift along the chains, maintaining their shape through small motions of the C−Hs. Drift motion in the absence of an electric field has been observed in magnetic resonance experiments [74] and optical experiments [79]. Magnetic resonance techniques, specifically electron spin resonance spectroscopy (ESR), nuclear magnetic resonance (NMR) and dynamic nuclear polarization (DNP), gave extensive data on a spin moving one dimensionally, i.e. along the chains, in undoped *t*-PA [74]. After considerable controversy over the meaning of the data there emerged wide agreement that the moving spin is to be identified with the neutral soliton. Deduction from the experimental results of the magnitude of the diffusion coefficient $D_\parallel$, and its temperature variation was fraught with controversy, however. A major source of the difficulty is that a fraction of the solitons is trapped, that fraction varying with temperature, and the trapping has different effects on different types of magnetic resonance measurements. As pointed out by Clarke and Scott [75], the

crux of the problem of determining $D_\parallel$ lies in determining what fraction of the spins is trapped. We show in Figure 1.5 the data of two experimental groups who determined this fraction with some care. The sizable difference seen between their results could be due to insufficient accuracy in the determination of the fraction trapped, or to genuine differences in trapping between the samples.

The quantity actually measured in the magnetic resonance experiments is not $D_\parallel$, but the pumping frequency at which the NMR signal is enhanced, which corresponds to a frequency component of the motion of the electronic spin [74]. This frequency can be considered as a kind of hopping frequency of the solitons. To obtain the value of $D_\parallel$ in $cm^2 s^{-1}$ this frequency must be multiplied by the square of what would correspond to a hopping distance. Because a soliton can occupy only even sites or only odd sites, the hopping distance would be twice the distance between carbon atoms. The resulting experimental $D_\parallel$ at 300 K, as seen in Figure 1.5, is $2–4 \times 10^{-2}$ $cm^2 s^{-1}$. It is also seen that this value of $D_\parallel$, as well as its temperature dependence, is in reasonable agreement with the results of a theoretical calculation, to be described in Section 2.4.2.

Another value of $D_\parallel$ for *t*-PA was obtained by Vardeny *et al.* [79]. They measured, with a polarized probe laser, the change in absorption parallel, $\Delta\alpha_\parallel$, and perpendicular, $\Delta\alpha_\perp$, to the pump induced by a pump laser. The *t*-PA sample was unoriented, thus macroscopically isotropic. It was found, however, that for a time, approximately nanoseconds, the induced change in absorption $\Delta\alpha_\parallel$ was greater than $\Delta\alpha_\perp$. The induced dichroism was attributed to the absorption of solitons created by the pump pulse in regions where the chains were parallel to the pump polarization. The decay, i.e. the loss of polarization memory, was attributed to the motion of photogenerated solitons along the chains; the distance travelled was sufficient for the chain direction to rotate by $\pi/4$. Averaging over all chain directions with respect to the pump polarization in a planar geometry, they concluded [79] that the data for the decay of induced dichroism gave $D_\parallel = 2.2 \times 10^{-2}$ $cm^2 s^{-1}$ at 300 K, in good agreement with the value cited above. The moving excitations were charged in this case rather than neutral and, as will be discussed in the next section, might have been polarons rather than solitons. It should not be possible to distinguish between polarons and solitons at room temperature in any case because it is predicted that their diffusion rates are comparable [80]. Irrespective of whether the excitations were

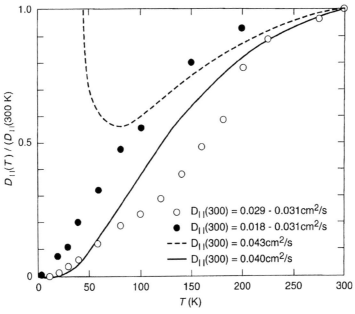

Figure 1.5. Intrachain diffusion coefficient of neutral solitons in *trans*-polyacetylene (normalized to 300 K value) vs. temperature. Closed circles: data of ref. 74; open circles: data of 76; — theoretical results from refs. 77 and 78 for phonon scattering only; – – theoretical results from refs. 77 and 78 for phonon scattering plus barriers of 0.01 eV. (Adapted by permission from S. Jeyadev and E.M. Conwell, *Phys. Rev.* **B36**, 3284, 1987).

solitons or polarons the background of ions was apparently small enough in the pristine samples used that they were not bound as envisioned by Kivelson [65].

Given the Einstein relation, equation (1.1), $\mu_\parallel$ for drift along the chains can be deduced from $D_\parallel$ given above. With the results of Figure 1.5 $\mu_\parallel$ vs. T for solitons is as shown in Figure 1.6. Although there is considerable scatter it is clear that the soliton $\mu_\parallel \simeq 1$ cm^2 V^{-1} s^{-1} from 300 K down to ~ 50 K, below which it decreases rapidly. Polaron μ at room temperature should be comparable [80].

2.4.1 Picosecond photoconductivity

What happens subsequent to the creation of an electron–hole pair by a photon has been much studied since the early days of conducting polymer research. For *t*-PA it was predicted that, when the electron and hole are created on the same chain, distortion of the chain would result in their being transformed into an S^+S^- pair within a few hundred femtoseconds [82]. When the electric vector of the light is perpendicular to the chains the electron and hole could be created on separate chains [83]. In that case, and in the case where an electron–hole pair is separated by the electron or

hole hopping within femtoseconds to a separate chain, distortion of the chains would result in a pair of polarons. These predictions were verified with sub-picosecond experiments by Rothberg *et al.* [84,85]. They found that for the electric vector of the light parallel to the chains, solitons were indeed created, with a quantum yield near unity, in the subpicosecond time frame as predicted. However, they decayed rapidly, victims of geminate recombination, to less than 5% of their original number within ~ 2 ps [85]. For the electric vector perpendicular to the chains, about 60% of the excitations were S^+S^- pairs that decayed rapidly; however the remaining $\sim 40\%$ of the excitations had a lifetime greater than 300 ps and a somewhat different absorption spectrum than the solitons [85]. Rothberg *et al.* identified the $\sim 40\%$ long-lived excitations as polarons [85].

Many papers have been written on how the solitons and polarons whose generation has just been described contribute to photoconductivity. Picosecond photoconductivity perpendicular to the chain is clearly due to polarons; this has been discussed in Sections 2.1.2 and 2.2.1. This section considers picosecond photoconductivity parallel to the chains. On the basis of the mobility values obtained in the last section, parallel photoconductivity in the first picoseconds after excitation should

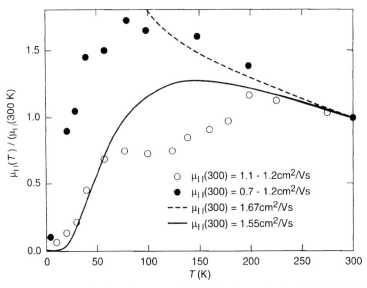

Figure 1.6. Intrachain mobility of charged solitons in *trans*-polyacetylene (normalized to 300 K value) calculated from $D_{\parallel}e/kT$; the $D_{\parallel}$ value is taken from Figure 1.5. Closed circles: $\mu_{\parallel}$ from data of ref. 74; open circles: $\mu_{\parallel}$ from data of ref. 77; - - - - - theoretical results from refs. 78 and 79 for phonon scattering only; ----- theoretical results from refs. 78 and 79 for phonon scattering plus barriers of 0.01 eV. (Adapted by permission from E.M. Conwell and S. Jeyadev, *Mol. Cryst. Liq. Cryst.* **160**, 443, 1988).

represent carrier drift along the chains. Given an average conjugation length of ~ 40 Å in *t*-PA, the average distance a carrier could drift before being stopped by the end of a chain is ~ 20 Å. For $\mu \sim 1$ cm^2 V^{-1} s^{-1}, as discussed in the last section, and a typical electric field of 2×10^4 V cm^{-1}, the average time the carrier could drift is 10 ps. After that time the carrier is either trapped or must hop to continue moving.

Early experiments on *t*-PA had a time resolution of ~ 100 ps and could not separate out the drift motion. Sinclair *et al.*, in the first experiments, found a large photoconductivity (typically ~ 0.3 ohm^{-1} cm^{-1}) that decayed with a lifetime of 300 ps [86]. Reichenbach *et al.* [87] showed by deconvolution of data taken with a resolution of ~ 50 ps [88] that the picosecond photoconductivity could be fitted with a sum of short relaxation time, ~ 10 ps, and a long time, ~ 170 ps. A more accurate result for the short time (4.5 ps) was obtained with an experimental technique that had highly improved temporal resolution [87]. From considerations similar to those above, Reichenbach *et al.* identified this time as the average time for a carrier to drift to the end of a chain [87]. To identify the carriers and determine their mobility Reichenbach *et al.* used the results of Rothberg *et al.*, refs. 84 and 85. For excitation by photons with an electric vector perpendi-

cular to the chains they took the carriers to be polarons, created by 40% of the incident photons. With this result, knowledge of the incident phonon flux, and the total charge flow during the drift time of 4.5 ps, Reichenbach *et al.* obtained $\mu \sim 1.3$ cm^2 V^{-1} s^{-1} with an uncertainty of about a factor two [87]. As will be seen in the next section, the value of polaron mobility calculated by Jeyadev and Conwell, was ~ 3 cm^2 V^{-1} s^{-1} (relatively good agreement) although the calculated value was admittedly more uncertain than that for the soliton mobility [80]. Extension of this analysis to the case of excitation polarized parallel to the chains led to difficulties. According to ref. 85, the parallel excitation photophotogeneration gave only solitons (although generation of a small polaron concentration could not be ruled out) of which less than 5% remained 2 ps after the pump pulse. This would lead to a minimum soliton μ of 10 cm^2 V^{-1} s^{-1}, considerably larger than that seen in other experiments or obtained by calculation. This discrepancy was a major reason for the suggestion that the 40% of the excitations with a long lifetime created by perpendicularly polarized excitation are not free polarons, but polarons bound in P$^+$P$^-$ pairs on adjacent chains by their Coulomb attraction [89]. Such pairs (exciplexes) have been more solidly established for PPV and PT, which also show long-lived excitations and a photoconductivity essentially independent of pump polarization, by the fact that the long-lived excitations are seen only in thin films but not in dilute solutions of these polymers [89]. If the long-lived excitations in t-PA are indeed polaron pairs that do not contribute to photoconductivity, then it is not clear whether the photocurrent is due to solitons or polarons or a mixture.

An interesting characteristic of parallel pump-induced photoconductivity, seen in Figure 1.1, is that it appears essentially constant over a wide temperature range, $300 > T > 10$ K [86]. The temperature independence was taken as evidence that the carriers are 'hot' [86]. When excited by photons with an energy well above the absorption edge the carriers are, as of course occurs in all semiconductors, created with high kinetic energy. This energy can be dissipated by phonon emission. A calculation of the rate of energy loss to acoustic phonons of hot solitons and polarons, using the scattering times to be given in the next section, shows that the thermalization times are fractions of a picosecond [90]. Given a typical time resolution of the experiments of 50–100 ps, it is apparent that hot carriers would not be seen. In any case, as shown in Figure 1.6, μ vs. T obtained from diffusion rate measurements of neutral solitons and from the theory [79,80] shows little change with temperature from 300 K down to 75 K. The rapid decrease in μ with T below 75 K is undoubtedly due to defect scattering or trapping. Of course this is sample dependent and might conceivably not be occurring in the particular samples measured in refs. 86 and 88. However, it has been pointed out that, particularly at low temperatures, the energy of the laser pulses is sufficient to cause considerable sample heating, especially for the parallel pump case [87,91].

Picosecond photoconductivity has been investigated in other conducting polymers, principally PPV and PT. In the latter two cases it was expected that, in contrast to t-PA, the principal excitations photogenerated would be excitons. Of course, if the binding energy of the excitons were $\leqslant kT$ they would not be bound; thus the principal excitations would be free electrons and holes, as in the usual three-dimensional (3D) semiconductors at room temperature and well below. One group of investigators claims this is indeed the case [92]. An argument for their position is that the threshold for photoconductivity in PPV and PT [93] coincides with the absorption edge. There are many arguments against their point of view, however. Theoretical calculations and experiments give the exciton binding energy in PPV as some tenths of an electron volt [94–98]. The

reason for the large binding energy, compared to that of the 3D semiconductors, is the ID nature of the exciton in these polymers, as well as the relatively low dielectric constant. The coincidence of the onset of photoconductivity and the absorption edge could be the result of exciton dissociation, which could occur at a surface, for example. It must be remembered that only $\sim 1\%$ of the photons create photocarriers in both t-PA [86] and PPV [99].

In contrast to expectations, extensive study of the effects of light on PPV has led to the conclusion that only 10–20% of the excitations are excitons, in which the bound electron and hole are on the same chain. The remaining 80–90% are the polaron pairs with the electron and hole on adjacent chains described in the last section [100,101]. Both of these excitations are neutral and cannot, as noted above, contribute to photoconductivity unless dissociated. There is good evidence that polaron pairs are created in considerable numbers also in PT [89,102,103]. Nevertheless the picosecond photoconductivity of PPV and PT is similar in magnitude to that of t-PA and has a similar decay time [99,93]. On the premise that only $\sim 1\%$ of the photons yield photocarriers, Bradley et al. have estimated the drift mobility of the carriers along the chains in PPV as $\sim 8 \times 10^{-2}$ cm^2 V^{-1} s^{-1} [99]. The carriers are polarons because it is not likely that bipolarons, the other possibility, would have had time to form. In fact it has been questioned whether a bipolaron is stable, despite the repulsion of the two like-charge polarons of which it is constituted, without the stabilizing effect of nearby ions of the opposite sign that would be present if the bipolaron resulted from doping [104]. Bipolarons have been invoked to account for the very long persistence (minutes to hours) of photoconductivity in PPV because they would be expected to have particularly low hopping mobility, which would slow the process of recombination [105]. However the slow decay could be attributed to the electrons being trapped by, e.g. the carbonyl traps referred to in Section 2.2.4.

2.4.2 Theory of drift mobility in t-PA

The mobility of solitons and polarons is determined by phonon scattering and by defect scattering or trapping. The former is expected to be more important at high temperatures, the latter at low temperatures. Because optical phonons in t-PA have high energy (~ 0.2 eV) the important phonon scattering at 300 K and below must be due to the lower energy acoustic phonons. In pristine material the soliton and polaron densities are not high and their energies may be taken to have a Maxwell–Boltzmann distribution. In the absence of scattering they have constant wavevectors given by $k = mv/\hbar$ where m is the soliton or polaron mass, introduced in Section 1.1, and v its velocity. It was assumed that the internal state of the soliton or polaron is unchanged by the scattering, which only causes a change in momentum. The soliton or polaron differs from an electron or hole in a 3D conductor, however, in that it has a limiting velocity, a consequence of the fact that its motion entails motion of the atoms on the chain. Calculations gave the limiting velocity of the soliton, denoted c, as $2.7 v_s$, where v_s is the sound velocity [106,107]. The existence of a limiting velocity indicates that the mass is a function of v, approaching infinity as $v \rightarrow c$. Such 'relativistic' behavior of mass with velocity is familiar for solitons obtained for other 1D systems, e.g. from sine-Gordon and Φ^4 field theories [108].

Jeyadev and Conwell [77,78,80] have set up a theoretical treatment of the phonon scattering along lines used for electron or hole scattering by phonons in a 3D semiconductor. Their theory gave results in good agreement with experimental results as seen in Figure 1.5. The change in mass with velocity was incorporated, but is not a big effect for the soliton. We consider first the case of the soliton, assuming its mass is a constant, m_s^0.

The conditions of conservation of energy and momentum lead to the possible values of phonon wavevector q or emission or absorption. In the range of soliton energy where the mass is constant at m_s^0,

$$q = 2(|k| - k_s) = \frac{2m_s^0}{\hbar}(|v| - v_s) \quad \text{emission} \quad (1.20)$$

where

$$k_s = m_s^0 v_s / \hbar \quad (1.21)$$

It is seen from equation (1.20) that phonon emission is not possible unless the soliton velocity is greater than the sound velocity. Solitons with $v < v_s$ may still be scattered by absorbing a phonon. The absorption process may take place either with the q vector of the phonon absorbed in the same direction as k or in the opposite direction to k. Conservation of energy and momentum for these cases leads to

$$q = 2(k_s \pm |k|) = \frac{2m_s^0}{\hbar}(v_s \pm |v|) \quad \text{absorption} \quad (1.22)$$

where the upper sign holds when the velocity after scattering is opposite in direction to the initial velocity,

and the lower sign holds when the velocities are in the same direction. From equations (1.20) and (1.22) it is possible to evaluate the wavelength $\lambda_p = 2\pi/q$ of the phonons important in scattering. The average kinetic energy of a freely moving soliton at the low concentrations we are concerned with is $\frac{1}{2}kT$. Thus the average velocity $\langle v \rangle = (kT/m_s^0)^{1/2}$. With $m_s^0 = 5.45 \, m_e$ [2], for $T = 100$ K $\langle v \rangle = 1.67 \times 10^6$ cm s^{-1}. The velocity of sound in polyacetylene has been measured as 1.5×10^6 cm s^{-1} [109]. Thus at 100 K, $\langle v \rangle$ is already larger than v_s. It is still sufficiently smaller than the limiting soliton velocity $c = 2.7 \, v_s = 4 \times 10^6$ cm s^{-1}, however, to justify taking $m_s = m_s^0$ as was done above in calculating $\langle v \rangle$. With $\langle v \rangle > v_s$ the average soliton will relax by emission, which is more important than absorption, particularly at low temperatures [78]. For $|v| = \langle v \rangle$ at 100 K, λ_p obtained from equation (1.20) is equal to a few hundred lattice constants. To obtain λ_p for a temperature of 300 K it is necessary to take into account the increase in mass of the soliton with $\langle v \rangle$. The result is that the wavelength of the phonon involved in the scattering (emission) process is about 40 lattice constants [77]. Thus the wavelengths of the acoustic phonons responsible for scattering in the range 100–300 K which, as will be seen, is the range in which phonon scattering determines $D_{\parallel}$, are, at the least, several times the length of the soliton.

The fact that the important phonons for scattering have wavelengths many times the soliton length means that the scattering may be thought of as due to the compression or expansion of the soliton by the phonon or sound wave. The compression or expansion changes the energy E_s of the soliton by changing the lattice constant a. The energy change due to a strain $\delta a/a$ is given by

$$\delta E_s = (\partial E_s / \partial a)\delta a = (a \partial E_s / \partial a)(\delta a / a) \quad (1.23)$$

The quantity $(a \partial E_s / \partial a)$ is a deformation potential, to be denoted E_D. To evaluate E_D for this case we use for E_s the creation energy of the soliton, equal to π^{-1} times the energy gap $2\Delta_0$ of t-PA [2]. With the dependence of Δ_0 on a taken from the continuum version of the SSH Hamiltonian [2] we obtain [77]

$$E_D = -(8\alpha a/\pi)(1 + 1/2\lambda)e^{-1/2\lambda} \quad (1.24)$$

where α is the change in the electronic coupling or transfer integral t_0 between neighboring n-orbitals per unit bond length change and λ is the dimensionless electron–phonon coupling, equal to $2\alpha^2/\pi K t_0$, K being the stiffness constant due to the σ-electrons. For the usual parameter values equation (1.24) gives $E_D = 3.3$ eV.

The deformation potential E_D is similar in spirit to the deformation potential introduced by Bardeen and Shockley [110] for long wavelength acoustic phonons in semiconductors. The importance of that deformation potential arises from the proof by Bardeen and Shockley that the matrix element for scattering of electrons from the state $\mathbf{k}$ to $\mathbf{k}'$ by one of these phonons is the matrix element between these states of $E_D(\partial y(x)/\partial x)$, where E_D is the shift in electron energy per unit strain, $\partial y(x)/\partial x$ is the strain and $y(x)$ is the displacement at the point x due to the scattering phonon. A similar theorem may be proved for solitons scattered by long wavelength phonons. The matrix element for scattering from $\mathbf{k}$ to $\mathbf{k}'$ is then

$$M_{k'k} = \langle k', N'(q)|E_D(\partial y(x)/\partial x)/k, N(q)\rangle \qquad (1.25)$$

where $N(q)$ and $N'(q)$ are, respectively, the number of phonons with wavevector q before and after the collision. With $y(x)$ given by [111]

$$y(x) = \sum_q (\hbar/2MNv_s q)(a_q e^{iqx} + a_q^\dagger e^{-iqx}) \qquad (1.26)$$

where M is the mass of a CH group and N is the number of sites on the chain, the absolute square of the matrix element for the case $m_s = m_s^0$ takes the familiar form [80]

$$|M_{k'k}|^2 = (\hbar/2MNv_s)E_D^2 q \begin{cases} N(q)\delta_{k',q\mp k} & \text{absorption} \\ (N(q)+1)\delta_{k',k-q} & \text{emission} \end{cases}$$

$$(1.27)$$

With equation (1.27) and first order perturbation theory the scattering time τ_{tot} including emission and absorption processes is readily obtained. It is found that τ_{tot} is inversely proportional to the soliton mass, the square of the deformation potential and the total number of phonons with wavevectors that satisfy conservation of energy and momentum for the three different scattering processes [77]. The diffusion coefficient is obtained from $\tau_{tot}(k)$ by means of

$$D_\parallel = \langle v(k)^2 \tau_{tot}(k)\rangle \qquad (1.28)$$

where the average is taken over the velocity distribution of the solitons.

Above 100 K the increase of the soliton mass must be taken into account. To do this a phenomenological dispersion relation was introduced [77,80]:

$$E_k = (\hbar^2 c^2 k^2 + (m_s^0)^2 c^4)^{1/2} + \bar{E} \qquad (1.29)$$

where $\bar{E} = E_s - m_s^0 c^2$. In the limit $c \to \infty$, equation (1.29) reduces to the usual dispersion relation, with the kinetic energy given by $\hbar^2 k^2/2m_s^0$. With the dispersion given by equation (1.29), equations (1.20) and (1.22) are no longer valid. The resulting expressions are complicated [80] and must be integrated numerically.

The resulting $D_\parallel$ vs. T is shown as a dashed line in Figure 1.5. It is seem that from ~ 200 K to 300 K $D_\parallel$ is in good agreement with the measured values, perhaps better than could be expected because of the scatter of the experimental values and the uncertainty in some of the parameters involved in both the theoretical and experimental determinations. It is noteworthy that there are no arbitrary parameters involved in this fit. Reasonable agreement persists down to perhaps 100 K; however, below that the calculated $D_\parallel$ increases rapidly, indicating that phonon scattering is ineffective. That is not unexpected; at low temperatures the number of phonons excited is very low, making absorption events rare, and phonon emission is no longer allowed for most solitons because $\langle v \rangle < v_s$. Thus at low temperatures the scattering must be due primarily to defects or traps in the materials, as is usually the case for electrons in semiconductors.

Because a soliton cannot, in principle, leave a chain (except for the improbable event of two nearby solitons jumping simultaneously) any defect on the chain presents a barrier to soliton motion. A barrier of height V_0 would essentially block the motion of solitons with energy less than V_0. To obtain the effect of a barrier on the diffusion rate, the lower limit of the integration (1.28) for $D_\parallel$ could be chosen as V_0 rather than zero. The solid line in Figure 1.5 shows the fit to the experimental data obtained by choosing $V_0 = 0.01$ eV. This value was chosen because of an earlier suggestion by Gibson et al. [112] that remanent cis-polyacetylene linkages in trans-polyacetylene chains present a barrier to soliton motion that they estimated as 0.01 eV. Further, they demonstrated, by measurements of infrared absorption, that there is always a minimum of 5% of cis linkages in trans-polyacetylene obtained by the Shirakawa process. There is thus in principle a sufficient density of these barriers to trap all the neutral solitons. Because the decrease in the theoretical $D_\parallel$ with decreasing T is quite similar to the decreases found experimentally, and the experimental samples were made by the Shirakawa process, it does appear that cis inclusions are responsible for the trapping below ~ 150 K. More work, both theoretical and experimental, on better defined samples, remains to be done in this area, but it appears that the basic processes determining the transport of neutral solitons in pristine polyacetylene are understood. The mobility of charged solitons derived from $D_\parallel$, shown in Figure 1.6, is not as certain; the motion of charged solitons may be affected by defects that do not affect neutral solitons, such as impurity ions. Thus the mobility shown in Figure 1.6 could be applicable only to photogenerated charged solitons in pristine t-PA.

The method of calculation described above for $D_\parallel$ and $\mu_\parallel$ of the solitons should also be valid for polarons because the wavelengths of the scattering phonons are also at least several times the length of the polaron [77,80]. The deformation potential E_D for the polarons is larger than that given in equation (1.24) for the solitons by the factor $\sqrt{2}$ because the creation energy is larger by that factor. Thus E_D for polarons is $3.3\sqrt{2} = 4.7$ eV. The maximum velocity c has not been calculated for polarons but is expected to be about the same as c for solitons. Calculations of $D_\parallel$ and $\mu_\parallel$ for polarons scattered by phonons, along the lines already described for solitons, lead to slightly larger values than for solitons. This is because, although E_D is larger, the low energy mass of the polarons is smaller than that of the solitons by a factor greater than 5 [77,80]. At 300 K $\mu_\parallel \simeq 2$ cm V^{-1} s^{-1}, about the same value as obtained for the soliton. As noted in Section 2.4 these values are in good agreement with experimental results. The smaller mass also results in the effects of mass variation being larger over all, and starting at lower temperatures, for the polarons. Of course the actual $D_\parallel$ and $\mu_\parallel$ may be smaller than those calculated, particularly at low temperatures, due to scattering by defects. As in the case of solitons, the calculations apply only to those polarons present in low concentration that drift along a chain in the absence of impurities or other pinning centers.

Calculations of $D_\parallel$ and $\mu_\parallel$ have not been carried out for other polymers. It is expected that these quantities will be smaller for other polymers because there are more atoms per unit cell, and thus more phonon branches to scatter the polarons. Consistent with this, the value of $\mu_\parallel$, estimated for PPV from the photoconductivity, was an order of magnitude smaller than that calculated for t-PA.

3 TRANSPORT IN DOPED NON-METALLIC SAMPLES

This section is devoted to non-metallic samples. Usually these have doping in the range of ~ 0.5–5% but there are non-metallic samples with higher doping. A wide range of behavior is found, depending on the polymers, the nature of the dopant and the amount. Typically conductivity increases rapidly with temperature, although the temperature dependence is by no means the same for all non-metallic samples. Figure 1.7 shows plots of conductivity vs. temperature for iodine-doped and Fe(Cl)$_3$-doped t-PA. In the latter case the most heavily doped samples are metallic, or close to it, and will be discussed in Section 4.

A prerequisite for understanding transport in doped samples is knowledge of the way the dopants go into the samples. The section starts by summarizing the available information on the structure of doped samples. This is followed by a section discussing the nature of the excitations that result. The data and the hopping theories that have been invoked to explain the transport data on non-metallic samples are then described.

3.1 Distribution of dopant ions in polymer samples

Most of the earlier polyacetylene samples were made by the Shirakawa process [114], which produces *cis*-polyacetylene, c-PA. Microscopic examination showed them to be a tangled mass of thin fibrils 5 to 50 nm in diameter, depending on the synthesis conditions, which fill only one-third of the total volume. The fibrils are highly crystalline, 75 to 90% according to X-ray studies [115], with a transverse coherence length estimated as ~ 8 nm [116]. The coherence length along the chain has been estimated by Fincher *et al.* [117] to have a lower limit of 10 nm, or ~ 80 C–H groups. Conversion of c-PA to t-PA by heating causes some loss of coherence, partly because some *cis*-linkages remain on the chain.

The Durham procedure for making PA, which involves using a precursor polymer [70], if the material is stretched during the transformation process, results in highly oriented, homogeneous, non-fibrillar and fully dense t-PA [118,119]. Coherence lengths are apparently somewhat smaller, however, than for Shirakawa t-PA [120]. More recently modifications of the Shirakawa technique by Naarmann [121] led to much higher conductivity on iodine doping, $\sim 10^5$ ohm^{-1} cm^{-1} rather than $\sim 10^3$ ohm^{-1} cm^{-1}. However, the samples, denoted N-t-PA, still showed fibrillar morphology and X-ray studies found them to be indistinguishable from Shirakawa material [121]. Further changes in the Shirakawa synthesis by Tsukamoto *et al.* [122] showed further improvements, with the metallic behavior of t-PA maintained down to very low temperatures. The

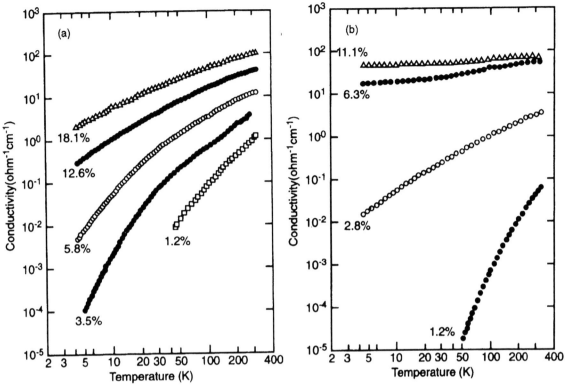

Figure 1.7. Variation of dc conductivity with temperature for (a) iodine-doped and (b) FeCl$_3$-doped t-PA. (Adapted by permission from K. Ehinger and S. Roth, *Phil. Mag.* **B53**, 301, 1986 [113]).

resulting t-PA, denoted v-t-PA, was found to be composed of a densely packed globular structure, more dense than N-t-PA [122]. X-ray studies showed no particular difference from S-t-PA, although v-t-PA is undoubtedly better ordered. Still all PA films, fibrillar or lamellar, are polycrystalline in nature with small polymer crystallites embedded within less ordered regions.

All of the more dense forms of t-PA are harder to dope than S-t-PA. Thus most of the information on the distribution of dopant ions comes from S-t-PA. Commonly used donor ions are Li$^+$, Na$^+$ and K$^+$. Commonly used acceptor ions are AsF$_5^-$, I$_3^-$, ClO$_4^-$ and FeCl$_4^-$. The dopant ions go into the voids between the chains. Typically it has been found that, beyond a dopant concentration of the order of 0.3%, doping is non-uniform, with a mixture of very lightly doped regions and heavily doped regions with a more or less regular arrangement (coherence lengths are never more than ~ 100 Å) of impurity ions [123]. For alkali metal doping the ions are arranged in channels parallel to the polymer chain axes. Small alkali metals (Li[123], Na in

PA [124], Na or K in PPV [125]) form structures in which each channel of ions is surrounded by three chains in the doped polymer. Larger alkali metals, such as potassium in PA, result in square lattices in which each channel is surrounded by four chains [127–129]. Most of the references suggest highly symmetric structures. A number of different stoichiometric phases were found. These phases could be described by the formula $[(C_nH_n)_mM]_x$, where m denotes the number of polymer chains per column of alkali metal ions (M) and n the number of C–H units between adjacent ions in the column. Experimentally it was found that $3 \leqslant n \leqslant 5$ and $2 \leqslant m \leqslant 4$ when M = Na$^+$ or K$^+$ [123,127]. The first ordered phase for sodium-doping, corresponding to $n = 5$, $m = 3$, has a concentration $y = 1/(5 \times 3) = 0.0667$. This concentration is significant because the rapid increase of susceptibility with doping, to be discussed in Section 4, occurs at $y \simeq 0.06$. More recent work on potassium doped PA [128] makes a case for lower symmetry than assumed in ref. 123 and other work on the structure. The details of the asymmetry will strongly affect interchain interactions, overlap integrals etc., but are not yet definitive.

Iodine doping results in more disordered PA. According to Winokur *et al.* [129] iodine doping appears to proceed through a two step process with the dopant initially forming a structure in which there are polyiodide channels, each surrounded by six PA chains. Subsequent doping requires dopant insertion into the host matrix a layer at a time [129]. Heavily doped material consists of layers of PA, layers in which every other PA chain is replaced by a polyiodide column and layers of all polyiodide columns [129,130]. The columns consist of mixtures of I_3^- and I_5^-. There is considerable evidence that iodine doping is non-uniform, particularly at low doping where it is more concentrated in the surface region [131].

For regions of low doping, perhaps less than a few tenths of a per cent, no definitive structure is reported even for relatively ordered polymers. It is not likely that doping is completely random, however, even in such regions, because the size of most dopants, particularly acceptors, is larger than the voids in the polymers; the opening of a channel by one large ion is likely to result in other ions going into that channel. This will probably hold also for highly disordered polymers.

3.2   Nature of excitations in doped non-metallic samples

Determination of the nature of the excitations resulting from doping has relied largely on magnetic susceptibility, χ, data. In interpreting these data it has been usual to assume that, after subtraction of the core susceptibility, χ is a sum of a Curie term $\chi_c = C/T$, due to localized spins, and a temperature independent, so called Pauli term, χ_p, due presumably to non-localized electrons or holes.

For materials in this doping range χ has generally been found to be small, particularly for alkali metal-doped *t*-PA with a few per cent doping [132,133]. On the basis of the small χ for *t*-PA it is concluded that the electrons or holes donated by the doping ions are largely in charged soliton states. In polymers with non-degenerate ground states there are, of course, no solitons. It is concluded for this case that the predominate excitations are singlet bipolarons. A Pauli component can arise from non-uniform doping producing metallic islands within the polymer. This could be the case, for example, for iodine-doped *t*-PA in this doping range [131]. However, there are cases where the so-called Pauli susceptibility increases with T

[131]. Also, there are cases where an apparent Pauli susceptibility is found in highly disordered material where extended wavefunctions are not likely. A clearcut case of this kind has been reported by Chauvet *et al.* [134]. Studying electrochemically grown poly-polypyrrole films heavily doped with large polyanions (highly disordered material) they found χ to deviate from $1/T$ variation for $T > 100$ K, growing progressively larger than χ_c as the temperature increased. Also, they found low spin susceptibility and very low spin to charge ratios ($\sim 1/400$). These features are surprising because about half of the chains or conjugation lengths should have an odd number of excess electrons or holes, and thus at least one polaron and one resulting spin. This should give rise to a large number of spins because conjugation lengths are typically five to 10 monomers [135]. Chauvet *et al.* suggest that the small χ indicates an efficient pairing of polarons between different chains [134]. They call the resulting spinless species transverse or interchain bipolarons. Bussac and Zuppiroli have shown that these bipolarons are stabilized by the doping ions, for reasonable values of the attractive potential of an ion, the on site repulsion U for a second electron and the lattice deformation energy [136]. Earlier work by other authors showed that it is likely that the wavefunction of a polaron created by an impurity ion is spread over the chains surrounding the ion [137]. Some possible configurations for such transverse polarons and bipolarons are shown schematically in Figure 1.8 for the randomly distributed dopant ions. It is seen that the bipolarons may connect two or three chains. The location of the doping ions essentially determines the wavefunctions of the polarons making up the bipolaron.

Bussac and Zuppiroli also suggest that the 'Pauli susceptibility' of the disordered polypyrrole samples is due to thermal activation of the singlet bipolarons into their triplet state [138]. It is usually assumed that the energy required to effect this is much greater than kT. They find that although the singlet state is much lower in energy when the two polarons (now assumed on the same chain) are within a few polaron lengths of each other (the polaron length is ~ 4 pyrrole monomers [139], at larger distances the attractive potential of the dopants and the Coulomb repulsion between the polarons stabilize a weakly bound bipolaron where the triplet energy is $\sim kT$ above the singlet. Assuming a random distribution of dopants, Bussac and Zuppiroli have calculated a spin susceptibility vs. T that is similar to experimental data for disordered PPy and disordered PANI [138].

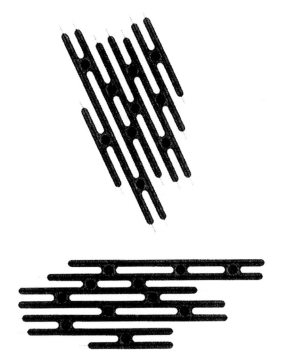

Figure 1.8. Schematic representation of polaronic filamentary clusters in a conducting polymer. The dopant centers are represented by black dots. Various possible configurations of a bipolaron are shown. (Adapted by permission from L. Zuppiroli, M.N. Bussac, S. Paschen, O. Chauret and L. Forro, *Phys. Rev.* **B50**, 5196, 1994.)

The new excitations—the transverse polarons and bipolarons and weakly bound bipolarons—discussed in refs. 134–138 clearly have important effects on transport as well as magnetic properties. These will be discussed at the end of the next section. Although refs. 134–138 claim only that these excitations occur in highly disordered PPy and PANI, it seems likely that they occur to some extent in all but the most ordered conducting polymer samples.

3.3 Hopping in moderately doped *t*-PA

In general, at any given T, conductivity of *t*-PA with a given dopant increases monotonically with the percentage of doping y, as is illustrated in Figure 1.7. Thermopower, on the other hand, decreases monotonically with increasing y, as shown in Figure 1.9.

As noted in the last section, the small spin susceptibility of *t*-PA samples doped in the range ~ 0.5 to $\sim 5\%$ has been taken to indicate that the

electrons and holes donated by the doping ions are essentially all in charged soliton states. Even if this were true, it need not mean that conduction does not involve polarons. It is to be expected that carriers enter the polymer one at a time, which means that a large percentage become polarons. It is straightforward to visualize that in an electric field, conduction takes place by an injected polaron entering one end of a chain and emerging at the other end. The polaron is able to go through solitons. If it were to be trapped, particularly in the lightly doped regions, transport could proceed by the type of hopping discussed in the sections on transport in pristine polymers. Nevertheless, almost all theories of transport in this doping range have assumed that it involves only solitons. According to a couple of theories transport takes place through drifting solitons [141], although no suggestion is made as to how the solitons overcome the strong pinning to the donor or acceptor ions. Chance *et al.* have suggested that interchain transport takes place by the simultaneous jump of two solitons [142], but this is an unlikely event.

Only theories based on hopping have led to any significant agreement with experimental results. The most extensively used theory is that of variable range hopping. This theory applies to systems with a large number of localized states occupying a small range of energies that includes the Fermi energy. In describing the transport it is useful to distinguish between two temperature ranges: (1) high T, where kT is large enough so that an electron or hole can usually find a phonon of sufficient energy to jump to its nearest neighbor; (2) low T, where kT is small enough so that the electron or hole will preferentially jump to a further site if that jump requires less energy. The latter process is called variable range hopping (VRH) because the average hopping distance, or range R, increases as T decreases. A simple calculation, making use of the assumption that the density of states $N(\varepsilon)$ is constant, N_0, over the range kT at the Fermi energy, ε_F, results in the VRH conductivity [143]

$$\sigma = 2(9\xi N_0/8\pi kT)^{1/2} v_{\text{ph}} \exp[-(T_0/T)^{1/4}] \quad (1.30a)$$

where

$$T_0 = 16/kN_0\xi^3 \quad (1.30b)$$

Here ξ is the distance from the site for the electron wavefunction to decay to $1/e$ of its value and v_{ph} is of the order of a phonon frequency. The significant feature of σ is its dependence on $\exp(-T_0/T)^{1/4}$. When the arbitrary assumption that the density of states is constant around ε_F is dropped, σ no longer varies with

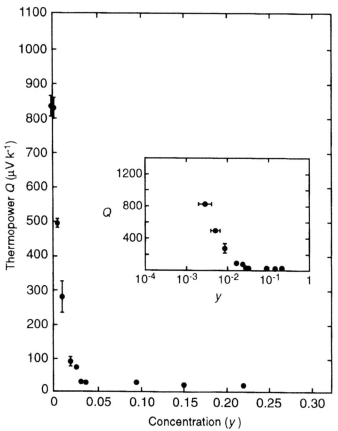

Figure 1.9. Thermopower of iodine-doped *t*-PA at room temperature as a function of iodine concentratin *y*. The inset shows the data with *y* on a log scale; the value for undoped *t*-PA is indicated with an arrow. (Adapted from Y.W. Park, A. Denenstein, C.K. Chiang, A.J. Heeger and A.G. MacDiarmid, *Solid State Commun.* **29**, 747, 1979, with kind permission from Elsevier Science Ltd, The Boulevard, Langford Lane, Kidlington, OX5 1GB, UK [140]).

T as $\exp[-(T_o/T)^{1/4}]$. For quadratic dependence of $N(\varepsilon)$ on ε, for example, σ varies with *T* as $T \exp[-(T_0'/T)^{1/2}]$, where [144]

$$T_o' = 2^{7/3}(18/\pi N_2 k^3 \xi^3)^{1/3} \qquad (1.31)$$

The thermoelectric power for variable range hopping is given by Mott and Davis [143] as

$$Q = (k^2/2e)(T_0 T)^{1/2}(\mathrm{d}\ln N(\varepsilon)/\mathrm{d}\varepsilon)\varepsilon_F \qquad (1.32)$$

A systematic study including structural, optical, magnetic, dc conductivity and thermopower measurements over a wide temperature range was carried out by Epstein *et al.* [145] for iodine-doped samples. Because most of the iodine is present as I_3^- it is convenient to characterize the doped material as $[CH(I_3)_y]x$. The samples studied by Epstein *et al.* had $y \simeq 0.017, 0.033,$ 0.042 and 0.048. Infrared absorption for the latter three

samples showed neither the interband transition at ~ 1.4 eV nor a strong midgap absorption. Instead there was a single intense absorption at 3000 to 3500 cm^{-1}, plus the 900 and 1370 cm^{-1} infrared absorption of vibration (IRAV) peaks associated with the solitons. The intensity of the latter two peaks was found to increase linearly with *y*. This fact, coupled with the small magnetic susceptibility, indicates that the charge transmitted from the iodine to the polymer chains in the indicated range of *y* goes into soliton states. Presumably the soliton band is quite wide, due both to overlap at the large soliton concentrations involved and to disorder resulting from the large amount of iodine. The 300 to 350 cm^{-1} peak presumably results from transitions between the soliton band and the conduction and valence bands.

The temperature dependence of σ in the range 10 to 300 K was found by Epstein *et al.* [145] to be well fitted by $T^{1/2} \exp[(T_o/T)^{1/4}]$ predicted for variable range hopping with constant $N(\varepsilon_F)$. T_o, taken as a parameter, was found to be $\sim 10^6$ K for samples with $y = 0.017$ and 0.333, and $\sim 10^5$ K for $y = 0.042$ and 0.048. As shown in equation (1.30b), T_o depends on the density of states at ε_F. N_o was determined from the temperature-independent part of the susceptibility χ_p by using the free electron equation

$$\chi_p = 2\mu_B^2 N_0 \qquad (1.33)$$

where μ_B is the Bohr magneton. The resulting N_o values were in the range 7 to 35×10^{20} states eV^{-1} cm^{-3}. Combined with the T_o values required for a fit, these N_o values led to ξ values, from equation (1.30b), of 0.5 ± 0.2 nm. These values for ξ are in reasonable agreement with $\xi = (\xi_\parallel \xi_\perp^2)^{1/3} = 0.36$ nm calculated by Kivelson [65]. Having determined N_o and ξ in this way, and taking $v_{ph} \sim 3.6 \times 10^{13}$ s^{-1}, an optical phonon frequency, Epstein et al. calculated the dc σ from equation (1.30a). The resulting σ was too small by a factor of 10^2 to 10^3; this was found also for VRH in amorphous semiconductors [143]. Values of σ in good agreement with experimental results could be obtained by replacing v_{ph} with $v_{ph} \exp[2R/\xi]$ according to a suggestion of Colson and Nagels [146]. It should be noted that somewhat different temperature-dependence for samples doped with iodine in this concentration range has been reported by others, e.g. behavior as $\exp-[T_o/T]^{1/4}$ at low temperatures only and falling below this at high temperatures [113]. Also it has been reported that the temperature-dependence of similar samples is fitted better by $\exp[-(T_o^2/T)^{1/3}]$ [147].

For the less heavily doped samples the temperature-dependence of the thermopower measured by Epstein et al. [145] is in agreement with the prediction of Q for VRH [equation (1.32)] provided d $\log[N(\varepsilon)]/d\varepsilon$ is independent of T. The value of d $\log[N(\varepsilon)]/d\varepsilon$ required for a quantitative fit to the data was 0.4 eV^{-1}. This was considered to represent slow variation of $N(\varepsilon)$ with ε, as required for the validity of the $\exp[-(T_o/T)^{1/4}]$ temperature dependence for σ, equation (1.30a). For the more heavily doped samples the temperature-variation of Q was not in accord with equation (1.32), Q increasing linearly with T for $y = 0.042$ and having a dependence between $T^{1/2}$ and T for $y = 0.048$. Since linear dependence of Q on T is characteristic of metallic behavior, the behavior of the more heavily doped samples was taken to be consistent with their being closer to the metallic state.

Although, according to the discussion above, reasonably good agreement has been demonstrated between experimental data for iodine-doped t-PA and VRH theory, it has been objected that the theory is not valid over a good part of the temperature range (10 to 300 K) in which it has been applied. Ehinger et al. [148] point out that the validity of the theory requires that, for T_o given by equation (1.30b), $(T_o/2T)^{1/4} > 10$. For the lowest values of T_o found by Epstein et al. [145], 3.8 and 4.6×10^5 for the more heavily doped samples, the theory is then not valid above about 20 K. Indeed, particularly for the thermopower, it was seen that agreement with VRH theory is not good for these samples. Even for the lowest doped sample, $(y = 0.017)$ the condition for validity is not met above 95 K. Ehinger et al. [148] have fitted the conductivity data that they obtained for iodine-doped t-PA samples with a more general hopping model based on the extended pair approximation. $N(\varepsilon_F)$ was not taken from the measured χ_p according to equation (1.33), on the grounds that it is not clear how well a free electron model works for electrons bound in soliton states, but was treated as a parameter. A good fit was obtained by Ehinger et al. [148] to the experimental σ vs. T for a density of states that varies considerably with ε around ε_F and is concentration dependent, as might reasonably be expected. It is clear that the t-PA data in this doping range can be fitted, more or less, with different models and parameters.

The temperature dependence of the conductivity of AsF$_5$-doped t-PA with $y = 0.005$ was found also to be fitted well by $\exp[-(T_o/T)^{1/4}]$ [149]. However, T_o was only $\sim 2 \times 10^4$ K, which makes the use of the VRH theory questionable.

Millimeter wave and far-infrared conductivity of iodine-doped polyacetylene were investigated by Genzel et al. [150]. They found that $\sigma(\omega)$ for samples with 1.1% of I$_3$ or AsF$_5$ decreased as the frequency went from 10 to $\sim 10^2$ GHz, after which it increased with further increase in ω. Because $\sigma(\omega)$ must increase with ω for hopping conduction, Genzel et al. proposed that the decrease in $\sigma(\omega)$ with ω arises from conduction by free carriers, holes in this case. Above $\sim 10^2$ GHz this contribution to $\sigma(\omega)$ has apparently decreased to the extent that hopping again dominates. Comparing their experimental results with a simple Drude theory, Genzel et al. found that at 300 K the free carrier concentration is $\sim 10^{15}$ cm^{-3} and the mobility μ is between 4×10^3 and 1.5×10^4 cm^2 V^{-1} s^{-1}. The mobility values are much higher than could reasonably be expected, casting doubt on the validity of the analysis.

For ClO$_4^-$-doped t-PA χ_p, the 'Pauli' susceptibility, is considerably lower for $y < 0.04$ than that for I$_3$-

doped t-PA, and increases with doping rather than occupying a plateau as in the case of I_3^-. Also χ_p for ClO_4^- doping is found to vary considerably with T as well as with doping. These features were taken to indicate an energy-dependent density of states. The simplest energy-dependence would be

$$N(\varepsilon) = N_o + (N_2/2)\varepsilon^2 \qquad (1.34)$$

On the assumption of free electron behavior of χ_p, (1.34) leads to

$$\chi_p = 2\mu_B^2[N_o + (\pi^2/6)(kT)^2 N_2] \qquad (1.35)$$

N_o and N_2 may be determined as functions of y by comparing (1.35) with χ_p vs. T for different y values. Detailed analysis of σ_{dc} vs. T was carried out by Epstein et al. [151] for only one ClO_4^- doped sample, with $y = 0.015$. They noted that σ_{dc} vs. T could be fitted with $\exp[-(T_o/T)^{1/4}]$ if, consistent with that temperature dependence, N_2 was taken equal to zero and the value of N_o obtained for χ_p by equation (1.33); however the resulting N_o and T_o led to σ_{dc} values five orders of magnitude too small. Also the experimental Q for this sample was constant for $T > 150$ K, rather than being proportional to $T^{1/2}$ as predicted by equation (1.32) for a constant density of states. However σ_{dc} fitted well with the temperature dependence predicted for $N(\varepsilon) \propto \varepsilon^2$, $\exp[-(T_o'/T)^{1/4}]$, from ~ 70 to 300 K. The value of ξ obtained with this fit from was 1.4 nm. This is several times the size expected for a soliton. However, it appears to be in agreement with the structural studies of ClO_4^--doped samples which show clusters of four ClO_4^- units between layers of t-PA chains [152]. With the values of T_o', N_2 and ξ deduced as described, Epstein et al. [151] found that the calculated σ_{dc} was in reasonable agreement with the experimental value.

X-ray and electrochemical studies on alkali metal-doped t-PA indicate that the ordered phases for these dopants are more extensive than those found for other dopants. For sodium-doped samples, as noted earlier, in the concentration range up to $\sim 6\%$ sodium, the ordered phase has doping of 6.67% while the surrounding disordered phase has doping of $\sim 0.4\%$. The regular arrangement of sodium ions results in a regular arrangement of solitons, or a soliton lattice in the ordered phase [127]. Consistent with this is the fact that the number of measured spins is three orders of magnitude smaller than the number of electrons donated [153]. In fact the susceptibility is so small that the use of the relations (1.33) or (1.35) leads to a number of neutral solitons, or empty states into which the electrons can hop, so small that it is not possible to

account for the observed σ of sodium-doped t-PA with VRH [132,154]. This casts serious doubt on the VRH model described above for iodine-doped t-PA [145] and ClO_4^--doped t-PA [151]. Since the iodine-doped and ClO_4^--doped t-PA are more disordered than the sodium-doped t-PA it is in fact possible that χ_p, for the two former cases, represents polarons rather than the neutral solitons required as hopping sites for electrons.

Another serious shortcoming of the considerations described above for transport in the moderate doping range is that they ignore the inhomogeneity of the doping. It is to be expected that the highly doped, ordered regions have a different conduction mechanism and higher conductivity than the poorly doped, disordered regions. Whatever the conduction mechanism within the ordered regions, the measured resistance must be dominated by the poorly doped regions. Undoubtedly conductivity in the latter regions takes place by hopping. Shacklette and Toth have applied VRH theory to the poorly doped regions [127]. With two phase material the average hopping distance R should reflect only the average spacing of sites in the poorly doped region and therefore be independent of y, at least until the heavily doped regions come close to filling the sample. This would appear to lead to σ being essentially linearly dependent on y because the main effect of doping would be to decrease the volume of poorly doped material. In actuality, σ increases much more rapidly with y than linearly. Shacklette and Toth were able to duplicate the rapid variation of σ with y by assuming that $R \propto y^{-0.33}$ as though the dopant were uniformly spread through the sample, and the resulting charged soliton states could somehow accept another electron. It is hard to justify these assumptions.

Little effort has been expended in consideration of the mechanism for transport in the heavily doped ordered regions. One suggestion is that this transport is due to free carriers. Calculations for a soliton lattice at $T = 0$ give, for 6.6% concentration, the Fermi energy ε_F 0.57 eV above midgap and a gap of 1.56 eV between the valence and conduction bands, compared with 1.4 eV in undoped material [155]. The larger gap results from the removal of states from the conduction and valence bands to form the soliton states. This calculation neglects the ions, but the fact that they are in a regular arrangement, and that each soliton is under the influence of many ions, may make it reasonable to think of them as a smeared-out background of positive charge. For $T \neq 0$ the Fermi energy would be expected to be somewhat higher than the $T = 0$ value; calculations by the method of Conwell [156] for a 6.67% concentration at 300 K lead to $\varepsilon_F = 0.64$ eV above midgap. With this value there is a conduction electron concentration of 10^{18} cm^{-3} in the heavily doped phase. The mobility due to phonon scattering for electrons in the conduction band has been calculated as 600 cm^2 V^{-1} s^{-1} [157], which gives $\sigma = 10^2$ Ω^{-1} cm^{-1}, well above that measured for sodium-doped t-PA in this doping range. As discussed earlier for iodine-doped samples, conduction due to such electrons might be seen in ac fields although it could not be seen directly in dc fields due to the confinement of the electrons by a barrier.

For the relatively undoped regions, with sodium concentration of $\sim 0.4\%$, ε_F would be lower than it is in the highly doped regions. The requirement that ε_F be the same throughout the sample results in a barrier between adjacent regions of high and low doping. Calculations using the results of Conwell [156] lead to a barrier height of 0.25 eV at 300 K [158]. In the presence of an electric field the barrier for electrons between the two regions would be decreased and a net number of electrons could go over it into a low doped region. Those that do would undoubtedly drop rapidly into polaron levels. The resulting polarons would be available to carry current.

The polaron levels, are calculated to be 0.2 eV below the band edge, at least in lightly doped material [159]. This suggests that conduction electrons from a heavily doped region could also tunnel into polaron levels, and then also carry current as polarons. This mechanism would lead to σ increasing linearly with y, contrary to experimental results, unless the new ordered regions resulting from an increase in y, rather than being distributed randomly, are created close to the original ones. This is plausible and could lead to a lowering of the barriers between the highly doped and less doped regions.

3.4 Hopping in moderately doped non-degenerate ground state polymers

A great deal of work has been done on transport in polyaniline (PANI) and its derivatives. This polymer differs from the others that have been discussed in that doping is usually accomplished by varying the number of protons on the polymer chain (specifically adding protons to the $-N=$ sites, while the number of electrons is held constant) [160]. The protonation level is characterized by $x = [H^+]/[N]$, the maximum possible being 0.50 for the emeraldine form of PANI, since half the Ns already have protons in the undoped state. (See Table 1.1.) In early work with unstretched and

presumably quite disordered samples it was found that in the highly doped state the polymer exhibits a number of metallic properties, i.e. a temperature-independent component of χ [160], a linear relation between thermopower and temperature [161] and a free carrier absorption typical of a metal [162]. Nevertheless the conductivity of these specimens was quite low, with a maximum of ~ 1 ohm^{-1} cm^{-1} at full protonation. Conductivity increased with temperature as has been seen for the various hopping models discussed [161]. It was proposed that phase segregation of doped and undoped regions results in PANI being a granular metal [160,162]. Agreeing with the granular metal picture, Mizoguchi et al., on the basis of NMR and ESR experiments, suggested that a 'grain' or conducting island consisted of a single conducting chain [163]. X-ray structural studies, however, demonstrated that the Pauli susceptibility was observed only when 3D crystalline regions were formed [164]. The average size of the islands increases with doping, reaching values of ~ 50 Å in parallel and perpendicular directions for full doping ($x \sim 0.5$) [164].

Conductivity variation with temperature was best fitted, for doped samples with $0.1 < x \leqslant 0.5$, by $\sigma = \sigma_0 \exp[-(T_0/T)^{1/2}]$, with T_0 decreasing with increasing χ [161]. It was suggested that the conductivity is determined by charging-energy-limited tunneling between metallic particles, in accordance with the theory of refs. 165 and 166. Later work on microwave transport in protonated emeraldine necessitated modification of this picture [167]. The microwave dielectric constant K was found to increase with protonation and with temperature, this increase being more rapid for more heavily protonated samples. On a model for the microwave dielectric constant involving ID motion of the carriers on interrupted strands of metallic wire [168], the value of κ required the carriers to be quite localized at low temperatures [167]. The picture of protonated emeraldine was then modified to include barriers within the metallic islands, which were then described as 'textured metallic islands' [167]. There is no doubt that doped PANI has islands of greater conductivity than the surrounding material but it is not clear that they are metallic. In any case, according to the work of Bussac and Zuppiroli discussed in Section 3.2, a ID model may not be appropriate for calculating K.

For more lightly protonated emeraldine, as x increases from 0 to 0.08 the dc conductivity varies from 10^{-10} to 10^{-6} ohm^{-1} cm^{-1}. It varies with temperature as $\exp[-(T_0/T)^{1/4}]$ with T_0 decreasing with increasing x [169]. Study of this material led to the conclusion that the temperature dependence does not fit a Kivelson-type mechanism [65] for hopping between polaron and bipolaron sites. It was proposed that there are a few bipolaron sites (with two holes) among the polaron sites (one hole) and σ is determined by holes hopping between them in the primarily unprotonated regions of the polymers [169].

A quite different model was invoked by Wang et al. [170] to explain σ vs. T and other data obtained on the protonated methyl ring-substituted derivative of PANI, poly-o-toluidine, abbreviated POT. As in the more heavily protonated PANI samples [161], σ_{dc} was found to vary with T as exp $[-(T_o/T)^{1/2}]$, with $T_o \sim 3 \times 10^4$ K, much larger than 6×10^3 K found typically for PANI. The hopping model found to fit σ_{dc}, thermopower, dielectric and magnetic data for POT was quasi-1D VRH [171,172] rather than charging energy-limited tunneling, which had been suggested to explain the PANI data [161]. Addition of the methyl groups results in a greater interchain separation and also decreased interchain coherence because there is randomness in the location of the methyl groups [170]. These effects decrease the interchain coupling in POT as compared with PANI. For weak interchain coupling the carriers are localized with the localization length perpendicular to the chains, b, equal to the interchain separation and the localization length along the chains $4l_i$, where l_i is the mean free path along the chains [170]. The calculated interchain hopping conductivity perpendicular to the chains is [171]

$$\sigma_\perp = \frac{e^2 N(\varepsilon_F) b^2 v_{ph}}{A} (t_\perp \tau_i / \hbar)^2 \exp[-(T_o/T)^{1/2}] \quad (1.36)$$

where v_{ph} is an attempt frequency, $t_\perp$ the interchain coupling, τ_i the mean free scattering time along the chain, A the average cross-sectional area of the chain and

$$kT_o = 4/N(\varepsilon_F)l_i z \quad (1.37)$$

z being the number of nearest neighbor chains. For the component of interchain hopping conductivity parallel to the chain direction, b is replaced by the most probable hopping distance to give [171]

$$\sigma_\parallel = \frac{4e^2 l_i v_{ph}}{kTzA} (t_\perp \tau_i / \hbar)^2 \exp[-(T_o/T)^{1/2}] \quad (1.38)$$

For weak interchain coupling ($t_\perp \tau_i / \hbar \ll 1$), the interchain conductivity [the sum of (1.36) and (1.38)] is much smaller than intrachain conductivity. Wang et al. state that they expect the observed temperature dependence of the conductivity to be determined by interchain hopping [170]. In agreement with this they found the

value of l_i obtained from the measured T_o and $N(\varepsilon_F)$ (determined from χ_p) to be of the same order of magnitude as the values they deduced from analysis of electron paramagnetic resonance spectroscopy (EPR) data and their microwave dielectric constant measurements. Also their measured thermopower varied with temperature as $Q_o + Q_1/T$, where Q_o and Q_1 are constants, in agreement with theoretical expectations for a sum of interchain and intrachain hopping contributions. Surprisingly, they found an electric field dependence of σ for the protonated POT that agrees approximately with the Poole–Frenkel formula, equation (1.2) provided the change in σ is due to a change in μ rather than the carrier concentration. However the fields required for the change were a couple of orders of magnitude lower than those required for the change in μ, as described in Section 2.1.

Despite some qualitative agreements found between quasi-1D VRH and the experimental data on POT, it is difficult to believe that the theory is applicable, i.e. that the rate limiting step in the transport is interchain hopping. Although this was not documented specifically for doped POT, it is to be expected that, just as in PANI [164], there are well protonated regions or conducting islands, separated by non-protonated or lightly protonated regions. The rate limiting step is, reasonably, hopping between the conducting islands. Further, for the quasi-1D VRH theory to be applicable to interchain transitions in doped polymers it would, at least, have to be modified to allow for the higher value of $t_\perp$ where a doping ion connects two chains. If the suggestion of Bussac and Zuppiroli concerning the formation of transverse bipolarons [136] is correct, interchain hopping in doped polymers is not at all described by the theory of refs. 171 and 172.

Failure to take into account the non-uniform distribution of dopants makes the applicability of almost all of the theoretical treatments of hopping used so far to account for transport in doped polymers very doubtful. One of the few treatments of the hopping conductivity that does take the non-uniformity into account is that of Zuppiroli et al. [173]. Their treatment also does not assume that the conducting islands are metallic. That assumption may be true for highly ordered material, as shown by Wang et al. [174] for oriented (stretched) highly doped PANI, but is not likely to be true for disordered material.

According to the treatment of Zuppiroli et al., the hops that determine the conductivity are those between doped regions such as the polaronic clusters shown in Figure 1.8. Motion within the clusters, which is identified as adiabatic hopping [173], is clearly much faster. The inter-cluster hopping rate is taken from equation (1.11). W_{ij}, the overlap term, is assumed, as usual, to be of the form $\exp[-2s/a]$, where s is the width of the gap between clusters, and a, the monomer size, is taken as the characteristic decay distance of the polaronic wavefunctions at the edge of the cluster [173]. $R(\Delta_{ij})$ is taken in the Miller–Abrahams form, equation (1.13), which is shown to be applicable for hopping between large size clusters. Δ in equation (1.13) is taken to be half the charging energy E_c. E_c is obtained by electrostatic theory as [173]:

$$E_c = \frac{e^2}{2\pi Kd} \quad \text{with } K = \kappa_0 \kappa_r (1 + d/2s) \quad (1.39)$$

where d is the diameter of a cluster (assumed spherical). The dielectric constant $\kappa_0 \kappa_r$ of the system is estimated by relating it to the on-site repulsion U. The repulsion of two electrons sitting at the distance a is $U = (1/4\pi\kappa_0\kappa_r)(e^2/a)$. Substituting for $\kappa_0\kappa_r$ in (1.39) we obtain

$$E_c = 2U \frac{a}{d(1 + d/2s)} \quad (1.40)$$

With (1.40) the logarithm of the jump rate is given by [173]

$$\ln \nu = \frac{U}{kT} \frac{a}{d(1 + d/2s)} - \frac{2s}{a} + \ln \nu_0 \quad (1.41)$$

To simplify this Zuppiroli et al. obtained a relation between d and s by assuming that within a cluster of size d the average distance between dopants δ is independent of d. This leads to

$$\frac{\delta}{\bar{\delta}} = \frac{d}{d + s} \quad (1.42)$$

where $\bar{\delta}$ is the distance between dopants averaged over the entire sample. The hopping conductivity is calculated by looking for the pairs of clusters that contribute the most to the conductivity because they have the highest hopping rates. Optimizing ν with respect to d/a, using relation (1.42), Zuppiroli et al. obtain [173]

$$\sigma = \sigma_0 \exp[-(T_0/T)^{1/2}] \quad (1.43)$$

where

$$T_0 = \frac{8U}{k} \frac{[(\bar{\delta}/\delta) - 1]^2}{[(\bar{\delta}/\delta) - \frac{1}{2}]} \quad (1.44)$$

With U taken from other experiments (~ 1 to 2 eV [138]) T_0 depends only on $\bar{\delta}/\delta$, or the degree of doping. Values of T_0 ranging from $1\,000$ to $10\,000$, as found

experimentally, are obtained for $\bar{\delta}/\delta$ values ranging from ~ 1.06 to ~ 1.2. In agreement with experimental results T_0 is smaller the larger the doping.

Thus theory based on the dopants being in separate clusters, which need not be metallic, leads to the observed temperature dependence of σ for POT and well protonated PANI. This theory is also in agreement with the temperature dependence of σ for highly disordered samples of polypyrrole doped with polyanions [173].

4 TRANSPORT IN METALLIC SAMPLES

4.1 Origin of the metallic state

As discussed in Sections 3.1 and 3.2, the low susceptibility of t-PA samples with a few percent doping, and the regular arrangement of ions (particularly in sodium or potassium doped samples) suggests the existence of a soliton lattice. The soliton states may occupy a wide band, but for low doping there is clearly a gap between that band, and the valence and conduction bands. For sodium-doped samples, when the doping concentration increases beyond $\sim 6\%$ the susceptibility begins to grow to a metallic value [132,133]. Although it is difficult to determine how sharp the transition to metallic susceptibility is with increasing doping, because the amount of doping cannot be precisely determined, it was argued that this represents a first order phase transition to a metallic state, the metallic state being a polaron lattice [133,175]. There are strong arguments, however, against the metallic state being a polaron lattice, for example, the linear increase to 18% potassium doping of the IRAV made infrared-active by the dopant ions [176,177]. The increase in susceptibility starting at $\sim 6\%$ sodium doping can be attributed to the doping beginning to create a new phase in which the solitons are spaced more closely and overlap sufficiently for the electrons to move freely along the chains [178,179]. Below $\sim 6\%$, the phase being built up as doping increases is the 6.67% phase. It has a spacing of 15 sites between soliton centers, which is large enough so that the electron wavefunctions on adjacent solitons overlap little. Around 6% this phase would more or less have filled the parts of the sample available to doping; thus a new phase is begun in which ion spacing in the columns is smaller, probably four sites, which leads to 12 sites between solitons and considerably greater overlap [178].

There is other evidence that the soliton lattice persists above the concentration where the metallic susceptibility is seen [177]. For doping just above 6% there is considerable evidence, from optical properties, for a pseudo gap at the Fermi energy [180,181]. This is what would be expected for weak overlap of the soliton band with the valence band (acceptor doping) or conduction band (donor doping) [177,179]. Photoemission data for t-PA [182], polypyrrole [183] and poly-(3-hexylthiophene) [184] doped just above the insulator–metal transition show ε_F lying in a tail of the density of states rather than corresponding to a sharp Fermi surface. The soliton model predicts that for doping close to the transition there can be samples with a small gap ε_g of about a few kT at 300 K [178]. Such samples would appear metallic at room temperature but become semiconducting at temperatures where $kT \ll \varepsilon_g$. Evidence for this behavior could be the finding (in a sample with 6% I_3^-) of a narrow free carrier absorption peak at 300 K, which was no longer visible below 200 K, indicating carrier freezeout [185,177]. As will be seen later, however, a different interpretation of this behavior is possible.

A number of calculations for the band structure of a soliton lattice in sodium- and potassium-doped t-PA have been carried out [178,179,186]. Calculations based on the SSH Hamiltonian with the usual approximations have not been able to eliminate the gap between the filled soliton band and the conduction band although it becomes small for high doping concentrations. The calculation by Stafström [186] came the closest, bringing the gap for 10% doping down to the order of the level spacing, which is still ~ 0.1 to 0.2 eV at this concentration for a chain length of ~ 100 sites. The feature of Stäfstrom's calculation that appears to have brought the gap down was the incorporation of interchain interactions due to the ions. However the treatment was basically ID and involved various unknown parameters. What appears to be required is a 3D treatment of the entire structure of polymer chains and ions.

For doping well above the transition in well ordered samples the soliton lattice might no longer exist, the solitons having lost their identity due to strong overlap. There is no experimental evidence on this point. The discussion of this section has so far been devoted to t-PA and the soliton lattice. Very similar considerations hold for conducting polymers with non-degenerate ground state and their characteristic bipolaron lattice [187].

In all of the above discussion a perfect lattice has been assumed. Even with the big improvements in synthesis that have been achieved recently this assumption is not justified. In practice two samples that have nominally undergone the same synthesis and doping procedures may end up with σ different by orders of magnitude at low temperature, one being metallic and the other insulating. This is a clear indication that in practice it is disorder that is determining whether an insulator–metal transition takes place in heavily doped samples. In the following sections we will describe what has been found in studies of conductivity, optical properties and thermopower of heavily doped samples.

4.2   Non-metallic aspects of conductivity in heavily doped t-PA samples

Although, as will be discussed, there are highly doped polymer samples that show conductivity comparable to that of copper, thermopower $Q \propto T$, negative microwave dielectric constant, etc., no sample has shown all of the properties usually associated with the metallic state. In particular, no sample has shown conductivity that either remains more or less constant (at low temperatures) or decreases with increasing temperature. Nevertheless it has been shown that, despite an increase of σ with T over some ranges of temperature there are samples that show truly metallic behavior at low temperatures. In this section the behavior of σ in the range where it is increasing with T is reviewed. The next section will be devoted to the discussion of σ for metallic samples in the low temperature limit.

The first samples to stay metallic in the low temperature limit were AsF_5-doped t-PA [188] made from Shirakawa polyacetylene. For these early samples, with 10% doping, room temperature σ was only $\sim 10^3$ ohm^{-1} cm^{-1}. Resistivity ρ increased with increasing T only above ~ 250 K, at which temperature it was a minimum. It then increased by a factor of two as T decreased to ~ 10 K, but below that was flat to 0.04 K (the lowest T at which ρ was measured [188]. In measurements to still lower temperatures Gould et al.. found that below 0.03 K ρ increased logarithmically down to ~ 0.001 K [189]. Behavior similar to that of AsF_5-doped t-PA has been seen in $FeCl_3$-doped t-PA, with the minimum ρ at ~ 200 K [190]. Metallic behavior for $FeCl_3$ doping is seen for the heavily doped samples of Figure 1.7b, although the logarithmic scale does not permit seeing what for this plot would be a shallow maximum in σ. It is reasonable to interpret the increase in ρ with T above the minimum as due to phonon scattering while the behavior below the minimum is due to defects or disorder. This was supported by Kaneko and Ishiguro [191] who showed that aging of the sample, which is known to increase ρ by introducing disorder, raised the temperature of the minimum.

In contrast to the situation for AsF_5-doped t-PA, ρ of iodine-doped Shirakawa t-PA increased strongly with decreasing T at low temperatures. Naarmann t-PA doped with iodine, although it provided samples with very high σ (2×10^4 to 10^5 ohm^{-1} cm^{-1} at room temperature [192,193]), also did not prove metallic at low temperatures. On being cooled from 300 K to 3 K, σ for such a sample decreased by a factor of between 2.8 and 4 [193]. Below that σ decreased more rapidly still with decreasing T. For high conductivity Tsukamoto t-PA iodine doping did produce samples with weak temperature dependence at low temperatures, even below 0.001 K [194]. The difference is attributed to better ordering in the Tsukamoto samples, due to stretchability greater by a factor of two to three and slow doping with iodine gas. The better ordering is seen in X-ray diffraction [194]. Significantly, polarized resonant Raman scattering has shown that in fully doped Tsukamoto samples the ratio of I_3^- to I_5^- is greater than 20 to one, while in Shirakawa t-PA the I_5^- species dominates in heavily doped material.

Even for Tsukamoto samples made apparently the same way, with doping to saturation, room temperature σ could be different by an order of magnitude, presumably due to defects or disorder. The lower the room temperature conductivity, the more rapidly did σ decrease on going below room temperature [194]. This is seem in Figure 1.7, for example. Only the highest σ samples showed the weak temperature dependence down to millikelvins.

Metallic behavior has also been found in iodine-doped t-PA samples [195] prepared by the non-solvent method described by Akagi et $al.$ [196]. This method of synthesis has also yielded metallic potassium-doped t-PA [197].

There have been a few theoretical efforts to understand the high conductivity of the highly doped polymers and why that conductivity increases with increasing T. Kivelson and Heeger pointed out that the high value of σ in ordered samples can be attributed to the quasi-ID nature of the polymers [198]. One important result of the ID nature is that the dopant ions cannot scatter the carriers. Because a carrier is confined to a chain, in going through an attractive potential (such as that due to a dopant ion which is off the chain) the carrier will be accelerated at first, but

after having gone by the ion it will eventually return to its original velocity [199]. Another important result of the ID nature is that the important phonon scattering for a degenerate case is from one side of the Fermi surface to the other, involving $2k_F$ phonons. Because these phonons have high energy not many are excited at room temperature. The large bandwidth of t-PA is another factor favoring high σ. Of course the quasi-ID nature could result in localization of the carriers rather than high σ but for the case of high molecular weight and few serious defects even relatively weak interchain coupling is sufficient to avoid localization [198]. Treating t-PA as a normal quasi-ID metal, Kivelson and Heeger calculated σ at 300 K along the chain direction to be greater than that of copper [198].

For iodine-doped t-PA samples with conductivity in the metallic range at room temperature many workers have found that Sheng's theory of thermal fluctuation induced tunneling [200], for particular choices of the parameters, gives an excellent fit to σ vs. T. (See for example ref. 193.) This theory is based on tunneling between metallic particles embedded in an insulator. According to the theory the tunneling that determines σ in the limit of 0 K increases with T because the thermal motion of the metallic electrons effectively decreases the barriers. As pointed out by Schimmel et $al.$ [193], Nogami et $al.$ [201] and others, however, the values of the parameters required for the fit to experimental data are not reasonable for t-PA and not consistent with Sheng's theory.

Another characteristic found for near metallic iodine-doped t-PA is an increase of σ with electric field E starting at fields as low as fractions of V cm^{-1} [202,201]. The effect was particularly large at low temperatures where the conductivity had decreased substantially from the room temperature value. Epstein et $al.$, with Shirakawa samples, found the increment in σ to vary as $\ln E$ [202], while Nogami et $al.$, with Tsukamoto samples found it to vary more or less linearly with E [201]. Nogami et $al.$ [201] showed that the field dependence at different temperatures could not be described consistently by the Sheng model.

The sample-dependent increase of σ with E starting at quite low fields suggests thermal activation over barriers, some of which may be quite small [203]. It was acknowledged by Sheng that thermal activation over barriers gives the same exponential dependence on $1/T$ at high temperatures as the thermal fluctuation-induced tunneling model [200]. One source of barriers is the undoped parts of samples, e.g. amorphous regions and very short chains or conjugation lengths. It has been estimated from electrochemical data on

Shirakawa samples that $\sim 10\%$ of a sodium- or potassium-doped sample remains undoped [127]. The percentage could be higher for iodine-doped samples, which are less ordered. Evidence for undoped regions in heavily iodine-doped and sodium-doped samples has been obtained by X-ray studies, [204,205]. In addition to semiconducting regions that result from parts of a sample being undoped, as noted earlier, chains with a short doping length, less than ~ 100 sites, still have a gap though doped. A barrier must arise between a semiconducting region, whatever the size of the gap, and a metallic region due to the requirement that ε_F must be constant across the sample in thermal equilibrium. Some of the barriers may be so large and extended as to effectively cut off parts of the sample from conduction. Others, arising perhaps from short semiconducting inclusions, may permit tunneling. Still others may separate regions with a small enough conductivity difference that the barrier between them could be surmounted with phonon assistance giving, as suggested above, σ increasing with T. In the case of a semiconducting inclusion with a small gap, such as a length of chain not quite long enough to be metallic, the resistance of the inclusion and thus the barrier height could decrease markedly with increasing T. If electrons cannot go through or surmount a particular barrier they may be able to go over to another chain without a barrier in that neighborhood and continue to contribute to $\sigma_\parallel$ on that chain. The process of going over to another chain may require phonon assistance, because the existence of barriers redistributes the potential drop along a chain [203]. Thus the presence of barriers gives rise to (1) σ increasing with increasing T and (2) a mixing of $\sigma_\parallel$ and $\sigma_\perp$, the admix depending on the defects in the sample [203].

Although part of the low temperature conductivity in metallic t-PA might be due to tunneling, another reason the Sheng model is not applicable is that it neglects the scattering within the metallic regions due to chain breaks, conjugation defects such as sp^3 [206] and carbonyl [206] inclusions and phonons. The wide variety of possible scatterers and barriers implies a small likelihood of σ being correlated with the concentration of a particular defect, e.g. sp^3, as has sometimes been claimed [192].

The above discussion suggests that the resistivity in the chain direction of metallic and neat metallic samples in the range where σ is increasing as T increases could be written as a sum of contributions representing really metallic regions, barrier regions in which hopping or tunneling occurs, a phonon contribu-

tion and residual resistance [203]. A program of this kind was carried out by Kaiser and Graham [207]. Simplifying the situation somewhat, they represented the polymer by metallic regions in series with regions consisting of a parallel combination of a barrier region in which hopping or tunneling occurs and a disordered metallic region. The latter was required to keep σ from vanishing in the low T limit. With the hopping conductivity taken in the form of equations (1.30) or (1.43), with T_0 as a parameter, and four other parameters, Kaiser and Graham were able to obtain good fits to σ vs. T for a set of some 20 t-PA samples, metallic and non-metallic, over the full temperature range of the data [207].

Prigodin and Efetov suggested that the insulator–metal transition in t-PA is the result of its fibrillar structure [208]. In their model each fibril is considered to be a weakly disordered metallic wire randomly crosslinked to other fibrils. In the absence of the junctions all of the electronic states in the wires are localized. The states become extended over the whole network of fibrils only for strong enough interwire electron transfer, i.e. a large enough concentration of interwire links or junctions [208]. This suggestion does not appear to apply, however, to the insulator–metal transition in t-PA or the other conducting polymers. When Shirakawa t-PA, in which the fibrillar structure is well developed, is compressed to higher density the conductivity increases by only the same factor as the increase in density [209]. Also, studies of the fibrillar structure in the interior of Naarman t-PA and its correlation with conductivity led to the conclusion that a lower concentration of interfibrillar crosslinks led to high conductivity on doping [210]. It was concluded by Tsukamoto that the higher conductivity of his t-PA may be attributed to its higher bulk density (as well as its better ordering) and that the contribution from interfibrillar processes to the dc conductivity is not large [211].

4.3   Metallic conductivity

A number of studies of heavily doped polymers have concluded that at low temperatures they are close to a disorder-induced metal–insulator transition. It was found by Thummes *et al.* [212,213] that for heavily potassium-doped t-PA, $\sigma = \sigma_0 + CT^{1/2}$ over a wide range, characteristic of a disordered 3D metal with strong electron–electron correlation. A rapid decrease of $\sigma(T)$ below 1.5 K was attributed to localization [213]. The $T^{1/2}$ dependence over a wide range was not found by others, however, for heavily potassium-doped t-PA [214] nor for heavily iodine-doped t-PA [215,216]. Interpretation of the later data did, however, frequently follow the theory of disordered metals undergoing Anderson localization.

It was noted that the quality of a sample, i.e. the degree of order, is well characterized by the ratio of ρ at some high T (about room temperature) and ρ at some arbitrarily chosen low T. Thus, for example, Reghu et al. [216] characterized iodine-doped t-PA samples by the ratio $\rho_r = [\rho(1.2 \text{ K})/\rho(260 \text{ K})]$. Their best samples, made by the method of Akagi et al. [196], and subsequently stretched to 15 times the original length, had $\rho_r < 4$. Nevertheless these samples did not show a positive temperature coefficient.

In the critical regime of the metal–insulator transition due to Anderson localization, theory suggests that ρ follows a power law dependence on T, $\rho(T) \propto T^{-\beta}$, where β is constant [217,218]. To find the critical regime it is convenient to plot the reduced activation energy function $W(T) = -T[d(\ln \rho)/dT]$ vs. T on a log–log scale. The critical region is identified as the region of constant $W = \beta$. For samples with ρ_r close to 3 Reghu et al. found $W(T)$ to be constant from 180 to 60 K [216]. Although the ratio of σ parallel to the chains to σ perpendicular to the chains is 100, $W(T)$ was found to be identical for the two directions of σ. At a pressure of 8 kbar both ρ_r and the anistropy have decreased due to the chains being closer together. At this pressure $W(T)$ is smaller [216], indicating that the system is becoming more metallic.

The log T dependence of ρ reported for iodine-doped t-PA samples by Ishiguro et al. [194] was correct for samples with $\rho_r \sim 14$; it did not fit samples with $\rho_r < 4$ well [216]. Reghu et al. [216] compared σ vs. T for samples with $\rho_r \sim 3$ with the theoretical form for 3D disordered metals [219]:

$$\sigma(T) = \sigma(0) + mT^{1/2} + BT^{p/2} \quad (1.45)$$

Here the $T^{1/2}$ term results from electron diffusion in states near ε_F, reduced by the electron–electron interaction, and the $T^{p/2}$ term is a contribution due to localization effects. The value of p is determined by the temperature dependence of the mean free scattering time. For electron–photon scattering $p = 3$; for inelastic electron–electron scattering $p = 2$ in the clean limit, $3/2$ in the dirty limit [219]. For the iodine-doped samples with $\rho_r \sim 3$ Reghu et al. found that $T^{3/4}$ provides a better fit, indicating that electron–electron interactions are important [216]. These conclusions must be considered tentative, however, because a 3D model is not really applicable [215].

As noted above, application of higher pressure, in addition to decreasing ρ_r and the anistropy, decreases $W(T)$, making the system look more metallic. Thus pressure may be thought of as a means of tuning through the critical regime of the insulator–metal transition [220]. Another means for tuning through the critical region is application of a magnetic field [220]. The effect of the magnetic field is to move up the position of the mobility edge relative to the Fermi energy [221,220], tending to make the system less metallic. These effects were demonstrated with log–log plots of $W(T)$ vs. T. As discussed above, the critical region is identified as the region of constant W. Application of pressure made W increase over some ranges of T, and thus ρ grow more rapidly with decreasing T. These effects were demonstrated for iodine-doped and potassium-doped t-PA, and also for samples of PPy and PANI in the critical region [220].

A wide range of behavior is found in magnetoresistance (MR) of heavily doped t-PA, including both positive and negative values. For well ordered samples with $\rho_r < 3$ MR is negative (magnetoconductance positive) for magnetic field parallel or perpendicular to the chains. This has been interpreted in terms of weak localization and electron–electron interaction in a disordered system [189,212,215,222]. Nogami et al. compared the observed magnitude of the magnetoconductance with 3D theory for localization and electron interaction contributions and concluded that the prediction of the 3D theory is too small by a couple of orders of magnitude for high conductivity samples [215]. For samples with somewhat greater disorder ($\rho_r \approx 3$–6) the sign of MR is negative (positive) for the field perpendicular (parallel) to the chains above 2 K [216]. Also the magnitude of the negative MR decreases gradually as T is lowered from 10 to 2 K [216]. These results indicate that the weak localization contribution dominates at high temperature, and the contribution from electron–electron interaction grows more important at lower temperatures [215,216], consistent with the earlier results cited for temperature dependence of σ.

More recently highly doped, highly conducting samples of PPy and PANI have shown similar phenomena to those discussed above for highly doped t-PA. The PPy films were prepared by anodic oxidation, and doped with PF_6. The room temperature conductivity was in the range 200–500 ohm^{-1} cm^{-1}, increasing to ~ 1000 ohm^{-1} cm^{-1} after tensile drawing [223, 224,194]. Carrier concentration was estimated as about one per three rings, a little lower than the one carrier per 10–12CH units estimated for iodine-doped t-PA [194].

The lower σ need not mean greater disorder, however; the mobility should be lower because of the larger number of normal modes of lattice vibration.

X-ray studies indicate that $PPy(PF_6)$ films are $\sim 50\%$ crystalline, with a coherence length of 20 Å [225]. The metallic properties are hard to understand given the small coherence length. J [226]. A prime source of disorder in PPy is the lack of symmetry of a pyrrole unit about the *para* positions. The local steric potential depends on the dihedral angle between the nitrogens on adjacent pyrrole rings. When the dihedral angle is $\sim 180°$ the chain will tend to be planar or rod-like. If the dihedral angle is $\sim 0°$, the polymer chain will tend to coil to minimize intrachain steric repulsion. Rod-like conformations are known to produce more crystalline polymer solids, while coiled regions will be disordered [226].

Evidence for the metallic nature of the PPy doped with PF_6 was the increase of ρ with increasing T in the range 10–20 K [194, 221]. A more detailed study of heavily doped $PPy(PF_6)$ was reported recently by Yoon *et al.* [227]. Characterizing their samples by $\rho_r = [\rho(1.4\ K)/\rho(300\ K)]$, they define them as being on the metallic side of the insulator–metal transition for $2 < \rho < 6$, where $\rho(T)$ remains finite as $T \to 0$, but ρ still increases with decreasing T for all temperatures. They define samples with $\rho_r < 2$ as being in the metallic regime (ρ decreasing with decreasing T at low temperatures). The properties in the various regimes could be analyzed in a similar fashion to those of t-PA. One difference was that only positive magnetoresis-magnetoresistance was seen, whereas both signs of magnetoresistance are seen in t-PA.

With the introduction of functionalized sulfonic acids to protonate PANI and the resulting solubility in the conducting form, the homogeneity and order of PANI samples were greatly improved [228,229]. With these improvements came the possibility of observing metallic features in transport. Critical behavior of the electrical conductivity near the disorder-induced metal–insulator transition was observed in PANI doped with camphor sulfonic acid (CSA) [230]. With ρ_r defined by $\rho(1.4\ K)/\rho(300\ K)$, it was found that the critical region was given by $2 < \rho_r < 6$; samples with $\rho_r < 2$ were said to be in the metallic region, and samples with $\rho_r > 6$ in the insulating region. Values of ρ_r as small as 1.6 were found. The critical region is identified by a power law variation of ρ, $\rho \propto T^{-\beta}$ [218], which typically extended from 40 K to below 2 K. β values were found typically in the range $0.26 < \beta < 0.4$ [230]. For samples on the metallic side of the transition the reduced activation energy W decreases but was not

observed to change sign, i.e. real metallic behavior has not been observed for PANI (CSA) in any temperature range. The temperature dependence of σ was found to be quite weak for samples with low ρ_r. For samples with $\rho_r = 1.6$, σ vs. T could be fitted by equation (1.45) with $p = 1$ or $B = 0$. Consistent with this, magnetoresistance was found to be positive [230].

4.4 Optical studies in the metallic region

The spectral response of metallic and near metallic polymers for low frequencies can be attributed to free carrier absorption. However the complex conductivity $\sigma^*(\omega)$ does not show normal Drude-like behavior, characterized by

$$\sigma^*(\omega) = \frac{\Omega_p^2 \tau}{4\pi(1 + i\omega\tau)} \qquad (1.46)$$

Here $\Omega_p^2 = 4\pi ne^2/m$, the square of the plasma frequency, m is the effective mass, and τ is the scattering time. According to equation (1.46) the real part of $\sigma^*(\omega)$ should increase monotonically to $\sigma_{dc} = ne^2\tau/m$ as $\omega \to 0$. This behavior is not found for highly doped conducting polymers. In t-PA, for example, Kim and Heeger [180] found Re $\sigma^*\omega$) suppressed below the Drude extrapolation as $\omega \to 0$ with a peak around 1.2 eV [231]. This was interpreted by them in the polyacetylene case as indicating a pseudo gap in the density of states [180], which was subsequently attributed to poor overlap of the soliton band and valence band [177–179]. Lee *et al.* [231], however, interpreted the suppression of the Drude tail for $\omega \to 0$ as arising from disorder-induced localization in PANI (CSA) and compared it to a localization-modified Drude model of Mott and Kaveh [232]. They found this model to be in good agreement with the data, leading to a mean free path of 7 Å, comparable to the unit cell dimension along the chains.

Hasegawa *et al.* [233] took $\sigma^*(\omega)$ data down to lower frequencies and analyzed it differently. They deconvoluted the data into a number of different contributions. The lowest frequency contribution (decreasing monotonically with increasing ω from $\sim 80 \sim 200\ cm^{-1}$) they identified as the Drude contribution. This contribution was followed, with further increase in ω, by a rise (peaking at $\sim 400\ cm^{-1}$) and then a decrease, constituting, according to the interpretation of Hasegawa *et al.* [233], a band due to ID localization. Analysis of the Drude contribution, under the assumption that the effective mass equals the free electron mass, led to carrier concentrations of 2.8×10^{20} and 7.2×10^{20} cm^3 for samples doped with 11% and 16% iodine, respectively. The values of τ obtained from the analysis were 4.5×10^{-14} and 1.0×10^{-13} s, respectively. These values led to quite reasonable mobility values, 80 and 180 cm^2 V^{-1} s^{-1}, respectively.

Equation (1.46) predicts a negative dielectric function, which in the low frequency region $\omega\tau \ll 1$ is $\kappa \simeq -\Omega_p^2\tau^2$. The first observation of the predicted negative κ was at microwave frequencies in PANI doped with CSA and processed in m-cresol [234]. Processing in m-cresol results in greater crystallinity than processing in CHCl$_3$ because of conformation changes to more rod-like material in the former case from coil-like material in the latter case. The microwave frequency dielectric constant κ_{mw} was found to be negative from room temperature down to 4.2 K, with a minimum value of $\sim -10^5$ at ~ 200 K where σ_{mw} had its maximum. Below 200 K κ_{mw} increased with decreasing T to $\sim -10^4$ at 4 K [234]. The plasma frequency was found to be ~ 0.015 eV from the relation, $\Omega_p \sim 4\pi\sigma_{mw}/(-\kappa_{mw})^{1/2}$ for $\omega\tau \ll 1$. This led to an anomalously large value of τ ($\sim 1.2 \times 10^{-11}$ s) at room temperature. For PANI samples other than those doped with CSA and processed in m-cresol κ_{mw} was positive and σ_{dc} was smaller and decreased monotonically below room temperature, approximately according to $\sigma(T) \propto \exp[-(T_o/T)^{1/2}]$. Below 50 K there was a linear relation between the value of κ_{mw} for a given sample and the square of its coherence length determined by X-rays. The coherence length is presumably a measure of the delocalization length, or the size of the metallic islands [235]. In some PANI (CSA) samples processed in m-cresol κ_{mw} was found to cross over from very large room temperature negative values to large positive ones with decreasing T below ~ 20 K [236]. This was attributed to the transition from a phonon-controlled metallic state to a non-metallic state with hopping transport. In the localized state there was an increase in Q and a linear decrease in W, the reduced activation energy, with decreasing T. The model predicted by Larkin and Khmelnitskii [218] does not fit this case. Joo $et\ al.$ suggest that the appropriate model is not their model of homogeneous 3D disorder, but rather a 3D network of highly conducting islands connected by single chains passing less coherently through the disordered regions.

A Drude metallic response was also found in the highly conducting state of PPy [236]. κ vs. ω showed two crossings from negative to positive values, or

plasma frequencies. The one in the far infrared, at ~ 100 cm^{-1} is attributed to the really delocalized electrons. For this plasma frequency n is quite small. If m^* is taken to be equal to the free electron mass m_e, $n/m^* \sim 10^{-4}n_o/m_e$, where n_o is the number of electrons added to the chain at full doping (one per three rings repeat). Thus only a very small fraction of the total number of added electrons is sufficiently delocalized to participate in the free electron Drude response. The value of τ was estimated as $\sim 3 \times 10^{-11}$ s. It was suggested that this anomalously long scattering time is due to an open Fermi surface, as expected for highly anisotropic materials [236]. The second plasma frequency, in the near infrared, is attributed to the remaining, confined conduction electrons. From the reflectance data it is deduced that the electrons contributing to this plasma frequency are bound by ~ 0.2 eV [236].

4.5 Thermoelectric power

Although conductivity vs. temperature for heavily doped polymers has yet to show metallic temperature dependence except for relatively few samples over restricted temperature ranges, the thermopower, in contrast, often shows metallic size and temperature dependence for heavily doped samples. Kaiser [237] has considered the effect on thermopower Q of sample inhomogeneity, pointing out that in a heterogeneous sample the contribution made by different regions to Q may be weighted differently from their corresponding contributions to the total conductivity. Thus, for example, for a fibril consisting of metallic regions, with characteristic thermopower Q_1 and thermal resistance W_1, separated by thin resistive regions or barriers, with characteristic thermopower Q_2 and thermal resistance W_2,

$$Q = (W_1/W)Q_1 + (W_2/W)Q_2 \qquad (1.47)$$

where W is the total thermal resistance. As pointed out by Kaiser, this situation can result in Q close to Q_1, but σ determined by σ_2, if $W_1 \gg W_2$, i.e. the thermal current carried by the phonons is less impeded by thin barriers than the electric current [237]. If transport in the resistive region is determined by variable range hopping, for which, according to equation (1.32), $Q \propto T^{1/2}$, equation (1.47) leads to

$$Q = AT + BT^{1/2} \qquad (1.48)$$

where A and B are constants. This expression was found by Kaiser [237] and by Park $et\ al.$ [238] to give a reasonably good fit for many polyacetylene samples; B is larger for samples with a few per cent doping and A increases as the doping increases.

For the heavily doped samples where σ remains finite as $T \rightarrow 0$ the shape of Q vs. T becomes more linear, as expected for metallic conductivity. Sometimes there is an increase in the slope below ~ 50 K (seen in $FeCl_4$ and AsF_5 doped polyacetylene) similar to the behavior seen in many disordered metals [237]. There it is ascribed to the electron–phonon enhancement of diffusion thermopower at low temperatures. The phonon dragpeak, which has the dominant effect on the temperature dependence of Q below room temperature in most good crystals, is suppressed by disorder [239]. This makes the enhancement effect in diffusion thermopower visible as an increase in the slope as T decreases and the enhancement increases below the Debye temperature [237]. Kaiser finds good agreement between Q vs. T for PA samples with 10% AsF_5 doping and 6% $FeCl_4$ doping, respectively, and the theory including the electron-phonon interaction [237]. Note, however, that some heavily doped t-PA samples show highly linear Q vs. T [240].

Until recently almost all thermopower data were taken on acceptor-doped, or p-type samples. Measurements on n-type samples, are more difficult because of their extreme sensitivity to air and moisture. Recently measurements were made for the first time on heavily alkali-doped t-PA, and yielded the surprising result that Q is positive [240]. This result was obtained for doping with sodium, potassium or rubidium. The magnitude of Q is as expected for highly doped material ($\sim 8 \mu V K^{-1}$ at room temperature) and it decreases more or less linearly with decreasing T [240]. If the alkali-doped samples were exposed to air, Q was found to decrease with time, becoming negative eventually. Park $et al.$ cite the positive Q as evidence for a soliton lattice. As discussed in Section 4.1, with increasing doping the soliton states broaden into a band. The usual picture is that at some concentration this band overlaps the valence band (p-type) or conduction band (n-type) allowing metallic conduction. In this picture the Fermi energy is in the region of overlap; here the density of states is relatively small, in agreement with photoemission results on metallic materials [182–184]. The thermopower would then be positive for p-type material, negative for n-type. To account for the result of Park $et al.$ [241] one must postulate that in the metallic state of the alkali-doped t-PA samples with $Q > 0$ the soliton band created by background acceptor impurities. One point in favor of this hypothesis is that the alkali-doped metallic samples are probably more

ordered than the acceptor-doped samples because the alkali ions are smaller, similar in size to the vacant spaces in the t-PA lattice. However if this hypothesis is correct it remains to be explained why heavily doped potassium-doped samples behave quite similarly to say, iodine-doped samples in regard to the magnitude of Q, temperature dependence of σ vs. T, etc.

5   Conclusions

Surveying the mass of experimental data and analysis, I have formed some conclusions about transport in the conducting polymers.

- There is no good evidence that dark transport in pristine or low doped $trans$-polyacetylene is due to solitons. I believe that the role of polarons in dark transport has been underestimated. Solitons probably do contribute to photoconductivity however.

- For non-degenerate semiconductors there is no good evidence that bipolarons are important in photoconductivity.

- It is essential to take into account the non-uniform distribution of impurities even for pristine and moderately doped samples. With rare exceptions this has not been done. Particularly for moderately doped samples, the rate limiting step in transport is likely to be hopping or tunneling between doped regions.

- As emphasized by Zuppiroli and colleagues [135], the role of impurity ions in transport has not been properly considered. At the least allowance must be made for the fact that the transfer integral $t_\perp$ is increased in the neighborhood of an impurity ion site. More work must be done on the effect of the ions on the excitations.

- There are apparently some heavily doped samples that are sufficiently homogeneous that a 3D model of the Anderson disorder driven metal–insulator transition is applicable. Agreement of experimental results with theory is probably helped by the fact that contribution of the most poorly conducting regions may freeze out at low temperatures. Despite the success of this disorder model for some samples, the model of conducting islands separated by poor conductivity regions is appropriate for other heavily doped samples [203,207,231]. It has been well demonstrated by structural studies that disorder in heavily doped PANI-CSA samples is highly inhomogeneous [234].

6 ACKNOWLEDGEMENT

I am grateful to Dr Yu Gartstein for a careful reading of the manuscript.

7 REFERENCES

1. E.M. Conwell and H.A. Mizes, *Phys. Rev B* **51**, 6953 (1995).
2. W.-P. Su, J.R. Schrieffer and A.J. Heeger, *Phys. Rev. B* **22**, 2099 (1980).
3. P. Vogl and D.K. Campbell, *Phys. Rev. B* **41**, 12797 (1990).
4. H.A. Mizes and E.M. Conwell, *Phys. Rev. Lett.* **70**, 1505 (1993).
5. D. Emin and K.L. Ngai, *Jour. de Phys. Colloq.* C **3**, 471 (1983).
6. P. Brendel, A. Grupp, M. Mehring, R. Schenck, K. Müllen and W. Huber, *Synth. Met.* **45**, 49 (1991).
7. A.J. Heeger, S. Kivelson, J.R. Schrieffer and W.-P. Su, *Rev. Mod. Phys.* **60**, 781 (1988).
8. E.M. Conwell, *IEEE Trans.* El. Ins., **El-22**, 591 (1987).
9. E.M. Conwell and H.A. Mizes, *Handbook on Semiconductors*, (ed. T.S. Moss) Vol. North Holland, Amsterdam 1992, pp 583–625.
10. W. Warta and N. Karl, *Phys. Rev. B* **32**, 1172 (1985).
11. H. Bässler, *Phys. Stat. Sol.* (B) **175**, 15 (1993).
12. M. Abkowitz, H. Bässler and M. Stolka, *Phil. Mag. B* **63**, 201 (1991).
13. T. Takiguchi, D.H. Park, H. Ueno and K. Yoshino, *Synth. Met.* **17**, 657 (1987).
14. J. Obrzut, M.J. Obrzut and F.E. Karasz, *Synth. Met.* **29**, E103, (1989).
15. P. Strohriegel and D. Haarer, *Makromol. Chem. Makromol. Symp.* **44**, 85 (1991).
16. B.R. Hsieh, H. Antoniadis, M.A. Abkowitz and M. Stolka, *Polymer Preprints* **33**, 414 (1992).
17. S. Karg, V. Dyakonov, M. Meier, W. Riess and G. Paasch, *Synth. Met.* **67**, 165 (1994).
18. H. Antoniadis, B.R. Hsieh, M.A. Abkowitz, S.A. Jenekhe and M. Stolka, *Mol. Cryst. Liq. Cryst.* **256**, 381 (1994).
19. M. Gailberger and H. Bässler, *Phys. Rev. B* **44**, 8643 (1991).
20. P. Delannoy, G. Horowitz, H. Bouchriha, F. Deloffre, F.-L. Fave, F. Garnier, R. Hajlaoui, M. Heyman, F. Kouki, J.-L. Monge, P. Valat, V. Wintgens and A. Yassar, *Synth. Met.* **67**, 197 (1994).
21. R.N. Marks, J.J.M. Halls, D.D.C. Bradley, R.H. Friend and A.B. Holmes, *J. Phys. Condens. Matter* **6**, 1379 (1994).
22. S. Zhou, H. Hong, Y. He, D. Yang, Z. Jin, R. Qian, T. Masuda and T. Higashimura, *Polymer* **33**, 2189 (1992).
23. A. Walser, A. Seas, R. Dorsinville, R.R. Alfano and R. Turbino, *Solid State Commun.* **67**, 333 (1988).
24. E.M. Conwell, H.-Y. Choi and S. Jeyadev, *Synth. Met.* **49–50**, 359 (1992).
25. S. Jeyadev and E.M. Conwell, *Phys. Rev. B* **25**, 6253 (1987).
26. E.M. Conwell and S. Jeyadev, *Synth. Met.* **28**, D439 (1989).
27. X.Z. Peng, G. Horowitz, D. Fichou and F. Garnier, *Appl. Phys. Lett.* **57**, 2013 (1990).
28. A. Dodabalapur, L. Torsi and H.E. Katz, *Science* **268**, 270 (1995).
29. F. Garnier, G. Horowitz, X. Peng and D. Fichou, *Adv. Mater.* **2**, 592 (1990).
30. F. Garnier, A. Yassar, R. Hajlaoui, G. Horowitz, F. Deloffre, B. Sevret, S. Ries and P. Alnot, *J. Am. Chem. Soc.* **115**, 8716 (1993).
31. F. Garnier, R. Hajlaoui, A. Yassar and P. Strivastava, *Science* **265**, 1684 (1994).
32. K. Waragi, H. Akimichi, S. Hotta, H. Kano and H. Sakaki, *Synth. Met.* **55–57**, 4053 (1993).
33. H. Akimichi, K. Waragai, S. Hotta, H. Kano and H. Sakaki, *Appl. Phys. Lett.* **58**, 1500 (1991).
34. G. Horowitz, X. Peng, D. Fichou and F. Garnier, *Synth. Met.* **51**, 419 (1992).
35. A. Tsumara, K. Hoezuka and T. Ando, *Synth. Met.* **25**, 11 (1988).
36. H. Koezuka, A. Tsumara, H. Fuchigami and K. Kuramoto, *Appl. Phys. Lett.* **62**, 1794 (1993).
37. A. Assadi, C. Svensson, M. Wilander and O. Inganäs, *Appl. Phys. Lett.* **53**, 195 (1988).
38. J. Paloheimo, P. Kuivalainen, H. Stubb, E. Vuorimaa and P. Yli-Lahti, *Appl. Phys. Lett.* **56**, 1157 (1990).
39. H. Fuchigami, A. Tsumura and H. Koezuka, *Appl. Phys. Lett.* **63**, 1372 (1993).
40. J.H. Burroughes, C.A. Jones and R.H. Friend, *Nature* **335**, 137 (1988).
41. G. Horowitz, D. Fichou, X. Peng and P. Delannoy, *J. Phys. France* **51**, 1489 (1990).
42. S. Jeyadev and J.R. Schrieffer, *Phys. Rev. B* **30**, 3620 (1984).
43. H.-Y. Choi and E.M. Conwell, *Synth. Met.* **41–43**, 3667 (1991); *Mol. Cryst. Liq. Cryst.* **194**, 23 (1991).
44. A. Terai and Y.J. Ono, *J. Phys. Soc. Japan* **55**, 213 (1986).
45. E.M. Conwell, H.-Y. Choi and S. Jeyadev, *J. Phys. Chem.* **96**, 2827 (1992).
46. Z.-B. Su and Y. Lu, *Phys. Rev. B* **27**, 5199 (1983).
47. T. Holstein, *Ann. Phys. (NY)* **8**, 343 (1959).
48. H.J. de Wit, *Phillips Res. Rep.* **23**, 449 (1968).
49. S.D. Phillips and A.J. Heeger, *Phys. Rev. B* **38**, 6211, (1988).
50. K. Pichler, D.A. Halliday, D.D.C. Bradley, P.L. Burn, R.H. Friend and A.B. Holmes, *J. Phys. Condens. Matter* **5**, 7155 (1993).
51. D. Chen, M.J. Winokur, M.A. Masse and F.E. Karasz, *Polymer* **23**, 3116 (1992).
52. D. Emin, *Adv. Phys.* **24**, 305 (1975).
53. R.A. Marcus, *Ann. Rev. Phys. Chem.* **15**, 155 (1964).

54. A. Miller and E. Abrahams, *Phys. Rev.* **120**, 745 (1960).

55. D. Belijonne, Z. Shuai, R.H. Friend and J.L. Brédas, *J. Chem. Phys.* **102**, 2042 (1955).

56. Yu.N. Gartstein and E.M. Conwell, *J. Chem. Phys.* **100**, 9175 (1994).

57. B.I. Shklovskii and A.L. Efros, *Electronic Properties of Doped Semiconductors*, Springer-Verlag, Berlin, 1984.

58. Yu.N. Gartstein and E.M. Conwell, *Phys. Rev. B* **51**, 6947 (1995).

59. H.A. Mizes and E.M. Conwell, *Synth. Met.* **68**, 145 (1995).

60. C. Zhang, D. Braun and A.J. Heeger, *J. Appl. Phys.* **73**, 5177 (1993).

61. P.L. Burn, A.B. Holmes, A. Kraft, D.D.C. Bradley, A.R. Brown and R.H. Friend, *J. Chem. Soc. Chem. Commun.* **32** (1992).

62. See, for example, S. Karg, W. Reiss, M. Meier and M. Schwoerer, *Synth. Met.* **55–57**, 4186 (1993).

63. M. Yan, L.J. Rothberg, F. Papadimitrakopoulos, M.E. Galvin and T.M. Miller, *Phys. Rev. Lett.* **73**, 744 (1994).

64. F. Papadimitrakopoulos, M. Yan, L.J. Rothberg, H.E. Katz, E.A. Chandross and M.E. Galvin, *Mol. Crys. Liq. Cryst.* **256**, 663 (1994).

65. S. Kivelson, *Phys. Rev. B* **25**, 3798 (1982).

66. K.L. Ngai and R.W. Rendell, *Handbook of Conducting Polymers*, (ed. T.A. Skotheim), Vol. 2, Marcel Dekker, New York, 1986, pp. 967–1040.

67. A.J. Epstein, H. Rommelmann, M. Abkowitz and H.W. Gibson, *Phys. Rev. Lett.* **47**, 1549 (1981).

68. A.J. Epstein, *Handbook of Conducting Polymers*, (ed. T.A. Skotheim), Vol. 2, Marcel Dekker, New York, 1986. pp. 1041–1097.

69. S. Summerfield and J.A. Chroboczek, *Solid State Commun.* **53**, 129 (1985).

70. J.H. Edwards and W.J. Feast, *Polym. Commun.* **21**, 595 (1980).

71. R.H. Friend, D.C. Bott, D.D.C. Bradley, C.K. Chai, W.J. Feast, P.J.S. Foot, J.R.M. Giles, M.E. Horton, C.M. Pereira and P. Townsend, *Philos. Trans. R. Soc. London A* **314**, 37 (1985).

72. J.C. Scott and T.C. Clarke, *J. Phys. (Paris)* **44**, C3–365 (1983).

73. K. Ito, Y. Tanabe, K. Akagi and H. Shirakawa, *Phys. Rev. B* **45**, 1246 (1992).

74. M. Nechtschein, F. Devreux, F. Genoud, M. Guglielmi and K. Holczer, *Phys. Rev. B* **27**, 61 (1983).

75. T.C. Clarke and J.C. Scott, *Handbook of Conducting Polymers*, (ed. T.A. Skotheim), Vol. 2, Marcel Decker, New York 1986. pp. 1127–1156.

76. K. Mizoguchi, K. Kume and H. Shirakawa, Abstracts of International Conference on Science and Technology of Synthetic Metals, Kyoto 55, 1986.

77. S. Jeyadev and E.M. Conwell, *Phys. Rev. Lett.* **58**, 258 (1987).

78. S. Jeyadev and E.M. Conwell, *Phys. Rev. B* **36**, 3284 (1987).

79. Z. Vardeny, J. Strait, D. Moses, T.-C. Chung and A.J. Heeger, *Phys. Rev. Lett.* **49**, 1657 (1982).

80. S. Jeyadev and E.M. Conwell, *Phys. Rev. B* **35**, 6253 (1987).

81. E.M. Conwell and S. Jeyadev, *Mol. Cryst. Liq. Cryst.* **160**, 443 (1988).

82. W.-P. Su and J.R. Schrieffer, *Proc. Natl. Acad. Sci. USA* **77**, 5626 (1980).

83. (a) P.L. Danielsen, *Synth. Met.* **20**, 125 (1987).
(b) Y.N. Gartstein and A.A. Zakhidov, *J. Mol. Electron* **3**, 163 (1987).
(c) D. Baeriswyl and K. Maki, *Phys. Rev. B* **38**, 8135 (1988).

84. L.J. Rothberg, T.M. Jedju, S. Etemad and G.L. Baker, *Phys, Rev. Lett.* **57**, 3229 (1986).

85. L.J. Rothberg, T.M. Jedju, P.D. Townsend, S. Etemad and G.L. Baker, *Phys, Rev. Lett.* **65**, 100 (1990).

86. M. Sinclair, D. Moses and A.J. Heeger, *Solid State Commun.* **59**, 343 (1986).

87. J. Reichenbach, M. Kaiser and S. Roth, *Phys, Rev. B* **48**, 14104 (1993).

88. H. Bleier, S. Roth, Y.Q. Shen, D. Schäfer-Siebert and G. Leising, *Phys. Rev. B* **38**, 6031 (1988).

89. E.M. Conwell and H.A. Mizes, *Phys. Rev. B* **51**, 6953 (1995).

90. S. Jeyadev and E.M. Conwell, *Phys. Rev. B* **35**, 5917 (1987).

91. L. Rothberg, T.M. Jedju, S. Etemad and G.L. Baker, *Phys. Rev. B* **36**, 7529 (1987).

92. K. Pakbaz, C.H. Lee, A.J. Heeger, T.W. Hagler and D. McBranch, *Synth. Met.* **64**, 295 (1994).

93. G. Yu, S.D. Phillips, H. Tomozawa and A.J. Heeger, *Phys. Rev. B* **42**, 3004 (1990).

94. P. Gomes da Costa and E.M. Conwell, *Phys. Rev. B* **48**, 1993 (1993).

95. M.J. Rice and Yu.N. Gartstein, *Phys. Rev. Lett.* **73**, 2504 (1994).

96. Yu.N. Gartstein, M.J. Rice and E.M. Conwell, *Phys. Rev. B* **52**, 1683 (1995).

97. M. Chandross, F. Guo and S. Mazumdar, *Synth. Met.* **69**, 625 (1995).

98. R.N. Marks, J.J.M. Halls, D.D.C. Bradley, R.H. Friend and A.B. Holmes, *J. Phys. Condens. Matter* **6**, 1379 (1994).

99. D.D.C. Bradley, Y.Q. Shen, H. Bleier and S. Roth, *J. Phys. C* **21**, L515 (1988).

100. J.W.P. Hsu, M. Yan, T.M. Jedju, L.J. Rothberg and B.R. Hsieh, *Phys. Rev. B* **49**, 712 (1994).

101. M. Yan, L.J. Rothberg, F. Papadimitrakopoulos, J.E. Galvin and T.M. Miller, *Phys. Rev. Lett.* **72**, 1104 (1994).

102. G.S. Kanner, X. Wei, B.C. Hess, L.R. Chen and Z.V. Vardeny, *Phys. Rev. Lett.* **69**, 538 (1992).

103. B. Kraabel, D. McBranch, N. Sariciftci, D. Moses and A.J. Heeger, *Mol. Cryst. Liq. Cryst.* **256**, 733 (1994); *Phys. Rev. B* **50**, 18543 (1994).

104. Y. Shimoi and S. Abe, *Phys. Rev.* B **50**, 14781 (1994).
105. C.H. Lee, G. Yu and A.J. Heeger, *Phys. Rev.* B **47**, 15543 (1993).
106. A.R. Bishop, D.K. Campbell, P.S. Lomdahl, B. Horovitz and S.R. Phillpot, *Phys. Rev. Lett.* **52**, 671 (1984); *Synth. Met.* **9**, 223 (1984).
107. F. Guinea, *Phys. Rev.* B **30**, 1884 (1984).
108. A.R. Bishop and T. Schneider (eds.) *Solitons and Condensed Matter Physics,* Springer-Verlag, Berlin, 1978.
109. D. Moses, A. Denenstein, A. Pron, A.J. Heeger and A.G. MacDiarmid, *Solid State Commun.* **36**, 219 (1980).
110. J. Bardeen and W. Shockley, *Phys. Rev.* **80**, 72 (1950).
111. C. Kittel, *Introduction to Solid State Physics*, John Wiley Sons, New York, 1976.
112. H.W. Gibson, R.J. Weagley, R.A. Mosher, S.B. Kaplan, W.M. Prest, Jr and A.J. Epstein, *Phys. Rev.* B **31**, 2328 (1985).
113. K. Ehinger and S. Roth, *Phil. Mag.* B **53**, 301 (1986).
114. H. Shirakawa and S. Ikeda, *Polymer* **J2**, 231 (1971).
115. J.P. Pouget in *Electronic Properties of Polymers and Related Compounds*, (eds. H. Kuzmany, M. Nehring and S. Roth) Springer-Verlag, Berlin, 1985, pp. 26–34.
116. P. Robin, J.P. Pouget, R. Comès, H.W. Gibson and A.J. Epstein, *J. Phys. (Paris)* **44**, C3–87 (1983).
117. C.R. Fincher, C.-E. Chen, A.J. Heeger, A.G. MacDiarmid and J.B. Hastings, *Phys. Rev. Lett.* **48**, 100 (1982).
118. G. Leiser, G. Wegner, R. Weizenhöfer and L. Brombacher, *Polymer Preprints* **25**, 221 (1984).
119. D. White and D.C. Bott, *Polymer Commun.* **25**, 98 (1984).
120. H. Kahlert, O. Leitner and G. Leising, *Synth. Met.* **17**, 467 (1987).
121. H. Naarmann, *Synth. Met.* **17**, 223 (1987).
122. J. Tsukamoto, A. Takahashi and K. Kawasaki, *Jpn. J. Appl. Phys.* **29**, 125 (1990).
123. R.H. Baughman, N.S. Murthy and G.G. Miller, *J. Chem. Phys.* **79**, 515 (1983).
124. N.S. Murthy, L.W. Shacklette and R.H. Baughman, *Phys. Rev.* B **40**, 12550 (1989).
125. M. Winokur, Y.B. Moon, A.J. Heeger, J. Barker, D.C. Bott and H. Shirakawa, *Phys. Rev. Lett.* **58**, 2329 (1987).
126. D. Chen, M.J. Winokur, M.A. Masse and F.E. Karasz, *Phys, Rev.* B **41**, 6759 (1990).
127. L.W. Shacklette and J.E. Toth, *Phys. Rev.* B **32**, 5892 (1985).
128. P.A. Heiney, J.E. Fischer, D. Djurado, J. Ma, D. Chen, M.J. Winokur, N. Coustel, P. Bernier and F.E. Karasz, *Phys. Rev.* B **44**, 2507 (1991).
129. M.J. Winokur, J. Maron, Y. Cao and A.J. Heeger, *Phys. Rev.* B **45**, 9656 (1992).
130. N.S. Murthy, G.G. Miller and R.H. Baughman, *J. Chem. Phys.* **89**, 2523 (1988).
131. A.J. Epstein, H. Rommelmann, M.A. Druy, A.J. Heeger and A.G. MacDiarmid, *Solid State Commun.* **38**, 683 (1981).
132. F. Moraes, J. Chen, T.-C. Chung and A.J. Heeger, *Synth. Met.* **11**, 271 (1985).
133. J. Chen and A.J. Heeger, *Synth. Met.* **24**, 311 (1988).
134. O. Chauvet, S. Paschen, L. Forro, L. Zuppiroli, P. Bujard, K. Kai and W. Wernet, *Synth. Met.* **63**, 115 (1994).
135. L. Zuppiroli, S. Paschen and M.N. Bussac, *Synth. Met.* **69**, 621 (1995).
136. M.N. Bussac and L. Zuppiroli, *Phys. Rev.* B **49**, 5876 (1994).
137. M. Seimiya and Y. Onodera, *J. Phys. Soc. Japan* **58**, 1676 (1989).
138. M.N. Bussac and L. Zuppiroli, *Phys. Rev.* B **47**, 5493 (1993).
139. J.-L. Brédas, J.C. Scott, K. Yakushi and G.B. Street, *Phys. Rev.* B **30**, 1023 (1984).
140. Y.W. Park, A. Denenstein, C.K. Chiang, A.J. Heeger and A.G. McDiarmid, *Solid State Commun.* **29**, 747 (1979).
141. T.-C. Chung, F. Moraes, J.D. Flood and A.J. Heeger, *Phys. Rev.* B **29**, 2341 (1984).
142. R.R. Chance, J.-L. Brédas and R. Silbey, *Phys. Rev.* B **29**, 4491 (1984).
143. N.F. Mott and E.A. Davis, *Electronic Processes in Non-Crystalline Materials*, Clarendon Press, Oxford, 1979.
144. E.M. Hamilton, *Phil. Mag.* **26**, 1043 (1972).
145. A.J. Epstein, H. Rommelmann, R. Bigelow, H.W. Gibson, D.M. Hoffman and D.B. Tanner, *Phys. Rev. Lett.* **50**, 1866 (1983).
146. R. Colson and P. Nagels, *J. Non-Cryst. Solids* **35**, 129 (1980).
147. C. Budrowski, J. Przyluski, A. Pron, K. Ehinger and S. Roth, *Synth. Met.* **16**, 117 (1986).
148. K. Ehinger, S. Summerfield, W. Bauhofer and S. Roth, *J. Phys. C* **17**, 3753 (1984).
149. M. Audenaert, *Phys. Rev* B **30**, 4609 (1984).
150. L. Genzel, F. Kremer, A. Poglitsch, G. Bechtold, K. Menke and S. Roth, *Phys. Rev.* B **29**, 4595 (1984).
151. A.J. Epstein, R.W. Bigelow, H. Rommelmann, H.W. Gibson, R.J. Weagley and A. Feldblum, *Mol. Cryst. Liq. Cryst.* **117**, 147 (1985).
152. J.P. Pouget, J.C. Pouxviel, P. Robin, R. Comès, D. Begin, D. Billaud, A. Feldblum, H.W. Gibson and A.J. Epstein, *Mol. Cryst. Liq Cryst.* **117**, 75 (1985).
153. J. Chen, T.-C. Chung, F. Moraes, and A.J. Heeger, *Solid State Commun.* **53**, 757 (1985).
154. F. Moraes, J. Chen, T.-C. Chung and A.J. Heeger, *Synth. Met.* **11**, 271 (1985).
155. M. Nakahara and K. Maki, *Phys. Rev.* B **24**, 1045 (1981).
156. E.M. Conwell, *Phys. Rev.* B **33**, 2465 (1986).
157. C. Menendez and F. Guinea, *Phys. Rev.* B **28**, 2183 (1983).

158. E.M. Conwell and S. Jeyadev, *Synth. Met.* **17**, 69 (1987).

159. D.K. Campbell and A.R. Bishop, *Phys. Rev.* B **24**, 4859 (1981).

160. J.M. Ginder, A.F. Richter, A.G. MacDiarmid and A.J. Epstein, *Solid State Commun.* **63**, 97 (1987).

161. F. Zuo, M. Angelopoulos, A.G. MacDiarmid and A.J. Epstein, *Phys. Rev.* B **36**, 3475 (1987).

162. A.J. Epstein, J.M. Ginder, F. Zuo, R.W. Bigelow, H.S. Woo, D.B. Tanner, A.F. Richter, W.-S. Huang and A.G. MacDiarmid, *Synth. Met.* **18**, 303 (1987).

163. K. Mizoguchi, M. Nechtschein, J.-P. Travers and C. Menardo, *Phys. Rev. Lett.* **63**, 66 (1989).

164. M.E. Jozefowicz, R. Laversanne, H.H.S. Javadi, A.J. Epstein, J.P. Pouget, X. Tang and A.G. MacDiarmid, *Phys. Rev.* B **39**, 12958 (1989).

165. B. Abeles, P. Sheng, M.D. Coutts and Y. Arie, *Adv. Phys.* **24**, 407 (1975).

166. P. Sheng, B. Abeles and Y. Arie, *Phys. Rev. Lett.* **31**, 44 (1973).

167. H.H.S. Javadi, K.R. Cromack, A.G. MacDiarmid and A.J. Epstein, *Phys, Rev.* B **39**, 3579 (1989).

168. M.J. Rice and J. Bernasconi, *Phys. Rev. Lett.* **29**, 113 (1972).

169. F. Zuo, M. Angelopoulos, A.G. MacDarmid and A.J. Epstein, *Phys. Rev.* B **39**, 3570 (1989).

170. Z.H. Wang, A. Ray, A.G. MacDiarmid and A.J. Epstein, *Phys. Rev.* B **43**, 4373 (1991).

171. E.P. Nakhmedov, V.N. Prigodin and A.N. Samukhin, *Sov. Phys.-Solid State* **31**, 368 (1989).

172. V.K.S. Shante, C.M. Varma and A.N. Bloch, *Phys. Rev.* B **8**, 4885 (1973).

173. L. Zuppiroli, M.N. Bussac, S. Paschen, O. Chauvet and L. Forro, *Phys. Rev.* B **50**, 5196 (1994).

174. Z.H. Wang, C. Li, E.M. Scherr, A.G. MacDiarmid and A.J. Epstein, *Phys. Rev. Lett.* **66**, 1745 (1991).

175. S. Kivelson and A.J. Heeger, *Phys. Rev. Lett.* **55**, 308 (1985).

176. D.B. Tanner, G.L. Doll, A.M. Rao, P.E. Eklund, G.A. Arbuckle and A.G. MacDiarmid, *Synth. Met.* **28**, D141 (1989).

177. See E.M. Conwell and H.A. Mizes, *Synth. Met.* **65**, 203 (1994) for further discussion.

178. E.M. Conwell, H.A. Mizes and S. Jeyadev, *Phys. Rev.* B **40**, 1630 (1989).

179. E.M. Conwell, H.A. Mizes and S. Jeyadev, *Phys. Rev.* B **41**, 5067 (1990).

180. Y.H. Kim and A.J. Heeger, *Phys. Rev.* B **40**, 8393 (1989).

181. S. Hasegawa, M. Oku, M. Shimizu and J. Tanaka, *Synth. Met.* **41**, 155 (1991).

182. K. Kamiya, H. Inokuchi, M. Oku, S. Hasegawa, C. Tanaka and J. Tanaka, *Synth. Met.* **41**, 155 (1991).

183. P. Bätz, D. Schmeisser and W. Göpel, *Solid State Commun.* **74**, 461 (1990).

184. M. Logdlund, R. Lazzaroni, S. Stäfstrom, W.R. Salaneck and J.-L. Brèdas, *Phys. Rev. Lett.* **63**, 1841 (1989).

185. H.S. Woo and D.B. Tanner, *Synth, Met.* **41–43**, 159 (1991).

186. S. Stäfstrom, *Phys. Rev.* B **47**, 12437 (1993).

187. E.M. Conwell and H.A. Mizes, *Phys. Rev.* B **44**, 937 (1991).

188. J.F. Kwak, T.C. Clarke, R.L. Greene and G.B. Street, *Solid State Commun.* **31**, 355 (1979).

189. C.M. Gould, D.M. Bates, H.M. Bozler, A.J. Heeger, M.A. Druy and A.G. MacDiarmid, *Phys. Rev.* B **23**, 6820 (1981).

190. Y.W. Park, C. Park, Y.S. Lee, C.O. Yoon, H. Shirakawa, Y. Suezaki and K. Akagi, *Solid State Commun.* **65**, 147 (1988).

191. H. Kaneko and T. Ishiguro, *Synth. Met.* **65**, 141 (1994).

192. N. Basescu, Z.-X. Liu, D. Moses, A.J. Heeger, H. Naarmann, N. Theophilou, *Nature* **327**, 403 (1987).

193. Th. Schimmel, G. Denninger, W. Riess, J. Voit and M. Schwoerer, *Synth. Met.* **28**, D11 (1989).

194. T. Ishiguro, H. Kaneko, Y. Nogami, H. Ishimoto, H. Nishimaya, J. Tsukamoto, A. Takahashi, M. Yamaura, T. Hagiwara and K. Sato, *Phys. Rev. Lett.* **69**, 660 (1992).

195. Reghu, M., K. Väkiparta, Y. Cao and D. Moses, *Phys. Rev.* B **49**, 16162 (1994).

196. K. Akagi, M. Suezaki, H. Shirakawa, H. Kyotani, M. Shimonura and T. Tanabe, *Synth. Met.* **28**, D1 (1989).

197. K. Väkiparta, M. Reghu, M.R. Anderson, Y. Cao, D. Moses and A.J. Heeger, *Phys. Rev.* B**47**, 9977 (1993).

198. S. Kivelson and A.J. Heeger, *Synth. Met.* **22**, 371 (1988).

199. E.M. Conwell, *Synth. Met.* **20**, 289 (1987).

200. P. Sheng, *Phys. Rev.* B **21**, 2180 (1980).

201. Y. Nogami, M. Yamashita, H. Kaneko, T. Ishiguro, A. Takahashi and J. Tsukamoto, *J. Phys. Soc. Japan* **62**, 664 (1993).

202. A.J. Epstein, H.W. Gibson, P.M. Chaikin, W.G. Clark and G. Grüner, *Phys. Rev. Lett.* **45**, 1730 (1980).

203. E.M. Conwell and H.A. Mizes, *Synth. Met.* **38**, 319 (1990).

204. P. Robin, J.P. Pouget, R. Comès, H.W. Gibson and A.J. Epstein, *Polymer* **24**, 1558 (1983).

205. S. Flandrois, C. Hauw and B. François, *Mol. Cryst. Liq. Cryst.* **117**, 91 (1985).

206. P. Sautet, O. Eisenstein and E. Canadell, *Chem. Mater.* **1**, 225 (1989).

207. A.B. Kaiser and S.C. Graham, *Synth. Met.* **36**, 367 (1990).

208. V.N. Prigodin and K.V. Efetov, *Phys. Rev. Lett.* **70**, 2932 (1993).

209. J. Plocharski, *Mater. Sci. Forum* **42**, 17 (1989).

210. D. Glaser, T. Schimmel, M. Schwoerer and H. Naarmann, *Makromol. Chem.* **190**, 3217 (1989).

211. J. Tsukamoto, *Adv. Phys.* **41**, 509 (1992).

212. G. Thummes, U. Zimmer, F. Körner and J. Kötzler, *Jpn. Appl. Phys.* **26–3**, 713 (1987).

213. G. Thummes, F. Körner and J. Kötzler, *Solid State Commun.* **67**, 215 (1988).

214. K. Väkiparta, Reghu, M., M.R. Anderson, Y. Cao, D. Moses and A.J. Heeger, *Phys. Rev.* **B47**, 9977 (1993).
215. Y. Nogami, H. Kaneko, H. Ito, T. Ishiguro, T. Sasaki, N. Toyota, A. Takahashi and J. Tsukamoto, *Phys. Rev.* B **43**, 11829 (1991).
216. Reghu, M., K. Väkiparta, Y. Cao and D. Moses, *Phys. Rev.* B **49**, 16162 (1994).
217. W.L. McMillan, *Phys. Rev.* **B24**, 2739 (1981).
218. A.I. Larkin and D.E. Khmelnitskii *Zh. Eksp. Teor. Fiz.* **83**, 1140 (1982) [*Sov. Phys.* **JETP56**, 647 (1982)].
219. P.A. Lee and T.V. Ramakrishrian, *Rev. Mod. Phys.* **57**, 287 (1985).
220. Reghu, M., C.O. Yoon, D. Moses, Y. Cao and A.J. Heeger, *Synth. Met.* **69**, 329 (1995).
221. D.E. Khmelnitskii and A. Larkin, *Solid State Commun.* **39**, 1069 (1981).
222. H.H.S. Javadi, A. Chakraborty, C. Li, N. Theophilou, D.B. Swanson, A.G. MacDiarmid and A.J. Epstein, *Phys. Rev.* B **43**, 2183 (1991).
223. M. Yamamura, T. Hagiwara and K. Iwata, *Synth. Met.* **26**, 209 (1988).
224. K. Sato, M. Yamamura, T. Hagiwara, K. Murata and M. Tokumoto, *Synth. Met.* **40**, 35 (1991).
225. Y. Nogami, J.P. Pouget and T. Ishiguro, *Synth. Met.* **62**, 257 (1994).
226. R.S. Kohlman, T. Ishiguro, H. Kaneko and A.J. Epstein, *Synth. Met.* **69**, 325 (1995).
227. C.O. Yoon, Reghu, M., D. Moses and A.J. Heeger, *Phys. Rev.* B **49**, 10851 (1994).

228. Y. Cao, P. Smith and A.J. Heeger, *Synth. Met.* **48**, 91 (1992).
229. Reghu, M., Y. Cao, D. Moses and A.J. Heeger, *Phys. Rev.* B **47**, 1758 (1993).
230. Reghu, M., C.O. Yoon, D. Moses, A.J. Heeger and Y. Cao, *Phys. Rev.* B **48**, 17685 (1993).
231. K. Lee, A.J. Heeger and Y. Cao, *Phys. Rev.* B **48**, 14, 884 (1993).
232. N.F. Mott and M. Kaveh, *Adv. Phys.* **34**, 329 (1985).
233. S. Hasegawa, M. Oku, M. Shimizu and J. Tanaka, *Synth. Met.* **38**, 37 (1990).
234. J. Joo, Z. Oblakowski, G. Du, J.-P. Pouget, E.J. Oh, J.M. Weisinger, Y. Min, A.G. MacDiarmid and A.J. Epstein, *Phys. Rev.* B **49**, 2977 (1994).
235. J. Joo, V.N. Prigodin, Y.G. Min, A.G. MacDiarmid and A.J. Epstein, *Phys. Rev.* **50**. 12226 (1994).
236. R.S. Kohlman, J. Joo, Y.Z. Wang, J.-P. Pouget, H. Kaneko, T. Ishiguro and A.J. Epstein, *Phys. Rev. Lett.* **74**, 773 (1995).
237. A.B. Kaiser, *Phys. Rev.* B **40**, 2806 (1989).
238. Y.W. Park, W.K. Han, C.H. Choi and H. Shirakawa, *Phys. Rev.* B **30**, 5847 (1984).
239. A.B. Kaiser, *Phys. Rev.* B **29**, 7088 (1984); B **35**, 4677 (1987).
240. Y.W. Park, C.O. Yoon, C.H. Lee, H. Shirakawa, Y. Suezaki and K. Akagi, *Synth. Met.* **28**, D27 (1989).
241. E.B. Park, J.S. Yoo, J.Y. Park, Y.W. Park, K. Akagi and H. Shirakawa, *Synth. Met.* **69**, 61 (1995).

CHAPTER 2

Charge Transport in Conducting Polymers

Reghu Menon*

Institute for Polymers and Organic Solids, University of California at Santa Barbara, USA

1 INTRODUCTION

The main objective of this chapter is to present an overview of the past few years, study of charge transport in doped conducting polymers. From the initial discovery [1] of nearly 12 orders of magnitude of enhancement in conductivity in doped polyacetylene in the year 1977, the study of charge transport in doped conducting polymers has made substantial progress [2].

The investigation of charge transport in conducting polymers was mainly focused on the hopping transport phenomena in earlier review articles [3], since the extent of disorder in previous generation materials was rather substantial and the intrinsic metallic features

were suppressed. However, the remarkable improvement in reducing the extent of disorder in doped conducting polymers in the past seven years has provided the opportunity to investigate the intrinsic metallic features in these systems. The disorder-induced metal–insulator transition (M–I transition) in new generation conducting polymers has been discussed extensively in a recent review article [2]. The main motivation for this review article is to attempt a comparative study of charge transport in conducting polymers with respect to other systems near the M–I transition. The anisotropy in electronic properties, electron–electron (e–e) correlation effects, disorder, etc. are some of the common features in these systems.

*Current address: CSIR, Regional Research Laboratory, Trivomdrum 695019, Kerala, India.
Handbook of Organic Conductive Molecules and Polymers: Vol. 4. Conductive Polymers: Transport, Photophysics and Applications.
Edited by H. S. Nalwa. © 1997 John Wiley & Sons Ltd

Hence, a comparative study could provide more insight into the similarities and differences of charge transport in doped conducting polymers with respect to other systems. Moreover, the universality of the current understanding of disorder-induced M–I transition could be verified during this process.

The role of disorder in the M–I transition in doped conducting polymers has been studied extensively during the last six years [2]. The charge transport properties as a function of temperature, pressure, magnetic field, etc. on a wide variety of samples of doped polyacetylene $(CH)_x$, polyaniline (PANI), poly-pyrrole (PPy), polythiophene (PT), poly(p-phenylene-vinylene) (PPV), etc. have been published in literature. Nevertheless, the theoretical understanding of charge transport phenomena is not yet completely developed in doped conducting polymers. The main bottlenecks were involved with the difficulty in quantifying the intrinsic and extrinsic parameters (e.g. the role of interchain interactions, intrinsic low dimensionality of the system, anisotropic diffusion coefficient of charge carriers, role of dopant ions, the extent of disorder, etc.) that influence the charge transport. The intrinsic and extrinsic contributions to charge transport are strongly intermixed and it is often not trivial to quantify the individual contributions from various parameters. For example, the 'intrinsic' vs. 'effective' electronic dimensionality of the system plays a crucial role in determining the microscopic transport properties. Thus several papers regarding the charge transport in doped conducting polymers have appeared with only a qualitative level discussion. Hence, the theoretical modeling of transport properties in conducting polymers is a challenging problem due to the extreme complexity of the system.

In general, polymeric materials were considered as insulators before the discovery of metallic poly(sulfur nitride) $[(SN)_x]$ in the beginning of 1970 [4], and the discovery of nearly 12 orders of magnitude enhancement in conductivity in doped $(CH)_x$ [1]. The theoretical possibility of conducting and superconducting organic materials was first proposed by Little in the year 1964 [5]. Prior to that, London [6] speculated the possibility of superconductivity in biomolecular systems as early as 1937. Later, Ladik *et al.* [7] and Pearlstein [8] pursued theoretical calculations in DNA and suggested the possibility of local superconductive-type enhancement in conductivity and the formation of a local Bose-condensate state.

The investigation of electronic properties of organic solid state materials slowly gathered momentum after the discovery of conducting charge transfer complexes in late 1960s (for e.g. TTF-TCNQ, etc.) [9,10]. The discovery of conductive polyconjugated systems [1] and organic photoconductive materials for xerography [11] in 1970s gave a tremendous boost to this field. Later, the observation of various collective ground state phenomena (e.g. superconductivity [10], charge density waver [12], spin density wave [13], bond-order wave [14], etc.), has provided additional stimulus for rapid developments in organic solid state materials. More recently, the non-linear optical properties [15] and electroluminiscence [16] in π-electron rich materials have drawn considerable attention towards potential applications.

The study of charge transport in systems similar to that of doped conducting polymers, for example, in molecularly doped polymers [17], organic charge-transfer complexes [10,18], carbon-black filled polymers [19], etc. has made substantial progress in the past many years. The attemps to extend some of those models applied to systems similar to that of conducting polymers have not been very successful. However, the recent study of disorder-induced M–I transition in new generation conducting polymers has unraveled the intrinsic metallic features [2].

Polyconjugated chains consist of alternating single (σ-bonds) and double bonds (π-bonds) $[(-C=C-C=C-C=C-)_n]$ [14,20,21]. The carbon atoms in single and double bonds are in sp^3 and sp^2 hybridization states, respectively. The π-electrons from p_z orbitals of sp^2 carbons are highly delocalized and easily polarizable, and this makes the electrical and optical properties of polyconjugated systems rather different from conventional systems. The extent of delocalization of π-electrons along the polymer backbone plays a dominant role in the electrical and optical properties of polyconjugated systems. The molecular-scale properties are mainly governed by the extent of intrachain delocalization of π-electrons; however, the extent of interchain delocalization plays a major role in the macroscopic scale solid state properties.

In general, doped conducting polymers belong to the category of doped semiconductors [14,20]. The conductivity of undoped polyconjugated systems is of the order of 10^{-6}–10^{-12} S/cm; hence it can be considered at the semiconductor–insulator boundary. The values of band gap in polyconjugated systems vary from 1 to 4 eV. The bandwidths parallel and perpendicular to the chain axis in $(CH)_x$ are of the order of 10 and 1 eV, respectively [14,21]. The comparative values of bandwidth in other organic systems are the following: organic charge transfer complexes along the π-orbital overlapping direction ≈ 0.5 eV [18], $C_{60} \approx$ 0.1 eV [22], etc. The charge carrier concentration in conducting polymers can be varied by many orders of magnitude (nearly 12 orders) by doping [14,20]. The carrier mobility in doped conducting polymers is lower with respect to that in inorganic semiconductors, and this is mainly due to the presence of relatively higher disorder in the former systems. However, high-quality materials of conducting polymers show the typical electronic properties of disordered metals [2]. In general, the transport properties are crucially dependent on the extent of disorder and the universal features of disorder-induced M–I transition are observed in doped conducting polymers.

The intrinsic quasi-one-dimensional (q-1D) nature of conducting polymer chains plays a major role in determining the structural and electronic properties of these materials. Polyconjugated chains can be considered as a q-1D metal with one charge carrier per carbon atom. It is well known that such a half-filled system is susceptible to Peierls instability by opening up an insulating gap at the Fermi level [10,23]. In general, the driving force of the broken symmetry ground state of low-dimensional systems is determined by the strength of various interactions, effective (electronic) dimensionality of the system, band filling, etc. Thus the competition among various parameters could create a complex phase diagram of the ground states or excited states of conducting polymers; however, the dominant role played by disorder smears out the rich features in real systems.

A rigorous understanding of the intrinsic/primary excitations in doped and undoped polyconjugated systems is still lacking [14,24–27]. The scenario does not look trivial in spite of the fact that a considerable amount of both theoretical and experimental work has been carried out by several groups focusing on this fundamental problem. The competition among various factors (electron–phonon interaction, e–e interaction, quantum lattice fluctuations, screening associated with interchain interactions and disorder, extent of both intra and interchain delocalization of electronic states, etc.) makes the situation so complex that the quantitative estimation of the contribution from individual parameters becomes a difficult task. Moreover, the extrinsic contributions from disorder, inhomogenous doping, etc. make it rather difficult to envisage a microscopic mechanism for charge transport in doped conducting polymers. Nevertheless, the significant improvement in reducing the extent of disorder has clarified many ambiguous notions regarding the transport properties in conducting polymers.

2 ELECTRONIC PROPERTIES OF LOW-DIMENSIONAL SYSTEMS

The intrinsic electronic properties of polyconjugated systems are governed by the physics of q-1D systems [24–27]. Thus in this section a brief overview of the electronic properties of low-dimensional systems is presented.

In general three types of M–I transitions can be expected in a 1D metal: (a) Peierls M–I transition which is mainly due to strong electron–phonon coupling [23], (b) Mott M–I transition due to e–e interactions [28] and (c) Anderson M–I transition due to random disorder potentials [29].

The pioneering works by Fröhlich [30] and Peierls [23] in the early 1950s have laid the foundation of the electronic properties of low-dimensional systems. The main finding is that a partially filled 1D system is unstable with respect to lattice modulation for any finite electron–phonon coupling. This lattice instability favors the opening up of a gap in the electronic spectrum at the Fermi level. The energy of the charge carriers is lowered by the opening up of the gap. Usually, the energy gained by the charge carriers is larger than the loss in elastic energy and this produces the lattice modulation. The associated modulation of the charge density and spin density favors the formation of a collective condensate named a charge density wave (CDW) [12] and spin density wave (SDW), respectively [13]. However, at finite temperatures the thermally excited carriers facilitates the lowering of the gap and it altogether vanishes above the transition temperature.

The Peierls instability for a 1D metal with a half-filled band consisting of a periodic array of atoms (lattice constant a) is shown in Figure 2.1 [10,23]. The periodic lattice distortion with a period ϕ is given by (π/k_F), where k_F is the Fermi wave vector. The density of states of undistorted lattice (metal) is shown in Figure 2.1a. In the Peierls insulator state, as shown in Figure 2.1b, the electronic energy of the system is lowered due to the opening of Peierls gap at the Fermi level and only the states up to $\pm k_F$ are occupied. The size of the single-particle gap (Δ) is proportional to the amplitude of lattice distortion (u). However, in a 3D array of 1D chains the quantum lattice fluctuations and the interchain coupling tend to reduce the Peierls gap.

In the Peierls transition, the increase of the elastic energy and the decrease of the electronic energy are proportional to u^2 and $u^2 \ln u$, respectively. Thus Peierls transition in low-dimensional systems is favored, since the total energy of the distorted state is lower than that of the undistorted metal. Near the transition tempera-

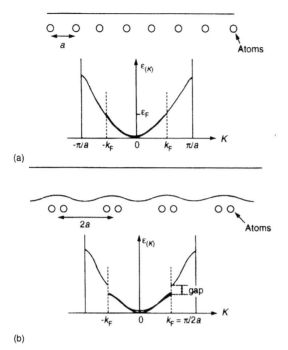

(a)

(b)

Figure 2.1. The Peierls instability for 1 D metal with a half-filled band consisting of a periodic array of atoms with lattice constant a: (a) undistorted (metal); (b) distorted lattice (insulator).

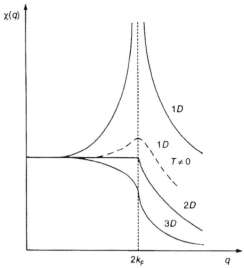

Figure 2.2. The $2k_F$ anomaly of electron gas in one, two and three dimensions. A plot of electronic susceptibility vs. the wavevector.

ture the anomaly in the phonon dispersion curve is exhibited by the sharp dip of the phonon frequency at $2k_F$, which is called the Kohn anomaly [31]. The dip in phonon dispersion (softening of the lattice) is caused by the strong coupling of $2k_F$ phonons with charge density.

The $2k_F$ anomaly of an electron gas at 1-, 2- and 3-dimensional systems can be easily identified from a plot of electronic susceptibility vs. wave vector as shown in Figure 2.2 [18,23]. The electronic susceptibility (χ_q) for a non-interacting electron gas is expressed by the Linhard function:

$$\chi_q = \sum[f(k) - f(k+q)]/[\varepsilon_{k+q} - \varepsilon_k] \quad (2.1)$$

where f(k) is the Fermi–Dirac distribution function, ε_k is the electronic energy of momentum $\hbar k$. The tendency for the divergence of (χ_q) at $2k_F$ upon reducing the dimensionality of the system is clearly shown in Figure 2.2. Thus the Kohn anomaly is related to the logarithmic singularity of (χ_q) at $2k_F$, and the $2k_F$ anomaly in 1D systems decreases as the temperature increases. This remarkable sensitivity of 1D metals makes them susceptible to large fluctuations in order parameters and the associated instabilities.

For a 1D metal with a half-filled band, all the states near the Fermi level differing by $2k_F$ are nearly degenerate, and the coupling between states at $+k_F$ and states at $-k_F$ occurs at the Peierls instability. Thus the energy for transferring the carriers from one side to other side of the Fermi distribution is rather small in 1D. The electrons and holes at $\pm k_F$ of the nested Fermi surface involves the formation of collective mode.

The collective condensate in the Peierls–Frohlich state is either commensurate or incommensurate with the underlying lattice depending up on the band filling [10,12]. In the case of arbitrary band filling the density wave is incommensurate (ϕ/a is irrational). The enhanced Frohlich conductivity due to the sliding motion of CDW in the Peierls state is rather sensitive to various extrinsic factors (disorder, etc), and it usually occurs at infrared frequencies. The sliding motion of the collective condenstate could be pinned by the commensurate lattice, disorder, etc. Considerable work regarding the pinning–depinning mechanisms of collective condensate and the associated non-linear processes have been reported in literature [12,13].

The dynamics of the collective state in 1D systems are mainly determined by interchain interactions, disorder, pinning-depinning processes, etc [12–14]. The interchain coupling stabilizes long-range order at finite temperature. The ratio of intrachain vs. interchain transfer integrals ($t_\perp/t_\parallel$) plays a decisive role in determining the transition temperature and the nesting of Fermi surface. As long as $t_\perp$ is less than the Fermi energy, the Fermi surface is open in the transverse direction. Moreover, the ratio of $t_\perp/t_\parallel$ plays a role in the phase and amplitude of the order parameter fluctuations near the transition temperature. The ground state of the collective mode is influenced by $t_\perp/t_\parallel$. This is vividly shown in the case of the pressure-induced transition from the CDW state to the superconducting state in niobium triselenide ($NbSe_3$) [12,32]. Thus the competition between the superconducting state and CDW/SDW state in q-1D systems is very much dependent on the ratio of $t_\perp/t_\parallel$.

Disorder is an important extrinsic parameter that tunes the forward and backward scattering process in 1D systems [10,14]. Usually, disorder tends to suppress or smear out the phase transition and the associated long-range ordering process. The phase of collective mode is rather easily pinned or modulated by the impurity potential. Thus, if the extent of disorder is large enough, it can completely suppress the dynamics of the collective mode. Similarly, the presence of commensurate potentials dampens the dynamics of the collective mode. However, any finite interchain interaction makes the q-1D system less sensitive to impurity potentials, and consequently the order parameter correlation tends to fall off algebraically, instead of exponentially.

It is well known that q-1D systems are prone to fluctuations and instabilities [10,14]. Moreover, the response of the collective mode to external perturbations is substantial. The extensive study of the frequency and electric field dependencies of complex conductivity in CDW systems is a typical example [10,12]. Various types of nonlinear processes (for e.g. mode-locking, bifurcations, chaos, etc.) have been observed in CDW systems [12,32]. Experimental studies about the static and dynamic properties of CDW [12] and SDW [13] systems in various organic [10] (tetrathiafulvalenium tetracyanoquinodimethane (TTF-TCNQ), bistetra-methyl-tetraselenafulvalene (TMTSF) salts, etc) and inorganic [32] ($NbSe_3$, TaS_3, $K_{0.3}MoO_3$ etc.) have been extensively reviewed.

Although the Peierls M–I transition due to electron–phonon interaction is the most common feature of 1D metals, the Mott M–I transition, due to e–e interaction, could also produce a gap at $T = 0$ K. Moreover, in a 1D Mott insulator the lattice instability occurs at $2\pi/2k_F$ [10]. Thus Mott and Peierls instabilities could coexist at certain band filling. Many TCNQ charge transfer complexes are Mott insulators, for example, K-TCNQ, Cs_2-TCNQ, etc. [18].

Although the Peierls and Mott M–I transitions are the two common intrinsic insulating ground state in 1D metals, the presence of any extrinsic disorder can easily

facilitate 1D localization. The case of Anderson localization in 1D systems is well known by the famous theorem that the eigenstates have nearly zero amplitudes on all of the sites in a 1D lattice with site-diagonal disorder; or in other words, the presence of any infinitesimal disorder induces exponential localization of all the electronic states in a 1D system [29]. However, recently Phillips and Wu [33] proposed that this situation is absent in some cases (e.g. the random dimmer model and replusive binary alloy). Moreover, they suggested that the random dimer model could be used to understand the metallic state of doped conducting polymers.

Recently, the investigation of the electronic properties of strongly correlated low-dimensional systems (e.g. high-temperature superconductors [34] has created a lot of interest in the search for alternate models to the conventional Fermi liquid theory. Clarke *et al.* [35] have estimated the interchain conductivity for a system of weakly coupled Hubbard chains. The low-energy state of such a system is identified as the spin–charge separated Luttinger liquid. The interchain single particle hopping could be coherent or incoherent (partially or completely) depending on the transfer integral. Moveover, the recent photoemission studies in organic conductors have suggested the possibility of the Luttinger liquid model in those systems [36,37]. The photoemission studies in (N, N'-dicyanoquinonediimine-Cu ((DMe-$DCNQI$)$_2Cu$) salts (a q-1D system) suggests a crossover from a 1D Luttinger liquid to a 3D Fermi liquid as a function of excitation energy, temperature and mass anisotropy [36]. Thus the preliminary studies of 1D metals from the point of view of the Luttinger liquid model is rather encouraging.

Although the above models have convincingly described many properties of various low-dimensional systems, none of these models was found to be quite satisfactory to explain all of the features of the metallic state in conducting polymers. The Su–Schrieffer–Heeger (or Huckel) model [14], the Pariser–Parr–Pople (PPP) model [38] and the Peierls–Hubbard model [39] are some of the models commonly used to investigate the physical properties of polyconjugated systems. These models are qualitatively described in Section 4.

3 ELECTRONIC PROPERTIES OF DISORDERED SYSTEMS

From the previous section it is known that the intrinsic q-1D nature of conducting polymers could induce M–I transition in these systems. However, the main source of M–I transition in conducting polymers is the disorder-induced localization [2]. Thus a brief introduction to the electronic properties of disordered systems is presented in this section, although this topic has been extensively reviewed by several authors.

The pioneering works by Mott [28] and Anderson [29] have laid the foundation for the study of M–I transition. In the Mott M–I transition the e–e correlation effects play a major role, while in the case of the Anderson transition the localization of electronic wavefunctions can occur if the random component of disorder potential is large enough with respect to the bandwidth.

In the Mott transition, the system undergoes a first-order phase transition from the metallic side, in which the Coulomb interaction is screened, to an insulating side when the screening breakdown. This occurs by varying either the interatomic distance (by chemical pressure or external pressure) or the charge carrier density (by doping). Consequently, the correlation effects split the band into filled and empty levels resulting in the formation of a gap at the Fermi level. In metals with weak correlation effects the onsite interaction energy of two electrons with opposite spins is not very large. Usually, in strongly correlated systems the band width is rather small. In such a case the electronic energy of upper and lower band levels (upper and lower Hubbard bands) is determined by the presence or absence of another electron on the same site. At the Mott transition the lower band is filled and the upper band is empty, and the system becomes an insulator. This type of Mott-Hubbard transition is observed in several oxides, charge transfer salts, etc [18]. The main difference between Mott and Anderson transitions is that the former occurs even in an ideal periodic structure due to e–e interaction, whereas the later is due to disorder in a single-particle (non-interacting) model. The Mott transition is usually achieved by varying the effective Coulomb interaction in the system, while the Anderson transition is induced by varying the extent of disorder and the resultant localization of scattered waves due to interference in random media.

The electronic dimensionality of the system plays a major role in Anderson localization. A 1D conductor becomes an insulator due to either site-energy fluctuations or correlation effects. The scaling theory of localization by Abrahams et al. [40] proposed that any arbitrarily weak disorder can localize all of the states in 2D (logarithmic function of temperature) and in 1D (exponential function of temperature). However, in 3D the electronic wave functions may remain extended or

localized in the entire system depending on the extent of disorder. The recent theoretical developments in Anderson localization for various models including both energy and positional site disorders have been reviewed by Skinner [41].

Although the wavefunction is locally perturbed near the fluctuating random potential it remains extended well beyond the mean free path in 3D systems. In the case of strong disorder the envelope of wave function decays exponentially as follows [28]:

$$|\psi(r)| \sim \exp\left(|r - r_0|/\xi\right) \qquad (2.2)$$

where ζ is the localization length of the localized state. The extent of disorder plays a decisive role in determining whether the wavefunction is localized in the vicinity of a site or if it remains extended in the whole system. Disorder tunes both the overlap and the phase of the wavefunctions. In the presence of strong disorder the overlap of the wavefunctions drops off exponentially and the phase of the wavefunctions varies randomly from site to site. In such a case the formation of a coherent state that remains delocalized over many sites is rather difficult. Consequently the overlap integral (nearest neighbor transfer matrix element, V) approaches zero; hence the spreading of the wavefunctions is suppressed and the tails of the wavefunctions overlap with exponentially small amplitude. Thus the temporal evolution of wavefunction in space remains localized on the same site. In this regime the coherent transport of charge carrier ceases and the system becomes an insulator at $T = 0$. On the other hand, if all the sites have nearly the same energy, then the wavefunctions of the Bloch states are extended over the entire system and the charge transport is ballistic.

The mean free path (λ) in a system with a bandwidth (the width of the uniform distribution of site energies) of W', overlap integral V and interatomic distance a is given by [28]

$$1/\lambda = 0.7(1/a)(V/W')^2 \qquad (2.3)$$

In the limit of weak disorder, $\lambda \sim (2V/W')^2$. The ratio of V/W' determines the transition from the coherent diffusive propagation of wavefunction (extended states) to the 'trapping' of wavefunctions in random potential fluctuations (localized states). If $V \gg W'$ then the transport is diffusive. If $W' \ll 2V$, then λ is nearly the size of the system and the electronic states are extended. In other words, the extended states are multiple-scattered plane waves that occupy some fraction of the system. The wave functions get localized in the disorder potential fluctuations as the value of W

increases; when λ becomes of the order of a (Ioffe–Regel limit [42]) the system is fully localized. Thus in 3D a continuous transition from extended to localized states is possible by tuning the ratio of V/W'.

Considerable numerical studies regarding this transition from extended to localized states have been carried out by various groups [28]. According to Anderson, localized states are formed throughout the band for a critical value of V/W' in 3D. When V/W' exceeds this critical value the wavefunctions fall off exponentially from site to site and the delocalized states can not exist in such a case. The states in the band tails are the first to be localized, since these are the first to lose the ability to tunnel resonantly as the extent of disorder increases. Moreover, the probability of finding the appropriate matching energy levels for the charge carriers to tunnel in the band tails also decreases upon increasing the extent of disorder.

The wavefunctions of localized states are of the form $e^{-r/Lc}\operatorname{Re}[\psi_o]$, where $\psi_o = \Sigma\, c_n \exp(i\phi_n)\psi_n$; ($c_n$ are the real coefficients, ϕ_n are the random phases), and the localization length (L_c) tends to infinity as (V/W') approaches the critical value [28].

If V/W' is below the critical value (W' is not very large) so that the entire band is not localized, then the extended states at the band center and the localized states at the band tails coexist. Mott name this critical energy that separates the localized states from the extended states as the mobility edge (E_c) [28]. The location of Fermi energy (E_F) with respect to the E_c depends on the extent of disorder, charge carrier density, etc. This could be experimentally identified from the temperature dependence of conductivity [$\sigma(T)$] at low temperatures ($T \approx 1$ K). When the extent of disorder is sufficiently large to induce the E_F to cross E_c (in other words, to place the Fermi energy in the region of localized states), then the transition from metallic state (finite value of the conductivity as T $\to$ 0) to a non-metallic state (the conductivity goes to zero as T $\to$ 0) takes place. Mott called this disorder-induced M–I transition by moving the E_F cross E_c as the Anderson transition [28]. The spread out of the wavefunctions of the localized states into the classically inaccessible states is determined by the energy difference ($E - E_c$). The localization length diverges as the Fermi energy approaches E_c and it becomes smaller as it moves to the center of the localized regime. If the localized and extended states are separated by a rather small energy difference, then L_c is large. Hence, $L_c \sim a[E_0/(E_c-E)]^\nu$; where $\nu \approx 1$, and E_0 is a constant.

According to the Ioffe–Regel criterion [42] the lower limit of the mean free path cannot be less than the de Broglie wavelength (the interatomic spacing) [28]. Based upon this fact, Mott proposed that the Anderson transition as $E_F \to E_c$ is discontinuous (first-order phase transition) and introduced the concept of minimum metallic conductivity ($\sigma_{\min}$). In 3D, $\sigma_{\min} = 0.03\, e^2/3\hbar a$. If the extent of disorder is not very large, then the conductivity follows the Drude equation ($\sigma = e^2 k_F^2 \lambda/3\pi^2\hbar$, where k_F is the Fermi wave vector). Thus in the vicinity of $\sigma_{\min}$, $k_F\lambda \approx 1$.

The scaling theory of localization by Abrahams *et al* [40] demonstrated that the M–I transition is continuous (second-order phase transition) in 3D. The conductivity of a metal goes smoothly to zero as $E_F \to E_c$, so that in general $\sigma_{\min}$ does not exist. Moreover, one could expect such a smooth transition due to the fact that in real systems the mobility edge is broadened by interaction with phonons or by e–e interactions. Another important fact is that, even at very low temperatures a charge carrier can jump to lower energy levels with the emission of a phonon; thus the localized states could be broadened by $\Delta E \sim \hbar/\tau$, where τ is the finite lifetime of localized state. Hence a localized state could be delocalized by phonons. All the above mentioned factors suggest that Mott's original proposal for a sharp discontinuous Anderson transition in 3D is not really valid.

Recently, Möbius [43] pointed out that the scaling theory does not completely disprove the existence of $\sigma_{\min}$, since the low-temperature conductivity near the M–I transition can be explained alternatively by a combination of $\sigma_{\min}$ on the metallic side and the Coulomb interaction contribution on the insulating side. However, the concept of $\sigma_{\min}$ is a useful parameter to identify systems with $\sigma \sim \sigma_{\min}$, which can be considered as near the M–I transition; systems with $\sigma < \sigma_{\min}$ can be considered as on the insulating side of the M–I transition.

In realistic disordered systems, however, the M–I transition is neither pure Anderson transition (localization due to quantum interference) nor Mott transition (Coulomb interaction) [44,45]. Thus neither model by itself is satisfactory to comprehend fully the complexity of M–I transition. For example, by increasing the extent of disorder the diffusion of the charge carriers decreases; hence they interact more strongly. Thus understanding the simultaneous roles of both localization and interaction near the M–I transition is a challenging task and there is no satisfactory solutions yet. However, the preliminary transport property measurements in oriented conducting polymers have shown that this is an ideal system for studying the interplay of both localization and interaction [2]. The anisotropic diffusion coefficient in oriented conducting polymers has provided an unique opportunity to investigate the rich interplay of both localization and interaction, when compared to other systems. The details about the interplay of localization and interactions in disordered metallic regime is discussed in Section 8.1.

Electronic properties on the insulating side of the M–I transition are described by various theoretical models, depending on the details of the system. For example, Mott's variable range hopping (VRH) [46], Fermi glass [47], Coulomb glass [48], superlocalization [49,50], mutifractal localization [51,52], etc. The $T^{-(D+1)}$ dependence of conductivity (where D is the dimensionality of the system) at low temperatures in various systems has provided ample evidence for Mott's VRH. In VRH, the carriers hop to sites with the least energy difference and the average hopping length increases as $T^{-(D+1)}$ at low temperatures. According to VRH theory [46],

$$\sigma(T) \propto \exp[-(T_0/T)^\gamma] \qquad (2.4)$$

where T_0 is the characteristic temperature that determines the thermally activated hopping among localized states. In conventional VRH, $\gamma = 1/(D+1)$. Mott's VRH has been extensively reviewed by various authors [28]. In highly disordered conducting polymers VRH behavior has been observed at low temperatures [3].

Anderson [47] proposed that a degenerate electron gas in random disorder potentials tends to localize if the magnitude of the disorder potential is large compared with the bandwidth. In such a case all the states become localized, and the system is a 'Fermi glass' which is an insulator with a continuous density of localized states (no energy gap) occupied according to Fermi statistics. The temperature dependence of conductivity at low temperatures in a Fermi glass system is due to phonon-induced hopping of localized quasiparticles and their polarization clouds. At high temperatures the conductivity shows activated behavior and at low temperatures it shows $T^{-1/4}$ dependence. This type of Fermi glass behavior has been observed in some doped PANI samples [2].

In strongly disordered systems, in which the Coulomb interaction is not very well screened, there is a possibility that a large number of charge carriers are subjected to long-range interactions. Such systems are considered as Coulomb glass [48]. Electronic and thermodynamic properties of Coulomb glass systems are mainly governed by the spectrum of low-energy many-particle excitations due to the long-range correlated displacement of many carriers. A simple phenom-

enological model of charge transport in Coulomb glass systems indicates that the hopping probability decreases exponentially with hopping distance, and the latter is proportional to the number of particles contributing to the correlated processes.

Levy and Souillard [49] have shown that in a fractal structure, the wavefunctions for states near the Fermi level are superlocalized, and decay as $\psi(r) \propto \exp[-(r/L_c)^\zeta]$, where L_c is the localization length and ζ is the superlocalization exponent which is greater than unity (in Anderson localization, $\zeta \approx 1$). Deutscher, Levy and Souillard predicted that the temperature dependence of electrical conductivity which results from VRH between superlocalized states would be of the form $\sigma(T) \propto \exp[-(T_0/T)^\gamma]$; where $\gamma = \zeta/(\zeta+D) \approx 3/7$. Harris and Aharony [53] have predicted that γ is approximately 0.38 and 0.35 for 2D and 3D, respectively. The theory of VRH among superlocalized states was generalized by van der Putten et al. [54] to include the Coulomb interaction; van der Putten et al. obtained $\gamma \approx 0.66$ and $\zeta \approx 1.94$ consistent with the experimental results in carbon black polymer composites. However, Aharony et al. [55] argued that $\zeta = 1.36$ for 3D and suggested that the generalized VRH equation should be used in the superlocalized regime:

$$\sigma(T) = \sigma_0(T_0/T)^s \exp[-(T_0/T)^\gamma] \qquad (2.5)$$

where s is the unknown exponent of the prefactor. However, due to the large uncertainty in the value of s it is not possible to determine the values of T_0 and γ unambiguously from equation (2.5). Moreover, Aharony et al. [55] pointed out that the a priori requisite for applying the fractal geometry near the percolation threshold is that the length scales satisfy the following condition: $a \ll L_c \ll r_h \ll \xi_p$, where r_h is the hopping length, L_c is the localization length and ξ_p is the percolation correlation length. The hopping transport in PANI–Camphorsulfonic acid (CSA)/PMMA [(poly(methylmethacrylate)) blends near the percolation threshold indicates the possibility of superlocalization in these systems [2].

The concept of fractal character of localized wavefunctions near the mobility edge has been proposed by various groups; for example, [56–58]. In weak (strong) disorder, wavefunctions are homogeneously extended over the whole sample (exponentially localized at a particular site of the sample). While approaching the mobility edge at the critical disorder W_c, the length scales of the self-similar fluctuations of the amplitude of the wavefunctions becomes larger than the lattice spacing. The nature of the wavefunction varies continuously from exponentially localized to fractally localized (power law behavior) to fractally extended to uniformly extended states [57]. The length scales of these multifractal fluctuations vary from the localization length of the localized states to the coherence length of extended states.

In the standard VRH theory [46], the total number of states involved in the hopping conduction is assumed to be x^D times the density of states per unit volume in a region of linear dimension x and $\sigma(T)$ is given by equation (2.4). Aoki [56] and Schreiber [57] have shown that near the mobility edge, the localized wavefunctions are fractal; due to the fractal nature of the wavefunctions, the number of states involved in hopping conduction behaves like x^d where d is the fractal dimensionality (rather than x^D). The conductivity resulting from VRH among such spatially fractal localized wavefunctions is expressed by

$$\sigma(T) \propto \exp[-(T_o/T)^{1/(d+1)}] \qquad (2.6)$$

Since $d < D$, γ is larger than the usual values (1/4 for 3D, etc). Moreover, calculations by Schreiber and Grussbach [59] have shown that D decreases significantly with increasing disorder, yielding a further increase in the exponent of $(1/T)$ for increasing localization. Although considerable theoretical work regarding the fractal nature of wavefunctions near the mobility edge has been carried out [56–59], the interpretation of experimental results by the use of this model has not made much progress.

The details about hopping transport are discussed in Section 10.

4 INTRINSIC EXCITATIONS (SOLITONS, POLARONS, BIPOLARONS AND EXCITONS) IN POLYCONJUGTED SYSTEMS

The role of intrinsic excitations in the charge transport mechanism of conducting polymers is not very well understood yet. Considerable theoretical and experimental work regarding the nature of intrinsic/primary excitations in polyconjugated systems has been carried out by several groups [14,38,60,61]. However, a common consensus regarding the dominant parameters that are involved in the generation of primary excitations in polyconjugated systems is lacking still. Hence, the relationship between intrinsic excitations and charge transport in conducting polymers is an issue yet to be fully resolved.

The issue of the nature of primary excitations is important not only from the point of view of optical properties but also in the context of charge transport, since charge carriers in polyconjugated systems can be generated by both photoexcitation and chemical doping [14,27]. Moreover, the similarities in the physical properties of carriers created by photoexcitation and chemical/electrochemical doping have made it even more significant to develop a comprehensive understanding of the nature of primary photoexcitations (charge carriers) in conducting polymers.

In 3D inorganic semiconductors the primary excitations are usually electrons and holes (solid state band model with a continuum of energy levels). Meanwhile in small organic molecules (for e.g. anthracene, etc) and also in polydiacetylene the primary excitations are excitons (molecular model with discrete energy levels) [27,60]. However, in polyconjugated systems this so-called band [14,61] vs. exciton [38,60] model issue is not very well understood. It seems that various intrinsic and extrinsic parameters, for example, interchain interaction, Coulomb interaction, screening, electron–phonon interaction, charge delocalization, extent of conjugation length, extent of disorder, quantum lattice fluctuations [62], lattice stiffness, etc., play crucial roles in determining the nature of primary excitations in these systems [14,38,61]. Thus theoretical modeling, by taking care of these parameters, is rather difficult, and the experimental results to a large extent depend on the nature and quality of the samples. This complex scenario has hindered the appropriate interpretation of the experimental results and the relationship between microscopic and macroscopic properties of these systems.

It is well known that the bond-altered structure of t-$(CH)_x$ is energetically favorable with respect to the structure with no bond alternation [(C–C $\approx$ 1.4 Å in t-$(CH)_x$, for comparison C–C $\approx$ 1.54 Å for polyethylene][14,21]. The bond-length modulation [difference between alternating long (1.44 Å) and short (1.36 Å) C–C bonds] in $(CH)_x$ is nearly 0.08 Å. The main reason for the stability of the bond-altered structure is the π-band, composed of unpaired electrons in the p orbitals perpendicular to the plane of the carbon backbone. The band structures of the symmetrical and bond-altered structures of t-$(CH)_x$ are shown in Figures 2.3 [21]. For the bond-altered structure, the energy levels of the half-filled π-band near the Fermi level at Z are shifted down and the unoccupied π^*-levels are shifted up with respect to the symmetrical structure. Hoffmann et al. [21] have systematically shown that this first-order energy splitting is most effective near Z, where the energy levels of the π- and π^*-bands are exactly or nearly degenerate. The stabilization of the bond-altered structure is due to the energy balance between

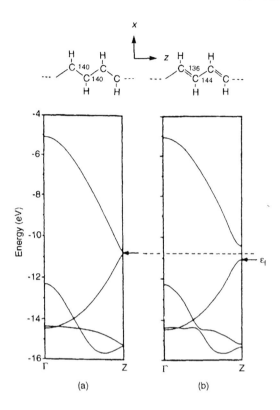

Figure 2.3. The band structures of t–(CH)x: (a) symmetrical and (b) bond–altered structures. (Reprinted with permission from ref. 21. Copyright 1991 American Chemical Society).

competing effects, consisting of the stabilization of the π and the destabilization of the σ-bonds.

The unaltered structure of t-(CH)$_x$ would have been a metal. However, Peierls distortion in lD systems with an incompletely filled band opens up a gap at the Fermi level; thus the system becomes a semiconductor. The intrinsic conductivity of undoped t-(CH)$_x$ is nearly 10^{-8} S cm^{-1} [14]. The Peierls distortion in solid state is analogous to the molecular Jahn–Teller distortion, in which a stabilizing deformation of the molecule destroys the degeneracy in the ground state [21]. Similarly, in t-(CH)$_x$ the degeneracy of the two orbitals formerly equal in energy is broken by a symmetry-lowering vibration [14,21].

The bond alternation defects are called 'misfits' by Pople and Walmsley [36]. The early quantum chemical work has considered the possibility of these mobile domain wall excitations and the formation of midgap states. However, the real breakthrough in the detailed understanding of these topological excitations was realized only after the seminal work of Su, Schrieffer,

and Heeger (SSH model) [14]. The basic physical properties of t-(CH)$_x$ could be understood within the framework of the SSH model. Although, later it was shown by various groups that the inclusion of Coulomb interaction and quantum lattice fluctuations in the SSH Hamiltonian are essential for the quantification of some experimental features [38,62].

The fundamental physical properties of solitons in t-(CH)$_x$ are the following [14]. The reversed spin-charge relationship is one of the most peculiar feature of the soliton in t-(CH)$_x$. The neutral soliton (radical) has no charge but has a spin. Meanwhile, by removing this defect state it becomes a spinless positively charged soliton (carbocation) and by the addition of another electron it becomes a spinless negatively charged soliton (carboanion). The soliton creation energy (E_s) is estimated to be nearly 0.6Δ (Δ is the band gap). Since E_s is less than Δ, the soliton is a stable excitation with respect to an electron or hole. In t-(CH)$_x$, $E_s \approx 0.42$ eV. The spatial extension of the soliton is determined by the relation $\{aW'/\Delta\}$ (where a is the lattice constant and W' is the bandwidth), which is nearly ($7a$) for t-(CH)$_x$. The bond alternation difference is nearly zero at the center of soliton and it gradually increases with distance from the center. This large spatial extension of the soliton and the large bandwidth (10 eV) along the chain direction for t-(CH)$_x$ leads to a rather small effective mass for the soliton, which is nearly six times the electron mass. The low effective mass of the soliton facilitates it to diffuse at nearly half of the velocity of sound.

Various optical and spin measurements confirmed the physical properties of solitons in t-(CH)$_x$ and more recently in another degenerate ground state conjugated polymer poly(1,6-heptadiyne) [64]. The lD spin diffusion of solitons and the associated spin dynamics are verified from electron spin resonance (ESR), light-induced ESR, nuclear magnetic resonance (NMR), and electron nuclear double resonance (ENDOR) measurements [14]. Solitons can be created either by low-level charge transfer doping or by photoexcitations. The experimental observation of photoexcitation-induced infrared-active vibrational modes, midgap absorption, etc. have been associated with the dynamic properties of solitons. The fast photogeneration of soliton–antisoliton pairs has been verified from the subpicosecond photoinduced absorption, and transient photoconductivity measurements [14,65,66]. The charged soliton–antisoliton pairs are photogenerated with 10^{-13}s. Recently, the enhancement of nonlinear optical properties ($\chi 3$) due to soliton–antisoliton pairs in degenerate ground state conjugated polymers has been observed [67]. The picosecond photoconductivity measurement in t-$(CH)_x$ [68] has shown that the photogenerated solitons undergo geminate decay within 1 picosecond and polarons are carrying the picosecond photocurrent. The details of transient photoconductivity measurements in t-$(CH)_x$ and polydiacetylene have been reviewed by Moses and Heeger [65]. The photoelectron spectroscopy in short polyene molecules [69] (conjugated sequence of seven double bonds) and ESR measurements in doped β-carotene polyene molecules [70] (conjugated sequence of eleven double bonds) indicate that the charge storage mechanism is due to the formation of spinless soliton pairs. Further details regarding the physical properties of solitons have been extensively reviewed by Heeger et al [14]. However, more theoretical and experimental efforts are required to understand the roles of interchain interaction, Coulomb interactions, disorder and quantum lattice fluctuations in influencing the physical properties of soliton–antisolition pairs.

The role of solitons in charge transport in doped $(CH)_x$ is not fully understood. Kivelson has proposed an intersoliton hopping model in a lightly doped regime [71]. Experimental results of the frequency-dependent conductivity measurements in lightly doped t-$(CH)_x$ samples are in resonable agreement with the predictions of the intersoliton hopping model [72]. The charge storage mechanism and the charge transport process by hopping among the midgap soliton states have been observed in electrochemical voltage spectroscopy [73]. Further details regarding the transport properties of soliton are discussed in Section 7.

Although the concept of polaron and bipolaron was well known in systems with strong electron–phonon coupling, the polaronic excitation in discrete chains of $(CH)_x$ was first proposed by Su and Schrieffer [14,74]. The theoretical studies suggest that polarons and bipolarons are the intrinsic and energetically favorable excitations in non-degenerate conjugated polymers [25,38,75,76]. A susbstantial amount of theoretical and experimental work regarding the physical properties of these excitations in cis-$(CH)_x$, PT, PPy, PPV, etc. has been carried out by various groups [14].

The bond alternation in non-degenerate ground state polymers usually results in incremental difference in ground state energies. The typical example is the energy difference between benzoid and quinoid structures of polyheterocyclic systems, like PPy, PT, etc. [14]. In these systems, the polaron can be considered as a charged soliton bound to a neutral soliton, whereas the bipolaron can be considered as a confined pair of charged solitons. Bipolarons are spinless particles like charged solitons. One of the significant differences between the soliton and the polaron/bipolaron is that the soliton states are situated at the midgap, whereas the polaron/bipolaron states are located symmetric to the band edges. The effective mass and the spatial extension of the polaron/bipolaron depend on how far these states are located from the band edges. If the Coulomb interaction is comparable with respect to the electron–phonon interaction, then the polaron and bipolaron can be considered as polaron–exciton and bipolaron-exciton, respectively. The competition between polarons and bipolarons has been studied by Shimoi and Abe [77]. They suggests that the Coulomb interactions significantly suppress the stability of bipolarons, and that the charge carriers are most likely in the form of polarons at low doping levels. However, at high doping levels the carriers tend to form a bipolaron lattice. Their model predicts a doping-induced transition from polaron to bipolaron lattice. The energetics of polarons and bipolarons formation have been reviewed by Hegger et al. [14].

Numerous optical spectroscopy, in situ (while chemical/electrochemical doping or photogeneration of carriers takes place), ESR, magnetic susceptibility measurements, etc. have given ample evidence for the static and dynamic properties of polarons and bipolarons in non-degenerate polyconjugated systems [14]. However, the roles of disorder, interchain interactions, chain ends, dopant ions, etc. in pinning and stabilizing the polaron/bipolaron excitations are not precisely known. Nechtschein et al. [78] proposed that the reduction in the number of spins at higher doping level is due to the formation of energetically favorable bipolarons with respect to polarons. Nevertheless, an alternative interpretation proposed by Mizoguchi et al. [79] suggests that this could be as well due to a crossover from Curie to Pauli behavior at higher doping levels.

Similarly, the optical property measurements in new-generation polyconjugated systems with reduced disorder indicate that the results could be alternatively interpreted by the exciton model [60,80] instead of the polaron/biploaron model. Thus the uniqueness of the interpretation of the experimental results from various types of measurements is not yet fully developed. The main hindrance towards this is that the physical properties of these materials depend on various intrinsic and extrinsic parameters (disorder, conjugation length, interchain interaction, etc.). Thus more work is essential in the carefull characterization of various parameters and its relationship with the static and dynamic properties of intrinsic excitations in doped and undoped polyconjugated systems.

The role of polarons/bipolarons in the transport properties of conducting polymers is not precisely understood. Although polaron/bipolaron hopping models [81] have been proposed by various groups, the distinct features in polaron/bipolaron hopping with respect to conventional VRH are not very clear. Thus the interpretation of experimental results is not really unambiguous. For example, the bipolaron hopping could take place either by a single hopping process of the bipolaron as a whole, or by the hopping of virtual polarons from the dissociated bipolarons [82]. Although it is well known that the activation energy for various types of polaron/bipolaron hopping mechanisms varies substantially, the understanding of the experimental results is not really straightforward. Thus more work is essential to understand the role of the polaron/bipolaron in the transport properties of conducting polymers, if at all they play an important role.

Although it is well known that solitons, polarons, and bipolarons are energetically the most favorable excitations in intrinsically 1D polyconjugated chains, Vogl and Campbell [83] have proposed that these excitations may not be stable in the presence of strong interchain interaction. Rough estimates indicate that the interchain coupling effects of the order of 0.6 eV or higher would be sufficient to break up the solitons on single chains [84,85]. In real systems, the stabilization of these nonlinear excitations by lattice relaxation is very much dependent on the trade off among competing factors: disorder, defects, dopant ions and chain ends tend to stabilize and interchain interaction tends to destabilize these excitations. One of the conditions for the stability of polarons in polyconjugated systems is that the ratio $t_\perp/t_\parallel < 0.01$. In case of t-$(CH)_x$, $t_\parallel \approx 2.5$ eV and $t_\perp \approx 0.12$ eV [84,85,86]. However, the calculations by Mizes and Conwell [84] have shown that the chain endings and other conjugation defects can stabilize the polarons in real systems, in spite of the destabilization factor due to interchain coupling.

The interplay of the extent of disorder and interchain interaction determines the dimerization and the formation of the electronic gap [84,85]. If polaron/bipolaron binding energy is low, even a small interchain transfer integral is enough to destabilize these excitations. Moreover, if the energy gained by the lattice relaxation on a single chain is of the order of the transverse bandwidth, then the excitations are rather spread out. Thus to a large extent the fundamental excitations in both doped and undoped polyconjugated systems depend on various intrinsic and extrinsic parameters.

Although the earlier theoretical and experimental studies have provided enough evidence for the presence of solitons, polarons, and bipolarons as the most energetically favorable excitations in polyconjugated systems [14], the recent investigations in PPV indicate that the lattice stiffness and Coulomb interactions favor the presence of excitons with a binding energy of the order of 0.3 ± 0.2 eV [87,88,89]. However, the recent transient photoconductivity measurements in oriented PPV samples suggest a much lower value of the binding energy of excitons [90]. The earlier photophysical studies showed the presence of excitons in small organic molecules (anthracene, etc.) [91]. The controversy regarding polarons/bipolarons vs. polaron/bipolaron–excitons vs. intermediate size excitons as the most energetically favorable excitation in polyconjugated systems is a relevant issue in the optical properties [92]; it does not have any significant role in the charge transport properties of doped systems. In doped polyconjugated systems, the charge carriers are either delocalized or localized depending on the extent of disorder in the system [2]. This topic is discussed in detail in Section 7.

5 ROLES OF INTERCHAIN INTERACTION, DISORDER AND DOPANT ON THE TRANSPORT PROPERTIES OF CONDUCTING POLYMERS

Interchain interaction and disorder play a substantial role in the electrical and optical properties of doped and undoped polyconjugated systems [14,84,85]. In $(CH)_x$, the average intercarbon distance in the same chain is nearly 1.4 Å, while on adjacent chains it is nearly 3–4 Å. Usually, the presence of side groups attached to the polyconjugated backbone reduces the interchain interaction. This is clearly demonstrated from the studies of the electrical and optical properties of unsubstituted and substituted $(CH)_x$ [20]. For example, the electrical conductivity of substituted $(CH)_x$ is considerably lower with respective to unsubstituted $(CH)_x$ [20]. Similarly, the enhancement of metallic properties by increasing the interchain interaction at high pressures is well known [2]. The interchain transfer integral in q-1D polyconjugated chains determines the effective electronic dimensionality of the system. If the overlap of the wavefunctions of π-orbitals in the transverse direction to the chain axis is substantial, then the system behaves more like an anisotropic 3D system. Usually, in undoped polyconjugated systems, the effective overlap of the wavefunctions of π-orbitals in the transverse direction to the chain axis is not very large. The typical values of the perpendicular bandwidths at the valence band minimum and maximum are 0.3–0.82 and 0.1–0.5 eV, respectively [14,84,85]. The current estimates of the interchain hopping matrix element are around 0.15 eV, and the q-lD models are rather successful in understanding the physical properties of undoped systems.

An approximate estimate of interchain vs. intrachain delocalization of π-electrons can be obtained from the ratio of bandwidths parallel and perpendicular to the chain axis. In oriented $(CH)_x$, this ratio is early 20 [14,84,85]. It is interesting to compare this value with respect to other anisotropic systems. In organic charge transfer complexes like TTF-TCNQ, the bandwidth transverse to the molecular stacking is nearly two orders of magnitude lower than those along the stacks [10,18]. In graphite the π-bandwidths parallel and perpendicular to the a–b plane are 8–9 and 1–2 eV, respectively [93,94]. The values of $\sigma(300\ K)$ and anisotropy in dc conductivity of various metallic conducting polymers, $(SN)_x$ [4], TTF-TCNQ [10b], HMTSF-TCNQ [10b], $(DMe\text{-}DCNQI)_2Cu$ [36], $(ET)_2I_3$ [10c], highly oriented pyrolytic graphite (HOPG) [93,94], $YBa_2Cu_3O_{7-\delta}$ [95,96], etc. are shown in Table 2.1. This shows that the anisotropy in conducting polymers is not very large with respect to other systems.

The role of interchain interaction in coupled metallic chains has been studied by several groups. Prigodin et al. [97] suggested that a system consisting of metallic chains remains q-lD if the bandwidth in the transverse direction ($4t_\perp$, where $t_\perp$ is the interchain exchange integral characterizing the strength of interchain tunneling or the width of the energy spectrum in the transverse direction) is such that $t_\perp\tau < 0.3$ (where τ is the scattering time) and $t_\perp/\varepsilon_F \ll 1$, where ε_F is the Fermi energy for a single metallic chain. Thus the degree of q-one-dimensionality can be characterized and the extent of disorder is characterized by the factor $k_F\lambda t_\perp/\varepsilon_F \approx t_\perp\tau$, where k_F is the Fermi wavevector and λ is the mean free path. When $t_\perp/\varepsilon_F \ll 1$, the Fermi surface consists of two open corrugated planes and $t_\perp/\varepsilon_F$ is a measure of this corrugation. The system of uncoupled lD chains becomes more 3D by increase $t_\perp/\varepsilon_F$. According to Prigodin and coworkers [97,98] an abrupt transition from extended to localized states is expected at the critical interchain exchange integral, $t_{\perp c} \approx (3\hbar/\tau)$. Thus the localization in q-lD metallic chains disappears only if $t_\perp$ becomes larger than the threshold value, $t_\perp \sim 0.3/\tau$. The transport properties according to this model are described in Section 7.

Table 2.1. D.C. conductivity (order of magnitude estimates* in conducting polymers) and anisotropy of conductivity in various anisotropic systems

Sample	$\sigma_\parallel(300\ K)$ (S cm^{-1})	$\sigma_\perp(300\ K)$ (S cm^{-1})	$\sigma_\parallel\sigma_\perp$ (300 K)	ρ_r	$\sigma_\parallel(1.4\ K)$ (S cm^{-1})	$\sigma_\perp(1.4\ K)$ (S cm^{-1})
I–$(CH)_x$	10^5	10^2–10^3	100–250	2–3	40000	–
FeCl$_3$–$(CH)_x$	20 000	100–500	100–150	1.3–2	10^4	–
PPV–H$_2$SO$_4$	10 000	100–200	100–150	1–1.5	7000	–
PPy–PF$_6$	1000–2500	100–200	10–20	1.5–2	10^3	–
$(SN)_x$ Crystal	1500–2000	5	300–400	–	10^5	3
$(SNBr_y)_x$ Crystal	12500^{‡2}	7^{‡1}	–	–	–	–

	$\sigma_\parallel(300\ K)$ (S cm^{-1})	$\sigma_\parallel/\sigma_a$ (300 K)	$\sigma_\parallel/\sigma_c$ (300 K)	$\sigma_\parallel$(Residual) (S cm^{-1})
TTF–TCNQ	300–500	350–500	120	10^{-4}
HMTSF–TCNQ	1500–2000	30	450	10^2
$(DMe\text{-}DCNQI)_2Cu$	1000–2000	10–20	–	10^5

	$\sigma_{plane}(300\ K)$ (S cm^{-1})	$\sigma_\perp(300\ K)$ (S cm^{-1})	$\sigma_{plane}(4.2\ K)$ (S cm^{-1})	$\sigma_\perp(4.2\ K)$ (S cm^{-1})
$(ET)_2I_3$	150	0.04	5000	15($T_c \sim 3.6\ K$)
HOPG	20 000	10	5×10^5	5.5
$YBa_2Cu_3O_{7-\delta}$	9000#	220	$T_c \sim 90\ K$	$T_c \sim 90\ K$

* $\rightarrow 10^4$ S cm$^{-1} \sim 10000 \pm 3000$.
$\rho_r \approx \rho(1.4\ K)/\rho(300\ K)$ for both $\sigma_\parallel$ and $\sigma_\perp$ to chain axis. Systems with negative TCR.
$^{\ddagger1} \rightarrow (SNBr_{0.39})_x$ and $^{\ddagger2} \rightarrow (SNBr_{0.1})_x$. σ_{max} for $(SNBr)_x \sim 10^5$ S cm$_{-1}$. For $(SN)_x \approx T_c \sim 0.3\ K$.
$\rightarrow \sigma_b(\parallel$ to CuO chains) $\approx 14\,000$ S cm^{-1} $\sigma_a(\perp$ to CuO chains) ≈ 6000 S cm^{-1}; $\sigma_b/\sigma_a \sim 2$; $\sigma_b/\sigma_c \sim 65$; $\sigma_a/\sigma_c \sim 30$. Note that σ_c and $\sigma_\perp$ are the same.
The references of the data are mentioned in the text. HMTSF = hexamethylenetetraselenafulvalene and $(ET)_2I_3$ = [bis(ethylenedithiolo)-tetrathiafulvalene]$_2I_3$.

Stafström [85] has extensively studied the effects of interchain interaction on the electronic properties of doped $(CH)_x$. He suggested that the interchain interactions reduce the energy gap around the Fermi energy considerably. Moreover, the interchain interactions add up along the chains and give rise to a specific interchain ordering of the bond-length alternation pattern along the chains. The interchain interactions produce an ordering of the phase of dimerization-order-parameter between the chains and the dimerized lattice is stabilized due to interchain interactions. However, in highly disorders samples it is possible that a major part of the chains are aligned out of phase with respect to each other. In highly doped samples, the metallic state is not attainable if the interchain exchange integral is set to zero. Thus the inclusion of the interchain exchange is essential to explain the observed metallic nature of highly doped conducting polymers.

Stafström [85] has shown that by turning on the interchain hopping strength it is possible to observe the delocalization of the wavefunctions of the soliton band over more than one chain. Moreover, he found that the transition from interchain delocalization to localization is smooth around $t_\perp \approx 0.1$–0.24 eV. However, this is contrary to the abrupt transition suggested by Firsov [98]. At full interchain hopping, the density of states on a single chain is roughly the same for all of the three chains used in the simulation [85]. Hence, realistic values of interchain hopping lead to 3D delocalization of carriers and this explains the Drude metallic behaviour of heavily doped $(CH)_x$, etc. Recently, Stafström [99] has extended this model by using the multi-channel Buttiker–Landauer formalism [100]. In such a model the interchain interactions reduce the effect of backscattering and this facilitates the delocalization over length scales considerably larger than the length of individual chains.

Conwell *et al.* [101,84] have studied the effect of interchain coupling on the electronic structure of both doped and undoped t-$(CH)_x$. They suggested that the interchain coupling is energy dependent, decreasing monotonically from a maximum value at the bottom of the valence band. Although one could assume that $t_\perp$ is lower in doped samples due to the opening up of the lattice in order to accommodate the dopant ions, the change in $t_\perp$ is rather small with respect to the undoped samples, particularly at the Fermi level. Conwell *et al.* suggested that the interchain interaction by itself is not enough for the transition into the metallic state in doped samples. Although in principle the interchain coupling is sufficient to shrink the energy gap between the soliton band and the conduction band to zero, in reality

the Coulomb interaction with dopant ions plays a significant role in giving rise to the metallic density of states. The detailed theoretical modeling is carried out only in the case of t-$(CH)_x$, and it should be extended to other systems in order to develop a comprehensive picture about the role of interchain interactions in both doped and undoped systems.

Recently Prigodin and Efetov [102] have developed a theoretical model for M–I transition in a random network of coupled metallic chains. In this model, the intrachain disorder due to intrinsic defects and the randomness in the distribution of interchain contacts induce localization. They have shown that the M–I transition in such a system is determined by the critical concentration of interchain crosslinks, which in turn depends on localization lengths and interchain coupling. Thus the localization–delocalization transition is expected at some critical concentration of interchain crosslinks. Moreover, a metallic state can really exist in such a random network of coupled metallic chains only if the concentration of interfibril contacts is large enough to overcome the percolation threshold.

The renormalization group calculations by Salkola and Kivelson [103] indicate that the Peierls instability generates an energy gap in 1D systems, when the largest phonon energy is lower with respect to the single-particle gap energy, and the interchain bandwidth is much larger than the single-particle gap. However, when the single-particle gap becomes comparable with respect to interchain bandwidth, the system becomes 3D. Since the transverse bandwidth ($8t_\perp \sim 0.8$ eV) is large compared to the incommensurate band gap in t-$(CH)_x$, even the 3D effects due to quantum lattice fluctuations are enough to suppress the Peierls instability.

Although the present theoretical models could give a qualitative picture of the interrelationship between interchain interactions and physical properties of polyconjugated systems, it seems that more rigorous work is essential for a quantitative understanding of various experimental results.

As mentioned in previous sections, the interchain transfer integral plays a significant role in charge delocalization, screening and Coulomb interaction. This is particularly important in the context of the nature and stability of primary excitations in polyconjugated systems. The interchain interaction has three significant effects on the optical response of a system constituting of 1D chains: (a) the gradual suppression of the inverse-square-root singularity which is a unique feature of 1D systems, (b) the reduction of 1D bandgap, and (c) the appearance of an electronic transition moment perpendicular to the chain axis. Comoretto *et al* [104] have shown that both photoinduced absorption and photoconductivity exhibit higher intensity when the exciting light is polarized perpendicular to the chain axis in t-$(CH)_x$. Moreover, the perpendicular excitation is found to be larger (2–5 times) than the parallel excitation. Although this is not fully understood, it seems that the interchain interaction is playing a substantial role in the ratio of intrachain with respect to interchain photocarrier generation and their recombination mechanisms. The discrepancy between theoretical calculations and experimental results is that in real samples a large number of chains are misaligned and it is not easy to estimate their contribution.

The role of interchain interaction is rather well studied in the case of oriented t-$(CH)_x$ by several groups. Leising [105] has carried out extensive studies about the role of interchain interaction and anisotropy in influencing the optical properties of undoped and doped Durham–Graz type $(CH)_x$. It is well known that q-1D metallic systems always exhibit a finite gap around the Fermi energy and are therefore non-metallic. However, in the case of highly conducting t-$(CH)_x$ Leising has shown that this gap is closed. This is consistent with the metallic properties of highly doped t-$(CH)_x$. The optical absorption (α) of t-$(CH)_x$ shows an anisotropy of $\alpha_\parallel/\alpha_\perp \approx 26$ at the absorption maximum (1.9 eV) [105]. The shifting of the absorption maximum for parallel and perpendicular polarizations indicates that the band structure is anisotropic. The anisotropy of the refractive index for oriented t-$(CH)_x$ is nearly five. The polarized reflectance of highly doped t-$(CH)_x$ shows a free electron-like behavior for parallel polarization, and a semiconductor-type behavior in perpendicular polarization. Since the density of Durham–Graz $(CH)_x$ is close to the theoretical density 1.2 g ml^{-1}), one could expect that the interchain interaction is substantial. However, a quantitative estimation of the role of interchain coupling in electrical and optical properties of various types of oriented $(CH)_x$ and other conducting polymers is still lacking.

The semiconductor-type behavior in the perpendicular direction to the chain axis in optical properties implies that either the interchain interaction is not strong enough or the extent of disorder limits the interchain transport. This is in contradiction with the fact that the dc conducitivty behavior for Naarmann-type t-$(CH)_x$ samples is nearly identical for both the parallel and perpendicular directions to the chain axis, which implies that the interchain interaction is substantial [106]. Moreover, the magnetotransport,

which is a sensitive probe for microscopic transport properties (e.g. the scattering mechanism, etc.) is nearly identical in both the parallel and perpendicular directions to the chain axis in oriented t-$(CH)_x$ [106] and oriented PPV-H_2SO_4 [107]. Nevertheless, Mizoguchi [108] pointed out from spin diffusion studies that the intrinsic q-1D nature of conducting polymers is playing an important role in the anisotropic diffusion of carriers. Epstein *et al* [109] suggested the possibility of 3D delocalization in crystalline regions and 1D localization in amorphous regions. Thus more detailed investigation regarding the effective electronic dimensionality of doped polyconjugated systems is essential to clarify the contradictions among various experiments.

It is well known that disorder is an inherent feature of polymeric systems which are only partially crystalline, and which often exhibit complex morphologies. Disorder in conducting polymers can be controlled to some extent by varying the parameters involved in sample preparation and processing methods. The classical examples are doped PANI and PPy; the counterion-induced processibility of PANI [110] and the low-temperature electropolymerization of PPy [111] enhanced the structural order in these systems. Although it is desirable to incorporate structural and morphological order by self-assembly techniques during material preparation and processing, in practice is not very easy to achieve this goal. The morphological order can be enhanced by tensile drawing so that the chains extent and orient to some extent. Even then, the disorder introduced while doping can be substantial. X-ray studies of conducting polymers indicate that the root mean square (rms) displacement of dopant ions could be few angströms [112,113].

The chain interruptions caused by chain breaking, sp^3 defects, cross linking between chains, etc. are some of the common sources of disorder in polyconjugated systems [14]. In both fibrillar and globular morphology, a large number of chains are weakly coupled together in a random fashion. The extent of overlap of π-orbitals (delocalization of the electronic states) depends on interchain interaction, chain interruptions, defects, etc. At chain interruptions the forward scattering of charge carriers to the adjacent conjugated segment is rather weak [84,85]. Thus the carriers are backscattered at each point of chain interruptions, and in absence of interchain interaction this leads to severe localization in q-1D metallic chains. This subtle interplay of disorder and interchain interaction in conducting polymers provides an unique opportunity to investigate the localization phenomena in low dimensional systems.

Disorder effects are well known to produce tails in the bands of regular systems, and this usually reduces the energy gap between valence and conduction bands [85]. Recent theoretical studies indicate that quantum lattice fluctuations play a significant role in the dynamical properties of polyconjugated systems [62,85]. The lattice fluctuations can cause substantial irregularities in the intra/inter chain hopping. This also leads to band tailing and a corresponding reduction of the energy gap. However, the broadening of band edges due to interchain hopping could camouflage the band tailing due to disorder. In doped t-$(CH)_x$ the reduction of the energy gap due to interchain hopping is nearly 0.2 eV, and the variation in gap due to the random distribution of dopant ions is only 0.03 eV [85]. Thus the disorder due to dopant ion distribution is relatively less significant with respect to the disorder in interchain hopping. Moreover, extrinsic disorder-induced effects could further lower the energy gap, and consequently the transition to metallic state is achieved at lower doping levels than that expected for a perfectly well-ordered system.

The effect of disorder in interchain hopping has been investigated by Wolf and Fesser [114]. Their work suggests that dimerization decreases and the density of the states in the gap increases as a function of increasing disorder in interchain coupling. In their model, the perfectly dimerized Peierls ground state breaks down towards a 'metal-like' state at a critical random distribution of interchain hopping. The random interchain coupling changes the properties of band gap to a pseudogap, with a small but non-zero density of states.

Harigaya et al. [115] have done extensive theoretical calculations regarding the effect of random impurities on the electronic properties of conjugated polymers. The impurities modify the electronic transfer integral and site energy at the same time. They have shown that at low impurity concentrations, an isolated impurity band is formed above the top of the valence band or below the bottom of the conduction band according to the sign of the impurity strength. Moreover, when impurity concentration and strength increases, the order parameter and energy gap altogether vanishes. At low impurity concentrations, the formation of impurity bands is suppressed when the strength of the bond component is larger than that of the site component. However, at high concentrations the energy gap vanishes for strong site components. When the bond component stabilizes the dimerization, it becomes more difficult to disrupt the energy gap so as to gain the electronic energy. Harigaya et al. suggested that the

energy gap at the Fermi level may vanish even when the bond-alternation patterns persist, and it is enhanced by impurity potentials. This is in agreement with X-ray diffraction studies in highly doped samples. Although the electronic properties are not very much influenced by the range of impurity potentials, the lattice configuration patterns are affected by the impurity potential range.

Wada [116] has reviewed the doping-induced disorder in conducting polymers. His study indicates that there is no real evidence for (a, b) plane disorder or incommensurability at any doping level, and the commensurability between the ion–ion distance in column and repeat units in the polymer chain is often assumed and it has not been substantiated. Moreover, both experimental and theoretical studies suggest that it is difficult to reduce the Peierls gap to zero without taking account of the effects of disorder.

Although considerable amount of work about the role of disorder in the structural and electronic properties of conducting polymers has been reported, systematic investigation of the physical properties by controlling and varying different types of disorder has not been accomplished yet. Thus more extensive studies are essential for a quantitative level of understanding of the role of different types of disorder affecting the structural and electronic properties of doped conducting polymers.

It is well known that doping of conjugated polymers is rather different from that in inorganic semiconductors. In inorganic semiconductors the dopants like phosphorous, boron, etc. substitute the host atomic sites and electrons or holes are generated depending upon the valency of the dopant ion. In crystalline inorganic semiconductors the charge carriers freely diffuse through the 3D lattice and their mobility is considerably higher with respect to organic semiconductors. Whereas, in the case of conjugated polymers the dopants are not substitutional but interstitial. The X-ray studies of alkali metal-doped $(CH)_x$ indicate that the counterions are displaced from the polymer chains by 2.3 Å in the direction perpendicular to the chain axis and each column of counterions interacts with several polymer chains [116]. The doping mechanism is similar to the intercalation process in layered structures like graphite [93]. The dopants randomly diffuse in between the polymer chains. Doped conducting polymers often exhibit a variety of 2D superlattices and these structures may be commensurate, incommensurate, or disordered [116].

The conductivity of conjugated polymers increases by several orders of magnitude at doping levels as low as 1%. The maximum level of doping in conjugated polymers could be as high as 20–30%, which corresponds to one dopant ion per three to five monomer units. In general the doping process is inhomogenous and the distribution of dopant ions in the polymer matrix is not uniform. This is mainly due to the complex morphology of the polymer matrix, which consists of both crystalline and amorphous regions [117]. One could envisage the intercalation of dopant ions in the polymer matrix as follows. The dopants easily diffuse into vacant spaces in the amorphous regions. Once the amorphous regions are saturated with dopant ions, they slowly migrate into crystalline regions.

The diffusion of dopant ions in the crystalline region is more complex than that in the amorphous region. Dopants open up the interchain separation at isolated domains in the crystalline region and this facilitates the diffusion of dopants into those domains. At the same time the chains get more tightly packed together in the undoped domains in the crystalline region and this hinders the motion of dopants into those domains. This could be one of the important reasons for the formation of segregated doped and undoped regions. Secondly, in fibrillar morphology, homogenous doping predominantly occurs only at the surface of the fibrils and the diffusion of dopants into the core of the fibrils is an inhomogenous process. Thus the mechanism of the diffusion of dopant ions in various domains is an extremely complex process and this usually results in inhomogenously doped samples, even though the undoped sample could be fairly homogenous. However, the recent breakthroughs in counterion-induced processing of conjugated polymers [110] and low-temperature electrochemical polymerization [111] have considerably improved the homogenous distribution of dopant ions in the conjugated polymer matrix.

It is often assumed that dopants act as a passive electron/hole reservoir, and an integral number of electrons/holes is transferred between dopants and the conjugated polymer chain. The resulting charges on the conjugated polymer backbone lead to the formation of solitons, polarons, etc. Although conventional wisdom suggests that the impurity scattering at high dopant concentration should be substantial and it could adversely affect the metallic state of doped conducting polymers, the Kivelson and Heeger model [118] suggests that the dopant counter-ion impurities make a negligible contribution to scattering, even if they are arranged in a disordered fashion. This is mainly due to the interstitial nature of the dopants and as a result the backscattering rate along the chain direction is negligible.

The interchain and intrachain hopping parameters in the presence of the potential due to dopant ions have been estimated by Cohen and Glick [119]. They find that in the neighborhood of a dopant ion the intrachain hopping is enhanced by 10% and the interchain hopping is enhanced by 100%. Thus the dopants may act as 'bridges' for interchain transport. The interchain hopping strength at sites in the presence and absence of dopant ions are estimated to be nearly 0.17 eV and 0.10 eV, respectively. The ESR measurements in conducting polymers by Bernier et al. [120] have demonstrated that the dopants play an active role in the charge transport process. Bulka [121] has estimated the interaction of a short-range impurity potential to the adjacent carbon atoms in the conjugated polymer backbone and found that the gap remains open even at high doping levels.

The theoretical calculations by Springborg and Eriksson [122] regarding the electronic structure of heavily doped $(CH)_x$ suggest that the hopping integral between the polymer and the dopants is of the order of 0.5 eV. Moreover, the existence of non-negligible hopping integrals between the dopants and the polymer suggests that only a fractional electron transfer takes place upon doping the polymer, and the dopants assist in both intrachain and interchain transport. The dopants modify the material from being q-lD to more 3D through the orbital interactions, such that the localization of electrons due to disorder is reduced. This also explains the rapid increase in the conductivity at very low doping levels.

Salkola and Kivelson [103] suggest that the counterions affect the energy gap. If counterions are located alternatively next to odd and even sites, then the gap is enhanced; if counterions are located next to even (or odd) sites then the gap is reduced. Thus even if the average counterion density is homogenous in a given sample, the metallic and Peierls states can coexist due to subtle microscopic differences in the arrangement of counterions in different regions of the same sample, and also due to interchain coupling.

Recently Yamashiro et al. [123] have estimated the dopant–chain interaction and its role in interchain transfers. Usually the dopant ion is a closed shell entity. A charge carrier passing via the dopant ion cannot directly enter this region because of the Pauli exclusion principle. The lowest such orbitals are the valence orbitals of the dopant. In both n-and p-type dopants the size of the electron or hole accepting orbitals are large, covering at least three or more carbon atoms on a $(CH)_x$ chain. This indicates that the dopants mediate the largest interchain transfers. The carbon atom in a chain has interchain transfer of about 0.3–0.1 eV with five to seven carbon atoms in another chain contacting with a common dopant column. The interchain transfer via dopants has little effect on intrachain states but gives a modification of the orbital energy spectrum.

In summary, several groups have arrived at the same conclusion that interchain interaction, disorder and dopants play a significant role in the charge transport properties of conducting polymers. This is discussed in detail in Section 7.

6 DOPING-INDUCED INSULATOR–METAL TRANSITION IN CONDUCTING POLYMERS

It is well known from the pioneering work by Heeger and coworkers [14] that in the case of doping of $(CH)_x$ the conductivity increases rapidly, whereas the Pauli susceptibility is rather small and it is nearly independent of doping up to 4–7%. At doping levels above 4–7%, Pauli susceptibility increases rapidly; however, there is no sharp transition in dc conductivity. Although considerable amount of work has been focused on this problem by several groups, there is still a lack of consensus on the mechanism of this doping-induced insulator–metal transition in t-$(CH)_x$. Thus the doping-induced insulator–metal transition is one of the important unresolved problem in this field.

Several groups have proposed various models to understand this problem. The most detailed study of doping-induced insulator–metal transition is carried out only in the case of doped $(CH)_x$. Mele and Rice [124] suggested that this is due to two competing effects in a 1D chain. The Peierls dimerization favors the 1D chain to be in the insulating state. Meanwhile, the random potentials of dopants suppress the bond alternation to lower the energy of the system. At doping-induced insulator–metal transition the latter dominates and this gives rise to a non-vanishing density of states at the Fermi level. However, the bond alternation persists into the metallic state and this favors the formation of a gapless Peierls distortion. The band tailing due to disorder and 3D coupling could favor a transition from commensurate CDW to incommensurate gapless Peierls system. However, there is no clear experimental evidence that disorder is inducing this transition, since this transition exists even in highly ordered samples.

Kivelson and Heeger [125] proposed that this doping-induced insulator–metal transition, at around 6% doping, is due to the first-order phase transition from soliton lattice to polaron lattice. In this model the

metallic properties above 6% doping is attributed to a half-filled polaron lattice. Although this model could convincingly explain the sharp increase in Pauli susceptibility at the critical doping concentration ($\approx 6\%$), the linear increase of infrared-active vibrational (IRAV) modes up to the highest doping level suggests that this model cannot satisfactorily explain all of the features of the metallic state. Since the order parameter does not change sign for a polaron as in the case of a soliton, its variation with distance does not increase significantly above the critical doping concentration. Moreover, the energy distribution in the gap probed by electron energy-loss spectroscopy (EELS) [126] suggests that the energy levels are well spread across the gap, which is more consistent with the band structure of a soliton lattice than that of a polaron lattice.

Recently Salkola and Kivelson [103] suggested an alternative scenario in which the interchain coupling causes the system to undergo a phase transition from a correlated soliton lattice to a metallic state; the later can be described by a weakly interacting but highly anisotropic 3D Fermi liquid. Although the strong infrared activity (IRAV modes) has been suggested as evidence for the Peierls state and the associated localized states, Salkola and Kivelson suggest an alternative possibility due to the usual phonon-assisted absorption processes in dirty metals. Moreover, for realistic 3D band structures it seems that a semimetallic-type phase coexists with non-metallic (Peierls) and metallic (Fermi liquid) phases.

Conwell *et al.* [127] have carried out extensive theoretical calculations and modeling of the doping-induced insulator–metal transition phenomena in $(CH)_x$. They proposed that the soliton band merges with valence band (conduction band) in the p-type (n-type) system, and this induces the insulator–metal transition in doped $(CH)_x$. Thus this first-order transition can be considered as a Mott transition and it could be associated with the characteristic variations in the counterion distribution. For example, the sharp increase in Pauli susceptibility in Na-$(CH)_x$ at around 6–8% is probably due to the transition from lattice spacing $5a$(6.67% doping) to $4a$ (8.33% doping). Moreover, Conwell *et al.* pointed out that the linear increase of IRAV intensity with doping, the existence of pseudogap, the persistence of midgap absorption, small plasma frequency, the absence of Drude behavior at low temperatures, the tail in density of states, etc. are in favor of the soliton lattice of metallic state.

Recent theoretical work by Stafström [128] has shown that the geometrical structure of $(CH)_x$ chains corresponds to a soliton lattice at all doping levels. However, he cautions that the soliton lattice can explain the metallic state at doping levels above 6% only if the dimerization amplitude is very small. The optimized dimerization amplitude at higher doping levels is less than half of the dimerization amplitude of the undoped system, and this could be as well due to correlation effects and lattice fluctuations. Moreover, he suggests that the IRAV modes in heavily doped systems might as well be due to the presence of an undimerized lattice. For realistic values of interchain interaction the soliton wavefunctions are sufficiently delocalized to form 3D electronic states.

Jeckelmann and Baeriswyl [39] have used a 1D Peierls–Hubbard model to describe the physical properties of conjugated polymers. In this approach both electron–phonon and e–e interactions are taken into account. In contrast to the slow decrease of the lattice distortion amplitude for conventional SSH model, the rapid fall-off of the lattice distortion amplitude at the critical doping concentration in the Peierls–Hubbard model gives a different perspective of doping-induced insulator–metal transition in conjugated polymers. Jeckelmann and Baeriswyl [39] found that the lattice distortion amplitude becomes so small for doping levels above 7% in $(CH)_x$ that it is no longer possible to decide whether it is really finite or even zero. Moreover, the optical gap falls off strongly between 4 and 6% doping in the Peierls–Hubbard model and it reaches a very small value of the order of 0.05 eV. One of the reasons for this is that at half filling the on-site Coulomb interaction enhances both dimerization and the optical gap. According to this model, the on-site Coulomb interaction (U) in $(CH)_x$ is nearly 7 eV. The charge susceptibility is strongly enhanced as a function of U, and the electronic charge modulation around an impurity is expected to increase by correlation effects. Thus the distortion of the lattice in the neighborhood of a charged dopant ion can induce Friedel-type oscillations and local IRAV modes. Even though this model could successfully explain some of the experimental observations, one could assume that a different scenario is possible by taking account of 3D long-range Coulomb interactions and quantum lattice fluctuations. However, the above additional effects favor a metallic state in the Peierls–Hubbard model.

Phillips and Cruz [129] also used an extended Peierls–Hubbard model in their analysis to understand the metallic state of doped $(CH)_x$. They have shown that the wavefunctions extrapolate smoothly (second-order phase transition when compared to the first-order phase transition in the Kivelson–Heeger model) from soliton to polaron excitations when the total energy of the system is minimized with respect to on-site (U), nearest-neighbor (V) and bond replusion (W) Coulomb interactions. Phillips and Cruz suggested that the soliton lattice is 0.6 eV lower in energy with respect to the polaron lattice at the critical doping level, and the values of U and V are 4 and 0.4 eV, respectively. Moreover, within the soliton model the sharp increase in the Pauli susceptibility at the insulator–metal transition is possibly associated with the spreading of the bound-state soliton levels in the energy gap. This could provide a finite density of states at the Fermi level. When the upper and lower soliton levels narrow down to a small minimum value the Pauli susceptibility increases rapidly and it becomes temperature independent.

The polson theory of metallic state in doped $(CH)_x$ suggests a lattice of hybrid soliton–polaron pairs, instead of a soliton lattice [130]. The polson unit has an unpaired electron and it has two coordinated dopants on the same side of the polyene chain. Although this model could explain both the soliton characteristics and Pauli susceptiblity of the metallic phase, the linear increase of IRAV intensity over the entire range of doping could not be understood. The ultraviolet photoemission spectroscopy (UPS) of doping-induced insulator–metal transition by Tanaka et al. [130] indicates that the 3D interaction of polson states is significant for the appearance of the metallic state. Yamashiro et al. [123] suggested that the intrachain and interchain motions of electrons/holes with a correlation cloud may give rise to the metallic state in doped conducting polymers. According to their model the correlational effects are essential to achieve the metallic state.

All the above mentioned models are mainly focused on doping-induced insulator–metal transition in the degenerate ground state conjugated polymer [t-$(CH)_x$)]. Similar type of systematic theoretical and experimental work in non-degenerate ground state conjugated polymers is still lacking. A large number of in situ doping measurements in PPy, PT etc. indicates that polarons and bipolarons are the energetically favored excitations at low and high doping levels, respectively. Thus one could envisage that a polaron–bipolaron transition occurs at the doping-induced insulator–metal transition in these systems [78]. However, the same data could be interpreted as a transition from Curie to Pauli at higher doping levels, as pointed by Mizoguchi et al. [79].

The IRAV intensity grows linearly at all doping levels in the case of doped poly(3-methylthiophene) (P3MT) [131,132]. The UPS and EELS measurements show a non-vanishing density of states at the Fermi level for doped P3HT [133] and PPy [134], as expected by a metal. This metallic state has been interpreted as a polaron lattice and this is consistent with the high Pauli susceptibility of the metallic state. The Fermi energy was observed to be moving down towards the valence band for p-type doping in both P3HT and PPy. These observations are in agreement with the valence effective Hamiltonian calculations on a polaron lattice. However, according to Conwell and Mizes [135], the tailing-off of the density of the states above the Fermi level could be expected as well for a bipolaron band merging with valence band, Nevertheless, the recent electronic absorption spectra and Raman spectroscopic studies [136] suggest that the two absorption bands for doped PT, PPP and PPV must be assigned to polarons, instead of bipolarons.

Recently several groups have discussed the competition between polarons and bipolarons in doped non-degenerate conjugated polymers [77,135,137]. According to Conwell and Mizes [135], the IRAV modes at high doping levels suggest the evidence for a bipolaron lattice. Moreover, they argue that the density of the states at the Fermi level need not vanish for the case of the bipolaron lattice, since the valence-effective Hamiltonian calculations at high doping levels indicate that the bipolarons band is nearly merged with the conduction band or valence band. Contrary, the polaron lattice in the metallic state cannot be stabilized by interactions with other chains or dopants, since these interactions are negligible [135].

The theoretical calculations by Shimoi and Abe [77] suggest that the Coulomb interactions significantly suppress the stability of bipolarons. Their work is mainly focused on the relative stability of polarons and bipolarons as a function of on-site and long-range Coulomb interactions. The roles played by dopant potentials and interchain interactions are not taken into account in their model. They suggest that it is possible to induce a transition from polaron to bipolaron lattice at high doping levels. The Coulomb repulsion between two similar charge carriers confined in the lattice deformation of a bipolaron costs extra energy with respect to two spatially separated pair of polarons. Hence, the relative stability of bipolarons with respect to polarons depends on the effective Coulomb interactions. The overlap of the lattice deformations between adjacent polarons in a polaron lattice affects their relative stability. When the overlap among polarons becomes substantial the energy gain due to electron–lattice coupling for a polaron lattice reduces, and this favors the transition towards a bipolaron lattice at higher doping levels. This is consistent with the

maximum spin density from ESR measurements in doped PPy [138]. The maximum spin density was observed at a doping concentration of about one charge carrier per six pyrrole untis. This is not a sharp transition due to the thermal population of metastable states, disorder, inhomogeneity in doping, etc. However, the modulated absorption measurement [139] indicates that the confinement parameter of bipolarons in doped conjugated polymers is not large enough for the formation of the bipolaron lattice.

The recent theoretical calculations by Xie et al. [137] indicate that both interchain coupling and Coulomb interaction with dopant ions play important roles in the stability of polarons and bipolarons. They have shown that the width of the bipolaron decreases considerably due to the interaction with the counterion, and this results in the concentration of the charge density in the vicinity of counterion. Moreover, they have suggested that the bipolaron state is stable even in the presence of interchain coupling, and bipolarons are energetically more favorable with respect to singlet or triplet polarons. The process of bipolaron or polaron creation in a isolated chain depends on Coulomb correlation (U), especially when $U \sim 2t_\perp$. In the presence of strong Coulomb correlation ($U > 2t_\perp$) one could expect a bipolaron–polaron transition. The counterion pinning effect increases the localization of both polarons and bipolarons, and at the same time decreases their creation energy. This localization enhances the Coulomb interaction inside the excitation, which is particularly strong in bipolarons with respect to polarons. Xie et al suggested that both polarons and bipolarons can even coexist under certain conditions due to the competition among various factors. This type of coexistence has been observed in some ESR measurements of conductiong polymers [78]. Thus the interplay of polaron and bipolaron excitations in doped non-degenerate gound state conjugated polymers is possible, depending on the relative strength of the various types of interactions.

Although considerable amount of theoretical and experimental work has been carried out by several groups, the doping-induced insulator–metal transition in both degenerate and non-degenerate ground state conjugated polymers is not very well understood. This is mainly due to the difficulties involved in the quantification of various interaction and localization parameters that influence the doping-induced insulator–metal transition. Moreover, a detailed understanding of this phenomena is essential for a comprehensive understanding of the charge transport mechanism in doped conjugated polymers.

7 MODELS OF CHARGE TRANSPORT IN DOPED CONDUCTING POLYMERS

In previous sections the current understanding of structure, morphology, doping, fundamental excitations, dimensionality, disorder, etc. in conducting polymers is discussed. The consensus is that the metallic state exists in conducting polymers due to interchain coupling, dopant–polymer chain interactions, etc. Moreover, there is a continuous density of states as a function of energy with a well-defined Fermi energy in metallic conducting polymers [2]. Thus the fundamental electronic structure of heavily doped conducting polymers is that of a metal. However, whether the characteristic features of the metallic state in conducting polymers are in agreement with that of a Fermi liquid, marginal Fermi liquid, Luttinger liquid or chiral-spin liquid, etc. is yet to be understood. Moreover, the nature of the Fermi surface, the dynamic properties of quasiparticles, the details about elastic and inelastic scattering processes, effective electronic dimensionality, etc. are yet to be studied in detail.

Although Little [5] suggested as early as 1964 that the large polarizability of the side groups attached to a conjugated polymer backbone could induce exciton-mediated motion of charge carriers along the conducting spine, it is not yet realized in practice due to the difficulties involved in synthesizing such a material.

The experimental finding of metallic conductivity and even superconductivity in the crystalline inorganic polymer $(SN)_x$ in the year 1973 [4] has indeed shown that it is feasible to achieve metallic properties in a polymeric system.

In first-generation conducting polymers the electrical conductivity was limited $\{(CH)_x \approx 10^3 \text{ S cm}^{-1}$ [1], PPy [140] and PT [141] $\approx 10^2 \text{ S cm}^{-1}\}$ due to disorder. Moreover, $\sigma(T)$ was strongly temperature dependent and $\Delta \ln \sigma/\Delta \ln T$ showed a negative temperature coefficient typical to that expected for transport on the insulating side of the M–I transition. In general a finite value of dc conductivity at $T \to 0$ was not observed in previous generation conducting polymers, except in I-$(CH)_x$ [142]. Although thermopower showed a quasi-linear temperature dependence [143], the temperature dependence of magnetic susceptibility indicated the presence of a relatively large Curie contribution [144]. Thus in previous generation conducting polymers the presence of strong disorder masked the intrinsic metallic behavior and the observed features were typical of that expected in an inhomogeneous system. However, the existence of a finite Pauli spin susceptibility [144], the quasi-linear temperature dependence of thermopower [143] and a linear term in specific heat [145] have provided early evidence of a continuous density of states with a well-defined Fermi energy. Nevertheless, the real 'fingerprints' of metallic behavior [e.g. positive temperature coefficient of $W \sim \Delta \ln \sigma/\Delta \ln T$, metallic reflectivity (Drude) in the infrared, etc.] were not observed.

During the last seven years, improved homogeneity and reduction in the degree of disorder resulting from improved synthesis and processing of conducting polymers [110,111,146–148] has provided a new opportunity for investigating the nature of the 'metallic' state through transport and optical measurements [2]. The improvement in the quality of the 'new generation' of materials has enabled the observation of the typical metallic positive temperature coefficient of resistivity (TCR), at high temperatures, in specific systems [149,150]. This has resulted in a deeper understanding of the role of disorder as the limiting factor in transport and as the origin of the M–I transition in doped conducting polymers.

Recent improvements in the quality of doped conducting polymers have allowed the investigation of the disorder-induced M–I transition in these materials [2]. Disorder is an inherent feature of polymers that often exhibit complex morphology; such systems are often partially crystalline and partially amorphous in nature. Since disorder leads to qualitatively different charge transport mechanisms in the homogenous and inhomogenous limits, it is important to quantify these two limits. The critical parameter is the localization length (L_c): if L_c is greater than the structural coherence length (which characterizes the length scale of the crystalline regions and thus the length scale for inhomogeneity), the disorder can be viewed as homogeneous; *the system sees only an average.* On the other hand, when there are large-scale inhomogeneities, as in a granular metal, the disorder must be viewed as inhomogeneous.

In previous generation conducting polymers, inhomogeneities often dominated the transport properties, and 'metallic islands' models were constructed to handle such larger scale granularity. Thus the availability of homogenous and air stable metallic conducting polymers, in the last few years, has provided a new opportunity to invigorate the study of the metallic properties of these systems. The high electrical conductivities of doped $(CH)_x$ reported by Naarmann *et al.* in 1987 signaled the onset of a new generation of conducting polymers [151–153]. Conducitvites of the order of 10^4 S cm^{-1}, comparable to that of traditional metals like lead, were reported. By continuing to improve the material, Tsukomoto increased the conductivity by another order of magnitude, to 10^5 S cm^{-1} in 1990 [154]. Correspondingly, the $\sigma(T)$ in doped $(CH)_x$ consistently became weaker as the conductivity increased. Although values of conductivity, of the order of 10^4 S cm^{-1}, were observed for doped oriented PPV in 1990, no rigorous transport measurements were carried out [147]. In 1990, the low-temperature electropolymerization of PPy-PF$_6$, and subsequently the stretch-oriented samples yielded σ (300 K) $\sim 10^3$ S cm^{-1} [111]. In this material, for the first time in conducting polymers, a significant positive TCR was observed at temperatures below 20 K. In 1991, the development of the counterion-induced processibility of PANI enhanced the conducitivity of PANI to 300–400 S cm^{-1} [110]. These samples showed substantially weaker $\sigma(T)$ values compared to previous generation PANI [150]. Moreover, for the first time, a doped conducting polymer showed a significant positive TCR in the temperture range from 160–300 K [150]. In 1992, the synthesis of regio-regular polyalkylthiophenes (PAT) resulted in materials with substantially enhanced conductivity with values of the order of 10^3 S cm^{-1} [148].

This brief summary of the significant developments that have occurred within the last few years in the preparation and processing of doped conducting polymers indicates substantial progress; further progress is necessary, however, to reduce the microscopic and macroscopic disorder and thereby to bring out the intrinsic metallic features. The systematic improvements in material quality needs to be characterized by rigorous transport measurements in order to quantify the various parameters involved in the disorder-induced localization which leads to the M–I transition. The conclusions inferred from such experimental studies will provide a deeper understanding of the microscopic parameters involved in charge transport in these metallic polymers and consequently a deeper understanding of the requirements for further improvement in the quality of the materials.

An important parameter for characterizing the disorder is the product of Fermi wavevector (k_F) and the mean free path (λ), which is the order parameter for the disorder-induced M–I transition; for $k_F\lambda \sim 1$, $\sigma \sim \sigma_{min}$. Recent progress has resulted in conducting polymers on the metallic side of the M–I transition with $k_F\lambda \geqslant 1$ and $\sigma \geqslant \sigma_{min}$. In this situation, the lifetime broadening of the electronic states is less than the Fermi energy ($\varepsilon_F\tau > \hbar$), so that a band model can be used as a starting point [28,118].

In the strict 1D limit, all wavefunctions are localized in the presence of disorder. Interchain electron transfer suppresses this extreme tendency toward localization. For a given level of disorder, the strength of the interchain coupling needed to suppress the localization depends on the coherence length along the q-1D chains; an electron must be able to hop to an adjacent chain prior to the resonant backscattering which inevitably leads to localization in 1D. Thus, although conducting polymers are correctly referred to as q-1D electronic systems, the interchain coupling can be sufficiently large to enable the formation of 3D metals [118,155]. When such a material is oriented (e.g. by tensile drawing), the properties are those of an anisotropic 3D metal. Without macroscopic orientation, the macroscopic properties of conducing polymers are isotropic even though on a microscopic level, the electronic structure (intrachain vs. interchain) is highly anisotropic.

The transport properties of disordered systems are sensitive to the presence of both extended and localized states [28,44]. The extent of disorder determines the relative importance of the roles played by localization and by e–e interactions; the extent of disorder determines the screening length and scattering processes involved in the charge transport. Among various length scales, the correlation length on the metallic side, the localization length on the insulating side, the e–e interaction length, the thermal diffusion length, and the inelastic scattering length determine the dominant mechanisms involved in the transport.

In the classical definition, a metal should have a positive TCR and $k_F\lambda > 1$, where

$$k_F\lambda = [\hbar(3\pi^2)^{2/3}]/(e^2\rho n^{1/3}), \qquad (2.7)$$

ρ is the resistivity and n is the number of charge carriers [28]. However, the more precise experimental definition of a metal requires that there is a finite conductivity as $T \rightarrow 0$ and that W shows a positive temperature coefficient [156]. In new generation conducting polymers there are examples that exhibit a positive temperature coefficient for W and even exhibit a positive temperature coefficient for the resisitivity (TCR) at temperatures above 150 K [149,150]. On the insulating side of the M–I transition ($\sigma \rightarrow 0$ as $T \rightarrow 0$) W shows the typical negative temperature coefficient [2,156]. The exponential temperature dependence, $\ln \sigma(T) \propto T^{-x}$ with $x < 1$, on the insulating side of the M–I transition is typical of hopping transport. The exponent (x) is determined by the extent of disorder, by the dimensionality of the system, and by the morphology, granularity and microstructure (homo-genous or inhomogenous). Most of the transport measurements in previous generation conducting poly-mers were in the insulating regime [3].

The new generation of conducting polymers has made possible detailed studies of the critical regime of the M–I transition [2], previously unexplored in doped conducting polymers. When the extent of disorder is near the critical disorder of the Anderson transition, the temperature dependence of conductivity follows a power law over a substantial range of temperatures, $\sigma_{crit} \propto T^{-\beta}$ and W is temperature independent [156–158]; $W = \beta$. This power law behavior is universal near the critical regime of the M–I transition and it does not depend on the details of the system. Thus, the power law temperature dependence plays a key role in defining the critical regime. Tuning through the critical regime by varying the extent of disorder, by varying the inter-chain interaction through application of high external pressure and by shifting the mobility edge through application of high magnetic fields has provided insight into the general features of the transport near the M–I transition [2,159].

Although the above general overview gives some global aspects of the charge transport in conducting polymers, the subtle details of the mechanism of charge transport are yet to be sorted out. Firstly, the role of solitons, polarons and bipolarons in the charge trans-port is not very well understood. Although Kivelson [71] proposed a phonon-assisted hopping mechanism between bound soliton states in the case of lightly doped $(CH)_x$, the interpretation of the experimental data based on alternative models suggests that the inter-soliton hopping model [81] for charge transport in doped $(CH)_x$, PPP, etc. could not unambiguously rule out the possibilities of other models in interpreting the experimental data. Secondly, is there any significant difference between microscopic and macroscopic transport properties? For example, the microscopic transport properties governed by the intrinsic q-1D of the system, and do the macroscopic transport properties reveal more 3D features due to the contributions from interchain and counterion interactions? Thirdly, are the charge transport properties in the crystalline (homo-genous) and amorphous (inhomogenous) regions in the system drastically different, and if so, is it essential to use two different models in the amorphous and crystalline regimes? Fourthly, is it possible to determine the cut-off length scales in which discrete and band models are applicable? How far these questions are addressed in the models proposed by various groups is discussed below.

In Kivelson's model [71] he argued that the thermally activated conduction due to the free motion of charged solitons is difficult, since the binding energy of solitons is rather large and at reasonable temperatures the population of solitons is small. Thus the hopping of electrons or holes between soliton states is a less strongly activated process. For example, the activation energy for phonon-assisted hopping of an electron from a charged soliton to a neutral soliton is rather small if a negatively charged soliton bound to a positively charged impurity is situated near a neutral soliton on adjacent chain. The intersoliton hopping conductivity is determined by the rate at which an electron hops between a pair of solitons. Epstein et al. [72,160] observed that the temperature and frequency depen-dence of conductivity is in agreement with VRH among soliton like states in the lightly doped regime. Similar to Kivelson's model, Chance et al. [81] proposed that the interchain bipolaron hopping is important to account for the low magnetic susceptibilities in highly conduct-ing polymers.

Pietronero [161] suggested that the main contribution to resistivity in conjugated polymer or graphite intercalation systems is due to the scattering between the conduction electrons and phonons of the conjugated polymer chains or graphite layers, while the phonons of the intercalate layers contribute much less to the scattering process. In graphite intercalation compounds the electron–phonon scattering involves mainly the electronic states within the same Fermi surface pocket connected through low-energy phonons, and the strong suppression of scattering due to phonon freezing occurs only at temperatures below 100 K. In a 1D chain the only possible scattering is from $+ k_F$ to $- k_F$ involving high-energy [$\hbar\omega \sim 0.2$ eV for $(CH)_x$] $2k_F$ phonons. Due to phonon freezing effects even at room temperature, the first-order scattering should induce a strong enhancement in conductivity in 1D chains.

Pietronero [161] suggested that in carbon polymers, usually $\hbar\omega \gg k_B T$ and an inelastic treatment of scattering is important. In the limit of elastic scattering ($k_B T \gg \hbar\omega$), the conductivity in the chain direction in a 1D chain can be expressed as

$$\sigma_{\parallel} = (e^2 na/\pi\hbar)v_F\tau = (e^2 na^2/\pi\hbar)(\lambda/a) \qquad (2.8)$$

where n is the conduction electron density per unit volume, a is the carbon–carbon distance along the chain direction, $v_F = (2t_0 a/\hbar)$ is the Fermi velocity, $t_0 (\approx 2–3$ eV) is the π-electron hopping matrix element and τ is the backscattering lifetime. In the limit of elastic scattering for a half-filled band system, $\sigma_{\parallel}$

$(300 \text{ K}) \approx 1.6 \times 10^5 \text{ S cm}^{-1}$, which is nearly an order of magnitude less than that estimated for graphite intercalation systems. However, Pietronero [161] argues that this estimation is not realistic, since the main scattering in a conducting polymer chain involves only $2k_F$ phonons, i.e. $\hbar\omega_0 \gg k_B T$, where $\hbar\omega_0 \approx 2600 \text{ K}$. Hence, by including the inelastic scattering process below the characteristic temperature ($k_B T \sim \hbar\omega_0/4$ and $T \sim 600 \text{ K}$) another two orders of magnitude of enhancement in conductivity should be possible (i.e. $\sigma_\parallel (300 \text{ K})\sigma_\parallel \approx 1.4 \times 10^7 \text{ S cm}^{-1}$.

Kivelson and Heeger [118] have carried out detailed theoretical estimation of the intrinsic conductivity in conducting polymers. The intrinsic conductivity along the chain direction for 1D chain is given by

$$\sigma_\parallel = (e^2/4\pi\hbar a)na^3(M\omega_0 t_0^2/\alpha^2\hbar)\exp(\hbar\omega_0/k_B T) \tag{2.9}$$

where $\alpha \sim 4.1 \text{ eV Å}^{-1}$ is the electron–phonon coupling constant, ω_0 is the $2k_F$ phonon frequency ($\hbar\omega_0 = 0.12 \text{ eV}$) and M is the carbon mass. The backscattering rate of an electron near the Fermi energy due to the $2k_F$ phonon is given by

$$(1/\tau_{ph}) \approx [8\alpha^2/M\omega_0 t_0]\exp(-\hbar\omega_0/K_B T) \tag{2.10}$$

The large value of t_0 (2–3 eV), the small number $2k_F$ phonons and $\sigma \propto \exp(-\hbar\omega_0/k_B T)$ yields conductivity at room temperature of the order of 10^7 S cm^{-1}; the conductivity is expected to increase exponentially at low temperatures.

Since the charged dopant ions are spatially removed from the q-1D conduction path, the resistive back-scattering is suppressed. Thus the combined effect of the anisotropic screening and the motion of carriers along the polymer chain indicates that the conductivity is only sensitive to the $2k_F$ Fourier component of the scattering potential of disorder. Thus the counterion potentials are not expected to cause any significant backscattering. In fact structural defects and chain breaks could induce strong backward scattering and the counterions mainly introduce forward scattering. More-over, only a small number of thermally excited $2k_F$ phonons is relevant in the scattering process, which is at much higher frequencies in conjugated polymers with respect to conventional metals. This itself suggests that the intrinsic conductivity in conjugated polymers is expected to be very high at room temperature. Hence, in a q-1D polyconjugated system, the primary momentum relaxation is from the scattering with the modest population of $2k_F$ phonons at room temperature. Thus one could expect intrinsic conductivity of the order of 10^7 S cm^{-1} [118].

As mentioned in Section 5, even relatively weak interchain coupling ($t_\perp \sim 0.1 \text{ eV}$) is sufficient for 3D delocalization of charge carriers. Thus for finite values of $t_\perp$ the coherent interchain quantum diffusion of carriers becomes 3D as long as the mean separation between chain breaks is sufficiently great that the chain break concentration (x) is much less than ($t_\perp/t_0 \approx 0.03$), i.e. $x \ll (t_\perp/t_0)$. Conversely, if $x \ll (t_\perp/t_0)^2$, incoherent interchain hopping due to 1D localization will limit the transport [118]. Thus the conductivity should increase in proportion to the mean distance between chain interruptions, crosslinks, sp^3 defects, etc. When the concentration of chain interruptions is sufficiently high so that (l/x) is only a few lattice constants, then the wavefunctions will be localized. The temperature dependence of intrinsic conductivity is dominated by the backscattering from phonons along the polyene chain. All the above factors indicate that in high-quality conducting polymers the electronic mean free path could be much larger than the structural coherence length and real metallic features can be observed.

Prigodin and Firsov have proposed similar charge transport models based on the intrinsic q-1D of the system [97,98]. A brief introduction regarding the role of interchain coupling in this model is already mentioned in Section 5. Firsov [98] has shown that for a q-1D system, when the interchain exchange integral characterizing the strength of interchain tunneling is less than the critical value for the interchain exchange integral at which the M–I transition occurs ($t_\perp < t_{\perp c}$), the system behaves like a disordered 3D insulator near the mobility edge; and for ($t_\perp > t_{\perp c}$) the system behaves like a disordered 3D metal near the mobility edge. Nakhmdeov et al. [97b] have carried out theoretical calculations about hopping transport in q-1D systems near the M–I transition with weak disorder. They have determined the cut-off temperature regimes, in which band transport crosses over to hopping transport. In the high-temperature regime, the band transport is governed by phonon scattering and the motion of electron between inelastic collisions is ballistic. In the intermediate temperature regime between band and hopping transport the temperature dependence of conductivity, near the vicinity of the M–I transition, shows a scaling behaviour (power law behavior). In lower temperature regimes, hopping at a constant average length crosses over to Mott's VRH.

However, Nakhmdeov et al. [97b] have pointed out that the above mentioned temperatures regimes are slightly different in q-1D system with weak interchain coupling; in which the extent of disorder, density of states and the interchain hopping integral play important roles. In this case (weak interchain coupling), the temperature dependence of conductivity gradually varies from lD behavior at higher temperatures to that of a isotropic 3D behavior at lower temperatures (Mott's VRH, $T^{-1/4}$ law); and in the in-between temperature regime, the longitudinal (transverse) conductivity follows $\exp[-T_0/2T]$ ($\exp[-(T_0/2T)^{1/2}]$) dependence.

Joo et al. [162] extended the above q-lD models suggesting that the impurity backward scattering time is the only effective scattering time and the long-range random potentials in the polyconjugated backbone that produce the forward scattering are not effective. Moreover, the phonon forward scattering destroys the weak localization by breaking the phase coherence of the impurity scattering and when the phonon forward scattering time becomes comparable with the impurity scattering, the temperature dependence of conductivity exhibits a maximum at some temperature.

Epstein et al. [109,163] have widely used these q-lD models to interpret the transport properties in both metallic and insulating conducting polymer systems. According to this model, conducting polymers are considered as inhomogeneous systems consisting of partially crystalline and amorphous regions. In crystalline regions the overlap of π-orbitals could result in anisotropic delocalized states, whereas in amorphous regions the polymer chains are weakly interacting and the electronic states are subjected to lD localization. When the size (crystalline coherence length) and volume fraction (crystallinity) of crystalline region increases with respect to amorphous region, the system is expected to undergo a transition from insulator to metal. In metallic conducting polymers the crystallinity is nearly 50–80% and the crystalline coherence length is nearly 20–80 Å [117].

Epstein et al. [109,163] suggested that the charge carriers are subjected to lD localization when they pass through the amorphous regions in between the crystalline regions. Hence, in bulk transport measurements the charge carriers must traverse through these lD localized regions and this usually dominates over the transport through the 3D extended states in the crystalline regions. However, when the volume fraction of 3D extended states is above some critical value one could expect a 'percolative metallic' transport. In other words, the flow of charge carriers prefers the path of least resistance and one could expect a quasi-metallic behavior as the volume fraction of the crystalline region increases. Moreover, this metallic island (mesoscopic size) model suggests that the usual Anderson localization in the homogenous disorder limit is not appropriate for conducting polymers since the micro-

scopic transport properties are different in crystalline and amorphous phases. Another important feature in this model is that the Fermi surface of metallic conducting polymers is assumed to be 'open' as expected for highly anisotropic systems. Thus the forward scattering of conduction electrons may be ineffective and the backward scattering from k_F to $-k_F$ may be essential for momentum relaxation. This could result in anomalously long scattering time (10^{-11} s). The implications of this model are discussed in next section.

The M–I transition in a random network of metallic wires has been suggested by various groups [164,102]. Some of these models were used in the context of the M–I transition in doped conducting polymers, assuming that the fibrils of polymer chains can be considered as metallic wires. The role of interchain coupling in the Prigodin–Efetov model [102] is discussed in Section 5. The transport properties are rather similar to those found in the q-lD models suggested by Firsov [98] and Nakhmdeov et al. [97b].

Recently, Andrade et al. [165] reported computer simulations in a random resistor network model. In this phenomenological model they found some correlations between microstructure and transport properties. Baughman and Shacklette [166] proposed a random resistor network model to study the effect of conjugation length on hopping transport. They proposed that the exponential temperature dependence of conductivity results from the finite conjugation lengths and the nearest neighbor interchain hopping could be modeled by using a random resistor network with a wide distribution of activation energies. A correlation between the distribution of activation energies and the average conjugation length and its distribution has been proposed in this model. Roth has suggested a similar correlation between the conjugation length, conductivity, T_0, and the exponential temperature dependence of conductivity [3].

Movaghar and Roth [167] proposed that the transport properties of conducting polymers can be explained by a percolation-type model for inhomogenously disordered systems. They suggested a power law of $\sigma(T)$ due to phonon-assisted intergrain transport mediated through tunneling and activated processes over local mobility edges, and a spectrum of activation energies over parallel conduction paths due to the inhomogenity in the system. The thermal fluctuations tend to homogenize the system by smearing off the potential barriers and creating conducting pathway; this increases the diffusivity of charge carriers. The scattering from disorder may tend to average over the local mobility edges of the metallic grains to give rise to a global mobility edge, which is magnetic field dependent. When the Fermi energy moves with the magnetic field the charge transport spill over into new percolating channels. This could produce a strong negative magnetoresistance due to the energy renormalizations. The magnetic field is expected to move the Fermi energy up via orbit and Zeeman spin energy by redistributing the up and down states and thus connecting the metallic domains with each other.

The Sheng's fluctuation-induced tunneling (FIT) model was widely used in the earlier interpretation of charge transport in doped conducting polymers [168]. Although this model was originally developed for granular metals (metallic particles in an insulating matrix, e.g. carbon black particles in insulating polymers) in which one usually expects a strong temperature dependence of conductivity at low temperatures, the FIT model has been used even in the case of highly conducting $(CH)_x$. The volage across the insulating barriers is subjected to large thermal fluctuations. These voltage fluctuations could modulate the tunneling probability of the carriers passing through the insulating junctions. The tendency for the saturation of conductivity at low temperatures for metallic conducting polymers is contrary to the predictions of Sheng's model. This discrepancy is probably due to the underestimation of the number of parallel paths with low activation energy in Sheng's model. The volage fluctuations in the metal–insulator interface tend to change the shape of the potential barriers and also the activation energy of the excitations. Moreover, in Sheng's model the tunneling due to energy-independent overlap is considered to be more important than the phonon-assisted activated processes.

Voit and Buttner [169] have shown that the fitting parameters in FIT model are not consistent with the physical properties of metallic conducting polymers. Kaiser and Graham [170] extended the FIT model for heterogeneous systems by introducing geometric factors to the insulating barriers. The temperature dependence of the insulating barrier is retained with an additional amplification by the geometric factors depending on the size of the highly conducting regions. The insulating barriers are produced by voids, inter-fibrillar contacts, undoped regions, structural defects, etc. In addition to the tunneling transport across the insulating barrier, a parallel phonon-assisted hopping transport was included in Kaiser and Graham model [170]. Thus, in their model the bulk conductivity is a combination of the Kivelson–Heeger q-1D transport, the hopping/tunneling transport and the 3D disordered

metallic transport. This phenomenological model was found to be satisfactory for a qualitative level of explanation for charge trasnport in previous generation conducting polymers. According to the heterogeneous model proposed by Kaiser [143], it is possible that the systems that are well into the insulating regime, having a negative temperature coefficient of W, could display a metallic quasi-linear temperature dependence of thermopower. Since the thermal current carried by phonons is less impeded by the thin insulating barriers than the electrical current carried by electrons or holes, the temperature dependence of thermopower is not expected to be as sensitive as the temperature dependence of conductivity and the latter is rather limited by the scattering processes and mean free path.

Paasch [171] has criticized the Kaiser–Graham model [170] by suggesting that this model is a combination of many transport mechanisms, and the seven (independent) parameters in this model may give a smooth fit to any data. He has proposed a slightly modified version of the FIT model. In this heterogeneous model, one of the main barrier for the tunneling process is the chain segments with residual dimerization. These segments exhibit a dimerization gap which acts as a tunneling barrier for the charge carriers. The corresponding barrier height is nearly half of the energy gap (1.5 eV/2). In general, however, the mutiple-parameter fitting procedure in FIT models has not been able to provide a satisfactory physical understanding of charge transport in conducting polymers.

Nogami *et al.* [172] have carried out detailed measurements to verify the tunneling/hopping models. Temperature dependence of conductivity by itself is not enough to verify the tunneling/hopping models. In fact, the electric field dependence of conductivity at different temperatures is more sensitive to check the appropriateness of tunneling/hopping models. Nogami *et al.* have cross-checked the temperature and electric field dependence of conductivity with numerical simulations. Their rigorous study has shown that the tunneling/hopping models are not consistent with the experimental data for metallic conducting polymers.

Conwell and Mizes [173] suggested that the conduction mechanism is not due to FIT in conducting polymers, since the metallic regions in the FIT model have negligible temperature dependence of conductivity. Moreover, the phonon scattering and the scattering effects due to chain breaks, conjugation defects, inclusion of semiconducting regions, defects within the metallic regions, etc. are neglected in Sheng's model. The conductivity is mainly limited by conjugation defects and chain breaks within the metallic regions. The increase in conductivity with temperature is due to thermal activation (the charge carriers absorb phonons and are thereby activated over the barriers). Moreover, Mizes and Conwell [174] suggested that the interchain coupling is large enough to destabilize the polaron; however, the presence of chain endings and conjugation defects can stabilize the polaron in real systems. The chains are fairly long and the conjugation defects could be eliminated, then the electron and hole transport in t-$(CH)_x$ would be like that in a silicon [174].

Conwell and Mizes [173] have used the method suggested by Soda et $al.$ [175] to determine the band motion vs. diffusive hopping transport in oriented metallic conducting polymers, and its comparison with respect to TTF-TCNQ-type systems. This is especially important to understand the effective electronic dimensionality of the system and its role in transport properties. A brief description for this analysis is as follows.

Consider that at time zero an electron is on chain 1 (wavefunction ψ_1) and $t_\perp$ induces a transition to chain 2 (wavefunction ψ_2). The wavefunction at time t is given by $\psi(t) = \psi_1 \cos(t_\perp t/\hbar) + \psi_2 \sin(t_\perp t/\hbar)$. The process of oscillating coherently between the two chains is interrupted by the scattering of the electron moving along the chain after an average time $\tau_\parallel$. If $t_\perp\tau_\parallel/\hbar \gg 2\pi$ the wavefunction oscillates between the chains many times before a phase-changing collision occurs and can be considered to be a coherent superposition of ψ_1 and ψ_2. If, on the other hand, $t_\perp\tau_\parallel/\hbar \ll 1$, $\psi(t)$ has not time to build up on chain 2 before its phase is destroyed and the motion between the chains is diffusive. Thus the criterion for band motion is $(t_\perp\tau_\parallel/\hbar) \gg 1$ and that for diffusive motion is $(t_\perp\tau_\parallel/\hbar) \ll 1$ [173].

The value for $\tau_\parallel$ in $(CH)_x$ can be estimated from $\sigma_\parallel = ne^2\tau_\parallel/m^*$, (where $\sigma_\parallel \approx 10^5$ S cm^{-1}, $n \approx 10^{22}$ cm^{-3} and m^* is the free electron mass), which is nearly 10^{-14} s [173]. Thus $t_\perp\tau_\parallel/\hbar \approx 1$ is the limiting value for this case. If the system consists of high conducting regions mixed with low conducting regions, then $\sigma_\perp$ is a mixture of band motion and hopping. Thus when the conductivity is 10^4 S cm^{-1} or lower, $\sigma_\perp$ would be more hopping. Soda et al [175] have derived the probability of transverse hopping per unit time, which is given by $\tau_\perp^{-1} = (2\pi/\hbar)(t_\perp^2)(\tau_\parallel/\hbar)$. Then the diffusion rate is given by $D = l^2\tau_\perp^{-1}$, where i is the interchain distance and $\sigma_\perp = ne^2l^2\tau_\perp^{-1}/k_BT$. Thus the anisotropy of conductivity is given by $\sigma_\parallel/\sigma_\perp = \hbar^2 k_BT/m^*l^2\tau_\perp^2$. According to this model, the anisotropy of conductivity is expected to be linear in temperature, if $\tau_\perp$ and $\sigma_\perp$ are not activated. However, the measurements of $\sigma_\parallel/\sigma_\perp$ as a function of temperature in oriented metallic $(CH)_x$ and TTF-TCNQ (above the Peierls transition) shows that $\sigma_\parallel/\sigma_\perp$ is nearly temperature independent, hence it is possible that $\sigma_\perp$ is diffusive hopping in both cases [173].

It is well known that the $transverse$ bandwidth in undoped $(CH)_x$ ($4t_\perp = 0.3$ eV) is only slightly smaller than the bandwidth $parallel$ to the TTF or TCNQ chains in the TTF-TCNQ system [10,173]. The values of $\sigma(300$ K), $\sigma_\parallel/\sigma_\perp$, etc of various anisotropic systems are shown in Table 2.1. Since TTF-TCNQ crystals exhibits positive TCR above the Peierls transition temperature, one could expect a positive TCR for both $\sigma_\parallel$ and $\perp$ in metallic $(CH)_x$ by reducing the disorder. Moreover, in the case of $(CH)_x$ the chains are not aligned as in the case of TTF-TCNQ-type systems (crystalline q-1D) and the misaligned chains could easily contribute a parallel component to $\sigma_\perp$. The experimental results indicate that the apparent ratio of $\sigma_\parallel/\sigma_\perp$ increases with $\sigma_\parallel$ [176] and the issue of whether $\sigma_\perp$ in highly oriented metallic conducting polymers is band motion or diffusive hopping is still open.

According to Stafström [99] the enhanced interchain interactions with increasing dopant concentration can induce 3D delocalization of the electronic states. Moreover, the calculations by using the multichannel Buttiker–Landauer formalism [100] indicate that both conductance and localization in a system of coupled polyconjugated chains are strongly dependent on the interchain hopping strength. Although Mizes and Conwell [173b] have used a Landauer-type transmission–reflection formalism for conduction in ladder polymers, the model proposed by Stafström is the first rigorous attempt to solve the problem of transport in conducting polymers by using the Landauer-type formalism. The multichannel Buttiker–Landauer conductance method is used to study the conductance of a system as a function of the length of the system. The channels consist of several conjugated segments that are separated by defects, chain interruptions etc. The interchannel hopping involves both intrachain hopping [2.5 eV for t-$(CH)_x$] and hopping between the two adjacent chain ends, i.e. interchain hopping [0.1 eV for t-$(CH)_x$] [99].

In the electron localization problem, the weak disorder in the site energies or interchain hopping is not as significant as chain interruptions. The interchain interactions are expected to reduce the effect of backscattering due to chain interruptions. The numerical simulations by Stafström show that the most relevant parameter that induces localization is the number of chain interruptions and not the channel length. Moreover, the numerical simulations show that the conductance is unaffected by chain interruptions up to a certain critical number. The critical chain length for which the conductance of the system starts to drop can be considered as the transition between diffusive and non-diffusive transport. The numerical simulations on the doped $(CH)_x$ system show that the coherent transport extends over a length scale larger than the average distance between two chain interruptions. This clearly explains the fact that the mean free path in high-quality metallic samples of $(CH)_x$ is of the order of 600 Å, which is considerably longer than the crystalline coherence length of the order of 100 Å [99]. Thus by using the multichannel Buttiker–Landauer formalism, Stafström concludes that the conductance properties of oriented metallic conducting polymers differ from both lD and 3D systems. It is possible that the conductance in some local region of the system could behave in accordance with the intrinsic q-lD nature of the conducting polymer and then it exponentially decays outside this local region.

Phillips *et al.* [33] proposed a random dimer model which has a set of delocalized conducting states, even in lD, that ultimately allow a particle to move through the lattice almost ballistically. They suggested that this interesting absence of Anderson localization in lD, despite the disorder, might be applicable to conducting polymers. Phillips *et al.* suggested that the disordered bipolaron lattice for doped PANI is a subset of the random dimer model.

In highly disordered conducting polymers, the usual exponential temperature dependence of conductivity is attributed to some type of phonon-assisted hopping mechanism or some type of tunneling mechanisms [for example FIT, the charging-energy-limited tunneling (CELT)] [168,177] or both [170]. Recently Zuppiroli *et al.* [178] proposed a polaron/bipolaron model involving correlated hopping and mutiphonon processes as an explanation for the exponential temperature dependence of conductivity. Their model is an extension of the work by Sheng *et al.* [168] and Ables *et al.* [177] on granular metals by incorporating the polaron/bipolaron clusters. In their model the charging energy is the main barrier for hopping transport. The formation of polaronic clusters, due to the fluctuations in dopant concentrations, functions as metallic grains in their 'granular-metal' hopping model. Zuppiroli *et al.* [178] suggest that the counterion potentials favor the possibility of interchain tunneling through dopants and as a result the interchain coupling is enhanced. When the size of polaronic clusters increases by improving the alignment of chains and the homogenous distribution of dopants, a crossover from non-adiabatic

to adiabatic hopping transport is expected; as a result long-range coherent transport is possible.

Among various models proposed by several groups for conducting polymers in the insulating side, the theoretical work by Ovchinnikov and Pronin [179] and Lewis [180] are slightly different from other models. Ovchinnikov and Pronin [179] proposed a q-lD percolation model for explaining the conductivity of conducting polymers in the insulating regime. In this model an impurity (e.g. acceptor) captures an electron from one of the adjacent chains and forms a charged impurity center. Such a carrier can detrap by an activated process and diffuse along the chain. This polaron can recombine with another impurity center near the chain and then escape to an arbitrary chain adjacent to the second impurity center. Thus, conduction by percolation is possible in such a system if an infinite cluster of chains can be connected by impurity centers.

Lewis [180] suggested that the charge transport occurs by tunnel transitions between localized states. Lattice fluctuations and electron–lattice coupling tend to broaden and reorganize the energy levels at each site. He has used a Redi–Hopfield [181] formalism to estimate the tunnel transmission probability across the potential barriers between localized states. Lewis has estimated the electric field, frequency and temperature dependences of charge transport in both conducting and non-conducting polymers. Moreover, another interesting suggestion in this model is that the localised states correspond to the Urbach states in amorphous systems, since the optical absorption from a distribution of states extends out from the fundamental absorption band edge.

The influence of molecular weight, orientation, conjugation length, crystallinity, mechanical properties, processing properties, etc. on conductivity and charge transport in conducting polymers has been investigated by the models proposed by Heeger and Smith [155], and Pearson *et al.* [182]. Heeger and Smith [155] have developed a phenomenological model based on the following parameters: carrier diffusion along the chain (D), the mean lifetime of the carrier (τ_i) on a given chain before it hops to another chain and the characteristic time (τ_c) for a carrier to diffuse along the backbone of the entire length of the chain. In the limit $\tau_i \gg \tau_c$ for an amorphous glass of random chains, the conductivity is expressed as $\sigma = ne^2 N\alpha^2/k_B T\tau_i$, where n is the concentration of carriers, e is the charge per carrier, N is the degree of polymerization, α is the persistence length of the chain and $N\alpha \sim L$ is the contour length of the macromolecule. In the case of an amorphous glass of random coils, the conductivity increases with increasing molecular weight as N^β with $1.0 < \beta < 1.2$. However, as N increases, eventually τ_c exceeds τ_i. In the limit $\tau_c \gg \tau_i$ for an amorphous glass of Gaussian chains, the conductivity is expressed as $\sigma = (ne^2\alpha/k_BT)(D/\tau_i)$; in this regime the conductivity is independent of the molecular weight. Hence, if the interchain transport occurs faster than the intrachain transport then the conductivity is limited by molecular weight and if the opposite occurs the conductivity is not limited by the molecular weight. Heeger and Smith found that in the case of oriented polymers the equations for conductivity are slightly different. A detailed analysis of experimental data regarding the relationship of conductivity with respect to the Young's modulus, tensile strength, tenacity, draw ratio, solution viscosity, etc. of several conjugated polymers has been discussed in the review by Heeger and Smith [155].

Pearson *et al.* [182] have used a similar phenomenological model to understand the effects of molecular weight and chain orientation on conductivity. According to their calculations, in the limit $\tau_i \gg \tau_c$, $\sigma = ne^2\alpha L/18k_BT\tau_i$ and in the limit $\tau_c \gg \tau_i$, $\sigma = (ne^2\alpha/6k_BT) \times (D/\tau_i)^{1/2}$. The crossover between the two regimes is smooth and it occurs when $\tau_c/\tau_i = 9/\pi^2 \approx 1$. This model predicts that the conductivity should increase with the square of the stretching ratio. The roles played by molecular weight distribution, defects, liquid crystallinity, etc. have been included in their model. The preliminary experimental results [183,184] are in agreement with the models proposed by both Heeger and Smith [115] and Pearson *et al.* [812].

The exponential temperature dependence of conductivity in the insulating regime of conducting polymers has been often attributed to VRH, as in equation (2.4). In conventional VRH, the exponents (γ) for one, two and three dimensions are 0.25, 0.33 and 0.5, respectively. However, a wide range of values (0.25–1) of the exponent has been observed in insulating conducting polymers [2,3]. Moreover, the $T^{-1/2}$ dependence of conductivity is often observed in granular metallic systems. Schreiber and Grussbach [59] suggested that the fluctuations in mesoscopic systems could give a wide range of values of the exponent due to the fractal nature of wavefunctions near the mobility edge. Moreover, in multifractal localization the nature of wavefunction varys continuously from exponentially localized to fractally localized in the insulating regime of the M–I transition [57].

Although the above theoretical models could shed some light on the transport mechanism in conducting polymers, the early low-temperature conductivity measurements in $(CH)_x$ have given some important clues for the identification of the transport mechanism in metallic systems [142,185]. The quantum corrections (contributions from weak localization and e–e interactions) to finite conductivity, at mK temperatures, have been observed in metallic $(CH)_x$. Later, Madsen *et al.* [186] have observed the e–e interaction contribution to conductivity ($\sigma \propto T^{1/2}$) in metallic PPV samples. Recently Ishiguro *et al.* [187] have observed temperature-independent conductivity, below 10 K, in high quality metallic $(CH)_x$ and PPy samples. The presence of large conductivity at mK temperatures clearly shows the intrinsic metallic nature of high-quality samples.

The possibility of superconductivity in conducting polymers has been investigated by some theoretical models [188,189]. Kivelson and Emery [188] have discussed the relationship between disorder, correlation effects and superconductivity for doped conducting polymers. Their schematic phase diagram of doped conducting polymers suggests that the intermediate doping level is highly favorable for observing a superconducting state in these systems. In an earlier model for superconductivity in conducting polymers, suggested by Voit [189], the strong coupling of intramolecular vibrations to electrons, high-frequency phonons, well-screened electronic interactions, etc. are considered as the essential ingredients for obtained a superconducting state in these systems. The polymer chain structure and fibrillar morphology are not detrimental to the process of obtaining a superconducting state, since inorganic polymer $(SN)_x$ is a superconductor at very low temperatures [4].

The stochastic network models and discrete models assume that doped conducting polymers are highly disordered and inhomogenous systems; this point of view is rather different from the Anderson localization in a homogeneously disordered system. The cut-off length scales for band vs. discrete models depend on the size of the extended states, the localization/correlation length, the inelastic scattering length, the Landu orbit, etc. The intragranular and intergranular disorder in highly disordered samples could smear off all the quantum features in the low-temperature transport. The implications of the effective Coulomb correlation energy and the long-range Coulomb interactions in charge transport are not yet fully understood. The details regarding the length scales and low-temperature transport are presented in next section.

8 TRANSPORT PROPERTIES OF CONDUCTING POLYMERS IN THE METALLIC REGIME

8.1 Conductivity

Although in the classical sense a metal should show positive TCR down to low temperatures, it is well known from the many years of work in disordered metallic systems that the extent of disorder in the system essentially determines the temperature dependence of conductivity at low temperatures [28,44]. The role of disorder in insulator–metal and insulator–superconductor transitions is very well demonstrated in metallic and insulating thin films, as shown in Figure 2.4 [190–192]. In the case of granular superconductors it is well known that a critical radius of 22 Å or about 750 electrons is needed to give rise to the formation of a superconducting state [193]. In other words, if the grain size is equal or less than the coherence length, then the thermal and quantum fluctuations inhibit the condensation of Cooper pairs. This subtle insulator–superconductor transition in disordered high $-T_c$ superconductors is shown in Figure 2.5 [194]. In granular aluminum films of 1 μm thickness (3D films) with a room temperature conductivity of 45 S cm^{-1} and $\rho(4.2$ K$)/\rho(300$ K$)\approx 4.6$, the signs of superconductivity have been observed below 1.4 K [195]. Moreover, the M–I transition marks the boundary between superconducting and non-superconducting behavior in disordered granular systems.

It is important to compare the behavior of the temperture dependence of conductivity of systems that are rather similar (in terms of low dimensionality, disorder, correlation effects, etc.) with respect to metallic conducting polymers. This might be helpful in developing a wider perspective for understanding the charge trasnport in conducting polymers. Hence the temperature dependence of conductivities of $(SN)_x$ [4], doped C_{60} [196], HOPG [94,197], TTF-TCNQ [10a] and BEDT-FFT salt, $(ET)_2I_3$ [10c] are shown in Figures 2.6–10. The conventional wisdom suggests that a low-room temperature conductivity $(k_F\lambda \approx 1)$ could adversely affect the typical metallic positive TCR. However, some of the above examples clearly rule out such constraints. Another classic example is $Li_{0.8}$ $Ti_{2.2}O_4$ which has a room temperature conductivity of 20 S cm^{-1}. This material shows a positive TCR and it becomes a superconductor at 10 K [198].

Although the room temperature conductivity of some of the above mentioned systems (see Table 2.1) is considerably lower with respect to metallic $(CH)_x$ PPV,

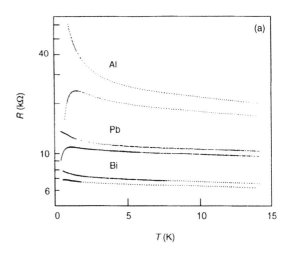

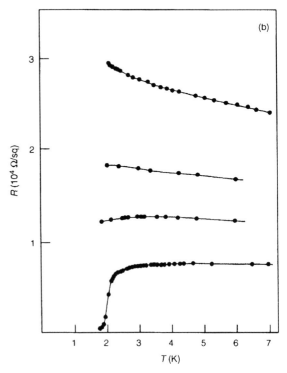

Figure 2.4. The role of disorder in insulator–metal and insulator–superconductor transitions in thin metal films: (a) sheet resistance R vs. T for the first superconducting and the last insulating films for bismuth, lead and aluminum (Reproduced by permission of the American Physical Society from ref. 192), (b) sheet resistance R vs. T for lead film [film thickness increases from 5 (top) to 20 (bottom) Å][190] (c) resistivity vs. T for granular niobium nitride thin films (8–7 nm) (Reproduced by permission of the *J. de Physique* from ref. 191).

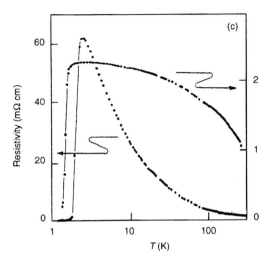

Figure 2.4. (*continued*)

PPy etc. and the anisotropy in conductivity in some of those systems is rather similar with respect to oriented metallic conducting polymers, the absence of a significant positive TCR in the latter indicates that disorder is playing a substantial role in limiting the metallic transport. The positive TCR and superconductivity in crystalline low-dimensional systems like $(SN)_x$ and BEDT-TTF indicates that the intrinsic low-dimensional nature of conducting polymers should not be restricting the metallic behavior. Moreover, the interchain interaction in conducting polymers is substantial with respect to charge transfer complexes due to larger interchain and polymer–dopant interactions. Thus, in principle one could expect real metallic behavior in high qualtiy conducting polymers.

Usually, in inorganic semiconductors the M–I transition occurs at carrier concentrations around 10^{18}–10^{19} cm^{-3} [28], and the M–I transition is mainly governed by the carrier density. In crystalline silicon doped with phosphorous, boron, etc. the conductivity near the M–I transition is around 50–100 S cm^{-1}. At temperatures below 4.2 K, the typical positive TCR for a good metal is observed for doped silicon samples on the metallic side of the M–I transition [199,200]. However, in doped silicon near or below the M–I transition, the TCR is negative at all temperatures. This change in the sign of the TCR at low temperatures (for samples on the metallic side of the M–I transition) is attributed to the breakdown of Thomas–Fermi screening as the M–I transition is approached.

This brief outline of the temperature dependence of conductivity of various systems might be helpful in

understanding the charge transport in conducting polymers. The typical examples of metallic conducting polymers are the following: p- and n-type doped oriented $(CH)_x$, p- and n-type doped oriented PPV, PPy-PF$_6$, PANI-CSA and ion-implanted PANI.

In previous generation conducting polymers hardly any positive TCR was observed below 300 K. Although a finite value of conductivity as $T \rightarrow 0$ K was observed in doped Shirakawa $(CH)_x$, the negative TCR was substantial due to disorder [142,185]. The first systematic observation of a positive TCR (in the temperature range 220–300 K) in FeCl$_3$-$(CH)_x$ was reported by Park et al. [149,201]. Later, a substantial positive TCR was observed down to 160 K in PANI-CSA [150]. Surprisingly, PPy-PF$_6$, [2,111], PPV-AsF$_5$ [186] and PPV-H$_2$SO$_4$ [107], ion-implanted PANI [202] and K-$(CH)_x$ [203] showed a *positive TCR at temperatures below 20 K*. Hence, in a nutshell, the real metallic features in conducting polymers are already observed in the temperature dependence of dc conductivity measurements and the details are given below.

The first criteria for the metallicity in any system is the existence of a finite conductivity at $T \rightarrow 0$ K, which implies that there is a finite density of states at the Fermi level. Although a positive TCR is desirable for a good metal, it is not an essential criteria for metallicity. The detailed study of disordered metallic systems in the past many years has shown that the presence of a weak negative TCR does not necessarily imply that the system is not metallic. However, in order to confirm the presence of a finite conductivity at $T \rightarrow 0$ K it is essential to have the $\sigma(T)$ data in the mK temperature range, which is not very easily accessible in many conducting polymer research laboratories. Hence a facile alternative method for characterizing the metallicity of the system, from the temperature dependence of conductivity, is essential.

From the study of $\sigma(T)$ in disordered semiconductors, Zabrodskii and Zeninova [156] have shown that the characteristic behavior of $\rho(T)$ can be understood in detail by defining the reduced activation energy (W) as the logarithmic derivation of $\sigma(T)$

$$W = -T\{d \ln \rho(T)/dT\} = -d(\ln \rho)/d(\ln T)$$
$$= d(\ln \sigma)/d(\ln T) \qquad (2.11)$$

where ρ is resistivity and σ is conductivity. Thus the best way to analyze the temperature dependence of conductivity is to plot W vs. T in a log–log scale [156]. This plot facilitates the identification of transport in various regimes. Of course, in metallic systems with

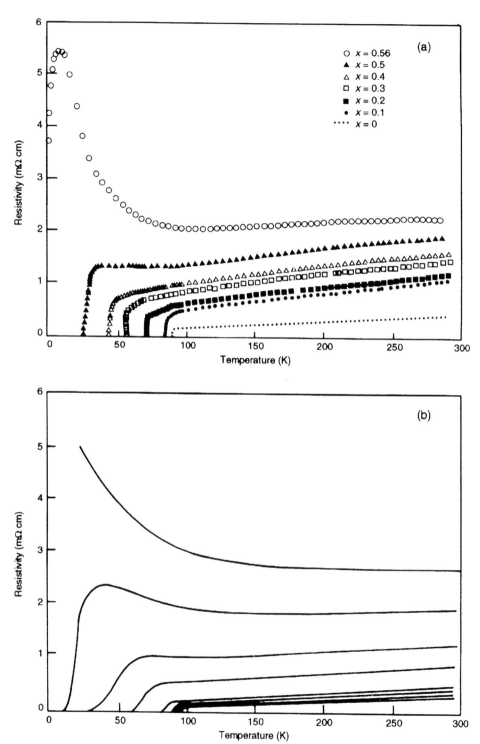

Figure 2.5. The insulator–superconductor transistion in high-T_c superconductors 194: resistivity vs. T (a) $Y_{1-x}PR_xBa_2Cu_3O_{7-\delta}$. (b) Ion damaged $YBa_2Cu_3O_{1-\delta}$ (Reproduced by permission of the American Physical Society from ref. 194).

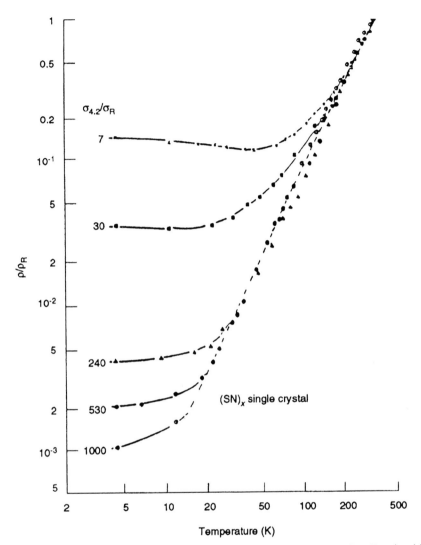

Figure 2.6. Normalized resistivity ρ/ρ_R [where $\rho_R \sim \rho(300\ K)$] vs. T for various $(SN)_x$ samples. (Reprinted by permission of Kluwer Academic Publishers from ref. 4).

positive TCR, W is negative. However, in systems with negative TCR, the appropriate regimes of charge transport in disordered systems can be identified as follows:

(1) If $W(T)$ has a positive temperature coefficient at low temperatures, then the system is on the metallic side of the M–I transition.
(2) If $W(T)$ is nearly temperature independent for a wide range of temperature, then the system is on the critical regime of the M–I transition.

(3) If $W(T)$ has a negative temperature coefficient at low temperatures, then the system is on the insulating side of the M–I transition.

This behavior of $W(T)$ is closely related to another important parameter, the resistivity ratio, $\rho_r \sim \rho(1.4\ K)/\rho(300\ K)$. In the case of crystalline-doped silicon near the M–I transition, ρ_r is nearly 1.2–3 [199,200] and for conducting polymers it is very much dependent on the quality of samples. Since the critical doping concentration for the M–I transition in conducting polymers is

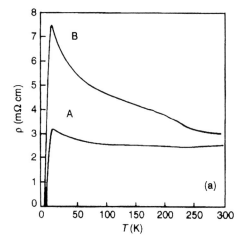

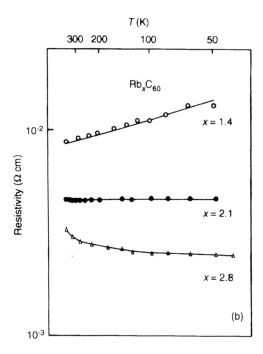

Figure 2.7. (a) resistivity vs. T for K_3C_{60} films [196 a]; (b) resistivity vs. T for Rb_xC_{60} films at various doping levels (x). (Reproduced by permission of the American Physical Society from ref. 196b).

not known as in the case of inorganic semiconductors, ρ_r is a useful parameter in characterizing the M–I transition in conducting polymers. The transport property measurements in conducting polymers indicate that the critical doping concentration near the M–I transition is around 10^{20} cm^{-3}. However, even in

the case of fully doped conducting polymers, the M–I transition is mainly governed by the extent of disorder, since conducting polymers are inherently partially crystalline and partially amorphous. Thus the extent of disorder plays an important role in controlling the M–I transition in conducting polymers.

In early publications [3], the data analysis was not focused on the precise identification of the metallic, critical and insulating regimes in conducting polymers. This somehow hampered the understanding of charge transport in conducting polymers. Only recently, the improvement in sample quality has substantially reduced the dominant role of disorder-induced localization (although it certainly remains important even in the best materials). Hence, the genuine metallic properties just begins to be observed in transport property measurements.

Usually, the conductivity in the disordered metallic regime at low temperatures is expressed by [199]

$$\sigma(T) = \sigma(0) + mT^{1/2} + BT^{p/2} \qquad (2.12)$$

where the second term ($T^{1/2}$) results from e–e interactions and the third term is the correction to $\sigma(0)$ due to localization effects. The value of p is determined by the temperature dependence of the scattering rate ($\tau^1 \propto T^P$) of the dominant dephasing mechanism. For electron–phonon scattering: $p = 3$; for inelastic e–e scattering: $p = 2$ in the clean limit or $p = \frac{3}{2}$ in the dirty limit. The calculation by Belitz and Wysokinski [204] give $p = 1$ very near to the M–I transition. In the disordered metallic regime, the conductivity depends on three length scales [28,44]: the correlation length L_c describing the M–I transition, the interaction length, $L_T = (\hbar D/k_B T)^{1/2}$, and inelastic diffusion length, $L_{in} = (D\tau_{in})^{1/2}$ (where D is the diffusion coefficient and τ_{in} is the inelastic scattering time). In practice, however, it is difficult to distinguish these contributions from $\sigma(T)$ alone; and the finer details of these contributions can be determined from magnetoconductance (MC) measurements.

Möbius [43] suggested that in the metallic regime the logarithmic derivative, $w(T) = d \ln \rho/d \ln T$ is far more senstive than the conductivity itself, expressed by equation (2.12). The logarithmic derivative of equation (2.12) is given by

$$w(T) = [0.5mT^{1/2} + (p/2)BT^{p/2}]/\sigma(T) \qquad (2.13)$$

This equation implies that $w(T)$ vanishes as $T \rightarrow 0$ for metallic samples.

In disordered metals, e–e interactions play an important role in the low temperature transport. The

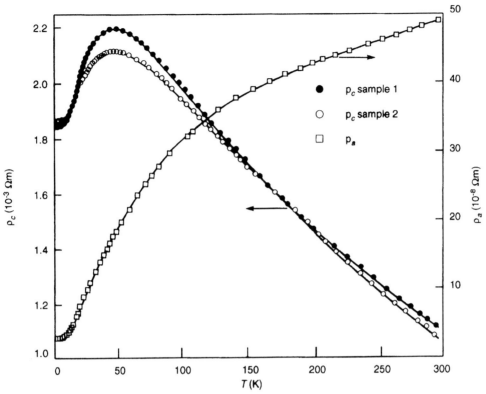

Figure 2.8. Resistivity vs. T for highly oriented pyrolytic graphite (HOPG): $\rho(c$ axis) on left hand side, $\rho_{(a-b\ \text{plane})}$ on right hand side [94].

e–e interaction contribution towards the low temperature conductivity $\sigma_1(T)$ can be expressed as [199]

$$\sigma_1(T) = \sigma(0) + mT^{1/2} \qquad (2.14)$$

where

$$m = \sigma[(\tfrac{4}{3}) - (3gF_\sigma/2)] \qquad (2.15)$$

$$\sigma = (e^2/\hbar)(1.3/4p^2)(k_B/2\hbar D)^{1/2} \qquad (2.16)$$

$$F_\sigma = 32[(1 + F/2)^{3/2} - (1 + 3F/4)]/3F \qquad (2.17)$$

The finite temperature correction term due to e–e interactions in equation (2.4) consists of exchange and Hartree contributions [28,44]. The sign of this correction depends on the relative size of the exchange and Hartree terms, which depend on the screening length. In doped semiconductors the sign of the finite temperature correction is related to various parameters, such as the degeneracy of the conduction band minima in k space (valleys), intervalley scattering, mass anisotropy, etc. The Hartree factor (F) is the screened

interaction averaged over the Fermi surface, α is a parameter depending on the diffusion coefficient (D) and γF_σ is the interaction parameter. The value of γ depends on the band structure [199]. The coefficient was found to change sign as a function of disorder [199], a change which can be interpreted as being due to a sign change in $[(\tfrac{4}{3})-(3\gamma F_\sigma/2)]$. Usually, the sign of m is negative when $\gamma F_\sigma > 8/9$.

Among various metallic conducting polymers, oriented I-(CH)$_x$ has been studied extensively by several groups [106,154,172,205,206]. The maximum room temperture conductivity parallel to the chain axis for the best quality oriented I-(CH)$_x$ is of the order of 10^5 S cm^{-1}, and the anisotropy is nearly 100 [154]. Shirakawa *et al.* [207] have observed that the stretchability and the maximum obtainable conductivity in I-(CH)$_x$ is very much dependent on the film thickness. The thinner the films, the higher the conductivity in both stretched and unstretched films. Recently Mizoguchi *et al.* [208] have shown that the main difference between Shirakawa and Naarmann (CH)$_x$ is the higher

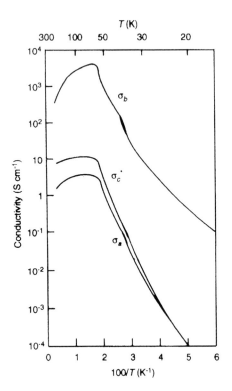

Figure 2.9. Conductivity vs. T for a TTF–TCNQ crystal in a, b and c^* planes [10a]. (Reproduced from *Organic Superconductors*, T. Ishiguro and K. Yamaji, p.26, Figure 2.12, 1990, copyright notice of Springer-Verlag).

density and the higher degree of chain orientation in the latter; the basic features of spin dynamics are identical in both.

The structural and physical properties of highly conducting $(CH)_x$ have been reviewed by Tsukamoto [154]. The average crystalline coherence lengths parallel and perpendicular to the chain axis are 120 Å and 50 Å, respectively. The number of charge carriers in heavily doped samples is of the order of 10^{22} cm^{-3} and the mean free path is approximately 500 Å. Epstein [206,209] has pointed out that in metallic $(CH)_x$ the elastic scattering length along the chain is nearly 30 Å. This indicates that a charge carrier can visit nearly 40 chains and travel an approximate distance of 2000 Å before undergoing any inelastic scattering collisions. The density of states at the Fermi level in fully doped samples is approximately 0.3 states $(eV - C)^{-1}$, which corresponds to 10^5 emu $-$ mole^{-1}C^{-1} [154].

Ishiguro and coworkers have reported extensive measurements of $\sigma(T)$ in iodine [172,187,205] and

FeCl$_3$ doped $(CH)_x$ [210] down to mK temperatures. Although the room temperature conductivity of FeCl$_3$-$(CH)_x$ is nearly an order of magnitude lower than that of I-$(CH)_x$, ρ_r for FeCl$_3$-$(CH)_x$ and I-$(CH)_x$ are similar: 1.3 and 2.8, respectively, as shown in Figure 2.11. Moreover, the positive TCR for FeCl$_3$-$(CH)_x$ above 200 K, indicates that the doping is more homogenous and the dopant-induced interchain transport is higher with respect to I-$(CH)_x$. In some I-$(CH)_x$ samples, a weak positive TCR has been observed below 4.2 K, as mentioned below. Usually, the conductivity is nearly temperature independent below 10 K, in high quality I-$(CH)_x$ samples, as shown in Figure 2.11a. However, the positive TCR near room temperature is absent in the case of I-$(CH)_x$, although its conductivity is nearly an order of magnitude higher at room temperature with respect to FeCl$_3$-$(CH)_x$. Although FeCl$_3$-$(CH)_x$ samples show a metallic positive TCR above 200 K, the low-temperature conductivity is not as weakly temperature dependent with respect to I-$(CH)_x$ samples.

Kaneko *et al.* [210] suggested that the mechanism for this weak positive TCR above 200 K in fresh FeCl$_3$-$(CH)_x$ samples is due to the phonon-induced non-metal–metal transition in q-lD systems. Ishiguro *et al.* [187] have observed a logarithmic temperature dependence of resistivity for metallic I-$(CH)_x$ samples. Although the understanding of this logarithmic temperature dependence of resistivity is not very clear, it has been interpreted to be due to low-energy excitations and the relaxations in molecular conformation at low temperatures. However, this temperature-independent conductivity (at low temperatures) in metallic $(CH)_x$ samples could due as well to some temperature-independence scattering mechanisms, competition among weak localization and e–e interaction contributions, etc. Since the temperature dependence of conductivity by itself is not enough to distinguish the most appropriate one among various models, the MC measurements (Section 8.2) are important in clearing the ambiguities.

A significant number of publications regarding $\sigma(T)$ of metallic $(CH)_x$ was mainly focused on the stretch-oriented $(\sigma_\parallel)$ direction. Although Schimmel *et al.* [153,211] reported the temperature dependence of both $(\sigma_\parallel)$ and $(\sigma_\perp)$ the ratio of $(\sigma_\parallel)/(\sigma_\perp)$ in their sample was only 25 and the tendency for the saturation of conductivity as $T \to 0$ K was not observed. Nevertheless, Schimmel *et al.* observed that the temperature dependences are nearly identical in both $(\sigma_\parallel)$ and $(\sigma_\perp)$. They have used the FIT model for the data analysis of $\sigma(T)$. Similarly Ahmed *et al.* [212] have used the FIT model for the data analysis of $\sigma(T)$ in I-$(CH)_x$ samples.

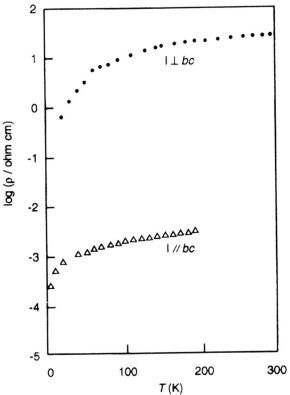

Figure 2.10. Log of resistivity vs. *T* for (ET)$_2$I$_3$ (both parallel and perpendicular to the *bc* plane) [10c]. (Reprinted from *Solid State Commun* **64**, K. Kajita *et al*, p. 1279, Copyright 1987, with kind permission from Elsevier Science Ltd, The Boulevard, Langford Lane, Kidlington OX5 1GB, UK).

Park *et al* [149,201] have carried out detailed study of $\sigma(T)$ and $(\sigma_{\parallel})/(\sigma_{\perp})$ in (CH)$_x$ samples, and they suggested a qualitative interpretation based on soliton condensation model.

The recent work by Reghu *et al.* [106] have shown that the mechanism of the temperature dependences of both $(\sigma_{\parallel})$ and $(\sigma_{\perp})$ are nearly identical. In high quality I-(CH)$_x$ samples with $(\sigma_{\parallel})/(\sigma_{\perp}) \approx 100$ and $\rho_r \approx 2$–3, the logarithmic dependence of $\sigma(T)$ was observed to be not satisfactory, as shown in Figure 2.12. $W(T)$ is nearly temperature independent from 200 to 60 K, at ambient pressure, as shown in Figure 2.13 and at 8 kbar $W(T)$ shows the typical metallic positive temperature coefficient below 200 K. The presence of the temperature-independent $W(T)$ regime (200–60 K) indicates that the system is just on the metallic side of the critical regime of M–I transition. The normalized conductivity, below this temperature-independent $W(T)$ regime, as a function of pressure and ρ_r is shown in Figure 2.14. These data clearly indicate that pressure increases the

interchain interaction and subsequently the metallic features of oriented conducting polymers are enhanced. Thus the interchain interaction plays a key role in the metallic properties of doped conducting polymers.

The $\sigma(T)$ in the temperature independent $W(T)$ regime follows the typical power law in the critical regime of the M–I transition [2]. In fact this temperature-independent $W(T)$ regime, at high temperatures, is a qualitative measure of the extent of inhomogenity in the system. This is explained in more detail in Section 9. The analysis of the temperature dependences [in the positive temperature coefficent $W(T)$ regime] of both $\sigma_{\parallel}$ and $\sigma_{\perp}$ is carried out by using the localization–interaction model for a 3D anisotropic system. Although Thummes *et al.* [185] have identified the interaction contribution in $\sigma(T)$ for metallic (CH)$_x$ samples [$\sigma(300 \text{ K}) \approx 100 \text{ S cm}^{-1}$], the quality of the samples used in these measurements was not high enough to distinguish the contributions from localization and interactions. In I-(CH)$_x$ samples with $\sigma_{\parallel}/\sigma_{\perp} \approx 100$ and

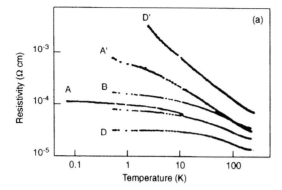

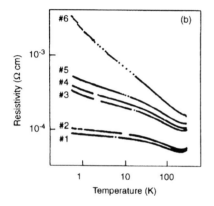

Figure 2.11. Resistivity vs. T for doped $(CH)_x$: (a) I–$(CH)_x$, $\rho_r \approx 2.8$–100 [205]; (b) $FeCl_3$–$(CH)_x$, $\rho_r \approx 1.3$–100 [210]. (Reproduced by permission of Taylor Francis from ref. 205). (Reproduced by permission of Elsevier Science SA from *Synth. Met.* **65**, 1994, 141).

$\rho_r \approx 2.5$–3, the dominance of the weak localization contribution at higher temperatures ($T \geqslant 4$ K) and the dominance of the e–e interaction contribution at lower temperatures ($T \leqslant 4$ K) could be clearly observed; the low-temperature conductivity follows [106]:

$$\sigma(T) = \sigma(0) + mT^{1/2} + BT^{3/2} \qquad (2.18)$$

The $T^{1/2}$ and $T^{3/4}$ fits (below 60 K), for both $\sigma_\parallel$ and $\sigma_\perp$, are shown in Figure 2.15. Although $\sigma_\parallel/\sigma_\parallel \approx 100$, these fits are identical for both $\sigma_\parallel$ and $\sigma_\perp$, indicating that an anisotropic 3D model is appropriate for highly oriented $(CH)_x$. The linearity of the $T^{3/4}$ fits is better than that of the $T^{1/2}$ fits ($T > 3$ K), implying that the contribution from the localization is dominant at higher temperatures [106]. The inelastic electron–phonon scattering ($p = 3/2$) is the dominant scattering mechanism when $\sigma \propto T^{3/4}$. However, when $\sigma \propto T^{1/2}$ ($T < 3$ K), e–e interactions are more dominant. The normalized conductivity for I–$(CH)_x$ samples ($l/l_o = 6$) as a function of $T^{1/2}$ (for both $\sigma_\parallel$ and $\sigma_\perp$,) at ambient pressure and 4 kbar is shown in Figure 2.16. A pressure-induced crossover from $T^{3/4}$ to $T^{1/2}$ is observed at low temperatures in $\sigma_\parallel$, as shown in Figure 2.17. Thus the localization and interaction effects dominate at high and low temperatures, respectively; and both could be fine-tuned by varying the interchain interactions with pressure, orientation, etc. This has been confirmed from MC measurements (Section 8.2) as described below.

The $\sigma(T)$ for [both $\sigma_\parallel$ and $\sigma_\perp$] for I–$(CH)_x$ samples at various stretching ratios (l/l_o) is shown in Figure 2.18. As l/l_o increases, the $\sigma_\parallel$ increases dramatically and $\sigma_\parallel(T)$ becomes weaker; whereas $\sigma_\perp$ gradually drops and $\sigma_\perp(T)$ remains more or less the same. This

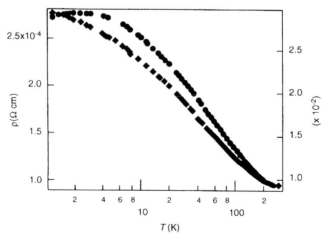

Figure 2.12. Resistivity [$\rho_\parallel$ (●) and $\rho_\perp$ (◆)] vs. T (log scale) for I–$(CH)_x$ ($\sigma_\parallel/\sigma_\perp \approx 100$ and $\rho_r \approx 2.5$–3). (Reproduced by permission of the American Physical Society from ref. 106).

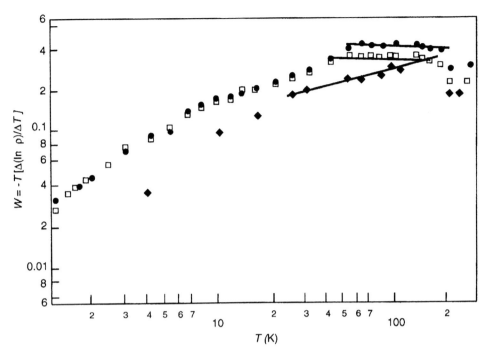

Figure 2.13. W vs. T for I–(CH)$_x$ ($\sigma_\parallel/\sigma_\perp \approx 100$ and $\rho_r \approx 2.5$–3) at ambient pressure $\sigma_\parallel$, ● and $\sigma_\perp$ □ and at 8 kbar (◆) for $\sigma_\parallel$. The lines are drawn to guide the eyes. (Reproduced by permission of the American Physical Society from ref. 106).

indicates that $\sigma_\parallel$ is rather sensitive to the alignment of the fibrils. The W vs. T plot for I-(CH)$_x$ samples (for $\sigma_\parallel$) as a function of ρ_r is shown in Figure 2.19. This consistently shows that the sample becomes more metallic as ρ_r decreases. Moreover, the temperature-independent $W(T)$ regime (200–30 K) for the sample with $\rho_r = 3.32$ shrinks to a smaller temperature range (120–70 K) as ρ_r decreases to 2.08. This clearly indicates that the system moves more into the metallic side of the critical regime of the M–I transition as ρ_r decreases; this could be due as well to the enhanced homogeneity of the system. Recently Mizoguchi et al. [108,208] have indirectly estimated the temperture dependence of microscopic conductivity in doped (CH)$_x$ samples from the NMR relaxation data. This shows a positive TCR, although the macroscopic $\sigma(T)$ shows the typical negative TCR.

Recently Bernier and coworkers [203] have observed a rather weak $\sigma(T)$ for K-(CH)$_x$ samples ($\rho_r < 2$); some samples even showed a positive TCR below 7 K. The room temperature conductivity of fully doped (16.1%) K-(CH)$_x$ samples is nearly 8000 S cm^{-1}. The $\sigma(T)$ of K-(CH)$_x$ samples at various doping level is shown in Figure 2.20. Moreover, they have observed a correla-

tion between staging-induced structural transitions and electronic properties in K-(CH)$_x$. However, more work is necessary to fully characterize the metallic regime in alkali metal-doped (CH)$_x$.

Madsen et al. [186] have shown that the $\sigma(T)$ of doped oriented PPV samples is considerably weak and this is the second best metallic conducting polymer, after doped (CH)$_x$. Surprisingly, much less work has been carried out in doped PPV samples, with respect to doped (CH)$_x$. The maximum $\sigma(300\ K)$ and the minimum ρ_r [note here, $\rho_r \approx \rho(4\ K)/\rho(300\ K)$] for oriented PPV-AsF$_5$ samples ($l/l_o \approx 10$) are 2360 S cm^{-1} and 1.28, respectively. Although Madsen et al. have observed a $T^{1/2}$ dependence of conductivity in PPV-AsF$_5$ samples at low temperatures, the interpretation of the data was only in a qualitative level. Ohnishi et al. [147a] reported that the $\sigma(300\ K)$ for PPV-H$_2$SO$_4$ samples ($l/l_o \approx 10$) is nearly 10^4 S cm^{-1}, which is of the same order of magnitude as that obtained for doped (CH)$_x$ samples.

Recently Ahlskog et al. [107] have carried out detailed $\sigma(T)$ measurements in PPV-H$_2$SO$_4$ samples. These showed extremely weak $\sigma(T)$ values with ρ_r as low as 1.15. This value of ρ_r is less than that reported

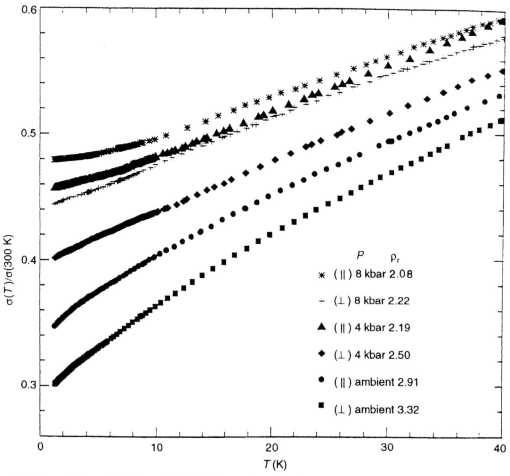

Figure 2.14. Normalized conductivity vs. T for I–(CH)$_x$ as a function of pressure and ρ_r.

for metallic (CH)$_x$ (see Table 2.1) [210]. As shown in Figure 2.21, the $\sigma(T)$ for both $\sigma_\parallel$ and $\sigma_\perp$ are nearly identical, which is same as that observed in doped (CH)$_x$ samples. The W vs. T plot for metallic PPV-H$_2$SO$_4$ samples is shown in Figure 2.22. The typical positive temperature coefficient of W indicates that the system is on the metallic side of the M–I transition. Moreover, $\sigma(T)$ follows equation (2.12) [107]. Thus in both metallic (CH)$_x$ and PPV-H$_2$SO$_4$ the localization–interaction model for 3D anisotropic system is observed to be valid.

Hagiwara *et al.* [111a] have reported that in oriented ($l/l_o \approx 2$) PPy-PF$_6$ samples, $\sigma_\parallel \approx 2500$ S cm^{-1} and in unstretched samples $\sigma(300$ K$) \approx 300$ S cm^{-1}. Sato *et al.* [111b] have observed rather weak $\sigma(T)$ in these PPy-PF$_6$ samples. The significant aspect of PPy-PF$_6$ is that it is the best metallic air stable conducting polymer,

because doped (CH)$_x$ and PPV are not stable in air for long time. The detailed $\sigma(T)$ measurements by Yoon *et al.* [213] have shown that high-quality PPy-PF$_6$ samples are on the metallic side of the M–I transition. The values of ρ_r at ambient pressure and 18 kbar for the best samples of PPy-PF$_6$ are 1.75 and 1.33, respectively. The $\sigma(T)$ of various samples of PPy-PF$_6$ are shown in Figure 2.23. The corresponding W vs. T, Figure 2.24, shows the typical positive temperature coefficient for samples on the metallic side of the M–I transition. The $\sigma(T)$ (for both $\sigma_\parallel$ and $\sigma_\perp$) for an oriented PPy-PF$_6$ sample is shown in Figure 2.25, the unstretched sample is shown for comparison. Although, for stretched PPy-PF$_6$ samples ($l/l_o \approx 2$) the value of $\sigma_\parallel(300$ K$)$ has enhanced by a factor of four, the behavior of $\sigma(T)$ is nearly identical to that of the unstretched samples. This indicates that the charge

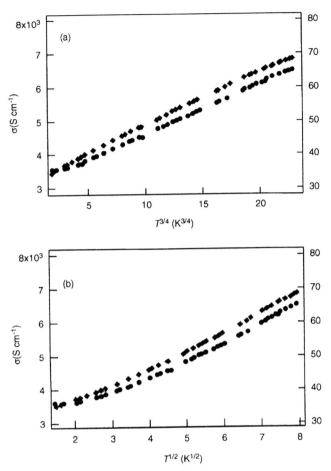

Figure 2.15. Conductivity $\sigma_\parallel$ ● and $\sigma_\perp$ ◆ vs. T: (a) $T^{3/4}$ and (b) $T^{1/2}$ fits for I–(CH)$_x$ ($\sigma_\parallel/\sigma_\perp \sim 100$ and $\rho_r \sim 2.5$–3). (Reproduced by permission of the American Physical Society from ref. 106).

delocalization does not improve significantly at low stretching ratios. Metallic PPy-PF$_6$ samples have large finite conductivity (≈ 100 S cm^{-1}) at mK temperatures, as shown in Figure 2.26. The behavior of low-temperature conductivity in metallic PPy-PF$_6$ follows the localization–interaction model, as shown below.

The surfactant counterion-induced processing of PANI has considerably improved the homogeneity in doping, morphology and carrier delocalization [110,150]. Although $\sigma(300\text{ K}) \approx 300$ S cm^{-1} for PANI-CSA, values of ρ_r as low 1.7 have been observed for samples in the metallic regime. The $\sigma(T)$ for PANI-CSA in the metallic regime is shown in Figure 2.27. The $\sigma(T)$ is in the metallic regime even at 8 T field. This indicates that the metallic feature is quite robust. The W vs. T plot, shown in Figure 2.28, confirms that

high-quality PANI-CSA is in the metallic regime. The field dependence of $\sigma \propto T^{1/2}$ (due to e–e interactions) at 0, 4 and 8 T is shown in Figure 2.29. The $\sigma(T)$ at mK temperatures, Figures 2.30, shows the presence of a finite value of conductivity in metallic PANI-CSA samples. The low-temperature data of $\sigma(T)$ for metallic PANI-CSA samples follow the localization–interaction model. The intrinsic conductivity of PANI-CSA along the chain axis is estimated to be greater than 10^4 S cm^{-1} at room temperature [150].

Recently Aleshin *et al.* [202] have observed that ion-implanted thin films of PANI are metallic. For ion-implanted PANI, $\sigma(300\text{ K}) \approx 100$–800 S cm^{-1} depending on the dosage of irradiation. Moreover, for metallic ion-implanted PANI samples the value of ρ_r is nearly 1.1. Aleshin *et al.* have observed that $\sigma(T)$ for metallic samples follow equation (2.12). This suggests that the

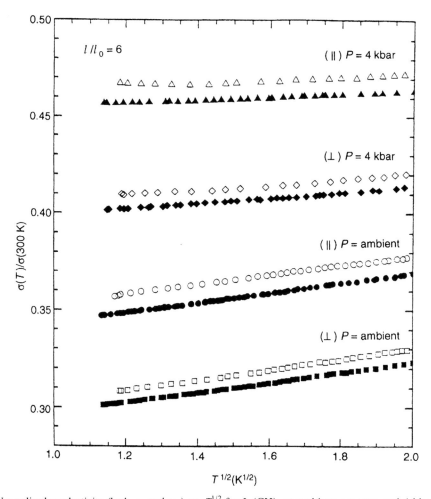

Figure 2.16. Normalized conductivity (both $\sigma_{\parallel}$ and $\sigma_{\perp}$) vs. $T^{1/2}$ for I–(CH)$_x$ at ambient pressure and 4 kbar.

localization–interaction model is applicable for metallic conducting polymers.

The *curious conductivity cusp* (crossover from negative TCR at higher temperatures to positive TCR at lower temperatures) at low temperatures has been reported by various groups in heavily doped (metallic) inorganic semiconductors for examples, Sasaki *et al* [214], Ootuka *et al.* [215], Rosenbaum *et al.* [216] etc. Only recently have such conductivity cusps been observed in metallic conducting polymers [PPy-PF$_6$, K-(CH)$_x$, I-(CH)$_x$, PANI-CSA, PPV-H$_2$SO$_4$, ion-implanted PANI]. This is a very important feature in the charge transport properties of metallic conducting polymers.

Sato *et al.* [111b] have reported the conductivity cusp in metallic PPy-PF$_6$ samples at low temperatures. This

cusp is sensitive to both pressure and magnetic field. It moves to higher temperatures as a function of pressure and it is either suppressed or shifted to lower temperatures as a function of the magnetic field [213,217]. This is shown clearly in Figures 2.31. Foxonet *et al.* [203] have observed the cusp at around 7 K in K-(CH)$_x$, as shown in Figure 2.20. The conductivity cusp and its field dependence in I-(CH)$_x$ are shown in Figure 2.32. It is very interesting to notice that the behavior of the cusp is strongly dependent on the direction of the field with respect to the chain axis. The cusp is nearly suppressed when the field is parallel to the chain axis and it remains the same (with a small enhancement in conductivity) when the field is perpendicular to the chain axis. Such anisotropic behavior of the conductivity cusp is observed for the

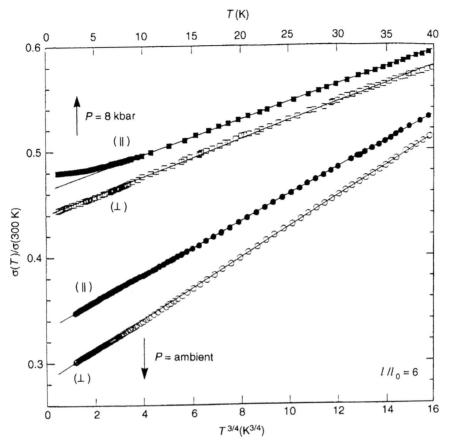

Figure 2.17. Normalized conductivity (both $\sigma_{\|}$ and $\sigma_{\perp}$) vs. $T^{3/4}$. Pressure–induced crossover (top curve) from $T^{3/4}$ to $T^{1/2}$ dependence for I–$(CH)_x$ paralled to chain axis at low temperatures. (Reproduced by permission of the American Physical Society from ref. 106).

first time in oriented I-$(CH)_x$ samples. In metallic PANI-CSA samples the cusp is observed around 150 mK, as shown in Figure 2.30. In PPV-H_2SO_4 samples the cusp is observed below 20 K, as shown in Figure 2.33. Aleshin *et al.* [202] have observed the cusp in ion-implanted PANI samples below 20 K, as shown in Figure 2.34.

Although the physical origin of this conductivity cusp in several metallic conducting polymers is not very clear, it could be due to the interplay of disorder, screening and interactions. In the case of PPy-PF_6 samples, this cusp region could be fitted to equation (2.12) as shown in Figure 2.35. The coefficient m in equation (2.14) changes sign as a function of disorder [213], and this could be due to the sign change in [(4/3)–(3γF_σ/2)]. Usually, the sign of m is negative when $\gamma F_\sigma > 8/9$. Although Bryksin [218] has proposed a

similar conductivity cusp due to the tunnel transport of polarons in insulating samples with activated transport, it seems that this model is not appropriate for samples in the metallic regime. However, if polarons are playing an active role in the low-temperature transport properties of conducting polymers, then one cannot completely rule out this possibility. This model has been used to interpret the conductivity cusp in some insulating PPy samples [219]. Nevertheless, this conductivity cusp is a very important feature in the search to unravel the mechanism for charge transport in these systems.

Although the recent improvement [148] in the chemical synthesis of structurally ordered regio-regular polyalkyl-thiophenes (PATs) has allowed conductivites of the order of 10^3 S cm^{-1} to be obtained, the behavior of $\sigma(T)$ is typical to that of an insulator [2]. This is probably due to the inhomogenous iodine doping of

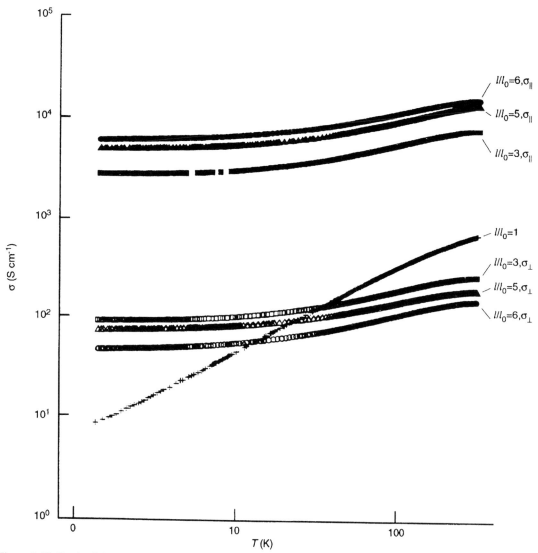

Figure 2.18. Conductivity (both $\sigma_\parallel$ and $\sigma_\perp$) vs. T for I–$(CH)_x$ at various stretching ratios.

PAT samples. However, the values of $\sigma(300\ K)$ are quite encouraging that it might be possible to observe the metallic features in PAT samples.

In summary, the presence of a finite value of conductivity at mK temperatures, the positive temperature coefficient of $W(T)$ in several metallic conducting polymers, etc. gives strong evidence that high-quality conducting polymers are indeed on the metallic side of the M–I transition. Thus, in spite of the molecular-scale inhomogenities due to structural, morphological and doping-induced disorders, the presence of robust metallic features at very low temperatures and high magnetic fields indicates that the possibility of any 1D localization due to the inherent q-1D nature of the polymer chains can be ruled out. Moreover, this gives clear evidence for the formation of 3D extended states due to interchain interactions and counterion-induced intra/interchain transport. This is further substantiated from the MC, thermpower and infrared reflectivity measurements as mentioned below.

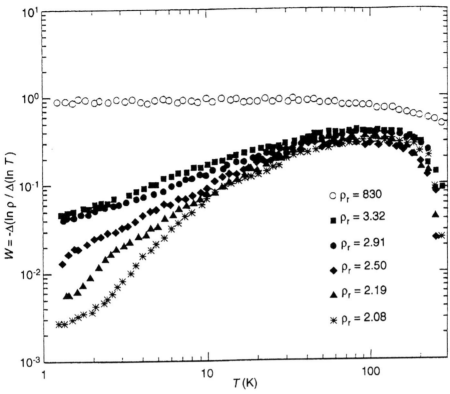

Figure 2.19. W vs. T for I–(CH)$_x$ (parallel to chain axis) as a function of ρ_r Unstretched I–(CH)$_x$, $\bigcirc$

8.2 Magnetoconductance

From the previous section it is evident that the main features of the disordered metallic state of conducting polymers are in agreement with the localization–interaction model. However, this does not unambiguously rule out the possibility that other models (as mentioned in Section 7) are less important in understanding the charge transport in metallic conducting polymers. Thus the information from $\sigma(T)$ by itself is not enough to fully understand the charge transport mechanism in metallic conducting polymers. Although one could use the frequency and electric field dependences of conductivity in order to check the validity of various models, the interpretation of experimental data from these measurements is not always straightforward. Since it is well known that MC is a sensitive local probe for investigating the scattering process in disordered metallic systems and it is closely related to $\sigma(T)$, it must be of considerable interest to know the behavior of MC in metallic conducting

polymers. Moreover, the quantitative level of understanding of MC in the localization–interaction model is rather helpful to check the appropriateness of this model for metallic conducting polymers. Finally, the detailed study of conventional disordered metallic systems in the past years has shown that the localization–interaction model is appropriate for such systems [44]; the internal consistency of using this model could be verified from $\sigma(T)$ and MC measurements.

It is well know that MC probes the local dynamics of charge transport in a conducting system [220,221]. The insight into the microscopic transport property parameters (the elastic and inelastic scattering length, scattering time etc.) is usually obtained from MC measurements. In crystalline 3D metals the dominant contribution to the weak negative MC ($-\Delta\sigma/\sigma \leqslant 5\%$) is due to the classical orbital motion. The classical transverse MC is mainly due to the bending of the charge carrier trajectory by Lorentz force. The classical negative MC is proportional to the square of the field ($-\Delta\sigma/\sigma = kH^2$, where the constant k is a function of the

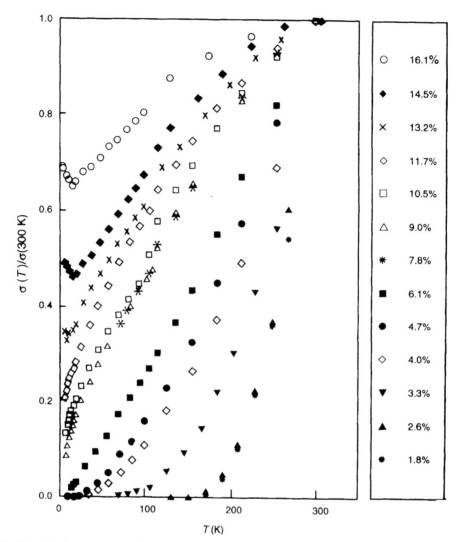

Figure 2.20. Normalized conductivity vs. T for K–(CH)$_x$ at various doping levels. (Reproduced by permission of Elsevier Publications from ref. 203).

charge transport scattering time). Moreover, MC gives information about the second derivative of the density of the states at the Fermi energy with respect to energy.

It is well known that in an ideal lD conductor the transverse orbital motion is restricted; thus the carriers cannot make circular motion in the presence of magnetic field [220,221]. Hence, there is hardly any MC in an ideal lD conductor. However, this scenario is not appropriate when the interchain transfer integral is turned on in a lD system, and in many q-lD systems some interchain coupling exists. In this section, the

significance of MC as a powerful tool to investigate the charge transport mechanism in metallic conducting polymers is shown vividly.

In disordered metallic systems it is well known that the quantum corrections due to weak localization and e–e interactions contribute to the MC at low temperatures [44,222]. The e–e interaction contribution towards the low-temperature conductivity in a zero magnetic field is given by equation (2.14). However, in the presence of fields sufficiently high that $g\mu_B H \gg k_B T$, the coefficient of the $T^{1/2}$ term (m) is altered to m' and

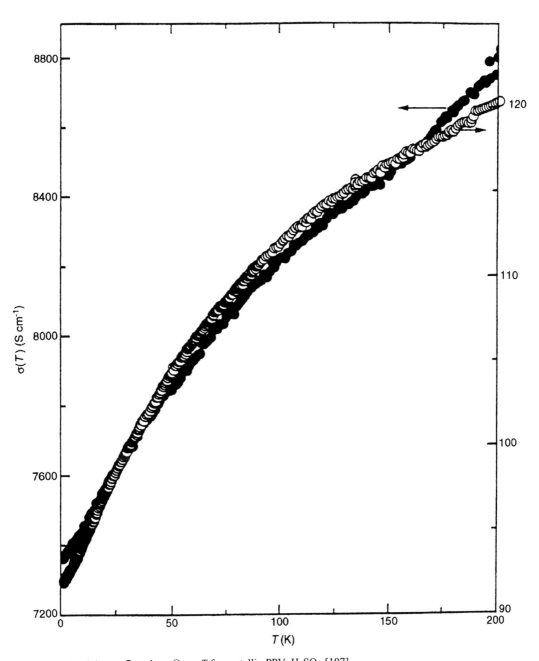

Figure 2.21. Conductivity $\sigma_\parallel$ ● and $\sigma_\perp$ ○ vs. T for metallic PPV–H$_2$SO$_4$ [107].

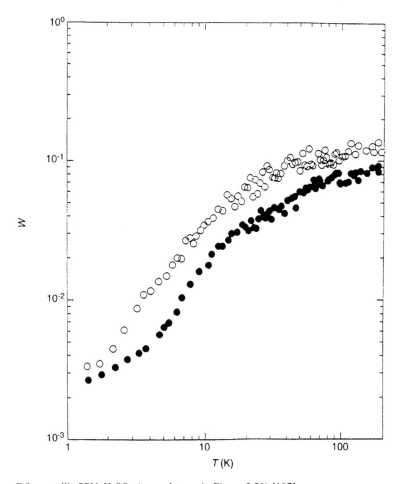

Figure 2.22. W vs. T for metallic PPV–H_2SO_4 (same data as in Figure 2.21) [107].

the interaction contribution of conductivity in the presence of a magnetic field is given by

$$\sigma_1(H, T) = \sigma(H, 0) + m'T^{1/2} \qquad (2.19)$$

$$\text{where } m' = \alpha[(\tfrac{4}{3}) - \gamma(F_\sigma/2)] \qquad (2.20)$$

Using equations (2.15) and (2.20)

$$\gamma F_\sigma = (\tfrac{3}{8})(m' - m)/(3m' - m) \qquad (2.21)$$

assuming that α, γ and F_σ are not dependent on the magnetic field [199,200]. Thus, the parameters, α and $\gamma F\sigma$, can be estimated from the values of m and $m' = m(H)$ obtained at $H = 0$ and at $H = 8$ T by using (2.15) and (2.20), respectively.

The MC at high fields arises mainly from the interaction contribution (weak localization contribution is less important in strong fields) [199]. In the free electron model, using the Thomas–Fermi approximation.

$$F_\sigma = x^{-1} \ln(1 + x) \qquad (2.22)$$

where $x = (2k_F\Lambda_s)^2$, and Λ_s is the Thomas–Fermi screening length. Equations (2.21) and (2.22) yield $0 < F_\sigma < 0.93$ and $0 < F_\sigma < 1$ (note that $F_\sigma \approx 1$ for short-range interactions and $F_\sigma \ll 1$ for long-range interactions). Decreasing γF_σ leads to a change in sign of m, corresponding to the divergence of screening length near the M–I transition, consistent with McMillan's prediction [157]. Kaveh and Mott [223]

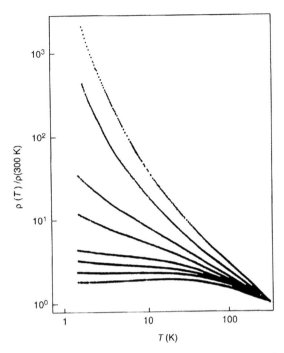

Figure 2.23. Normalized resistivity vs. T for several PPy–PF$_6$ samples. (Reproduced by permission of the American Physical Society from ref. 213).

suggested that the inelastic e–e scattering should dominate near the M–I transition.

The field dependence of the contribution to MC from the e–e interactions can be written as $\Delta\Sigma(H,T) = \sigma(H,T) - \sigma(0,T)$ [199]

$$\Delta\sum_{I}(H,T) = -0.041\alpha(g\mu_B/k_B)^2\gamma F_\sigma T^{-3/2}H^2$$

$$(g\mu_B H \ll k_B T) \quad (2.23)$$

$$\Delta\sum_{I}(H,T) = \alpha\gamma F_\sigma T^{1/2} - 0.77\alpha(g\mu_B/k_B)^{1/2}\gamma F_\sigma H^{1/2}$$

$$(g\mu_B H \gg k_B T) \quad (2.24)$$

Thus at low and high fields, $\Delta\Sigma_I(H,T)$ is proportional to H^2 and $H^{1/2}$, respectively.

For the low magnetic field regime, it is assumed that the contributions to $\Delta\Sigma(H,T)$, which arise from e–e interactions and weak localization, are additive [199,200]. Thus, the total low-field MC is given by the following:

(a) For weak spin–orbit coupling (positive contribution to MC),

$$\Delta\sum H, T = -0.141\alpha(g\mu_B/k_B)^2\gamma F_\sigma T^{-3/2}H^2$$

$$+ (1/12\pi^2)(e/c\hbar)^2 G_0(l_{in})^3 H^2 \quad (2.25)$$

(b) For strong spin–orbit coupling (negative contribution to MC),

$$\Delta\sum(H,T) = -0.041\alpha(g\mu_B/k_B)^2\gamma F_\sigma T^{-3/2}H^2$$

$$- (1/48\pi^2)(e/c\hbar)^2 G_0(l_{in})^3 H^2 \quad (2.26)$$

where $G_0 = (e^2/\hbar)$ and l_{in} is the inelastic scattering length. The first term on the right hand side of (2.25) and (2.26) is the contribution from e–e interactions (negative MC), and the second term on the right hand side is the contribution from weak localization (positive or negative MC). The first term can be estimated by using equations (2.23) and (2.24). Then, using the slope of $\Delta\sigma(H,T)$ vs. H^2 in the low-field region, the second term can be estimated. In this way, the value of the inelastic scattering length can be calculated at each temperature. Although the theoretical estimate for quantum corrections to MC in conventional metallic systems is below 2% (i.e. $\Delta\sigma/\sigma < 2\%$) [222], the observed MC in oriented metallic conducting polymers is slightly higher ($\Delta\sigma/\sigma \approx 2$–5%), which is probably associated with the anisotropic diffusion coefficient, anisotropic mass, etc.

The initial MC measurements, down to mK temperatures, in Shirakawa (CH)$_x$ by Gould et al. [142] has shown the presence of weak localization contribution to MC. Later, Thummes et al. [185] used the localization-interaction model for the analysis of MC data in both I-(CH)$_x$ and K-(CH)$_x$. Roth has reviewed the earlier MC work [3]. The detailed study of MC [106,205] in new types of (CH)$_x$ [Naarmann, Tsukamoto, Akagi, etc.] has confirmed the presence of quantum corrections to MC in metallic conducting polymers, inspite of the assumptions that the molecular-scale inhomogenities in the system could wash off the quantum effects. The recent detailed study of MC in metallic PPV samples by Ahlskog et al. [107] has reinforced this view. Thus in both metallic (CH)$_x$ and PPV, this conclusion holds well. However, in other metallic conducting polymers, like PPy-PF$_6$ and PANI-CSA, more work is essential to develop a complete understanding of the behavior of MC.

Ishiguro and coworkers have reported a detailed study of MC in I-(CH)$_x$ and FeCl$_3$-(CH)$_x$ samples [210,205]; the behavior of MC is rather different in

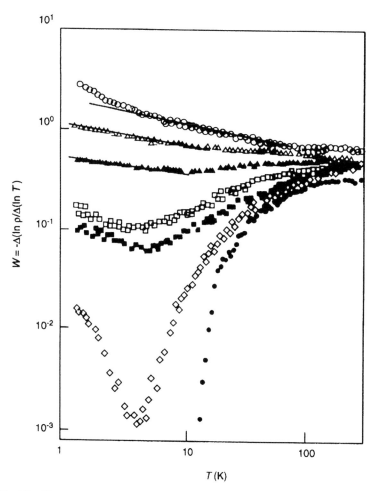

Figure 2.24. W vs. T for PPy–PF$_6$ samples (same data as in Figure 2.23). (Reproduced by permission of the American Physical Society from ref. 213).

both systems, as shown in Figure 2.36. The MC measurements in oriented I-(CH)$_x$ [$l/l_o \approx 5$–10, $\sigma_{\parallel}/\sigma_{\perp}$ (not known), $\rho_r \approx 3$–20] display a wide range of behavior, including both positive and negative MC. When ρ_r decreases, the sign of MC shifts from negative to positive. For samples with intermediate disorder ($\rho_r \approx 3$–6), the sign of MC was positive (negative) when the field was perpendicular (parallel) to the chain axis at temperatures above 2 K, as shown in Figure 2.37. In either cases (H parallel or perpendicular to chain axis), the magnitude of positive MC decreases gradually as the temperature is lowered from 10 to 1 K. This indicates that the weak localization contribution (positive MC) dominates at higher temperatures, and the contribution from e–e

interactions (negative MC) becomes increasingly important only at lower temperatures. However, in many samples with $\rho_r \leqslant 3$, Nogami et al. [205] have observed an additional positive contribution to MC at high fields ($H \geqslant 6$ T) and at very low temperatures ($T \geqslant 1$ K), which is probably not due to the usual weak localization. Although this additional contribution to positive MC at high fields is not substantial, it is important to know its origin. It is likely that some field-induced delocalization mechanism is taking place, especially when the magnetic length [$L_H = (c\hbar/eH)^{1/2}$; $L_H(10$ T$) \approx 80$ Å] and the interaction length [$L_T = (\hbar D/k_B T)^{1/2}$] becomes comparable. However, more theoretical and experimental work is essential for a complete understanding of MC in the entire range of temperature and field.

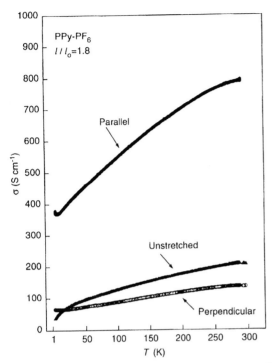

Figure 2.25. Conductivity vs. T for stretched PPy–PF$_6$. The data for an unstretched sample are shown for comparsion.

In the case of highly metallic I-(CH)$_x$ samples ($\rho_r \leqslant 3$), Nogami *et al.* [205] have observed that the positive contribution to MC dominates even at lower temperatures ($T \leqslant 1$ K), as shown in Figure 2.36a. However, in this case it seems that the dominance of the interaction contribution may appear at temperatures below 1 K. This suggests that the interplay of localization and interaction contributions to MC is sensitive to the extent of disorder in anisotropic systems. Moreover, it is surprising that the positive MC increases linearly up to a 14 T field, which is not fully consistent with the localization–interaction model. The additional positive contribution to MC above 8 T (Figure 2.36a) is playing a role in this high-field linear behavior in MC. Probably this is due to the microsocpic anisotropy of the system and the associated features in the anisotropic MC.

Nogami *et al.* [205] have observed that the positive contribution to transverse MC is nearly three times larger with respect to longitudinal MC in I-(CH)$_x$ samples ($\rho_r \approx 3$). Javadi *et al.* [206] have reported that the positive contribution to transverse MC is nearly 10 times larger with respect to longitudinal MC in I-(CH)$_x$ samples ($l/l_o \approx 5$, $\sigma_{\parallel}/\sigma_{\perp} \approx 30$, $\rho_r > 5$). This indicates

that the diffusion coefficent in parallel and perpendicular directions to the chain axis is strongly dependent on the angle between chain axis and field. In the case of a I-(CH)$_x$ sample with $\rho_r \approx 3$, the positive MC, $\Delta\sigma/\sigma$ at 15 T $\approx 6\%$ at 1.6 K, and this exceeds the theoretical estimate for weak localization contribution in isotropic systems. Nevertheless, such large values are possible if one takes into account the roles of the anisotropic mass ratio and anistropic diffusion coefficient in MC. Moreover, the anisotropy in MC is especially sensitive to the actual number of misaligned chains (anisotropy on the molecular scale) present in the system.

A complete study of the anisotropy in conductivity and MC (in both parallel and perpendicular directions to chain axis) of I-(CH)$_x$ samples ($l/l_o \approx 10$–14, $\sigma_{\parallel}/\sigma_{\perp} \approx 100$, $\rho_r \approx 3$) has been reported Reghu *et al.* [106]. The transverse and longitudinal MC in highly anisotropic I-(CH)$_x$ samples are shown in Figure 2.38. The MC has H^2 dependence at low fields. The temperature dependence of the inelastic scattering length is determined with the help of equations (2.23)–(2.26). The temperature dependence of the inelastic scattering length [estimated from $L_{in} = (D\tau_{in})^{1/2}$], for both parallel and perpendicular directions to the chain axis, is shown in Figure 2.39. At 1.2 K, the inelastic scattering lengths, parallel and perpendicular to the chain axis are 1163 Å and 210 Å, respectively. The $T^{-3/4}$ dependence of L_{in} is typical of inelastic e–e scattering in disordered metals [28]. This is in agreement with the $T^{3/4}$ dependence of conductivity (Figure 2.15a). This indicates that the behavior of conductivity and MC in I-(CH)$_x$ samples [$l/l_o \approx 10$, $\sigma_{\parallel}/\sigma_{\perp} \approx 100$, $\rho_r \approx 3$] is consistent with the localization–interaction model for anisotropic disordered metals.

It is significant to observe the sharp features in MC (maxima due to competing contributions from weak localization and e–e interactions), as shown in Figures 2.38. This (indirectly) indicates that the molecular-scale anisotropy of the samples used in ref. [106] is relatively higher with respect to those in refs. [205 and 206]. This is consistent with the stretching ratios of the samples used in these measurements ($l/l_o \approx 10$–14 in [106] and $l/l_o \approx 5$–10 in [205,206]); although this comparison between the anisotropic MC and the l/l_o ratio is not always true, because the molecular-scale anisotropy depends on various parameters. Firstly, in transverse MC the abrupt reduction of positive MC, at 1.3 K and 2 T, clearly indicates the onset of the dominance of e–e interactions. Secondly, in longitudinal MC there is hardly any positive MC. The substantial anisotropy in MC, involving a sign change from positive (transverse MC, H perpendicular to the chain axis) to negative

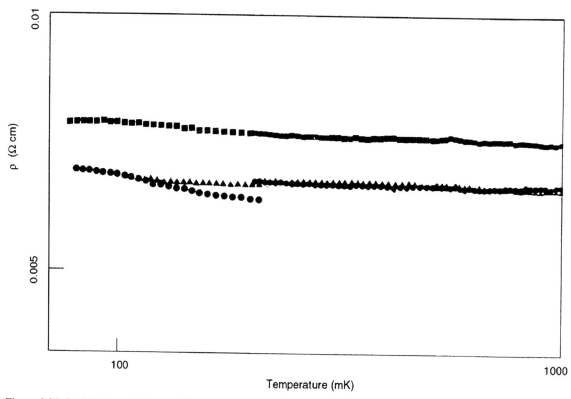

Figure 2.26. Resistivity vs. T for metallic PPy–PF$_6$ at $H = 0$ (●), 2 (▲) and 8 T (■) at mK temperatures. (Reproduced by permission of Elsevier Science SA from *Synth. Met.* **69**, 1995, 215) [224].

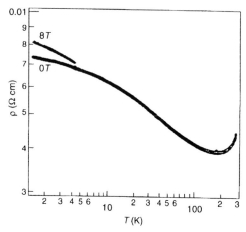

Figure 2.27. Resistivity vs. T for metallic PANI–CSA at 0 and 8 T. (Reproduced by permission of the American Physical Society from ref. 150).

(longitudinal MC, H parallel to the chain axis), is rather unusual. Nogami *et al.* have observed a rather similar feature, as shown in Figure 2.37, in one of their samples (samples C in ref. 205). It could be either due to the dominance of the interaction contribution over the localization contribution or due to the substantial reduction of the localization contribution in longitudinal MC, or both simultaneously. Although the physical understanding of this behavior is not known clearly, it seems that the anisotropy in MC is closely associated with the anisotropic mass ratio and anisotropic diffusion coefficient. In order to verify that this anisotropic feature in MC is not unique to metallic I-(CH)$_x$ samples, Ahlskog *et al.* [107] have carried out detailed MC measurements in metallic PPV samples.

In PPV-H$_2$SO$_4$ samples ($l/l_o \approx 10$, $\sigma_\parallel / \sigma_\perp \approx 100$, $\rho_r \approx 1.1$) Ahlskog *et al.* [107] have observed a rather similar anisotropic MC, as in I-(CH)$_x$ samples. The transverse and longitudinal MC in PPV-H$_2$SO$_4$ samples are shown in Figure 2.40. Thus the sign change from positive (transverse MC) to negative (long-

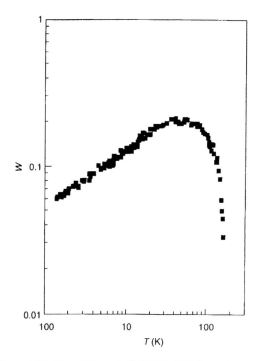

Figure 2.28. W vs. T for metallic PANI–CSA (same data as in Figure 2.27).

itudinal MC) is a general feature in highly anisotropic metallic conducting polymers. Moreover, the temperature dependence of the anisotropic features (in MC) in I-$(CH)_x$ and PPV-H_2SO_4 samples is similar to that observed in $(SN)_x$ crystals [4]. It is well known from MC measurements in $(SNBr_y)_x$, as a function of doping level and pressure, that the anisotropic features gradually decrease upon increasing the doping level or pressure [4]. The pressure dependence of MC in I-$(CH)_x$ also showed similar features [106]. In general, the positive contribution to MC increases at higher doping level (or at pressures) due to the enhanced interchain transport. This indicates that the negative (positive) contribution to longitudinal MC decreases (increases) when the interchain interactions in highly anisotropic metallic system increases. Hence, it seems that the anisotropic MC is closely related to the microscopic anisotropy in transport properties and that this is a general feature in all of the anisotropic systems. Thus the anisotropic MC can be used as a sensitive probe for investigating the intrinsic microscopic anisotropy of the electronic properties in anisotropic 3D or q-1D metallic systems.

Kaneko and Ishiguro [210] have observed a rather different behavior of MC in metallic $FeCl_3$-$(CH)_x$, with respect to I-$(CH)_x$ samples. The negative MC at higher temperatures gradually shifts to positive at lower temperatures, shown in Figure 2.36b. Although the exact mechanism for this special behavior of MC in $FeCl_3$-$(CH)_x$ is not understood, it indicates that some temperature-dependent competing contribution to MC is playing an important role. The main difference between I-$(CH)_x$ and $FeCl_3$-$(CH)_x$ is that in the latter the dopant ion is magnetic. Thus one could assume that this magnetic dopant is playing a subtle role in the MC in $FeCl_3$-$(CH)_x$. This is very interesting problem for future investigations.

The MC in metallic PANI-CSA and PPy-PF_6 samples is shown in Figures 2.41 and 2.42, respectively [150,213]. The low field-MC has a H^2 dependence and it follows equation (2.23). Since these systems are relatively less metallic with respect to I-$(CH)_x$ and PPV-H_2SO_4 samples, there is hardly any positive MC at temperatures above 1 K. The extent of disorder plays a major role in the negative MC $[-\Delta\sigma/\sigma(1.4\ \text{K}/8\ \text{T}) \approx 5$–$10\%]$ in metallic PANI-CSA and PPy-PF_6 samples. The negative MC decreases when ρ_r decreases. This is clearly shown for PPy-PF_6 samples in Figure 2.43. Although the origin for this relatively large negative MC is not known clearly, it seems that some contribution from hopping transport in disordered regions is playing a role. Hence, MC measurement in disordered metallic conducting polymers can be used as a sensitive probe for investigating the extent of disorder and its influence in the transport properties.

The MC measurements in metallic PPy-PF_6 samples at mK temperatures is shown in Figure 2.44 [224]. The positive MC is nearly 30% at 100 mK (2 T field). This large positive MC at low fields and its relatively weak field dependence at high fields is not consistent with the localization–interaction model, because the magnitude of positive MC is nearly 20 times larger than that usually observed due to quantum corrections in disordered systems. On the other hand, this large positive MC in metallic PPy-PF_6 samples (at mK temperatures) is reminiscent of the large positive MC observed in granular aluminum near the boundary of the superconducting–non-superconducting transition [195]. In granular aluminum film of 1 μm thickness [3D films, $\sigma(300\ \text{K}) \approx 45$ S cm^{-1} and $\rho(4.2\ \text{K})/\rho(300\ \text{K} \approx 4.6]$ the positive MC at 1 K and 9 T is nearly 20%; for another granular aluminum film $[\sigma(300\ \text{K}) \approx 16$ S cm^{-1} and $\rho(4.2\ \text{K})/\rho(300\ \text{K} \approx 20]$ the positive MC is nearly 100%, at 0.5 K and 9 T [195]. Thus this anomalously large positive MC in metallic PPy-PF_6 samples at very low temperatures is really an interesting problem for further detailed study.

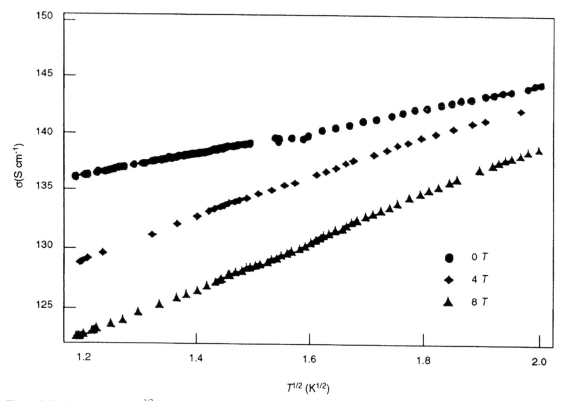

Figure 2.29. Conductivity vs. $T^{1/2}$ fit for metallic PANI–CSA at low temperatures 0, 4 and 8 T. (Reproduced by permission of the American Physical Society from ref. 150).

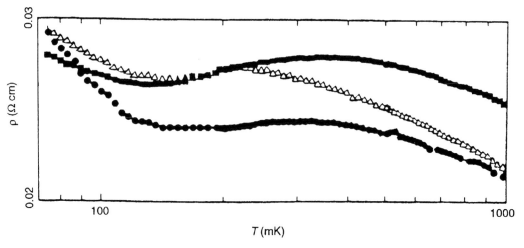

Figure 2.30. Resistivity vs. T for metallic PANI–CSA at 0 (●), 2 (△) and 8 T (■) at mK temperatures. (Reproduced by permission of Elsevier Science SA from *Synth. Met.* **68**, 1995, 215).

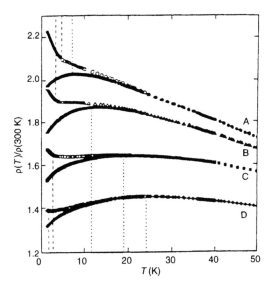

Figure 2.31. Normalized resistivity vs. T for metallic PPy–PF$_6$ at $H = 8$ T (open symbols) and $H = 0$ T (closed symbols). Samples A and B at ambient pressure and C and D at 10 kbar. The dotted lines show T_t for the conductivity cusp. (Reproduced by permission of Elsevier Science SA from *Synth. Met.* **64**, 1994, 53).

In summary, the sizable MC in metallic conducting polymers can be utilized as an important tool to investigate the microscopic charge transport mechanism in these systems. In oriented metallic conducting polymers, the sensitivity of MC to the angle between the chain axis and the field can give some information regarding the molecular-scale anisotropy in these systems. The MC in oriented metallic conducting polymers is in qualitative agreement with the localization–interaction model for disordered systems. In general, the weak localization (positive MC) contribution dominates at higher temperatures and lower fields and the e–e interaction contribution (negative MC) dominates at lower temperatures and higher fields. However, at very high fields ($H > 8$ T) and very low temperatures ($T < 1$ T) some sort of field-induced delocalization mechanism is occurring, which is not yet understood. In less metallic conducting polymers (PANI-CSA and PPy-PF$_6$) the magnitude of negative MC is in accordance with ρ_r, and this information is useful in characterizing the extent of microscopic disorder present in the system.

8.3 Thermopower

The thermopower (S) of doped conducting polymers has been studied for many years [143]. In doped conducting polymers, thermopower is not as sensitive to disorder as electrical conducitivity, because the latter is strongly dependent on various scattering and hopping processes involved in the charge transport in the disorder system. Moreover, the mean free path involved in electrical transport is rather sensitive to the slightest amount of disorder present in the system. Although the typical metallic positive TCR has not been observed in highly conducting polymers (except in PANI-CSA with a positive TCR in the range 350–160 K), the typical metallic temperature dependence of thermopower in a wide range of temperature (300–10 K) is usually observed in all conducting polymers. The quasi-linear temperature dependence of thermopower is observed to persist well into the insulating regime [2].

Park *et al.* [201] and Nogami *et al.* [205] have carried out detailed thermopower measurements in metallic (CH)$_x$, as shown in Figure 2.45. Remarkably linear $S(T)$ has not been observed even in crystalline metals, since the thermopower in crystalline metals peaks due to phonon drag contribution at low temperatures. This thermopower peak due to the phonon drag effect is usually suppressed in disordered system. Moreover, the diffusion thermopower shows deviation from linear behavior at low temperatures due to the enhancement in $S(T)$, which is usually attributed to the electron–phonon interaction. However, the predicted non-linearities in the diffusion thermopower due to electron–phonon enhancement is rather similar to that due to the hopping contribution, except at very low temperatures. Kaiser [143] suggested that the increasing slope of $S(T)$ below 50 K indicates the presence of electron–phonon enhancement of the diffusion thermopower at low temperatures. Moreover, the change in the slope of the thermopower indicates a variation in the number of carriers (usually an increase in the number carriers lowers the slope of the thermopower). The change in the slope of the thermopower can also occur due to the variation in scattering mechanisms, the energy dependence of scattering time, etc. Thus the interpretation of $S(T)$ involves some ambiguity, unless the band structure and the electronic structure of the system is well known.

The metallic diffusion thermopower is given by the general equation:

$$S_d(T) = \frac{1}{e} \int \left(-\frac{\partial f}{\partial E} \right) \frac{\sigma(E)(E - E_F)}{\sigma} \frac{1}{T} \equiv \frac{k_B}{e} \frac{\langle (E - E_F) \rangle}{k_B T}$$

$$(2.27)$$

where e is the electronic charge, $E - E_F$ is the energy of the carrier with respect to the Fermi energy, f is the

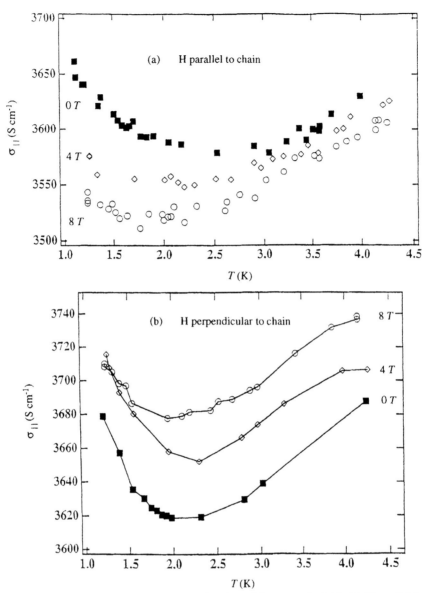

Figure 2.32. Conductivity cusp in metallic I–(CH)$_x$ as a function of the magnetic field: (a) H parallel to the chain axis; H perpendicular to the chain axis for the same sample.

Fermi function. The standard Mott equation for the metallic diffusion thermopower is given by [2,143]

$$S_{\mathrm{d}}(T) = (\pi^2/3)(k_{\mathrm{B}}/e)(k_{\mathrm{B}}T)[\mathrm{d}\ln\sigma(E)/\mathrm{d}E]_{E_{\mathrm{F}}} \quad (2.28)$$

or alternatively,

$$S_{\mathrm{d}}(T) = +(\pi^2/3)(k_{\mathrm{B}}/|e|)(k_{\mathrm{B}}T)(z/E_{\mathrm{F}}) \quad (2.29)$$

where the energy dependence of $\sigma(E)$ arises from a combination of the details of band structure and scattering mechanism and z is a constant (of order unity), again determined from the band structure and the energy dependence of the mean scattering time. Note that $S(T)$ is linear in temperature. Although one can use equation (2.28) to estimate the density of the states at the Fermi level, the band effects and energy-

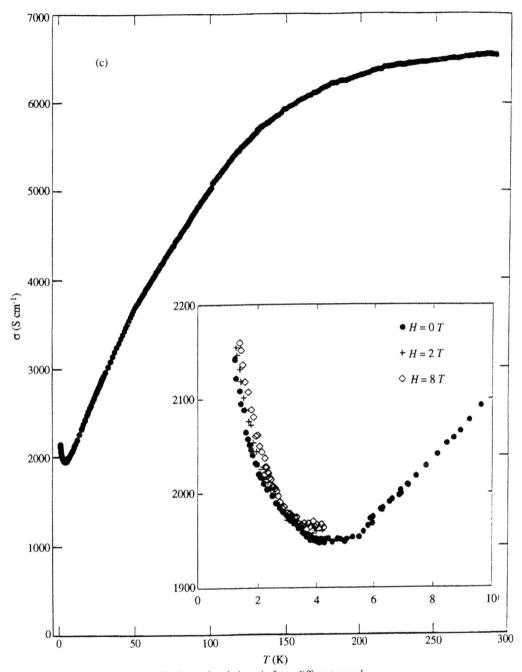

Figure 2.32. (*continued*) (c) *H* perpendicular to the chain axis for a different sample.

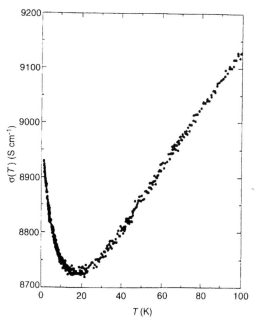

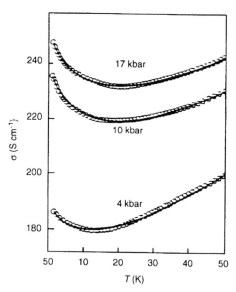

Figure 2.33. Conductivity cusp for metallic PPV–H_2SO_4 [107].

Figure 2.35. Conductivity cusp for metallic PPy–PF_6 fitted to $\sigma(T) = \sigma(0) + mT^{1/2} + BT^p/2$ [equation (2.12)] where $p = 2.50 \pm .04$ and $B = 0.4 \pm 0.01$. Note that $\sigma(0)$ and m depend on pressure.

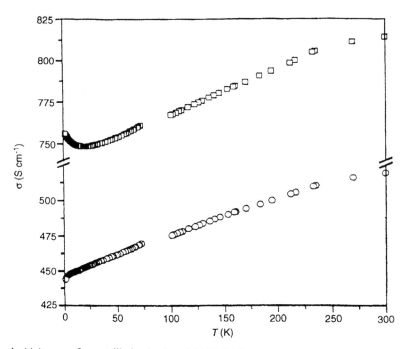

Figure 2.34. Conducitivity cusp for metallic ion-implanted PANI [202].

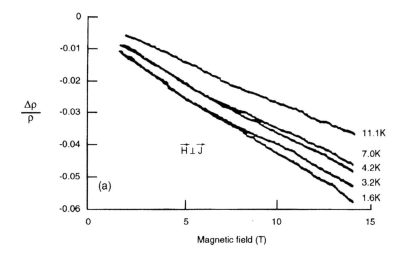

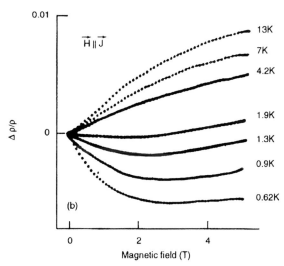

Figure 2.36. Magnetoresistance ($\Delta\rho/\rho$) vs. H for metallic $(CH)_x$: (a) I–$(CH)_x$, (b) $FeCl_3$–$(CH)_x$. (Reproduced by permission of Elsevier Science SA from *Synth. Met.* **65**, 1994, 141).

dependent scattering processes could modify the energy dependence of $\sigma(E)$ significantly and this could alter the estimated value of the density of the states at the Fermi level. The case with the sign and magnitude of thermopower is similar. As in equation (2.27), the thermopower gives the first derivative of the density of the states at the Fermi energy. Once again, the above mentioned factors indicate that the interpretation of thermopower is not straightforward.

Kaiser [143] has carried out detailed study of the behavior of thermopower in conducting polymers and compared it to conventional disordered metallic systems. In undoped conducting polymers $[\sigma(300\ K) < 10^{-6}\ S\ cm^{-1}]$ the thermopower at room temperature is nearly $10^3\ \mu V\ K^{-1}$. This large value of thermopower gradually decreases upon doping and in fully doped conducting polymers the thermopower is nearly $10\ \mu V\ K^{-1}$. Kaiser [143] suggested that the

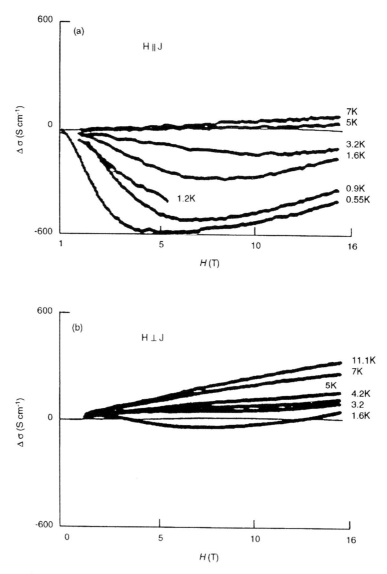

Figure 2.37. Magnetoconductance ($\Delta\sigma$) vs. H for metallic $(CH)_x$ with intermediate disorder ($\rho_r \approx 3$–6). The current is parallel to the chain axis. (a) H is parallel to both current and chain axis; (b) H is perpendicular to both current and chain axis. (Reproduced by permission of the American Physical Society from ref. 205).

thermopower in metallic $(CH)_x$ is larger than that usually observed in most metals and this indicates a relatively strong variation of $\sigma(E)$ in equation (2.27). Moreover, the remarkable linearity of $S(T)$ and the negligible non-linear contribution to thermopower in high quality metallic $(CH)_x$ indicates that the electron–phonon renormalization of diffusion thermopower, the phonon drag effects, etc. are not significant. This indirectly suggests that the electron–phonon interaction in metallic $(CH)_x$ is rather small and this could act favorably in achieving very high intrinsic conductivity in ordered metallic conducting polymers.

In less metallic conducting polymers [$\sigma(300\ K) < 100\ S\ cm^{-1}$] Kaiser has proposed a heterogeneous model [143]. In the heterogeneous model, the system consists of various types of intrafibril regions (for example inter-fibril barriers, undoped regions, chain interruptions, voids, etc.) that are connected in series

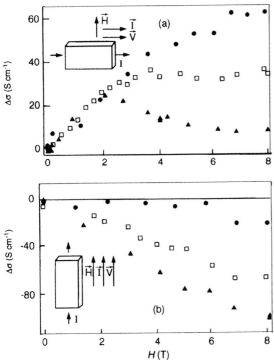

Figure 2.38. Magnetoconductance ($\Delta\sigma$) vs. H in I–(CH)$_x$ at 4.2 (●), 2 (□) and 1.2 K (▲). The current is parallel to the chain axis. (a) H is perpendicular to both current and chain axis; (b) H is parallel to both current and chain axis [106]. (Reproduced by permission of the American Physical Society from ref. 106).

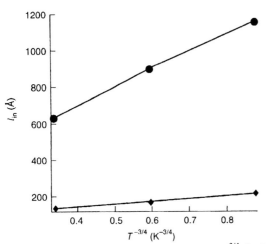

Figure 2.39. The inelastic scattering length vs. $T^{-3/4}$ for I–(CH)$_x$: (●) parallel and (◆) perpendicular to chain axis [106]. (Reproduced by permission of the American Physical Society from ref. 106).

with the conduction path. In such a model, the less conducting regions that limit the motion of charge carriers, determine the bulk transport properties. However, because thermopower is an intrinsic property of a system, the geometric factors involved in the electrical conductivity expression are absent. Thus the net thermopower is weighted in favor of the regions where the largest temperature gradient occurs. If the thermal current carried by phonons is less impeded by thin insulating barriers than the electrical current carried by electrons or holes, then the system indeed shows a metallic thermopower. In other words, one can assume that, most of the temperature gradient occurs across the highly conducting regions and most of the electrical potential drop occurs in the thin insulating barriers.

Kaiser [143] suggested that the thermopower from disordered regions is weighed more heavily at high temperatures than at low temperatures, and this leads to a reduction of the thermopower at low temperature towards the value characteristic of the crystalline regions. Moreover, one could expect a larger value of thermopower, if the charge carriers tend to carry more heat during the thermally assisted tunneling process. This also increases the slope of the thermopower as temperature increases. Because in metallic (CH)$_x$ the large enhancement of thermopower due to the tunneling contribution has not been observed, suggests that the barriers in the heterogeneous model contribute little to the thermopower.

Recently Yoon et al. [213,225,226] have carried out detailed study of the termopower in various conducting polymers. The thermopower data for various PANI-CSA samples are shown in Figure 2.46 [225]. For PANI-CSA samples, $S(300 \text{ K}) \approx 10 \pm 2 \ \mu\text{V K}^{-1}$. The positive sign of the thermopower is consistent with the calculated band structure of metallic emeraldine salt; a three-quarter-filled π-band with one hole per ($-$B$-$NH$-$B$-$NH$-$) repeat unit [225]. Although ρ_r for PANI-CSA samples varies by three orders of magnitude near the M–I transition, the quasi-linear thermopower for all the samples near the M–I transition is relatively insensitive to ρ_r. The linear temperature dependence of $S(T)$ corresponds to the diffusion thermopower [equation (2.29)]. The relatively large magnitude, $S(300 \text{ K}) = 8 \sim 12 \ \mu\text{V K}^{-1}$, indicates the diffusion of charge carriers in electronic states with relatively narrow bandwidth. Using $z = 1$, $E_F \approx 1$ eV [227] and $T = 300$ K, equation (2.28) yields $S(300 \text{ K}) \approx 7.5 \ \mu\text{V K}^{-1}$, close to the measured value. The density of the states estimated from the magnitude of $S(T)$ is 1.1–1.6 states per electron volt per two rings

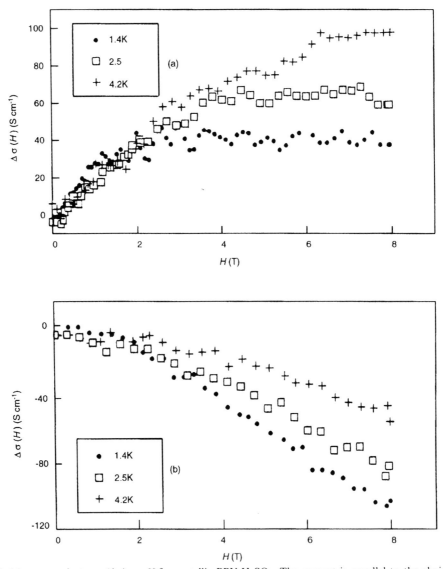

Figure 2.40. Magnetoconductance ($\Delta\sigma$) vs. H for metallic PPV-H_2SO_4. The current is parallel to the chain axis. (a) H is perpendicular to both current and chain axis; (b) H is parallel to both current and chain axis [107].

(assuming energy independent scattering) [225], consistent with the value of 1 state per electron volt per two rings obtained from magnetic susceptibility measurements [110,228].

The temperature dependences of $S(T)$ for various PPy-PF_6 samples near the M–I transition are shown in Figure 2.47 [213]. The room temperature value is positive, with $S(300\ K) = +(9–12)\ \mu\ V\ K^{-1}$; the magnitude decreases as ρ_r decreases. The relatively large magnitude of $S(T)$ again implies that the partially

filled π-band is relatively narrow; in this case less than 1 eV. The density of the states at the Fermi level, estimated from equation (2.28) is $N(E_F) \sim 1.0$–1.6 states per electron volt per four pyrrole units assuming the ideal doping level of one dopant per four rings. As for PANI-CSA, $S(T)$ for PPy-PF_6 is rather insensitive to ρ_r near the disorder-induced M–I transition.

However, the sign of thermopower in metallic conducting polymers is a mystery. Recently, Park *et al.* [229] have observed a positive thermopower for *n*-

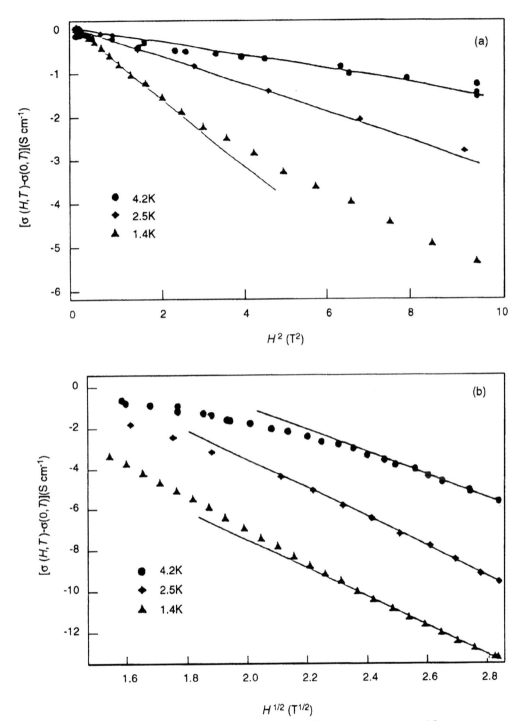

Figure 2.41. Magnetoconductance ($\Delta\sigma$) for metallic PANI–CSA (a) H^2 fit at low fields; (b) $H^{1/2}$ fit at high fields. (Reproduced by permission of the American Physical Society from ref. 150).

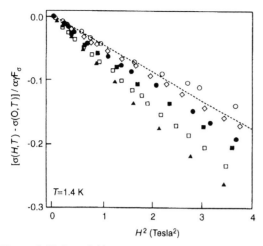

Figure 2.42. Low-field magnetoconductance normalized by $\delta\gamma F_\sigma$ plotted as a function of H^2 for various metallic PPy–PF$_6$ samples at 1.4 K. The dashed line is the theoretical estimate [equations (2.23) and (2.24)]. (Reproduced by permission of the American Physical Society from ref. 213).

doped (alkali metals) (CH)$_x$, the same as in p-doped materials. This is contrary to the conventional notion that the sign of the thermopower depends on the sign of the charge carrier. However, this is not really surprising, since the sign of the thermopower is determined by the details of the band structure, etc. Moreover, a discrepancy between the sign of the thermopower and the Hall coefficient has been observed in metallic PPy–PF$_6$ samples [230]. In this p-doped system, although the sign of the thermopower is positive as expected, the sign of the Hall coefficient is negative, as shown in Figure 2.48. Similar Hall–thermopower sign reversal

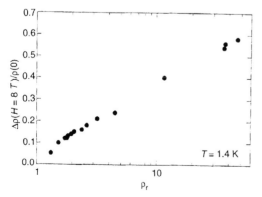

Figure 2.43. Magnetoresistance ($\Delta\rho/\rho$) vs. the resistivity ratio (ρ_r) for various PPy–PF$_6$ samples. (Reproduced by permission of the American Physical Society from ref. 213).

has been observed in several semiconducting systems like chalchogenide glasses, etc. [231]. However, in these low-mobility systems the charge carriers are mainly small polarons and the carrier concentration is rather low with respect to metallic PPy-PF$_6$ samples, in which there is a free carrier absorption in the infrared region and a carrier density of the order of 10^{22} cm^{-3} [232]. Since, only a small number of Hall coefficient measurements have been carried out in conducting polymers, the relationship between the sign of the thermopower, the Hall coefficient and the sign of the charge carrier is not very well known. Thus further theoretical and experimental studies are essential to explain these anomalies.

Various models have been proposed for the behavior of the thermopower in the deep insulating side of the M–I transition in conducting polymers. This is discussed in Section 10.

8.4 Microwave and optical conductivity

Epstein *et al.* [109] have used the dielectric constant measurement in the microwave frequency range to investigate the charge carrier localization–delocalization phenomena in conducting polymers. They have observed that the microwave dielectric constant (ε_{mw} at 6.5 GHz) is proportional to the square of the crystal-crystalline coherence length (ξ'). Moreover, ε_{mw} is independent of the orientation of the sample with respect to the electric field, which demonstrates that the charge is delocalized three-dimensionally within the crystalline regions. Epstein *et al.* [109] have used a simple metallic box model for the data analysis. The dielectric response is given by $\varepsilon_{mw} = \varepsilon_\infty + (2^{4.5}/\pi^3)e^2 N(E_F)L^2$, where ε_∞ is the core polarization, e is the electronic charge, $N(E_F)$ is the density of the states at the Fermi energy and L is the localization length. From the above expression, the value of $N(E_F)$ can be estimated by assuming that L is same as ξ' from X-ray diffraction. They have observed that this value of $N(E_F)$ is consistent with those obtained from magnetic susceptibility measurements [233].

The temperature dependence of microwave conductivity (6.5 GHz) of metallic (CH)$_x$, PPy-PF$_6$ and PANI-CSA is shown in Figure 2.49 [109]. Surprisingly, the temperature dependence of the microwave and dc conductivity are rather similar indicating that there is hardly any frequency dependence (up to 10 GHz) of the conductivity in metallic conducting polymers. The temperature dependence of the microwave dielectric

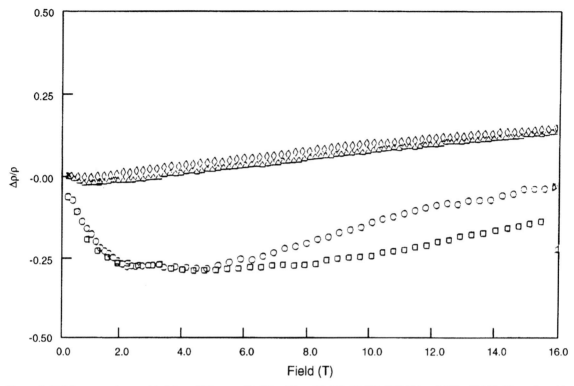

Figure 2.44. Magnetoresistance ($\Delta\rho/\rho$) vs. *H* for metallic PPy–PF$_6$ at 20 ($\square$), 70 ($\bigcirc$), 370 ($\triangle$) and 500 mK ($\diamondsuit$). (Reproduced by permission of Elsevier Science SA from *Synth. Met.* **69**, 1995, 215).

constant of metallic (CH)$_x$, PPy-PF$_6$ and PANI-CSA are shown in Figure 2.50 [109]. Epstein *et al.* [109,233] have pointed out that this large negative value of ε_{mw} is typical of that observed in conventional metallic systems. Hence, by using a Drude model for low frequencies ($\omega\tau \ll 1$), one obtains, $\varepsilon \cong -\omega_p^2\tau^2$ and $\sigma \cong (\omega_p^2/4\pi)\tau$, where τ is the relaxation time and ω_p is the plasma frequency. The values of ω_p and τ for metallic (CH)$_x$, PPy-PF$_6$ and PANI-CSA are 100–200 cm^{-1} (0.025–0.012 eV) and 10^{-11}s, respectively [109,233]. Thus both values of ω_p and τ are rather different from the usual Drude response in typical metals. This relatively low value of ω_p in metallic conducting polymers indicates that only a small fraction of the conduction band carriers are fairly delocalized and the low-frequency microwave response is mainly from these delocalized carriers. Epstein *et al.* [109,233] have observed two plasma frequencies, associated with the response from the 'most delocalized carrier' and from the 'entire carriers in the conduction band'. They conclude that the

presence of two 'plasma frequencies' suggests the evidence for inhomogenous disorder in metallic conducting polymers.

As pointed out in Section 7, Epstein *et al.* [109,233] suggested that this anomalously long τ arises from the open Fermi surface of highly anisotropic systems, because the essential mechanism for the relaxation of the carrier momentum is associated with the back-scattering from k_F to $-k_F$, rather than the forward scattering of charge carriers in lD systems. Moreover, the elastic backward scattering process is due to disorder, phonons, etc. residing outside the metallic chains, and therefore the role of random static scatterers in momentum relaxation is rather weak. Moreover, the freezing of high-frequency $2k_F$ phonons in lD systems results in weakening of the inelastic backward scattering. Since τ is nearly temperature independent above 100 K, the elastic backward scattering dominates the phonon backscattering and in turn determining the relaxation time even at high temperatures. Hence, the investigation of microwave

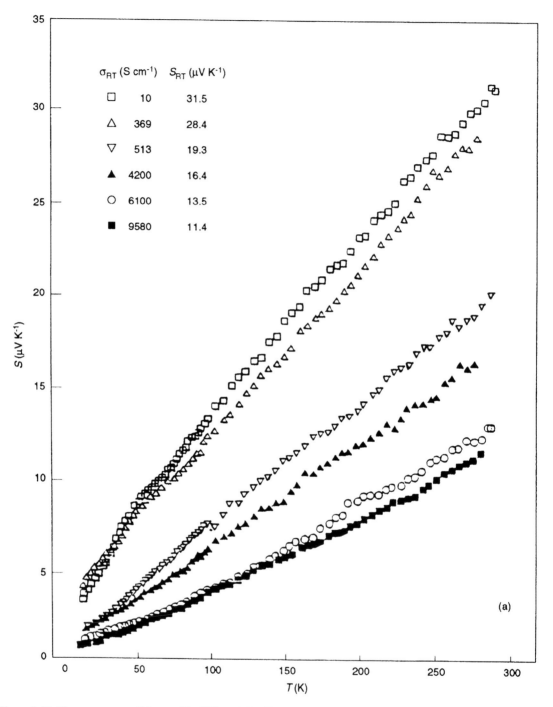

Figure 2.45. Thermopower vs. T for metallic $(CH)_x$: (a) $MoCl_5–(CH)_x$ [201]; (b) $I–(CH)_x$. (Reprinted from *Solid State Commun* **76**, Y. Nogami *et al.*, p.583, Copyright 1990, with kind permission from Elsevier Science Ltd, Kidlington OX5 1GB, UK).

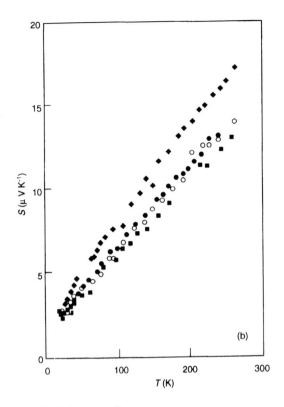

Figure 2.45. (*continued*)

dielectric response in metallic conducting polymers suggest that lD localization persists in these metallic systems, since the charges suffer lD localization when they move through uncorrelated chains through the disordered regions (amorphous domains) separating the crystalline domains [109,233]. The metallic state at high temperatures is due to the delocalization via phonons and this is expected to interrupt the quantum interference of the elastic scattering process [109].

However, the apparent contradiction with the above interpretations is as follows. Firstly, from dc conductivity measurements it is well known that metallic $(CH)_x$, PPy-PF$_6$ and PANI-CSA have large finite conductivity (above 100 S cm^{-1}) at mK temperatures. This clearly rules out the presence of any lD localization in these metallic systems. Secondly, the long relaxation time ($\tau \approx 10^{-11}$s) from the above data analysis [109] implies that the mean free path (λ) and $k_F\lambda$ are 1–100 μm and 10^5, respectively. While considering the extent of disorder in metallic $(CH)_x$, PPy-PF$_6$ and PANI-CSA, these numbers for λ and $k_F\lambda$ are not realistic. Even if one assumes that it is possible to achieve these numbers for λ and $k_F\lambda$ within the

crystalline domains, it is still not realistic since the crystalline coherence length ($\zeta'_\| \approx 50$–100 Å, $\zeta'_\perp \approx 20$–50 Å) is many orders of magnitude lower with respect to λ. Thirdly, the anomalously long relaxation time ($\tau \approx 10^{-11}$s) is based upon the argument that metallic conducting polymers are highly anisotropic q-lD systems with a open Fermi surface. However, the detailed study of various ideal q-lD systems [TTF-TCHQ, (TMTSF)$_2$PF$_6$, (SN)$_x$, NbSe$_3$, K$_2$Pt(CH)$_4$Br$_{0.3}$ H$_2$O, etc.] in which both the anisotropy in conductivity (see Table 2.1) and bandwidth are larger with respect to metallic conducting polymers, has shown that the value of τ in those ideal q-lD systems is around 10^{-14}–10^{-15} s [4,10,12]. This discrepancy suggests that the argument for the anomalously long relaxation time in metallic conducting polymers is not fully justified by relying on the assumption that metallic conducting polymers are highly anisotropic q-lD systems with an open Fermi surface. On the other hand, the value of τ obtained by various groups in highly doped $(CH)_x$ is 10^{-13}–10^{-15}s [105,234,235].

The detailed study of infrared reflectivity measurements in metallic conducting polymers has been carried out by Leising [105], Yang *et al.* [234], Hasegawa *et al.* [235], Lee *et al.* [236,232], Kohlman *et al.* [233], etc. Leising [105] has observed a free electron-like behavior for parallel polarization and a semiconductor-type behavior for perpendicular polarization in highly doped $(CH)_x$. The plasma energy is nearly 3.51 eV. The optical anisotropy [$(\alpha_\|/\alpha_\perp) \approx 26$] is nearly the same as that of undoped materials. Yang *et al.* [234] have reported that the band gap is not closed in fully doped $(CH)_x$ and the bond length distortion becomes incommensurate at doping levels above 0.04, and in presence of strong disorder the gapless Peierls state exists with a finite but reduced density of the states at the Fermi level.

Woo and Tanner [237] have ascribed the narrow Drude peak about 100 cm^{-1} to free-carrier absorption in heavily doped $(CH)_x$; this peak vanished below 200 K, which is assumed to be due to the freezeout of carriers at low temperatures. They have attributed the additional absorption at low temperatures due to the trapping of charge carriers by the Coulomb interaction with dopant ions. Conwell and Mizes [238] have pointed out that the progressive shift of absorption to lower frequencies, as temperature is lowered, is due to the progressive freezeout of the metallic phase in smaller gap chains. Kim and Heeger [239] have observed the features of a 'pseudo-gap' in heavily doped $(CH)_x$ at low temperatures. However, the recent infrared reflectivity measurements in metallic conduct-

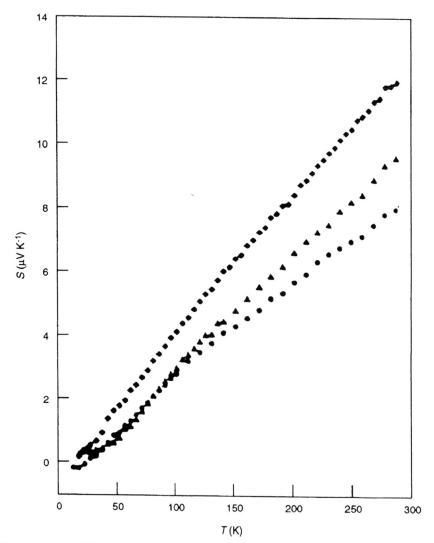

Figure 2.46. Thermopower vs. T for various PANI–CSA (Reproduced by permission of the American Physical Society from ref. 225).

ing polymers by Lee *et al.* [236,232] suggested that this anomalous features at low temperatures could be understood by using the localization-modified Drude model.

Hasegawa *et al.* [235] have observed free carrier creation and new band formation above the critical doping concentration, which is consistent with the observation of Pauli susceptibility above the critical doping concentration, in $(CH)_x$. They have observed that the relaxation time of free carriers in metallic $(CH)_x$ is nearly 10^{-13}–10^{-15} s. Conwell and Mizes [238] have analyzed the optical absorption data and they find

that the plasma frequencies for 11 and 16% iodine-doped $(CH)_x$ are 0.34 and 0.48 eV, respectively. The values for carrier concentration and τ for 11% (16%) I-$(CH)_x$ are 2.8×10^{20} cm^{-3} (7.2×10^{20} cm^{-3}) and 4.5×10^{-14} s (1.0×10^{-13} s), respectively. The theoretical estimate made by Stafström [85] for τ in highly doped $(CH)_x$ is of the order of 10^{-15}. More-over, Javadi *et al.* [206] obtained $\tau \approx 2.7 \times 10^{-15}$ s in highly doped $(CH)_x$ from transport property measurements.

Epstein *et al.* [109,233] have reported the optical response in highly doped conducting polymers. They

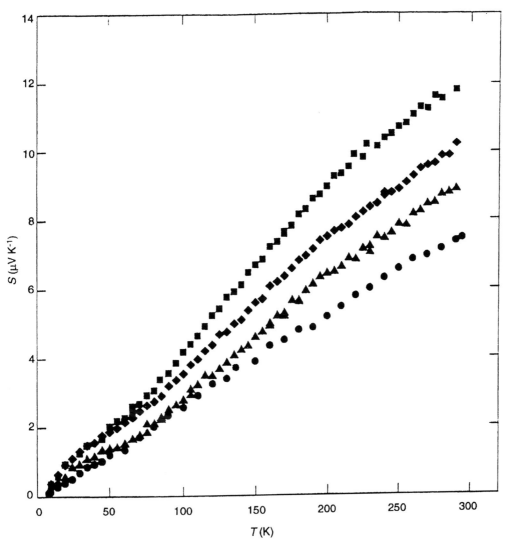

Figure 2.47. Thermopower vs. T for various PPy-PF$_6$ samples near the M–I transition. (Reproduced by permission of the American Physical Society from ref. 213).

have observed a crossover from a positive to a negative dielectric constant as the frequency is lowered through the effective screened plasma-frequency response of the entire conduction band. The dielectric constant deviates from Drude behavior below a characteristic frequency. This negative dielectric constant below the second plasma frequency (from infrared response) is similar to that observed in the microwave response. In some conducting polymers, Epstein *et al.* [109,233] have observed two or three zero crossings of the real part of the dielectric function as the frequency is lowered. Meanwhile, for conducting polymers in the insulating side of the M-I transition, $\varepsilon(\omega)$ remains positive for the entire frequency regime. From the above results, Epstein *et al.* [109,233] inferred that metallic conducting polymers are inhomogenously disordered systems, consisting of a composite of metallic islands (crystalline regions, $\tau \approx 10^{-11}$ s) embedded in a semiconducting (amorphous regions) matrix, in which the volume fraction of metallic islands is above the percolation threshold.

However, the percolative transport through the connected network of metallic islands, having a mean free path (λ) of the order of 1–100 μm, is really

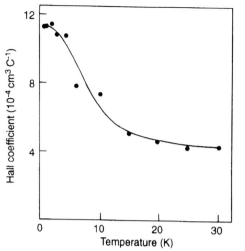

Figure 2.48. The Hall coefficient vs. T for metallic PPy–PF$_6$ [230].

surprising [109,233]. If the volume fraction of metallic islands is below the percolation threshold, then the system is on the insulating side of the M–I transition. If this model is really true, one could expect quantum-size effect in these materials, since the metallic islands are quantum-size (nanoscopic-scale) grains. However, so far no such behavior has been observed. As mentioned before, *metallic samples of K-(CH)$_x$, PANI-CSA, PPV-H$_2$SO$_4$, PPy-PF$_6$ and ion-implanted PANI show a positive TCR below 20 K,* which one could hardly expect in percolative transport through the connected network of metallic islands. Thus once again, these experimental observations are not consistent within the framework of percolative transport through the connected network of metallic islands.

Recently Lee *et al.* [232,236] have carried out detailed measurements of the temperature dependence of infrared reflectivity in metallic PANI-CSA and PPy-PF$_6$ down to 10 K. Lee *et al.* [232,236] have used the localization-modified Drude model,

$$\sigma_{LD}(0) = \sigma_{Drude}\left[1 - \frac{C}{(k_F l)^2}\right] \quad (2.30)$$

in which the frequency-dependent Drude conductivity $[\sigma_D(\omega) = (\omega_p)^2\tau/\{4\pi(1 + \omega^2\tau^2)\}]$ is modified by introducing a first order correction due to weak disorder-induced localization. In this correction term, L_ω is the distance over which a charge carrier diffuses within an optical period and C is of the order of unity, $L_\omega = (D/\omega)^{1/2}$, where $D = (\lambda^2/3\tau)$ is the electronic diffusion constant. In this localization-modified Drude model, the extent of disorder is assumed to be within the

homogenous limit (correlation length/localization length is larger than the crystalline coherence length) for metallic conducting polymers. This assumption is justified from the large finite conductivity observed at mK temperatures, the observation of quantum corrections in dc conductivity/MC, the absence of any frequency dependence of conductivity in the KHz–GHz range, etc.

According to this localization-modified Drude model [equation (2.30)] the best fit for reflectance data, for metallic PPy-PF$_6$, is shown in Figure 2.51. The fit directly yields the value of $k_F\lambda \approx 1.6$, which is identical to that obtained (independently) from dc conductivity measurements. Moreover, this analysis yields $\tau \approx 10^{-15}$ s, which is considerably different from that obtained by Kohlman *et al.* [233]. The values of τ and $k_F\lambda$ for metallic PANI-CSA are rather similar [236]. In high-quality PPy-PF$_6$ samples, Lee *et al.* [232,236] have observed a narrow peak in $\sigma(\omega)$ below 100 cm^{-1} at low temperatures, which implies that metallic PPy-PF$_6$ is a conductor with a gap in the carrier excitation spectrum. This type of behavior is usually observed in systems in which the formation of some collective mode occurs at low temperatures. Work is in progress to fully understand this phenomena.

In summary, the preliminary results of microwave and optical conductivity in metallic conducting polymers have shown that this is a powerful technique to investigate both the static and dynamic properties of charge carriers in these systems. However, more work is essential to improve the understanding of the experimental results.

8.5   Magnetic susceptibility

From early magnetic susceptibility ($\chi = \partial M/\partial H$, where M is the magnetization) studies in doped (CH)$_x$ it is well known that susceptibility measurements are important in understanding the charge storage mechanism and the spin–charge relationship in conducting polymers [14,144]. The inverse spin–charge relationship from susceptibility measurements provided an important clue for solitons in lightly doped (CH)$_x$. Similarly, the susceptibility measurements indicated the presence of spin 1/2 polarons and spinless bipolarons in non-degenerate ground state conducting polymers. The temperature dependence of susceptibility measurements are useful in identifying whether the charge carriers are localized or delocalized. The presence of localized spins can be easily identified from the well know l/T dependence of $\chi(T)$, namely the

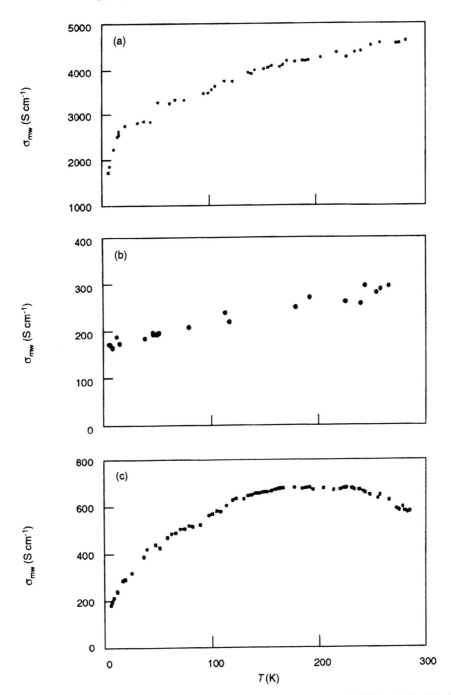

Figure 2.49. Microwave conductivity (6.5 GHz) vs. T for metallic conducting polymers: (a) $(CH)_x$; (b) $PPy-PF_6$ (c) PANI–CSA. (Reproduced by permission of Elsevier Science SA from *Synth. Met.* **65**, 1994, 149).

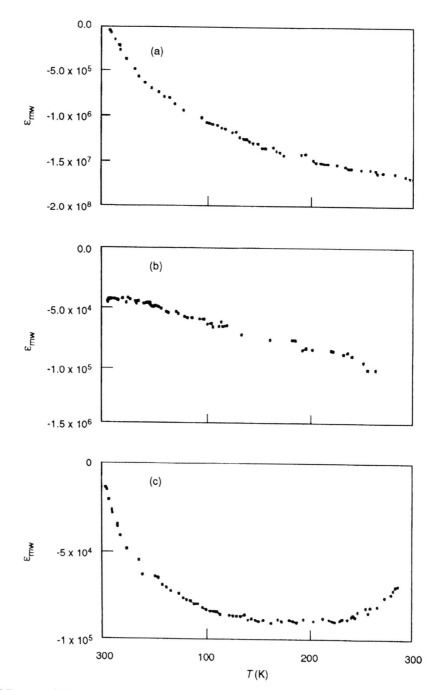

Figure 2.50. Microwave dielectric constant (6.5 GHz) vs. *T* for metallic conducting polymers: (a) $(CH)_x$; (b) PPy–PF$_6$, (c) PANI–CSA. (Reproduced by permission of Elsevier Science SA from *Synth. Met.* **65**, 1994, 149).

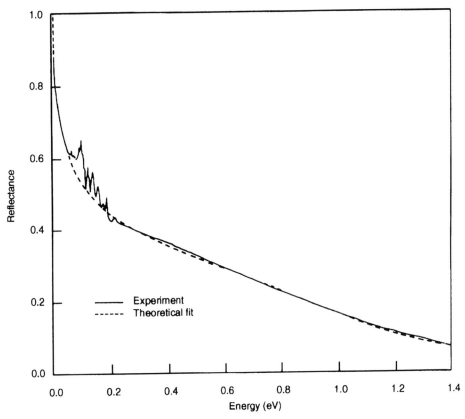

Figure 2.51. Reflectance vs. energy: the fit for the reflectance data in the localization-modified Drude model [equation (30)] for metallic PPy–PF$_6$. (Reproduced by permission of the American Physical Society from ref. 236).

Curie law behavior ($\chi^c = N_c \mu_B^2 / k_B T$, where N_c is the number of localized spins and μ_B is the Bohr magneton). In metallic systems, the temperature-independent Pauli susceptibility (χ^P) is a characteristic feature for delocalized carriers. χ^P is directly proportional to the density of states at the Fermi level through the relation, $\chi^P = \mu_B^2 N(E_F)$. Thus one obtains $N(E_F)$ directly from the temperature-independent χ^P for metallic systems. In disorder systems, the measured $\chi(T)$ is the sum of both the Curie and Pauli terms; thus one could estimate the extent of disorder in the system. Apart from susceptibility measurements, the Korringa relation from NMR relaxation is also important in characterizing the metallic state in disordered systems. For example, recently Kolbert et al. [240] have observed a modified Korringa relation from ^{13}C NMR for metallic PANI-CSA samples.

Nogami et al. [205] have reported the evolution of susceptibility as a function of doping in Tsukamoto (CH)$_x$. The values of $N(E_F)$ for iodine (9.4% I$_3^-$ units)

doped Tsukamoto (CH)$_x$ are 0.22 (unstretched) and 0.31 (stretched) states/eV C. The value of χ^P for metallic (CH)$_x$ is 7×10^{-6} emu (mol C)$^{-1}$. The $\chi(T)$ shows a Curie-type behavior below 100 K, as shown in Figure 2.52. It is surprising to observe such a large Curie contribution in metallic (CH)$_x$ samples [$\sigma(300\ K) \approx 8 \times 10^4$ S cm^{-1}], which indicates that the presence of localized spins due to impurities, defects, neutral solitons in undoped regions, etc. are substantial.

Although the room-temperature conductivity of metallic PANI-CSA and PPy-PF$_6$ samples [$\sigma(300\ K) \approx 300$ S cm^{-1}], is nearly two orders of magnitude less than that of metallic (CH)$_x$, the $\chi(T)$ measurements in these systems showed a substantialy weaker Curie contribution at low temperatures, as shown in Figures 2.53 and 2.54, respectively. Adams et al. [241] have reported a weak $\chi(T)$ for PANI-HCl samples, although the $\sigma(T)$ of these samples showed the typical insulating behavior. A weak Curie contribution

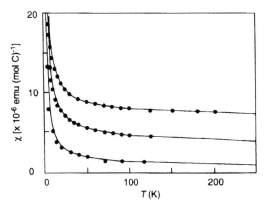

Figure 2.52. $\chi(T)$ vs. T for metallic I–(CH)$_x$. (Reprinted from *Solid State Commun* **76**, Y. Nogami *et al.* p. 583, Copyright 1990, with kind permission from Elsevier Science Ltd, Kidlington OX5 1GB, UK).

in high-quality PANI and PPy-PF$_6$ samples appears below 50 K. The approximate values of $N(E_F)$ for metallic PANI-CSA [228] and PPy-PF$_6$ [242] are one (states/eV/two rings) and three (states/eV/four rings), respectively. The approximate values of χ^p for PANI-CSA, PANI-HCl and PPy-PF$_6$ samples are around 2×10^{-5} emu (mol C + N)$^{-1}$, 4×10^{-5} emu (mol C + N)$^{-1}$ and 7×10^{-6} emu (mol C)$^{-1}$, respectively. In a recent review Kohlman *et al.* [233] have tabulated the reported values of both χ^p and $N(E_F)$ for various conducting polymers.

Although considerable work regarding the behavior of magnetic susceptibility and $N(E_F)$ as a function of doping, temperature and magnetic field has been reported, the finer details are yet to be understood. Earlier studies in doped PPy and PT indicated that the reduction of spins at higher doping level is associated with the transition from spin 1/2 polarons to spinless bipolarons. Recently Mizoguchi *et al.* [79,108] pointed out that this could be due as well to the transition from localized carriers (Curie type) at lower doping levels to

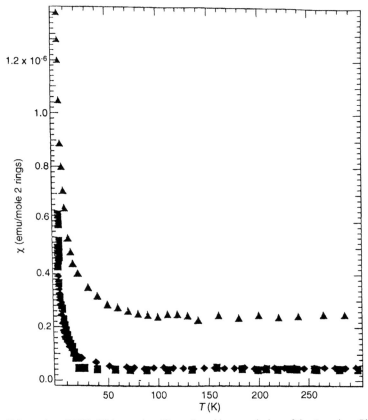

Figure 2.53. $\chi(T)$ vs. T for various PANI–CSA samples. (Reproduced by permission of the American Physical Society from ref. 228).

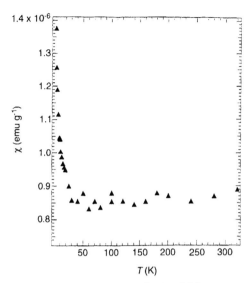

Figure 2.54. $\chi(T)$ vs. T for metallic PPy–PF6.

delocalized carriers (Pauli type) at higher doping levels. This also explains the $\chi(T)$, since some carriers are localized at defect/impurity sites to show a Curie-type behavior at low temperatures. However, Le Guennec *et al.* [78b] suggested that the effect of disorder and interaction between quasi-particles varies the polaron–bipolaron statistics significantly. Thus the difference in the behavior of χ as a function of doping level and temperature is probably due to the variations in disorder, undoped regions, defects, interchain interactions, etc. In some systems, although an insulating behavior was observed in $\sigma(T)$, the $\chi(T)$ value was found to be of a Pauli type instead of the expected Curie behavior in insulating systems at low temperatures. Thus a consistent pattern of behavior in $\sigma(300\ K)$, $\sigma(T)$ $\chi(3000\ K)$, $\chi(T)$ and $N(E_F)$ in various conducting polymers is yet to emerge.

9 TRANSPORT NEAR THE CRITICAL REGIME OF THE METAL–INSULATOR TRANSITION IN CONDUCTING POLYMERS

As mentioned in Section 8.1, for systems near the critical regime of the M–I transition, $W(T)$ is nearly temperature independent for a wide range of temperature. For a 3D system in the critical regime of the M–I transition, the correlation length is large and has a power law dependence on $\delta = |\,E_F - E_c/E_F\,| < 1$ with a

critical exponent v, $L_c = a\delta^{-1/v}$ where a is a microsopic length, E_F is the Fermi energy and E_c is the mobility edge [157,158]. In this critical region, the resistivity is not activated, but rather follows a power law dependence on the temperature as shown by McMillan's scaling theory [157] and the model proposed by Larkin and Khmelnitskii [158]:

$$\rho(T) \approx (e^2 p_F/\hbar^2)(k_B T/E_F)^{-1/\eta} \propto T^{-\beta} \qquad (2.31)$$

where p_F is the Fermi momentum and e is the electron charge. The predicted range of validity includes $1 < \eta < 3$; i.e. consistent with the observed values $0.33 < \beta < 1$.

This power law is universal and requires only that the disordered system be in the critical regime where $\delta \ll 1$. A value of $\eta > 3$ indicates that the system is just on the metallic side of the M–I transition. Although $\eta > 3$ is above the theoretical limit for the power law dependence, values for η as large as 4.5 have been reported for *n*-doped germanium near the critical regime [156]. At mK temperatures, however, the system becomes either a metal or an insulator depending on the extent of disorder; extension of the power law dependence to $T = 0$ requires that the system be *precisely* at the critical point. Conducting polymers are particularly interesting for investigating the critical behavior near the M–I transition since a wide range of parameters (e.g. carrier concentration, interchain interaction and extent of disorder) are involved in this phenomena.

This scaling theory for the temperature dependence of resistivity $[\rho(T) \propto T^{-\beta}]$ near the critical regime of the M–I transition is fully valid for $0.33 < \beta < 1$, although values of $\beta \approx 0.2$ have been observed in doped PPV samples [107]. The continuous variation of β from 0.2 to 1 may be due to the continuous variation of the nature of the wavefunctions near the mobility edge, from fractally localized (power law behavior) to fractally extended (power law behavior), as the system approaches the M–I transition [57]. The physical origin of the power law behavior of $\rho(T)$ near the critical regime of the M–I transition is not very well understood. The typical example of various conducting polymers in the critical regime of the M–I transition is shown in Fig. 2.55 [2,245]. The sample-to-sample variation of the critical exponent (η) in the power law temperature dependence correlates with ρ_r; samples with smaller (larger) ρ_r give smaller (larger) values of η. In the power law regime [temperature-independent $W(T)$ regime] $W = \beta$ and $\rho(T) \propto T^{-\beta}$. Thus both the temperature range and the value of the exponent β in

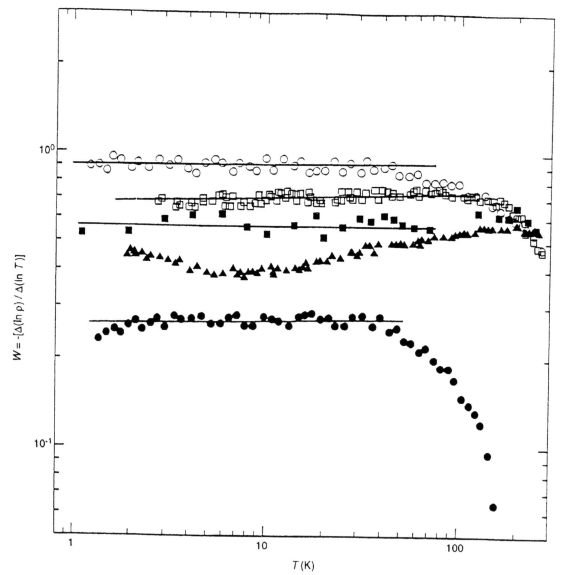

Figure 2.55. W vs. T for various conducting polymers in the critical regime of the M–I transition: I–(CH)$_x$ (O); oriented I–(CH)$_x$ (□); oriented K–(CH)$_x$ (■); PPy–PF$_6$ (▲); PANI–CSA (●). The straight lines are drawn to guide the eye. Reproduced by permission of Elsevier Science SA from *Synth. Met.* **69**, 1995, 329).

the power law regime can be verified from the log–log plots of W vs. T and ρ vs. T.

The log–log plot of W vs. T, as a function ρ_r, for both stretched and unstretched I-(CH)$_x$ is shown in Figure 2.19. For unstretched I-(CH)$_x$ samples, $W(T)$ is nearly temperature independent from 300–1.4 K and $\beta \approx 1$. Another log–log plot of W vs. T for I-(CH)$_x$ samples, as a function of doping level and pressure, is shown in Figure 2.56. For I-(CH)$_x$ sample with $\rho_r \approx 3$, $W(T)$ is

nearly temperature independent from 180–60 K, at ambient pressure. However, at 8 kbar, this temperature-independent $W(T)$ regime undergoes a transition to a positive temperature coefficient of the $W(T)$ regime, as shown in Figure 2.56 [2]. This clearly indicates that the system has become more metallic due to enhanced interchain interactions at high pressure. In slightly less doped samples, the system moves more towards the critical regime of the M–I

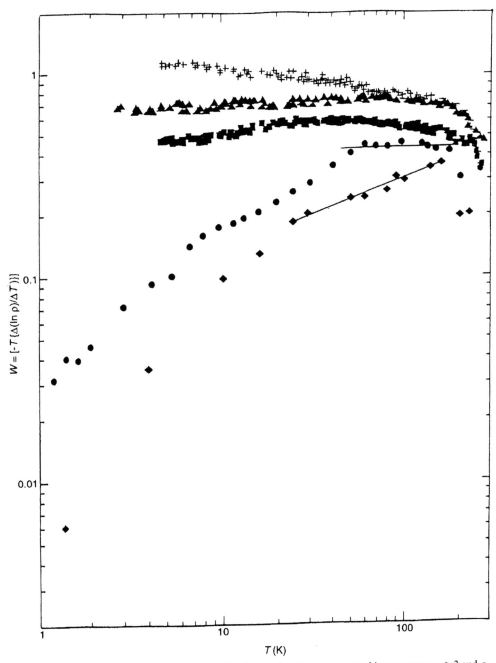

Figure 2.56. W vs. T for I–(CH)$_x$ samples as a function of doping level and pressure: at ambient pressure, $\rho_r \approx 3$ and $\sigma_{RT} \approx 11\ 000$ S cm^{-1} (●); at 8 kbar, $\rho_r \approx 2$ and $\sigma_{RT} \approx 9500$ S cm^{-1} (◆); at ambient pressure, $\rho_R \approx 9$ and $\sigma_{RT} \approx 7000$ S cm^{-1} (■); at ambient pressure, $\rho_r \approx 17$ and $\sigma_{RT} \approx 3500$ S cm^{-1} (▲); at ambient pressure, $\rho_r \approx 32$ and $\sigma_{RT} \approx 2450$ S cm^{-1} (+) [243]. The straight lines are drawn to guide the eye. Reproduced by permission of Elsevier Science SA from *Synth. Met.* **69**, 1995, 329).

transition as the temperature-independent $W(T)$ regime extends down to lower temperatures. For I-$(CH)_x$ samples with $\rho_r \approx 32$, $W(T)$ shows a negative temperature coefficient, typical to that of an insulator. This clearly demonstrates that highly doped conducting polymers are near the M–I transition, and the transition from the metallic side to the insulating side is continuous via the critical regime. This can be vividly observed from W vs. T plots.

The pressure-induced transition to the metallic side and magnetic field-induced transiton to the insulating side for PPy-PF_6 samples in the critical regime is shown in Figure 2.57 [243]. In this case, the negative temperature coefficient of $W(T)$ at 8 T field shows the transition from the critical regime to the insulating regime. Although in inorganic semiconductors several groups have observed the critical behavior by tuning the interaction length by magnetic field and the critical carrier density by pressure [244,245], it is somewhat different from that observed in conducting polymers. In inorganic semiconductors, the field-induced transition

was observed from the $T^{1/2}$ dependence of conductivity [$\sigma(T) = \sigma(0) + mT^{1/2}$, due to e–e interaction] in the disordered metallic regime to VRH [$\ln \rho \propto (T_0/T)^x$] in the insulating regime at mK temperatures. However, the field-induced transition in conducting polymers is from the critical regime with a power law behavior [$\rho(T) = zT^{-\beta}$, where z is a constant] to the insulating regime with an exponential law behavior [i.e. VRH, $\ln \rho \propto (T_0/T)^x$]. This field-induced transition for PANI-CSA sample from critical to insulating regime is clearly shown in Figure 2.58.

Since an ideal lD system should not be sensitive to a magnetic field, the magnetic field-induced transition from the critical regime to the insulating regime provides unambiguous evidence for the importance of interchain transport in conducting polymers. Khmelnitskii and Larkin [246] have presented a scaling argument which indicates that the mobility edge can be shifted with respect to the Fermi energy by a magnetic field. Schreiber and Grussbach [52] have suggested a field-induced multifractal localization model. There is,

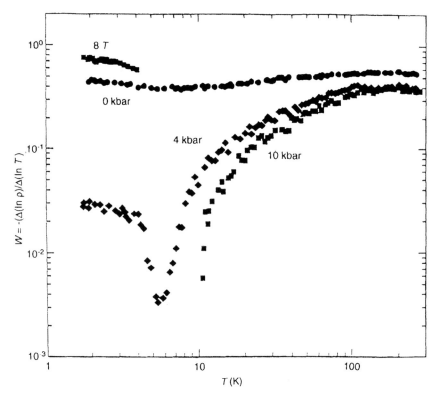

Figure 2.57. W vs. T for a PPy–PF_6 sample in the critical regime: the pressure-induced transition to the metallic side and the field-induced transition to the insulating side. Reproduced by permission of Elsevier Science SA from *Synth. Met.* **69**, 1995, 329).

however, no detailed microscopic theory for describing the effect of a magnetic field on the M–I transition, particularly in an anisotropic (q-1D) system. The overlap of the wavefunctions in a q-1D metal is rather sensitive to a magnetic field, since the field tends to shrink the wavefunctions [247]. The field-induced crossover from the critical regime to VRH among localized states is consistent with a field-induced shift of the mobility edge with respect to the Fermi level. The crossover occurs when the localization length and magnetic length $[(\hbar c/eH)^{1/2}]$ become comparable.

The positive magnetoresistance (MR) in the critical regime is substantially higher ($\Delta\rho/\rho > 5\%$) with respect to that in the metallic regime. As ρ_r increases, when the system moves from the metallic to the insulating regime, the positive MR increases substantially. This is shown clearly in the case of various PPy-PF$_6$ samples, as in Figure 2.43. The MR for a PANI-CSA sample in critical regime, with $\rho_r \approx 4$ is shown in Figure 2.59. In the critical regime, the overlap of the wavefunctions near the mobility edge is rather sensitive to magnetic field. Hence, slight shrinkage of the overlap of the wavefunctions in the presence of a magnetic field easily shifts the mobility edge towards the insulating side of the M–I transition.

Thus the fine tuning of the critical regime by pressure and magnetic field is well observed in $\sigma(T)$ and MR in conducting polymers. However, temperature depen-

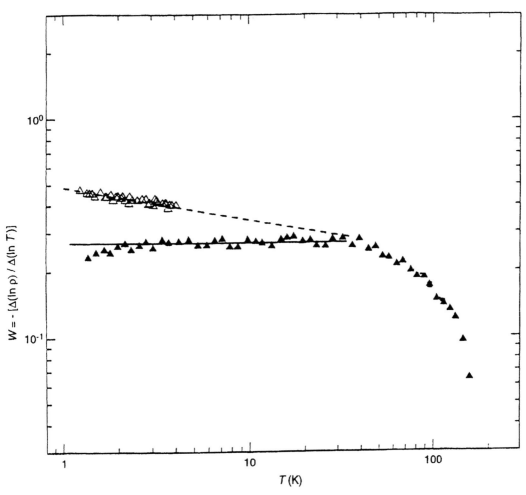

Figure 2.58. W vs. T for PANI–CSA in the critical regime: the field-induced transition to the insulating side. $H = 0$ (▲) and 8 T (△). Reproduced by permission of Elsevier Science SA from *Synth. Met.* **69**, 1995, 329).

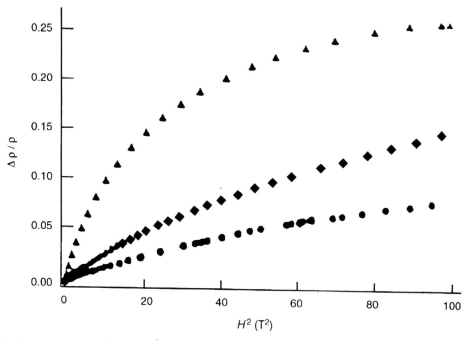

Figure 2.59. Magnetoresistance ($\Delta\rho/\rho$) vs. H^2 for PANI–CSA in the critical regime of the M–I transition: 4.2 (●), 2.5 (◆) and 1.4 K (▲). (Reproduced by permission of the American Physical Society from ref. 150).

dence of the thermopower in the critical regime shows the typical quasi-linear behavior in the metallic regime. The frequency and electric field response of conductivity near the critical regime of the M–I transition is not very well studied yet. In general, the power law response of systems near the critical regime of a phase transition (here the M–I transition) is very clearly observed for the first time in $\rho(T)$ in conducting polymers. This has provided an unique opportunity to investigate the static and dynamical properties of wavefunctions near the mobility edge in systems near the critical regime of the M–I transition.

10 HOPPING TRANSPORT IN THE INSULATING REGIME OF CONDUCTING POLYMERS

In the insulating regime, transport occurs through VRH among localized states as described by Mott [46] for non-interacting carriers and by Efros–Shklovskii (ES) [247] when the Coulomb interaction between the

electron and the hole left behind is dominant. For Mott VRH conduction in 3D

$$\ln \rho \propto (T_0/T)^{1/4} \tag{2.32a}$$
$$T_0 = 18/k_B L_c{}^3 N(E_F) \tag{2.32b}$$

where k_B is the Boltzmann constant, L_c is the localization length and $N(E_F)$ is the density of the states at the Fermi level. In the Efros–Shklovskii limit

$$\ln \rho \propto (T_0'/T)^{1/2} \tag{2.33a}$$
$$T_0' = \beta_1 e^2/\varepsilon k_B L_c \tag{2.33b}$$

where e is the electron charge, ε is the dielectric constant, and $\beta_1 = 2.8$ (a numerical constant).

The crossover from Mott to ES VRH has been observed in insulating samples of PPy-PF$_6$ and PATs near the M–I transition [226]. The experiments have also shown that equation (2.33) holds in a strong magnetic field ($H = 8$ T) where $T_0'(H)/T_0'(0) \propto H^p$ with $p = 1.0 \sim 1.2$.

The crossover from equation (2.32) to equation (2.33) occurs when the mean hopping energies (Δ_{hop})

in the Mott and ES limits are comparable [248,249]. The mean hopping energy from each theory is given by

$$\Delta_{hop} = 1/4(k_BT)(T_0/T)^{1/4}, \quad (Mott, x = \tfrac{1}{4}) \quad (2.34a)$$

$$\Delta'_{hop} = 1/2(k_BT)(T'_0/T)^{1/2}, \quad (ES, x = \tfrac{1}{2}) \quad (2.34b)$$

The Efros–Shklovskii VRH theory [247] predicts a power law energy dependence in the density of the states near the Fermi level which occurs with the Coulomb gap, Δ_c:

$$\Delta_C = e^3N(E_F)^{1/2}/\varepsilon^{3/2} \quad (2.35)$$

where $N(E_F)$ is the unperturbed density of states at the Fermi level (i.e. in the absence of the Coulomb gap) and ε is the dielectric constant. Castner [159] pointed out that near the M–I transition, the dielectric constant can be expressed as

$$\varepsilon = \varepsilon'_\infty + 4\pi e^2N(E_F)L_c^2 \quad (2.36)$$

where ε'_∞ is the core dielectric constant and the second term results from the polarizability of the localized states. Note that L_c diverges as $\delta \to 0$. Thus, if the system is not too far from the M–I transition, then the second term is dominant, and $\varepsilon \approx 4\pi e^2N(E_F)L_c^2$. Assuming that the above approximation is valid, Castner has noted the following relations [159,248]:

$$T_0/T'_0 = 18(4\pi)/\beta_1 = 81 \quad (2.37)$$

$$\Delta_C \approx k_BT_0/18(4\pi)^{3/2} \approx k_BT'_0/\beta_1(4\pi)^{1/2} \quad (2.38)$$

$$T_{cross} = 16(T'_0)^2/T_0 \quad (if\ \Delta_{hop} \approx \Delta'_{hop}) \quad (2.39)$$

The localization length can be estimated from the expression for the weak magnetic field dependence of the VRH resistivity [247]

$$\ln[\rho(H)/\rho(0)] = t(L_c/L_H)^4(T_0/T)^{3x} \quad (2.40)$$

where $t = 0.0015$ for $x = 1/4$, $t = 0.0035$ for $x = 1/2$, and $L_H = (c\hbar/eH)^{1/2}$, the magnetic length [247].

The experimental ratio, $T_0/T'_0 = 85 \sim 115$, is close to the value predicted by Castner [159]; and $\Delta_c = 0.3–0.6$ meV. The experimental values of T_{cross} are in good agreement with those estimated from the theory [226]. The validity of the approximation $[\varepsilon \approx 4\pi e^2N(E_F)L_c^2]$ is tested from the fact that $4\pi e^2N(E_F)L_c^2/\varepsilon = 0.70–0.95$. This indicates that ε'_∞ makes a relatively small contribution to ε.

For samples farther into the insulating regime ($\rho_r > 10^3$), two different kinds of materials are known, homogeneous and inhomogeneous [226].

(1) For 'homogeneous' systems, the localization length is greater than the disorder length scale, e.g. greater than the structural coherence length (ζ) in a polymer which has both crystalline and amorphous regions ($L_c \geqslant \zeta$). Since Mott's $T^{-1/4}$ law is usually observed in a homogenously disordered insulating system, the observation of a similar temperature dependence of conductivity at low temperatures in conducting polymers can be considered as a verification of the presence of homogenous disorder. Moreover, in the homogeneously disordered regime, thermopower has a linear relationship with temperature.

(2) For 'inhomogeneous' (inhomogenous doping, phase segregation of doped and undoped regions, partial dedoping and large-scale morphological disorder, etc.) systems $L_c \leqslant \zeta$. In such systems, a granular metallic behavior is indicated by the $\ln \rho \propto (T'_0/T)^{1/2}$ behavior [250–252]. Although the factors leading to the $T^{-1/2}$ fit for granular metals are not completely understood, the recent theoretical work by Cuevas et al. [251] has shown that the low-temperature transport properties can be dominated by the long-range Coulomb interaction rather than by charging effects (as previously believed). Moreover, the hopping contribution to $S(T)$ is substantial at low temperatures, as give below.

The VRH hopping contribution to thermoelectric power depends on the details of the hopping mechanism [46,253].

$$S_{hop}(T) = \tfrac{1}{2}(k_B/e)(\Delta_{hop}^2/k_BT)[d\ln N(E)/dE]_{E_F} \quad (2.41)$$

where Δ_{hop} is the mean hopping energy. From equations (2.34a) and (2.34b), we have $S_{hop} \propto T^{1/2}$ for $x = 1/4$ and $S_{hop} = $ constant for $x = 1/2$ [226]. Thus, in a Fermi glass with a finite density of states at E_F, $S(T)$ should have contributions from both $S_d(T)$ and $S_{hop}(T)$. One finds that the hopping contribution to the total thermoelectric power of PANI-CSA samples fits the empirical formula well

$$S(T) - AT = BT^{\frac{1}{2}} + C \quad (2.42)$$

where A is the linear slope of $S(T)$, and B and C are fitting parameters. The magnitude of the hopping thermoelectric power increases with ρ_r. The origin of the positive or negative sign for hopping contributions is not understood (generally the sign depends on asymmetry corrections to the density of the states with respect to the Fermi level [253]).

Assuming energy-independent scattering for $S_d(T)$, the magnitude of the hopping contribution can be estimated by using equations (2.29) and (2.41).

$$S_{hop}/S_d \approx (3/2\pi^2)(\Delta_{hop}/k_B T)^2 = (3/2\pi^2)W^2 \quad (2.43)$$

where $W = \Delta_{hop}/k_B T = x(T_0/T)^x$ is the reduced activation energy. For $W < 1$, $S_{hop}/S_d \ll 1$, and the hopping contribution to the thermoelectric power is insignificant. For $x = 1/4$, equation 2.43 becomes $S_{hop}/S_d \approx (\lambda T_0/T)^{1/2}$ where $\lambda \sim 10^{-3}$; for $x = 1/2$, $S_{hop}/S_d \approx (\lambda' T_0'/T)$ where $\lambda' \sim 10^{-1}$ [226]. The condition for the hopping thermoelectric power to be comparable to the diffusion thermoelectric power is, for example

$T_0 \geqslant 10^5$ K or $T_0' \geqslant 10^3$ K at 100 K. In the homogenous limit ($\rho_r < 10^2$, $x = 1/4$), $T_0 < 10^3$ K, implies that the hopping contribution to $S(T)$ is negligible. In the inhomogeneous limit ($\rho_r > 10^3$, $x = 1/2$), the hopping thermoelectric power contributions from the large values of $T_0 > 10^4$ and $T_0' > 10^2$ become important, and the temperature dependence of the resistivity is $\ln \rho \propto T^{-1/2}$.

The crossover from Mott to ES at low temperatures is clearly shown from the log–log plots of W vs. T for both PPy-PF$_6$ and iodine-doped polyalkylthiophene (I-PAT) samples, as shown in Figures 2.60 and 2.61, respectively [226]. The temperature regimes for $T^{-1/4}$ and

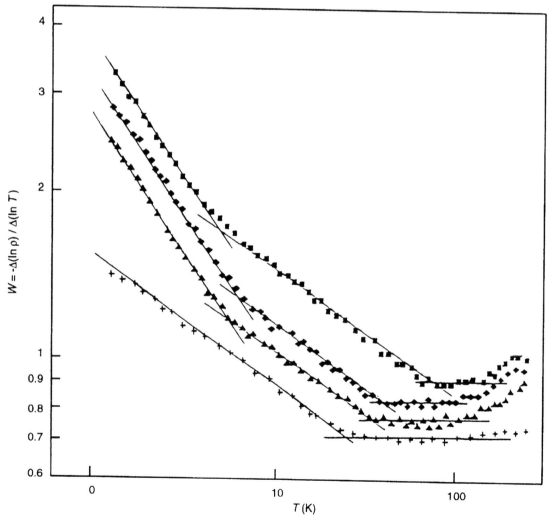

Figure 2.60. W vs. T for various insulating PPy–PF$_6$ samples. The slope change indicates the crossover from Mott ($T^{-1/4}$) to ES ($T^{-1/2}$) at low temperatures. From ref. 226.

$T^{-1/2}$ behavior are distinct from the slopes in W vs. T plots. In one PAT (polytetradecylthiophene) sample the $T^{-1/2}$ dependence of conductivity is observed from 3 to 100 K, as shown in Figure 2.61. This is a typical example for $\rho(T)$ in a granular system.

The field dependence of $\rho(T,H)$ for magnetic fields up to $H=8$ T for PPy-PF$_6$ and iodine-doped POT samples is shown in Figure 2.62 [226]. Although strong magnetic fields significantly alter the localized electronic wavefunctions, decreasing the overlap and increasing the hopping length, the $\ln \rho \propto (T_0'/T)^{1/2}$ law remains valid, but with increased T_0'. Various VRH exponents ($x=3/5$, $1/2$ and $1/3$) have been suggested as appropriate in a strong magnetic field in the presence of the Coulomb gap [247,254]; however, the data in Figure 2.62 clearly indicate $x=1/2$. Moreover, the data indicate $T_0'(H)/T_0' \propto H^p$ with $p=1.0$–1.2, consistent with Shklovskii's theory [247].

The typical H^2 dependence of MR, at low fields, in the VRH regime for PPy-PF$_6$ samples is shown in Figure 2.63. The large positive MR in the VRH regime is attributed to the shrinkage of the overlap of the tails of the wavefunctions [247]. Thus hopping becomes more difficult in the presence of the magnetic field, and as a result $\rho(H)$ increases significantly.

Recently Yoon et al. [226] observed that the gradual change of $S(T)$ from the positive linear temperature dependence to the negative U-shaped behavior is

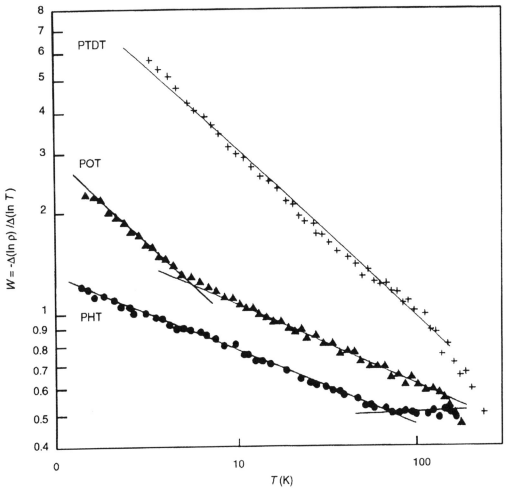

Figure 2.61. W vs. T for various insulating I–PAT samples (H-hexyl, O-octyl, TD-tetradecyl). The slope change indicates the crossover from Mott ($T^{-1/4}$) to ES ($T^{-1/2}$) at low temperatures. From ref. 226.

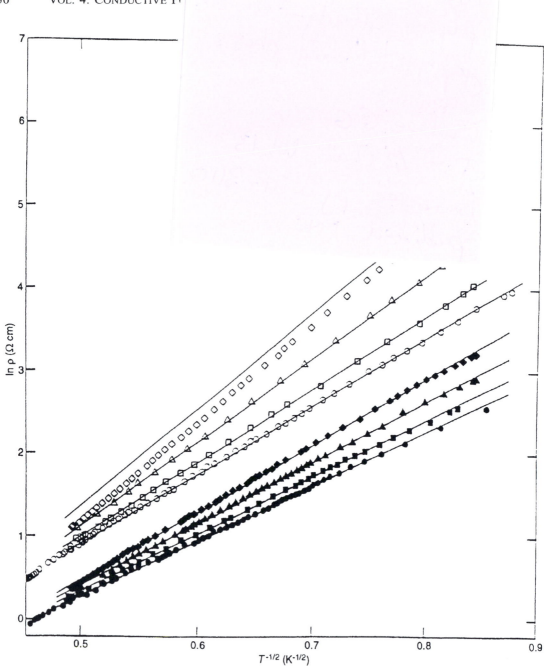

Figure 2.62. Ln ρ vs. $T^{-1/2}$ for insulating samples: PPy–PF$_6$ ($\rho_r = 734$, solid symbols) and I–POT ($\rho_r = 1640$, open symbols): $H = 0$ ($\bullet$, $\bigcirc$), 2 ($\blacksquare$, $\square$), 5 ($\blacktriangle$, $\triangle$) and 8 ($\blacklozenge$, $\diamondsuit$). From ref. 226.

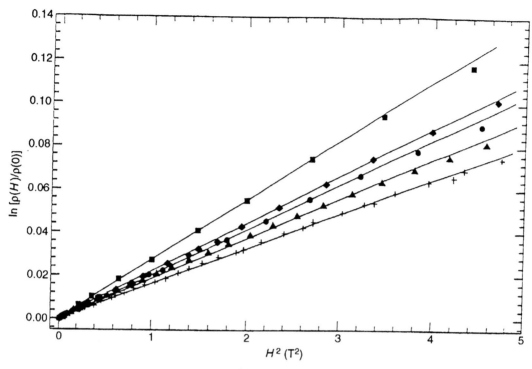

Figure 2.63. Low-field magnetoresistance ($\Delta\rho/\rho$) vs. H^2 for insulating PPy–PF$_6$ samples.

correlated with microstructure, and this is indicative of negative hopping contributions in addition to the metallic diffusion thermoelectric power. The hopping contribution to thermopower for various doped PANI samples is shown in Figure 2.64.

In summary, all features of hopping transport is observed in $\rho(T)$, MR and thermopower in conducting polymers on the insulating side of the M–I transition.

11 CHARGE TRANSPORT IN CONDUCTING POLYMER BLENDS

In recent years various types of conducting polymer blends have emerged [255,256]. In conventional conducting polymer blends, conducting materials like carbon black powder, metal particles, carbon or metal fibres, etc. are mechanically blended with insulating polymers like polyethylene, etc. [19]. The percolation threshold (f_c) and conducting properties of such systems are very much dependent on the processing conditions, size of particles, the aspect ratio (the ratio of length to diameter) of wires, etc. However, the processing problems and the cost effectiveness for obtaining high-quality homogenous conducting poly-

mer blends are still challenging issues for many potential applications. The increasing need of new materials for antistatic applications, electromagnetic interference (EMI) shielding, transparent electrical conductive coatings, etc. has stimulated substantial effort in this field. Another class of charge-transporting materials from molecularly doped polymers (MDP) is widely used in organic photoreceptors [17]; and more recently has found attractive applications in organic electroluminescent devices, photorefractive devices, synapse bond devices, etc. In this section the main focus is on conducting polymer blends prepared by mixing intrinsically conducting polymers (ICP) in a conventional insulating polymer matrix.

Although classical percolating systems (where $f_c \approx 16\%$ by volume fraction, for globular conducting objects dispersed in an insulating matrix) have been studied in detail for many years [257], the transport properties of conducting polymer blends consisting of a network of fibrillar conducting objects are in the early stages of investigation [256,258]. From previous experimental and theoretical studies of polymer composites filled with metal [259] or carbon fibres [260], it is known that the percolation threshold decreases when

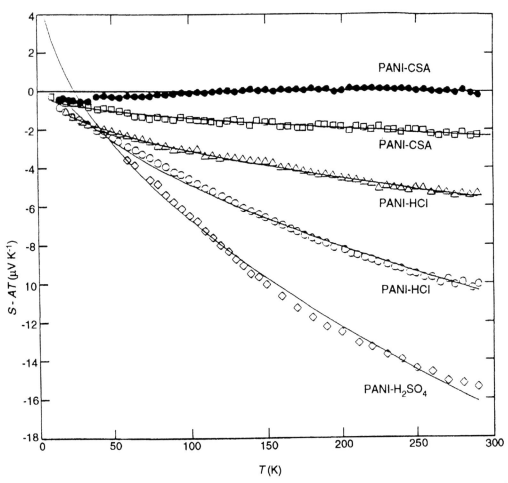

Figure 2.64. Hopping contribution to thermopower for various doped PANI samples [see equation (2.34)]. From ref. 226.

the aspect ratio of conducting object increases [258]. Although a percolation threshold as low as 0.1–0.05 vol. % has been reported for polymer composites consisting of nanoparticles of carbon black [54] or PANI [261], the conductivity near the percolation threshold for those samples is less than 10^{-6} S cm^{-1}. Moreover, the processibility, homogeneity, reproducibility and economic viability for mass producing the above systems (nanoparticulate composites) are challenging issues.

Recently, 'reticulate-doped' (*in situ* crystallization of a soluble organic electroconductor under non-equilibrium conditions in a vitrifying polymer matix) polymer composite films and interpenetrating network of ICP in insulating polymer matrices have made significant breakthroughs.

(1) Highly *conductive and transparant* films ($\sigma = 1$–5 S cm^{-1}) consisting of networks of tetraselenotetracene chloride radial cation salt [(TSeT)$_2$Cl, $\sigma(300 \text{ K}) = 2500$ S cm^{-1}, needle-like crystallites, aspect ratio $\sim 1000{:}1$] in a polymer matrix have been reported by Finter *et al* [262]. In this reticulate-doped system the percolation threshold is below 0.5% w/w. This is the first conductive, transparent and flexible polymer composite system with a *metallic conductivity* (positive TCR) in the range 400–60 K.

(2) Another reticulate-doped system consisting of 2 wt % of (BEDT-FFT)$_2$I$_3$ in a polycarbonate matrix shows a room-temperature conductivity around 1 S cm^{-1} [263]. These flexible and semitransparent reticulate-doped polymer films show a *metallic*

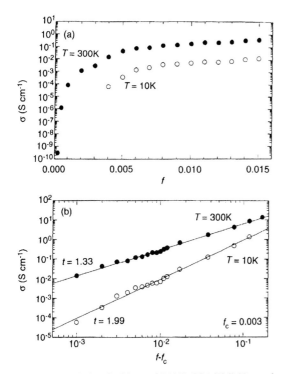

Figure 2.65. (a) Conducitivty of PANI–CSA/PMMA vs. the volume fraction of PANI–CSA; (b) the scaling fit [equation (44)] for PANI–CSA/PMMA blends. $f_c = 0.3 \pm 0.5\%$ and $t = 1.00 \pm 0.04$. (Reproduced by permission of the American Physical Society from ref. 252).

conductivity (positive TCR) in the range 4–300 K, and some samples even exhibited a precursor of superconductivity below 10 K.

(3) The self-assembled interpenetrating fibrillar network of PANI-CSA in a polymer matrix has a percolation threshold around 0.3 % w/w, and the room-temperature conductivity is around 10^{-3} S cm^{-1} [252,264]. Flexible, semi-transparent and homogenous films of any size and shape can be easily fabricated either by co-dissolving the conducting PANI-CSA and a suitable matrix polymer in a common solvent and casting onto a substrate or by melt processing the blend. This multiply-connected molecular-scale nanocomposite is the first real example of a ICP-polyblend.

In this section the main focus is on PANI-CSA blends in PMMA [252,264]. The details about the preparation of these blends, transmission electron microscopy (TEM), morphology, solution properties, transport properties, etc. are widely studied. The main

results of the transport property measurements are presented in this review.

The conductivity of PANI-CSA/PMMA blends is shown in Figure 2.65. The data fit to the scaling law of percolation [257]:

$$\sigma(f) \approx \sigma_T |f - f_c|^t \qquad (2.44)$$

where $\sigma_T \approx (r_h)\zeta_R \Sigma(r_h)$, which is interpreted as the conductance for each basic unit; t is the critical exponent ($t = 1$ in 2D and $t = 2$ in 3D); ζ_R is the resistivity scaling exponent ($\zeta_R = 0.975$ in 2D and 1.3 in 3D); r_h is the hopping length. The fit to equation (2.44) is shown in Figure 2.65b. For PANI-CSA/PMMA blends, $f_c = 0.3 \pm 0.05\%$ and $t = 1.99 \pm 0.04$, in agreement with the predicted universal value of $t = 2$ for percolation in 3D [257]. At room temperature, however, $t = 1.33 \pm 0.02$ (and $f_c = 0.3$). The smaller value of the exponent at room temperature arises from thermally induced hopping transport between disconnected (or weakly connected) parts of the network.

The temperature dependence of resistivity of PANI-CSA/PMMA blends is shown in Figure 2.66. The

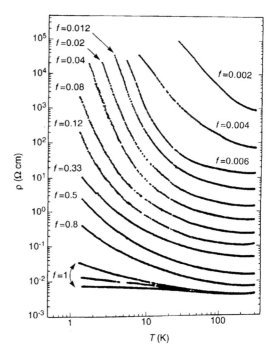

Figure 2.66. Resistivity vs. T for PANI–CSA/PMMA blends. (Reproduced by permission of Elsevier Science SA from *Synth. Met.* **69**, 1995, 255).

corresponding W vs. T plot is shown in Figure 2.67. The temperature dependence of the resistivity of PANI-CSA/PMMA blends can be classified in three different categories [252].

(1) $0.01 \leqslant f \leqslant 1$: the VRH exponent x increases systematically from 0.25 to 1.
(2) $0.006 \leqslant f \leqslant 0.01$: the VRH exponent, $x \approx 1$;
(3) $0.002 \leqslant f \leqslant 0.006$: $x \approx 1/2$.

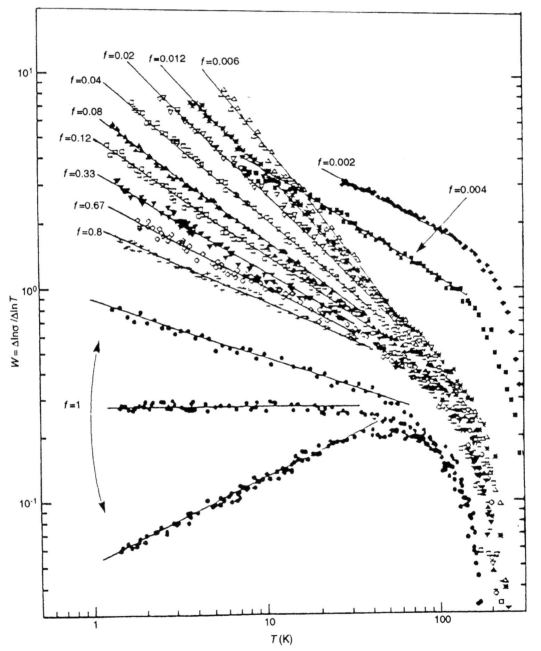

Figure 2.67. W vs. T for PANI–CSA/PMMA blends (same data as in Figure 2.66). (Reproduced by permission of Elsevier Science SA from *Synth. Met.* **69**, 1995, 255).

The systematic increase of the VRH exponent x upon decreasing the volume fraction of PANI-CSA is probably due to superlocalization in a fractal network [54,55]. Another interpretation of the systematic increase of x is possible if the VRH is occurring among localized wavefunctions with a fractal character as in equation (2.6) [56]. Activated conductivity ($\ln \rho \propto 1/T$ dependence) is usually observed when the dominant contribution to charge transport takes place by nearest neighbor hopping. This could occur for samples $0.006 \leqslant f \leqslant 0.01$, if the hopping length becomes comparable to the diameter of the fibrillar links [252].

The exponent, $x(f)$, goes through a maximum at the percolation threshold; for samples containing volume fractions of PANI-CSA below 0.5%, the exponent decreases rapidly from 1 to 0.45 ± 0.05, as shown in the W vs. T plot (Figure 2.67). Dramatic change in the transport properties occurs near the percolation threshold, where the connectivity of the PANI-CSA network breaks up. When the volume fraction of PANI-CSA decreases below 0.5%, the fibrillar diameter of the links between multiply-connected regions decreases until the connected network cannot be sustained. Precisely at the point where the morphology changes, the charge transport undergoes a transition to that typical of granular metallic systems. The $\ln \sigma \propto T^{-1/2}$ dependence for samples containing volume fractions of PANI-CSA below the percolation threshold is typical of granular metals [252].

The behavior of MR for PANI-CSA/PMMA blends is shown in Figure 2.68. The magnitude of positive MR shows a temperature-dependent maximum upon decreasing the volume fraction of PANI-CSA. Above 4.2 K the MR is rather low for both 100% PANI-CSA and blends. At 4.2 K, the MR is maximum at 1.5% PANI-CSA, and at 1.4 K, the MR is maximum at 8% PANI-CSA. The increase in MR upon dilution is consistent with the VRH model since the overlap of the wavefunctions of the localized states decreases due to the superlocalization of the wavefunctions on the fractal network upon decreasing the volume fraction of PANI-CSA. It seems that the maximum in MR as a function of f occurs when the hopping length and the diameter of the links becomes rather similar. At temperatures below 4.2 K, the hopping length continues to increase, and the maximum in MR shifts to higher volume fractions of PANI-CSA (larger diameter of the fibrillar links) both of which are consistent with the observations at 1.4 K. This is also consistent with the $\ln \rho \sim 1/T$ dependence for samples containing volume fractions of PANI-CSA from 1 to 0.5%. The rapid decrease in MR on

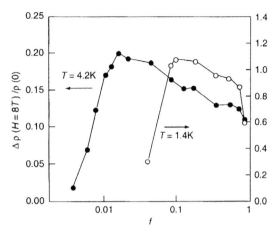

Figure 2.68. Magnetoresistance $[\Delta\rho(H=8t)/\rho(0)]$ vs. the volume fraction (f) of PANI-CSA in PMMA at 4.2 and 1.4 K for $1 < f < 0.004$. (Reproduced by permission of the American Physical Society from ref. 252).

approaching the percolation threshold is in agreement with effective medium theory [265]

Although the temperature dependences of electrical conductivity and MR in PANI-CSA/PPMA blends are sensitive to dilution, the temperature dependence of thermopower remains linear at high temperatures (metallic behavior), as shown in Figure 2.69 [225,264]. However, below 80 K there is slight deviation from linearity and the magnitude of this deviation increases as f decreases. The observation of the weak U-shaped contribution to $S(T)$ at low volume fractions of PANI-CSA is consistent with the existence of large-scale inhomogeneity (metallic islands) in PANI-HCl; the U-shape results from a hopping contribution to $S(T)$ [226].

In summary, the self-assembled interpenetrating network of PANI-CSA results in a low percolation threshold (0.3% w/w) with rather high conductivity at the threshold in comparison with other percolating systems. The conductivity near percolation threshold is 0.003 S cm^{-1} at room temperature. The value of x in the $\ln \sigma \propto T^{-x}$ dependence increases systematically, from 0.25 to 1, upon dilution. this suggests that the exponent depends on the complex morphology of the network, perhaps due to superlocalization on the fractal network near the percolation threshold. In the disconnected regime below the percolation threshold, the $\ln \sigma \propto T^{-1/2}$ dependence is typical of granular metals. The positive MR increases upon decreasing the volume fraction of PANI-CSA, and when the fibrillar diameter and the hopping length become comparable, the MR decreases rapidly. Although thermopower remains

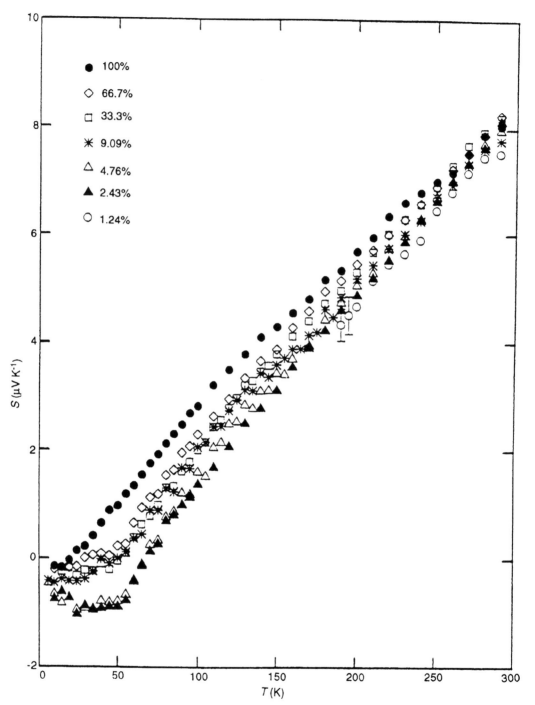

Figure 2.69. Thermopower vs. *T* for PANI–CSA/PMMA blends. (Reproduced by permission of Elsevier Science SA from *Synth. Met.* **69**, 1995, 255).

linear at temperatures above 100 K upon dilution, the U-shaped feature at lower temperatures for concentrations near the percolation threshold indicates the importance of the contribution from hopping transport. Various applications of this self-assembled network of conducting PANI such as transparent conducting films, carrier injecting electrodes for LEDs and polymer grid triodes, etc. have been reviewed by Heeger [256].

12 CONCLUDING REMARKS

Considerable progress has been achieved in understanding the electronic properties of conjugated polymers in the past few years [266]. This is mainly due to the structural and morphological improvement in the materials.

The transport studies indicate that conducting polymers are metallic, however, the disorder-induced localization plays a major role in limiting the metallic properties. The metallic nature of high-quality conducting polymers is shown from the following experimental facts: large finite conductivity at mK temperatures, linear thermopower, quantum corrections (weak localization and e–e interaction) to MC, Pauli susceptibility, a linear term in the specific heat, large negative dielectric constant in the microwave frequencies, plasma edge in the infrared, metallic (Drude-like) reflectivity in the infrared. The following conducting polymers show some of the above mentioned metallic features: p and n-type doped $(CH)_x$, p and n-type doped PPV, PPy-PF_6, PANI-CSA and ion-implanted PANI.

Although in doped $(CH)_x$ and PPV samples the room temperature conductivity along the chain axis is greater than 10^4 S cm^{-1} the absence of a positive TCR (except in PANI-CSA and $FeCl_3$-$(CH)_x$ samples above 180 K) indicates that both charge transport perpendicular to the chain axis and disorder are the main factors that limit the positive TCR. If this behavior of negative TCR in metallic conducting polymers is compared to positive TCR observed in reticulate-doped polymeric systems [2% w/w $(TSeT)_2Cl$ or $(BEDT-TTF)_2I_3$ in polycarbonate matrix [262,263], with a room temperature conductivity of 1–20 S cm^{-1}], then it is quite obvious that it is the disorder in both crystalline and amorphous phases of the conducting polymers that really limits the positive TCR.

The large finite conductivity in metallic conducting polymers at mK temperatures suggests that these systems are not susceptible to the pathologies of 1D systems (for example, Peierls transition, Anderson localization in 1D, etc.). This clearly shows that the interchain interaction, counterion-mediated transport, long-range Coulomb interactions, etc, are making the system more 3D. In this sense metallic conducting polymers are very much different from other low-dimensional systems. Indeed, the 3D nature of charge transport in oriented metallic conducting polymers ($\sigma_\parallel/\sigma_\perp \approx 100$) can be inferred from the close similarities of $\sigma(T)$ and MC in both parallel and perpendicular directions to the chain axis. The similarities in MC is especially important, since it is a local probe for charge transport and scattering mechanisms.

A semiquantitative level of understanding of $\sigma(T)$ and MC in metallic conducting polymers can be made within the framework of the weak localization/e–e interaction model for disordered metallic systems. However, the consequences of anisotropic carrier mass, anisotropic diffusion coefficient, etc. in this model are not known very well. Moreover, this model is used under the assumption that the Fermi liquid theory is valid for the metallic state of conducting polymers. Thus more detailed work is essential to prove conclusively that this approach is fully valid. The anomalous sign of the thermopower and Hall coefficient in some of the metallic conducting polymers may give special insight into this problem. Moreover, some anomalies have been observed in the infrared reflectance of metallic conducting polymers.

A positive TCR below 20 K has been observed in metallic samples of K-$(CH)_x$, I-$(CH)_x$, PANI-CSA, H_2SO_4-PPV and ion-implanted PANI. The exact reason for this anomalous conductivity cusp is not very well know. In the localization–interaction model, this change in TCR is due to the sign change in $[(4/3)-(3\gamma F_\sigma/2)]$. Usually, the sign of m is negative when $\gamma F_\sigma > 8/9$. Alternatively, in a polaronic metallic system, a similar positive TCR at low temperatures is possible due to the intersite tunnel percolation of polarons. The conductivity cusp is suppressed by magnetic field and it is enhanced by pressure. This conductivity cusp could provide some important clues about the metallic state of conducting polymers.

The linear $S(T)$ in conducting polymers near the M–I transition is typical of that expected due to metallic diffusion thermopower. However, as the extent of the disorder increases and the system moves well into the insulating regime, the hopping contribution to the thermopower appears at low temperatures. The temperature-independent Pauli susceptibility for metallic conducting polymers indicates the presence of the finite density of the states at the Fermi level. As the extent of disorder increases, a small Curie contribution appears at low temperatures. The metallic reflection, plasma edge and free carrier absorption in the infrared shows the presence of the metallic state in high-quality conducting polymers. Moreover, the observation of the modified Korringa relation from ^{13}C NMR in PANI-CSA samples gives evidence for the metallic state in high-quality samples. All the above mentioned experimental results really substantiate the metallic state in conducting polymers. However, more work is essential to understand the precise nature and the characteristic features of this metallic state.

Epstein et al. [109] have observed two 'plasma frequencies' associated with the response from the 'most delocalized carrier' and from the 'entire carriers in the conduction band' in conducting polymers. This is considered as evidence for the inhomogenous disorder in conducting polymers. The relatively low value of ω_p [100–200 cm^{-1} or 0.025–0.012 eV] in metallic conducting polymers indicates that only a small fraction of the conduction band carriers are fairly delocalized and the low-frequency microwave response is mainly from these highly delocalized carriers. Epstein et al. suggested that the anomalously long time for τ (10^{-11} s) arises from the open Fermi surface of highly anisotropic systems, since the essential mechanism for the relaxation of carrier momentum is associated with the backscattering from k_F to $-k_F$. However, Lee et al. [236] suggested that the localization-modified Drude model can explain the static and dynamic properties of the infrared response in metallic conducting polymers. All these suggest that the infrared response of metallic conducting polymers is a very interesting probe to unravel the mysteries of the metallic state in these materials.

The W vs. T plots have considerably helped to understand the behavior of $\sigma(T)$; they have been especially useful in precisely identifying the metallic [positive temperature coefficient of $W(T)$], critical [temperature-independent $W(T)$] and insulating [negative temperature coefficient of $W(T)$] regimes. Tuning the critical regime of the M–I transition by pressure and magnetic field has provided an unique opportunity to investigate the nature of the wavefunction of localized states near the mobility edge, which is not easily accessible in other systems. The field-induced transitions from a critical to an insulating regime and from a metallic to a critical regime are very interesting. These subtle tuning effects on wavefunctions near the mobility edge have not been observed in any other systems before. However, the theoretical understanding of this tuning near the critical regime of the M–I transition by pressure and magnetic field is not very well understood.

The behavior of charge transport in the insulating regime is consistent with the VRH model. For systems in the insulating regime near the M–I transition, a crossover from $T^{-1/4}$ to $T^{-1/2}$ has been observed below 10 K, which indicates the presence of a Coulomb gap. Whereas, the observation of a $T^{-1/2}$ dependence (without any crossover) in a wide range of temperature indicates the typical charge transport in granular metallic systems. This is clearly shown in case of PANI-CSA/PMMA blends. This behavior is also observed in highly disordered and inhomogenously doped conducting polymers.

In the insulating regime the positive MR increases as ρ_r increases. This is mainly due to the shrinkage of the overlap of the wavefunctions, and consequently the probability for hopping transport decreases. The negative U-shaped behavior in low-temperature thermopower is indicative of the negative hopping contribution in addition to the weak metallic diffusion thermoelectric power in the insulating regime. The Curie contribution to the temperature dependence of susceptibility is enhanced when the system becomes more insulating. However, both $S(T)$ and $\chi(T)$ are not as sensitive as $\sigma(T)$ and MR when the system moves to the insulating regime.

The intrinsically conducting polymer blends (ICP) are exciting novel systems for both fundamental studies and various applications [267]. The fractal morphology of self-assembled PANI-CSA networks and the charge transport in such fractal networks are interesting topics for further investigations. For PANI-CSA/PMMA blends, the percolation threshold is rather low (0.3% w/w) and the conductivity at percolation threshold is relatively high (10^{-3} S cm^{-1}). These interesting features have made these systems very attractive for various applications.

13   ACKNOWLEDGEMENTS

I thank Professor A.J. Heeger for reading this manuscript. Moreover, I very much value his suggestions and discussions over the last four years. Thus, I acknowledge all his support in the completion of this review article. I thank all my past and present colleagues for their help in various levels of work over the last four years. I especially acknowledge Professor F. Wudl, Dr D. Moses, Dr K. Vakiparta, Dr Yong Cao, Dr C.O. Yoon, Dr N.S. Sariciftci, Dr Kwanghee Lee, Dr C.Y. Yang, Dr M. Ahlskog. My special thanks go to Professor T. Ishiguro for sending a copy of his paper before publication and for some discussions, Dr A.

Aleshin for sending me a figure of his most recent work before publication, and Professor Patrick Bernier for some discussions regarding doping in conducting polymers. I also thank NSF (MRL: DMR-9123048) for supporting the transport measurements in our laboratory at UCSB.

14 REFERENCES

1. H. Shirakawa, E.J. Louis, A.G. MacDiarmid, C.K. Chiang and A.J. Heeger, *J. Chem. Soc., Chem. Commun.*, 578 (1977); C.K. Chiang, C.R. Fincher, Y.W. Park, A.J. Heeger, H. Shirakawa, E.J. Louis, S.C. Gua and A.G. MacDiarmid, *Phys. Rev. Lett.*, **39**, 1098 (1977); C.K. Chiang, M.A. Druy, S.C. Gua, A.J. Heeger, H. Shirakawa, E.J. Louis, A.G. MacDiarmid and Y.W. Park, *J. Am. Chem. Soc.*, **100**, 1013 (1978).

2. Reghu M. C.O. Yoon, D. Moses and A.J. Heeger, in *Handbook of Conducting Polymers,* 2nd Ed. (Eds. T.A. Skotheim, R.L. Elsenbaumer and J.R. Reynolds), Dekker, New York, 1997.

3. S. Roth and H. Bleier, *Adv. Phys.,* **36**, 385 (1987); S. Roth, in *Hopping Transport in Solids,* (Eds M. Pollak and B. Shklovskii) North-Holland, Amsterdam 1991, p. 377, and references therein; *Conducting Polymers— Transport Phenomena* (Eds. J. Przyluski and S. Roth), Trans Tech Pub., Brookfield, 1993; S. Roth, *One Dimensional Metals,* VCH, New York 1995.

4. For a review see K. Kaneto, K. Yoshino and Y. Inuishi, in *Electronic Properties of Inorganic Quasi-One-Dimensional Materials,* Vol II (Ed. P. Monceau), D. Reidel, Dordrecht, 1985 pp. 69–109, and references therein.

5. W.A. Little, *Phys. Rev.* **134**, A1416 (1964); *Sci. Am.*, **212**, 21 (1965).

6. (a) F.J. London, *J. Phys. Radium* **8**, 397 (1937).
   (b) R. Pethig, *Dielectric and Electronic Properties of Biological Materials* John Wiley & Sons, New York 1979, p.340.

7. J. Ladik, G. Biczo and J. Redly, *Phys. Rev.* **188**, 710 (1969).

8. R.M. Pearlstein, *Phys. Rev. Letts.* **20**, 594 (1968).

9. J.T. Devreese and V.G. van Doren (eds.) Papers in *Highly Conducting One-Dimensional Solids Plenum 1979.*

10. (a) T. Ishiguro and K. Yamaji, *Organic Superconductors* Springer, Berlin 1990 and references therein.
    (b) For a reveiw see D. Jerome and H.J. Schulz, *Adv. in Phys.* **31**, 299 (1982).
    (c) K. Kajita, Y. Nishio, S. Moriyama, W. Sasaki, R. Kato, H. Kobayashi and A. Kobayashi, *Solid State Commun.* **64**, 1279 (1987).

11. For a reveiw see D.M. Pai and B.E. Springett, *Rev. of Mod. Phys.* **65**, 163 (1993) and references therein.

12. For a review see G. Gruner, *Rev. of Mod Phys.* **60**, 1129 (1988) and references therein.

13. For a review see G. Gruner, *Rev. of Mod. Phys.* **66**, 1, (1994) and references therein.

14. For a review see A.J. Heeger, S. Kivelson, J.R. Schrieffer and W.P. Su, *Rev. Mod. Phys.* **60**, 781 (1988) and references therein; A.J. Heeger, in *Handbook of Conducting Polymers*, (Ed. T.A. Skotheim), Dekker, New York, p. 729, 1986.

15. H.S. Nalwa, *Ad. Mat.* **5**, 341 (1993); H.S. Nalwa and S. Miyata, eds, *Nonlinear Optics of Organic Molecules and Polymers*, CRC Press, Boca Raton, Florida, 1997.

16. Papers in *Synth. Met.* Vol. **65**, July 1994 (NEOME Polymer LED Minisymposium Sept. 1993, Eindhovan, Netherlands); S. Miyata and H.S. Nalwa, eds, *Organic Electroluminescent Materials and Devices*, Gordon & Breach, 1997.

17. (a) R.H. Young, J.A. Sinicropi, J.J. Fitzgerald, *J. Phys. Chem.* **99**, 9497 (1995).
    (b) H. Bassler, *Phys. State. Sol. (b)* **175**, 15 (1993).
    (c) P.M. Borsenberger and D.S. Weiss, *Organic Photoreceptor for Imaging Systems*, Dekker, New York 1993.

18. A. Graja, *Low-Dimensional Organic Conductors*, World Scientific, Singapore 1992.

19. E.K. Sichel, *Carbon Black–Polymer Composites*, Dekker, New York 19982.

20. T.A. Skotheim (ed.), Papers in *Handbook of Conducting Polymers*, Vols. 1 and 2, Dekker, New York 1986.

21. R. Hoffmann, C. Janiak and C. Kollmar, *Macromolecules* **24**, 3725 (1991).

22. (a) H. Ehrenreich and F. Spaepen (eds), *Solid State Properties of Fullerenes,* Academic, Boston 1994.
    (b) H. Kuzmany, J. Fink, M. Mehring and S. Roth (eds.), Papers in *Progress in Fullerene Research*, World Scientific, Singapore, 1994.

23. R.E. Peierls, *Quantum Theory of Solids*, Oxford University Press, London 1955.

24. W.P. Su, in *Handbook of Conducting Polymers* Vol. 2 (Ed. T.A. Skotheim) Dekker, New York p. 757. 1986; R.R. Chance, D.S. Boudreaux, J.L. Bredas and R. Silbey, ibid p.825; D.K. Campbell, A.R. Bishop and M.J. Rice, ibid p.937.

25. S.A. Brazovskii and N.N. Kirova, *Sov. Sci. Rev. A. Phys.* **5**, 99 (1984) and references therein.

26. Y. Lu (ed.), *Solitons and Polarons in Conducting Polymers*, World Scientific, Singapore, 1988.

27. H.G. Kiess (ed.), *Conjugated Conducting Polymers*, Springer Series in Solid State Sciences, Vol. 102, Springer, Berlin, 1992.

28. N.F. Mott, *Metal–Insulator Transition*, 2nd ed., Taylor Francis, London, 1990 and references therein.

29. P.W. Anderson, *Phys. Rev.* **109**, 1492 (1958).

30. J. Fröhlich, *Proc. R. Soc.* A, **223**, 296 (1954).

31. W. Kohn, *Phys. Rev. Lett* **2**, 393 (1959).

32. P. Monceau, in *Electronic Properties of Inorganic Quasi-One-Dimensional Materials,* (Ed. P. Monceau,

Vol. II, Reidel, Dordrecht pp. 139–257 and references therein.

33. P. Phillips and H.L. Wu, *Science* **252**, 1805 (1991); P.Phillips, *Annu. Rev. Phys. Chem.* **44**, 115 (1993); D.H. Dunlap, H.L. Wu and P. Phillips, *Phys. Rev. Lett* **65**, 88 (1990).

34. Proc. of the Int. Conf. on Materials and Mechanisms of High T_c Superconductors, *Physica C,* Vol. **235–240**, 1994.

35. D.G. Clarke, S.P. Strong and P.W. Anderson *Phys. Rev. Letts.* **74**, 4499 (1995).

36. A. Sekiyama, A. Fujimori, S. Aonuma, H. Sawa and R. Kato, *Phys. Rev.* B**51**, 13899 (1995) and references therein.

37. J. Voit, *Synth. Met.* **70**, 1015 (1995).

38. For a review see D. Baeriswyl, D.K. Campbell and S. Mazumdar in *Conjugated Conducting Polymers* Vol. 102, (Ed. H.G. Kiess), Springer, Berlin, and references therein.

39. E. Jeckelmann and D. Baeriswyl, *Synth. Met.* **65**, 211 (1994) and references therein.

40. E. Abrahams, P.W. Anderson, d.C. Licciardello and T.V. Ramakrishnan, *Phys. Rev. Lett.* **42**, 695 (1979).

41. J.L. Skinner, *J. Phys. Chem.* **98**, 2503 (1994).

42. A.F. Ioffe and A.R. Regel, *Prog. Semicond.* **4**, 237 (1960).

43. A. Möbius, *J. Phys. C: Solid State Phys.* **18**, 4639 (1985); *Phys. Rev.* B**40**, 4194 (1989).

44. P.A. Lee and T.V. Ramakrishnan, *Rev. Mod. Phys.* **57**, 287 (1985) and references therein.

45. D. Belitz and T.R. Kirkpatrick, *Rev. Mod. Phys.* **66**, 261 (1994).

46. N.F. Mott and E.A. David, *Electronic Process in Noncrystalline Materials*, Oxford University Press, Oxford, 1979.

47. P.W. Anderson, *Comments Solid State Phys.* **2**, 193 (1970).

48. K. Tenelsen and M. Schreiber, *Phys. Rev.* B**49**, 12662 (1994).

49. (a) Y.E. Levy and B. Souillard, *Europhys. Lett.* **4**, 233 (1987).
(b) G. Deutscher, Y.E. Levy and B. Souillard, ibid. **4**, 577 (1987).

50. A. Ahorony and A.B. Harris, *Physica* **163**, 38 (1990); ibid. **205**, 335 (1994).

51. A. Aharony and A.B. Harris, *Physica* A**191**, 365 (1992).

52. M. Schreiber and H. Grussbach, *Mod. Phys. Lett* B**6**, 851 (1992); H. Grussbach and M. Schreiber, *Physica* A**191**, 394 (1992); H. Grussbach and M. Schreiber, *Chem. Phys.* **177**, 733 (1993); *Phys. Rev. Lett* **51**, 663 (1995).

53. A.B. Harris and A. Ahorony, *Europhys. Lett.* **4**, 1355 (1987).

54. van der Putten, J.T. Moonen, H.B. Brom, J.C.M. Brokken-Zijp and M.A.J. Michels, *Phys. Rev. Lett* **69**, 494 (1992); ibid. **70**, 4161 (1993)

55. A. Aharony, O. Entin-Wohlman and A.B. Harris, *Physica* A**200**, 171 (1993); *Phys. Rev. Lett.* **70**, 4160 (1993).

56. H. Aoki, *J. Phys. C: Solid State Phys.* **16**, L205 (1983); *Phys. Rev.* B**33**, 7310 (1986).

57. M. Schreiber, *Phys. Rev.* B**31**, 6146 (1985).

58. C.M. Soukoulis and E.N. Economou, *Phys. Rev. Lett.* **52**, 565 (1984).

59. M. Schreiber and H. Grussbach, *Philos. Mag.* B**65**, 707 (1992).

60. H. Bässler, M. Deußen, S. Heun, U. Lemmer and R.F. Mahrt, *Zeit. fur Phys. Chemie,* **184**, 233 (1994); V.I. Arkhipov, H. Bässler, M. Deussen, E.O. Gobel, R. Kersting, H. Kurz, U. Lemmer and R.F. Mahrt, *Phys. Rev.* B**52**, 4932 (1995).

61. K. Pakbaz, C.H. Lee, A.J. Heeger, T.W. Hagler and D. McBranch, *Synth. Met.* **64**, 295 (1994).

62. A. Takahashi, *Phys. Rev.* B**51**, 16479 (1995); ibid, **46**, 11550 (1992); *Prog. Theor. Phys.* **81**, 610 (1989).

63. J.A. Pople and S.H. Walmsley, *Mol. Phys.* **5**, 15 (1962).

64. K. Pakbaz, R. Wu, F. Wudl and A.J. Heeger, *J. Chem. Phys.* **99**, 590 (1993).

65. D. Moses and A.J. Heeger, in *Relaxation in Polymers* (Ed. T. Kobayashi), World Scientific, Singapore, 1993.

66. L. Rothberg, T.J. Jedju, S. Etemad and G.L. Baker, *Phys. Rev.* B**36**, 7529 (1987); L. Rothberg, T.J. Jedju, P.D. Townsend, E. Etemad and G.L. Baker, *Phys. Rev. Lett.* **65**, 100 (1992).

67. C. Halverson, T.W. Hagler, D. Moses, Y. Cao and A.J. Heeger, *Chem. Phys. Lett.* **200**, 364 (1992); *Synth. Met.* **49**, 49 (1992).

68. J. Reichenbach, M. Kaiser and S. Roth, *Phys. Rev.* B**48**, 14104 (1993).

69. M. Logdlund, P. Dannetun, S. Stafström, W.R. Salaneck, M.G. Ramsey, C.W. Spangler, C. Fredriksson and J.L. Brédas, *Phys. Rev. Lett.* **70**, 970 (1993).

70. E. Ehrenfreund, D. Moses, A.J. Heeger, J. Cornil and J.L. Brédas, *Chem. Phys. Lett.* **196**, 84 (1992).

71. S. Kivelson, *Phys. Rev. Lett.* **46**, 1344 (1981); *Phys. Rev.* B**25**, 3798 (1982).

72. A.J. Epstein, in *Handbook of Conducting Polymers*, Vol.2, (Ed. T.A. Skotheim), Marcel Dekker, New York, p.1041.

73. J.H. Kaufman, J. Kaufer, A.J. Heeger, R. Kaner and A.G. MacDiarmid, *Phys. Rev.* B**26**, 4 (1983); J.H. Kaufman, T. C. Chung and A.J. Heeger, *J. Electrochem. Soc.* **131**, 2847 (1984).

74. W.P. Su and J.R. Schrieffer, *Proc. Natl. Acad. Sci. USA* **77**, 201 (1980).

75. S.A. Brazovskii and N.N. Kirova, *Pis'ma Zu. Eksp. Teor. Fiz.* **33**, 6 (1981) [*JETP Lett.* **34**, 4 (1981)].

76. D.K. Campbell, A.R. Bishop and K.Fesser, *Phys. Rev.* B**26**, 6862 (1982); K. Fesser, A.R. Bishop and D.K. Campbell, ibid, **27**, 4808 (1983); D.K. Campbell, A.R. Bishop and M.J. Rice, *Handbook of Conducting Polymers* Vol 2, (Ed. T.A. Skotheim), Dekker, New York, p.937.

77. Y. Shimoi and S. Abe, *Phys. Rev.* **B50**, 14781 (1994).
78. (a) M. Nechtschein, R. Devreux, F. Genoud, E. Vieil, J.M. Pernaut and E. Genies, *Synth. Met.* **15**, 59 (1986); (b) P. Le Guennec, M. Nechtschein and J.P. Travers, *Synth. Met.* **55–57**, 630 (1993).
79. K. Mizoguchi, T. Obana, S. Ueno and K. Kume, *Synth. Met.* **55–57**, 601 (1993).
80. M. Chandross, S. Mazumdar, S. Jeglinski, X. Wei, Z.V. Vardeny, E.W. Kwock and T.M. Miller, *Phys. Rev. B* **50**, 14702 (1994).
81. (a) R.R. Chance, J.L. Bredas and R. Silbey, *Phys. Rev.* **B29**, 4491 (1984); J.L. Bredas and G.B. Street, *Ac. Chem. Res.* **18**, 309 (1985). (b) P. Kuivalainen, H. Stubb, H. Isotalo, P. Yli-Lahti and C. Holmström, *Phys. Rev.* **B31**, 7900 (1985).
82. H. Bottger and V.V. Bryksin, *Hopping Conduction in Solids*, Deerfield, Florida, 1985; V.V. Bryksin and V.S. Voloshin, *Fiz. Tverd. Tela* **26**, 2357 (1984) [*Sov. Phys. Solid State* **26**, 1429 (1984)].
83. P. Vogl and D.K. Cambell, *Phys. Rev.* **B41**, 12797 (1990).
84. H.A. Mizes and E.M. Conwell, *Phys. Rev.* **B43**, 9053 (1991).
85. S. Stafström, *Phys. Rev.* **B47**, 12437 (1993).
86. P.M. Grant and I.P. Batra, *Solid State Commun.* **29**, 225 (1979).
87. U. Lemmer, S. Karz, M. Scheidler, M. Deussen, W. Rieß, B. Cleve, P. Thomas, H. Bassler, M. Schwoerer and E. Gobel, *Synth. Met.* **67**, 169 (1994); R. Kersting, U. Lemmer, M. Deussen, H.J. Bakker, R.F. Mahrt, H. Kurz, V.I. Arkhipov, H. Bassler and E. Gobel, *Phys. Rev. Lett.* **73**, 1440 (1994).
88. E.M. Conwell and H.A. Mizes, *Phys. Rev.* **B51**, 6953 (1995).
89. I.J. Campbell, T.W. Hagler, D.L. Smith and J.P. Ferraris *Phys. Rev. Lett.* **76**, 1900, 1996.
90. D. Moses, C.H. Lee, A.J. Heeger, T. Noguchi and T. Ohnishi, *Phy. Rev.* **B54**, 4748, 1996.
91. M. Pope and C.E. Swenberg, *Electronic Process in Organic Crystals,* Clarendon Press, Oxford, 1982.
92. Proc. of the 2nd Inter. Conf. on Optical probes of Conjugated Polymers and Fullerens, *Mol. Cry. Liq. Cryst.* **256**, (1994).
93. (a) H. Zabel and S.A. Solin (eds.), *Graphite Intercalation Compounds II,* Springer, Berlin, 1992; (b) P. Bernier (ed.), Papers in *Chemical Physics of Intercalation II,* Plenum, New York, 1993.
94. G.J. Morgan and C. Uher, *Phil. Mag.* **B44**, 427 (1981).
95. (a) T.A. Friedmann, M.W. Rabin, J. Giapintzakis, J.P. Rice and D.M. Ginsberg, *Phys. Rev.* **B42**, 6217 (1990). (b) U. Welp, S. Fleshler, W.K. Kwok, J. Downey, Y. Fang, G.W. Crabtree and J.Z. Liu, *Phys. Rev.* **B42**, 10189 (1990).
96. (a) Z. Schlesinger, R.T. Collins, F. Holtzberg, C. Field, S.H. Blanton, U. Welp, G.W. Crabtree, Y. Fang and J.Z. Liu, *Phys. Rev. Lett.* **65**, 801 (1990).
(b) J. Schutzmann, S. Tajima, S. Miyamoto and S. Tanaka, *Phys. Rev. Lett.* **73**, 174 (1994).
97. (a) V.N. Prigodin and Y. A. Firsov, *Pisma Zh. Eksp. Teor. Fiz.* **38**, 241 (1983) [JETP Lett. **38**, 284 (1983). (b) E.P. Nakhmedov, V.N. Prigodin and A.N. Samukhin, *Fiz. Tverd. Tela* **31**, 31 (1989) [*Sov. Phys. Solid State* **31**, 368 (1989)].
98. Y.A. Firsov in *Localization and Metal–Insulator Transition* (Eds. H. Fritzsche and D. Adler,) Plenum, New York, 1985, p.477.
99. S. Stafström, *Synth. Met.* **65**, 185 (1994); ibid, **69**, 667 (1995); *Phys. Rev.* **B51** 4137 (1995).
100. M. Buttiker, Y. Imry, R. Landauer and S. Pinhas, *Phys. Rev.* **B31**, 6207 (1985).
101. E.M. Conwell and H.A. Mizes, *Synth. Met.* **55–57**, 4284 (1993); E.M. Conwell H.Y. Choi and S. Jeyadev, *Synth. Met.* **49–50**, 359 (1992).
102. V.N. Prigodin and K.B. Efetov, *Synth. Met.* **65**, 195 (1994).
103. M.I. Salkola and S.A. Kivelson, *Phys. Rev.* **B50**, 13962 (1994).
104. D. Comoretto, G. Dellepiane, G.F. Musso, R. Tubino, R. Dorsinville, A. Walser and R.R. Alfano, *Phys. Rev.* **B46**, 10041 (1992); D. Comoretto, R. Tubino, G. Dellepiane, G.F. Musso, A. Borghsei, A. Piaggi and G. Lanzani, *Phys. Rev.* **B41**, 3534 (1990).
105. G. Leising, *Phys. Rev.* **B38**, 10313 (1988); *Synth. Met.* **28D**, 215 (1989).
106. Reghu M., K. Vakiparta, Y. Cao and D. Moses, *Phys. Rev.* **B49**, 16162 (1994); C.O. Yoon, Reghu M., A.J. Heeger, E.B. Park, Y.W. Park, K. Akagi and H. Shirakawa, *Synth. Met.* **69**, 70 (1995).
107. M. Ahlskog, Reghu M, A.J. Heeger, T. Noguchi and T. Ohnishi, *Phy. Rev.* **B53**, 15529, 1996.
108. K. Mizoguchi, *Jpn. J. Appl. Phys.* **34**, 1 (1995) and references therein.
109. A.J. Epstein, J. Joo, R.S. Kohlman, G. Du, A.G. MacDiarmid, E.J. Oh, Y. Min, J. Tsukamoto, H. Kaneko and J.P. Pouget, *Synth. Met.* **65**, 149 (1994); Z. Wang, A. Ray, A.G. MacDiarmid and A.J. Epstein, *Phys. Rev.* **B43**, 4373 (1991).
110. Y. Cao, P. Smith and A.J. Heeger, *Synth. Met.* **48**, 91 (1992); Y. Cao and A.J. Heeger, ibid. **52**, 193 (1992); Y. Cao, J.J. Qiu and P. Smith, ibid, **69**, 187, 191 (1995); Y. Cao, P. Smith and A.J. Heeger, in *Conjugated Polymeric Materials: Opportunities in Electronics, Optoelectronics and Molecular Electronics* (Eds. J.L. Bredas and R.R. Chance, Kluwer, Dordrecht, 1990.
111. (a) T. Hagiwara, M. Hirasaka, K. Sato and M. Yamaura, *Synth. Met.* **36**, 241 (1990). (b) K. Sato, M. Yamaura, T. Hagiwara, K. Murata and M. Tokumoto, ibid. **40**, 35 (1991).
112. D. Chen, M.J. Winokur, Y. Cao, A.J. Heeger and F.E. Karasz, *Phys. Rev.* **B45**, 2035 (1992); M.J. Winokur, J. Maron, Y. Cao and A.J. Heeger, ibid. **45**, 9656 (1992); T.J. Prosa, M.J. Winokur, J. Moulton, P. Smith and A.J. Heeger, ibid. **51**, 159 (1995); P. Papanek, J.E. Fisher,

J.L. Sauvajol, A.J. Dianoux, G. Mao, M.J. Winokur and F.E. Karasz, ibid. **50**, 15668 (1994).

113. R.J. Baughman, N.S. Murthy and H. Eckhardt, *Phys. Rev.* B**45**, 10515 (1992); N.S. Murthy, R.H. Baughman and L.W. Shacklette, ibid. **41** 3708 (1990).

114. M. Wolf and K. Fesser, *J. Phys. Condens. Matt.* **3**, 5489 (1991).

115. K. Harigaya, A. Terai and Y. Wada, *Phys. Rev..* B**43**, 4141 (1991); K. Harigaya and A. Terai, ibid. **44**, 7835 (1991).

116. Y. Wada, in New Horizons in Low-Dimensional Electronic Systems, (Eds. H. Aoki, M. Tsukada, M. Schluter and F. Levy), Kluwer, Dordrecht, 1992.

117. J.P. Pouget, Z. Oblakowski, Y. Nogami, P.A. Albouy, M. Laridjani, E.J. Oh, Y. Min, A.G. MacDiarmid, J. Tsukamoto, T. Ishiguro and A.J. Epstein, *Synth. Met.* **65**, 131 (1994); Y. Nogami, J.P. Pouget and T. Ishiguro, ibid. **62**, 257 (1994).

118. S. Kivelson and A.J. Heeger, *Synth. Met.* **22**, 371 (1989).

119. R.J. Cohen and A.J. Glick, *Phys. Rev..* B**42**, 7659 (1990) and references therein.

120. P. Bernier, A. El-Khodary, F. Rachdi and C. Fite, *Synth. Met.* **17**, 413 (1987).

121. B.R. Bulka, *Synth. Met.* **24**, 41 (1988).

122. M. Springborg and L.A. Eriksson, *Synth. Met.* **57**, 4302 (1993); *Phys. Rev.* B**46**, 15833 (1992).

123. A. Yamashiro, A. Ikawa and H. Fukutome, *Synth. Met.* **65** 233 (1994).

124. E.J. Mele and M.J. Rice, *Phys. Rev.* B**23**, 5397 (1981).

125. S.A. Kivelson and A.J. Heeger, *Phys. Rev. Lett.* **53**, 308 (1985).

126. J. Fink, N. Nucker, B. Scheerer, A.V. Felde and G. Leising, in *Electronic Properties of Conjguated Polymers* (Eds. H. Kuzmany, M. Mehring and S. Roth), Springer Series in Solid State Sciences, Vol. 76, Springer, Berlin, p.84, 1987.

127. E.M. Conwell, H.A. Mizes and S. Jeyadev, *Phys. Rev.* B**40**, 1630 (1989); ibid **41**, 5067 (1990).

128. S. Stafström, *Phys. Rev.* B**43**, 9158 (1991).

129. P. Phillps and L. Cruz, *Synth. Met.* **65**, 225 (1994).

130. J. Tanaka, C. Tanaka, T. Miyamae, M. Shimizu, S. Hasegawa, K. Kamiya and K. Seki, *Synth. Met.* **65**, 173 (1994).

131. S. Hasegawa, K. Kamiya, J. Tanaka and M. Tanaka, *Synth. Met.* **18**, 225 (1987).

132. Y.H. Kim, S. Hotta and A.J. Heeger, *Phys. Rev.* B**36**, 7486 (1987).

133. M. Logdlund, R. Lazzaroni, S. Stäfstrom, W.R. Salaneck and J.L. Brédas, *Phys. Rev. Lett.* **63**, 461 (1989).

134. P. Bätz, S. Schmeisser and W. Göpel, *Solid State Commun.* **74**, 461 (1990); *Phys. Rev.* B**43**, 9178 (1991).

135. E.M. Conwell and H.A. Mizes, *Phys. Rev.* B**44**, 937 (1991).

136. Y. Furukawa, *Synth. Met.* **69**, 629 (1995).

137. S. Xie, L. Mei and D.L. Lin, *Phys. Rev.* B**50**, 13364 (1994).

138. F. Genoud, M. Guglielmi, M. Nechtschein, E. Genies and M. Salmon, *Phys. Rev. Lett.* **55**, 118 (1985).

139. A.J. Barssett, N.F. Colaneri, D.D.C. Bradley, R.A. Lowrence, R.H. Friend, h. Murata, S. Tokito, T. Tsutsui and S. Saito, *Phys. Rev.* B**41**, 10586 (1990); K.E. Ziemelis, A.T. Hussain, D.D.C. Bradley, R.H. Friend, J. Ruhe and G. Wegner, *Phys. Rev. Lett* **66**, 2231 (1991).

140. G.B. Street, in *Handbook of Conducting Polymers* (Ed. T.A. Skotheim), Dekker, New York 1986 p.265.; P. Pfluger, G. Weiser, J. Campbell Scott, G.B. Street ibid. Vol.2, p.1369.

141. G. Tourillon, in *Handbook of Conducting Polymers* Vol.1 (Ed. T.A. Skotheim) Dekker, New York, 1986 p.293; A.O. Patil, A.J. Heeger and F. Wudl, *Chem. Rev.* **88**, 183 (1988); J. Roncali, ibid. **92**, 711 (1992).

142. C.M. Gould, D.M. Bates, H.M. Bozler, A.J. Heeger, M.A. Dury and A.G. MacDiarmid, *Phys. Rev.* B**23**, 6820 (1980).

143. A.B. Kaiser, *Phys. Rev.* B**40**, 2806 (1989) and references therein; *Synth. Met.* **45**, 183 (1991); *Electronic Properties of Conjugated Polymers IV.* (eds. H. Kuzmany, M. Mehring and S. Roth), Springer Series in Solid State Sciences, Vol. 76, Springer, Berlin, 1991.

144. (a) P. Bernier, in *Handbook of Conducting Polymers* Vol.2 (Ed. T.A. Skotheim) Marcel Dekker, New York, 1986 p.1099 and references therein.
(b) H.S. Nalwa, *Phys. Rev.* B**39**, 5964 (1989).

145. D. Moses, A. Denenstein, A. Pron, A.J. Heeger and A.G. MacDiarmid, *Solid State Commun.* **36**, 219 (1980).

146. J. Tsukamoto, A. Takahashi and K. Kawasaki, *Jpn. J. Appl. Phys.* **29**, 125 (1990).

147. (a) T. Ohnishi, T. Noguchi, T. Nakano, M. Hirooka and I. Murase, *Synth. Met.* **41–43**, 309 (1991).
(b) F.E. Karaz, J.D. Capistran, D.R. Gagnon and R.W. Lenz, *Mol. Cry. Liq. Cryst.* **118** 327 (1985).
(c) I. Murase, T. Ohnishi, T. Noguchi and M. Hirooka, *Synth. Met.* **17**, 639 (1984).

148. (a) R.D. McCullough and R.D. Lowe, *J. Org. Chem.* **70**, 904 (1993).
(b) T.A. Chen and R.D. Rieke, *J. Am. Chem. Soc.* **114**, 10087 (1992).

149. Y.W. Park, C. Park, Y.S. Lee, C.O. Yoon, H. Shirakawa, Y. Suezaki and K. Akagi, *Solid State Commun.* **65**, 147 (1988).

150. Reghu M, Y. Cao, D. Moses and A.J. Heeger, *Phys. Rev.* B**47**, 1758 (1993); Reghu M, C.O. Yoon, D. Moses, A.J. Heeger and Y. Cao, ibid **48**, 17685 (1993).

151. H. Naarmann, *Synth. Met.* **17**, 223 (1987); H. Naarmann and N. Theophilou, ibid. **22**, 1 (1987).

152. N. Basescu, Z.X. Liu, D. Moses, A.J. Heeger, H. Naarmann and Theophilou, *Nature* **327**, 403 (1987).

153. T. Schimmel, D. Glaser, M. Schwoerer and H. Naarmann, in *Conjugated Polymers*, (Eds. J.L. Bredas and R. Silbey), Kluwer, Dordrecht, 1991 p.49.

154. J. Tsukamoto, *Adv. Phys.* **41**, 509 (1992) and references therein.

155. A.J. Heeger and P. Smith in *Conjugated Polymers* (Eds. J.L. Bredas and R. Silbey, Kluwer, Dordrecht, 1991, p.141.

156. A.G. Zobradskii and K.N. Zeninova, *Zh. Eksp. Teor. Fiz.* **86**, 727 (1984) [*Sov. Phys. JETP* **59**, 425 (1984); A.G. Zabrodskii, *Fiz. Tekh. Poluprovodn.* **11**, 595 (1977) [*Sov. Phys. Semicond.* **11**, 345 (1977)].

157. W.L. McMillan, *Phys. Rev* B**24**, 2739 (1981).

158. A.I. Larkin and D.E. Khmelnitskii, *Zh. Eskp. Teor. Fiz.* **83**, 1140 (1982) [*Sov. Phys. JETP.* **56** 647 (1982)].

159. T.G. Castner, in *Hopping Transport in Solids* (Eds. M. Pollak and B.I. Shklovskii, North-Holland, Amsterdam, 1990, and references therein.

160. A. J. Epstein, R.W. Bigelow, A. Feldblum, H.W. Gibson, D.M. Hoffman and D.B. Tanner, *Synth. Met.* **9**, 155 (1984).

161. L. Pietronero, *Synth. Met.* **8**, 225 (1983).

162. J. Joo, V.N. Prigodin, Y.G. Min, A.G. MacDiarmid and A.J. Epstein, *Phys. Rev. B* **50**, 12226 (1994); J.Joo, Z. Oblakowski, G. Du, J.P. Pouget, E.J. Oh, J.M. Wiesinger, Y. Min, A.G. MacDiarmid and A.J. Epstein, *Phys. Rev.* B**49**, 2977 (1994); J.Joo and A.J. Epstein, *Rev. Sci. Instrum.* **65**, 2653 (1994); J.Joo, G. Du. V.N. Prigodin, J. Tsukamoto and A.J. Epstein, *Phys. Rev.* **52**, 8060 (1995).

163. A.J. Epstein, J.M. Ginder, F. Zuo, H.S. Woo, D.B. Tanner, A.F. Richter, M. Angeloupolos, W.S. Huang and A.G. MacDiarmid, *Synth. Met.* **21**, 63 (1987); A.J. Epstein, J.Joo, C.Y. Wu, A. Benatar, C.F. Faisst, Jr., J. Zegarski and A.G. MacDiarmid, in *Intrinscially Conducting Polymers: An Emerging Technology* (Ed. M. Aldissi), Kluwer, Dordrecht, 1993 p. 165 and references therein.

164. P.M. Bell and A. MacKinnon, *J. Phys. Condens. Matter* **5**, 8337 (1993); B. Kramer and A. MacKinnon, *Rep. Prog. Phys.* **56**, 1469 (1993).

165. J.S. Andrade, Jr., J. Shibusa, Y. Arai and A.F. Siqueira, *Synth. Met.* **68**, 167 (1995).

166. R.J. Baughman and L.W. Shacklette, *Phys. Rev.* B**39**, 5872 (1989); *J. Chem. Phys.* **90**, 7492 (1989).

167. B. Movaghar and S. Roth, *Synth. Met.* **63**, 163 (1994).

168. P. Sheng, *Phys. Rev.* B**21**, 2180 (1980); P. Sheng and J. Klasfter, ibid. **27**, 2583 (1983).

169. J. Voit and H. Buttner, *Solid State Commun.* **67**, 1233 (1988).

170. A.B. Kaiser and S.C. Graham, *Synth. Met.* **36**, 367 (1990).

171. G. Paasch, *Synth. Met.* **51**, 7 (1992).

172. Y. Nogami, M. Yamashita, H. Kaneko, T. Ishigura, A. Takahashi and J. Tsukamoto, *J. Phys. Soc. Jpn.* **62**, 664 (1993).

173. E.M. Conwell and H.A. Mizes, *Synth. Met.* **38**, 319 (1990); *Phys. Rev.* B**44**, 3963 (1991).

174. H.A. Mizes and E.M. conwell, *Phys. Rev. Lett* **70**, 1505 (1993); *Synth. Met.* **55–57**, 4284 (1993).

175. G. Soda, D. Jerome, M. Weger, J. Alizon, J. Gallice, H. Robert, J.M. Fabre nd L. Giral,. *J. Phys. (Paris)* **38**, 931 (1987)

176. Y. W. Park, C.O. Yoon, C.H. Lee, H. Shirakawa, Y. Suezaki and K. Akagi, *Synth. Met.* **28**, D27 (1989).

177. B. Abele, P. Sheng, M.D. Coutts and Y. Arie, *Adv. Phys.* **24**, 407 (1975).

178. L. Zuppiroli, M.N. Bussac, S. Paschen, O. Chauvet and L. Forro, *Phys. Rev.* B **50**, 5196 (1994); M.N. Bussac and L. Zuppiroli, ibid. **49**, 5876 (1994); ibid. **47**, 5493 (1993); O. Chauvet, S. Paschen, L. Forro, L. Zuppiroli, P. Bujard, K. Kai and W. Wernet, *Synth. Met.* **63**, 115 (1994).

179. A.A. Ovchinnikov and K.A. Pronin, *Synth. Met.* **41–43**, 3373 (1991); *Solv. Phys. Solid State* **28**, 1666 (1986) [*Fiz. Tverd. Tela* **28**, 2964 (1986)].

180. T.J. Lewis, *Faraday Discuss. Chem. Soc.* **88**, 189 (1989).

181. R. Redi and J.J. Hopfield, *J. Chem. Phys.* **72**, 6651 (1980).

182. D.S. Pearson, P.A. Pincus, G.W. Haffner, and S.J. Dahman, *Macromolecules* **26**, 1570 (1993); G.W. Haffner, S.J. Dahman, D.S. Pearson and C.L. Gettinger, *Polymer* **34**, 3155 (1993).

183. S. Tokito, P. Smith and A.J. Heeger, *Polymer* **32**, 464 (1991).

184. Y. Cao, P. Smith and A.J. Heeger, *Polymer* **32**, 1210 (1991).

185. G. Thummes, U. Zimmer, F. Korner and J. Kotzler, *Solid State Commun.* **67**, 215 (1988).

186. J.M. Madsen, B.R. Johnson, X.L. Hua, R.B. Hallock, M.A. Masse and F.E. Karasz, *Phys. Rev.* B**40**, 11751 (1989).

187. T. Ishiguro, H. Kaneko, Y. Nogami, H. Ishimoto, H. Nishiyama, M. Yamaura, T. Haiwara and K. Sato, *Phys. Rev. Lett.* **62**, 660 (1992).

188. S.A. Kivelson and V.J. Emery, *Synth. Met.* **65**, 249 (1994).

189. J. Voit, *Phys. Rev. Lett.* **64**, 323 (1990).

190. Y. Imry and M. Strongin, *Phys. Rev.* B**24**, 6453 (1981).

191. R. Cabanel, J. Chaussy, J. Mazuer and J.C. Villegier, *J. De. Physique* **49**, 795 (1988).

192. Y. Liu, D.B. Haviland, B. Nease and A.M. Goldman, *Phys. Rev.* B**47**, 5931 (1993).

193. E. Simanek, *Inhomogenous Superconductors* Oxford University Press, New York, 1994.

194. A.G. Sun, L.M. Paulius, D.A. Gajewskii, M.B. Maple and R.C. Dynes, *Phys. Rev.* **50**, 3266 (1994).

195. H.K. Sin, P. Lindenfield and W.L. McLean, *Phys. Rev.* B**30**, 4067 (1984).

196. (a) Z.H. Wang, A.W.P. Fung, G. Dresselhaus, M.S. Dresselhaus, K.A. Wang, P. Zhou and P.C. Eklund, *Phys. Rev B*, **47**, 15354 (1993)
(b) F. Stepniak, P.J. Benning, D.M. Poirier and J.H. Weaver, *Phys. Rev.* **48**, 1899 (1993).

197. M.S. Dresselhaus and G. Dresselhaus, *Adv. Phys* **30**, 139 (1981) and references therein.

198. D.C. Johnston, *J. Low. Temp. Phys.* **25**, 145 (1976).
199. P. Dai, Y. Zhang and M.P. Sarachik, *Phys. Rev.* B**45**, 3984 (1992); ibid. **46**, 6724 (1992).
200. T.F. Rosenbaum, R.M.F. Milligan, G.A. Thomas, P.A. Lee, T.V. Ramakrishnan and R.N. Bhatt, *Phys. Rev. Lett.* **47**, 1758 (1981); T.F. Rosenbaum, R.F. Milligan, M.A. Paalanen, G.A. Thomas, R.N. Bhatt and W. Lin, *Phys. Rev.* B**27**, 7509 (1983).
201. Y.W. Park, C.O. Yoon, C.H. Lee and H. Shirakawa, *Makromol. Chem. Macromol. Chem. Macromol. Symp.* **33**, 341 (1990); Y.W. Park, C.O. Yoon, B.C. Na, H. Shirakawa and K. Akagi, *Synth. Met.* **41–43**, 27 (1991); C.O. Yoon, PhD Thesis, Dept. of Physics, Seoul National Univ., Korea, 1992.
202. A.N. Aleshin, N.B. Mironkov, A.V. Suvorov, J.A. Conklin, T.M. Su and R.B. Kaner, *Phy. Rev. B* (in press).
203. N. Foxonet, P. Bernier and J. Voit, *J. de Chime. Phy. et de Phy-Chimie Bio.* **89**, 977 (1992); N. Coustel, P. Bernier and J.E. Fisher, *Phys. Rev.* B**43**, 3147 (1991); D. Bormann and P. Bernier, personal communication.
204. D. Beliz and K.I. Wysokinski, *Phys. Rev.* B**36**, 9333 (1987).
205. Y. Nogami, H. Kaneko, H. Ito, T. Ishiguro, T. Sasaki, N Toyota, A. Takahashi and J. Tsukamoto, *Phys. Rev.* B**43**, 11829 (1991); Y. Nogami, H. Kaneko, T. Ishiguro, A. Takahashi, J. Tsukamoto and N. Hosoito, *Solid State Commun.* **76**, 583 (1990).
206. H.H.S. Javadi, A. Chakraborty, C. Li, N. Theophilou, D.B. Swanson, A.G. MacDiarmid and A.J. Epstein, *Phys. Rev.* B**43**, 2183 (1991).
207. H. Shirakawa, Y.X. Zhang, T. Okuda, K. Sakamaki and K. Akagi, *Synth. Met.* **65**, 93 (1994).
208. K. Mizoguchi, H. Sakurai, F. Shimizu, S. Masubuchi and K. Kume, *Synth. Met.* **68**, 239 (1995); K. Mizoguchi, S. Masubuchi, K. Kume, K. Akagi and H. Shirakawa, *Phys. Rev.* B**51**, 8864 (1995).
209. General discussion on charge transport properties in conducting polymers, *Faraday Discuss. Chem. Soc.* **88**, 239 (1989).
210. H. Kaneko and T. Ishiguro, *Synth. Met.* **65**, 141 (1994); H. Kaneko, T. Ishiguro, J. Tsukamoto and A. Takahashi, *Solid State Commun.* **90**, 83 (1994).
211. Th. Schimmel, G. Denninger, W. Riess, J. Voit, M. Schwoerer, W. Schoepe and H. Naarmann, *Synth. Met,* **28D**, 11 (1989).
212. M.T. Ahmed, A.B. Kaiser, S. Roth and M.D. Migahed, *J. Phys. D: Appl. Phys.* **25**, 79 (1992); R. Zuzok, A.B. Kaiser, W. Pukachki and S. Roth, *J. Chem. Phys.* **95**, 1270 (1991).
213. C.O. Yoon, Reghu M., D. Moses and A.J. Heeger, *Phys. Rev.* B**49**, 10851 (1994).
214. W. Sasaki and R. De Bruyn, *Physica* **27**, 877 (1961); C. Yamanouchi, K. Mizuguchi and W. Sasaki, *J. Phys. Soc. Jpn* **22**, 859 (1967).
215. Y. Ootuka, S. Kobayashi, S. Ikehata, W. Sasaki and J. Kondo, *Solid State Commun.* **30**, 169 (1979).

216. T.F. Rosenbaum, K. Andres, G.A. Thomas and P.A. Lee, *Phys. Rev. Lett.* **46**, 568 (1981).
217. Reghu M., C.O. Yoon, D. Moses and A.J. Heeger, *Synth. Met.* **64**, 53 (1994).
218. V.V. Bryksin, *Fiz. Tverd. Tela* **28**, 2981 (1986) [*Sov. Phys. Solid State* **28**, 1676 (1986)]; ibid. **32**, 343 (1990) [*Sov. Phys. Solid State* **32**, 197 (1990)].
219. Reghu M. and S.V. Subramanyam, *Solid State Commun.* **72**, 325 (1989).
220. A.B. Pippard, *Magnetoresistance in Metals,* Cambridge University Press, New York, 1989.
221. C. Kittel, *Quantum Theory of Solids*, John Wiley & Sons New York, 1972.
222. A.L. Efros and M. Pollak (eds.), *Electron–Electron Interactions in Disordered Systems,* North-Holland, Amsterdam, 1985.
223. M. Kaveh and N.F. Mott, *Philos. Mag.* B**55**, 1 (1987).
224. J.C. Clark, G.G. Ihas, A.J. Rafanello, M.W. Meisel, Reghu M., C.O. Yoon, Y. Cao and A.J. Heeger, *Synth. Met.* **69**, 215 (1995).
225. C.O. Yoon, Reghu M., D. Moses, A.J. Heeger and Y. Cao, *Phys. Rev.* B**48**, 14080 (1993).
226. C.O. Yoon, Reghu M., D. Moses, A.J. Heeger, Y. Cao, T.A. Chen, X. Wu and R.D. Rieke, *Synth. Met.* **75**, 229, 1995.
227. S. Stafström, J.L. Brédas, A.J. Epstein, H.S. Woo, D.B. Tanner, W.S. Huang and A.G. MacDiarmid *Phys. Rev. Lett.* **59**, 1464 (1987); D.S. Boudreaux, R.R. Chance, J.F. Wolf, L.W. Shacklette, J.L. Brédas, B. Themans, J.M. Andre and R. Silbey, *J. Chem. Phys.* **85**, 4584 (1986).
228. N.S. Sariciftci, A.J. Heeger and Y. Cao, *Phys. Rev.* B**49**, 5988 (1994).
229. E.B. Park, Y.S. Yoo, J.Y. Park, Y.W. Park, K. Akagi and H. Shirakawa *Synth. Met.* **69**, 61 (1995).
230. T.H. Gilani, T. Masui, G. Y. Logvenov and T. Ishiguro *Synth. Met.* **78**, 327, 1996.
231. (a) E.J. Yoffa and D. Adler, *Phys. Rev.* B**15**, 2311 (1977).
    (b) D. Emin, C.H. Seager and R.K. Quinn, *Phys. Rev. Lett.* **28**, 813 (1972).
232. K. Lee, Reghu M, E.L. Yuh, N.S. Sariciftci and A.J. Heeger, *Synth. Met.* **68**, 287 (1995).
233. R.S. Kohlman, J. Joo and A.J. Epstein, in *Physical Properites of Polymers Handbook,* (Ed. J. Mark), AIP Press, in press.
234. X.Q. Yang, D.B. Tanner, A. Feldblum, H.W. Gibson, M.J. Rice and A.J. Epstein, *Mol. Cryst. Liq. Cryst.* **117**, 267 (1985); D.M. Hoffman, D.B. Tanner, A.J. Epstein and H.W. Gibson, *Mol. Cryst. Liq. Cryst.* **83**, 143 (1982).
235. S. Hasegawa, M. Oku, M. Shimizu and J. Tanaka, *Synth. Met.* **38**, 37 (1990); K. Kamiya, H. Inokuchi, M. Oku, S. Hasegawa, C. Tanaka and J. Tanaka, *Synth. Met.* **41**, 155 (1991).
236. K. Lee, A.J. Heeger and Y. Cao *Phys. Rev.* B**48**, 14884 (1993); *Synth. Met.* **72**, 25 (1995); K. Lee, Reghu M.,

C.O. Yoon and A.J. Heeger, *Phys. Rev.* B**52**, 4779 (1995).

237. H.S. Woo and D.B. Tanner, *Synth. Met.* **41–43**, 281 (1991).

238. E.M. Conwell and H.A. Mizes, *Synth. Met.* **65**, 203 (1994).

239. Y.H. Kim and A.J. Heeger, *Phys. Rev.* B**40**, 8393 (1989).

240. A.C. Kolbert, S. Caldarelli, K.F. Thier, N.S. Sariciftci, Y. Cao and A.J. Heeger, *Phys. Rev.* B**51**, 1541 (1995).

241. P.N. Adams, P.J. Laughlin, A.P. Monkman, and N. Bernhoeft, *Solid State Commun.* **91**, 875 (1994); J.P. Travers, P. Le Guyadec, P.N. Adams, P.J. Laughlin and A.P. Monkman, *Synth. Met.* **65**, 159 (1994).

242. N.S. Sariciftci, Reghu M., *et al.* unpublished work.

243. Reghu M., C.O. Yoon, D. Moses, Y. Cao, and A.J. Heeger, *Synth. Met.* **69**, 329 (1995); Reghu M., K. Vakiparta, C.O. Yoon, Y. Cao, D. Moses and A.J. Heeger, *Synth. Met.* **65**, 167 (1994).

244. G. Biskupski, A. El Kaaouachi and A. Briggs, *J. Phys. Condens. Matter* **3**, 8417 (1991).

245. D.J. Newson and M. Pepper, *J. Phys. Condens. Matter* **19**, 3983 (1986).

246. D.E. Khmelnitskii and A.I. Larkin, *Solid State Commun.* **39**, 1069 (1981).

247. B.I. Shklovskii and A.L. Efros, *Electronic Properties of Doped Semiconductors,* Springer, Heidelberg 1984.

248. R. Rosenbaum, *Phys. Rev.* B**44**, 3599 1991.

249. Y. Zhang, P. Dai, M. Levy and M.P. Sarachik, *Phys. Rev. Lett* **64**, 2687 (1990); A. Aharony, Y. Zhang and M.P. Sarachik, *Phys. Rev. Lett.* **68**, 3900 (1992).

250. M. Pollak and C.J. Adkins, *Philos. Mag.* B**65**, 855 (1992).

251. E. Cuevas, M. Ortuno and J. Ruiz, *Phys. Rev. Lett.* **71**, 1871 (1993).

252. Reghu M., C.O. Yoon, C.Y. Yang, D. Moses, P. Smith, A.J. Heeger and Y. Cao, *Phys. Rev.* B**50**, 13931 (1994).

253. I.P. Zvyagin, in *Hopping Transport in Solids* (Eds. M. Pollak and B.I. Shkovskii), North-Holland, Amsterdam, 1990.

254. I. Shlimak, M. Kaveh, M. Yosefin, M. Lea and P. Fozooni, *Phys. Rev. Lett.* **68**, 3076 (1992).

255. M. Aldissi (ed. *Intrinsically Conducting Polymers: An Emerging Technology,* Kluwer, Dordrecht, 1993.

256. A.J. Heeger, *TRIP* **3** 39 (1995).

257. A. Aharony and D. Stauffer, *Introduction to percolation Theory,* 2nd end. Taylor Francis, London, 1993, and references therein.

258. S.H. Munson-McGee, *Phys. Rev.* B**43**, 3331 (1991).

259. B. Bridge and H. Tee, *Int. J. Electronics* **6**, 785 (1990).

260. F. Carmona, *Physica A* **157**, 461 (1989) and references therein.

261. P. Banerjee and B. Mandel, *Macromolecules* **28**, 3940 (1995).

262. J. Finter, C.W. Mayer, J.P. Ansermet, H. Bleier, B. Hilti, E. Minder and D. Neuschafer, *Synth. Met.* **41–43** 951 (1991); H. Bleier, J. Finter, B. Hilti, W. Hofherr, C.W. Mayer, E. Minder, H. Hediger and J.P. Ansermet, *Synth. Met.* **55–57**, 3605 (1993).

263. J. Ulanski, J.K. Jeszka, P. Polanowski, I. Glowachki, M. Kryszewskii, D. Staerk and J.W. Helberg, *Acta Phy. Polonica A* **87**, 899 (1995); J. Ulanski, J.K. Jeszka, A. Tracz, I. Glowachki, M. Kryszewskii and E.E. laukhina, *Synth. Met.* **55–57**, 2011 (1993).

264. Reghu M., C.O. Yoon, C.Y. Yang, D.Moses, A.J. Heeger and Y. Cao, *Macromolecules,* **26**, 7245 (1993); *Synth. Met.* **63**, 47 (1994); *Synth, Met.* **69**, 255 (1995).

265. D. Stroud and F.P. Pan, *Phys. Rev.* B**13**, 1434 (1976).

266. Y.W. Park and H. Lee (eds.), Proc. of the International Conference of Science and Technology of Synthetic metals (ICSM '96), Seoul, Korea, *Synth. Met.* **69**, (1995).

267. Neste Conductive Polymer Technical Brochure, Neste Oy, Finland, 1995.

CHAPTER 3

Electronic Structure of π-Conjugated Polymers

Miklos Kertesz
Georgetown University, Washington DC 20057–1227, USA

1 INTRODUCTION

Conjugated π-electron polymers have attracted much interest, especially since the discovery of highly conducting doped polyacetylene [1]. Since then, many π-conjugated polymers have been synthesized, characterized and fabricated into useful or potentially useful devices [2–6]. In recent years much research has been focused on a number of novel properties such as nonlinear optics, electroluminescence, photovoltaics, etc. that will be amply reviewed in many chapters of this book.

The electronic structural theory of π-conjugated polymers has helped substantially to formulate a microscopic conceptual framework within which the physical properties of these organic materials are being interpreted and understood. Several research groups have contributed to this progress, which will be mentioned in the chapter. A number of excellent reviews have been written on the subject [7–18].

Some of the reviews have focused on techniques for calculating total energy and conjugated structures of π-conjugated polymers [7–10]. Other reviews focused on polyacetylene, and its bond length alternating ground state as opposed to a higher energy metallic state [8,11,12] which gives rise to a degeneracy of the ground state and the existence of soliton defects [13,14]. The difficult subject of accurately describing

Handbook of Organic Conductive Molecules and Polymers: Vol. 4. Conductive Polymers: Transport, Photophysics and Applications.
Edited by H. S. Nalwa.

local defects, such as solitons, polarons and bipolarons has also been extensively reviewed [15]. Studies on the excited states of conjugated polymers primarily based on highly correlated semi-empirical models were reviewed by various authors [16–18]. The book by André, Delhalle and Bredas reviewed some of the electronic structure theory on π-conjugated polymers including polyaniline and non-linear optics [10]. The area of interchain interactions has not been extensively reviewed yet. The computational techniques that account for intrachain interactions reliably, are possibly less accurate, when applied to interchain interactions. Some progress has been made [19,20], but there is still a need for more reliable interchain calculations.

Figure 3.1 illustrates a few π-conjugated polymers that will be discussed in this review. They have been chosen to include several main categories:

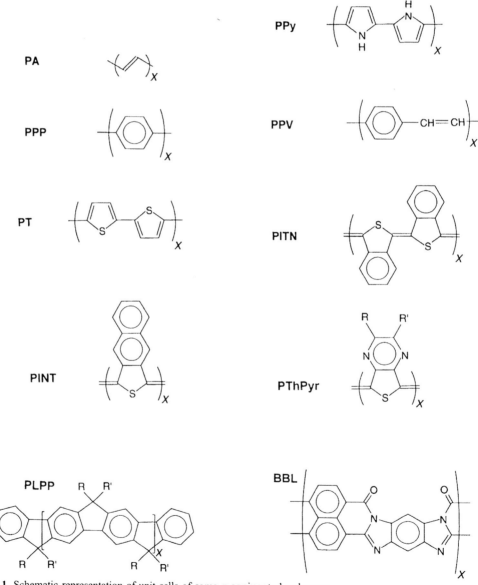

Figure 3.1. Schematic representation of unit cells of some π-conjugated polymers.

- -polyacetylene (PA)
- -polyphenylenes [PPP: poly(*p*-phenylene); PLPP: planarized polyphenylene, PPV: poly(*p*-phenylene-vinylene)]
- -heterocyclic polymers polymerized along a single main chain [PPy: polypyrrole, PT: polythiophene; PITN: poly(isothianaphthene); PINT: poly(iso-naphthothiophene): PThPyr: poly(R,R'-thienopyra-zine)]
- -ladder polymers [BBL: poly(benzimidazobenzophe-nanthroline)].

Even though there are many similarities between π-conjugated and σ-conjugated polymers, such as poly-silanes, as pointed out by Soos and Hayden [21], the latter will not be covered in this chapter. Polyanilines [10] will not be covered either. They represent a borderline case, because the degree of π-conjugation in the various forms of polyanilines is directly related to their degree of coplanarity. Calculations alone are not sufficiently accurate to determine the torsional angles that are present in the condensed phase samples of polyanilines. More experimental structural work is needed to ascertain the degree of coplanarity in the various forms of polyaniline.

This review is based on the *molecular orbital viewpoint* [22]. Thus, this chapter focuses on chemical *trends* among series of similar polymers, discusses the effects of heteroatomic substitutions, side groups and connectivity, the central concept of the amplitude mode, some computational aspects and the electronic structures of specific examples and groups of π-conjugated polymers are discussed.

2 AMPLITUDE MODE, BAND-GAP AND THE PEIERLS DISTORTION IN PA

A common and central feature of π-conjugated polymers is that their electronic structures are closely coupled to certain geometrical degrees of freedom, related to bond length alternation [23], which go by the name of the amplitude mode [24], ja-mode [25] or simply Peierls mode. (Figure 3.2) The latter term refers to a theorem by Peierls [26,27] which states, that one-dimensional strings cannot be metals. As applied to polyacetylene [23–28], this theorem implies that structure A, or the equivalent structure B are the stable form in Figure 3.3, and not structure C. (The Peierls distortion breaks the symmetry and doubles the size of the repeat unit in this case.) A and B are equivalent,

constituting an example of a so-called degenerate ground state polymer. (Some rather large unit cell ladder polymers also belong to this category.) In structure C all C–C bonds are equivalent, the electronic structure has a zero bandgap and should therefore be metallic. However, according to Peierls' theorem [26], or a number of calculations [8–11,23–27] A (or B) is more stable than C. The bond length altering structure (A or B) has the equilibrium alternation value of approximately $r_{long} - r_{short} = 0.1$ Å, as found by various experiments [29] and an associated bandgap around 1.9 eV [30].

In this review, the experimental bandgap is the value that corresponds to the longest wavelength absorption maximum, λ_{max}. This choice has some disadvantages in that it does not account for the possibly excitonic nature of the transition [31], nor for the effect of three-dimensional couplings on the line shape of the absorption [30]. However, in many instances the data available do not allow for the delineation of these details, since the extent of π-conjugation or the morphology of the polymer is frequently not precisely defined and known. On the other hand, the first absorption maximum is experimentally easily accessible. Since this chapter is concerned about the *chemical trends* for a series of similar systems, this definition of the experimental bandgap is very fruitful [32].

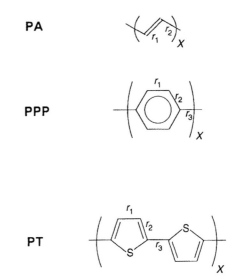

Figure 3.2. Unnormalized amplitude modes (q_{AM}) in polyacetylene, polythiophene and poly(*p*-phenylene) are defined as $r_2 - r_1$ for PA, $4r_2 - 2r_1 - r_3$ for PPP and $r_1 + r_3 - 2r_2$ for PT.

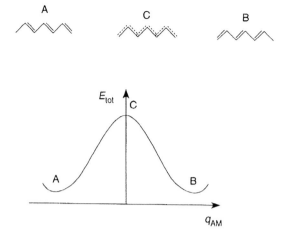

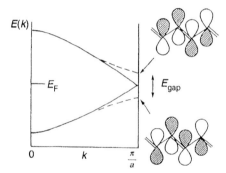

Figure 3.3. Peierls distortion of polyacetylene. Top: three alternative structures. Middle: total energy as a function of q_{AM}. Bottom: splitting of the degenerate orbitals at the Fermi level.

The fact that structure A (or B) should be more stable than C can be easily understood by using a simple orbital argument concerning the energy of the highest occupied level. The advantage of this argument is that it can be easily extended to other polymers, including ladder-type polymers, or heterocyclic polymers, as illustrated in Figure 3.3. The point simply is that as the structure is being changed from C to A, along a generalized coordinate

$$q_{AM} = q_P = a(r_s - r_d)$$

the highest occupied orbital (HOO) will be stabilized both on account of gaining bonding character in the double bond region (r_d) and losing antibonding character in the single bond region (r_s). (The lowest unoccupied orbital, LUO, undergoes a similar perturbation in the opposite direction, leading to a sizeable bandgap, $E_g = E_{LUO} - E_{HOO}$.) The Peierls coordinate,

q_p, provides a linear interpolation between a geometrical configuration with two equivalent bonds ($a = 0$) and one with the fully developed single and double bonds ($a = 1$). This general scheme does not account for the equilibrium value of q_{AM}^{equ}, or r_s^{equ} or r_d^{equ} after the bond alternation has been introduced, nor for the amount of the associated energy gain. Likewise, the magnitude of orbital stabilization or the bandgap value need to be determined by detailed calculations, which ought to be accurate and also account for the other details of the geometry, such as bond angles and other bond distances (and torsional angles). Within non-SCF theories, such as simple Hückel theory, the bandgap is proportional to the amplitude mode, [26–28]

$$E_g = k|r_s^{equ} - r_d^{equ}| = ka^{equ}(r_s - r_d) = k|q_{AM}^{equ}| \quad (3.1)$$

showing how fundamental the connection is between certain aspects of the geometry and the electronic structure for polyacetylene (q_{AM}^{equ} is the equilibrium value of the amplitude mode). This connection, although significantly modified, remains a characteristic feature of most π-conjugated polymers.

3 COMPUTATIONAL TECHNIQUES

Ideally the computational techniques that one would like to employ for the study of the electronic structure of π-conjugated polymers should be

(a) applicable to such complex systems as listed in Figure 3.1;
(b) amenable to geometry optimization *and* should yield good calculated geometrical parameters;
(c) able to produce good electronic structure characteristics, such as bandgap, ionization potential (IP), electron anninity (EA), charge distributions and wavefunctions in general;
(d) capable of yielding good results for other calculated physical observables, such as force constants, transition dipoles, etc.

Different research groups in the field use various approaches that vary with the focus of the applications even within research groups [32–37]. Generally, no single technique has evolved so far that satisfies all criteria above. While the focus of this review is elsewhere, these criteria are discussed briefly in turn.

Item a   It is customary to assume that the polymer in question is long enough, so that an *infinite periodic model* provides a useful model. Within such a model the language of *energy bandstructure* theory is applicable [7–12,22]. This amounts to the introduction of reciprocal or *k* vectors, leading to the *k*-dependent energy levels of the system (energy bands). The calculations require the evaluation of interactions over a certain range of the polymers and should correspond to a converged result with respect to this range (lattice sum). While this represents certain convergency problems, such calculations have been available within several of the techniques (Hamiltonians) that have their origins in the corresponding quantum chemical method. These include band theoretical versions of non-self-consistent field (non-SCF) techniques such as Hückel theory, HT [27–34], extended Hückel theory, EHT [22,32], valence effective Hamiltonian theory, VEH [10,33,37], or SCF techniques, such as Pariser–Parr–Pople, PPP [16,38], modified neglect of diatomic overlap MNDO (including the more popular AM1 and PM3 parametrizations) [33,37,39] or *ab initio* Hartree–Fock [7,11], density functional theory, DFT [35], all of which have been applied to polymeric electronic structure calculations.

An alternative approach, which is free of any lattice sum approximation, is based on a series of calculations performed on a sequence of longer and longer oligomers, consisting of *m* repeat units, (*m*-mers) that in principle leads to the polymer as $m \to \infty$. In practice, *m* may be very limited. Still, this approach offers various advantages [40,41]; it permits the use of well-tested molecular computer codes and it produces information on oligomers that are often accessible experimentally and are of interest in their own right. Another advantage of the oligomer technique is that for non-degenerate ground state polymers, it allows the calculation of the energy difference between structure A and structure Q, where such structural alternatives make chemical sense. (Structure A is usually referred to as the aromatic or benzenoid, structure Q as the quinonoid one, see Figure 3.4.

This energy difference is calculated according to the following scheme. An *m*-mer and an (*m* + 1)-mer are considered, with terminal groups appropriate for the aromatic structure (e.g. –H). The approximate per repeat unit (pru) energy representing the energy of the inserted middle unit is

$$E_{\text{pru}}^{\text{A}}(m) = E^{\text{A}}(m+1) - E^{\text{A}}(m) \qquad (3.2)$$

where $E^{\text{A}}(m)$ is the total energy of the aromatic *m*-mer, including the end group as calculated with the given Hamiltonian. Then, a quinonoid *m*-mer and an (*m* + 1)-

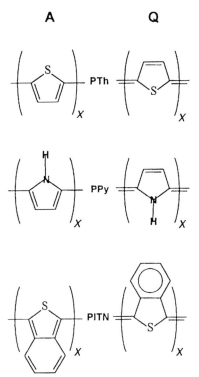

Figure 3.4. Aromatic (A) and quinonoid (Q) structures of three selected π-conjugated polymers. Terminal groups used in the oligomer calculations are not shown.

mer are considered, terminated with proper end group such as –CH$_2$ [40]. The difference is the corresponding approximate per repeat unit energy.

$$E_{\text{pur}}^{\text{Q}}(m) = E^{\text{Q}}(m+1) - E^{\text{Q}}(m) \qquad (3.3)$$

Then, the difference of the differences is taken [40,41], which is a good predictor of the relative stability of the two (A and Q) forms.

$$\Delta E_{\text{pru}}^{\text{AQ}}(m) = E_{\text{pru}}^{\text{A}}(m) - E_{\text{pru}}^{\text{Q}}(m) \qquad (3.4)$$

Results for three polymers based on this simple straightforward approach are summarized in Table 3.1–3.3. Fast convergence as a function of *m* is noteworthy. As expected, the energy difference values scatter over a wide range depending on the Hamiltonian. However, the signs indicating the relative stability of the A vs. Q forms are independent of the technique used.

Item b Geometry optimization of organic molecules has been the staple of quantum chemistry in the last two decades. Techniques at semi-empirical and primarily *ab initio* levels are well known and wide spread; the errors

Table 3.1. Relative stability of the A and Q forms of polythiophene as shown by the calculated values of $\Delta E_{\text{pru}}^{\text{AQ}}(m)$, as defined in equation (4) [40]. Negative value indicates a more stable aromatic structure

m	3-21G*	PM3	LHS
2	-7.90	-5.84	-7.65
3	-8.01	-5.77	
4		-5.76	

Table 3.2. Relative stability of the A and Q forms of polypyrrole as shown by the calculated values of $\Delta E_{\text{pru}}^{\text{AQ}}(m)$, as defined in equation (4) [40]. Negative value indicates a more stable aromatic structure

m	3-21G*	PM3	MNDO
2	-16.09	-15.29	-12.30
3	-15.97	-14.65	11.84
4		-14.49	-11.80

Table 3.3. Relative stability of the A and Q forms of poly(isothianaphthene) as shown by the calculated values of $\Delta E^{\text{AQ}}_{\text{pru}}(m)$, as defined in equation (4) [40]. Negative value indicates a more stable aromatic structure

m	3-21G'	PM3	PM3 (non-planar)	LHS
2	9.20	7.48	4.73	3.04
3		7.85	5.14	

of predictions associated with particular basis sets and standard correlation theories are well under control [42]. Usually for closed shell systems, the Hartree–Fock formalism is used. Predictions using the band-structure formalism usually achieve similar accuracy as the respective molecular techniques [40,41]. (For instance, MNDO overestimates the equilibrium internal torsion angle, in biphenyl, and so does the polymer variant for poly(p-phenylene). The PM3 parametrization on the other hand gives a more realistic torsional angle for both.) Obviously *ab initio* molecular computer codes beyond the level of the Hartree–Fock or density functional theory with gradient corrections in combination with oligomer extrapolation are the most reliable if affordable for a given series of oligomers. Correlation corrections at the *ab initio* level in conjunction with such oligomer to polymer extrapolations have unfortunately been reported for simple systems only [41]

Item c Unfortunately, Hartree–Fock (HF) grossly overestimates the bandgap of polymers [8], while many practical computer realizations of the DFT underestimate it [43]. (This HF problem occurs at all levels of approximation, such as *ab initio*, CNDO, MNDO, etc.)

These artifacts of the HF and DFT methods can be overcome in practice by turning to one of the techniques, usually of the tight binding form, which produces bandstructures that are free from this gap problem. A number of such tight binding methods have been developed [29,32–37,39], all of which are non-self-consistent. The calculations often produce very good correlation between the calculated and experimental bandgap and ionization potential data for a wide series of conjugated polymers.

Item d In the conjugated polymers theory literature that concentrates on properties other than total energy, geometry or electronic structure, the following properties have attracted much attention: non-linear optics, vibrational spectra and the related IR and Raman intensities, and transport properties. The calculations of these properties are not reviewed here, but *ab initio* techniques (HF or DFT) are becoming prevalent. To indicate the level of difficulty, the outstanding issue of the exciton binding energy should be mentioned; it should be amenable to theoretical calculations, but remains controversial [31,44,45].

The typical compromise for electronic structure calculations therefore consists of the following steps:

(1) Geometry optimization by either periodic boundary condition or the oligomer technique.
(2) Non-SCF tight binding band calculation.

4   TRENDS IN THE ELECTRONIC STRUCTURES OF CONJUGATED POLYMERS: DERIVATIVES OF POLYTHIOPHENE, POLY(p-PHENYLENE), POLYPYRROLE AND THEIR COPOLYMERS WITH VINYLENES AND METHENYLS

Following the scheme of electronic structure calculations described at the end of Section 3, several groups have calculated the electronic structures of families of conjugated polymers, including substituted derivatives and various copolymers. Due to the strong coupling of the electronic structures to the geometry, it is essential that the electronic structure calculations be preceded by a geometry optimization *and* an analysis of the stability of competing structural alternatives. For instance, based on the overwhelming stability of the A form of polythiophene, relative to the Q form, one has to base the electronic structure calculation of thiophene on the structure of the more stable A form, and similarly, on the Q form of PITN, etc. The analysis of both the relative stabilities of these forms, as well as the general trends in the electronic structures themselves is greatly aided by the analysis of the orbital patterns, especially those of the frontier HO (highest occupied) and LU (lowest occupied) crystal orbitals (CO) [7–10,22].

As an illustration of this concept, consider the frontier orbitals of polythiophene. This analysis refers to one of the infinite possible k values (between 0 and π/a) that characterize the crystal orbitals of an idealized infinite periodic model of a polymer. For the analysis $k=0$ is chosen, which corresponds to the $e^{ika}=1$ phase factor that determines the orbital coefficients in the unit cells outside the reference cell by the following rule (Bloch's theorem [22]): coefficients in a cell that is j cells apart from the reference cell are related to those in the reference cell by the phase factor e^{ijka}. As can be shown by detailed calculations, the HOCO–LUCO gap corresponds to precisely the special value of $k=0$. Figure 3.5 shows the corresponding orbitals within one

repeat unit for the two forms of polythiophene. According to Bloch's theorem, the coefficients in all the other unit cells are identical to those in the reference cell at $k=0$, making the intercell C–C bonds bonding for the HOCO and antibonding for the LUCO.

An amplitude mode-like coordinate, q, connects the two configurations on the opposite sides of the level crossing that are obviously not equivalent, unlike PA (cf. Figure 3.3). Still, some connection with polyacetylene remains, and the analysis leads to an approximate formula for the bandgap that contains a term linear in the average alternation. However, the structure with $E_g=0$ is now determined by an accidental degeneracy as opposed to the symmetry related to the zero bandgap of equidistant polyacetylene, and therefore this $E_g=0$ structure does not play a special role in deriving the most stable structure. Nevertheless, there is a substantial contribution coming from bond alternation along the polyacetylenic backbone of the polymer,

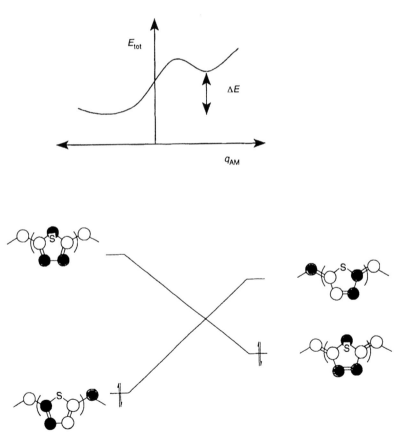

Figure 3.5. Top: Schematic change of the total energy per unit cell. The general *direction* of the connecting coordinate is approximately q_{AM} as defined in equation (5). Bottom: highest occupied and lowest unoccupied crystal orbitals (at $k=0$) of polythiophene along the same direction.

which is approximately linear in q_{AM}^{equ}, as defined as the average equilibrium bond length alternation [26]

$$q_{AM}^{equ} = (\sum r_{even} - \sum r_{odd})/n \qquad (3.5)$$

where r_{even} and r_{odd} refer to the optimized C–C bond distances along the carbon skeleton of the polymer and n is the number of terms in one of the sums in equation (5). This analysis has led to the following perturbation theory formula [32,46]

$$E_g = k|q_{AM}^{equ}| + E_g^{heter} + E_g^{torsion} + E_g^{hn} + corrections \qquad (3.6)$$

where the first term is a Peierls-like term, followed by a heteroatomic correction, and a term that is applicable only if the torsion of the neighboring rings is small (which is a precondition for π-conjugation), hn refers to higher neighbors, meaning direct π-overlap terms between atoms on the chain that are not first neighbors. This approximate formula helps to break down the calculated bandgaps into different components, but is applicable only for polymers, for which a polyacetylenic backbone can be identified (e.g. not for PPP). This analysis also permits observation of the trends and understanding of the role of the first term. Figure 3.6 shows this relationship among several conjugated polymers by plotting calculated E_g values as a function of the average bond length alternation, q_{AM}. It is clear from the figure that while the bandgaps follows a linear trend if the heteroatoms are omitted from the electronic structure (but not from the geometry optimization), E_g^{hetero} can be a dominant contribution. This explains,

for instance, why E_g(polypyrrole) $> E_g$(polythiophene) even though the former has a smaller alternation ($|q_{AM}|$ value), than the latter.

This relationship also explains the relatively small calculated bandgap of methenyl derivatives of polythiophene or PPy in a natural way. Figure 3.7 displays the HOCO and LUCO of the simplest such copolymer. Due to the nodal structure of these orbitals, the heteroatomic contribution is nearly equal for both, and they are both antibonding. As a consequence, in this case $E_g^{hetero} \approx 0$ [47]. Thus the bandgap is almost entirely determined by the $k|q_{AM}|$ term. Similar conclusions have been obtained for the polypyrrole-methylene copolymer [48] and other analogous π-conjugated polymers [47].

The correlation between E_g and q_{AM} is profoundly modified if additional rings are fused to the π-rings constituting the π-conjugated polymer. Figure 3.8 displays such a series based on PT, together with the corresponding HOCO-LUCO levels (again at the appropriate $k = 0$ value). Here the trends again can be well understood based upon the bonding and antibonding characteristics of the frontier orbitals at $k = 0$ [34]. However, as f (the number of fused rings perpendicular to the main chain) is increasing, the connection to the polyacetylene-based perturbation formula becomes remote, unless the rings attached to the thiophene unit are also treated as the heteroatomic perturbations. The trends found in Figure 3.8 and the corresponding calculated bandgap values have been confirmed experimentally, as shown in Figure 3.9.

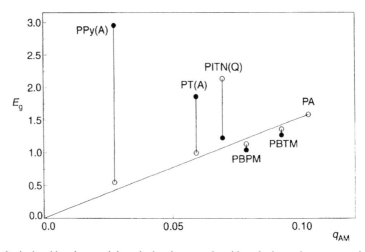

Figure 3.6. Correlation of calculated bandgap and the calculated average bond length alternation, q_{AM}, as defined in equation (5). The two values for each system correspond to one calculation without (○) and another with (●) the inclusion of heteroatoms.

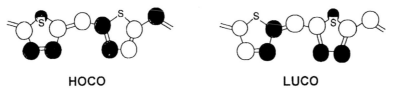

HOCO   **LUCO**

Figure 3.7. Highest occupied and lowest unoccupied crystal orbitals in a methenylene derivative of polythiophene. Both are pushed up relative to the polyacetylene orbitals, but the effect of sulfur nearly cancels the determination of E_g.

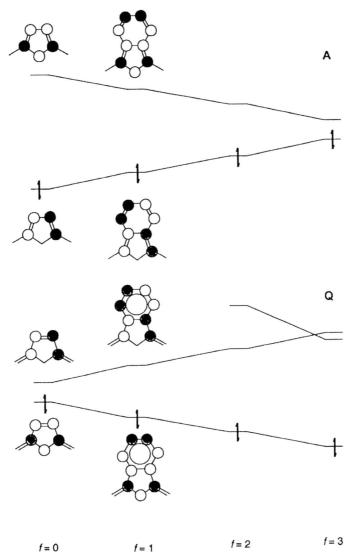

$f = 0$ $f = 1$ $f = 2$ $f = 3$

Figure 3.8. Node pattern of the frontier crystal orbitals in a series of polythiophene-related polymers ($k = 0$). Sulfur atoms are not shown; f designates the length of the oligoacenic fragment that is fused to each thiophene unit. $f = 0$: PT, $f = 1$: PITN, $f = 2$ PINT as defined in Figure 3.1. A: aromatic, Q: quinonoid structures.

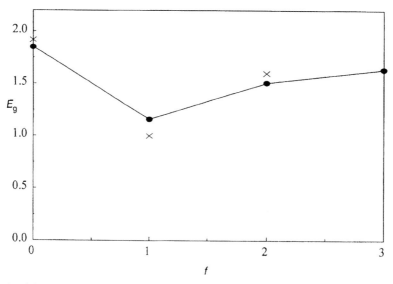

Figure 3.9. Calculated (o) vs. experimental (optical, x) bandgaps in the series shown in Figure 3.8. (See text.)

5   DESIGN OF SMALL BAND-GAP POLYMERS

The motivation behind the efforts represented by the results in Figure 3.9 on both the experimental and theoretical front, was to design new conjugated polymers with small E_g, a property that is deemed to be desirable for ease of acceptor and donor doping, high non-linear optical (NLO) response, possible transparency in the visible region of the absorption spectrum, etc. [10,46,49]. The design of such a conjugated system has been a challenge in its own right, regardless of potential utility. In addition to the systems mentioned in the previous section, a few others have been proposed as good candidates for small E_g. Several alternative ideas have been tried [46–59]. (Figure 3.10)

(a) Copolymers of ring systems with methenyl linkages $(-T_A-CH=T_Q-CH-)_x$.
(b) Copolymers of two different units $(\cdots S_A \cdots T_Q \cdots)_x$.
(c) Similar to (b) except the ratio of the two different repeat units (S and T) is not equal to 1.
(d) Conjugated ladder polymers (these will be discussed in the next section).

Copolymers made according to strategy (a) can be viewed as systems on which the less stable units (e.g. Q in the case of thiophene) are forced to co-exist with the more stable kind (in the thiophene example A). In cases (b) and (c) the inter-ring connecting bonds are being

forced to assume a length that is intermediate between the one that would occur if both units had the same aromatic structure, and the one that would occur if both were quinonoid. Table 3.4 compares a few such calculated values with experimental results. With the exception of poly(ITN-T$_2$), whose short chain segments may dominate the optical E_g measurement [58], the predictions are in remarkably good agreement with the experiments.

It should be emphasized that in achieving small E_g it is not the quinonoid structure itself, but the rather delicate balance between competing structures that leads to this desired property. Hong *et al.* have recently shown that for several $(ST)_x$ copolymers the bandgap can be well approximated by the formula:

$$E_{g,copolymer} \approx |\, w_A E_g{}^S{}_A + w_Q E_g{}^T{}_Q \,|, \qquad (3.7)$$

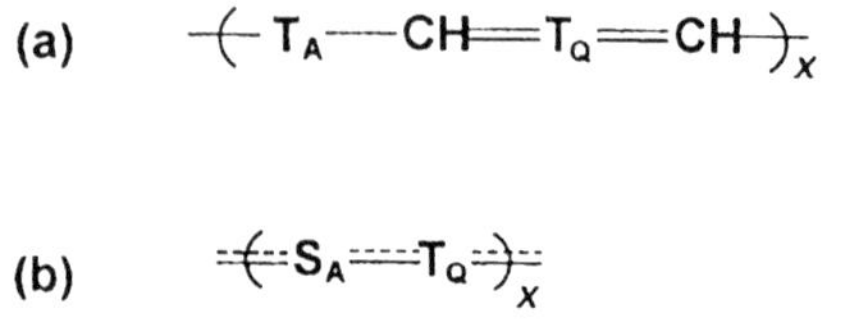

(a) $-\!\!\left(T_A\!-\!\!-\!CH\!=\!T_Q\!=\!\!CH\right)_{\!\! x}$

(b) $\cdots\!\!\left(S_A\!\cdots\!\!-T_Q\!\cdots\right)_{\!\! x}$

Figure 3.10. Schematic representation of two design strategies towards small bandgap-conjugated polymers: (a) copolymer of a ring system with (−CH−) methine groups. T_A and T_Q designate the aromatic and quinonoid structures of the *same* ring system, respectively. (b) Copolymer of two *different* ring systems, S_A and T_Q, where the homopolymer poly(S_A) would be aromatic and poly(T_Q) would be quinonoid in their respective ground undoped states.

Table 3.4. Correlation of experimental vs. theoretical E_g values for a series of thiophene-based small intrinsic E_g polymers (in eV)*

Polymer	E_g (experiment)	E_g (theory) [49]
Poly(T)	2.1 [55]	1.95
Poly(ITN)	1.0 [56]	0.88
Poly(ITN-T$_2$)	1.6 [58]	0.63
Poly(INT-T)	–	0.54
Poly(INT-T$_2$)	0.6 [59]	0.42
Poly(ThPyr)	1.0 [57]	0.91

*Poly(A–B$_n$) is a copolymer with the sequence of –A–B$_n$–A–B$_n$– etc. See Figure 3.1 for the various A and B units used.

where w_A and w_Q are the stoichiometric weights, $E_g{}^S{}_A$ and $E_g{}^T{}_Q$, are the bandgaps of the A-type and Q-type homopolymers, respectively. (Negative values for E_g are permitted here, referring to the reversal of the orbital order in the aromatic to quinonoid 'transition'.)

It should be noted, however, that the concepts of 'quinonoid' and 'aromatic' are very approximate and intermediate notions, and that neither are being fully realized. Ultimately, it is the nature of the orbitals that determines the properties of these systems.

6 LADDER POLYMERS

Fully conjugated ladder polymers have become experimentally available only recently [61–67]. Figures 3.11–3.15 show several main categories of these fascinating systems. Many of these new macromolecules have either five-membered carbon rings or seven-π-electron six-membered rings. An abstract transport calculation by Mizes and Conwell [68] has confirmed the expectation that the ladder topology is favorably facilitating transport along the chain, in the sense that even if one 'leg' of the ladder is broken, transport is still possible. Another intuitive expectation has been that the ladder topology, corresponding to more two-dimensional-like conjugation, ought to be favorable for delocalization. It will be seen that this latter expectation does hold for some ladder polymer, but not for all of them [69]. In fact, it turns out that many ladder polymers have an intrinsically well-localized electronic structure, with large bandgap values and rather narrow energy bands. At the same time, some members of this broad and growing class of conjugated polymers have small bandgaps, wide bands and rather delocalized defect (bipolaron) states that are favorable for high intrachain conductivity.

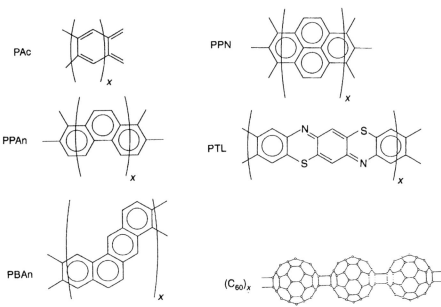

Figure 3.11. Selected ladder polymers discussed in this review. PAc: polyacene; PPAn: poly(phenanthrene); PBAn: poly(benzanthrene); PN; poly(perinaphthalene); PTL: poly-(phenothiazine); (C$_{60}$)$_x$: linear ladder polymer of buckminsterfullerenes.

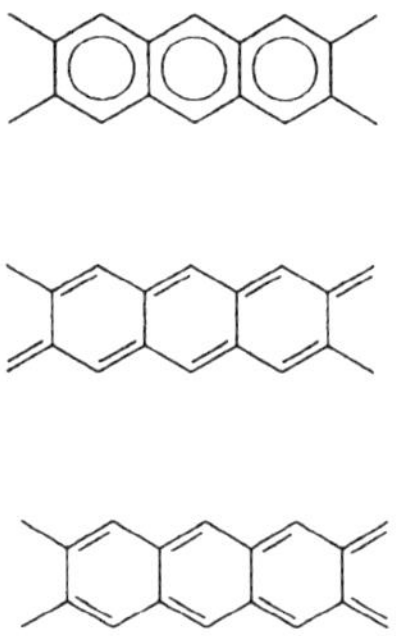

Figure 3.12. Polyacene models. Top: high symmetry, metallic (A); middle: *trans*-bond-localized (B); and bottom: *cis*-bond localized (C) structures.

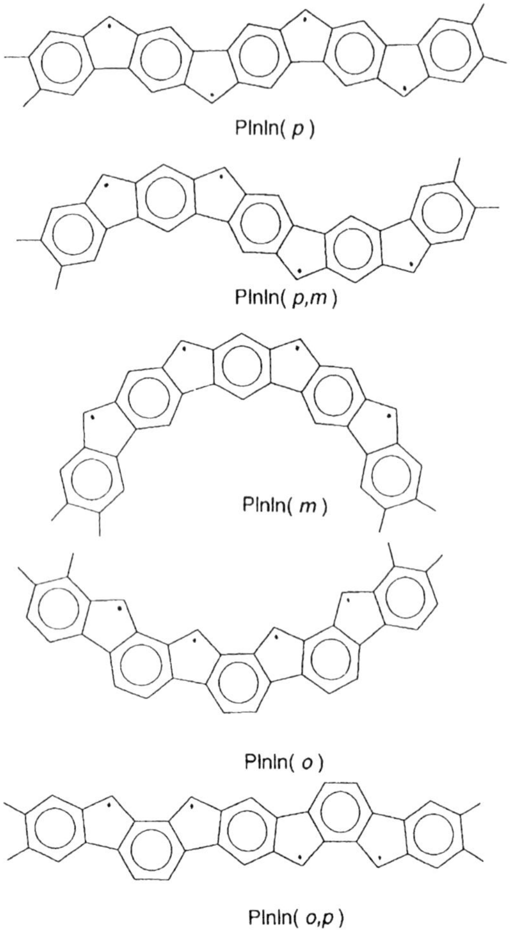

PInIn(*p*)

PInIn(*p,m*)

PInIn(*m*)

PInIn(*o*)

PInIn(*o,p*)

Figure 3.13. Poly(indenoindene), PInIn, isomers: Dots indicate the odd-electrons, side groups are not shown.

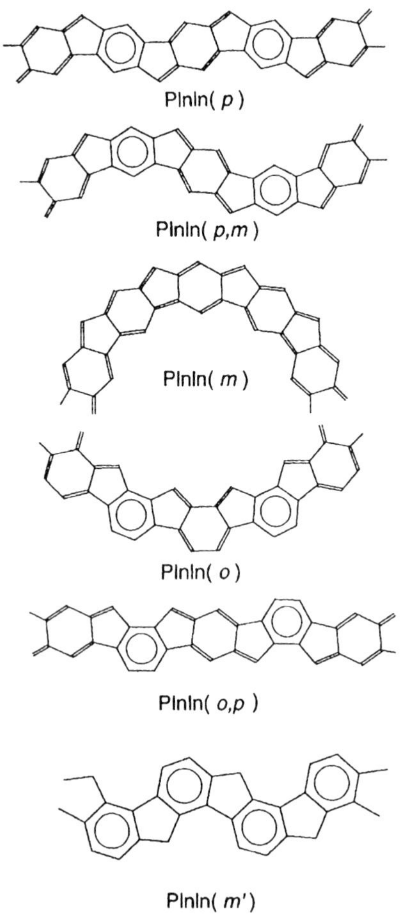

PInIn(*p*)

PInIn(*p,m*)

PInIn(*m*)

PInIn(*o*)

PInIn(*o,p*)

PInIn(*m'*)

Figure 3.14. Bond-localized forms of the isomers of poly(indenoindene) polymers. In the *m'* form bond localization is not possible [82].

Ladder polymers cover a wide variety of intrinsic electronic structures [68–76] ranging from very small bandgap systems to semimetals, to semiconductors, to large bandgap insulators. In this section, the electronic structures of various hydrocarbon and carbon systems, and heteroconjugated systems will be reviewed.

6.1 Polyacene

Polyacene is an ideal target for theoretical studies although it has not been made yet. The issue of the bond localization problem as illustrated in Figure 3.12 is different from that of polyacetylene in that there are two lower symmetry structures (B,C) into which the metallic structure (A) can develop. Most calculations,

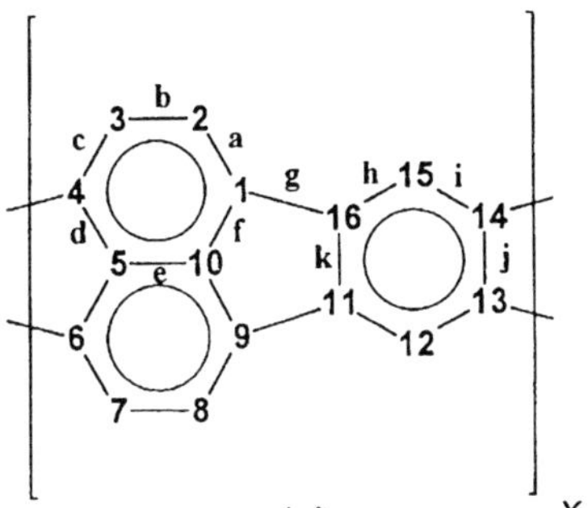

(a)

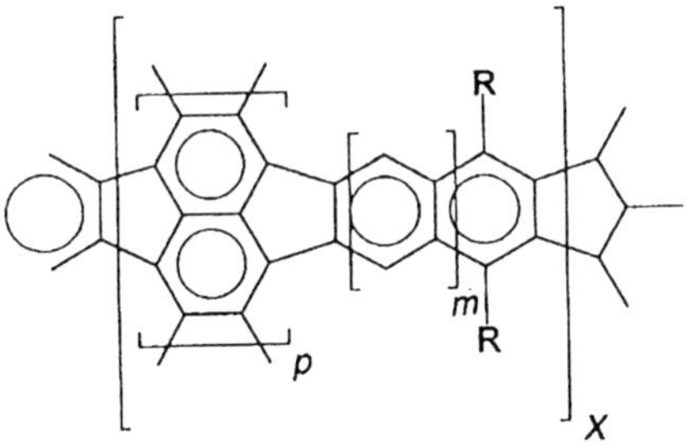

(b)

Figure 3.15. Polyfluoranthene, PFA, polymers. (a) $a = 2$, $m = 0$ case; (b) general formula. (p is the number of benzene rings in the polyacenic fragments perpendicular to the chain, ($m + 1$) is the same for those parallel to the chain.)

except LHS [27,28] predict that one of the non-metallic structures is slightly more favorable by a few tenths of a kcal mol^{-1} [75,76].

Predicted bandgap values based on the MNDO optimized geometry, followed by ETH bandstructure calculations, are in qualitative agreement with the extrapolated value based on the experimental $\pi\pi^*$ transitions of finite oligoacenes [77]. As a result, it seems that the bandgap of the hypothetical system of PAc has been established around 0.3–0.5 eV. Hetero-substituted PAc ladder polymers have been made, however, and will be discussed in Section 6.8.

6.2 Other fully conjugated planar ladder hydrocarbons based on six-membered rings

Several theoretical studies have been reported on fused ring polymers other than polyacene [75,78,79–81].

Most of these calculations have been done without the benefit of detailed comparison with experimental results, although most recently significant progress in synthesis has been made [67]. Figure 3.11 shows the schematic connectivity of three such polymers: poly(phenanthrene), poly(benzanthracene) and poly-(perinaphthalene). The non-geometry-optimized calculations of Liegener *et al.* [78] indicates, that the bandgap of poly(phenanthrene) and poly(benzanthracene) should be much larger than that of polyacene. This is in qualitative agreement with the geometry-optimized results of Tanaka *et al.* [75] for poly(phenanthrene), although both calculations suffer from the usual gross overestimation error of the bandgap by Hartree-Fock [10]. (The Hartree-Fock bandgap of polyacene by Tanaka *et al.* [75] has a very realistic value of 0.5 eV, however.) The optimized geometry of poly(phenanthrene) by Tanaka *et al.* [75] indicates strong bond localization. Toussaint and Bredas have arrived at both optimized geometries (based on oligomer calculations) and realistic calculated bandgaps (2.86 eV) for poly(benzanthracene). The experimental value is very similar, 2.79 eV [67]. These results indicate a relatively large degree of π-electron localization.

As a number of 'legs' of the ladder increases, the material becomes more and more similar to graphine [72] (one sheet of graphite), which is semimetallic [70].

6.3 Poly(indenoindene)s: π-ladder polymers with six- and five-membered rings

Scherf and Müllen [61] have achieved a synthetic breakthrough in synthesizing a new planarized poly-(phenylene) (PLPP, see Figure 3.1), from which, by dehydrogenation, a new type of ladder polymer incorporating indenoindene-like condensed six- and five-membered rings has been created. Figure 3.13 shows various possible isomers of these poly(indenoindene), PInIn, polymers. The m, o, and p indices refer to the relative positions of the methine carbon of the adjacent six-membered ring. So far the (p,m) isomer has been made [61]. The methine carbons contribute a single π-electron, making these systems possess an odd number (seven) of π-electrons per chemical repeat unit. Formally, a Peierls distortion is expected to occur, accompanied by a unit cell doubling (14 electrons) and bond localization according to the bond localization pattern shown in Figure 3.14.

The situation is more complicated, however. Peierls' theorem is only applicable, if the bandgap is zero before the symmetry breaking distortion is taken into account. A simple Hückel calculation on the (p,m) isomer produces a bandgap of 0.59 eV [82,83]. Even though this is not a Peierls system, geometry relaxation is still significant, as has been stressed in connection with many other examples of non-degenerate ground state polymers (see Section 4). If the geometrical relaxation is also taken into account, the calculated bandgap becomes 0.98 eV (LHS) or 1.19 eV (PM3 optimized geometry, bandstructure by LHS) [82,83]. These calculated values compare well with the experimental value of 0.7 to 0.9 eV [84]. In general, the bandgap values can be analyzed by the following generalization of equation (6) [80,82,83].

$$E_g = E_g^{topol} + E_g^{relax} + \text{corrections} \qquad (3.8)$$

where E_g^{topol} is the topologically determined bandgap, obtained from a Hückel-type calculation without geometry relaxation (the 0.59 eV value in the above case of the (p,m) isomer). The Peierls' theorem is applicable when $E_g^{topol} = 0$; E_g^{relax} becomes the Peierls term, $k \, | q_{AM}^{equ} |$ in the class of polymers discussed around equation (3.6).

The relaxation term is significant for the PInIn (p,m) polymer. For example in the LHS model, $E_g^{relax} \approx E_g^{calc} - E_g^{topol} = 0.98 - 0.59$ eV $= 0.39$ eV, which is a very significant contribution. The optimized geometry of PInIn (p,m), both as based on LHS or PM3, displays an alternation of benzenoid and quinonoid six-membered rings, indicating a significant degree of π-electron localization. The largest bond distance difference between adjacent bonds due to symmetry breaking in this system is at the two sides of the methine carbon and is 0.15 Å (by LHS) or 0.11 Å (by PM3).

By inspecting the symmetry of the other PInIn isomers in Figure 3.13, it becomes apparent that with the exception of the (p,m) isomer all of the other isomers considered have zero topological bandgaps. Therefore, a Peierls distortion should occur for all these other isomers. However, their bandgaps, after geometry optimization, fall into three groups as follows (in eV).

- First group $(p) - 0.55$, $(o) - 0.88$, $(po) - 0.55$.
- Second group: $(m) - 0.12$, $(p, m,m) - 0.13$.
- Finally, the third group has only one member (m') which does not undergo (non-magnetic) Peierls distortion; the bandgap remains zero even if geometry relaxation is permitted in the calculation.

Ukrainskii [85] has shown that the Peierls argument is also applicable to systems in which the unit cell has an even number of electrons, but a reduced cell with odd number of electrons exists, provided that the whole polymer can be built up from the reduced cell by a screw axis of symmetry. The example used by Ukrainskii was polyacetylene, but the argument is more generally valid [85]. This explains why the following isomers have a zero topological bandgap: (p), (m), (m'), (o).

A simplified analysis of these three types of isomers has been given in terms of the electronic structure of poly(arylenemethine)s, PAMs, that can be generated by severing, for the sake of the analysis, those C$-$C bonds that directly connect neighboring six-membered rings [82,83]. Accordingly, the Peierls distortion of those PInIn isomers that have bandgaps in the 0.55 to 0.88 eV (by LHS) region can be accounted for by an ordinary orbital analysis of the related PAM polymer. However, the second group of isomers ($E_g = 0.12$ to 0.13 eV) distort only slightly, in concordance with the almost complete localization in the respective isomers of the odd electrons on the methine linking π-centers. Finally, the (m') isomer does not distort at all at the LHS level of theory. Parenthetically, it should be noted that these latter isomers are potential candidates for polymers with ferri-magnetic properties [86]. The 'frustration' provided by the topology of the (m') structure leads to the impossibility to develop aromatic and quinonoid substructures, leading to delocalized π-electrons and zero bandgap. As mentioned above, in the largest bandgap isomer consider, the (p,m) isomer, the six-membered rings attain an alternating aromatic-quinonoid structure leading to appreciable π-electron localization.

6.4 Polyfluoranthenes: buckyboard polymers

Another group of innovative ladder polymers incorporating five-membered rings and benzene and/or oligoacenic fragments has been made recently in a remarkable synthetic effort by Schlüter et al. [62,63]. Figure 3.15 shows a family of such polymers. These polyfluoranthene, PFA, polymers consist of cumulated oligoacenic fragments connected by five-membered rings, and resemble two-dimensional, flattened and polymerized buckyballs, or 'molecular boards'. Each polymer is characterized by a p value (number of benzene rings in the perpendicular oligoacenic fragments) and an m value ($m + 1$ is the number of benzene rings in the parallel oligoacenic fragments). Even though only a few members (all with $p = 2$) have been made, the analysis provided by the variation of both p and m allows a better understanding of the degree of electron localization and the analysis of selected electronic structure parameters [69].

Below recent calculations [69] on the structural and electronic structure features of these polymers as a function of p and m are summarized. These studies allow the following: the question of how the ladder topology helps in maintaining delocalization under the influence of breakages in one 'leg' of the ladder structure, and the issue concerning the type of localized defect structures that are generated if additional charges are transferred to the ladder polymer.

Table 3.5 compares some typical C–C bond distances in fluoranthene and polyfluoranthene ($p = 2$, $m = 0$), and further compares them with the experimental values for fluoranthene [87]. The difference between molecular fluoranthene and the polymer is quite small, aside from the obvious change of symmetry. This rather small difference between the polymer and monomer geometry is indicative of the significant degree of electron localization to the $4n + 2$ membered benzenoid and the oligoacenic fragments in this polymer.

Figure 3.16 shows the dependence of the calculated bandgap as a function of p. The topological bandgap values follow a similar trend as the fully relaxed values, as given in Table 3.6, even though the relaxation itself provides a significant contribution. Both trends display

a minimum gap, although this minimum occurs at a slightly different value of p. Consequently, the most attractive target for synthesizing a reduced bandgap member of the whole series correspond to $m = 0$, $p = 4$ and 6.

The V-shaped dependence of E_g on p can be readily understood by inspecting the frontier orbitals of the $m = 0$ series. A level crossing occurs between $p = 4$ and 6, as shown in Figure 3.17. Such a crossover is the consequence of the topology of the system and occurs already for the non-relaxed, 'topological' calculation.

The optical absorption edge for the simplest polyfluoranthene ($p = 2$, $m = 0$) has been measured by Schlüter et al. [62,63]. The absorption spectrum has an ill-defined extended peak in the 550 to 600 nm region (~ 2.2 eV). Due to possible chain length or conjugation length limitations of the samples, it is reasonable to conclude that the calculated bandgap of 1.63 eV for the infinite PFA ($p = 2$, $m = 0$) should be lower than the experimental transition energy, and therefore it is in qualitative agreement with the available experimental data.

The geometry of PFAs as a function of p displays a very interesting trend, as illustrated for the $m = 0$ case in Figure 3.18. (The trend for the $m = 1$ case is similar, and is now shown here.) For $p = 2$, $m = 0$ (one of the experimentally synthesized polymers), the connecting five-membered ring has the longest C-C bond, indicating that the electrons in the two kinds of oligoacenic fragments are localized to some extent. This π-electron localization on the $4n + 2$ Hückel-type substructures, i.e. in the oligoacenic fragments, is characteristic for the $p = 2$ series. This explains not only the occurrence of

Table 3.5. Bond distances (in Å) for fluoranthene molecule (FA) and polyfluoranthene (PFA)

Bond type (Figure 3.15)	X-ray for FA Hazell et al [87]	FA LHS optimized geometry	PFA
g	1.476	1.472	1.471
a	1.372	1.390	1.393
b	1.407	1.420	1.416
i	1,383	1.398	1.405
h	1.388	1.407	1.405

Table 3.6. Variation of the fully relaxed bandgap (LHS), and the topological bandgap, as a function of p, the number of 'vertical' polyacenic rings in the unit cell of PFA (in eV)

p	$m = 0$		$m = 1$	
	LHS	topological	LHS	topological
2	1.637	1.062	1.633	1.066
4	0.601	0.086	0.760	0.264
6	0.118	0.285	0.293	0.051
8	0.401	0.454	0.098	0.199
10	0.527	0.478	0.245	0.277

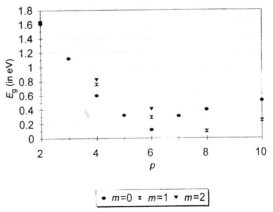

• $m=0$ × $m=1$ ▼ $m=2$

Figure 3.16. Energy bandgap, E_g of PFAs, as a function of p (p and m are defined in Figure 3.15b). (Reprinted with permission from ref. 69, Copyright 1996 American Chemical Society)

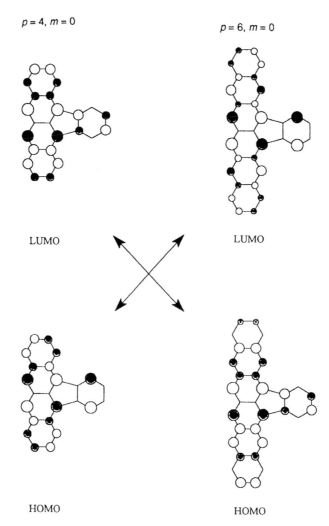

Figure 3.17. Frontier orbitals of the repeat units of PFA showing a level crossing from the polymer $p=4$ to $p=6$ ($m=0$). (Reprinted with permission from ref. 69, Copyright 1996 American Chemical Society)

the longest bonds in the five-membered rings, but also the trends of the high degree of localization of the bipolarons. However, with increasing p values the longest C–C bond (g) decreases, indicating a better delocalization across the five-membered rings. Around the crossover point, discussed above, π-conjugation increases and the bandgap decreases.

Another tool that is available for bandstructure 'engineering' of conjugated polymers is provided by heteroatomic substitutions. According to calculations performed on five different substituted polyfluoranthene polymers [69], pyridine-type nitrogen substitutions produce shifts in the $+0.2$eV to -0.15eV range. These substitutional effects follow the patterns

of the frontier orbitals; the smallest bandgap corresponds to the substitution at the C_{15} carbon site and is 1.5 eV for the $p=2$, $m=0$ case. The combination of the three effects: topology, geometrical relaxation and heteroatomic substitutions, allows fine-tuning of the electronic structure that has not been fully explored so far, either experimentally, or theoretically.

6.5 Bipolarons and solitons in ladder polymers

In this section the recent results on LHS-type calculations concerning polarons and solitons on ladder-type

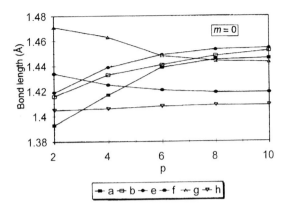

Figure 3.18. Variation of the C–C bond distances as a function of p, the number of rings in the perpendicular oligoacenic fragments of polyfluoranthene polymers. ($m = 0$ series) (See Figure 3.15a for the notation of bonds.)

polymers are summarized, and the differences and similarities in comparison with single main chain, i.e. non-ladder, polymers are emphasized. For comparative purposes consider some features of similar calculations on charged poly(p-phenylene), with 15 phenylene units constituting the supercell (both $+2$ and -2). The results are shown in Figure 3.19. These show the following trends.

(1) There is full symmetry between the positively and negatively charged bipolarons, in agreement with the fact that PPP is an alternant hydrocarbon and the LHS model in such a case displays perfect electron-hole symmetry.

(2) The results are in very good agreement with the best available *ab initio* calculations [88] in terms of the size of the bipolaron (depending on the definition, five or seven unit cells), although the center of the bipolaron has a somewhat more uniform charge distribution according to the *ab initio* calculations [88].

(3) The C–C bond distances by LHS also agree very well with the large double zeta type *ab initio* calculations. The extent of the bipolaron can be also identified from Figure 3.19b, which shows the inter-ring C–C bond distance along the supercell of 15 phenylene rings. Again, the bipolaron extends over five to seven rings; the shortest and longest bond length values are 1.409Å and 1.435Å, respectively. The corresponding *ab initio* values are 1.395Å and 1.445Å, as estimated from the paper of Karpfen and Ehrendorfer [88]. The *ab initio* calculation shows a slight asymmetry with respect to the sign of the bipolaron, while the LHS model calculation should not and does not.

The results on the PFA Schlüter polymers are shown in Figure 3.20. These display the characteristics of a localized bipolaron, as can be seen from the pattern of bond distances. The bipolarons clearly show localization over three unit cells (six groups in terms of those separated by five-membered rings), with very minor perturbations extending over one further neighbor on each side. The degree of localization of the bipolaron for PFA is very similar to that of PPP if the fact that the unit cell in the latter case is much larger is taken into account.

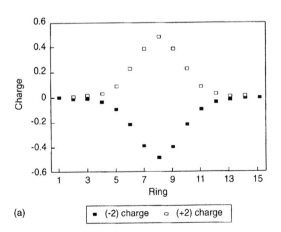

(a)

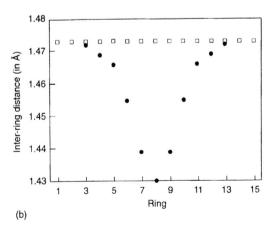

(b)

Figure 3.19. Bipolarons in poly(p-phenylene), PPP. Calculations refer to both positive and negative bipolarons. (a) Charges along the supercell of 15 rings; (b) inter-ring bond distances; the upper data points indicate the value without the bipolaron (in Å).

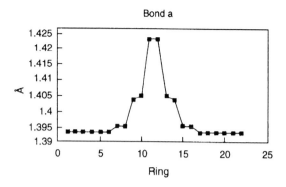

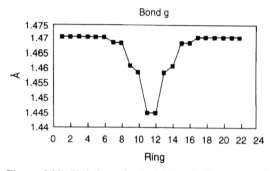

Figure 3.20. Variation of selected bond distances in the bipolaronic lattic calculation of PFA ($p=2$, $m=0$). Charge is (-2) for the whole supercell. The bipolaron is in the middle of a supercell containing 11 chemical repeat units.

Another way to look at the bipolaron is to plot the charge distribution along the polymer. In Figure 3.21 charges for each unit are shown along the chain. Accordingly, the bipolaron extends over five to seven oligoacenic groups (benzene and naphthalene groups), although the charge distribution is more peaked around the center than in PPP. As one moves away from the center of the defect, the charges of these units converge to the same value independently of the sign of the charge of the bipolaron, clearly indicating that the size of the bipolaron is about seven units (three and a half chemical repeat units). The symmetry with respect to the sign of the biplaron has been lost relative to PPP, since PFAs are non-alternant hydrocarbons.

What is the influence of the bandgap on the degree of localization of the bipolaron? It follows from very general consideration that the bipolaron should be completely delocalized for zero bandgap and should be completely localized to one unit cell for a molecular species for which there is no appreciable coupling between the unit cells. This question can be assessed

more quantitatively, by plotting the bipolaron bond distance distribution for the $p=4$, $m=0$ polymer, which has a bandgap of less than half that of the $p=2$, $m=0$ system. The corresponding bond distances are plotted in Figure 3.22, showing a defect extending over about seven to nine chemical units, i.e. a defect that is, by about a factor of two, more delocalized than in the $p=2$, $m=0$ case, as can be seen also by making a comparison with Figure 3.20. Since the amount of charge is the same, the deformations of the bonds relative to the uncharged polymer are smaller. This indicates that the geometrical relaxation of the small bandgap polymers will be less significant and more delocalized as compared to their large bandgap relatives.

6.6 Conjugation interruption sp³ defects in a ladder polymer

The role of sp³-type carbon defects may be less significant in PFA polymers than in single strand polymers, given their ladder topology. Mizes and Conwell have shown [68] that as long as one of the two 'legs' of the ladder topology remains available for π-conjugation, the on-chain conductivity may be maintained. It is relatively straightforward to model such defects in the LHS scheme. For instance, C_{15} can be transformed into an sp₃-type carbon by eliminating bonds H and I in the unit cell (numbering refers to Figure 3.15). Such a change leads to a polaronic defect once the lattice is allowed to relax. The charge distribution and bond relaxation pattern, corresponding to such a defect for the $p=2$, $m=0$ PFA polymer is shown in Figure 3.23. The charge distribution extends over only four rings, and is therefore somewhat more localized than the bipolaron (Figure 3.20 and 3.21) in the same system.

An important issue is, how does the makeup of the repeat unit influence this localization. The comparison of the charge distribution of the polaron for two cases, ($p=2$, $m=0$ vs $p=4$, $m=0$) shows that the latter has a much smaller bandgap, and that the corresponding polaron extends over seven or more units. This conclusion is very similar to the dependence of the size of the bipolaron on the bandgap. In both cases, small bandgaps tend to produce less localized defects. This provides a mechanism for reduced scattering of sp³ defects in ladder polymers, which should enhance their electrical conductivities.

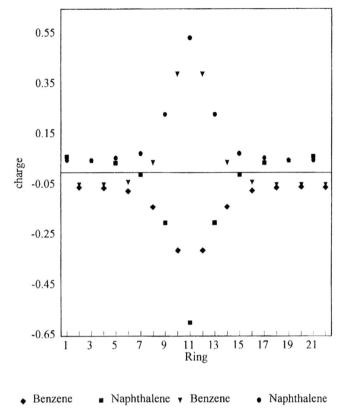

Figure 3.21. Variation of charge distribution among naphthalenic and benzenoid units in the bipolaronic lattice calculation for PFA. The bipolaron is in the middle of a supercell containing 11 chemical repeat units. ($p = 2$, $m = 0$ case). Both positive (filled symbols) and negative (empty symbols) bipolarons are given. (Reprinted with permission from ref. 69, Copyright 1996 American Chemical Society)

6.7 Carbon nanotubes

Carbon nanotubes [89–92] are the ultimate realizations of multiladder polymers. They can be viewed as the wrapping of a graphite sheet to form a tube with a diameter of the order of a nanometer. Many such wrappings are possible, and some of the most interesting theoretical work [90–92] has focused on establishing the connection of the energy bandstructure of a honeycomb graphite layer (called graphine) to those of a multitude of possible wrappings, many of which produce chiral tubes [90,91]. Special wrappings correspond to non-chiral tubes, some of which can be viewed as analogs of the extension of the series that one obtains starting from polyacetylene and polyacene, poly(phenanthrene), and continuing to poly(peri-naphthalene) and so on.

The most fascinating aspect of this mapping is that it can be used to rationalize large-scale sophisticated DFT calculations on carbon nanotubes. Accordingly [90], there are two distinct groups of periodic nanotubes, as far as the energy gap is concerned. In one group, the bandgap is significantly different from zero (of the order of 1 to 0.3 eV for a radius up to 1.5 nm), and the bandgap is approximately inversely proportional to the radius of the tube. In the other group the bandgap is very small, and the tube does not undergo a Peierls distortion at normal temperatures [92]. An analysis of these results along the lines of equation (3.8) might shed light on the absence of this Peierls distortion. While the chemical processes for making nanotubes are very different from those of conjugated hydrocarbon polymers, their existence and interesting properties imply that further efforts to make new

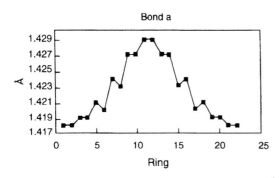

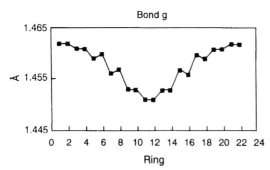

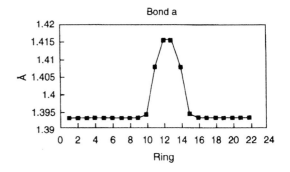

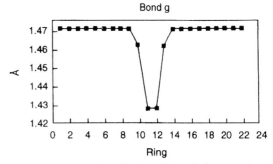

Figure 3.22. Variation of selected bond distances in the bipolaronic lattice calculation for PFA. ($p = 4$, $m = 0$ case). The (2-) bipolaron is in the middle of a supercell containing 11 chemical repeat units. (Connecting lines are provided to guide the eye; bonds are defined in Figure 3.15a.) (Reprinted with permission from ref. 69, Copyright 1996 American Chemical Society)

Figure 3.23. Variation of selected bond distances in the polaronic lattice calculation for PFA. The polaron is in the middle of a supercell containing 11 chemical repeat units. ($p = 2$, $m = 0$ case), the charged (spinless) polaron arises due to an sp^3 defect in the polyacenic fragments at C_{15}. (numbering refers to Figure 3.15a).

ladder-type conjugated polymers might be very fruitful and attainable.

6.8 Heteroconjugated ladder polymers with degenerate ground state: PXL

Three isoelectronic ladder polymers have been made by Kim [93] and Dalton *et al.* [94] that can be derived from a polyacene framework by substituting the apical carbons in every other six-membered ring by an imine nitrogen and another group, X. The generic name of PXL (L for ladder) will be adopted here [93]. Depending on this other group or atom, one obtains:

- for X = NH: poly(phenoquinoxaline), (PQL) [94]
- for X = S: poly(phenothiazine), (PTL) [93]
- for X = O: poly(phenoxazine), (POL) [93].

From the point of view of π-electron counting, all three X groups above provide two π-electrons, while

the imine nitrogen provides one. Therefore, the chemical repeat unit of each contains eight π-centers (six carbons, one imine nitrogen and one X) and nine π-electrons. [95]. Formally, this would imply four and one half occupied energy bands and therefore a metallic ground state (zero bandgap). Of course, the Peierls distortion intervenes, and the structure distorts by unit cell doubling, leading to a unit cell of 16 π-centers and 18 π-electrons (nine occupied bands) with a non-zero bandgap. Figure 3.24 illustrate one of these polymers before and after the Peierls distortion has been taken into account. In the distorted structure, PTL(II), the distortion breaks the symmetry and instead of all six-membered rings being equivalent, as in PTL(I), an alternation of benzenoid (aromatic) and quinonoid rings occurs, in close analogy to the alternation of long and short bonds in polyacetylene. The ground state is degenerate, because the energy of the system is independent from whether the even-numbered or the odd-numbered rings are quinonoid (AQAQ sequence vs QAQA sequence).

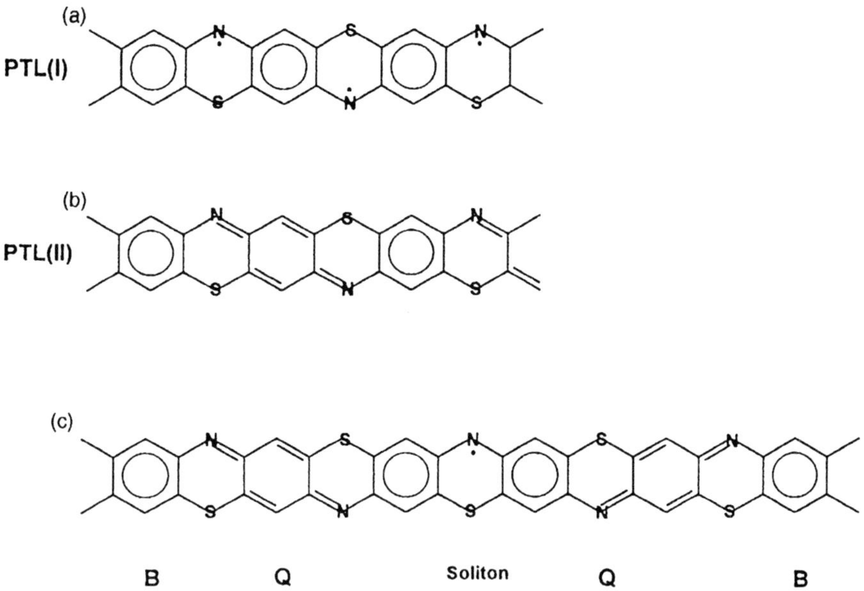

Figure 3.24. Structure of the unit cell of PTL: (a) before the Peierls distortion is taken into account: PTL (I); (b) including the Peierls distortion: PTL (II); (c) solitonic defect in PTL (spin = $\frac{1}{2}$, charge = 0).

Since the electronic structure of the PXL polymers can be derived from a half-filled band case, the same type of local defects can be obtained as in the case of other half-filled band polymers, such as polyacetylene. Such a solitonic defect is illustrated in Figure 3.24c, incorporating a transition region that occurs as the ...AQAQ... sequence is turning into the ...QAQA... sequence. A major difference with respect to poly-acetylene is expected to occur, however. The extent of the Peierls distortion, and therefore the size of the resulting bandgap should be strongly affected by the more rigid ladder topology. Calculations bear out these qualitative points [95]. The MNDO band theory calculations with optimized geometries show that the type II structures are more stable than type I structures by 7 to 8 kcal mol^{-1} for all three PXL systems. The lower symmetry type II structures display the expected ...AQAQ... alternation, although the extent of the alternation is not sufficiently large to call the Q structures purely quinonoid. The largest distortion occurs around the imine nitrogens, where the difference between the two bonds is 0.10 to 0.11 Å.

The AQ structural alternation leads to calculated bandgaps that are in the range of a few tenth of an eV [95], and in the same range as the bandgap of polyacene [76] as obtained by the same techniques (EHT band theory preceded by MNDO band theoretical geometry optimization). These calculated bandgap values (0.32 to

0.46 eV for the three PXL polymers) might explain the broad peak in the IR spectrum of PTL that has been observed [93] at 2600 cm^{-1} (0.3 eV). Other features of the calculated bandstructure are also in qualitative agreement with the UV/vis spectra [93]. However, it must be noted that the similarity in the calculated bandgaps of polyacene and the three PXLs is coincidental. The origin for the bandgap is very different: in PXL it is due to a Peierls distortion of a half-filled band, while in polyacene its origin is a second-order Peierls distortion.

6.9   Heteroconjugated ladder polymers with non-degenerate ground state: BBL

Rigid rod polymers constitute an important class of conjugated polymers [96], primarily because of their thermal stability and mechanical properties. They have been investigated in various groups as conductors and high NLO materials [97]. One such polymer is BBL, poly(benzimidazobenzophenanthroline), a genuine ladder polymer (see formula of the repeat unit in Figure 3.1). Other members of this family of polymers are occasionally being referred to as ladder-type, but in these other polymers, the ladder topology is occasionally interrupted, and in going from one unit to the next at least one of the connections is a single strand. In

BBL, just like in the other genuine ladder polymers discussed in this review, there are always at least two connections between each neighboring units.

BBL has a very robust geometrical and electronic structure [98]. The electrons are well localized in the aromatic naphthalenic and the benzenoid units that are separated by the imine groups. As a consequence, the system has a relatively large bandgap (1.99 eV by the modified EHT based on MNDO geometry optimization [98], 2 to 2.2 eV by optical absorption [93,97].

BBL behaves differently upon doping than other conducting polymers, in that it does not display a red shift. This has been attributed to the small geometrical relaxation of BBL upon doping and to the very small energy dispersion of the conduction band. The small overall geometry relaxation upon protonation is due to the localized nature of the effect of protonation not affecting the aromatic nature of the naphthalene groups very much. According to this interpretation [98] no bipolarons are formed in this non-degenerate ground state polymer upon the doping process.

7   EFFECT OF CHARGE TRANSFER ON THE GEOMETRY OF CONDUCTING POLYMERS

Charge transfer upon doping modifies the electronic properties of conjugated polymers enormously, often making them highly conducting [1]. The mechanism of this doping process is extremely complex both structurally and electronically, and the nature of the highly conducting metallic-like state is still being intensively studied [99]. The effect of the dopants, their positions and distributions, the creation of solitons and polarons (bipolarons) as a result of lattice relaxation of the polymer are not fully understood. This review will summarize some results on modeling one important aspect of these complex and interrelated

issues the effect that large amount of charge transfer has on the geometry of conjugated polymers. In this discussion, the following assumptions are made:

(1) The counterions are not explicitly taken into account. This assumption allows us to focus on the relaxation of the polymer without the inhomogeneous field of the counterions, which is likely to contribute to the localization of the transferred charges in the vicinity of these counterions. However, the location of these ions is not known experimentally.

(2) The transferred electrons are uniformly delocalized over the polymer. This assumption is not necessary, but allows the use of band theory. It excludes the possibility of the relaxation into solitons, bipolarons, etc. and is certainly *not* valid in the low doping regime, where these localized defects can and will form freely. At dopant concentrations high enough to produce significant overlap between the defect states, band theory becomes more applicable. This is the regime where the calculations reviewed in this section can be relevant. In particular, they ought to be relevant in describing average changes and trends, as various groups of highly doped conducting polymers are compared.

A measure of the strength of the geometry relaxation, albeit a somewhat crude one, is the critical amount of charge transfer at which the symmetry breaking distortion disappears in the calculation. Table 3.7 displays the calculated critical charge transfer values for a few systems with degenerate ground states including ladder and single strand polymers. Polyacetylene converts to the non-alternating structure at a much larger dopant concentration than the ladder polymers. The quite low critical doping value for PInIn(p) calculated in this oversimplified manner can be traced back to a similar behavior of the corresponding PAM(p)

Table 3.7 Effect of charge transfer on bond length alternation using band theory

Polymer	Critical charges (in e) at which alternation vanishes	Geometry optimized by
(CH)$_x$ [96]	0.1 /carbon	MNDO
[82]	0.1 /carbon	LHS
PTL [95]	0.8 /unit cell $= 0.05$ /π-atom	MNDO
PAM(p), no closed five-membered ring [82]	0.25 /unit cell $= 0.017$ /carbon	LHS
PInIn(p) [82]	0.18 /unit cell $= 0.012$ /carbon	LHS

polymers. The fact that this critical concentration is very low makes the ladder polymers attractive, because they may be doped into the highly conductive regime at relatively low dopant concentrations.

Doping also changes the size of the unit cell, a property that perhaps may be utilized in electrical actuators [100]. Figure 3.25 shows the dependence of the size of the unit cell as a function of dopant concentration for polyacetylene in two models: in one, the length of the unit cell, C, is plotted against the

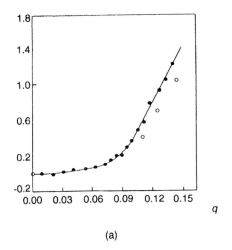

(a)

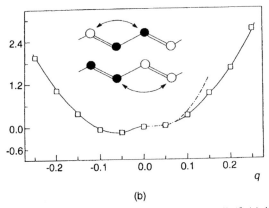

(b)

Figure 3.25. Change in the size of the unit cell ($\delta c/c$) in doped polyacetylene, $(CH^q)_x$ as a function of doping level, q. (a) soliton lattice [101]; (b) periodic (C_2H_2) chain model ([96]. Experimental data from M.J. Winokur *et al. Solid State Commun.* **68**, 1055 (1988), are represented by ● in (a) and by $\cdots\cdots\cdots$ in (b). The insert in (b) shows the HOCO and LUCO orbitals that are both antibonding in the second neighbor sense, causing the lack of symmetry with respect to the sign of the doping. (Reprinted with permission from the American Institute of Physics.)

charge transfer where soliton relaxation has been taken into account [101]. The results agree remarkably well with the results of the oversimplified energy band model in which the periodicity of two CH units has been artificially maintained [96]. This latter model does not include the relaxation into a soliton lattice; nevertheless the two models agree quite well with each other and with the available experimental data. It seems that the overall change upon doping is well characterized on average by the fundamental nodal properties of the wavefunctions around the Fermi level, which are slightly antibonding. As a result, even without taking the soliton relaxation into account, changes upon doping in polyacetylene are such that donor doping expands and acceptor doping shrinks the lattice.

8 CONCLUDING REMARKS

In this section a few exciting new systems will be mentioned for which theoretical work has not been undertaken or has only just began. More theoretical work will lead to a better understanding of the properties of these promising new materials and to the design of new ones.

Polymers including directly connected C_{60} units have been suggested to exist by several groups [64]. Theoretical work has attempted to describe the properties of these polymers, but can do so only by assuming a certain connectivity of the buckyballs. There are several alternative ways to make those connections; the connections may have very unusual bonding [64] and no consensus has yet emerged among competing theoretical models even for the simplest $2+2$ cycloaddition model of a buckyball polymer [102,103]. Tanaka *et al.* [102] suggested other conjugated polymers in which the C_{60} units are not directly connected, but are joined by various conjugated groups, showing that in some cases the delocalization is severed by the alternation of C_{60} with the conjugated linking group.

Experimental work in the direction of making new unsaturated ladder polymers is producing further new π-conjugated polymers [104]. New heteroconjugated polyacenic systems are also being made [105], which await more physical characterization and theoretical interpretation.

9 ACKNOWLEDGMENT

This work has been partially supported by NSF through a grant # DMR-9115548.

10 REFERENCES

1. C.K. Chiang, C.R. Fincher, Y.W. Park, A.J. Heeger, H. Shirakawa and E.J. Louis, *Phys. Rev. Lett.* **39**, 1098 (1977).
2. T.A. Skotheim (ed.) Handbook of Conducting Polymers, Vols. 1 and 2, Marcel Dekker, New York, 1986.
3. J.L. Bredas and R.R. Chance (eds.) *Conjugated Polymeric Materials*, Kluwer Academic, Dordrecht, 1990.
4. J.L. Bredas and R. Silbey (eds.) Conjugated Polymers, Kluwer Academic, Dordrecht, 1991.
5. H.G. Kiess (ed.) Conjugated Conducting Polymers, Springer-Verlag, Berlin, 1992.
6. W.R. Salaneck, I. Lundstrom and B. Ranby (eds.) Conjugated Polymers and Related Materials, Oxford University Press, 1993.
7. J.M. Andre, *Adv. Quantum Chem.* **12**, 65 (1980).
8. M. Kertesz, *Adv. Quantum Chem.* **15**, 161 (1982).
9. J.J. Ladik, *Quantum Theory of Polymers and Solids*, Plenum Press, New York, 1988.
10. J.M. André, J. Delhalle and J.L. Bredas, *Quantum Chemistry Aided Design of Organic Polymers*, World Scientific, Singapore 1991.
11. A. Karpfen, in *Polydiacetylenes* (eds. D Bloor and R.R. Chance), NATO ASI Series E, Vol. 102, Nijhoff, Dordrecht, 1985.
12. A.A. Ovchinnikov and I.I. Ukrainskii, *Sov. Sci. Rev. B. Chem.* **9**, 125 (1987).
13. D.K. Campbell, A.R. Bishop and M.J. Rice, in *Handbook of Conducting Polymers* (ed. T.A. Skotheim), Vol. 2, Marcel Dekker, New York, 1986, p. 937.
14. A.J. Heeger, S. Kivelson, J.R. Schrieffer and W.P. Su, *Rev. Mod. Phys.* **60**, 781 (1988).
15. R.R. Chance, R.S. Boudreaux, J.L. Bredas and R. Silbey, in *Handbook of Conducting Polymers* (ed. T.A. Skotheim), Vol. 2, Marcel Dekker, New York, 1986, p. 825.
16. Z.G. Soos and G.W. Hayden, in *Excited States in Conjugated Polymers* (ed. T.A. Skotheim), Marcel Dekker, New York, 1988, pp. 197-266.
17. B.E. Kohler, in *Conjugated Polymers* (eds. J.L. Bredas and R. Silbey, Kluwer Academic, Dordrecht, 1991, p. 405.
18. D. Baeriswyl, D.K. Campbell and S. Mazumdar, in *Conjugated Conducting Polymers* (ed. H.G. Kiess), Springer-Verlag, Berlin, 1992, p. 7.
19. P. Vogl and D.K. Campbell, *Phys. Rev. Lett.* **62**, 2012 (1989); P.G. daCosta *et al. Synth. Met.*, **57**, 4320 (1993), C. Ambroschdraxl *et al. Synth. Met.* **69**, 411 (1995).
20. E.M. Conwell and H.A. Mizes, *Phys. Rev. B* **51**, 6953 (1995) and references therein.
21. Z.G. Soos and G.W. Hayden, *Chem. Phys.* **143**, 199 (1990).
22. See for example, R. Hoffmann, *Solids and Surfaces: A Chemist's View of Bonding in Extended Structures*, VCH Publishers, New York, 1988.
23. W.P. Su, J.R. Schrieffer and A.J. Heeger, *Phys. Rev. Lett.* **42**, 1698 (1979), M.J. Rice, *Phys. Lett.* **A71**, 1521 (1979).
24. B. Horowitz, Z. Vardeny, E. Ehrenfreund and U. Brafman, *Synth. Met.* **9**, 215 (1984).
25. G. Zerbi, M. Gussoni and C. Castiglioni, in *Conjugated Polymers* (eds. J.L. Bredas and R. Silbey), Kluwer Academic, Dordrecht, 1991, p. 435; C. Castiglioni, M. Del Zoppo and G. Zerbi, *J. Raman Spectrosc.* **24**, 485 (1993).
26. R. Peierls, *Quantum Theory of Solids*, Oxford University Press, 1954, p. 108.
27. H.C. Longuet-Higgins and L. Salem, *Proc. Roy. Soc. London A* **251**, 172 (1959).
28. More recent application of the LHS [27] Hamiltonian include M. Kertesz, P.R. Surján, *Solid State Commun.* **39**, 611 (1981); D.S. Boudreaux, R.R. Chance, J.E. Frommer, J.L. Bredas and R. Silbey, *Phys. Rev. B* **31**, 652 (1985); K. Kürti and P.R. Surján, *J. Chem. Phys.* **92**, 3247 (1990).
29. C.R. Fincher, C.E. Chen, A.J. Heeger, A.G. MacDiarmid and J.B. Hastings, *Phys. Rev. Lett.* **48**, 100 (1982); C.S. Yannoni and T.C. Clarke, *Phys. Rev. Lett.* **51**, 1191 (1983); H. Kahlert, O. Leitner and G. Leising, *Synth. Met.* **17**, 467 (1987); Y.B. Moon, M.J. Winokur, A.J. Heeger, J. Barker and D.C. Bott, *Macromolecules* **20**, 2457 (1987); M.J. Winokur, Y.B. Moon, A.J. Heeger, J. Barker and D.C. Bott, *Solid State Commun*, **68**, 1055 (1988).
30. T.C. Chung, F. Moraes, J.D. Flood and A.J. Heeger, *Phys. Rev. B* **29**, 2341 (1984); S. Etemad *et al. Mol. Cryst. Liq. Cryst.* **117**, 275 (1985).
31. See for example K. Pakbaz, C. Lee, A.J. Heeger, T. Hagler and D. McBranch, *Synth. Met.* **64**, 295 (1994).
32. S.Y. Hong and D. Marynick, *J. Chem. Phys.* **96**, 5497 (1992).
33. J.L. Bredas, B. Themans, J.P. Fripiat, J.M. Andre and R.R. Chance, *Phys. Rev. B* **29**, 6761 (1984).
34. Y.S. Lee and M. Kertesz, *Int. J. Quantum Chem. Symp.* **21**, 163 (1987).
35. J.W. Mintmire, C.T. White and M.L. Elert, *Synth. Met.* **25**, 109 (1988).
36. J. Kürti and P.R. Surjan, *J. Chem. Phys.* **92**, 3247 (1990).
37. J.M. Toussaint and J.L. Bredas, *Synth. Met.* **69**, 637 (1995).
38. Y. Shimai and S. Abe, *Synth. Met.* **69**, 687 (1995).
39. Y.S. Lee, M. Kertesz and R.L. Elsenbaumer, *Chem. Mater.* **2**, 625 (1990).
40. A. Karpfen and M. Kertesz, *J. Phys. Chem.* **93**, 7681 (1991).
41. J. Cioslowski, *Chem. Phys. Lett.* **153**, 446 (1988).

42. W.J. Hehre, L. Radom, P.v.R. Schleyer and J.A. Pople, Ab initio *Molecular Orbital Theory*, John Wiley and Sons, New York, 1986.

43. J.W. Mintmire and C.T. White, *Phys. Rev. Lett.* **50**, 1010 (1982).

44. N. Colaneri, D. Bradley, R.H. Friend, P. Bara, A. Holmes and C. Spangler, *Phys. Rev. B* **42**, 1170 (1990).

45. J. Leng, S. Jeglinski, X. Wei, R. Benner, Z. Vardeny, F. Guo and S. Mazumdar, *Phys. Rev. Lett.* **75**, 156 (1994).

46. J.M. Toussaint, B. Themans, J. M. Andre and J.L. Bredas, *Synth. Met.* **28**, C205 (1989).

47. Y.S. Lee and M. Kertesz, *J. Chem. Phys.* **88**, 2609 (1988).

48. M. Kertesz and Y.S. Lee, *J. Phys. Chem.* **91**, 2690 (1987).

49. J. Kürti, P.R. Surjan and M. Kertesz, *J. Am. Chem. Soc.* **113**, 9865 (1991).

50. F. Wudl, M. Kobayashi and A.J. Heeger, *J. Org. Chem.* **49**, 3381 (1984).

51. M. Kobayashi, N. Colaneri, M. Boysel, F. Wudl and A.J. Heeger, *J. Chem. Phys.* **82**, 5717 (1985).

52. J.L. Bredas, *Mol. Cryst. Liq. Cryst.* **118**, 49 (1985).

53. S.A. Jenekhe, *Nature*, **322**, 345 (1986); R. Becker, G. Blöchl and H. Bräunling, in *Conjugated Polymeric Materials* (eds. J.L. Bredas and R.R. Chance) Kluwer Academic, Dordrecht, 1990, p. 133.

54. M. Karikomi, C. Kitamura, S. Tanaka and Y. Yamashita, *J. Am. Chem. Soc.* **117**, 6791 (1995); C. Kitamura, S. Tanaka and Y. Yamashita, *Chem. Lett.* **1996**, 63 (1996).

55. T.C. Chung, J.H. Kaufman, A.J. Heeger and F. Wudl, *Phys. Rev. B* **30**, 702 (1984).

56. F. Wudl, M. Kobayashi and A.J. Heeger, *J. Org. Chem.* **36**, 3382 (1984).

57. M. Pomeranz, B. Chaloner-Gill, L.O. Harding and W.J. Pomeranz, *Chem. Commun.* 1672 (1992).

58. J.P. Ferraris, A. Bravo, W. Kim and D.C. Hrncir, *Chem. Comm.* 991 (1994).

59. D. Lorcy and M.P. Cava, *Adv. Mater.* **4**, 562 (1992).

60. S.Y. Hong, *Bull. Korean Chem. Soc.* **16**, 845 (1995). S. Y. Hong, S.J. Kwon and S.C. Kim. *J. Chem. Phys.* **105**, 1301 (1996).

61. U. Scherf and K. Müllen, *Polymer*, **33**, 2443 (1992); U Scherf and K. Müllen, *Synthesis*, 23, (1992).

62. A.D. Schlüter, M. Loffler and V. Enkelman, *Nature* **368**, 831 (1994).

63. M. Loffler, A.D. Schlüter, K. Gessler, W. Saenger, J.M. Toussaint and J.L. Bredas, *Angew. Chem. Int. Ed. Engl.* **33**, 2209 (1994).

64. P.C. Ecklund *et al.*, *Science* **259**, 955 (1993); P. Zhou *et al. Chem. Phys. Lett.* **211**, 337 (1993); S. Pekker *et al. ibid.* **265**, 1077, (1994); P.W. Stephens *et al. Nature* **3701**, 636 (1994); M. Nunez-Regueiro, L. Marques, O. Hodeau, J.L. Bethoux and M. Rerroux, *Phys. Rev. Lett.* **74**, 278 (1995); J.E. Fisher, *Science* **246**, 1548 (1994).

65. J.J.S. Lamba and J.M. Tour, *J. Am. Chem. Soc.* **117**, 11723 (1994).

66. M.B. Goldfinger and T.M. Swager, *J. Am. Chem. Soc.* **116**, 7895 (1994).

67. K. Chmil and U. Scherf, *Macromol. Chem. Rapid Commun.* **14**, 217 (1993).

68. H. Mizes and E.M. Conwell, *Phys. Rev.* B **44** 3963 (1993).

69. M. Kertesz and A. Ashertehrani, *Macromolecules*, **29**, 940 (1996).

70. M.H. Whangbo, R. Hoffmann and R.B. Woodward, *Proc. Roy. Soc. London*, A **366**, 23 (1979).

71. T. Yamabe, K. Tanaka, K. Ohzeki and A. Yata, *Solid State Commun*, **44**, 823 (1982).

72. M. Kertesz and R. Hoffmann, *Solid State Commun*, **47**, 97 (1982).

73. Other theoretical papers on ladder polymers include W.A. Switz, D.J. Klein, T.G. Schmalz and M.A. Garcia-Bach, *Chem. Phys. Lett.*, **115**, 139 (1985); J.M. Toussaint, J.L. Bredas, *Synth. Metals.* **46**, 325 (1992); A.K. Bakhshi, C.M. Liegener, J.J. Ladik, *Chem. Phys.* **173**, 65 (1993); S. Karabunarliev, L. Gherghel, K H. Koch, M. Baumgarten, *Chem. Phys.*, **189**, 53 (1994).

74. J.P. Lowe, S.A. Kafafi and J.P. LaFemina, *J. Phys. Chem.*, **90**, 6602 (1986).

75. K. Tanaka, K. Ohzeki, S. Nankai, T. Yamabe and H. Shirakawa, *J. Phys. Chem. Solids* **44**, 1069 (1983).

76. M. Kertesz, Y.S. Lee and J.J.P. Stewart, *Int. J. Quantum Chem.* **35**, 305 (1989).

77. A.M. Gyul'maliev, *Zh. Fiz. Chim.* **60**, 1682 (1986).

78. C.M. Liegener, A.K. Bakshi and J. Ladik, *Chem. Phys. Lett.* **199**, 62 (1992).

79. J.M. Toussaint and J.L. Bredas, *Synth. Met.* **46**, 325 (1992).

80. M. Baumgarten, S. Karabunarliev, K.H. Koch, K. Müllen and N. Tyutyulkov, *Synth. Met.* **47**, 21, (1992).

81. J.L. Bredas, R.R. Chance, R.H. Baughman and R. Silbey, *J. Chem. Phys.* **76**, 3673 (1982).

82. M. Kertesz, *Macromolecules*, **28**, 1475 (1995).

83. M. Kertesz and T.R. Hughbanks, *Synth. Met.* **69**, 699 (1995).

84. U. Scherf and K. Müllen, *Macromol. Chem. Macromol. Symp.* **69**, 23 (1993).

85. I.I. Ukrainskii, *Theor. Chim. Acta (Berl.)* **38**, 139 (1975); C.X. Cui and M. Kertesz, *J. Am. Chem. Soc.* **111**, 4217 (1989); K. Yoshizawa and R. Hoffmann, *J. Am. Chem. Soc.* **117**, 6921(1995).

86. N. Mataga, *Theor. Chim. Acta (Berl.)* **10**, 372 (1968). A few selected later references on magnetic ground states in polymers are N. Tyutyulkov, P. Schuster and O. Polansky, *Theor. Chem. Acta (Berl.)* **63**, 291 (1983); D.A. Dougherty, *Mol. Cryst; Liquid Cryst.* **176**, 25 (1989); J.S. Miller and A.J. Epstein, *Angew. Chem. Int. Ed. Engl.* **33**, 385 (1994).

87. A.C. Hazell, D.W. Jones and J.M. Sowden, *Acta Cryst.* **33**, 1516 (1977).

88. C. Ehrendorfer, A. Karpfen, *J. Phys. Chem.* **99**, 10196 (1995).

89. S. Ijima, *Nature*, **354**, 56 (1991).

90. J.W. Mintmire and C.T. White, *Carbon*, **33**, 893 (1995).

91. M.S. Dresselhaus, G. Dresselhaus and R. Saito, *Carbon*, **33**, 883 (1995).

92. J.W. Mintmire, B.I. Dunlap and C.T. White, *Phys. Rev. Lett.* **68**, 631 (1992).

93. O.K. Kim, *Mol. Cryst. Liq. Cryst.* **105**, 161 (1984); O.K. Kim, *J. Polym. Sci. Polym. Lett. Ed.* **23** 137 (1985).

94. L.R. Dalton, J. Thomson and H.S. Nalwa, *Polymer*, **28**, 543 (1987).

95. S.Y. Hong, M. Kertesz, Y.S. Lee and O.K. Kim, *Chem. Mater.* **4**, 378 (1992).

96. S.Y. Hong, M. Kertesz, *Phys. Rev. Lett.* **64**, 3031 (1990).

97. F. Coter, Y. Belaish, D. Davidov, L.R. Dalton, E. Ehrenfreund, M.R. McLean and H.S. Nalwa, *Synth. Met.* **29**, E471 (1989).

98. S.Y. Hong, M. Kertesz, Y.S. Lee and O.K. Kim, *Macromolecules*, **25**, 5424 (1992).

99. See for example J. Paloheimo, J. von Boehm, *Solid State Commun.* **87**, 487 (1993).

100. R.H. Baughman, *Synth. Met.* **78**, 339 (1996).

101. R.H. Baughman, N.S. Murthy, H. Eckhardt and M. Kertesz, *Phys. Rev B* **46**, 10515 (1992).

102. K. Tanaka, Y. Matsuura, Y. Oshima and T. Yamabe, *Chem. Phys. Lett.* **237**, 127 (1995); K. Tanaka, Y. Matsuura, Y. Oshima, T. Yamabe, Y. Asai and M. Tokumoto, *Solid State Commun.* **93**, 163 (1995).

103. P.R. Surjan and K. Nemeth, *Solid State Commun.* **92**, 407 (1994); S.C. Erwin, G.V. Krishna and E.J. Mele, *Phys. Rev. B* **51**, 7345 (1995); M. Springborg, *Phys. Rev. B* **52**, 2935 (1995).

104. B. Schlicke, H. Schirmer and A.D. Schluter, *Adv. Mater.* **7**, 544 (1995).

105. T. Freund, U. Scherf and K. Mullen, *Angew. Chem.* **33**, 2424 (1995).

CHAPTER 4
Photochemical Processes of Conductive Polymers

Mohamed S. A. Abdou and Steven Holdcroft

Simon Fraser University, Burnaby, Canada

1 INTRODUCTION

Unique microelectronic [1,2] and photonic applications [3,6] are emerging in which π-conjugated polymers and oligomers complement, or even replace, conventional inorganic and metallic components (Figure 4.1). The relationship between the chemical and physical properties has become a subject of intense study because of the potential impact on novel and existing technologies. For many of these applications, the effect of light on π-conjugated polymers is a critical issue which determines whether they are ever likely to play a significant role in specialized technical innovations such as optoelectronic and photoconduction displays, electro-

luminescent displays, photovoltaic cells, photolithography, and electronically conducting polymer wires. However, until recently very few studies existed in which chemical reactions resulting from the interaction of light with π-conjugated polymers were investigated. A growing interest in photochemical processes in π-conjugated polymers has occurred since the emergence of polymer-based electroluminescent display technologies. For these devices to be commercially feasible, the polymers must be photochemically stable. The aim of this chapter is to summarize the present status regarding photochemical processes of π-conjugated polymers. It includes sections on general photochemistry, photochemical doping and de-doping, photoelectrochemistry,

Handbook of Organic Conductive Molecules and Polymers: Vol. 4. Conductive Polymers: Transport, Photophysics and Applications.
Edited by H. S. Nalwa. © 1997 John Wiley & Sons Ltd

photoelectron transfer, and an extensive review of photolithography of conductive polymers. This chapter does not report on the large body of work conducted on the photophysics of conductive polymers nor on solid state photoelectron transfer or photoconduction properties, except where they are directly relevant to photochemical processes.

2 PHOTOPROCESSES

2.1 Photophysics

No photochemical reaction can occur unless a photon of light is absorbed. For polymers, two types of chromophore are important: intrinsic chromophores, such as functional groups: and extrinsic chromophores arising from impurities and additives. The principal laws governing the absorption of light by materials are Lambert's and Beer's laws [7,8]. The first of these states that 'the proportion of light absorbed by a transparent medium is independent of the intensity of the incident light and that each successive layer of the medium absorbs an equal fraction of the light passing through it'. Mathematically, this leads to the expression

$$\log_{10}\left[\frac{I_0}{I}\right] = Kl \qquad (4.1)$$

where I_0 and I are the intensity of the incident and transmitted light, respectively, l is the thickness of the absorbing material (cm), and K is the penetration depth of light and is defined as the distance over which the incident light attenuates e times. Beer's law incorporates a concentration variable, and states that 'the amount of light absorbed is proportional to the number of molecules of absorbing substance through which the light passes'. It is defined mathematically as

$$\log_{10}\left[\frac{I_0}{I}\right] = \varepsilon c l \qquad (4.2)$$

where ε is the molar extinction coefficient (1 mol^{-1}cm^{-1}), and c is the concentration of chromophores (mol l^{-1}). For convenience, the term $\log_{10}$ (I_0/I) is usually replaced by absorbance, A.

Absorption of light creates an excited state analog of the ground state from which photochemical reaction can occur. Various photophysical routes for deactivation of the excited state are available. The rate of deactivation determines the inherent lifetime of the excited state and the time period over which photochemical reaction must proceed. Macromolecules share many fundamental photophysical processes with monomeric analogs but subtle differences exist due to the contiguous nature of polymer chains and the relatively high local concentration of chromophores. For detailed accounts on the photophysics of polymers, the reader is referred to one of a number of excellent monographs on the subject [7,8].

Key photophysical processes related to polymers are shown in Figure 4.2. Absorption of a photon promotes an electron from the ground state (M) to a higher energy state, usually singlet in nature (^{1}M*). Fluorescent transitions (Fl) occur following excitation between states of the same multiplicity ($\tau \sim 10^{-12}$–10^{-8} s), while a phosphorescent transition (Ph) between states of different multiplicity occurs on a much longer lifetime ($\tau \sim 10^{-6}$–10^2 s). Non-radiative transitions of internal conversion (IC) and inter-system crossing (ISC) compete with radiative processes. IC is relatively fast ($\tau \sim 10^{-12}$ s) and occurs via vibronic coupling between an upper excited state to the lowest excited or ground state. ISC is a spin-forbidden

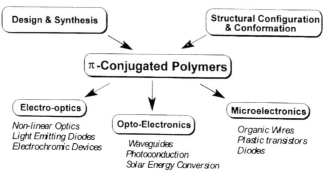

Figure 4.1. Applications of π-conjugated polymers.

radiationless transition ($\tau \sim 10^{-11}$–10^{-6} s) by which triplet states are usually populated.

The nature and local concentration of the chromophore determines whether deactivation occurs *unimolecularly* or *bimolecularly*. When the local density of chromophores is high, such as in the case of π-conjugated polymers, interaction can exist between the excited species and its nearest ground state neighbors. When this interaction occurs within the lifetime of the excited state, bimolecular mechanisms dominate deactivation processes. Quenching (Figure 4.2, paths 1 and 4), migrative energy transfer (paths 2 and 5), and annihilation (path 3) are bimolecular processes. Annihilation is rare for singlet states but can occur with triplets to yield delayed fluorescence. Alternatively, an excited singlet state can interact with a ground state molecule of the same species to yield an excimer ($^1D^*$), characterized by a broad, structureless fluorescence, which is red-shifted with respect to monomer fluorescence. Many excimers have very low quantum yields of luminescence. The ground state dimer (D) is unstable and readily dissociates. Complex formation between excited state species and ground states of different molecules, gives rise to an exciplex. Emission characteristics are similar to those associated with excimers. Typical emission characteristics of lumiphore-bearing polymers are shown in Figure 4.3.

Both radiative and internal conversion processes, whether originating from individual excited state molecules or excited state complexes, generally serve to deactivate the excited state with respect to the photochemical reaction. However, in some instances, non-radiative processes can enhance photochemical reaction. In the first instance, energy transfer, as defined as the donation of excitation energy from one molecule to another, can serve to funnel excitation energy into a reactive center. The 'antennae effect' has been well documented for lumiphore-bearing polymers [7,8]. Secondly, oxygen is an efficient quencher of both

singlet and triplet excited states, forming singlet oxygen ($^1O_2$) in the process. Singlet oxygen has a sufficiently long lifetime and is sufficiently reactive to initiate reactions with most unsaturated hydrocarbons.

2.2 Photochemical processes relevant to conductive polymers

Relatively little information is available concerning the photochemistry of π-conjugated polymers. In order to gain insight into possible photochemical processes of the latter, we draw on the wealth of information which exists concerning the photochemistry of polymers in general. In this section, common photochemical reactions of polymers are described. These reactions by no means cover all of the photochemical reactions likely to be observed for π-conjugated polymers. Key photochemical processes related to polymers are schematically shown below in Figure 4.4.

2.2.1 Photooxidation

The process of photooxidation of polymers can be rationalized in terms of a free radical chain reaction comprised of initiation, propagation and termination [7,9,10]. The mechanism of initiation of photooxidation of conventional polymers such as polyethylene, polyvinyl chloride and polystyrene, by light of wavelength > 300 nm is complicated due to the fact that these polymers are transparent and have no allowed electronic transitions in that region of the spectrum. Photoinitiation is not an intrinsic property of these materials. Extrinsic chromophores such as transition metal impurities incorporated during polymerization of the material, and peroxide or carbonyl groups created during polymerization or processing of the polymer are chromophores which might initiate photooxidation. Transition metal impurities usually absorb light in the UV vis region and have high extinction coefficients

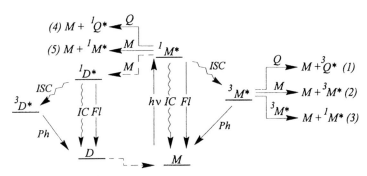

Figure 4.2. Photophysical processes of polymers.

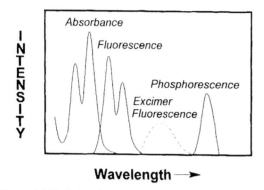

Figure 4.3. Emission characteristics of luminescent polymers.

$(\varepsilon_{max} \sim 10^4 \text{L mol}^{-1} \text{ cm}^{-1})$. $M^{n+}X^-$ ion pairs, where M^{n+} is a transition metal ion and X^- is a halogen or hydroxyl ion, have broad electronic absorption bands extending over a wide wavelength range and often undergo photoreduction to produce free radicals $X^{\bullet}$ [equation (4.3)]. The resulting radicals ($X^{\bullet}$) are very reactive and capable of abstracting hydrogen from most hydrocarbons (RH) thus initiating photodegradation reactions [equation (4.4)].

$$M^{n+}X^- \xrightarrow{h\nu} [M^{(n-1)+}X^{\bullet}] \longrightarrow M^{(n-1)+} + X^{\bullet} \quad (4.3)$$
$$RH + X^{\bullet} \longrightarrow R^{\bullet} + HX \quad (4.4)$$

The H–X bond strength affects the ability of $X^{\bullet}$ to abstract a hydrogen atom [11]. The bond energy of H–Cl is 102 kcal mol^{-1}, and thus chlorine atoms are relatively non-selective species and attack aliphatic hydrocarbons in a rather indiscriminate manner. In contrast, the bromine atom (bond energy H–Br = 86 kcal mol^{-1}) selectively abstracts benzylic or allylic hydrogen atoms (bond energy ~ 82 kcal mol^{-1}) over more strongly bound aliphatic hydrogen atoms (bond energy ~ 95 kcal mol^{-1}).

Alkyl radicals are reactive and form peroxy radicals in the presence of oxygen [equation (4.5)].

$$R^{\bullet} + O_2 \longrightarrow ROO^{\bullet} \quad (4.5)$$

The resulting radical, although more stable than the parent alkyl radical, can also abstract a hydrogen atom and regenerate $R^{\bullet}$ [equation (4.6)].

$$ROO^{\bullet} + RH \rightarrow ROOH + R^{\bullet} \quad (4.6)$$
$$O_2$$

Peroxides and hydroperoxides absorb light in the region 300–360 nm ($\varepsilon \sim 10$–100 l mol^{-1} cm^{-1}). Absorption of light cleaves the O–O bond to yield hydroxy ($^{\bullet}$OH) and alkoxy (RO$^{\bullet}$) radicals according to equation (4.7).

$$ROOH \xrightarrow{h\nu} RO^{\bullet} + {}^{\bullet}OH \quad (4.7)$$

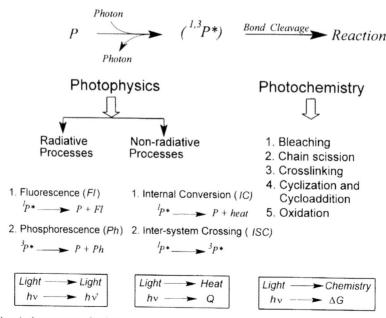

Figure 4.4. Photochemical processes of polymers.

The quantum yield of the photochemical decomposition of t-butyl peroxide, for example, in solution at 313 nm is unity [12]. The radical pairs can initiate degradation reactions.

Termination of free radical chain reactions occurs by several processes as listed below.

(1) Radical coupling wherein coupling of alkyl radicals leads to crosslinking of the polymer [equation (8)]. The reactions, however, must be sufficiently rapid ($> 10^9 \ l \ mol^{-1} \ s^{-1}$) to compete with the addition of oxygen [equation (4.5)].

$$2R^{\cdot} \longrightarrow R{-}R \qquad (4.8)$$

Coupling of alkoxy ($RO^{\cdot}$) or peroxy radicals ($ROO^{\cdot}$) does not terminate the propagation process since the products are photochemically active and result in free radicals upon further photolysis.

(2) Hydrogen abstraction by an alkoxy radical to yield an alcohol.

$$RO^{\cdot} + RH \longrightarrow ROH + R^{\cdot} \qquad (4.9)$$

(3) β Scission of the alkoxy radical to yield a carbonyl group.

$$\underset{RCH{-}CH_2 \backsim}{\overset{O^{\bullet}}{|}} \rightarrow \underset{RCH}{\overset{O}{\|}} + {}^{\bullet}CH_2 \backsim \qquad (4.10)$$

(4) Reaction of hydroperoxides resulting in ether linkages and crosslinking of the polymer

$$2ROOH \longrightarrow ROR + O_2 + H_2O \qquad (4.11)$$

Photooxidation reactions such as those described above might be applicable to π-conjugated polymers which possess alkyl side chain substituents. In fact the presence of a π-delocalized system next to an aliphatic group enhances hydrogen atom abstraction from the aliphatic substituent by stabilizing the resulting free radical. Such an effect is observed for polystyrene, for which hydrogen atom abstraction occurs primarily at the α substituent (Figure 4.5). A similar effect will be described later for the poly(3-alkylthiophenes).

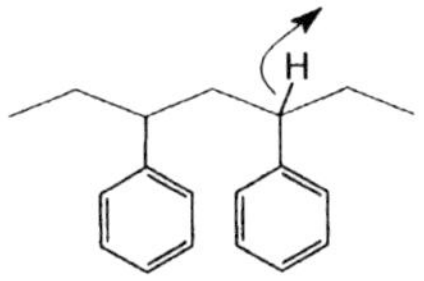

Figure 4.5. Hydrogen atom abstraction from polystyrene.

2.2.2 Singlet oxygen photosensitization

In addition to molecular oxygen playing an important role in free radical photochemistry of polymers, oxygen is an efficient quencher of both singlet and triplet excited states (Figure 4.6) [13,14]. Formation of $^1O_2$ is often restricted to the triplet state annihilation processes (T–T annihilation). Its formation is usually thermodynamically feasible because the energy gap between the ground state and the first excited singlet state of molecular oxygen $(3(\Sigma)O_2 \rightarrow {}^1\Delta_gO_2)$ is only 22.5 kcal mol^{-1}. In some instances, molecular oxygen enhances intersystem crossing of the singlet excited state thereby increasing the triplet concentration, which in turn leads to an enhancement of triplet–triplet annihilation [15]. Singlet oxygen is extremely reactive. Reaction with aromatic compounds often yields dioxetanes and endoperoxides as shown in Figure 4.7 [15,16].

In contrast, olefins, which contain allylic hydrogen atoms, react with $^1O_2$ to form peroxides via the *ene* reaction [17].

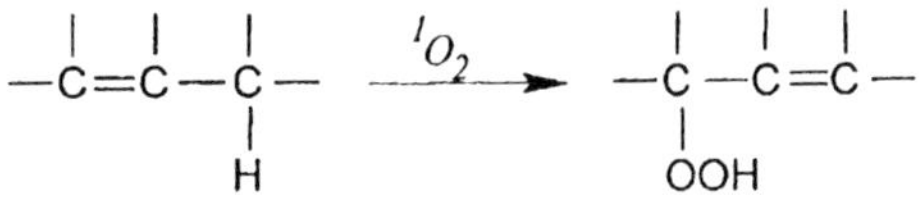

$$\underset{H}{\overset{|}{-C{=}C}}\overset{|}{-C}\overset{|}{-} \quad \xrightarrow{{}^1O_2} \quad \underset{OOH}{\overset{|}{-C}}\overset{|}{-C}{=}C\overset{|}{-} \qquad (4.12)$$

Direct reaction of singlet oxygen with saturated hydrocarbons is not observed because the reaction is endothermic. Polypropylene, for example, has been reported to be unreactive to singlet oxygen after 51 hours of exposure [18].

Molecular oxygen possesses a relatively low formal potential, $E'_o = -0.82 \ V$ (SCE) [19], and can form a ground state charge transfer complexes (CTC) with alkanes, olefins and aromatic hydrocarbons [20]. An additional contribution to the photosensitization of singlet oxygen is photolytic dissociation of an oxygen CTC. Recently, Ogilby *et al.* illustrated photogeneration of singlet oxygen in polystyrene by irradiation of

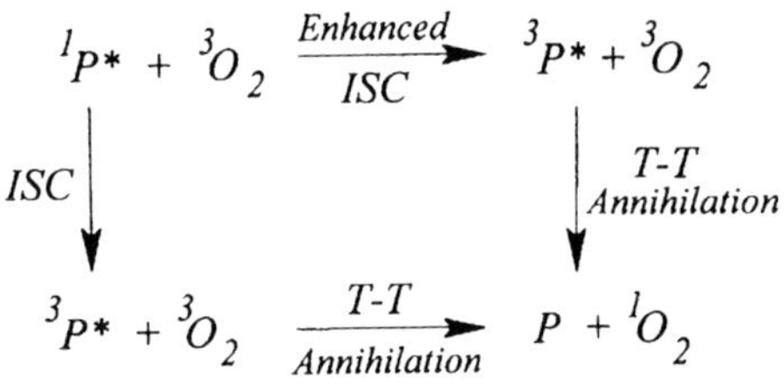

Figure 4.6. Photosensitization of singlet oxygen. P represents polymer.

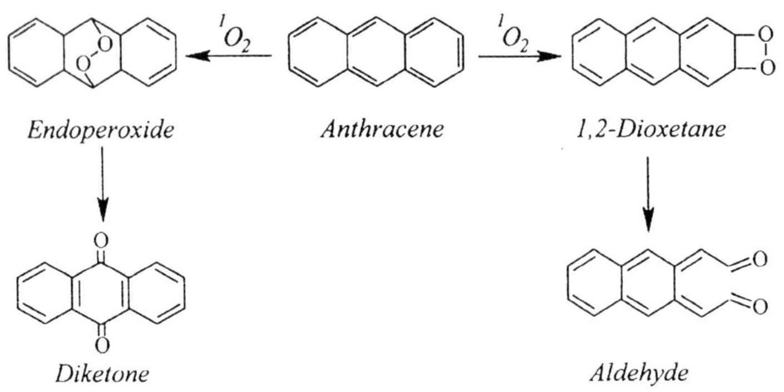

Figure 4.7. Reaction of $^1O_2$ with anthracene.

the oxygen–polystyrene charge transfer band [equation (4.13)] [21]. Since π-conjugated polymers exhibit much lower ionization potentials than polystyrene it appears likely that such polymers are also susceptible to CTC formation.

$$P_0 + {}^3O_2 \rightleftharpoons {}^3[P..O_2]/{}^3CTC$$

$$^3CTC \xrightarrow{h\nu} {}^3[CTC] \rightleftharpoons {}^3[CTC] \longrightarrow P_0 + {}^1O_2 \qquad (4.13)$$

2.2.3 Carbonyl-containing polymers

Ketone functionality holds historical importance in the photochemistry of polymers. The most frequently observed electronic transition in ketones is the n-π^* transition, which forms the excited singlet state. The lifetime of the singlet state of aliphatic ketones, is typically 10^{-9} s, and quantum yields of luminescence are $c.$ 0.01. Excited state ketones can undergo a variety of reactions. The most common are α cleavage (Norrish type 1 reaction) and hydrogen atom abstraction (e.g. Norrish type II reaction):

$$R-\overset{O}{\underset{}{C}}-R' \xrightarrow{h\nu} R-\overset{O}{\underset{}{C}}{}^{\bullet} + R'^{\bullet} \xrightarrow{\Delta} R^{\bullet} + CO + R'^{\bullet} \quad (4.14)$$

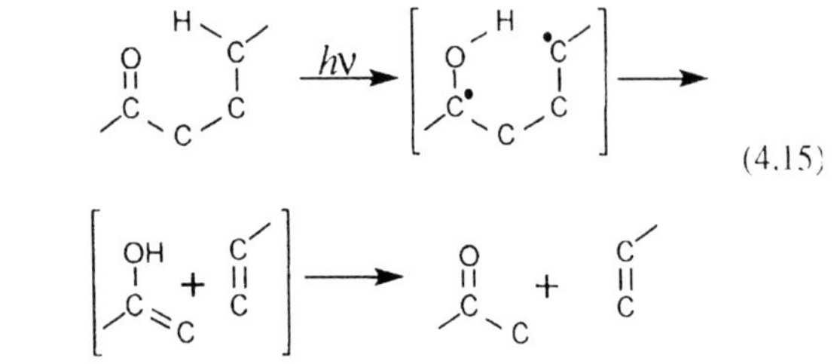

$$(4.15)$$

The Norrish type I reaction [equation (4.14)] yields a primary radical, the fate of which depends on the local chemical environment. The most common fate is abstraction of a hydrogen atom from a surrounding hydrocarbon, but at elevated temperature, the acyl radical can eliminate carbon monoxide. The resultant alkyl radical can also participate in hydrogen atom abstraction reactions. The rate of this reaction depends on the stability of the resultant alkyl radical. If resonance-stabilizing aromatic or t-butyl functionality are attached α- to the alkyl group then the rate increases dramatically.

The Norrish type II reaction [equation (4.15)] is a special case of hydrogen atom abstraction. It occurs when a hydrogen atom is abstracted from the carbon situated γ to the carbonyl group. Cleavage of the α–β carbon–carbon bond and tautermerization of the resultant enol yields a ketone in addition to a terminal olefinic group.

Since both Norrish I and II reaction pathways result in cleavage of C–C bonds, polymers possessing ketone groups in the main chain undergo chain scission as demonstrated below for ethylene–carbon monoxide polymers [equations (4.16) and (4.17)]. In this example, quantum yields are $c.$ 0.001 and 0.02 for type I and type II reactions, respectively.

For a ketone functionality situated in the side chain, the quantum yields of the type I reaction are somewhat larger due to rapid diffusion of the small molecule radical out of the primary recombination cage. Photolytic studies of poly(methyl vinyl ketone) solutions, for example, exhibit quantum yields of 0.04 and 0.025 for type I and type II reactions, respectively [equations (4.18) and (4.19)] [7].

When considering the role of these mechanisms in the photochemistry of π-conjugated systems it must be borne in mind that the above have been demonstrated in

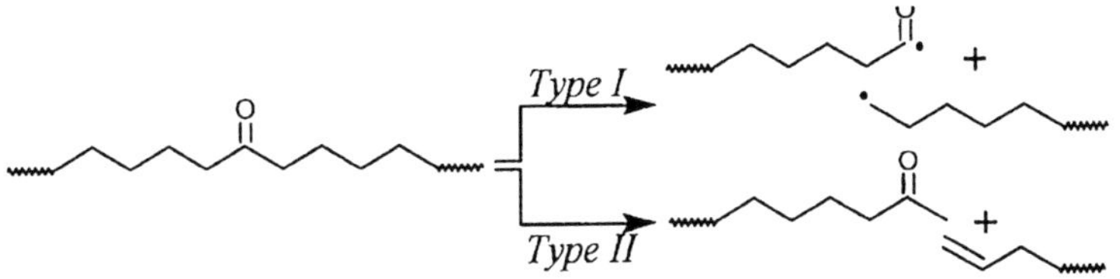

$$(4.16)$$

$$(4.17)$$

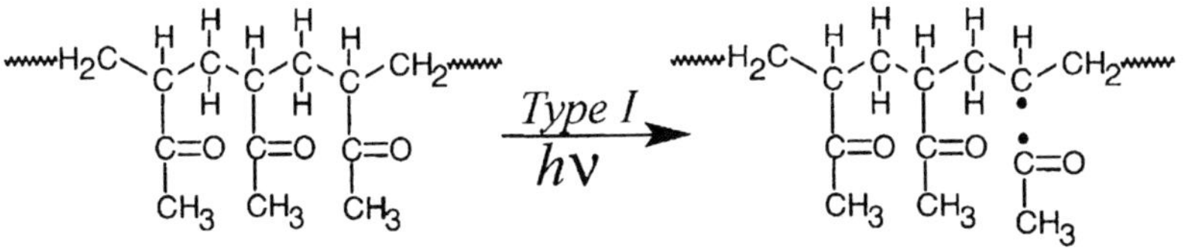

$$(4.18)$$

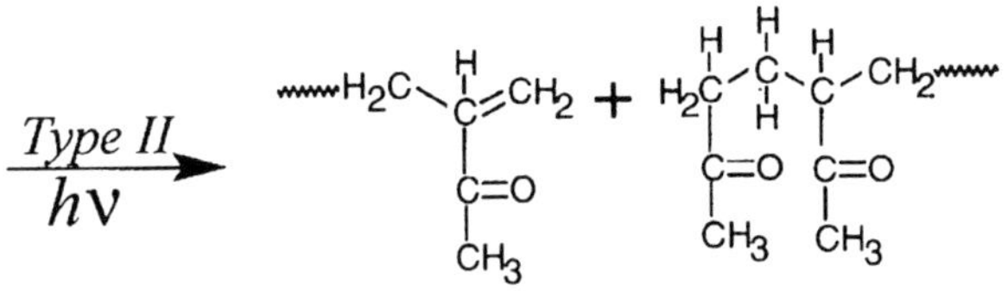

$$(4.19)$$

aliphatic ketones where the dominant chromophore is the carbonyl group itself. The absorption tail associated with the n–π transition extends to c. 350 nm but its extinction coefficient is < 100 mol^{-1} cm^{-1}. Thus, for π-conjugated polymers, the π-system associated with the backbone will usually dominate the absorption process and prevent direct excitation of the carbonyl functionality. However, many π-conjugated polymers exhibit an absorption window between 300 and 350 nm, and therefore the wavelength of the incident irradiation must be considered before ruling out direct excitation.

A number of π-conjugated polymers exist in which a ketone functionality is placed α- to the π-conjugated backbone. The first point to be made about the photochemistry of aromatic ketones is, in contrast to aliphatic ketone, the singlet excited state of aromatic ketones undergoes rapid intersystem crossing to the triplet due to the presence of low-lying π–π^* states. Thus the dominant reaction is often Norrish type II in nature. For aryl phenyl ketones the quantum yield of this reaction approaches unity in polar solvents. For compounds of the structure C$_6$H$_5$(C=O)R where R is an alkyl group longer than propyl, quantum yields are c. 1.0.

In addition to Norrish I and II reactions described above, a number of other photochemical reaction pathways have been observed for carbonyl-containing polymers; these include cyclization of the biradical formed by hydrogen atom abstraction to form cyclo-

butanols; photocycloaddition of exciplexes consisting of carbonyl compounds and olefins; photoreduction of carbonyl species; elimination of heteroatom species on the carbon situated α to the carbonyl. The relative importance of these and other reactions depends on the inherent nature of the species under investigation. The reader should consult literature specific to the chemical species present in their polymer system.

Another important photochemical reaction of the carbonyl species is the photo-Fries reaction of aromatic esters. This is a general reaction [equation (4.20)] which occurs with high efficiency in polymeric systems [equation (4.21)]. In the example shown below, quantum yields are as high as 0.4 in both solution and solid state.

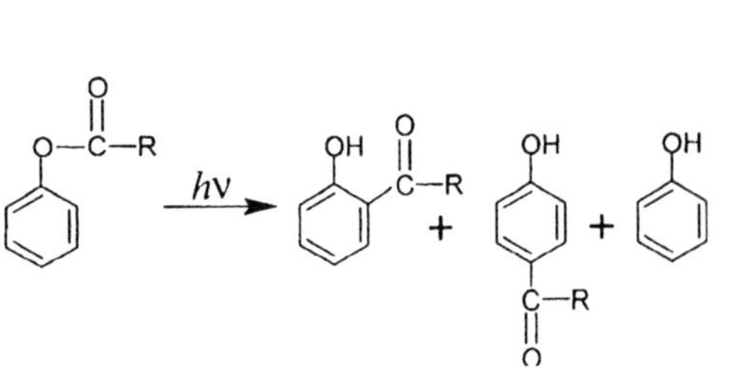

$$(4.20)$$

2.2.4 Photocyclization

Photocyclization can occur in polymer systems for which the position and nature of the chromophore is

$$(4.21)$$

such that the excited state of a molecule reacts to form two new sigma bonds (usually C–C bonds). Photocyclization of small molecule aromatic compounds is well documented [15]. The most common examples are [4 + 4] cycloadditions, as exemplified by the photocyclodimerization of anthracene

$$(4.22)$$

No evidence has been reported concerning the photophotocyclodimerization of benzene although there is evidence for photocyclization of 2-thenylethylenes [22]. Furthermore, aromatic molecules are known to undergo a variety of photocycloadditions in the presence of olefins [15], as exemplified by the [2 + 2] cycloaddition between benzene and tetramethyl ethylene [equation (4.23)]. The reactions of benzene appear to go through the singlet state, possibly through a charge transfer complex or exciplex formation. Reactions such as these are not restricted to benzene and may well be present in a number of π-conjugated systems.

$$(4.23)$$

2.2.5   Chain scission and crosslinking

Many photochemical reactions involve the formation of free radicals along a polymer chain and cleavage of C–C bonds. Thus it is not uncommon to find that irradiation of polymers results in a decrease in molecular weight (MW) due to chain scission, or insolubilization due to intrachain coupling of radicals (crosslinking) (Figure 4.8). The descriptions of these processes at both the molecular and macromolecular

level have been well documented because of their importance in the microchip fabrication industry.

The relative extents of chain scission to crosslinking (the sol-gel fraction) following irradiation determines whether a polymer is insolubilized upon irradiation. Charlesby and Pinner developed a theory relating the number of crosslinks and chain scission events as a function of irradiation dose to estimated changes in MW and solubility of the polymer [23]. The theory assumes that both crosslinking and scission occur randomly, and are proportional to the radiation dose. For a polymer having a random molecular weight distribution, the change of the weight-average molecular weight, M_W, with irradiation dose, r, follows the expression

$$\frac{1}{M_w} = \frac{1}{M_{w0}} + \left[\frac{\Phi_s}{2} - 2\Phi_x\right]r \qquad (4.24)$$

where ϕ_s and ϕ_x are the quantum yields of chain scission and crosslinking, respectively. This equation implies that gel formation occurs when $4\,\phi_x > \Phi_s$. In the unlikely event that $\Phi_s = 4\phi_x$, chain scission fully compensates crosslinking and the molecular weight of the polymer is independent of the irradiation dose.

The solubility of the polymer (sol fraction, S) can be determined as a function of the irradiation dose by solvent extraction. The solubility equation is given by

$$S + S^{1/2} = \frac{p_o}{q_o} + \frac{1}{q_o \overline{DP_n}} \frac{1}{r} \qquad (4.25)$$

where p_o and q_o represent the fracture and crosslinking densities per unit radiation dose, and $\overline{DP_n}$ is the number-average degree of polymerization ($\overline{DP_n} = M_n/$

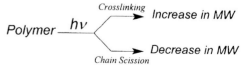

Figure 4.8. Effect of irradiation on the molecular weight of polymers.

M_o). The term p_o/q_o is the relative extent of scission to crosslinking as a function of irradiation. There are two limiting values for p_o/q_o at infinite r: $p_o/q_o \geqslant 2$ which yields a value for S of 1 at infinite radiation dose and implies that the polymer will remain completely soluble, independent of the radiation dose; and $p_o/q_o < 2$ which implies that crosslinking is the dominant photochemical reaction. The dose of incipient gelation, r_{gel}, under this condition, will be governed by the relative ratio of p_o to q_o. Typical changes of the sol fraction of a polymer with irradiation dose are shown in Figure 4.9.

For polymers which undergo efficient photocrosslinking, such as negative-tone photoresists, photochain scission can be neglected. The quantum yield of crosslinking (gelation) is defined as [24–26].

$$\Phi_x = \frac{\text{No. of crosslinks}}{\text{No. of quanta absorbed}} \quad (4.26)$$

The number of crosslinked units formed as a result of radiation dose r, is given by

$$\text{No. of crosslinked units} = qN \quad (4.27)$$

where q is the crosslinking density, and N is number of repeating units per unit area as defined by

$$N = \frac{l\rho}{M_o} N_A \quad (4.28)$$

l and ρ are the thickness and density of the polymer, M_o is the molecular weight of the repeating units, and N_A is Avogadro's number. The number of crosslinks is thus given by

$$\text{No. of crosslinks} = \frac{N}{2} q = \frac{l\rho}{2M_o} \quad (4.29)$$

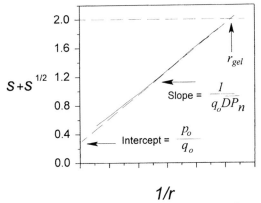

Figure 4.9. Solubility function of polymers $(S+S^{1/2})$ as a function of irradiation dose.

since each crosslink involves the linking together of two chains.

The number of photons absorbed is given by

$$\text{No of quanta absorbed} = \phi\, rN_A \quad (4.30)$$

where ϕ is the fraction of incident radiation density r (Einsteins cm^{-2}) absorbed by the material ($\phi = 1 - 10^{-A}$, where A is the optical density of the polymer film). Thus the quantum yield of crosslinking is given by

$$\Phi_x = \frac{l\rho q}{2M_o \varphi r} \quad (4.31)$$

The gel dose (D_g^i or r_{gel}), as defined above, is the minimum exposure required for incipient gel formation. This corresponds to an average crosslinking density of one crosslink per chain, which can be expressed quantitatively by the Stockmeyer's rule as [27].

$$q = \frac{M_o}{M_w} \quad (4.32)$$

The gel dose, r_{gel}, is obtained by combining equations (4.31) and (4.32):

$$r_{gel} = \frac{l\rho}{2M_w \varphi \Phi_x} \quad (4.33)$$

For optically thin polymer films, the fraction of light absorbed (ϕ) is adequately approximated by

$$\phi = 2.303\ A = 2.303\varepsilon cl \quad (4.34)$$

where ε is the extinction coefficient of the chromophore, and c is the concentration of the chromophore in the solid film. Equation (4.33) now takes the form

$$r_{gel} = \frac{\rho}{2(2.303M_w \varepsilon c\Phi_x)} \quad (4.35)$$

This is a fundamental equation for photo-induced insolubilization and is often used for evaluating photoresists. It relates the quantum yield of crosslinking to the experimentally measurable parameter, gel dose (r_{gel}). It shows that r_{gel} depends on both the molecular weight and density of the polymer. r_{gel} is, however, independent of the polymer film thickness, assuming optically thin films.

3 PHOTOPHYSICS OF CONJUGATED POLYMERS

Detailed photophysical studies of π-conjugated polymers began in the 1970s with poly(phenylacetylene) when it was realized that energy migration between pendant phenyl groups attached to a polyconjugated chain was extremely efficient due to strong interactions between the chromophore and the unsaturated backbone [28]. Photophysical studies have since been used to provide fundamental information regarding the electronic structure of π-conjugated polymers. Their photophysical properties have also achieved significant technical interest because of potential applications in optoelectronics and electro-optics, particularly in the area of electroluminescent displays. Demonstrations of lasing, photoconduction, photodegradation and photo-imaging of conjugated polymers also draws attention to the fate of excited states of these materials.

Many of this class of polymer fluoresce because of the delocalized π-conjugated system. For π-conjugated polymers other than the parent polyacetylene, two non-degenerate configurations are possible: the benzenoid and quinoid forms. Usually, the former has the lower ground state energy, but the reverse is the case in the excited state. Photoexcitation results in formation of a 'polaron–exciton'. There is a tendency for this excitation to become self-trapped and confined to the polymer chain because of the quasi-one-dimensional nature of the chain. Dissociation of the excited state species results in photoconductivity and non-radiative decay whereas recombination can yield luminescence. The excited state configuration and the corresponding electronic energy diagram of a typical conjugated polymer, poly(p-phenylene), is illustrated in Figure 4.10

Of the conjugated polymers, photoluminescence of polyacetylenes, poly(phenylene)s poly(aromatic vinylene)s and polythiophenes have been extensively examined. Many related polymers, including polypyrroles and polyanilines, are not considered photoluminescent because of competitive and efficient non-radiative processes. The photophysical properties of conjugated polymers are similar because of their common π-structure but variations exist in the color of emission, radiative lifetime and quantum efficiency of luminescence. These attributes are dependent on the molecular architecture which determines the band-gap energy. π–π^* transition energies (absorption and emission) decrease as the extent of conjugation increases according to the 'particle in a box' theory. The photophysics also depends on the conformation of the polymer. In the solid state the polymer takes up a rod-like conformation which imparts a red shift to the fluorescence compared to the coil-like conformation adopted in solution. Thus, the wavelength of maximum emission exhibits a blue shift of up to 100 nm upon dissolution of the polymer due to interannular bond rotation. Similarly, the presence of large bulky substituents attached to the conjugated backbone prevents the polymer from achieving a planar configuration and increases the π–π^* transition energy. Noticeably, both absorption and emission spectra are broad due to the conformational statistics of the polymer coil which segmentalizes π-conjugated lengths into a range of persistence lengths. Stoke's shifts are usually large due to relaxation of the structure in the excited state.

The triplet state of conjugated polymers has not been studied in much detail, largely due to its very weak phosphorescence. As a result, there is considerable uncertainty about the role of triplets in the electronic structure of the ground and excited states of conjugated polymers. It appears the triplet may be weakly populated which means that photochemistry is not likely to take place from this longer-lived state. A heavy-aton effect has also been employed to induce more efficient intersystem crossing due to enhanced spin–orbit coupling.

4 PHOTOCHEMISTRY OF CONDUCTIVE POLYMERS

4.1 Polyacetylenes

Polyacetylene is the parent π-conjugated polymer and the first to undergo detailed photochemical study. In 1982, oxidation of polyacetylene was reported in the

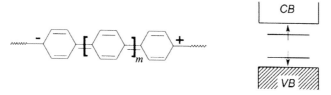

Figure 4.10. Excited state of poly(p-phenylene).

presence and absence of fluorescent lighting [29]. In the absence of light, air oxidation was reported to occur with psuedo-first-order kinetics due to the reaction of molecular oxygen with existing free radicals on the polymer chain. Under fluorescent lighting a similar rate order was observed but the kinetics of reaction were faster. Details of the photochemistry were not reported but the authors speculate on reactions involving singlet oxygen. More detailed photochemical studies were reported for a series of substituted polyacetylene films [30]. Photobleaching, chain scission and crosslinking were observed during photolysis in air. The rate of these processes depended on the pendant substituent. Many degraded rapidly, whereas polymers possessing pendant aromatic substituents were reported to be relatively stable. A list of substituted polyacetylenes studied by Tsuchihara *et al.* is given below (Table 4.1). The absorption maxima of these polymers varied between 235 and 540 nm. A few examples of absorption spectra are given in Figure 4.11

These authors monitored the change in molecular weight of polymer samples during irradiation with a 200 W high pressure mercury lamp using gel permea-

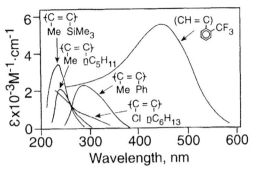

Figure 4.11. Absorption spectra of substituted polyacetylenes. Reprinted with permission from ref. 30.

tion chromatography. The variation in the stability of the polymers is demonstrated in Figure 4.12. An estimation of the relative probabilities of chain scission (β_s) and crosslinking (β_x) was obtained from these data using equations derived from radiolysis [equations (4.36) and (4.37)]. The rate of change in molecular weight decreased with irradiation time due to the changing chemical structure of the polymer during photolysis. The relative probabilities of chain scission (β_s) and crosslinking (β_x) are shown in Table 4.1.

$$\beta_s - \beta_x = \frac{1}{DP_n} - \frac{1}{DP_{n,0}} \qquad (4.36)$$

$$(\beta_s - 4\beta_x)/2 = \frac{1}{DP_w} - \frac{1}{DP_{w,0}} \qquad (4.37)$$

β_s values were generally 3–9 times greater than β_x, indicating a predominance for the chain scission reaction. Three classes of substituted polyacetylenes were found: aliphatic disubstituted polymers were the most photosensitive; heteroatom-containing and aromatic disubstituted polyacetylenes were of medium photosensitivity; and aromatic monosubstituted polyacetylenes were amongst the most stable. Poly[o-(trifluoromethyl) phenylacetylene] was found to be the most stable of all the polymers examined.

Upon further photochemical study of one of the polymers, poly[1-(trimethylsilyl)-1-propyne], it was concluded that no main chain scission occurs in vacuum and that chain scission in the presence of air is accompanied by photooxidation. On the basis of these and other results the authors offered the following mechanisms of photochemical degradation. For polyacetylenes possessing allylic hydrogens, the mechanisms involves C–H bond scission at the allylic position, followed by transfer of the radical to the main chain, formation of hydroperoxide by its reaction with

Table 4.1. Photochain scission and photocrosslinking of substituted polyacetylenes [30]

Polymer #	$(CR = CR')_n$ R	R'	β_s ($\times 10^4$)	β_x ($\times$)10^4
1	Me	n-C$_3$H$_7$	580	81
2	Me	n-C$_5$H$_{11}$	490	56
3	Me	n-C$_7$H$_{15}$	470	100
4	Me	SiMe$_3$	30	7.7
5	Me	S-nBu	15	2.5
6	Me	Ph	67	11
7	Cl	Ph	36	9.1
8	H	⟨CF$_3$-phenyl⟩	1.8	0.4
9	H	⟨F-substituted phenyl, nBu⟩	4.5	0.9
10	H	⟨SiMe$_3$-phenyl⟩	—	—
11	Cl	n-C$_4$H$_9$	—	—
12	Cl	n-C$_6$H$_{13}$	—	—
13	Cl	nC$_8$H$_{17}$	—	—

Me, methyl group; Bu, butyl group; Ph, phenyl group.

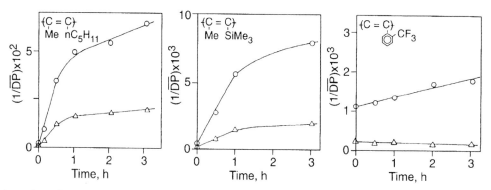

Figure 4.12. Plots of 1/(degree of polymerization) vs. irradiation time for selected substituted polyacetylenes: ($\bigcirc$, $1/\overline{DP}_n$, ($\triangle$), $1/\overline{DP}_w$. Reprinted with permission from ref. 30.

oxygen, and subsequent ketone formation (Figure 4.13). In the absence of allylic hydrogens, excitation of the double bond and subsequent opening of the double bond was hypothesized. The authors did not speculate on the formation and reaction of singlet oxygen, although this would appear to be an additional route for photooxidation of the polymers. Another factor may be the nature of the transition metal catalyst used in the preparation of the polymers. Such catalysts are difficult to completely remove and often initiate photooxidation.

Feast *et al.* examined the photochemistry of polyacetylenes and precursors to polyacetylene [31,32]. A scheme depicting the reaction pathways relevant to this study is shown in Figure 4.14, wherein the tricyclic molecule **I** is ring-opened to give the precursor polymer **II**. Elimination of hexafluoroorthoxylene yields polyacetylene with a large percentage of *cis* couplings **III**. Thermal treatment yields the *trans* analog **IV**. These authors found that elimination of hexafluoroorthoxylene from II can be controlled so that copolymer **V** can be obtained. It is the latter, together

with *trans*-polyacetylene, for which the photochemistry has been reported.

Films of **V** in which one in three of the repeat units of **II** had undergone elimination were irradiated under vacuum with a 1000 W medium pressure mercury lamp. The resultant change in the UV vis absorption spectra showed only a slight increase in the degree of conjugation as evidenced by a slight red shift in the spectra. This film could be subsequently thermally treated to yield polyacetylene. Photolysis of **V** in air, however, resulted in photobleaching and insolubilization. IR analysis of samples photolyzed in air gave absorption bands consistent with the formation of hydroxyl or hydroperoxide groups, carbonyl groups and other species associated with photooxidation. Also notable were the decrease in bands associated with olefinic C−H bonds. When the photolyzed precursor V was heated to attempt complete elimination of the volatile products, hexafluoroorthoxylene, water and carbon dioxide were detected. The latter two species originated from photooxidation products. CPMAS ^{13}C NMR indicated the presence of carboxylic acid, ketone

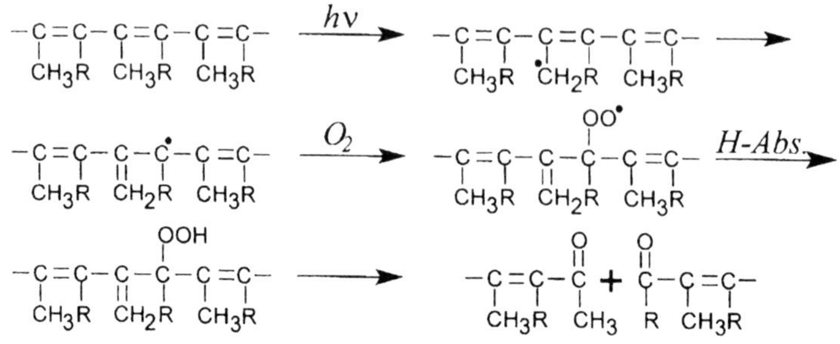

Figure 4.13. Proposed mechanism for the photodegradation of polyacetylenes possessing allylic hydrogens. Reprinted with permission from ref. 30.

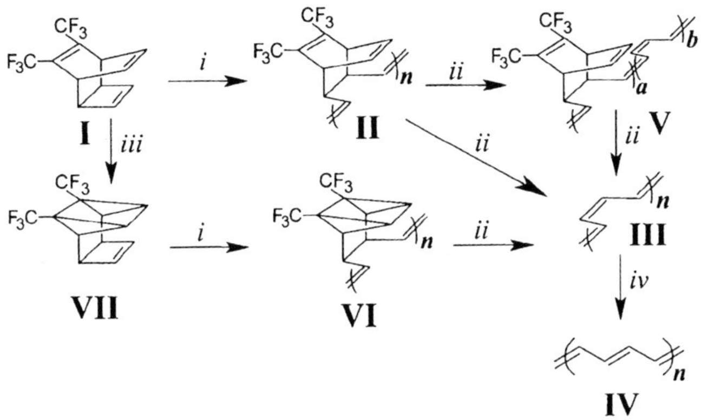

Figure 4.14. The precursor route for the preparation of polyacetylene. Reprinted with permission from ref. 31.

carbonyl, olefinic carbons, saturated carbons possessing one oxygen atom and an allylic carbon.

A free standing film of polyacetylene prepared by the precursor route, and exposed to laboratory lighting over a period of two years, exhibited similar IR features as those described above, i.e. IR bands due to $-O-H$, $C=O$ and carboxylic units. CPMAS ^{13}C NMR provided data which were consistent with those observed for photolyzed and thermalized precursor **V**, confirming the similarity of degradation mechanisms between the two polymers. Thermal analysis was consistent with extensive oxidation. Elemental analysis yielded 49.6% carbon and 4.7% hydrogen, from which a $> 45\%$ oxygen content was inferred. The authors conclude by suggesting that photooxidation involves singlet oxygen generation, and reaction of single oxygen to form an oxidized, crosslinked network.

γ-Irradiation of polyacetylene has been investigated [33]. It was reported that γ- or electron beam irradiation of *cis*-polyacetylene prepared by the Shirakawa route retarded isomerization to the *trans*-configuration, thereby enhancing the stability of polyacetylene samples (Table 4.2). The *trans* form of the pristine polymer was found to undergo rapid degradation in storage as determined by IR analysis and cyclic voltammetry in

Table 4.2. *cis*-Content of polyacetylenes following γ-irradiation after various periods of storage. Reprinted with permission from ref. 33

Storage time (days)	cis-Content of polyacetylene	
	Non-irradiated sample	Irradiated with 75 Mrad
—		
0	97	90.5
100	75.5	89.5
170	66	88.0

non-aqueous electrolytes. In contrast, irradiated films stored for several months were found to exhibit reversible electrochemistry similar to the pristine polymer.

4.2 Polythiophenes

Although a great deal of research on conjugated polymers has been performed on polythiophenes and their soluble analogs, poly(3-alkylthiophene)s, there has been little attention addressed to their photochemical properties until recently. It has since been shown that poly(3-alkylthiophene)s (P3ATs) undergo facile photodegradation whereas the parent polythiophene does not [34–39]. Irradiation of poly(3-hexylthiophene) in air-saturated CHCl$_3$ results in a decrease in the optical density (OD) (Figure 4.15) and a blue shift in γ_{max}. Photobleaching was attributed to disruption and shortening of the π-conjugated segments. In addition to photobleaching, photochain scission was observed (Figure 4.15, inset). The degree of polymerization $(\overline{DP}_n)$ decreased from > 50 to values as low as ~ 8. FTIR vibrational spectroscopy of photolyzed polymers was used to determine the presence of characteristic hydroxyl and hydroperoxide functionality (Figure 4.16). Keto and/or aldehydic groups, and sulfine residues $(C-S^+-O^-$ or $C-S-O)$ were observed, together with a loss of the alkyl side chain, a decrease of the absorption bands characteristic of inter-annular stretching modes and the disappearance of aromatic $C-H$ stretch and $C-H$ out-of-plane deformation modes. Absorption wavenumbers and the structural assignments for new IR bands are listed in Table 4.3. IR data for pristine polymers are listed for comparison.

Table 4.3. FTIR data of P3HT following irradiation. Reprinted with permission from ref. 34.

Pristine polymer		Photolyzed polymer	
cm^{-1}	Assignment	cm^{-1}	Assignment
3055	C–H str.(aromatic)	3580	OOH str.
2955	CH_3 asym. str.	3450	OH str.
2926	CH_2 in-phase vib.	2955 ⎫	
2870	CH_3 sym. str.	2926 ⎬	C–H str. (aliph.)
2856	CH_2 out-of-phase vib.	2857 ⎭	
1655	Overtone of thiophene ring	1765 ⎫	CHO str.
1560 ⎫		1740 ⎭	
1512 ⎬	C_2–C_3 and C_4–C_5 antisym. and sym.	1710 ⎫	C=O str. (ketone)
1460 ⎭	str. modes of thiophene ring	1685 ⎭	
1377	Methyl def.	1650 ⎫	C=C str.
1260	C–C inter-ring bond str.	1625 ⎭	
1190 ⎫	Wagging and twisting of	1431	C–H in-plane def. (olefin)
1155 ⎭	methylene groups	1395 ⎫	sym. and asym. bending of OH
1090	C–H in-plane bending of	1370 ⎭	groups
	thiophene ring	1253 ⎫	C=S=O
823	C–H out-of-plane	1266 ⎭	
	def. of thiophene ring	1090 ⎫	$C=S^+=O^-$
725	Rocking mode of methylene group	1920 ⎭	
		924	C–O str.
		790	C–H def. (olefin)

Films cast on KBr disks.

^{1}H NMR of photolyzed polymers in solution was used to detect olefinic protons, C<u>H</u>(OH), hydroperoxide protons, and α- and β-unsaturated aldehydic protons. A decrease in the aliphatic proton resonance signals, and a decrease in the aromatic proton signal indicating ring opening, were also observed, consistent with changes in the IR spectra. ^{13}C NMR analysis indicated a decrease in aromatic and aliphatic carbon signals, the appearance of primary and secondary alcohols, and the formation of olefinic structure.

Thiophene oligomers and polythiophene derivatives are efficient sensitizers of singlet oxygen ($^1\Delta_gO_2$)

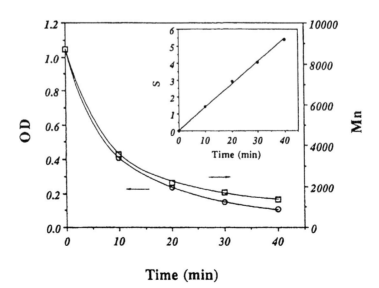

Time (min)

Figure 4.15. Decrease in the optical density of P3HT solution (λ, 435 nm) and decrease in M_n with irradiation. Inset: number of chain scissions, S, vs. irradiation time. Reprinted with permission from ref. 34.

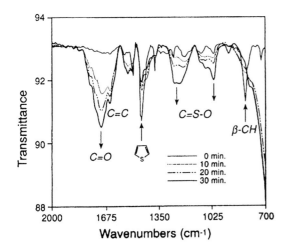

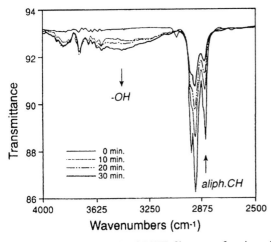

Figure 4.16. FTIR analysis of P3HT films as a function of time of irradiation (photolyzed in CHCl₃). Reprinted with permission from ref. 34.

[34,40–43]). $^1O_2$ photosensitized by P3ATs was trapped using anthracene. GC–MS, IR and UV vis spectroscopic analysis confirmed the loss of anthracene and the build up of the product, anthraquinone. In addition, the presence of the singlet oxygen trap decreased the rate of photobleaching of the polymer. However, the relative rates of photochain scission in the presence or absence of a quencher were unchanged, indicating that chain scission is unaffected by $^1O_2$. This was confirmed by introducing singlet oxygen generated *ex situ* into polymer solutions whereupon it was found that the optical density of the polymer solution rapidly decreased in contrast to the molecular weight of the polymer which remained unchanged (Figure 4.17).

FTIR analysis of the polymer following exposure to $^1O_2$ showed signals due to sulfine, keto and olefinic residues. No hydroxyl functionalities or disappearance of the aliphatic C–H stretching band was observed. The following mechanism was proposed to account for photobleaching (Figure 4.18). Endoperoxide residues (Figure 4.18, **2**) were envisaged as intermediates in the formation of the sulfine, but these primary oxygenation products can only be inferred because endoperoxides are unstable.

The mechanism of photosensitization of singlet oxygen generation was initially postulated to result from triplet–triplet annihilation according to Figure 4.19. However, this appears unlikely, given the recent phosphorescence and transient spectroscopic data which indicates a very low quantum yield of triplet formation of polythiophene [44]. An alternative explanation for photosensitization of singlet oxygen is based on the dissociation of an excited state charge transfer complex (CTC) arising from association of the polymer (electron donor, D) with oxygen (electron acceptor, A) [equation (4.38)]

$$D + A \rightleftharpoons [D^{\delta+} \cdot A^{\delta-}] \xrightarrow{h\nu} [D^{+\bullet}, A^{-\bullet}] \quad (4.38)$$

The relationship between the energy of the charge transfer transition, E_{CTC}, the ionization potential of the donor, IP_D, and the electron affinity of the acceptor, EA_A, is given by [45]

$$E_{CTC} = IP_D - EA_A - W \quad (4.39)$$

where W is the Coulombic attraction energy of the complex. E_{CTC} can be estimated with reasonable accuracy using the empirical relation [46,47]

$$E_{CTC}(eV) = E_{1/2}^D - E_{1/2}^A + 0.15(\pm 0.10) \quad (4.40)$$

$E_{1/2}^D$ and $E_{1/2}^A$ represent the half wave potentials of the donor and acceptor, respectively. The 0.15 component is an empirical factor determined from the linear correlation of the CTC energy with the energy calculated via $E_{1/2}^D$-$E_{1/2}^A$. Using measured $E_{1/2}$ values of -0.82 and $+0.94$ vs. SCE in CH₃CN/tetraethylammonium perchlorate (TEAP) for oxygen and P3HT respectively, E_{CTC} was estimated to be $\sim 1.9 \pm 0.1$ eV (648 ± 33 nm) [48].

In order to detect this absorption, films of P3HT were subjected to variable pressures of oxygen and the visible spectrum was recorded *in situ*. Films of P3HT possessed a broad absorption spectrum with a maximum absorption of ~ 505 nm and an absorption tail which extends to 620 nm. The presence of oxygen was found to perturb the optical density of the absorption

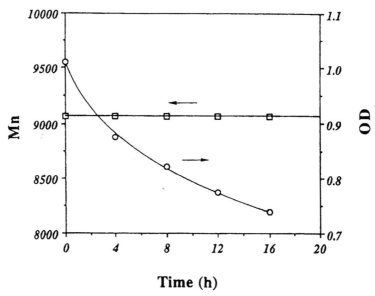

Figure 4.17. Decrease in the optical density (λ 435 nm) and change in M_n of P3HT in CHCl$_3$ during reaction with chemically generated singlet oxygen. Reprinted with permission from ref. 34.

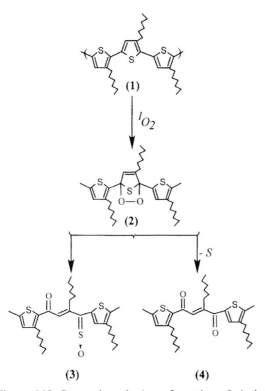

Figure 4.18. Proposed mechanism of reaction of singlet oxygen with P3AT. Reprinted with permission from ref. 34.

tail of P3HT. The difference spectra of the polymer in the presence and absence of oxygen shows a structureless band at ~ 630 nm (Figure 4.20) [48]. The proximity of this band to the predicted value for E_{CTC}, and the fact that the absorbance of the charge transfer complex was proportional to the pressure of oxygen were considered as strong evidence for a charge transfer complex between molecular oxygen and P3HT.

The structure proposed for the P3HT–O$_2$ CTC is depicted in Figure 4.21

It has been demonstrated that photogeneration of singlet oxygen in solid films of polystyrene can occur from irradiation of oxygen–polystyrene charge transfer complexes [21]. A similar mechanism can be anticipated for P3ATs. Indeed, singlet oxygen emission at 1270 nm was detected following irradiation of solutions of P3HT in the presence of oxygen. The following mechanism is offered to explain this photosensitization process (Figure 4.22). In this scheme, the ground state P3HT–O$_2$ CTC is a triplet due to multiplicity multi-

$$
\begin{aligned}
{}^1P_0 \quad & \xrightarrow{\ h\nu\ } \quad {}^1P* \\
{}^1P* \quad & \xrightarrow{\ ISC\ } \quad {}^3P* \\
{}^3P* + {}^3O_2 \quad & \longrightarrow \quad {}^1P_0 + {}^1O_2
\end{aligned}
$$

Figure 4.19. Singlet oxygen photosensitization by triplet–triplet annihilation. P represents polymer.

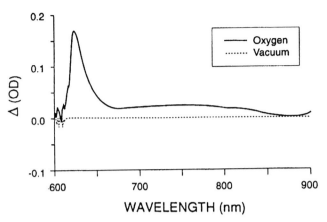

Figure 4.20. UV vis absorption difference spectra of a P3HT. P3HT thin film (20 μm) was in contact with oxygen (10 atm) and *in vacuo*. Reprinted with permission from ref. 48

plication of the singlet ground state of P3HT and the triplet ground state of molecular oxygen. Excitation of the $P3HT-O_2$ CTC yields a triplet excited state. Typically, this is very close in energy to the excited singlet state. Dissociation of the latter yields singlet oxygen. Further photophysical investigation is necessary to clarrify this mechanism. The formation of a charge transfer complex with π-conjugated polymers is anticipated to be a general effect. The degree of charge transfer and the magnitude of the equilibrium constant of complex formation will certainly be a function of the ionization potential and morphology. Those polymers possessing low ionization potentials will be particularly susceptible to charge transfer complex formation in the presence of oxygen.

In order to explain the origin of photochain scission, a classical photooxidative route was proposed [34]. Since photooxidation of polyolefins is known to be initiated by photosensitization of free radicals by catalyst impurities, experiments were designed to elucidate the role of free radicals in the chemistry of P3HT. An independent source of radicals, namely the thermolysis of *t*-butyl peroxide, was used to initiate successfully chain scission in the absence of light. Molecular weights of P3HT in refluxing benzene/*t*-butyl peroxide, at 160°C decreased from 8500 to 5000

over a period of 45 hours representing an average of ~ 0.8 chain scissions. In the absence of the free radical source no decrease in molecular weight was found. Over the same period of time the optical density of the polymer decreased by only a negligible amount ($7 \sim 3\%$).

Pristine P3HT was found to contain residual iron impurities even after extensive purification [34]. The impurities have been shown to be octahedral Fe(III) hydroxychloro-complexes ($[Fe(H_2O)_4(OH)_2]X$) where X is a chloride or hydroxyl ion. These salts have broad electronic absorptions which extend from the UV to the visible region of the spectrum (~ 250–450 nm) [49]. $Fe^{3+}-X^-$ undergoes photoreduction to yield iron(II) and a free radical $X^\bullet$

$$Fe^{3+}X^- \xrightarrow{h\nu} [Fe^{2+}X^\bullet] \longrightarrow Fe^{2+} + X^\bullet \qquad (4.41)$$

The resulting radicals ($X^\bullet$) are sufficiently reactive to abstract hydrogen atoms from most hydrocarbons (RH), thus initiating photodegradation reactions

$$RH + X^\bullet \longrightarrow R^\bullet + HX \qquad (4.42)$$

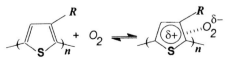

Figure 4.21. Proposed structure for the $P3HT-O_2$ charge transfer complex. Reprinted with permission from ref. 48

Figure 4.22. Singlet oxygen photosensitization by photodissociation of polymer (P)–oxygen charge transfer complex.

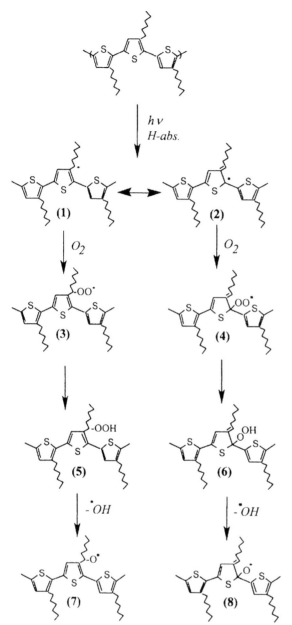

Figure 4.23. Proposed mechanism of photooxidation of P3AT. Reprinted with permission from ref. 34.

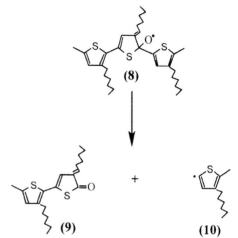

Figure 4.24. Proposed mechanism of chain scission of P3AT. Reprinted with permission from ref. 34.

oxygen with free radicals is known to be diffusion controlled. The following mechanism of photooxidation was proposed [34] (Figure 4.23). In this scheme, the resulting P3HT radical is resonance stabilized by the π-system to yield radicals **1** and **2** which rapidly react with oxygen to produce peroxides **3** and **4**. These abstract a hydrogen from the polymer chain, or solvent, to yield hydroperoxides **5** and **6**. Cleavage of the

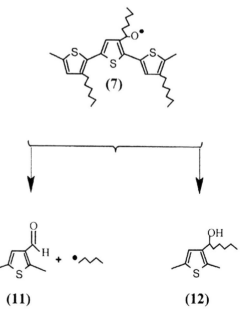

Figure 4.25. Reactions of thienyl alkoxy radicals. Reprinted with permission from ref. 34

The primary point of hydrogen abstraction from P3AT is the α-carbon atom of the alkyl group due to the lower dissociation energy of the α-C-H bond. Abstraction of a hydrogen atom was presumed to be followed by the addition of oxygen to the polymer-bound radical and formation of hydroperoxide. The rate of reaction of

hydroperoxide group by light results in alkoxy radicals **7** and **8**. In this mechanism only rearrangement of the alkoxy radical **8** can cause chain scission, forming ketones **9** and thienyl radicals **10**, (Figure 4.24). The alkoxy radical **7** (Figure 4.23), on the other hand, can undergo several of the following possibilities as shown in Figure 4.25: (1) β-scission to yield an aldehyde **11**, and volatile hydrocarbons; (ii) hydrogen abstraction to give a secondary alcohol **12**, (iii) radical coupling to form a crosslinked polymer. The latter possibility which is suppressed in solution was proposed viable in solid films.

FTIR measurements on thin solid polymer films were used to determine the photochemical processes in the solid state [50]. Figure 4.26 shows FTIR spectra of thin films of the polymer before and after irradiation in air using polychromatic light (> 300 nm). FTIR bands indicated the formation of similar functionality as found in the case of polymer solutions. Notably, a decrease in the C−H stretching band of the alkyl chain was observed together with the formation of sulfine residues (C−O, C=C and C−S$^+$O$^-$), as a result of a 1,4 Diels-Alder addition of photosensitized singlet oxygen.

Kinetic studies established that these absorption bands evolve immediately upon irradiation and that all three bands increase in intensity at the same proportional rate, indicating their common reaction pathway (Figure 4.27).

Irradiation of films of P3HT caused both photo-bleaching and chain scission [50]. The former was proposed to involve photosensitization and reaction of singlet oxygen, and the latter, a classical photooxidation route. This was confirmed by reacting polymer films with singlet oxygen ($^1O_2$) generated *ex situ*. The optical density of the film at 500 nm was reported to decrease by ∼ 20%, and the polymer remained soluble, indicating the absence of crosslinking. FTIR analysis of the polymer following exposure to $^1O_2$ showed strong signals due to sulfine, keto and olefinic residues, but no hydroxyl or ether functionalities were observed. The reaction products are consistent with the reported mechanisms observed for the reaction of $^1O_2$ with P3ATs in solution, i.e., Figure 4.23.

Photo-induced crosslinking appears to involve photo-oxidation of polymers and was rationalized in terms of a free radical chain reaction comprised of initiation, propagation and termination steps possibly involving Fe(III) impurities. As in the solution case, the primary point of hydrogen abstraction from P3HT was proposed to be the α-carbon atom of the alkyl group, followed by addition of oxygen to the polymer-bound radical and the formation of hydroperoxide. Cleavage of the peroxide by the absorption of light, or by reaction with Fe(II) species, produces alkoxy radicals which can couple with polymer-bound radicals to form ether linkages (Figure 4.28). Hydroxyl radicals are produced in subsequent reactions. These propagate the photo-oxidation process by participating in additional hydrogen atom abstraction reactions.

When thin P3HT films were irradiated in an argon atmosphere, the films remained soluble in toluene and

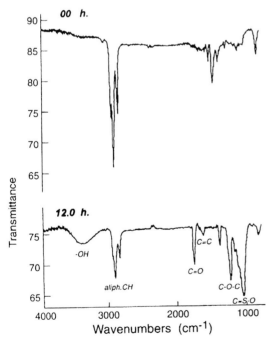

Figure 4.26. FTIR spectra of P3HT films before and after 12 h irradiation. Irradiation in air with a polychromatic light (> 300 nm). Reprinted with permission from ref. 50.

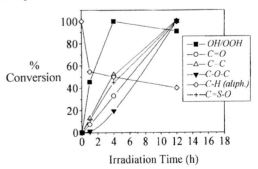

Figure 4.27. Time-dependent evolution of functional groups during irradiation of P3HT in air. Reprinted with permission from ref. 50.

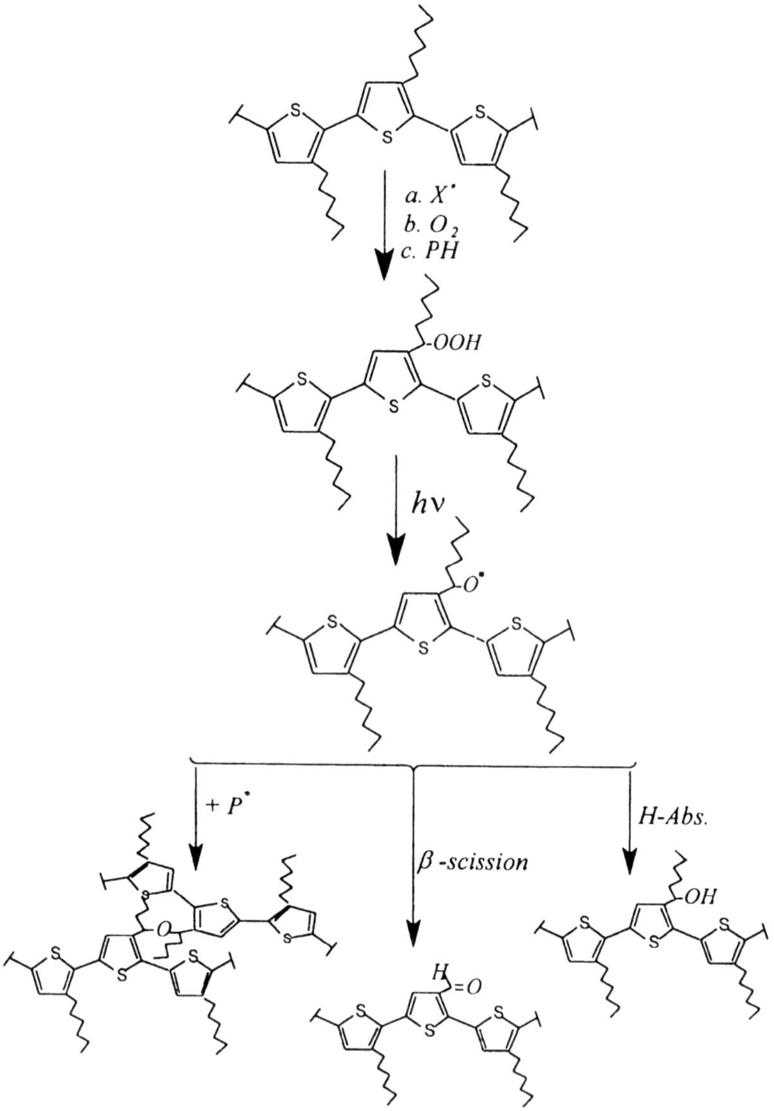

Figure 4.28. Proposed mechanism of photo-induced crosslinking of P3AT. Reprinted with permission from ref. 50.

changes in the IR spectra were negligible, even after 24 h of photolysis. Furthermore, parallel studies with side chain-free, α-hexathiophene and polythiophene revealed the absence of photooxidation under the same experimental conditions used for P3HT, thus corroborating the hypothesis that alkyl side chains are involved in the photooxidation process. Photobleaching was still observed for these systems since photosensitization and the reaction of singlet oxygen is independent of the presence or absence of the alkyl side chain.

It is useful to draw upon thermal oxidation studies of P3ATs. Films heated *in vacuo* at 200°C for 15 h remained completely soluble in organic solvents and FTIR spectra of the polymer before and after thermal treatment were identical [50]. However, when polymer films were heated in the presence of an oxygen atmosphere, the films were rendered insoluble and exhibited FTIR spectra very similar to those produced by irradiation of P3HT in oxygen, i.e. loss of C−H, and an increase in −OH and C−O−C. A sulfine

(C=S$^+$-O$^-$) residue was absent in the thermally treated sample since this is a photochemically generated product. The similarity between the thermal and photochemical products of P3HT in the presence of oxygen infers similar chemical routes. It is well known from studies of poiyolefins that thermal treatment in the presence of oxygen results in a free radical chain reaction initiated by hydrogen abstraction, followed by the formation of hydroperoxide, peroxide bond cleavage and crosslinking via ether bond formation [50]. The same chemical events appear to occur in the photooxidative process, the difference being that in one case the initial hydrogen abstraction is thermally driven, while in the other, it is photochemically driven.

The relative efficiencies of both photo-induced crosslinking and chain scission were evaluated by subjecting the polymers to a sol-gel analysis using the Charles–Pinner Theory [23]. Figure 4.29 shows a solubility plot of P3HT films following irradiation with 435 nm light. This graph was constructed by plotting the soluble fraction $S + S^{1/2}$ of P3HT films as a function of $1/r$ according to equation (4.25). The intercept of the linear plot at infinite exposure dose provides the ratio of photochain scission to photocrosslinking, p_o/q_o. A value of 0.23 was reported indicating that films of P3HT undergo simultaneous crosslinking and chain scission. The crosslinking density, q_o, was found to be ~ 4.5 times greater than the fracture density, p_o. The energy required for incipient gelation of the polymerise, r_{gel}, calculated from extrapolation of the linear plot to a sol fraction $S + S^{1/2}$, was 49 mJ cm^{-2}, or 1.78×10^{-7} Einstein cm^{-2}. The quantum yield of crosslinking, ϕ_x, at 435 nm was calculated to be 2.37×10^{-3} mol Einstein^{-1}. Using the ratio p_o/q_o, the quantum yield of main chain scission, Φ_s, was estimated to be 5.45×10^{-4} mol Einstein^{-1}. Main chain scission was indeed observed experimentally as evidenced by

a decrease in M_n of the soluble extractable component of the polymer following irradiation and development. However, the relative efficiencies of these processes was proposed to be dependent on the concentration of radicals and hence the irradiation intensity, so that quantum yield of crosslinking obtained at lower incident intensifies may be substantially less.

The photochemistry of water soluble conjugated analogs, i.e. poly[ω-(3'-thienyl)alkanesulfonate]s (P3TASs) (Figure 4.30) in the solid state and in aqueous solution has been investigated in order to determine whether similar photochemical mechanisms apply [39].

Solutions of sodium poly[ω-(3'-thienyl)alkanesulfonate]s were found to undergo a relatively small degree of photobleaching upon exposure to UV or visible light and a blue shift in the absorption spectrum upon prolonged photolysis. This is consistent with a shortening of the π-conjugation length due to singlet oxygen sensitization and its subsequent 1,4 Diels–Alder addition to thienyl units as previously outlined. FTIR spectroscopic analysis was consistent with the proposed photochemical mechanism of degradation of non-suifonated P3ATs. However, aqueous solutions of the suifonated polymers were reported to be considerable more photostable than their non-aqueous analogs. A slight dependence of the alkanesulfonate chain length on the rate of photobleaching was observed. Photobleaching of polymer films exposed to ambient atmosphere was very much slower compared to polymer solutions, and much slower than films of their non-sulfonated analogs, as shown in Figure 4.31. Films of the sodium salt of P3TASs were also found to be more stable than their solutions. However, films of sodium poly [6-(3'-thienyl)alkanesulfonate]s (P3TASNa) exhibited enhanced photostability over P3ATs when irradiated in ambient air. This was interpreted on the basis of the hygroscopic nature of these sulfonate-based polyelectrolytes, which take up moisture from the atmosphere, and which lead to efficient quenching of photosensitized singlet oxygen and thus lower rates of photobleaching. The presence of moisture was confirmed by FTIR spectroscopy.

The sodium salt of the polymers can be converted to the acid form (P3TASH) by passing solutions of the

$S + S^{1/2}$

2.0
1.6
1.2
0.8
0.4
0.0

0.000　　0.008　　0.016

$1/r$ (cm mJ^{-1})

Figure 4.29. Solubility curve of P3HT following irradiation with 435 nm light. Reprinted with permission from ref. 50.

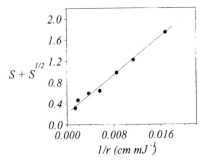

Figure 4.30. Poly(ω-(3'-thienyl)alkanesulfonate)s.

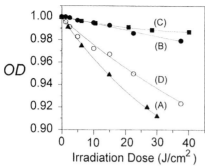

Figure 4.31. Change in optical density (at λ_{max}) of polymer films during photolysis in ambient air (unless stated): (A) poly(3-hexylthiophene); (B) sodium poly(6-(3'-(thienyl))-hexanesulfonate) P3THSNa; (C) poly(6-(3'-(thienyl))hexanesulfonic acid) P3THSH; (D) anhydrous P3THSNa in dry air in dry air. Reprinted with permission from ref. 39.

polymers through a cation exchange column. Films of the sulfonic acid form of these polymers exposed to ambient atmosphere were found to be even more photostable than their sodium salt analogs (Figure 4.31). Remarkably only a small change in optical density ($\sim 1\%$) was observed upon prolonged exposure (40 J cm^{-2}).The enhanced stability of the acid form of the sulfonated polymers might be due to a combination of moisture uptake, enhanced rigidity of the polymer backbone leading to efficient internal quenching of excitation via lattice relaxation, and the presence of polaronic and bipolaronic charges on the polymer backbone which serve as quenching centers for the deactivation of excitation.

Unlike non-sulfonated P3ATs, P3TAS films exposed to ambient atmosphere remained water soluble even after 2 h of irradiation. This was presumed due to residual water interfering with the crosslinking process. In support of this it was found that anhydrous films were rendered completely insoluble upon exposure to moderate doses of UV vis irradiation [39]. In addition, photochemical chain scission has been used to render non-soluble P3ATs soluble [51]. In this work, P3HT was grown electrochemically to yield a non-soluble polymer film on the electrode. Following irradiation of the film in an appropriate oxygenated solvent with an argon ion laser, or halogen lamp, the polymer became soluble. The bandgap of the soluble portion was slightly larger than the pristine polymer indicating a shorter conjugation length and/or lower molecular weight. IR analysis of the photoiyzed polymers indicated the formation of carbonyl species. Of note is the observation that films of poly(3-ethy]thiophene) could not be solubilized. In a similar experiment, films of P3HT were irradiated in chloroform solution under

an atmosphere of carbon dioxide [52]. In addition to the presence of carbonyl groups, a C–O stretch mode was observed in the IR region. The authors speculated the formation of chemical structures associated with these reactions (Figure 4.32).

Crosslinking of soluble P3ATs has also been induced by γ-irradiation [53]. The resultant material is a gel which can be processed because of its fusibility. Figure 4.33 shows the dependence of gel formation on the irradiation dose. Gelation begins after 20 Mrad. The mechanism of crosslinking was not addressed in detail but the authors inferred that the presence of the alkyl side chain was critical for crosslinking.

In another photochemical study of polythiophenes, films of poly(3-octylthiophene) were reported to be photochemically stable [54]. The discrepancy between this and previous observations [34,50] might be due to the lower irradiation doses employed and the fact that a low pressure mercury lamp lwas employed. Low pressure mercury lamps emit light mainly of wavelength < 366 nm with a maximum intensity at 254 nm. In the former studies, a high pressure mercury lamp was used in which the emission was much more likely to be absorbed by impurities and the $\pi-\pi^*$ system.

Photochemistry of P3ATs possessing a photolabile functionality attached to the side chain are attracting increasing attention due to applications in photoimaging. Photosolubilization of P3AT has already been addressed. Kaeriyama *et al.* have reported photochemical studies of copolymers of poly[3-(4'-butyl-4'-methyl-4'-silaotcyl)-thiophene-co-(3-nonylthiophene)] (Figure 4.34) and concluded that the extent of photooxidation of the latter is more pronounced that the corresponding homopolymer, poly(3-nonylthiophene) [55].

Figure 4.35 shows the affect of irradiation of poly(3-nonylthiophene) (P3NT) and the copolymer with a high pressure mercury lamp (250 mW cm^{-2}) The copolymer shows a more significant blue shift than the homopolymer indicating more efficient disruption of conjugation. The authors suggest that accelerated photooxidation is due to the formation of side chain

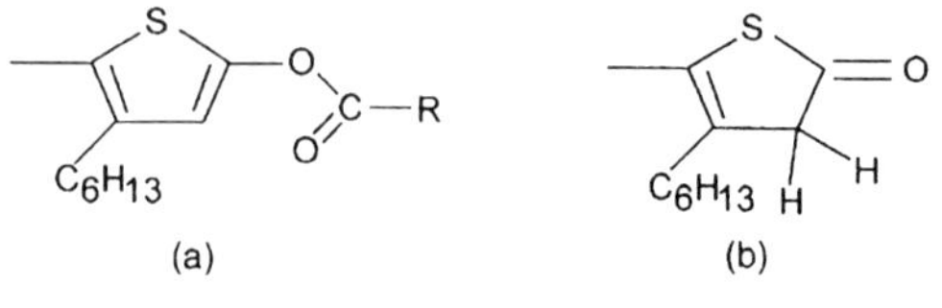

Figure 4.32. Speculated chemical structures resulting from photo-induced solubilization of insoluble P3HT: (a) under CO_2 and (b) under O_2. Reprinted with permission from ref. 52.

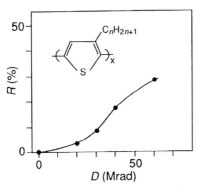

Figure 4.33. Dependence of gel formation (R) on the γ-irradiation of P3HT. Reprinted with permission from ref. 53

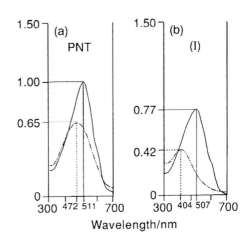

Figure 4.35. UV vis spectra of (a) P3NT and (b) poly-[3-(4'-butyl-4'-methyl-4'-silaotcyl)-thiophene-co-(3-nonylthiophene)](I) films before (——) and (after) (-··-) irradiation. Reprinted with permission from ref. 55.

radicals associated with decomposition of the silyl-containing alkyl group. The photolyzed copolymer could not be oxidatively doped with ferric chloride to give conductive films, whereas photolyzed P3NT could. The effect of irradiation of the π-electron system was demonstrated using cyclic voltammetric analysis of polymer films. Photolyzed P3NT was still electrochemically active, albeit to a lesser degree than the pristine polymer, whereas the copolymer irradiated with an equivalent irradiation dose was completely inactive.

In a similar vein, the synthesis, photochemistry and photoimaging of π-conjugated polymers and copolymers based on poly[3-(2-methacryloyloxyethyl)-thiophene] (Figure 4.36) has been reported [56].

The methacrylate functionality attached to the thienyl group is light sensitive. FTIR studies showed a decrease in the vinylidene band and a shift in the carbonyl stretch towards a higher frequency. The latter was attributed to a decrease in α, β-conjugation associated with the carbonyl and the double bond in the methacrylate side chain. Functional groups including hydroxyls, hydroperoxides, sulfine residues, all products of reaction with photosensitized singlet oxygen, were also observed since polythiophenes are efficient singlet oxygen

photosensitizers. Methacrylate-functionalized polythiophenes were rendered insoluble upon irradiation with much greater efficiency than P3ATs. As a result, only low irradiation doses were required to completely insolubilize the film, during which time only small losses in the optical density due to singlet oxygen reactions were observed. Thus, high conductivities of photolyzed films were observed upon oxidative doping because π-conjugation was retained during photoinsolubilization. FTIR studies indicated that insolubilization was most likely the result of crosslinking via free

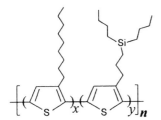

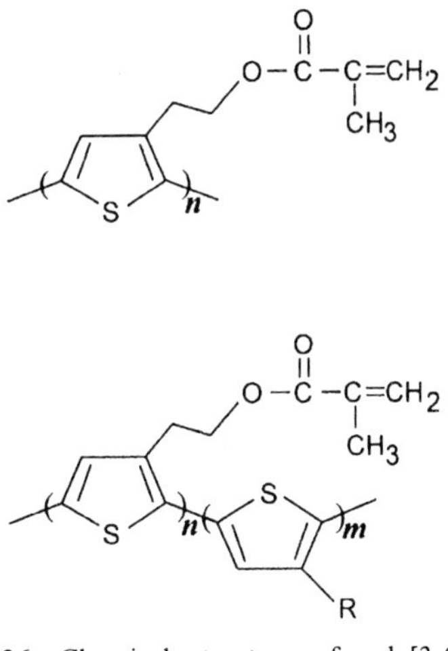

Figure 4.34. Chemical structure of poly[3-(4'-butyl-4'-methyl-4'-silaotcyl)-thiophene-co-(3-nonylthiophene)]. Reprinted with permission from ref. 55.

Figure 4.36. Chemical structure of poly[3-(2-methacryloyloxyethyl)thiophene] (PMET) and PMET-co-(3-hexylthiophene). Reprinted with permission from ref. 56.

radical-induced coupling of the methacrylate groups. More details of the lithographic process are given in Section 8.

4.3   Poly(*p*-phenylene sulfide)

A laser flash photolysis study of oligomers corresponding to poly(*p*-phenylene sulfide) was reported in 1993 [57]. In this study, eight model compounds and two aromatic disulfides were studied by nanosecond flash photolysis (Table 4.4). Only weak luminescence was observed despite the short lifetimes of the singlet state ($< 10^{-9}$ s). Upon photolysis in solution, short-lived triplets were detected. Figure 4.37 shows the transient absorption spectra for the dimer 0.1 and 0.6 μ s following the laser pulse. The signals at 450 and 630 nm are due to triplets, which rapidly decay to yield a longer-lived species. The triplets could be quenched by a variety of species. The authors suggest that rapid quenching by oxygen ($c.\ 10^{10}\ M^{-1}\ s^{-1}$) indicates that a charge transfer complex with oxygen plays a major role in the quenching process. The residual spectrum is associated with C$-$S bond cleavage and the formation of thiophenoxy radicals ($\Phi_\phi \sim\ < 0.05$). This was confirmed from the photolysis of disulfides which yielded thiophenoxy radicals exhibiting similar transient absorption spectra. It was suggested that photocleavage occurred from the singlet state. Similar processes were found for longer oligomers. Photo-induced cleavage of C$-$S bonds was inefficient due to intersystem crossing to the triplet state. This trend is expected to continue for the analogous polymers, but has yet to be proved. The participation of triplets in the photosensitization of singlet oxygen and subsequent reactions with the oligomers or polymers was suggested but no experimental evidence was presented.

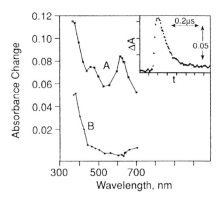

Figure 4.37. Transient absorption spectra for C$_6$H$_5$SC$_6$H$_5$ in acetonitrile (A) 0.1 μs and (B) 0.6 μs following the laser pulse. Inset shows the rate of decrease of the triplet at 635 nm. Reprinted with permission from ref. 57.

4.4   Poly(*p*-phenylenevinylene)

Poly(*p*-phenylenevinylene) (PPV) has attracted considerable attention because of its luminescence properties and its applications in electroluminescent displays. The usual synthetic route for this polymer involves synthesis of the polymeric precursor which often contains the pendant tetrahydrothiophenium functionality. Thermal elimination of tetrahydrothiophene yields the corresponding π-conjugated polymer. Photoelimination of precursors has been investigated in order to achieve a low temperature procedure [58] (Figure 4.38).

Thin films of the precursor polymer were irradiated with a 200 W xenon lamp *in vacuo* and the absorption spectra were monitored. Figure 4.39 shows the spectra before and after elimination. The increased absorption at *c.* 3 eV is evidence of increased π-conjugation and hence photoelimination of tetrahydrothiophene. The onset of optical absorption of the resultant polymer was

Table 4.4. Model compounds of poly(*p*-phenylene sulfide) studied by laser flash photolysis. Reprinted with permission from ref. 57

Polymer #	Formula*	Polymer #	Formula*
1	C$_6$H$_5$SC$_6$H$_5$ (dimer)	6	*p*BrC$_6$H$_4$SC$_6$H$_4$Br-*p*
2	C$_6$H$_5$SC$_6$H$_4$SC$_6$H$_5$ (trimer)	7	*p*-PhCOC$_6$H$_4$SC$_6$H$_4$COPh-*p*
3	C$_6$H$_5$SC$_6$H$_4$SC$_6$H$_4$SC$_6$H$_5$ (tetramer)	8	*p*-CH$_3$C$_6$H$_4$SC$_6$H$_4$SC$_6$H$_5$CH$_3$-*p*
4	*p*-ClC$_6$H$_4$SC$_6$H$_5$	9	C$_6$H$_5$SSC$_6$H$_5$
5	*p*-ClC$_6$H$_4$SC$_6$H$_4$Cl-*p*	10	*p*-ClC$_6$H$_4$SSC$_6$H$_4$Cl-*p*

* Phenylene groups ($-$C$_6$H$_4-$) are linked at the *p*-positions.

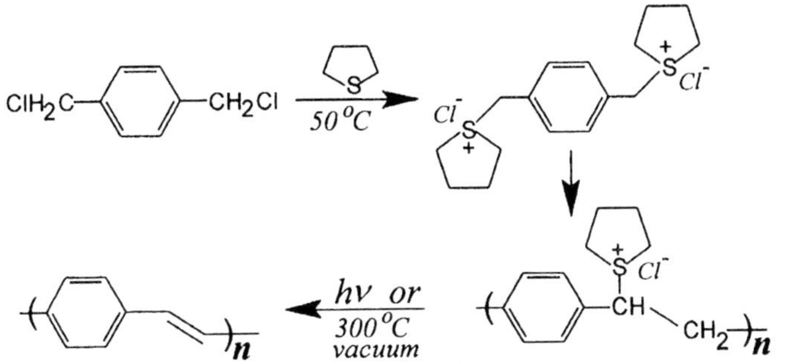

Figure 4.38. Scheme for the thermal or photochemical preparation of PPV from the precursor polymer. Reprinted with permission from ref. 58.

c 2.52 eV, which is slightly higher in energy than the thermally eliminated polymer (2.34–2.44 eV).

IR spectroscopy confirmed the emergence of vinylene units upon photolysis. However, photoelimination also led to unwanted side reactions which hindered conversion by subsequent heat treatment. Thus the resultant films continued to exhibit strong absorptions between 3.5 and 4.0 eV. The nature of the side reactions was not determined, although the possibility of the reaction of the partially converted polymer with methanol or chlorine molecules was ruled out. Nevertheless, when the photoelimination was performed at elevated temperature (120°C) the reaction proceeded smoothly and the photoconverted polymer exhibited an absorption curve more akin to thermally-prepared PPV (Figure 4.40).

A similar procedure has been employed using Ar$^+$ laser irradiation of precursor polyelectrolyte films of both unsubstituted PPV (H-PPV), and its 2,5- dimethoxyphenylene derivative (DMEO-PPV) (Figure 4.41) [59].

Figure 4.42 shows the UV vis absorption spectra of the precursor to H-PPV in relation to the Ar$^+$ laser line.

The absorption spectra of cast films indicate that partial elimination occurs during casting. Even so, the laser emission does not overlap with that of the absorption band of the precursor. Thus even laser irradiation with 350 W cm^2 did not affect photoelimination. In contrast, the absorption spectra of the corresponding dimethoxy derivative is blue-shifted so that irradiation of partially eliminated thin films caused a color change from orange to red indicating photoelimination (Figure 4.43). FTIR analysis of the latter indicated a substantial change in the ratio of aromatic ring C–H vibrations to C–H stretching vibrations of the vinylene groups. A characteristic *trans*-vinylene C–H out-of-plane bending band was observed. Surprisingly, very little carbonyl formation was observed even though irradiation was carried out in ambient air.

In order to facilitate absorption of the argon ion laser, photosensitizers were impregnated into precursor films prior to irradiation. Photoeffects of the sodium salt of 4-(2-hydroxy-1-naphthylazo)-l-naphthalenesulfonic acid (commercial name, Acid Red 88) on the photoelimination of precursor polymers were reported in detail [59].

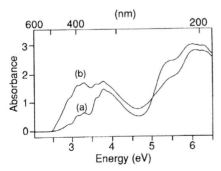

Figure 4.39. Optical absorption spectra of precursor to PPV (a) before and (b) after irradiation with UV light at room temperature *in vacuo*. Reprinted with permission from ref. 58.

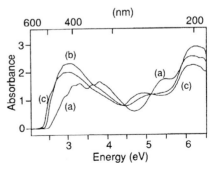

Figure 4.40. Optical absorption spectra of precursor PPV (a) before and (b) after irradiation with UV light at 120°C *in vacuo*. (c) Thermally prepared PPV. Reprinted with permission from ref. 58.

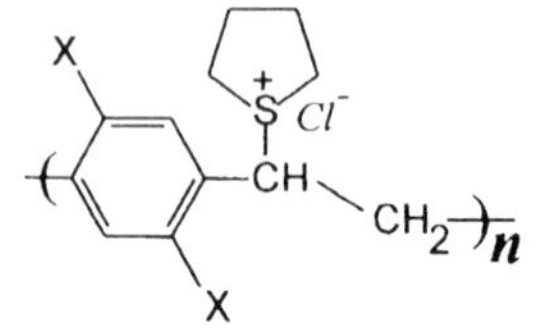

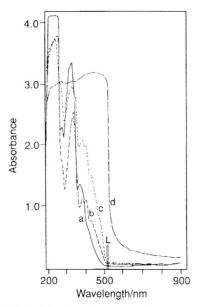

Figure 4.41. Chemical structures of H-PPV (x = –H) and its 2,5-dimethoxyphenylene derivative, DMEO-PPV (x = –OCH₃). Reprinted with permission from ref. 59.

An anionic sensitizer was chosen in order to obtain uniform dispersions in the cationic polyelectrolyte. This dye exhibited a strong absorption band at the laser emission wavelength and effectively sensitized the photoelimination of precursor films of both the H-PPV and dimethoxy derivatives. The mechanism of sensitization was not clear but it was suggested that the absorption and dissipation of irradiation could indirectly lead to a thermal elimination process.

It has been found that fluorescence from PPVs decreases in intensity when irradiated with visible light in air [60]. Under nitrogen, irradiation does not affect the photoluminescence yield. Photooxidation appears responsible for the decrease in luminescence. FTIR indicated the emergence of carbonyl groups and the disappearance of the vinyl functionality (Figure 4.44).

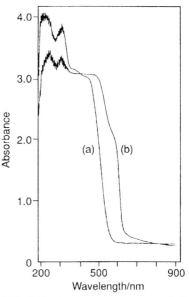

Figure 4.43. UV vis absorption spectra of films of the precursor to DMEO-PPV(a) before and (b) after irradiation. Reprinted with permission from ref. 59.

The photoluminescent yield was found to be inversely related to the concentration of carbonyl groups (Figure 4.44).

A detailed study of the mechanisms of photochemical degradation of PPV also confirms the photosensitivity of the vinyl group [61]. The authors provide evidence that the difference IR spectra of photolyzed PPV are similar to the small molecule analog, stilbene-4,4′-dialdehyde. The authors conclude that photochemical oxidation of PPV occurs at the vinylene bond and results in chain scission with terminal aldehyde formation.

The photochemistry of a soluble derivative of PPV, poly[2,5-bis(cholestanoxy)-1,4-phenylenevinylene] (BCHA-PPV) has been investigated on gold and aluminium substrates using a low pressure mercury irradiation source (Figure 4.45) [54]. The films were analyzed *in situ* using IR reflection spectroscopy (IRRAS) and *ex situ* X-ray photoelectron (XPS) and attenuated total reflection (ATR) spectroscopies. When films of BCHA-PPV were irradiated *in vacuo* no change in the IR spectra were observed even after prolonged exposure. When irradiated in air, a shift in the carbonyl peak and an attenuation of the signals associated with aromatic C−H bonds was observed. ATR indicated the disappearance of phenyl groups. Methylene and methyl groups associated with the side chain were also reported to decrease. An unassigned IR

Figure 4.42. UV vis absorption spectra of the precursor to H-PPV: (a) in solution, (b) and (c) with different degrees of elimination and (d) the fully conjugated polymer. The spectral position of the Ar⁺ laser line is shown as L. Reprinted with permission from ref. 59.

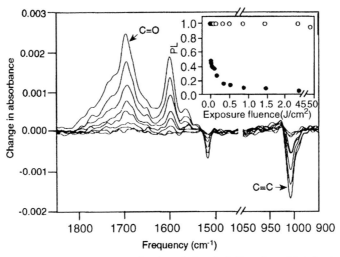

Figure 4.44. Differential FTIR spectroscopy of PPV as a function of photolysis time. Inset: Photoluminescence (PL) as a function of irradiation dose in air (●) and under nitrogen (○). Reprinted with permission from ref. 60.

band at 3400 cm^{-1} emerged. Volatile aldehydes were also detected by IRRAS, while XPS indicated the formation of C−O bonds. The proposed mechanism (Figure 4.46) is consistent with photooxidation of the vinylene bond in PPV [60] and involves formation of alkoxy radicals from the 1,2-cycloaddition of singlet oxygen. It was inferred that BCHA-PPV is an efficient photosensitizer of singlet oxygen which appears to be a reasonable assumption given the energy levels of the participant species. Hydrogen atom abstraction is presumed to follow the initial reaction with singlet

oxygen, thus initiating formation of hydroperoxides and carbonyl groups.

4.5 Polyanilines and Polypyrroles

Relatively little information regarding the photochemistry of nitrogen-containing conducting polymers, such as polyanilines and polypyrroles, has been reported. This most likely attests to the efficient non-radiative deactivation of the excited state. The various forms of

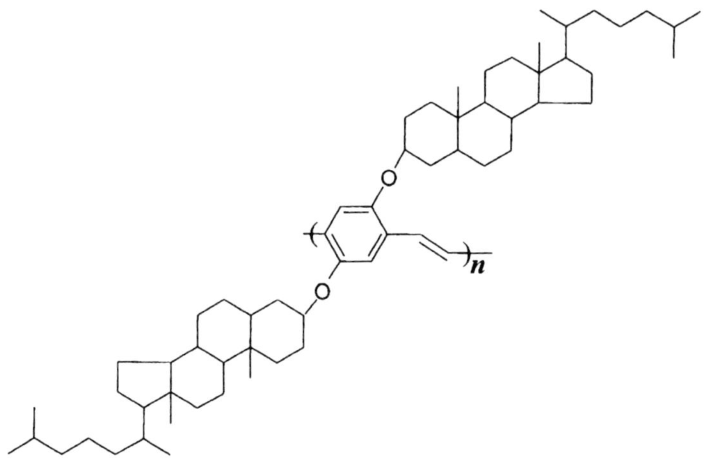

Figure 4.45. Structure of BCHA-PPV. Reprinted with permission from ref. 54.

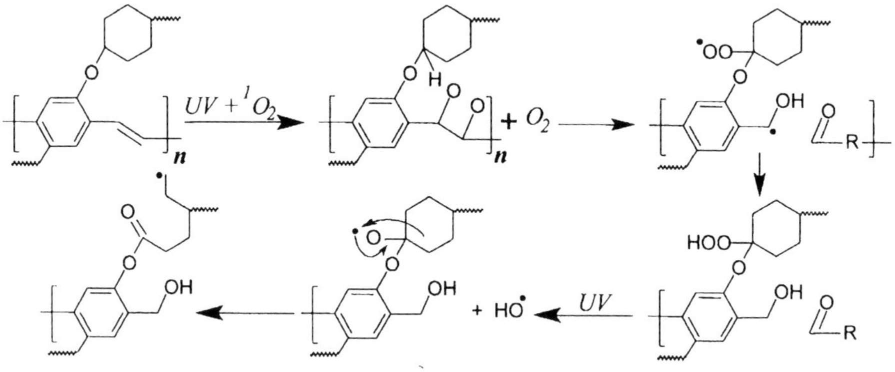

Figure 4.46. Proposed mechanism of photooxidation of BCHA-PPV. Reprinted with permission from ref. 54.

polyaniline discussed in this section are shown in Figure 4.47 The reduced form of polyaniline, leucoemeraldine, has been reported to undergo an enhanced rate of oxidation in N-methylpyrrolidone (NMP) under irradiation [62]. Figure 4.48 shows the UV vis absorption spectrum of leucoemeraldine in NMP and the absorption peak at 345 nm. The weak absorption signal at 635 nm increases with time, and is accelerated by irradiation with a 150 W xenon arc lamp. Eventually the absorbance ratio of the higher energy peak to the lower energy peak approaches that observed for the pure emeraldine form. With further irradiation, both these absorption bands exhibit a blue shift.

IR analysis indicates the formation of carbonyl structures and a progressive increase in the photooxidation of the aromatic rings. XPS indicates the possibility of solvent molecules being grafted onto the polymer chain. The emeraldine form is found to photodegrade, but at a much slower rate than the leucoemeraldine form. In the absence of dissolved oxygen, leucoemeraldine solutions are thermally stable but a slight blue shift in the 345 nm peak is observed under irradiation,

which indicates partial photodegradation. The 635 nm peak, however, is barely affected indicating that degradation takes the form of photooxidation and not simple electron transfer. These results stress the importance of oxygen in photochemical degradation. Figure 4.49 shows the effect of irradiation on leucoemeraldine solutions. The irradiated sample shows an increase in the 1600 cm^{-1}/1500 cm^{-1} ratio compared to the pristine sample, which indicates an increase in the quinoid-like content of the polymer (imine units) over the benzenoid content (amine units) as a result of irradiation. The emergence of a peak at 1380 cm^{-1} is also evidence of the increase in quinoid character. A small band at 1680 cm^{-1} was also observed, providing more evidence of photooxidation and the formation of carbonyl groups. XPS also confirmed this oxidation process.

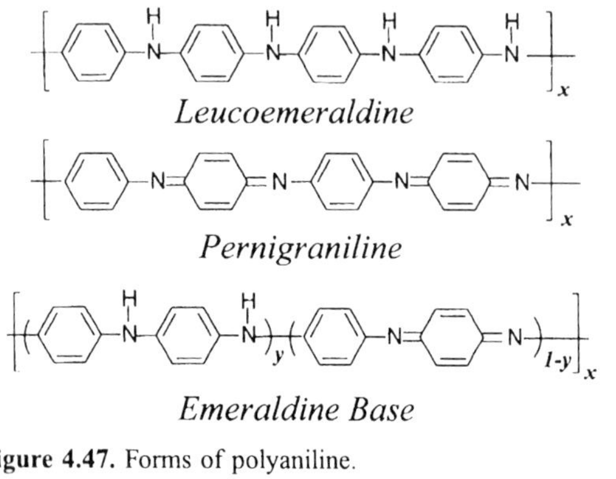

5   PHOTOCHEMICAL ELECTRON TRANSFER

Photoelectron transfer reactions are important photoprocesses of π-conjugated polymers. This section addresses photoelectron transfer processes of conjugated polymers which result in chemical change. These processes are distinguished from solid state photovoltaic or photogalvanic systems in which the absorbed photon causes current flow without chemical change. The latter will not be reviewed in this chapter.

Oligomers serve as model compounds for corresponding π-conjugated polymers. Scaiano and coworkers have demonstrated using transient spectroscopy that bithienyl and terthienyl oligomers undergo electron transfer reactions to form radical cations when irradiated in the presence of electron acceptors such as

Leucoemeraldine

Pernigraniline

Emeraldine Base

Figure 4.47. Forms of polyaniline.

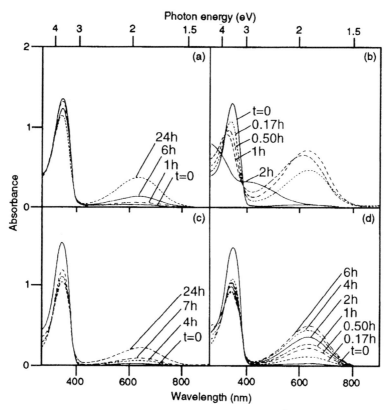

Figure 4.48. UV-vis absorption spectra of leucoemeraldine in NMP: (a) 2×10^{-4} M not irradiated; (b) 2×10^{-4} M irradiated; (c) 2×10^{-3} M not irradiated; (d) 2×10^{-3} M irradiated. Reprinted with permission from ref. 62.

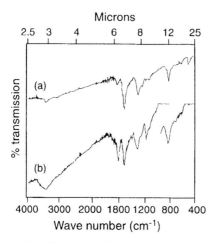

Figure 4.49. Effect of irradiation on IR spectra of solid leucoemeraldine following irradiation in NMP solutions: (a) pristine polymer and (b) following irradiation. Reprinted with permission from ref. 62.

tetracyanoethylene (TCNE) and methylviologen (mv) [41]. Detailed quenching studies are reported. Electron transfer apparently proceeds through the triplet state with rate constants c. 5–30×10^9 $M^{-1}s^{-1}$.

$$3T \xrightarrow{h\nu} {}^3(3T)^* \xrightarrow{h\nu, TCNE} T^{+\cdot} + TCNE^{-\cdot} \quad (4.43)$$

Similarly, Garnier *et al.* reported the photochemical generation of radical cations from thiophene oligomers (nT), for terthiophene (3T) to sexithiophene (6T), in dichloromethane solution using laser flash photolysis [63]. Triplet states are observed with lifetimes of tens of microseconds. The corresponding radical cations, nT$^{+\cdot}$, were reported to form with rate constants $> 10^{10}$ M^{-1} s^{-1} via the triplet state in the presence of an electron acceptor. Figure 4.50 shows the transient absorption spectra of a number of oligomeric thiophene radical cations.

In the absence of a good electron acceptor, excited state triplet oligomers and radical cations can still be observed. The former are reportedly quenched by

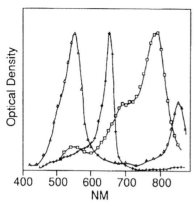

Figure 4.50. Transient absorption spectra of $\Delta 3T^{+\cdot} + 4T^{+\cdot}$, and $\square$ $6T^{+\cdot}$ in dichloromethane. Reprinted with permission from ref. 63.

oxygen but the presence of oxygen has no affect on the generation of $nT^{+\cdot}$. This observation, together with other quenching and sensitization experiments, led the authors to suggest that $nT^{+\cdot}$ formation proceeds via the singlet state [equation (4.44)] in contrast to the work of Scaiano *et al.*. The chlorinated solvent was suggested as the recipient of the ejected electron [equation (4.45)]. Also of note is the observation that $nT^{+\cdot}$ formation

increased by a factor of *c.* five upon increasing the conjugation length from $n = 3$ to $n = 6$, presumably due to a decrease in the electrochemical oxidation potential with increasing conjugation length. By extrapolation, polymeric thiophenes are expected to participate in photochemical electron transfer reactions. In this work, it was also recognized that the resultant radical cations participated in coupling reactions to yield higher order analogs, i.e. sexithiophene. This species was observed in the UV vis absorption spectra of photolyzed terthienyl.

$$3T \xrightarrow{h\nu} {}^{3}(3T)^{*} \longrightarrow T^{+\cdot} + e^{-} \qquad (4.44)$$

$$e^{-} + RCl \longrightarrow R^{\cdot} + Cl^{-} \qquad (4.45)$$

Oligothiophene radicals have also been formed photochemically in acidic solutions and acidic polymeric media such as Nafion [64]. Figure 4.51 shows the oligomeric thiophenes used in this study. The presence of terminal groups on the oligomers was intended to prevent α–α' coupling reactions upon oxidation.

A solution of compound 3Th-SMe (Figure 4.51) in methylene chloride/trifluoroacetic acid was reported to change color from yellow to green, and finally blue within 24 h in the presence of laboratory light. The presence of the radical cation in these solutions was

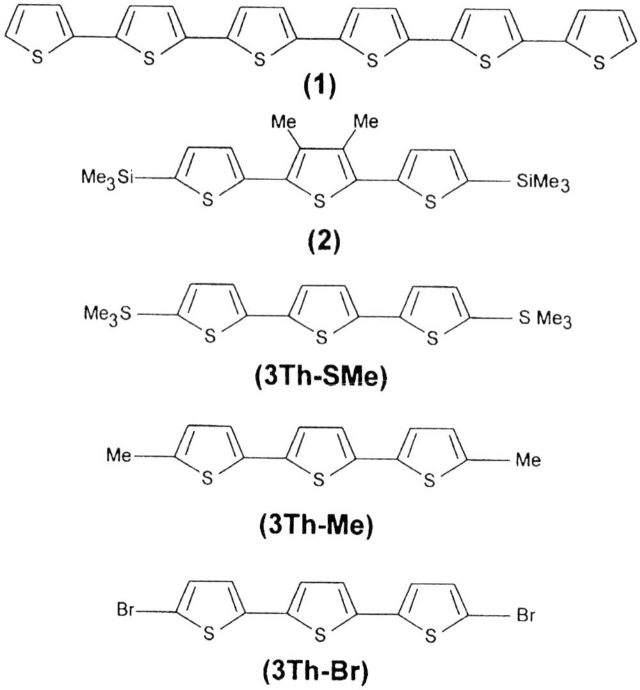

(1)

(2)

(3Th-SMe)

(3Th-Me)

(3Th-Br)

Figure 4.51. Oligomeric thiophenes used by Zinger *et al.* to prepare radical cations. Reprinted with permission from ref. 64.

confirmed by UV vis, NIR and electron paramagnetic resonance spectroscopies. The rate of oxidation increased by two orders of magnitude when irradiated with a UV vis source. No reaction occurred in the dark, nor did the reaction proceed in the presence of light when trifluoroacetic acid was replaced with acetic acid. Dinitrobenzene sulfonic acid and methanesulfonic acid also were found to be suitable media for photooxidation to occur. However, in concentrated acid solutions, stable dications were formed immediately upon dissolution of the oligothiophene. Irradiation of the same compound in acetonitrile solutions yielded both 3Th-SMe$^{+\cdot}$ and the cation radical dimer (3Th-7SMe$^{+\cdot}$)$_2$ due to the increase in polarity of the solvent. In contrast to 3Th-SMe, 3Th-Me$^{+\cdot}$ could not be formed from photolysis of 3Th-Me in CH$_3$CN/trifluoroacetic acid, and only decomposition products were observed. It was presumed that the thiomethyl substituent in 3Th-SMe stabilized the cation radical to a larger extent than the methyl substituent due to its electron-donating interaction with the π-system. Similarly, 3Th-Br was less prone to photooxidation due to its more positive electrochemical oxidation potential. The photochemical formation of radical cations of 3Th-SMe could also be achieved when the compounds were irradiated in Nafion impregnated with acetonitrile. Nafion is a super-acid.

Consistent with studies of the photochemical generation of radical cations from thiophene oligomers in non-aqueous solvent is the observation that P3TASs photolyzed in aqueous solution develop an increase in absorption intensity at 800 nm upon irradiation [39]. This can be observed in Figure 4.52 for sodium poly[8-(3-thienyl)octylsulfonate]. This observation suggests that the polymers are photochemically oxidized, generating polarons and bipolarons. Upon prolonged

irradiation, the absorption intensity at 800 nm decreases. This can be explained on the basis of disruption of the π-system, as evidenced by photobleaching of the π–π^* band and its corresponding blue shift, upon prolonged exposure.

Polyaniline in its reduced form, leucoemeraldine, undergoes an enhanced rate of oxidation in NMP solutions, to the emeraldine form, under irradiation [62]. While little information on the photoelectron transfer reactions of polyanilines in solution is available, there is extensive knowledge on photoelectron transfer reactions occurring at polymer films, i.e. photoelectrochemistry. This is presented in Section 7.

Photochemical reduction of oxidized colloidal polypyrrole by methylene blue (MB$^+$) using laser flash photolysis has been reported [65]. The absorption spectra of methylene blue and oxidized polypyrrole (ppy$^+$) and methylene blue/polypyrrole in ethanol are shown in Figure 4.53.

The corresponding absorption transients (photomultiplier response) at 420 nm, associated with MB$^+$, following laser flash photolysis with 694 nm light is shown in Figure 4.54. Following excitation of methylene blue, the emergence of transient signals at 420 nm and 820 nm were observed due to the formation of the triplet excited state, ^{3}(MB$^+$)*. As can be observed from the transient spectra, ^{3}(MB$^+$)* is quenched by the presence of ppy$^+$ (Figure 4.54).

Furthermore, the absorption spectra of photolyzed MB$^+$/ppy$^+$ indicate the formation of neutral polypyrrole and a decrease in the bands due to MB$^+$/ppy$^+$ (Figure 4.55). The absorption peaks corresponding to MB$^+$ reappeared after 2 h but the neutral polymer was not reoxidized. The authors postulate a reaction mechanism involving electron transfer via the triplet

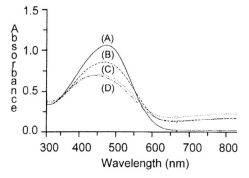

Figure 4.52. Photolysis of sodium poly[8-(3-thienyl)octylsulfonate] in aqueous solution: (A) before irradiation ($t = 0$ min.); (B) $t = 2$ min; (C) $t = 60$ min; (D) $t = 120$ min. Reprinted with permission from ref. 39.

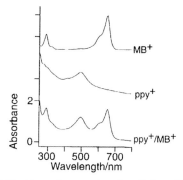

Figure 4.53. Absorption spectra of methylene blue (MB$^+$) and oxidized polypyrrole (ppy$^+$) and methylene blue/polypyrrole (MB$^+$/ppy$^+$) in ethanol. Reprinted with permission from ref. 65.

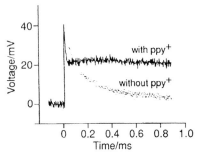

Figure 4.54. Absorption transient at 420 nm (photomultiplier response) of MB^+ following laser flash photolysis with 694 nm light. Reprinted with permission from ref. 65.

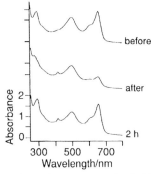

Figure 4.55. Absorption spectra of MB^+/ppy^+ before, immediately following and 2 h following laser flash photolysis experiments. Reprinted with permission from ref. 65.

state of methylene blue and subsequently reduction of the oxidized polymer.

Oligo(p-phenylene)s have been found to be photocatalytic towards the reduction of carbon dioxide to formic acid and carbon monoxide [66]. In the presence of triethylamine as a sacrificial electron donor, irradiation of oligo(p-phenylene)s (OPP) are reported to undergo reduction to the radical anion. Transient absorption spectra of the radical anions of oligo(p-phenylene)s of various length (OPP-n) are shown below (Figure 4.56).

Subsequent electron transfer from the radical anion of OPP to carbon dioxide results in the following reactions:

$$OPP^{-\bullet} + CO_2 \longrightarrow OPP + CO_2^{-\bullet} \qquad (4.46)$$

$$CO_2^{-\bullet} \xrightarrow{H^+} HCO_2^{\bullet} \xrightarrow{e^-} HCO_2^- \qquad (4.47)$$

$$CO_2^{-\bullet} \xrightarrow{CO_2, e^-} CO + CO_3^{2-} \qquad (4.48)$$

Of the oligomers studied, the trimer and tetramer were found to be the most effective catalysts. Quantum yields of HCO_2^- formation were c. 0.08. The corresponding pentamer, hexamer and poly(p-phenylene) were found to be inefficient photocatalysts for this reaction. The authors inferred that this is due to the less negative redox potential of the longer chain analogs, $E_{OPP/OPP}^{-\bullet}$;

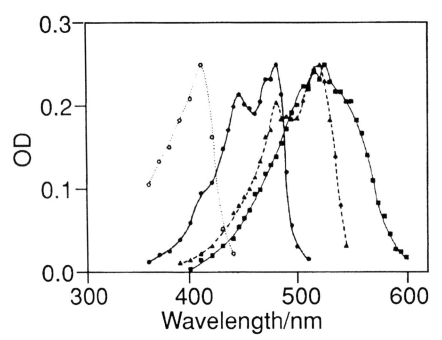

Figure 4.56. Transient absorption spectra of the radical anions of OPP-n. Left to right: $n = 2$, 3, 4 and 5. Reprinted with permission from ref. 66.

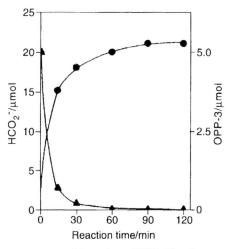

Figure 4.57. Rate of formation of $HCO_2^{-\bullet}$ and consumption of the OPP-3 Δ during photocatalysis of CO_2 in DMF/triethylamine. Reprinted with permission from ref. 66.

which results from a larger degree of π-conjugation. During photocatalysis, the oligomers were found to undergo photodegradation possibly through a mechanism involving protonation and disproportionation to yield dihydro derivatives according to the Birch reduction. Residual water or oxidation of triethylamine was suggested as the source of protons. The rate of formation of HCO_2^- and rate of consumption of the oligomer are shown below for photocatalysis using the trimer (Figure 4.57).

Electron transfer reactions between conjugated polymers, including PPV and P3AT, as electron donors and C_{60} as the electron acceptor have been well documented.

6 PHOTOCHEMICAL DE-DOPING

Significant effort has been made towards understanding the instability of conducting polymers in their oxidized, conducting form. The reversion of the conductive form to the neutral, semiconducting counterpart is commonly termed de-doping. The majority of studies to date have been concerned with thermal de-doping but it has recently been shown however that de-doping of oxidized P3ATs can occur photolytically [67,68]. Indeed, for thin films of $P3AT/FeCl_4^-$, photolytic de-doping dominates over thermal de-doping under ambient lighting [68]. Figure 4.58 shows pseudo-first-order kinetic plots for the decrease in conductivity of oxidized polymer films. The rate of decrease in conductivity in the presence of light is ~ 50 times faster than in the dark.

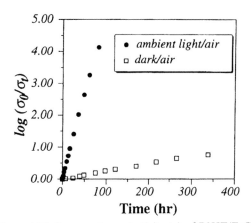

Figure 4.58. Decrease in conductivity (σ) of $P3HT/FeCl_4^-$ films during exposure to ambient light. Ambient laboratory conditions σ_0: ~ 6 S cm^{-1}. Reprinted with permission from ref. 66.

The photochemical stability of the polymer-counter ion pair was also wavelength dependent as shown in Figure 4.59. Irradiation with 580 nm did not lead to photodegradation.

UV vis NIR IR and Mössbauer spectroscopic analyses was used to explain that photodegradation was primarily due to excitation, and photoreaction, of the counter ion, $FeCl_4^-$. Photolysis in air and *in vacuo* showed similar mechanisms, i.e. reduction of the oxidized polymer, disappearance of $FeCl_4^-$ and loss of conductivity. The net reaction was proposed to involve photoreduction of $FeCl_4^-$, hydrogen atom abstraction from the side chain, and electron transfer

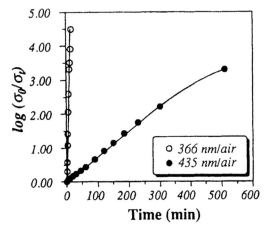

Figure 4.59. Effect of irradiation wavelength on the photochemical reduction of P3HT $FeCl_4^-$ films. Ambient laboratory conditions, σ_0: ~ 6 S cm^{-1}. Reprinted with permission from ref. 68.

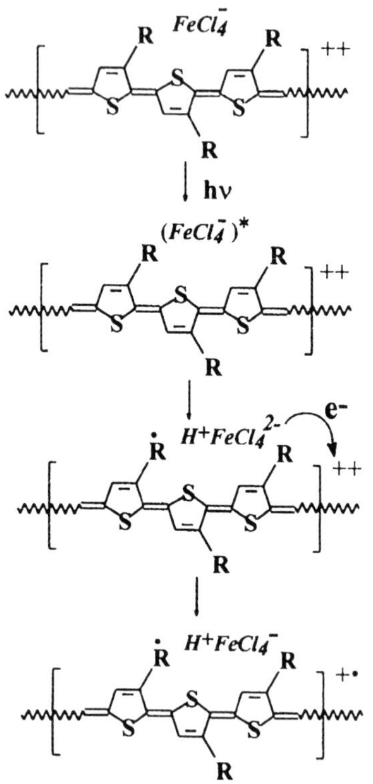

Figure 4.60. First stages of photochemical degradation of P3AT $FeCl_4^-$ conducting polymers in the absence of moisture. (R˙) represents an alkyl side chain radical linked to the polymer backbone. Reprinted with permission from ref. 68.

to the polymer as illustrated in Figure 4.60. Subsequent photoreduction led to the fully reduced polymer. When present, moisture was shown to be the predominant sacrificial electron donor, and the photochemical reaction was accelerated. Although photochemistry of oxidized P3HT containing various counter ions was reported [68], the majority of systems were insufficiently thermally stable to distinguish photochemical degradation from thermal degradation. However, $AuCl_4^-$, containing conducting polymers were found to be two orders of magnitude more photochemically stable than $FeCl_4^-$ doped polymers by virtue of the photochemical stability of $AuCl_4^-$.

7 PHOTOELECTROCHEMISTRY

An intense study of photoelectrochemical solar energy conversion occurred during the 1970s and 1980s. This was largely inspired by the perception that the semiconductor/electrolyte interface would experience fewer interface states than solid state analogs thus paving the way for large scale photoelectrochemical devices using less expensive polycrystalline photoelectrodes. Indeed, it has been found that photoelectrochemical cells can operate on polycrystalline and amorphous devices although high efficiencies ($> 15\%$) are typically reserved for highly crystalline electrodes. The photoelectrochemical process is shown schematically in Figure 4.61 for a p-type semiconductor.

In this scheme, irradiation with light of energy greater than the bandgap creates an electron–hole pair. Under the influence of an applied field electrons are driven towards the electrode/solution interface where electrochemical reduction of solution species occurs. Photogenerated holes are driven away from this interface and oxidize a solution species at the counter electrode. For a detailed discussion of photoelectrochemistry, the reader should consult one of the detailed reviews available on the subject [69–71].

Thin films of π-conjugated polymers have been investigated [72] as an alternative source for semiconductor electrodes for photoelectrochemistry for the following reasons.

(1) Evidence exists for the injection of electrons into solution, and the subsequent formation of cationic species, upon irradiation.
(2) In the solid state, such polymers exhibit semiconducting properties as demonstrated by solid state devices such as Schottky barrier devices, electroluminescent displays and field effect transistors.
(3) They can be readily processed in the form of thin films.

Polyacetylene was the first π-conjugated polymer to be investigated; it developed a small photovoltage (40–

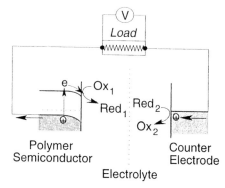

Figure 4.61. Photoelectrochemical reaction scheme.

100 mV) upon illumination in aqueous solutions [73]. The polymer served as a *p*-type semiconductor. Upon irradiation, electrons are injected into solution and a photocurrent was observed which decayed to a steady state value. When the light was removed, the current decreased to a negligible value. The cathodic photocurrent observed for *trans*-polyacetylene under a negative bias is show in Figure 4.62. The action spectrum of *trans*-polyacetylene is shown in Figure 4.63. The spectrum is somewhat different to the absorption spectrum of the polymer film. The origin of this discrepancy and the nature of the photoelec-photoelectrochemical reduction reaction was not discussed.

Similarly, photoelectrochemical photovoltaic cells have been fabricated using polyacetylene as the active photocathode and sodium polysulfide as the electrolyte [74]. An open circuit potential of *c.* 0.3 V and a short circuit current (I_{sc}) of 10 µA cm^2 were obtained with illumination of *c.* 1 Sun. Irradiation of polyacetylene under these conditions caused the reduction of sodium polysulfide

$$S_y^{2-} + 2e^- \longrightarrow S_{y-n}^{2-} + S_n^{2-} \qquad y \geq n+1 \quad (4.49)$$

S_y^{2-} is reoxidized at the counter electrode. The absolute quantum yield was reported to be *c.* 1% at 2.4 eV irradiation.

Polypyrrole was reported to be photoelectrochemically active in the presence of aqueous solutions of Cu^{2+} when irradiated with light and biased with a potential more negative than 0.3 V (SCE) [75].

$$Cu^{2+} + e^- \longrightarrow Cu^+ \qquad (4.50)$$

The action spectrum of polypyrrole in acetonitrile solutions using a bias of -1.0 V is shown in Figure 4.64. The formal quantum efficiency coincides with absorption spectrum of the polymer indicating that excitation of the $\pi-\pi^*$ transition of the polymer is responsible for the photocurrent. The photocurrent showed a linear dependence on light intensity. Photo-

currents were reported to decrease by 15% over a period of 2 h.

Formation of a liquid junction between polypyrrole and aqueous solutions and subsequent photoelectrochemistry upon irradiation was reported by Yamada *et al.* [76]. The photocurrent reached 50 µA cm^2 under an applied bias of -1.0 V (vs. SCE). It was suggested that water participated in the electrochemistry but no details regarding this reaction were provided. Prolonged photolysis resulted in the emergence of an absorption peak at 650 nm and a decrease in the 400 nm absorption band associated with the $\pi-\pi^*$ transition of the polymer film, indicating photochemical oxidation/doping of the polymer. Photoelectrochemistry of a fully oxidized, polypyrrole film yielded a higher photocurrent than the undoped film. It was speculated that this was due to photochemically induced de-doping of the oxidized polymer. The dependence of the photocurrent on the applied potential on the doped and undoped polypyrrole films is shown in Figure 4.65.

Photoelectrochemistry of polyaniline has been studied in aqueous and non-aqueous solution. In aqueous solution, oxidized polyaniline, prepared by electropolymerization of aniline, was reported to form a liquid junction when immersed in an aqueous solution containing 0.1 M LiClO$_4$ [77]. The illuminated electrode supported a cathodic photocurrent of 21 µA cm^{-2} under an applied bias of -0.5 V (vs. Ag/AgCl). Evolution of gas at the irradiated electrode was not observed and only a slight dependence of photocurrent on pH was found. The authors concluded that reduction of protons was not significant in the photoelectrochemical process.

MacDiarmid *et al.* noted that *NN'*-diphenyl-*p*-phenylene diamine, which corresponds to the leuco-emeraldine form of polyaniline, undergoes photo-oxidation in ethanol/acetic acid media to form the blue-colored semiquinone analog via electron injection to the solvent [equation (4.51)] [78]. These authors studied the photoelectrochemistry of polyanilines to determine whether similar photoelectron transfer reactions occur at polyaniline films.

$$\text{(4.51)}$$

$$\text{(4.52)}$$

$$\tfrac{1}{2}O_2 + H_2O + 2e^- \longrightarrow 2OH^- \qquad (4.53)$$

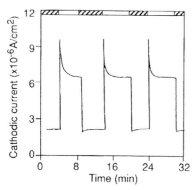

Figure 4.62. Cathodic photocurrent observed for *trans*-polyacetylene under a negative bias using intermittent irradiation in aqueous solution. Reprinted with permission from ref. 73.

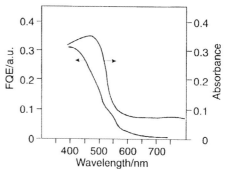

Figure 4.64. Action spectrum of polypyrrole in 0.1 M tetrabutylammonium perchlorate/acetonitrile. FQE represents the formal quantum current efficiency. Reprinted with permission from ref. 75.

The dark current and photoelectrochemical response of the polymer in 0.1 M $ZnSO_4/H_2SO_4$ as a function of oxidation state is shown in Figure 4.66.

The polymer was initially in its emeraldine state. The background current decay is due to the reduction of the polymer to the leucoemeraldine form. The current spikes upon irradiation are associated with photoelectrochemistry of the reduced (leucoemeraldine) units. As shown, photoelectrochemistry is associated with oxidation of the leucoemeraldine form. This was confirmed by the observation that films of the leucoemeraldine form yielded a photocurrent of > 100 µA cm^2 under a negative bias. The following mechanism was proposed

A fast photoresponse to a NIR light pulse has been observed for polyaniline films in HF and acetonitrile solutions [79]. A cyclic voltammogram of a polyaniline film in NH_4F/HF during 50 µs flashes is shown in Figure 4.67. The response of the electrode is extremely rapid and believed due to creation and separation of

electron–hole pairs. However, current flows only in one direction suggesting the presence of a Schottky barrier. The effect of UV vis irradiation was reported to be very small. Similarly, the effect of continuous irradiation had a much less dramatic affect. With continuous irradiation the photoreduction of chemical species in solution could be achieved. This was demonstrated with the photoreduction of chloral (CCl_3CHO) to the corresponding alcohol.

The effect of redox couples in solution have been shown to have a dramatic effect on the magnitude and stability of the photocurrent generated at polyaniline photocathodes. In one study, polyaniline films were electrodeposited from solutions of aniline and dilute sulfuric acid, and irradiated with a 400 W tungsten–halogen lamp [80]. The photocurrents observed when polyaniline films are irradiated in 1 M sulfuric acid, ferricyanide/ferrocyanide, and KI/I_2 solutions are shown in Figure 4.68.

In sulfuric acid alone, photoanodic current was observed at the most positive bias employed. As the

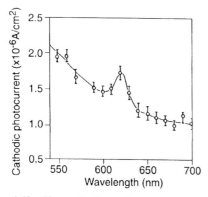

Figure 4.63. Photocathodic action spectrum of *trans*-polyacetylene. Reprinted with permission from ref. 73.

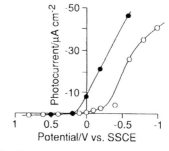

Figure 4.65. Dependence of the photocurrent on applied potential for doped (●) and undoped (○) polypyrrole films in 0.050 M KCl/0.01 M HCl. Reprinted with permission from ref. 76.

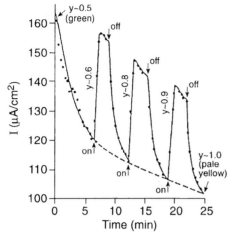

Figure 4.66. Photoelectrochemical response of the polyaniline in 0.1 M $ZnSO_4/H_2SO_4$ as a function of oxidation state. y represents the fraction of reduced (leucoemeraldine) units in the polymer. Reprinted with permission from ref. 78.

bias was made more negative the magnitude of the photoanodic current decreased and eventually was observed only transiently. At more negative biases, photocathodic current was observed. The same trend was observed with other redox couples. However, the bias at which the photoanodic current was switched to photocathodic current was dependent on the nature of the redox couple. The presence of an active redox couple significantly increased the magnitude of the photocurrent.

Photoelectrocatalysis was demonstrated with the reduction of peroxodisulfate as shown in Figure 4.69. The onset of reduction of peroxodisulfate at illuminated polyaniline is shifted c. 1 V compared to the reduction at bare platinum. In the absence of light, no current associated with the reduction of peroxodisulfate was observed. Mechanisms of photoelectrochemical reactions were not described.

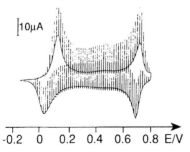

Figure 4.67. Cyclic voltammogram of a polyaniline film in $NH_4F/2.3$ M during 50 μs flashes of NIR irradiation. Reprinted with permission from ref. 79.

Further studies of the photoelectrochemistry of polyaniline have been reported by Peter *et al.* [81]. These authors observe a small photocathodic current when films are irradiated in 1 M H_2SO_4 and much larger photocurrents under potentiodynamic conditions, i.e. cyclic voltammetry. It was found that maximum photocurrent was attained when the leucoemeraldine form was partially oxidized to the emeraldine form. Quantum yields for photocurrent were reported to be < 1% indicating that deactivation of the excited state and/or recombination of the carriers are efficient pathways. An action spectrum correlating the absorption spectrum of leucoemeraldine with the wavelength-dependent photocurrent indicates that the photocurrent is associated with the π–π^* transition of the polymer. Furthermore, the maximum photocurrent was observed with a film thickness which corresponded well with the penetration depth of light (c. 60 nm) in the film. The authors suggest that the polymer does not act like a classical p-type semiconductor but that the partially oxidized units serve as the current collector. The following generalized mechanism was proposed:

Photoexcitation	P $\xrightarrow{h\nu}$ P*	
Electron transfer quenching	P* + A $\longrightarrow$ P$^+$ + A$^-$	
Reduction of P$^+$	P$^+$e$^-$ $\longrightarrow$ P	(4.54)

P$^+$ is a polaron, or an emeraldine unit, and A is an electron acceptor, in some instances, oxygen. However, the authors report that the photocurrent decreased when the solutions were oxygenated. These systems deserve further investigation in order to clarify the model.

Polythiophenes exhibit interesting photoelectrochemical properties with power efficiencies up to 4.7%. Poly(3-methylthiophene) shows a significant stable photocurrent when irradiated in acetonitrile solutions [82,83]. The current–voltage curves for poly(3-methylthiophene) in $CH_3CN/LiClO_4$ are shown in Figure 4.70. Photocurrents up to 230 μA cm^2 were obtained. Maximum photocurrents were reported to occur at irradiation wavelengths of 480 nm, consistent with the wavelength of the maximum absorption of the polymer.

Photocurrents were reported to be dependent on the nature of the counter ion used during electropolymerization of the polymer. This was explained on the basis of the structural integrity/regularity of the polymer. The photoelectrochemical system was found to be relatively stable over a period of hours following an initial decrease in photocurrent. The electrochemical reactions leading to photocurrent were not reported. The position of the energy bands of the polymer, flat band potentials

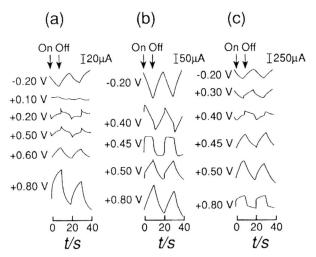

Figure 4.68. Photocurrent observed at polyaniline films under intermittent light: (a) 1 M H_2SO_4; (b) 1 M $H_2SO_4 + 0.01$ M $Fe(CN)_6^{3-}/Fe(CN)_6^{4-}$ (C) 1 M $H_2SO_4 + 0.02$ M $KI/0.01$ M I_2. Photocathodic current ↑. Reprinted with permission from ref. 80.

and the density of the charge carriers was determined by capacitance measurements. It was determined that platinum and gold form ohmic contacts with the polymer whereas indium and aluminum, which possess lower work functions, form rectifying contacts.

Photoelectrochemical hydrogen evolution at thin films of the processable polymer, P3HT, has been investigated in aqueous solutions [84]. Figure 4.71 shows thermodynamic potentials of the H_2O/H^+ and O_2/H_2O redox couples with respect to the energy bands

of the polymer. This diagram indicates that the electrons generated via excitation of the polymer have sufficient energy to reduce protons in solution. In the dark, the polymer film passivates the underlying carbon electrode to hydrogen evolution. Upon irradiation with light of wavelength 435 nm, a photocurrent is observed at potentials more negative than $+0.3$ V (SCE). The onset of the photocurrent correlates with the flat band potential. The effect of pH on the photocurrent is shown in Figure 4.72. The onset potential remains constant at $+0.3$ V (SCE) which indicates that the energy bands are not susceptible to Fermi level pinning.

Since the onset potential remains constant and the redox potential of the H_2O/H^+ couple moves negative with increasing pH, the photovoltage (extrapolated to

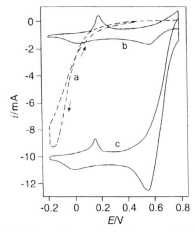

Figure 4.69. Photoelectrocatalysis of peroxodisulfate reduction at polyaniline electrodes: Cyclic voltammetry of polyaniline films in 1 M $H_2SO_4 + K_2S_2O_8$ (0.1 M) (a) platinum electrode; (b) polyaniline film in the dark; (c) irradiated polyaniline film. Reprinted with permission from ref. 80

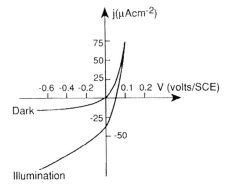

Figure 4.70. Current–voltage curves for P3MT in $CH_3CN/LiClO_4$. Reprinted with permission from ref. 83.

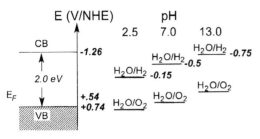

Figure 4.71. Thermodynamic redox potentials of water with respect to the energy bands of P3HT. Reprinted with permission from ref. 84.

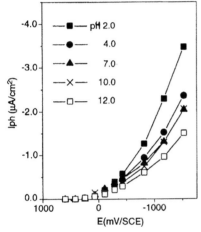

Figure 4.72. Effect of pH on the photocurrent at P3HT photocathodes. Reprinted with permission from ref. 84.

zero current) increases to values of c. 1.0 V at the highest pH. Furthermore, while the hydrogen ion concentration decreases by a factor of 10^{10} from pH 2 to 12, the photocurrent varies much less implying that the photocurrent is not limited by an electrochemical kinetic event. The quantum efficiencies are low due to geminate electron-hole recombination. Strategies leading to enhanced electron-hole separation will enable higher quantum efficiencies.

8 PHOTOLITHOGRAPHY OF π-CONJUGATED POLYMERS

8.1 Introduction

Conjugated polymers have been important targets for processing due to the rapidly growing interest in

constructing polymer-based microelectronics and optoelectronics. For many of these applications, it is essential that the conjugated polymer be deposited in a controlled fashion. An emerging approach to patterning these materials is photolithography. A brief discussion of microlithography of polymers is presented, followed by a review of microlithography of conducting polymers.

Microlithography of polymers refers to a process in which a thin polymer film is patterned, using a radiation source, with micrometer and submicrometer resolution [85–87]. The polymeric patterns are often called resists because of their ability to resist semiconductor processing chemicals. In traditional semiconductor devices, the resist does not play an *active role* in the electronic operation of the circuit. It is used only for pattern transfer onto the underlying substrate, and is usually removed after use. Conjugated polymer patterns, however, might be employed as an *integral* component of the device because of their unique electrical properties.

The basic processes of photolithography are shown in Figure 4.73. When the polymer resist is exposed to light through a photomask, or exposed directly with a focused point source, chemical reactions take place which result in a change in the polymer's solubility. Utilizing the solubility difference between the exposed and unexposed area, a pattern image can be developed

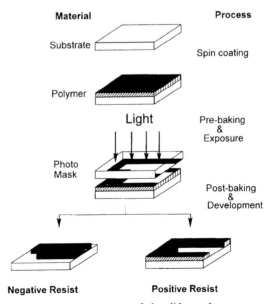

Figure 4.73. Basic processes of photolithography.

using an appropriate development technique. Three main photochemical processes are employed to photochemically modify the solubility of the polymer photoresists. These are crosslinking, chain scission, and changes in hydrophobicity/hydrophilicity. Crosslinking occurs in the exposed area resulting in decreased solubility and negative-tone resists. Chain scission results in an increase in the solubility of the exposed regions and forms the basis for positive-tone resists. Changes in polarity can be achieved photochemically, resulting in either a decrease or increase in solubility of the photoresist.

Photoreactivity of the resist materials is expressed in terms of three lithographic parameters. These are the interface gel dose (D_g^i or r_{gel}) and contrast (g_n) For negative photoresists, D_g^i is defined as the minimum radiation dose required to induce a reaction such that an insoluble gel of polymer remains after development. Sensitivity is signified by the exposure dose required to yield a particular response in the resist polymer after development. It is most commonly defined as the dose required to leave 50% of the original thickness and is represented by $D_g^{0.5}$ (Figure 4.74). In order to construct a sensitivity curve, the normalized film thickness remaining (l/l_0) after photolysis is plotted against exposure dose. l_0 and l represent the original and residual film thicknesses of the photoresist, respectively. The contrast of the photoresist, γ_n, is defined as the rate of insolubilization at constant input energy, and is given by

$$\gamma_n = \frac{1}{\log D_g^0 - \log D_g^i} = \left[\log \frac{D_g^0}{D_g^i} \right]^{-1} \qquad (4.55)$$

where D_g^0 is the dose required to produce 100% gel formation of the original film, and is determined by extrapolation of the linear portion of the sensitivity plot to a dose value equivalent to the normalized film thickness.

The basic aspects of positive photoresists are similar to those of negative resists except that photolysis enhances the rate of dissolution. Gel doses can also be determined from a plot such as that shown in Figure 4.74, where D_d^i and D_d^0 represent the exposure dose of incipient and complete dissolution of the resist, respectively.

The resolution of a lithographic resist is a critical processing parameter but is difficult to quantify. It has been found that the resolution is often related to the contrast of the resist. Resolution is also related to the wavelength of the incident light due to diffraction of the incident light by the photomask. As a consequence, research into deep-UV lithography is currently active [88].

8.2   Photolithographic chemistry

Several strategies have been used to photochemically modify the solubility of conjugated polymers for photolithographic applications. The majority of studies so far have focused on chemistry which decreases the solubility of the exposed regions. These include photocrosslinking, the use of acid photogenerators and photopolymerization. Crosslinking between polymer chains reduces the solubility of the polymer and leads to a negative image of the photomask. Photo-

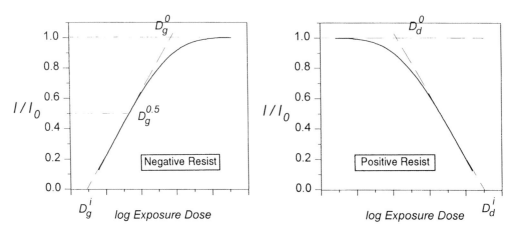

Figure 4.74. Sensitivity curves of negative and positive resists.

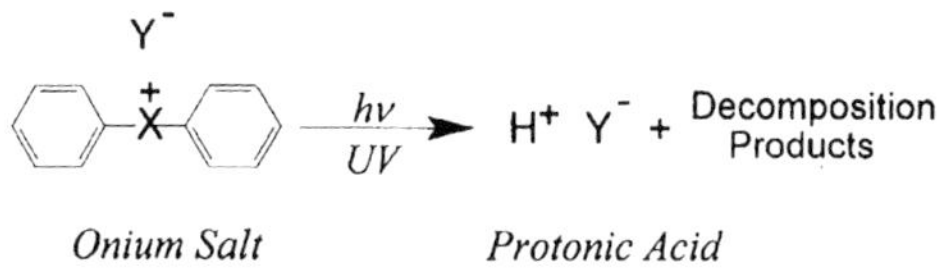

Figure 4.75. Acid photogenerators based on onium salts; X = lidide or an aryl sulfide.

chemically induced crosslinking is often initiated using free radical photosensitive initiators. Photoinitiators usually possess high molar absorptivities at the wavelength of interest, high quantum yields for the desired photochemical reaction, are thermally stable under processing conditions, and are chemically compatible with the polymer resist [89,90]

Acid photogenerators have found a wide application as photoinitiators for acid-catalyzed reactions in polymeric systems. Acid-catalyzed deprotection of functionalized polystyrenes, for example, by irradiation of suitable onium salts has been extensively used in chemically-amplified photoresist technology [91]. Figure 4.75 depicts the basic process of acid generation upon photolysis of onium salts.

Organic, non-ionic acid photogenerators based on 2-nitrobenzyl sulfonate esters are also available and generate sulfonic acid upon photolysis (Figure 4.76). The acid photogenerated from these compounds is sufficiently strong to effect doping of conjugated polymers such as polyaniline [92]. Acid photogenerators have also been employed to effect acid-cataiyzed elimination and deprotection reactions for the conversion of precursor polymers to their fully conjugated form [93].

Another strategy used for the imaging of conjugated polymer resists is photopolymerization, in which

relatively low-molecular-weight monomers undergo photointiated, and photoassisted, polymerization. The initiation process can be photoactivated using a photosensitizer. Photosensitization is useful when the molecule of interest possesses a low optical density at the desired wavelength. Initiation can also be achieved photoelectrochemically using inorganic semiconductors [71]. In this technique, the energy of the incident light should be greater than the bandgap energy of the semiconductor in order to generate electron–hole pairs. For the polymerization to take place, the electron (or hole) must be sufficiently energetic to initiate the polymerization process, i.e. the conduction band energy of the n type semiconductor must be more negative than the redox potential of the monomer. Miscellaneous patterning strategies include light-induced reduction of doped polymers in which oxidized, conductive polymers are photochemically reduced to their neutral form [94]; *in situ* polymerization on preformed catalyst patterns in which a polymerization catalyst is first patterned using conventional microlithography, prior to *in situ* polymerization on the catalyst [95–97]; and laser ablation which uses laser light to photothermally deposit materials of controlled dimensions [98].

8.3 π-Conjugated polymer resists

8.3.1 *Polyacetylenes*

Intractability has been a major obstacle in processing polyacetylene. The precursor, or Durham route, was developed to overcome this problem. The precursor polymer is soluble in many polar solvents. It is an insulator and non-conjugated, but can be converted into

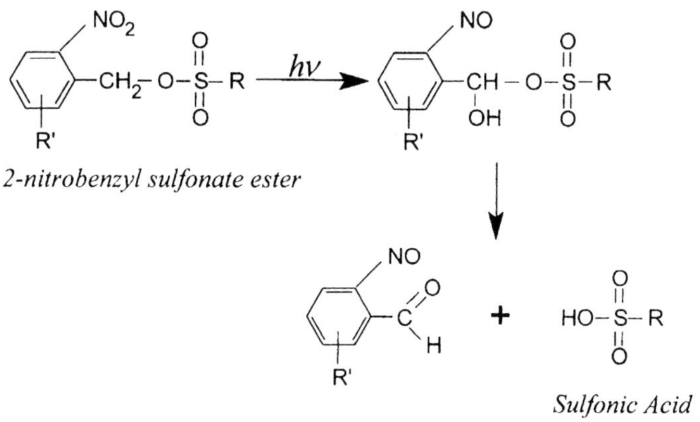

Figure 4.76. Photolysis of non-ionic acid photogenerators based on 2-nitrobenzyl sulfonate esters.

the conjugated form using acid-catalyzed or thermal elimination reactions. Patterning of the soluble Durham precursor was demonstrated using the crosslinking chemistry [31,32]. The strategy employed in the patterning process is shown schematically in Figure 4.77.

The soluble precursor is spin-cast onto a substrate from a butanone solution, exposed through a photomask, and heated under nitrogen to effect elimination of hexafluoroxylene. The exposed regions undergo photophotobleaching and insolubilization reactions and yield an insulating pattern, even after thermal treatment. Thermal treatment at ~ 150–$200°C$ was reported to convert only the unexposed regions into polyacetylene. Conductive patterns of polyacetylene were thus generated in an insulating polymer matrix. A resolution of 0.3 μm was obtained using contact printing lithography. The mechanism of photobleaching was explained in terms of the reactions of singlet oxygen with the polymer.

Enhancing the solubility of polyacetylene was also achieved by structural modification of the parent polymer with long side chains. Poly(bis-thioalkyl-acetylenes), for example, are completely soluble (Figure 4.78). Irradiation of thin films of this polymer rendered it insoluble. The conductivity of the resultant material ($c.$ 10^2 S cm^{-1}) was 16 orders of magnitude greater than the unexposed polymer [99,100]. The electron spin density increased from 2.7×10^{17} spins g^{-1} upon exposure, indicating an increase in the free carrier density. The formation of a π-conjugated carbon skeleton linked by sulfur was proposed to explain the increase in conductivity. Direct writing of

R
|
S
R = Methyl or Ethyl
S
|
R

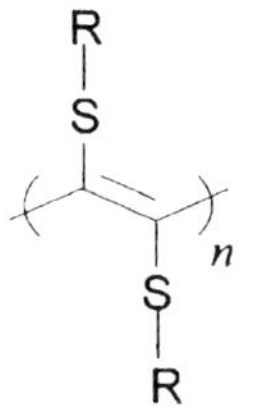

Figure 4.78. Chemical structure of poly(bis-thioalkyl-acetylenes). Reprinted with permission from ref. 100.

thin films of poly(bis-thioalkyl-acetylenes) using a focused argon laser (488 nm) was employed to fabricate conducting polymer electronic patterns.

8.3.2 Poly(3-alkylthiophene)s

Films of P3ATs undergo crosslinking, becoming insoluble, when irradiated with UV vis light [36–38,50]. These characteristics make P3ATs interesting candidates for photolithographic technology since polymer films can be readily spin-cast onto solid substrates, irradiated through a photomask, and unexposed polymer removed by dissolution. Oxidation of the crosslinked polymer results in an electronically conducting pattern. Inorganic photoinitiators have been postulated as photoinitiators of crosslinking [34,50]. In the case of P3ATs prepared by oxidative coupling using ferric chloride the impurities have been shown to be octahedral Fe(III) hydroxychlorocomplexes, e.g. $Fe(H_2O)_4(OH)_2Cl$, and exist at concentrations > 0.05 wt%, even after extensive purification [49]. Irradiation of residual iron impurities has been reported to lead to the formation of free radicals which are capable of abstracting hydrogen atoms from the alkyi side chain. A mechanism proposed for the crosslinking of P3ATs is shown in Figure 4.79

Lithography of P3ATs was established using P3HT. Photoimaging of P3HT was demonstrated using both conventional photolithography with a focused, high intensity UV vis light (> 300 nm), and by laser, direct-write microlithography [36–38]. In the latter, relief images were formed in thin polymer films on various surfaces by tracing a pattern onto the surface of the film with a focused laser beam. Immersion of the relief pattern in a suitable solvent dissolved the unexposed regions leaving only exposed polymer. The residual pattern was intensely red-colored signifying the persistence of extensive π-conjugation. Figure 4.80 shows an optical micrograph of P3HT wires on a silicon wafer.

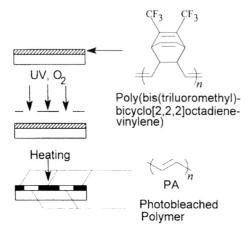

CF$_3$ CF$_3$

UV, O$_2$

Poly(bis(triluoromethyl)-bicyclo[2,2,2]octadiene-vinylene)

Heating

PA

Photobleached Polymer

Figure 4.77. Photoimaging of polyacetylene via the Durham precursor route. Reprinted with permission from refs 31,32.

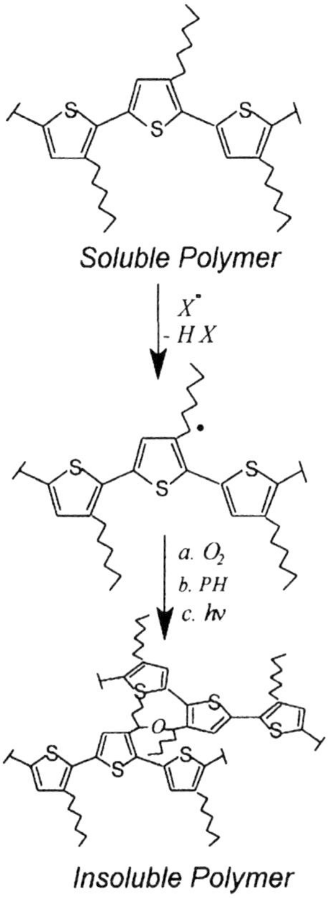

Soluble Polymer

$$\downarrow \begin{array}{l} X^{\cdot} \\ -HX \end{array}$$

$$\downarrow \begin{array}{l} a.\ O_2 \\ b.\ PH \\ c.\ h\nu \end{array}$$

Insoluble Polymer

Figure 4.79. Proposed mechanism for photocrosslinking of P3AT. Reprinted with permission from ref. 50.

The resolution of the fine features is 4 μm. Structures of 1 μm resolution were also reported.

The lithographic parameters of P3HT at 435 nm were determined from the sensitivity curve shown in Figure 81. The curve was constructed by plotting the remaining thickness of polymer films, after solvent development, against exposure dose. The gel content (film thickness) increased with increasing exposure dose until the, thickness remaining was equal to the original film thickness. Values of gel dose, r_{gel} or D_g^i, sensitivity, $D_g^{0.5}$, and contrast, γ, were calculated to be 38, 190 mJ cm^{-2} and 0.8 respectively. The energy required for incipient gelation, r_{gel}, of P3HT was calculated to be 49 mJ cm^{-2} from a solubility plot of $S + S^{1/2}$ vs. irradiation dose. Accordingly, the quantum yield of crosslinking, ϕ_x, at 435 nm was calculated to be 2.37×10^{-3} mol Einstein^{-1} [49].

The effect of irradiation on the conjugation length and conductivity of P3HT was reported since a significant decrease in π-conjugation would impair the electrical properties of the material. π-Conjugation was monitored by the change optical density of the film with irradiation. Irradiation of thin films of P3HT with 435 nm light in ambient air caused the optical density of the film to decrease as shown in Figure 4.82. Neutral, undoped polymer film has a strong absorption band at 500 nm indicating a significant degree of π-conjugation along the polymer backbone. However, the effective conjugation length of the polymer did not change appreciably with irradiation doses <300 mJ cm^{-2} [49]. This observation illustrates that the polymer films can be photochemically insolubilized, yet still possess a high degree of conjugation.

Neutral irradiated, and pre-irradiated, polymer films have low conductivity ($\sim 10^{-7}$ Ω^{-1} cm^{-1}). Upon oxidative doping, using anhydrous acetonitrile solutions of nitrosonium tetrafluoroborate (NOBF$_4$) or ferric chloride, the films turned blue in color and were found to be electronically conductive by the four-point probe technique. The initial conductivity of doped P3HT was ~ 5 Ω^{-1}cm^{-1}. In order to determine the effect of light on the conductivity of the films, several neutral films were irradiated, each to a different exposure dose. Following this, each film was oxidized and the conductivity measured by the four-point probe technique. The conductivity of the film decreased in a nonlinear fashion with irradiation dose (Figure 4.83) and fell by several orders of magnitude after prolonged photolysis. However, when the irradiation dose was kept low (<300 mJ cm^{-2}), the change in electronic conductivity was negligible. The conductivities of oxidized polymer wires were determined by depositing five parallel polymer channels, each 40 μm wide, 200 nm long, 90 nm thick and separated by 40 μm, onto glass slides as shown schematically in Figure 4.84 [37]. The polymer was oxidized with nitrosonium tetrafluoroborate and exhibited a resistance of 150 kΩ per channel, yielding a bulk conductivity of ~ 5 Ω^{-1} cm^{-1}. The environmental stability of the oxidized form of the polymer is critical for practical applications. The oxidized, conductive patterns of P3HT are relatively stable under a nitrogen blanket but revert back to their neutral, semiconducting state shortly after exposure to the atmosphere. The process is considerably more rapid with micrometer–sized images compared to relatively thick free-standing films due to the small mass of the former. Investigation of the mechanism of these de-doping processes and controlling lfactors continues to be an area of active interest [10].

The vinyl moiety is well known to undergo efficient crosslinking or polymerization under illumination. This chemistry has been used in negative-tone conjugated

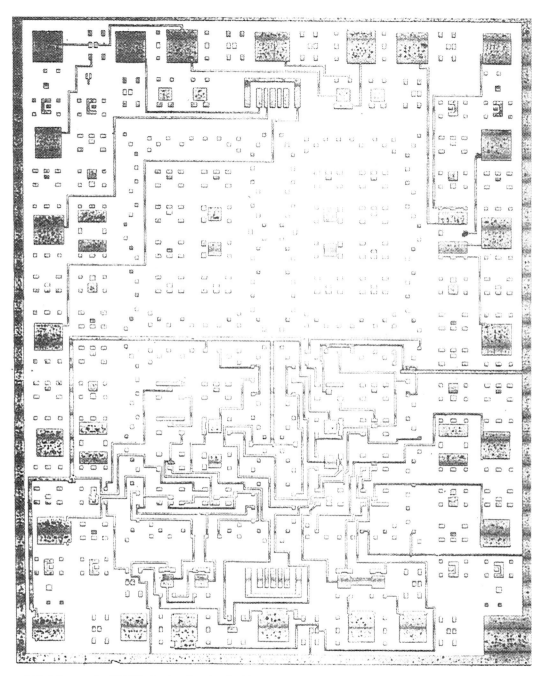

Figure 4.80. Micrograph of P3HT circuitry. Reprinted with permission from Elsevier, ref. 37.

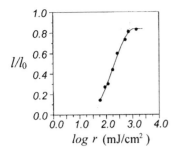

Figure 4.81. Sensitivity plot for P3HT using 435 nm light. Reprinted with permission from ref. 50.

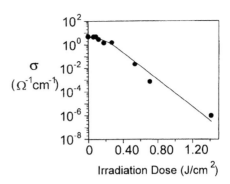

Figure 4.83. Conductivity of oxidized P3HT vs. irradiation dose. Reprinted with permission from ref. 50.

polymer photolithography. In poly(2-(3-thienyl)ethyl methacrylate) (PTEM) crosslinking occurs through the unsaturated side chain [38,56]. The chain mechanism of this reaction leads to highly photosensitive materials. D_g^i, and $D_g^{0.5}$ were determined to be 1.0 and 3.7 mJ cm^{-2}, respectively, using 313 nm irradiation [56]. The corresponding copolymer, poly[2-(3-thienyl)ethyl methacrylate-co-(3-hexylthiophene)] (PTEM-co-3HT) (Figure 4.85) was approximately six times lower in sensitivity than PTEM (D_g^i 2.6 mJ cm^{-2}; $D_g^{0.5}$, 24 mJ cm^{-2}) but enhanced resolution was obtained due to the lesser extent of post-irradiation free radical coupling reactions.

Figure 4.86 shows a comparison between the sensitivity plots of P3HT and PTEM-co-3HT. PTEM-co-3HT can be clearly crosslinked with greater efficiency than P3HT due to the presence of the methacrylate functionality. The photosensitivity of these vinyl-containing polymers has served in reducing the losses in π-conjugation during irradiation. Micrographs of images obtained using this polymer are shown in Figure 4.87.

Another example of a crosslinkable polythiophene-based polymer is poly[3-octylthiophene-co-N-(3-thienyl)-4-amino-2-nitrophenol] functionalized with a cin-

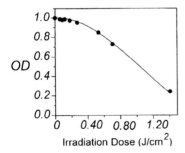

Figure 4.82. Relative optical density (at $\lambda = 500$ nm) for P3HT films vs. irradiation dose. Reprinted with permission from ref. 50.

namoyl group (Figure 4.88) [102]. Cinnamoyl chromophores were crosslinked via [2 + 2] cycloaddition using a 254 nm light.

The use of bis-arylazide chemistry is another strategy used to enhance the photosensitivity of P3ATs [103–105]. These compounds are sensitive to light and decompose to give nitrene intermediates. Insertion of the nitrene into carbon–hydrogen bonds leads to crosslinking and insolubilization of the material. Perfluorophenyl bisazides have been used to design deep-UV and electron-beam P3AT-based negative-tone resists as illustrated in Figure 4.89. In this method, a mixture of the polymer and the crosslinking agent in xylene was spin cast onto a silicon wafer, baked at 60°C for 30 min, and exposed to UV or electron beam radiation. The exposed regions became insoluble allowing pattern generation. The mechanism of insolubilization is believed to occur via a two-step process involving hydrogen abstraction and radical combination. Using IR spectroscopy, it has been shown that the crosslinking reaction occurred via CH insertion along the aikyl side chains rather than the thiophene rings.

In contrast to UV light, electron beam radiation was found to effect crosslinking and insolubilization of P3ATs in the absence of the crosslinking agent bisazide. Crosslinking occurred with exposure doses of ~15 μC cm^{-2}, slightly larger than in the presence of the crosslinking agent. Although the crosslinking agent enhances the sensitivity of the material it results in patterns with poor resolution due to the propagation of the generated free radicals beyond the boundary of the exposed regions. An image resolution of 0.2 μm was achieved in the absence of bisazide. Figure 4.90 shows a micrograph of a P3AT conducting pattern obtained by electron beam lithography.

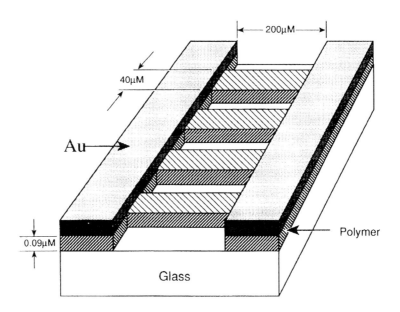

Figure 4.84. Schematic diagram of a microstructure used for measuring the conductivity of polymer wires. Reprinted with permission from ref. 37.

A third technique employed to image P3ATs is photo-induced doping using acid photogenerators [106]. Due to the poor mixing between organic-soluble P3ATs and ionic onium salts, non-ionic sulfoxyimide acid photogenerators were employed. Large quantities of the acid photogenerators (~ 11 wt %) and high doses of deep-UV or electron beam irradiation were required to photochemically dope the material due to the high absorption cross-section of P3ATs and decomposition of a large fraction of the acid photogenerator.

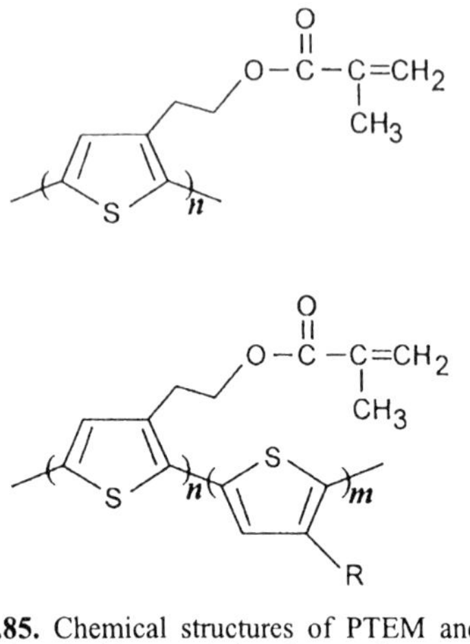

Figure 4.85. Chemical structures of PTEM and PTEM-co-3HT. Reprinted with permission from ref. 56.

Poly(thienylvinylene) (PTV) and poly(phenylene-vinylene) (PPV) polymer precursors are examples of conjugated polymer precursor resists that have been processed using the photoelimination chemistry [58,59,107,108]. Irradiation of the thin films of the precursors with visible light was reported to promote elimination of the pendant group and lead to a reduction in the solubility of the irradiated regions. Solvent development of the unexposed areas resulted in partially conjugated polymer patterns [107]. A fully conjugated material was obtained after heating under an inert atmosphere at 100–400°C. This heat treatment process was accelerated by carrying out the process under acid atmosphere such as HCl [107]. The process is shown schematically is Figure 4.91.

Photo-induced de-doping was employed to generate non-conducting polymer patterns in a conducting matrix. In studies concerning the photochemical stability of $FeCl_4^-$-doped P3ATs, it was discovered that this dopant–polymer system was susceptible to photochemical reduction, forming neutral, de-doped material [68]. In another study, it was found that the oxidation of neutral films of P3ATs using nitromethane solutions of $AuCl_3$ resulted in $AuCl_4^-$-doped polymer and electroless deposition of metallic gold [101,109]. These two phenomena were coupled to selectively deposit gold on conducting polymers [110]. $FeCl_4^-$-doped P3AT films were irradiated through a photomask

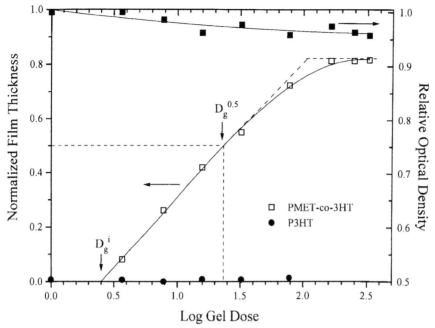

Figure 4.86. Gel dose curve and relative optical density of λ_{max} of PMET-co-3HT and P3HT as a function of exposure dose. Reprinted with permission from ref. 56.

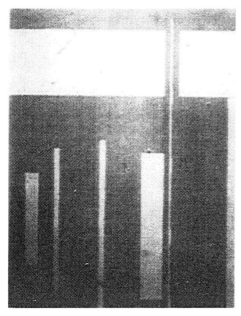

Figure 4.87. Photolithography of PMET-co-3HT(II) on silicon. Polymer thickness ~75 nm. Narrowest line width = 3 μm. Reprinted with permission from The American Chemical Society, ref. 56.

using UV vis light. The exposed regions became neutral while the unexposed areas remained oxidized. Upon exposure to AuCl$_3$ solution in nitromethane, gold(O) was selectively deposited on the neutral regions resulting in gold-decorated conducting polymer wires. Conducting polymer wires was achieved using contact printing lithography [110]. Photoselective electroless deposition was also achieved using photo-induced doping using acid photogenerators. In this case, films of P3AT–acid photogenerator composites were irradiated through a photomask, followed by exposure to AuCl$_3$ solution in nitromethane [110] Au(O) was selectively deposited onto nexposed regions since the exposed regions were photochemically doped. Au(O) was selectively deposited onto the unexposed regions since the exposed regions were photochemically doped. Au(O) deposition of pre-made polymer patterns was also examined. The following schemes outline the photoselective deposition process using both photo-induced doping and de-doping processes (Figure 4.92).

Conventional, non-conjugated polymers with pendant oligothiophenes have been patterned using crosslinking chemistry [111,112]. In addition to cross-linking of the hydrocarbon backbone, it has been proposed that the terthiophene side chains underwent a

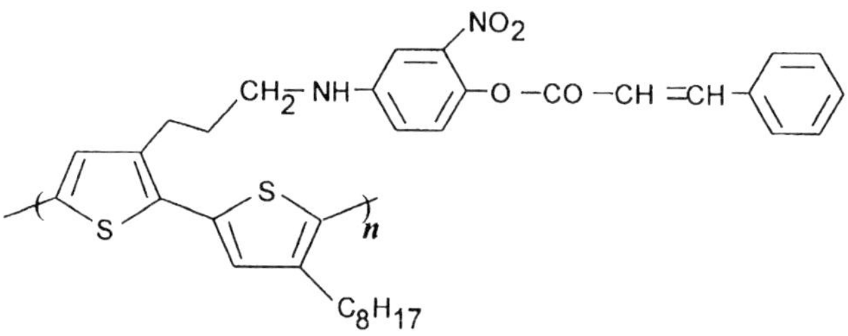

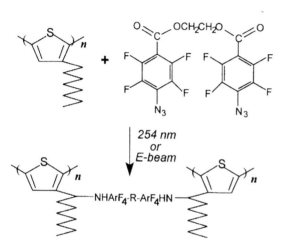

Figure 4.88. Poly[3-octylthiophene-co-*N*-(3-thienyl)-4-amino-2-nitrophenol] functionalized with a cinnamoyl group. Reprinted with permission from ref. 102.

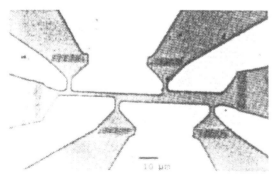

Figure 4.89. Mechanism of deep-UV and electron beam lithography of P3AT using perfluorophenyl bisazides. Reprinted with permission from ref. 103.

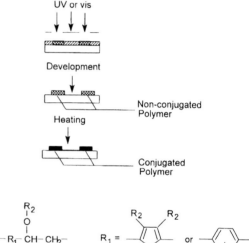

Figure 4.91. Photolithography of poly(aromatic vinylenes) from precursor polymers. Reprinted with permission from ref. 107.

Figure 4.90. Micrograph showing a conducting wire structure defined by electron beam lithography. Reprinted with permission from ref. 103.

coupling reaction to yield the corresponding sexithienyl residues (Figure 4.93). This assertion was supported using UV vis spectroscopy which showed a new peak at ~450 nm following irradiation.

The photochemistry and photolithography of water soluble polythiophenes, namely P3TASs have been studied [39]. It was found that P3TASs can be processed as both negative and positive resist depending on the water content of the material. In the presence of moisture, P3TAS films remained soluble in water even after extensive irradiation. Photochain scission was proposed as a possible rationale for the enhanced solubility of the polymer after irradiation. In contrast, anhydrous films of P3TAS s became insoluble upon irradiation. Enhanced dissolution of the material after irradiation was used to generate positive resists. Water

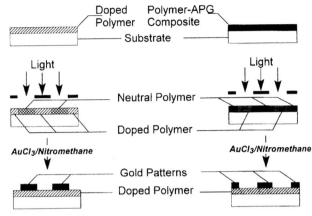

Figure 4.92. Photoselective deposition of gold on P3AT.

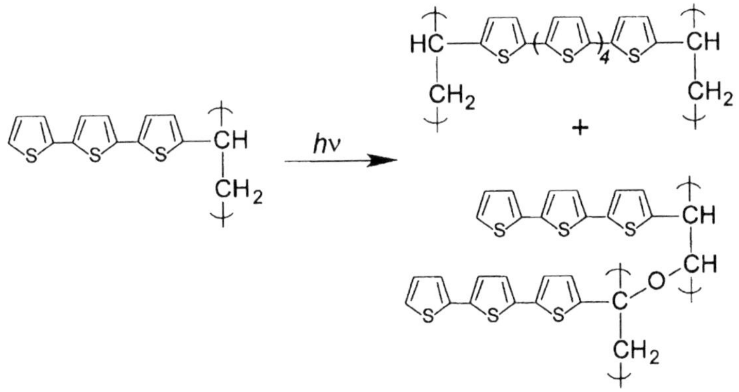

Figure 4.93. Photocrosslinking of non-conjugated polymers possessing pendant oligothiophenes. Reprinted with permission from ref. 111.

soluble P3TAs were prepared by oxidative coupling using ferric chloride as oxidative catalyst/agent, a recipe which is similar to that used in the preparation of organic soluble P3ATs. The concentration of the residual iron content was reportedly ~ 10–20 ppm. A very low concentration was found to be insufficient to photoinitiate crosslinking of the material. It was necessary to incorporate ferric chloride as a photo-initiator through a conventional doping process. $FeCl_4^-$-doped P3ATs are light sensitive and undergo facile photochemistry leading to de-doping of the material [68]. $FeCl_4^-$-doped P3TASs undergo photo-chemical reduction reactions upon exposure to UV or vis irradiation affording the construction of positive or negative-tone conjugated polymer resist. A positive image was obtained when films of poly[3-(3-thienyl)-propanesulfonic acid] P3TPSH were doped with

$FeCl_4^-$, irradiated, and developed with water. In this case, photolysis converted the oxidized polymer to its neutral form. The dissolution rate of neutral polymers in water is faster than the oxidized material and hence image transfer was achieved. In contrast, when development was carried out using acetic acid solution a negative resist was obtained. This was attributed to changes in the polarity of the material following exposure. Figure 4.94 shows a micrograph of positive- and negative-tone resist of P3TASs. Negative-resist P3TASs exhibit conductivity in the order of 0.01–0.1 S cm^{-1} following doping in nitromethane solutions of $FeCl_3$. Positive-resist P3TASs exhibit similar conductivity due to the self-doping nature of sulfonated polythiophenes.

Water soluble polythiophenes were also patterned using radiation-induced doping. Electron beam litho-

Figure 4.94. Photolithography of P3TASs: top, negative image of P3TOS; bottom, positive image of P3TPS. Resolution of the image is 0.5 mm. Reprinted with permission from ref. 39.

graphy was employed using sulfoxyimide, non-ionic compounds as the acid photogenerator [106].

8.3.3   Poly(p-phenylene sulfide)

Photoimaging of poly(p-phenylene sulfide) (PPS) has relied on photodoping of chemically modified PPS [113]. The technique is based on the difference in solubility created between an exposed (photodoped) and unexposed (neutral) polymer. S-phenylated PPS yields a soluble derivative of PPS and contains triaryl sulfonium residues in the main chain. The polymer serves as both the conjugated material and the acid photogenerator. Under UV irradiation, the polymer yields radical-coupled products and a Brønsted acid according to Figure 4.95. Concurrent doping of PPS occurs via the photogenerated acid. S-phenylated PPS is soluble in acetone whereas the photolyzed product is not. Thus negative-tone images can be achieved with this solvent. Alternatively, the use of dimethylsulfoxide as the developer, for which the photolyzed product is soluble, affords a positive-tone image.

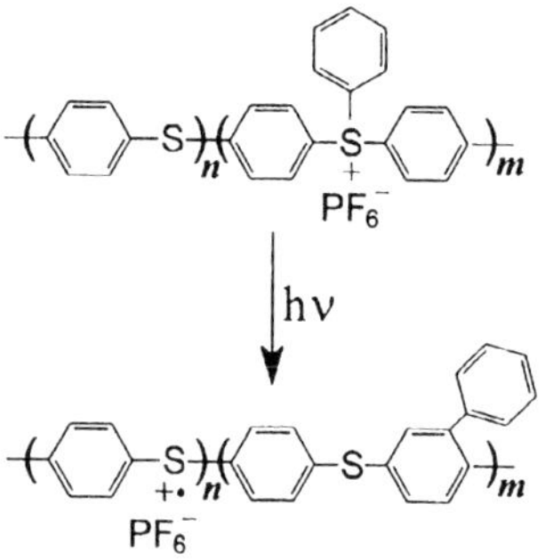

Figure 4.95. Photolysis of S-phenylated PPS. Reprinted with permission from ref. 113.

8.3.4   Poly(p-phenylenevinylene)

Acid-catalyzed elimination reactions of precursor polymers has been employed in the microlithography of PPV [93]. Precursors to PPV polymers are soluble in polar solvents and can be blended with acid photogenerators to yield light-sensitive materials. Upon exposure to UV light, acid is generated which causes an elimination reaction and converts the soluble precursor into the insoluble πconjugated polymer (Figure 4.96). Dissolution of the unexposed regions generates the conjugated polymer pattern.

Direct photoconversion of the polymer precursor into PPV has also been investigated as a lithographic process. Conversion of polymer precursors derived from 1,4-bis(tetrahydrothiophenium salts) of benzene and its 2,5-derivatives were examined using Ar⁺ laser irradiation [59,108]. No elimination was observed when films were irradiated with light of wavelength 514 or 488 nm even after prolonged irradiation, due to the low absorption coefficients of the polymer at these wavelengths. Irradiation with light of wavelength 254 nm, a wavelength at which the polymer precursor is highly absorbing, resulted in only poor conversion [59]. Prolonged irradiation introduced new functional groups, including carbonyl and hydroxyl groups, indicating photodegradation of the material [108].

An azo-sulfonic dye was used to photosensitize the reaction and enhance the efficiency of elimination with light of wavelength [59]. The dye was introduced into the polymer by either immersing the water-insoluble polymer films in solutions of the dye in water–alcohol mixtures, or by osmosis, in which the polymer film was placed between two compartments containing solutions of different concentrations of the dye. Impregnation of the dye was confirmed by UV vis spectroscopy. Although the elimination was improved in the presence

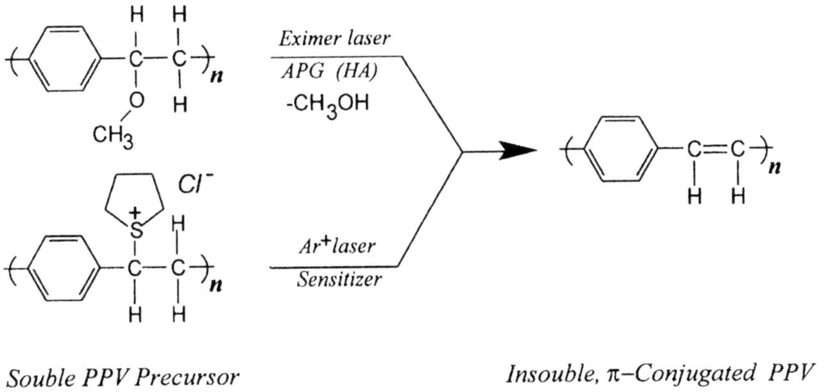
Eximer laser
APG (HA)
-CH3OH
Cl-
Ar+laser
Sensitizer
Souble PPV Precursor
Insouble, π–Conjugated PPV

Figure 4.96. Photoconversion of precursor polymers to PPV. Reprinted with permission from ref. 93.

of the sulfonic acid dye, the overall process was not sufficiently efficient to convert the precursor into a fully conjugated polymer. The mechanism of elimination was proposed to be photothermal rather than photochemical [59]. However, a fully conjugated material was obtained when irradiation was carried out at 120°C under vacuum.

8.3.5 Polyanilines

Photolithography of polyanilines (PANI) has attracted attention among conjugated polymers because of its solution processability and relative stability. Radiation-induced doping using acid photogenerators has been employed in the patterning of polyanilines [92,114–116]. When a film of polyaniline-containing dispersed triarylsulfonium or diaryliodonium salts is irradiated, the film becomes conducting. This process is depicted in Figure 4.97

Electronic absorption spectra of the composite film following irradiation with light of wavelength 240 nm are shown in Figure 4.98. The pristine film exhibits absorption peaks at 2 and 4.1 eV, characteristic of the undoped form of the polymer. Upon exposure, the 2 eV peak decreases and two new peaks appear at 1.5 and 2.8 eV due to electronic absorptions of the polaronic species indicating the formation of the doped form (emeraldine salt). The conductivity of the composite was reported to be a function of the concentration of the

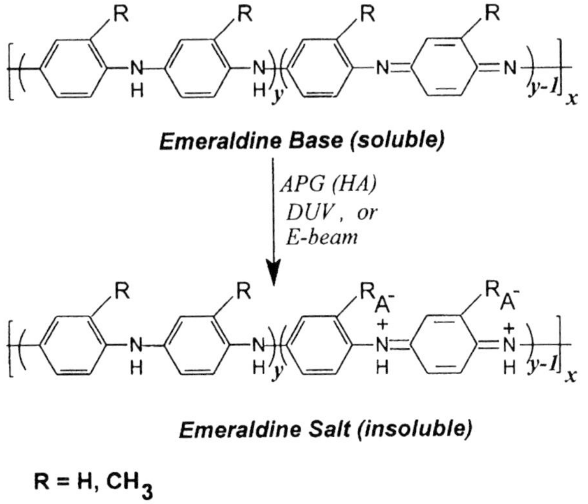
Emeraldine Base (soluble)
APG (HA)
DUV, or
E-beam
Emeraldine Salt (insoluble)
R = H, CH3

Figure 4.97. Photodoping of polyaniline using acid photogenerators (APG).

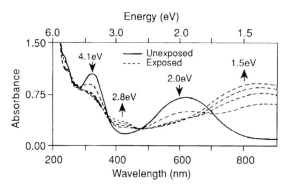

Figure 4.98. Optical absorption spectra for a polyaniline/ onium salt film before (solid line) and after exposure (dotted lines). Reprinted with permission from ref. 116.

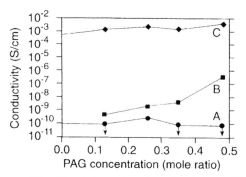

Figure 4.99. Electrical conductivity of poly(*o*-toluidine)/ nitrobenzylsulfonate ester films as a function of the molar ratio of the polymer to acid photogenerator (PAG) . Reprinted with permission from ref. 117.

onium salt and exposure dose. Electrical conductivities were measured to be *c.* 0.1 S cm^{-1} for onium salt concentrations ranging from 10 to 50 mol % and exposure doses up to 300 mJ cm^{-2}. Electron beam irradiation was also found to effect doping of the polymer–acid photogenerator composite films. The pristine polymer is soluble in *N*-methylpyrrolidinone but the photochemically doped polymer is insoluble. The differential solubility was employed to generate conducting polymer patterns. Images possessing resolutions of 0.5 and 0.25 µ were attained by deep-UV and electron beam lithography respectively.

Ionic acid photogenerators, such as nitrobenzylsulfonate esters, which produce sulfonic acid when irradiated with UV light, have also been used to fabricate submicrometer patterns of polyanilines derivatives [117]. The changes in the electronic absorption spectra of films of poly(methylaniline) containing the acid photogenerator accompanying irradiation with light of

wavelength 264 nm are analogous to those observed with polyaniline–onium salt composite films, indicating photo-induced doping. The conductivity of the film was monitored as a function of both the irradiation dose and the molar ratio of polymer to acid photogenerator. Figure 4.99 shows the electrical conductivities of films possessing different molar ratios when exposed to 1000 mJ cm^{-2} of light of wavelength 254 nm. At a mole ratio of 0.48, the conductivity of the film increased from *c.* 10^{10} to 10^{-6} S cm^{-1} following exposure. The relatively small increase in conductivity of the film was attributed to incomplete decomposition of the acid photogenerator due to an inner filter effect caused by a significant absorption of light by the polymer. Upon exposure to HCl vapor, the conductivity of the photochemically doped film was significantly increased to 0.01 S cm^{-1}.

Although photo-induced doping was not very effective increasing the conductivity of the material, a sufficient difference in solubility was obtained, between exposed and unexposed areas to achieve photolithography. It was shown that thin films (0.1–1.0 µm) of poly(methylaniline)–acid photogenerator mixtures irradiated with 500–3000 mJ cm^{-2} of 248 nm light of wavelength through a projection mask, and developed with chloroform resulted in dissolution of the unexposed regions leaving the exposed photochemically doped polymer intact. A resolution of 0.8 µm was achieved in this study. Figure 4.100 shows a scanning electron micrograph of these patterns.

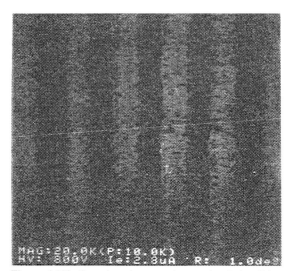

Figure 4.100. Scanning electron micrograph of photolithographed films of poly(*o*-toluidine)/nitrobenzylsulfonate ester. Reprinted with permission from ref. 117.

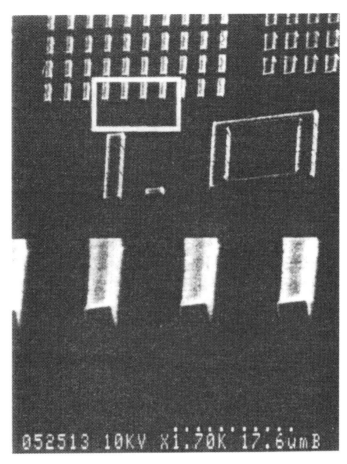

Figure 4.101. Micrograph of 1.0 μm conducting polyaniline lines in a 0.75 μm thick film prepared from Panaquas™ using electron beam irradiation. Reprinted with permission from ref. 118.

Chemical changes accompanying irradiation of these polymer composite films were monitored using optical spectroscopy. Infrared spectra of thin films exposed to different exposure doses of irradiation of wavelength 254 nm indicate no major changes occurred to the chemical structure of poly(methylaniline).

More recently, water soluble polyanilines (Pan-Aquas™) containing a crosslinkable functionality have been developed by IBM [118]. Conducting patterns of PanAqua™ have been constructed using electron beam lithography. Figure 4.101 shows a micrograph of 1 μm polyaniline patterns constructed using electronic beam lithography.

Photocrossiinkable polyaniline was also reported to be achieved by the covalent attachment of vinyl substituents [119]. Copolymers of aniline and allyiani-

line were prepared as potential materials for negative-tone lithographic resists (Figure 4.102). It has been suggested that exposure of films of this copolymer to UV or electron beam radiation should crosslink the allyl groups and hence insolubilize the material, but no lithographic studies have been reported to date.

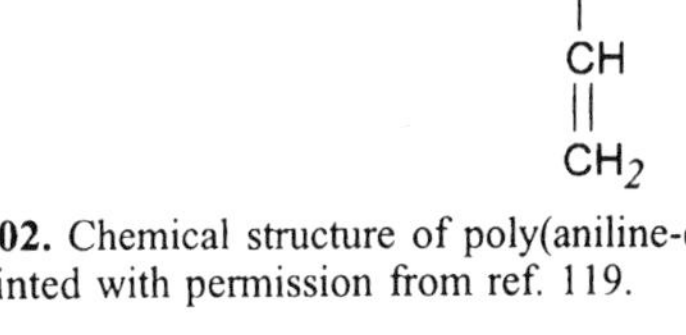

Figure 4.102. Chemical structure of poly(aniline-co-allylani-line). Reprinted with permission from ref. 119.

Photolithography of polyaniline has also been demonstrated via light-induced reduction of the doped, insoluble form to the neutral form at semiconductor electrodes [94]. In this system, inorganic semiconductor photosensitizers possessing a conduction band energy more negative than the formal electrochemical potential of polyaniline, i.e. -0.1 V (SCE), and the valence band energy must be more positive than the formal electrochemical potential of a sacrificial reducing agent employed. A wide bandgap semiconductor,

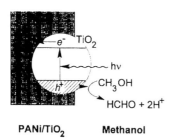

PANi/TiO₂   Methanol

Figure 4.103. Light-induced reduction of doped polyaniline. Reprinted with permission from ref. 94.

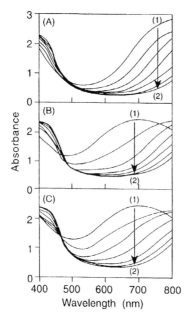

Figure 4.104. Changes in the electronic absorption of the doped polyaniline/TiO₂ system under open circuit conditions following irradiation. (1) and (2) represent before and after photolysis. (A) 1 M HCl, (B) pH 7, (C) methanol. Reprinted with permission from ref. 94.

TiO_2, was employed as a photosensitizer and methanol as the sacrificial reducing agent as depicted in Figure 4-103.

Changes in the electronic absorption of the doped polyaniline/TiO_2 system under open circuit conditions following irradiation are shown in Figure 4.104. A decrease in optical density c. 800 nm is observed together with a corresponding change in the color of polyaniline from blue-green (doped polymer) to yellow (neutral, deprotonated polymer) [94].

Laser-assisted electropolymerization has been used to deposit polyaniline patterns [120]. Aniline can be polymerized electrochemically at potentials more positive than $+0.9$ V (SCE). Under laser irradiation, the electropolymerization potential is reduced to 0.65 V (SCE) thus allowing laser-directed electrodeposition of polyaniline on metal electrodes. Thermal effects due to the laser beam was proposed as the reason for the decrease in the potential required for electropolymerization.

8.3.6   Polypyrrole

Polypyrrole was the first conjugated polymer to be imaged photochemically. Initial efforts involved photoelectrochemical polymerization of pyrrole on semiconductor electrodes as illustrated in Figure 4.105 [121–124].

Irradiation of a semiconductor electrode with a photon of energy greater than the bandgap resulted in the generation of electron–hole pairs. Using n-type TiO_2, a bias of -0.6 V (vs. SCE) was necessary to initiate polymerization of pyrrole on the illuminated electrodes as shown in Figure 4.106 [121,122]. This voltage bias is less positive than the potential required for the polymerization of pyrrole at metal electrodes

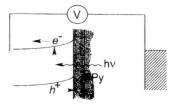

n-TiO₂   Pyrrole   Metal

Figure 4.105. Photoelectrochemical polymerization of polypyrrole.

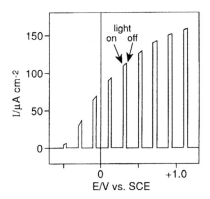

Figure 4.106. Current–voltage characteristics for the polymerization of pyrrole on n-TiO$_2$ semiconductor electrodes under irradiation and in the dark. Reprinted with permission from ref. 121.

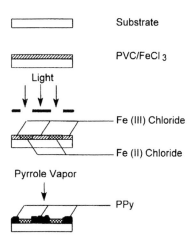

Figure 4.108. Photoselective polymerization of polypyrrole on insulating substrates. Reprinted with permission from ref. 97.

leading to photoselective deposition of polypyrrole on illuminated regions of the electrode.

Polypyrrole was selectively deposited on the illuminated areas of ZnO and n-type silicon [123,124], in addition to TiO$_2$, but the technique suffers from the disadvantage that it is a 'wet' process and thus is incompatible with solid state processing technology. Furthermore, patterns were reported to be of poor resolution due to the relatively large diffusion length of the photogenerated holes in the semiconductor.

Deposition of polypyrrole on insulating substrates using photosensitized polymerization of pyrrole has also been demonstrated (Figure 4.107) [125]. Tris(2,2'-bipyridine) ruthenium(II) [Ru(bpy)$_3$$^{2+}$] was used as a photosensitizer and penta-aminechlorocobalt(II) ([Co(NH$_3$)$_5$Cl]Cl) was employed as a sacrificial oxidizing agent. Selective deposition of polypyrrole on the surface of the Nafion membranes was also reported.

A mixture of Cu(I) diphenylphenanthroline and p-nitrobenzylbromide has also been used to effect the photochemical deposition of polypyrrole patterns [126]. In the presence of p-nitrobenzylbromide, the Cu(I)

complex is photochemically oxidized to Cu(II). Cu(II) is an effective polymerization agent of pyrrole. When photolysis was performed through a photomask, selective photodeposition of polypyrrole was achieved.

Similarly, photoselective polymerization of polypyrrole on various substrates has been achieved by photochemical patterning of an oxidant, followed by *in situ* polymerization of pyrrole from the vapor phase [96,97,127]. Potential initiators include salts of Cu(II), Fe(III) and Ce(IV), onium salts and peroxides. Patterns of Fe(II) salts, for example, were generated by selective irradiation of iron(III) compounds incorporated in thin polyvinyl chloride (Figure 4.108) [97]. Upon irradiation, Fe(III) is converted to Fe(II), which is ineffective for the polymerization of pyrrole. *in situ* polymerization of pyrrole generates the polypyrrole pattern. SEM micrographs of polypyrrole deposited onto Cu(II) patterns are shown in Figure 4.109 [127]. This technique apparently suffers from uncontrolled polymerization which leads to ill-defined polymer patterns.

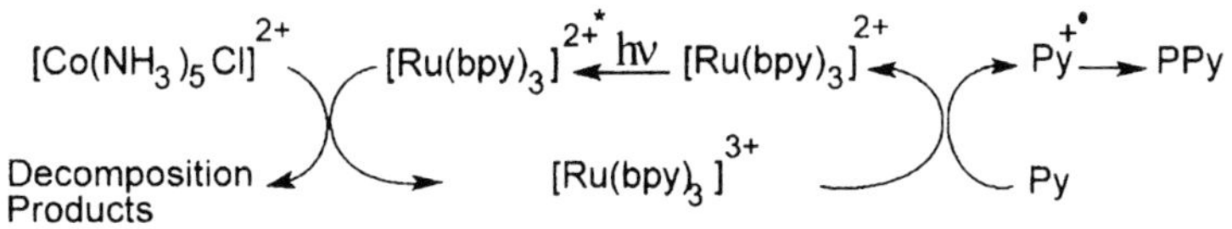

Figure 4.107. Photosensitized polymerization of polypyrrole. Reprinted with permission from ref. 125.

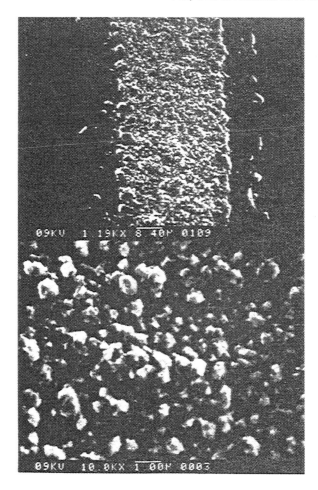

Figure 4.109. SEM micrograph of polypyrrole deposited onto Cu(II) salt. Reprinted with permission from ref. 127.

Laser ablation of polypyrrole by irradiation of thin polymer films through a mask with pulsed UV laser radiation renders pattern generation [98]. This technique leads to damage of the polymer surrounding the irradiated spot and produces patterns of low resolution.

9   SUMMARY

Photochemical processes of conjugated polymers have only recently been subject to detailed study. There are considerable differences in both the efficiency and mechanism of photochemical reactions between the various classes of conjugated polymers. All of the polymers, however, appear to undergo photooxidation in the presence of oxygen. Carbonyl group formation and disruption of the π-conjugated system are common features during irradiation. The mechanism of photooxidation may vary considerable between polymers but only two or three classes of conjugated polymers have been examined to the extent that a photochemical pathway can be proposed with reasonable certainty. Nitrogen-containing polymers are distinctly more photochemically stable than polyacetylenes, polythiophenes, and poly(aromatic vinylenes). For the latter polymers, the reaction of singlet oxygen across double bonds has been clearly demonstrated. The mechanism of singlet oxygen is not clear at present. The usual triplet–triplet annihilation route might not be viable if the triplet population is formed at very low concentration. Alternatively, singlet oxygen might arise from photodissociation of a charge transfer complex, as suggested for poly(3-alkylthiophenes). These issues require further study.

An emerging area of interest is photolithography. The use of conjugated polymers in this regard is in its infancy. Both negative- and positive-tone images have been demonstrated using photocrosslinking, photochain scission and photoacid generation chemistry. Innovative photochemical pathways, unique to conjugated polymers, are expected to be developed. There is considerable room for improving photosensitivity, contrast and resolution. In order to deposit electronically conducting polymer wires, the deposited polymer must be stable in its oxidized state. This is not a trivial issue. The development of conducting polymer circuitry will go hand-in-hand with development of stable conducting polymer systems. A related property which has not yet been examined is the resistance of conjugated polymers to microchip fabrication chemicals. Photolabile conjugated polymers may well prove to be alternatives to conventional photoresists.

The degree of photochemical reaction of a polymer is inversely proportional to the sum of all of the other rates of deactivation of the excited state. Photophysical properties play a major role in the efficiency of photochemical reactions. For small molecule photochemistry, mechanisms can be accurately modeled through knowledge of photochemical reactions and non-photochemical pathways. Evaluating rate constants of photochemical reactions for conjugated polymers is difficult because of rapid excitonic migration along the polymer chain. The difficulty is increased in the solid state due to the influence of the morphological state of the polymer and its effect on the rate of photophysical processes.

Most conjugated polymers serve as sources of electrons in photoelectron transfer reactions. The resultant polymer is in its partially oxidized state. The sacrificial oxidizing agent is reduced. In the form of solid films on electrodes, photogenerated electron–hole pairs can be separated under the influence of an applied field. The separation of charges means conjugated polymers can be employed as photocathodes in photoelectrochemical cells for continuous electrochemical reduction of solution species. Studies to date have focused on phenomenological relationships. Photoelectrochemical efficiencies are quite low. In order to increase this efficiency the effect of chemical structure and morphology on the efficiency of electron–

hole separation needs to be addressed. It is reasonable to assume that the efficiencies can be increased with a suitable choice of polymer and careful control of experimental conditions. The employment of electron acceptors, such as C_{60} may well prove to be useful in separating electron–hole pairs, as demonstrated in solid state devices. Photochemical stability will be of a concern and more information regarding polymer photodegradation, and its prevention, is required

As the importance of photonic applications increases, so does the need to amass information on photochemical properties. General concepts pertaining to photochemical stability or photochemical lability are expected to emerge concurrently as a consequence of their technological applications.

10 REFERENCES

1. H. Koezuka, A. Tsumara and T. Ando, *Synth. Met.* **18**, 699, (1987)
2. J.H. Burroughes, A.C. Jones and R.H. Friend, *Synth. Met.* **28**, C735, (1989).
3. A.O. Patil, A.J. Heeger and F. Wudl, *Chem. Rev.* **88**, 183 (1988).
4. H.S. Nalwa and S. Miyata, eds. *Nonlinear Optics of Organic Molecules and Polymers*, CRC Press, Boca Raton, Florida, 1997.
5. J.H. Burroughes, D.D.C. Bradley, A.R. Brown, R.N. Marks, K. Mackay, R.H. Friend, P.L. Burns and A.B. Holmes, *Nature* 539, (1990).
6. D. Moses, *Appl. Phys. Lett.* **60**, 3215, (1992).
7. J.E. Guillet, *Polymer Photophysics and Photochemistry*, Cambridge University Press, New York, 1985.
8. D. Phillips (ed.), *Polymer Photophysics*, Chapman & Hall, New York, 1985.
9. B. Ranby and J.E. Rabek, *Photodegradation, Photooxidation and Photostabilization of Polymers*, John Wiley & Sons, Chichester, 1975.
10. J.F. McKeller and N.S. Allen, *Photochemistry of Man-Made Polymers*, Applied Science Publishers, London, 1979.
11. K.U. Ingold, *Pure Appl. Chem.* **15**, 49 (1967.
12. F.R. Mayo, *Acc. Chem. Res* **1**, 193 (1968).
13. J.B. Birks, *Photophysics of Aromatic Molecules*, Wiley & Sons, London, 1970.
14. F. Wilkinson, W.P. Helman and A.B. Ross, *J. Phys. Chem. Ref. Data* **22**, 113, 1993.
15. N.J. Turro, *Modern Molecular Photochemistry*, Benjamin/Cumkings Pub., California, 1978.
16. K. Gollnick and G.O. Schenck, in *1,4-Cycloaddition Reactions*, (ed. J. Hamer,) Academic Press, New York, Chapter 10.

17. K. Gollnick, in *Singlet Oxygen-Reaction with Organic Compounds and Polymers*, (eds. B. Ranby and L.F. Rabek) John Wiley & Sons, Chichester, 1978.

18. F.H. Winslow and A.M. Trozzolo, *Macromolecules* **1** 98 (1968).

19. D.T. Sawyer, *Oxygen Chemistry*, Oxford University Press, New York, 1991.

20. H. Tsubomura and R.S. Mulliken, *J. Am. Chem. Soc.* **82**, 5966 (1960).

21. P.R. Ogilby, M.P. Dillon, M. Kristiansen and R.L. Clough, *Macromolecules* **25**, 3399 (1992).

22. C.E. Loader and C.J. Timmons, *J. Chem. Soc. (C)* 1677 (1967).

23. A. Charlesby and S.H. Pinner, *Proc. R. Soc.* A **249**, 367 (1958).

24. A. Charlesby, *Atomic Radiation and Polymers*, Pergamon Press, New York, 1960.

25. A. Reiser, *Photoactive Polymers*, John Wiley & Sons, New York, 1989.

26. H.H.G. Jellinek (ed.) *Aspects of Degradation and Stabilization of Polymers*, Elsevier Scientific, Amsterdam, 1978.

27. W.H. Stockmayer, *J. Chem. Phys.* **12**, 125 (1944).

28. A.M. North and D.A. Ross, *J. Polym. Sci., Polym. Symp.* **55**, 259 (1976).

29. H.W. Gibson and J.M. Pochan, *Macromolecules* **15**, 242 (1982).

30. K. Tsuchihara, T. Masuda and T. Higashimura, *J. Polym. Sci., Polym. Chem. Ed.* **29**, 471 (1991).

31. P.I. Clemonson, W.J. Feast, M.M. Ahmad, P.C. Allen, D.C. Bott, C.S. Brown and L.M. Connors, *Polymer* **33**, 4711 (1992).

32. P.C. Allen, D.C. Bott, C.S. Brown, L.M. Connors, S. Gray, N.S. Walker, P.I. Clemonson and W.J. Feast, *Electronic Properties of Conjugated Polymers III*, (eds. H. Kuzmany, M. Mehring and S. Roth, Springer-Verlag, Berlin, 1989.

33. L.I. Tkachenko, G.I. Kozub, A.Ph. Zueva, O.S. Roschupkina, O.N. Efimov and M.L. Khidekel, *Synth. Met.* **40**, 173 (1991).

34. M.S.A. Abdou and S. Holdcroft, *Macromolecules* **26**, 2954 (1993).

35. S. Holdcroft, *Macromolecules* **24**, 4834 (1991).

36. M.S.A. Abdou, M.I. Arroyo, G. Diaz-Quijada and S. Holdcroft, *Chem. Mater.* **3**, 1003 (1991).

37. M.S.A. Abdou, Z.W. Xie, A. Leung and S. Holdcroft, *Synth. Met.* **52**, 159 (1992).

38. M.S.A. Abdou, Z.W. Xie, J. Lowe and S. Holdcroft, *Advances in Resist Technology and Processing* SPIE, **2195**, 756 (1994).

39. M.I. Arroyo, G. Diaz-Quijada, M.S.A. Abdou and S. Holdcroft, *Macromolecules* **28**, 975 (1995).

40. D. McLachlan, T. Arnason and J. Lam, *Photochem. Photobiol.* **39**, 177 (1984).

41. C. Evans, D. Weir, J.C. Scaiano, A. MacEachern, J.T. Arnason, P. Morand, B. Hollebone, L.C. Leitch and

B.J.R. Philogene, *Photochem. Photobiol.* **44**, 441 (1986).

42. J.P. Reyftmann, J. Kagan, R. Santos and P. Moliere, *Photochem. Photobiol.* **41**, (1985).

43. C. Evans and J.C. Scaiano, *J. Am. Chem. Soc.* **112**, 2694 (1990).

44. B. Xu and S. Holdcroft, *J. Am. Chem. Soc.* **115**, 8447 (1993).

45. R.S. Mulliken and W.B. Person, *Molecular Complexes*, John Wiley & Sons, London, 1969.

46. A. Weller, in *The Exciplex*, (eds. M. Gordon and W.R. Ware), Academic Press, New York, 1975.

47. D. Rehm, and A. Weller, *Z. Phys. Chem.* **69**, 193 (1970).

48. M.S.A. Abdou, F. Orfino, Z.W. Xie, M.J. Deen and S. Holdcroft, *Adv. Mater.* **6**, 838 (1994).

49. M.S.A. Abdou, X. Lu, Z.W. Xie, F. Orfino, M.J. Deen and S. Holdcroft, *Chem. Mater.* **7**, 631 (1995).

50. M.S.A. Abdou and S. Holdcroft, *Can. J. Chem.* **73**, 1893 (1995).

51. K. Yoshino, T. Kuwabara, Y. Manda, S. Nakajima and T. Kawai, *Jpn. J. Appl. Phys.* **29**, L1716 (1990).

52. T. Kawai, T. Kuwabara and K. Yoshino, *Jpn. J. Appl. Phys.* **31**, L49 (1992).

53. K. Yoshino, K. Nakao and M. Onoda, *J. Phys. Condens, Matter* **2**, 2857 (1990).

54. B.H. Cumpston and K.F. Jensen, *Synth. Met.* **73**, 195 (1995).

55. M. Sandberg, S. Tanaka and K. Kaeriyama, *Synth. Met.* **60**, 171 (1993).

56. J. Lowe and S. Holdcroft, *Macromolecules* **28**, 4608 (1995).

57. P.K. Das, P.J. DesLauriers, D.R. Fahey, F.K. Wood and F.J. Cornforth, *Macromolecules* **26**, 5024 (1993).

58. J. Bullot, B. Dulieu and S. Lefrant, *Synth. Met.* **61**, 211 (1993).

59. A. Torres-Filho and R.W. Lenz, *J. Polym. Sci., Polym., Phys. Ed.* **31**, 959 (1993).

60. M. Yan and L.J. Rothberg, F. Papadimtrakopoulos, M.E. Galvin, T.M. Miller, *Phys. Rev. Lett.* **73**, 744 (1994).

61. F. Papadimtrakopoulos, M. Yan, L.J. Rothberg, H.E. Katz, E.A. Chandross and M.E. Galvin, *Mol. Cryst. Liq. Cryst.* **256**, 663 (1994).

62. K.G. Neoh, E.T. and K.L. Tan, *Polymer* **33**, 2292 (1992).

63. V. Wintgens, P. Valat and F. Garnier, *J. Phys. Chem.* **98**, 228 (1994).

64. B. Zinger, K.R. Mann, M.G. Hill and L.L. Miller, *Chem. Mater.* **4** 1113 (1992).

65. D. Matthews, A. Altus and A. Hope, *Aust. J. Chem.* **47**, 1163 (1994).

66. S. Matsuoka, T. Kohzuki, C. Pac, A. Ishida, S. Takamuku, M. Kusaba, N. Nakashima and S. Yanagida, *J. Phys. Chem.* **96**, 4437 (1992).

67. M. Sandberg, S. Tanaka and K. Kaeriyama *Synth. Met.* **55–67**, 3587 (1993).

68. M.S.A. Abdou and S. Holdcroft, *Chem. Mater.* **6**, 962 (1994).

69. A.J. Bard, *Science* **207**, 139 (1980).

70. M.S. Wrighton, *Pure Appl. Chem.* **57**, (1985).

71. A.J. Nozik, *Annu. Rev., Phys. Chem.* **29**, 189 (1978).

72. T.J. Skotheim (ed.) *Handbook of Conducting Polymers*, Vols. 1 and 2, Marcel Dekker, New York, 1986.

73. H. Shirakawa, S. Ikeda, M. Aizawa, J. Yoshitake and S. Suzuki, *Synth. Met.* **4**, 43 (1981).

74. S.N. Chen, A.J. Heeger, Z. Kiss, A.G. MacDiarmid, S.C. Gau and D.L. Peebles, *Appl. Phys., Lett.* **36**, 96 (1980).

75. T. Inoue and T. Yamase, *Bull. Chem. Soc. Jpn* **56**, 985 (1983).

76. M. Keneko, K. Okuzumi and A. Yamada, *J. Electroanal. Chem.* **183**, 407 (1985).

77. M. Keneko and H. Nakamura, *J. Chem. Soc. Chem. Commun*, 346 (1985).

78. M.X. Wan, A.G. MacDiarmid and A.J. Epstein, in *Electrical Properties of Conjugated Polymers*, Springer-Verlag, Berlin, *76*, 216 (1987).

79. E.M. Genies and M. Lapkowski, *Synth. Met.* **24**, 69 (1988).

80. P.K. Shen and Z.Q. Tian, *Electrochem. Acta* **34**, 1611 (1989).

81. M. Kalaji, L. Nyholm, L.M. Peter and A.J. Rudge, *J. Electroanal. Chem.* **310**, 113 (1991).

82. S. Glenis, G. Horowtz, G. Tourillon and F. Garnier, *Thin Solid Films* **111**, 93 (1984).

83. S. Glenis, G. Tourillon and F. Garnier, *Thin Solid Films* **122**, 9 (1984).

84. O. El-Rashiedy and S. Holdcroft, unpublished results.

85. W.M. Moreau, *Semiconductor Lithography*, Plenum Press, New York, 1988.

86. W.S. DeForest, *Photoresist*, McGraw-Hill, New York 1975.

87. L.F. Thompson, C.G. Wilson and M.J. Bowden, (eds.) *Introduction to Microlithography*, ACS Symposium Series 219, American Chemical Society, Washington, 1983.

88. S.R. Turner, K.D. Ahn and C.G. Willson, *Proc. ACS Division Polymer Material Science and Engineering*, American Chemical Society, Washington, 1986.

89. S.P. Pappas, *Radiation Curing*, Plenum Press, New York, 1992.

90. B.M. Monroe and G.C. Weed, *Chem. Rev.* **93**, 435 (1993).

91. S.A. MacDonald, C.G. Willson and J.M.J. Frechet, *Acc. Chem. Res.* **27**, 151 (1994).

92. M. Angelopoulos, J.M. Shaw, W.-S. Huang and R.D. Karplan, *Mol. Cryst. Liq. Cryst.* **189**, 221 (1990).

93. S. Taguchi and T. Tanaka, *Eur Patent Appl.* EP 261, 991, 1988.

94. S. Kuwabata, N. Takahashi, S. Hirao and H. Yoneyama, *Chem. Mater.* **5**, 437 (1993).

95. Y.-M. Wuu, F.F. Fan and A. Bard, *J. Electrochem. Soc.* **136**, 885 (1989).

96. A. Nannini and G. Serra, *J. Molec. Electron* **6**, 81 (1990).

97. W. Behnck and J. Bargon, *Synth. Met.* **54**, 223 (1993).

98. L.S. Van Dyke, C.J. Brumlik, C.R. Martin, Z. Yu and G.J. Collins, *Synth. Met.* **52**, 299 (1993).

99. H.-K. Roth, H. Gruber, E. Fanghanel, A.M. Richter and W. Horig, *Synth. Met.* **37**, 151 (1990).

100. J. Bargon and R. Baumann, in *SPIE Proceedings, Vol. 1910* (eds. E.M. Conwell , M. Stolka and M.R. Miller), 1993, p.92.

101. M.S.A. and S. Holdcroft, *Synth. Met.* **62**, 213 (1993).

102. K.G. Chittibabu, L. Li, M. Kamath, J. Kummar and S.K. Tripathy, *Chem. Mater.* **6**, 475 (1994).

103. S.X. Cai, J.F.W. Keana, J.C. Nobity and M.N.Wybourne, *J. Molec. Electron.* **7**, 63 (1991).

104. S.X. Cai, M. Kanskar, J.C. Nobity, J.F.W. Keana and M.N. Sybourne, *J. Vac. Sci. Technol.* B **10**, 2589 (1992).

105. S.X. Cai, M. Kanskar, M.N. Wybourne and F.W. Keana, *Chem. Mater.* **4**, 879 (1992).

106. W.-S. Haung, *Polymer* **35**, 4057 (1994).

107. T. Tanaka and S. Doi, *US Patent 4*, **988**, 608 (1991).

108. M. Onoda, H. Nakayama and K. Yoshino, *IEEE Trans. Elect. Ins.* **27**, 636 (1992).

109. M.S.A. Abdou and S. Holdcroft, *Chem. Mater.* **8**, 26 (1996).

110. M.S.A. Abdou and K. Bartle and S. Holdcroft, unpublished results.

111. R.K. Khanna and H. Cui, *Macromolecules* **26**, 7076 (1993).

112. R.K. Khanna and N. Bhingare, *Chem. Mater.* **5**, 899 (1993).

113. B.M. Novak, E. Hagen, A. Viswanathan and L. Magde, *ACS Polymer Preprints* **31**, 482 (1990).

114. M. Angelopoulos, R.D. Kaplan and S. Perreault, *J. Vac. Sci. Technol* B **7**, 1519 (1989).

115. M. Angelopoulos, J.M. Shaw, K.-L. Lee, W.-S. Huang, M.-A. Lecorre and M. Tissier, *J. Vac. Sci. Technol.* B **9**, 3428 (1991).

116. M. Angelopoulos, J.M. Shaw, K.-L. Lee, W.-S. Huang, M.-A. Lecorre and M. Tissier, *Mol. Cryst. Liq. Cryst.* **189**, 221 (1990).

117. G. Venugopal, X. Quan, G.E. Johnson, F.M. Houlihan, E. Chin and O. Nalamasu, *Chem. Mater.* **7**, 271 (1995).

118. M. Angelopoulos, N. Patel, J.M. Shaw, N.C. Labianca and S.A. Rishton, *J. Vac. Sci. Technol.* B **11**, 2794 (1993).

119. L.H. Dao, M.T. Nguyen and T.N. Do, *ACS Polymer Preprints* **33(2)**, 408 (1992).

120. J.Y. Jin, N. Teramae and H. Haraguchi, *Chem. Lett.* 101 (1993).

121. M. Okano, K. Itoh, A. Fujishima and K. Honda, *Chem. Lett.* 469 (1986).

122. M. Okano, K. Itoh, A. Fujishima and K. Honda, *J. Electrochem. Soc.* **134**, 837 (1987).

123. H. Yoneyama, K. Kawai and S. Kuwabata, *J. Electrochem. Soc.* **135**, 1699 (1988).

124. M. Okano, E. Kikuchi, K. Itoh and A. Fujishima, *J. Electrochem. Soc.* **135**, 1641 (1988).

125. H. Segawa, T. Shimidzu and K. Honda, *J. Chem. Soc. Chem. Commun.* 132 (1989).

126. J.-M. Kern and J.-P. Sauvage, *J. Chem. Soc. Chem. Commun.* 657 (1989).

127. A. Nannini and G. Serra, *J. Molec. Electron.* **6**, 81 (1990).

CHAPTER 5
Photorefractive Polymers

Luping Yu*, Wai Kin Chan, Zhonghua Peng, Wenjie Li and Ali R. Gharavi
University of Chicago, Chicago, Illinois, USA

1 INTRODUCTION

Photorefractive (PR) materials are multifunctional materials which combine photoconductivity and electro-optic (E-O, Pockel effect) activity to manifest a photorefractive effect [1]. In these materials, the indices of refraction can be modulated by light via their E-O effect and a photoinduced space charge field. It is generally believed that there are four basic processes involved in the photorefractive effect: the photocharge generation upon absorbing photons; the transportation of the generated charge carriers either by thermal diffusion or electric drifting; the trap of the charge carriers in trapping centers which results in the charge separation and formation of the space charge field; and the formation of a phase grating due to the space charge field modulation of the refractive index via the linear

electro-optical effect [1]. Figure 5.1 illustrates these basic processes. Notice that two light beams are needed to generate the oscillating space charge field. A unique feature of the space charge field is that it has a phase shift from the interference pattern. This feature results in a special property for photorefractive materials: the optical energy exchange of the two incident beams occurs in an asymmetric way, one of the beams loses energy while the other beam gains energy.

The driving force for the studies of the photorefractive effects is their potential applications, such as in three-dimensional holographic light processing, phase conjugation and the handling of large quantities of information in real time [1]. From Figure 5.1, at least two of the applications can be visualized. First, if the material is a good insulator, the photoinduced charge separation can be maintained for a long time in the dark and can therefore be used to store information. Second, if one of the incident beams is a weak optical signal

---

*Author to whom all correspondence should be addressed.

Handbook of Organic Conductive Molecules and Polymers: Vol. 4. Conductive Polymers: Transport, Photophysics and Applications.
Edited by H. S. Nalwa. © 1997 John Wiley & Sons Ltd

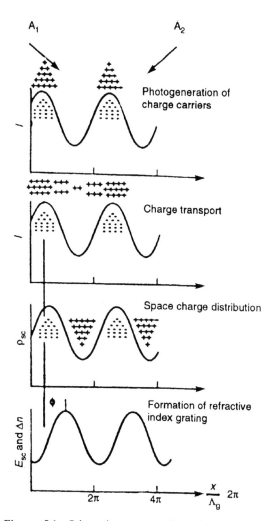

Figure 5.1. Schematic representation of the processes involved in the photorefractive effect. Two laser beams were applied.

properties such as ease of preparation for large area samples, low dielectric constants, high optical performance and low cost [5].

In the past several years, two approaches have been developed to prepare or synthesize PR polymeric materials, namely, the composite material approach and the fully functionalized polymer approach [5–16]. Both approaches have enjoyed success in identifying PR polymers. The materials' performances have been greatly enhanced, for example, a nearly 85% diffraction efficiency and $200\,cm^{-1}$ optical net gain was observed in a composite material [17]. Prototype devices based on these materials were also attempted [18]. The composite systems exhibit a major advantage: they make it easier to prepare different materials and to survey different compositions quickly. However, the composite systems also exhibit several serious problems, such as their intrinsic instability due to phase separation and the low glass transition temperature. Although fully functionalized polymer systems exhibit several disadvantages, such as time-consuming chemical synthesis and difficulty in rational design, these polymers offer advantages that the composite materials do not possess, such as the stable transparency, minimum phase separation and high stability in dipole orientation. Therefore both approaches are necessary for the development of PR polymers. Judging from the history of the development of second order nonlinear optical (NLO) polymers, the functionalized polymer approach will be the preferred one [19,20]. This chapter is devoted to a brief survey on the rational design and preparation or synthesis of PR polymers, including both composite materials and fully functionalized PR polymers. Detailed discussions will be given towards the fully functionalized polymer systems. An excellent review on composite polymeric PR materials can be found in ref. 5.

Despite the exciting progress, we must emphasize that PR polymer studies are still in their infancy. There are numerous issues which prevent these materials from practical application. The concluding remarks will discuss the problems existing in this area.

2 BRIEF DESCRIPTION OF THE FOUR PROCESSES OF THE PHOTOREFRACTIVE EFFECTS

As discussed in the introduction, the photorefractive effect involves four complicated processes. The most important process is the establishment of the space charge field. Based upon a band transport model for

beam, it can be amplified by using a strong pump beam due to the asymmetric energy exchange.

Photorefractive phenomena have been studied for almost thirty years since its discovery in ferroelectric single crystals ($LiNbO_3$) in 1966 [2]. Before 1989, research on the photorefractive effect was exclusively focused upon inorganic materials, such as ferroelectric crystals ($LiNbO_3$, $LiTaO_3$, $BaTiO_3$), semiconductors ($GaAs$, InP, CdF_2) and sillenites ($Bi_{12}SiO_{20}$) [1,2]. Recently, organic single crystal and PR polymers have emerged as new forms of optical materials [3,4]. It is expected that PR polymers will possess unique properties supplemental to their inorganic counterparts:

inorganic single crystal materials, Kukhtarev *et al.* were able to obtain the lowest Fourier component of the space charge field in a material illuminated by a sinusoidal interference pattern [21]. Several sets of equations are needed to describe the process of the formation of the space charge field (a slight modification was made to accommodate the case of polymeric materials):

$$\frac{\partial n_h}{\partial t} + \frac{\partial J_h}{e\,\partial x} = (S_h I + \beta_h)N - n_h N^- \mu - \mu n_h M + \gamma_h M^+$$
(5.1)

$$J_h = e\mu_h n_h E - k_B T \mu_h \frac{\partial n_h}{\partial x}$$
(5.2)

$$\frac{\partial N^-}{\partial t} = (s_h I + \beta_h)N - \mu n_h N^-$$
(5.3)

$$\frac{\partial M^+}{\partial t} = \mu n_h M - \gamma_h M^+$$
(5.4)

$$\frac{\partial E}{\partial x} = \left(\frac{e}{\varepsilon\varepsilon_0}\right)(n_h - N^- + M^+)$$
(5.5)

$$N_0 = N + N^-$$
(5.6)

$$M_0 = M + M^+$$
(5.7)

$$\int_0^L E\,\mathrm{d}z = V$$
(5.8)

where n_h is the hole density, s_h is the photoionization constant for holes, β_h is the thermal generation rate for holes, N^- is the ionized acceptor density, N is the acceptor density, N_0 is the initial acceptor density, μ is the carrier's mobility, e is the elementary charge, T is the temperature, γ_h is the detrapping rate, k_B is Boltzman's constant, ε is the dielectric constant, ε_0 is the vacuum permittivity, M is the trap density, M^+ is the ionized trap density, M_0 is the initial trap density, n is the index of refraction of the PR material, and E_t is the total electric field. Equation (5.1) describes the rate of hole generation, assuming the hole is the dominant charge carrier in most of the polymeric materials. Equation (5.2) is the expression for the current, and equations (5.3) and (5.4) are the rates of formation of the ionized sensitizer and ionized trappers. Equation (5.5) is Poisson's equation, describing the relationship between the charges and the field. Equations (5.6) and (5.7) are the mass balance equations for the sensitizer and trap species. Equation (5.8) is required if an external field is applied through the sample with a thickness of L.

These are nonlinear equations and their exact solutions are formidable even for the most simple situation in inorganic PR single crystals where the following assumptions were made: only one charge carrier and one charge trapping level exist, and the quantum yield of photogeneration of the carriers is independent upon both the light intensity and space charge field. However, after several further assumptions, these equations became solvable for inorganic single crystal materials [21]. These assumptions include small optical modulation, $m \ll 1$; small dark conductivity, $\beta \ll sI$; and low free carrier density. A detailed description of these mathematical manipulations is outside the scope of this chapter. We only cite the expression for the space charge field under the steady-state condition [21,22].

$$E_{sc} = E_q \sqrt{\left[\frac{E_0 2 + E_d^2}{E_q^2 + (E_d + E_d)^2}\right]}$$
(5.9)

where $E_d = k_B TK/e$ is the diffusion field, $E_q = eN_e/\varepsilon_0\varepsilon_K$ is the trap-limited space charge field, E_0 is the external electric field, ε is the dielectric constant, and N_e is the effective density of the empty trap centers. This expression predicts that the space charge field is affected by the external field and the maximum achievable space charge field is the trap-limited field E_q.

For polymeric materials, the situation becomes much more complicated. Several assumptions made in the case of inorganic single crystals are not valid for polymeric materials. For example, the trapping level in polymeric materials has a wide distribution, as indicated by a long tail beyond the absorption edge; the quantum efficiency for the photogeneration of carriers strongly depends upon the applied field and the light intensity; the bandwidth in a disordered polymer is very narrow and the charge transport is most likely to occur via a hopping mechanism. To appreciate this complication, it is helpful to discuss some of the features in the photogeneration and transport of charge carriers in polymeric materials.

3 PHOTOGENERATION OF THE CHARGE CARRIERS

When a polymer absorbs a photon, an excited state is formed, which can be deactivated by several possible mechanisms [23]. It can either relax internally to the ground state within the molecule via the nonradiative or radiative decay processes or it can 'thermalize' to form a bound electron–hole pair where the electron and the hole are localized on the same or different molecules. The hole and the electron are separated by a distance of r_0, which is called the thermalization distance. The bound electron–hole pair can recombine due to Coulombic attraction (germinate recombination), or it can further dissociate into free charge carriers. There are numerous mechanisms for the photogeneration of the charge carriers from an exciton, such as spontaneous or thermal ionization of an exciton, exciton dissociation due to the interaction of excitons and trapped carriers, collisions of an exciton with a dissociation center and so on. All of these mechanisms are facilitated under the assistance of thermal motion or an external field. Therefore, the dissociation has a strong dependence upon the external electric field and temperature. Since the polymeric materials exhibit very small dielectric constants, long distance charge interaction exists, which is obviously dependent upon the light intensity. The higher the light intensity, the larger the number of generated carriers and thus the higher the possibility of charge recombination.

The quantum yield of the charge carrier formation, η, defined as the number of free electrons and holes formed per absorbed light quantum, is usually used to characterize the photogeneration processes. It can be determined from the photocurrent measurements, which are a function of the field and temperature. The relationship between the photocurrent and the quantum efficiency is given by the following equation:

$$\eta = \frac{I_{ph}}{eP[1 - \exp(-\alpha L)]} \quad (5.10)$$

where P is the power of the incident light, and α is the absorption coefficient of the material. From the photocurrent measurements, photoconductivity can also be deduced, which equals the current density divided by the applied electric field strength,

$$\sigma = J/E = e(n\mu_n + p\mu_p) \quad (5.11)$$

where J is the current density, E is the applied electric field, e is the charge of the electron, n, p are the number of electrons and holes (the charge carriers) per unit volume, and μ_n, μ_p are the mobilities of the electron and the hole, respectively.

It is usually found that the quantum yield in polymeric photoconductors is very low due to many deactivation mechanisms and it exhibits a superlinear relationship with the applied electric field. A theory developed by Onsager a long time ago is used to explain the above phenomena [24]. This theory assumed that after the formation of short-lived excited states by the absorption of a photon, a coulombically bound electron–hole pair is generated with a probability η_0 and an initial pair distance of r_0. The quantum efficiency of the pair dissociation, $\eta(E, T)$, can be expressed as

$$\eta(E, T) = \eta_0 \int p(r, \theta, E, T) g(r, \theta) d^3r \quad (5.12)$$

where η_0 is the initial yield of the thermalized bound pairs and is assumed to be independent of the electric field; $p(r, \theta, E, T)$ is the probability of dissociation of the thermalized pairs as a function of the initial separation, field and temperature; and $g(r, \theta)$ is the spatial distribution of the initial pairs. It is commonly assumed that $g(r, \theta)$ is isotropic and all pairs have the same thermalized distance r_0. Integration of equation (5.12) gives

$$\eta(r_0, E) = \eta_0 \left[1 - \left(\frac{kT}{eEr_0} \right) \sum_{g=0}^{\infty} I_g \left(\frac{e^2}{\varepsilon_0 kTr_0} \right) I_g \left(\frac{eEr_0}{kT} \right) \right] \quad (5.13)$$

where $I_g(x)$ is a recursive function which is equal to $I_{g-1}(x) - [\exp(-x)x^{g/g'}]$ and $I_0(x) = 1 - \exp(-x)$. Equation (5.13) predicts that under a low field condition $(E < 3 \times 10^4 \text{ V cm}^{-1})$, the relationship between quantum yield and the field will be linear:

$$\eta(r_0, E) = \eta_0 \left(1 + \frac{e^2 E}{2\varepsilon(kT)^2} \right) \exp\left(-\frac{r}{r_0} \right) \quad (5.14)$$

If the quantum yield is plotted against an electric field, the thermalization yield η_0, and the initial distance r_0 can be obtained from curve fitting based on equation (5.13). These relationships were indeed observed in PR polymers. For example, Figure 5.2 shows the results of the quantum yield as a function of the field for a conjugated PR polymer (CP polymer **V**) and the

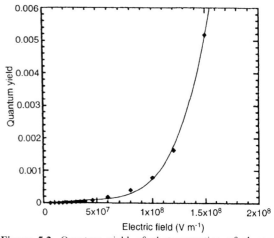

Figure 5.2. Quantum yield of photogeneration of charge carriers as a function of electric field for conjugated photorefractive CP polymer **V** (see Scheme 5.2 for its structure).

theoretical fit (solid line) based upon Onsager's model, from which r_0 and Φ_0 values of 15 Å and 0.0026 were obtained [9].

For most of the photoconductive polymers, Onsager's model gave an excellent fit. However, as pointed out by Mort and Pfister, the uncertainty about the description of the photogeneration process remains [25]. For example, before the bound electron–hole pair dissociate, an exciton migration might occur, which can not be accounted for by Onsager's model.

4 CHARGE TRANSPORTATION

Upon the formation of charge carriers (either a hole or an electron, depending on the nature of the materials), they will travel in the material in the direction of the external field. In semiconducting polymeric solids or molecularly doped polymers, the charge carriers propagate by a hopping mechanism which is largely different from the single crystal inorganic materials where a band transport prevails. Let us consider the hole transport process as in the case of the PVK system. Under the influence of an externally applied electric field, neutral molecules will repetitively transfer electrons to their neighboring cations. The net result of this process is the motion of a positive charge across the bulk of the sample film.

For a hole transport system, the charge transporting agent is an electron donor in its neutral state. For electron transport, the electrons hop from the radical anions to their neighboring neutral molecules, so the charge transporting agent is an electron acceptor in its neutral state. We should note that both hole and electron transport can occur in the same sample, depending upon the relative amount of hole and electron carriers. For example, hole transport occurs by hopping through the uncomplexed carbazole in the poly(vinyl-carbazol) (PVK): 2,4,7-Tri-nitrofluorenone (TNF) system when the PVK concentration is high [25]. Addition of TNF to PVK removes uncomplexed carbozole units, resulting in a rapid decrease in hole mobility. As a result, the charge transport processes occur through the electrons when the TNF density is high.

One parameter to describe the rate of charge transport is the drift mobility of the charge, which is defined as the velocity of charge per unit field strength:

$$\mu = L/Et_{tr} = L^2/Vt_{tr} \qquad (5.15)$$

where L is the distance between electrodes and V is the voltage applied across the electrode.

Furthermore the expression of the photocurrent can be expressed as

$$I_{ph} = e(F\mu\tau/L^2)V \qquad (5.16)$$

where 2 is the mean life time of the change carriers, F is the generation velocity of change carriers.

It can be clearly seen that the photocurrent is a linear function of the voltage and so is the quantum yield, if μ and τ do not depend on the voltage. This relationship is true when the electric field strength is low. At a higher electric field, a deviation from linearity is observed due to many factors, such as the field-assisted thermal ionization of the ion pair, called the Poole–Frenkel effect, the field dependence of the mobility and the charge carrier injection from the electrodes (see Figure 5.3). [23].

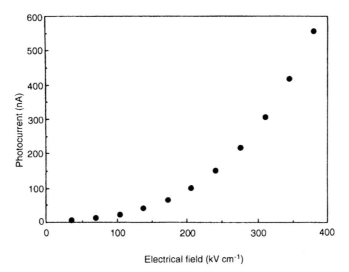

Figure 5.3. Photocurrent as a function of the electric field for CP polymer **V**.

Since the local environments of the individual transport molecules are different from each other, the electronic states of these species undergo fluctuation and the charge carriers thus experience different energetic environments at each hopping step. The charge transport in these disordered materials is called a dispersive transport. Furthermore, when the charge carriers move to a site, the surrounding lattice will be deformed to form polaronic states which gain stabilization energy. The charge transporting process or carrier mobility is thus temperature dependent. Based upon the hopping mechanism in a disordered material, a phenomenological equation accounting for the relationship between mobility and other factors was derived [25,26]

$$\mu \sim \frac{1}{E}\sinh\left[\frac{e\rho E}{2kT}\right]^{1/\alpha}\exp\left(\frac{-\Delta_0}{kT}\right) \qquad (5.17)$$

where ρ is the hopping distance (related to the concentration of the transporting species), α is the parameter characterizing the effect of the disorder on the transport process (the more disordered the system, the smaller the α value), Δ_0 is the zero-field activation energy (related to the concentration of the transporting species and the nature of the polymers), and E the electric field. This equation again points out the complicated nature of charge transport in polymer photoconductors.

There are two methods to measure the mobility of the charge carriers, namely the time-of-flight experiment and the xerographic discharge experiment. The time-of-flight experiment is performed by injecting a thin carrier sheet from the surface of the sample film with a highly absorbing wavelength of light [25]. The transient time, t_{tr}, for the sheet of charge drifting across the film under an electric field is then measured. The polarity of the applied field determines the sign of the carriers which drift across the sample.

The drifting charge is manifested as a constant current such that

$$i = q\mu E/L \qquad (5.18)$$

and the current abruptly drops to zero when the carrier sheet arrives at the other side of the electrode at a time of $t_{tr} = L/(\mu E)$ [equation 5.15]. q is the total charge injected into the sample by the incident light flash F (absorbed photons per second). In disordered solids, the current pulse often deviates from the ideal rectangular shape due to the dispersive nature of the charge transport (see Figure 5.4).

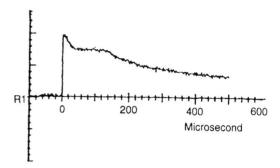

Figure 5.4. Time-of-flight photocurrent trace for CP polymer V.

The second type of technique is called xerographic discharge which has the advantage of having one surface of the sample that does not have to be electroded [25]. In this experiment, a thin film is cast onto a conducting surface which is subsequently charged with a corona surface potential V_0. After a flash of light, the change in surface potential $V(t)$ is monitored.

For a sufficiently low light intensity, the change in surface potential is small compared to V_0. The discharge rate at $t = 0$ is given by

$$(\mathrm{d}V/\mathrm{d}t)_0 = \phi eF/C \qquad (5.19)$$

where ϕ is the quantum efficiency, and C is the capacitance of the sample. It can be seen that the measurement of the initial decay, $\mathrm{d}V/\mathrm{d}t$, gives the value of the quantum efficiency. Therefore, the mobility can be deduced by using equations (5.10) and (5.16).

5   CHARGE TRAPPING

The charge trapping in polymeric materials is difficult to study mainly because of the wide range of charge trapping levels residing in the forbidden energy gap. This is manifested in a long absorption tail beyond the absorption band-edge. A variety of trapping centers can exist in a polymer, such as impurities, the structural defects. They exert a profound effect on the charge carriers mobility and hence the photorefractive performance. Their detailed structures, however, are usually unknown which clearly makes theoretical work on the photorefractive effect of polymers more difficult.

These three processes related to the photoconduction demonstrate the complication in the PR polymer systems. Further complication was introduced due to the existence of the E-O chromophores which are species with large ground state dipole moments and which have to be aligned under an external field to manifest an E-O effect. This effect is usually viewed not to interweave with the photoconduction process and is treated separately. However, it is very easy to imagine that the dipole-oriented chromophores will definitely affect the local dielectric properties of the polymers. If the NLO chromophore is utilized to play a role in photoconduction, either in photogeneration or transport, the process becomes much more complicated and it is almost formidable to treat it theoretically.

Despite these difficulties, Schildkraut *et al.* attempted to numerically simulate space charge field distribution in photoconductive electro-optic polymers [28]. In their simulation, the field dependence of the quantum yield of the photgeneration assumed a simple relationship ($\phi = E^p$, p being the constant to be determined from the experiment). An exponential relationship between the mobility and electric field was also used ($\mu = \exp[C(E^{1/2} - 1)])$. After judiciously selecting proper experimental data, they were able to reveal some of the features of the space charge field distribution. This chapter will not review the detailed work; readers who are interested are referred to the original papers [28].

6   ELECTRO-OPTIC EFFECT

The electro-optic effect, a second order nonlinear optical effect, is another required property for the photorefractive effect [1]. It is known that when a medium is subjected to an intense electric field such as an intense laser pulse, the polarization response of the material become nonlinear which is usually described by [29]

$$P = \chi^{(1)}E + \chi^{(2)}EE + \chi^{(3)}EEF + \cdots. \quad (5.20)$$

The coefficients $\chi^{(1)}$, $\chi^{(2)}$, $\chi^{(3)}$ are called first-order, second-order and third-order nonlinear susceptibilities of the medium, respectively. If we consider a single molecule instead of a bulk material, the molecular polarization can be written as

$$p = \alpha E + \beta EE + \gamma EEE + \cdots. \quad (5.21)$$

where α is the linear molecular polarizability, and β and γ are the first and second hyperpolarizability, respectively. Symmetry arguments require all of the even-ordered terms in the electric field to vanish if the molecule possesses an inversion center [29]. Similarly in the bulk material, the second order term becomes

zero if the material has an inversion center, e.g. amorphous material or crystal with a centro-symmetric structure.

In an organic polymer with a nonlinear optical chromophore, the molecular nonlinear hyperpolarizability β and the macroscopic $\chi^{(2)}$ are related by the equation [30]:

$$\chi^{(2)} = Nf\langle\beta\rangle \quad (5.22)$$

where N is the number density of the nonlinear optical chromophore in the material, f is the product of different local field factors which is a correction of the external field due to the sum of the dipole field from all of the polarizable particles, and $\langle\beta\rangle$ is the statistical average of β over all orientations of the chromophore. The orientations of the chromophores in a polymer matrix are random; therefore, the nonlinear optical response will be zero. In order to make a polymer nonlinear optically active, a dipole has to be aligned by an external field (or anchored to a substrate in an asymmetric way).

All second order NLO materials exhibit the electro-optic effect. In the presence of a dc or nearly dc field, the refractive index of a second order nonlinear material can be modified, resulting in a change that is linearly proportional to the electric field. To describe this effect, another tensor, r_{ijk}, is defined according to equation (5.23) where the ijth component of the change in the dielectric constant is given by

$$\Delta\left(\frac{1}{\varepsilon}\right)_{ij} \equiv \frac{1}{\varepsilon_{ij}} - \frac{1}{\varepsilon_{ij}(0)} \approx -\frac{\Delta\varepsilon_{ij}}{\varepsilon_{ij}^2} = r_{ijk}E_k \quad (5.23)$$

where $\varepsilon_{ij}(0)$ is the ijth component of the dielectric constant in the absence of the external dc field and the approximation is valid in the limit where the field-induced changes in the dielectric constant are small. The electro-optic coefficient can be related to the corresponding susceptibility tensor elements by

$$\chi_{ijk}^{(2)}(-\omega;\omega,0) = -(\tfrac{1}{2})\varepsilon_{ii}(\omega)\varepsilon_{jj}(\omega)r_{ijk}(-\omega;\omega,0) \quad (5.24)$$

For the material with ∞mm symmetry the independent components are r_{13}, r_{33} and r_{51}. Since $\varepsilon = n^2$, equation (5.25) implies

$$\Delta n = -\tfrac{1}{2}n^3 r_{\text{eff}}E \quad (5.25)$$

where n is the index of refraction, and r_{eff} is the effective E-O coefficient.

7 GRATING FORMATION AND BEAM COUPLING

7.1 Two-beam coupling

After the space charge field is formed, the refractive index of the materials can be modulated according to equation (5.25). As discussed in the introduction, if two beams were used, an oscillating index grating is formed, which exhibits a phase shift compared to the interference pattern formed between the two intersected beams. The index grating with a phase shift will cause beam coupling, leading to asymmetric energy exchange [1]. The process is characterized by an exponential gain factor Γ, which is related to the EO coefficient and the space charge field by [31]

$$\Gamma = p\frac{2\pi}{m\lambda}n^3 r_{\text{eff}} E_{\text{sc}} \tag{5.26}$$

where λ is the wavelength of the writing beams, p is a projection factor, r_{eff} is the effective E-O coefficient, m is the modulation depth, and E_{sc} is the space charge field. The optical gain coefficient can be experimentally determined by monitoring the optical energy exchange. The following equation is usually used to deduce the optical gain coefficient [1]

$$\Gamma = \frac{1}{L}\ln\left(\frac{1+a}{1-\beta a}\right) \tag{5.27}$$

where $a = \Delta I_s(L)/I_s(L)$, $\Delta I_s(L)$ is the intensity change of the signal beam, $I_s(L)$ is the total signal beam intensity emerging from the sample, L is the optical path length of the writing beam, and $\beta = I_s(0)/I_p(0)$ is the ratio of the intensity of the incident beams. The experimental set up for two beam coupling is shown in Figure 5.5.

7.2 Four-wave mixing experiment

In a PR material, an index grating derived from the interference of pumping beams 1 and 2 can diffract a reading beam 3, which has a much lower intensity and is propagated in the opposite direction to beam 1. This is the four-wave mixing (FWM) experiment, the configuration of which is shown in Figure 5.6. From this experiment, the efficiency of diffraction, defined as the ratio of the intensities of diffracted beam 4 to the input beam 3, can be measured. If the writing beams are both s-polarized, the diffraction efficiency of a s-polarized reading beam can be written as [32].

$$\eta_s = \sin^2\left(\frac{\pi n^3 r_{13} E_{\text{sc}} d \sin\theta_G}{2\lambda(\cos\theta_1\cos\theta_2)^{1/2}}\right) \tag{5.28}$$

and for a p-polarized reading beam

$$\eta_p = \sin^2\left(\frac{\pi n^3 r_{\text{eff}}^{\text{P}} E_{\text{sc}} d \cos 2\theta_0}{2\lambda(\cos\theta_1\cos\theta_2)^{1/2}}\right) \tag{5.29}$$

where E_{sc} is the space charge electric field, d is the sample thickness, θ_0 is the full angle between the writing beams, and θ_G is the angle between the index grating wave vector and the sample surface plane (see Figure 5.6). The electro-optic coefficient, $r_{\text{eff}}^{\text{p}}$, in equation (5.30) is given by [5]

$$r_{\text{eff}}^{\text{p}} = r_{13}[\cos\theta_1\sin(\theta_2+\theta_G) + \sin\theta_1\cos\theta_2\cos\theta_G]$$
$$+ r_{33}\sin\theta_1\sin\theta_2\sin\theta_G \tag{5.30}$$

The ratio η_p/η_s predicts the polarization anisotropy in a PR material. Even though measurement of the polarization anisotropy suggests the formation of a PR grating, care must be taken when interpreting this data. It is still possible that other grating formation mechanisms such as photochromic grating, thermal or mechanical gratings may also produce the anisotropy. It should be noted that in order to conclude that a material is photorefractive, the asymmetric optical

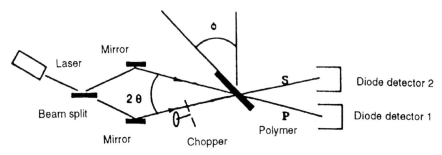

Figure 5.5. The experimental setup for two-beam coupling, where S and P are the signal and pump beams, respectively.

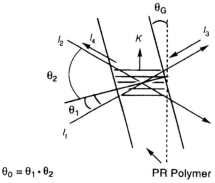

$\theta_0 = \theta_1 \cdot \theta_2$ PR Polymer

Figure 5.6. Schematic representation of four-wave mixing in a photorefractive material.

energy exchange and a non-zero degree phase shift of index grating from the two-beam coupling (2BC) experiment must be observed. Although the FWM experiment will give the data about the diffraction efficiency of the grating, the FWM results alone are not conclusive.

8 RESPONSE TIMES

Apart from the beam coupling and the diffraction efficiency, several other physical parameters are used in characterizing the PR performances of polymers, including dielectric response time, figure of merit and photorefractive sensitivity. The dielectric response time relates to the time constant to build up a grating through a space charge field and the E-O optic effect, and is defined as

$$\tau_{\text{di}} = \frac{\varepsilon \varepsilon_0}{\sigma(I_0)} \qquad (5.31)$$

where $\sigma(I_0)$ is the photoconductivity at the average intensity [1]. The expression for the time constant to build up a grating can be derived from the equations of Kukhtarev *et al.* for inorganic single crystal materials [21]. It becomes complicated for polymeric materials because $\sigma(I_0)$ not only depends on the light intensity it also depends on the electric field and the time. The treatments for inorganic materials can no longer be applied to the polymeric materials due to the reasons stated in the above sections. Generally, the response time is measured in the diffraction efficiency during erasure of the grating $\eta_e(t)$ and the diffraction efficiency during the writing of the grating $\eta_w(t)$ are given by

$$\eta_e(t) \propto [\exp -t/\tau]^2 \qquad (5.32)$$

$$\eta_w(t) \propto [1 - \exp -t/\tau]^2 \qquad (5.33)$$

where the time constant τ can be determined from the data fitting [33]. In the experiments on polymeric materials, the results can accidentally be fitted well with this model. The detailed reason is unknown.

Photorefractive sensitivity is defined as the refractive index change per unit absorbed energy and volume at the initial stage of photorefractive recording [34].

$$S = \frac{\partial n}{\partial w} = \left(\frac{n^3}{2} \frac{r_{\text{eff}}}{\varepsilon_{33}\varepsilon_0} \right) \left(\frac{\Phi}{h\nu} \right) (eL_{\text{eff}}) m \qquad (5.34)$$

where $w = \alpha W_0$ the energy absorbed by the materials, L_{eff} is the charge transport length which is a function of the materials' properties, applied field and the incident angles of the two incident beams, and Φ is the quantum yield. This expression indicates the way to improve the photorefractive sensitivity, namely, to enhance the E-O coefficient, the quantum yield for the charge photogeneration and the charge transport length. Another frequently used figure of merits is $n^3 r_{\text{eff}}/\varepsilon_{33}$, which is just a partial expression from the above relationship, where the relevant parameters are easier to obtain from experimental results.

After discussing some of these basic concepts and the photorefractive principles, the design idea for PR polymers becomes clear: two physical properties, photoconductivity and electro-optic effect, have to be brought together to manifest the photorefractive effect. The photoconductivity is required to achieve the charge separation and the electro-optic activity is needed to modify the index of refraction of the materials. This can be achieved by either the composite materials approach or by a fully synthetic approach. In the next section, we will review the general strategy of preparing these composite materials. Our major focus, however, will be on reviewing the detailed work of fully functionalized PR polymers.

9 REVIEW OF ORGANIC PHOTOREFRACTIVE MATERIALS

9.1 Organic crystalline materials

The first organic PR material reported in 1989 by Gunter *et al.* was 2-cyclooctylamino-5-nitropyridine (COANP1) single crystal doped with 7,7,8,8-tetra-cyanoquinodimethane (TCNQ2) [4]. The COANP has an asymmetric crystal structure and is second order nonlinear optically active. The doped TCNQ formed a charge transfer complex with COANP and extended the photosensitive region to a lower wavelength. A

COANP
(1)

TCNQ
(2)

MNBA
(3)

diffraction efficiency of 0.1% was found by a FWM experiment. The physical properties of this material are summarized in Table 5.1. Although it exhibited photorefractive properties, its diffraction efficiency was low and the preparation of these charge transfer complex crystals is difficult. These problems clearly limited their potential for future development.

More recently, Sutter et al. reported the photorefractive observation in a pure organic single crystal of 4'-nitrobenzylidene-3-acetamino-4-methoxyaniline MNBA3) [35]. This single crystal had a strong absorption at 515 nm (60 cm^{-1}). A maximum photorefractive gain of 2 cm^{-1} was observed under a zero electric field. The response time of the photorefractive effect was found to be in the order of 10 to 100 s for writing intensities about 100 mW cm^{-2}.

9.2 Composite photorefractive polymers

In general, there are three approaches to prepare composite PR polymers: (1) based on the NLO polymers, (2) based on the photoconducting polymers and (3) based on the inert host polymers. In the first approach, a second order NLO polymer is doped with charge sensitizers and charge transporting molecules. In the second approach, a photoconducting polymer is doped with second order NLO molecules and charge sensitizers. The third approach utilizes inert polymers as the host matrix for NLO chromophores, charge generating and transporting species.

9.2.1 Photorefractive polymers based on NLO polymers

The first PR polymer reported in 1991 was based on a second order NLO polymer (bisA-NPDA4) doped with

a hole transporting agent, diethylaminobenzaldehyde diphenylhydrazone (DEH5)[5]. Photorefractive measurements were done at 647 nm using a Kr$^+$ ion laser. From the 2BC and FWM experiments, the diffraction efficiency η, refractive index change Δn, phase shift between the intensity and index gratings $\Delta\Phi_p$, and optical gain Γ were measured to be 5×10^{-5}, 4.5×10^{-6}, $90°$ and 0.33 cm^{-1}, respectively [36]. This is a significant work in the sense that it demonstrated the possibility to prepare polymeric PR materials. However, the performance of the materials was obviously poor. The diffraction efficiency was low, which was attributed to the short optical path length and low electro-optic coefficient of the system. Since the NLO chromophore was utilized for the charge sensitizer, the photochromic effect certainly contributed to the diffraction efficiency. The polymer host, bisA-NPDA, had a glass transition temperature (T_g) of 65°C. It was a prepolymer which could undergo thermal crosslinking to increase the stability of the second order optical nonlinearity. However, phase separation was observed when this polymer was cured at high temperatures. As a result, the optical quality of the polymer film deteriorated due to the aggregation of DEH. Even at room temperature, the photoconductivity decreased by 50% in two weeks due to the decrease in the number of hopping sites, which was also due to the phase separation.

The photorefractive effect was observed in a copolymer 6 of poly(methylmethacrylate) (PMMA) and methylsulphonylazobenzene (MSAB) derivative doped with borondiketone (BDK7) and tri-p-tolylamine (TTA) as a sensitizing and charge transporting agent, respectively. The ratio of BDK:TTA:polymer was 0.3:320:710 [37]. The polymer film could be poled at 70°C without phase separation and the r_{33} value was found to be 1.28 pm V^{-1}. The quantum yield data for

Table 5.1. Photorefractive properties of COANP–TCNQ crystal

d_{33} (pm V^{-1})	r_{eff} (pm V^{-1})	σ (Ω^{-1}cm^{-1})	Δn	$\Delta\Phi_a$	$\Delta\Phi_p$
13.7	10	1.7×10^{-1}	1.8×10^{-6}	$0 \pm 20°$	$20–70 \pm 20°$

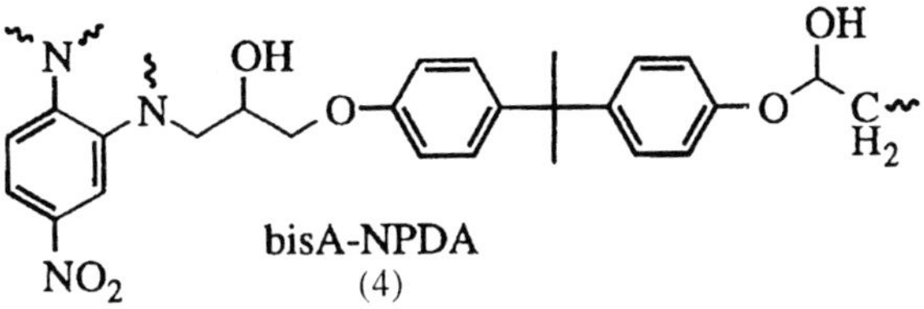

bisA-NPDA (4)

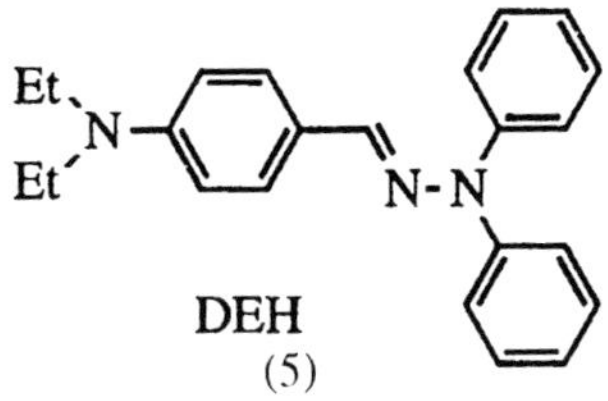

DEH (5)

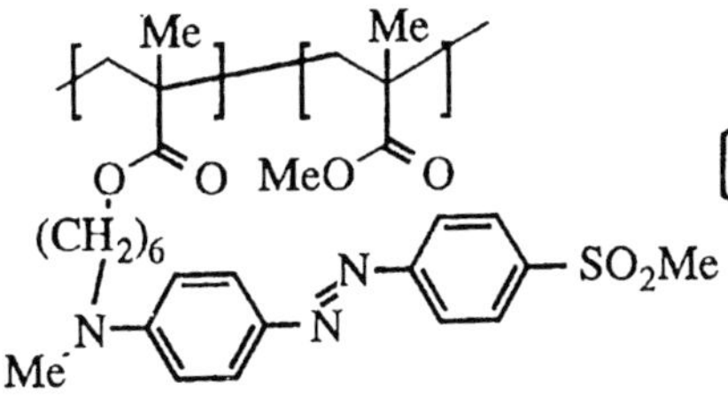

PMMA–MSAB (6)

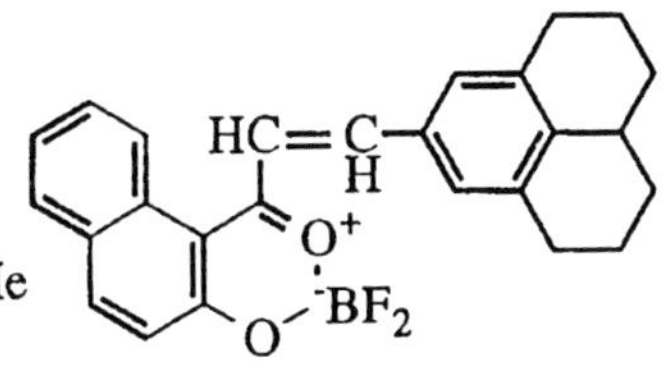

BDK (7)

the charge photogeneration was fitted to the Onsager model with good agreement. A diffraction efficiency of 3.2×10^{-5} was observed from the degenerate four-wave-mixing experiment. However, the value of the optical gain and the details about index grating were not reported. The evidence for the photorefractive index grating is thus not conclusive.

In a later report, C_{60} was used as a sensitizer in a composite PR polymer (also see **8** for the polymer's structure) [38]. The utilization of the C_{60} as the charge generator in composite PR materials was stimulated by the report that fullerenes function as a hole generator in polyvinylcarbazole systems and result in a high photoconductivity [39]. The C_{60} (0.2 wt %) was doped into a copolymer of poly(methylmethacrylate-co-p-nitrophyenylaminomethyl methacrylate) (PMMA-PNA**8**). DEH (30 wt %) was used as the charge transporting agent. Compared with the undoped sample, the enhancement in sensitization at longer

PMMA–PNA (8)

wavelength can be seen from the increase in absorption in the UV/vis spectrum.

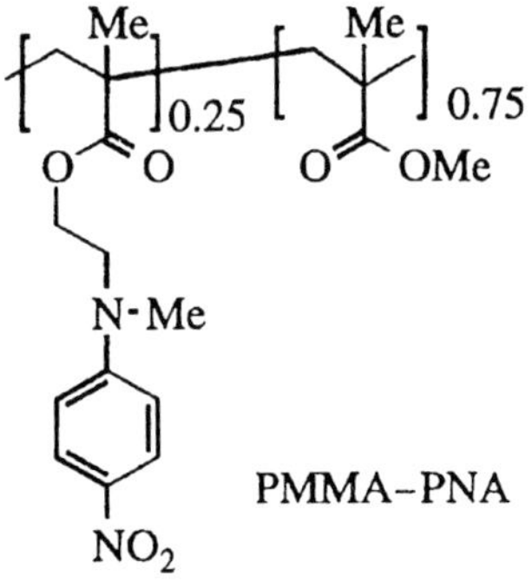

After doping with C_{60}, there was an increase in the photoconductivity, the grating growth rate and the diffraction efficiency. The η and Γ values were found to be 4.8×10^{-5} and $0.6\,cm^{-1}$, respectively. These values are very similar to those in the bisA-NPDA:DEH system. The low photorefractivity might be due to the small electro-optic coefficient, since a NLO chromophore (p-nitroaniline) with a small β value was used. Under an externally applied electric field, the $n^3 r_e$ value at 632 nm was found to be only $0.3\,pm\,V^{-1}$ and was independent of the C_{60} concentration. The sensitization process was limited by the concentration of fullerene, which usually has very low solubility in organic polymers. Further addition of more C_{60} molecules may result in the phase separation.

Another NLO polymer, bisphenol A 4,4'-nitroaminostilbene (BisA-NAS**9**), was utilized to prepare a PR composite after doping with DEH (29 wt %) [40]. The polymer also had a low glass transition temperature ($T_g = 60°C$). The stilbene dye acted as both a sensitizer and a NLO chromophore. Under an external field, the Γ value increased with the field and a value as high as $56\,cm^{-1}$ was observed under a field of $50\,V\,\mu m^{-1}$. Reversing the field polarity lead to a negative Γ value, i.e. the direction of the energy exchange was reversed. No net gain was observed at the visible region because of a large absorption coefficient, α, value. The author suggested that by extrapolation, a net gain would be obtained at 710 nm. But no follow-up experimental results were reported.

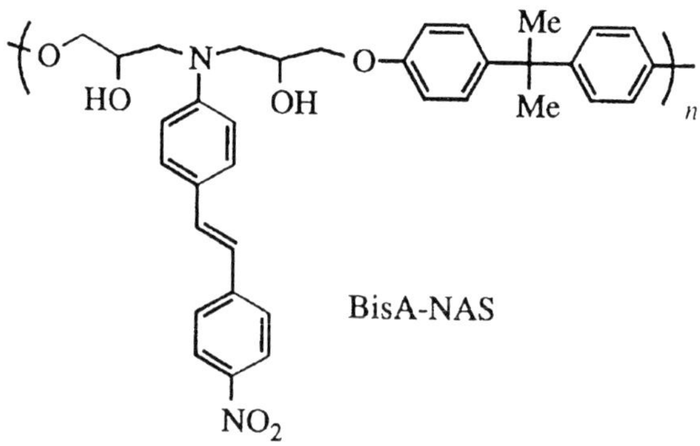

(9)

BisA-NAS

Further studies on the photoconductive properties of this polymer indicated a reduction of approximately 30–75% in photocurrent following a rapid rise after illumination by light [41]. This 'photocurrent fatigue' was attributed to the presence of two types of traps for the charge carriers although the detailed structures of the traps were not identified. It may have also been due to the *cis–trans* isomerization of the chromophore. Moreover, the photosensitivity and dark conductivity decreased steadily with the sample age. This photoconductivity decay seemed to relate to the crystallization of the DEH dopant within the polymer host after a long period of time. The powder X-ray diffraction indeed showed the evidence of crystallization. These results revealed a dilemma in designing the composite photorefractive systems. In order to maintain a high charge carrier mobility and large E-O effect, the charge carrier molecules and the NLO chromophores had to be kept at a high concentration. However, the incompatability between the guest molecules and the polymer host imposed a limit on the amount of guest molecules in the matrix.

9.2.2 Photorefractive polymers based on photoconducting polymers

Another approach to prepare PR composite materials is based on a photoconducting polymer doped with a NLO chromophore and a photosensitizer. Polyvinylcarbazole is the most popularly used host polymer. For example, Prasad *et al.* prepared the composite using polyvinylcarbazole (PVK **10**) doped with diethylamino-β-nitrostyrene (DEANST **11**) as the NLO chromophore and C_{60} as the sensitizer [42]. The concentration of C_{60} was kept very low (PVK:DEANST:C_{60} = 560:265:16) in order to avoid the strong absorption and the phase separation.

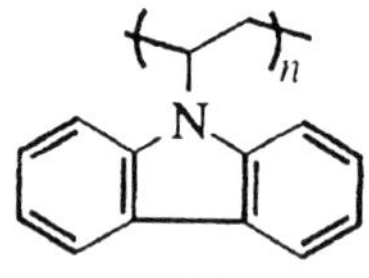

PVK
(10)

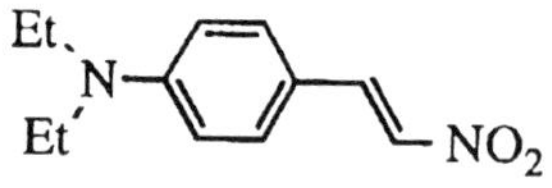

DEANST
(11)

After poling between a silver and an indium tin oxide (ITO) electrode, the E-O coefficient, r_{eff} was measured to be 4 pm V^{-1} in a sample with an unknown thickness. A maximum diffraction efficiency of 2×10^{-5} was observed from FWM experiments, which was found to be strongly field dependent. No 2BC results were reported.

Moerner *et al.* reported a similar PVK-based system by using 3-fluoro-4-*N,N*-diethylamino-β-nitrostyrene (F-DEANST, 33 wt %) as the NLO chromophore and 2,4,7-trinitrofluorenone (TNF, 1.3 wt %) as the photosensitizer [42]. The glass transition temperature of the resulting composite was low, *c.* 40°C, a dramatic decrease from that of pure PVK ($T_g = 212°C$). The TNF was used to form a charge transfer complex with PVK so that the absorption region could be extended to a longer wavelength. The photoconductivity was measured to be 2.0×10^{-12} (Ω^{-1} cm^{-1})/(W cm^{-2}). The E-O effect was characterized by using the Mach–Zehnder interferometric technique, a $n^3 r_{13}$ value of 2.4 pm V^{-1} at E = 40 V μm^{-1} and $\lambda = 830$ nm was found. 2BC experiments showed that the net optical gain (7.2 cm^{-1}) was observed at $\lambda = 753$ nm when an external field of 40 V μm^{-1} was applied (the Γ value was 8.6 cm^{-1}, while the α value was 1.4 cm^{-1}). At shorter wavelengths, the Γ values were higher but the α values also increased and no net gain was observed. This was the first example of an organic PR polymer composite exhibiting a net optical gain. FWM showed that at 647 nm, the η value was 1.2%, which was higher than that at longer wavelengths due to higher sensitivity. It was also pointed out that the NLO chromophore accounted for 10% absorption at this wavelength. Therefore, the absorption grating may have also contributed to the diffraction of the reading beam in the FWM experiment.

It was also noticed that the performance of the above polymer system is too good to be explained by the E-O photorefractive effect alone. The orientational enhancement of the photorefractive performance was invoked [44]. It was proposed that in addition to the normal E-O effect, the dipoles of the NLO chromophore will respond to the oscillating space charge field and thus introduce the birefringence both in the E-O coefficient and in the index of refraction.

The photorefractive performances of the PVK-based materials were affected by numerous factors, such as the nature of the NLO chromophores and the photosensitizer. It was found that the variation of the PVK:TNF composites by doping different NLO chromophores (different aminonitrobenzene derivatives) resulted in a wide range of optical gain values, from 1.2 to 8.0 cm^{-1} [45]. The photorefractive responses for these polymers were not satisfactory. The detailed reasoning for their poor performance was not given. However, the comparison between the chromophore structures and the photorefractive performance strongly indicated the role of the E-O molecules. A trend was that the photorefractive response of the material decreased as the size of the chromophore increased, which is attributed to increasing hindrance towards the dipole orientation caused by both the poling field and the space charge field. It is not difficult to imagine that the NLO chromophore well affect all four of the processes in the photorefractive effect. Different molecules possess different molecular and electronic structures and it becomes extremely difficult to assess the detailed roles that NLO chromophores play. More systematic work on this aspect is needed.

A different version of the PVK:TNF composite was obtained by using 2,5-dimethyl-4-(p-nitrophenylazo)-anisole (DMNPAA**12**) as the NLO chromophore [46]. It was reported that this azo dye did not exhibit any photisomerization properties.

The polymer composite showed a diffraction efficiency of 5% with a field of 40 V μm^{-1} with a film thickness of 105 μm. By using a diode laser (674 nm), a gain coefficient of $\Gamma = 30$ cm^{-1} was measured at a field of 40 V μm^{-1}, which resulted in a net gain of 6 cm^{-1} [46]. In a later report, the photorefractive properties of this polymer were further dramatically improved by adding N-ethylcarbazole (ECZ) to decrease the glass transition temperature [17]. The composition of the polymer was PVK:DMNPAA:ECZ:TNF = 33:50:16:1 wt %. It contained a very high percentage of NLO chromophore. This composite showed a very high optical gain ($\Gamma = 220$ cm^{-1} at a field of 90 V μm^{-1}), which gave a net gain of 207 cm^{-1}. It was suggested that the large photorefractivity was due to a high chromophore density and the periodic orientation of the chromophore by the internal space charge field [44]. When the polymer has a low T_g, the chromophore is more mobile. Therefore both the E-O coefficient and the birefringence of the material are modulated by the space charge field and results in a larger photorefractive effect. However, a system like this is obviously not a thermodynamically stable system. It can be viewed as the small molecules' mixture doped with PVK polymers. Phase separations due to crystallization of small molecules are unavoidable. Further studies are needed to elucidate the true mechanism in this particular composite system.

A similar strategy was applied by Prasad et al. to modify their PVK:DEANST:C$_{60}$ system [47]. The glass transition temperature of the polymer was lowered by adding 20 wt % of dibutyl phthalate (DBP) as the plasticizer to form the new composite containing DEANST (25 wt %), C$_{60}$ (0.56 wt %). Due to its low T_g, the electro-optic coefficient was dependent on the external field. At 40 V μm^{-1}, the r_{33} value was determined to be 10 pm V^{-1}. The maximum optical gain of 4 cm^{-1} (50 V μm^{-1}) was deduced. The FWM experiments indicated fast index writing and erasure processes. The rise time and the erase time of the diffracted beam signal were of the order of milliseconds, which were also dependent on the intensity of the writing/erasure beams.

DMNPAA (12)

Further depression in T_g, of the above PVK:DEANST system was achieved by using another plasticizer, tricresyl phosphate (TCP) [48]. The polymer composite consisted of PVK, DEANST (3.75 wt %), TCP (36 wt %) and C_{60} (0.22 wt %) and exhibited a T_g below $14°C$. Although the chromophore content was lower than the previous system, a higher E-O coefficient was observed ($r_{33} = 37.6$ pm V^{-1} at 140 V μm^{-1}). From the FWM and 2BC experiments, the maximum values of η and Γ were found to be 33% and 133.6 cm^{-1} (both at $E = 110\,V$ μm^{-1}), respectively. This corresponds to a net gain of 116.6 cm^{-1}. It was interesting to observe that the photorefractive properties changed tremendously just by altering the chromophore density or lowering the T_g. The exact reasons for these changes in property are not understood. One certain thing is that it must not totally be due to the space charge induced photorefractive effect. The orientation-enhanced effect may explain some of the facts. However, more contrasting experiments are needed to fully understand this phenomenon. For example, the temperature dependence of the optical gain or diffraction efficiency should be carefully studied. Samples with similar T_g but without E-O chromophores should be prepared to preform similar experiments in order to shed more light on these complicated systems.

In a composite of PVK:TNFdoped with disperse red-1 (DR1), a dual grating formation, a photorefractive and a photosensitizing grating, was observed [49]. It was suggested that under the illumination of light, the azo dye would undergo *cis–trans–cis* photoisomerization which was accompanied by reorientation until the molecule was perpendicular to the polarization of light. Therefore when two *s*-polarized laser beams are writing a grating in the polymer, the azo dye molecules in the bright region tend to align in a plane perpendicular to the polarization of light. A reading beam with an *s*-polarization would read the photoisomerization grating, while the reading beam with a *p*-polarization would read the photorefractive one. Studies of this dual grating were achieved by using two different reading beams with *s*- or *p*-polarization. It was observed that the *s*-polarized reading beam was field independent and the buildup time was much slower than that of the photorefractive grating. The *p*-polarized reading, on the other hand, showed a field dependent diffraction efficiency with a maximum η value of 1.6×10^{-5}.

Malliaras *et al.* prepared a PVK:TNF:*N*,*N*-diethyl-*p*-nitroaniline (EPNA) composite [50]. Unlike other reports, the optical gain results were obtained by applying an external field through corona discharging to a 65 μm-thick polymer film. The optical gain was dependent on the voltage applied to the corona discharging needle. The highest voltage applied was 10 kV at which a net gain of 18 cm^{-1} was demonstrated. By this corona discharge method, although a higher electric field could be applied to the polymer film surface, the exact field strength was unknown. The effect of the surface charge can not be evaluated. The situation becomes more complicated if other physico-chemical phenomena related to the corona discharging were considered, such as surface etching, electrostriction etc.

The above PVK:TNF:EPNA system was further studied by adding a different amount of DEH into the composite [51]. It was attempted to study the role of charge trap species. The experiments showed that at a very low concentration of DEH (0.1% relative to carbazole), it served as the hole trap and the grating was increased (inverse erase time constant $\tau^{-1} = 0.21$ s^{-1}). When the amount of DEH was increased, the DEH began to act as the hole transporter. The value of τ^{-1} increased to 8.0 s^{-1} when DEH was 20% relative to PVK. However, before the role of the intrinsic trapping centers (such as the impurity and defects) is fully understood, an explanation for the above observations needs to be cautious. It is unclear whether small amounts of DEH plays the role of trap center through its redox process or simply through perturbation of the polymer host lattice.

Polysilanes are other types of photoconducting polymers which have also been used as polymer hosts in the composite PR materials [52]. For example, poly(4-*n*-butoxyphenylethylsilane) (PBPES**13**) was doped with two different kinds of NLO chromophores: the laser dye Coumarin-153 (C-153**14**, 20 wt %) or (*E*)-*β*-nitro-(*Z*)-*β*-methyl-3-fluoro-4-*N*,*N*-diethyl-aminostyrene (FDEAMNST**15**, 40 wt %). Either C_{60} or TNF (0.2 wt %) was used as the sensitizer. Three different composites were prepared; the PBPES:FDEAMNST:C_{60} showed a η value of 0.1% and exhibited the largest photorefractivity of the three composites. Under an external field of 11.4 V μm^{-1}, an optical gain of 1.7 cm^{-1} was observed, which exhibited a net gain of 0.7 cm^{-1}. The charge transport studies in those polymers indicated that the PBPWS:C-153:C_{60} composite had a very high mobility (10^{-4} cm^2 (Vs)$^{-1}$ at 11.4 V μm^{-1}, which was significantly higher than other DEH-doped polymer composites. Since C-153 was not known as a charge transporter, the role of C-153 in the enhancement of charge mobility was not known.

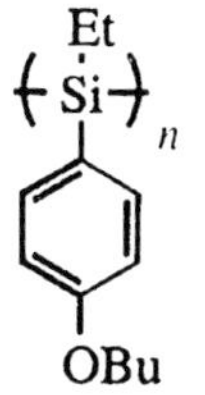

PBPES (13)

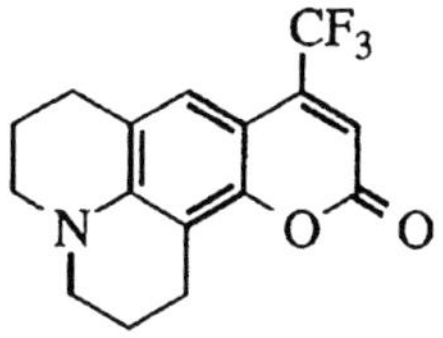

C-153 (14)

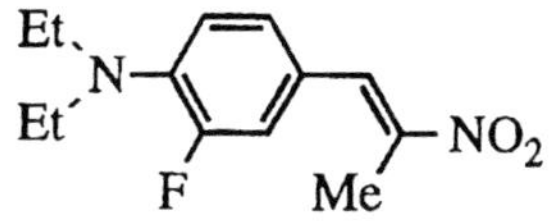

FDEAMNST (15)

9.2.3 Photorefractive polymers based on inert host polymers

The third approach to prepare the composite PR materials is to mix all of the species necessary for the photorefractive effects with an inert polymer host. Yokoyama *et al.* prepared a PMMA host composite which contained DEH (30 wt %), (S)-(− **16**, 30 wt %), and squarylium dye (SQ**17**, 0.1 wt %) as the charge transporter, NLO chromophore and sensitizer, respectively [53].

This PMMA:DEH:SQ:NPP composite, PR_{SQ}, had a very low T_g (18.2°C). The ionization potentials of each dopant were estimated from cyclic voltammetry (CV). The half-wave oxidation potential, $E_{1/2}$, of NPP (> 1 V) and SQ (0.8 V) were higher than that of DEH (0.63 V). This indicated that NPP and SQ did not act as deep traps during the charge transporting process, and DEH was the major hole transporting species. However, the photoconductivity and quantum yield of this polymer were quite low (8.2×10^{-14} S cm^{-1} and 0.1% respectively, at 633 nm and under a field of 230 kV cm^{-1}). Due to its low T_g, the 2BC experiment was performed under a high field. At 230 kV cm^{-1}, the values of η, Δn, and $\Delta\Phi$ were determined to be 4.0×10^{-6}, 7.2×10^{-6} and 90°, respectively. These results indicated that the material's performances were poor compared to other composite systems. Neither the absorption coefficient not optical gain results were reported.

The above polymer system was modifed by using TCNQ (0.2 wt %) as a sensitizer, which forms a charge transfer complex with DEH [54]. This polymer composite showed a larger photoconductivity and photorefractive response than the previous one, but no quantitative results were given. There was one intriguing result worth mentioning. A polymer composite was prepared by mixing polycarbonate (PC), NPP (30 wt %), and DEH (50 wt %). When this mixture was heated to 250°C during the film formation, the DEH molecules decomposed thermally. Surprisingly, this decomposed polymer system exhibited the largest photorefractive response of the three polymer composites. Under a field of 34.9 V μm^{-1}, the diffraction efficiency and optical gain were determined to be 1.1% and 10 cm^{-1} respectively. Although a possible reason was given to explain the above observation, this is clearly not the approach one should take to prepare PR polymers. After the material decomposed, the structural information was lost and became extremely difficult to decipher. Thus all of the physical measurements lost the structural bases to interpret and became less significant.

Another example of such a polymer composite is composed of PMMA:DTNBI (1,3-dimethyl-2,2-tetra-methylene-5-nitrobenzeimidazoline **18**):C$_{60}$ [55]. The dopant DTNBI served as both a NLO chromophore and a hole transporting reagent. It was found that the steady-state diffraction efficiency could be as high as 7% under a high electric field. The α and Γ values were determined to be 12 and 28 cm^{-1}, respectively. A net

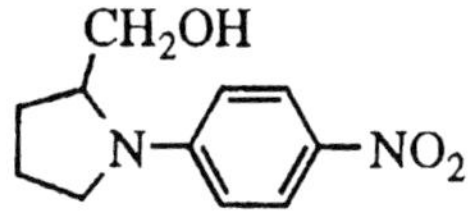

NPP (16)

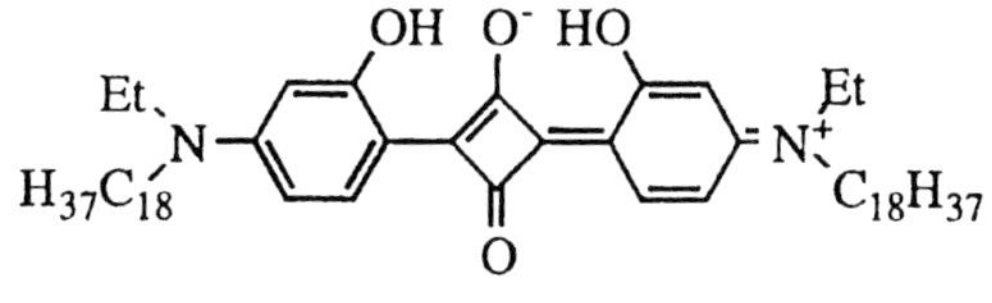

SQ (17)

gain of $16 \, \text{cm}^{-1}$ was observed. Moreover, in the FWM experiment, the intensity of the output reading beam was kept stable for a long time ($> 24 \, \text{h}$), provided that the intensity of the reading beam was low. This is an interesting phenomenon, called quasi-nondestructive readout. Phenomenon like this has been extensively studied in inorganic materials for the purpose of developing volume holographic memory. This is the first time that such an observation was made in polymeric materials. A two-trap level model was proposed to qualitatively mimic the intensity dependent decay rate and the transition to quasi-nondestructive reading.

(18)

DTNBI

It was also noticed that when the polymer film was illuminated uniformly with a fluence before writing the grating, the diffraction efficiency was enhanced [56]. This preirradiation process also increased the writing time of the grating and decreased the decay rate of η. A 2BC experiment showed that after preradiation, the optical gain coefficient Γ increased. This was attributed to the optical trap activation, in which preradiation produces deep trapping sites in the polymer.

This polymer composite was utilized to fabricate a prototype device element, a stratified volume holographic optical element (SVHOE) structure [18]. It was prepared by sandwiching two or four layers of polymer films between ITO electrodes. FWM experiments were then carried out on these SVHOE structures. It was observed that the diffraction efficiency increased with the square of the active layer thickness. The increase in diffraction efficiency was explained by the coherent addition of the light diffracted from each layer. This experiment demonstrated that it is possible to utilize PR polymers in true devices.

9.3   Fully functionalized polymers

The composite materials approach offers advantage to quickly survey different compositions. But the intrinsic instability of these materials in the photorefractive performance makes it desirable to synthesize polymers with all of the four functionalities covalently attached to

the polymer backbone. Moreover, this multifunctional polymer approach offers further opportunity to explore new structures rationally for the photorefractive effects, which is difficult in composite systems.

9.3.1   Polyurethanes [6–8,10]

The idea to synthesize fully functionalized PR polyurethanes was simple: attach four different species onto the polymer backbone. The design principle is based on the general view that in a PR material, four functional species exist simultaneously: photocharge generator, charge transporter, charge trap and second order nonlinear optical moiety. As pointed out in the introduction, the trapping centers naturally exist in the polymer sample due to defects and conformational disorders. In the first such polymer, the charge trapping centers were intentionally not incorporated; only the NLO chromophore, the charge generator and the transporting compound were covalently linked to the polyurethane backbone. There are numerous compounds which can be chosen to play these three roles. When selecting different species, an important consideration is the absorption windows of different species. The charge generator was chosen to absorb at long wavelengths (say, above $500 \, \text{nm}$) so that a normal laser light could be used to excite the species. The charge transporting compound and the NLO chromophore were selected to absorb at short wavelength. It was designed to avoid the excitation of the other species when the charge generator is excited, so that the physical properties of the charge transporter and the NLO chromophore would not change. Scheme 5.1 shows the polymers' structures.

The synthesis of polyurethanes was straightforward (see Scheme 5.1 for the polymers' compositions). After polymerization, the maximum absorption wavelength of the charge generators did not change significantly, which indicated that there were no charge transfer complexes formed between the charge generator and the various other species. The absorptions of the NLO chromophore and the transporting compound were similar to those of the corresponding monomers.

All of the polymers I–III were found to be photoconductive. A maximum photocurrent of $1.5 \, \mu\text{A}$ was observed for polymers I and II with an applied electric field of $100 \, \text{V}$ and a laser intensity of $0.4 \, \text{W}$ cm^{-2} (at $632 \, \text{nm}$). A photoconductivity of $1.3 \times 10^{-13} \, \Omega^{-1} \, \text{cm}^{-1}$ for polymer III was obtained under a field strength of $290 \, \text{kV} \, \text{cm}^{-1}$ and a laser intensity of $270 \, \text{mW} \, \text{cm}^{-2}$ (at $690 \, \text{nm}$), corresponding

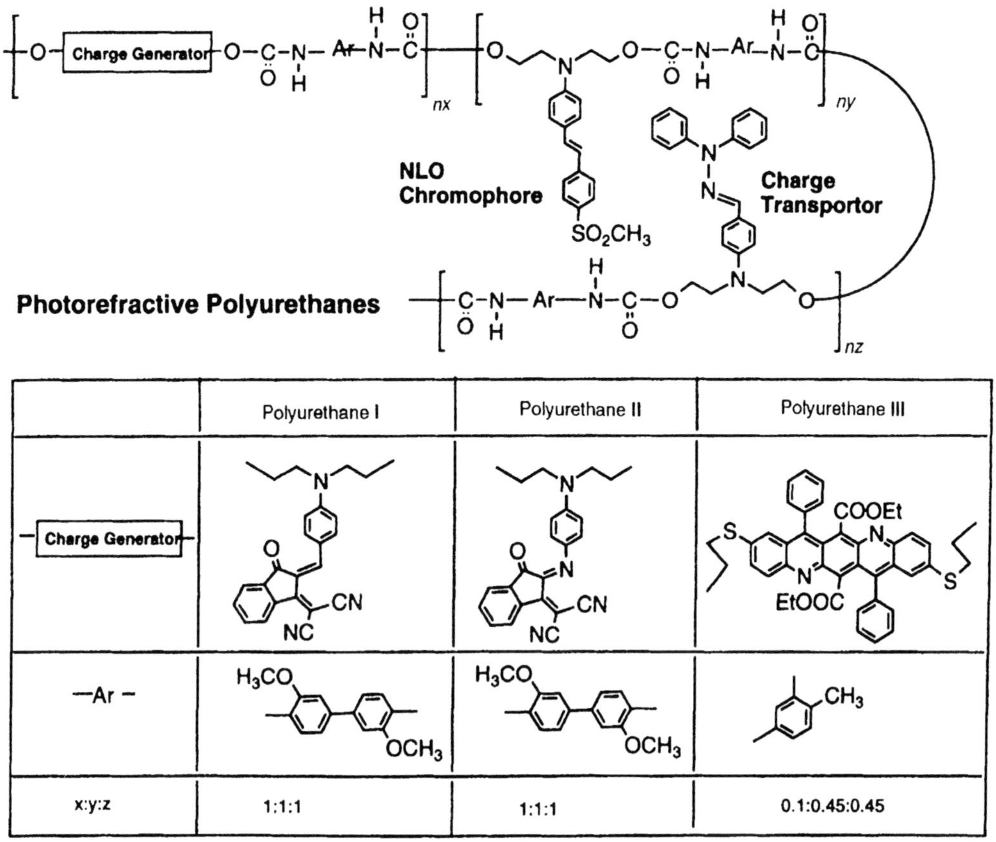

Photorefractive Polyurethanes

NLO Chromophore

Charge Transportor

	Polyurethane I	Polyurethane II	Polyurethane III
Charge Generator			
—Ar —			
x:y:z	1:1:1	1:1:1	0.1:0.45:0.45

Scheme 5.1. Synthesis and structures of photorefractive polyurethanes.

to a quantum efficiency of charge generation, c. 4.3×10^{-7}. Like other photoconductive polymers, the photocurrents of these polymers were both field and intensity dependent, i.e. as the field increased, the photoconductivity increased [8]. A linear relationship was found between the photocurrent and the incident laser intensity, indicating that bimolecular recombination of the charge carrier is absent in this polymer system (Figure 5.7).

The pristine polymer thin films which have a random dipole orientation and exhibit no electro-optic activity, were poled under an external electric field by using a corona discharging method when they were heated near their glass transition temperatures. The materials became asymmetric and the E-O coefficients, r_{33}, of 12.2 and 13.0 pm V^{-1} were detected for polymers **I** and **II**, respectively, at 632 nm. The linear E-O coefficient, r_{33}, of polymer **III**, was determined to be 4.0 pm V^{-1} (thickness 0.8 μm) at 690 nm. One of the advantages of these polymers is that their E-O

coefficients were quite stable due to hydrogen bonding. For example, a r_{33} value of 4 pm V^{-1} was detected three months after poling for polymer **I** (Figure 5.8).

The photorefractive effect was studied by two-beam coupling in a setup shown in Figure 5.5 where the polymer film was tilted by an angle of ϕ in order to have a nonzero effective E-O coefficient value, r_{33}. For example, polymer **III** was studied by using a diode laser (690 nm, s-polarized) as the light source with the sample being tilted 30° [10]. The effect of the asymmetric optical energy exchange between the two beams was clearly demonstrated (see the inset of Figure 5.9). When the two writing beams, beam S (signal beam) and beam P (pump beam), were overlapped in the sample, beam S (curve b) lost optical energy and beam P (curve c) gained optical energy. After the sample was rotated 180° around the rotation axis in the sample plate, the phenomena became reversed; beam S (curve d) gained optical energy and beam P (curve e) lost energy. Unpoled samples showed

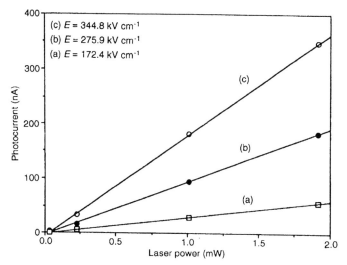

Figure 5.7. Photocurrent as a function of the light intensity of polyurethane **III**.

no such phenomena. This experimental result indicated that this polymer was indeed photorefractive.

The optical gain results were measured as a function of the grating period ($\Lambda = (\lambda/2 \sin \theta)$ in Figure 5.9. Two parameters characterizing the PR polymer, the effective density of the empty trap centers (N_e) and the maximum refractive index change (Δn), could be extracted. According to band transport theory the value of grating spacing corresponding to the maximum optical gain is equal to the Debye screening length $L_t = (4\pi^2 \varepsilon_0 \varepsilon k_B T / e^2 N_e)^{1/2}$, where k_B is Boltzman's constant [21]. Thus, from the results of Figure 5.9, an N_e

value of c. 2.3×10^{14} cm^{-3} was obtained. It is known that the maximum optical gain is related to Δn by $\Gamma_{max} = 2\pi\Delta n/\lambda \cos \theta$, where θ is the half interaction angle between two writing beams [1]. The value of Γ_{max}, 0.88 cm^{-1}, from Figure 5.9, lead to an Δn value of 1×10^{-5}.

For polymers **I** and **II**, the 2BC experiments were carried out in a waveguide structure using a HeNe laser. In this experiment, the E-O coefficient tensor component, r_{51} was utilized [1b]. The largest optical gain coefficient under a zero external field was 2.3 cm^{-1} at a grating wavenumber of 0.6×10^5 cm^{-1}.

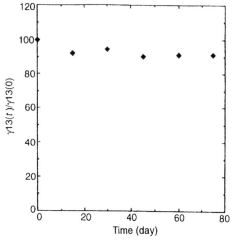

Figure 5.8. Electro-optic coefficient as a function of time at room temperature for polyurethane **I**.

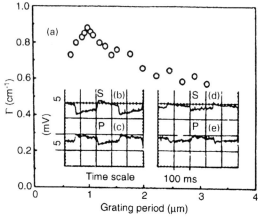

Figure 5.9. Optical gain coefficient, Γ, as a function of the grating period for polyurethane **III**. The inset is the asymmetric energy exchange signals.

It is worth mentioning that although these polymers are indeed photorefractive, the observed optical gain under a zero field condition was not a net gain because of the strong absorption.

9.3.2 Conjugated photorefractive polymers (CP polymers)

There are several common features in those functionalized polyurethanes [6–8,10]. First, a comonomer (diisocyanate in this case) was used to link the different species; therefore, the density of the different species was limited to a low level. To optimize the photorefractive effect, the densities of the charge transporter and the NLO chromophore should be optimized. In the above polymer systems, this could not be done simultaneously. Second, the polymer backbones were polyurethanes that cannot effectively transport the photocharge carriers. Third, since they are copolymers with three functional species, their molecular weights are very difficults to optimize. Any impurities or weighing errors in the monomers could have lead to the mismatch of the monomers' stoichiometry which lowered the molecular weight. These were the reasons why the second type of PR polymers, which contain a conjugated backbone and a second order NLO chromophore, were designed [9,11,13]. Figure 5.10 shows the schematic structure of this polymer.

The rationale for designing these new materials was that the conjugated backbone absorbs photons in the visible region and plays the triple role of charge generator, charge transporter and backbone. It is known that conjugated polymers have relatively high photogenerated carrier mobilities (10^{-3}–10^{-5} cm^2 V^{-1}s^{-1}) [57]. Thus, the four functions necessary to manifest the photorefractive effect exist simultaneously in a single polymer. Furthermore, the use of conjugated backbones enhances the density of the NLO chromophores.

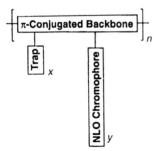

Figure 5.10. Schematic structure of the conjugated PR polymer.

However, in order to synthesize these polymers, a new polymerization approach had to be developed because these polymers contain many functional groups which may not be tolerated in many polymerization processes, such as Zigler–Natta polymerization [58], electrochemical polymerization and the oxidative coupling reaction [59]. Fortunately, the Stille coupling reaction, a palladium-catalyzed reaction between organic halides (or triflate) and organotin compounds, offers a solution to this problem [60–62]. The reaction requires very mild conditions and can tolerate different substituents in the reactant, such as amines, esters, ethers, etc. allowing one to introduce different functionalities into the polymer backbone. The structures of the synthesized conjugated PR polymers are shown in Scheme 5.2 (polymers **IV** to **VII**) where the dihydropyrrolopyrroldione (DPPD) compound is introduced as the photosensitizer to extend the photosensitive region into a longer wavelength [63]. These types of compounds have a strong absorption in the visible region.

The polymerization went smoothly while a typical Stille catalyst system was used. Either Pd(PPh$_3$)$_4$ with LiCl or Pd(PPh$_3$)$_2$Cl$_2$ alone can be used as the catalyst. Four polymers were synthesized with various compositions. These polymers were fully characterized by various spectroscopic and other techniques. The results were all consistent with the proposed polymer structures.

As the composition changed, the UV/vis spectral features of the polymers also change (Figure 5.11). It should be noted that CP polymer **VII** has an absorption at c. 390 nm, mainly due to the absorption of the NLO chromophore. CP polymers **IV** and **VI**, however, showed absorptions at 573 nm, which were dramatically shifted from the absorption of ditriflate DPPD monomer (483 nm). This illustrates clear evidence for the incorporation of the DPPD unit into the polymers. The absorption strength correlated well with the concentration of the DPPD untis (Figure 5.11). These results indicate that the absorption strength of polymers at specific regions can be controlled. This is very important for the design of PR polymers. In order to demonstrate the photorefractive effect, the materials must have certain but small absorptions at the wavelength of a working laser. Otherwise, either the charge carriers can not be generated or the materials are too absorbing. For this reason, CP polymer **V** has been studied in detail.

Photoconductivity studies of these polymers showed a strong dependence on the external field strength. At the beginning, as the electrical field was increased, the

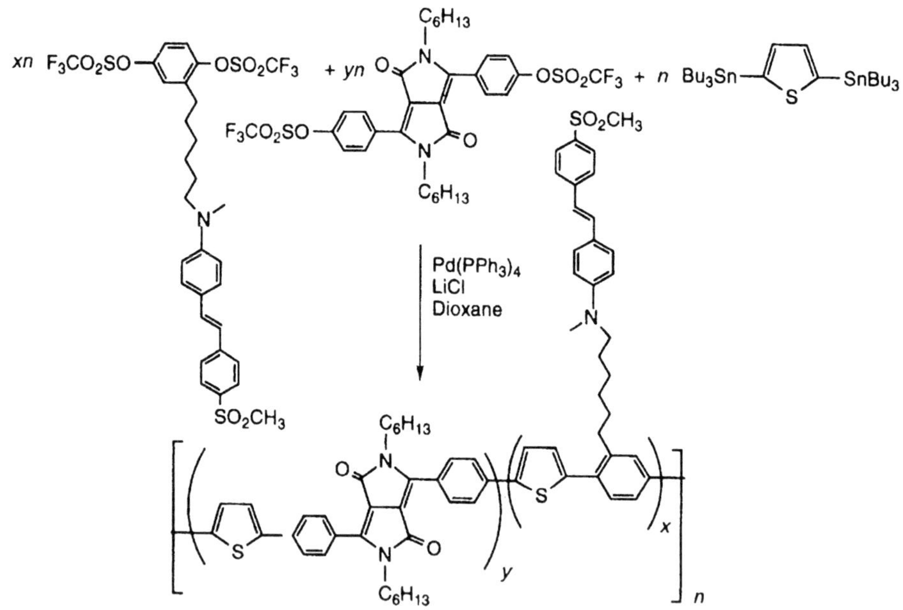

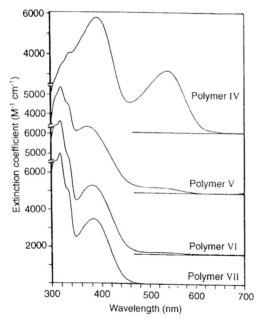

Scheme 5.2. Synthetic scheme of CP polymers **IV** to **VII** utilizing the Stille coupling reaction.

CP polymer **IV**: $y = 0.41$, $x = 0.59$
CP polymer **V**: $y = 0.5$, $x = 0.95$
CP polymer **VI**: $y = 0.1$, $x = 0.99$
CP polymer **VII**: $y = 0$, $x = 1$

Figure 5.11. UV/vis spectra of CP polymers **IV** to **VII** in THF.

photocurrent increased linearly (ohmic behavior), and then almost quadratically (Figure 5.3). A typical photocurrent response time of $c.$ 100 ms was estimated. A photoconductivity of 1.8×10^{-11} Ω^{-1} cm^{-1} was obtained for CP polymer **V** under a field strength of 1500 kV cm^{-1} and a laser intensity of 311 mW cm^{-2}. The photoconductivities for CP polymers **IV** and **VI** were found to be $c.$ 8×10^{-11} and 4×10^{-11} s cm^{-1}, respectively. These values are comparable to those of well-known conjugated polymers, such as poly(phenylene-vinylene) [57,64]. The photocurrent of CP polymer **V** was also measured as a function of several wavelengths of excitation at the same laser intensity (48 mW cm^{-2}) and electric field (400 kV cm^{-1}). It was found that the spectral dependence of the photocurrent had a similar shape to the absorption spectrum of the conjugated PR polymer. This seemed to indicate that the optical excitation of the conjugated backbone was the origin of the photocharge generation.

The quantum yield of the photogeneration of charge carriers deduced from the photocurrent results showed an electric field dependence which can be theoretically simulated by Onsager's model of the geminate-pair dissociation [25]. The results are shown in Figure 5.2 where the r_0 and η_0 values can be abstracted. The r_0 value was also known to relate to the activation energy E_A at low field conditions [25]

$$r_0 = e^2/4\pi\varepsilon_0\varepsilon E_A \qquad (5.35)$$

Temperature dependent measurements of the carrier mobility indicated an activation energy of 0.16 eV for this polymer. The r_0 value was then calculated to be 18 Å, in agreement with experimental fitting results (15 Å).

The charge carrier mobility (μ) was determined by using the time-of-flight (TOF) technique. The results showed a dispersive charge transportation as indicated by the initial decay and the tail of the transient signals. The temperature dependent measurements of the charge carrier's mobility indicated that the charge

transporting process was thermally activated with an Arrhenius activation energy of 0.16 eV and 0.23 eV for CP polymers **V** and **IV**, respectively. The field dependence of the carrier mobility was anomalous; as the field increased, the mobility decreased. Similar phenomena were observed in the poly(2-phenyl-1,4-phenylenevinylene) systems [64], and in a few composite photoconductive systems [65]. It was attributed to the random walk of the charge carrier within a random potential field [64]. The randomness of the potential field in this polymer was manifested by the broadened absorption spectrum (Figure 5.11).

2BC experiments were performed to characterize the photorefractivity of the polymers. The asymmetric optical energy exchange is shown in Figure 5.12. The resulting Γ values for CP polymer **V** are plotted as a function of the grating wavenumber

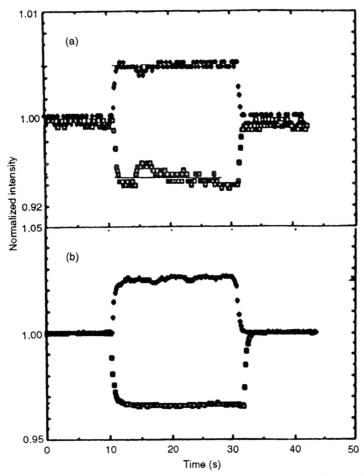

Figure 5.12. Asymmetric energy exchange for CP polymer **V** under (a) zero external field, (b) 600 kV cm^{-1} external field.

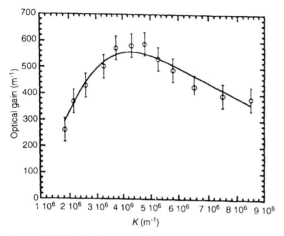

Figure 5.13. Optical gain coefficient, Γ, as a function of the grating wavenumber for CP polymer **V**.

$(K = 4\pi \sin \theta / \lambda)$ in Figure 5.13. Similar results were obtained for CP polymers **IV** and **VI**. Two parameters characterizing the CP polymer **V**; the effective density of the empty trap centers (N_e) and the maximum refractive index change (Δn), extracted from Figure 5.13, are *c.* 2×10^{14} cm^{-3} and 4×10^{-5}, respectively. For polymer **IV**, an N_e of 1.9×10^{15} cm^{-3} and a Δn of 3×10^{-5} were deduced.

The refractive index change is related to the space charge field by $\Delta n = n^3 r_{eff} E / 2$, where r_{eff} is the effective electro-optic coefficient. The large index change ($\Delta n \sim 7.1 \times 10^{-5}$) under zero field conditions implied either a large space charge field (~ 100 kV cm^{-1}) or a large electro-optical coefficient (200 pm V^{-1}). Both implications were in contrast with theoretical expectations and with the results obtained by independent measurements. According to the photorefractive theory based on the band transport model, the largest space charge field which can be achieved under a zero field condition is the thermal diffusion field $E_d = k_B TK / e$ (~ 1.37 kV cm^{-1}). The experimental value of the E-O coefficient for this polymer was about 2–4 pm V^{-1} (Electrode poling). These discrepancies implied that there were other factors responsible for such a large optical gain. A reasonable assumption is that an internal field exists which assists the charge separation and enhances the photorefractivity under a zero field. This assumption has a solid base because of the orientation of the dipoles in the nonlinear chromophore after electric poling.

Further evidence was obtained from the field dependent studies of the optical gain. It was found that the optical gain decreased when increasing the field in the region of 0–100 kV cm^{-1}, and there existed a valley near the field of 10^5 V cm^{-1} (Figure 5.14). The optical gain increased as the field further increased. The results clearly indicated that the internal field was about 10^5 V cm^{-1} in a direction opposite to the poling field.

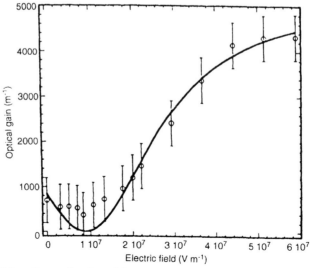

Figure 5.14. Optical gain coefficient, Γ, as a function of the external electric field for CP polymer **V**.

Further theoretical fitting was carried out based on the relationship between the optical gain and the space charge field (E_{sc}) [1]:

$$\Gamma = \frac{\pi n^3 \sin \phi r_{eff} E_{sc}}{\lambda m \cos \theta}$$

$$= \frac{\pi n^3}{\lambda \cos \theta} \sin \phi r_{eff} E_q \sqrt{\left[\frac{E_0^2 + E_d 2}{E_0^2 + (E_d + E_q)^2} \right]} \quad (5.36)$$

with

$$\phi = \tan^{-1} \left[\frac{E_d}{E_0} \left(1 + \frac{E_d}{E_q} + \frac{E_0^2}{E_d E_q} \right) \right] \quad (5.37)$$

where ϕ is the phase shift between the interference pattern and the index grating, r_{eff} is the effective E-O coefficient, $E_d = k_B T K / e$ is the diffusion field, $E_q = e N_e / \varepsilon_0 \varepsilon K$ is the limiting space charge field, E_0 is the external electric field, ε is the dielectric constant (4.85), and N_e is the effective density of the empty trap centers. By incorporating the internal field term, the E_0 term in the above equation becomes $\sin \theta (E^* + E_i)$, where θ is the angle between the grating wave vector and the sample plane (30°), E^* is an applied field (0–588 kV cm^{-1}), and E_i is an internal field. The simulation of the field dependent data according to equations (5.12) and (5.13) yielded $E_q = 1.0 \times 10^5$ V cm^{-1}, $E_i = 8.8 \times 10^4$ V cm^{-1} and $r_{eff} = 1.0$ pm V^{-1}. The extracted value of E_i from the fitting was very interesting and was in reasonable agreement with the value theoretically in other polar polymer systems [66].

It is known that there are many different types of conjugated polymers which can be explored for photorefractive studies. An obvious extension is to utilize other conjugated polymer backbones, such as poly(phenylenevinylene) (PPV) which exhibits high photoconductivity. Studies by us and other groups also indicate that PPV backbones are easy to functionalize by utilizing the Heck coupling reaction [67]. Two dibromobenzene derivatives substituted with a NLO chromophore and an alkoxy group were used as monomers. These monomers were copolymerized with p-divinylbenzene to give the resulting copolymers (Scheme 5.3). However, detailed photorefractive studies revealed that although the copolymers possessed both photoconductivity and electro-optic activity, the polymer systems exhibited no photorefractive phenomenon. This is one example which shows the complication in designing PR polymers; the existence of two necessary physical properties in a single polymer does not sufficiently ensure that the polymer is photorefractive.

9.3.3 Functionalized polyimides [12]

This photorefractive system was designed based upon the fact that porphyrin–electron acceptor (quinones or imide moieties) systems are well-known model compounds for photosynthetic processes and exhibit very interesting charge transfer properties [68]. A high quantum yield of charge separation can be achieved in these systems. Polyimides were found to be photoconductive and to allow charge transportation [69]. Furthermore, polyimides possess high glass transition temperatures and therefore the electric field-induced dipole orientation can be fixed afte imidization [70].

As shown in Scheme 5.4, polymerization was carried out in aprotic polar solvents, such as DMAC or NMP under room temperature. Spectroscopic studies sup-

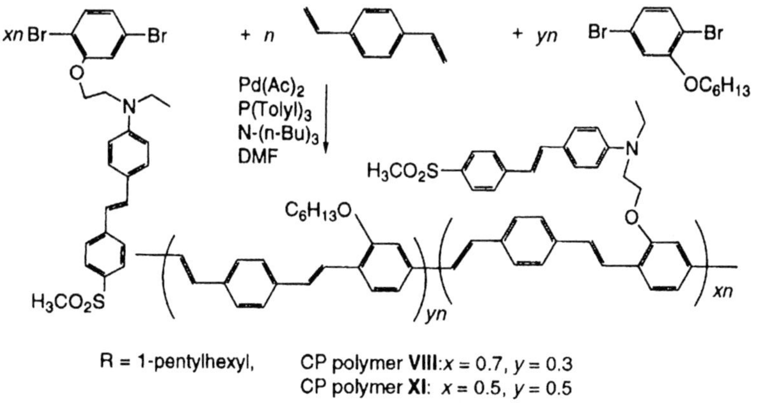

R = 1-pentylhexyl,   CP polymer **VIII**: x = 0.7, y = 0.3
                     CP polymer **XI**:   x = 0.5, y = 0.5

Scheme 5.3. Synthetic scheme of functionalized poly(phenylenevinylene) (CP polymers **VIII** and **IX**) utilizing the Heck coupling reaction.

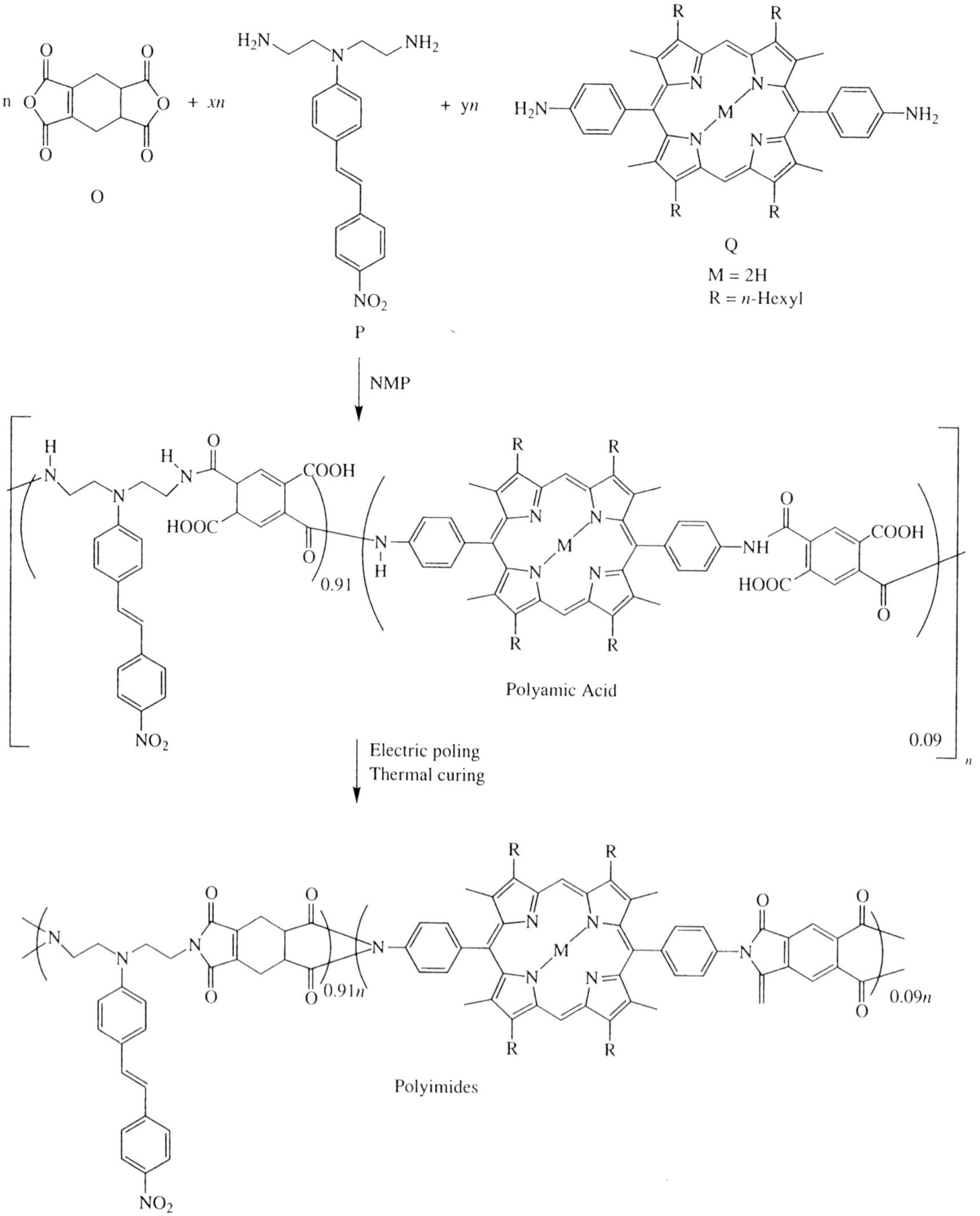

Scheme 5.4. Synthesis of photorefractive polyimide containing porphyrin moieties.

ported the structure of the polymer. The differential scanning calorimetry (DSC) studies showed that the polyamic acid has a glass transition temperature at c. 90°C and starts to imidize around 160°C; the polyimide exhibited a much higher glass transition temperature (250°C). The absorption coefficient at 690 nm was determined to be 1260 cm^{-1}.

This polyimide was photoconductive and second order nonlinear optically active. The photoconductivity of the polyimide was determined to be 1.1×10^{-12} Ω^{-1}cm^{-1} under an external field of 1500 kV cm^{-1} using a diode laser ($\lambda = 690$ nm) as the light source ($I = 5.9$ mW cm^{-2}). After the sample was poled using a corona poling technique, second harmonic generation measurements revealed a sizable d_{33} value (c. 110 pm V^{-1} at 1064 nm). This PR polyimide exhibited very high temporal stability in dipole orientation at elevated temperatures. There was no significant decay in the d_{33} value at 90°C and 150°C. Long term stability was observed even at 170°C; the initial d_{33} value of 80% was retained after 120 h (Figure 5.15).

This polyimide demonstrated interesting photorefractive effects. For instance, an asymmetric optical energy exchange in the 2BC experiment has been observed. For an unpoled sample, no such phenomenon was observed. A very large optical gain coefficient (22.2 cm^{-1}) under zero field condition was detected. This unusually large optical gain originated from the existence of an internal field in the poled polymer samples.

9.3.4 Other functionalized photorefractive polymers

Several examples of functional PR polymers have appeared in literature other than those discussed above. Peyghambarian et al. [15] prepared a copolymer of polyacrylate which contained carbazole and tricyanovinylcarbazole groups as the side chain. A plasticizer (benzylbutylphthalate) was added to lower the glass transition temperature so that the polymer film could be poled at room temperature. A small E-O coefficient (0.25 pm V^{-1}) under an external electric poling field (26 V μm^{-1}) was detected. FWM experiments revealed a small diffraction efficiency, 3×10^{-7}, with an applied field of 45 V μm^{-1}.

Another example was reported by Sansone et al. [16]. A PMMA polymer backbone was functionalized with dialkylaminonitrostilbene. The NLO chromophores in these polymers served as both the charge sensitizer and the charge transporter. FWM experiments indicated a diffraction efficiency of the order of 10^{-4}. However, since the physical studies were carried out at a wavelength of 633 nm, which is in the absorbing region of the NLO chromophore, photoinduced cis–trans isomerization of the stilbene is likely to have occurred. This photochemical transformation certainly affected the NLO properties and refractive index of the polymer. The grating observed probably originated from the photochromic effect, as pointed out correctly by these authors. In both papers, only the field

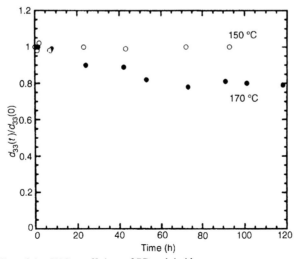

Figure 5.15. Temporal stability of the SHG coefficient of PR polyimide.

dependent diffraction efficiencies and electro-optic activities were studied. No information about the 2BC experiment was given. Therefore, the evidence is not conclusive to support the photorefractive nature of these polymers.

Recently Prasad *et al.* synthesized a PMMA derivative containing a NLO chromophore and carbazole moieties (19) [71]. This polymer exhibited a low T_g and was found to be photorefractive after a small amount of C_{60} (0.33 wt %) was doped as the photosensitizer. A E-O coefficient of $9 \, pm \, V^{-1}$ under a field of $135 \, V \, \mu m^{-1}$ was obtained. FWM experiments indicated a diffraction efficiency of 0.9% under a field of $115 \, V \, \mu m^{-1}$. 2BC studies showed an optical gain coefficient of $7 \, cm^{-1}$ under a field of $100 \, V \, \mu m^{-1}$ ($\alpha \sim 25 \, cm^{-1}$). a grating write time of $0.5 \, s$ was observed at a power level of $250 \, mW \, cm^{-2}$ and an electric field of $100 \, V \, \mu m^{-1}$.

(19)

$$x:y:z = 0.17:0.53:0.3$$

10   CONCLUDING REMARKS

Syntheses and studies on the photorefractive effects of multifunctional polymers are a new research area which is filled with surprises and challenges. The past five years has witnessed rapid progress in this area, especially in composite polymer systems [5]. Some of the characteristic parameters in these systems, such as optical gain and diffraction efficiency, have reached extraordinarily high values. However, the challenges we are facing are also overwhelming: very little knowledge has been gained on the mechanism for these PR materials. Photoconductive polymers have been studied for many decades and their mechanism for photo-

conduction is still not well understood. In a PR polymer, the introduction of electro-optic species makes the system tremendously more complicated. Further complications in composite systems arise from the fact that they suffer from phase separation and low glass transition temperatures. So far, only those composite systems with glass transition temperatures below room temperature have given rise to net optical gains. Our fully functionalized polymers exhibiting higher glass transition temperatures did not show any net optical gains although the two necessary physical properties were comparable or even better than some of these composite systems. This raises a question about the true mechanism of the photorefractivity in polymeric materials.

Theoretical works on the photorefractive effect in organic polymer systems are extremely difficult due to both their complication and the lack of data. At the present stage, the band transport theory developed for inorganic single crystals was applied in interpreting the experimental results [29]. Although this theory has worked well so far, its validation is not justified. For example, it is not proper to describe the energetic structure of the amorphous polymer according to the band theory. The charge transportation in polymeric materials is dispersive and a wide distribution of the charge carrier's mobility occurs most probably via a hopping mechanism. A wide range of trapping centers exist in polymeric materials as evidenced by a long tail in the absorption spectrum. However, despite the complication in these polymer materials, the potential application in photonic devices, the advantages of these materials over their inorganic counterparts, the challenge in synthesis, physical studies and fundamental understanding justify the pursuit of research in this area. Further developments in both synthetic chemistry and physics of conductive molecules and polymers will definitely add strength to the growth of this area.

11   REFERENCES

1. (a) P. Gunter and J.P. Huignard, (Eds.) *Photoretractive Materials and their Applications*, Vols. 1 and 2 Springer-Verlag, Berlin 1988.
   (b) A. Yariv, *Optical Electronics*, 4th edn, Harcourt Brace Jovanovich, Orlando, 1991.
2. A. Ashkin, G.D. Boyd, J.M. Dziedzic, R.G. Smith, A.A. Ballmann and K. Nassau, *Appl. Phys. Lett.* **9**, 72 (1966); F.S. Chen, *J. Appl. Phys.* **38**, 3418 (1967); F.S. Chen, J.T. LaMacchia and D.B. Fraser, *Appl. Phys. Lett.* **13**, 223 (1968);
   F.S. Chen, *J. Appl. Phys.* **40**, 3389 (1969).

3. K. Sutter, J. Hulliger and P. Gunter, *Solid State Commun.* **74**, 867 (1990); K. Sutter and P. Gunter, *J. Opt. Soc. Am.* B, **7**, 2274 (1990).

4. S. Ducharme, J.C. Scott, R.J. Tweig and W.E. Moerner, *Phys. Rev. Lett.* **66**, 1846 (1991).

5. W.E. Moerner and S.M. Silence, *Chem. Rev.* **94**, 127 (1994).

6. L.P. Yu, W.K. Chan, Z.N. Bao and S. Cao, *Polymer Preprints* **33** (2), 398 (1992).

7. L.P. Yu, W.K. Chan, Z.N. Bao and S. Cao, *J. Chem. Soc. Chem. Commun.* 1735 (1992).

8. L.P. Yu, W.K. Chan, Z.N. Bao and S. Cao, *Macromolecules* **26** 2216 (1993).

9. W.K. Chan, Y.M. Chen, Z.H. Peng and L.P. Yu *J. Am. Chem. Soc.* **115**, 11735 (1993).

10. Y.M. Chen, Z.H. Peng, W.K. Chan and L.P. Yu *Appl. Phys. Lett.* **64**, 1195 (1994)

11. L.P. Yu, Y.M. Chen, W.K. Chan and Z.H. Peng *Appl. Phys. Lett.* **64**, 2489 (1994)

12. Z.H. Peng, Z.N. Bao, Y.M. Chen and L.P. Yu *J. Am. Chem. Soc.* **116**, 6003 (1994).

13. L.P. Yu, Y.M. Chen and W.K. Chan, *J. Phys. Chem.* **99**, 2797 (1995)

14. W.K. Chan, and L.P. Yu, *Macromolecules* **28**, 2410 (1995).

15. B. Kippelen, K. Tamura, N. Peyghambarian, A.B. Padias and K.H. Hall, Jr, *J. Appl. Phys.* **74**, 3617 (1993); B. Kippelen, K. Tamura, N. Peyghambarian, A.B. Padias and K.H. Hall, Jr, *Phys. Rev. B* **48**, 10710 (1993);

16. M.J. Sansone, C.C. Teng, A.J. East and M.S. Kwiatek, *Opt. Lett.* **18**, 1400 (1993).

17. K. Meerholz, B.L. Volodln, Sandalphon, B. Kippelen, and N. Peyghambarian, *Nature* **371**, 497 (1994).

18. J.J. Stankus, S.M. Silence, W.E. Moerner and G.C. Bjorklund, *Opt. Lett.* **19**, 1480 (1994).

19. D.M. Burland, R.D. Miller and C.A. Walsh, *Chem. Rev. B* **94**, 31 (1994);

20. T.J. Marks and M.A. Ratner, *Angew. Chem. Int. Ed. Engl.* **34**, 155 (1995).

21. N.V.Kukhtarev, V.B. Markov, S.G. Odulov and M.S. Vinetskii, *Ferroelectrics* **22**, 949 (1979).

22. Ref. 1a, Chapter 3 or Ref. 1b Chapter 18.

23. K.C. Kai and W. Hwang, *Electrical Transport in Solids*, Pergamon Press, Oxford, 1981.

24. L. Onsager, *Phys. Rev.* **54**, 554 (1938). A. Mozumder, *J. Chem. Phys.* **60**, 4300 (1974).

25. J. Mort and G. Pfister, *Electronic Properties of Polymers* John Wiley and Sons, New York, 1982.

26. G. Pfister, *Phys. Rev. B* **16**, 3676 (1977); M. Silver, G. Schonherr and H. Bassler, *Phil. Mag.* **43**, 943 (1981).

27. R.G. Kepler, *Phys. Rev.* **119**, 1226 (1960).

28. J.S. Schildkraut and A.V. Buettner, *J. Appl. Phys.* **72**, 1888 (1992); J.S. Schildkraut and Y. Cui, *J. Appl. Phys.* **72**, 5055 (1992).

29. Y.R. Shen, *The Principles of Nonlinear Optics*, John Wiley Sons, New York, 1984.

30. K.D. Singer, M.G. Kuzyk and J.E. Sohn, *J. Opt. Soc. Am.* B, **4**, 968 (1987).

31. T.J. Hall, R. Jaura, L.M. Connors and P.D. Foote, *Progress Quantum Electron.* **10**, 77 (1985).

32. H. Kogelnik, *Bell Syst. Tech. J.* **48**, 2909 (1969).

33. G.C. Valley and M.B. Klein, *Opt. Eng.* **22**, 705 (1983).

34. P. Guter, *Phys. Reports* **93**, 199 (1982).

35. K. Sutter, J. Hullinger, R. Schlesser and P. Gunter, *Opt. Lett.* **18**, 778 (1993).

36. C.A. Walsh and W.E. Moerner, *J. Opt. Soc. Am. B*, **9**, 1642 (1992).

37. Y. Cui, Y. Zhang, P.N. Prasad, J.S. Schildkraut and D.J. William, *Appl. Phys. Lett.* **61**, 2132 (1992)

38. S.M. Silence, C.A. Walsh, J.C. Scott and W.E. Moerner, *Appl. Phys. Lett.* **61**, 2967 (1992)

39. Y. Wang, *Nature* **356**, 585 (1992).

40. M. Liphardt, A. Goonesekera, B.E. Jones, S. Ducharme, J.M Takacs and L.Zhang, *Science* **263**, 367 (1994).

41. B.E. Jones, S. Ducharme, M. Liphardt, A. Goonesekera, J.M Takacs, L. Zhang and R. Athalye, *J. Opt. Soc. Am. B*, **11**, 1064 (1994).

42. Y. Zhang, Y. Cui and P.N. Prasad, *Phys. Rev. B* **46**, 9900 (1992).

43. M.C.J.M. Donckers, S.M. Silence, C.A. Walsh, F. Hache, D.M. Burland, W.E. Moerner and R.J. Tweig, *Opt. Lett.* **18**, 1044 (1993).

44. W.E. Moerner, S.M. Silence, F. Hache and G.C. Bjorklund, *J. Opt. Soc. Am. B*, **11**, 320 (1994).

45. S.M. Silence, M.C.J.M. Donckers, C.A. Walsh, D.M. Burland, R.J. Tweig and W.E. Moerner, *Appl. Opt.* **33**, 2218 (1994).

46. B. Kippelen, Sansalphon, N. Peyghambarian, S.R. Lyon, A.B. Padias and H.K. Hall, *Electronics Lett.* **29**, 1873 (1993).

47. M.E. Orczyk, J. Zieba and P.N. Prasad, *J. Phys. Chem.* **98**, 8699 (1994).

48. M.E. Orczyk, B. Swedek, J. Zieba and P.N. Prasad, *J. Appl. Phys.* **76**, 4990 (1994).

49. Sansalphon, B. Kippelen, N. Peyghambarian, S.R. Lyon, A.B. Padias and H.K. Hall, *Opt. Lett.* **19**, 68 (1994).

50. G.G. Malliaras, V.V. Krasnikov, H.J. Bolink and G. Hadziloannou, *Appl. Phys. Lett.* **65**, 262 (1994).

51. G.G. Malliaras, V.V. Krasnikov, H.J. Bolink and G. Hadziloannou, *Appl. Phys. Lett.* **66**, 1038 (1995).

52. S.M. Silence, J.C. Scott, F. Hache, E.J. Ginsburg, P.K. Jenkner, R. Miller, R.J. Tweig and W.E. Moerner, *J. Opt. Soc. Am. B*, **10**, 2306 (1993).

53. K. Yokoyama, K. Arishima, T. Shimada and K. Sukegawa, *Jpn. J. Appl. Phys.* **33**, 1029 (1994).

54. K. Yokoyama, K. Arishima and K. Sukegawa, *Appl. Phys. Lett.* **65**, 132 (1994).

55. S.M. Silence, R.J. Tweig, G.C. Bjorklund and W.E. Moerner, *Phys. Rev. Lett.* **73**, 2047 (1994).

56. S.M. Silence, G.C. Bjorklund and W.E. Moerner, *Opt. Lett.* **19**, 1822 (1994).

57. T.A. Skotheim (Ed.) *Handbook of Conducting Polymers*, Marcel Dekker, Basel, 1989; T.A. Skotheim (Ed.)

Electroresponsive Molecular and Polymeric Systems, Vol. 2, Marcel Dekker, Basel, 1991.

58. G. Ordian, *Principles of Polymerization*, 2nd edn, John Wiley Sons, New York, 1981.

59. J. Roncali, *Chem. Rev.* **92**, 711 (1992).

60. J.K. Stille, *Angew. Chem. Int. Ed. Engl.* **25**, 508 (1986).

61. A.M. Echavarren and J.K. Stille, *J. Am. Chem. Soc.* **109**, 5478 (1987).

62. Z.N. Bao, W.K. Chan and L.P. Yu, *Chem. Mater.* **6**, 2 (1993).

63. T. Potrawa and H. Langhals, *Chem. Ber.* **120**, 1075 (1987).

64. M. Gailberger and H. Bassler, *Phys. Rev B* **44**, 8643 (1991).

65. P.M. Boresenberger, L. Pautmeier and H. Bassler, *J. Chem. Phys.* **94**, 5447 (1991).

66. N. Minami, K. Sasaki and K. Tsuda, *J. Appl. Phys.* **54**, 6764 (1983).

67. R.F. Heck, *Organic Reactions* **27**, 345 (1982). Examples of Heck coupling reaction utilized in polymer synthesis: W. Heitz, W. Brugging, L. Freund, M. Gailberger, A. Greiner, H. Jung, U. Kampschulte, N. Neibner and F. Osan, *Makromol. Chem.* **189**, 119 (1988); Greiner and W. Heitz, *Makromol. Chem.* **192**, 967 (1991); H.P. Weitzel and K. Mullen, *Makromol. Chem.* **191**, 2837 (1990); Z. Bao, Y. Chen, R. Cai and L. Yu, *Macromolecules* **26**, 5281 (1993).

68. M.R. Wasielewski, *Chem. Rev.* **92**, 435 (1992).

69. K. Iida, T. Nohara, S. Nakamura and G. Sawa, *Jpn. J.Appl. Phys.* **28**, 1390 (1989).

70. Z.H. Peng and L.P. Yu, *Macromolecules* **27**, 2638 (1994).

71. C.F. Zhao, C.K. Park, P.N. Prasad, Y. Zhang, S. Ghosal and R. Burzynski, *Chem. Mater.* **7**, 1237 (1995).

CHAPTER 6

Nonlinear Optical Properties of π-Conjugated Materials

Hari Singh Nalwa

Hitachi Research Laboratory, Hitachi Ltd, Ibaraki, Japan

1 INTRODUCTION

In the 1970s, studies on nonlinear optics were focused on inorganic materials which lead to the development of quartz, potassium dihydrogen phosphate, lithium niobate, and semiconductors such as cadmium sulfide, selenium, tellurium, cadmium germanium arsenide. Recent interests in conjugated organic molecular and polymeric materials were stimulated due to their intrinsic large optical nonlinearities. The strong second-harmonic generation in organic molecules originated from the fact that they possess electron donor and acceptor groups attached to a π-conjugated system. On the other hand, large third-order optical nonlinearity are observed in organic materials possessing highly delocalized π-conjugated systems. Therefore, current research activities are progressing on study of organic molecular and polymeric materials for second-order and third-order nonlinear optics, optical limiting and photorefractive effects. The importance of organic molecules and polymers stems from their promise of large nonlinear optical response, architectural flexibility, fast optical response on a subpicosecond scale, high laser damage thresholds, and ease of processing and fabrication. In the past decade, the research activities have intensified on the search for new organic molecular and polymeric materials that exhibit interesting nonlinear optical properties which can be applied in future photonic technology. Nonlinear optical materials have been recognized as the key photonic elements to future optical communication technologies where interesting photonic functions can be integrated with electrical and magnetic components. This chapter is

Handbook of Organic Conductive Molecules and Polymers: Vol. 4. Conductive Polymers: Transport, Photophysics and Applications.
Edited by H. S. Nalwa. © 1997 John Wiley & Sons Ltd

an attempt to give a systematic review of π-conjugated materials for nonlinear optics from the point of view of the chemical physicist and polymer chemist. Here the emphasis will be on structure–property relationships that provide a basic understanding on how these speciality materials behave under different nonlinear optical (NLO) processes and how to control their optical nonlinearities by using molecular design and synthetic strategies. This chapter aims to present the description of computational approaches for calculating polarizabilities, techniques of measuring hyperpolarizabilities and NLO susceptibilities of organic π-conjugated molecular and polymeric materials. In particular there is great emphasis on organic materials that have been investigated for second- and third-order nonlinear optics, electro-optics, optical limiting, and photorefractive effect and on their potential applications.

2   Nonlinear Optical Phenomena

A quantitative description of the nonlinear optical processes are well documented in several textbooks; here emphasis is on those features important to the understanding of the NLO properties of π-conjugated materials. The fundamental aspects of nonlinear optics are briefly discussed here to explain the relationship between molecular polarizabilities and chemical structures. A laser is a source of light. When the electromagnetic field of a laser beam is illuminated on an atom or a molecule, interesting optically nonlinear properties are introduced. The polarization p induced in a molecule under the applied electric field E of an incident electromagnetic wave at frequency ω can be expressed by [1–8]

$$p_i\omega_1 = \sum_j \alpha_{ij}(-\omega_1; \omega_2)E_j(\omega_2)$$
$$+ \sum_{jk} \beta_{ijk}(-\omega_1; \omega_2, \omega_3)E_j(\omega_2)E_k(\omega_3)$$
$$+ \sum_{jkl} \gamma_{ijkl}(-\omega_1; \omega_2, \omega_3, \omega_4)E_j(\omega_2)E_k(\omega_3)E_l(\omega_4)$$
$$+ \cdots \tag{6.1}$$

where $p_i(\omega_1)$ is the induced polarization in a microscopic medium at a laser frequency ω_1 along the ith molecular axis, α is the polarizability tensor, β and γ are the first hyperpolarizability and the second hyperpolarizability tensors, respectively, and E_j is the applied electric field component along the jth direction. Analogously, the macroscopic polarization induced in bulk media by a

laser beam can be expressed in a power series as

$$P_i(\omega_1) = \sum_J \chi_{IJ}^{(1)}(-\omega_1; \omega_2)E_J(\omega_2)$$
$$+ \sum_{JK} \chi_{IJK}^{(2)}(-\omega_1; \omega_2, \omega_3)E_J(\omega_2)E_K(\omega_3)$$
$$+ \sum_{JKL} \chi_{IJKL}^{(3)}(-\omega_1; \omega_2, \omega_3, \omega_4)$$
$$\times E_J(\omega_2)E_K(\omega_3)E_L(\omega_4)$$
$$+ \cdots \tag{6.2}$$

here $\chi^{(1)}$, $\chi^{(2)}$ and $\chi^{(3)}$ are the first-, second-, and third-order nonlinear optical susceptibilities, respectively. Therefore, $\chi^{(n)}$ is the $(n+1)$ rank tensor associated with the nonlinear optical response of a medium. Both the field E and polarization P are vectors while the nonlinear coefficients are tensors which depend on the frequency and are generally written as functions of the frequency of the electromagnetic field. The even-order tensors vanish in a centrosymmetric medium while the odd-order tensors have no symmetry requirements. For example, the second-order NLO effects appear only in the noncentrosymmetric media whereas the third-order NLO processes can be observed in any media, such as solids, liquids, and gases. Third-order NLO susceptibility is displayed by all crystal classes unlike second-harmonic generation which is only exhibited by the noncentrosymmetric materials. The macroscopic optical nonlinearities are correlated to the corresponding microscopic terms α, β, and γ by the following expressions [2]:

$$\chi_{IJ}^{(1)} = N\alpha_{ij}f_{Ii}f_{Ji} \tag{6.3}$$
$$\chi_{IJK}^{(2)} = N\beta_{ijk}f_{Ii}f_{Ji}f_{Kk} \tag{6.4}$$
$$\chi_{IJKL}^{(3)} = N\gamma_{ijkl}f_{Ii}f_{Ji}f_{Kk}f_{Li} \tag{6.5}$$

where N is the molecular number density and f is the local field factors due to intermolecular interactions. By knowing the magnitude of microscopic counterparts, a general trend of their corresponding macroscopic NLO coefficients can be estimated. The orientationally averaged first and second hyperpolarizabilities can be written as

$$\beta = \beta_{xxx} + \beta_{yyy} + \beta_{zzz} \tag{6.6}$$
$$\langle \gamma \rangle = \tfrac{1}{5}(\gamma_{xxxx} + \gamma_{yyyy} + \gamma_{zzzz} + 2\gamma_{xxyy} + 2\gamma_{xxzz} + 2\gamma_{yyzz}) \tag{6.7}$$

The individual tensor components for second-harmonic generation (SHG) and third-harmonic generation (THG) processes discussed here follow the work of Bloembergen [9,10] and Ward [11]. The quantum mechanical formula derived by time-dependent perturbation theory can be expressed as follows:

$$\beta_{ijk}(-2\omega; \omega, \omega) =$$

$$-\frac{e^3}{8\hbar^2}\left[\sum_{\substack{n\neq g \\ n'\neq n}}\sum_{\substack{n'\neq g}}\left\{\begin{array}{l}(r^j_{gn'}r^i_{n'n}r^k_{gn} + r^k_{gn'}r^i_{n'n}r^j_{gn})\left(\dfrac{1}{(\omega_{n'g}-\omega)(\omega_{ng}+\omega)} + \dfrac{1}{(\omega_{n'g}+\omega)(\omega_{ng}-\omega)}\right) \\[2ex] + (r^i_{gn'}r^j_{n'n}r^k_{gn} + r^i_{gn'}r^k_{n'n}r^j_{gn})\left(\dfrac{1}{(\omega_{n'g}+2\omega)(\omega_{ng}+\omega)} + \dfrac{1}{(\omega_{n'g}-2\omega)(\omega_{ng}-\omega)}\right) \\[2ex] + (r^j_{gn'}r^k_{n'n}r^i_{gn} + r^k_{gn'}r^j_{n'n}r^i_{gn})\left(\dfrac{1}{(\omega_{n'g}-\omega)(\omega_{ng}-2\omega)} + \dfrac{1}{(\omega_{n'g}+\omega)(\omega_{ng}+2\omega)}\right)\end{array}\right\}\right.$$
$$\left. + 4\sum_{n\neq g}\left\{[r^j_{gn}r^k_{gn}\Delta r^i_n(\omega^2_{ng}-4\omega^2) + r^i_{gn}(r^k_{gn}\Delta r^j_n + r^j_{gn}\Delta r^k_n)(\omega^2_{ng}+2\omega^2)]\times\frac{1}{(\omega^2_{ng}-\omega^2)(\omega^2_{ng}-4\omega^2)}\right\}\right]$$

$$(6.8)$$

where r^i_{gn} and $r^j_{n'n}$ are matrix elements of the ith components of the dipole operator for the molecule between the unperturbed ground and excited states and between two excited states, respectively, ω is the incident laser frequency, $\hbar\omega_{ng}$ corresponds to the excitation energy from ground state ($|g\rangle$) to excited state ($|n\rangle$). The individual tensor components for third-harmonic generation can be expressed using perturbation theory [9–12]:

$$\gamma_{ijkl}(-3\omega; \omega, \omega, \omega) = k_{THG}\frac{P_{jkl}}{6}\left(\frac{e^4}{8\hbar^3}\right)$$

$$\times\sum_{n,n'm}\left[\begin{array}{l}r^i_{gn'}r^j_{n'm}r^k_{mn}r^l_{ng}\left(\dfrac{1}{(\omega_{n'g}+3\omega)(\omega_{mg}+2\omega)(\omega_{ng}+\omega)} + \dfrac{1}{(\omega_{n'g}-3\omega)(\omega_{mg}-2\omega)(\omega_{ng}-\omega)}\right) \\[2ex] + r^j_{gn'}r^k_{n'm}r^l_{mn}r^i_{ng}\left(\dfrac{1}{(\omega_{n'g}-\omega)(\omega_{mg}-2\omega)(\omega_{ng}-3\omega)} + \dfrac{1}{(\omega_{n'g}+\omega)(\omega_{mg}+2\omega)(\omega_{ng}+3\omega)}\right) \\[2ex] + r^l_{gn'}r^i_{n'm}r^j_{mn}r^k_{ng}\left(\dfrac{1}{(\omega_{n'g}-\omega)(\omega_{mg}+2\omega)(\omega_{ng}+\omega)} + \dfrac{1}{(\omega_{n'g}+\omega)(\omega_{mg}-2\omega)(\omega_{ng}-\omega)}\right) \\[2ex] + r^k_{gn'}r^l_{n'm}r^i_{mn}r^j_{ng}\left(\dfrac{1}{(\omega_{n'g}-\omega)(\omega_{mg}-2\omega)(\omega_{ng}+\omega)} + \dfrac{1}{(\omega_{n'g}+\omega)(\omega_{mg}+2\omega)(\omega_{ng}-\omega)}\right)\end{array}\right]$$

$$(6.9)$$

Here all other definitions are similar to the SHG process except that $k_{THG}=4$, the degeneracy factor for THG. $\hbar\omega_{mg}$ refers to the excitation energy from the ground state ($|g\rangle$) to excited singlet state ($|m\rangle$). The nonlinear optical processes associated with second- and third-order nonlinear optical susceptibilities are shown in

No.	Input Frequency	Output Frequency	Susceptibility $\chi^{(n)}$	Operating Optical Processes
1.	ω	ω	$\chi^{(1)}(-\omega; \omega)$	Linear absorption
2.	ω	2ω	$\chi^{(2)}(-2\omega; \omega, \omega)$	Second harmonic generation
3.	$\omega, 0$	ω	$\chi^{(2)}(-\omega; 0, \omega)$	Linear electro-optical effects (Pockels effect)
4.	ω	0	$\chi^{(2)}(0; -\omega, \omega)$	Inverse linear electro-optical effects (Optical rectification)
5.	ω_1, ω_2	ω_3	$\chi^{(2)}(-\omega_3; \omega_2, \omega_1)$	Sum frequency generation
6.	ω	3ω	$\chi^{(3)}(-3\omega; \omega, \omega, \omega)$	Third harmonic generation
7.	$\omega_1, \omega_2, \omega_3$	ω_4	$\chi^{(3)}(-\omega_4; \omega_3, \omega_2, \omega_1)$	Four-wave sum frequency mixing
8.	ω_1, ω_2	ω_3, ω_4	$\chi^{(3)}(-\omega_3; \omega_4, \omega_1, \omega_2)$	Four-wave difference frequency mixing (four-wave parametric mixing, amplification, oscillation)
9.	ω_1, ω_2	ω_1, ω_2	$\chi^{(3)}(-\omega_2; \omega_1, \omega_1, \omega_2)$	Nondegenerate optical Kerr effect
10.	ω	ω	$\chi^{(3)}(-\omega; \omega, \omega, -\omega)$	Degenerate four-wave mixing, optical field induced birefringence, self-focusing, optical Kerr effect
11.	ω	ω	$\chi^{(3)}(-\omega; \omega, \omega, -\omega)$	Degenerate two-photon absorption
12.	ω	ω	$\chi^{(3)}(-\omega; \omega, \omega, -\omega)$	Absorption saturation
13.	$\omega, 0$	ω	$\chi^{(3)}(-\omega; 0, 0, \omega)$	dc Kerr effect
14.	$\omega, 0$	2ω	$\chi^{(3)}(-2\omega; 0, \omega, \omega)$	Electric field induced second harmonic generation
15.	ω_1	ω_2	$\chi^{(3)}(-\omega_2; -\omega_1, \omega_1, \omega_2)$	Spontaneous Raman scattering
16.	ω_1	ω_2	$\chi^{(5)}(-\omega_2; \omega_1, \omega_1, \omega_2, -\omega_1, \omega_2)$	Hyper-Raman scattering

Figure 6.1. Linear and nonlinear optical processes related to the input and output frequencies for measuring nonlinear optical susceptibilities.

Figure 6.1. Optical phenomena, such as SHG, linear electro-optic (EO) or Pockel's effect, sum frequency generation, optical rectification, etc. arise from $\chi^{(2)}$. For SHG, the NLO coefficient d_{ijk} is related to second-order NLO susceptibility $\chi^{(2)}$ and now the polarization can be expressed by:

$$P_i^{2\omega} = d_{ijk}(-2\omega; \omega, \omega)E_j(\omega)E_k(\omega) \quad (6.10)$$

By interchanging j and k in the above equation, one can replace the subscripts kj and jk by the contracted indices [13]. The contracted d_{ij} tensor obeys the symmetry restrictions as does the piezoelectric tensor; hence they are reduced to only a few independent elements but have the same form in a given point-group symmetry.

Third-harmonic generation (THG), degenerate four-wave mixing (DFWM), self-focusing, optical Kerr effect, electric field-induced second-harmonic generation (EFISH), four-wave sum frequency mixing, four-wave difference mixing (four-wave parametric mixing, oscillation), nondegenerate optical Kerr effect, dc Kerr effect, spontaneous Raman scattering, degenerate two-photon absorption and absorption saturation arise from $\chi^{(3)}$.

The components of the third-order optical susceptibility are related to the third-order nonlinear optical coefficient (C) as follows [14]:

$$\chi_{jklm}^{(3)}(-3\omega; \omega, \omega, \omega) = 4C_{jklm}(-3\omega; \omega, \omega, \omega) \quad (6.11)$$

Both incident field and polarization are vector quantities and $\chi^{(n)}$ is a $(n+1)$ rank tensor.

The development of the material and device design is an interdisciplinary approach. A systematic approach in this modern computer age involves several step-by-step strategies; the definition of the requirements of a specialty material, the design of the products by molecular modelling, the establishment of the structure–property relationship, molecular engineering to optimize the desired functions, and the conversion of the product to a device format. The systematic approach of the material's development and design process can be seen through the various chapters in this book. The fields of electronics and nonlinear optics are examples of truly interdisciplinary areas of research where expertise in molecular modelling, chemical syntheses, molecular engineering, quantitative measurements, materials processing and device fabrication are needed.

The basic concepts that describe electronic and photonic properties of organic materials are rather different than those of inorganic semiconductors and metals. Of particular interest are piezoelectrics and π-conjugated materials that share common goals in the fields of nonlinear optics and electronics. Tremendous opportunities exist in studying the electrical and nonlinear properties of organic, organometallic and polymeric materials. The discovery of new high performance electronic materials should direct us to new technological innovations. Since there exists a wide range of electronic and photonic processes, there are numerous fields of possible technological applications.

3 SECOND- AND THIRD-ORDER NONLINEAR OPTICAL PROPERTIES OF π-CONJUGATED MATERIALS

Organic NLO materials are emerging at an alarming rate because of rapidly broadened research activities in this field. The most important advantages of organic materials are that they offer tremendous tailoring possibilities yielding a wide variety of chemical species possessing desired electronic and photonic functions. The organic molecular and polymeric NLO materials that have been investigated, can be categorized into single crystals, Langmuir–Blodgett (LB) films, polar polymers, guest–host systems, NLO-chromophore functionalized polymers, self-assembled systems, liquid crystals and biomaterials [15]. Another important class of NLO materials is organometallic compounds, where metal-to-ligand or ligand-to-metal charge transfer substantially contributes in enhancing optical nonlinearities. The interactions between the metal d orbitals and the π-electron orbitals of the conjugated ligands foster fine-tuning of optical nonlinearity through the maneuvering of metal ions with diverse oxidation states and ligand environments [16]. Conjugated molecules, well-defined oligomers and polymers have been recognized as playing a vital role in developing structure–property relationships and such study is a central theme of this chapter. Therefore, this review is necessarily selective describing the nonlinear optical properties of π-conjugated materials. As a framework, nonlinear optical phenomena and their associated measurements techniques are briefly discussed. In particular, discussions will be focused on second- and third-order NLO properties of fullerenes, phthalocyanines, porphyrins and π-conjugated polymers. The optical limiting

properties of fullerenes, phthalocyanines, porphyrins are also described.

Attempts have been made to establish relationships between the chemical structures and various nonlinear optical processes by evaluating the current trends in the design and synthesis of new organic molecular and polymeric materials through computational and experimental probes. Important inorganic materials and organic molecules are discussed briefly in parallel to the polymers to evaluate their electro-optic, photorefractive, optical limiting, second- and third-order nonlinear optical properties. The state-of-the-art of nonlinear optical π-conjugated materials in view of their potential advantages in integrated optic devices is critically evaluated.

3.1 Second-order nonlinear optical effects

The powder technique of Kurtz and Perry [17] is one of the most convenient, frequently used method for rapidly screening the second-order NLO activity of powdered materials. This also facilitates the categorization of NLO materials into three different classes: (1) SHG active phase-matchable; (2) SHG active nonphase-matchable; and (3) SHG inactive centrosymmetric. The phase-matching materials can be identified from the particle size dependence. The SHG intensity of a nonphase-matchable material is expressed by the following approximation:

$$I(2\omega) \propto d^2 l_c^2 / 2\langle r \rangle$$

$$(\langle r \rangle > l_c; \text{nonphase-matchable})$$

The SHG intensity of a nonphase-matchable material decreases with increasing averaging particle size where $\langle r \rangle$ is larger than the coherence length l_c. On the other hand, the SHG intensity of a phase-matchable materials does not decrease when $\langle r \rangle$ is larger than l_c. The second-harmonic power $P(2\omega)$ generated by a single-mode Gaussian beam of power $P(\omega)$ incident on a plane parallel slab of a NLO crystal is given by [18]:

$$P(2\omega) = \frac{128\pi^2\omega^2 L^2 P(\omega)^2 d_{ijk}^2}{c^3 w_0^2 n_{(\omega)}^2 n_{(2\omega)}} \frac{\sin^2(\frac{1}{2}L\Delta k)}{(\frac{1}{2}L\Delta k)^2} \quad \text{(in CGS units)}$$

$$(6.12)$$

where ω is the angular frequency of the fundamental wave, w_0 is the spot radius of the fundamental beam, d_{ijk} is the NLO coefficient, $n_{(\omega)}$ and $n_{(2\omega)}$ are the refractive indices of the material at the fundamental and the SH wavelengths respectively, c is the velocity of light in a vacuum, $\Delta k = k_{(2\omega)} - 2k_{(\omega)} = 2\omega/c(n_{(2\omega)} - n_{(\omega)})$, the phase mismatch between wave-vectors at the harmonic and fundamental wavelengths, and L is the optical path length of material. The above expression indicates that the SH power undergoes periodic oscillation as a function of thickness and is called Maker-fringe when Δk is not zero. This period is called a coherence length (l_c) and is important in generating the SH power. The experimental setup for the Maker-fringe method is shown in Figure 6.2. The d^2_{eff}/n^3 is called the figure of merit for SHG.

Nonlinear optical properties such as second-harmonic generation and electro-optic (EO) Pockel's effect are closely associated with the chemical and crystalline structures of the polymeric materials. Of the noncentrosymmetric classes, 20 exhibit piezoelectric behavior; hence they are inherently active to the nonlinear optical SHG and EO Pockel's effects. In molecules where the dipoles are aligned in such a fashion that self-cancellation does not occur, then spontaneous polarization appears because of its unique chemical and morphological structures. Polarization can also be induced by applying a poling technique, though that polarization is electrically quite different from spontaneous polarization. Based on the dielectric properties of a poled polymer system, both SHG and EO Pockel's effects are also influenced by its elastic behavior because these NLO parameters increase rapidly near the glass transition temperature (T_g). It is now possible

to generate SHG and EO effects by electrical poling in organic dyes dispersed in a polymer matrix. Therefore it is rather important to study the effect of the poling field and the temperature on the SHG and EO properties of polymers and optimize them, as they strongly depend on poling conditions.

This chapter deals with poled polymers for SHG activity and electro-optic effects. Poled polymers have a mm2 point-group symmetry; the d_{ij} tensor can be written as:

$$\begin{pmatrix} 0 & 0 & 0 & 0 & d_{15} & 0 \\ 0 & 0 & 0 & d_{24} & 0 & 0 \\ d_{31} & d_{32} & d_{33} & 0 & 0 & 0 \end{pmatrix} \qquad (6.13)$$

There are five nonzero coefficients in a poled polymer system, three of which are independent [19]. According to Kleinman symmetry [20], these tensor components are interrelated by $d_{31} = d_{32}$; $d_{24} = d_{15}$. Now considering $\omega = 2\omega$ and $\omega_1 = \omega_2 = \omega$; the nonlinear polarization is given by:

$$P_{NL} = \begin{pmatrix} 2d_{15}(-2\omega; \omega, \omega)E_1(\omega)E_2(\omega) \\ 2d_{15}(-2\omega; \omega, \omega)E_2(\omega)E_3(\omega) \\ d_{31}(-2\omega; \omega, \omega)E_1^2(\omega) \\ \quad +d_{31}(-2\omega; \omega, \omega)E_2^2(\omega) \\ \quad +d_{33}(-2\omega; \omega, \omega)E_3^2(\omega) \end{pmatrix} \qquad (6.14)$$

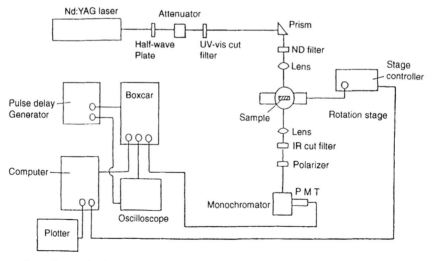

Figure 6.2. An experimental setup for SHG measurement by the Maker-fringe method. (Reprinted from ref. 73, by courtesy of Marcel Dekker Inc. 1995)

where d_{15}, d_{31} and d_{33} refer to the second-order NLO coefficient and $E = (E_1, E_2$ and $E_3)$ refer to the optical field at a fundamental frequency ω. Far away from the resonance, Kleinman symmetry applies, so that $d_{15} = d_{31}$; therefore there are only two independent components for a poled polymer system. The relationship between hyperpolarizability β and NLO coefficients can be given as [21]:

$$d_{33}(-2\omega; \omega, \omega) = \tfrac{1}{2} N f(\omega) f(\omega) f(2\omega) \beta_{zzz}(-2\omega; \omega, \omega)$$
$$\times \langle \cos^3 \theta \rangle \qquad (6.15)$$

where N is the number density of NLO chromophores, β_{zzz} the dominant hyperpolarizability component in the z molecular axis, θ is the angle between the permanent dipole moment and electric poling field, f is the corresponding local field factors, and the orientational average $\langle \cos^3 \theta \rangle$ is given as:

$$\langle \cos^3 \theta \rangle = \frac{\int_0^\pi \sin \theta \cos^3 \theta F(\theta, E_p) d\theta}{\int_0^\pi \sin \theta F(\theta, E_p) d\theta} \qquad (6.16)$$

With Boltzmann's law, the $d_{33}(-2\omega; \omega, \omega)$ can be now written as:

$$d_{33}(-2\omega; \omega, \omega) = \tfrac{1}{2} N f(\omega) f(\omega) f(2\omega) \beta_{zzz}(-2\omega; \omega, \omega)$$
$$\times \frac{\mu_z E_p}{5kT} \varepsilon \frac{(n^2 + 2)}{(n^2 + 2\varepsilon)} \qquad (6.17)$$

where

$$\langle \cos^3 \theta \rangle = \frac{\mu_z E_p}{5kT} \varepsilon \frac{(n^2 + 2)}{(n^2 + 2\varepsilon)} \qquad (6.18)$$

ε is the dielectric constant of the medium, n is the refractive index at ω, μ_z is the dipole moment of the NLO chromophore along the dominant axis, E_p is the poling field, T is the poling temperature, k is Boltzmann's constant, and the local field factor $f(\omega)$ is Onsager's type [22]. The local field factors for optical fields by Lorentz–Lorenz type relations is given by [23]:

$$f(\omega) = \frac{(n^2 + 2)}{3} \qquad (6.19)$$

The second-order NLO d_{ijk} coefficient is related to the second-order nonlinear optical susceptibility $\chi^{(2)}$ as follows:

$$d_{ijk}(-2\omega; \omega, \omega) = \tfrac{1}{2} \chi^{(2)}_{ijk}(-2\omega; \omega, \omega) \qquad (6.20)$$

The change of refractive index caused by the Pockel's effect is directly associated with the second-order nonlinear optical susceptibility $\chi^{(2)}$. The linear electro-optical coefficient r_{ijk} of a poled polymer and the second-order NLO susceptibility $\chi^{(2)}$ are related through the following equation [24]:

$$r_{ijk}(-\omega; \omega, 0) = -(2/n^4)\chi^{(2)}_{ijk}(-\omega; \omega, 0) \qquad (6.21)$$

here n is the refractive index of the NLO material. Another relationship between the electro-optic coefficient (r) and the NLO coefficient (d) can be written through the two-level model [25]:

$$r_{ijk}(-\omega; \omega, 0) = -\frac{4d_{kij}}{n_i^2(\omega)n_j^2(\omega)} \frac{f_{ii}^\omega f_{JJ}^\omega f_{kk}^0}{f_{kk}^{2\omega} f_{ii}^{\omega'} f_{jj}^{\omega'}}$$
$$\times \frac{(3\omega_0^2 - \omega^2)(\omega_0^2 - \omega'^2)(\omega_0^2 - 4\omega'^2)}{3\omega_0^2(\omega_0^2 - \omega^2)^2}$$
$$(6.22)$$

here ω_0 is the frequencies of the first strongly absorbing electronic transition in the molecule, and ω' and ω are the fundamental wavelength in second-harmonic generation and for electro-optic coefficient measurements, respectively. One can roughly estimate the electro-optic coefficients from the SHG coefficients. The electro-optic tensors follow the same symmetry rules as the piezoelectric tensors and the d_{ij} tensors. An excellent description of electro-optic coefficients and nonlinear optical coefficients in organic single crystals has been provided by Bosshard et al. [26]. The electro-optic tensor elements of poled polymeric systems can be written as follow:

$$\begin{bmatrix} 0 & 0 & r_{13} \\ 0 & 0 & r_{23} \\ 0 & 0 & r_{33} \\ 0 & r_{42} & 0 \\ r_{51} & 0 & 0 \\ 0 & 0 & 0 \end{bmatrix} \qquad (6.23)$$

After applying Kleinman symmetry, $r_{13} = r_{23} = r_{42} = r_{51}$ it leaves two independent components; r_{13} and r_{33}. The electro-optic component r_{33} is parallel to the polar axis whereas r_{13} is perpendicular to the polar axis.

During the poling process, an external electric field is generally applied at temperatures higher than the T_g of the polymer systems and then the field is switched off at around room temperature to freeze the induced orientation of dipoles. Like thermoelectrics, the NLO properties of amorphous polymers also depend on the poling conditions, such as poling field and temperature. Now it has become a well-known fact that poling field strength significantly affects the magnitude of SHG in NLO polymers. Polar order and relaxation behavior of SHG activity as a function of the poling electric field in amorphous polymers have been studied [27,28]. Under low poling conditions if the poling field is greater than the threshold field, the SHG intensity was found to be proportional to E_p^2. The lengthening of the orientational relaxation time occurs as the poling field strength increases and the relaxation time is proportional to the polar order initially induced by the poling field.

3.1.1 Origin of second-order optical nonlinearity

The origin of second-order NLO effects in organic molecules is association with strong donor–acceptor intramolecular interactions. A typical SHG active molecule should have an electron acceptor group and an electron donor group attached to a π-conjugated system as depicted below, though it must lack a centre of symmetry. The symmetry requirements eliminates many materials from being SHG active if they are centrosymmetric. Therefore one has to consider either chemical or physical means of introducing noncentrosymmetry in the molecular structures.

The π-conjugated systems could be either a benzene, azobenzene, stilbene, tolan, biphenyl, benzylidene, heterocycle or polyene backbone. The electron acceptor and donor groups that can be attached to a π-conjugated system are as follows [15].

1 Acceptor groups:
NO$_2$, NO, CN, COOH, COO$^-$, CONH$_2$, CONHR, CONR$_2$, CHO, SSI, SO$_2$R, SO$_2$C$_3$F$_7$, SO$_2$CH$_3$, COR, COCF$_3$, CF$_3$, COCH$_3$, CH=C(CN)$_2$, C$_2$(CN)$_3$, SO$_2$NH$_2$, N$_2^+$, NH$_3^+$, N(CH$_3$)$_3^+$ and aromatic* (R is an alkyl group).

2 Donor groups:
NH$_2$, NHCH$_3$, N(CH$_3$)$_2$, NHR, N$_2$H$_3$, F, Cl, Br, I, SH, SR, OR, CH$_3$, OH, NHCOCH$_3$, OCH$_3$, SCH$_3$, OC$_6$H$_5$, C(CH$_3$)$_3$, COOCH$_3$, O$^-$, S$^-$, and aromatic* (R is an alkyl group).
(*an Aromatic group is capable of both kinds of effects)

Both the increasing π-conjugation length or strength of donor acceptor groups influence the optical absorption band, which will have either a bathochromic (red shift toward a longer wavelength) or hypsochromic (blue shift toward to shorter wavelength) features. Therefore, the position of an absorption band could be remarkably changed among stilbene, azobenzene, benzylidene and tolan π-backbones. A significant effect in the optical spectrum can also be observed for organometallic compounds due to the charge transfer between conjugated system and ligands. Electron donor–acceptor groups tend to cause bathochromic shift with an increase in the intensity of absorption band. A donor group can provide additional electrons into the π-conjugated system giving rise to strong interactions. The relative positions of donor–acceptor groups at the π-conjugated system also plays a vital role. More complex NLO chromophores can be envisaged using multiple donor and acceptor groups linked to different π-conjugation either by length or elements. Now three decades after the discovery, this simple notion that the large second-order optical nonlinearity originates from π-conjugated molecules with an electron acceptor group at one end and a donor group at the opposite end of a benzene ring has given birth to field of nonlinear optics. The field of nonlinear optics is uniquely concerned with the molecular design of functional dyes where NLO response is governed largely by the chromophores involving interactions with laser light.

Lalama and Garito [29] made detailed calculations of the β-component from the all-valence-electron self-consistent-field linear-combination-of-atomic-orbitals procedure, which includes molecular orbital configuration interaction. A relationship between the excited states of p-nitroaniline with β_{ijk} was described taking into account the interactions between the NO$_2$ and NH$_2$ groups with a benzene ring. P-Nitroaniline has a planar

$$\boxed{\text{Donor} \longrightarrow \boxed{\pi\text{-Conjugated system}} \longrightarrow \text{Acceptor}}$$

structure and belongs to the point group C_{2v}, the nonzero tensor components of β_{ijk} are β_{xxx}, β_{xyy}, β_{zxx}, β_{yxy}, β_{yyz}, β_{zxz} and β_{zzx}, where $\beta_{yyx} = \beta_{yxy}$ and $\beta_{zzx} = \beta_{zxz}$ for second-harmonic generation; therefore after Kleinman symmetry consideration, only three nonzero components β_{xxx}, β_{xyy} and β_{xzz} remain. Molecular orbital calculations of p-nitroaniline indicate a transfer of electron density from the NH_2 group across the benzene ring to the NO_2 group with the involvement of neighboring carbons. In p-nitroaniline, β_x is mainly influenced by β_{xxx}, β_{xyy} and β_{yyx} whereas β_{xzz} and β_{zzx} have a negligible effect. As a result of the charge transfer, p-nitroaniline has a one-dimensional (ID) character and only nonzero tensor components contribute to the molecular hyperpolarizability. Besides 1D charge transfer molecules, two-dimensional (2D) and octupolar species are of considerable interest. For example, methyl-(2,4-dinitrophenyl)-aminopropanoate (MAP) molecule as a 2D character was described by Zyss and Oudar [30]. After Kleinman symmetry relations, only four tensor components, viz. β_{xxx}, β_{xyy}, β_{yyy} and β_{yxx} remain. Nalwa $et\ al.$ [31,32] were first to investigate 2D materials based on 1,5-diamino-2,4-dinitrobenzene which are characteristically different because they show a very large off-diagonal β-component. In these and analogous 2D charge transfer molecules, the off-diagonal orientation plays an important role in stabilizing second-order optical nonlinearity both in LB films and poled polymer systems [33]. In a series of papers, Zyss [34–36] provided an excellent description of octupolar compounds for second-harmonic generation.

3.1.2 Optimization of second-order optical nonlinearity

The trade-off between optical transparency and SHG efficiency is important from an application viewpoint. The first hyperpolarizability of an organic molecule is related to the charge transfer characteristics eventually governed by the π-conjugation length and the strength of donor and acceptor groups. The large β-values can be obtained from highly polar conjugated molecules but in turn a loss in optical transparency occurs. Both β- and optical transparency are affected by the nature of the π-conjugated bonds, π-conjugation length, strength of electron donor and acceptor substituents, heteroatoms, dimensionality and conformation. Some important factors that affect the magnitude of second-order optical nonlinearity are discussed below.

3.1.2.1 Effect of π-conjugation length

A systematic study on a series of disubstituted conjugated molecules was carried out by Huijts and Hesselink [37] to establish a relationship between the conjugation length and first hyperpolarizability. EFISH studies were performed on p-methoxynitrobenzene, α-p-methoxyphenyl-ω-p-nitrophenyl polyene (MPNP, with $n = 1, 2, 3, 4$, and 5 for ethene to decapentaene). These molecules have the methoxy (CH_3O) group as a donor and the nitro (NO_2) group as an acceptor. The length of conjugation between the CH_3O and NO_2 groups varies from two to nine π-bonds where the phenyl ring was counted as two. The $\mu\beta$ varies with a 3.4 power of the number of π-bonds, while β_0, corrected for resonance effects, varies as the third power of the length over the entire range of measurements. The length of π-conjugation has a very significant effect on the magnitude of hyperpolarizability. For example, the β value of compound 7 is more than two orders of magnitude larger than that of compound 1. From the viewpoint of efficiency–transparency trade-off, though β increases remarkably as the length of π-conjugation increases but the increased conjugation length leads to a loss in optical transparency.

The π-conjugation length dependence of β was also studied in push-pull polyenes and carotenoids [38–40]. The donor groups were dimethylaminophenyl (8a through 8e), benzodithia (9a through 9e), julolidinyl (10a through 10e) and ferrocenyl (11a through 11d) and the acceptor groups were formyl (CHO), dicyanovinyl [$CH=C(CN)_2$], p-cyanophenylvinyl ($CH=CHC_6H_4CN$) and p-nitrophenylvinyl ($CH=CHC_6H_4NO_2$).

The $\mu\beta$ values were measured by the EFISH technique in chloroform while $\mu\beta(0)$ values were calculated from the two-level model. The β value increases significantly with increasing conjugation length of the polyene chain (Table 6.1). The off-resonance $\mu\beta(0)$ values of 8 and 9 molecules are 50 times larger than that of p-nitroaniline. The compound 8e with a dicyanovinyl group showed a $\mu\beta(0)$ value 17 times larger than that of 4-(N,N-dimethylamino)-4'-nitrostilbene (DANS). A triple bond in the middle of the polyenic bridge reduces $\mu\beta$ values as shown by the p-cyanophenyl and p-nitrophenyl derivatives. The effect of the triple bond depends on the end group which may be useful from a transparency point of view. This is somewhat visible in the push–pull carotenoid-containing formyl acceptor group. The benzodithia group seems less effective at enhancing β than the dimethylamino group. The β does not level-off for a length of up to 30 Å for these push–pull substituted polyenes.

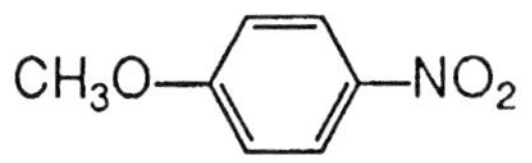

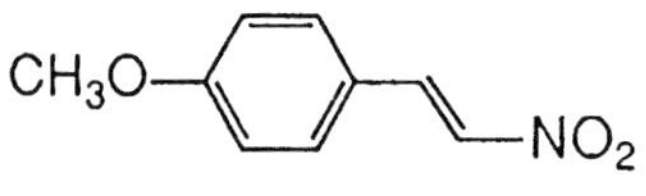

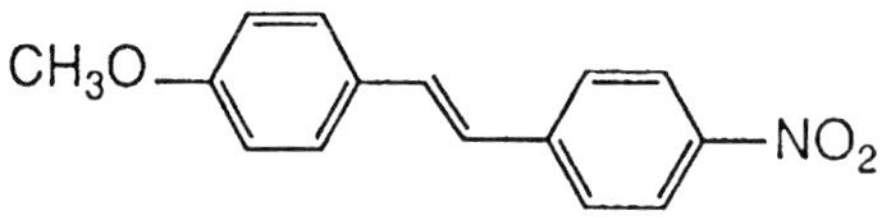

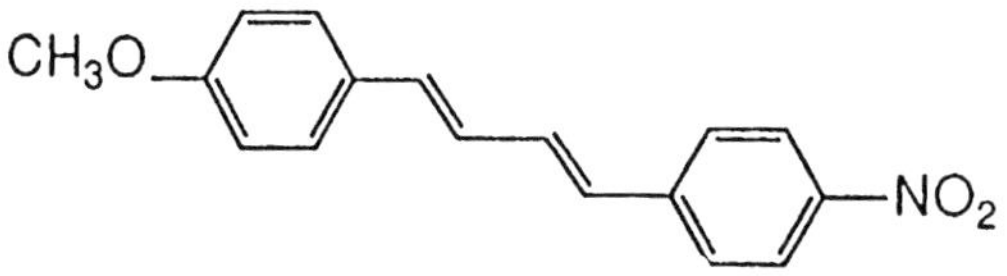

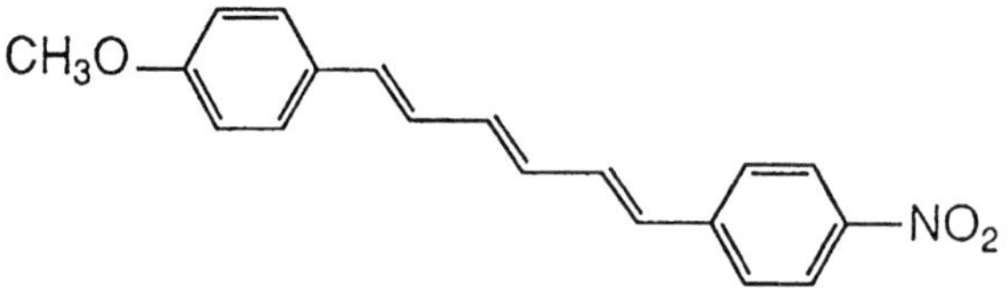

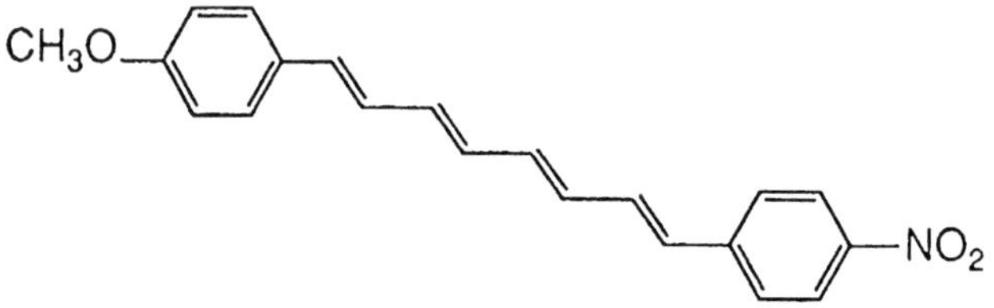

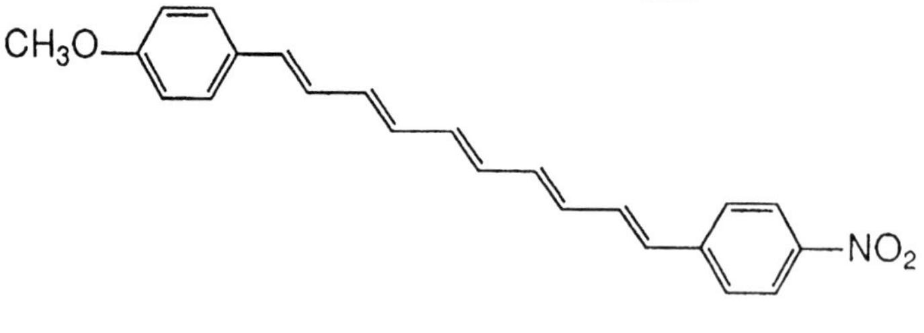

The conjugation length dependences of molecular hyperpolarizabilities for donor–acceptor substituted π-conjugated systems were also reported by Chang et al. [41]. Table 6.2 lists the dipole moment, α, β and γ values of para-disubstituted benzenes, α-phenylpolyene, and α,ω-diphenylpolyene charge transfer oligomers. The β values increase as the length of π-conjugation increases. For polyphenyls, β optimizes

with $n = 2$–3 and biphenyl induces coplanarity, and a good combination of donor–acceptor groups gives best coplanarity. The α-phenylpolyene and α, ω diphenylpolyene show significantly larger β values.

Morley et al. [42] calculated the hyperpolarizabilities of polyphenyls and polyenes containing nitro and dimethylamino groups at opposite ends of the conjugated systems. For polyphenyls, β increased slowly with increasing number of phenyl units at zero frequency $(\omega, 0)$ and at 1.17 eV. β_x increased up to

$(CH_3)N$ — [phenyl] — CH=CH — Acceptor   **8a**

$(CH_3)N$ — [phenyl] — polyene chain — Acceptor   **8b**

$(CH_3)N$ — [phenyl] — polyene chain — Acceptor   **8c**

$(CH_3)N$ — [phenyl] — polyene chain — Acceptor   **8d**

$(CH_3)N$ — [phenyl] — polyene chain — Acceptor   **8e**

[benzodithiole] = CH — Acceptor   **9a**

[benzodithiole] = polyene chain — Acceptor   **9b**

[benzodithiole] = polyene chain — Acceptor   **9c**

[benzodithiole] = polyene chain — Acceptor   **9d**

[benzodithiole] = polyene chain — Acceptor   **9e**

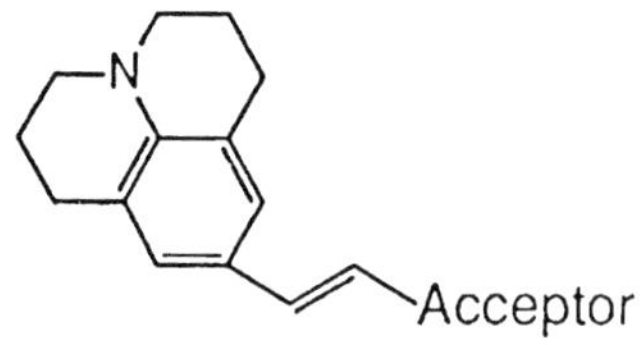

six phenyl rings though the effect per unit volume was optimized for 4-dimethylamino-4'-nitroterphenyl. On the other hand, Change *et al.* [43] pointed out that the β value of polyphenyls peaked for two phenyl units and decreases as the number of phenyl rings increases. Both theoretical and experimental data support the fact that larger β values are difficult to obtain in polyphenyls. β increased rapidly with increasing ethenyl units for polyenes with the same donor and acceptor groups and the effect per unit volume was optimized for 20 units. Polyenes show β values 20 times larger than that of the polyphenyls.

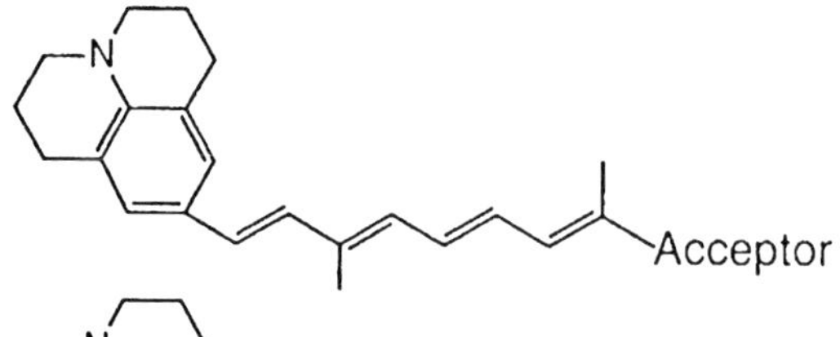

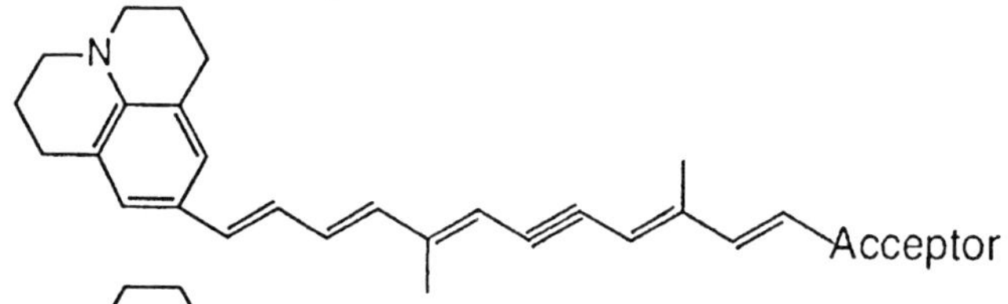

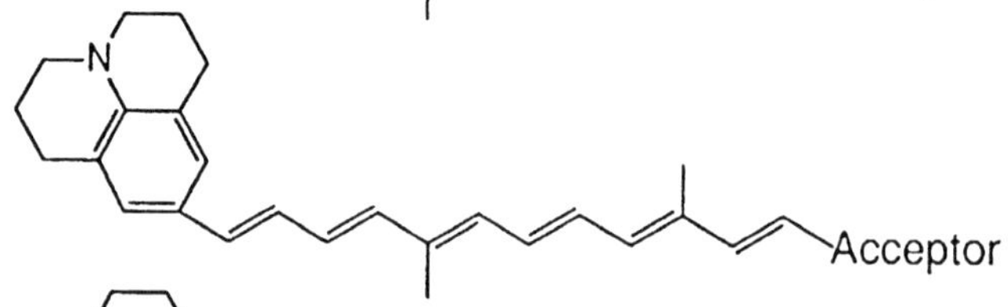

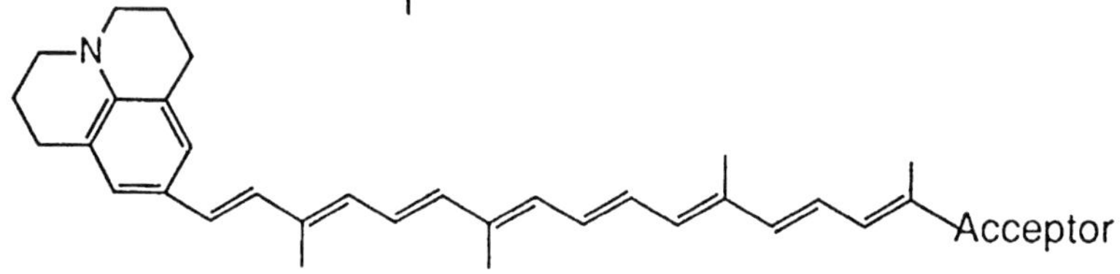

Morley [44] calculated the hyperpolarizabilities of *trans* donor–acceptor substituted cumulenes and compared them with those of corresponding polyenes and polyynes. The number of conjugated carbon atoms (N) between the donor and acceptor groups varied from four to 44. Both dipole moments and hyperpolariz-

abilities of cumulenes, polyenes and polyynes increased sharply with increasing π-conjugated chain length. The β_0 values calculated from empirical structures for eight conjugated carbon atoms were 16.5×10^{-30} cm^5 esu^{-1} for cumulene, 82.3×10^{-30} cm^5 esu^{-1} for polyene and 48.5×10^{-30} cm^5 esu^{-1} for polyyene. The β_0 values for 36 conjugated carbon atoms were 3834.4×10^{-30} cm^5 esu^{-1} for cumulene, 2239×10^{-30} cm^5 esu^{-1} for polyene and 124.1×10^{-30} cm^5 esu^{-1} for polyyne. Both dipole moment and β values of donor–acceptor cumulenes were substantially larger than the corresponding polyenes and polyynes at extended π-conjugated length. However, the molecular hyperpolarizability of polyenes was found to be geometry dependent.

The static β_{xxx} component for cumulene ($n = 5$) was estimated as 156.1×10^{-30} cm^5 esu^{-1} from singly excited configurations (SC1), 197.1×10^{-30} cm^5

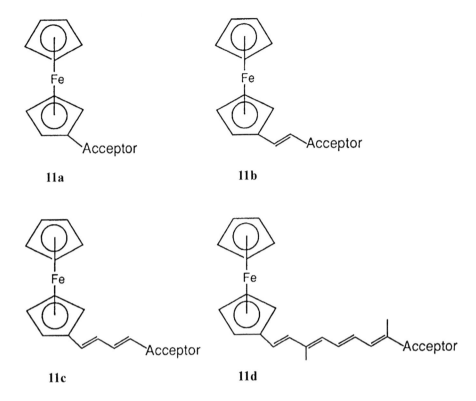

esu^{-1} from singly and doubly excited configurations (SDC1), and 373×10^{-30} cm^5 esu^{-1} from time-dependent Hartee–Fock (TDHF) method. The static β_{xxx} component (largest component of β) for polyene ($n=6$) was 138.7×10^{-30} cm^5 esu^{-1} from singly excited configurations (SC1), 142.2×10^{-30} cm^5 esu^{-1} from singly and doubly excited configurations (SDC1), and 220.1×10^{-30} cm^5 esu^{-1} from the

TDHF method. For the similar chain length ($n=6$), the static β_{xxx} component for polyyne were estimated as 81.8×10^{-30} cm^5 esu^{-1} from singly excited configurations (SC1), 88.0×10^{-30} cm^5 esu^{-1} from singly and doubly excited configurations (SDC1), and 392.7×10^{-30} cm^5 esu^{-1} from the TDHF method. For polyyne ($n=6$), the β_{xxx} component changed from 107×10^{-30} cm^5 esu^{-1} to 120×10^{-30} cm^5 esu^{-1} at

Cumulenes **12**

Polyenes **13**   Polyynes **14**

Table 6.1. Absorption maximum wavelength (λ) recorded in chloroform, hyperpolarizabilities measured by EFISH of polyene compounds showing the effect of π-conjugation length and donor–acceptor groups on β (after ref. 38–40.)

Compound	Acceptor	λ_{max} (nm)	$\mu\beta(2\omega)$ (10^{-48} esu)	$\mu\beta(0)$ (10^{-48} esu)	Measurement wavelength (μm)
8a	CHO	372	30	20	1.34
	CH=C(CN)$_2$	446	460	230	1.34
		446	230	170	1.91
	CH=CHC$_6$H$_4$CN	410	250	140	1.34
	CH=CHC$_6$H$_4$NO$_2$	452	1000	480	1.34
8b	CHO	456	1200	570	1.34
	CH=C(CN)$_2$	562	2850	1700	1.91
	CH=CHC$_6$H$_4$CN	465	1950	900	1.34
	CH=CHC$_6$H$_4$NO$_2$	488	2200	900	1.34
8c	CHO	466	2200	1000	1.34
	CH=C(CN)$_2$	540	3700	2300	1.91
	CH=CHC$_6$H$_4$CN	457	1600	750	1.34
	CH=CHC$_6$H$_4$NO$_2$	467	1660	750	1.34
8d	CHO	485	2700	1100	1.34
8e	CHO	500	7250	2800	1.34
9a	CHO	384	320	200	1.34
	CH=C(CN)$_2$	489	1030	710	1.91
9b	CHO	450	2000	1000	1.34
	CH=C(CN)$_2$	560	5500	3300	1.91
9c	CHO	461	4200	1950	1.34
	CH=C(CN)$_2$	538	4300	2700	1.91
9d	CHO	474	4300	1900	1.34
	CH=C(CN)$_2$	574	9400	5450	1.91
9e	CHO	498	8900	3400	1.34
	CH=C(CN)$_2$	588	13400	7500	1.91
10a	CH=C(CN)$_2$	520	1500	980	1.91
10b	CH=C(CN)$_2$	570	5500	3200	1.91
10c	CH=C(CN)$_2$	536	5800	3650	1.91
10d	CH=C(CN)$_2$	572	11000	6400	1.91
10e	CH=C(CN)$_2$	572	13700	8000	1.91
11a	CH=C(CN)$_2$	320,626	90	60 ~ 80	1.91
11b	CH=C(CN)$_2$	370,556	420	250 ~ 340	1.91
11c	CH=C(CN)$_2$	456,568	1120	660 ~ 875	1.91
11d	CH=C(CN)$_2$	458	4600	3300	1.91

0.656 eV with doubly excited states. The cumulene ($n = 5$) showed remarkable change when the β_{xxx} component changed from 313×10^{-30} cm^5 esu^{-1} to 469×10^{-30} cm^5 esu^{-1} at 0.656 eV in moving from singly excited configurations to doubly excited states. Polyenes showed larger β values at short chain lengths of up to eight carbons than those of cumulenes; however this trend was reversed as the chain length further increased i.e. both showed larger β values than those of the corresponding polyynes. This study demonstrated the importance of donor–accepted cumulenes for second-order nonlinear optics. These studies are evidence that n-conjugation length significantly affects second-order optical nonlinearity.

3.1.2.2 Strength of donor and acceptor groups

Though there are numerous reports, the excellent studies to develop the relationship between the molecular structures and intrinsic molecular optical nonlinearities have been conducted by the DuPont group. They analyzed the effects of the strength of donor and acceptor groups, nature of the conjugated systems, and the π-conjugated length on polarizabilities and hyperpolarizabilities by performing solution-phase dc EFISH and THG measurements on benzene, stilbene, styrene, biphenyl, tolane and fluorene derivatives [41,43]. Table 6.3 lists the μ, α, β and γ values of

Table 6.2. Absorption maximum, dipole moment, α, β and γ values (in esu) of *para*-disubstituted benzenes (after ref. 41, American Chemical Society, 1991.)

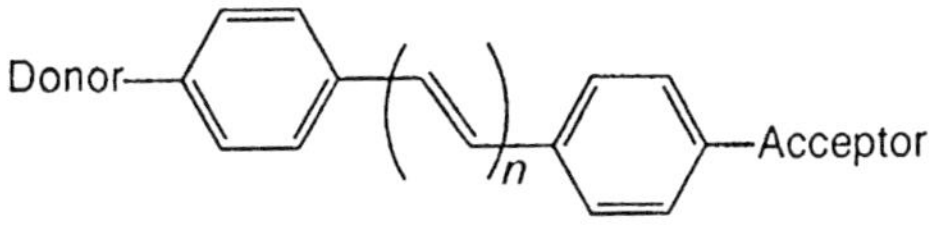

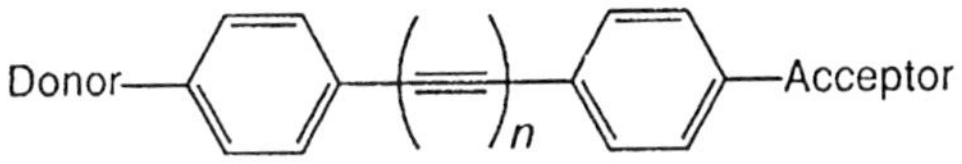

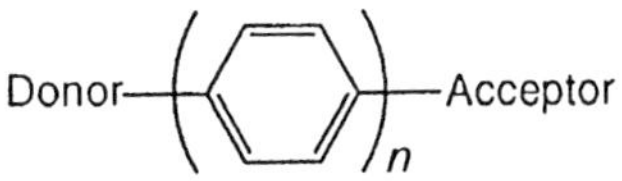

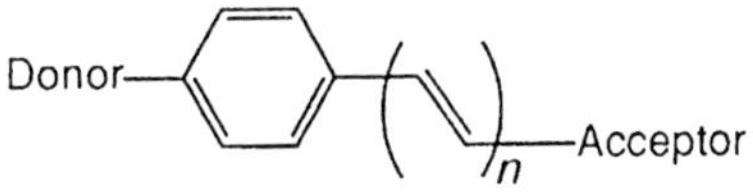

Acceptor	Donor	Solvent	n	λ_{max} (nm)	μ (10^{-18})	α (10^{-23})	β (10^{-30})	γ (10^{-36})
				Donor—⬡—(=)ₙ—⬡—Acceptor				
CN	OCH_3	chloroform	1	340	3.8	3.4	19	54
			2	360	4.5	3.8	27	122
			3	380	4.4	4.4	40	234
NO_2	OCH_3	chloroform	1	376	4.5	3.4	34	93
			2	397	4.8	4.0	47	130
			3	414	5.1	4.2	76	230
			4	430	5.8	4.8	101	
NO_2	$N(CH_3)_2$	chloroform	1	430	6.6	3.4	73	225
			2	442	7.6	4.0	107	
			3	458	8.2	4.2	131	
			4	464	9		190	
				Donor—⬡—(≡)ₙ—⬡—Acceptor				
CN	SCH_3	chloroform	1	333	4.0	3.5	15	35
			2	330	4.7	3.8	17	42
NO_2	SCH_3	chloroform	1	362	4.0	3.8	20	95
			2	338	3.9	4.0	17	61
				Donor—(⬡)ₙ—Acceptor				
NO_2	NH_2	NMP	1	370	7.8	1.6	10	21
			2	372	7.8	2.6	24	96
			3	360	7.6	3.5	16	124
			4	344	10	3.3	11	133
No_2	OCH_3	p-dioxane	1	302	4.6	1.5	5.1	10
			2	332	4.5	2.8	9.2	39
			3	340	5.0	3.8	11	
				Donor—⬡—(=)ₙ—Acceptor				
CHO	OCH_3	chloroform	1	318	4.0	2.5	12	28
			2	350	4.3	3.0	28	43
			3	376	4.6	3.5	42	120
CHO	$N(CH_3)_2$	chloroform	1	384	5.6	2.6	30	63
			2	412	6.0	3.3	52	140
			3	434	6.3	4.0	88	257
$CHC(CN)_2$	$N(CH_3)_2$	chloroform	1	486	8.4	3.2	82	
			2	520	9.0	3.6	163	

Table 6.3. Absorption maximum, dipole moment, α, β and γ values (in esu) of *para*-disubstituted benzenes (after ref. 41 and 43, American Chemical Society, 1991)

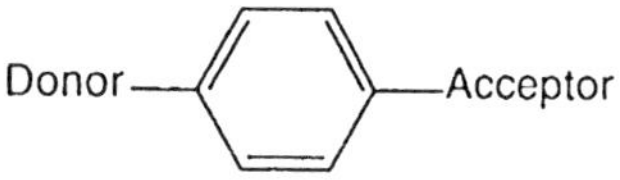

Donor —⟨benzene ring⟩— Acceptor

Acceptor	Donor	Solvent	λ_{max} (nm)	μ (10^{-18})	α (10^{-23})	β (10^{-30})	γ (10^{-36})
SO$_2$CH$_3$	OH	p-dioxane	290	3.4	1.7	1.3	3
CN	CH$_3$	neat		4.4	1.5	0.7	5
CN	Cl	p-dioxane		2.3	1.6	0.8	5
CN	Br	p-dioxane		2.4	1.8	1.1	7
CN	OC$_6$H$_5$	p-dioxane		4.1	2.6	1.2	9
CN	COH$_3$	p-dioxane	248	4.8	1.7	1.9	4
CN	SCH$_3$	p-dioxane		4.4	2.0	2.8	9
CN	NH$_2$	p-dioxane	270	5.0	1.6	3.1	6
CN	N(CH$_3$)$_2$	p-dioxane	290	5.6	2.1	5.0	10
CHO	CH$_3$	neat		3.0	1.6	1.7	7
CHO	OC$_6$H$_5$	neat	269	2.8	2.5	1.9	12
CHO	OCH$_3$	neat	269	3.5	1.7	2.2	8
CHO	SCH$_3$	neat	310	3.1	1.9	2.6	13
CHO	N(CH$_3$)$_2$	p-dioxane	326	5.1	2.0	6.3	18
i-SO$_2$C$_3$H$_7$	OCH$_3$	chloroform	290	5.4	2.7	3.3	5
COCF$_3$	OCH$_3$	p-dioxane	292	3.5	2.9	3.6	12
COCF$_3$	OC$_6$H$_5$	p-dioxane	292	4.0	2.0	3.6	7
COCF$_3$	N(CH$_3$)$_2$	p-dioxane	356	5.9	2.4	10	16
NO	N(CH$_3$)$_2$	p-dioxane	407	6.2	2.1	12	
NO$_2$	CH$_3$	p-dioxane	272	4.2	1.6	2.1	8
NO$_2$	Br	p-dioxane	274	3.0	1.8	3.3	
No$_2$	OH	p-dioxane	304	5.0	1.5	3.0	8
NO$_2$	OC$_6$H$_5$	p-dioxane	294	4.2	2.6	4.0	9
NO$_2$	OCH$_3$	p-dioxane	302	4.6	1.5	5.1	10
NO$_2$	SCH$_3$	p-dioxane	322	4.4	1.9	6.1	17
NO$_2$	N$_2$H$_3$	p-dioxane	366	6.3	1.8	7.6	9
NO$_2$	NH$_2$	acetone	365	6.2	1.7	9.2	15
NO$_2$	N(CH$_3$)$_2$	acetone	376	6.4	2.2	12	28
NO$_2$	CN	p-dioxane		0.9	1.7	0.6	7
NO$_2$	CHO	p-dioxane	376	2.5	1.7	0.2	7
CHC(CN)$_2$	OCH$_3$	p-dioxane	345	5.5	2.4	9.8	30
CHC(CN)$_2$	N(CH$_3$)$_2$	chloroform	420	7.8	2.8	32	
CHC(CN)$_2$	julolidine	CH$_2$Cl$_2$	458	8.0	3.0	44	
C$_2$(CN)$_3$	NH$_2$	CH$_2$Cl$_2$	498	7.8	3.4	39	
C$_2$(CN)$_3$	N(CH$_3$)$_2$	CH$_2$Cl$_2$	516	8.2	3.7	50	
C$_2$(CN)$_3$	julolidine	CH$_2$Cl$_2$	556	8.5	3.9	60	

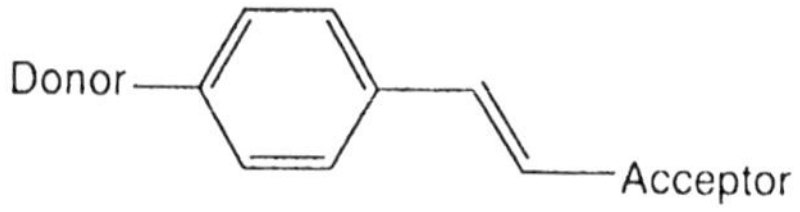

Donor —⟨stilbene⟩— Acceptor

Acceptor	Donor	Solvent	λ_{max} (nm)	μ (10^{-18})	α (10^{-23})	β (10^{-30})	γ (10^{-36})
CN	OCH$_3$	chloroform	304	4.2	2.3	7.0	11
CN	N(CH$_3$)$_2$	chloroform	364	6.0	2.8	23	29
CHO	Br	chloroform	298	2.0	2.3	6.5	26
CHO	OCH$_3$	chloroform	318	4.2	2.5	11	28
CHO	N(CH$_3$)$_2$	chloroform	384	5.6	2.6	30	63
NO$_2$	OH	chloroform	312	5.1	2.4	18	52
NO$_2$	OCH$_3$	chloroform	352	4.6	2.6	17	35
NO$_2$	N(CH$_3$)$_2$	chloroform	438	6.5	3.2	50	

Table 6.3. (*continued*)

Acceptor	Donor	Solvent	λ_{max} (nm)	μ (10^{-18})	α (10^{-23})	β (10^{-30})	γ (10^{-36})

Donor—⟨⟩—=—⟨⟩—Acceptor

Acceptor	Donor	Solvent	λ_{max} (nm)	μ (10^{-18})	α (10^{-23})	β (10^{-30})	γ (10^{-36})
$SO_2C_6F_{13}$	OCH_3	p-dioxane	347	7.8	4.8	14	93
$COCF_3$	OCH_3	p-dioxane	368	4.2	3.9	16.4	83
CN	OH	p-dioxane	344	4.5	3.2	13	52
CN	OCH_3	chloroform	340	3.8	3.4	19	54
CN	$N(CH_3)_2$	chloroform	382	5.7	3.9	36	125
NO_2	OH	p-dioxane	370	5.5	3.3	17	104
NO_2	OCH_3	p-dioxane	364	4.5	3.4	28	79
		chloroform	370	4.5	3.4	34	93
NO_2	SCH_3	chloroform	380	4.3	3.8	34	100
NO_2	$N(CH_3)_2$	chloroform	427	6.6	3.4	73	225
NO_2	julolidinamine	chloroform	438	7.0	4.5	96	

Donor—⟨⟩—=—⟨⟩—Acceptor

Acceptor	Donor	Solvent	λ_{max} (nm)	μ (10^{-18})	α (10^{-23})	β (10^{-30})	γ (10^{-36})
SO_2CH_3	NH_2	chloroform	338	6.5	3.4	13	59
CO_2CH_3	NH_2	chloroform	332	3.8	3.7	15	62
$COCH_3$	SO_2CH_3	chloroform	334	3.3	3.2	12	29
CN	NH_2	chloroform	342	5.2	3.2	20	55
CN	$NHCH_3$	chloroform	358	5.7	3.4	27	90
CN	$N(CH_3)_2$	chloroform	372	6.1	3.7	29	99
NO_2	OCH_3	p-dioxane	356	4.4	3.9	14	52
NO_2	SCH_3	chloroform	362	4.0	3.8	20	95
NO_2	NH_2	chloroform	380	5.5	3.2	24	120
		NMP	410	5.5	3.6	40	140
NO_2	$NHCH_3$	chloroform	400	5.7	4.0	46	130
NO_2	$N(CH_3)_2$	chloroform	415	6.1	4.1	46	151

NMP = N-methyl-2-pyrrolidone

donor–acceptor substituted π-conjugated systems. The experimental data demonstrate that (1) the efficacies of acceptor groups increase in the order of SO_2CH_3, CN, CHO, $COCF_3$, NO, NO_2, $CHC(CN)_2$ and $C_2(CN)_2$; (2) with a nitro acceptor group, the relative effectiveness of the various donor groups in an increasing order were $OCH_3 < OH < Br < OC_6H_5 < OCH_3 < SCH_3 < N_2H_3 < NH_2 < N(CH_3)_2$ and the julilodine amine; (3) the magnitude of the optical nonlinearities is governed by the strength of the donor–acceptor groups and the best combination can provide an enhancement of about 10 times; (4) the β values are quite large for disubstituted benzenes with $CHC(CN)_2$, $C_2(CN)_2$, dimethylamino and julolidine groups; (5) the β value of a donor–acceptor substituted tolan is about half that of the respective stilbenes because acetylene introduces strong bond alteration and there is a lack of conjugation

between donor and acceptor groups, (6) the β values of donor–acceptor substituted styrenes fall between those of benzene and stilbenes derivatives. Donor–acceptor groups play an important role in governing optical nonlinearities.

3.1.3 π-conjugated materials

A very wide variety of organic materials have been investigated for second-order nonlinear optics that includes second-harmonic generation and the electro-optic effect (Pockel's effect). Second-order NLO material can be classified into several different categories depending upon their nature and processing characteristics [15].

Here poled polymers can be divided into NLO dye-functionalized polymers, cross-linked polymers and ferroelectric polymers categories; they are structurally different. LB films, self-assembled films and liquid crystals are characteristically different due to their chemical structures and the way of processing. A great emphasis have been focused on poled polymeric systems both for SHG and EO studies. There are a few reports where second-order NLO properties of π-conjugated materials, such as fullerenes, phthalocyanines, porphyrins, and NLO dye-functionalized π-conjugated polymers have been reported.

3.1.3.1   Fullerenes

Highly stable icosahedral-cage molecule C_{60} fullerene is a new form of carbon, which has a soccer ball structure. C_{60} molecule has a highly aromatic character where all the atoms are connected by sp^2 bonds and the remaining 60 π-electrons are distributed [45–47]. C_{60} is an ideal candidate for third-order nonlinear optics because of its interesting π-electron conjugation. Despite the symmetry, there are also some reports on SHG activity of C_{60} from vapor-deposited thin films and LB films.

Wang *et al.* [48] prepared 100 Å $\sim$ 2 nm thick films of C_{60} on a glass substrate by thermal sublimation. These yellow films were shiny and smooth. The transmission SHG measurements on the films were performed at 1.064 μm in the p-polarized geometry. The transmission SHG intensity for different incident angles was measured and the ratio of $\chi^{(2)}_{zyy}/\chi^{(2)}_{zzz} = 0.48$, was in agreement with the reflection SHG measurements. Also the square root of the SHG intensity increased linearly as a function of the film thickness. A $\chi^{(2)}_{zzz}$ of 2.1×10^{-9} esu was measured for a C_{60} thin film at room temperature, which is about 1.5 times larger than that of quartz under similar conditions. The SHG signal slowly decayed with time on exposure to air. The transmission SHG was recorded for corona poled films heated from room temperature to 170°C. Figure 6.3 shows the temperature dependence of SHG for a C_{60} film poled under $+6$ kV potential. The SHG was found to be temperature dependent; as it started rising at 90°C, reached a maximum at 140°C and then

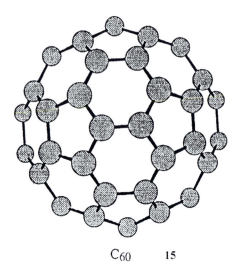

C$_{60}$ **15**

$C_{60}(C_6H_9NO_4)$ **16**

$C_{60}(C_8H_{13}NO_4)$ **17**

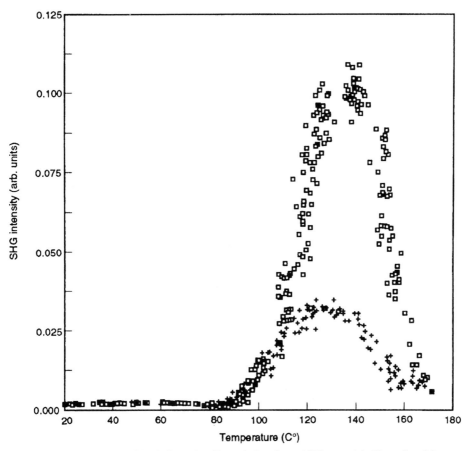

Figure 6.3. Temperature dependence of SHG for a C_{60} film poled under $+6$ kV potential. (Reproduced by permission of the American Institute of Physics from ref. 48).

decreased with further increasing temperature. The $\chi^{(2)}$ value was 1.0×10^{-8} esu at about 140°C, which is about 10 times larger than that at room temperature. SHG followed the same pattern during cooling and it appeared again after reheating the cooled films. The possible origin of SHG was considered to be electric quadrupole or magnetic dipole contributions, which are allowed in centrosymmetric materials. A similar type of mechanism has been proposed for phthalocyanines thin films as will be discussed later.

Kajzar and coworkers [49] measured SHG from C_{60} thin films at 1.064 µm by comparing it with an α-quartz single-crystal plate under similar experimental conditions. A resonant $\chi^{(2)}$ value of 3.8×10^{-9} esu (1.6 pm V^{-1}) was estimated for C_{60} thin films. This $\chi^{(2)}$ value is about two times larger than that reported by Wang *et al.* at room temperature. The incidence angle dependence of SHG showed a ratio of

$\chi_{113}^{(2)}/\chi_{333}^{(2)} = 0.53$, almost close to that reported by Wang and coworkers. [48].

There have been a few reports on SHG properties of LB films of C_{60} derivatives. Gan *et al.* [50] reported SHG from the Langmuir–Blodgett films of $C_{60}(C_4H_8N_2)$. The $C_{60}(C_4H_8N_2)$ derivative was synthesized by reacting C_{60} with 2-aminoethyl hydrogen sulfate in a multisolvent system toluene/ethanol/water. The Langmuir films were deposited from the spreading situation on a pure water subphase. The solvents were allowed to evaporate for 15–20 min and then the films were compressed at a rate of 20 mm min^{-1}. The monolayer was transferred as a Z type onto hydrophilic fused quartz plate. The $C_{60}(C_4H_8N_2)$ films showed three bands at 220, 270 and 328 nm. Assuming a refractive index of 1.90 and a film thickness of 1.5 nm, $\beta = 3.6 \times 10^{-29}$ esu and $\chi^{(2)} = 1.8 \times 10^{-7}$ esu were calculated. The relative SHG intensities were 1.0, 1.2

and 2.0 for one, three and five layers, respectively. The SHG showed a subquadratic dependence.

Zhou *et al.* [51] prepared both Langmuir and LB films of two C_{60}-glycine ester derivatives. The $C_{60}(C_6H_9NO_4)$ and $C_{60}(C_8H_{13}NO_4)$ were synthesized by photochemical reaction between C_{60} and glycine methyl or ethyl ester. The monolayers of both compounds were transferred as Z type onto a hydrophilic fused quartz plate. The observed limiting area was 0.67 nm²/molecule for $C_{60}(C_6H_9NO_4)$ and 0.75 nm²/molecule for $C_{60}(C_8H_{13}NO_4)$.

Optical absorption spectrum showed three peaks: 227, 278 and 337 nm for $C_{60}(C_6H_9NO_4)$ and 225, 275 and 336 nm for $C_{60}(C_8H_{13}NO_4)$. $\beta = 2.3 \times 10^{-29}$ esu and $\chi^{(2)} = 1.5 \times 10^{-7}$ esu were estimated for $C_{60}(C_6H_9NO_4)$ at 1.064 µm. The relative SHG of this compound was nine times larger for three layers than that of one layer. $\beta = 1.9 \times 10^{-29}$ esu and $\chi^{(2)} = 1.1 \times 10^{-7}$ esu were measured for $C_{60}(C_8H_{13}NO_4)$ films.

Leigh *et al.* [52] reported formation and SHG from LB films of amphiphilic fullerene-aza-crown ethers and their potassium ion complexes. The chemical structures of three derivatives are shown below. The deposition of the films was carried out for both with and without K^+ ions in the subphase using the same conditions. It was found that Langmuir films with significantly higher areas/molecule were obtained by applying a high concentration of K^+ ions in the subphase than those with no ions in the subphase. The pressure–area (π–A) isotherms yielded areas/molecule of about 91, 94 and 90 Å² for compounds **18–20** with no ions in the subphase, respectively.

The SHG activity of monolayers were measured by a ND:YAG laser at 1.064 µm and the corresponding $\chi^{(2)}$ values are shown along with chemical structures. The difference in $\chi^{(2)}$ values was considered to be associated to variations in film quality rather than the

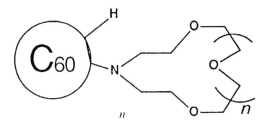

18	C_{60}-1-aza-12-crown-4 ($n=1$)	$\chi^{(2)} = 2.3$ pm V^{-1}
19	C_{60}-1-aza-15-crown-5 ($n=2$)	$\chi^{(2)} = 3.6$ pm V^{-1}
20	C_{60}-1-aza-18-crown-6 ($n=3$)	$\chi^{(2)} = 3.2$ pm V^{-1}

size of the attached crowns. All compounds yielded a $\chi^{(2)}$ of about 1 pm V^{-1} with K^+ ions. The lower $\chi^{(2)}$ values of films with K^+ ions were considered to be related to either tilting of the magnetic dipoles, the lower packing density due to KCl intercalation or poor film quality.

3.1.3.2   Conjugated organometallics

Like fullerenes, metallophthalocyanines are not expected to exhibit SHG due to their symmetry. However the observation of SHG in phthalocyanines has become possible through fabrication techniques by breaking the symmetry. In particular, SHG activity has been reported by several research groups in copper phthalocyanine (CuPc) (M = Cu, R = None) (**21**) which possesses a centrosymmetric D_{4h} structure. Chollet *et al.* [53] reported SHG for the first time from vacuum-deposited thin films of CuPc.

Second-order NLO properties of CuPc were analyzed using the Langmuir–Blodgett film approach and poled polymer film approach. Using the LB approach with a

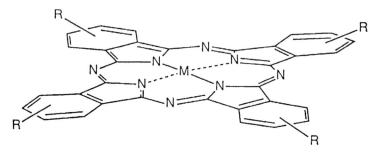

random orientation in the substrate plane, the second-order nonlinear optical coefficients were estimated from the following relationships:

$$d_{pp} = d_{33}(\sin^3 \theta_\omega \cos^3 \varphi + \tfrac{3}{2}\sin \theta_\omega \cos^2 \theta_\omega \sin^2 \varphi \cos \varphi) \quad (6.24)$$

$$d_{sp} = \tfrac{1}{2}d_{33}\sin^2 \varphi \cos \varphi \sin \theta_\omega \quad (6.25)$$

The estimated d_{33} coefficients were 11.5×10^{-9} esu for p–p polarization and 65×10^{-9} esu for s–p polarization for a 162 nm thick CuPc film. This approach was not found to be as suitable because of a large discrepancy that was observed between the d_{33} values determined from p–p and s–p polarization configurations; both configurations are expected to yield the same value of d_{33}. Using the poled polymer films approach, the second-order nonlinear optical coefficients were estimated from the following relationships:

$$d_{pp} = (d_{33} \sin^2 \theta_\omega + d_{31} \cos^2 \theta_\omega) \sin \theta_{2\omega} + 2d_{31} \cos \theta_\omega \sin \theta_\omega \cos \theta_\omega \quad (6.26)$$

$$d_{sp} = 2d_{31} \sin \theta_\omega \cos \theta_\omega \cos \theta_\omega \cos \theta_{2\omega} \quad (6.27)$$

From these equations, the d_{33} coefficients were 31×10^{-9} and 11.3×10^{-9} esu for p–p polarization for 162 and 78 nm thick CuPc films, respectively. The d_{31} coefficients were estimated as 11.4×10^{-9} esu for s–p polarization both for 162 and 78 nm thick CuPc films. The d_{31} coefficients were 3.76×10^{-9} esu for the p–p polarization for a 78 nm thick film. The ratio d_{31}/d_{33} was found to be 0.75 from the poled polymers approach instead of theoretically $1/3$ value.

Kumagai et al. [54] also observed strong SHG from evaporated CuPc thin films with thicknesses between 40 to 200 nm. The SHG intensity was found to decrease around 100 nm due to the phase mismatch between the fundamental beam and the generated SH beam. The second-order nonlinear optical susceptibility $\chi^{(2)}_{ZYY}$ of CuPc film was estimated to be 4.51×10^{-8} esu where Z and Y are the perpendicular and parallel directions to the film surface, respectively. The authors postulated that the SHG in CuPc thin films originates from the slightly deformed crystal structure. The SHG activity was increased due to a resonance of the SH light with an electronic transition. The origin of SHG in vacuum-deposited CuPc thin films was analyzed by Yamada et al. [55]. They investigated in situ SHG from CuPc during the evaporation process as a function of film thickness. Figure 6.4 shows the thickness dependence SH intensity of CuPc films in transmission geometry. The SH intensity shows a

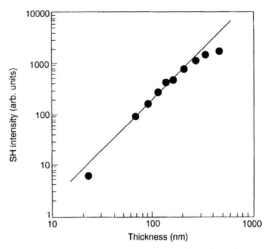

Figure 6.4. Thickness dependence SH intensity of copper phthalocyanine (CuPc) films in transmission geometry. (Reproduced by permission of Gordon and Breach Publishers from ref. 57).

quadratic dependence in the 70 to 200 nm region but it deviates for thinner and thicker regions. Kumagi et al. [54] also observed a deviation from the quadratic dependence in thicker regions in a reflection geometry. These authors suggested the origin of SHG in CuPc vacuum-deposited films from the electric quadrupolar mechanism or preferably from a magnetic dipole mechanism. Takezoe et al. [56] reported strong SHG from vacuum-evaporated films of cobalt phthalocyanine (CoPc), zinc phthalocyanine (ZnPc) and metal-free phthalocyanine (H$_2$Pc) which is in contrast to the results of Chollet et al. [53] where no SHG was observed in these phthalocyanines. The authors proposed that SHG of CoPc, ZnPc and H$_2$Pc originates from bulk optical nonlinearity, irrespective of the central metal, though these phthalocyanines are centrosymmetric. Thickness-dependence SHG was measured in H$_2$Pc thin films. The SHG spectrum showed a narrow resonance peak at about 555 nm which corresponds to the edge of the linear absorption band. The origin of the resonant SHG was considered to be the quadrupole mechanism.

Hoshi et al. [57,58] reported SHG in oxovanadium phthalocyanine (VOPc) (M = VO, R = none) **21** thin films. Thin films of VOPc of different thicknesses were epitaxially grown on a KBr(100) surface by the molecular beam epitaxy (MBE) method. The incident angle dependence of SH intensity for three films of different thicknesses are shown in Figure 6.5. VOPc thin films showed no SHG at normal incidence;

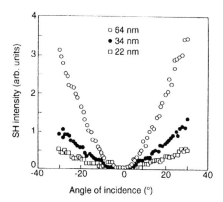

Figure 6.5. Incident angle dependence of SH intensity of oxovanadium phthalocyanine (VOPc) for three films of different thickness. (Reproduced by permission of the Japanese Journal of Applied Physics from ref. 58).

however SHG activity started appearing as the films were rotated. The thickness dependence SH intensity at an incident angle of 20° was studied. The SH intensity increases quadratically with increasing film thickness up to at least 64 nm and showed a saturation behavior at about 100 nm. The origin of SHG has been considered a polar VOPc orientation in the epitaxially grown structure. The epitaxial polar structure was found to change into a nonpolar bulk crystal structure at about 80 nm film thickness. The structural difference between thinner and thicker VOPc films was confirmed by the infrared and UV-vis absorption spectroscopy, X-ray diffraction and scanning electron microscopy.

Liu et al. [59] reported SHG from the Langmuir-Blodgett films of an asymmetrically substituted metal-free phthalocyanine; nitro-tri-tert-butylphthalocyanine ($M = H_2$ and $R_3 = C(CH_3)_3$, $R = NO_2$) (**21**). The absorption spectrum of the LB film (six layers) showed the Q band at 630 nm which indicated an one-dimensional linear stacking of phthalocyanine molecules. SHG was measured from the monolayer and multilayers. The dependence of SH intensity on the number of layers was not found to be quadratic. The SH intensities of the monolayer were $4.0 10^{-7}$ and 7.5×10^{-8} relative to a Z-cut quartzwedge for p–p and s–p configurations, respectively. The β value of $2–3 \times 10^{-30}$ esu and $\chi^{(2)}$ of $20–30 \times 10^{-9}$ esu was estimated for the monolayer. Neuman et al. [60] reported SHG from nickel tetra(cumylphenoxy) phthalocyanine $NiPc(CP)_4$ ($M = Ni$, $R = OC_6H_4–C(CH_3)_2–C_6H_5$) films developed by the LB technique. Multilayer (35-layer) LB films from an equimolar mixture of $NiPc(CP)_4$ and octadecanol were deposited onto a quartz substrate. The SH intensity was found to

increase remarkably when the substrate was kept within a few degrees of the optimal orientation. The SH signal varied with the angle of rotation and a maximum SH intensity was observed at around 90°. Two factors: the alignment of the LB film covered quartz substrate and the angle of incidence were critical for SHG activity. The 35-layer LB films of $NiPc(CP)_4$ and octadecanol showed a nonlinear grating behavior, which had the same periodicity and groove depth over the entire surface.

Hoshi et al. [61] reported SHG in ultrathin films of fluoro-bridged aluminium phthalocyanine polymer $(AlPcF)_n$. The SHG intensity from $(AlPcF)_n/KBr$ was found to be 23.8 times larger than that of $(AlPcF)_n/$silica for the same thickness (20 nm) film. The behavior of SHG was described by point groups, 4 mm for $(AlPcF)_n/KBr$ and $C_{\infty V}$ for $(AlPcF)_n/$silica, which support the existence of asymmetry along the direction of the thickness.

The above results indicate that SHG activity in centrosymmetric phthalocyanines can be insinuated by breaking inversion symmetry through physical means. In particular, the film growth processes of MBE and Langmuir–Blodgett techniques proved important. The SHG in CuPc is a deformation-induced property and the possibility of quadrupolar and magnetic dipole mechanisms can not be ruled out. On the other hand, a polar orientation of VOPc molecules in epitaxial films holds the key to SHG appearance. The SG intensity data on $(AlPcF)_n$ films suggest the role of a substrate to be a polar orientation that leads to SHG activity. SHG in phthalocyanines seems to be a highly surface-selective optical phenomenon that can be induced from the break in symmetry during fabrication processes.

Suslick et al. [62] measured the hyperpolarizabilities of porphyrins with donor and acceptor substituents using the EFISH technique at 1.19 µm for chloroform solutions. Porphyrins are also two-dimensional π-conjugated materials like phthalocyanines. The dipole moment and β values were found to be affected by the position of the donor and acceptor groups. The β value of porphyrin with $R_1 = R_2 = R_3 = NO_2$, $R_4 = NH_2$ was $\geq 10 \times 10^{-30}$ esu. In this case, only the amino group pushes charge into the macrocycle while the three nitro groups drain it, which reduces the β value. A β value of 30×10^{-30} esu and a dipole moment of 7×10^{-18} esu was measured for porphyrin with $R_1 = R_2 = NO_2$, $R_3 = R_4 = NH_2$. The porphyrin with $R_1 = NO_2$, $R_2 = R_3 = R_4 = NH_2$ exhibited a β value of 20×10^{-30} esu and dipole moment of 5×10^{-18} esu. In this case, the charge is pushed into the

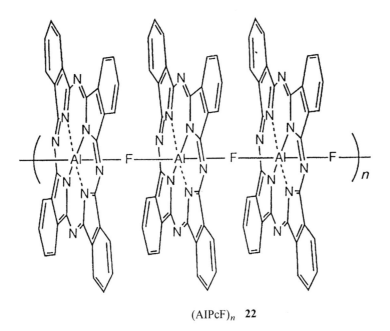

(AlPcF)$_n$ **22**

π-conjugated ring by three amino groups. The charge transfer being well aligned yielded a large β value.

Peng *et al.* [63] synthesized a multifunctional polyimide with a metal-free porphyrin (M = H$_2$) (**25**) into a polyimide backbone as well as a NLO chromophore side chain. The photoconductivity arises from conjugated porphyrin units while SHG originates from the pendant NLO chromophore. This polyimide showed a T_g of 250°C. The absorption spectra exhibited bands at 510, 540 and 630 nm for porphyrin rings and 458 nm for the NLO chromophore. This polyimide is photoconductive as well as SHG active. The photo-photoconductivity of 1.1×10^{-12} S cm^{-1} was determined under an external field of 1500 kV cm^{-1} at 690 nm. A d_{33} value of 110 pm V^{-1} at 1064 nm was measured for polymer film after corona poling. No decay of the d_{33} value was observed at 90 and 150°C. The d_{33} of 80% was retained even at 170°C after 120 h.

Peng *et al.* [64] also reported a new photorefractive polyimide with zinc porphrin (**27**) in the main chain ($m = 0.005$) and a NLO chromophore side chain ($n = 0.995$). This fluorinated polyimide exhibited a T_g of 228°C and thermal stability up to 350°C. Its photoconductivity was highly dependent on the electric field. The SHG showed good thermal stability and remained unchanged up to 180°C.

SHG properties of one-dimensional nitrido-conjugated polymers were reported by Pollagi *et al.* [65]. The [N≡M(OR)$_3$]$_n$ (**28**) structure has triple and single

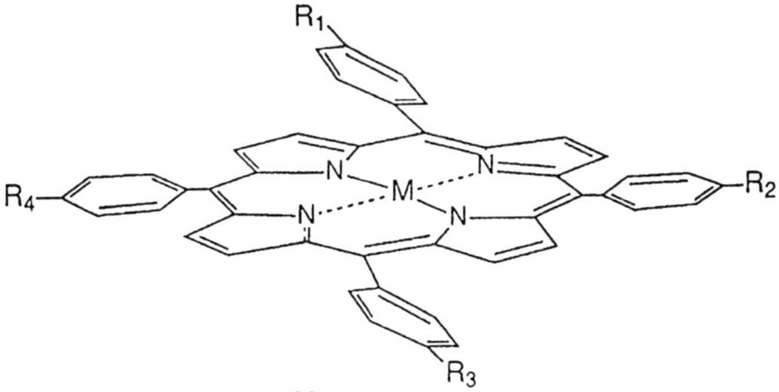

23

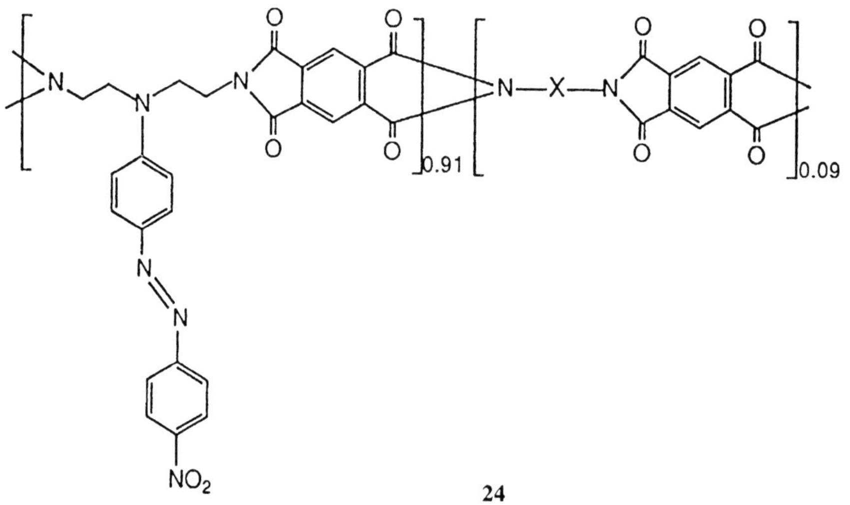

24

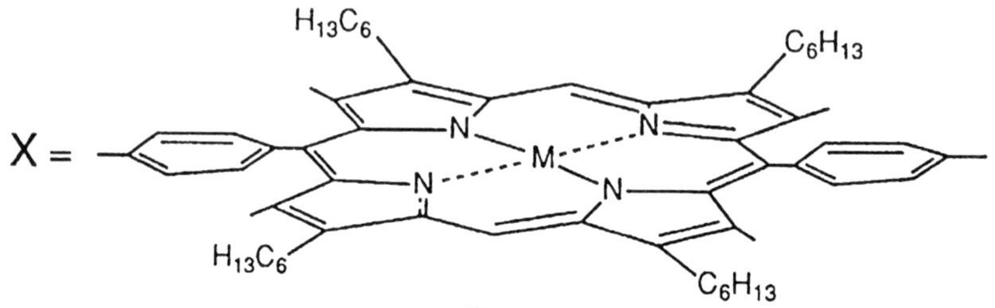

25

bonds where $M = Mo$, W and $R = Me_3$, $R = CMe_3CF_3$. The $[MoN(OCMe_3)_3]_n$ showed SHG efficiency 0.25 times that of urea and it crystallizes into P6$_3$ cm space group. The SHG efficiencies of $[WN(OCMe_3)_3]_n$ and $[MoN(OCMe_2CF_3)_3]_n$ were 0.20 and 0.35 times that of urea at 1.064 µm, respectively. No SHG was observed for $[MoN(OCMe_2Et)_3]_n$. The SHG efficiency and emission energy were found to be affected by the nature of alkoxide ligands and the effective conjugation length of the $[MN]_n$ backbone.

3.1.3.3 NLO chromophore functionalized conjugated polymers

Like piezoelectric properties, the absence of a center of symmetry is a prerequisite for the appearance of second-order NLO effects. A non-centrosymmetric structure is required for the display of piezoelectricity, second-harmonic generation, and Pockel's effect. Several chemical and physical strategies have been applied to generate noncentrosymmetry [15]. The poling technique is one of the most important methods. The NLO chromophores can be covalently attached to the linear chain of a polymer by the simple chemical processes. The most promising are polystyrene and poly(methylmethacrylate) because they provide high optical transparency over a broad frequency range as well as high glass transition temperature, low dielectric constant and good processability. NLO chromophore functionalized polymers constitute a major class of second-order NLO materials on which current research and development efforts are focused. Organic polymers are considered the most valuable specialty materials for the next generation of photonic devices because

(1) they provide tremendous architectural flexibility to tailor the desired photonic functions and material performance;
(2) they can be economically produced on a large scale;
(3) they provide ease of processing and fabrication;
(4) they can be prepared into any desired shape and size;

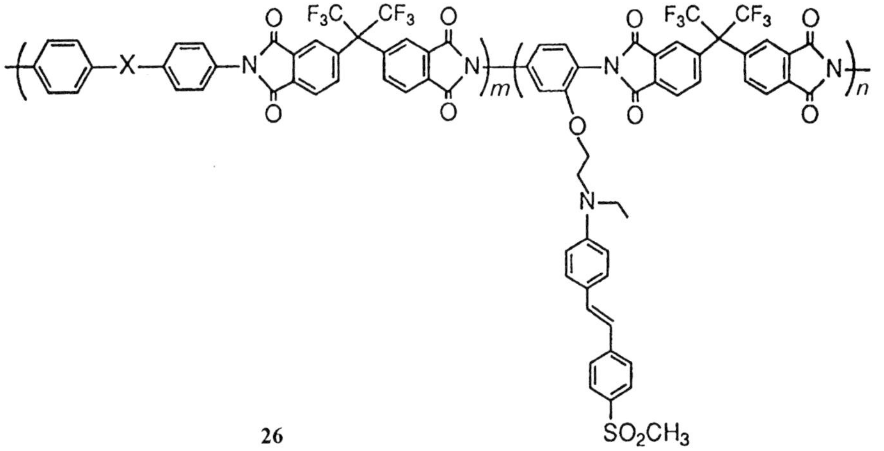

26

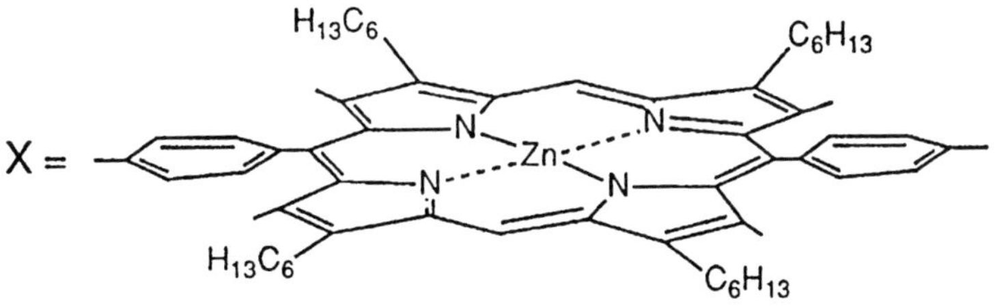

$X =$

27

(5) their chemical modification facilitates the generation of new properties;

(6) they are compatible to metals, glasses, ceramics and other substrates;

(7) they are environmentally stable and robust materials;

(8) they are lightweight and tough materials.

Polymers are the only materials that can combine all these properties in order to be successfully applied in integrated optics. The versatility of organic polymers is responsible for the recent rapid growth of research interest in the field of photonics. Just to introduce readers to the aspect of poled NLO polymers for second-order nonlinear optics, a few selective examples are provided.

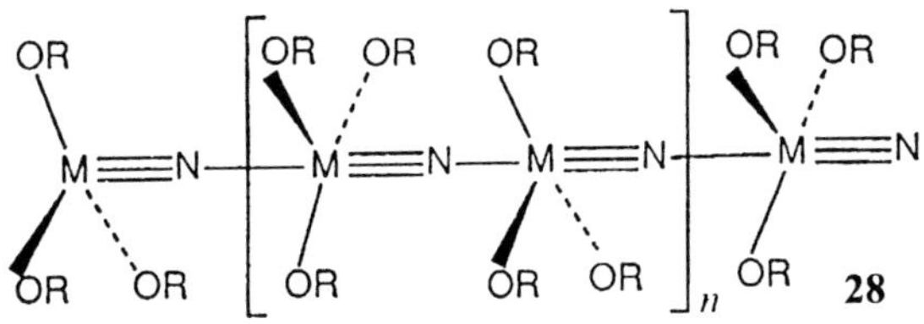

1. NLO chromophore functionalized side-chain polymers In these polymers, NLO chromophores are covalently attached as the side chains. A NLO chromophore may be covalently bound as a pendant group to the polymer backbone directly or through a spacer group. Typical examples are side-chain polymers containing Disperse Red (DR1) dye as a NLO chromophore, such as those reported by Chen *et al.* [66]. The side-chain NLO polymer (**29**) has a T_g of 120 and a d_{33} value of 250 pm V^{-1}. Its largest d_{33} value appears to be due to higher chromophore density and resonant enhancement and it retains 80–90% of the stabilized d_{33} values.

A very wide variety of NLO chromophore functionalized side-chain polymers have been reported in the literature [15]. Another example to develop the structure–property relationship is that of aromatic polyimides containing hexafluoroisopropylidene, reported by Yu *et al.* [67]. The d_{33} coefficients, measured at 1064 nm, were 169, 146 and 103 pm V^{-1} and r_{33} coefficients at 780 nm were 27, 46 and 24 pm V^{-1} for

28

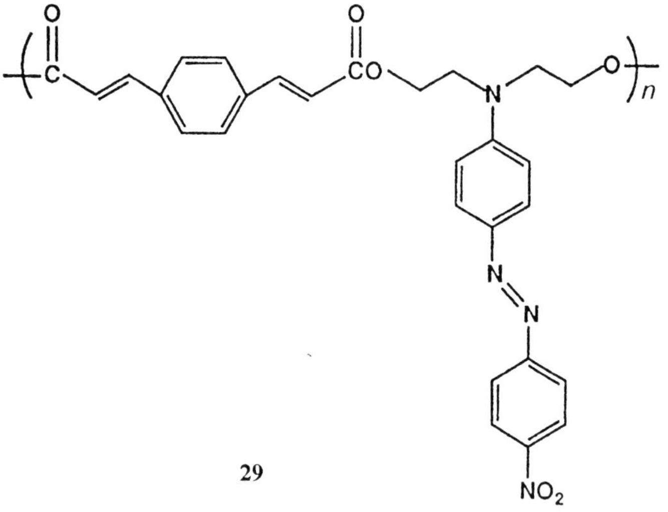

29

polyimides **30–32**, respectively. The T_g values of polyimides **30–32** were 215, 217 and 227°C, respectively. The large d_{33} values of these polyimides are due to resonance enhancement as they show optical absorption between 400 and 477 nm. It is apparent here that thermal, optical and NLO properties of these polyimides were tailored by incorporating a different π-conjugated moiety between two phenyl rings. Polyimide **32** retained 75% of the initial values in air at 150°C over a period of more than 800 h. The SHG intensity decays near its T_g (227°C) and disappears when the temperature was above T_g. Interestingly, a SHG signal

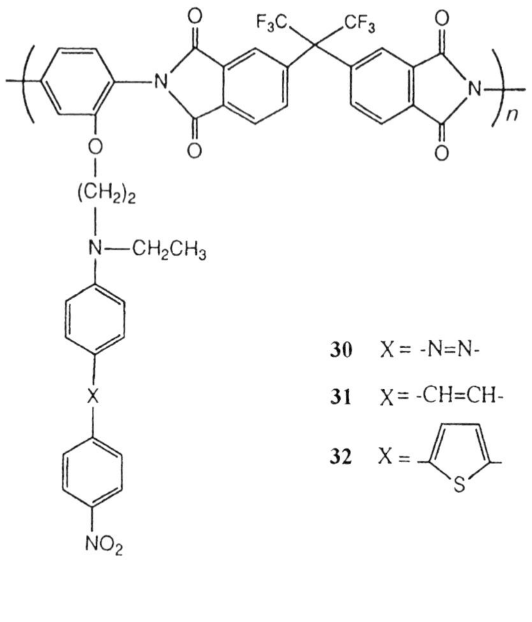

of the same magnitude was recovered when it was repoled by corona discharge, which indicates that the NLO chromophore was not damaged after applying high temperatures. NLO chromophore functionalized polymides are good choices for applications in photonic devices because of their very high thermal stability and high second-order optical nonlinearities, many such polymides have been reported [15].

2. NLO chromophoric main-chain polymers The NLO chromophores are imparted into the polymer main chain either in head-to-tail or tail-to-head configuration. Xu *et al.* [68] reported main-chain polymers that consist of amino-sulfone azobenzene chromophores. The polymer MC1 has a polyurethane backbone (**33**) while MC2 has polyester backbone (**34**).

The MC2 polymers showed a glass transition temperature of 114°C and a d_{33} value of 125 pm V^{-1}. MC1 exhibited a T_g of 62°C and a d_{33} value of 60 pm V^{-1}. Polymer MC1 shows a more rapid decay than polymer MC2 due to the long flexible segments between the rigid chromophores. An initial decay of 25% of the initial d_{33} value was observed for uncrosslinked film whereas no decay was noticed for the crosslinked film; this also suggests that crosslinking can provide temporal NLO stability to poled polymers.

3. Chromophoric main-chain polymers with NLO dye side chains In these polymers, the NLO chromophores can be attached as a side chain as well as incorporated into the main-chain polymer backbone. Such chemical strategy should allow a very high density of NLO moieties. Aromatic polyureas with pendant NLO chromophores are such examples [69,70].

30 X = -N=N-

31 X = -CH=CH-

32 X =

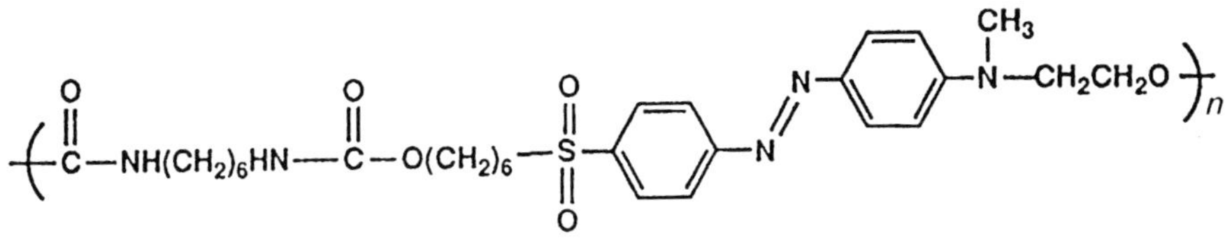

Polymer MC1 **33**

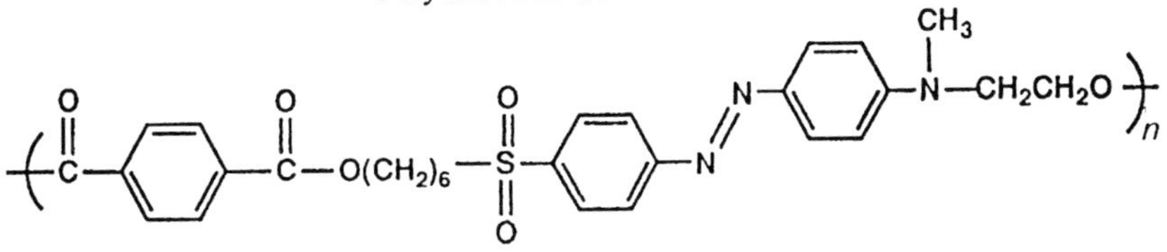

Polymer MC2 **34**

4. Crosslinked NLO polymers The above classes of NLO polymers, either with side-chain NLO chromophores or chromophoric main-chain polymers, can be transformed into a cross linked network polymer using crosslinking agents or thermal curing techniques. Boogers *et al.* [71] reported such crosslinked polyurethanes (**35**). A d_{33} value of 60 pm V^{-1} at 1064 nm was measured for this crosslinked polymeric system. Such a cross linked system provides SHG stability and similar results were observed at 70°C where the d_{33} value decreased to 40 pm V^{-1} after *ca.* 800 h.

5. Ferroelectric polymers Piezoelectric polymers are inherently active to second-order NLO effects.

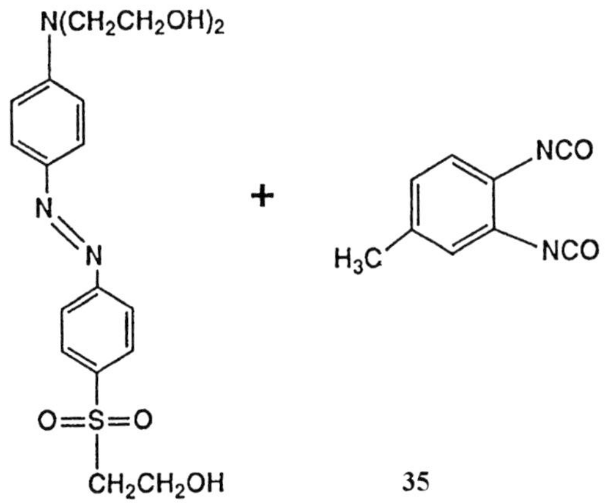

Poly(vinylidene fluoride) PVDF, copolymers of PVDF, cyano-polymers, ferroelectric liquid crystal polymers, odd-numbered nylons and polyureas are examples of such polymers (**36**) [72,73]. The SHG properties of an aromatic polyurea were reported by Nalwa *et al.* [74,75]. This ferroelectric polymer is transparent up to 300 nm and has a d_{33} value of 5.5 pm V^{-1}. It showed very little temporal decay because the d_{33} value decreased to 5.0 pm V^{-1} over a period of 1000 h at room temperature.

6. Guest–host systems Numerous guest–hosts system consisting of NLO dyes and host polymers have been studied because of their versatility and the ease with which they can be processed and fabricated. Watanabe *et al.* [76] reported the SHG properties of a *p*-nitroaniline (p-NA) and poly(oxyethylene) (POE) system. The system showed a SHG intensity 20 to 30 times that of urea depending upon the p-NA concentration. For example, the maximum SHG was observed at 35 wt % –p-NA and it decreased at higher concentrations due the formation of SHG inactive p-NA. More interesting was p-NA/poly(ε-caprolactone) (PCL) where large SHG was observed even without the poling process [77].

NLO polymers offer several advantages, including ease of processing in thin films onto desired substrates, quasi-phase matching, multilayers formation, easy integration with optical components and possible tailoring of NLO properties and material performance

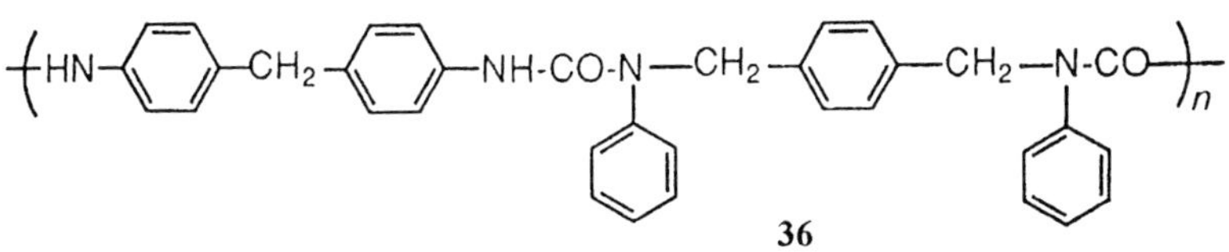

35

36

via chemical modifications. In this chapter, discussion is restricted to NLO chromophore functionalized π-conjugated polymers which are not the conventional type. The SHG and EO properties of π-conjugated oligomers and polymers such as polythiophenes, poly(p-phenylene-vinylenes) (PPV), poly(1,6-hepta-diynes), poly(p-phenylenes) and polydiacetylenes are discussed here. These π-conjugated polymeric systems can be viewed as mutlifunctional materials because they not only show interesting second- and third-order NLO properties but electronic properties as well.

Polyarylenes
Rao *et al.* [78–80] reported the synthesis, electronic absorption and first hyperpolarizabilities of two series of donor–acceptor substituted conjugated thiophene compounds. These thiophene compounds show two absorption bands; an intense low absorption band in the visible region associated with the intramolecular charge transfer transition and second weak absorption band in the UV vis region between 250–700 nm. The lowest absorption band is solvatochromic in nature and denotes a large dipole moment change between the ground and excited states. The lowest absorption charge-transfer bands of various compounds and their chemical structures are listed in Table 6.4.

The β-values of thiophene compounds were measured by EFISH technique in a 1,4-dioxane at 1.907 μm. The compounds with a dicyanovinyl group showed larger $\beta\mu$ values than those containing nitro groups as it is a strong electron acceptor. The replacement of one benzene does not affect the $\beta\mu$ values so much whereas the replacement of both benzene rings with thiophene rings leads to a two-fold increase in $\beta\mu$. A similar trend was observed for

dialkylamino-dicyanovinyl stilbene compounds. A drastic increase of the β value was noticed with the replacement of both benzene rings with thiophenes in dicyanovinyl and nitro-substituted stilbene compounds. The $\beta\mu$ value was found to be quite sensitive to the conjugating moieties connecting the donor and acceptor groups. The $\beta\mu$ values of thiophene compounds are 1.8 to 2.5 times larger than the corresponding benzenoid stilbene compounds. Furthermore, a red shift of about 100 nm in the absorption spectra was observed for a thiophene ring. The plot of $\beta\mu$ against the number of conjugated bonds for thiophene substituents shows no saturation and the linear fit of the data yields a power law exponent of 1.8 for the dicyanovinyl group series and 2.0 for the nitro group series. A series of dithiolyldinemethyl derivatives was reported by Jen *et al.* [81] with an effort to improve electron–donor activity and thermal stability. The presence of similar substituents leads to greater donor ability of the dithiolyldinemethyl group with larger $\mu\beta$ values. The incorporation of a benzenoid ring in the dithiolyldinemethyl group causes a decrease in $\mu\beta$ values. The thiophene ring seems quite effective in enhancing the β value because it introduces more electron delocalization in donor–acceptor compounds than a benzenoid ring. Bithiophene compounds show better thermal stability (at least higher by 50°C) than the stilbene compounds.

Migani *et al.* [82] reported the synthesis and first hyperpolarizabilities of new thiophene compounds. Donor and acceptor groups were attached to the α-positions of thiophene in order to achieve large second-order optical nonlinearity. Two thiophene derivatives (**37** and **38**) exhibited large β values, similar to that of DR1. A comparative study of thiophene compounds with benzene analogs indicates that the thiophene

λ_{max}=510 nm $\beta\mu$=604×10^{-48} esu   **37**

λ_{max}=499 nm $\beta\mu$=260×10^{-48} esu   **38**

Table 6.4 The $\beta\mu$ values measured by the EFISH technique at 1.907 μm and the absorption maxima recorded in dioxane for some thiophene-based donor–acceptor compounds and their benzene analogs (after ref 78–81)

Compound	λ_{max} (nm)	$\beta\mu$ (10^{-48} esu)
Me_2N—⟨benzene⟩—NO_2	370	110
Me_2N—⟨benzene⟩—CH=CH—⟨benzene⟩—NO_2	424	580
Et_2N—⟨benzene⟩—CH=CH—⟨thiophene⟩—NO_2	478	600
Et_2N—⟨benzene⟩—(CH=CH—⟨thiophene⟩)$_2$—NO_2	506	1400
(piperidine)N—⟨thiophene⟩—CH=CH—⟨benzene⟩—NO_2	460	660
(pyrrolidine)N—⟨thiophene⟩—⟨thiophene⟩—NO_2	500	563
Et_2N—⟨thiophene⟩—CH=CH—⟨thiophene⟩—NO_2	516	1040
Et_2N—⟨benzene⟩—CH=C(CN)—CN	419	300
Et_2N—⟨benzene⟩—CH=CH—⟨benzene⟩—CH=C(CN)—CN	468	1100
Et_2N—⟨benzene⟩—CH=CH—⟨benzene⟩—C(CN)=C(CN)—CN	594	2700
Et_2N—⟨benzene⟩—CH=CH—⟨thiophene⟩—CH=C(CN)—CN	513	1300

(*continued overleaf*)

Table 6.4 (*continued*)

Compound	λ_{max} (nm)	$\beta\mu$ (10^{-48} esu)
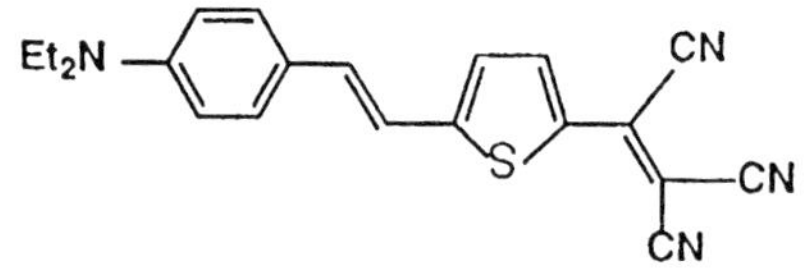	640	6200
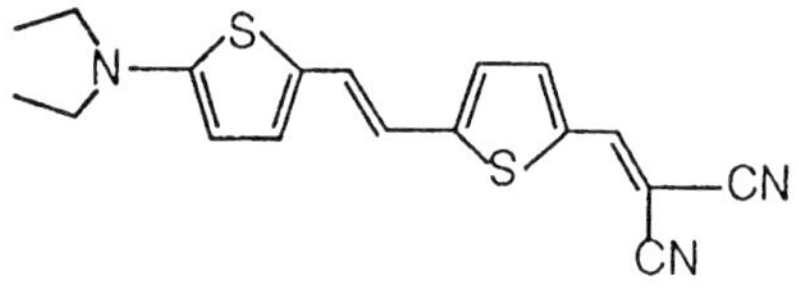	584	2600
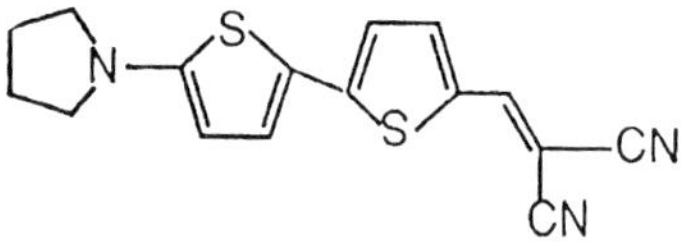	558	1250
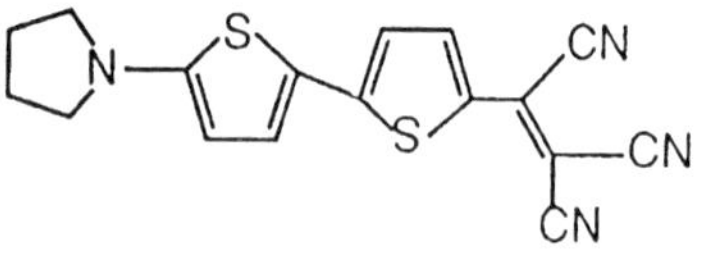	650	2850
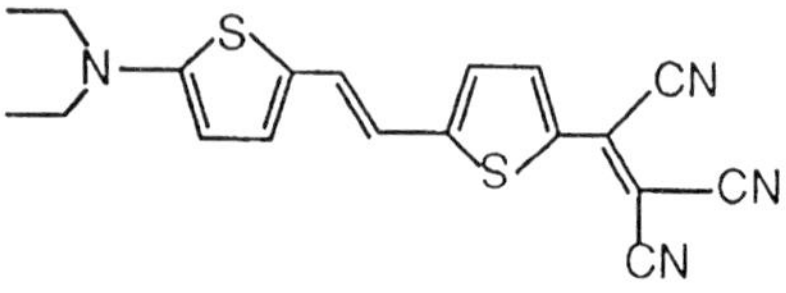	718	6900
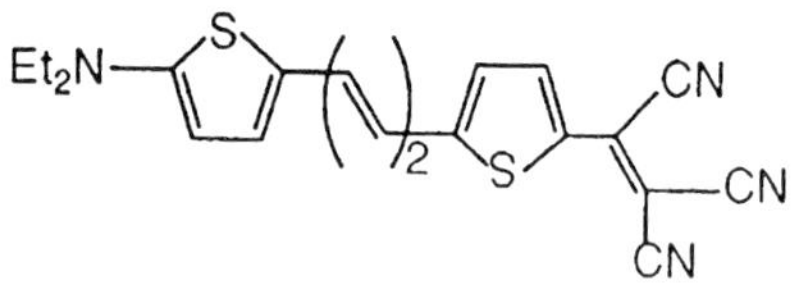	662	9100
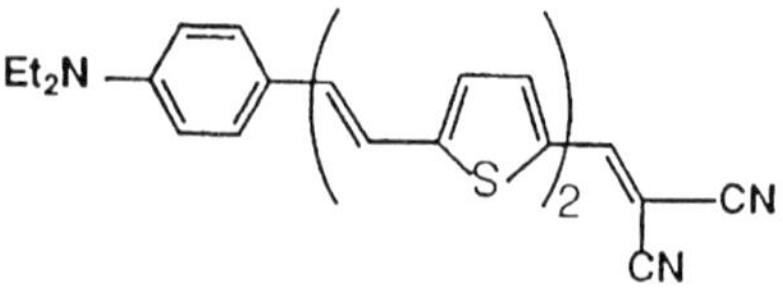	547	2300
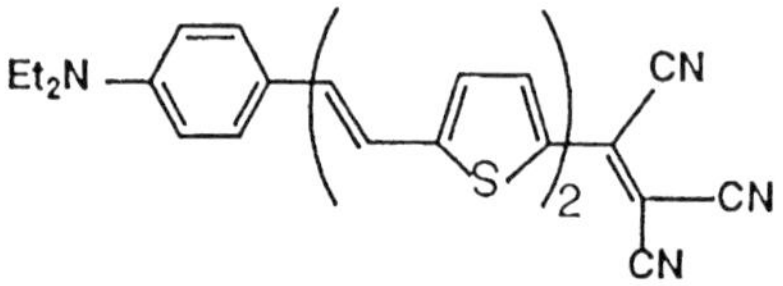	653	7400
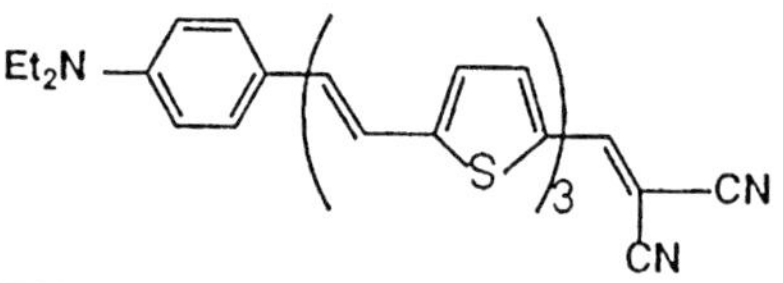	556	3800

Table 6.4 (*continued*)

Compound	λ_{max} (nm)	$\beta\mu$ (10^{-48} esu)
	556	950
	594	1500
	625	1600
	660	3800
	604	1350
	635	3300
	638	2400

moiety provides a good π-electron delocalization pathway. Low-molecular-weight donor–acceptor group substituted thiophenes have emerged as a new class of NLO chromophores with large second-order optical nonlinearity.

Cai and Jen [83] reported a highly thermally stable guest–host system that comprised of polyquinoline (PQ-100) (**39**) and diethyl-amino-tricyanovinyl substituted cinamyl thiophene (RT-9800). This composite system showed a nonresonant r_{33} coefficient of 45 pm V^{-1} at 1.3 μm. An initial decay of r_{33} of about 40% was observed for 100 h and the EO coefficient r_{33}

of 26 pm V^{-1} remain over a period of more than 2000 h.

Chittibabu *et al.* [84] reported the synthesis of poly(3-octylthiophene-co-*N*-(3-thienyl)-4-amino-2-nitrophenol (POMDT) (**41**), which has a side-chain NLO chromophore. The second-order NLO properties of this copolymer were studied when further functionalized with a photo-crosslinkable cinnamoyl group and doped with a photo-crosslinkable NLO chromophore which has large molecular hyperpolarizability. A small SHG signal was detected for the poled and photo-crosslinked cinnamoylated POMDT. This co-

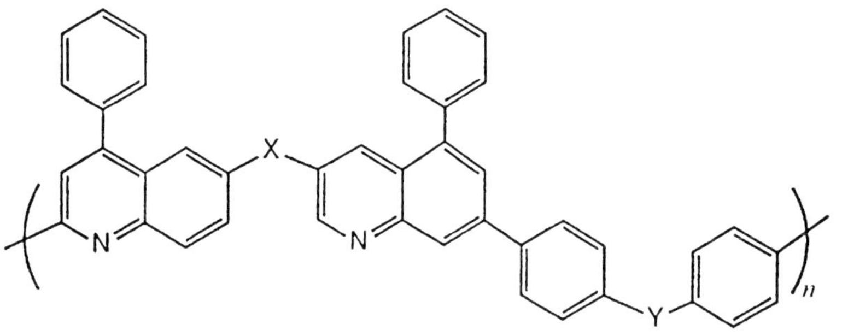

PQ-100 **39**

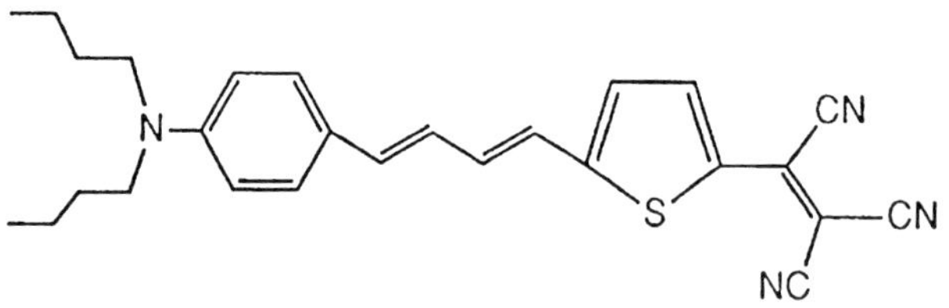

RT-9800 **40**

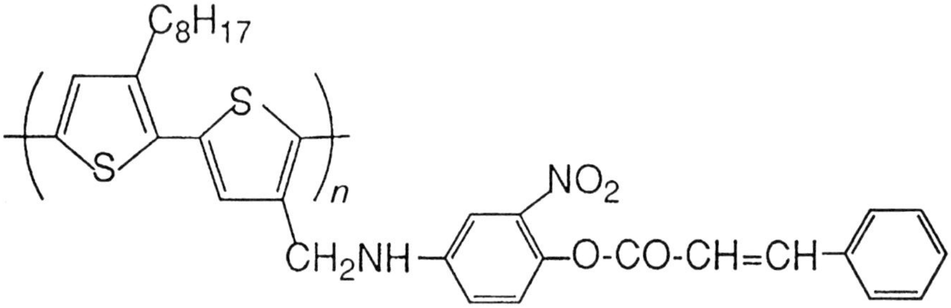

Cinnamoylated POMDT21 **41**

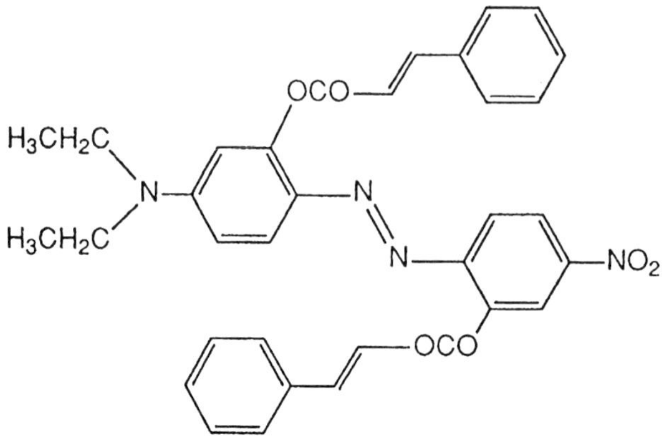

CNNB-R dye **42**

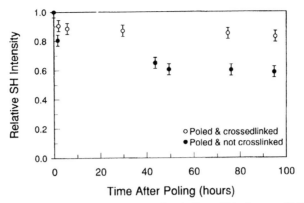

Figure 6.6. Temporal decay of SHG for poled cinnamoylated poly(3-octylthiophene-co-N-(3-thienyl)-4-amino-2-nitrophenol (POMDT) films with and without photo-crosslinking. (Reprinted with permission from ref. 84. Copyright 1994 American Chemical Society).

polymer was doped with 17 wt % CNNB-R dye, **42**. The poled, photo-crosslinked and CNNB-R dye-doped copolymer showed d_{33} values of 18.6 pm V^{-1} at 1064 nm and 3.3 pm V^{-1} at 1542 nm. Figure 6.6 shows the temporal decay of SHG for poled polymer films with and without photo-crosslinking. a significant enhancement in SHG stability was observed for photo-crosslinked polymer films than uncrosslinked films. The NLO coefficient of the photo-crosslinked copolymer was stable at room temperature for 100 h.

Abe *et al.* [85] reported the SHG properties of poly(p-phenylene) derivatives containing the 1-carbonyl-4-(N,N-dimethyl)aminobenzene chromophore, (**43** and **44**). The glass transition temperatures of Poly-X-800 and Poly-X-810 were 222 and 232°C respectively. SHG from cast film after corona poling at about 120°C

was measured at 1.064 μm. The absorption peak of Poly-X-800 was shifted from 340 to 320 nm with an intensity decrease after poling. This was considered an indication of side chains oriented by corona poling and the twisted poly(p-phenylene) main chain. Poly-X-800 corona-poled films showed a d_{33} value of 15.2 pm V^{-1} and d_{31} of 2.5 pm V^{-1} after 24 h of poling, and after a month, a d_{33} value of 11.2 pm V^{-1} and a d_{31} value of 1.9 pm V^{-1} was measured. The d_{33} value of 6.8 pm V^{-1} and d_{31} of 1.9 pm V^{-1} was obtained for Poly-X-810 after 24 h of poling and there was a slight decrease after one month. The high SHG stability of poly(p-phenylene) derivatives was caused by the thermal stability and rigidity of polymer main chains. The temperature dependence of SHG showed a maxima around 130°C for both polymers.

Poly-X-800 **43**

Poly-X-810 **44**

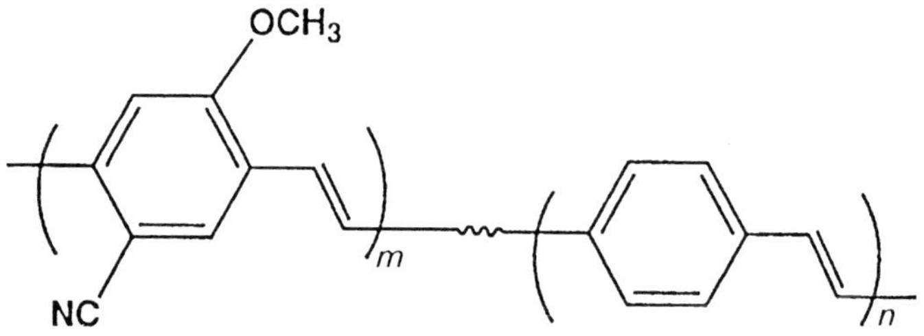

P(CMPV-co-PV)   **45**

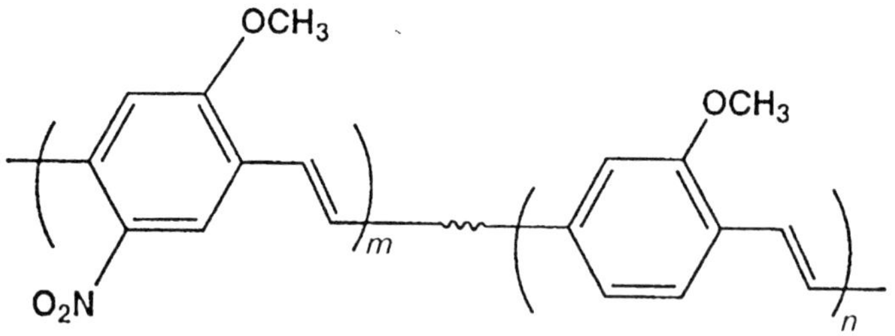

P(MNPV-co-MPV) **46**

Kim *et al.* [86,87] reported the second-order NLO properties of poly(2-cyano-5-methoxy-1,4-phenylene)-vinylene (PCMPC) and *p*-phenylenevinylene copolymers [P(CMPV-co-PV) **45**]. The precursor polymers were prepared from the copolymerization of the bis-sulfonium salt monomers. The conjugated copolymer was obtained by thermal heating of precursors. The EO coefficient of the poled films was measured at

632.8 nm using a Michelson interferometer. The EO coefficient of poled P(CMPV-co-PV) at 1 MV cm^{-1} increased linearly up to 35 wt % of the CMPV contents in the copolymer and then remained unchanged at higher concentrations. A maximum EO coefficient of the copolymer was 1.2 pm V^{-1}. Thermal aging of the poled copolymer films showed stability up to 100°C for several hours when poled during the

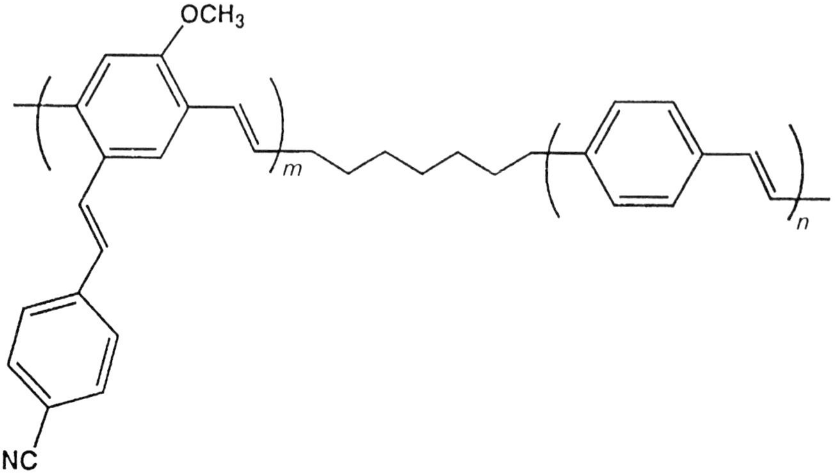

P(CEMPV-co-PV) **47**

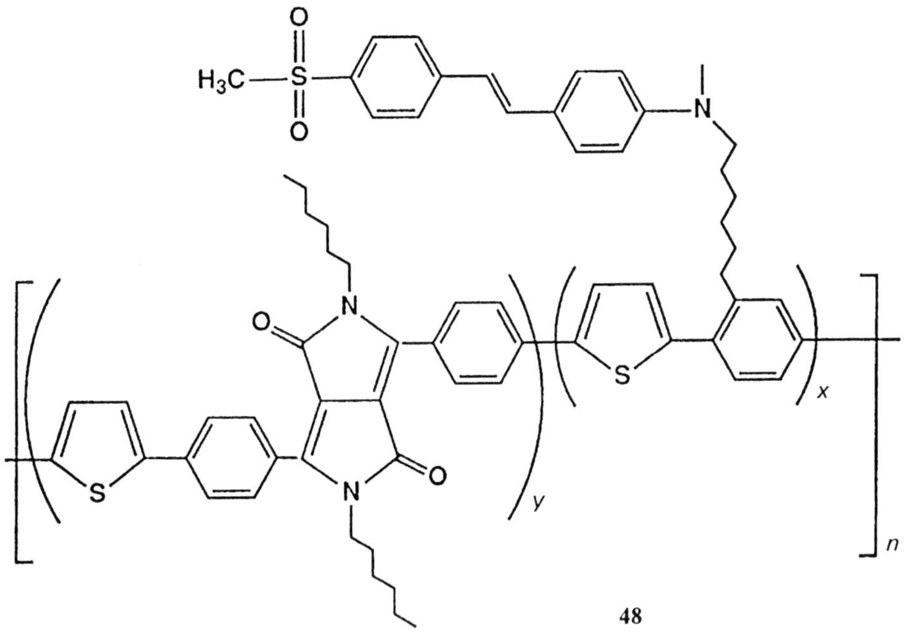

48

elimination process. The SHG of poly[2-methoxy-5-nitrol,4-phenylene-vinylene-co-(2-methoxy-1,4-phenylene)vinylenes) [P(PMNP-co-MPV)] was reported by Jin and Lee [88]. P(MNPV-co-MPV) (46) copolymer containing 67.5 mol% of the MNPV units showed $\chi^{(2)}$ of 1.0×10^{-8} esu. Stable SHG was observed even for thin films heated to 100°C.

Hwang et al. [89] reported second- and third-order NLO properties of poly[2-(2-(4-cyanophenyl)ethenyl)-5-methoxy-1,4-phenylene]vinylene (PCMPV) and a series of copolymers containing 1,4-phenylenevinylene (PV) units. The EO coefficient of corona-poled copolymer films containing 32 mol % of the CEMPV (45) units was found to be 1.2 pm V^{-1}. This EO coefficient is the same as that reported for P(CMPV-co-PV) copolymer by Kim et at [86,87]. The temporal decay of the r_{33} coefficient at room temperature, ambient conditions was studied and no significant decay was observed over a period of 2600 h.

Yu et al. [90,91] reported a multifunctional conjugated photorefractive polymer with a conjugated backbone and a NLO chromophore (48). This polymer ($x = 0.59$ and $y = 0.41$) showed a large d_{33} value of 54 pm V^{-1} at 1064 nm and a r_{33} value of 4.26 pm V^{-1}. This has a 32% weight density of NLO chromophore. This polymer showed the highest SHG stability due to its high T_g. Two-beam experiments

showed large optical gain, 5.7 cm^{-1}, under zero-field conditions. The d_{33} value increased with the increasing content of the NLO chromophore (48). Polymers containing 54% weight ($y = 0.05$, x-0.95) and 57% weight ($y = 0$, $x = 1$) density of NLO chromophores showed d_{33} values of 89 and 94 pm V^{-1} at 1064 nm, respectively.

Polydiacetylenes and polyacetylenes
Kim et al. [92–95] synthesized novel soluble polydiacetylenes (PDAs) consisting of a polarizable side group at one end to enlarge the hyperpolarizability and a high entropy side group to introduce solution processability. The hydrogen-bonded network of poly[(8-butoxycarbonyl)methyl urethanyl) 1-(5-pyrimidyl)-octa-1,3-diyne] (BPOD) (49) represents such an example of a self-assembled film. The spin-coated thin films of poly(BPOD) showed effective SHG coefficients (d_{eff}) of 5.57, 4.43 and 3.32 pm V^{-1} at 1.064 μm for sample thicknesses of 0.31, 0.25 and 0.19 μm, respectively. The SHG of poly(BPOD) remained unchanged upon repetitive heating and cooling though it dropped abruptly above 130°C. Poly(BPOD) with an asymmetric pyrimidyl ring as a weak electron acceptor and flexible urethane side groups showed SHG in both spin-coated and LB films without electrical poling. The 19-multilayer Z-type LB film showed

Poly(BPOD)   **49**

PHD-1 (*x*=3)   PHD-2 (*x*=6)   **50**

PHD-3   **51**

d_{33} of 1.52 pmV^{-1}. Poly(NBPD) and poly (NABD) having NLO chromophores; $-CH_2OCONH-C_6H_4-C_6H_4-NO_2$ and $-CH_2OCONH-C_6H_4-N=N-C_6H_4-NO_2$ showed d_{33} coefficients of 12 pmV^{-1} and 23 pmV^{-1} from LB films, respectively [95]. The noncentrosymmetric alignment responsible for the appearance of SHG resulted from the combined effect of the asymmetry of the two side groups and spontaneous alignment of the urethane groups via hydrogen bonding.

Lee *et al.* [96] investigated the electro-optic properties of poly(1,6-heptadiyne) derivatives bearing pendant NLO chromophores. The fully π-conjugated backbone of poly(1,6-heptadiyne) (PHD) (50) gives rise to third-order NLO properties whereas pendant NLO chromophores exhibit second-order NLO effects. Therefore they are multifunctional polymers. NLO polymers were obtained by homopolymerization and copolymerization of corresponding chromophore monomers with ethyl dipropargyl(diethoxyphosphoryl) acetate using a metathesis catalyst. The resulting polymers were amorphous and thermally stable up to 250°C. Soluble copolymers have a concentration of chromophore monomer up to 50%. These polymers showed two absorption peaks around 390 and 550 nm due to the pendant NLO chromophore and π-conjugated polyene backbone.

The electro-optic coefficients of poled thin films of PHDs (51) were measured by reflection technique at 1.3 μm. The r_{33} coefficients of PHD-1 with a 29 and 43% mole fraction of the chromophoric monomer in the copolymer were 1.7 and 5.2 pm V^{-1}. The r_{33} coefficients for PHD-2 with a 28 and 49% mole fraction of the chromophores in the copolymer were 4.6 and 10.1 pm V^{-1}. PHD-3 with a 20% mole fraction of monomer chromophore in the copolymer showed a r_{33} coefficient of 3.2 pm V^{-1}. The r_{33} coefficient increased as the concentration of NLO chromophore increased in PHD-1 and PHD-2 copolymers. PHD-2 copolymer showed no significant decay of orientational relaxation within 60 days after poling.

3.2 Third-order nonlinear optical effects

3.2.1 Evaluation of third-order optical nonlinearities

3.2.1.1 Theoretical methods

Many quantum chemistry methods and experimental techniques have been used to evaluate the microscopic optical nonlinearities of organic molecules [3,97–99].

Third-order optical nonlinearities of well-defined molecular systems have been calculated using the following approaches.

(1) Empirical methods
    (a) Pariser–Parr–Pople (PPP) method
    (b) The extended Hückel (EH) method
(2) Semiempirical methods
    (a) Complete neglect of differential overlap (CNDO) method
    (b) Modified neglect of diatomic overlap (MNDO) method
    (c) Intermediate neglect differential overlap (INDO) method
(3) *Ab initio* method

The magnitude of the polarizabilities (α) and second hyperpolarizabilities (γ) for a single molecule can vary from one approach to another because of the different molecular considerations. Theoretical methods that are applied to identify useful chemical species can be tailored for third-order nonlinear optics. Computational methods can provide first hand information on the polarizabilities and hyperpolarizabilities of organic materials, though a precise evaluation to match exactly experimental results has not yet been reached. These theoretical approaches have been applied to defined chain of oligomers of polyenes, polyynes, polyazines, polyphenylenes, poly(diacetylenes), polyazomethines as well as to single molecules such as fullerenes. More details of these theoretical methods can be found in ref. 15.

3.2.1.2 Measurement techniques

Measurement techniques such as THG, degenerate four-wave mixing (DFWM), EFISH, optical Kerr gate (OKG), and Z-scan methods are generally used to evaluate the second hyperpolarizability, nonlinear refractive index and third-order NLO susceptibility of organic materials. A brief description of the different techniques is presented in order to understand the origin of various third-order nonlinear optical processes.

Third-harmonic generation (THG)
The THG technique has been frequently used to evaluated the $\chi^{(3)}(-3\omega; \omega, \omega, \omega)$ of organic materials either as thin films or in solutions. A method developed by Kajzar *et al.* [100,101] is discussed

here. The nonlinear polarization for THG process in a medium under an electric field E given by

$$P_{NL}(3\omega) = \frac{\varepsilon_0}{4}\chi^{(3)}(-3\omega; \omega, \omega, \omega)E_\omega^3 \qquad (6.28)$$

THG measurements are done by transmission as a function of the incident light wavelength and the sample (polymer films) is rotated along the axis perpendicular to the beam propagation direction and parallel to the incident light polarization. The harmonic intensity measured as a function of the incident angle (θ) is given by the following equation:

$$
\begin{aligned}
I_{3\omega}(\theta) = {} & \frac{64\pi^2}{c^2}\left(\frac{\chi^{(3)}}{\Delta\varepsilon}\right)_s^2 |\exp[i(\phi_\omega^s + \phi_{3\omega}^p)] \\
& \times \{T_1[1 - \exp(-i\Delta\phi_s)] \\
& + \rho T_2 \exp(i\phi)[\exp(i\Delta\phi_p) - 1]\}|^2 I_\omega^3 \quad (6.29)
\end{aligned}
$$

T_1 and T_2 are factors from the transmission and boundary conditions, I_ω is the incident light intensity, $\Delta\varepsilon = \varepsilon_\omega - \varepsilon_{3\omega}$ is dielectric constant dispersion. The phase mismatch $\Delta\phi$ between fundamental (ω) and harmonic (3ω) frequencies is written as

$$\Delta\phi = \phi_\omega - \phi_{3\omega} = \frac{6\pi}{\lambda}(n_\omega \cos\theta_\omega - n_{3\omega}\cos\theta_{3\omega})l \qquad (6.30)$$

where θ_ω and $\theta_{3\omega}$ are the propagation angles at ω and 3ω frequencies in a given medium respectively, l is sample thickness and ρ is written as

$$\rho = \frac{\left(\dfrac{\chi^{(3)}}{\Delta\varepsilon}\right)_p}{\left(\dfrac{\chi^{(3)}}{\Delta\varepsilon}\right)_s} \qquad (6.31)$$

where the subscripts s and p refer to the substrate (such as fused silica) and the polymer film. The refractive index of a polymer film at harmonic frequency 3ω is complex:

$$n_{3\omega} = n_{3\omega}^r + i\kappa_{3\omega} \qquad (6.32)$$

The harmonic intensity resulting from the polymer film and substrate are calibrated to the 1 mm thick silica plate reference measured under identical conditions and is given by

$$I_{3\omega}^s = \frac{256\pi^4}{c^2}\left(\frac{\chi^{(3)}}{\Delta\varepsilon}\right)_s^2 T^2 \sin^2\left(\frac{\Delta\phi^s}{2}\right)I_\omega^3 \qquad (6.33)$$

where T is a transmission factor associated to the Fresnel transmission factors and boundary conditions at the front and back faces of the plate. The measured

$\langle\chi^{(3)}(-3\omega; \omega, \omega, \omega)\rangle$ is averaged over all polymer chain distributions. The $\chi^{(3)}$ tensor component along the polymer chain direction is enhanced and can be given by

$$\chi_{xxxx}^{(3)} = \langle\chi^{(3)}(-3\omega; \omega, \omega, \omega)\rangle/\langle\cos^4\theta\rangle \qquad (6.34)$$

θ as defined earlier is the angle between the polymer chain and the incident light polarization direction. For a three-dimensional chain disorder, $\langle\cos^4\theta\rangle$ is $1/5$; while when considering all polymer chains parallel to the substrate where disorder is two dimensional, $\langle\cos^4\theta\rangle$ is $3/8$ [102]. $\langle\cos^4\theta\rangle$ is 1 when all the polymer chains are parallel to a given direction. This indicates an increase of a factor of five in $\chi^{(3)}$ and a factor of at least 25 in efficiency depending on the third-order NLO process while going from a completely disordered system to a mono-oriented system such as a single crystal [103]. If the fundamental measuring wavelength corresponds to 3ω in the absorption band then the polymer would have a three-photon resonance enhancement, in that case optical nonlinearity is given by

$$|\chi^{(3)}(-3\omega; \omega, \omega, \omega)| \propto \frac{A}{[(E_{ng} - 3\omega)^2 + \Gamma^2]^{\frac{1}{2}}} \qquad (6.35)$$

where A is a parameter that depends on the frequency and oscillator strength, E_{ng}(eV) is the location of the resonance and Γ is the phenomenological damping term.

Degenerate four-wave mixing (DFWM)

DFWM, one of the most important technique, provides information on the magnitude and the response of third-order nonlinearity. This technique can also be applied to study the photorefractive effects. Many reports on the aspects and analysis of the DFWM technique have been published from the late 1970s. Several excellent reviews and monographs on DFWM and its related nonlinear optical processes have been written [104–108]. The nonlinear polarization for a DFWM process is written as

$$P_{NL}(\omega) = {}^3/_4\chi^{(3)}(-\omega; \omega, \omega - \omega)E_1(\omega)E_3(\omega) \quad (6.36)$$

In the DFWM process, all pump and probe beams have, at the same frequency ω, [109,110]:

$$|\chi^{(3)}(-\omega; \omega, \omega - \omega)| = \frac{4\alpha_0 c^2 n^2 \varepsilon_0}{3\omega I_s^0} \qquad (6.37)$$

where the absorption is given as

$$\alpha_0 = \frac{\Delta N_0 T_2 \omega \mu^2}{2\varepsilon_0 hnc} \qquad (6.38)$$

where n is the linear refractive index, ω is the frequency, I is the pump intensity, T is the transmission at I, and ΔN_0 is the population difference. In the DFWM process, three input beams of intensity $I_1(\omega)$, $I_2(\omega)$ and $I_3(\omega)$ and equal frequency ω interact in a medium where two of them, i.e. $I_1(\omega)$ and $I_2(\omega)$ counterpropagate and the third $I_3(\omega)$; the probe beam crosses at a small angle θ. A beam of intensity $I_4(\omega)$ and the same frequency is generated by the third-order NLO interaction $\chi^{(3)}$ counterpropagates relative to the $I_3(\omega)$ beam. $I_4(\omega)$ is related to $\chi^{(3)}$ by the following equation:

$$I_4(\omega) = \left(\frac{\omega}{2\varepsilon_0 cn_0^2}\right)^2 (\chi^{(3)})^2 l^2 I_1 I_2 I_3 \qquad (6.39)$$

where n_0 is the linear refractive index, l is the interaction length, and I_1, I_2 and I_3 are the pump beam intensities. By knowing $I_4(\omega)$ at the incident fields, the tensor component $\chi^{(3)}_{1111}(-\omega; \omega, \omega, -\omega)$ of the $\chi^{(3)}$ can be determined. $\chi^{(3)}$ for an absorbing medium is given by [109]

$$\chi^{(3)}_{sample} = \chi^{(3)}_{ref}$$

$$\times \frac{\left(\dfrac{n_{sample}}{n_{ref}}\right)^2 \left(\dfrac{l_{ref}}{l_{sample}}\right)\left(\dfrac{I_{sample}}{I_{ref}}\right)^{1/2} \alpha l_{sample}}{\exp\{(-\alpha l_{sample}/2)[1 - \exp(-\alpha l_{sample})]\}} \qquad (6.40)$$

where n is the refractive index, l is the interaction length in the sample and reference, and α is the linear absorption coefficient of the sample. The reference is carbon disulfide (CS_2). It is assumed that the reference CS_2 does not absorb at the measurement wavelength. For nonabsorbing media, the equation is modified as follows [8]:

$$\chi^{(3)}_{sample} = \chi^{(3)}_{ref}\left(\frac{n_{sample}}{n_{ref}}\right)^2 \left(\frac{l_{ref}}{l_{sample}}\right)\left(\frac{I_{sample}}{I_{ref}}\right)^{\frac{1}{2}} \qquad (6.41)$$

The second hyperpolarizability (γ) which is related to $\chi^{(3)}$ can be calculated by the following expression:

$$\langle \gamma_{xxxx} \rangle = \frac{\chi^{(3)}_{xxxx}}{L^4 N} \qquad (6.42)$$

where L is the local field factor $|(n^2 + 2)/3|$ and N is the number of density of the sample.

Generally the $\chi^{(3)}$ value obtained from the DFWM method is enhanced compared to that obtained with THG because of the degeneracy of the macroscopic nonlinearity, changes in the microscopic mechanism, thermal grating and increased resonant enhancement.

Electric field-induced second-harmonic generation (EFISH)

Both first hyperpolarizability (β) and second hyperpolarizability (γ) of organic materials can be evaluated using solution-phase dc EFISH. Initially, Levine and Bethea [111] used dc EFISH technique for measuring hyperpolarizabilities in solutions. Both the modulus and phase of $\chi^{(3)}(-2\omega; \omega, \omega, 0)$ can be evaluated from the EFISH technique by performing measurements as a function of solute concentration. In centrosymmetric medium, the nonlinear polarization at 2ω results from the term $\chi^{(3)}(-2\omega; \omega, \omega, 0)$ while in the non-centrosymmetric medium both first and second hyperpolarizabilities contribute to the $\chi^{(3)}(-2\omega; \omega, \omega, 0)$. The effective γ can be written as

$$\gamma^{EFISH} = \gamma(-2\omega; \omega, \omega, 0) + \frac{\mu\beta}{5kT} \qquad (6.43)$$

where the μ is the permanent dipole moment and β is the first hyperpolarizability. For π-conjugated polymers, β is zero due to their centrosymmetry; therefore the $\gamma(-2\omega; \omega, \omega, 0)$ term contributed and is directly comparable to the γ involved in THG. The γ-term has both electronic and vibronic second hyperpolarizabilities and their sum is related to γ-THG. For long π-conjugated molecules end-capped with donor–acceptor groups, the $\gamma(-2\omega; \omega, \omega, 0)$ contribution is significantly larger than the $\mu\beta/5kT$ term. Contrary to this, if the $\mu\beta/5kT$ term is much larger than $\gamma(-2\omega; \omega, \omega, 0)$ contribution then one may neglect this part. Both the γ and $\chi^{(3)}$ values of thin films and solutions or organic materials can be determined using the EFISH technique.

It can be seen that different nonlinear optical processes are involved with different measurement techniques. For example, the third-order optical nonlinearity of $\chi^{(3)}(-3\omega; \omega, \omega, \omega)$, $\chi^{(3)}(-\omega; \omega, -\omega, \omega)$, $\chi^{(3)}(-2\omega; \omega, \omega, 0)$ $\chi^{(3)}(-\omega; 0, 0, \omega)$ are measured by the THG technique, DFWM method, EFISH generation, and the dc Kerr effect, respectively.

Nonlinear refractive index

The figure of merit (FOM) for $\chi^{(3)}$ can be written as $\chi^{(3)}/\alpha\tau$ where α is the absorption coefficient and τ is the response speed. In the case of one-photon absorption, the FOM for the NLO device can also be expressed as $n_2(\lambda)/\alpha\lambda$ where n_2 is the nonlinear refractive index, λ is the wavelength of the light and α is the absorption coefficient at λ. The intensity-dependent refractive index of the OKE is expressed as

$$n = n_0 + n_2 I \qquad (6.44)$$

where n_0 is the linear refractive index and n_2 is the nonlinear refractive index. n_2 and $\chi^{(3)}$ are related to each other as [8]

$$n_2 = \frac{12\pi^2}{cn_0^2}\chi^{(3)}(-\omega; \omega, \omega, -\omega) \qquad (6.45)$$

and when both are in esu units

$$n_2 = \frac{5.26 \times 10^{-6}\chi^{(3)}}{n_0^2} \qquad (6.46)$$

where n_2 is in MKS (m^2 W^{-1}) units and $\chi^{(3)}$ is in esu units. The conversion of n_2 to $\chi^{(3)}$ can be done as follows:

$$n_2(m^2/v^2) = \frac{3}{8n}\chi^{(3)}(m^2/v^2) \qquad (6.47)$$

and

$$\chi^{(3)}(m^2/v^2) = \frac{4\pi}{9 \times 10^8}\chi^{(3)} \text{ (esu)} \qquad (6.48)$$

The sign and the magnitude of the real and the imaginary parts of the excited state optical nonlinearities can be evaluated by the Z-scan technique. The details of this technique can be found in refs. 112–114. In the case of the saturation absorption (SA), n_2 has both a real and an imaginary part [115]. The real part of $\chi^{(3)}$ is related to second hyperpolarizability γ as

$$\text{Re } \chi^{(3)} = \frac{cn_0^2}{16\pi^2}\gamma \qquad (6.49)$$

and the imaginary part of $\chi^{(3)}$ is related to the change of the absorption $\Delta\alpha$ as follows:

$$\text{Im } \chi^{(3)} = \frac{cn_0^2}{32\pi^2}\frac{c\alpha_0}{\omega I_s} \qquad (6.50)$$

When the pump frequency reaches the single-photon transition frequency in the NLO materials, then the resonant part of $\chi^{(3)}(-\omega; \omega, -\omega, \omega)$ is dominated by the following terms [8]:

$$\chi_R^{(3)} \propto i\left[\frac{1}{(\omega_{gi} - \omega + i\Gamma_{gi})^2} - \frac{1}{(\omega_{gi} - \omega)^2 + \Gamma_{gi}^2}\right] \qquad (6.51)$$

This represents the two-level system with ^1A_g as the ground state g and ^1B_u as the excited state i, and Γ_{gi} is the damping term.

3.2.2 Optimization of third-order optical nonlinearity

Both theoretical and experimental studies on organic materials demonstrate that third-order optical nonlinearity can be significantly altered by factors such as the length of π-conjugation, atoms and bonds, π-bonding sequence, confirmation, dimensionality, donor–acceptor substituents and charge transfer complex formation. A detail description of these factors have been provided by Nalwa [116]. The impact of dimensionality and orientation on third-order optical nonlinearities of conjugated polymers has been discussed by Kajzar [117]. Ducing [105] ascertained that the second hyperpolarizability should decrease as the square root of the increasing dimensionality. In this regard, third-order optical nonlinearity should follow an order: 1D (polyenes, polyynes) > 2D (phthalocyanines, porphyrins) > 3D (fullerenes). To visualize the effect of various factors on third-order optical nonlinearity of two-dimensional materials, such as phthalocyanines and porphyrins, would not be as straightforward as in one-dimensional conjugated systems. In the case of 1D π-conjugated materials, the two most important factors that govern the magnitude of third-order optical nonlinearity are (1) the length of π-conjugation and (2) the nature of the π-bonding sequence.

3.2.2.1 Length of π-conjugation

Many systematic studies for developing relationships between the π-conjugation length and third-order optical nonlinearities have been performed by taking into account the model compounds. A few examples substantiating the effect of length of π-conjugation on third-order nonlinearity of π-conjugated polymers are provided here. The following relationships between the delocalization length L and the number of electrons N was developed by Ducing [118]:

$$\alpha \propto L^4/N \qquad (6.52)$$
$$\gamma \propto L^{10}/N^3 \qquad (6.53)$$

For the one-dimensional system, $\alpha \propto L^3$, and $\gamma \propto L^7$ since $N \propto L$. For a two-dimensional system, $\alpha \propto L^2$, and $\gamma \propto L^4$ since $N \propto L^2$. This indicates that both the

Table 6.5. Averaged γ values (10^{-36} esu) and VEH band gap (E_g) for polyenes ranging in size from $N = 4$ to 30. (After ref. 122.)

N	Band gap (eV)	Shuai and Bredas [122]		Kurtz [123]	Hurst et al. [124]		
		VEH	SSH-SOS	MNDO-FF	STO-3G	6-31G	6-31G + PD
4	5.41	0.51	1.36	2.14	0.25	0.55	7.48
6	4.13	6.42	14.20	15.18	2.74	4.97	17.69
8	3.49	36.70	59.34	52.06	11.33	20.54	41.41
10	3.08	127.50	163.42	125.45	31.12	57.73	89.87
12	2.78	333.55	349.16	243.30	66.69	127.85	174.13
14	2.59	721.20	628.90	408.15	120.90	293.94	303.98
16	2.43	1358.62	1003.40	618.27	197.47	407.40	491.71
18	2.32	2305.96	1464.36	869.25	290.29	619.66	
20	2.22	3608.00	1998.52	1155.95	405.38	896.76	
22	2.14	5295.17	2590.92	1472.84	526.39	1198.92	
24	2.08	7374.19	3227.44	1814.86			
26	2.03	9838.30	3896.22	2178.10			
28	1.98	12664.93	4587.58	2558.32			
30	1.93	15826.55	5294.38	2952.42			

delocalization length and the dimensionality of the molecules have a significant effect on polarizability and second hyperpolarizability. Rustagi and Ducuing [119] estimated a relationship for the one-dimensional free-electron gas model as $\alpha \propto L^3$, and $\gamma \propto L^5$; this demonstrated the overestimated calculations in the unsolved approximation. A weaker dependence ($\gamma \propto L^6$) of the delocalization length was demonstrated by a tight-bonding approximation. Agrawal *et al.* [120,121] suggested that $\chi^{(3)}$ in the optical transparency region follows a sixth-power dependence on the π-electron conjugation length where the following expression is a good approximation for one-dimensional π-conjugated polymers:

$$\chi^{(3)} \propto (N_d)^6 \qquad (6.54)$$

where N_d is the π-electron delocalization length at the Brillouin edge. One-dimensional tight-bonding theory for infinite chain conjugated polymers predicted the following relationship between the energy gap E_g and $\chi^{(3)}$:

$$\chi^{(3)} \propto (1/E_g)^6 \qquad (6.55)$$

The above studies showed that the π-conjugated polymers with small band gap are useful materials for third-order nonlinear optics.

Shuai and Bredas [122] used the valence effective Hamiltonian (VEH) method to calculate the static and dynamic γ tensors of oligomers of polyacetylene, poly(p-phenylenevinylene) (PPV), and poly(thienylene vinylene) (PTV) and compared their data with SSH-SOS, MNDO finite field by Kurtz [123] and Hartee–

Fock *ab initio* calculations by Hurst *et al.* [124]. Table 6.5 compares the averaged γ values for polyenes ranging in size from $N = 4$ to 30. Hurst *et al.* [124] applied coupled-perturbed Hartree–Fock (CPHF) theory to calculate α and γ values of all-*trans* polyene molecules in the series C_4H_6 to $C_{22}H_{24}$. The γ value of the $C_{20}H_{22}$ polyene was estimated to be an order of magnitude larger than that of the $C_{10}H_{12}$ polyene by them. The static $\chi^{(3)}$ values for the polyacetylene chain

$n = 1–8$ **52**

53

54

55

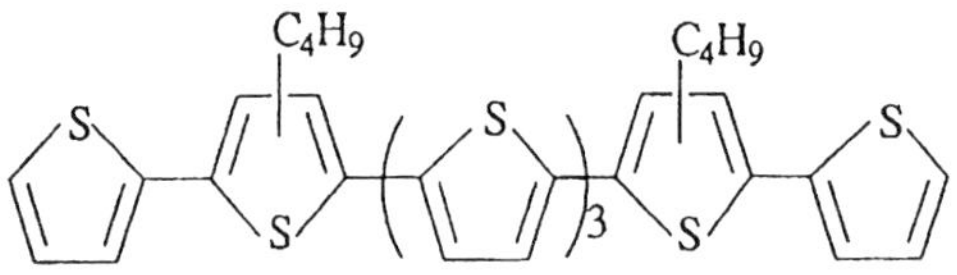

56

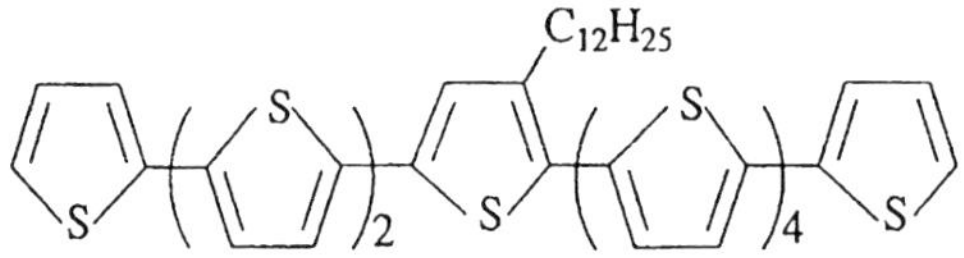

57

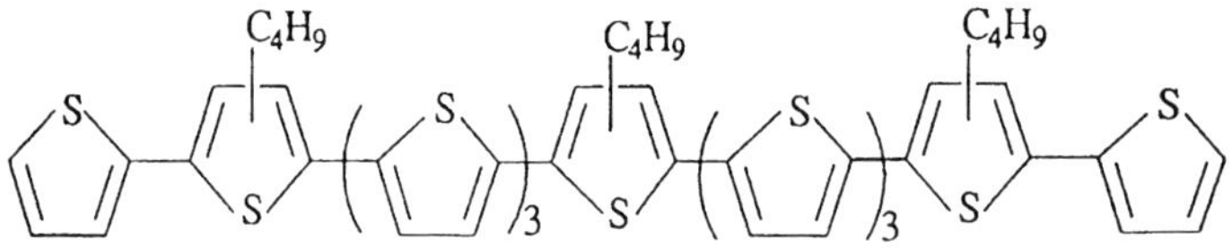

58

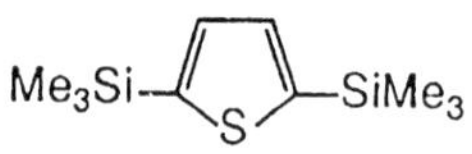

59

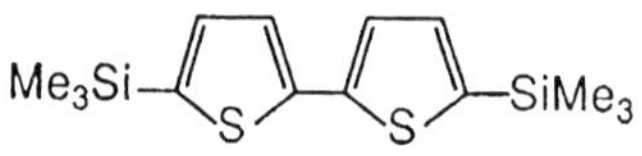

60

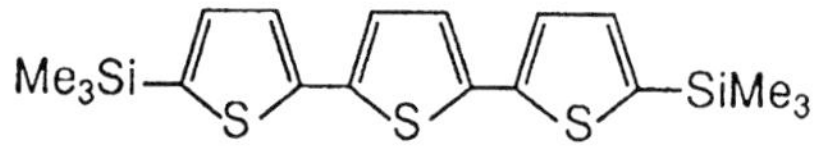

61

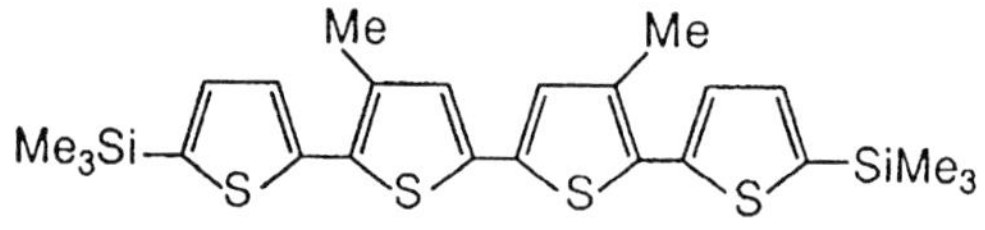

62

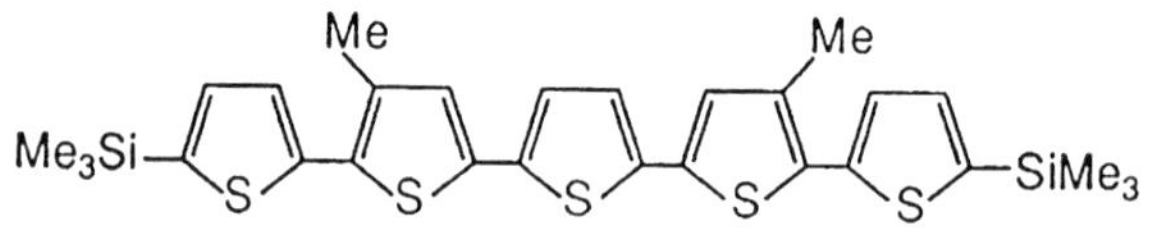

63

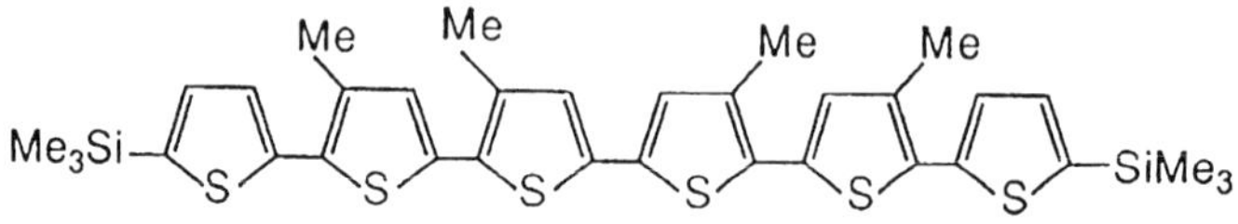

64

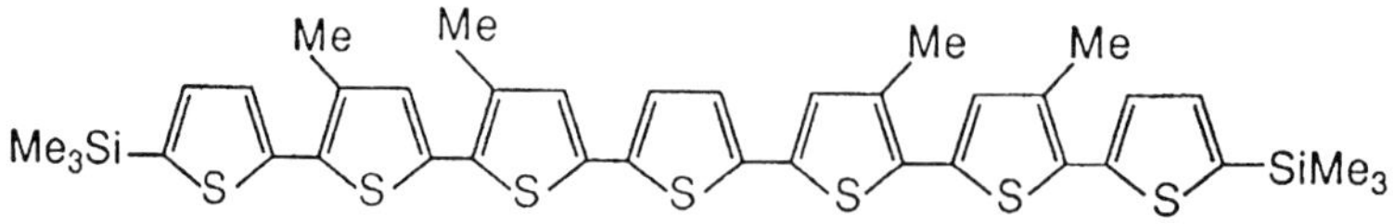

65

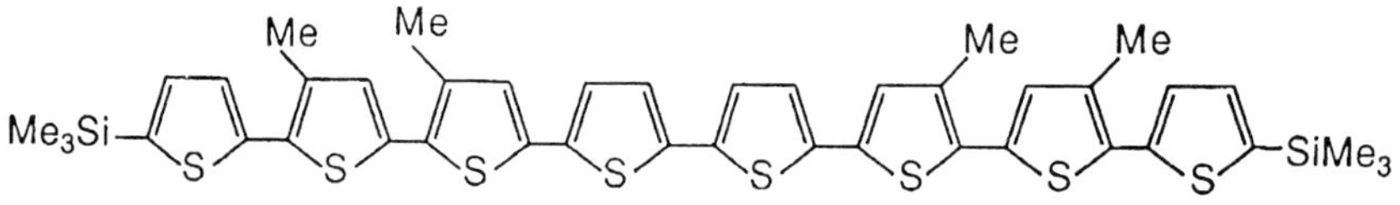

66

was estimated to be 4.2×10^{-12} esu from γ-values obtained from the 6-31G basis set.

The γ values calculated by the VEH method showed a rapid increase for large-sized N values, though their magnitude was within the order of 6-31G *ab initio* values. On the other hand, the SSH-SOS values were close to the *ab initio* 6-31G+PD for short and intermediate polyene chain lengths. For $N = 200$ polyene, the γ_{xxxx} value was estimated as 3.75×10^{-31} esu and the $\chi^{(3)}$ value as 9.84×10^{-11} esu. The value of $\chi^{(3)}$ corresponds to 1.85×10^{-10} esu for a well-oriented polyacetylene chain where a factor of $^3/_8$ was applied. The features of the VEH-SOS theoretical THG spectra of PPV for $N = 1$ and 4 oligomers were found to be similar to those of the $N = 6$ and 30 polyenes. The VEH-SOS α and γ values for phenyl-capped PPV and PTV oligomers were compared and the γ values for the polyene chain were 16 times larger than those for PPV oligomers and five times larger than those for PTV oligomers. The magnitude of the static γ values was also larger for PTV oligomers than for the PPV oligomers due to the delocalization charge which is more facilitated in PTV outside of the thiophene ring along the chain than benzene.

Andre *et al.* [125,126] studied the influence of the π-conjugated chain length on the dipole polarizability for *trans*-1,3,5-hexatriene, *trans*-1,3,5,7,9-decapentaene 1,5-hexadiene-3-yne, and 1,5,9-decatriene-3,7-diyne by *ab initio* CPHF using the STO-3G basis set. The α values increased with increasing chain length. Garito and coworkers [127,128] reported a remarkable increase in γ values with increasing π-conjugation chain length for polyenes.

The nonlinear optical properties of a series of α-conjugated thiophene oligomers (α-nT) were reported by Fichou *et al.* [129]. This study constitutes an excellent example of the understanding of the structure–property relationship. The α-conjugated thiophene oligomers consisting of repeat thiophene units $n = 3,4,5,6$, and 8 were prepared as the model compounds. α-3T, α-4T, α-5T, α-6T are soluble while α-8T is an insoluble material. The thin films of

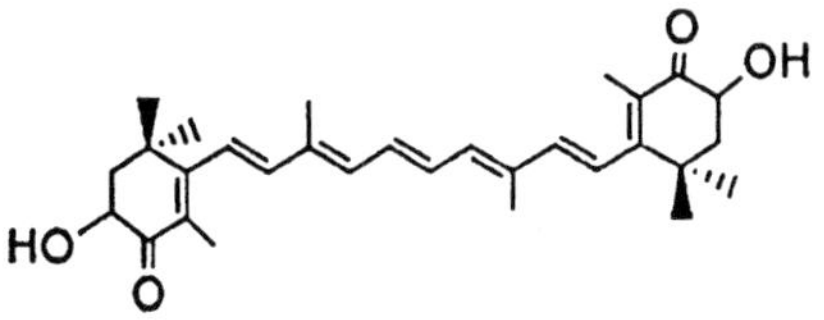

C30 astaxanthin **68**

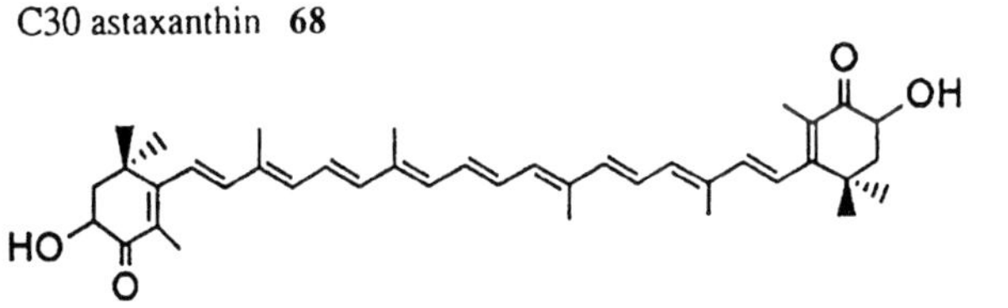

C40 astaxanthin **69**

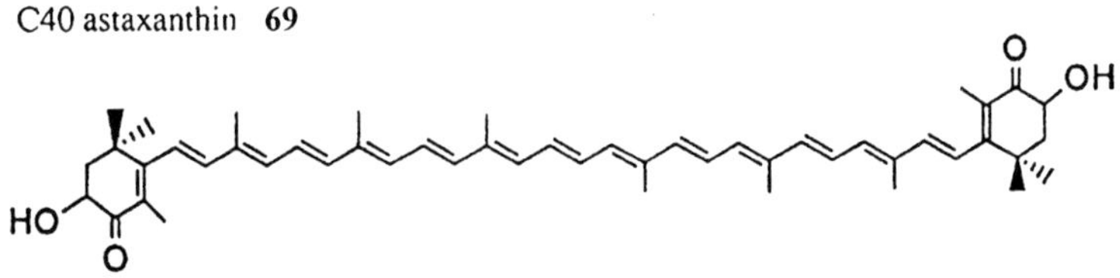

C50 astaxanthin **70**

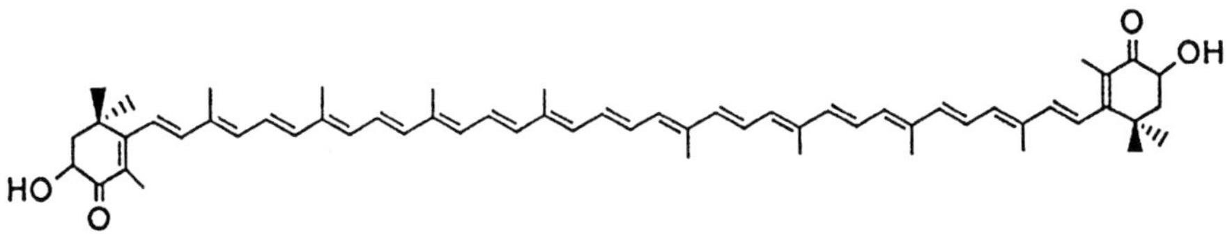

C60 astaxanthin **71**

thicknesses 2000 to 9000 Å of these α-nT oligomers were vacuum evaporated onto a glass slide by heating the powdered materials under suitable conditions of pressure. The $\chi^{(3)}$ values of the thin films were determined from THG using the Maker-fringe method by rotating the films around their vertical axis. The variation of $\chi^{(3)}$ of α-3T and α-6T as a function of the fundamental wavelength over the 0.8 to 1.9 μm measurement wavelength region was studied. α-6T showed a $\chi^{(3)}$ value of 2.38×10^{-12} esu at 1.907 μm. The $\chi^{(3)}$ spectrum has three peaks at around 1.06, 1.35 and 1.91 μm. In a benzene solution, α-6T shows an absorption peak at 432 nm, while this peak was shifted to 415 nm and accompanied by several small shoulders in the 398 to 600 nm region. The first two peaks arise from three-photon resonance and correspond to α-6T conformers absorbing in the 350 to 450 nm region with two to three planar thiophene units of equivalent conjugation length. The third peak may also be due to three-photon resonance in highly conjugated α-6T conformers with six planar thiophene units of equivalent conjugation length. The large optical nonlinearity at 1.91 μm results from the increased planarity of the conjugated system which is a suitable condition for large $\chi^{(3)}$ values. The $\chi^{(3)}$ value of other low-molecular-weight oligomers was smaller than that of α-6T. For α-3T, the value of $\chi^{(3)}$ was about an order of magnitude smaller than that of α-6T over all the wavelength measurement region. α-5T exhibits a $\chi^{(3)}$ value of 1.88×10^{-12} esu at 1.907 μm.

The γ values of α-nT (where n, the number of thiophene units, is 4, 6, 8, 10, 12 and 14) were calculated using self-consistent CNDO computation of the excited states energies and transition dipole moments. In these theoretical computations, mono-excited states and configuration interactions were also taken into consideration. The calculated γ and $\chi^{(3)}$ values as a function of the number of thiophene repeat units were determined. There was a continuous increase of nonlinear optical parameters as the thiophene units (n) reaches to 14. On the other hand, the energy of the first allowed transition showed no significant change after eight thiophene units (n = 8). Both THG experimental results and computational data demonstrate that the nonlinear optical properties of oligothiophenes are strongly influenced by both the conjugation length and the spatial conformation. Both γ and $\chi^{(3)}$ values of the α-nT oligomers increased significantly up to N = 14 while the optical gap remains almost constant after N = 8.

Thienpont et al. [130] synthesized alkyl-substituted oligothiophenes with a repeat unit (n) of 3, 4, 5, 7, 9

and 11 (53–58). The second hyperpolarizabilities of these well-defined oligothiophenes were measured by the EFISH technique at the fundamental wavelength of 1.064 μm. The spin-coated thin films of these materials on indium tin oxide (ITO) glass substrate were prepared by dispersing a few weight per cent of sample in a poly(methyl methacrylate) PMMA matrix using chlorobenzene as a solvent. The maximum concentrations were 5 wt % for n = 3 to 0.8 wt % for n = 11. The dc electric field, up to 1.5 MV cm^{-1} was applied in a corona poling technique. A y-cut crystal quartz plate was used as a reference to calibrate the second-harmonic intensity. The concentration and electric field dependence yield the hyperpolarizability parameters of the oligothiophenes. A strong dependence of conjugation chain length on the γ values was observed for $n \geq 7$; it then starts slowing down as the π-electron system increases further. For example, the γ value changes by two orders of magnitude from n = 3 to 7. There was also a significant decrease in the band gap, which also shows some preliminary saturation behavior at n = 9. The power law expressed by the extent of conjugation dependence on α and γ gave an exponent of 2.4 and 4.6. Fichou et al. [129] point out that a smaller optical nonlinearity would arise if the thiophene ring rotates around the single bonds connecting the thiophene units. Thienpont et al. [130] suggested that the alkyl-substituted oligothiophene molecules are not fully planar in a PMMA matrix at room temperature and a large γ value can be anticipated if the molecules are planarized for long conjugated oligothiophenes. Therefore conformation should play a great role in the optimization of third-order optical nonlinearity.

Cheng et al. [131] reported γ values of trimethylsilyl end-capped α-thiophene oligomers (59)–(66) up to the octamer measured by the THG technique at 1.91 μm in tetrahydrofuran (THF) solution. These trimethylsilyl end-capped oligomer have red-shifted absorption maxima when compared with unsubstituted thiophenes. The λ_{max} values measured in chloroform solution were at 248, 320, 368, 396, 418, 430, 448 and 458 nm for n = 1, 2, 3, 4, 5, 6, 7 and 8, respectively where n is the number of thiophene units. Figure 6.7 shows γ vs. the number of thiophene units; it compares the data from Thienpont et al. [130] and Cheng et al. [131]. The $\gamma(-3\omega: \omega, \omega, \omega)$ values were 6, 15, 41, 142, 249, 320, 450 and 620×10^{-36} esu for n = 1, 2, 3, 4, 5, 6, 7 and 8, respectively. The γ value increases rapidly as the length of the π-conjugation in the oligomers increases. The rate of increase of the γ values was significantly smaller than those reported by

Table 6.6 Polarizability components of oligoazines obtained from 6-31G basis set. The α is in Å^3 and γ is in 10^{-36} esu units. The orientationally averaged values are $\langle\alpha\rangle = \frac{1}{3}(\alpha_{xx} + \alpha_{yy} + \alpha_{zz})$ and $\langle\gamma\rangle = \frac{1}{5}(\gamma_{xxxx} + \gamma_{yyyy} + \gamma_{zzzz} + 2\gamma_{xxyy} + 2\gamma_{xxzz} + 2\gamma_{yyzz})$. The polarizability axes x and z are perpendicular to the π-conjugated backbone while the y axis is along the periodicity direction. The $\chi^{(3)}(-3\omega;\omega,\omega,\omega)$ of oligoazines was measured by THG at 1.50 μm. (After ref. 131–133)

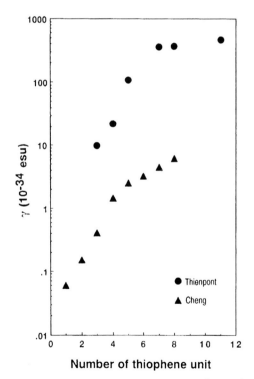

(67)

Repeat unit (n)	Polarizability tensors (R = H)			$\chi^{(3)}(10^{-13}$ esu) (R = CH$_3$)
	$\langle\alpha\rangle$	γ_{yyyy}	$\langle\gamma\rangle$	
1	7.99	15.22	3.34	1.39
2	16.32	79.00	19.73	3.89
3	25.60	228.38	57.19	8.67
4	35.30	390.40	112.03	–
5	45.18	592.88	175.91	15.3
7	–	–	–	23.8

Thienpont *et al.* from EFISH measurements [130]. The heptamer shows a THG γ value which is seven times lower than the EFISH value. The scaling law yields a value of 2.8, remarkably lower than the 4.54 obtained from EFISH measurements and no saturation of γ was noticed up to octamer, which is also in disagreement with the results of Thienpont *et al.* [130], though in good agreement to that reported by Fichou *et al.* [129]. It may be concluded that third-order optical nonlinearity of polythiophenes can not be optimized for short-chain oligomers and much larger π-conjugation is required. This is similar to that shown experimentally for polyenes.

Nalwa *et al.* [131–133] calculated α and γ values of the oligoazines with the chemical formula $H_2N-[N=CH-CH=N]_n-NH_2$ where n is 1, 2, 3, 4, and 5 using *ab initio* CPHF with the 6-31G basis set. Table 6.6 lists the components of α and γ for oligoazine model compounds. The $\langle\gamma\rangle$ values increase by a factor of six in going from the monomer to the dimer and a factor of more than 53 from the monomer to the pentamer. The exponents for α and γ are in good agreement with other π-conjugated systems [124–129, 134]. The $\chi^{(3)}$ values of oligoazines thin films measured by the THG technique at 1.5 μm are also listed in Table 6.6. The $\chi^{(3)}$ values of the pentamer is more than an order of magnitude higher than that of the monomer. Both *ab initio* calculations and THG

Figure 6.7. Second hyperpolarizability (γ) vs. the number of thiophene units. The plots compare data from refs. 130 (Thienpont *et al.*) and 131 (Cheng *et al.*).

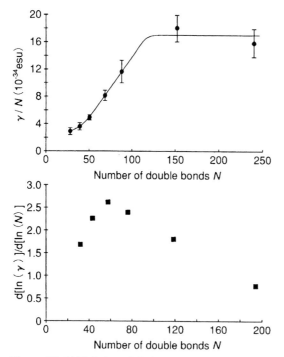

Figure 6.8. (a) Variation of $\gamma(-3\omega; \omega, \omega, \omega)/N$ as a function of the number of double bonds N for polyene oligomers measured by THG at 1.907 μm. (b) Plot of $p = d[\ln(\gamma)]/d[\ln(N)]$ as a function of π-conjugated chain length for the data in (a). (Reprinted with permission from I.D.W. Samuel *et al.*, *Science* **265**, 1070 1994. Copyright 1994 American Association for the Advancement of Science).

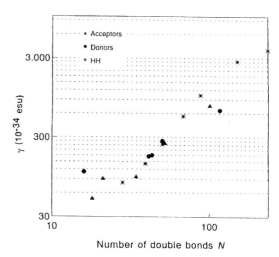

Figure 6.9. Variation of THG-measured $\gamma(3\omega; \omega, \omega, \omega)/N$ as a function of the number of double bonds N for push–push [R = N(CH₃)₂], pull–pull (R = CN) and non-substituted polyene oligomers. (Reproduced by permission of Gordon & Breach Publishers from I.D.W. Samuel et al., *Nonlinear Optics* Vol. 10, 263, 1995).

measurements showed a significant increase in third-order optical nonlinearity with increasing π-conjugation chain length, a trend similar to oligothiophenes [129,130] and polyenes [135].

The BASF research team [136,137] reported the $\chi^{(3)}$ values of thin films of astaxanthene molecules (**68**)–

(**71**) of increasing π-conjugation chain length. The $\chi^{(3)}$ values of C₃₀ ($n = 1$); C₄₀ ($n = 2$); C₅₀ ($n = 3$); and C₆₀ ($n = 4$) astaxanthene molecules were measured as a function of the wavelength between 1.0 to 2.5 μm. $\chi^{(3)}$ values of 2.0×10^{-11} esu for C₃₀, 4.4×10^{-11} esu for C₄₀, 8.2×10^{-11} esu for C₅₀ and 1.2×10^{-10} esu for C₆₀ astaxanthene were measured. These experimental results show that the $\chi^{(3)}$ increases with increasing conjugation length of the astaxanthene molecules. MNDO calculations also supported the linear increase of hyperpolarizability value with increasing conjugation length. The extrapolation of $\chi^{(3)}$ against the conjugation length predicted 300–600 effective conjugated double bonds for aligned polyacetylene taking

72

into account the $\chi^{(3)}$ value of 5×10^{-9} and 1×10^{-8} esu This predicted value is remarkably larger than theoretical numbers as discussed further.

Theoretical studies have predicted the saturation of third-order optical nonlinearity for double bonds ranging between 20–50 depending upon the calculation methods. Garito *et al.* [127,128] suggested that the γ values of polyenes should begin to saturate at some length shorter than 60 Å, which corresponds to 50 carbon sites for *trans*-polyenes. Bratan *et al.* [138] demonstrated a rapid increase in γ value up to 10–15 repeat units and then a more slow increment up to 40 repeat units for *trans*-polyenes. The computational studies on a variety of subsequently build oligomers have supported the notion that an intermediate π-electron delocalization length should be sufficient for optimizing third-order optical nonlinearity rather than the infinite polymer chain. Unfortunately earlier experimental studies were restricted to oligomers with fewer than 20 double bonds though γ/N showed no saturation [135,139,140].

The first experimental systematic study of the third-order optical nonlinearity of model polyene oligomers (**72**) with double bonds from 28 to 240 was conducted by Samuel *et al.* [141,142]. The polyene synthesized in living polymerization yielded soluble long chains with

up to 240 double bonds. The γ values of these soluble long-chain model polyene oligomers were measured in THF solution by THG at 1.9 µm. Figure 6.8a shows the variation of THG-measured $\gamma(3\omega; \omega, \omega, \omega)/N$ as a function of the number of double bond N for polyene oligomers (R = H). The γ value increases significantly as the number of double bonds increased, reaching as high as 3.79×10^{-31} esu for $N = 240$ double bonds. The γ/N plot as a function of N shows the onset of saturation around $N = 120$, which is remarkably higher than that predicted theoretically. The rate of increase of γ with N was also slower than that reported for shorter-chain molecules. A clear saturation is apparent in Figure 6.8b showing $p = d[\ln(\gamma)]/d[\ln(N)]$ vs. N where a maximum occurs for $N = 60$ double bonds. p has a maximum slightly above 2.5 for $N = 60$ and approaches 1 for $N = 200$. Mukamel and Wang [143] pointed out a maximum p for $N = 5$ which approaches 1 for $N = 21$ using a Hubbard potential $U = 11.26$ eV in a PPP model. Spano and Soos [144] predicted a saturation for short chains and a maximum p at N = 15 in Hückel theory. Figure 6.8 shows the variation of THG-measured $\gamma(-3\omega;\omega,\omega,\omega)/N$ as a function of the number of double bond N for push–push [R = N(CH$_3$)$_2$], pull–pull (R = CN) and non-substituted polyene oligomers. The effect of electron donor or acceptor groups on γ values with varying N is not evident here because unsubstituted polyenes have larger γ values than push–push or pull–pull polyenes. Probably the end groups dilutes the effect for longer-chain molecules. Puccetti *et al.* [145] measured γ values of various α,ω-disubstituted polyenes of increasing π-conjugation length and bearing push–push, pull–pull end groups at 1.34 µm by the EFISH technique. No saturation in the γ value occurred up to 40 Å and the steepest increase in the γ value was observed for polyenes with electron donors. This study does indicate that donor–acceptor endgroups have an effect on third-order optical nonlinearity for short conjugation length. This was not so visible for long conjugation length polyenes [141].

Useful attempts, made on oligothiophenes, oligoazines and polyenes, demonstrate that the length of the π-electron conjugated system is a major contributing factor for large and fast nonlinear optical responses. The scaling law exponents have been calculated for polyenes, PPV, polyazomethine, polyazine, polyaromatics and their push–push or pull–pull, push–push substituted derivatives. Theoretical studies predict a power law dependence of α with an exponent ranging from 1.3 to 3.0 and of γ with an exponent from 3.2 to 5.2; π-conjugated polymers somewhat follow this trend.

73

74

75

76

77

Series I (monomer) $n = 1$ for (1), (2), (3), (4), and (5)
Series II (dimmer) $n = 2$ for (6), (7), (8), (9), and (10)
Series III (trimer) $n = 3$ for (11), (120, (13), (14), and (15)

Table 6.7. Polarizabilities in Å^3 and second hyperpolarizabilities in 10^{-39} esu of series I, II and III π-conjugated systems end-capped with amino groups. (After ref. 152. Copyright American Chemical Society)

Molecules	α_{yy}	$\langle \alpha \rangle$	γ_{yyyy}	$\langle \gamma \rangle$
series I				
I-1	12.37	7.84	14 677.46	3232.52
I-2	12.00	7.42	4720.50	1029.72
I-3	12.84	7.63	9557.36	2016.50
I-4	13.13	8.51	11 818.31	2370.13
I-5	15.43	6.90	10 310.00	2384.90
series II				
II-1	26.54	16.07	77 003.16	19 264.77
II-2	27.25	15.49	58 627.10	12 956.63
II-3	30.36	15.80	83 911.54	17 425.71
II-4	32.08	18.81	141 048.05	33 155.04
II-5	39.07	15.59	114 950.11	23 450.98
series III				
III-1	42.76	25.26	207 564.91	56 011.50
III-2	45.19	24.62	199 785.61	46 476.15
III-3	52.10	24.92	280 522.87	56 998.07
III-4	55.88	31.16	568 879.74	143 678.37
III-5	68.37	26.16	445 537.66	89 636.07

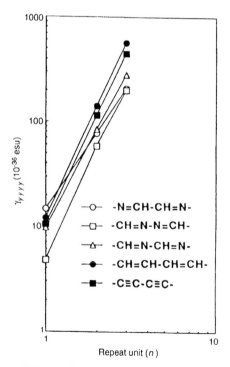

Figure 6.10. Log–log plot for γ_{yyyy} as a function of repeat unit for five different types of well-defined π-conjugated organic systems. (Reprinted with permission from ref. 152. Copyright 1994 American Chemical Society).

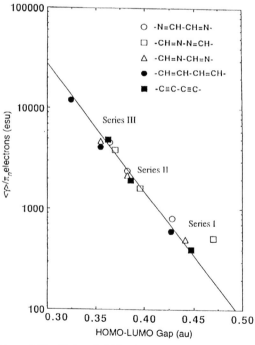

Figure 6.11. Plots of $\langle \gamma \rangle / \pi$-electrons as a function of HOMO–LUMO gaps for five π-bonding sequence. (Reprinted with permission from ref. 152. Copyright 1994 American Chemical Society).

It can be concluded that extended π-electron delocalization is required for optimizing third-order optical nonlinearity in 1D polyenes.

3.2.2.2 Nature of the π-bonding sequence

The nature of π-conjugation also plays an important role in determining the magnitude of third-order optical nonlinearity. Bodard et al. [146,147] indicated that larger polarizabilities of a polyene chain (CH=CH) than that of a polyyne (C≡C) chain result from the more homogeneous bonding sequence. This was also supported by the CNDO/S-Cl calculations of hyperpolarizabilities for polyenes, polydiacetylenes and other related π-conjugated molecules [134]. Nalwa et al. [148] reported that the α and γ values of polyazine and polyene chains are quite different for the same π-electron chain length. The γ values of polyene derivatives were also estimated as an order of magnitude larger than those of polyacene systems by Meyers and Bredas [149]. Hamada and coworkers [150,151] reported that polyenes show the highest γ values, polyacenes the lowest, and polydiacetylenes and

polyynes show intermediate values. This further supported the role of the π-bonding sequence for second hyperpolarizability.

A systematic study on the effect of the π-bonding sequence on third-order optical nonlinearity was carried out by Nalwa and coworkers [152–154] by employing polyenes, polydiacetylenes, polyacenes, polyazine, polyazomethine, and other π-conjugated systems. The static α and γ values were calculated via ab initio CPHF with the HONDO-7 program using STO-3G basis set and GAUSSIAN 92. Polarizabilities and hyperpolarizabilities of series I, II and III molecules are listed in Table 6.7. α_{yy}, which is the dominant component of polarizability, increases with increasing π-conjugation length and decreases in the order I-5 > I-4 > I-3 > I-1 > I-2 for series I. For series II and II, the polarizability of 2 is greater than that of I. The mean polarizability $\langle \alpha \rangle$ shows a slightly different trend than the α_{yy} component for all these series. The N=CH–CH=N bond showed the most effective delocalization of π-electrons whereas the CH=N–CH=N is the least polarizable of the five π-bonding sequences. The dominant component of the static polarizability (α_{yy}) of polyyne molecules I-5, II-5 and III-5 are larger than those of polyene molecules I-4,

Table 6.8 The $\chi^{(3)}$ values of polyacetylene, polydiacetylne, polyazine and polyazomethine measured by THG technique

π-polymer	$\chi^{(3)}(-3\omega; \omega, \omega, \omega)$ (esu)	Wavelength (μm)	Ref.
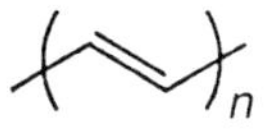 (polyacetylene structure)	1.3×10^{-9}	1.907	156
	2.7×10^{-8}	1.907	157
	1.8×10^{-7}	0.715	158
(polydiacetylene structure) R=CH$_2$OSO$_2$C$_6$H$_4$CH$_3$	8.5×10^{-10}	1.89	159
	2.0×10^{-10}	1.90	160
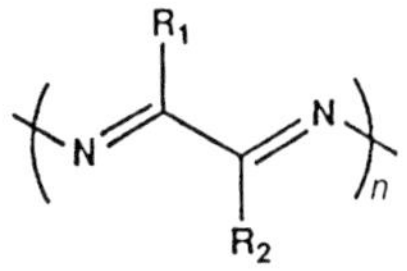 (polyazine structure) R$_1$=R$_2$=CH$_3$	9.0×10^{-12}	1.50	133
R$_1$=CH$_3$ R$_2$=CH$_2$CH$_2$CH$_3$	8.0×10^{-12}	1.50	133
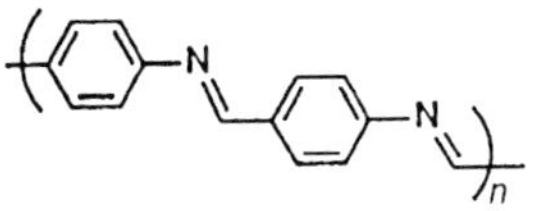 (polyazomethine structure)	1.6×10^{-12}	2.38	1.61

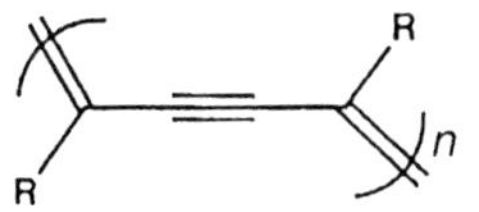

II-4 and III-4. The second hyperpolarizabilities of series I ($n = 1$), II ($n = 2$), and III ($n = 3$) molecules are also listed in Table 6.7.

Figure 6.10 shows the log–log plot for γ_{yyyy} as a function of repeat unit for five different types of π-conjugated organic systems. γ_{yyyy} increases as the length of π-conjugation increases and decreased in an order I-1 > I-5 > I-4 > I-3 > I-2 for series I (monomeric model molecules where $n = 1$). The polyazine chain with the N=CH−CH=N bond has the largest γ values because both $\langle \gamma \rangle$ and γ_{yyyy} of the N=CH−CH=N bond are about 3 and 1.5 times larger than that of CH=N−N=CH and CH=N−CH=N bonds, respectively. $\langle \gamma \rangle$ and γ_{yyyy} of model compounds I-4 with a CH=CH bond and I-5 with a C≡C bond compete with each other, though the polyene chain is somewhat better than the polyyne chain. The $\langle \gamma \rangle$ value of I-1 was found to be about 1.4 times larger than that of I-4 indicating that the polyazine π-conjugated backbone is more promising than the polyene chain. For series II ($n = 2$) $\langle \gamma \rangle$ decreases in the order II-4 > II-5 > II-1 > II-3 > II-2, and an almost similar trend was noticed with the series III model molecules ($n = 3$); III-4 > III-5 > III-3 > III-1 > III-2 for $\langle \gamma \rangle$. It is apparent from these data that in series II and III, molecules with polyene (CH=CH) and polyyne (C≡C) bonds become more important that the polyazine (N=CH−CH=N) bond due to their larger second hyperpolarizability. $\langle \gamma \rangle$ and γ_{yyyy} of the CH=CH bond in series II are more than 1.72 and 1.83 times larger than that of the N=CH−CH=N bond, respectively. Likely, $\langle \gamma \rangle$ and γ_{yyyy} of the CH=CH bond in series III are 2.57 and 2.74 times larger than that of the N=CH−CH=N bond, respectively. Pariser–Parr–Pople calculations of Albert et al. [155] and Hamada [147,148] also showed that the second hyperpolarizabilities of polyenes are larger than those of polyynes, which is in agreement to ab initio results. The simple power law for $n = 3$ showed exponents of 0.94, 0.82, 0.84, 0.92 and 0.93 for the $\langle \alpha \rangle$ values of the N−CH=CH−N, CH=N−N=CH, CH=N−CH=N, CH=CH−CH=CH and C≡C−C≡C systems, respectively. For $\langle \gamma \rangle$, the exponents were 2.59, 3.3, 3.73, 3.48, and 3.04 for N−CH=CH−N, CH=N−N=CH, CH=N−CH=N, CH=CH−CH=CH and C≡C−C≡C systems, respectively. Theoretically the exponents for the first three NLO coefficients are 1.3 to 3.0 for α, 3 for β and 5.0 for γ as derived from the free electron theory [2]; the exponent of π-conjugated polymers depends upon the chemical species and appears around these range.

Figure 6.11 shows the plots of $\langle \gamma \rangle$/π-electrons as a function of the highest occupied molecular orbital-

lowest unoccupied molecular orbital HOMO–LUMO gaps for five π-bonding sequence. It is apparent that the HOMO–LUMO gap decreases and the γ values increase with the increase of the conjugation length. The gap between the HOMO and LUMO decreases with increasing number of π-electrons. The polyacetylene has the smallest values of the HOMO–LUMO gap among all π-bonding sequences and the largest α and γ values. The HOMO–LUMO gap of the π-conjugated systems containing a nitrogen atom is significantly larger than that of the polyacetylene backbone.

The effect of the π-bonding sequence on third-order optical nonlinearity can further be supported by examining the experimental data of $\chi^{(3)}$ of different π-conjugated polymers reported in the literature. Table 6.8 lists $\chi^{(3)}$ values of four different type π-conjugated polymers: polyacetylene, polydiacetylene, polyazine and polyazomethine which have a somewhat identical π-bonding sequence to the 1D conjugated system theoretically studied by ab initio [152]. Only THG measured $\chi^{(3)}(-3\omega; \omega, \omega, \omega)$ values are compared to avoid confusion with DFWM, where generally large $\chi^{(3)}(-\omega; \omega, \omega, -\omega)$ values are observed due to one-photon resonance enhancement. Though the comparison here is crude due to different measurement conditions such as wavelength, the state of the materials absorption, etc., it can be seen that polyacetylene has the largest $\chi^{(3)}(-3\omega; \omega, \omega, \omega)$ value (of the order of 10^{-8} esu). While polydiacetylene–toluene sulfonate (PDA–PTS) have a value of 10^{-10} esu. Both polyazines and polyazomethines compete with each other and have nonresonant $\chi^{(3)}(-3\omega; \omega, \omega, \omega)$ values in the range 10^{-11} to 11^{-12} esu. However a slightly higher $\chi^{(3)}(3\omega; \omega, \omega, \omega)$ value was measured for polyazine than polyazomethine. The $\chi^{(3)}(-3\omega; \omega, \omega, \omega)$ values show a decreasing order of polyacetylene > polydiacetylene > polyazine ⩾ polazomethine; this is in good agreement with the theoretical results at $\omega = 0$ reported by Nalwa et al. [152]. This comparison should be considered rather crude as both experimental $\chi^{(3)}(-3\omega; \omega, \omega, \omega)$ values and theoretical γ value of different π-conjugated 1D systems are compared. The π-bonding sequence of a 1D system plays an important role in third-order optical nonlinearity and should be considered when designing new materials for applications in third-order nonlinear optics. Similar may hold true for the 2D and 3D π-conjugated materials.

Nalwa and Shirk [162] compiled data on a very large variety of metallophthalocyanines to establish structure–property relationships. Experimental observations on the variation in the magnitude of the $\chi^{(3)}$ values indicates that in metallophthalocyanines factors, such as (I) metal atom, (ii) axial substitution, (iii) peripheral substitution, (iv) length of π-electron conjugation, (v) crystal structure and (vi) fabrication, have a remarkable effect on third-order optical nonlinearity.

3.2.3 π-conjugated materials

Materials where symmetry requirements are lifted can be used for third-order nonlinear optics, and π-conjugated polymers are such materials, as they have a dual role by exhibiting large $\chi^{(3)}$ values and electrical conductivity. Third-order NLO effects have been investigated in a variety of organic molecular and polymeric systems such as liquids, dyes, charge transfer complexes, π-conjugated polymers, NLO dye-grafted polymers, organometallic compounds, composites and liquid crystals [15]. Polymeric materials are expected to possess desirable properties to show their potential for third-order NLO devices. The importance of organic polymers has been realized because of the promise of large nonlinear optical figures of merit, high optical damage thresholds, ultrafast optical responses, architectural flexibility and ease of fabrication. Third-order optical nonlinear susceptibilities determined by, for example THG, DFWM, and self-focusing techniques greatly differ from each other due to the distinct nonlinear optical processes and because of the applied experimental conditions such the measurement wavelength, environment and material states. Third-order optical nonlinearity values are often quoted as resonant and nonresonant values resulting from their wavelength dispersion within or far from the optical absorption regions of a nonlinear material. The resonant $\chi^{(3)}$ values can be several orders of magnitudes larger than that of the non resonant value.

3.2.3.1 Fullerenes

Fullerenes are conjugated π-electron systems. C_{60} is a highly aromatic molecule that have all the atoms connected by sp^2 bonds; therefore it is perfectly suited for third-order nonlinear optics. Table 8.9 summarizes the γ values of fullerenes calculated using quantum chemistry approaches. Li et al. [163] calculated $\langle \gamma \rangle$ $(-\omega;\omega,-\omega,\omega)$ of 7.30×10^{-34} esu at $\omega = 1.064$ μm, 3.11×10^{-34} esu at $\omega = 0.532$ μm) and $\langle \gamma \rangle$ $(-2\omega; \omega, \omega, 0)$ of 6.90×10^{-34} esu at ω-1.9 μm using the

INDO/SC1 method. Shuai and Bredas [165] calculated the average α values as 1.54 Å^3 for C_{60} and 214 Å^3 for C_{70} using the valence-effective-Hamiltonian-sum-over-states (VEH-SOS) approach, the averaged γ values of C_{70} was more than four times larger than that of C_{60}. Karna and Wijekoon [172] calculated the α and γ values of C_{60} using ab initio, semiempirical INDO-SOS and CNDO-SOS calculations at different frequencies. The γ values were estimated in EFISH $(\gamma)-2\omega;\omega,\omega,0)$ THG $\gamma(-3\omega; \omega, \omega, \omega)$, DFWM $\gamma(-\omega;\omega,\omega,-\omega)$ and electric-field-induced Kerr effect $\gamma(-\omega;0,0,-\omega)$ methods at 1.064, 1.370, 1.50, 1.907 μm by the CNDO and INDO techniques. The semiempirical results showed agreement with experimental data while the γ values obtained by ab initio were two orders of magnitude smaller compared to experiments. Theoretically calculated γ values for C_{60} were found to vary in an order: $\gamma(-3\omega;\omega,\omega,\omega) > \gamma(-2\omega;\omega,\omega,0) > \gamma(-\omega; \omega, \omega, -\omega)$ $> \gamma(-\omega; 0, 0, \omega) > \gamma(-0; 0, 0, 0)$. The magnitude of the γ value depends upon the theoretical methods and optical frequencies. Harigaya and Abe [173] calculated the frequency dependence of $\chi^{(3)}$ of C_{60} with tight binding, and $\chi^{(3)}$ values of 1.22×10^{-12} esu near zero frequency and about 10^{-11} esu around 2.5 eV were estimated. The larger value near peak at 2.5 eV is due to a three-photon resonance enhancement.

Table 6.10 lists the experimentally measured γ and $\chi^{(3)}$ values of fullerenes, chemically modified fullerenes and composites of fullerenes with conjugated materials. The third-order optical nonlinearities of fullerenes vary by as much as three orders of magnitude depending upon the chemical compositions, material state (solution and thin films), measurement wavelengths, techniques and resonance contributions. The $\chi^{(3)}$ value of C_{60} ranges from 10^{-10} to 10^{-12} esu depending upon the resonance and nonresonance contributions. The $\chi^{(3)}$ value of C_{70} ranges from 10^{-9} to 10^{-12} esu, its larger $\chi^{(3)}$ value results from it having more quasi-free electrons than C_{60}. The research team at Naval Research Laboratory [191, 196] measured the $\chi^{(3)}$ values of thin films of C_{60} and its composite with poly[2-methoxy-5-(2'-ethylhexyloxy)p-phenylenevinylene](MEH-PPV) using DFWM and nonlinear transmission at 590.5 nm. The charge transfer complex between C_{60} and MEH-PPV showed an enhancement of an order of magnitude in the $\chi^{(3)}$ value and the figure of merit relative to either of the components. Some enhancement in the $\chi^{(3)}$ values were also observed for photopolymerized C_{60}, and for C_{60} and meso-tetraphenyl porphyrin (TPP) composites. One, two- and three-photon resonances have been observed in both C_{60} and C_{70} through THG and

Table 6.9. Second hyperpolarizabilities (γ) of fullerenes calculated by different theoretical approaches

Fullerene (C_n)	Method	Technique	Wavelength (μm)	γ (esu)	Ref.
C_{60}	SOS-INDO/S	DFWM	$\omega = 0.532$	3.11×10^{-34}	163
		DFWM	$\omega = 1.064$	7.3×10^{-34}	163
		DFWM	$\omega = 1.91$	6.9×10^{-34}	163
	MNDO-PM$_3$		$\omega = 0$	2.39×10^{-36}	164
	VEH-SOS	THG	$\omega = 0$	2.02×10^{-34}	165
	LDA-DFT		$\omega = 0$	7.0×10^{-36}	166
	TB		$\omega = 0$	9.0×10^{-36}	167
	SOS-CNDO/S	THG	$\omega = 0$	4.58×10^{-34}	168
	SOS		$\omega = 0$	1.06×10^{-35}	169
	TBM		$\omega = 0$	5.08×10^{-34}	170
	PPP-FF		$\omega = 0$	5.87×10^{-36}	171
	ab initio	EFISH/DFWM	$\omega = 0$	1.36×10^{-36}	172
	CNDO-SOS	EFISH/DFWM	$\omega = 0$	8.26×10^{-36}	172
	INDO-SOS	EFISH/DFWM	$\omega = 0$	6.92×10^{-36}	172
	ab initio	EFISH	$\omega = 1.064$	1.61×10^{-36}	172
		DFWM	$\omega = 1.064$	1.55×10^{-36}	172
	CNDO-SOS	EFISH	$\omega = 1.064$	6.25×10^{-35}	172
		DFWM	$\omega = 1.064$	4.54×10^{-35}	172
	INDO-OS	EFISH	$\omega = 1.064$	4.09×10^{-35}	172
		DFWM	$\omega = 1.064$	2.73×10^{-36}	172
	CNDO-SOS	EFISH	$\omega = 1.907$	1.21×10^{-35}	172
		THG	$\omega = 1.907$	1.84×10^{-35}	172
		DFWM	$\omega = 1.907$	1.08×10^{-35}	172
	CNDO-SOS	EFISH	$\omega = 1.500$	1.62×10^{-35}	172
		THG	$\omega = 1.500$	3.43×10^{-35}	172
		DFWM	$\omega = 1.500$	1.34×10^{-35}	172
	CNDO-SOS	EFISH	$\omega = 1.370$	1.93×10^{-35}	172
		THG	$\omega = 1.370$	5.18×10^{-35}	172
		DFWM	$\omega = 1.370$	1.54×10^{-35}	172
	INDO-SOS	EFISH	$\omega = 1.907$	9.44×10^{-36}	172
		THG	$\omega = 1.907$	1.41×10^{-35}	172
		DFWM	$\omega = 1.907$	8.4×10^{-36}	172
	INDO-SOS	EFISH	$\omega = 1.500$	1.25×10^{-35}	172
		THG	$\omega = 1.500$	2.60×10^{-35}	172
		DFWM	$\omega = 1.500$	1.03×10^{-35}	172
	INDO-SOS	EFISH	$\omega = 1.370$	1.48×10^{-35}	172
		THG	$\omega = 1.370$	3.88×10^{-35}	172
		DFWM	$\omega = 1.370$	1.18×10^{-35}	172
C_{70}	VEH-SOS	THG	$\omega = 0$	4.09×10^{-34}	165
	MNDO-PM3		$\omega = 0$	4.52×10^{-36}	164
	SOS-CNDO/S	THG	$\omega = 0$	8.57×10^{-34}	168
	SOS		$\omega = 0$	1.19×10^{-35}	169
C_{76}	SOS-CNDO/S		$\omega = 0$	1.21×10^{-33}	168
	SOS		$\omega = 0$	1.91×10^{-35}	169
C_{84}	SOS-CNDO/S	THG	$\omega = 0$	1.81×10^{-33}	165
$C_{60}H_{62}$	VEH-SOS	THG	$\omega = 0$	8.0×10^{-32}	165
$C_{59}B$	SOS		$\omega = 0$	1.4×10^{-36}	169
$C_{59}N$	SOS		$\omega = 0$	2.22×10^{-35}	169

DFWM methods. The large variation in $\chi^{(3)}$ occurs due to the resonance effects, where off-resonance optical nonlinearities are smaller than those of resonant ones.

Figure 8.12 shows the wavelength dependence $\chi^{(3)}(-3\omega;\omega,\omega,\omega)$ of C_{60} and C_{70} thin films. The $\chi^{(3)}(-3\omega;\omega,\omega,\omega)$ spectrum of C_{60} films shows two resonance enhancements: the first is located at 1.06 μm and the second is located at 1.3 μm. On the other hand, the $\chi^{(3)}(-3\omega;\omega,\omega,\omega)$ spectrum of C_{70} films showed only one broad resonance around 1.4 μm and a shoulder at 1.064 μm. Kajzar [197] interpreted that, in C_{60}, the peak at 1.06 μm is due to a three-photon

Table 6.10. Third-order nonlinear optical susceptibilities and hyperpolarizabilities of fullerenes

Fullerene (C_n)	Measurement technique	Wavelength (μm)	$\chi^{(3)}$ (esu)	γ (esu)	Ref.
C_{60} (benzene)	DFWM	1.064	1×10^{-10}	1.1×10^{-30}	174
C_{60} (benzene)	SA	0.440	6.8×10^{-12}		175
		0.520	3.2×10^{-11}		175
		0.580	1.7×10^{-10}		175
C_{60} (toluene)	DFWM	0.532	2.0×10^{-10}	1.83×10^{-27}	176
		0.532	$\sim 10^{-13}$	$\sim 10^{-31}$	177
		0.532	1.4×10^{-13}	3.3×10^{-31}	178
C_{60} (benzene)	DFWM	0.532	1.2×10^{-13}	3.5×10^{-31}	178
C_{60} (styrene)	DFWM	0.532	1.2×10^{-13}	2.2×10^{-31}	178
C_{60} (xylene)	DFWM	0.532	0.8×10^{-13}	3.5×10^{-31}	178
C_{60} (film)	DFWM	1.064	7.0×10^{-12}	3.0×10^{-34}	179
		0.633	2.0×10^{-10}		180
	THG	1.064	2.0×10^{-10}		181
		1.064	2.0×10^{-11}		48
		1.32	3.0×10^{-11}		182
		1.064	1.4×10^{-11}		182
		1.91	9.0×10^{-12}		182
		1.91	9.0×10^{-12}		182
		2.37	4.0×10^{-12}		182
		0.85	1.5×10^{-11}		183
		1.06	8.7×10^{-11}		183
		1.32	6.1×10^{-11}		183
		1.907	3.2×10^{-11}	1.1×10^{-32}	183
C_{60} (toluene)	THG	1.064	7.2×10^{-11}	1.3×10^{-33}	184
		1.50	3.0×10^{-11}	3.1×10^{-33}	184
		2.0	3.7×10^{-11}	1.6×10^{-33}	184
		2.37	4.0×10^{-12}		184
	EFISH	1.91	1.6×10^{-11}	7.5×10^{-34}	185
C_{60} (film)	DFWM	0.597	3.8×10^{-10}		186
		0.675	0.82×10^{-10}		186
$C_{60} + C_{70}$ (toluene)	DFWM	1.064	3.3×10^{-9}	1.6×10^{-31}	187
C_{60}/DEA	THG	1.064		6.7×10^{-29}	182
$C_{60} + C_{70}$/PMMA	DFWM	0.608	$\sim 10^{-10}$	5.0×10^{-30}	188
C_{60}-polymer	DFWM	0.602	1.4×10^{-14}		189
C_{60} (film)	DFWM	0.745	5.11×10^{-12}		190
		0.875	1.58×10^{-11}		190
		0.5905	5.2×10^{-10}		191
		1.675	8.2×10^{-11}		192
Poly-C_{60} (film)	DFWM	0.5905	8.4×10^{-10}		191
		0.675	8.7×10^{-11}		192
$C_{60} - O_2$ (film)	DFWM	0.5905	4.3×10^{-10}		192
		0.675	9.5×10^{-11}		192
C_{60}/TMPD	DFWM	0.675	8.5×10^{-13}		192
C_{60}:TPP(film)	DFWM	0.5905	7.2×10^{-10}		191
C_{60}:MEH-PPV (10:90)	DFWM	0.5905	1.0×10^{-8}		191
(20:80)		0.5905	1.2×10^{-8}		191
(30:70)		0.5905	6.5×10^{-9}		191
(40:60)		0.5905	6.1×10^{-9}		191
(50:50)		0.5905	2.7×10^{-9}		191
C_{70} (benzene)	SA	0.520	2.7×10^{-10}		175
		0.580	2.0×10^{-12}		175
		0.610	5.1×10^{-11}		175
C_{70} (toluene)	DFWM	1.064	5.6×10^{-12}	1.2×10^{-30}	193
		0.532	0.9×10^{-13}	7.3×10^{-31}	178
	EFISH	1.91	4.4×10^{-33}	1.3×10^{-33}	185

(*continued overleaf*)

Table 6.10. (*continued*)

Fullerene (C_n)	Measurement technique	Wavelength (µm)	$\chi^{(3)}$ (esu)	γ (esu)	Ref.
		1.83	9.6×10^{-12}	7.5×10^{-34}	185
	THG	1.064	1.4×10^{-9}	5.0×10^{-32}	184
		1.5	5.4×10^{-10}	2.6×10^{-32}	184
		2.0	9.1×10^{-11}		184
C_{70} (film)	DFWM	0.633	3.0×10^{-10}		180
		0.597	2.1×10^{-9}		186
		0.675	6.4×10^{-10}		186
		1.064	1.2×10^{-11}	5.0×10^{-34}	194
	THG	1.064	2.6×10^{-11}		183
		1.4	9.0×10^{-11}		195
		1.907	2.4×10^{-11}	1.1×10^{-32}	195

DEA = N,N'-diethylaniline;   TMPD = $N,N,N'.N'$-tetramethyl-1,4-phenylenediamine;   MEH-PPV = poly[2-methoxy-5-(2'-ethylhexyloxy)p-phenylenevinylene]; TPP = *meso*-tetraphenyl porphyrin

resonance with one photon allowed T_{1u} excited state, while the peak at 1.3 µm is due to a two-photon resonance with one photon forbidden band located around 1.9 eV. The resonance, in C_{70} films, around 1.4 µm could be a strong three-photon resonance and shoulder as a weaker two photon resonance with the two phonon states lying slightly below the one photon allowed state. Two- and three-photon resonances in C_{60} and C_{70} have been shown by THG while one-photon resonance has been shown by DFWM. Third-order optical nonlinearity measurements and theore-

tical calculations indicate that fullerenes are promising materials for third-order nonlinear optics.

3.2.3.2   *Phthalocyanines*

Organometallic materials are attracting a great deal of attention in the field of nonlinear optics. In particular, metallophthalocyanines, metalloporphyrins, metallocenes, metal–polyynes and polysilanes have been studied for third-order nonlinear optics. The large molecular hyperpolarizability in organometallic com-

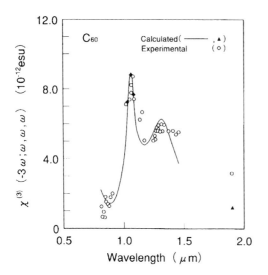
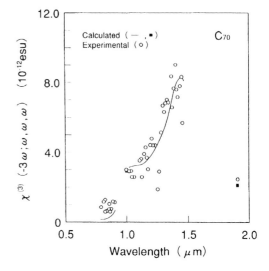

Figure 6.12. Wavelength dependnce $\chi^{(3)}(-3\omega; \omega, \omega, \omega)$ of C_{60} and C_{70} thin films shown by circle (O). Solid lines, rectangles and triangles represent the calculated values. (Reproduced by permission of the Society of Photo-Optical Instrumentation Engineers (SPIE) from F. Kajzar, *SPIE Proc.* **2284**, 1994).

pounds arises from the metal–ligand bonding due to the transfer of electron density between the metal atom and the conjugated ligand systems. The diversity of central metal atoms, oxidation states, their size and the nature of the ligands helps to tailor the NLO properties of organometallic materials by optimizing the charge transfer interactions. The first review article on organometallic materials for nonlinear optics was published by Nalwa in 1991 [16]. Recently more progress has been made in developing organometallic materials for photonics technologies. Metallophthalocyanines have been considered the most valuable materials for the development of nonlinear optical devices because:

(1) about 70 different metal atoms can be incorporated into the phthalocyanine ring;
(2) they provide tremendous architectural flexibility for tailoring physical, electrical, magnetic and photonic functions;
(3) they can be economically produced on a large scale;
(4) they are thermally and environmentally stable robust materials;
(5) they are nontoxic biocompatible materials;
(6) they provide ease of processing and fabrication;
(7) they are highly compatible with metals, ceramics and other substrates and can be prepared into any desired shape and size;
(8) their chemical modification either by central metals or axial and peripheral substituents, facilitates the generation of new chemical species;
(9) they show high laser damage thresholds;
(10) they are multifunctional materials with applications in catalysis, photodynamic therapy, printers, memory disks, dyes, pigments, lubricants, semiconductors.

Metallophthalocyanines are the only materials that can combine various chemical and physical properties into a single system which can be successfully applied in integrated optics. Metallophthalocyanines are not true organometallic materials owing to the absence of metal-to-carbon bonding in the ring; instead they have metal-to-nitrogen bonding. The metal-to-ligand bonding in MPcs is considered to give rise to large molecular hyperpolarizability because of the transfer of the electron density between the central metal atom and ligands. Furthermore larger intramolecular interactions in MPcs complexes can be expected due to the overlapping of the electron orbitals of the ligands with the metal ion orbitals.

Nalwa and Shirk [162] have reviewed second- and third-order nonlinear optical properties, as well as the optical limiting properties of MPcs including the various measurement techniques used for analyzing these systems and the experimental results. The phthalocyanine ring has a two-dimensional conjugated π-electron system [198,199], and the third-order NLO susceptibilities of many metallophthalocyanines (MPcs) have been measured by THG, DFWM, Z-scan, two-photon absorption (TPA) and EFISH techniques. Third-order NLO properties of the following phthalocyanine systems were compiled to establish the relationships between the chemical structures and various nonlinear optical effects by evaluating MPcs through computational and experimental probes:

(1) metallophthalocyanines;
(2) metallonaphthalocyanines;
(3) bis-phthalocyanines and oligomers;
(4) polymer-functionalized metallophthalocyanines;
(5) porphyrazines.

The magnitude of third-order nonlinear optical susceptibility $\chi^{(3)}$ varies by several orders depending on the metal atom substitution [200–212]. The effect of factors such as central metal atoms, peripheral groups, axial substituents; crystal structure, π-conjugation and fabrication process were found to affect significantly the third-order nonlinear optical properties of phthalocyanines.

Third-order NLO properties of MPcs are enhanced by as much as one order of magnitude by polymorphism and an example is provided here. For MPcs, it is well known that their photoconductivity is significantly affected by crystal structures, such as that for TiOPc and H_2Pc. TiOPc, which exists in three different polymorphs α, β and Y, is the best example of such a relationship.

The polymorphs of TiOPc were prepared from vacuum-deposited thin films following solvent treatment. First, TiOPc was evaporated in a vacuum of $10^{-5} \sim 10^{-6}$ Torr and thin films were deposited on fused silica substrate at room temperature. Figure 6.13 shows the processes of solvent treatment to transform TiOPc into different polymorphs. The α-TiOPc was prepared by exposing amorphous TiOPc to tetrahydrofuran vapors while β-TiOPc was obtained by treating amorphous TiOPc to xylene vapors. The Y-TiOPc was obtained from amorphous TiOPc by exposing deposited thin films to the vapors of a chlorobenzene–water (10:1) mixture. Solvent vapor exposure was carried out for 13–15 h at room temperature in all cases. The

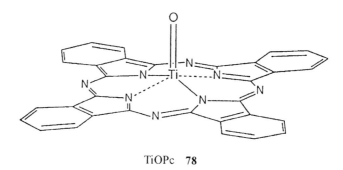

TiOPc **78**

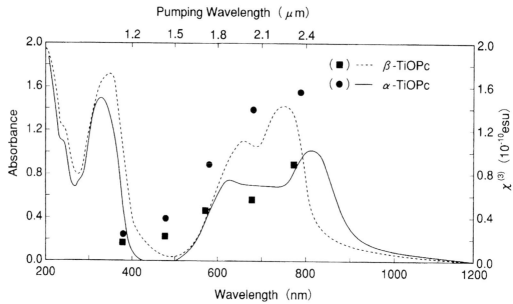

Figure 6.13. Processes of treatment to transform TiOPc into the respective polymorphs. (Reprinted with permission from ref. 212. Copyright 1994 American Chemical Society).

structures of the TiOPc polymorphs were confirmed by powder X-ray diffraction patterns.

Figure 6.14 and 6.15 show the dispersion of $\chi^{(3)}$ values of different polymorphs of TiOPc measured by the THG technique in the wavelength region 1.2 to 2.5 μm along with their optical absorption spectra. The magnitude of the $\chi^{(3)}$ value varied in the order $\alpha > Y > \beta >$ amorphous at a wavelength of 2.43 μm where $\chi^{(3)}$ of α-form was enhanced by a factor of 3.3 compared to the β-form. The value of $\chi^{(3)}$ is maximum at 2.43 μm because it is influenced by the three-photon (3ω) resonance with the Q-band and decreases toward lower wavelengths. At lower wavelengths, the resonance effect is smaller and hence partially affects the

Figure 6.14. A comparison of optical absorption spectra and the dispersion of $\chi^{(3)}(-3\omega; \omega, \omega, \omega)$ of α and β polymorphs of TiOPc.

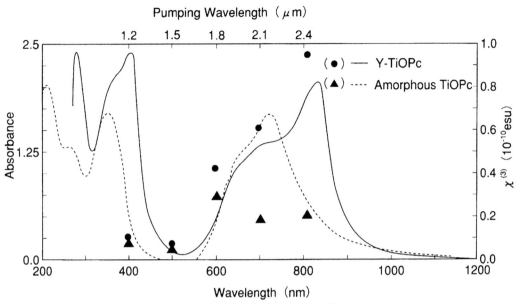

Figure 6.15. A comparison of optical absorption spectra and the dispersion of $\chi^{(3)}(-3\omega;\omega,\omega,\omega)$ of amorphous and Y-polymorphs of TiOPc.

$\chi^{(3)}$ values. The α-TiOPc shows the largest $\chi^{(3)}$ value over the whole wavelength region and the largest resonance enhancement. The dispersion of the third-order optical nonlinearity of TiOPc polymorphs over the fundamental wavelength range of $1.2 \sim 2.43$ μm was similar to other organic molecular and polymeric materials. The large third-order optical nonlinearity observed in α- and Y-TiOPc polymorphs results from the highly ordered structures. The strong intermolecular interactions in TiOPc molecules in the α-form give rise to a larger $\chi^{(3)}$ value compared to other polymorphs. These results therefore establish the relationship between the crystal structure and the third-order optical nonlinearity of TiOPc polymorphs where $\chi^{(3)}$ can vary by an order of magnitude depending upon the crystal structure. The polymorphs of VOPc [213], CuPc [214] and H_2Pc derivatives [215] also show the similar effect of crystal structure on third-order optical nonlinearity.

The magnitude of third-order optical nonlinearity can be enhanced by enlarging the π-electron delocalization length as discussed earlier for well-defined chain length oligomers of polyenes, polythiophenes, polyazines. Nalwa *et al.* investigated metallonaphthalocyanines (MNcs) to enlarge third-order optical nonlinearity because the MNc system has an extended π-electron delocalization over the phthalocyanine macrocycle. Third-order NLO properties of soluble MNc derivatives

containing metal atoms: Cu, Zn, Pd, Ni, InCl, Ru, Rh and VO were found to be remarkably altered by a change of the central metal atom [208–210]. In a further study, the third-order NLO properties of the central metal: Si, Ge, Sn, Al, Mn and VO of the MNc derivatives were investigated by THG between 1.05 to 2.10 μm and DFWM at 800 nm [216, 217]. Table 6.11 lists the chemical structures of MNc derivatives: (I) $NcMX_1X_2R_4$ derivates, where both axial groups are identical ($X_1 = X_2$), M is either Si, Ge, Sn or Mn, and R refers to the peripheral groups; (ii) a $NcAlX_1X_2R_4$ derivative with different axial groups, where the $X_1 = Cl$ and $X_2 = OH$, and R is $C(CH_3)_3$; and (iii) a $NcVX_1X_2R_4$ derivative where $X_1 = O$ (there is no X_2 group) and R is $C(CH_3)_3$. The $\chi^{(3)}$ values of the MNc derivatives at 1.05 μm range from 10^{-11} to 10^{-12} esu, where SiNc derivatives show the largest $\chi^{(3)}$ values, about an order of magnitude higher than that of other MNc derivatives. The $\chi^{(3)}$ values measured at 1.50 μm were even larger. Two of the SiNc derivatives show $\chi^{(3)}$ values on the order of 10^{-10} esu which arise from the strong resonant contributions at the Soret band edge. A three-photon resonance (3ω) of the 350 nm Soret band occurs for the fundamental wavelength of 1.05 μm; this is one-third of the fundamental measurement wavelength and therefore a three-photon resonance leads to the larger $\chi^{(3)}$ value of compound 6, which is more than 4.2 times that of 7 and 35 times that

Table 6.11. Chemical structures and the $\chi^{(3)}(-3\omega; \omega, \omega, \omega)$ values of the spin-cast thin films of MNc derivatives measured between 1.05 µm to 2.10 µm wavelength region After ref. 216.

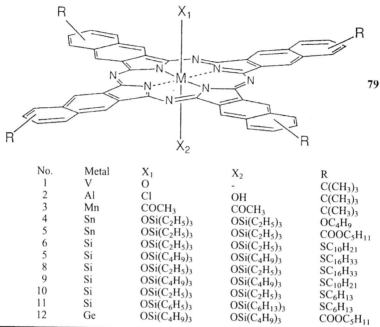

79

No.	Metal	X_1	X_2	R
1	V	O	-	$C(CH_3)_3$
2	Al	Cl	OH	$C(CH_3)_3$
3	Mn	$COCH_3$	$COCH_3$	$C(CH_3)_3$
4	Sn	$OSi(C_2H_5)_3$	$OSi(C_2H_5)_3$	OC_4H_9
5	Sn	$OSi(C_2H_5)_3$	$OSi(C_2H_5)_3$	$COOC_5H_{11}$
6	Si	$OSi(C_2H_5)_3$	$OSi(C_2H_5)_3$	$SC_{10}H_{21}$
5	Si	$OSi(C_4H_9)_3$	$OSi(C_4H_9)_3$	$SC_{16}H_{33}$
8	Si	$OSi(C_2H_5)_3$	$OSi(C_2H_5)_3$	$SC_{16}H_{33}$
9	Si	$OSi(C_4H_9)_3$	$OSi(C_4H_9)_3$	$SC_{10}H_{21}$
10	Si	$OSi(C_2H_5)_3$	$OSi(C_2H_5)_3$	SC_6H_{13}
11	Si	$OSi(C_6H_5)_3$	$OSi(C_6H_{13})_3$	SC_6H_{13}
12	Ge	$OSi(C_4H_9)_3$	$OSi(C_4H_9)_3$	$COOC_5H_{11}$

MNc compound no.	Film thickness (nm)	$\chi^{(3)}(10^{-12}$ esu) at Wavelength (µm)							
		1.05	1.20	1.35	1.50	1.65	1.80	1.95	2.10
1	94	2.45	1.71	2.71	3.41	4.23	4.60	4.40	4.0
2	168	1.79	1.35	2.20	1.41	2.21	4.27	3.82	2.60
3	181	1.80	4.08	3.77	3.32	2.53	2.84	2.33	2.03
4	153	1.60	160	2.30	2.07	2.61	2.71	3.16	2.51
5	178	1.66	2.07	2.32	2.40	2.40	2.44	1.85	-
6	47	53.2	44.4	34.4	56.0	31.7	19.4	23.5	25.2
7	30	12.6	7.0	9.50	10.6	13.3	13.8	13.0	9.70
8	117	17.8	70.0	47.5	107.0	37.3	20.5	10.9	16.3
9	67	24.2	92.6	104.0	136.0	52.0	13.7	7.07	14.2
10	168	38.0	31.8	37.6	42.6	31.4	19.7	23.7	19.3
11	110	31.8	38.6	40.0	31.3	19.1	7.60	3.71	6.01
12	97	18.0	24.3	29.0	18.6	12.0	7.62	5.71	4.70

of compound 4. The GeNc derivative showed the same order of $\chi^{(3)}$ to those of the SiNc derivatives.

Figure 6.16 shows the optical absorption spectrum of compound 6 and its $\chi^{(3)}(-3\omega;\omega,\omega,\omega)$ values measured in the 1.05 to 2.1 µm wavelength region. The absorption peak in the Soret band is at 260 nm while in the Q band it is at 740 and 821 nm. The $\chi^{(3)}$ values are largest at 1.05 and 1.50 µm and decreases toward higher wavelengths. $\chi^{(3)}(-3\omega;\omega,\omega,\omega)$ at 1.05 µm is enhanced by a three-photon resonance; however the

increase at 1.5 µm could be from a two-photon resonance lying just above the one-photon state because this proximity does not seem to be strongly three-photon resonant. Compounds 8, 9 and 10 show the similar two-photon resonance effect. DFWM measurements on compound 6 yielded a $\chi^{(3)}(-\omega; \omega, \omega, -\omega)$ value of 5.0×10^{-7} esu at 800 nm owing to one-photon resonance enhancement and the corresponding figure of merit $\chi^{(3)}/\alpha$ is 4.2×10^{-12} esu cm, which is significantly larger and on the same order of

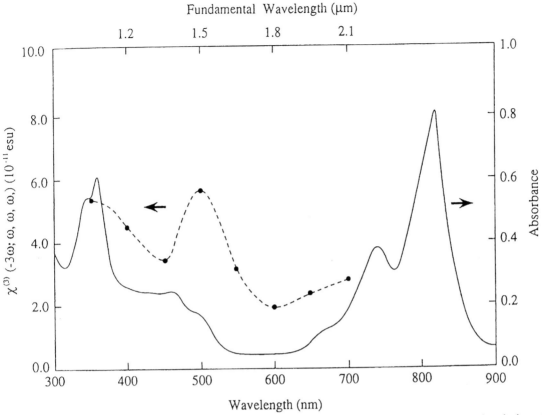

Figure 6.16. Optical absorption spectrum of a SiNc derivative (compound 6 in Table 6.11) and its wavelength dependent $\chi^{(3)}(-3\omega; \omega, \omega, \omega)$ measured between 1.05 to 2.1 μm. After ref. 216.

magnitude as reported for C_{60}-MEH-PPV systems [195] and highly conjugated astaxanthene molecules [139]. The large variation between the $\chi^{(3)}(-3\omega; \omega, \omega, \omega)$ and $\chi^{(3)}(-\omega; \omega, \omega, -\omega)$ values of compound 6 is due to the different resonance contributions, because the one-photon (ω) resonance enhancement in the DFWM measurements greatly exceeds those of the two-photon (2ω) and three-photon (3ω) resonances in the THG measurements. The $\chi^{(3)}(-\omega; \omega, \omega, \omega)$ values of 2.09×10^{-11} esu at 598 nm from the DFWM technique for a 30% silicon naphthalocyanine (SiNc)/PMMA guest–host system [206], and as high as 10^{-6} esu, for another silicon naphthalocyanine derivative, were deduced from the nonlinear refractive index measured by the SA technique [218]. The large difference in the $\chi^{(3)}$ values of SiNc derivatives originates from the different nonlinear optical processes involved with the DFWM, THG and SA techniques. a resonant $\chi^{(3)}(-\omega; \omega, \omega, -\omega)$ value as large as 10^{-7} esu for oligomeric $[t\text{-Bu}_4\text{PcRu(dib)}]_n$

thin films was measured by the DFWM technique [207]. Therefore the extended π-electron conjugation in MNcs and oligomeric MPcs could be recognized as the origin of the larger $\chi^{(3)}$ values. Unlike 1D π-conjugated systems, the effect of enlarged π-electron conjugation on the third-order NLO properties of phthalocyanines is rather complicated as they are 2D systems.

Like phthalocyanines, metalloporphyrins have a two-dimensional π-electron conjugated system. Metalloporphyrins also offer tremendous opportunities to incorporate a variety of central metal atoms into the ring structures and modifications at the peripheral sites of the rings. Third-order NLO properties of metalloporphyrins have been compiled by the author [15]. Metalloporphyrins containing central metal atoms: Zn, Co, Ni, Cu, VO, Mg, Mn or metal-free show $\chi^{(3)}$ values between 10^{-8} to 10^{-12} esu depending upon the chemical composition and measurement conditions.

3.2.3.3 π-conjugated polymers

In the 1980s, conjugated polymers emerged as a new class of electronic materials that exhibit large electrical conductivity when doped with electron acceptor and electron donor species [219]. The importance of π-conjugated polymers in the field on nonlinear optics was realized when large third-order optical nonlinearity was observed in polydiacetylene p-toluene sulfonate (PDA-PTS) [220]. This view was further extended to other π-conjugated polymers and it became a fact that the π-electron backbone holds the key to large third-order nonlinear optical effects. As a consequence, these polymers attracted the attention of the scientific community as new materials for third-order nonlinear optics. Both theoretical and experimental results demonstrate that large third-order optical nonlinearity is displayed by the highly conjugated π-electron system. With this view, organic π-conjugated polymers attracted much attention in the field on nonlinear optics, where significant research activities are still progressing.

Polydiacetylenes and Polyacetylenes
Polydiacetylenes are prepared by the solid-state polymerization of diacetylene monomers with a structural formula, $R_1-C\equiv C-C\equiv C-R_2$, where R_1 and R_2 refer to the substituent side groups. PDAs are interesting materials because they offer dual advantages; the π-conjugated carbon backbone play an important role in unique electronic and NLO properties, whereas the side groups provide tremendous possibilities in structure control and material processing. Some of the representative examples of the diacetylene polymers $(R_1-C\equiv C-C\equiv C-R_2)_n$ which have been studied for third-order nonlinear optics are as follows.

Third-order NLO properties of a large variety of PDAs have been investigated by a number of research groups, in the form of single crystals, solutions, thin films and LB films because of their ease of availability in different forms. In 1976, Sauteret et al. [220] reported esu $\chi^{(3)}$ values of 8.5×10^{-10} and 0.7×10^{-10} esu at 1.89 µm for PDA-PTS and PDA-TCDU, respectively. PDA-PTS shows a laser damage threshold of about 50 GW cm^{-2} in the picosecond regime and a fast response time of 10^{-14} seconds. The time and wavelength resolved nonlinear optical spectroscopy of PDA-PTS was studied by Carter et al. [221]. The $\chi^{(3)}$ value of PDA-PTS was found to be very sensitive to the measurement wavelength and varies from 9×10^{-10} esu at 6515 Å to 5×10^{-10} esu at 7015 Å. DFWM measurements demonstrated the excited-state life time of PDA-PTS as 1.8 ps. Kajzar and Messier [222] investigated the $\chi^{(3)}$ values of PDAs Langmuir–Blodgett films at various different wavelengths. The THG measurements on LB multi-layers of PDAs showed two resonances in the $\chi^{(3)}$ values in the wavelength range 0.8 to 1.9 µm; one at about 1.35 µm due to a two-photon resonance and another at 1.907 µm due to a three-photon resonance. Tripathy and coworkers [92–95] reported $\chi^{(3)}$ of 4.0×10^{-8} esu at 532 nm for the poly (BPOD) LB

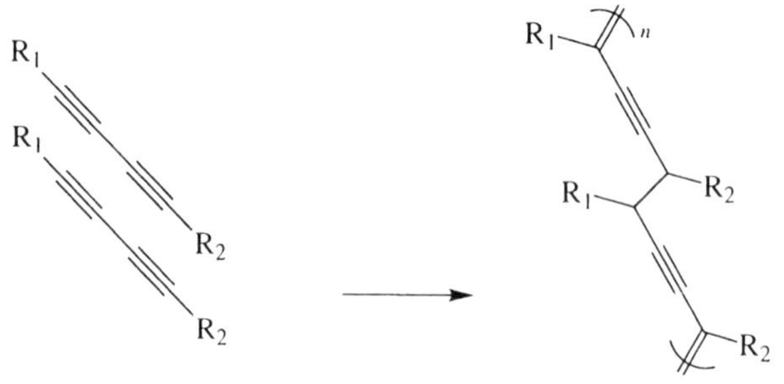

$R_1 = R_2 = -CH_2-O-SO_2-C_6H_4-CH_3$ (PDA-PTS)

$R_1 = R_2 = -(CH_2)_4-O-SO_2-C_6H_4-CH_3$ (PDA-PTS-12)

$R_1 = R_2 = -(CH_2)_4-O-CO-NH-C_6H_5$ (PDA-TCDU)

$R_1 = R_2 = -(CH_2)_3-O-CO-NH-CH_2-COO-C_nH_{2n+1}$ (PDA-nBCMU)

$R_1 = R_2 = CH_3-(CH_2)_{15}-C-C\equiv C-C-(CH_2)_8-COO-$ (PDA-AFA)

$R_1 = R_2 = (CH_2)_4-O-CO-NH-(CH_2)_{n-1}-CH_3$ (PDA-C$_4$UC$_n$)

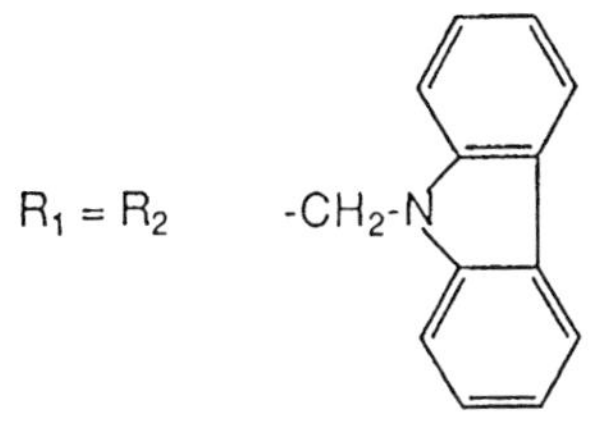

$R_1 = R_2$ $-CH_2-N$⟨carbazole⟩ PDA-DCHD **81**

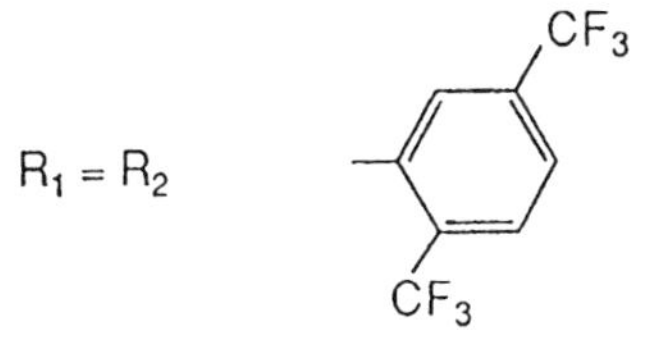

$R_1 = R_2$ ⟨phenyl with CF$_3$, CF$_3$⟩ PDA-DFMP **82**

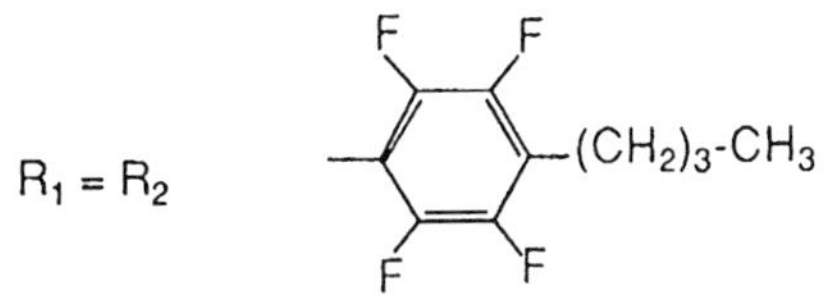

$R_1 = R_2$ ⟨tetrafluorophenyl⟩$-(CH_2)_3-CH_3$ PDA- BTFP **83**

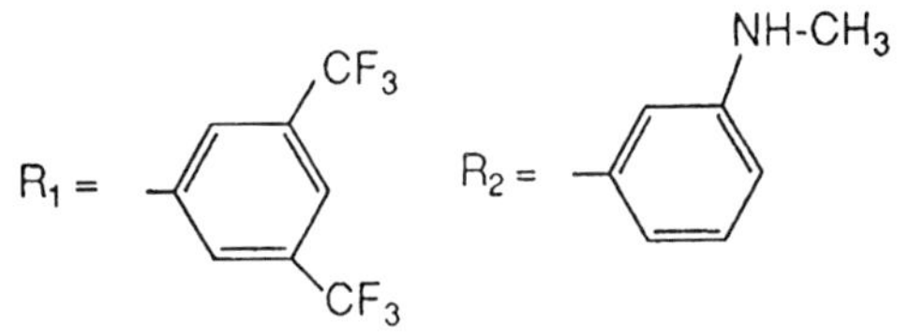

$R_1 =$ ⟨phenyl with CF$_3$, CF$_3$⟩ $R_2 =$ ⟨phenyl with NH-CH$_3$⟩ PDA-MADF **84**

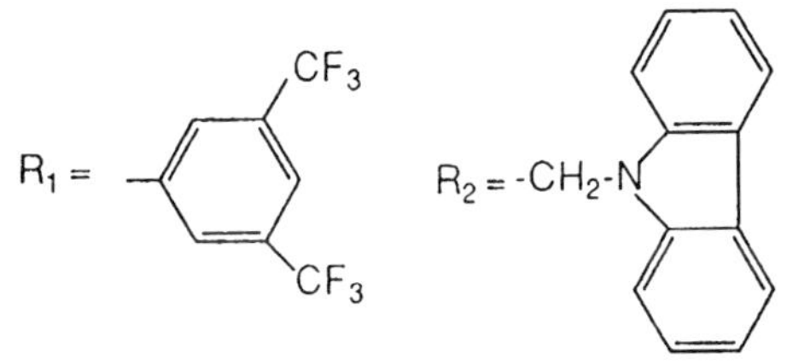

$R_1 =$ ⟨phenyl with CF$_3$, CF$_3$⟩ $R_2 = -CH_2-N$⟨carbazole⟩ PDA-DFCP **85**

films parallel to the dipping direction and 5.9×10^{-9} esu vertical to the dipping direction, measured by DFWM.

Table 6.12 lists the $\chi^{(3)}$ values of several PDAs, which vary by about seven orders of magnitude from 10^{-5} to 10^{-12} esu depending upon the chemical structure and resonance contributions. Greene *et al.* [223] reported a $\chi^{(3)}$ of 2×10^{-5} esu for PDA-PTS. The imaginary $\chi^{(3)}$ values up to 2×10^{-5} esu have been deduced from a saturation of the absorption on resonance at ~ 2 eV from PDA-PTS by Bolger *et al.* [224]. The Im$\chi^{(3)}$ and $|\chi^{(3)}|$ values in the range of 3×10^{-7} to 2×10^{-9} esu have been measured between 1.92 to 1.71 eV. The resonant $\chi^{(3)}$ values are increased by the linear absorption coefficient. In particular, polydiacetylenes have a strong one-photon absorption at 1.99 eV associated to a 1B$_u$ state with excitonic character. Around this region, very large $\chi^{(3)}$ values of PDAs, due to strong resonance enhancement, have been observed from DFWM and saturable absorption measurements. An enhancement in the $\chi^{(3)}$ values of PDAs occurs from the one-photon, two-photon and three-photon resonances. However, very large $\chi^{(3)}$ values of PDAs were also reported away from resonance [243]. Nonlinear optical spectroscopy of PDAs was analyzed by Etemad and Soos [244]. Figure 6.17 shows the linear absorption spectrum of PDA-PTS thin films [245], two-photon resonance enhancement in the THG spectrum of a PDA-PTS crystal [246], and the electroabsorption signal at 1.75 eV of a 0.35 mm thick PDA-PTS single crystal [247]. Comparing these data, it was concluded that the lowest-energy two-photon state lies below the lowest-energy one-photon

Table 6.12 The $\chi^{(3)}$ data on polydiacetylenes and polyacetylenes

Polymer		$\chi^{(3)}(10^{-10}$ esu$)$	Wavelength (μm)	Experimental technique	Ref.
PDA-PTS (single crystal)		8.5	1.89	THG	220
PDA-PTS (thin films)		90	0.6515	DFWM	221
PDA-PTS (single crystal)		2×10^5	0.623	SA	223
PDA-PTS-12 (solution)		5.0	1.064	EFISH	225
		0.11	1.064	THG	225
PDA-TCDU (single crystal)		0.7	1.89	THG	220
PDA-AFA (LB film–blue)		0.56	1.2	THG	222
PDA-AFA (LB film–red)		0.08	0.38	THG	226
PDA-4-BCMU (red gel)		0.13	1.064	DFWM	227
PDA-4-BCMU (thin film)		1.0	1.96	THG	228
PDA-4-BCMU (thin film)		2.0	0.620	DFWM	229
PDA-5-BCMU (thin film)		2.4	2.1	THG	228
PDA-3-BCMU (thin film)		9.0	1.064	DFWM	227
ODA-DCH (single crystal)		2×10^4	0.637	DFWM	230
PDA-C$_4$UC$_3$ (oriented film)		3.8	1.90	THG	231
PDA-C$_4$UC$_4$ (oriented film)		2.9	1.90	THG	231
PDA-C$_4$UC$_5$ (oriented film)		2.1	1.90	THG	231
PDA-BTPF (thin film)		7.7	1.83	THG	231
PDA-DFMP (thin film)		1.7	1.83	THG	231
PDA-DFCP (thin film)		1.1		THG	231
PDA-MADF (thin film)		8.0	1.90	THG	232
PDA-BPOD (LB film)		400	0.532	DFWM	92–95
Polyacetylene (Shirakawa)		13	1.907	THG	233
		1.0	1.064	THG	233
	(Durham)	270	1.907	THG	234
	(Naarmann)	0.2	2.0	THG	235
	(*trans*)	1600	2.25	TPA	236
		56.0	2.070	THG	237
	(oriented)	1800	2.2	THG	238
		170.0	1.907	THG	239
		90.0	1.064	THG	239
	(*trans*)	500.0	0.620	DFWM	240
		990.0	0.720	DFWM	241
PPA R = H		0.025	1.064	THG	242
R = CH$_3$		0.093	1.064	THG	242
R = C$_2$H$_5$		0.071	1.064	THG	242
R = C$_8$H$_{17}$		0.064	1.064	THG	242
R = Si(CH$_3$)$_3$		0.130	1.064	THG	242

state. Polydiacetylenes are one of the best examples of how NLO properties can be altered by variations in the polymer backbone.

Polyacetylene **86** represents an excellent example of a 1D highly conjugated π-electron system consisting of an array of single and double bonds in a carbon chain. It can be prepared by several synthetic routes. It exists in *cis*- and *trans*-isomers and doping with electron acceptor or donor species raises its electrical conductivity as high as 10^4 S cm^{-1} at room temperature [248].

Heeger's group made extensive study on the third-order optical nonlinearity of *cis*- and *trans*-polyacetylenes (PA) using THG [238,249] and two-photon absorption spectroscopy [236]. The nonresonant $\chi^{(3)}$ value of *trans*-PA (more than 10–20 times larger than that of the *cis*-isomer) has been reported for the same Shirakawa polyacetylene sample due to the fundamental change in polymer symmetry [249]. The *trans*-polyacetylene shows a $\chi^{(3)}$ value of an order of magnitude larger over the entire spectral range 0.4–

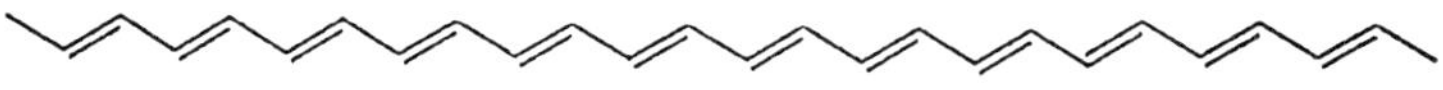

trans-polyacetylene **86**

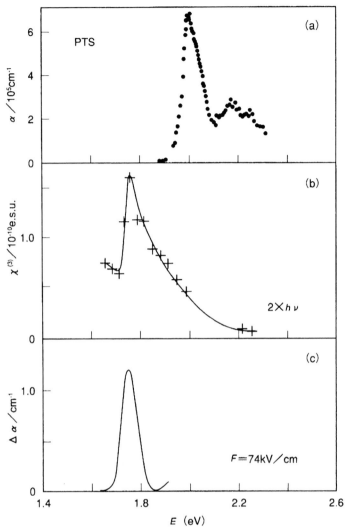

Figure 6.17. (a) Linear absorption spectrum of PDA-PTS thin films (Adapted from ref. 245), (b) two-photon resonance enhancement in the THG spectrum of a PDA-PTS crystal (Adapted from ref. 246), and (c) the electroabsorption signal at 1.75 eV for a 0.35 mm thick PDA-PTS single crystal (Adapted from ref. 247).

1.1 eV, and two peaks, at 0.6 and 0.9 eV were observed [237]. Figure 6.18 shows the THG-measured $\chi^{(3)}$ values of *cis*- and *trans*-polyacetylene over the range 0.5 eV to 1.3 eV. The $\chi^{(3)}$ values of *trans*-polyacetylene is five times larger than the *cis*-isomer through its three-photon resonance is an order of magnitude narrower. The $\chi^{(3)}$ values of oriented *trans*-polyacetylene (draw ratio 10) was found to be a factor of 40 larger than that of non-oriented sample near the two- and three-photon resonances owing to the orientation of the polymer chains and the high

molecular density attained by the material. The $\chi^{(3)}$ values of oriented *trans*-polyacetylene were as high as 1.8×10^{-7} esu at 0.55 eV. The peaks in the $\chi^{(3)}$ spectrum were interpreted as a three-photon resonance to the $1B_u$ state and a two-photon resonance to the $2A_g$ state. The $\chi^{(3)}$ values of polyacetylene are listed in Table 6.12; they vary by a factor of 100 depending upon the material, measurement technique and wavelength.

Interesting third-order NLO properties have been observed for substituted polyacetylenes. For example, the $\chi^{(3)}$ values of poly(phenyl acetylene) PPA change

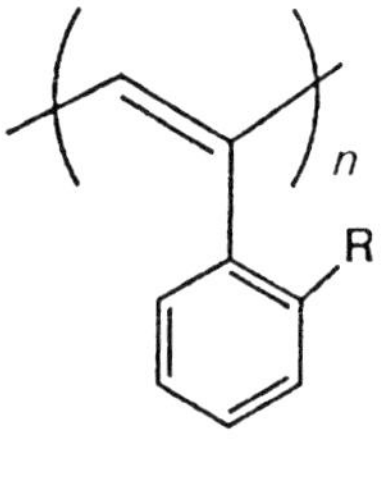

PPA **87** 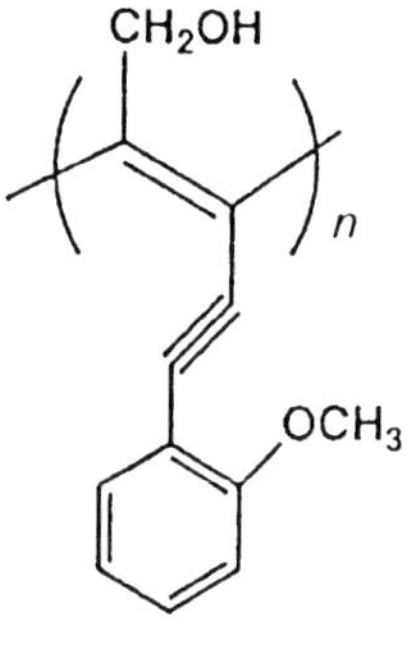   PMDO **88**

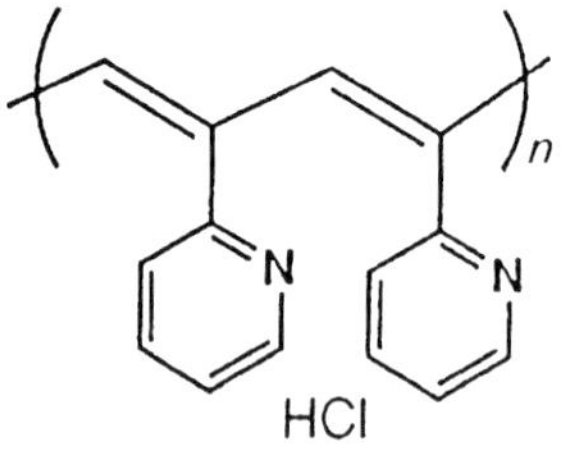

HCl
89

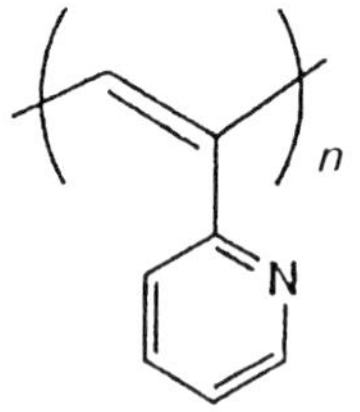

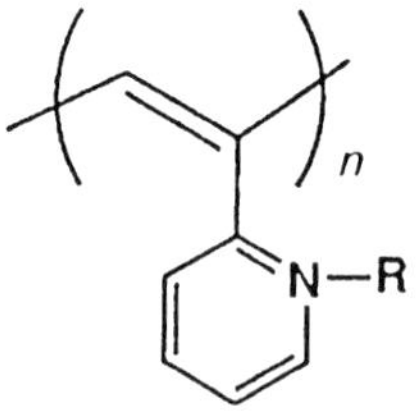

by an order of magnitude depending upon the phenyl ring groups [242]. The size of the substituents lowers the average $\chi^{(3)}$ values by diminishing the density of the polymer chain; this was clearly shown for PPA containing methyl, ethyl and octyl hydrocarbon chains (R). The $\chi^{(3)}$ of poly[1-(2-methoxyphenyl)penta-1,3-dyn-5-ol] (PMDO) was measured as 1.03×10^{-13} esu at 1.9 µm by THG [250]. $\chi^{(3)}$ values on the same order of magnitude were observed for poly(2-pyridy-lacetylene)s [251]. The $\chi^{(3)}$ values of poly(2-ethynyl-pyridine hydrochloride-co-2'-ethylnylpyridine), poly(2-ethynylpyridine), poly(2-N-octylpyridinium)acety-lene bromide) and poly(2-N-stearylpyridinium)acety-lene bromide) were 5.7×10^{-13}, 1.5×10^{-13}, 2.6×10^{-13} and 2.0×10^{-13} esu, respectively. Poly-

mers with HCl and bromide groups show larger third-order optical nonlinearity.

Lee *et al.* [96] measured the $\chi^{(3)}$ values in the range of 10^{-11} esu at 1.907 µm from the THG technique for poly(1,6-heptadiyne) (PHD) derivatives with pendant NLO chromophores. The $\chi^{(3)}$ values of PHD-2 with 28 and 49% mole fraction of chromophores in the copolymer were 2.1×10^{-11} and 2.7×10^{-11} esu respectively. PHD-3 with 20% mole fraction and PHD-1 with 45% mole fraction of monomer chromo-phore in the copolymer showed $\chi^{(3)}$ values of 3.3×10^{-11} and 2.6×10^{-11} esu, respectively. The $\chi^{(3)}$ values of polyethyl dipropargyl(diethoxy-phosphoryl)acetate (PTDPA) thin films were estimated as 3.4×10^{-11} esu. The concentration of the pendant

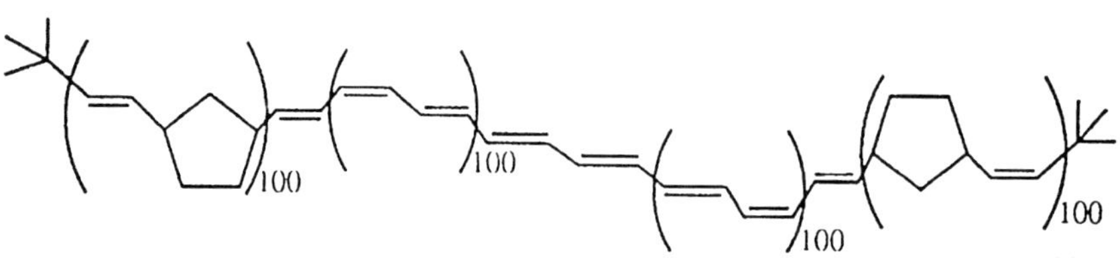

90

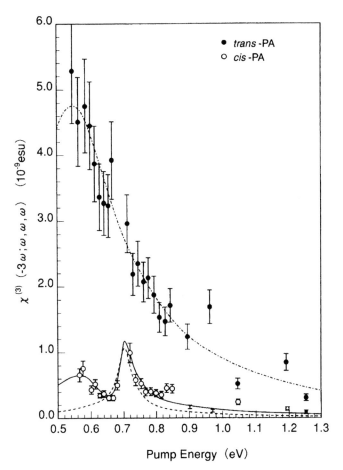

Figure 6.18. $\chi^{(3)}(-3\omega; \omega, \omega, \omega)$ values of *cis*- (open circle) and *trans*-polyacetylene (PA) (solid points) over the range 0.5 eV to 1.3 eV. The solid curves represent the data analyzed in terms of the effective medium theory $[\chi^{(3)}(3\omega)) = (1-f)$ $\chi^{(3)}(3\omega)l_{cis} + f\chi^{(3)}(3\omega)l_{trans}]$ where f is the volume fraction of *trans*-PA in the *cis*-PA. The dashed and dot-dashed curves represent pure *cis*-PA and *trans*-PA, respectively. (Reprinted from Holvorson *et al.*, *Synth. Met.*, 55–57, 3961, 1993 with kind permission from Elsevier Science SA, Lausanne, Switzerland).

NLO chromophore has little effect on third-order optical nonlinearity.

Craig *et al.* [252] reported $\chi^{(3)}$ values of spin-coated films of triblock copolymer containing Durham PA. The copolymers are represented by $[2]_x$-$[ene]_{2y}$-$[2]_x$ where 2 is either norbornene (2a) or methyltetracyclodocene (2b) and the subscripts x and y range from 50 to 100 and from 10 to 200, respectively. Table 6.13 lists THG and DFWM measured $\chi^{(3)}$ values of $[2a]_{100}$-$[ene]_n$-$[2]_{100}$ before and after isomerization. The $\chi^{(3)}$ values of triblock copolymers increase with increasing π-conjugation length, for example, the $\chi^{(3)}(-\omega; \omega, \omega, -\omega)$ value of $[2a]_{100}$-$[ene]_{400}$-$[2a]_{100}$ is 1.3×10^{-8} esu, about 80 times larger than $[2a]_{100}$-$[ene]_{20}$-$[2a]_{100}$.

The $\chi^{(3)}$ values are also affected by resonance contributions. There is significant difference in the $\chi^{(3)}$ values of triblock copolymers containing polynorbornene and polymethyltetracyclodocene segments. For example, the $\chi^{(3)}(-\omega; \omega, \omega, -\omega)$ at 715 nm and $\chi^{(3)}(-3\omega; \omega, \omega, \omega)$ values at 1.90 μm of $[2a]_{100}$ $[ene]_{400}$-$[2a]_{100}$ is about 6 and 4.8 times larger than $[2b]_{100}$-$[ene]_{400}7$-$[2b]_{100}$, respectively. Moreover, $\chi^{(3)}(-3\omega; \omega, \omega, \omega)$ values are larger at 1.65 μm than at 1.90 μm and thermal isomerization of the polyacetylene to all-*trans* led to lower $\chi^{(3)}$ values due to degradation. The $\chi^{(3)}(-\omega; \omega, \omega, -\omega)$ values were significantly increased at 715 nm and were almost constant between 726 to 806 nm.

Table 6.13. Third-order nonlinear optical susceptibility of triblock copolymers containing Durham PAs (Adapted from ref. 252 American Chemical Society.)

Copolymer	$\chi^{(3)}(10^{-10}\text{esu})$		
	THG method		DFWM Method
	1.650 μm	1.90 μm	0.715 μm
$[2a]_{100}\text{-}[ene]_n\text{-}[2a]_{100}$			
$n = 20$	0.026		1.60
(all-*trans*)	0.031		4.0
$n = 100$	0.60	0.50	38.0
(all-*trans*)	0.53	0.32	32.0
$n = 200$	2.0	1.80	110.0
(all-*trans*)	2.20	1.20	90.0
$n = 400$	3.20	2.50	130.0
(all-*trans*)	3.10	1.50	120.0
$[2b]_{1-}\text{-}[ene]_n\text{-}[2b]_{100}$			
$n = 100$	0.32	0.18	10.0
(all-*trans*)	0.28		4.30
$n = 200$	0.90	0.18	15.0
(all-*trans*)	0.60		10.0
$n = 400$	2.20	0.52	22.0
(all-*trans*)	2.0		27.0

Polyarylenes

Poly(arylenevinylene)s (PAV) consists of a π-conjugated backbone like PA. The effect of increasing π-electron conjugation on $\chi^{(3)}$ of poly(p-phenylenevinylene) (PPV) was studied by Bradley and Mori [253]. $\chi^{(3)}$ values of 1.6×10^{-11} and 7.5×10^{-11} esu were recorded after thermal conversion of the sulfonium polyelectrolyte precursor to fully conjugated PPV at 100 and 300°C under vacuum for a period of 3 h, respectively. This demonstrates that $\chi^{(3)}$ is very sensitive to the extent of π-electron conjugation. Kamiyama *et al.* [254] reported that the TH intensity in the direction parallel to the orientation axis was approximately 15 times larger than that in the direction perpendicular to the orientation axis in the highly oriented poly(2,5-dimethoxy-p-phenylenevinylene) (MO-PPV) LB films. $\chi^{(3)}$ values in the direction parallel and perpendicular to the orientation axis were

2.9×10^{-9} and 0.75×10^{-10} esu at the fundamental wavelength of 1.064 μm, respectively. Like PA, the importance of orientation was also noticed for MO-PPV because the $\chi^{(3)}$ value for an unoriented film was about 10 times smaller than that of an oriented MO-PPV film.

Mathy *et al.* [255] investigated the third-order optical nonlinearity of PPV and well-defined chain length oligo-p-phenylenevinylenes **93** where n is 3,4, and 5 using THG. The $\chi^{(3)}(-3\omega;\omega,\omega,\omega)$ spectrum of PPV thin films showed two resonance maxima, a major one at 1.336 μm with a $\chi^{(3)}$ value of 1.6×10^{-10} esu and a smaller one at 1.155 μm. PPV has λ_{max} at 458 nm; therefore $\chi^{(3)}$ values at 1.336 μm are increased by a three-photon resonance. In the case of oligomers, λ_{max} increases with increasing π-conjugation length; it is located at 383 nm for $n = 3$, at 394 nm for $n = 4$ and at 406 nm for $n = 5$. The $\chi^{(3)}$ values measured at

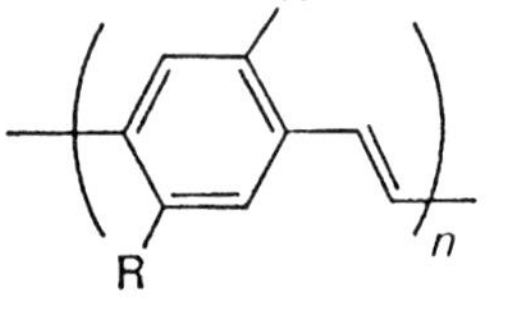

PPV where R=H; MO-PPV where R=OCH₃

91

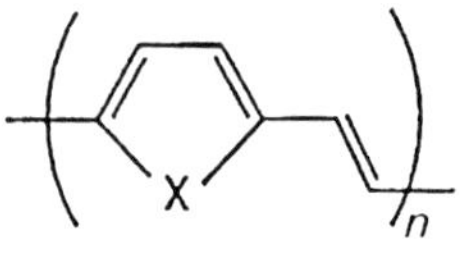

PTV where X=S; PFV where X=O

92

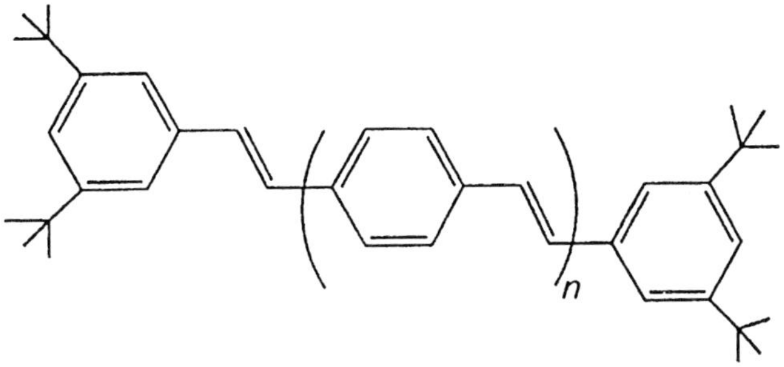

Oligo-*p*-phenylene vinylene where *n*=3,4 and 5   **93**

1.064 μm were on the order of 10^{-11} esu but they did not change much with increasing conjugation length. However the $\chi^{(3)}$ values measured in the range 1.15 to 1.22 μm showed some variation with increasing conjugation. This wavelength region is at $\lambda_L = 3\lambda_{max}$ for these oligomers; hence $\chi^{(3)}$ values have a three-photon resonance contribution.

Tsutsui and coworkers [256] investigated the $\chi^{(3)}(-3\omega;\omega,\omega,\omega)$ values of four PAV thin films, PPV, MO-PPV, poly(2,5-thienylenevinylene) PTV and poly(2,5-furylenevinylene) PFV. The optical band gaps determined from the lower energy edges of the π–π* transitions were 2.4 eV for PPV, 2.1 eV for MO-PPV, and 1.8 eV for both PTV and PFV. The $\chi^{(3)}(-3\omega;\omega,\omega,\omega)$ values measured in the 1.475 to 2.10 μm region showed a peak of $\chi^{(3)}$ at harmonic wavelengths of 540 nm in MO-PPV and 650 nm in PTV. The $\chi^{(3)}(-3\omega;\omega,\omega,\omega)$ values around the three-photon resonance region were 4.5×10^{-10} esu at 1.95 μm for PTV and 1.6×10^{-10} esu for MO-PPV. Like PPV, the $\chi^{(3)}(-3\omega;\omega,\omega,\omega)$ values of PTV films also increased with thermal conversion [256]. The unstretched PTV films showed three-photon resonances at 1.6 and 1.95 μm while in uniaxially stretched films, the resonance peak at 1.95 μm was enlarged, obscuring the peak at 1.6 μm.

Polythiophene is another π-conjugated polymer which has created much interest because of its good environmental stability compared to PAs. Third-order nonlinear properties of polythiophene (pT), polythieno(3,2-*b*)-thiophene (pTT), polydithieno(3,2-*b*,2',3'-*b*)thiophene (pDTT) were reported by Dorsinville *et al.* [257]. Figure 6.19 shows the wavelength dependence of $\chi^{(3)}(-\omega;\omega,\omega,-\omega)$ values for four polythiophenes; pT, pTT, pDTT and poly(3-dodecylthiophene)(pDDT) thin films. $\chi^{(3)}$ values of 5×10^{-10} esu at 602 nm and 4.0×10^{-11} esu at 705 nm were measured for pDDT [258]. The pT, pTT and pDTT thin films showed a $\chi^{(3)}$ value of 3×10^{-11} esu at 1.06 μm. The $\chi^{(3)}$ value measured at 532 nm was about two orders of magnitude larger due to the resonance enhancement. The $\chi^{(3)}$ values of all these polythiophenes were resonantly enhanced in the vicinity of 500 to 600 nm because they strongly absorb in this region.

In order to introduce solubility into the parent polythiophene, alkyl chain (C_nH_{2n+1}) substituted poly(3-alkylthiophene) (P3ATs) were studied. The $\chi^{(3)}$ values of P3ATs with an alkyl chain from butyl (CH_3) to tetradecyl ($C_{14}H_{29}$) have also been measured by various researchers. The $\chi^{(3)}$ values of 1.2×10^{-11} esu for poly(3-hexylthiophene), 5.6×10^{-12} esu for poly

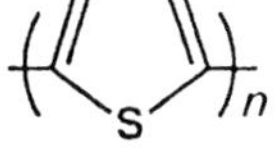

pT **94** 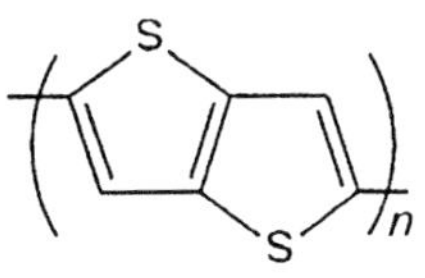   pTT **95** 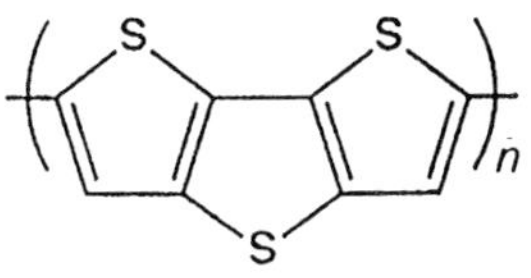   pDTT **96**

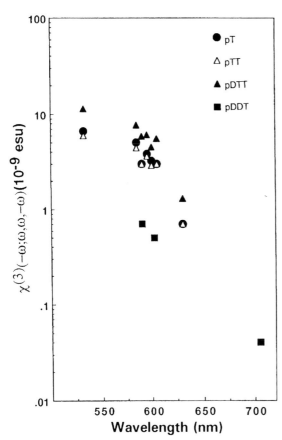

Figure 6.19. Wavelength dependence of $\chi^{(3)}(-\omega; \omega, \omega, -\omega)$ data for polythiophene (pT), polythieno(3,2-*b*)-thiophene (pTT), polydithieno(3,2-*b*,2',3'-*b*)thiophene (pDTT) and poly(3-dodecylthiophene) p-DDT thin films. The data for pT, pTT and PDTT are from ref. 257, and those for p-DDT are from ref. 258.

(1) $R_1 = C_4H_9$ $R_2 = H$ poly(3-butylthiophene)
(2) $R_1 = C_6H_{13}$ $R_2 = H$ poly(3-hexylthiophene)
(3) $R_1 = C_{10}H_{21}$ $R_2 = H$ poly(3-decylthiophene)
(4) $R_1 = C_{12}H_{25}$ $R_2 = H$

poly(3-dodecylthiophene)

(5) $R_1 = C_{14}H_{29}$ $R_2 = H$

poly(3-tetradecylthiophene)

(6) $R_1 = OC_8H_{17}$ $R_2 = CH_3$

poly(3-octyloxy-4-methylthiophene)

(7) $R_1 = OC_4H_9$ $R_2 = OC_4H_9$

poly(3,4-dibutoxythiophene)

(8) $R_1 = H$ $R_2 = (CH_2)_3(CF_2)_5CF_3$

poly[3-(1,2,3,4,5-perfluorononyl)thiophene]

(9) $R_1 = H$ $R_2 = C_6H_4\text{-}p\text{-}OC_8H_{17}$

poly[3-(*p*-octyloxyphenyl)thiophene]

(10) $R_1 = H$ $R_2 = C_6H_4\text{--}O\text{--}OC_8H_{17}$

poly[3-(*o*-octyloxyphenyl)thiophene]

Jenekhe *et al.* [260] first reported large third-order optical nonlinearities of mixed aromatic-quinoid polythiophenes. The $\chi^{(3)}(-\omega;\omega,\omega,-\omega)$ values measured at 532 nm in a dichloromethane solution by the DFWM technique were found to be the largest. The optical nonlinearity was resonantly enhanced because the measurement wavelength of 532 nm lies in the absorbing region of the optical spectrum. From an effective two-level system, the resonantly enhanced factor is written as

$$\omega - \omega_{ng} + i\Gamma_{ng}/\omega_{nr} - \omega_{ng} + i\Gamma_{ng} \qquad (6.56)$$

where ω_{ng} is absorption peak frequency at the absorption maximum of 451 nm, Γ_{ng}, the bandwidth, is $10^{11}s^{-1}$, ω and ω_{nr} are the nonresonance and near resonance measurement frequencies, respectively. From this assumption, the resonance enhancement was determined to be less than a factor of five. Poly(α-[5,5'-bithiophenediyl]benzylidene-block-α-[5,5'-bithiophenequinodi-methanediyl] (PBTBQ) showed an extremely large $\chi^{(3)}$ value of 2.7×10^{-7} esu at the same wavelength [261 Braunling *et al.* [262] reported $\chi^{(3)}$ values of 1.4×10^{-9} esu for PTM, 9.4×10^{-9} esu for PBTBQ and 21.4×10^{-9} esu for PBTSBQ from the saturable absorption measurement at 633 nm. The γ values were 1.2×10^{-31}, 1.6×10^{-29} and 3.7×10^{-30} esu for PMTBQ, PBTBQ and PBTABQ,

(3-tetradecylthiophene), 6.0×10^{-12} esu for poly(3-octyloxy,4-methylthiophene), 7.5×10^{-12} esu for poly(3-dibutoxythiophene), 7.5×10^{-13} esu for poly(3-(p-octyloxyphenylthiophene), 1.4×10^{-12} esu for poly[3-(*o*-octyloxyphenyl)thiophene] and 2.3×10^{-12} esu for poly[3-(1,2,3,4,5-perfluorononyl)thiophene] thin films were reported by Callender *et al.* [259]. The absorption maxima of poly[3-(*o*-octyloxphenyl]thiophene (POOPT) and poly[3-(*p*-octyloxyphenyl)thiophene] (PPOPT) increased compared to P3ATs. The $\chi^{(3)}(-3\omega;\omega,\omega,\omega)$ values of these polythiophenes ranged between 1.2×10^{-11} and 0.75×10^{-12} esu showing the minor effect of substitution on the $\chi^{(3)}$ values. The $\chi^{(3)}(-3\omega;\omega,\omega,\omega)$ value of poly(3-hexylthiophene) was twice that of poly(3-tetradecylthiophene).

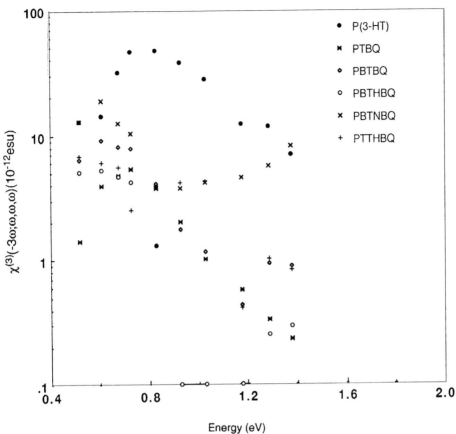

Figure 6.20. Dispersion of $\chi^{(3)}(-3\omega; \omega, \omega, \omega)$ in the range of 0.52 eV to 1.38 eV for P3HT, PTBQ, PBTBQ, PBTHBQ and PTTHBQ (Plotted from *Synth. Met.* **49**, J.S. Meth *et al.*, 59, 1992, with kind permission from Elservier Science SA, Lausanne, Switzerland).

respectively. Meth *et al.* [263] studied the dispersion of $\chi^{(3)}(-3\omega;\omega,\omega,\omega)$ in the range 0.5 to 1.4 eV for poly(3-hexylthiophene) (P3HT), polymers derived from precursors poly(α-[5,5′-bithiophenediyl]benzylidene (PBTB), poly(α-[5,5′-bithiophediyl]*p*-nitrobenzylidene (PBTNB), poly(α-[5,5′-bithiophene-diyl]*p*-heptyloxy-benzylidene (PBTHB), poly(α-[5,5′-terthiophene-diyl]*p*-acetoxybenzylidene (PTTHB); their conjugated derivatives are PBTBQ, PBTNBQ, PBTHBQ, and PTTHBQ. Figure 6.20 shows the dispersion of $\chi^{(3)}(-3\omega; \omega, \omega, \omega)$ in the range of 0.52 to 1.38 eV for P3HT, PTBQ, PBTBQ, PBTHBQ and PTTHBQ. The $\chi^{(3)}(-3\omega; \omega, \omega, \omega)$ values are more resonantly enhanced than those measured in the non resonant regions. The $\chi^{(3)}(-3\omega;\omega,\omega,\omega)$ value of P3HT was described by the two-level model. As can be seen here there is a difference of about five orders of magnitude between $\chi^{(3)}(-3\omega;\omega,\omega,\omega)$ and $\chi^{(3)}(-\omega;\omega,\omega,-\omega)$, the

former being remarkably smaller. It seems that an aromatic and quinoid arrangement in both the ground and excited states is a favorable condition to nonlinear optical process.

Callender *et al.* [259] measured the $\chi^{(3)}$ values of copolymers from 3-decylthiophene and 3-methylthiophene. The P(DT-3MeT) copolymer **99** in the ratio of 50:50 showed a $\chi^{(3)}$ value of 6.6×10^{-12} esu at 1.064 μm, about twice that of the 65:35 copolymer (3.6×10^{-12} esu). Bao *et al.* [264] synthesized an alternating copolymer of benzene and thiophene **100** by the Stille coupling reaction. The band gap of this copolymer is 2.38 eV (520 nm); it lies between the band gap of poly(*p*-phenylene) 3.0 eV (413 nm) and polythiophene 2.1 eV (590 nm). The chloroform solution of the copolymer has a $\chi^{(3)}$ value of 1.77×10^{-13} esu with a concentration of 0.527 gl^{-1} as measured from the DFWM technique at 532 nm.

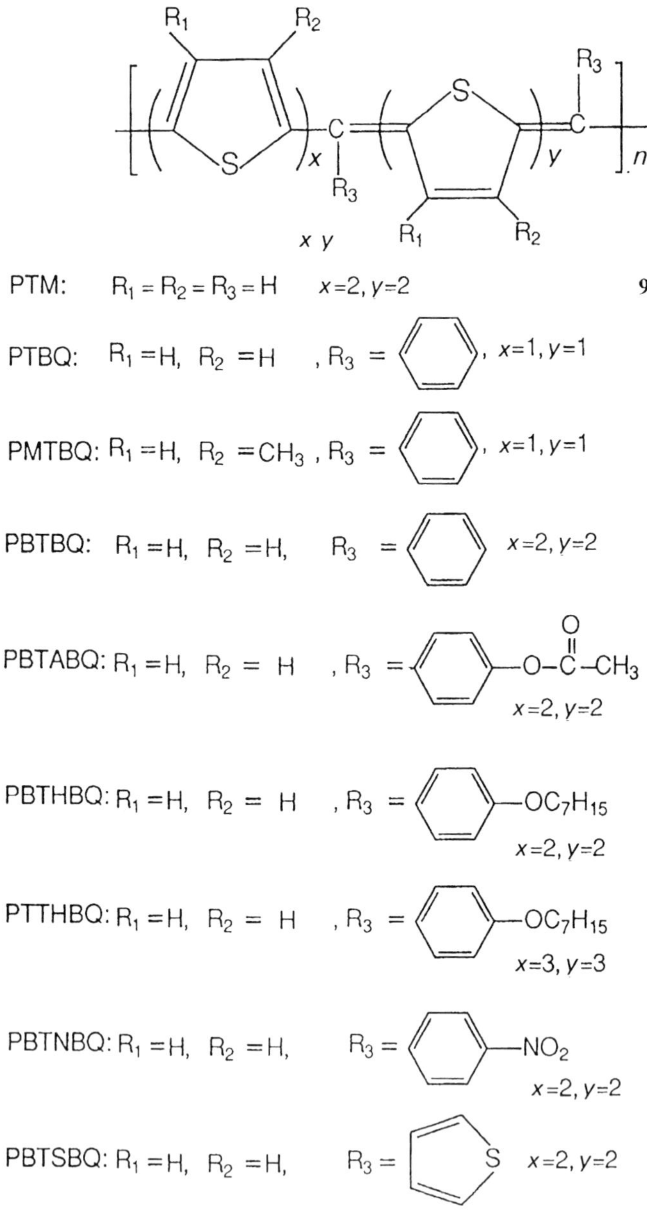

PTM: $R_1 = R_2 = R_3 = H$ $x=2, y=2$ **98**

PTBQ: $R_1 = H$, $R_2 = H$, $R_3 = $ [phenyl] , $x=1, y=1$

PMTBQ: $R_1 = H$, $R_2 = CH_3$, $R_3 = $ [phenyl] , $x=1, y=1$

PBTBQ: $R_1 = H$, $R_2 = H$, $R_3 = $ [phenyl] $x=2, y=2$

PBTABQ: $R_1 = H$, $R_2 = H$, $R_3 = $ [phenyl]$-O-\overset{\overset{\displaystyle O}{\|}}{C}-CH_3$ $x=2, y=2$

PBTHBQ: $R_1 = H$, $R_2 = H$, $R_3 = $ [phenyl]$-OC_7H_{15}$ $x=2, y=2$

PTTHBQ: $R_1 = H$, $R_2 = H$, $R_3 = $ [phenyl]$-OC_7H_{15}$ $x=3, y=3$

PBTNBQ: $R_1 = H$, $R_2 = H$, $R_3 = $ [phenyl]$-NO_2$ $x=2, y=2$

PBTSBQ: $R_1 = H$, $R_2 = H$, $R_3 = $ [thienyl, S] $x=2, y=2$

Polythiophenes are intensely colored materials; therefore, the third-order susceptibility has a resonant contribution due to the high absorption. In an attempt to overcome this problem, Nalwa [265] prepared two random copolymers of 3-methylthiophene and methyl methacrylate 101; one is yellow while the other one is red. Both copolymers are solution processable and show good environmental stability. The main advantage of the copolymer is its optical quality and good solution processibility compared with the parent polythiophenes. Thin films prepared by the solution casting technique show different optical absorption characteristics. It was demonstrated that by maneuvering the ratio of conjugated to nonconjugated segments in the polymer strand, the electrical conductivity of 3-methylthiophene and methyl methacrylate copolymers can be tailored by five orders of magnitude up to 6.5 S cm^{-1} on iodine doping [264,265]. The increased

99

100

components of 3-methylthiophene π-conjugation give rises to larger electrical conductivity where the increased components of methylmethacrylate segments assist in tailoring the optical transparency. The $\chi^{(3)}$ values of thin films of 3-methylthiophene and methylmethacrylate copolymers were measured using the DFWM and THG techniques [266,267]. The red copolymer showed resonant $\chi^{(3)}(-\omega;\omega,\omega,-\omega)$ values on the order of 8.7×10^{-11} esu about 12 times larger than that of a yellow copolymer determined using the DFWM technique at 0.602 μm. The THG-measured $\chi^{(3)}(-3\omega;\omega,\omega,\omega)$ values were 5.0×10^{-13} esu for the yellow copolymer and 1.3×10^{-11} esu for the red copolymer at 1.860 μm. The large resonant $\chi^{(3)}$ values of the red copolymer are attributed to the highly delocalized π-electron system of the 3-methylthiophene segments whereas the increased nonconjugated methyl-methacrylate segments significantly reduces the magnitude of $\chi^{(3)}$ in the yellow copolymer. The copolymers of 3-methylthiophene and methylmethacrylate are some of the best examples of where manipulation of π-conjugated and nonconjugated segments in a single polymer backbone through polymerization, yields novel polymers with excellent thermal stability, desired optical transparency, high electrical conductivity and third-order NLO susceptibility ($\chi^{(3)} = \sim 10^{-10}$ esu).

101

The $\chi^{(3)}$ value of polythiophenes, P3ATs and copolymers can be compared. The difference in $\chi^{(3)}$ as a function of the alkyl chain length occurs to the dilution of π-conjugation with increasing chain length

which also cause variations in the refractive indices. PT and P3ATs absorb in the vicinity of 450–600 nm with λ_{max} around 550 nm; hence the $\chi^{(3)}(-3\omega; \omega, \omega, \omega)$ values can be enhanced several times at fundamental frequencies around 1.35–1.80 μm with peaking around 1.65 μm due to three-photon resonance. On the other hand, one-photon resonance can give rise to significantly larger $\chi^{(3)}(-\omega; \omega, \omega,-\omega)$ values in the DFWM measurement if performed in the vicinity of 450–600 nm. For example, Kajzar et al. [268] investigated by $\chi^{(3)}(-3\omega; \omega, \omega, \omega)$ value of pTT, poly[1,4-di(2-thienyl)benzene] (PDTB) in the fundamental wavelength range 1.064 to 1.907 μm; a three-photon resonance was observed in the $1.25 \sim 1.45$ μm region. For PDTB, the three-photon resonance band was rather narrow ($\Omega = 780$ cm^{-1}). The π-π^* transition in PDTB occurs at an energy 3 eV different to that of PTT. A non planar geometry of the PDTB backbone caused by the steric hindrance of the hydrogen atoms of the adjacent benzene and thiophene ring, results in a high-energy π-π^* transition. The possible difference in the optical nonlinearity of pTT and PDTB occurred due to the different conjugation lengths Torruellas et al. [269a] measured the $\chi^{(3)}(-3\omega;\omega,\omega,\omega)$ of poly(3-butylthio-phene) and poly(3-decylthiophene) in the range 0.65 to 1.3 eV and evaluated difference resonances. The magnitude of $\chi^{(3)}$ is also significantly influenced by the π-delocalization length, donor–acceptor functional-ities, chain orientation and conformation in polythio-phenes and the measurement techniques. Yang et al. [270] measured the sign and magnitude of real and the imaginary parts of $\chi^{(3)}$ of polythiophene thin films by the Z-scan method. The real $\chi^{(3)}$ of -6.1×10^{-9} esu and Im$\chi^{(3)}$ of -1.9×10^{-9} esu were estimated at 532 nm; here both the real and the imaginary part of $\chi^{(3)}$ are negative. The absolute $\chi^{(3)}$ calculated from $|(\text{Re}\chi^{(3)})^2 + (\text{Im}\chi^{*(3)})^2|^{1/2}$ was 6.30×10^{-9} esu for polythiophene which is comparable to the DFWM $\chi^{(3)}$ value of 5.9×10^{-9} esu at 532 nm [257]. The agreement between third-order optical nonlinearity measured by Z-scan and DFWM techniques indicates the same response time < 30 ps. As can be seen by comparing $\chi^{(3)}$ data measured by the DFWM and THG techniques the large difference between the $\chi^{(3)}$ values occurs because of the measurement wavelength and NLO process associated with the measurement techniques.

Polyanilines show interesting electrical properties which are governed by addition or removal of electrons or protons from the polymer backbone, derivatization of the *p*-phenylene rings and by various forms [271]. The resonant $\chi^{(3)}$ of 10^{-8} esu and a similar value for the emeraldine salt were reported by Ginder *et al.* [272]. Wong *et al.* [273] measured the $\chi^{(3)}$ of emeraldine-based polyaniline as 8×10^{-10} esu at 602 nm by the DFWM technique; it has the nonlinear figure of merit ($\chi^{(3)}/\alpha$) of 2×10^{-14} esu cm. The transient $\chi^{(3)}$ response for nonresonant excitation was < 100 fs. Callender *et al.* [259] reported a $\chi^{(3)}(-3\omega; \omega, \omega, \omega)$ value of 1.6×10^{-12} esu at 1.064 μm for poly(2-decyloxyaniline). This polymer has a λ_{max} at 610 nm and a refractive index of 1.66 at 1053 nm. The phase of $\chi^{(3)}(-3\omega; \omega, \omega, \omega)$ for poly(2-decyloxyaniline) is 25°; therefore the multiphoton processes are different from those of poly(thiophenes) and the possibility of the presence of a two-photon resonance was indicated.

MacDiarmid and Epstein [274] reported that the electrical conductivity of the emeraldine base increases by as much as 13 orders of magnitude from 10^{-10} S cm^{-1} to 10^{3} S cm^{-1} upon protonation showing an insulator-to-metal transition. Halvorson *et al.* [275] reported dispersion of the $\chi^{(3)}(-3\omega; \omega, \omega, \omega)$ value over the range 0.55 to 1.3 eV for three base forms of

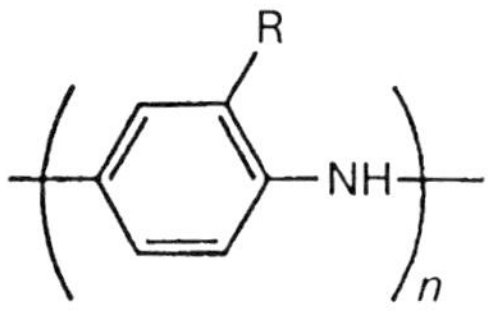

102

Polyaniline where R=H
Poly(2-decylocyaniline) where R=OC$_{10}$H$_{21}$

polyaniline: leucoemeraldine (fully-reduced form), emeraldine (semioxidized form) and pernigraniline (fully oxidized form). The $\chi^{(3)}(-3\omega; \omega, \omega, \omega)$ value of the emeraldine base was found to be the largest, about 4.4×10^{-10} esu at 0.55 eV on three-photon resonance, and about 3.4 times larger than pernigraniline form. The $\chi^{(3)}(-3\omega; \omega, \omega, \omega)$ values of leucoemeraldine were the lowest over the entire spectral range and about ten times smaller than that of emeraldine and pernigraniline forms. The leucoemeraldine showed a $\chi^{(3)}(-3\omega; \omega, \omega, \omega)$ value 1.8×10^{-11} esu at 1.18 eV due to three-photon resonance and an off-resonance value of 4.5×10^{-12} esu at 0.75 eV. The $\chi^{(3)}(-3\omega; \omega, \omega, \omega)$ value increased by an order of magnitude by converting the leucoemeraldine form (π–π^* energy gap=3.7 eV) to the emeraldine base (π–π^* energy gap=2.2 eV).

Chen *et al.* [276] measured $\chi^{(3)}(-3\omega; \omega, \omega, \omega)$ values of poly(4,4'-diphenylimine methine) PDPIM and poly(4,4'-diphenylimine *p*-heptyloxybenzylidene) PDPIHB thin films in the wavelength range of 0.9 to 2.37 μm. Thin films of PDPIHB showed larger

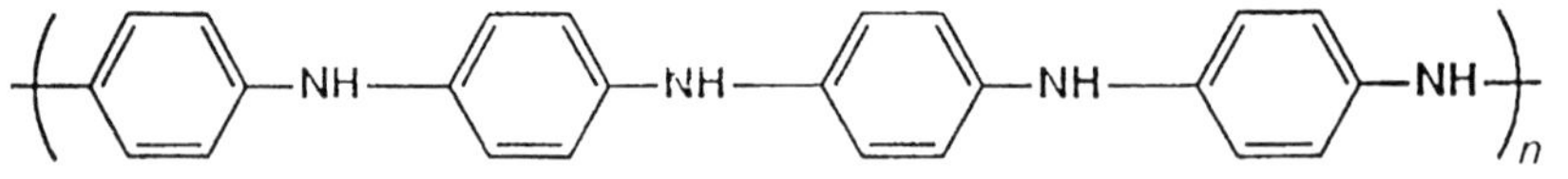

Leucoemeradline form

103

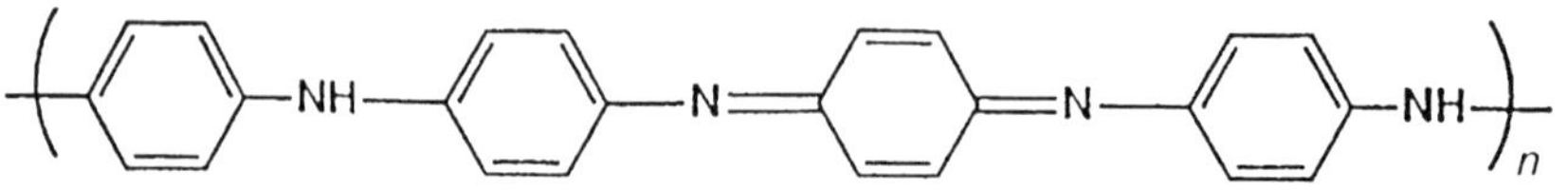

Emeraldine form **104**

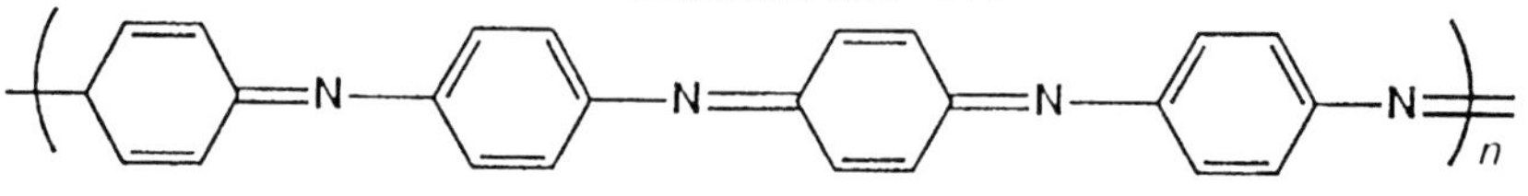

Pernigraniline form **105**

PDPIM   **106**

PDPIHB   **107**

$\chi^{(3)}(-3\omega; \omega, \omega, \omega)$ values than PDPIM over the entire range. The increase in $\chi^{(3)}(-3\omega; \omega, \omega, \omega)$ for PDPIHB results from the attached pendant *p*-heptyloxyphenylene groups and larger π-electron delocalization and oscillator strength. The substitution of the *p*-heptyloxyphenylene group at the bridge methine carbon also leads to a lower refractive index of 1.59 at 2.1 µm for PDPIHB compared to that of 1.81 at 2.1 µm for PDPIM. The $\chi^{(3)}(-3\omega; \omega, \omega, \omega)$ spectrum of PDPIHB showed three resonance peaks located at 1.05, 1.5 and 1.83 µm which correspond to absorption maxima at 363, 507 and 660 nm in the optical absorption spectrum. From the viewpoint of the chemical structure, the replacement of one of the sp^2 nitrogen atom in polypernigraniline with an sp^2 carbon atom changes both the linear and third-order NLO properties of these derivatives.

Polyazines offer tremendous architectural flexibility; therefore their molecular structures can be modified easily in order to minimize the absorption losses due to the optical beam. Nalwa *et al.* [133] have studied oligomeric and polymeric azine compounds. Polyazine is isoelectronic with polyacetylene. Both polyacetylene and polyazine have a simple linear chain of atoms with alternating single and double bonds, but polyazine consists of pairs of nitrogen atoms substituted for pairs of carbon atoms in a polyacetylene chain. This arrangement of carbon and nitrogen atoms in the polyazine π-conjugated backbone offers environmental stability and leads to better optical transparency. Nalwa *et al.* [153,154] extended their studies to novel polyazine materials in which different aromatic units were imparted into the π-conjugated backbone. This chemical modification provides novel polyazines ran-

ging from yellow to black and offers solution processability and environmental stability. The novel polyazines were prepared by either reacting diamino compounds with 2,3-hexanedione or diacetyl compounds with 2,3-hexanedione dihydrazone in dimethylsulfoxide (DMSO) using glacial acetic acid as a catalyst. The molecular structure of the conjugated polyazine backbone was modified by incorporating aromatic units of pyridine (C13PY), ferrocene and of polyazomethines by amino-1,3,5-triazine (C13TMT), tetrafluorobenzene (C13F4) and octafluorobiphenyl (C13F8).

The $\chi^{(3)}(-3\omega; \omega, \omega, \omega)$ values of polyazines and polyazomethines were measured by the THG technique between a fundamental wavelength 1.05 and 1.95 µm [277]. Polyazines showed $\chi^{(3)}(-3\omega; \omega, \omega, \omega)$ values in the range of 10^{-11} to 10^{-12} esu depending on the polyazine structures and measurement wavelengths. For example, the $\chi^{(3)}(-3\omega; \omega, \omega, \omega)$ values of 7.7×10^{-12} esu at 1.5 µm and 2.2×10^{-11} esu at 1.8 µm were observed for polyazines containing pyridine (C13PY) and ferrocene (C13Fc) moieties, respectively. The larger $\chi^{(3)}$ value for the ferrocene moiety arises from metal–ligand bonding. Figure 6.21 shows the optical absorption spectrum and the wavelength dependence of $\chi^{(3)}(-3\omega; \omega, \omega, \omega)$ for thin films of polyazine with the pyridine unit (C13PY) in the 1.05 to 1.95 µm region. The larger $\chi^{(3)}$ of 7.8×10^{-12} esu at the $\lambda_{\max}$ occurs due to a three-photon resonance. The conjugation breaking due to two α–α' coupling in pyridine may be a factor for lower $\chi^{(3)}$ of pyridine containing polyazine. The results indicated that the incorporation of aromatic units in the π-conjugated polyazine backbone affects both the magnitude of $\chi^{(3)}$ and the optical transparency.

Relatively larger $\chi^{(3)}(-3\omega; \omega, \omega, \omega)$ values were measured for the polyazomethines: C13AB and C13OX. C13AB showed a $\chi^{(3)}(-3\omega; \omega, \omega, \omega)$ value of 6.85×10^{-12} esu and here the magnitude of $\chi^{(3)}$ is enhanced due to the near resonance contribution, because thin films show an absorption peak at 408 nm. The $\chi^{(3)}$ values of 1.15×10^{-11} esu for C13OX and 9.43×10^{-12} esu for C13Fc were measured at 1.05 µm. The $\chi^{(3)}$ values increased near the band edge as has been reported for the polyacetylene conjugated backbone. Therefore a comparison of the optical spectrum with 3ω reveals the resonant origin of the $\chi^{(3)}(-3\omega; \omega, \omega, \omega)$ value in polyazines and polyazomethines.

Benzimidaxobenzophenanthroline-type BBL and BBB ladder polymers are another class of π-conjugated polymers known for their excellent environmental and thermal stability [278]. The polymer chain linkage indicates a complete ladder structure for BBL and a semiladder structure for BBB polymer. The off-resonant $\chi^{(3)}$ values of 1.5×10^{-11} esu for the BBL polymer and 5.5×10^{-12} esu for the BBB polymer at 1.064 μm from DFWM measurements were reported by Lindle *et al.* [279]. BBL showed a resonant $\chi^{(3)}$ value of 2.0×10^{-9} esu at 0.532 μm, about two orders of magnitude larger than that of the off-resonant value. The $\chi^{(3)}$ of the BBL polymer electrochemically doped with tetrabutylammonium ions (BF_4^-) was enhanced by about 30%. Jenekhe *et al.* [280] measured $\chi^{(3)}$ values of 2.90×10^{-11} and 0.96×10^{-11} esu for BBL and BBB at 1.05 μm, respectively.

Heterocyclic rigid-rod polymers such as poly(*p*-phenylenebenzobisoxazole) (PBO), poly(*p*-phenylene-benzobisthiazole) (PBZT) and poly(*p*-phenylenebenzo-bisimidazole) (PBZI) are also well known for their high thermal stability and mechanical strength. Jenekhe *et al.* [281] measured the $\chi^{(3)}$ of PBO and PBZT thin films by a THG technique between 0.9 and 2.4 μm. The PBO showed a strong three-photon resonance of $\chi^{(3)}$ at 1.2 μm while PBZT showed at around 1.3 μm. These resonances are in accordance with the optical spectrum where PBO shows a blue shift. Away from the resonance at 2.4 μm, the $\chi^{(3)}$ values of both PBO and PBZT were about an order of magnitude lower than that of the resonance values. Both PBO and PBZT have $\chi^{(3)}$ values of a similar magnitude in the resonant and off-resonant region. The $\chi^{(3)}$ value is on the order of 10^{-11} esu at 1.064 μm for a rigid-

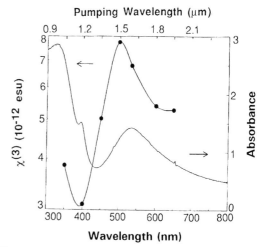

Figure 6.21. Optical absorption spectrum (solid curve) and the wavelength dependence of $\chi^{(3)}(-3\omega;\omega,\omega,\omega)$ for thin films of polyazine with pyridine units (C13PY) [277].

rod polymer, poly(phenothiazinobisthiazole) (PPT); it has an analogous chemical structure to PBZT and was measured using THG by Kistenmacher *et al.* [282]. The magnitude of the $\chi^{(3)}$ values for PPT, PBO and PBZT around the 1.06 μm region is almost the same. This also supported the fact the heteroatoms have no significant effect on third-order optical nonlinearity.

C13PY **108**

C13Fc **109**

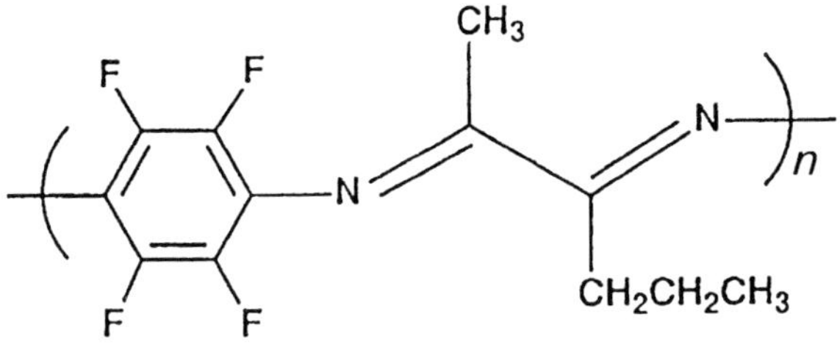

C13F4 **110**

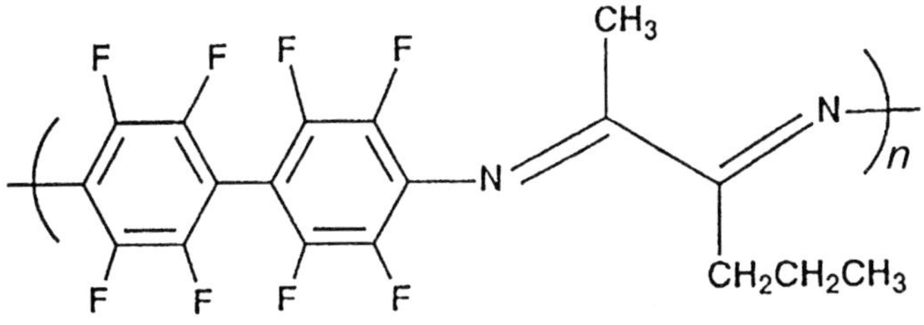

C13F8 **111**

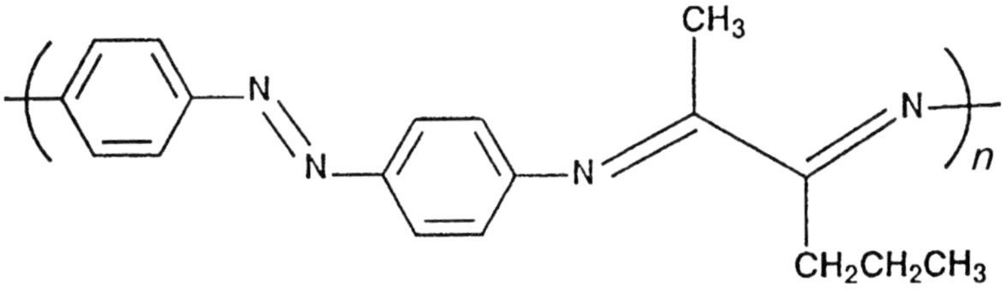

C13AB **112**

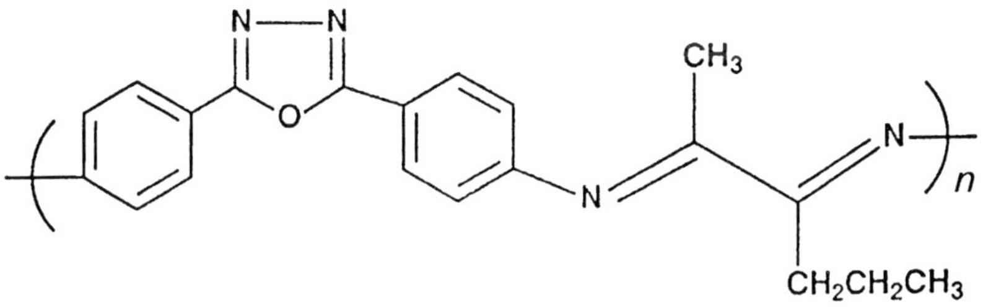

C13OX **113**

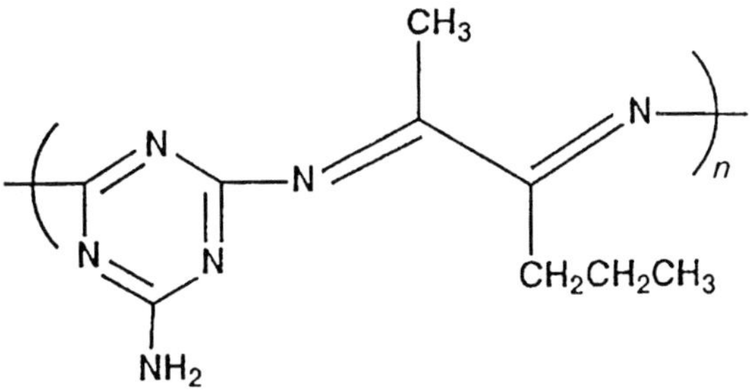

C13TMT **114**

The $\chi^{(3)}$ values of different polyarylenes measured by THG and DFWM at various wavelengths are listed in Table 6.14. The $\chi^{(3)}$ values change by three orders of magnitude from 10^{-9} esu to 10^{-12} esu depending upon the π-conjugated polyarylene structures, resonance contributions and measurement techniques. The $\chi^{(3)}$ values obtained by DFWM are comparatively larger than those measured by THG. In the case of polyazines, introduction of a ferrocene moiety into the π-conjugated backbone gives rise to larger third-order optical nonlinearity than those with and without aromatic moieties. Protonation has a significant effect on the $\chi^{(3)}$ of polyanilines. Moreover, Langmuir–Blodgett films of MO-PPV have significantly large

$\chi^{(3)}(-3\omega;\omega,\omega,\omega)$ because of high orientation. Another important feature is that incorporation of heteroatoms in π-backbone such as polyazines, polyazomethines, polythiophenes, ladder polymers do not noticeably result in larger $\chi^{(3)}$ values compared to polyacetylene and poly(p-phenylene vinylene).

Bubeck [287] attempted to develop a structure–property relationship between chemical structures, linear and third-order NLO properties of π-conjugated polymers by plotting $\chi^{(3)}(-3\omega;\omega,\omega,\omega)$ data, measured at 1.064 μm, as a function of λ_{max} of π-conjugated materials as depicted in Figure 6.22. The $\chi^{(3)}-3\omega;\omega,$ $\omega,\omega)$ values of polyacetylene (PA) [233–237]. Poly-(phenylacetylenes) PPA [242], poly(3-decylthiophene)

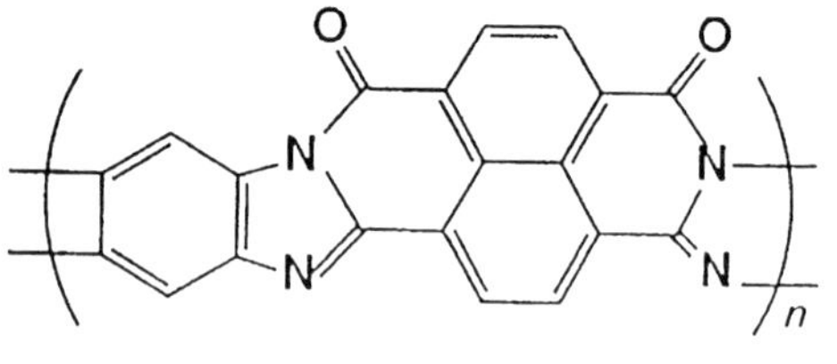

BBL polymer **115**

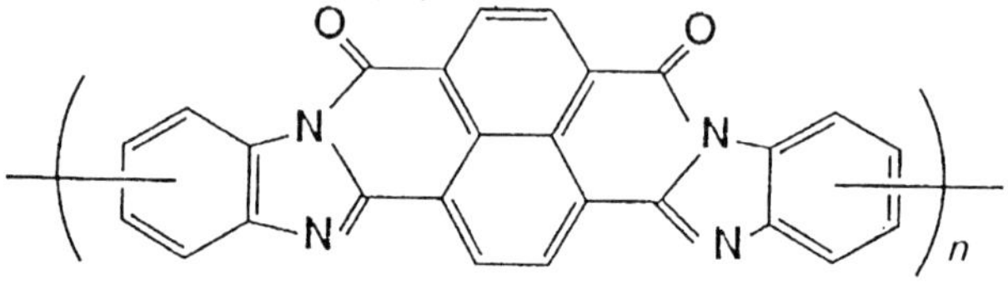

BBB polymer **116**

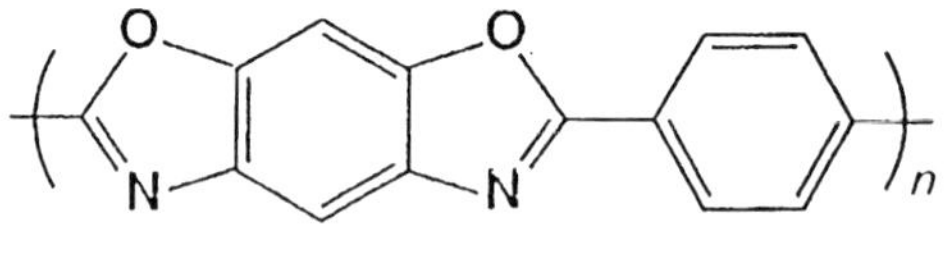

PBO **117**

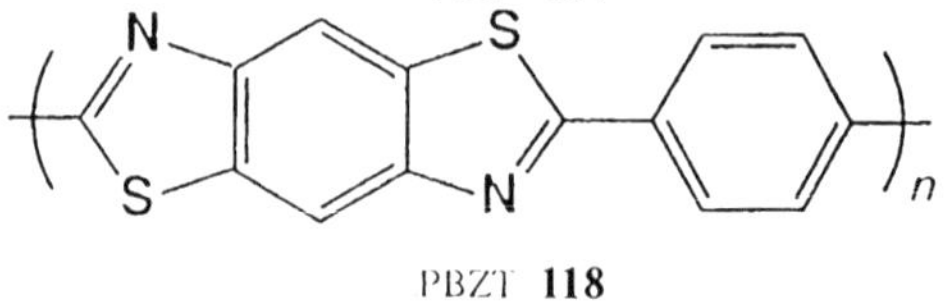

PBZT **118**

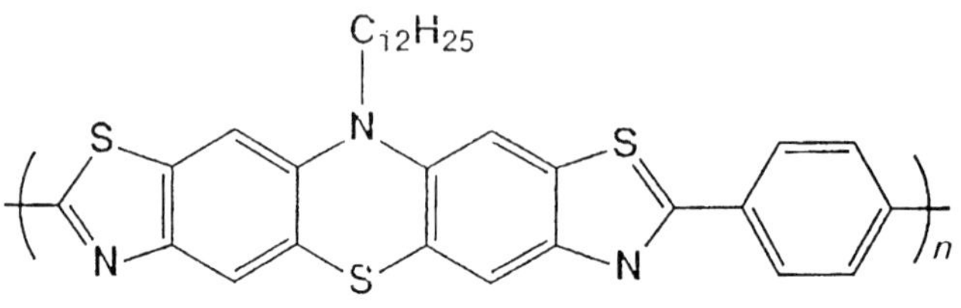

119

Table 6.14. The $\chi^{(3)}$ data of polyarylenes π-conjugated polymers

Polymer	$\chi^{(3)}$ (10^{-10} esu)	Wavelength (µm)	Experimental technique	Ref.
Polythiophene (pT)	66	0.532	DFWM	257
pTT	59	0.532	DFWM	257
pDDT	113	0.532	DFWM	257
PBTBQ	2700	0.532	DFWM	261
P(3-MeTH/MMA)	0.87	0.602	DFWM	267
	0.13	1.86	THG	267
Poly(3-decylthiophene)	0.118	1.064	THG	269
Poly(3-dodecylthiophene)	5.0	0.602	DFWM	258
	7.0	0.590	DFWM	258
	0.4	0.705	DFWM	258
Propylmethylpolyazine	0.08	1.50	THG	133
Permethylpolyazine	0.09	1.50	THG	133
Polyazine-pyridine (C13PY)	0.038	1.05	THG	276
	0.053	1.95	THG	276
Polyazine-ferrocene (C13Fc)	0.094	1.05	THG	276
	0.21	1.95	THG	276
Polyazomethine (C13F4)	0.012	1.05	THG	276
	0.017	1.95	THG	276
(C13F8)	0.052	1.05	THG	276
	0.051	1.95	THG	276
(C13TMT)	0.024	1.05	THG	276
	0.040	1.95	THG	276
(C13AB)	0.068	1.05	THG	276
	0.042	1.95	THG	276
(C13OX	0.115	1.05	THG	276
	0.078	1.95	THG	276
BBL	1.5	1.064	DFWM	283
BBB	0.55	1.064	DFWM	283
PBT	1.09	0.602	DFWM	284
PBZT	0.83	1.2	THG	281
	0.06	2.4	THG	281
PBO	0.7	1.2	THG	281
	0.081	2.4	THG	281
PDPIM	0.084	0.9	THG	276
	0.19	1.2	THG	276
	0.089	1.68	THG	276
	0.089	2.37	THG	276
PDPIHB	0.60	0.9	THG	276
	0.138	1.05	THG	276
	0.175	1.5	THG	276
	0.232	1.83	THG	276
	0.209	2.37	THG	276
Polyquinoline	0.11	1.2	THG	285
Polypernigraniline	0.14		THG	286
Poly(o-toluidine)	0.18		THG	286
Polyemeraldine	0.37		THG	286
Poly(2-methoxyaniline)	0.39		THG	286
Poly(thiophenevinylene)	4.5	1.950	THG	256
Poly(p-phenylenevinylene)	0.836	1.064	THG	255
	1.60	1.336	THG	255
	0.321	1.512	THG	255
Oligo(p-phenylenevinylene) ($n=3$)	0.161	1.064	THG	255
	0.152	1.155	THG	255
	0.032	1.440	THG	255
($n=4$)	0.17	1.064	THG	255
	0.191	1.155	THG	255
	0.04	1.485	THG	255
($n=5$)	0.172	1.064	THG	255
	0.36	1.222	THG	255
	0.081	1.485	THG	255
MO-PPV (LB film)	29.0	1.064	THG	254

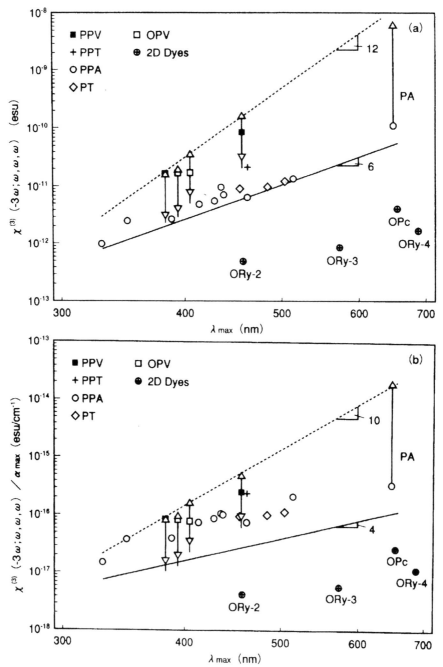

Figure 6.22. $\chi^{(3)}(-3\omega; \omega, \omega, \omega)$ data, measured at 1.064 μm, are plotted as a function of λ_{max} of π-conjugated materials. (a) Plot of figure of merit $\chi^{(3)}(-3\omega; \omega, \omega, \omega)/\alpha_{max}$ vs. λ_{max}. (b) The $\chi^{(3)}(-3\omega; \omega, \omega, \omega)$ values of polyacetylene (PA) [242,243], poly(phenylacetylenes) (PPA) [249], poly(3-decylthiophene) (PT) [269], poly(p-phenylenevinylene) (PPV) [256], poly-(phenothiazinobisthiazole (PPT) [286] oligo-(p-phenylenevinylene) (OPV) [255], [t-Bu$_4$PcRu(dib)]$_n$ oligomer (OPc) [288] were measured at 1.064 μm. (Reproduced by permission of Gordon & Breach Publishers from C. Bubeck, *Nonlinear Optics* Vol. **10**, p. 13, 1995).

PT [269], poly(phenothiazinobisthiazole) (PPT) [282], poly(p-phenylenevinylene) (PPV) [255], oligo-(p-phenylenevinylene) OPV [255], [t-Bu$_4$PcRu(dib)]$_n$ oligomer which is axially linked by a p-diisocyano-benzene (dib) bridging ligand [288] measured at 1.064 μm vs. λ_{max} are shown. Another plot shows figure of merit $\chi^{(3)}(-3\omega; \omega, \omega, \omega)/\alpha_{max}$ vs. λ_{max}. In these plots, the upper triangles refer to $\chi^{(3)}$ at the three-photon resonance with $\lambda_L = 3\lambda_{max}$ while the lower triangles refer to the low-resonant $\chi^{(3)}$ values at λ_L-3λ_0 where λ_0 is the cut-off wavelength with zero absorption. $\chi^{(3)}$ increases with increasing λ_{max}. The values of the exponent, according to the power law, are six and four for the solid lines for the off-resonance $\chi^{(3)}$ values and 10 and 12 for the dashed lines for the three-photon resonant $\chi^{(3)}$ values. The author concluded that the one-dimensional π-conjugated polymers are better candidates for achieving large off-resonance $\chi^{(3)}$ values than two-dimensional π-conjugated systems. The off-resonance $\chi^{(3)}$ values are limited to $\leqslant 10^{-10}$ esu for a π-conjugated system with $\lambda_{max} < 700$ nm.

The third-order optical nonlinearities can be significantly enhanced such as by increasing the π-electron density and by molecular orientation. The NLO studies on polydiacetylene, polyacetylene and PPV show that chain orientation plays an effective role in enhancing the third-order optical nonlinearity. The factors such as π-electron delocalization length, donor–acceptor groups, chain orientation and packing density, conformation, dimensionality and charge transfer complex formation also significantly affect the magnitude of $\chi^{(3)}$ as have been observed in different classes or organic materials.

3.2.4 Resonance phenomena and ultrafast response in $\chi^{(3)}$

The wavelength spectrum showing resonance effects and multiphoton resonances in $\chi^{(3)}$ have been discussed in a variety of π-conjugated materials such as fullerenes, phthalocyanines and conjugated polymers. Two important features for determining the resonant and nonresonant origins of the optical nonlinearity are the fundamental radiation of a laser system used to measure $\chi^{(3)}$ and the linear optical absorption characteristics of the individual material used. $\chi^{(3)}$ should have either a nonresonant value in the case there is no absorption in a medium at fundamental and harmonic wavelengths or a resonant value if the nonlinear medium is absorbing close to these wavelengths, and as a consequence the magnitude of the $\chi^{(3)}$ values may

Table 6.15. Resonances in organic molecules and polymers

Materials	Measurement technique	$\chi^{(3)}$ (10^{-12} esu)	Resonances	Ref.
C$_{60}$ (thin film)	DFWM	200	one photon (ω)	183
	THG	61	two photon (2ω)	182
	THG	87	three photon (3ω)	187
C$_{70}$ (thin film)	DFWM	300	one photon (ω)	180
	THG	26	two photon (2ω)	183
	THG	90	three photon (3ω)	183
α-TiOPc (thin film)	THG	159	three photon (3ω)	212
α-VOPc (thin film)	THG	99.5	three photon (3ω)	212
Y-TiOPc (thin films)	THG	98	three photon (3ω)	212
SiNc derivative (thin film)	THG	24.2	three photon (3ω)	216
	THG	136	two photon (2ω)	216
	SA	6.0×10^6	one photon (ω)	218
PDA-PTS (single crystal)	THG	850	three photon (3ω)	220
PDA-TCDU (single crystal)	THG	160	three photon (3ω)	220
PDA-C$_4$UC$_n$ (thin film)	THG	1700	three photon (3ω)	230
trans-Polyacetylene (thin film)	THG	5600	three photon (3ω)	237
trans-PA (oriented film)	THG	17000	three photon (3ω)	239
Astaxanthene (thin film)	THG	120	three photon (3ω)	139
Polythiophene (thin film)	DFWM	6600	one photon (ω)	257
Poly(thienylvinylene) (thin film)	THG	450	three photon (3ω)	256
Poly(3-dodecylthiophene) (thin film)	DFWM	700	one photon (ω)	258
PBO (thin films)	THG	70	three photon (3ω)	281
PBZT (thin films)	THG	83	three photon (3ω)	281

PA = polyacetylene; PBO = poly(p-phenylenebenzobisoxazole); PBZT = poly(p-phenylenebenzobisthiazole).

fluctuate by folds due to the nonresonant and resonant contributions. The $\chi^{(3)}$ may have both a resonant and a nonresonant contribution;

$$\chi^{(3)} = \chi^{(3)}\text{resonant} + \chi^{(3)}\text{nonresonant} \qquad (6.57)$$

Single-photon, two-photon, and three-photon resonances can be observed while measuring $\chi^{(3)}$ as a function of wavelength. The origin of single photon (ω) resonance lies in the coinciding of incident frequency with an allowed dipole transition from the ground state. Two photon (2ω) resonance results from the sum of difference of two input frequencies, while a three-photon (3ω) resonance involves a combination of three incident frequencies. One, two- and three-photon resonance effects can be observed depending upon the number of photons required to coincide with an excited level of an individual system. In particular, the value of $\chi^{(3)}$ is remarkably enhanced by a one-photon resonance relative to other resonance effects. The resonance effects can be investigated from THG, DFWM, EFISH and other measurement techniques. The dispersion of $\chi^{(3)}$ measured by THG provides information on two- and three-photon resonances, Byrne and Blau [289] discussed multiphoton nonlinear interactions in π-conjugated polymers such as poly-diacetylenes and polythiophenes. Charra *et al.* [290] also investigated the influence of conformation on two-photon spectra of P3ATs. Different experimental techniques used to measure $\chi^{(3)}$ in a variety of π-conjugated materials have shown the presence of resonances as listed in Table 6.15.

The recent advances in information processing technology have created a great demand of photonic materials with ultrafast response time because such developments are aimed to substitute the electronic functions with photonics, where the speed of electricity is surpassed by the speed of light. The applications of ultrafast technology include large-scale information processing and communication technology, short optical pulse generation, optical switching and the study of ultrafast dynamics in organic materials. In particular, nonlinear optical materials have been considered the key photonic elements for optical computing systems with a rate of processing as high as terabit per second. The response time of organic materials vary from millisecond to femoseconds. The dynamics of third-order optical nonlinearity have been evaluated in fullerenes, metallophthalocyanines, polydiacetylenes, poly(3-alkylthiophenes) and copolymers by using techniques such as DFWM, OKE, pump and probe, saturation absorption and photoluminescence. An excellent review on the ultrafast response of π-conjugated polymers with large third-order optical nonlinearity has been written by Kobayashi in this book [291]. The author [291,293] has studied the ultrafast nonlinear optical responses of polythiophenes. The formation times were 70 ± 50 and 100 ± 50 fs for electrochemically prepared poly(3-methylthiophene) and poly(3-dodecylthiophene) respectively, while the decay time of self-trapped excitons was in the picosecond range. Braunling *et al.* [262] reported relaxation times in the range of 10 ps for poly(thienylmethylidenes) from saturable absorption and between 50 to 60 ps from the DFWM technique. Nisoli *et al.* [294] reported response times of 90 fs and 700 fs for poly(3-pentoxythiophene vinylene), 140 fs and 0.97 ps for poly(4,4'-dipentoxy-2,2-bithiophene) and 110 fs and 0.92 ps for poly(4,4''-dipentoxy-2,2':5,2''-terthiophene) using the pump–probe method. The DFWM measurements showed a fast relaxation time of 70 ps in [t-Bu$_4$PcRu(dib)]$_n$ oligomer thin films at a laser [288]. The dynamics studies of NLO-active MPcs derivatives shows that the MPcs have a nonlinear response on a subpicosecond scale indicating that they may be good candidates for ultrafast information processing though they strongly absorb in the Q-band spectral region.

4 OPTICAL LIMITING PROPERTIES OF π-CONJUGATED MATERIALS

4.1 Optical limiting phenomena

Optical limiting is an important nonlinear optical property studied for applications in passive solid-state sensors and human eye protection from high-intensity visible light sources [295–297]. The simplest explanation of optical limiting properties can be provided with the help of Beer's law which postulates that the ratio of the transmitted intensity to the incident intensity through a medium is a constant. However some materials deviate from this behavior as the ratio of the transmitted to the incident light intensity changes. If the transmission increases with increasing incident fluence, the material bleaches and this phenomena is called as saturable absorption. On the other hand, if the transmission of a material decreases compared to a simple linear behavior with increasing incident light intensity, the phenomena is known as reverse saturable absorption (RSA). Figure 6.23 shows a schematic diagram of the response of an optical limiter to increasing input energy against output energy. The

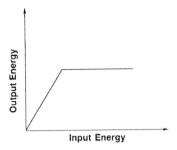

Figure 6.23. A simple schematic diagram of the response of an optical limiter to increasing input energy against output energy.

output energy of an ideal limiter rises linearly with input energy until a threshold is reached and then changes gradually by limiting output energy until a threshold is reached by limiting output energy to some specific value. The figure of merit for optical limiters can be expressed using the ratio of the effective excited-state absorption cross-section σ_e to ground-state absorption cross-section σ_g and with parameters such as the saturation fluence (F_s) and the optical damage threshold (F_d) [298,299].

The RSA was first discovered by Giuliano and Hess in 1967 while studying indathrone and Sudenshwarz B compounds [300]. Since then optical limiting has been observed in many materials such as organometallic clusters [297,298], cyanine dyes [301], indanthrone and its derivatives [302] and others [303,304]. The basic requirements for the applications of a optical limiting material include high linear transmission, low limiting threshold, low saturated output, large dynamic range, fast response time and broad spectral bandwidth. Optical limiting in organic materials is associated with the molecular structure. In past few years, optical limiting properties of fullerenes, porphyrins, and phthalocyanines have attracted much attention; they are discussed here to show a relationship between the chemical structures and optical limiting properties.

4.2 Optical limiting materials

4.2.1 Fullerenes

Fullerenes are multifunctional NLO materials because of their very interesting second and third-order NLO properties as well as optical limiting properties. Tutt and Kost [305] first reported the optical limiting properties of solutions of C_{60} and C_{70} in methylene chloride and toluene using 8 ns optical pulses at

532 nm. The transmittance of C_{60} toluene solution started decreasing at an incident intensity of about 100 mJ cm^{-2}, and a similar optical limiting response was observed for C_{60} in toluene and methylene chloride. A 70% C_{70} solution in toluene showed the output limiting fluence of about 350 mJ cm^{-2}. The optical limiting response of 70% transmitting solutions of C_{60}, chloroaluminum phthalocyanine (CAP), inda-throne and metal clusters at 532 nm was compared. C_{60} solution showed optical limiting with the lowest input threshold. Misra et al. [306] measured the optical properties of C_{60} and C_{70} solutions in toluene using 30 ns pulses at 527 nm. Besides absorption, the possibility of nonlinear refraction and scattering to optical limiting in C_{60} solution was proposed. The variation of output influence with input influence of nearly the same concentration of C_{60} in toluene, pyridine and carbon disulfide showed the origin of scattering and reflection due to thermal effects. Optical limiting of C_{60} solutions for 13 ns pulses at 532 nm, 25 ns pulse at 694 ns and 10 ns optical pulses at 514 nm were measured by Kost et al. [307]. The optical limiting of the C_{60} toluene solution for the input influence of less than 50 J cm^{-2} was due to a combination of RSA and self-defocusing. The relationship between optical fluence limiting and temperature dependence of the refractive index (dn/dT) was evaluated at 532 nm for solutions of C_{60} in 3-methylthiophene, tetraline, 1,2-dichlorobenzene and 1-chloronaphthalene; the fluence limiting data were nearly the same as that of the C_{60} toluene solution. The optical fluence limiting with the C_{60} toluene solution at 694 nm was clamped at a value 20 times larger than that at 532 nm. The C_{60} toluene solution for microsecond pulses at 514 nm was found to be a better optical limiter than carbon-black suspension. Hood et al. [308] reported that the solutions of C_{60} in toluene and chloronaphthalene limited strongly compared with carbon-black suspension at 694 nm. The transmitted fluence of the chloronaphthalene solution was clamped to a lower level than that for the toluene solution. Nonlinear absorption of solutions of C_{60} have also been reported by Henri et al. [175], Brandelik et al. [309] and McLean et al. [310].

Kost et al. [311] carried out RSA and optical limiting studies of C_{60} in PMMA matrix and compared the results with the measurements of C_{60} in solutions. The C_{60}:PMMA composite also showed the magenta color of C_{60} solutions. The C_{60} toluene solution was found to be a better limiter by about a factor of two compared to the C_{60}:PMMA sample due to nonlinear scattering at large angles. a comparison of the optical limiting

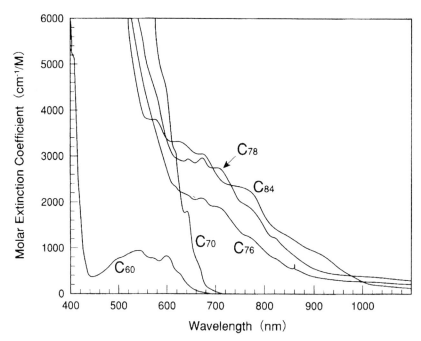

Figure 6.24. Molar extinction coefficients for C_{60}, C_{70}, C_{76}, C_{78} and C_{84} in tetrahydronaphthalene solutions. (Reproduced by permission of the Society of Photo-Optical Instrumentation Engineers (SPIE) from A. Kost et al., *SPIE Proc.* **2284**, 208, 1994).

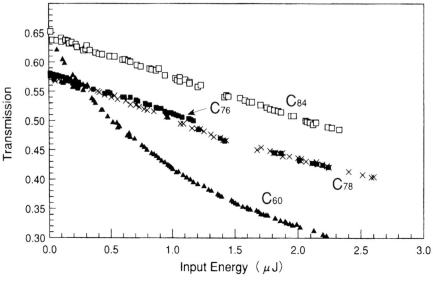

Figure 6.25. Transmission through the C_{60}, C_{76}, C_{78} and C_{84} solutions as a function of input energy at 532 nm. (Reproduced by permission of the Society of Photo-Optical Instrumentation Engineers (SPIE) from A. Kost et al., *SPIE Proc.* **2284**, 208, 1994).

performance at 532 nm of C_{60}:PMMA with CAP, N-methylthioacridine, King's complex and ruthenium King's complex in PMMA showed C_{60} to be the most promising reverse saturable absorber. Both CAP:PPMA and N-methylthioacridine:PMMA samples also showed good optical limiting properties. The King's complex behaved as a saturable absorber in PMMA and as an optical limiter in methylene chloride solution. Benti-vegna *et al.* [312] measured reverse saturable absorption of C_{60} in solid xerogel matrix. McBranch *et al.* [313] used the sol-gel process to study optical limiting of C_{60} in silicon dioxide sonogels and pointed out that C_{60}/SiO_2 gels exhibit optical limiting similar to the C_{60} solution in toluene and C_{60}:PMMA composites films.

Kost *et al.* [314] compared the optical limiting of C_{60} tetrahydronaphthalene solution with C_{76}, C_{78} and C_{84} tetrahydronaphthalene solutions and C_{60} proved to a better limiting material than higher fullerenes. Figure 6.24 shows the molar extinction coefficients for C_{60}, C_{76}, C_{78} and C_{84} solutions. C_{60} has no absorption at wavelengths higher than 680 nm while C_{70} absorbs at 700 nm and C_{76}, C_{78} and C_{84} all absorb at 1.064 μm and beyond. Transmission through the C_{60}, C_{76}, C_{78} and C_{84} solutions as a function of input energy at 532 nm is shown in Figure 6.25. C_{76} and C_{78} solution have the same limiting properties while C_{60} tetrahydronaphthalene solution is the most promising optical limiter compared with the higher fullerenes. Comparison of optical spectra suggest that fullerenes can be used in different wavelength regions where C_{76}, C_{78} and C_{84} could show optical limiting in the near-infrared region. Interestingly, the C_{84} tetrahydro-naphthalene solution also showed optical limiting at 1.064 μm though its mechanism was not clear. Kojima *et al.* [314b] reported optical limiting of the poly-styrene-bound C_{60}, about five times larger than that of a C_{60} solution in styrene.

4.2.2 Phthalocyanines

The RSA and optical limiting properties of a variety of metallophthalocyanines have been investigated by the research teams at the Jet Propulsion Laboratory and Naval Research Laboratory; their results along with others are described here. To understand the optical limiting behavior of these materials, it would be more appropriate to outline first the energy-level scheme of the MPc molecule associated with optical limiting. Figure 6.26 shows the energy-level diagram for optical limiting in metallophthalocyanines [315]. The MPc molecule has two energy-level systems which involves

singlet–singlet and triplet–triplet transitions with excited-state cross-sections of σ_s and σ_t, respectively. Here S_0, S_1 and S_n are singlet states and T_1 and T_n are triplet states. The incident photons with the same frequency can be simultaneously absorbed by MPc molecules in the ground state as well as in both singlet and triplet excited states. In MPc molecules, the excited-state cross-sections σ_s and σ_t are much larger than the ground-state absorption cross-section σ_g at 532 nm, and σ_t is larger than σ_s. It is the ratio of the excited-state absorption cross-section σ_e to ground-state absorption cross-section σ_g that determines the strength of RSA for optical limiting in the MPc molecules; therefore the material with a larger σ_e/σ_g ratio is a better optical limiter. The strong singlet–singlet (S_1–S_n) and triplet–triplet (T_1–T_n) absorptions contribute to the optical limiting behavior of MPc molecules.

Coulter *et al.* [304] in 1989 reported optical limiting in solutions of MPcs and naphthalocyanines. The limiting measurements performed at 532 nm using nanosecond pulses showed a limiting throughput of $1 \sim 10$ millijoules with alcohol solutions of 30–70% nominal transmission. The chloro-aluminium phthalo-cyanine (CAP) **120** solutions showed limiting through-puts of about 60 μj with nominal transmission of 58% from 30 ps pulses. For ClAlPc, the ratio of the excited-state to ground-state extinction coefficients was determined to be between 10–50 at 532 nm. Chloro-indium phthalocyanine, chloro-aluminium t-butylna-phathlocyanine (ClAltbuNc), and zinc t-butylnaphath-locyanine showed limiting throughputs of about 2.5 and 1 mJ respectively. Li *et al.* [316] reported RSA

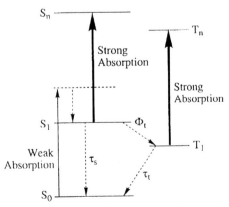

Figure 6.26. Energy-level diagram for optical limiting in metallophthalocyanines. (Reproduced by permission of the Society of Photo-Optical Instrumentation Engineers (SPIE) from K. Mansour et al., *SPIE Proc.* **2143**, 239, 1994)

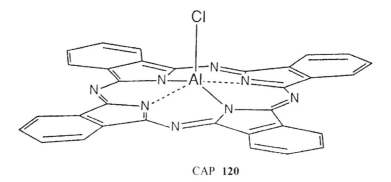

CAP **120**

and optical limiting in CuPc solutions. The CuPc showed a ratio of 10 compared to three for C_{60}.

Helfin *et al.* [317] studied the dispersion of optical limiting in the reverse saturable absorbers silicon naphthalocyanine (SiNc) and CAP over the wavelength range of 543 to 651 nm and compared with C_{60}. Figure 6.27 shows the wavelength dependence effective σ_e/σ_g ratio for SiNc, CAP and C_{60}. Improved optical limiting was observed from 543 to 568 nm for CAP and there was no limiting at 611 nm. The large ground-state cross-section σ_g caused significant decrease in the RSA of CAP. The effective value of the ratio of excited- to ground-state absorption cross-sections (σ_e/σ_g)eff in CAP decreased monotonically from 543 to 589 nm and there was no limiting at longer wavelengths. For SiNc, the (σ_e/σ_g)eff showed a peak around 610 and then decreased sharply at longer wavelengths. CAP and SiNc were less efficient limiters at around 600 and 640 nm, respectively. Contrary to this, C_{60} become the most efficient limiter at the longer wavelengths (at 641 nm) due to larger excited-state and smaller ground-state absorption cross-sections. The remarkable difference in the optical limiting of these three materials occurs due to the dispersion in the ground- and excited-state absorption cross-sections.

Wie *et al.* [318] reported the values of the excited-singlet state σ_s/σ_g as 10.5 for CAP and 14.0 for [SiNc(OSi(hexyl)$_3$)$_2$] **121** utilizing picosecond experiments. Perry *et al.* [319] reported excited-state absorption and optical limiting in solutions of MPcs and

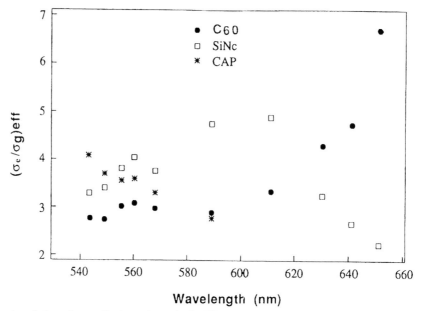

Figure 6.27. Wavelength dependence effective σ_e/σ_g ratio for SiNc, CAP and C_{60} (Reprinted with permission from ref. 317. Copyright 1994 Americal Chemical Society).

MNcs. SiNc showed stronger excited-state absorption at 532 nm than CAP. A threshold limiting energy of ~ 1.6 µJ was measured for a solution of CAP for a 30 ps, 532 nm pulse whereas for the SiNc solution, it was 0.45 µJ with the same linear transmission. The characteristics of CAP and SiNc solutions limiters were compared with tandem dye solutions/ZnSe hybrid limiters. The threshold limiting energies E_L for CAP/ZnSe and SiNc/ZnSe limiters were half that of CAP and SiNc solutions. The damage thresholds E_D for CAP/ZnSe and SiNc/ZnSe hybrid limiters were 50 and 80 µJ, respectively. The dynamic range defined as E_D/E_L was 63 for the CAP/ZnSe and 320 for the SiNc/ZnSe limiters, which is significantly larger than that of the wide-gap semiconductor ZnSe alone. The hybrid optical limiters of tandem combinations of CAP and SiNc solutions and ZnSe slab showed thresholds in submicrojoule range. A low threshold limiting energy may be achieved by utilizing the dye solution in front of ZnSe slab, and it would protect the semiconductor resulting in an increased dynamic range.

Shirk *et al.* [320] studied the optical limiting properties of lead tetrakis(cumylphenoxy phthalocyanine) PbPc(CP)₄. The PbPc(CP)₄/CHCl₃ showed the limiting threshold of 8 nJ which is more than an order of magnitude lower than that of an equivalent thermal limiter or a carbon-black suspension limiter, and the high transmission intensity and its dynamic range exceeded three orders of magnitude. For a PbPc(CP)₄ f/5 optical limiter, the output energy was <0.25 J for incident energies of 20 µJ. The efficient optical limiting in PbPc(CP)₄ is attributed to its large nonlinear absorption coefficient. A thin film of this material showed the threshold for limiting as 25 nJ and a dynamic range of about 35. The threshold of the thin film is about a factor of three larger than for an equivalent limiter based on a PbPc(CP)₄ solution. George *et al.* [321,322] reported the synthesis

of tetrakis (3-[N,N-bis(3-phenoxy-2-hydroxypropylether)]-aminophenoxy) lead phthalocyanine abbreviated as PbPc(PGE3Ap)₄. The NLO properties of this glassy PbPc were determined with picosecond time-resolved DFWM and transient absorption measurements. The excited-state absorption cross-section was found to be larger than that of the ground state from 430 to >580 nm. The DFWM measurements performed at 590 nm showed a rise time of less than 20 ps for the excited state and a decay time of more than 5 ns. For both thin film and CHCl₃ solution, the wavelength dependence of the nonlinear absorption and the dynamics of the excited state decay were similar. The large excited-state absorption and excited-state lifetime indicated its usefulness as an optical limiter.

Perry *et al.* [323–325] reported RSA and the optical limiting properties of phthalocyanines with central metal atoms from groups IIIA (Al, Ga, In) and IVA (Si, Ge, Sn, Pb) to examine the effect of the heavy metal atom. The chemical structures of AlPc, GaPc and InPc of the form [tri-*n*-hexylsiloxy] MPc and Si, Ge and Sn of the form bis[tri-*n*-hexylsiloxy] MPc and PbPc with tetra(t-butyl) groups are shown below (**122–124**).

The triplet quantum yield (Φ_t) increased while the first excited-singlet lifetime (τ_s) decreased with the heavier central metal atom. The cross-section ratio σ_e/σ_g for SiPc, GePc, SnPc and PbPc was between 10 to 18 whereas for AlPc, GaPc and InPc it was between 10 to 16. However the ratio of effective σ_e/σ_g for phthalocyanines containing heavy-metal atoms such as In, Sn and Pb was about a factor of two larger than those phthalocyanines containing lighter metals, such as Al and Si. The optical limiting response in an f/8 geometry of PbPc, SiNc in toluene and CAP in methanol at 532 nm is shown in Figure 6.28. The output energy of PbPc is 1.5 and 4 times lower than that of SiNc and CAP, respectively. The homogeneous

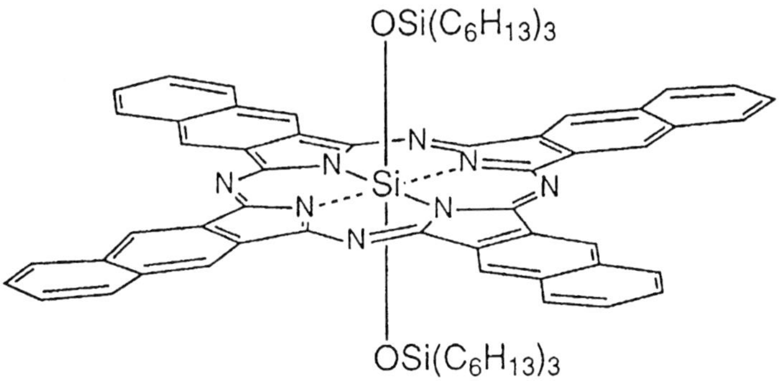

121

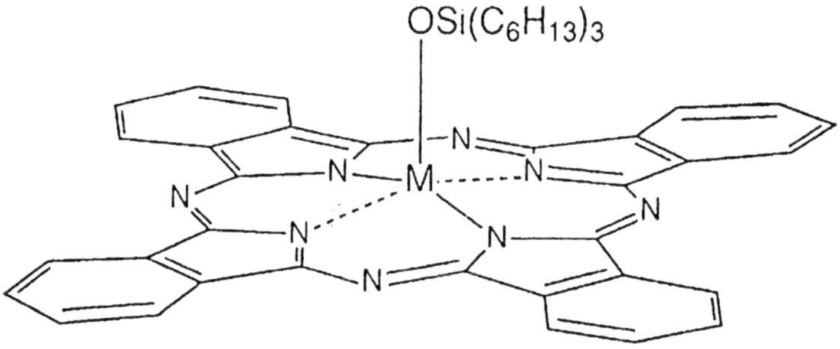

$OSi(C_6H_{13})_3$

M= Al, Ga and In   **122**

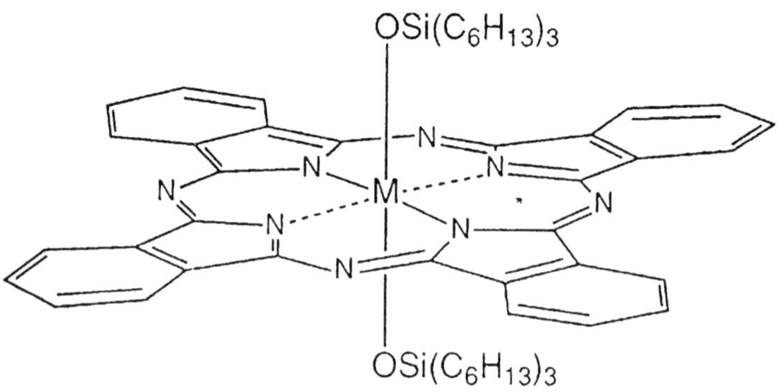

$OSi(C_6H_{13})_3$

$OSi(C_6H_{13})_3$

M= Si, Ge, and Sn   **123**

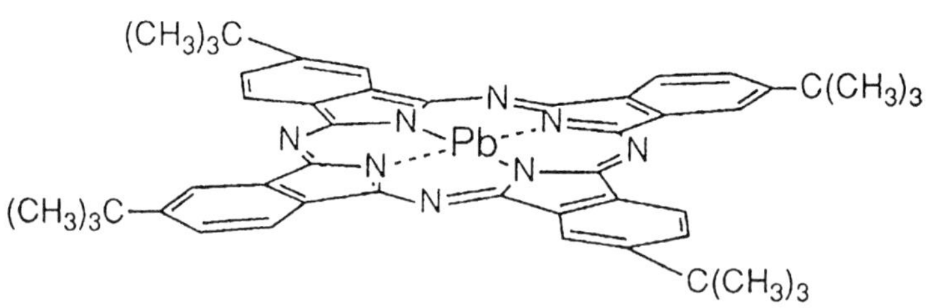

$(CH_3)_3C$ $C(CH_3)_3$

$(CH_3)_3C$ $C(CH_3)_3$

124

solutions of heavy-metal phthalocyanines with a linear transmission of 30% in an $f/8$ optical geometry allowed $\leqslant 3$ µJ output energy for incident energies as high as 800 µJ. The above studies demonstrated the enhanced optical limiting performance by the use of the heavy metal atom in phthalocyanine ring.

Mansour *et al.* [325] measured solid-state optical limiting in the above described MPcs doped PMMA, and originally modified silica sol-gels (ORMOSILS) and compared them with MPcs in toluene solutions. The excited-state properties such as the triplet–triplet absorption spectrum, the triplet quantum yields of the MPcs doped into PMMA and ORMOSILS were nearly

similar to those of toluene solutions. The ratio of σ_e/σ_g varied between 10 to 30 depending on the metal atom though they were same in PMMA, ORMOSIL or toluene. The saturation fluence (F_s) values were in the range of 0.2 to 0.5 J cm^{-2} where MPcs in PMMA or ORMOSIL showed slightly larger values than in solution. The authors indicated that the MPcs doped PMMA or sol-gel materials are driven to strong saturation prior to damage. The optical limiting performance of the MPcs containing a heavy-metal atom is approaching that required for the desired devices.

The characteristics of MPcs at 532 nm and the requirements for bottleneck optical limiters have been

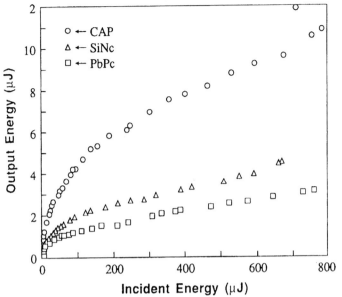

Figure 6.28. Optical limiting response in an $f/8$ geometry of PbPc, SiNc in toluene and CAP in methanol at 532 nm. The solutions had a linear transmittance of 0.3 at 532 nm and the concentrations were 7×10^{-4} M for SiNc, 9×19^{-4} M for CAP and 1.3×10^{-3} M for PbPc. (Reproduced by permission of the Optical Society of America from J.W. Perry et al., *Opt. Lett.* **19**, 625 1994).

evaluated [326]. Miles [327] outlined three figures of merit for saturated excited-state absorptive limiters. The transmitted energy for an optical system depends inversely on the product $(\Phi\sigma_g)^3(Cv\Delta T_m)^2$ where $Cv\Delta T_m$ is the heating limit of the matrix materials and $\Phi\sigma_g$ the effective ground-state absorption cross-section. The third figure of merit is determined from the ration σ_e/σ_g as defined earlier. The McPs have pulse suppression ratios larger than 10^3 at 532 nm for $T = 0.7$. Miles [237] indicated that for SnPc pulse-energy suppression ratios above 10^4 together with a linear transmittance more than 70% for pulse lengths in excess of the singlet–triplet transfer time of 4 ns should be possible. MPcs seem to be the most promising materials for optical limiters.

4.2.3 Miscellaneous compounds

Shi *et al.* [328,329] reported nonlinear excited-state absorption in a series of metallo-texaphyrin **125** compounds in acetonitrile solutions using 8 ns and 23 ps optical pulses at 532 nm. The texaphyrin compounds show reverse saturable absorption only at low fluences and saturable absorption at high fluences for picosecond pulses; however only reversible saturable absorption for nanosecond optical pulses was

shown. A six-level model was presented to explain the nonlinear absorption effects. Cadmium texaphyrin showed strong RSA and nonlinear absorption while other derivatives were very weak.

Staromlynska *et al.* [330] reported optical limiting from a platinum ethynyl compound **126** at 532 nm using nanosecond optical pulses and pointed out that nonlinear absorption is a dominant limiting mechanism below 6 J cm^{-2} while nonlinear scattering dominates above this input fluence.

Hochbaum *et al.* [331] pointed out that the limiting threshold decreases with increased conjugation length, planarity and substitution of donor/acceptor end groups for polyphenyl derivatives. Wood and Mott [332] evaluated nonlinear material requirements for passive optical limiting while considering intensity-dependent $\chi^{(3)}$. The authors indicated that the maximum permissible exposure for the eye at short pulses in the visible region is $\leqslant 0.2$ μJ and to ensure the output energy below this level, the smallest nonlinear refractive index n_2 is -5.5×10^{-8} esu where n_2 is the real part of $\chi^{(3)}$. Excellent reviews on organic reverse saturable absorbers for optical limiters have been written by Perry [333] and on their applications by van Stryland *et al.* [334]. Hagan *et al.* [298] pointed out that reverse saturable absorption dyes with high saturation fluence and high optical damage thresholds are required.

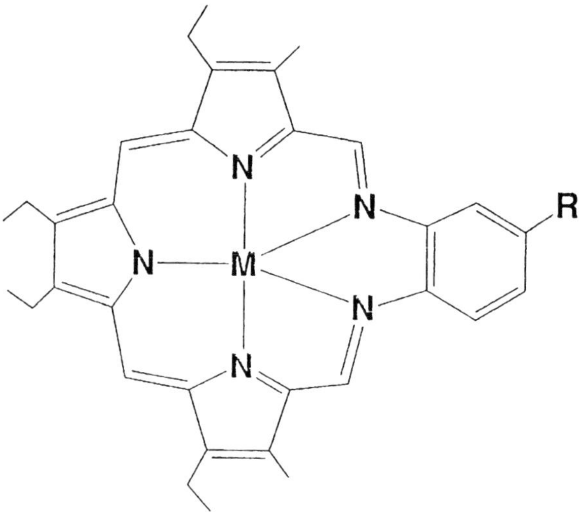

(i) M=Cd   R=H. Cl and CH$_3$

(ii) M=Sm   R=H

(iii) M=Gd   R=COONa

125

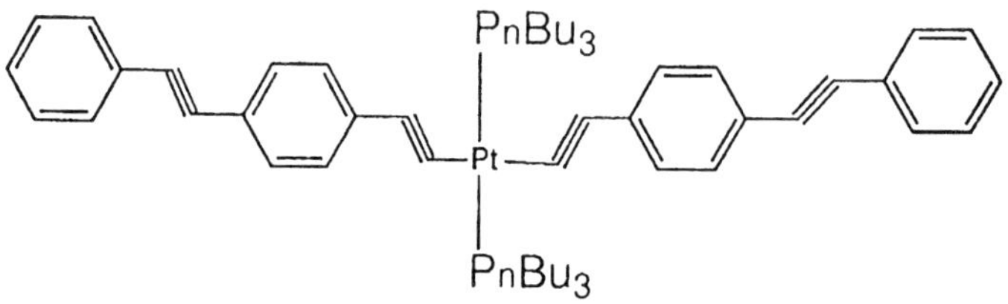

126

Optical limiters have applications in sensors susceptible to high intensity laser sources [335,336].

5 PHOTOREFRACTIVE EFFECTS IN π-CONJUGATED MATERIALS

A photorefractive material combines three features: photoconductive, electro-optic and defects in their lattice that can be optically ionized. Therefore, a photorefractive material exhibits photoconducting behavior, possesses photocarrier trap sites and does not have an inversion center of symmetry leading to a Pockel's linear EO effect. A characteristic of photorefractive materials is the change of the refractive index of a nonlinear optical medium as a function of static electric field strength. Therefore the photorefractive effects is a nonlinear optical phenomenon caused by the transport of photocarriers [337]. Photocarriers are generated by photogeneration of electron–hole pairs across the band gap or by photoionization of defects. The static electric fields are generated by the nonuniform charge distributions trapped at the defect sites resulting from the nonuniform light intensity. The resulting space-charge is associated with a spatially varying electric field that modifies the refractive index by the EO effect. Therefore the resulting static electric fields play an important role in the photorefractive effects [338]. Inorganic photorefractive materials can be divided into three classes: (I) oxygen octahedra ferroelectrics such as barium titanate ($BaTiO_3$), lithium niobate ($LiNbO_3$), lithium tantalate ($LiTaO_3$), barium sodium niobate ($Ba_2NaNb_5O_{15}$) BBN, $KTa_{1-x}Nb_xO_3$ (KTN), $Pb_{1-x}La_xZr_yTiO_3$ (PLZT), etc; (ii) sillenites such as $Bi_{12}SiO_{20}$(BSO), $Bi_{12}GeO_{20}$ (BGO), $Bi_{12}TiO_{20}$ (BTO); and (iii) compound semiconductors such as GaAs, GaP, InAs, InP, and InSb (III-V compounds),

CuCl, CdS, CdSe, CdTe, ZnS, ZnSe, ZnTe, HgSe, HgS and HgTe (II-VI compounds), InP:Fe, AlGaAs, $Hg_{1-x}Cd_xTe$ and CdZnTe. An excellent review on photorefractive crystals and their applications has been written by Roy and Singh [339]. The photorefractive materials show promise for applications in optical computing, telecommunication, image processing, phase conjugation, filters, optical interconnects, and modulators; therefore the research activities in new photorefractive materials, associated techniques and their based devices are growing very rapidly. More details of photorefractive nonlinear optics are documented in refs. 340–351. The important parameters that have to be considered when selecting photorefractive materials for different applications are the following, as pointed out by Gunter and Huignard [352]:

1. photorefractive sensitivity;
2. dynamic range (maximum refractive index change);
3. laser wavelength for inducing refractive index change;
4. photorefractive recording and erasing time;
5. phase shift between refractive index and light intensity distribution;
6. spatial frequency dependence;
7. external electrical field dependence;
8. signal-to-noise ratio;
9. resolution;
10. room temperature operation.

These requirements are important for evaluating novel materials for photorefractive effects. It is clear that any medium that fulfills the requirements of a photoconductor and displays an EO effect fits into the category of photorefractive material. As a result a variety of inorganic ferroelectric materials have emerged as the good candidates for studying photorefractive effects. The studies on organic photorefractive materials are rather new. In 1990, Sutter et al. [353] from the Swiss Federal Institute of Technology, Zurich, reported photorefractive effects for the first time in 2-cyclooctylamino-5-nitropyridine (COANP) organic single crystals doped with an electron acceptor 7,7,8,8-tetracyanoquinodimethane (TCNQ). The COANP:TCNQ complex showed diffraction efficiencies up to 0.1%. This report initiated interest in the search for organic photorefractive materials. Photorefractive polymeric materials are gaining much attention; they have been produced either by doping a photoconductive polymer with a NLO dye and then utilizing a sensitizer or synthesizing a multifunctional photo-

refractive polymer. The first photorefractive polymeric system was developed by the IBM research team by doping a NLO polymer with a charge transporting molecule [354]. Since then many polymeric photorefractive materials have been developed with EO polymers either with a sensitizer or a carrier transport agent [355–357].

The IBM group has made significant progress by doping photoconductive polymers with a variety of NLO dyes and sensitizing agents. The two-beam coupling measurements showed internal gain by doping poly(N-vinylcarbazole) (PVK) with the NLO dye 3-fluoro-4-N,N-diethylamino-β-nitrostyrene (FDEANST) and sensitizing it for charge generation with 2,4,7-trinitro-9-fluoronone (TNF) [358]. An 125 μm thick film of the composite exhibits a diffraction efficiency of more than 1% with a grating growth time of ~ 100 ms at 1 W cm^{-2}. It is most likely that a variety of novel photorefractive polymeric materials were prepared by doping PVK with NLO dyes and sensitizing agents including C_{60} [359]. A grating growth time as short as 39 ms at 0.547 μm, $E_o = 11.4$ V μm^{-1}, and an intensity of 1 W cm^{-2} were obtained by doping poly(4-n-butoxyphenyl)ethysilane] PBPES with NLO dye and sensitizing it with agents such as TNF and C_{60} [360]. The mobility of the doped PBPES polymer was 10^{-4} V cm^{-1} μm^{-1}, two to three orders of magnitude higher than other photorefractive polymers.

Theoretical analysis demonstrated that the good performance of new photorefractive appears to be due to a new orientational enhancement mechanism [361]. Moerner and Silence [362] reviewed the progress and developments of polymeric photorefractive materials. The effect of various sensitizing agents, including fullerenes, was investigated for the PMMA-PNA:DEH system. Silence et al. [363,364] incorporated TNF, squaryllium dye, the charge transfer complex anthracene-TCNQ, C_{60} and N,N'-bis(2,5-di-tert-butylphenyl)-3,4:9,10-perylenebis(dicarboximide) (p-dci). C_{60} and p-dci showed the largest enhancement of the diffraction efficiency and C_{60} appeared to be the most promising sensitizing agent for the PMMA-PNA:DEH system. C_{70} showed a similar sensitizing capability.

Kippelen et al. [365] reported two different types of erasable holograms using polarization and field-dependent four-wave mixing experiments for a new doped polymer composite comprised of PVK/TNF/DR1. The field-dependent quantum efficiency of this composite was 1% and the mobility $μ$ was 7×10^{-6} cm^2 V^{-1}s^{-1} at an applied voltage of 100 V μm^{-1}. The linear EO coefficient measured by the Mach–Zehnder interfero-

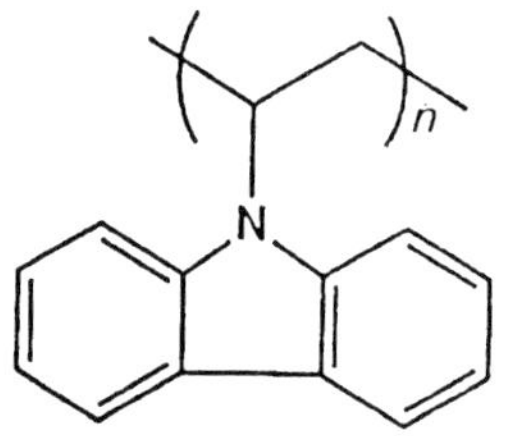

PVK   127

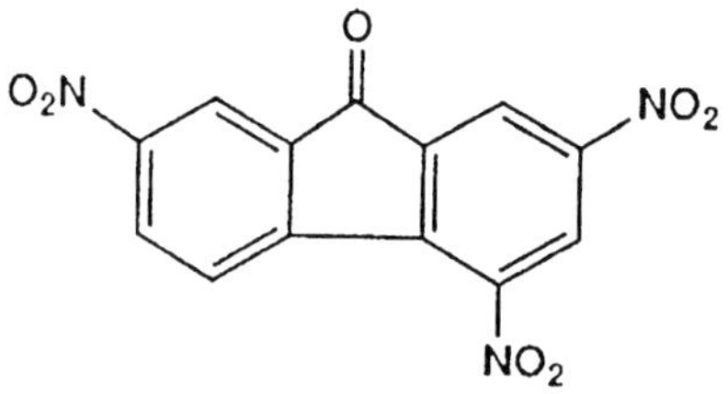

TNF   128

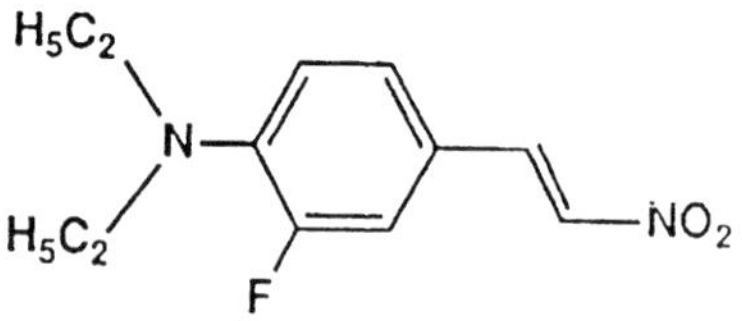

FDEANST   129

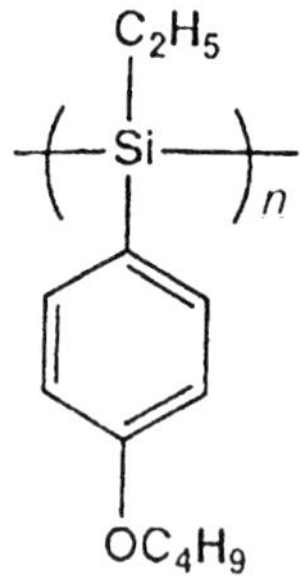

PBPES   130

metric technique was $n^3 r_{33} = 9$ pm V^{-1} at 0.674 µm with a poling field of 50 V µm^{-1}. The composite showed no photorefractive signal without an applied field, though the photorefractive grating starts to grow as the field is switched on at $t = 70$ s. The photorefractive diffraction efficiency was found to be field-dependent and the photorefractive signal began to increase as the applied field increases from 15 to 20 V µm^{-1} at $t = 135$ s. The photorefractive signal disappears completely as the applied poling field is removed at $t = 200$ s, though the photorefractive effect was fully reversible. The composite showed the longest storage time of the photorefractive holograms, and 70% of the diffraction efficiency 5 min after the end of the writing of the hologram was observed. A new highly efficient photorefractive polymer composite of PVK/TNF containing N-ethylcarbazole (ECZ) and 2,5-dimethyl-4-(p-nitrophenylazo)anisole (DMNPAA) has been reported [366]. A diffraction efficiency of 5% and a gain coefficient of 30 cm^{-1} was obtained with a field of 40 V µm^{-1} in a 105 µm thick film.

Tamura et al. [367] synthesized the first fully functionalized photorefractive polymer 133 that can be used without using carrier transport agents of sensitizers. The new methacrylic ester polymer contains the tricyanovinylcarbazole groups with an alkylene

spacer. This single-component polymer showed an coefficient r_{33} of 6.1 pm V^{-1} and high photoconductivity at 514 nm.

Yu et al. [368] reported a multifunctional single-component photorefractive polymer with a NLO chromophore, charge-transporting moiety, and charge-generating molecule covalently attached to the same polymer backbone. This multifunctional photorefractive polymer showed an EO coefficient r_{33} of 12 pm V^{-1} at 632 nm and a photocurrent quantum yield of 2.6×10^{-3}. Since then, Yu's research team [63,64,369] has developed many multifunctional photorefractive polymers.

Until now the milestone discovery is the demonstration of a diffraction efficiency of nearly 100% in a composite DMNPAA:PVK:ECZ:TNF (50:33:16:1 wt %) from a research team at the University of Arizona [370,371]. The 2,5-dimethyl-4-(p-nitrophenylazo)-anisole (DMNPAA) is an electro-optic chromophore, N-ethylcarbazole (ECZ) is the plasticizing agent that helps in easy alignment of chromophores under applied electric field, PVK the photoconductive polymer and TNF is a strong electron acceptor that increases photosensitivity in the visible. PVK and TNF form a charge-transfer complex. Figure 6.29 shows the electric field dependence of the DFWM diffraction efficiency for the DMNPAA:PVK:ECZ:TNF composite with s-

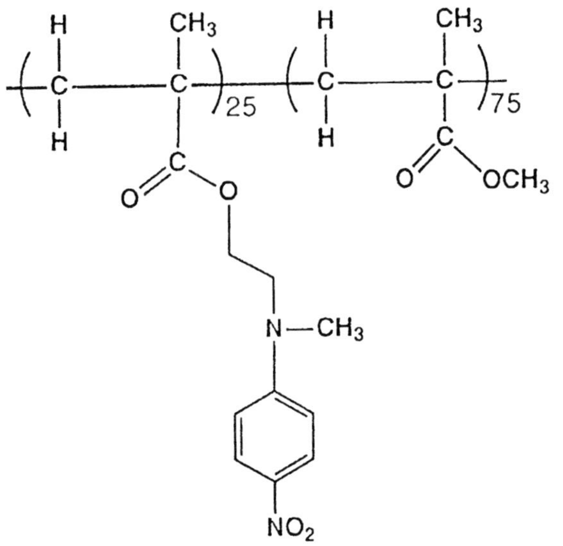

PMMA-PNA **131**

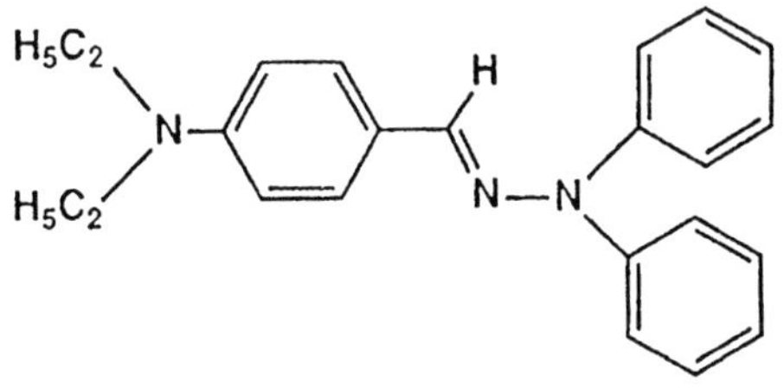

DEH **132**

133

polarized writing beam and a p-polarized reading beam. The diffraction efficiency increases with the applied electric field and reaches a maximum of 86% at 60 V/µm for DMNPAA:PVK:ECZ:TNF composite. The diffraction efficiency was smaller for s-polarized readout and no maximum was observed for fields up to 90 V/µm. Authors stated that under their experimental condition, the diffraction efficiency rises to about 95% of the maximum value within 100 ms and reaching a steady-state value after 10 s. When all beams and the electric fields were switched off, the efficiency of grating dropped to 15% of the maximum value within 24 h. The maximum optical gain of 220 cm^{-1} at an electric field of 90 V/µm was obtained. This is remarkable development because the inorganic crystals show an optical gain coefficient of 10–50 cm^{-1}.

Cox *et al.* [372] reported a photorefractive composite consisting of PVK, TNF and 1-(2′-ethylhexyloxy-2,5-dimethyl-4,(4′-nitrophenylazo)benzene (EHDNPB) which showed 60% device diffraction efficiency at 676 nm and 120 cm^{-1} two-beam coupling gain.

Zobel *et al.* [373] reported a new photorefractive polymer based on polysiloxane backbone. The multi-component system containing polysiloxane (56 wt %), the NLO dye DMNPAA (43 wt %) and sensitizer TNF (1 wt %) showed diffraction efficiency and optical gain comparable to those reported by Meerholz *et al.* [370]. Interestingly no plasticizer was used in the multi-component system because of the low intrinsic T$_g$ of polysiloxane (51°C). This polysiloxane-based photo-refractive polymer was further investigated by Poga *et al.* [374] by using a different NLO dye, 3-fluoro-4-

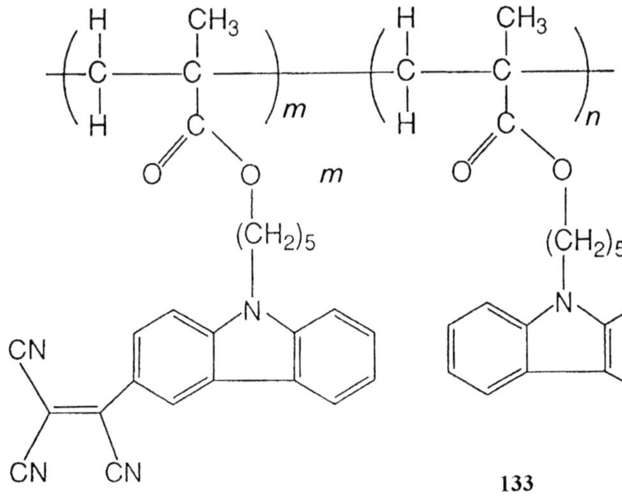

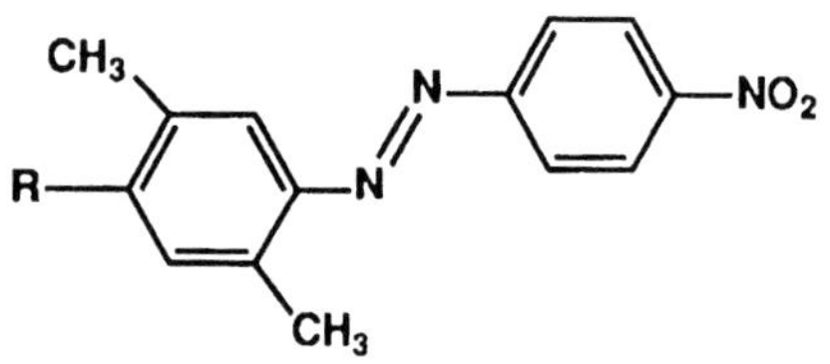

R = OCH₃   **(DMNPAA)**
R = O-CH₂-CH(C₂H₅)-(CH₂)₃-CH₃   **(EHDNPB)**

134

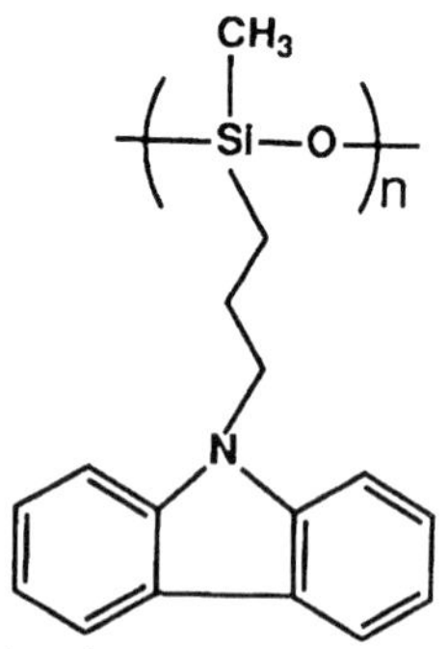

(carbazole substituted polysiloxane)

135

N,N'-diethylamino-β-nitrostyrene (FDEANST). The multicomponent system containing polysiloxane (66 wt %), FDEANST (33 wt %) and TNF (1 wt %) was used for holographic storage measurements. The diffraction efficiency was 35% ($\Delta n = 1.4 \times 10^{-3}$) at an electric field of 77 V/μm while the optical gain was $\Gamma = 98$ cm^{-1} at an electric field of 80 V/μm. The

64 Kbit holograms recorded in an area of 0.12 cm² yielded in a recording density of 0.52 Mbit/cm² and reading holograms back without error. The application of this photorefractive system was demonstrated by storing and reading high-density digital data.

Wang *et al.* [375] reported photorefractive effect in a photoconductive carbazole trimer sensitized with TNF. In the trimer, the two cyanovinylcarbazoles act as NLO moiety while carbazole rings act as a carrier transporting agent. The 0.06 wt % doped carbazole trimer has a T_g of 29°C, refractive index of 1.687 at 633 nm, electro-optic coefficient r_{33} of 1.12 pm/V and absorption coefficient α of 5.78 cm^{-1}. The diffraction efficiency of 18.3% was recorded at an electric field of 30.6 V/μm while the optical gain of 76 cm^{-1} was obtained in a 0.06 wt % TNF doped carbazole trimer.

Photorefractive polymers have great potential in holography storage. High diffraction efficiency, optical gain, data storage time and temporal stabilities are some of the properties required for a high-performance material that can be really applied to optical data storage devices. Photorefractive polymers, either composite types or multifunctionalized polymers could achieve the target but much work need to be done on developing new high-performance photorefractive materials.

6   PHOTONIC APPLICATIONS

The main goal of any scientific discovery is to create new immunities for society, and research on organic electronics and nonlinear optical materials pursue the same motivations. One of the main advantages of

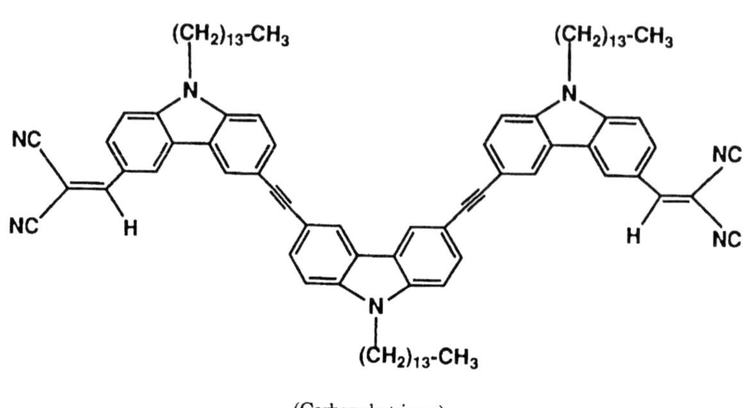

(Carbazole trimer)

136

organic materials over inorganic materials is their architectural flexibility to design and chemically modify materials according to the requirements of a particular device. Opportunities to synthesize novel organic materials to establish structure–property relationships are nearly unlimited. Recent scientific discoveries have created a multidisciplinary atmosphere where the needs of different scientific backgrounds are essential to achieve a fruitful impact on new emerging technologies. The applications of electronic polymers range from solid-state technology to biomedical engineering. The applications of NLO materials are widespread. Figure 6.30 lists the promising applications of electronics and photonic organic materials in industries.

Though a very wide range of physical properties exists, only three types of organic polymers, viz ferroelectric, conducting, nonlinear optical are taken into account. These are considered as active polymers since they take an active part in the functioning of the device. The active polymers are emerging as specialty

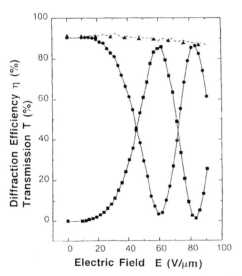

Figure 6.29. Electric field dependence of the DFWM diffraction efficiency and transmission for the DMNPAA: PVK:ECZ:TNF composite with s-polarized writing beams (power density ~ 1 W/cm^2, beam intensity ratio $b = 1.3$) and a p-polarized reading beam (power density 0.35 mW/cm^2). The squares represent the diffraction efficiency η. The circle and triangle refer to the transmission in the presence and absence of the writing beam, respectively. The dashed line is the sum of the diffracted and transmitted intensity in the presence of the writing beams, showing that the maximum achievable diffraction efficiency is limited by absorption and refraction losses ($\sim 12\%$). The solid lines are guides to the eye. (Reproduced from ref. 371, with permission from CRC Press, Boca Raton, Florida, 1997).

materials for electronic, photonic and molecular electronics. The passive optical polymers have potential as optical fibers, data communications, sensors, patterning, holography and waveguides. Nonlinear optical polymers have been considered as substitutes for electronic components where their photonic function will provide similar features in a cleaner and faster manner. The density of memory chips per year is increasing and efforts to have electronic and photonic switches on a molecular scale has been targeted using organic molecular and polymeric materials.

The potential applications of second-order NLO materials are in new lasers, displays, optical disks, optical telecommunication devices, printers and laser processing equipments and optical telecommunication systems. Harmonic generation is the most simplest application, where a SHG material can convert at 1064 nm Nd–YAG laser beam into a 532 nm intense green laser beam and a THG material can convert the same 1064 nm near-infrared laser beam into a 355 nm ultraviolet laser beam; hence NLO materials can be used as frequency doublers and frequency triplers respectively. New laser frequencies can be generated either by interacting two laser beams of the same frequency or differing frequencies. In particular, a great potential has been realized for second-order NLO polymers in optical communication systems, frequency doubling, modulation, switching, directional couplers, voltage sensors, parametric oscillator, electromagnetic radiation detector and other photonic devices. EO materials can be used in similar or related applications. Lipscomb et al. [376,377] suggested applications of EO polymers in photonic large-scale integration (PLSI) by hybrid integration of electronic and photonic devices. For example, the applications of optical interconnections include telecommunication, box-to-box, board-to-board and chip-to-chip interconnects. Other potential applications of EO polymers were indicated in optical multichip modules, reconfigurable optical connectors, high-speed multiplexers and demultiplexers, high-speed switching networks, high-speed digital and analog-to-digital modulator arrays, high-speed analog-to-digital convertors, two-dimensional optical source arrays.

The device concepts associated with third-order NLO materials are at the early stage. The $\chi^{(3)}$ of organic polymers range by more than eight orders of magnitude between 10^{-5} and 10^{-13} esu at the resonance and off-resonance regions. Third-order NLO materials with large figures of merit have potential applications in optical fiber communication, data storage, optical computing, image processing, optical bistables, display, printers, dynamic hologra-

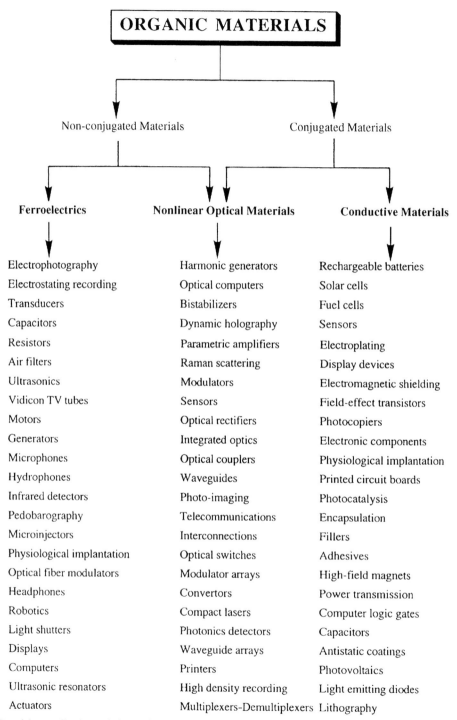

Figure 6.30. Promising applications of electronic and photonic organic materials.

and optical switching. Several prototype devices have already been demonstrated using organic third-order NLO materials. All-optical bitability devices were fabricated from polydiacetylene Langmuir–Blodgett films [374]. A directional coupler from PDA-4-BCMU thin films has also been fabricated [375]. Norwood *et al.* [380] developed the nonlinear Bragg mirror using ultrathin layers of the saturably absorbing copolymer of silicon phthalocyanine and methylmethacrylate. The all-optical phase modulation in waveguides from the single-crystal film of a PDA-PTS was demonstrated [381]. Third-order NLO materials at shorter wavelengths are highly desirable for laser frequency conversion in the fabrication of high-power laser sources that could be used in ultraviolet and near-infrared frequency regions.

The materials with interesting RSA may have potential photonics applications in optical memories, optical limiters, switches, modulators, communications, computing, filters and optical logics. Applications of photorefractive polymeric materials are in holographic data storage, security systems, biomedical technologies, filters, image processing, associative memories, neural networks, etc.

7 CONCLUDING REMARKS

Nonlinear optical properties of a wide variety of π-conjugated materials have been reviewed. A lot of research work has been carried out on π-conjugated materials which now constitute an important class of nonlinear optical media. Metallophthalocyanines are centrosymmetric materials though SHG can be generated from the break in symmetry through the fabrication processes. SHG in MPcs seems to be a highly surface-selective optical phenomenon. Thin films of fullerenes also show interesting SHG properties. The π-conjugated polymers with pendant NLO chromophores have been explored as a new class of multifunctional materials that exhibit both second- and third-order NLO properties. The magnitude of the SHG and EO coefficients in π-conjugated materials is rather low compared to NLO chromophore functionalized polymers.

The π-conjugated polymers show resonant $\chi^{(3)}$ values as large as 10^{-5} esu though their nonresonant values are in the range 10^{-11} to 10^{-12} esu. For all-optical switching, off-resonance $\chi^{(3)}$ on the order of 10^{-8} esu could be sufficient coupled with low optical losses. Therefore more work is needed for a trade-off between $\chi^{(3)}$ and the optical absorption losses. The $\chi^{(3)}$ values of π-conjugated materials such as MPcs,

porphyrins, fullerenes, polyacetylenes, polydiacetylenes, polythiophenes, poly(p-arylenevinylenes), polyanilines, heteroaromatic ladder polymers, polyazines and polyazomethines, are significantly enhanced through one-photon (ω), two-photon (2ω) and three-photon (3ω) resonances and can be tailored through strategic chemical modifications. For π-conjugated materials, a response time τ at the subpicosecond scale is achievable, but the remaining questions are high $\chi^{(3)}$ and low optical losses for technological applications. Fullerenes and MPcs have emerged as the most interesting optical limiting materials. Among higher fullerenes, C_{60} is the best optical limiter. MPcs and MNcs with metal atoms such as lead have good optical limiting properties compared to other MPcs. The application of MPcs could be as optical limiters. Photoconductive polymer composites and multifunctional single-component polymers show a large diffraction efficiency and a good temporal stability. The composite DMNPAA:PVK:ECZ:TNF showed a diffraction efficiency of nearly 100%. Organic materials offer tremendous architectural flexibility for generating numerous interesting chemical species for applications in photonics. The development of new high-performance materials is a challenging task for scientists involved in this fastest growing field of technical endeavour.

8 CONVERSION FACTORS

The following conversion factors of atomic units (au) to electrostatic units (esu) and international system (SI) units should be considered.

Polarizability (α)
$1\ \text{au} = 1.481 \times 10^{-25}\ \text{esu} = 1.648 \times 10^{-41}\ \text{C}^2\ \text{m}^2\ \text{J}^{-1}$
First hyperpolarizability (β)
$1\ \text{au} = 8.639 \times 10^{-33}\ \text{esu} = 3.206 \times 10^{-53}\ \text{C}^3\ \text{m}^3\ \text{J}^{-2}$
Second hyperpolarizability (γ)
$1\ \text{au} = 5.037 \times 10^{-40}\ \text{esu} = 6.235 \times 10^{-65}\ \text{C}^4\ \text{m}^4\ \text{J}^{-3}$
Second-order nonlinear optical susceptibility $\chi^{(2)}$
$1\ \text{m V}^{-1} = 2.387 \times 10^3\ \text{esu}$
Third-order nonlinear optical susceptibility $\chi^{(3)}$
$1\ \text{m}^2\ \text{V}^2 = 7.162 \times 10^7\ \text{esu}$
Non linear refractive index n_2 (MKS) $= 5.26 \times 10^{-6}$
$\chi^{(3)}/n^2$(esu)

9 FREQUENTLY USED SYMBOLS

E	External electric field
E_g	Band gap
f	Local field factor

P Polarization
P_i Induced polarization
L Length
L_π Delocalization length
N_π Number of electrons
μ Dipole moment
α Polarizability
β First hyperpolarizability
γ Second hyperpolarizability
$\chi^{(1)}$ First-order nonlinear optical susceptibility
$\chi^{(2)}$ Second-order nonlinear optical susceptibility
$\chi^{(3)}$ Third-order nonlinear optical susceptibility
$\chi^{(n)}$ nth-order nonlinear optical susceptibility
θ Angle of incidence
λ Wavelength
λ_{max} Wavelength of absorption maximum
λ_c Cutoff wavelength
ε Dielectric constant
ω Frequency
2ω Frequency of two-photon resonance
3ω Frequency of three-photon resonance
τ Decay time
d Nonlinear optical coefficient
r Electro-optical coefficient
T Temperature
N Number density of chromophores
E_p Electric field of polarization
T_p Temperature of polarization
t_p Time of polarization
T_g Glass transition temperature
n Index of refraction
n_2 Nonlinear refractive index
η Photorefractive efficiency

10 REFERENCES

1. N. Bloembergen, *Nonlinear Optics*, Benjamin, New York, 1965.
2. D.S. Chemla and J. Zyss, (eds.), *Nonlinear Optical Properties of Organic Molecules and Crystals*, Academic Press, Orlando, 1987.
3. H.S. Nalwa and S. Miyata (eds.), *Nonlinear Optics of Organic Molecules and Polymers*, CRC Press, Boca Raton, 1997.
4. J. Zyss, (ed.), *Molecular Nonlinear Optics*, Academic, New York, 1994.
5. H.S. Nalwa, *Adv. Mater.* **5**, 341 (1993).
6. S.R. Marder, J.E. Sohn and G.D. Stucky, (eds.), *Materials for Nonlinear Optics*, ACS Symposium Series 455, American Chemical Society, Washington DC, 1991.
7. N. Bloembergen, H. Logan and R.T. Lynch, *Indian J. Pure Appl. Phys.* **16**, 151 (1978).

8. Y.R. Shen, *The Principle of Nonlinear Optics*, John Wiley & Sons, New York, 1984.
9. J.A. Armstrong, N. Bloembergen, J. Ducuing and P.S. Pershan, *Phys. Rev.* **127**, 1918 (1962).
10. N. Bloembergen and Y.R. Shen, *Phys. Rev.* **133**, A37 (1964).
11 (a) J.F. Ward, *Phys. Rev.* **37**, 1 (1965).
11 (b) J.B. Orr and J.F. Ward, *Mol. Phys.* **20**, 513 (1971).
12. D. Li, T.J. Marks and M.A. Ratner *J. Phys. Chem.* **96**, 4325 (1992).
13. A. Yariv and P. Yeh, *Optical Waves on Crystals*, John Wiley & Sons, New York, 1983.
14. S. Singh, in *Handbook of Laser Science and Technology*, Vol III (ed. M.J. Weber) CRC Press, Boca Raton, 1986, pp. 3–228.
15. H.S. Nalwa, in *Nonlinear Optics of Organic Molecules and Polymers* (eds. H.S. Nalwa and S. Miyata) CRC Press, Boca Raton, 1997, Chapter 4 pp.89–350 and Chapter 11, pp. 611–797.
16. H.S. Nalwa, *Appl. Organometal. Chem.* **5**, 349 (1991).
17. S.K. Kurtz and T.T. Perry, *J. Appl. Phys.* **39**, 3798 (1968).
18. P.D. Maker, R.W. Terhune, M. Nisenhoff and C.M. Savage, *Phys. Rev. Lett.* **8**, 21 (1962).
19. J.F. Nye, *Physical Properties of Crystals*, Clarendon, Oxford, 1957.
20. D.A. Kleinman, *Phys. Rev.* **126**, 1977 (1962).
21. K.D. Singer, M.K. Kuzyk and J.E. Sohn, *J. Opt. Soc. Am.* **B 4**, 968 (1987).
22. K.D. Singer and A.F. Garito, *J. Chem. Phys.* **75**, 3572 (1981).
23. C.J. Boettcher, *Theory of Electric Polarization*, Elsevier, Amsterdam, 1952.
24. I.P. Kaminov, *An Introduction to Electrooptic Devices*, Academic Press, New York, 1974, Chapter 2, p.59.
25. K.D. Singer, S.J. Lalama, J.E. Sohn and R.D. Small, in ref. 2, Vol. 1, p. 437.
26. Ch. Bosshard, K. Sutter, R. Schlesser and P. Gunter, *J. Opt. Soc. Am.* B **10**, 867 (1993).
27. C.H. Wang, *J. Chem. Phys.* **98**, 3457 (1993).
28. C.H. Wang, S.H. Gu and H.W. Guan, *J. Chem Phys.* **99**, 5597 (1993).
29. S.J. Lalama and A.F. Garito, *Phys. Rev. A* **20**, 1179 (1979).
30. J. Zyss and J.L. Oudar, *Phys. Rev. A* **26**, 2028 (1982).
31. H.S. Nalwa, K. Nakajima, T. Watanabe, K. Nakamura, A. Yamada and S. Miyata, *Jpn. J. Appl. Phys.* **30**, 983 (1991).
32. H.S. Nalwa, T. Watanabe and S. Miyata, *Opt. Mater.* **2**, 73 (1993).
33. H.S. Nalwa, T. Watanabe and S. Miyata, *Adv. Mater.* **7**, 1991 (1995).
34. J. Zyss, *Nonlinear Opt.* **1**, 3 (1991).
35. J. Zyss, *J. Chem. Phys.* **98**, 6583 (1993).
36. J. Zyss and I. Ledoux, *Chem. Rev.* **94**, 77 (1994).
37. R.A. Huijts and G.L. J. Hesselink, *Chem. Phys. Lett.* **156**, 209 (1989).

38. M. Barzoukas, M. Blanchard-Desce, D. Josse, J.M. Lehn and J. Zyss, *Chem. Phys.* **133**, 323 (1989).

39. M. Blanchard-Desce, *Condensed Matter News* **2**, 12 (1993).

40. M. Blanchard-Desce, V. Bloy, J.M. Lehn, C. Runser, M. Barzoukas, A. Fort and J. Zyss, *SPIE Proc.* **2143**, 20 (1994).

41. L.T. Chang, W. Tam, S.R. Marder, A.E. Stiegman, G. Rikken and C.W. Spangler, *J. Phys. Chem.* **95**, 10643 (1991).

42. J.O. Morley, V.J. Docherty and D. Pugh, *J. Chem. Soc. Perkin Trans. 2* 1351 (1987).

43. L.I. Chang, W. Tam, S.H. Stevenson, G.R. Meredith, G. Rikken and S.R. Marder, *J. Phys. Chem.* **95** 10631 (1991).

44. J.O. Morlay, *J. Phys. Chem.* **99**, 10166 (1995).

45. H.W. Kroto, A.W. Allaf and S.P. Balm, *Chem. Rev.* **91**, 1213 (1991).

46. R.C. Haddon, *Acc. Chem. Res.* **25**, 127 (1992).

47. R. Taylor and D.R.M. Walton, *Nature* **363**, 685 (1993).

48. X.K. Wang, T.G. Zhang, W.P. Lin, S.Z. Liu, G.K. Wong, M.M. Kappes, R.P.H. Chang and J.B. Ketterson, *Appl. Phys. Lett.* **6**, 810 (1992).

49. F. Kajzar, C. Taliani, R. Zamboni, S. Rossins and R. Danieli, *Synth. Met.* **54**, 21 (1993).

50. L.B. Gan, D.J. Zhou, C.P. Luo, C.H. Huang, T.K. Li, J. Bai, X.S. Zhao and X.H. Xia, *J. Phys, Chem.* **98**, 12459 (1994).

51. D. Zhou, L. Gan, C.P. Luo, H. Tan, C. Huang, Z. Liu, Z. Wu, X. Zhao, X. Xia, S. Zhang, F. Sun and Y. Zou, *Chem. Phys. Lett.* **235**, 548 (1995).

52. D.A. Leigh, A.E. Moody, F.A. Wade, T.A. King, D. West and G.S. Bahra, *Langmuir* **11**, 2334 (1995).

53. P.A. Chollet, F. Kajzar and J.L. Moigne, *SPIE Proc.* **1273**, 87 (1990).

54. K. Kumagai, G.Mizutani, H. Tsukioka, T. Yamauchi and S. Ushioda, *Phys. Rev. B* **48**, 14488 (1993).

55. T. Yamada, H. Hoshi, K. Ishikawa, H. Takezoe and A. Fukuda, *Jpn. J. Appl. Phys.* **34**, L299 (1995).

56. T. Yamada, T. Manaka, H. Hoshi, K. Ishikawa, H. Takezeo and A. Fukuda, *Nonlinear Opt.* **15**, 193 (1996)

57. H. Hoshi, T. Yamada, K. Kajikawa, K. Ishikawa, H. Takezoe and A. Fukuda, *Mol. Cryst. Liq. Cryst.* **267**, 1 (1995).

58. H. Hoshi, K. Hamamoto, T. Yamada, K. Ishikawa, H. Takezoe, A. Fukuda, S. Fang, K. Kohama and Y. Maruyama, *Jpn. J. Appl. Phys.* **33**, L1555 (1994).

59. Y. Liu, Y. Xu, D. Zhu, T. Wada, H. Sasabe, L. Liu and W. Wang, *Thin Solid Films* **244**, 943 (1994).

60. R.D. Neuman, P. Shah and U. Akki, *Opt. Lett.* **17**, 798 (1992).

61. H. Hoshi, N. Nakamura and Y. Maruyama, *J. Appl. Phys.* **70**, 7244 (1991).

62. K.S. Suslick, C.T. Chen, G.R. Meredith and L.T. Cheng, *J. Am. Chem. Soc.* **114**, 6928 (1992).

63. Z. Peng, Z. Bao and L. Yu, *J. Am. Chem. Soc.* **116**, 6003 (1994).

64. Z. Peng, A. Gharavi and L. Yu, *Polym. Preprints* **36**, 41 (1995).

65. T. P. Pollagi, T.C. Stoner, R.F. Dallinger, T.M. Gilbert and M. Hofkins, *J. Am. Chem. Soc.* **113**, 703 (1991).

66. M. Chen, L. Yu, L.R. Dalton, Y. Shi and W.H. Steier, *Macromolecules* **24**, 5421 (1991).

67. D. Yu, A. Gharavi and L. Yu, *Polym. Preprints* **36**, 39 (1995).

68. C. Xu, B. Wu, M.W. Becker, L.R. Dalton, P.M. Ranon, Y. Shi and W.H. Steier, *Chem. Mater.* **5**, 1439 (1993).

69. H.S. Nalwa, T. Wattanabe, A. Kakuta and S. Miyata, *Nonlinear Opt.* **8**, 157 (1994).

70. H.S. Nalwa, T. Watanabe, A. Kakuta, A. Mukoh and S. Miyata, *Electron. Lett.* **28**, 1409 (1992).

71. J.A.F. Boogers, P.T. Klasse, J.J.D. Vlieger and A.H.A. Tinnemans, *Macromolecules* **27**, 205 (1994).

72. H.S. Nalwa (ed.) *Ferroelectric Polymers*, Marcel Dekker, New York, 1995.

73. T. Watanabe, S. Miyata and H.S. Nalwa, in *Ferroelectric Polymers* (ed. H.S. Nalwa) Marcel Dekker, New York, 1995, Chapter 12, p. 611.

74. H.S. Nalwa, T. Watanabe, A. Kakuta, A. Mukoh and S. Miyata, *Synth. Met.* **57**, 3895 (1993).

75. H.S. Nalwa, T. Watanabe, A. Kakuta, A. Mukoh and S. Miyata, *Appl. Phys. Lett.* **62**, 3223 (1993).

76. T. Watanabe, K. Yoshinaga, D. Fichou and S. Miyata, *J. Chem. Soc. Chem. Commun.* 250, (1988).

77. T. Miyazaki, T. Watanabe and S. Miyata, *Jpn. J. Appl. Phys.* **27**, L1724 (1988).

78. V.P. Rao, A.K. Jen, K.Y. Wong, K.J. Drost and R.M. Minimi, *SPIE Proc.* **1775**, 32 (1992).

79. K.Y. Wong, A.K. Jen, V.P. Rao, K.J. Drost and R.M. Minini, *SPIE Proc.* **1775**, 74 (1992).

80. A.K. Jen, V.P. Rao, K.Y. Wong and K.J. Drost, *J. Chem. Soc. Chem. Commun.* 90 (1993).

81. A.K. Jen, V.P. Rao, K.J. Drost, Y.M. Cai, R.M. Mininni, J.T. Kenney, JE.S. Binkley, L.R. Dalton and S.R. Marder, *SPIE Proc.* **2143**, 30 (1994).

82. G. Mignani, F. Leising, R. Meyrueix and H. Samson, *Tetrahedron Lett.* **31**, 4743 (1990).

83. Y.M. Cai and A.K. Jen, *Appl. Phys. Lett.* **67**, 299 (1995).

84. K.G. Chittibabu, L. Li, M. Kamath, J. Kumar and S.K. Tripathy, *Chem. Mater.* **6**, 475 (1994).

85. H. Abe, Y. Kiyohara, M. Rikukawa, K. Sanui and N. Ogata, *Nonlinear Opt.* **15**, 259 (1996).

86. J.J. Kim, S.W. Kang, D.H. Hwang and H.K. Shim, *Synth. Met.* **57**, 4024 (1993).

87. J.J. Kim, D.H. Hwang, S.W. Kang and H.K. Shim, *Mater. Res. Soc. Symp. Proc.* **277**, 229 (1992).

88. J.I. Jin and Y.H. Lee, *Mol. Cryst. Liq. Cryst.* **247**, 67 (1994).

89. D.H. Hwang, J.I. Lee, H.K. Shim, W.Y. Hwang, J.J. Kim and J.I. Jin, *Macromolecules* **27**, 6000 (1994).

90. W.K. Chan, Y.M. Chen, Z.G. Peng and L.P. Yu, *J. Am. Chem. Soc.* **115**, 11735 (1993).

91. L.P. Yu, Y.M. Chen, W.K. Chan and Z.H. Peng, *Appl. Phys. Lett.* **64**, 2489 (1994).

92. W.H. Kim, N.B. Kodali, J. Kumar and S.K. Tripathy, *Macromolecules* **27**, 1819 (1994).

93. S.K. Tripathy, W.H. Kim, B. Bihari, D.W. Cheong and J. Kumar, *Mat. Res. Soc. Symp. Proc.* **328**, 433 (1994).

94. W.H. Kim, B. Bihari, R. Moody, N.B. Kodali, J. Kumar and S.K. Tripathy, *Macromolecules* **28**, 642 (1995).

95. S.K. Tripathy, W.H. Kim, G. Masse, X. Jiang and J. Kumar, *Nonlinear Opt.* **15**, 111 (1996).

96. H.J. Lee, S.J. Kang, H.K. Kim, H.N. Cho, J.T. Park and S.K. Choi, *Macromolecules* **28**, 4638 (1995).

97. D.P. Shelton and J.E. Rice, *Chem. Rev.* **94**, 3 (1994).

98. D.R. Kanis, M.A. Ratner and T.J. Marks, *Chem. Rev.* **94**, 195 (1994).

99. J.L. Bredas, C. Adant, P. Tackx, A. Persoons and B.M. Pierce, *Chem. Rev.* **94**, 243 (1994).

100. F. Kajzar and J. Messier, *Phys. Rev. A* **32**, 2352 (1985).

101. F. Kajzar, J. Messier and C. Robilio, *J. Appl. Phys.* **60**, 3040 (1986).

102. F. Kajzar, J. Messier, C. Sentein, R.I. Elsenbaumer and G.G. Miller, *SPIE Proc.* **1147**, 36 (1989).

103. F. Kajzar, *Nonlinear Opt.* **5**, 329 (1993).

104. A. Yariv, *J. Opt. Soc. Am.* **66**, 310 (1976).

105. A. Yariv, *IEEE J. Quantum Electron* **QE-14**, 650 (1978).

106. R. W. Hellwarth, *J. Opt. Soc. Am* **67**, 1 (1977).

107. J.F. Reintjes, in *Nonlinear Optical Parametric Process in Liquids and Gases*, Academic, New York, 1984, Chapter 5, pp. 327–418.

108. R.A. Fisher, *Optical Phase Conjugation* Academic, New York, 1984.

109. R.G. Caro and M.C. Grower, *IEEE J. Quantum Electron* **QE-18**, 1376 (1982).

110. C. Maloney, H. Byrne, W.W. Dennis, W. Blau and J.M. Kelly, *Chem. Phys.* **121**, 21 (1988).

111. B.F. Levine and C.G. Bethea, *Appl. Phys. Lett.* **24**, 445 (1974).

112. M. Sheik-bahabe, A.A. Said and E.W. Van Stryland, *Opt. Lett.* **14**, 955 (1990).

113. M. Sheik-bahabe, A.A. Said and E.W. Van Stryland, *IEEE J. Quantum Electron.* **26**, 760 (1990).

114. E.W. Van Stryland, M. Sheik-bahabe, A.A. Said and D.J. Hagan, *Prog. Crystal Growth and Character* **27**, 279 (1993).

115. L. Yang, R. Dorsinville, Q.Z. Wang, P.X. Ye, R.R. Alfano, R. Zamboni and C. Taliani, *Opt. Lett.* **17**, 323 (1992).

116. H.S. Nalwa, in *Nonlinear Optics of Organic Molecules and Polymers* (eds. H.S. Nalwa and S. Miyata), CRC Press, Boca Raton, 1997, Chapter 9, p. 515.

117. F. Kajzar, *Nonlinear Opt.* **5**, 329 (1993).

118. J. Ducuing, *Optical Nonlinearity in Conjugated One-dimensional Systems* (ed. N. Bloembergen), North Holland, Amsterdam, 1977, p. 276.

119. K.C. Rustagi and J. Ducuing, *Opt. Commun.* **10**, 258 (1974).

120. G.P. Agrawal and C. Flytzanis, *Chem. Phys. Lett.* **44**, 366 (1976).

121. G.P. Agrawal, C. Cojan and C. Flytzanis, *Phys. Rev. B* **17**, 776 (1978).

122. Z. Shuai and J.L. Bredas, *Phys. Rev. B* **46**, 4395 (1992).

123. H.A. Kurtz, *Int. J. Quantum Chem., Quantum Chem. Symp.* **24**, 791 (1990).

124. G.J.B. Hurst, M. Dupuis and E. Clementi, *J. Chem. Phys.* **89**, 385 (1988).

125. J. Andre, C. Barbier, V. Bodart and J. Delhalle, in Ref. 2, Vol. 2, Chapter III-3, P. 137.

126. J. Andre and J. Delhalle, *Chem Rev.* **91**, 843 (1991).

127. A.F. Garito, J.R. Heflin, K.Y. Wong and O. Samani-Khamiri, in *Organic Materials for Nonlinear Optics* (eds. R.A. Hann and D. Bloor) Royal Society of Chemistry, Special Publication No. 69, 1989, p. 116.

128. J.R. Helfin and A.F. Garito, in *Electroresponsive Molecular and Polymeric Systems* (ed. T.A. Skothein), Marcel Dekker, New York, 1991, Vol. 2, pp. 1–48.

129. D. Fichou, F. Garnier, F. Charra, F. Kajzar and J. Messier, in *Organic Materials for Nonlinear Optics* (eds. R.A. Hann and D. Bloor), Royal Society of Chemistry, Special Publication No. 69, 1989, p. 176.

130. H. Thienpont, G.L.J.A. Rikken, E.W. Meijer, W. Ten Hoeve and H. Wynberg, *Phys. Rev. Lett.* **56**, 2141 (1990).

131. L.T. Cheng, J.M. Tour, R. Wu and P.V. Bedworth, *Nonlinear Opt.* **6**, 87 (1993).

132. H.S. Nalwa, A. Kakuta and A. Mukoh, *Jpn. J. Appl. Phys.* **32**, L193 (1993).

133. H.S. Nalwa, T. Hamada, A. Kakuta and A. Mukoh, *Nonlinear Opt.* **7**, 157 (1993).

134. M. Nakano, M. Okumara, K. Yamaguchi and T. Fueno, *Mol. Cryst. Liq. Cryst.* **182A**, 1 (1990).

135. J.P. Hermann and J. Ducuing, *J. Appl. Phys.* **45**, 5100 (1974).

136. K.-H. Haas, A. Ticktin, A. Esser, H. Fisch, J. Paust and W. Schrof, *J. Phys. Chem.* **97**, 8675 (1993).

137. A. Esser, H. Fisch, H.K. Haas, E. Hadicke, J. Paust, W. Schrof and A. Ticktin, *SPIE Proc.* **1775**, 349 (1992).

138. D.N. Bratan, J.N. Onuchi and J.W. Perry, *J. Chem. Phys.* **91**, 2696 (1987).

139. J.F. Ward and D.S. Elliot, *J. Chem. Phys.* **69**, 5438 (1978).

140. M. Blanchard-Dasache, J.M. Lehn, M. Barzoukas, I. Ledoux and J. Zyss, *Chem. Phys.* **181**, 281 (1994).

141. I.D.W. Samuel, I. Ledoux, C. Dhenaut, J. Zyss, H.H. Fox, R.R. Schrock and R.J. Silbey, *Science* **265**, 1070 (1994).

142. I.D.W. Samuel, I. Ledoux, C. Dhenaut, J. Zyss, H.H. Fox, R.R. Schrock and R.J. Silbey, *Nonlinear Opt.* **10**, 263 (1995).

143. S. Mukamel and H.X. Wang, *Phys. Rev. Lett.* **69**, 65 (1992).

144. F.C. Spano and Z.G. Soos, *J. Chem. Phys.* **99**, 9265 (1993).

145. G. Puccetti, M. Blanchard-Desce, I. Ledoux, J.M. Lehn and J. Zyss, *J. Phys. Chem.* **97**, 9385 (1993).
146. V.P. Bodart, J. Dehalle, J.M. Andre and J. Zyss, *Can. J. Chem.* **63**, 1631 (1985).
147. V.P. Bodart, J. Dehalle, J.M. Andre and J. Zyss, in *Polydiacetylens: Synthesis, Structure and Electronic Properties* (eds. D. Bloor and R.R. Chance), Nijhoff, Dordrecht, 1984, p. 125.
148. H.S. Nalwa, T. Hamada, A. Kakuta and A. Mukoh, *Nonlinear Opt.* **7**, 193 (1994).
149. F. Meyers and J.L. Bredas, *Synth. Met.* **49**, 181 (1992).
150. T. Hamada, 66th Fall Meeting of Chemical Society of Japan, Nishinomiya, Hyougo, Japan, 27–30 September, p. ID1310 (1993).
151. A. Kakuta, Y. Imanishi, H.S. Nalwa, T. Hamada, S. Ishihara and S. Hattori, *Proceedings of International Symposium on Nonlinear Photonics Materials* 24–25 May, 1994, Tokyo, p. 263.
152. H.S. Nalwa, J. Mukai and A. Kakuta, *J. Phys. Chem.* **99**, 10766 (1995).
153. H.S. Nalwa, Y. Imanishi and S. Ishihara, *Nonlinear Opt.* **15**, 227 (1996).
154. H.S. Nalwa, Y. Imanishi and S. Ishihara *Extended Abstracts of Second International Conference on Organic Nonlinear Optics* (ICONO'2), 23–26 July, 1995, Gumna, Japan, p. 49.
155. I.D.L. Albert, D. Pugh, J.O. Morley and S. Ramasesha, *J. Phys. Chem.* **96**, 10160 (1992).
156. F. Kajzar, S. Etemad, G.L. Baker and J. Messier, *Solid State Commun.* **63**, 1113 (1987).
157. M.R. Druy, *Solid State Commun.* **68**, 417 (1988).
158. C. Halvorson, T.W. Hagler, D. Moses, Y. Cao and A.J. Heeger, *Synth. Met.* **57**, 3961 (1993).
159. C. Sauteret, J.P. Herman, R. Fey, F. Pradere, J. Ducuing, R.H. Baughman and R.R. Chance, *Phys. Rev. Lett.* **36**, 956 (1976).
160. H. Nakanishi, H. Matsuda, S. Okada and M. Kato, *Polym. Adv. Technol.* **1**, 75 (1990).
161. S.A. Jenekhe, C.J. Yang, H. Vanderzeele and M.S. Meth, *Chem. Mater.* **3**, 985 (1991).
162. H.S. Nalwa and J.S. Shirk, in *Phthalocyanines*, Vol. 4 (eds. C.C. Leznoff and A.B.P. Lever) VCH Publishers, New York, Chapter 3, pp. 79–181.
163. J. Li, J. Feng and J. Sun, *Chem. Phys. Lett.* **203**, 560 (1993).
164. N. Matsuzawa and D.A. Dixon, *J. Phys. Chem.* **96**, 6241 (1992).
165. J. Shuai and J.L. Bredas, *Phys. Rev.* **46**, 17135 (1992).
166. A.A. Quong and M.R. Pederson, *Phys. Rev. B* **46**, 12906 (1993).
167. Y. Wang, G.F. Vertsch and D.Z. Tomanek, *Phys. D.* **25**, 181 (1993).
168. M. Fanti, G. Orlandi and F. Zerbetto, *J. Am. Chem. Soc.* **117**, 6101 (1995).
169. K.C. Rustagi, L.M. Ramaniah and S.V. Nair, *SPIE Proc.* **2284**, 90 (1994).
170. K.C. Rustagi, L.M. Rumaniah and S.V. Nair, *Int. J. Mod. Phys. B* **6**, 3941 (1992).
171. F. Willaime and L.M. Falicov, *J. Chem. Phys.* **98**, 6369 (1993).
172. S.P. Karna and W.M.K.P. Wijekoon, *SPIE Proc.* **2284**, 111 (1994).
173. K. Harigaya and S. Abe, *Jpn. J. Appl. Phys.* **31**, L887 (1992).
174. W.J. Blau and D.J. Cardin, *J. Mod. Phys. Lett. B* **22**, 1351 (1992).
175. F. Henari, J. Callaghan, H. Stiel, W. Blau and D.J. Cardin, *Chem. Phys. Lett.* **199**, 144 (1992).
176. R. Vijaya, Y.V.G.S. Murti, G. Sundarajan, C.K. Mathews and P.R.V. Rao, *Opt. Commun.* **94**, 353 (1992).
177. Z. Zhang, D. Wang, P. Ye, Y. Li, P. Wu and D. Zhu, *Opt. Lett.* **17**, 973 (1992).
178. R.J. Aranda, D.V.G.L.N. Rao, J.F. Roach and P. Tayebati, *J. Appl. Phys.* **73**, 7949 (1993).
179. Z.H. Kafafi, J.R. Lindle, R.G.S. Pong, F.J. Bartoli, L.J. Lingg and J. Milliken, *Chem. Phys. Lett.* **188**, 492 (1992).
180. M.J. Rosker, H.O. Marcy, T.Y. Chang, J.T. Khoury, K. Hansen and R.L. Whetten, *Chem. Phys. Lett.* **196**, 427 (1992).
181. H. Hoshi, N. Nakamura, Y. Maruyama, T. Nakagawa, S. Suzuki, H. Shiromaru and A. Achiba, *Japan J. Appl. Phys.* **30**, L1397 (1991).
182. J.S. Meth, H. Vanherzeele and Y. Wang, *Chem. Phys. Lett.* **197**, 26 (1992).
183. (a) F. Kajzar, C. Taliani, R. Zamboni, S. Rossini and R. Danieli, *Synth. Met.* **54**, 21 (1993). (b) F. Kajzar, *SPIE Proc.* **2025**, 352 (1993).
184. D. Neher, G.I. Stegeman, F.A. Tinker and N. Peyghambarian, *Opt. Lett.* **17**, 1491 (1992).
185. Y. Wang and L.T. Cheng, *J. Phys. Chem.* **96**, 1530 (1992).
186. S.R. Flom, R.G.S. Pong, F.J. Bartoli and Z.H. Kafafi, *Phys. Rev. B* **46**, 15598 (1992).
187. Q. Gong, Y. Sun, Z. Xia, Y.H. Zou, Z. Gu, X. Zhou and D. Qing, *J. Appl. Phys.* **71**, 3025 (1992).
188. W. Ji, S.H. Tang, G.Q. Xu, H.S.O. Chan, S.C. Ng and W.W. Ng, *J. Appl. Phys.* **74**, 3669 (1993).
189. M. Berrada, T. Watanabe and S. Miyata, *Sen-Gekkai Preprints* 1994 p. G121, (in Japanese).
190. F.P. Strohkendl, R.J. Larsen, L.R. Dalton, R.W. Hellwarth, H.W. Sarkas and Z.H. Kafafi, *SPIE Proc.* **2284**, 78 (1994).
191. Z.H. Kafafi, S.R. Flom, H.W. Sarkas, R.G.S. Pong, C.D. Merritt and F.J. Bartoli, *SPIE Proc.* **2284**, 134 (1994).
192. S.R. Flom, R.G.S. Pong, F.J. Bartoli and Z.H. Kafafi, *Nonlinear Opt.* **10**, 183 (1995).
193. S.C. Yang, Q. Gong, Z. Xia, Y.H. Zou, Y.Q. Wu, D. Qiang, Y.L. Sun and Z.N. Gu, *Appl. Phys. B* **55**, 51 (1992).

194. J.R. Lindle, R.G.S. Pong, F.J. Bartoli and Z.H. Kafafi, *Phys. Rev. B* **48**, 9447 (1993).
195. F. Kajzar, C. Taliani, R. Danieli, S. Rossini and R. Zamboni, *Chem. Phys. Lett.* **217**, 418 (1994).
196. S.R. Flom, H.W. Starkas, R.G.S. Pong, F.J. Bartoli and Z.H. Kafafi, ACS *Polymer Preprints* **35**, 110 (1994).
197. F. Kajzar, *SPIE Proc.* **2284**, 57 (1994).
198. F.H. Moser and A.L. Thomas, *The Phthalocyanines* CRC Press, Boca Raton, (1983).
199. C.C. Lenznoff and A.B.P. Lever, *Phthalocyanines: Properties and Applications* VCH Publishers, Weinheim, Vols. 1–4, 1989, 1993, 1996.
200. Z.Z. Ho, C.Y. Ju and W.M. Hetherrington III, *J. Appl. Phys.* **62**, 716 (1987).
201. Z.Z. Ho and N. Peyghambariam, *Chem. Phys. Lett.* **148**, 107 (1988).
202. N.Q. Wang, Y.M. Cai, J.R. Helfin and A.F. Garito, *Mol. Cryst. Liq. Cryst.* **189**, 39 (1990).
203. J.S. Shirk, J.R. Lindle, F.J. Bartoli, C.A. Hoffman, Z.H. Kafafi and A.W. Snow, *Appl. Phys. Lett.* **55**, 1287 (1989).
204. J.S. Shirk, J.R. Lindle, F.J. Bartoli and M.F. Boyle, *J. Phys. Chem.* **96**, 5847 (1992).
205. H.S. Nalwa, M.K. Engel, M. Hanack and H. Schultz *Appl. Organometal. Chem.* **10**, 661 (1996).
206. R.A. Norwood and J.R. Sounik, *Appl. Phys. Lett.* **60**, 295 (1992).
207. A. Grund, A. Kaltbeitzel, A. Mathy, K. Schwarz, C. Bubeck, P. Vermehren and M. Hanack, *J. Phys. Chem.* **96**, 7450 (1992).
208. H.S. Nalwa, A. Kakuta and A. Mukoh, *J. Phys. Chem*, **97**, 1097 (1993).
209. H.S. Nalwa, A. Kakuta and A Mukoh, *Phys. Chem. Lett.* **203**, 109 (1993).
210. H.S. Nalwa, S. Kobayashi and A. Kakuta, *Nonlinear Optics* **6**, 169 (1993).
211. H.S. Nalwa and A. Kakuta, *Thin Solid Films* **254**, 218 (1994).
212. H.S. Nalwa, T. Saito, A. Kakuta and T. Iwayanagi, *J. Phys. Chem.* **97**, 10515 (1993).
213. M. Hosoda, T. Wada, A. Yamada, A.F. Garito and H. Sasabe, *Jpn. J. Appl. Phys.* **30**, L1486 (1991).
214. K. Ishikawa, M. Kajita, T. Koda, H. Kobayashi and K. Kubodera, *Mol. Cryst. Liq. Cryst.* **218**, 123 (1992).
215. H. Matsuda, S. Okada, A. Masaki, H. Nakanishi, Y. Suda, K. Shigehara and A. Yamada, *SPIE Proc.* **1337**, 105 (1990).
216. H.S. Nalwa and S. Kobayashi, *J. Porph. Phthalocy.* 1997, in press.
217. H.S. Nalwa, Y. Imanishi and S. Ishihara. Second International Conference on Organic Nonlinear Optics (ICONO'2) 23–26 July 1995, Gumna, Japan, p.49.
218. J.W. Wu, J.R. Helfin, R.A. Norwood, K.Y. Wong, O. Zamani-Khamiri, A.F. Garito, P. Kalyanaraman and J. Sounik, *J. Opt. Soc. Am. B* **6**, 707 (1989).
219. T.A. Skotheim, *Handbook of Conducting Polymers* Marcel Dekker, New York, 1986.
220. C. Sauteret, J.P. Herman, R. Frey, F. Pradere, J. Ducuing, R.H. Baughman and R.R. Chance, *Phys. Rev. Lett.* **36**, 956 (1976).
221. G.M. Carter, J.V. Hryiewicz, M.K. Thakur, Y.J. Chen and S.E. Mayler, *Appl. Phys. Lett.* **49**, 998 (1986).
222. F. Kajzar and J. Messier, *Thin Solid Films* **132**, 11 (1985).
223. B.I. Greene, J. Orenstein, R.R. Millard and L.R. Williams, *Phys. Rev. Lett.* **58**, 2750 (1987).
224. J. Bogler, T.G. Harvey, W. Ji, A.K. Kar, S. Molyneux, B.S. Wherrett, D. Bloor and P. Norman, *J. Opt. Soc. Am. B* **9**, 1552 (1992).
225. P.A. Chollet, F. Kajzar and J. Messier, in *Nonlinear Optics of Organics and Semiconductors* (ed. T. Kobayashi) Springer, Berlin, 1988, p. 171.
226. F. Kajzar and J. Messier, *Polym. J.* **19**, 275 (1987).
227. J.M. Nunzi and D. Grec. *J. Appl. Phys.* **62**, 2198 (1987).
228. T. Doi, S. Okada, H. Matsuda, A. Masaki, N. Minami, H. Nakanishi and K. Hayamizu, Paper presented at the 52nd Meeting of the Japan Society of Applied Physics, 9–12 October 1991, Okayama, Japan.
229. K.S. Wong, S.G. Han and Z. Vardeney, *Synth. Met.* **41**, 3209 (1991).
230. T. Kanetake, K. Ishikawa, T. Hasegawa, T. Koda, K. Takeda, M. Hasegawa, K. Kubodera and H. Kobayashi, *Appl. Phys. Lett.* **54**, 2287 (1989).
231. H. Nakanishi, H. Matsuda, S. Okada and M. Kato, *Polym. Ad. Technol.* **1**, 75 (1990).
232. M. Ohsugi, S. Takaragi, H. Matsuda, S. Okada, A. Masaki and H. Nakanishi, *SPIE Proc.* **1337**, 162 (1990).
233. F. Kajzar, S. Etemad, G.L. Baker and J. Messier, *Solid State Commun.* **63**, 1113 (1987).
234. M.R. Drury, *Solid State Commun.* **68**, 417 (1988).
235. A. Ticktin, *Proc. of the 2nd Symposium on Photonic Materials, Basic Technologies for Future Industries*, 22–23 October 1991, Tokyo, Japan.
236. C. Halvorson and A.J. Heeger, *Chem. Phys. Lett.* **216**, 488 (1993).
237. W.S. Fan, S. Benson, J.M.J. Madley, S. Etemad, G.L. Baker and F. Kajzar, *Phys. Rev. Lett.* **62**, 1492 (1989).
238. C. Holvorson, T.W. Hagler, D. Moses, Y. Cao and A.J. Heeger, *Synth. Met.* **57**, 3961 (1993).
239. F. Krausz, E. Wintner and G. Leising, *Phys. Rev. B* **39**, 3701 (1989).
240. K.S. Wong and Z. Vardeny, *Synth. Met.* **49**, 13 (1992).
241. W. Schrof, J.W. Wunsch, A. Esser, K.H. Haas and H. Naarmann, *Nonlinear Opt.* **10**, 69 (1995).
242. D. Neher, A. Kalbeitzel, A. Wolf, C. Bubeck and G. Wegner, *J. Phys. D., Appl. Phys.* **24**, 1193 (1991).
243. D.M. Karol and M. Thakur, *Appl. Phys. Lett.* **56**, 1406 (1990).
244. S. Etemad and Z.G. Soos, in *Spectroscopy of Advanced Materials* (ed. R.J.H. Hester) John Wiley & Sons, New York, 1991, Chapter 2, p. 87.
245. G.J. Blanchard, J.P. Heritage, G.L. Baker and S. Etemad, *Chem. Phys. Lett.* **158**, 329 (1989).

246. G.L. Baker, S. Etemad and F. Kajzar, *SPIE Proc.* **824**, 102 (1987).

247. L. Sebastian and G. Wieser, *Chem. Phys.* **62**, 447 (1981).

248. H. Naarmann and H. Thiophilou, *Synth. Met.* **17**, 22 (1987).

249. M. Sinclair, D. Moses, K. Akagi and A.J. Heeger, *Phys. Rev. B* **38**, 10724 (1988).

250. H.J. Lee, M.C. Suh and S.C. Shim, *Synth. Met.* **73**, 141 (1995).

251. H. Okawa, K. Kurosawa, T. Wada and H. Sasabe, *Synth. Met.* **71**, 1657 (1995).

252. G.S.W. Craig, R.E. Cohen, R.R. Schrock, A. Esser and W. Schrof, *Macromolecules* **28**, 2512 (1995).

253. D.D.C. Bradley and Y. Mori, *Jpn. J. Appl. Phys.* **28**, 174 (1989).

254. K. Kamiyana, M. Era, T. Tsutsui and S. Saito, *Jpn. J. Appl. Phys.* **29**, L840 (1990).

255. A. Mathy, K. Ueberhofen, R. Schenk, H. Gregorius, R. Garay, K. Mullen and C. Bubeck, to be published.

256. (a) H. Murata, N. Takeda, T. Tsutsui, S. Sato, T. Kurihara and T. Kaino, *J. Appl. Phys.* **70**, 2915 (1991). (b) T. Tsutsui, H. Murata, J.C. Kim and S. Saito, in *Nonlinear Optics* (ed. S. Miyata), Elsevier Science Publishers, Amsterdam, 1992, p. 311.

257. R. Dorsinville, L. Yang, R.R. Alfano, R. Zamboni, R. Danieli, G. Ruini and C. Taliani, *Opt. Lett.* **14**, 1321 (1989).

258. B.P. Singh, M. Samoc, H.S. Nalwa and P.N. Prasad, *J. Chem. Phys.* **92**, 2756 (1990).

259. C.L. Callender, L. Robitallie and M. Leclerc, *Opt. Eng.* **32**, 2246 (1993).

260. S.A. Jenekhe, S.K. Lo and S.K. Flom, *Appl. Phys. Lett.* **54**, 2524 (1989).

261. S.A. Jenekhe, W.C. Chen, S.K. Lo and S.R. Flom, *Appl. Phys. Lett.* **57**, 126 (1990).

262. H. Braunling, R. Becker and Blochl, *Synth. Met.* **55**, 833 (1993).

263. J.S. Meth, H. Vanherzeele, W. Chen and S.A. Jenekhe, *Synth. Met.* **49**, 59 (1992).

264. Z. Bao, W. Chan and L. Yu, *Chem. Mater.* **5**, 2 (1993).

265. (a) H.S. Nalwa, *Synth. Met.* **35**, 387 (1990). (b) H.S. Nalwa, *Polymer* **32**, 745 (1991).

266. H.S. Nalwa, *J. Phys. D., Appl. Phys.* **23**, 745 (1990).

267. H.S. Nalwa, *Solid Thin Films* **235**, 175 (1993).

268. F. Kajzar, G. Ruani, C. Taliani and R. Zamboni, *Synth. Met.* **37**, 223 (1990).

269. (a) W.E. Torruellas, D. Neher, R. Zamoni, G.I. Stegeman, F. Kajzar and M. Leclerc, *Chem. Phys. Lett.* **175**, 11 (1990). (b) D. Neher, A. Wolf, M. Leclerc, A. Kaltbeitzel, C. Bubeck and G. Wegner, *Synth. Met.* **37**, 249 (1990).

270. L. Yang, Q.Z. Wang, R. Dorsinville, R.R. Alfano, R. Zamboni and C. Taliani, *Opt. Lett.* **17**, 5 (1992).

271. A.G. MacDiarmid, J.C. Chang, M. Hatpern, W.S. Wang, S.L. Mu, N.L.D. Somasiri, W. Wu and S.I. Yaniger, *Mol. Cryst. Liq. Cryst.* **121**, 173 (1985).

272. J.M. Ginder, A.J. Epstein and A.G. MacDiarmid, *Snyth. Met.* **29**, E-395 (1989).

273. K.S. Wong, S.H. Han and Z.V. Vardeny, *J. Appl. Phys.* **70**, 1896 (1991).

274. A.G. MacDiarmid and A.J. Epstein, *J. Faraday Discuss. Chem. Soc.* **88**, 317 (1989).

275. C. Halvorson, Y. Cao, D. Moses and A.J. Heeger, *Synth. Met.* **57**, 3941 (1993).

276. W.C. Chen, S.A. Jenekhe, J.S. Meth and H. Vanherzeele, *J. Polym. Sci., Polym. Phys. Ed.* **32**, 195 (1994).

277. H.S. Nalwa, *Extended Abstract of the 4th Pacific Polymer Conference*, Kauai, Hawaii, 12–16 December 1995, p. 206

278. H.S. Nalwa, *Polymer* **32**, 802 (1991) and references therein.

279. J.R. Lindle, F.J. Bartoli, C.A. Hoffman, O.K. Kim, Y.S. Lee, J.S. Shirk and Z.H. Kafafi, *Appl. Phys. Lett.* **56**, 712 (1990).

280. S.A. Jenekhe, M. Roberts, A.K. Agrawal, J.S. Meth and H. Vanherzeele, *Mater. Res. Soc., Symp. Proc.* **214**, 55 (1991).

281. S.A. Jenekhe, J.A. Osaheni, J.S. Meth and H. Vanherzeele, *Chem. Mater.* **4**, 683 (1992).

282. A. Kistenmacher, T. Soczka, U. Baier, K. Ueberhofen, C. Bubeck and K. Mullen, *Acta Polymerica* **45**, 228 (1994).

283. J.S. Meth, H. Vanherzeale, S.A. Jenekhe, M.F. Roberts, A.K. Agarwal and C.J. Yang, *SPIE Proc.* **1560**, 13 (1991).

284. H. Vanherzeele, J.S. Meth, S.A. Jenekhe and M.F. Roberts, *J. Opt. Soc. Am. B* **9**, 524 (1992).

285. S.A. Jenekhe, J.A. Osaheni, H. Vanherzeele and J.S. Meth, *Chem. Mater.* **4**, 683 (1992).

286. J.A. Osaheni, S.A. Jenekhe, H. Vanzerheele, J.S. Meth, Y. Sun and A.G. MacDiarmid, *J. Phys. Chem.* **96**, 2830 (1992).

287. C. Bubeck, *Nonlinear Opt.* **10**, 13 (1995).

288. A. Grund, A. Kaltbeitzel, A. Mathy, R. Schwarz, C. Bubeck, P. Vermehren and M. Hanack, *J. Phys. Chem.* **96**, 7450 (1992).

289. (a) H.J. Byrne and W. Blau, *Synth. Met.* **27**, 231 (1990). (b) H.J. Byrne and W. Blau, *Synth. Met.* **32**, 229 (1990).

290. F. Charra, J. Messier, C. Sentein, A. Pron and M. Zagorska, in *Organic Molecules for Nonlinear Optics and Photonics* NATO-ASI Series E, Dordrecht, 1991.

291. T. Kobayashi, in *Handbook of Organic Conductive Molecules and Polymers* Vol. 4, (ed. H.S. Nalwa) John Wiley & Sons, Chichester, 1997. Chapter, 7. p.365.

292. T. Kobayashi, M. Yoshizawa, U. Stamm, M. Taiji and M. Hasegawa, *J. Opt. Soc. Am. B* **7**, 1558 (1990).

293. T. Kobayashi, M. Yoshizawa, U. Stamm, M. Taiji and M. Hasegawa, *SPIE Proc.* **1775**, 85 (1992).

294. M. Nisoli, V. Pruneri, S. De Silvestri, V. Magni, A.M. Gallazzi, R. Romanoni, G. Zerbi and G. Zotti, *Chem. Phys. Lett.* **220**, 64 (1994).

295. S.W. McCahon and M.B. Klein, *SPIE Proc.* **1105**, 119 (1989).

296. K. Mansour, E.W. Van Stryland and M.L. Soileau, *SPIE Proc.* **1105**, 91 (1989).

297. L.W. Tutt, S.W. McCahon and M.B. Klein, *SPIE Proc.* **1307**, 315 (1990).

298. D.J. Hagan, T. Xia, A.A. Said and E.W. Stryland, *SPIE Proc.* **2229**, 179 (1994).

299. K. Mansour, P. Fuqua, S.R. Marder, B. Dunne and J.W. Perry, *SPIE Proc.* **2143**, 239 (1994).

300. C.R. Guiliano and L.D. Hess, *J. Quant. Electron.* **QE-3**, 338 (1967).

301. S. Hughes, G. Spruce, B.S. Wherrett, K.R. Welford and A.D. Lloyds, *Opt. Commun.* **100**, 113 (1993).

302. R.C. Hoffman, K.A. Stetyick, R.S. Potember and D.G. McLean, *J. Opt. Soc. Am.* B **6**, 772 (1989).

303. D.J. Hagan, E.W. Van Stryland, M.J. Soileau and Y.Y. Wu, *Opt. Lett.* **13**, 315 (1988).

304. D.R. Coulter, V.M. Miskowski, J.W. Perry, T. Wei, E.W. Van Stryland and D.J. Hagan, *SPIE Proc.* **1105**, 42 (1989).

305. L.W. Tutt and A. Kost, *Nature* **356**, 225 (1992).

306. S.R. Misra, H.S. Rawatt, M.P. Joshi, S.C. Mehendale and K.C. Rustagi, *SPIE Proc.* **2284**, 220 (1994).

307. A. Kost, J.E. Jensen, M.B. Klein, S.W. McMahon, M.B. Haeri and M.E. Ehritz, *SPIE Proc.* **2229**, 78 (1994).

308. P.J. Hood, B.P. Edmonds, D.G. McLean and D.M. Brandelik, *SPIE Proc.* **2229**, 91 (1994).

309. D. Brandelik, D. McLean, M. Scmitt, B. Epling, C. Colcalsure, V. Tondiglia, R. Pacher, K. Obermeirer and R.L. Crane, *MRS Proceedings* **247**, 361 (1992).

310. D.G. McLean, R.L. Sutherland, M.C. Brant, D.M. Brandelik, P.A. Fleitz and T. Pottenger, *Opt. Lett.* **18**, 858 (1993).

311. A. Kost, L. Tutt, M.B. Klein, T.K. Dougherty and W.E. Elias, *Opt. Lett.* **18**, 334 (1993).

312. F. Bentivegna, M. Canva, P. Georges, A. Brun, F. Chaput, L. Mailer and J.P. Boilot, *Appl. Phys. Lett.* **62**, 1721 (1993).

313. D. McBranch, B.R. Mattes, A. Koskelo, J.M. Robinson and S.P. Love, *SPIE Proc.* **2284**, 15 (1994).

314 (a) W.A. Kost, J.E. Jensen, M.B. Klein, J.C. Withers, R.O. Loufty, M.B. Haeri and M.E. Ehritz, *SPIE Proc.* **2284**, 208 (1994).

314 (b) Y. Kojima, T. Matsuoka, H. Takahashi and T. Kurauchi, *Macromolecules*, **28**, 8868 (1995).

315. K. Mansour, P. Fuqua, S.R. Marder, B. Dunne and J.W. Perry, *SPIE Proc.* **2143**, 239 (1994).

316. C. Li, L. Zhang, H. Wang and Y. Wang, *Phys. Rev. A* **49**, 1149 (1994).

317. J.R. Helfin, S. Wang, D. Marciu, J.W. Freeland and B. Benkins, *Polym. Preprints* **35**, 238 (1994).

318. T.H. Wie, D.J. Hagan, M.J. Sence, E.W. Van Stryland, J.W. Perry and D.R. Coulter, *Appl. Phys.* B **54**, 46 (1992).

319. J.W. Perry, L.R. Khundar, D.R. Coulter, D. Alvarex, Jr, S.R. Marder, T.H. Wei, M.J. Sence, E.W. Van Stryland and D.J. Hagan, in *Organic Molecules for Nonlinear Optics and Photonics* (ed. J. Messier), Kluwer, Dordrecht, 1991, p. 359.

320. J.S. Shirk, R.G.S. Pong, F.J. Bartoli and A.W. Snow, *Appl. Phys. Lett.* **63**, 1880 (1993).

321. R.D. George, A.W. Snow, J.S. Shirk, S.R. Flom and R.G.S. Pong, *ACS Polym. Preprints.* **35**, 236 (1994).

322. S.R. Flom, R.G.S. Pong, J.S. Shirk, R.D. George and A.W. Snow, *ACS Polym. Preprints* **35**, 240 (1994).

323. J.W. Perry, K. Mansour, S.R. Marder, K.J. Perry, D. Alvarez and I. Choong, *Opt. Lett.* **19**, 625 (1994).

324. K. Mansour, D. Alvarez, Jr, I Choong, K.J. Kelly, S.R. Marder and J.W. Perry, *Proceedings of Conference on Lasers and Electro-Optics*, Technical Digest Series, Vol. 11, p. 614, 1993.

325. K. Mansour, D. Alvarez, Jr, K.J. Perry, I. Choong, S.R. Marder and J.W. Perry, *SPIE Proc.* **1853**, 132 (1993).

326. P.A. Miles, *SPIE Proc.* **2143**, 251 (1994).

327. P.A. Miles, *Appl. Opt.* **33**, 6965 (1994).

328. J. Shi, M. Yang, Y. Wang, L. Zhang and C. Li, *Appl. Phys. Lett.* **64**, 3083 (1994).

329. J. Shi, M. Yang, Y. Wang, L. Zhang, C. Li, D. Wang, S. Dong and W. Sun, *Opt. Commun.* **109**, 487 (1994).

330. J. Staromlynska, P.B. Chapple, J.R. Davy and T.J. McKay, *SPIE Proc.* **2229**, 59 (1994).

331. A. Hochbaum, Y.Y. Hsu and J.L. Fergason, *SPIE Proc.* **2229**, 48 (1994).

332. G.L. Wood and A.G. Mott, *SPIE Proc.* **2229**, 158 (1994).

333. J.W. Perry, in *Nonlinear Optics of Organic Molecules and Polymers* (eds. H.S. Nalwa and S. Miyata) CRC Press, Boca Raton, 1997, Chapter 13, p. 813

334. E.W. Van Stryland, D.J. Hagan, T. Xia and A.A. Said, in *Nonlinear Optics of Organic Molecules and Polymers* (eds. H.S. Nalwa and S. Miyata) CRC Press, Boca Raton, 1997, Chapter 14, p. 841

335. L.W. Tutt and S.W. McMahon, *Opt. Lett.* **15**, 700 (1990).

336. S.W. McMahon, L.W. Tutt, M.B. Klein and G.C. Valley, *SPIE Proc.* **1307**, 304 (1990).

337. P. Gunter and J.P. Huignard (eds.) *Photorefractive Materials and Their Applications I and II.* Topics in Applied Physics, Vol. 61 and 62, Springer-Verlag, Berlin, 1988.

338. D. Nolte, *Condensed Matter News* **1**, 17 (1992).

339. A. Roy and K. Singh, *Atti. Fond. G. Ronchi* **48**, 327 (1993).

340. J.O. White, M. Cronin-Gollomb, B. Fischer and A. Yariv, *Appl. Phys. Lett.* **40**, 450 (1982).

341. J. Feinberg, *Phys. Today* **41**, 46 (1988).

342. D.D. Nolte, D.H. Olson, G.E. Doran, W.H. Know and A.M. Glass, *J. Opt. Soc. Am.* B **7**, 2217 (1990).

343. D.M. Pepper, J. Feinberg and N.V. Kukhtarev, *Scientific American* October, p. 34 (1990).

344. Q.N. Wang, D.D. Nolte and M.R. Melloch, *Appl. Phys. Lett.* **59**, 256 (1991).

345. W.A. Schroeder, T.S. Stark, M.D. Dawson, T.F. Boggess and A.L. Smirl, *Opt. Lett.* **16**, 159 (1991).

346. W.S. Robinovich and B.J. Feldman, *Opt. Lett.* **16**, 708 (1991).

347. H. Knog, C. Wu and M. Cronin-Golomb, *Opt. Lett.* **16**, 1183 (1991).

348. M. Cronin-Golomb, B. Fischer, J.O. White and A. Yariv, *IEEE J. Quantum Electron.* **QE-20**, 12 (1984).

349. P. Gunter, *Phys. Rep.* **93**, 199 (1982).

350. D.M. Pepper, in *Laser Handbook*, Vol. IV (eds. M.L. Stitch and M. Bass), North-Holland, Amsterdam, 1985, p. 333.

351. P. Yeh, *IEEE J. Quantum Electron.* **QE-25**, 484 (1989).

352. P. Gunter and J.P. Huignard, in *Photorefractive Materials and Their Applications I* (eds. P. Gunter and J.P. Huignard) Topics in Applied Physics, Springer-Verlag, Berlin, 1988, Vol. 61, Chapter 2, p. 7.

353. K. Sutter, J. Hulliger and P. Gunter, *Solid State Commun.* **74**, 867 (1990).

354. S. Ducharme, J.C. Scott, R.J. Twieg and W.E. Moerner, *Phys. Rev. Lett.* **66**, 1846 (1991).

355. J.S. Schildkraut, *Appl. Phys. Lett.* **58**, 340 (1991).

356. C.A. Walsh and W.E. Moerner, *J. Opt. Soc. Am.* B **9**, 1642 (1992).

357. S.M. Silence, C.A. Walsh, J.C. Scott, T.J. Matray, R.J. Twieg, F. Hache, G.C. Bjorklund and W.E. Moerner, *Opt. Lett.* **17**, 1107 (1992).

358. M.C.J.M. Donekers, S.M. Silence, C.A. Walsh, F. Hache, D.M. Burland, W.E. Moerner and R.J. Twieg, *Opt. Lett.* **18**, 1044 (1993).

359. S.M. Silence, M.C.J.M. Donekers, C.A. Walsh, D.M. Burlnad, R.J. Twieg and W.E. Moerner, *Appl. Opt.* **33**, 2218 (1993).

360. S.M. Silence, J.C. Scott, F. Hache, E.J. Ginsburg, P.K. Jenkner, R.D. Miller, R.J. Twieg and W.E. Moerner, *J. Opt. Soc. Am.* B **10**, 2306 (1993).

361. W.E. Moerner, S.M. Silence, F. Hache and G.C. Bjorklund, *J. Opt. Soc. Am.* B **11**, 320 (1994).

362. W.E. Moerner and S.M. Silence, *Chem. Rev.* **94**, 127 (1994).

363. S.M. Silence, C.A. Walsh, J.C. Scott and W.E. Moerner, *Appl. Phys. Lett.* **61**, 2967 (1992).

364. S.M. Silence, F. Hache, M.C.J.M. Donekers, C.A. Walsh, D.M. Burlnad, G.C. Bjorklund, R.J. Twieg and W.E. Moerner, *SPIE Proc.* **1852**, 253 (1993).

365. B. Kippelen, Sandalphon, N. Peyghambarian, S.R. Lyon, A.B. Padias and H.K. Hall, Jr, *Organic Thin Films for Photonic Applications*, 1993 Technical Digest Series Vol. 17, p. 228.

366. B. Kippelen, Sandalphon, N. Peyghambarian, S.R. Lyon, A.B. Padias and H.K. Hall, Jr, *Electron. Lett.* **29**, 1873 (1993).

367. K. Tamura, A.B. Padias, H.K. Hall and N. Peyghambarian, *Appl. Phys. Lett.* **60**, 1803 (1992).

368. L. Yu, W. Chan, Z. Bao and S.X.F. Cao, *Macromolecules* **26**, 2216 (1993).

369. L. Yu, W.K. Chan, Z. Peng, W. Li and A.R. Gharavi, in *Handbook of Organic Conductive Molecules and Polymers Vol. 4* (ed. H.S. Nalwa), John Wiley & Sons, 1997, Chapter 5, p. 233.

370. K. Meerholz, B.L. Volodin, Sandalphon, B. Kipperlen and N. Peyghambarian, *Nature* **371**, 497 (1994).

371. B. Kipperlen, K. Meerholz and N. Peyghambarian, in *Nonlinear Optics of Organic Molecules and Polymers* (eds. H.S. Nalwa and S. Miyata), CRC Press, Boca Raton, 1997, Chapter 8, p. 465.

372. A.M. Cox, R.D. Blackburn, D.P. West, T.A. King, F.A. Wade and D.A. Leigh, *Appl. Phys. Lett.* **68**, 2801 (1996).

373. O. Zobel, M. Eckl, P. Strohriegl and D. Haarer, *Adv. Mater.* **11**, 911 (1995).

374. C. Poga, P.M. Lundquist, V. Lee, R.M. Shelby, R.J. Twieg and D.M. Burland, *Appl. Phys. Lett.* **69**, 1047 (1996).

375. L. Wang, Y. Zhang, T. Wada and H. Sasabe, *Appl. Phys. Lett.* **69**, 728 (1996).

376. G.F. Lipscomb and R. Lytel, *Nonlinear Opt.* **3**, 41 (1992).

377. G.F. Lipscomb, R. Lytel and A.J. Ticknor, *Nonlinear Opt.* **10**, 421 (1995).

378. K. Sasaki, K. Fujii, T. Tomioka and T. Kinoshita, *J. Opt. Soc. Am.* B **5**, 457 (1988).

379. P.D. Townsend, J.L. Jackel, G.L. Baker, J.A. Shelburne and S. Etemad, *Appl. Phys. Lett.* **55**, 1829 (1989).

380. R.A. Noorwood, J.R. Sounik, D. Holcomb, J. Popolo, D. Swanson, R. Spitzer and G. Hensen, *Opt. Lett.* **17**, 577 (1992).

381. M. Takur in *Polymers for Lightwave and Integrated Optics* (ed. L.A. Hornak) Marcel Dekker, 1992, Chapter 22, p. 667.

CHAPTER 7

Ultrafast Responses in π-Conjugated Polymers with Large Optical Nonlinearity

Takayoshi Kobayashi

University of Tokyo, Bunkyo-ku, Tokyo, Japan

Handbook of Organic Conductive Molecules and Polymers: Vol. 4. Conductive Polymers: Transport, Photophysics and Applications.
Edited by H. S. Nalwa. © 1997 John Wiley & Sons Ltd

1 INTRODUCTION

The latter half of the twentieth century is marked by the development of electronics based on semiconductor industry and technology. However, the speed and memory of computers will reach their limit and these will not be sufficient in the near future. Since the bandwidth of an optical wave is much broader than that available with electronics, the next century is expected to be an 'optopia', where extensive development of optronics and/or photonics will be undertaken. The industry of the early stage of optronics or photonics, namely, opto-electronics, has been developed during the last ten years, and the gross products of this industry in 1992 in the author's country, Japan, is reported to have exceeded five trillion yen. In order for the industry of optronics or photonics to realize the optopia, the search for nonlinear optical materials is of key importance [1]. The properties of the materials used in information processing and memory storage should satisfy some or all of the following requirements: (1) large third-order optical nonlinearity; (2) ultrafast nonlinear response; (3) low absorption, reflection and scattering loss at operation wavelengths; (4) possibility of manufacturing waveguide; (5) long-term stability of nonlinear susceptibility and dynamics; (6) stability and absence of degradation at the operational level of lasters. From these viewpoints there are few groups of materials that interest scientists in the field.

Among others, conjugated polymers seem to be promising candidates for future practical applications in nonlinear optical devices. Because of their often large nonlinear optical susceptibilities and their role as model compounds for quasi-one-dimensional semiconductors, considerable interest has been directed to the electrical and optical properties of polymers with long conjugated chains such as polyacetylene (PA), polydiacetylene (PDA), polythiophene (PT), polypyrrole, polythienylenevinylene (PTV), polyphenylene (PP) and polyphenyl-

enevinylene (PPV) [1–23]. Figure 7.1 shows the molecular structures of these polymers. The major differences in the electronic and optical properties of these polymers are due to the backbone structure. For example, PA possesses the conjugated polyene type $(-CH=CH-)_n$ main chain, PDAs have the diacetylene-type configuration $(=CR-C\equiv C-CR=)_n$ with both

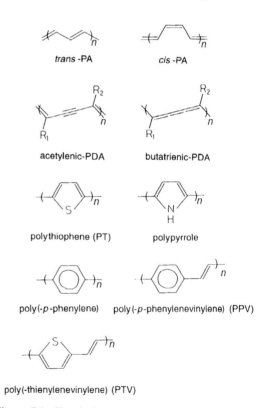

Figure 7.1. Chemical representations of the backbones in *trans*-polyacetylene, *cis*-polyacetylene, acetylenic-polydiacetylene (PDA), butatrienic-polydiacetylene, polythiophene, polypyrrole, polyphenylene and polyphenylenevinylene. R_1 and R_2 represent the side groups of PDA.

double and triple bonds, and PTs, PTV, PP and PPV contain ring structures in their repeating units.

PA attracted attention first because of high electric conductivity introduced by chemical doping, close to that of cupper. Its mechanism is discussed in terms of solitons. It is a prototype of the conjugated polymers since it has the simplest structure and is known to have quasi-one-dimensional conductive and optical properties, even though aligned samples have become available only recently. Both *trans* and *cis* forms of PA display large optical nonlinearities. The relaxation dynamics of the photoexcitations have been extensively investigated in *trans*-PA because this compound has the simplest structure of the polyenes and related conjugated polymers, and has solitons [4–9,11].

The relaxation processes in *trans*-PA are summarized as follows. Theoretical work predicts that an electron–hole pair generated in a single chain is subsequently converted in 10^{-13} s to a charged soliton–antisoliton pair induced by strong coupling between the C–C and C=C stretching modes and the electronic excitation [4]. However, the theoretical calculation cannot determine definitely the formation time; it simply gives the order of magnitude. It also does not take into account the flat potential minimum of the free exciton, which will be discussed later in the chapter. The formation process of the soliton could not be resolved in time by femtosecond spectroscopy of 150 fs resolution. The soliton pair disappears by intrachain recombination in the picosecond region [5–7]. A polaron pair is created from an electron–hole pair which is photogenerated in neighboring chains by perpendicularly polarized light, and the creation process is also associated with a similar geometrical relaxation to that in soliton–pair formation. After the creation of the polarons, the interchain relaxation processes of the polarons occur after a duration of the order of nanoseconds to microseconds [8,9,11].

Nonlinear optical properties for PDAs are the most extensively studied of the various polymers, because they have the following remarkable characteristics (see Figure 7.1). The properties of PDAs with different substituent groups, R_1 and R_2, differ considerably from each other. The polymerization of a large number of diacetylenes with various side groups has been reported [12–15], and PDAs with various side chains [16] (for instance, TS, DFMP, TCDU, ETCD and 1OH) poly-1,6-di(N-carbazoly)-2,4-hexadiyne (DCHD) can be obtained in the form of highly ordered single crystals [16]. See ref. 16 for the full names of these polymers. PTS is probably the most extensively studies PDA because large, single crystals of a high quality are available. There is a series of soluble PDAs, some of

which are referred to as PDA-nBCMU (n-butoxycarbonylmethylurethane). The yellowish orange color of the solutions, in a good solvent, of PDA-3BCMU ($n = 3$) and PDA-4BCMU ($n = 4$) changes to blue and red, respectively, when a poor solvent is added. PDA-3BCMU and PDA-4BCMU are sometimes referred to as poly-3BCMU and poly-4BCMU. A rod-to-coil transition has been reported as the origin of the color change for PDA-4BCMU, with the use of quasi-elastic light scattering [18,19]. Since the relatively small hydrodynamic radius increases abruptly when the molar fraction of the poor solvent (C_5H_{14}) in the $CHCl_3/C_5H_{14}$ mixed solvent is increased, it was concluded that PDA-4BCMU has a coil-like structure in the $CHCl_3$-rich solution, and a rigid rod-like structure in the C_5H_{14}-rich solution. PDAs also show similar dramatic color changes due to phase transitions induced by pressure and heat [20–22].

The color changes had been supposed to be due to two possible chemical structures of the main chain of PDAs (Figure 7.1): acetylenic (type A) and butatrienic (type B) until several years ago. Most PDAs are known to have the acetylenic structure, but some of them had been considered to have the butatrienic structure. PDA-3BCMU was accordingly thought to have four phases (BC, BR, AC and AR) where A, B, C and R denote acetylenic, butatrienic, coil-like and rod-like, respectively, from phase diagrams with various solvent compositions and temperatures [22], while PDA-4BCMU was considered to have only two phases (BC and BR). Later on, the structures of PDAs with a so-called butatriene backbone were re-examined in more detail by X-ray diffraction and it was found that the structure is not butatrienic but acetylenic [24].

Although the acetylenic and butatrienic structures are not appropriate for representing the phases of PDAs in the ground state, they can be used to describe the structure of elementary excitations in these compounds. The electronic ground state in PDAs is nondegenerate; consequently, these compounds cannot have solitons but have self-trapped (ST) excitons together with polarons. Bipolarons are formed also in PAs because the electronic excitation is strongly coupled with phonons (as will be discussed later) (Figure 7.2). Following the conventional description, positively and negatively charged polarons are regarded as butatrienic cation and anion radicals, respectively. Polarons are generated by electron–hole pair creation in neighboring chains by excitation with light polarized perpendicularly to the main chains. A doubly charged positive or negative bipolaron can be formed by t he fusion of two positive or negative polarons as shown in Figure 7.3.

ST excitons can be regarded as neutral bipolarons formed by a bound pair of positively and negatively charged polarons. After the photoexcitation to free excitons, the geometrical relaxation from free excitons to ST excitons takes place. The 'ST excitons' are sometimes called 'neutral bipolarons' or 'exciton polarons', since they are neutral in charge and 'deformed' by excitons–phonon coupling.

PDAs exhibit various colors depending on the side groups, degree of polymerization, morphology and phase. For example, PDA-3BCMU films are metallic dark blue or black, while PDA-4BCMU films are red. The absorption spectra have been measured by Chance *et al.* [20] between −150°C and 140°C, from lower to higher temperatures. At higher temperatures the spectra were shifted to higher frequencies; in PDA-3BCMU films the color change was reversible many times, while PDA-4BCMU films have poorer reversibility. It was concluded that the thermally induced color change is due to hydrogen bonding between intramolecular side groups, as determined by Fourier transform infrared (FTIR) spectrometry [20].

PDAs are reported to have remarkably large third-order nonlinear susceptibility coefficients and ultrashort time constants of the phase relaxation T_2 [25,26]. Short time constants of the energy relaxation, T_1 are reported extending from nanoseconds to femtoseconds [16,27–32]. Thus it is expected that PDAs will find applications in devices such as optical shutters and optical switches using bistability induced by the optical Kerr effect. The switching time of the bistable system using polydiacetylene-*p*-toluene sulfonate (PDA-TS) as a non-linear material was reported to be of the order of 0.1 ps [33].

Fluorescence of PDAs has been observed in red (rod-like polymers) and yellow (coil-like polymers) phase solutions [34]; the fluorescence decay is not exponential and the $1/e$ lifetime was in the order of a few hundred ps [35]. The decay kinetics of the fluorescence have also been studied in other PDAs in the red and yellow phases and the lifetimes obtained were in the range of 9–12 ps [27]. In contrast, no fluorescence from PDA-3BCMU and PDA-MADF (1-(3-methoxylanino)-phenyl)-4-(3,5-bis(trifluoromethyl) phenyl)-1,3-buta-diyne) in the blue phase has been detected. The nonradiative decay in the blue phase is expected to be much faster than in the red phase because the fluorescence quantum efficiency from the blue phase PDAs is lower than 10^{-5}.

PTs possess a simple backbone geometry resembling that of *cis*-PA. The *cis*-like structure is stabilized by the sulfur atom which is known to interact only weakly with the π-electron system of the backbone [36]. Thus polythiophenes can be considered to be pseudo-polyenes [37]. Most knowledge of elementary excitations in PTs has been gained from steady-state photoluminescence and photoinduced absorption spectroscopies, and from light-induced electron-spin resonance experiments [37–45]. It is established that polarons and bipolarons are formed in PTs by photoexcitation or electrochemical doping [37–43]. The appearance of bipolarons is due to the non-degenerate ground state in PTs, which leads to the confinement of soliton–antisoliton pairs. The bipolaron

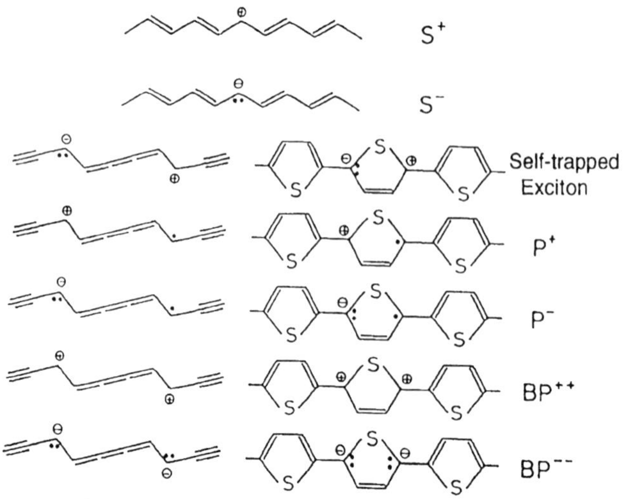

Figure 7.2. Structures of solitons (S^+ and S^-) in *trans*-polyacetylene and self-trapped exciton, polarons (P^+ and P^-) and bipolarons (BP^{2+} and BP^{2-}) in polydiacetylene and polythiophene.

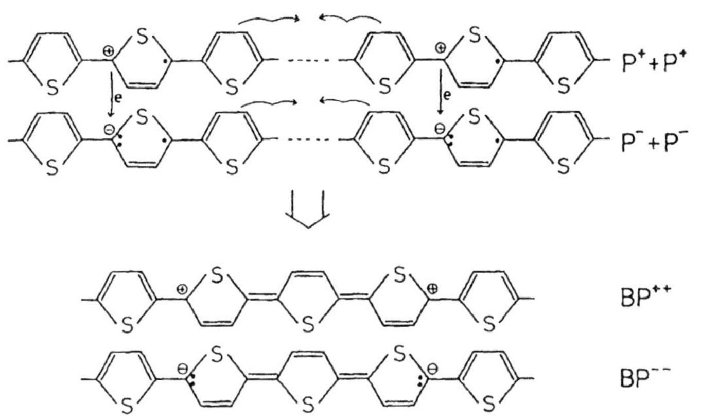

Figure 7.3. The formation of bipolarons (BP^{2+} and BP^{2-}) by the fusion process of two polarons ($P^+ + P^+$ and $P^- + P^-$) after the random walk on a chain.

formation is known to give rise to localized vibrational modes connected with the local structural distortions [7,40,43] and to the appearance of symmetric gap states [37–39,41,42]. There is evidence that bipolarons rather than polarons are the dominant charge storage configurations in PTs after steady-state photoexcitation or electrochemical doping, and thus the energy of the bipolaron state is lower than that of two polarons.

The photoluminescence of PTs has recently been reported [38,41] to be due to either free charge carriers or excitons. It has been suggested that the generation of photoluminescent excitons may be independent of the formation of polarons or bipolarons [41]. From time-resolved fluorescence experiments, it is known that PT shows a fast fluorescence decay; PT has been estimated to possess a decay time shorter than 9 ps [45].

PTV and PPV belong to the poly(arylenevinylene) group of polymers with the chemical structure of $C+X-CH=CHC+_n$, where X is a thienylene or phenylene group, respectively. A feature of poly-(arylenevinylene)s is that they can be prepared by an organic solvent-soluble precursor route, which facilitates their fabrication into high-optical-quality thin films. After fabrication, PTV can be fully converted into the conjugated form by a thermal process. PTV is a member of the polyheteroaromatics, and its main chain is a copolymer of PT and PA. Therefore in the same way as the backbone of PT resembles *cis*-polyacetylene, PTV has a backbone geometry similar in part to *trans*-polyacetylene and in part to *cis*-polyacetylene. The electronic structure is modified in PTV by the insertion of the vinylene linkage in PT. From this viewpoint it is of great interest to compare the electronic structure in the exciton state of these polymer compounds and their relaxation dynamics. It is also interesting to study PTV

in comparison with PPV, which is another poly(arylene-vinylene). PPV does not contain heteroatoms and the conjugated π-electronic system may be quite different from that of PTV because of the phenyl rings in the main chain.

The absorption edge of the lowest photon energy transition at 1.77 eV in PTV is found to overlap with the stepwise change of the photoconductivity signal, which is a characteristic of an interband transition [46]. However, the classification of the π–π* transition in PPV as either an interband or an exciton transition is not yet known [47]. In comparison with the structurally similar polymer PPV, the band gap of PTV was reported to be 25% lower in energy [48]. Additionally, the photoluminescence quantum yield in PTV was observed to be extremely weak (10^{-5}) in comparison with PPV, which has a quantum yield of about 0.08. Recently an extremely high fluorescence quantum efficiency (0.6) of a PTV derivative has been reported, and lasing was observed in a Hänsch-type cavity configuration using the polymer derivative in solution as a gain medium. A previous femtosecond investigation of PTV exhibited the photoinduced transient bleaching response only at a single probe photon energy of 1.97 eV [49].

It is of great interest to study the kinetics of excitations in PDAs, PTs and PTV and compare the results with those of PA, especially the formation and decay processes of solitons in PA and ST excitons in PDAs and PTs. The kinetics of photoexcitations have been studied in a PDA-TS single crystal. The population decay time was measured to be less than 2 ps by transient absorption spectroscopy [31] and by degenerate four-wave mixing (DFWM) in a PDA-TS single crystal [29]. In nanosecond time-resolved reflection spectroscopy studies, decay kinetics represented by $\text{erf}[(t/\tau)^{-1/2}]$ with $\tau = 1.0\ \mu s$ have been reported [23,50]. The decay kinetics were explained in terms of the geminate recombination of a pair of excitations after random walks in one-dimensional chains [51]. The photoexcitations in the nanosecond region were assigned to photoinduced polarons. The long-lived component observed in these time-resolved spectroscopies [29,51,52] when pumped with visible light was assigned to the triplet excitons formed by 'stepwise' two-photon absorption [53].

In this study, the relaxation dynamics of photoexcitations of soluble polydiacetylenes, PDA-3BCMU and PDA-4BCMU in cast films, PDA-3BCMU epitaxially grown in KCl single crystal, PDA-DFMP (1,4-bis (2,4-bis(trifluoromethyl)(phenyl)-1-3-butadiyne single crystal, and electrochemically prepared films of poly(3-methylthiophene) (P3MT) and poly(3-dode-cylthiophene) (P3DT), were investigated by femtosecond absorption spectroscopy. The third-order nonlinear optical susceptibilities of various processes were determined for both PDA-3BCMU and P3MT films. The fluorescence decay kinetics of the PDA-4BCMU films were also investigated using a single-photon counting system.

2 ULTRAFAST SPECTROSCOPY: TECHNIQUES AND POLYMERIC MATERIALS

2.1 Femtosecond Pump–probe experimental system

A block diagram of the apparatus used for femtosecond time-resolved absorption spectroscopy is shown in Figure 7.4. The cavity configuration of the colliding-pulse mode-locked (CPM) ring dye laser pumped by a CW (continuous wave) argon-ion laser (Spectra Physics, model 2030) was changed from that used in our earlier study [54] to compensate for the group-velocity dispersion using pairs of prisms [55]. All the cavity mirrors were single-stack coated and the reflectance of the outcoupling mirror was 95%. The gain medium was a 2.0×10^{-3} mol l^{-1} rhodamine 6G solution in ethylene glycol pumped with a power of 4–6 W at 514.5 nm. The saturable absorber was a 1.5×10^{-3} mol l^{-1} DODCI (diethylthiadicarboxyl chloride) solution in ethylene glycol. The thickness of the saturable absorber jet was about 55 μm. The full width at half maximum (FWHM) of the CPM laser pulse measured using the background-free autocorrelation scheme with a 0.22 mm thick KDP (KH_2PO_4)

crystal was 54 fs assuming sech^2 pulses. The center wavelength of the pulse was 630 nm and the average power was 20 mW at 100 MHz.

The output of the CPM laser was amplified 10^6 times in a four-stage dye amplifier system pumped by the second harmonic (532 nm, 150 mJ, 5–7 ns) of a Q-switched Nd:YAG laser (Quanta-Ray, DCR-1) at 10 Hz. In the early stage of experiment, the first three stages and the fourth stage were pumped transversely and longitudinally, respectively. Recently the system was modified in such a way that all the stages are sidepumped and new data were taken using the amplifier pumped in this configuration. The first two dye-amplifier stages are dye solutions and ordinary rectangular cells, while the third and the fourth are a Bethune cell and an axicon cell, respectively. Rhodamine 640 in water, with 4% ammonyx LO was used in the first stage and sulforhodamine was used in the other stages. Three thin jets of saturable absorber (malachite green in ethylene glycol) were inserted between adjacent stages of the amplifier to suppress the amplified spontaneous emission (ASE) and to reduce the intensity of the leading edge of the pulse [56]. The amplified pulses were compressed by group-velocity dispersion compensation using two pairs of prisms of high-index glass SF10. The typical duration, wavelength, and energy of the amplified pulses after the compressor were 80–90 fs, 625–630 nm and 0.2 mJ, respectively. Recently the compression system was also changed to a grating pair and a typical pulse width was 80–100 fs. The amplified femtosecond pulses were split into two beams for the pump and probe/reference by a partially reflecting mirror. Both the former and the latter passed through variable and fixed optical delay lines, respectively. The pump was the fundamental (630 nm, 1.97 eV) or the second harmonic (315 nm, 3.94 eV) of the amplified CPM laser. White continuum extending between near-ultraviolet (400 nm) and near-infrared (1000 nm) was generated by self-phase modulation in a 37 mm CCl_4 cell and was split into probe and reference beams. Both the pump and probe beams were focused on the sample and overlap with each other, and the spot size of the pump pulses was adjusted to be two or three times larger than that of probe pulses. The probe and reference spectra were measured by two coupled sets, each consisting of a multichannel photodiode and a polychromator (Unisoku). The spectral region of width 240 nm was averaged at one experimental run to cover the whole visible spectral region and near-UV down to 400 nm and near-IR up to 1000 nm, whereby four different spectral regions were measured with some overlap between each region. Over 100–200 pulses of the probe and reference spectra were averaged by a microcomputer (Epson, PC286V). The detection system was modified in later work. The reference and probe beams were collected by lenses into two fiber bundles connected to the entrance slit of another polychromator (Jarrel-Ash, Monospec 27) combined with a CCD (charge coupled device) sensor.

The time separation between the pump and probe pulses was changed by a power translation stage controlled by the microcomputer. Temporal overlap of the pump and probe pulses was measured by a cross-correlation method using a KDP crystal. The time resolution of the present system was about 100 fs and 300 fs for the fundamental and second-harmonic excitations in the whole spectral region (400–1000 nm). The chirp of the continuum was typically about 50 fs/100 nm and larger than 200 fs/100 nm at wavelengths longer than 700 nm and shorter than 500 nm, respectively, and was corrected using the cross-correlation data measured at more than ten wavelengths.

A femtosecond resonant Kerr experiment was performed by inserting a set consisting of a polarizer and an analyzer before and after the sample, respectively, into the pump/probe experimental apparatus in Figure 7.4.

2.2 Fluorescence measurements

Figure 7.5 is a block diagram of the single-photon-counting system used for the study of fluorescence decay kinetics. The excitation source was a rhodamine 6G laser (Spectra Physics, model 375) synchronously pumped by a CW mode-locked argon-ion laser (Spectra Physics, model 171) or by the second harmonic of a CW mode-locked Nd:YAG laser (Spectra Physics, model 3000). The dye laser was cavity dumped using a modulator (Spectra Physics, model 344) at 4 MHz.

The fluorescence from a sample was detected by a combined system of cut-off filters, a 50 cm focal-length monochromator (Ritsu, model MC-50), and a single-photon-counting photomultiplier (Hamamatsu Photonics, R1564U) with two microchannel plates. The output of the photomultiplier was amplified by a home-

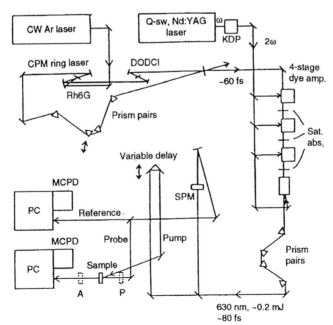

Figure 7.4. Block diagram of the femtosecond spectroscopy apparatus. Output of a colliding-pulse mode-locked (CPM) ring laser was amplified by a four-stage dye amplifier (dye amp.) pumped by the second harmonic of a Q-switched (Q-sw.) Nd : YAG laser. The dye amplifier consisted of four dye cells and three saturable absorber (sat. abs.) jets. The amplified femtosecond pulses were used as pump and probe/reference pulses. The probe/reference pulses were white continuum generated in a self-phase modulation (SPM) cell and detected by two coupled sets of a multichannel photodiode (MCPD) and a polychromator (PC). P and A are a polarizer and an analyzer, respectively, for the resonant Kerr experiment (Adapted from ref. 51.).

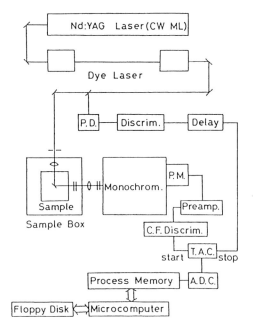

Figure 7.5. Block diagram of a single-photon-counting system. PM, photomultiplier; PD, PIN diode; TAC, time-to-amplitude converter; ADC, analog-to-digital converter; CF, constant fraction [51].

2.3 Electromodulation measurements

The high-voltage supply consisted of an oscillator (Torio, AG-203) and a power amplifier (Matsusada Precision, HEOP-2B2-TU) with two output terminals for a high-voltage output and a monitor. The monitor output voltage is $1/200$ of the high-voltage output. The output for the monitor was connected to an oscilloscope (Iwatsu, SS5415A) to allow the voltage to be viewed. The signal from the oscillator was also used as a reference for a lock-in amplifier (EG&G, model 5101) which was used to detect the field-induced change. A sinusoidal field at a frequency of 1 kHz was applied to the sample. The peak amplitude of the voltage was 2 kV; hence the maximum field was 507 kV cm^{-1}

The light from the tungsten lamp (Kondo Sylvania, model DRA, 120 V, 300 W) was passed through a monochromator (Ritsu, MC-25N) and then linearly polarized by a Glan–Thompson prism. The beam was directed to the area of the sample between the electrodes. The absolute value of the wavelength of the spectrometer was $+0.5$ nm. Polarization dependence of the electroabsorption experiment was measured with linearly polarized light with the electric field (E), which was parallel ($E \parallel F$) or perpendicular ($E \parallel F$) to the applied electric field (F). While the electric field amplitude was modulated, the intensity of the light transmitted through the sample was detected by a signal photodiode, a preamplifier (NF, model L176), and the lock-in amplifier. The output of both the lock-in amplifier (ΔI_T) and the preamplifier (I_T) was concurrently digitized with a digitizer (Autonics, model s121), and stored by a microcomputer (NEC, PC-8001mkII). Since the ratio ($\Delta T/T$) of transmittance change with the applied field (ΔT) to that without applied field (T) was less than 10^{-3}, $\Delta T/T$ was given by $\Delta I_T/I_T$. By moving the sample out of the optical pass, the incident light intensity (Ip) was also measured with the same signal photodiode used to measure the transmitted light. During the measurement of Ip and I_T, 15% of the incident beam from the lamp was directed to a reference photodiode with a beam splitter placed between the monochromator and the Glan–Thompson prism. The signal of the reference photodiode was amplified and digitized with the output of the signal photodiode, and they were recorded with the microcomputer. The intensity fluctuation of the light from the lamp during the measurement was corrected by the intensity of the split reference beam.

The reflection and electroreflection spectra of the single crystals of PDA-CPDO (5-19-carbazolyl)-2,4-

made wideband amplifier, discriminated by a constant-fraction discriminator (Ortec, model 583), and then applied to the start input of the time-to-amplitude converter (TAC) (Ortec, model 467). A part of the excitation pulse was split by a glass plate for reference and detected by a PIN photodiode (Hamamatsu Photonics, S1188). The output was amplified by another home-made wideband amplifier, selected by a discriminator (Ortec, model 934), and delayed to the stop input of the TAC.

To obtain the time characteristics of the emitted light, the output of the TAC was analyzed by the analog-to-digital-converter (ADC) (NAIG, model E-551), and accumulated in a multichannel analyzer (NAIG, model E-562). The data thus obtained were transferred to a microcomputer, and then stored on floppy disks. The difference in delay time between two neighboring channels was 41 ps, and the resolution time of the total system was 0.2 ns.

The fluorescence spectra were measured with a 25 cm focal-length polychromator of a modified Ebbert mounting and a multichannel photodiode (MCPD) (Unisoku). The excitation light source was the dye laser mentioned above or a CW argon-ion laser at 455 nm.

pentadiyn-1-ol in the visible region (1.13–2.50 eV) were measured using the apparatus described above.

Aluminum coplanar electrodes were deposited on the PDA-CPDO crystal. The electric field (F) was applied along the sample surface [(100) face] and parallel to the main chain ($F \parallel b$). To achieve this field application configuration the parallel edges of the electrodes were aligned perpendicular to the main chain. The measurement was conducted with the polarization of the incident light both parallel ($E \parallel b$) and perpendicular ($E \perp b$) to the direction of the modulation electric field at 297 ± 1 K. The peak amplitude of the voltage of 1 kV corresponding to the peak field strength of 25 kV cm^{-1} was applied.

2.4 Polymer samples

For femtosecond transient absorption spectroscopy, cast films of PDA-3BCMU and -4BCMU, and electrochemically prepared films of P3MT and P3DT, were used as samples. For the study of fluorescence decay kinetics, fully polymerized and partially polymerized chloroform cast films of PDA-4BCMU were investigated.

Films of PDA-3BCMU and PDA-4BCMU were prepared from CHCl$_3$ solution by drying under high vacuum and keeping in a vacuum flask [57]. The molecular weight of fully polymerized PDA-4BCMU in CHCl$_3$ cast film was about 3.4×10^5 as determined from gel permeation chromatography (GPC) data using the polystyrene standard. The number of repeat units in PDA-4BCMU cast film was about 670. The thickness of the sample for fluorescence measurement was about 0.1 mm.

Highly oriented films of PDA-4BCMU, which have aligned main chains, were prepared from the diacetylene monomers by the following procedure [34]: a vacuum deposition of a thin primary monomer layer on a glass substrate, a photopolymerization, a mechanical rubbing process, second vacuum deposition of a thicker monomer layer and final photopolymerization.

Red-phase oriented films of PDA-4BCMU were prepared by the thermal annealing of the blue-phase oriented films.

In the study of various nonlinear optical processes in PDA, described in Sections 3.8 and 3.9, PDA-3BCMU bi-oriented small-size crystals on a KCl substrate were used; the polymerization used UV light. A single crystal of KCl was cleaved in air and heated at 150°C for 30 min in a vacuum chamber before deposition. The deposition rate of the monomer was controlled at

0.3 nm min^{-1} under the pressure of 3×10^{-5} Torr. The substrate was kept at 45°C during the deposition. The monomer crystals grown on the KCl surface have a square shape of about 2×2 μm. From an analysis of the electron diffraction pattern, the crystals were bi-oriented and monomer molecules were perpendicular to the substrate surface stacking along the KCl[110] or [1̄10] direction. The thickness of the crystal was 0.1–0.5 μm. The monomer crystals were irradiated by UV light (254 nm) to polymerize.

A partially polymerized film was also prepared from CHCl$_3$ solution for the fluorescence decay study. UV irradiation, was used to obtain partially polymerized PDA-4BCMU, since it is readily converted to the fully polymerized form by γ-ray irradiation. The conversion rate of PDA-4BCMU was controlled by the duration of the UV irradiation. Single crystals of PDA-DFMP were also used.

PDA-MADF [(=CR−C≡C−CR′−)$_n$: R = C$_6$H$_3$(CF$_3$)$_2$ and R′ = C$_6$H$_4$NHCH$_3$)] which has aromatic rings directly attached to the main chain and is asymmetrically substituted with the side groups, was used for the triplet exciton formation study.

The monomer of PDA-MADF was synthesized and then polymerized according to the procedure described by Okada et al. [58]. After synthesis monomer crystals were recrystallized from ethanol solution by slow evaporation. The monomer single crystals were cut perpendicular to the long axis for easy handling. The typical dimension of the cut monomer crystals was 5 mm × 0.8 mm × 0.2 mm. The monomers were then polymerized by γ-ray irradiation of 380 Mrad in the solid state. The percentage of monomer to polymer conversion was more than 80%. The color of the surface of the PDA-MADF crystal was red, and a metallic luster could be observed.

Oriented films of blue-phase PDA-C$_4$UC$_4$- [(=CR−C≡C−CR′−)$_n$: R=R′=−(CH$_2$)$_4$OCONH (CH$_2$)$_3$CH$_3$] are also prepared by the procedures in ref.34: vacuum deposition of a thin primary monomer layer on a glass substrate, photopolymerization, mechanical rubbing process, second vacuum deposition of a thicker monomer layer, and final photopolymerization. The thickness of the sample film is 200 ± 10 nm. The absorption spectrum of the blue-phase film has a peak at 1.94 eV and an optical density of 1.7 for the light polarized along the rubbing direction. The absorption spectrum also has a broad phonon side band around 2.1 eV. The dichroic ratio $A \parallel A \perp$ is about 10 at the peak photon energy (1.94 eV).

The films of P3DT were synthesized electrochemically on an In−Sn oxide (ITO) conducting glass substrate. The obtained neutral films were peeled off ITO glass and stretched uniaxially up to twice the original length [38,59]. The thickness of the P3DT films used in this study was between 0.1 and 2.0 μm.

PTV is prepared from methoxy pendant-type precursors, as previously described by Murata et al. [46]. The precursor polymer is dip-coated onto a fused silica substrate and allowed to dry. The thermal conversion of the precursor to the PTV sample is performed between 80 and 170°C under a stream of nitrogen with a small amount of HCl. The film thickness of PTV sample used in this experiment is about 80 nm.

3 ANALYSIS OF ULTRAFAST RESPONSES IN π-CONJUGATED POLYMERS

3.1 Electromodulation spectra of PDA-3BCMU and PDA-CPDO

The visible absorption spectrum and electroabsorption of a cast film of PDA-3BCMU were measured. The absorption spectrum can be used to explain the signals of the electroabsorption spectrum.

Figure 7.6 shows the absorption spectra at 305 ± 1 and 30 ± 0.1 K; they are essentially the same as in literature [20]. At 305 K there is an absorption peak at

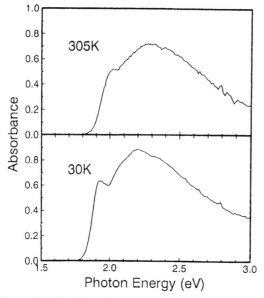

Figure 7.6. The absorption spectra of PDA-3BCMU at 305 and 30 K.

2.0 eV, and at energies higher than the peak, there is broad, large and featureless absorption. The lowest optically allowed transition of PDA-3BCMU is thought to be excitonic [22]. The line width of the transition at 2.0 eV was approximately 0.2 eV. This value is much larger than those of other polydiacetylenes such as PDA-TS ($\Delta = 0.11$ eV) [60] and PDA-DCHD ($\Delta E = 0.14$ eV) [61]. This broadening is probably due to inhomogeneous disorder in the polymer backbone. The absorption coefficient of this peak is 3.4×10^4 cm^{-1}. This value is much less than those of PDA-TS (7.3×10^5 cm^{-1}) [60,] and PDA-DCHD (5.0×10^5 cm^{-1}) [61]. One reason for the reduction of absorption coefficient of a cast film of PDA-3BCMU is inhomogeneous broadening of the rod-like conformation of the polymer in the blue phase. Two other reasons can be given as follows. The cast film of the PDA-3BCMU consists of microcrystalline and/or amorphous polymers; hence, the orientation of the main chains is randomly distributed in the plane of the substrate. The absorption coefficient of the exciton peak parallel to the chain direction of completely aligned PDA-3BCMU is calculated as 6.8×10^4 cm^{-1} from the absorbance of the cast film sample. Soluble polydiacetylenes, such as PDA-3BCMU and PDA-4BCMU, in chloroform/n-hexane mixtures show different features in the absorption spectra with different mole fractions of chloroform and different temperatures [22]. Since the absorption spectrum of the cast film of PDA-3BCMU seems to consist of these different spectra, the reduction of the lowest optical absorbance may be partly due to the coexistence of the polymers in different phases and configurations, such as the rod-like and coil-like configurations. The absorption spectra of molecules with configurations other than rod-like should contribute to the broad large peak in the energy region where phonon sidebands have been observed in other single crystals of polydiacetylene [60,61], except for PDA-CPDO. At 30 K, transition energies in the visible absorption spectrum shifted to lower energies by 0.08 eV compared to those at room temperature, and structures other than the lowest absorption peak could be observed. The structures at 2.11 and 2.20 eV are phonon sidebands. Se [62] has experimentally measured by Raman spectroscopy, the phonon transition energies in the ground state of PDA-3BCMU as 1461 cm^{-1} (0.1810 eV) and 2079 cm^{-1} (0.2578 eV), which correspond to the C=C and C≡C stretching modes, respectively, in the main chain. These transition energies are consistent with the observed photon energies of two features, shoulder and peak in the absorption spectrum. Therefore, we assign the absorption peaks at 2.11 and 2.20 eV are assigned to the double and triple bond-stretching energies.

The electroabsorption spectrum of PDA-3BCMU, ($\Delta T/T$) obtained at 30 ± 0.1 K and 50 KVcm^{-1} maximum field strength recorded with linearly polarized light parallel to the applied field ($E \parallel F$) is shown in Figure 7.7, where E and F are the electric field vectors of the polarized light and the applied field, respectively. Also shown in Figure 7.7 is the first derivative curve of the transmittance signal with respect to the probe photon energy, $(\mathrm{d}T/\mathrm{d}E_{ph})/T$, where E_{ph} is a photon energy. At the energy region of the lowest observed optical transition (1.7–2.3 eV), the electroabsorption spectrum can be approximated by the first derivative of the absorption spectrum. The observed red shift of the transition energy in the absorption spectrum is a dominant mechanism of the electroreflectance signal. The red shift of the transition energy in the visible region due to the applied electric field has been previously observed in the absorption and reflection spectra among single-crystal polydiacetylenes such as PDA-TS [63,64], PDA-DCHD [63,65], PDA-TCDU (poly-5,7-decadiyne-1,12-diol-bis-phenylurethane) [66] and PDA-DFMP (poly-2,2,5,5-tetraxis(trifluoromethyl)diphenyl-1,3-butadiyne) [67] and the vacuum-deposited film of PDA-C₄UC₄[R=R'=(CH₂)₄–OCONH–C₄H₉] [68]. In the energy region from 2.4 to 2.8 eV a field-induced

absorption can be seen. A structure at this energy has been previously found in the electromodulation spectra of several polydiacetylenes listed above.

The anisotropy (polarization dependence) in the electroabsorption signal recorded at 305 K and 50 kV cm^{-1} maximum field strength is displayed in Figure 7.8. The signal was also detected when the linear polarization of light was fixed perpendicular to the applied electric field ($F \perp E$), because the main chains of the polymer film were not oriented with respect to one another.

Figure 7.9 shows the field-strength dependence of the electroabsorption spectrum of PDA-3BCMU in the energy region of the lowest observed optical transition. The signal is found to vary quadratically with increasing field strength, and the line shape does not change with field strength. In Figure 7.10, the field-strength dependence of the maximum amplitude of the electroabsorption signal is displayed. The dashed line represents the line of slope 2. At electric fields below 50 kV cm^{-1}, it is obvious that the signal changes quadratically with increasing field strength. The same quadratic field dependence of the electroreflectance signal has been found in PDA-DCHD [63].

Figure 7.11 shows the electroreflectance spectrum ($\Delta R/R$) of the single crystal of PDA-CPDO obtained at 297 ± 1 K with linearly polarized light oriented both parallel and perpendicular to the polymer backbone, and with a maximum strength of the applied electric field of 25 kV cm^{-1}. There is a field-induced reflection minimum at 1.9 eV when measured with linearly polarized light parallel to the polymer backbone ($E \parallel b$), while there is no apparent signal for the perpendicular case ($E \perp b$), where E is the electric field vector of the linearly polarized light. This structure

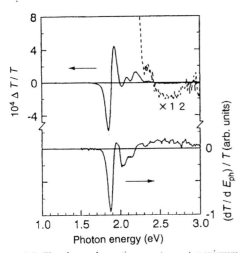

Figure 7.7. The electroabsorption spectrum at maximum field strength of 50 kV cm^{-1} and the first derivative of the transmittance signal of PDA-3BCMU at 30 K. The dashed curve represents the electroabsorption curve measured from 2.4 to 3.0 eV, the amplitude of which is expanded by a factor of 12.

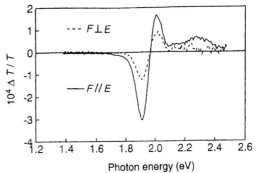

Figure 7.8. The polarization dependence in the electroabsorption signal at 305 K for the polarization of light (E) parallel ($F \parallel E$, solid curve) and perpendicular ($F \perp E$, dashed curve) to the applied electric field (F).

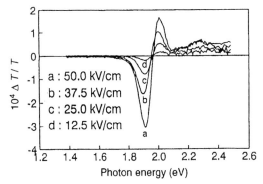

Figure 7.9. The electric field strength dependence of the electroabsorption spectrum of PDA-3BCMU at room temperature measured at 12.5, 25.0, 37.5 and 50.0 kV cm−1.

corresponds to the reflection peak at 1.89 eV [69]. In Figure 7.11 the second derivative curve $(\mathrm{d}^2R/\mathrm{d}E_{ph}^2)/R$, with respect to the photon energy (E_{ph}) is shown below the electroreflection curve. It is known that when a peak in an optical spectrum is slightly broadened by a perturbation, the difference between the perturbed and the original unperturbed spectrum of the peak is approximated by the second derivative of that original peak. This approximation holds on the condition that the oscillator strength of the peak is conserved. On the other hand the electroreflection curve is well approximated by the second derivative of the reflection spectrum with respect to the photon energy. Therefore, the main mechanism for the electroreflection signal of PDA-CPDO in the visible region is the broadening of the reflection peak. The electroreflection spectra of other polydiacetylenes such

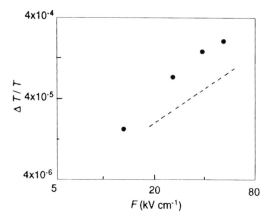

Figure 7.10. A log–log plot of the maximum amplitude of the electroabsorption spectrum of PDA-3BCMU vs. the strength of the applied electric field. The dashed line represents the line of slope 2.

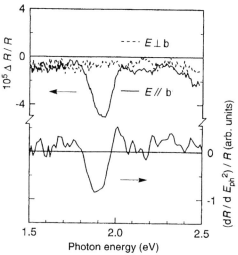

Figure 7.11. The electroreflection spectra of PDA-CPDO at 297 K with linearly polarized light oriented both parallel $(E \parallel b)$ and perpendicular $(E \perp b)$ to the main chain. The maximum strength of the applied electric field (F) was 25 kV cm^{-1}, and the electric field was oriented parallel to the main chain $(F \parallel b)$. The lower curve represents the second derivative curve of the zero-field reflection spectrum.

as PDA-TS [63,64], PDA-DCHD [63,64], PDA-TCDU [66] and PDA-DFMP [67] have been reported. In all of these polydiacetylenes, the features in the electroreflection spectra at approximately 2 eV, which correspond to the lowest optically observed transition, were found to be due exclusively to the red shift of the reflection spectra with the applied fields. In contrast to other polydiacetylenes only PDA-CPDO exhibits broadening in the energy region of the lowest optical transition.

The electroabsorption spectrum of the cast film of PDA-3BCMU is well represented by the first derivative curve of the absorption spectrum. Therefore, the effect of an applied electric field results in the red shift of the spectrum. The observations of red shifting and quadratic field-strength dependence for the amplitude of the electroreflection signal are characteristic of the spectral response of the ^1B_u exciton state previously seen in polydiacetylenes. Tokura *et al.* studied the electroreflectance of single crystals of PDA-DFMP [67], PDA-TCDU-A, and PDA-TCDU-B (PDA-TCDU in the red phase is called PDA-TCDU-B) [66]. They attributed the signals, which occurred at 2.25 eV, 1.90 eV and 2.3 eV in these polydiacetylenes to the quadratic Stark effects in the lowest ^1B_b exciton transition.

Let us consider the electroabsorption signal in PDA-3BCMU from the viewpoint of the quadratic Stark effect on the ^1B_b exciton. When the polarization of light is parallel to the polymer backbone, the absorbance change ΔA_c for one polymer chain is given by

$$\Delta A_c(\theta) = AF^2 \cos^2(\theta) \qquad (7.1)$$

where A is a proportional constant, F is the strength of the applied field and θ is the angle between the direction of the field and the main chain, because only the field projected onto the main-chain direction is efficiently coupled to the transition moment [18,63,65,70]. The absorbance change for an unoriented film detected by linearly polarized light is obtained by averaging the contributions in all possible orientations of the polymer backbone. Therefore, the anisotropy parameter in the electroabsorption in the cast film is given by

$$\frac{\Delta A_\perp}{\Delta A_\parallel} = \frac{\int_0^{2\pi} \sin^2(\theta) \Delta A_c(\theta) d\theta}{\int_0^{2\pi} \cos^2(\theta) \Delta A_c(\theta) d\theta} = \frac{1}{3} \qquad (7.2)$$

where $\Delta A_\perp$ and $\Delta A_\parallel$ are absorbance changes measured with linearly polarized light parallel $(E \parallel F)$ and perpendicular $(E \perp F)$, respectively, to the applied field direction (F). If the absorbance change is much less than unity, a transmittance change can be considered to be proportional to the absorbance change. The experiment performed here satisfied this condition. The value of the experimental anisotropy was found to be 0.39 in the cast film of PDA-3BCMU. The anisotropy experimentally determined is close to the calculated value of 0.33. The small discrepancy between them may suggest that a polymer in the cast film is not perfectly one dimensional. If the observed amplitude of the electroabsorption signal to the parallel case $(F \parallel E)$ is modified by dividing by the factor

$$\frac{1}{2\pi} \int_0^{2\pi} \cos^4(\theta) d\theta = \frac{3}{8} \qquad (7.3)$$

the amplitude of electroabsorption signal can be estimated if all he chains are aligned in the same direction (parallel to the applied field and light polarization direction) within a cast film. In this way the Stark shift of an oriented sample of 30 K and at a field strength of 50 kV cm^{-1} is estimated to be approximately 8×10^{-5} eV. This value is comparable to that of PDA-TCDU-A, which is reported to be 4.8×10^{-5} eV at 77 K and $F = 60$ kV cm^{-1} [66].

At 30 K a field-induced absorption in PDA-3BCMU was observed at 2.6 eV. Other polydiacetylenes such as PDA-TS, PDA-DCHD, PDA-TCDU and PDA-DFMP also reveal a field-induced absorption at approximately 0.5 eV above the energy of the lowest ^1B_u exciton transition. This field-induced transition has been assigned to the first-order dipole-forbidden ^1A_g exciton transition [66,67]. Sebastian and Wiser performed the electroreflection spectroscopy of PDA-TS [63] and PDA-DCHD [65] and have interpreted the structure which occurs 0.5 eV higher than the ^1B_u exciton state as the reflectance change due to a field-modulated interband transition by the Franz–Kedkysh effect. Their interpretation was based on the observation that the photocurrent could only be found at energies higher than 2.3 eV [71], which corresponded to the energy at which the electroreflection signal was observed. In addition, the field-strength dependence of the amplitude and the width of the reflectance change was found to be consistent with the Franz–Keldysh broadening. However, it has been suggested that the ^1A_g state also contributes to the photoconduction response of polydiacetylenes and that the field-strength dependence observed in PDA-DCHD can also be explained in terms of the ^1A_g exciton state mixed with the lower-energy ^1B_u exciton by an electric-field perturbation [66]. Hasegawa et al. have measured the third-order nonlinear optical susceptibilities $|\chi^{(3)}|$ and electroabsorption spectra of PDA-C$_4$UC$_4$ in both the blue and red phases [68]. They have also demonstrated the importance of the ^1A_g state in the nonlinear properties of polydiacetylenes. They showed that a calculation employing a three-level model which consists of the ground, the ^1B_u and the ^1A_g levels gives a quantitative explanation about the relative magnitude of $|\chi^{(3)}|$ of PDA-C$_4$UC$_4$ in both phases. Therefore, it is reasonable to assign the electric-field-induced structure of polydiacetylenes that occurs at approximately 0.5 eV above the lowest optically allowed transitions to the ^1A_g exciton transition at present.

From the results of the absorption and electroabsorption measurement of the cast film of PDA-3BCMU the following has been shown. In the zero-field absorption spectrum at visible photon energies, the observed lowest energy features are the ^1B_u exciton at 2.0 eV and the phonon side-bands at 2.1–2.3 eV. The field-induced absorption at 2.6 eV is assigned to the ^1A_g exciton, by analogy to previous observations of field-induced absorption in other polydiacetylenes, such as PDA-TCDU, PDA-DFMP and PDA-C$_4$UC$_4$. It has also been found that the optical response of the absorption band, due to the ^1B_u exciton, under an applied electric field was a red shift. In addition, a quadratic dependence of the field strength was observed. It is important to note that the agreement between all of these observations and the previous spectroscopic measurements of other polydiacetylenes is significant for the following reason. All the samples, in previous investigations using field-induced spectroscopies, were either single crystals or highly oriented films. In the present study, the cast film of PDA-3BCMU contains several phases, or different geometric configurations in the sample. This 'disordering' of the sample has the effect that the zero-field ^1B_u exciton absorption peak was broad $(\Delta E = 0.2\ \text{eV})$ and weak $(a = 3.4 \times 10^4\ \text{cm}^{-1})$ compared to other polydiacetylenes, such as PDA-TS and PDA-DCHD. In this 'disordered' system it is expected that the length of the rigid segments along the main chains is shorter and the variance of the distribution of the conjugation length should be larger than for other polydiacetylenes. However, even though the cast film is 'disordered', the transition in the visible region was observed to shift to red with the applied field, and no significant broadening was observed. This observation strongly suggests that the effect of field-induced broadening, which was observed in PDA-CPDO, can not be attributed to any 'disorder' in the polymer.

The electroreflection spectrum of the single crystal of PDA-CPDO, measured with the linear polarizations of light and the applied electric field both oriented parallel to the main chain $(E \parallel b$ and $F \parallel b)$ can be well fitted by the second-derivative curve of the zero-field reflection spectrum, as shown in Figure 7.11. Therefore, an applied field resulted in a spectral broadening of the observed optical transition in the reflection spectrum of the crystal. This type of field-induced response is different from the red-shift response for the lowest optical transition found in other polydiacetylenes. In the present study, it has been found that even the cast film of PDA-3BCMU, which does not have a highly ordered polydiacetylene structure and was observed to have a small exciton peak that is inhomogeneously broadened, exhibited a red shift and no significant broadening when modulated by an applied electric field. Therefore, the nature of the lowest transition of PDA-CPDO is not consistent with excitonic behaviour.

The Franz–Keldysh effect is well known as a broadening effect for interband transitions. The observations of this effect have been made in band-to-band transitions in silicon and germanium [72]. Hence the result of electroreflection measurement of the lowest-energy observed transition of PDA-CPDO is probably due to the interband transition. The interpretation that the band-gap energy of PDA-CPDO is approximately 1.6 eV was supported from the experimental results.

3.2 Transient absorption spectra of a cast film of PDA-3BCMU

Absorption spectra of a 100 nm-thick film of PDA-3BCMU at 10 and 290 K are shown in Figure 7.12. Comparing the absorption spectra with those of PDA-3BCMU solution [22], a 1.93 eV absorption peak at

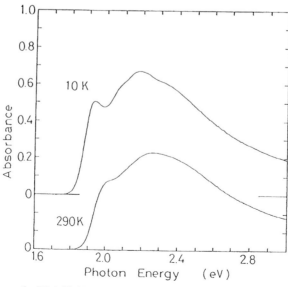

Figure 7.12. Absorption spectra of a PDA-3BCMU cast film at 10 K and 290 K (Adapted from ref. 51.).

10 K and a 2.0 eV shoulder at 290 K are assigned to the lowest singlet (^1B_u) excitons in the blue-phase PDA.

Since the photon energy of the fundamental femtosecond pulse (1.97 eV) for the pump–probe experiment is very close to the absorption peak of the ^1B_u excitons in the blue phase and smaller than the absorption edge in the red phase (2.2 eV) [22], the 1.97 eV pump pulse excites the ^1B_u excitons in the blue phase selectively. Electron–hole pairs are generated by the second-harmonic pulse (3.94 eV) with higher photon energy than the band gaps of PDAs in both the blue and red phases, which are reported to be between 2.4 and 3.1 eV depending on the side groups [23,53,73].

Figure 7.13 shows the spectra of the transient photoinduced absorption in the PDA-3BCMU cast film at 10 K and the femtosecond pump pulse. The excitation photon density is 9.6×10^{14} photons cm^{-2} with 1.97 eV [52]. The polarization of the pump and probe pulses are parallel to each other.

At −0.2 ps, a small but reproducible oscillating structure around 1.97 eV and a broad bleaching above 1.85 eV are observed. The oscillatory structure is most probably due to the perturbed free-induction decay observed also in semiconductors and organic molecules, when the probe pulse precedes the pump pulse [74,75]. The sharp bleaching peak at 1.97 eV and the

two small peaks at 1.79 and 1.71 eV are observed only between −0.1 ps and 0.1 ps, i.e. the pump and probe pulses overlap each other in time at the sample. The sharp peak of bleaching at 1.97 eV is due to hole burning. The two small minima at 1.79 and 1.71 eV are considered to be due to Raman gain. The corresponding Raman shifts are 1450 and 2100 cm^{-1}, which are assigned to the C=C and C≡C stretching modes, respectively [76]. This means that the C=C and C≡C stretching modes are strongly coupled with the excitons. Recent experiments in which PDA-TS was excited at photon energies below the excitonic resonance have proved the importance of the inverse Raman effect, which has been called phonon-mediated optical nonlinearity [77]. The same vibrational modes were found to be strongly coupled with the excitons [77,78]. The result of the Raman gain offered here corresponds to the inverse process of the inverse Raman effect. In this case two pump photons with higher energy and one probe photon with lower energy lead to the amplification of the probe light, while in the experiment of Blanchard et al. [77] two pump photons with lower energy and one probe photon with higher energy result in the loss of the probe light.

The transient absorption spectra induced by the excitation photon density of 3.8×10^{15} photons cm^{-1} are 10 K are shown in Figure 7.14. The absorbance changes have the same structures as the spectra induced by the weak photoexcitation shown in Figure 7.13 except for two features: first, the oscillations around 1.97 eV cannot be observed at −0.2 ps because of the larger bleaching peak due to the burned hole; and second, the asymmetry of the bleaching around 1.97 eV becomes more prominent and the bleaching above 2.0 eV at 0 ps becomes larger. The bleaching around 2.1 eV may be due to the phonon side holes of the C=C and C≡C stretching modes and the asymmetry may be due to the optical Stark effect.

The optical Stark effect has been widely studied in various materials [79–81] and has been discussed also for PDA [51,52,77,78,82]. The Stark effect in conjugated polymers is complicated because of competition with many other nonlinear optical phenomena such as induced Raman processes. All the experiments have been performed when the photon energy of the pump laser has been smaller than that of the absorption edge of the exciton transition. The pump photon energy in this study is slightly higher than that of the absorption peak of the excitons. Then the transition energy shift due to the optical Stark effect is to lower energy and the absorbance change has a maximum at 1.95 eV just below the pump photon energy.

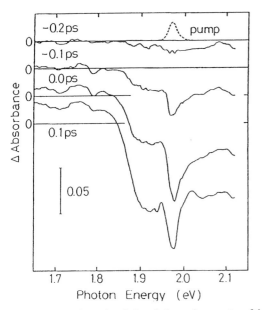

Figure 7.13. Transient photoinduced absorption spectra of the PDA-3BCMU film at 10 K and the spectrum of the pump pulse (dashed curve). The excitation photon density is 9.5×10^{14} photons cm^{-2} [51].

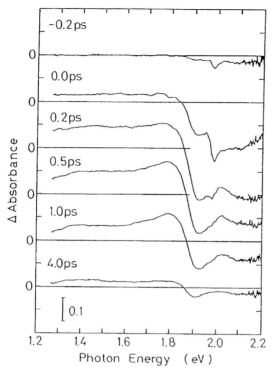

Figure 7.14. Transient absorption spectra of the PDA-3BCMU film induced by the 1.97 eV pump pulse at 10 K. The excitation photon density is 3.8×10^{15} photons cm^{-2} [51].

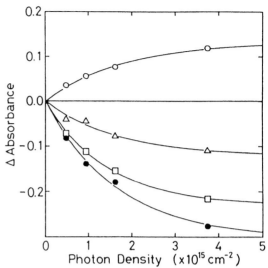

Figure 7.15. The excitation photon density dependence of the absorbance changes in the PDA-3BCMU film. Open circles, closed circles and open squares are the absorbance changes at 1.80 eV, 1.92 eV and 1.97 eV, respectively. Open triangles represent the Raman gain at 1.79 eV. The curves are best fitted with equation (4) [51].

The photoinduced bleaching at 0.5 ps has a peak at 1.92 eV and the shape is similar to the absorption spectrum of the blue-phase solution. It is due to the absorption saturation of the ^1B_u excitons in the blue-phase PDA. The photoinduced absorption below 1.85 eV is not observed at -0.2 ps and appears more slowly than the bleaching. At 0.0 and 0.2 ps the absorption spectrum is flat except for two small minima at 1.79 and 1.71 eV due to the Raman gain, while it has a maximum at 1.80 eV and drops below 1.35 eV at delay times slightly longer than 0.5 ps. This indicates that the photoinduced absorption shifts to a higher energy from 0.0 to 0.5 ps.

The dependence of the induced absorbance change on the excitation power density is shown in Figure 7.15. Because of the ground-state depletion, the absorbance at the pump photon energy decreases and the absorbance changes are saturated in the excitation intensity region in this study. Since the exciton decay time of about 2 ps is much longer than the duration of

the pump pulse, the saturated absorbance change ΔA can be approximated as

$$\Delta A = \Delta A_s[1 - \exp(-q/q_s)] \qquad (7.4)$$

All the absorbance changes in Figure 7.15 seem to have almost the same saturation photon density q_s (within experimental error). It is roughly estimated to be 1.5×10^{15} photon cm^2.

The intensity-dependent complex refractive index n_{NL} can be represented by

$$n_{NL} = n_0 + i\kappa_0 + (n_2 + i\kappa_2)I \qquad (7.5)$$

where n_0 and κ_0 are real linear refractive index and extinction coefficient, and n_2 and κ_2 correspond to nonlinear coefficients, respectively. $\kappa_2 = \kappa_0 \Delta a/2I$ can be obtained from the change in absorption coefficient, Δa, and the intensity, I. The corresponding imaginary part of the third-order susceptibility $\chi^{(3)}(-\omega; \omega, \omega, -\omega)$ can be calculated as -2.6×10^{-10} esu at the exciton photon energy $\hbar\omega = 1.097$ eV using the equation

$$\mathrm{Im}[\chi^{(3)}_{1111}(-\omega; \omega, -\omega, \omega)] = \frac{n_0^2 c \kappa_2}{16\pi^2} \qquad (7.6)$$

where c is the velocity of light in a vacuum.

The dynamics and high-energy shift of the absorbance change can be explained by the ST exciton. A model of the relaxation kinetics and potential surface of the ^1B_u exciton is shown in Figure 7.16. The binding energies of the excitons in the blue phase have been calculated by Tanaka *et al.* [83] as 1.33 eV for ^1B_u excitons, and as 0.57 eV for ^1A_g excitons. The band gap of the PDA-3BCMU cast film was estimated to be 3.1 eV from the excitation spectra of the triplet excitons observed by nanosecond spectroscopy [53] because the triplet excitons are not generated via the exciton but via the electron–hole pairs in the case of low excitation density. The transition energy from the ground state to the ^1B_u exciton is calculated to be $3.1 - 1.33 = 1.77$ eV and agrees well with the observed stationary absorption edge at 1.8 eV. The observed photoinduced absorption is assigned to the ST excitons formed from the free ^1B_u excitons initially photogenerated in polydiacetylene. The transition energies from the free ^1B_u excitons to the m $^1A_g (m > 2)$ excitons and to the electron–hole pairs are calculated from the binding energies as $1.33 - 0.57 = 0.76$ eV and 1.33 eV [83], respectively. The 0.76 eV transition is outside of the observed spectral region. The discussion on the assignment of the transient absorption will be made later in this paper.

The time-dependent absorbance change is fitted to the observed data as follows. The free ^1B_u excitons (F) are created by the pump pulse with a duration of τ_p. Then the free excitons relax to form the ST excitons (S) with a time constant τ_{FS}. Since the excitons are strongly coupled with the C=C and C≡C stretching modes, the strongly coupled coordinate is associated with a change in the bond alternation. Finally, the ST excitons relax into the ground state (G) with a time constant τ_{SG}. The kinetics of the excitations are then described by

$$\frac{dn_F}{dt} = W(t)(n_G - n_F) - \frac{n_F}{\tau_{FS}} \qquad (7.7)$$

$$\frac{dn_S}{dt} = \frac{n_F}{\tau_{FS}} - \frac{n_S}{\tau_{SG}} \qquad (7.8)$$

$$n_0 = n_G + n_F + n_S \qquad (7.9)$$

where n_G, n_F, and n_S are the densities of the ground state, free excitons and ST excitons, respectively; and n_0 is the saturation density of the excitons. The pump rate $W(t) = \sigma_{GF} I_p(t)/\hbar\omega_p$ is calculated using the absorption cross-section σ_{GF}. For both the pump, $I_p(t)$, and probe, $I_{pr}(t)$, the pulse shapes are assumed to be $sech^2(2.269t/t_p)$ with the same width of $\tau_p = 150$ fs, determined by the auto- and cross-correlation measurements.

From the time dependence of the absorbance changes at 1.92 and 1.77 eV in Figure 7.17 the lifetime and formation time of the ST exciton were determined as $\tau_{SG} = 2.0 \pm 0.1$ ps and $\tau_{FS} = 150 \pm 50$ fs, respectively. The curves in Figure 7.17 show the calculated time dependence of the absorbance changes at 1.92 and 1.77 eV, respectively, by

$$\Delta A_a(t) \propto \int_{-\infty}^{\infty} [n_G(t') - n_0 - n_F(t')] I_{pr}(t' - t) dt' \qquad (7.10)$$

and

$$\Delta A_b(t) \propto \int_{-\infty}^{\infty} n_S(t') I_{pr}(t' - t) dt' \qquad (7.11)$$

In equation (7.10) the bleaching includes not only the depletion of the ground state (n_G) but also the transition from the free excitons (n_F) to the ground state.

The kinetics of the absorbance around 1.8 eV does not change over the excitation photon-density range between 5×10^{14} and 1.5×10^{16} photons cm^{-2}, while those above 1.85 eV and around 1.5 eV start to show a longer-lived component of greater than several tens of picoseconds at densities higher than 1.5×10^{15} photons cm^{-2}. The absorbances of the component are proportional to the square of the pump photon density and are attributed to the triplet excitons formed from electron–hole pairs generated by the two-photon absorption, as in the case of PDA-TS single crystals [31]. The lifetime of the triplet excitons in PDA-3BCMU was determined as 18 μs by nanosecond spectroscopy [53].

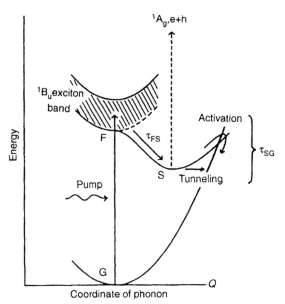

Figure 7.16. A model of the relaxation kinetics and potential surface of 1^1B_u excitons in polydiacetylene [51].

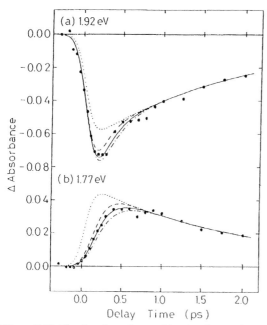

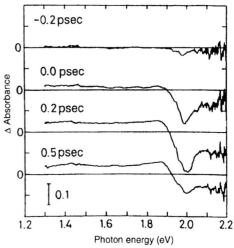

Figure 7.18. The transient absorption spectra of PDA-3BCMU induced by the 1.97 eV pump pulse at 290 K. The excitation photon density is 2.0×10^{15} photons cm^{-2} [51].

Figure 7.17. The time dependence of the absorbance changes in PDA-3BCMU at 10 K. The excitation photon density of the 1.97 eV pump pulse is 4.8×10^{14} photons cm^{-2}. The curves are the calculated transient absorbance changes with the formation times (τ_{FS}) of 0 fs (dotted curves), 100 fs (dashed curves), 150 fs (solid curves) and 200 fs (dash-dotted curves). The pulse duration (t_p) and the decay time (τ_{SG}) are 150 fs and 2.0 ps, respectively [51].

Figure 7.18 shows the photoinduced absorption at 290 K by the 1.97 eV pump pulse at 3×10^{15} photons cm^{-2}. The broad-band absorption below 1.9 eV and the bleaching of the ground state absorption are observed. The bleaching appears within the pulse duration, while the formation of the absorbance change at 1.80 eV is slower and the time constant is 150 ± 50 fs, which is equal to the time constant at 10 K, and is consistent with the barrierless potential of a ST exciton in one-dimensional systems [84]. The decay time of the absorbance change is 1.5 ± 0.2 ps, which is slightly shorter than at 10 K.

The lifetime of the ST exciton is PDA-3BCMU is much shorter than the radiative life estimated from the oscillator strength of the ^1B_u exciton transition. This means that the relaxation is mainly dominated by a radiationless process. The small difference in the lifetime between 10 and 290 K indicates that the thermal activation process required to overcome the potential barrier is not dominant in the relaxation process of the ST exciton. Excitons in many inorganic semiconductors have lifetimes between several hundred

picoseconds and a few nanoseconds, and those in various organic molecular crystals are usually of the order of nanosecond or several tens of nanoseconds at low temperatures. The activation potential is very often observed in most excitons in semiconductors, organic molecular crystals and also in ionic crystals. The lifetimes of these excitons are sensitive to temperature. In many cases, more than one order of magnitude differences in the lifetime are found between low temperatures, for example 10 K, and room temperature. The lifetime of ST excitons in PDA-3BCMU and also in PDA-4BCMU and P3MT is relatively insensitive to temperature as described later. The property is closely related to their unusually short lifetime.

These phenomena can be explained as follows. The potential curve of the ST exciton starts from the minimum of the potential curve of the free exciton in such one-dimensional systems as PDAs and PTs, and the curve is substantially lower in energy than that of the free exciton potential because of very strong electron–phonon coupling. As a result, the potential curves of the ST exciton and the ground state intersect at a point elevated only slightly above the minimum of the ST exciton potential curve. This intersection results in an efficient radiationless relaxation even at low temperatures.

The absorption spectra of PDA-3BCMU at 10 K induced by a 3.94 eV pump pulse with a density of 5×10^{14} photons cm^{-2} are shown in Figure 7.19. The polarizations of the pump and the probe pulses are parallel to each other. The ST exciton is formed within 300 fs from the electron–hole pair created by a 3.94 eV pump. The decay time of the absorbance change at 1.80 eV is 2.0 ± 0.4 ps. At 10 ps, the longer-lived component around 1.5 eV is clearer than the 1.97 eV photoexcitation shown in Figure 7.14. It is consistent with the assignment that the longer-lived component is due to the triplet excitons.

The self-trapping of the excitons is induced by strong electron–phonon coupling. After tunneling through or crossing over a barrier between the potential minima of the free exciton and the ST exciton in ordinary three-dimensional systems, the excitons are relaxed by the emission of lattice phonons with times of the order of the phonon cycle. Since the formation process has no barrier in the one-dimensional system [85], the formation of the relaxed ST excitons is expected to take place within the period of the coupled phonon cycle $T_M = 2\pi/\omega_M$. The experimental results of the excitation-intensity dependence of the Raman gain, the resonance Raman scattering [76] and the phon-mediated optical nonlinearity [77,82] show that the singlet excitons in polydiacetylenes are strongly coupled with the C–C stretching modes of double and triple bonds. The frequencies of the strongly

coupled stretching modes are about 1500 and 2100 cm^{-1} and correspond respectively to oscillation periods of 15 and 20 fs, which are much shorter than the 150 fs self-trapping time. There are three possible explanations for the slow self-trapping of the excitons.

(1) Su and Schrieffer investigated the formation of a soliton–antisoliton pair and a polaron in *trans*-PA by molecular dynamics [4]. When a soliton–antisoliton pair is created by an electron–hole pair, the binding energy is released to the kinetic energies of the soliton and antisoliton. Within a time of the order of the phonon cycle period, 36 fs, in *trans*-PA there is a well-formed soliton–anti-soliton pair separating at a velocity comparable to the speed of sound. When a polaron is generated from an electron–hole pair, the binding energy remains as the kinetic energy of lattice vibration and the polaron cannot relax to its ground state until the energy is removed.

The slow formation of the ST excitons in PDA can then be explained by the difference between the soliton–antisoliton pair and the ST exciton. Since the ST exciton is regarded as a fused polaron (neutral bipolaron), the self-trapping mechanism is expected to be similar to the polaron formation, in which the binding energy remains as the kinetic energy of the lattice oscillation. Therefore, the formation time of the ST exciton is not determined by the dissociation of the domain walls as in the case of soliton pairs, but by the energy redistribution from the strongly coupled phonon to the other phonon modes.

The time needed for the energy redistribution by phonon propagation in the chain is estimated to be $\chi/v_s = 500$ fs, where $v_s = 5.5 \times 10^3$ m s^{-1} is the velocity of sound along the PDA chain direction [86] and $\chi = 25$–30 Å is the exciton length [83]. It is much longer than the observed formation time. Therefore, the vibrational modes of the side groups are considered to play an important role in the formation of the ST excitons. The effect of the side groups will be clarified by studies of PDAs with different side groups.

(2) Jortner reported that the cooling rate of hot ST excitons in solid rare gases is given by $\omega_M \exp(-\omega_M/\omega_D)$, where ω_M and ω_D are the frequencies of the strongly coupled phonon model and the Debye frequency [87]. Hence the rate can be much smaller than ω_M. Since for a PDA-TS single crystal the Debye temperature θ_D is reported to be 45.6 K [88], the value evaluated for the

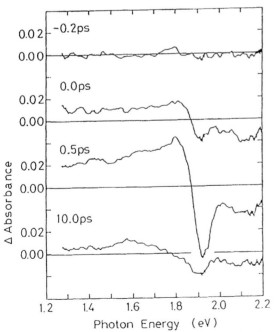

Figure 7.19. The transient absorption spectra of PDA-3BCMU induced by the 3.94 eV pump pulse at 10 K. The excitation photon density is 5×10^{14} photons cm^{-2} [51].

cooling rate of hot ST excitons is much slower than in the present experiment. This may be to the much higher vibrational mode density in polymers than in solid rare gases.

(3) Calculations of the exciton relaxation show that the time dependence of the average energy has a plateau at zero delay time [89]. Therefore it takes time for the phonon wave packet to experience the potential gradient force near the minimum of the free exciton potential and to initiate the sliding motion on the potential surface.

3.3 Transient absorption spectra of a cast film of PDA-4BCMU

The relaxation kinetics of photoexcitations have also been investigated in PDA-4BCMU. The absorption spectrum of PDA-4BCMU has a peak at 2.3 eV and the absorption edge is 2.1 eV. The second harmonic (3.94 eV) of the femtosecond pulse was used as the pump pulse, because the absorption edge is higher than the fundamental photon energy. The transient absorption spectra in PDA-4BCMU film between 1.25 and 2.60 eV were studied.

The features of the absorbance changes in PDA-4BCMU are similar to those in PDA-3BCMU excited by the 3.94 eV pump pulse. The bleaching due to the ground-state depletion has a peak at 2.35 eV at 290 K. The photoinduced absorption appears below 2.22 eV and is very broad and flat in the observed spectral region down to 1.25 eV. This broad spectrum is explained to be due to transition from 1^1B_u exciton to several $m^1A_g (m > 1)$ ST excitons. At 10 K the difference absorption peak shifts to lower energy. The bleaching peak is observed at 2.30 eV and the broad absorption appears below 2.15 eV. The absorbance changes due to Raman gain, hole burning, and other nonlinear optical effects, which were observed in PDAs and PTs excited by the 1.97 eV pulse, cannot be observed, because the pump photon energy is far away from the probe photon energies.

When a pump pulse of 3.94 eV, which is higher than the band gap is used, electron–hole pairs are photo-generated. The electron–hole pairs then relax to both singlet 1^1B_u and triplet excitons. The formation processes of the ST state of singlet excitons cannot be resolved by using second-harmonic pulses because of the broader pulse width of the second harmonic than the ST exciton formation time. Both the bleaching around 2.3 eV and the absorbance change below 2.2 eV decrease with time from 0.5 to 10 ps. The transient

absorption due to the exciton exhibits a substantial deviation from an exponential decay for a duration of a few picoseconds. The later decay becomes slower than the initial decay. The lifetimes of the ST excitons have been defined from the initial slope of the semiloga-rithmic plot of the transient absorbance. The decay times thus determined are 3.0 ± 0.3 ps and 2.1 ± 0.2 ps at 10 and 290 K, respectively. At delay times longer than 10 ps, the absorbance change around 2.0 eV disappears, while the bleaching and the absorption below 1.8 eV remain. The long live component lasts for a much longer time than 100 ps; this component is also observed in PDA-3BCMU and is assigned to the triplet excitons.

3.4 Transient absorption spectra of a PDA-DFMP single crystal

The absorption spectra of a single crystal of PDA-DFMP obtained from the measured transmission and reflection spectra, assuming negligible multiple reflection, recorded at a temperature of 297 K is shown in Figure 7.20. The spectrum shows a peak at 2.27 eV (FWHM of 0.12 eV) due to the 1^1B_u exciton transition.

Femtosecond pump–probe absorbance and reflectance-change signals of PDA-DFMP are performed

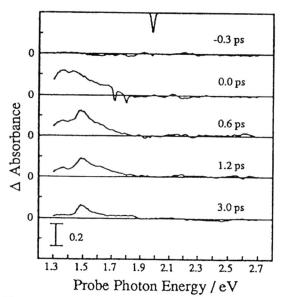

Figure 7.20. Absorption spectrum of a single crystal of PDA-DFMP obtained from the measured transmission and reflectance spectra at 297 K, assuming negligible multiple reflection.

with a pump photon energy at 1.99 eV, detuned from the peak of the ^1B_u exciton transition at 2.27 eV. Both the photoinduced absorption and reflection spectra at the probe photon energies from 1.3 to 2.0 eV and from 1.5 to 2.5 eV, respectively, are recorded with linearly polarized laser beams parallel to the polymer backbone, at pump–probe delay times from −1 to 80 ps at a temperature of 297 K.

At a delay time of $t = -0.12$ ps, when the probe pulse precedes the leading edge of the pump, a positive absorbance change signal is observed with the maximum absorbance change at 1.34 ± 0.02 eV. At this delay time an absorbance change signal is not detected at higher photon energies than 1.457 eV. This broad photoinduced absorption peak has been attributed in other PDAs to the upward transition from the free $1\,^1B_u$ exciton state to either high lying $m\,^1A_g(m > 1)$ exciton states and/or the edge of the continuum. Additionally, stimulated emission processes may also contribute to the observed absorbance change signal. At a longer delay time of 0.08 ps, the absorbance change signal can be observed in the spectral region from 1.3 to 1.6 eV with the maximum absorbance change at 1.4 eV. As the delay time is increased from $t = -0.12$ to 0.08 ps, not only does the overall absorbance change increase, but also the absorbance change spectrum shifts to higher photon energies. The time-varying spectral shift, with a time constant of 100 ± 30 fs, can be quantatively determined by the fitting of the rise time constant of the photoinduced transient absorption, as discussed later. This rapid spectral shift has been observed in other quasi-one-dimensional conjugated polymers, such as PDA-3BCMU, PDA-4BCMU, mentioned before, and polythiophene, to be described later, and is interpreted using the self trapped exciton (STE) model to be due to the formation of the STE state and phonon-emission process.

The decay kinetics of the photoinduced absorption is obtained from the plot of the transient absorbance responses at five different photon energies obtained with a pump photon density of 2×10^{15} photons cm^{-2}. The decay curves are well described by numerical fit by the convolution of a single exponential decay with a variable Gaussian pulse width and a constant long-time component, where time zero is treated as a free parameter. The single-exponential decay time constants, of the photoinduced absorbance changes in the region from 1.3 to 1.7 eV are 1.2 to 1.3 ps, while the decay time at a photon energy of 1.7 eV is 1.6 ps. For the purpose of the comparison of decay time among different photon energies, the transient decay curves near the long-lived state centered at 1.49 eV are

excluded from this analysis and are modelled using a different kinetic scheme.

The decreasing decay rate constants with increasing photon energies strongly suggest a distribution of the relaxation rates of the decay process from the STE to the ground state, which depends on the energy of the STE. The distribution of the rate constants may depend on the extent of vibrational population among highly vibrational states of the STE, which varies with time. Photoinduced absorption changes which at low (i.e. < 1.70 eV) photon energies monitor decay processes primarily due to vibrationally hot or unrelaxed states within the STE potential energy surface. Photoinduced absorption changes at relatively high (i.e.1.7 eV) photon energies monitor decay processes primarily due to vibrationally cool or relaxed STE states. The small decrease in the decay rate with larger photon energy may indicate that vibrationally hot STE relax more efficiently to the ground state than states which are lower in the STE potential-energy surface.

In the kinetic scheme of STE relaxation, the formation time from free exciton to ST excitons and decay time from the ST exciton to the ground state is described by the rate constants τ_{FS} and τ_{SG}. The time dependence of the absorbance changes at 1.82 eV with a pump photon density of 2×10^{-10} photons cm^{-2} is numerically fitted with the kinetic rate equations obtained from the above model. A linear dependence for the amplitude of the absorbance change as the pump photon density is increased from 1 to 3×10^{15} photons cm^{-2} is observed. For this nonfluorescent polymer, the single-exponential decay kinetics are similar to previous observations of blue-phase PDAs in which ST excitons are believed to relax to the ground state mainly via tunneling with significant thermalization processes.

3.5 Transient absorption spectra of an electrochemically prepared film of P3MT

Figure 7.21 displays a series of absorption difference spectra of P3MT at 10 K measured at various delay times between the excitation and probe pulses [51,90]. When the probe pulse arrives at the sample before the peak of the excitation pulse, a clear oscillating structure in the absorption difference spectra is observed around the photon energy of the excitation pulse. This oscillation, also observed for semiconductors near the band edge [74,91–94], is due to the coherent interaction of the weak probe pulse with the polarization induced by the intense pump pulse in the P3MT film, as is expected from theory [75,94]. The coherent oscillation vanishes when the maximum of the probe pulse arrives at the sample later than the pump pulse.

A striking feature of the photoinduced absorption spectra at early delay times (pronounced at -0.1 and 0 ps) is the appearance of a strong minimum at 1.8 eV and a weaker one at 1.62 eV. The minimum at 1.8 eV at early times reaches negative absorbance values, while there is no absorption of the ground state at 1.8 eV. The minima have only been observed when the excitation and the probe pulses overlap. Therefore this minimum is ascribed again to the Raman gain with no doubt. This assignment is also supported by the resonance Raman study of P3MT, in which a single dominant mode in the Raman gain spectra suggests that only one phonon mode with an energy of about 0.18 eV is much more strongly associated with the excitation of P3MT than any other modes [95].

In the same way as for PDA-3BCMU, the third-order susceptibilities corresponding to the absorption saturation and the Raman gain have been determined; $\mathrm{Im}[\chi^{(3)}(-\omega;\ \omega,\ -\omega,\ \omega)] = -3.5 \times 10^{-10}$ esu for $\hbar\omega = 1.97$ eV, and $\mathrm{Im}[\chi^{(3)}(-\omega_2;\ \omega_2,\ -\omega_1,\ \omega_1)] = -1.5 \times 10^{-10}$ esu for $\hbar\omega_1 = 1.97$ eV and $\hbar\omega_2 = 1.79$ eV. The smaller values are due to the lower absorption coefficient at 1.97 eV in P3MT than in PDA-3BCMU.

The other features of the spectra displayed in Figure 7.21 are now discussed. The strong bleaching around the excitation energy results from the generation of excitations near the band edge as well as from the ground-state depopulation. The sharp line bleached at 0 ps is due to nonthermal free excitons. The bleached spectral region at later times after thermalization becomes much broader. On the lower-energy side, a broad absorption band arises during excitation caused by the onset of a strong excited-state absorption. There is no clear difference in the transient absorption spectrum and the dynamics between parallel and perpendicular polarizations of the pump and probe pulses. Thus further discussion is restricted to parallel polarization.

The transient behavior of the photoinduced absorbance change is discussed in more detail using Figure 7.22. The time constant of the induced absorption at 1.86 eV is estimated to be 70 ± 50 fs. The variance is large because of the limited time resolution and signal-to-noise ratio.

As discussed for PDA-3BCMU, the observed fast relaxation corresponds to the exciton self-trapping, for

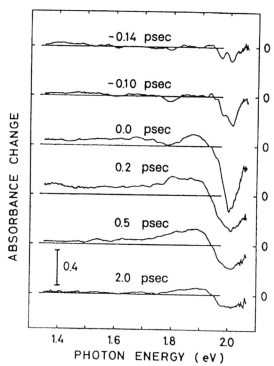

Figure 7.21. The differential absorption spectrum of P3MT at various delay times at 10 K. The photon energy of the excitation pulse was 1.98 eV (Adapted from ref. 90.).

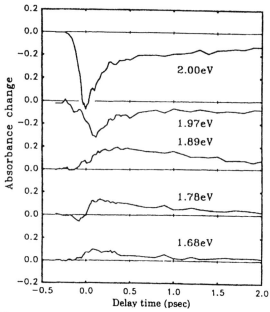

Figure 7.22. The time dependence of the photoinduced absorption of P3MT at 10 K for various photon energies of the probe pulse [90].

which the time constant was 150 ± 50 fs in PDA-3BCMU [32,51,52]. The formation time of the STE in PDA-3BCMU is longer than that of P3MT. This can be explained by the difference in the chemical structures of these polymers, i.e. P3MT has a rigid structure of thiophene rings conjugated with each other. The substituent methyl group does not efficiently contribute to the electronic structure of the exciton and is also not expected to affect the relation due to strong electron–phonon coupling. Then the relaxation is limited to the conformational change in thiophene rings. Therefore, the energy dissipation of the induced C=C stretching motion by the electronic excitation takes place within a few cycles of the oscillation. The vibrational energy of C=C stretching is transferred to the other modes such as C–C–C and C–S–C bending modes and C–H stretching mode to achieve a new stable configuration in the STE state, where the methyl substituent groups are relatively free from the main thiophene rings.

On the other hand PDA-3BCMU has bulky substituent groups attached to the main PDA chain. They extend a few tens of angströms from the backbone and form a sheet in the backbone plane with the help of hydrogen bonding. When electronic excitation occurs, the whole side-group structure must be reorganized in such a way that they can fit a new stable configuration in the ^1B_u excitation state. This change must be associated with the conformational reorganization of the whole substituent side groups in the polymer, and this process is expected to take a longer time than in P2MT. From the experimental results, the reorganization of the whole configuration takes place after seven to eight oscillation periods (10–20 fs) induced by the electronic excitation in PDA-3BCMU.

The relaxation of the ST exciton is now discussed. Just at excitation (0 ps), broad featureless absorption spectra have been observed in the photon energy region of 1.35–1.9 eV in both P3MT and PDA. The common feature indicates that the absorption is due to the same transition i.e. the free exciton to higher exciton states. At delay times between 0.2 and 0.5 ps, the spectrum of the ST exciton can be seen with a peak in the spectral region 1.8–1.95 eV, which corresponds to the transition to the excitons of higher energies or to the conduction band. A second fast component with a time constant of 800 ± 100 fs is apparent in the bleaching recovery. The same decay time was found in the broad band excited-state absorption. Thus, this decay indicates the relaxation of ST excitons to the ground state or other long-lived species.

The time-resolved fluorescence measurement of PT has been reported by Wong et al. [45]. They could only

determine the upper limit of the fluorescence lifetime to be 9 ps, because of limited temporal resolution. This corresponds to fluorescence in PT, since the excited-state absorption has only one component with a shorter life than 9 ps. Since the Stokes shifts of the transient fluorescence from PT and the steady-state fluorescence from P3MT are similar, they are emitted from the ST excitons with similar stabilization energy. The steady-state fluorescence of P3MT has a peak around 1.9 eV (Figure 7.23). The stabilization energy due to the self-trapping is estimated to be 0.1 eV, by assuming the same curvature of the potential for the ST exciton of the ground state and a free exciton energy of 2.1 eV.

After the 800 fs relaxation, a long-remaining absorbance change has been observed. Its time constant is longer than a few tens of picoseconds if exponential decay is assumed, but it may obey nonexponential decay kinetics. The induced absorption may be considered to be due to triplet excitons, polarons and/or bipolarons [51,52]. In the case of higher excitation pulse energy, the electron–hole pair is generated by two-photon absorption. Therefore the induced absorption in the region of 1.35–1.8 eV has some contribution from the triplet excitons in some case [51,52]. However, other species also seem to exist, since the broad absorption spectrum, which is different from the triplet exciton absorption, can still be observed even under weak excitation. It should have some contributions from the polarons or the bipolarons since the small absorbance change observed with a high pulse energy of a nanosecond laser indicates a small number of polarons or bipolarons. The injection of a

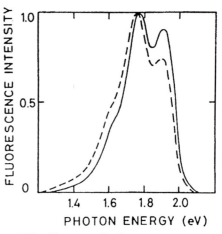

Figure 7.23. The normalized steady-state spontaneous fluorescence spectrum (solid line) and the calculated spectrum of the stimulated emission (dashed line) of P3MT [90].

single charge always occurs via the formation of polarons which can build up charged bipolarons. However, the time for the formation of bipolarons should be dependent on the density of excitations generated in the material. As known from the steady-state photoinduced absorption spectra measured during electrochemical doping, the formation of bipolarons in P3MT gives rise to the appearance of an absorption band around 1.6 eV and a strong one with a peak around 0.65 eV [38,39]. If polarons remain, an additional peak of about 1.2 eV is expected to be observed. The spectrum observed by the present femtosecond spectroscopy is so broad and featureless that it is difficult to determine which is the major long-lived species, triplet exciton, bipolaron or polaron.

It is also noted that neither the decay kinetics nor the spectrum measured at 295 K and 10 K show substantial differences. At room temperature, again the 800 fs decay component followed by a long remaining absorption and bleaching around the band-edge energy is found. This leads to the conclusion that the involved relaxation channel is temperature insensitive. No clear temperature dependence in the formation of the ST exciton is consistent with the barrierless potential of the ST exciton on a one-dimensional chain due to strong electron-phonon coupling. The temperature-insensitive ultrafast decay of the ST exciton in both PDA and P3MT of the order of one picosecond can be explained by assuming that the position of the crossing point of two potential surfaces between the ground state and the ST exciton is located below or only slightly higher than the bottom of the free exciton.

The femtosecond time-resolved resonant Kerr experiment has been performed by inserting a set consisting of a polarizer and an analyzer before and after the sample, respectively, into the pump/probe experimental apparatus. This experiment offers the time dependence of the anisotropy of absorbance. Since it is a null method, the signal-to-noise ratio is much better than the pump/probe experiment. The data are shown in Figure 7.24. At 1.77 and 2.00 eV, the signal has a very rapid decaying component due to Raman gain and hole burning, respectively. Both processes contribute to the signal with a similar dependence to the pump/probe pulses. At 1.88 eV the signal follows a power-law decay, t_d^{-a}, where t_d is the delay time and the exponent a is estimated to be about one. This indicates that anisotropy $\Delta\chi$ varies as $t_d^{-0.5}$ and that the anisotropy disappears by a geminate recombination type process following the one-dimensional random walk. This process can be explained in terms of the fusion process $P^{\pm} + P^{\pm} \rightarrow BP^{2\pm}$ after the random walk of $P^{\pm}$ as

shown in Figure 7.3, since the bipolaron (BP) is more stable than two polarons (P + P) in P3MT.

From the transmitted light intensity, the birefringence of the sample film induced by femtosecond pump pulse has been determined. From the birefringence value, the anisotropy of $\chi^{(3)}$, i.e. $|\Delta\chi^{(3)}(-\omega_2; \omega_2, -\omega_1, \omega_1)| = |\chi^{(3)}_{1111}(-\omega_2; \omega_2, \omega_1, -\omega_1) - \chi^{(3)}_{1122}(\omega_2; \omega_2, -\omega_1, \omega_1)| = 3.9 \times 10^{-11}$ esu has been obtained for $\hbar\omega_1 = 1.97$ eV and $\hbar\omega_2 = 1.88$ eV.

3.6 Transient absorption spectra of an electrochemically prepared film of P3DT

Figure 7.25 shows the stationary and transient absorption spectra of P3DT film at 290 K and the pump pulse spectrum. The bleaching due to the depletion of the ground state and the photoinduced absorption below the absorption edge are observed also in P3DT. The absorbance change at 0.0 ps has a structure at the pump photon energy, which is probably due to the induced frequency shift of the probe pulse as observed in the red-phase PDA-4BCMU. A small minimum at 1.80 eV is due to the Raman gain. The corresponding Raman mode is assigned to the stretching vibration of the C=C bond. The bleaching peak at 2.14 eV is due to the 0–1 transition of the same phonon mode. At 0.5 ps the absorbance change around 2 eV becomes positive and the photoinduced absorption has a peak at 1.88 eV. If excitons are assumed to be photoexcited in P3DT by the 1.97 eV pump pulse, this spectral change can be explained by the formation of the relaxed STEs in the same way as PDA-4BCMUs. The time constant of the formation is estimated as 100 ± 50 fs.

P3DT has similar decay kinetics to the red-phase PDA-4BCMU. The decay kinetics have short- and long-lived components. The time constant of the long-lived component is longer than 100 ps. The short-lived component can be fitted to biexponential functions. The sets of time constants at 290 K are summarized in Table 7.1. The time constant τ_1 is shorter than 0.5 ps and is expected to correspond to the decay of the free excitons and thermalization of the self-trapped exciton. The time constant τ_2 corresponds to the decay of the relaxed STEs and is estimated as 4.7 ± 1.2 ps.

If the decay curves are fitted to a power-law decay and long-lived component, the power a at 290 K is obtained as 1.02 ± 0.23 at 2.21 eV, 0.67 ± 0.05 at 1.88 eV and 0.61 ± 0.06 at 1.63 eV. The power-law decay was reported also in other PTs and the value of a was 0.37 in poly(3-hexylthiophene) at 1.17 eV, 0.22 in

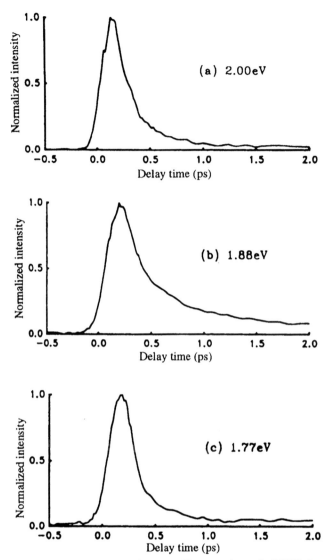

Figure 7.24. The results of the femtosecond time-resolved resonant Kerr experiment for P3MT observed at (a) 2.00, (b) 1.88 and (c) 1.77 eV. The results in (a) and (c) suffer from the DFWM and Raman gain and show very fast relaxation near 0.0 ps [51].

poly(3-octylthiophene) at 1.17 eV and 0.9 in polythiophene at 2 eV [96,97]. The dependence of the power a on the probe photon energy and the spectral change at the delay time from 0.5 to 5 ps are difficult to be explained by a simple recombination model.

The time dependence of the absorbance changes at 1.50 eV induced by the pump pulse polarized parallel ($\parallel$) and perpendicular ($\perp$) to the polymer chains was studied. The short-lived component due to the STEs decreases rapidly and the long-lived component remains much longer than 100 ps. It was clearly seen

that the long-lived component is induced more efficiently by the perpendicular pump pulse. This mean the long-lived component is not due to triplet excitons but due to either polarons or bipolarons generated by the interchain photoexcitation. A polaron pair is formed from an electron–hole pair excited by a single photon, while a bipolaron is generated by a collision of two polarons with a charge of the same sign. The observed intensity dependence shows that the long-lived component in P3DT increases proportionally to the pump photon density up to 2×10^{16} photons cm^{-2}. Hence,

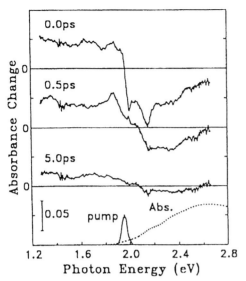

Figure 7.25. Transient absorption spectra of a P3DT film at 290 K induced by the 1.97 eV pump pulse. The excitation photon density is 7.5×10^{15} photons cm^{-2}. The dotted curve is the stationary absorption spectrum.

the long-lived component in P3DT is concluded to be due to polarons.

3.7   Transient absorption spectra of a dip-coated film of PTV

The femtosecond photoinduced absorption spectra are recorded at probe photon energies from 1.15 to 2.55 eV. A pump pulse of 100 fs width and with a photon energy of 1.97 eV measured at 297 K (and 1.96 eV at 77 K) is resonant with the π–π^* transition of PTV as shown in Figure 7.26. As the temperature is decreased, the edge of the absorption band, at 1.77 eV, shifts to lower energies and phonon features are found to be enhanced [48]. The change of the absorbance signals are recorded with a linearly polarized pump and probe

Table 7.1. Decay time constants of absorbance changes in a P3DT film at 290 K

Probe photon energy (eV)	Time constants (ps)	
	τ_1	τ_2
2.21	0.3 ± 0.1	4.7 ± 1.6
1.88	0.5 ± 0.1	4.7 ± 1.2
1.63	0.3 ± 0.1	7.9 ± 1.8
1.55	0.3 ± 0.1	7.9 ± 2.3

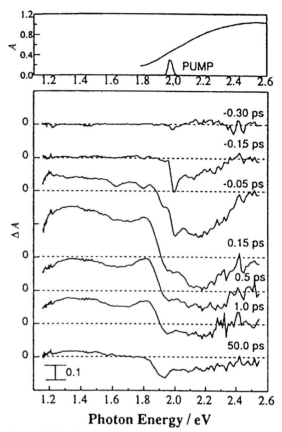

Photon Energy / eV

Figure 7.26. Photoinduced absorption, ΔA, vs. the probe photon energy for seven pump–probe delay times for a PTV film after excitation with a 1.97 eV pump pulse at 297 K. The excitation photon density is 2.4×10^{15} photons cm^{-2}. The absorption spectrum and the pump pulse spectral profile are shown for comparison.

beams parallel to each other at pump–probe delay times from $t_d = -1$ to 125 ps.

The ultrafast response of PTV near 0 ps delay time reveals the following three spectral features: (1) bleaching signal and derivative structure in the vicinity of the center photon energy of the pump and probe lasers at 1.97 eV; (2) spectrally broad bleaching signals in the region from 2.0 to 2.2 eV; (3) a negative dip in the absorbance change signal at 1.80 eV.

At delay times between -0.30 and $+0.05$ ps the photoinduced absorption spectra recorded at 297 K, shown in Figure 7.26, show a spectral structure which results from the combination of both a bleaching signal and a derivative structure near the spectral peak of the pump pulse at 1.97 eV. This spectral feature is accompanied by a maximum and minimum structures

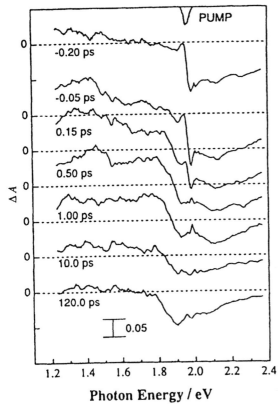

Figure 7.27. Photoinduced absorption, ΔA, vs. probe photon energy for seven pump–probe delay times for a PTV film induced by a 1.96 eV pump pulse at 77 K. The excitation photon density is 5.9×10^{14} photons cm^{-2}. The pump pulse spectral profile is shown for comparison.

probe free-induction decay, can contribute to the absorbance-change signal. Derivative structures near the pump photon energy, which are also observed in red-phase polydiacetylene, PDA-4BCMU, and P3DT were attributed to the effect of IPM.

A spectrally broad negative absorbance change signal is observed in the region from 2.0 to 2.2 eV at delay times between $t = -0.05$ and 0.15 ps recorded at both 297 and 77 K, as displayed in Figures 7.26 and 7.27, respectively. Population changes induced by the pump pulse at positive delay times produce a spectral bleaching signal due to the depopulation of the ground state. The bleaching observed until a delay time of about 50 fs is also due to the population of the free exciton, which is the final state of the relevant transition. Even though the stationary absorption spectrum is very broad and featureless, and laser spectrum is located at the tail of the spectrum, there is a very clear bleaching structure observed at 50 ps delay at room temperature, as shown in Figure 7.26. This indicates that the very broad stationary spectrum is mainly due to inhomogeneous broadening caused by various perturbation and defects in the sample. Hence, from now on the relaxation processes after photoexcitation of the exciton picture in terms not the band picture [47]. The transient response of the photoinduced bleaching signals at a probe photon energy of 2.4 eV, recorded with a photon density of 2.4×10^{15} photons cm^{-2} at 297 K, is displayed in Figure 7.28. The bleaching recovery time constant of $\tau = 0.48 \pm 0.05$ ps at 2.14 eV is obtained from the numerical fit to a single exponential decay function plus a constant component. As the pump laser intensity is increased from 5.9×10^{14} to 2.4×10^{15} photons cm^{-2} the signal amplitude is found to be proportional to the laser

at 1.97 and 2.00 eV, respectively. At delay times near $t = 0.30$ ps, since the probe pulse proceeds the rising edge of the pump pulse, the signals are not due to population changes. The small signal observed near the pump photon energy may arise from the process of pump perturbed probe-free induction decay discussed in three dimensions. If the pump excitation is slightly detuned from the resonant transition energy a derivative structure may result from pump perturbed probe free-induction decay. Near the delay time of $t = 0.05$ ps, nonlinear optical processes, such as pump polarization coupling and optical Stark effect [51,98–102] can contribute to the bleaching signals, and induced phase modulation (IPM) of the probe spectrum may be responsible for the derivative structures. At positive delay times, within the temporal overlap of the pump and probe pulses, all of the previously mentioned nonlinear optical processes, except pump perturbed

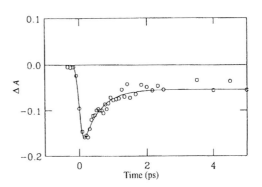

Figure 7.28. Time dependence of the absorbance changes, ΔA, corresponding to bleaching in PTV at 297 K recorded at photon energies of 2.14 eV. The excitation photon density is 2.4×10^{15} photons cm-2.

intensity, and only a reduction of the decay time constant smaller than 20% is observed.

The absorbance change spectra exhibit a peak minimum near 1.80 eV at delay times between $t = -0.15$ and $+0.10$ ps as shown in Figure 7.26. In contrast to the bleaching signals observed near 2.157 eV which are found to persist at delay times longer than 150 fs, the peak minimum signal at 1.80 eV can only be observed during the temporal overlap of the pump and probe beams. This signal is due to the Raman gain process taking place at the Stokes frequency of about 1410 cm^{-1}. The resonant spontaneous Raman spectrum of PTV was observed, with an excitation photon energy of 1.96 eV, to have an intense peak with a shift of 1409 cm^{-1}, and was assigned to the vinylene C=C stretching mode [103,104].

The imaginary part of the third-order susceptibility $\chi^{(3)}(-\omega_2; \omega_2, -\omega_1, \omega_1)$, where the subscripts 1 and 2 represent the pump and probe, respectively, is $\mathrm{Im}[\chi^{(3)}] = 1.9 \times 10^{-10}$ esu, for $\hbar\omega_1 = 1.97$ eV and $\hbar\omega_2 = 1.80$ eV. Raman gain signals have been observed in the femtosecond photoinduced absorption spectra of several polydiacetylenes and polythiophenes [51,52,98–100]. The occurrence of this gain signal indicates that the ground-state vinylene C=C stretching mode is significantly coupled to the π–π^* transition. Another minimum peak is also found to occur around time zero near 1.62 eV. This peal can be observed only when the sample cryostat is not evacuated and its assignment is presently unknown.

Kinetic fitting of the photoinduced transient absorption response, as discussed below, reveals an extremely rapid rise time component which cannot be temporally resolved and is faster than the 100 fs pulse duration of the laser. This initial rise time component is associated with the formation of the free exciton state. A second slightly slower rise time constant is found by examination of the positive absorbance changes at two different photon energies, as follows. In Figure 7.26 at a delay time of $t = 0.05$ ps, the amplitude of the positive absorbance change signal near 1.8 eV is small compared to the signal at lower photon energies near 1.4 eV. The broad absorption spectrum is attributed to transitions from the excited state to either higher excited states of $m^1 A_g (m > 2)$ symmetry and/or the continuum, i.e. the ionized state of the exciton. At the delay times from $t = 0.0$ to 0.3 ps, the ratio of the amplitudes of the absorbance-change signal at 1.8 eV to that near 1.4 eV increases with longer delay times. An accurate determination of the rise time constant of this spectrum is complicated by the contribution of the

pulse duration to the signal and is estimated to be 100–200 fs. This rapid spectral shift to higher photon energies has been observed in several one-dimensional conjugated polymers and is the signature of the geometrical relaxation to a neutral self-localized state called the STE state and includes a phonon emission process. Rise time constants of the transient absorbance changes measured in polydiacetylenes, PDA-3BCMU, and polythiophene, P3MT have been found to be 150 ± 50 and 70 ± 50 fs, respectively [51,52,90,98]. The absorbance change spectra measured at 77 K (Figure 7.27) are similar to those at 297 K (Figure 7.26) demonstrating that the rise time constant of the spectral shift of the photoinduced absorption does not significantly depend on the temperature as was found for several other polymers.

In Figure 7.29 the transient decay curve of the photoinduced absorption, recorded at six probe energies with a pump photon density of 2.4×10^{15} photons cm^{-2} at 297 K, is displayed as a semilogarithmic plot. In this plot a constant amplitude component, $\Delta A(l)$, which does not decay within 150 ps, has been subtracted. The slope of the semilogarithmic plot illustrates that there is significant deviation from linearity for the photon energies shown in the figure. Figure 7.30 shows that the transient decay curves can be numerically fitted to a biexponential decay function with an extremely long-lived component with constant absorbance change in the present observation time range, $\Delta A_1 \exp(-t/\tau_1) + \Delta A_2 \exp(-t/\tau_2) + \Delta A(l)$. The fitted decay constants τ_1 and τ_2 are shown in Table 7.2. Like the bleaching signals, the amplitudes of the photoinduced absorption

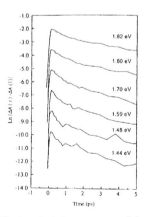

Figure 7.29. The transient decay curves of the photoinduced absorption for six different photon energies at 297 K displayed as a semilogarithmic plot; the long-time component has been subtracted.

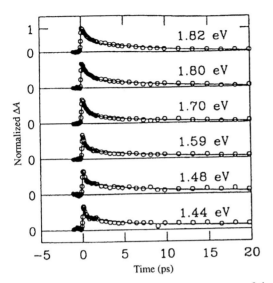

Figure 7.30. The normalized transient decay curves of the photoinduced absorption for six different photon energies at 297 K fitted to the biexponential decay function, $\Delta A_1 \exp(-t/\tau_1) + \Delta A_2 \exp(-t/\tau_2) + \Delta A(1)$. Here $\Delta A(1)$ is a constant amplitude absorbance change which does not decay within 150 ps.

signals are proportional to the pump laser intensity, which is varied from 5.9×10^{14} to 2.4×10^{15} photons cm^{-2}. Only a small decrease ($<20\%$) of the decay time constants is observed with increasing laser intensities, while the differences between τ_1 and τ_2 is much larger than 20% of τ_2. Thus, the dominant kinetic behavior is not a bimolecular decay process. Biexponential decay kinetics, also found in red-phase polydiacetylenes, PDA-4BCMU, and polythiophenes, P3MT and P3DT are interpreted according to a model of the STE relaxation process [51,52,90,98–100], as follows.

Table 7.2. The two decay constants, τ_1 and τ_2, are obtained from the biexponential decay fits of the transient absorbance changes recorded at six probe photon energies at 297 K. The uncertainties listed below represent one standard deviation error

Photon energy (eV)	Decay constant τ_1 (ps)	Decay constant τ_2 (ps)
1.82	0.86 ± 0.05	14.2 ± 1.4
1.80	0.73 ± 0.05	9.8 ± 1.0
1.70	0.56 ± 0.05	7.0 ± 1.0
1.59	0.36 ± 0.03	6.6 ± 0.9
1.48	0.21 ± 0.04	5.3 ± 0.8
1.44	0.16 ± 0.06	4.1 ± 0.9

The decay constant τ_1 is attributed to the relaxation of the hot STE to the ground state, including the emission of phonons occurring between vibrational levels strongly coupled to the excited transition. The slow decay constant τ_2 includes both the decay of thermalized STE states via tunneling to the ground-state potential-energy surface and a thermalization process. As seen in Table 7.2, both decay constants decrease with smaller probe photon energies. This is consistent with the STE relaxation model and suggests that there is a distribution of relaxation processes from the STE to the ground state which depends on the energy of the STE. Decay processes monitored at lower photon energies, corresponding to STE states with a significant population in high-lying vibrational states, may relax more efficiently into the ground-state potential-energy surface than population states which are lower in the potential-energy surface of the STE. The decay kinetics observed in PTV are quite similar to the previously observed dynamics in both polydiacetylenes and polythiophenes, with the following difference. Biexponential decay kinetics with $\tau_2 = 5 - 10$ ps have been reported in previous papers only in fluorescent polymers, such as polythiophenes and red-phase polydiacetylenes, while the slow decay constant in nonfluorescent polymers, such as blue-phase polydiacetylenes is 1.5–2.0 ps and the decay kinetics can be fitted to a single-exponential decay function. In contrast to these systems, the decay kinetics of the nonfluorescent polymer PTV, are not well-described by a single-exponential decay function, and can be fitted relatively well to a biexponential decay function.

In the photoinduced absorption spectra shown in Figures 7.26 and 7.28, a photoinduced bleaching signal is observed in the spectral region from 1.9 to 2.4 eV at $t > 5$ ps. The bleaching signal is observed at both the 297 and 77 K to have a lifetime much longer than 100 ps. The transient response of the bleaching signal at 1.90 eV, recorded at 77 K is displayed in Figure 7.31. The transient decay is numerically fitted to a biexponential decay function with a long-lived constant component. The fast decay component, $\tau_1 = 0.6 \pm 0.3$ ps, corresponds to the decay time constant of the initial bleaching signal. The slow component, $\tau_2 = 24 \pm 4$ ps, is the rise time constant of the long-lived bleaching signal. The formation time of this long-lived bleaching signal is discussed as follows. It is expected that some contributions of the long-lived induced absorption signal may be associated with polarons, and/or bipolarons. The fusion of two like-charge polarons can create a charged bipolaron. The photoinduced absorption spectrum recorded with chopped CW excitation at 488 nm, reported by Brassett *et al.*, revealed absorption peaks near 1.0 and 0.44 eV, which were attributed to sub-gap transitions associated with bipolarons [48]. Induced bleaching peaks were found near 2 eV. The lifetime of these features was estimated to be approximately 3 ms. Bleaching features in the electro-absorption measurements of PTV recorded at 100 kV cm^{-1} by Gelson *et al.*, were found in the same energy region as the bleaching structures in the CW photoinduced absorption spectrum [105]. The bleaching features from the electro-absorption spectrum were attributed to the electromodulation of the band edge due to local fields resulting from the presence of charged photoexcitations. The long-lived bleaching signal observed in the femtosecond experiment may result from the formation of long-lived excitations such as bipolarons and polarons, which are associated with the transfer of oscillator strength to these intragap levels. Thus, the time constant of 24 ps may correspond to the formation time of polarons and/or bipolarons by a polaron fusion process. Additional experimental studies of the polarization and intensity dependence of the long-lived bleaching are required in order to attribute the signal to either polaron or bipolaron formation. A derivative-structured spectral signal, which would be produced by electromodulation of the band edge from local fields due to bipolarons, is not readily observed in the long-lived bleaching signal, within the maximum of the recorded delay time; however, an electromodulation effect may contribute to the long-lived bleaching at much longer decay times.

3.8 Mechanism of exciton self-trapping in a one-dimensional system

Since there is no barrier in the potential curve between free and ST excitons in an ideal one-dimensional system as shown in Figure 7.32, the formation of the ST excitons is expected to take place within the period of the coupled phonon cycle $T_M = 2\pi/\omega_M$, the experimental results of the absorption spectra, the Raman gain [51,52,90], the resonance Raman scattering and the phonon-mediated optical nonlinearity show that the singlet excitons in polydiacetylenes are coupled with the C=C and C≡C stretching modes. The coupled stretching-mode frequencies are about 1500 cm^{-1} (C=C stretching) and 2100 cm^{-1} (C≡C stretching) and correspond to oscillation periods of 20 and 15 fs, respectively. Both of them are much shorter than the experimentally observed appearance times of ST excitons, which are 1507 fs in PDA-3BCMU, 100 fs in PDA-4BCMU, 70 fs in P3MT and 100 fs in P3DT.

In the following, the terminology 'self-trapping' is used in two different ways. One is the change in the potential curve of excitons which have the same energy as the free excitons. The self-trapping time estimated above from the oscillation period is used in this meaning. The other is the emission of phonons of the most strongly coupled mode to the excitonic transition. After this process the exciton energy is lower than that of the free exciton, and the population of the ST excitons is distributed near the bottom of the STE potential in Figure 7.32.

According to Jortner [84], the rate of the self-trapping process, of the second meaning, in rare gas

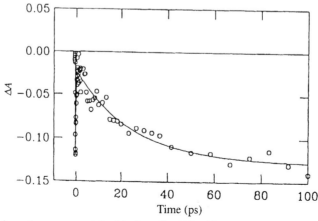

Figure 7.31. The transient absorption response, ΔA, of the long-lived bleaching signal recorded at a photon energy of 1.90 eV at 77 K. The fitted rise time constant is 24 ± 4 ps.

(a)   (b)

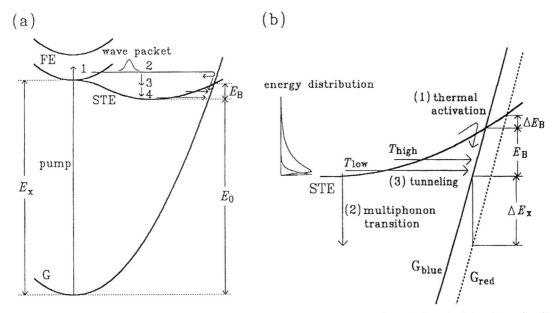

Figure 7.32. A model of the relaxation kinetics and the dependence on exciton energy shown in the potential surfaces of excitons in PDA. States 1, 2, 3 and 4 are free exciton (FE), nonthermal self-trapped exciton (STE), quasi-thermal STE and thermal STE, respectively. G_{blue} and G_{red} are the ground states of the blue-phase PDA and the red-phase PDA, respectively. Numbers (1)–(4) show the model decay kinetics of STE discussed in the text (Adapted from ref. 114.).

solids such as neon and argon at low temperatures is given by $k_{ST} = \omega_M \exp(-\omega_M/\omega_D)$, where ω_M and ω_D are the frequencies of the most strongly coupled mode and the Debye frequency, respectively. Using the values of ω_D in literature and ω_M given above, the calculated rate k_{ST} is several orders of magnitude smaller than the observed rate in the polymers [51,52]. This is because of the too-simple phonon structure in the model which is only suited to the rare gas solids. Much longer self-trapping time can be explained by the geometrical reorganization (relaxation) involving the lower-frequency modes than the oscillation period of the strongly coupled modes. The lower-frequency modes may include those associated with the structural change in the bulky side chains, which are lacking the rare gas solids.

There is difference in the formation time of ST excitons, namely the 'self-trapping' time of the second meaning, between PDAs and PTs. The formation times ar e 150 fs in PDA-3BCMU and 140 fs in PDA-4BCMU in the blue-phase and 70 fs in P3MT and 100 fs in P3DT. This can be explained by the difference in the chemical structures of these two groups of polymers. PDA-3BCMU and PDA-4BCMU have side groups with hydrogen bonds connecting neighboring side chains, while P3MT and P3DT have rigid

structures of thiophene rings conjugated with each other. The side chains of PTs have a helical noncoplaner structure due to the steric hindrance among the side groups [106,107]. The van der Waals interactions between the side chains, and between the main chain and the side chains are more intense in P3DT than in P3MT because of the longer side group in the former. This explains the longer formation time in the former.

The self-trapping time constant of excitons in P3MT is 70 fs, which only differs from the value (150 fs) for PDA-3BCMU by a factor of two; however, the differences in the structure and size of the side chains are very large. This may be due to the following three possible mechanisms. Mechanism (1): finite time is needed for the STE to slide down along the adiabatic potential because the potential curve of the STE deviates from the free-exciton potential bottom, which is flat with a zero derivative [89,108]. Mechanism (2): according to Sumi's theory concerning the self-trapping process in reduced dimensionalities [108], the spontaneous geometrical relaxation, which takes place from the infinitely extending free exciton to the STE, follows a reaction pathway with a very small slope along the reaction coordinate. Mechanism (3): because of the interchain interaction, the conjugated polymers may not be an ideal one-dimensional system. Then there may be a low but finite barrier between two minima in the potential curves of the ground state and the STE state.

3.9   Decay mechanism of self-trapped excitons in conjugated polymers

In this section the following properties are addressed concerning the excited state dynamics in conjugated polymers, especially polydiacetylenes.

(1) Many conjugated polymers are nonfluorescent or only weakly fluorescent. In particular the quantum efficiency of the nonfluorescent blue form of the polydiacetylenes is estimated to be lower than 10^{-5} and that of the fluorescent red-phase polydiacetylenes is only of the order of 10^{-4}

(2) There are common features in the transient absorption spectra of both PDAs (blue-phase PDA-3BCMU and PDA-4BCMU and red-phase PDA-4BCMU) and PTs either in random films or in oriented films. One feature is very broad absorption spectra in the 1.35–1.9 eV region observed just at excitation (0 ps).

(3) Triplet exciton cannot be formed directly from the singlet exciton. They are formed only at high-density excitation of exciton transition or by charge carrier creation by the interband transition.

(4) The formation and decay times and photoluminescence properties of the ST excitons in several PDAs and PTs studied are summarised in Table 7.3.

Figure 7.33 shows the time dependence of the absorbance change due to the STEs in several PDAs and PTs, after the component which does not decay within 150 ps has been subtracted. As can clearly be seen in the figure, the decay of nonfluorescent PDA-3BCMU in the blue phase and PDA-4BCMU in the blue phase are approximately given by an exponential function except for the early stage within 500 fs. The other fluorescent polymers have a nonexponential decay from a very early delay time until 30 ps. The initial decay times, defined by the slope of the semilogarithmic plot of the absorbance change against the delay time just after excitation, are determined as 890 ± 160, 620 ± 60 and 450 ± 50 fs in PDA-4BCMU, P3MT and P3DT, respectively, at 10 K. The decay becomes slower and slower at longer delay times because of the less efficient tunneling due to the thicker barrier width through which the population of excitons must tunnel to relax to the ground state.

The lifetime of the ST ^1B_u excitons in PDA-3BCMU is much shorter than the radiative life (a few ns) estimated from the oscillator strength of the transition. This means predominant radiationless relaxation in the exciton decay process. Excitons in many semiconductors and organic molecular crystals have lifetimes between several hundred picoseconds and several hundred nanoseconds at low temperatures. More than one order of magnitude differences in the lifetime of these excitons are usually found between low temperatures and room temperature. The activation potential barrier is very often invoked to explain the temperature dependence in most excitons in semiconductors, organic molecular crystals and also in ionic crystals.

Table 7.3. The formation time (τ_f), the decay time* (τ_d) of self-trapped (ST) excitons and the fluorescence of several polymers

Polymer	τ_f(fs) (290 K)	τ_d(ps) (10 K)	τ_d(ps) (290 K)	Fluorescence property[†]
PDA-3BCMU (blue, oriented film[‡])	150 ± 40	1.9 ± 0.2	1.6 ± 0.1	n
PDA-3BCMU (blue, cast film)	150 ± 50	2.0 ± 0.2	1.5 ± 0.2	n
PDA-DFMP (blue, single crystal)	100 ± 30	–	1.6 ± 0.1	n
PDA-4BCMU (blue, oriented film)	140 ± 50	2.1 ± 0.1	1.6 ± 0.1	n
PDA-4BCMU (red, oriented film)	120 ± 60	0.96 ± 0.09[§]	–	wf
PDA-4BCMU (red, cast)	< 200	0.88 ± 0.08[§]	–	wf
P3MT (electrochemical preparation)	70 ± 50	0.62 ± 0.07[§]	–	wf
P3DT (electrochemical preparation)	100 ± 50	0.45 ± 0.06[§]	–	wf

* The formation of the ST exciton corresponds to the emission process of a strongly coupled phonon to the excitonic transition.
[†] Polymers with about 10^{-4} and lower fluorescence quantum efficiency than 10^{-5} are indicated by wf (weakly fluorescent) and n (nonfluorescent), respectively.
[‡] The sample is prepared by the polymerization of an evaporated monomer film of PDA-3BCMU on a KCl crystal. The sample is composed of small crystalline and amorphous regions.
[§] This decay time is determined by the initial slope of the semilogarithmic plot in the early delay time shorter than 1 ps.

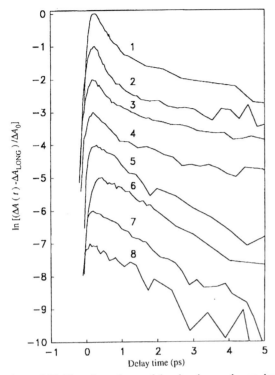

Figure 7.33. Time dependence of the absorbance change due to the ST excitons in several PDAs and PTs after the very long-lived component has been subtracted. The decay curves from the top to the bottom correspond to 1, P3MT (observed at 10 K at the probe photon energy of 1.63 eV); 2, P3DT (10 K, 2.34 eV); 3, PDA-4BCMU (red-phase cast film, 10 K, 1.77 eV); 4, PDA-4BCMU (blue-phase oriented film, 10 K, 1.77 eV); 5, PDA-4BCMU (blue-phase oriented film, 290 K, 1.77 eV); 6, PDA-3BCMU (blue-phase cast film, 290 K, 1.77 eV); 7, PDA-3BCMU (oriented single crystals and amorphous domains prepared by evaporation on a KCl single crystal surface (100), 290 K, 1.77 eV); and 8, PDA-DFMP (297 K, 1.98 eV). The excitation pulse width and energy are 100 fs and 1.97 eV, respectively. The samples of PDA-4BCMU in the red phase are excited by the two-photon process.

The relaxation of photoexcitations in polymers in general can be described as follows using PDAs and PTs as examples [51,52,90,100]. The differences in the relaxation kinetics of excitons in the nonfluorescent PDA-3BCMU and PDA-4BCMU in the blue-phase and fluorescent PDA-4BCMU in the red-phase, and P3DT and P3MT are discussed. The discussion is based on the Toyozawa theory [109,110]. Figures 7.34 and 7.35 show the potential curves of the ground-state (G) and self-trapped exciton (STE) and free-exciton (FE) band of the nonfluorescent (Figure 7.34) and fluorescent

(Figure 7.35) polymers, respectively. The curvature of the potential of STE is larger than those of the G and FE states in order to take into account the discussion on the reaction coordinate by Sumi *et al.* [108] and the small Stokes shift of fluorescent polymers. The electronic structure of the polymer systems are too complicated to be represented by the one-dimensional potential curve. The small Stokes shift may also be explained by the ground-state potential curve having two minima with a small barrier between two configurations corresponding to the acetylenic and butatrienic resonance structures, or to the aromatic and quinoid resonance structures in the cases of PDAs and PTs, respectively.

In Figures 7.34 and 7.35 the ST process from FE (indicated by the number 1 in the figures) to STE (indicated by 2) is the process 1→2, followed by the emission process (2→3) of a phonon of strongly coupled modes to the excitonic transition.

The STEs after the phonon emission process still remain in the nonthermal states (3). The nonthermal STEs (3) are coupled with phonon modes (intrachain vibrations) of low frequencies and thermalize to states, where temperatures of the intrachain vibration can be defined. The thermalization process (3→4) is observed as the spectral change at delay times from 0.5 to 5.0 ps. The time constant of the thermalization can be determined by the time dependence of absorbance change ratio among several different wavelengths. The obtained time constant of the thermalization process is about 17 ps in both red-phase PDAs and P3DT. However, the temperature of the STEs defined from equilibrated vibrational modes after this 1 ps thermalization process may be still higher than the bulk temperature of the polymer sample. Therefore, the cooling down process (4→5) with a time constant of several or a few tens of picoseconds is expected to take place after the observed thermalization process. However, the spectral change due to the cooling-down process of the STEs could not be clearly detected in this study because of the limited signal-to-noise ratio of the spectral data.

The change of the decay rate observed in the red-phase PDA-4BCMU and P3DT can be explained by the competition between the thermalization and tunneling processes. The STEs relax to the ground state mainly by tunneling through the barrier between the STEs and ground-state potentials. The initial fast decay is the relaxation from the nonthermal and quasi-thermal STEs (indicated by 3→4 in Figure 7.35) with a time constant of about 1 ps or slightly less. After thermalization, the STEs come down close to the bottom of the potential

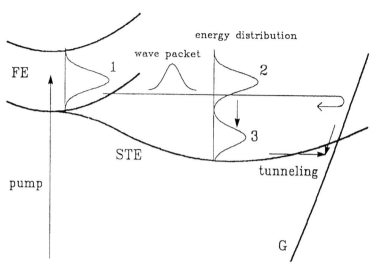

Figure 7.34. Potential curves of the ground state (G), free exciton (FE) and self-trapped exciton (STE) of the nonfluorescent PDAs. The meanings of numbers 1, 2 and 3 are the same as in Figure 7.32.

and the loss rate by the tunneling becomes slower because of the increase in the barrier height and thickness. The decay time constant of about 5 ps is due to the tunneling from the thermal STE (indicated by 4 in Figure 7.35). The thermalization process of the STE is estimated to be about 1 ps from the spectral change. The wavelength dependence of the decay curve can be explained by the thermalization process. The absor-

bance change at 2.00 eV in the red-phase PDAs is mainly due to the quasi-thermal STEs, and the thermal STEs have an absorption peak at 1.6 eV. Therefore, the decay curve at 2.00 eV can be fitted to a single-exponential function with a time constant of about 1 ps. The time constant of the absorbance change below 1.8 eV is longer than that of the bleaching, because the quasi-thermal STEs thermalize with a time constant of

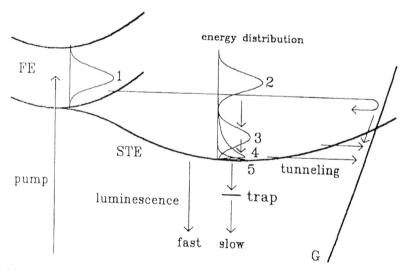

Figure 7.35. Potential curves of the ground state (G), free exciton (FE) band and self-trapped exciton (STE) of the fluorescent polymers. Numbers 1, 2 and 3 are the same as in Figure 7.32. Number 4 indicates the STE after thermal equilibration among intramolecular vibrational modes to a local temperature, which is higher than the bulk experimental temperature of the samples. Number 5 indicates the STE after the local temperature is lowered to the bulk experimental temperature.

1 ps and the absorption peak due to the thermal STEs appears at 1.6 eV. The observed data for P3MT and P3DT can be explained in the same way.

The decay of the nonthermal (but relaxed, namely after high-frequency phonon emission) STE in blue-phase PDAs is faster than in the red-phase, because the crossing point is lower than in red-phase PDAs. The major part of the STEs relaxes to the ground state before thermalization. The decay rate from the nonthermal STEs to the ground state is not much slower than the decay rate from the quasi-thermal STEs. Therefore, the decay curves in the blue-phase PDAs can be fitted to single-exponential functions. The wavelength dependence of the time constant is due to the mixed signals of the STE states. The bleaching signal exhibits unrelaxed and relaxed (i.e. after high-frequency phonon emission) but nonthermal STEs. while the absorption around 1.8 eV is due to the unthermalized STEs. Therefore, the observed decay time constant of the bleaching is shorter.

The difference between fluorescent and nonfluorescent polymers is in the time constants of the tunneling from the nonthermal and quasi-thermal STEs to the ground state. The time constants in the blue-phase are between 1 and 3 ps and they are shorter than that in the red-phase PDAs. Therefore the number of remaining STE at the bottom of the STE potential curves in the blue-phase PDA is smaller and the luminescence could not be detected. The fast component of the luminescence in the red-phase PDAs is considered to be due to the free excitons and/STEs and the slow component us due to the traps in the polymer chains. The detailed study using highly oriented samples is also consistent with the above model [96].

The proposed relaxation model predicts that the decay rate of the STE in conjugated polymers depends on the exciton energy. Conjugated polymers with large exciton energy are considered to have excitons with long lifetime and to be more fluorescent. For example poly(p-phenylenevinylene)(PPV), which has an absorption peak at 2.5 eV, has strong luminescence with a quantum yield of several per cent [111,112]. The much higher fluorescence quantum efficiency of PPV than red-phase PDAs may also be due to the absence of the decay channel to the lower-lying 2^1A_g state and due to non ideal one dimensionality in the polymer, caused by interchain interaction because of the small side groups.

Recently, a Langmuir–Blodgett (LB) film of PDA-(12,8), which has an exciton peak at 1.88 eV, has been studied and the decay time constant of the STE is estimated as 1.3 ± 0.1 ps at 290 K [98]. It is shorter than that in blue-phase PDAs, which have an exciton

peak at 1.97 eV. These results are consistent with the above proposed model. However, PDA under the hydrostatic pressure has different decay kinetics. When the pressure increases, the absorption edge of red-phase PDA-4BCMU shifts to lower energy but the decay kinetics becomes slower [113]. This pressure effect is probably due to a subsequent three-dimensional distortion of the polymer chain. The decay kinetics under pressure may be different from that in the one-dimensional system.

Thus low-fluorescence quantum efficiency of the one-dimensional polymers is systematically explained by the proposed model shown by Figures 7.34 and 7.35. In fluorescent PDA-4BCMU in the red-phase, and P3MT and P3DT, the crossing points are higher than in PDA-3BCMU and PDA-4BCMU in the blue phase, because of the higher energy of the free and ST excitons. However, in the red-phase PDA-4BCMU a larger fraction (probably 10 to several tens of per cent) of free excitons are relaxed to ST excitons with 3 ps lifetime at 10 K. This is shown in Figures 7.34 and 7.35 as fast tunneling through and passing through the crossing point of the potential curve between the ST exciton and the ground state. The ratio of the contribution of the tunneling and activation processes must be determined for each polymer by detailed experiments on the temperature dependence of the primary decay kinetics. The time constants of the various relaxation processes in three polymers are listed in Table 7.4.

3.10 Temperature dependence of decay kinetics

The photoinduced absorbance changes (ΔA) at 0.5 ps in PDA-(12,8) LB films are shown in Figure 7.36 with the pump spectra. Bleaching due to absorption saturation and broad photoinduced absorption below the absorption edge are observed. The photoinduced absorption has two peaks at 1.4–1.6 eV and just below the absorption edge. The peak around 1.4–1.6 eV and that near the absorption edge are assigned to the transitions from the lowest ^1B_u exciton to $m^1A_g(m > 1)$ excitons and to a composite exciton state, i.e. biexciton with ^1A_g symmetry, respectively [100].

The transient absorption spectra in PDA-(12,8)-PAA are shown in Figure 7.37. The sharp bleaching peak at 1.88 eV and associated two peaks at 2.04 and 2.18 eV are due to the saturation of the excitonic absorption and the phonon side bands, respectively. At 0.0 ps, three small minima due to Raman gain are clearly seen at

Table 7.4. Time constants of the relaxation processes of excitons at 290 K

Relaxation processes[*]	PDA-4BCMU (blue) (oriented film)	PDA-4BCMU (red) (oriented film)	P3DT
1–2 (self trapping)[†]	(10–20 fs)	(10–20 fs)	(10–20 fs)
2–3 (phonon emission)	140 ± 40 fs	120 ± 60 fs	100 ± 50 fs
3–4 (thermalization)	–[‡]	1.1 ± 0.1 ps	1.0 ± 0.2 ps
2–G (tunneling from unrelaxed STE)	< 1.0 ps	0.7 ± 0.1 ps[§]	0.3 ± 0.1 ps[§]
3–G (tunneling from nonthermal STE)	1.6 ± 0.1 ps		
4–G (tunneling from thermal STF)	¶	4.4 ± 0.5 ps[‖]	4.7 ± 1.2 ps[‖]

[*] 1–4 and G are the states shown in Figure 7.32.
[†] The time constants of the self-trapping processes could not be time resolved in this study.
[‡] The thermalization process could not be detected in the blue-phase PDA.
[§] The tunneling processes from the unrelaxed STE and from the nonthermal STE could not be separated.
¶ The tunneling from thermal STE in the blue-phase PDA-4BCMU could not be observed because the nonthermal STEs disappear before thermalization.
[‖] The decay times are determined by the absorbance change at longer delay time than 5 ps.

1.71, 1.62 and 1.44 eV. The corresponding Raman shifts, 2100, 2820 and 4270 cm^{-1}, are assigned to the stretching vibration of the C≡C bond and the overtones of the C=C and C≡C bonds, respectively. The coherent coupling between pump polarization and probe field which has been observed at the pump photon energy in blue-phase PDAs in previous papers [51.98] is not clear, because of the complexity introduced by the location of the pump photon energy between the lowest

exciton peak and the phonon side band. The photo-induced absorption at 0.0 ps is larger at lower probe photon energies down to 1.2 eV. Then the absorption has two peaks at 1.4 and 1.8 eV at delay times longer than 0.2 ps. The time constant of this spectral change is determined as 120 ± 50 fs from the transient response at 1.8 eV. A similar spectral change with a time constant of about 150 fs is also observed in blue- and red-phase PDA-(12,8)-Cd LB films and other PDAs [51,52,98,100,114].

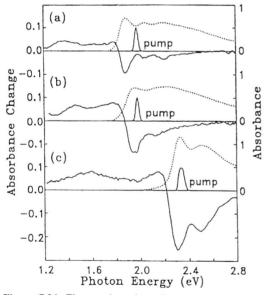

Figure 7.36. The transient absorption spectra at 0.5 ps in PDA LB films at 290 K. (a) PDA-(12,8)-PAA, (b) blue-phase PDA-(12,8)-Cd and (c) red-phase PDA-(12,8)-Cd. The absorption (dotted curves) and pump spectra are shown together. The pump photon densities are (a) 9.2×10^{14}, (b) 1.8×10^{15} and (c) 2.5×10^{15} photons cm^{-2} [114].

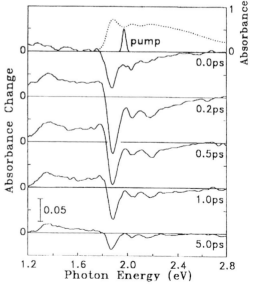

Figure 7.37. The transient absorption spectra of PDA-(12,8)-PAA at 290 K. The absorption (dotted curve) and pump spectra are shown together [114].

The transient responses of the bleaching in the PDA LB films are plotted in a semilogarithmic scale in Figure 7.38. Here, long-lived components (ΔA_c) due to triplet excitons are subtracted from the observed absorbance changes [114]. The decay curves do not fit a single-exponential function. The decay times defined by the initial slope of the semilogarithmic plot at the delays between 0.2 and 0.6 ps are shorter than 1 ps, and ΔA decays more slowly after this rapid decay. The time constants defined by the decay curves at delay times longer than 27 ps are summarized in

Table 7.5. The decay time constant becomes shorter at higher temperatures and in PDA with smaller exciton energy.

Since the ultrafast relaxation processes in PDAs are insensitive to the sample morphology [114] the exciton energy dependence of the decay kinetics can be discussed using a model shown in Figure 7.32. Here, the potential curves of the FE and STE are assumed to be independent of the exciton transition energy E_x. Polyenes have the 2^1A_g state below the 1^1B_u state, [115] but the 2^1A_g state in longer polyene systems is

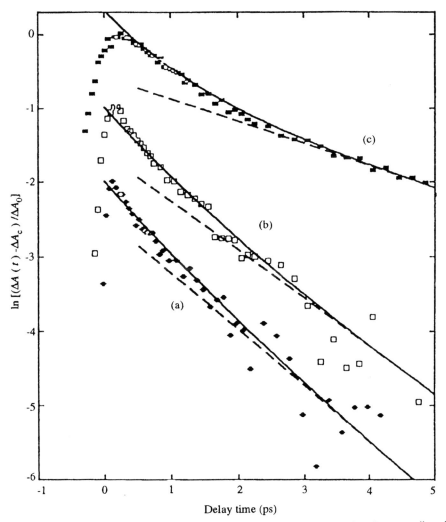

Figure 7.38. The time dependence of the absorbance changes in PDA LB films at 290 K plotted on a semilogarithmic scale. (a) PDA-(12,8)-PAA (closed circles), (b) blue-phase PDA-(12,8)-Cd (open squares) and (c) red-phase PDA-(12,8)-Cd (closed squares). The probe photon energies are (a) 1.88, (b) 1.94 and (c) 2.30 eV. The dashed lines are single-exponential functions obtained from the data at delay times longer than 2 ps for (a) and (b) and than 3 ps for (c). The solid curves are decay kinetics simulated using equations (7.19)–(7.21) with $T_0 = 3000$ K [114].

Table 7.5. Decay time constants obtained from the data at delay times longer than 2 ps in PDA-(12,8)-PPA and PDA-(12,8)-cd (blue) LB films and longer than 3 ps in PDA-(12,8)-Cd (red) LB films at several temperatures

PDA LB films	Exciton energy E_x (eV)	Decay time constant (ps)			
		10 K	100 K	200 K	290 K
PDA-(12,8)-PAA	1.88	1.8 ± 0.1	1.7 ± 0.1	1.5 ± 0.3	1.3 ± 0.1
PDA-(12,8)-Cd (blue)	1.97	2.1 ± 0.1	1.8 ± 0.1	1.6 ± 0.1	1.5 ± 0.1
PDA-(12,8)-Cd (red)	2.32	6.1 ± 1.2	4.2 ± 0.5	3.9 ± 0.6	3.3 ± 0.5

expected to separate to form two ^3B_u states [116]. The observed long-lived triplet excitons in PDAs are not created directly via the 1^1B_u exciton that is generated by single-photon excitation of the 1.97 eV pump pulse, but are created by fission of a higher excited singlet exciton generated by the Auger process of two 1^1B_u excitons [53]. Therefore, the relaxation rate from the 1^1B_u exciton to the 2^1A_g exciton is considered to be much smaller than the relaxation rate to the ground state. The 2^1A_g state in PDA can be neglected in the ultrafast relaxation of the 1^1B_u exciton.

The spectral change of the photoinduced absorption with a time constant of about 150 fs is due to the geometrical relaxation of the FE to the STE. Since the formation process of the STE has no barrier in the one-dimensional system, the photoexcited FE (state 1) is coupled with the C–C stretching modes within the phonon periods of 10–20 fs. However, the STE has not relaxed to the bottom of the potential curve and the stabilization energy remains as the kinetic energy of the lattice vibration [nonthermal STE (state 2)]. The nonthermal STE then relaxes to the quasi-thermal STE (state 3) coupled with vibrational modes within a single chain of polymer. The observed spectral change with a time constant of about 150 fs is considered to be due to this process. Slow spectral change with a time constant of 1.1 ± 0.1 ps, observed in red-phase PDA-4BCMU, corresponds to relaxation from the quasithermal STE to the thermal STE (state 4) [98]. There are three possible mechanisms of nonradiative relaxation in the decay kinetics from the STE to the ground state as shown in Figure 7.32b i.e. (1) thermal activation, (2) multiphonon transition and (3) tunneling in the configuration space.

(1) If the relaxation from STE to the ground state is mainly due to passing through the crossing point shown in Figure 7.32b(1), the temperature dependence of the decay rate is given by

$$R_a(T) = R_{a0} \exp(-E_B/kT) \qquad (7.12)$$

where E_B is the activation energy to pass through the crossing point and R_{a0} is the frequency of the activation process. Using the data in Table 7.5, E_B and R_{a0}^{-1} in PDA-(12,8) LB films are obtained as (a) 0.18 emV (2.1 K) and 1.5 ps, (b) 0.24 eV (2.8 K) and 1.6 ps, (c) 0.45 meV (5.2 K) and 3.7 ps, respectively. The weak temperature dependence gives a very small activation energy. The fastest decay rate in the activation process ($kT \gg E_B$) is R_{a0}, but the observed initial decay rate is faster than the obtained decay rate R_{a0}. Therefore, the relaxation via the crossing point is concluded not to be a major relaxation process.

(2) The ultrafast nonradiative decay in *cis*-polyacetey-lene was once explained by multiphonon transition [117]. The temperature dependence of the multi-multiphonon transition rate is approximately given by

$$R_m(T) \propto \left[\frac{1}{\exp(\hbar\omega/kT) - 1} + 1 \right]^{E_0/\hbar\omega} \qquad (7.13)$$

where $\hbar\omega$ is phonon energy and E_0 is the energy difference between the bottoms of the STE potential curve and the ground state [118]. Since the corresponding temperature of the phonon energy in PDA is much higher (0.19–0.26 eV = 2200–3000 K) than the experimental temperature, the difference between multiphonon transition rates at 10 and 290 K is very small and estimated to be less than 1%. Therefore, the multiphonon transition is not a main relaxation process.

(3) The main relaxation process from STE to the ground state is considered to be due to tunneling in the configuration space from the STE potential to the ground state [Figure 7.32b(3)] and the following phonon emission process from high vibrationally excited levels of the ground state. The nonradiative transition probability from the excited state $|j,x\rangle$ to the ground state $|j+p,g\rangle$ with the

same energy, where both j and $j+p$ specify the vibrational levels, was calculated considering the overlap of the vibrancy wavefunctions [119,120]. The transition rate was obtained as

$$W_j = C|\langle j, x|j+p, g\rangle|^2$$

$$= C \exp(-S)S^p \frac{j!}{(j+p)!}[L_j^p(S)]^2 \quad (7.14)$$

where C is a physical factor, $p = E_0/\hbar\omega$, $S = (E_x - E_0)/\hbar\omega$, and L_j^p is the Laguerre polynomial. Since S and j in PDA are expected to be much smaller than $p(S, j \ll p)$, the transition rate is approximately given by

$$W_j \sim W_0 \frac{(j+p)!}{j!p!} \sim W_0 p^j \quad (j = 0\text{-}1). \quad (7.15)$$

This is consistent with the tendency that the STE with a higher energy from the bottom of the potential curve has a faster tunneling rate because of a narrower barrier width. Since STEs in conjugated polymers are coupled with both intrachain and interchain phonon modes, the STE energies are expected to be distributed continuously. Therefore, the transition rate of STE with energy E is assumed as

$$W(E) = W_0 \exp(\alpha E) = W_C \exp[\alpha(E - E_B)]$$
$$(E < E_B)$$

$$W(E) = W_0 \exp(\alpha E_B \equiv W_C \quad (E > E_B)$$
$$(7.16)$$

where $\alpha = \log p/\hbar\omega$. W_C is the transition rate of STE with an energy larger than the crossing point and it is assumed to be constant and independent of the exciton transition energy. The energy distribution of the thermal STE is defined by the temperature. At higher temperatures the thermal STEs are distributed to higher energies and the transition rate is faster. The relaxation rate of STEs at temperature T is calculated as

$$R_t(T) = \frac{\int_0^\infty n(E)W(E)dE}{\int_0^\infty n(E)dE}$$

$$= W_C \frac{\exp(-\alpha E_B - \alpha kT \exp(-E_B/kT))}{1 - \alpha kT},$$
$$(7.17)$$

where the distribution of the exciton, $n(E)$, is assumed to be $\exp(-E/kT)$. Since the position of the ground state shown in Figure 7.32 depends on the exciton energy E_x, the height of the crossing point depends on the exciton energy. When the exciton energy increases by ΔE_x (solid curve G_{blue}

to dotted curve G_{red} in Figure 7.32b, the crossing point becomes higher by ΔE_b. When the change in the exciton energy is small, the barrier height is given by

$$E_B = \frac{\Delta E_B}{\Delta E_x}(E_x - E_{x0}) \quad (7.18)$$

The parameters are determined by least-square fitting using the data in Table 7.5 as $W_C^{-1} = 0.98 \pm 0.28$ ps, $\alpha^{-1} = 0.066 \pm 0.016$ eV, $\Delta E_B/\Delta E_x = 0.15 \pm 0.03$, and $E_{x0} = 1.62 \pm 0.14$ eV. E_{x0} is the exciton transition energy with zero barrier height. However, equation (7.18) is not a good approximation for $E_x \sim E_{x0}$. The actual exciton energy with zero barrier height is smaller than the E_{x0} obtained. The value of α^{-1} is calculated from the relation between equations (7.15) and (7.16) using $p = 10$ and $\hbar\omega = 0.19$ eV as 0.083 eV. This is consistent with the value obtained from the experimental results. The potential barrier height (E_B) and the relaxation time from the bottom of the STE potential curve (W_0^{-1}) in PDA LB films are obtained, respectively, as (a) 0.04 eV and 1.8 ps (b) 0.05 eV and 2.1 ps and (c) 0.11 eV and 5.2 ps. The decay time constant in yellow-phase PDA with an exciton energy of 2.6 eV is calculated as 6.2 ± 2.1 ps at 293 K. This is consistent with the fluorescence lifetime of 9 ± 3 ps obtained in PDA-3KAU(poly[4,6-decadiyne-1,10-diol-bis(carboxylmethylurethane)] yellow-phase solution [27].

The main relaxation process of STE is concluded to be due to the tunneling in the configuration space.

The time dependence of the bleaching shown in Figure 7.38 can be explained by the thermalization processes of the STE. Since the pump pulse excites the excitonic absorption peak, the distribution of the nonthermal STE is defined as $n_{nonth}(E) = \delta[E - (E_x - E_0)]$. Then the nonthermal STE relaxes to the quasithermal STE with time constant $\tau_{nonth} = 150$ fs. The temperature of the quasi-thermal STE decreases with time constant $\tau_{quasi} = 1.1$ ps, and the relaxation rate to the ground state becomes slower. Therefore, the rate equations for the populations and temperature change are given by

$$\frac{dN_{nonth}}{dt} = -\left[W(E_x - E_0) + \frac{1}{\tau_{nonth}}\right]N_{nonth} \quad (7.19)$$

$$\frac{dN_{quas}}{dt} = \frac{N_{nonth}}{\tau_{nonth}} - R_t(T)N_{quas} \quad (7.20)$$

$$T = T_0 \exp(-t/\tau_{quas}) + T_{therm} \quad (7.21)$$

where N_{nonth} and N_{quasi} are the populations of the nonthermal STE and the quasi-thermal STE respectively; T_{therm} is the experimental temperature, and the initial temperature increase; T_0, is assumed to be $(E_x - E_0)/k$ from the stabilization energy of the STE. Since the population of thermal STE corresponds to N_{quasi} at T_{therm}, the photoinduced bleaching signals are proportional to the sum of N_{nonth} and N_{quasi}. The calculations with $T_0 = 3000$ K give well-fit curves for each PDA, as shown in Figure 7.38. The stabilization energy $E_x - E_0$ is determined as 0.26 eV. It is larger than the potential barriers in both blue- and red-phase PDAs. PDAs in both phases correspond to type II in Toyozawa's classification [109] which is consistent with a very low quantum efficiency of fluorescence. Su

has calculated the stabilization energy of the triplet exciton polaron (triplet STE) in PDA as 0.46 eV [121].

To summarize, the exciton energy dependence of the ultrafast relaxation processes has been investigated in the PDA-(12,8) LB films with three exciton transition energies. The relaxation of the exciton is faster in PDA with a lower exciton energy. The decay kinetics are discussed using the model potential curves of FE, STE and the ground state. The main relaxation mechanism from STE to the ground state is the tunneling in the configuration space. The dependence of the decay kinetics on the exciton energy and temperature is well explained. The stabilization energy of STE and the barrier height between STE and the ground state are estimated.

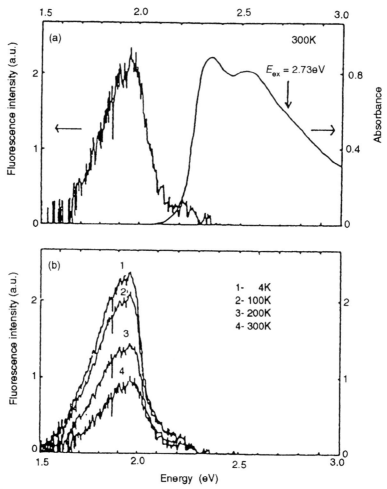

Figure 7.39. (a) The absorption and fluorescence spectra of a cast film of PDA-4BCMU at room temperature. (b) Temperature dependence of the fluorescence spectrum of the cast film of PDA-4BCMU. The energy of the excitation photons from the argon-ion laser is 2.73 eV [51].

3.11 Time-resolved fluorescence spectra of cast films of PDA-4BCMU

The fluorescence and absorption spectrum of a PDA-4BCMU cast film at room temperature is shown in Figure 7.39a; the fluorescence spectra at several temperatures between 4 and 300 K are shown in Figure 3.39b. At room temperature, the absorption spectrum of PDA-4BCMU film is broad with a peak of around 2.35 eV. The fluorescence spectrum, with a peak around 1.96 eV at all temperatures of measurement, is approximately a mirror image of the absorption spectrum. The excitation energy for a single-photon-counting measurement was 2.18 eV (570 nm), while the 2.73 eV (455 nm) CW line from an argon laser was used for a stationary spectrum.

The time dependence of the fluorescence intensity observed at 620 nm (photon energy = 2.0 eV) at 47 K is shown in Figure 7.40. The initial part of the decay curve is very steep in the linear plot of intensity shown in Figure 7.40a. In the semilogarithmic plot (b), the detectable signal exists even after 10 ns and the decay is not single exponential. A logarithmic plot (c) of the decay curve is close to being a straight line with a slope of about −2.

Figure 7.41 shows the time-resolved fluorescence spectra measured at 47 K by single-photon-counting at intervals of 2 nm. In the large-delay time region, the spectrum has a smaller signal-to-noise ratio, but it can be said that in each time region the fluorescence spectra are very similar to each other. At temperature below 200 K, the time dependence of fluorescence intensity is independent of the probe wavelength, but at room temperature it depends on the wavelength. The fluorescence decay below 200 K can be expressed in terms of a single exponential.

3.12 Models for the explanation of fluorescence decay kinetics

Many models have been proposed to explain the nonexponential relaxation of elementary excitations or excited species in solids, especially in amorphous solids or molecular systems. PDAs are known to have a quasi-one-dimensional structure. One-dimensional relaxation models are now considered. These are classified into two groups, single dimensional and fractal dimensional. There are two relaxation models of disappearance of the elementary excitations or excited species after one-dimensional diffusion due to random walk. One of these is a one-dimensional geminate recombination

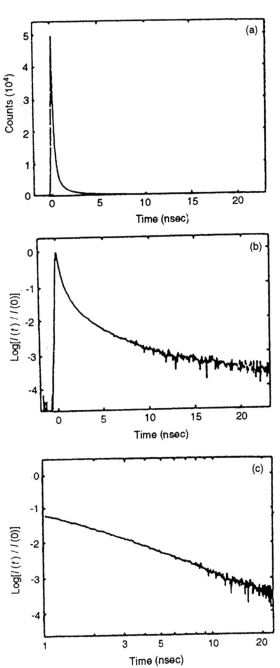

Figure 7.40. The typical time dependence of the fluorescence intensity from the PDA-4BCMU film at 4 K. Excitation photon energy and observed photon energy are 2.18 and 2.00 eV, respectively: (a) linear, (b) semilogarithmic and (c) logarithmic plot [51].

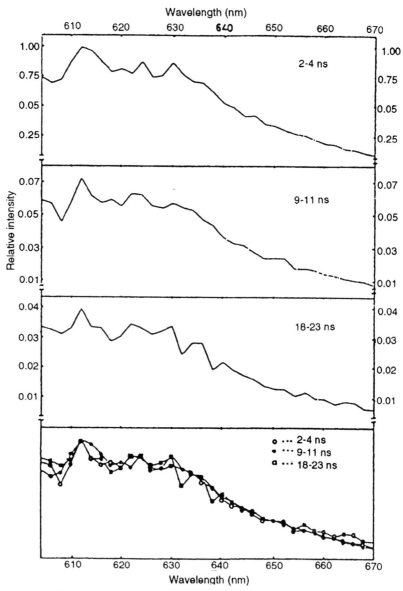

Figure 7.41. The time-resolved fluorescence spectra of the PDA-4BCMU film at 4 K. The excitation photon energy is 2.18 eV [51].

model, which was adopted to describe the decay of the photoinduced absorption (PIA) in *trans*-polyacetylene (PA) films in a sub-picosecond to picosecond time region [5,8]. Shank *et al.* [5] attributed the origin of the relaxation to geminate recombinations of solitons after one-dimensional random walk on OPA chains. The author [8,11] invoked geminate recombinations of two polarons generated on two neighboring chains to explain successfully the experimental results of photo-induced absorption in PA films in the nanosecond region at several temperatures. In the one-dimensional geminate recombination model, the decay function is of the form $\mathrm{erf}[(t/\tau)^{-\frac{1}{2}}]$, which is approximated very well by $t^{-\frac{1}{2}}$ for $t \gg \tau$. If the decay function has the form $t^{-\frac{1}{2}}$, in a logarithmic plot, the gradient of the decay curve must be $-\frac{1}{2}$. Since, in the PDA-4BCMU film, the gradient of the decay curve is about -2 as shown in Figure 7.40, the decay of the emission cannot be described by the one-dimensional geminate recombination.

The other one-dimensional diffusion model is a one-dimensional trap model. Movaghar et al. [113] obtained an exact analytic solution for the survival fraction in the one-dimensional diffusion problem with randomly distributed deep traps. According to their result the time dependence of fluorescence intensity $I(t)$ for long time periods at the small trap concentration is given by the following equation

$$I(t) = 8(4x^2 Wt/3\pi)^{1/2} \exp[-3(\pi^2 x^2 Wt/4)^{1/3}]$$
$$\text{for } t > W^{-1} \quad (7.22)$$

Here x is the trap concentration represented by the probability of finding a trap at each site in the single-dimensional PDA chains and W is the hopping rate. The asymptotic long-time behavior of the one-dimensional deep-trap problem can be approximated to $\exp[-(x^2 t/t_0)^{-1/3}]$. The decay of the photocurrent in a PDA-1OH film has been observed in the time range between 1×10^4 and 2×10^4 s [122]. The measurements are fitted extremely well by an $\exp[-(x^2 t/t_0)^{-1/3}]$ law. Bloor et al. [17] tried to fit the observed luminescence decay from PDA-1OH to $\exp[-(t/\tau)^{-1/3}]$. If the time dependence of the fluorescence from the PDA-4BCMU film measured in this study can also be represented by the same function $\exp[-(t/\tau)^{-1/3}]$ with an approximate τ, the decay curve must become straight in a $\ln[I(t)] - t^{-1/3}$ plot. However, the decay curve deviates slightly from a straight line in the region of observation as shown in Figure 7.42; this region is wider than that in Figure 7.40. The PDA-1OH samples used by Bloor et al. [17] are films of highly oriented polymer fibers. However the PDA-4BCMU films used in this work were $CHCl_3$ cast films. The X-ray diffraction photographs of the PDA-3BCMU/$CHCl_3$ cast film and the stretched film indicate that the former is not oriented, but that the latter is oriented.

Another model is the relaxation on a fractal structure. The annihilation reactions $A + A \rightarrow 0$ and $A + B \rightarrow 0$ after random walk on a fractal structure, where A and B are different elementary excitations or excited species in solids or molecular systems, have been examined by Blumen et al. [123]. According to them, the $A + A \rightarrow 0$ annihilation pattern varies as $1/S(t)$ with $S(t)$ given by

$$S(t) \propto t^{\tilde{d}/2} \; (\tilde{d} < 2)$$
$$S(t) \propto t^{\tilde{d}} \; (\tilde{d} > 2) \quad (7.23)$$

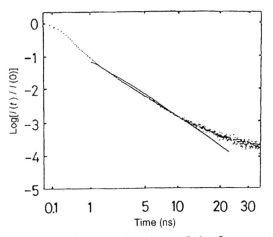

Figure 7.42. The time dependence of the fluorescence intensity from PDA-4BCMU at 4 K plotted as log(intensity) vs. $t^{1/3}$. The excitation photon energy and observed photon energy are 2.18 and 2.00 eV, respectively [51].

Here $\tilde{d}$ is the spectral dimension of the fractal. On the other hand the bimolecular reaction $A + B \rightarrow 0$ is dominated by density fluctuations. At long times, the decay is expected to follow a $t^{-\tilde{d}/4}$ law instead of a $t^{-\tilde{d}/2}$ behavior (for $\tilde{d} < 2$). However the gradient of the slope in Figure 7.40c is about -2. Therefore the relaxation in the PDA-4BCMU film in this study cannot be explained in terms of the above-mentioned bimolecular reaction models.

The time-dependent survival probability, $I(t)$, in disordered materials was explained by a random walk on the fractal. According to the authors, $I(t)$ is given in a short-time region by

$$I(t) = \exp[-\lambda a t^{\tilde{d}/2} + \lambda^2 b t^{\tilde{d}/2}] \quad (7.24)$$

where $\lambda = -\ln(1 - p)$, a and b are constants depending on the fractal, and p is the probability with which a site in the chain is occupied by a trap. Numerical calculations showed that equation (7.24) for $\tilde{d} = 1$, is a good approximation for the number of steps less than 1000 [124]. Long-time properties of trapping on fractals have been derived [125,126]. The asymptotic decay obeys the equation

$$I(t) = \exp[-C_3 p^{2/(\tilde{d}+2)} t^{\tilde{d}/(\tilde{d}+2)}] \quad (7.25)$$

Here C_3 is given by

$$C_3 = [(2/\tilde{d})^{\tilde{d}/(\tilde{d}+2)} + (\tilde{d}/2)^{2/(\tilde{d}+2)}]$$
$$\times C_{\tilde{d}}^{2/(\tilde{d}+2)} (\gamma_{\tilde{d}}/\tau_0)^{\tilde{d}/(\tilde{d}+2)} \quad (7.26)$$

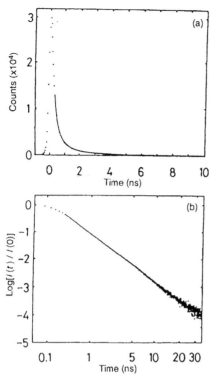

Figure 7.43. The time dependence of the fluorescence intensity from PDA-4BCMU at 4 K in (a) linear (b) log(intensity) vs. t^c ($c = 0.2$). The excitation photon energy and observed photon energy are 2.18 and 2.00 eV, respectively [51].

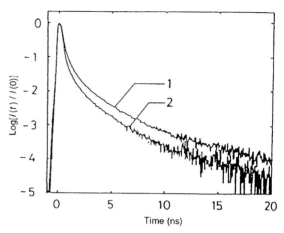

Figure 7.44. The temperature dependence of the time dependence of fluorescence intensity from the PDA-4BCMU film. Excitation photon energy and observed photon energy are 2.18 and 2.00 eV, respectively. 1: 4 K, 2: 200 K [51].

where $\bar{d}$ is the Hausdorff dimensionality, C_d is a number depending on the dimension $\bar{d}$ [$\lambda_2 = z_0^2$ with $z_0 = 2.4048$, the first zero of the Bessel function $J_0(z)$; $\gamma_3 = \pi^2$ etc.], and τ_0 is a jump time [127]. As shown in Figure 7.43 $\exp[-(t/\tau)^c]$ gives a good fit to the fluorescence decay. The decay of the fluorescence from the PDA-4BCMU film seems to be explained approximately by the trapping model for long durations. Constants c and τ, obtained by computer calculation, were 0.2–0.3 and 1×10^{-5}–1×10^{-4} s, respectively, and depend on the temperature. Figure 7.44 shows the temperature dependence of the time dependence of the fluorescence intensity. At lower temperatures, the decay curve has more contributions from slowly varying components. The spectral dimensionality $\tilde{d} = 2c/(1 - c)$ was obtained as a parameter by fitting the calculated curve with equations (7.25) and (7.26). The obtained value changes between 0.50 and 0.85 at temperatures between 4 and 200 K.

Equation (7.25) is a good approximation when $t \gg \tau/\lambda_{\bar{d}}$. In all cases, the maximum value of τ is 10 ps, and $\lambda_{\bar{d}}$ is of the order of about 10. Equation (7.25) is a good approximation to describe the fluorescence decay of the PDA-4BCMU film.

By applying the fractal-dimension model, spectral dimensions were obtained to be between 0.50 and 0.85. By the continuous random-walk theory for the space with Hausdorff dimension of $\bar{d}$ and spectral dimension of $\tilde{d}$, the squared average distance traveled in time t is given by [128]

$$\langle r(t)^2 \rangle \approx t^{\tilde{d}/\bar{d}} \tag{7.27}$$

And is well known, the squared average distance in the Euclidean lattice is given by

$$\langle r(t)^2 \rangle \approx t \tag{7.28}$$

On the assumption that the Hausdorff dimensionality of the PDA-4BCMU film is close to $5/3$ as determined from many other polymers, the squared average distance in the PDA-4BCMU film is given by

$$\langle r(t)^2 \rangle \approx t^c \qquad \text{with } 0.3 < c < 0.51 \tag{7.29}$$

If the PDA-4BCMU film is truly described by the fractal, the excited states in the film occupy a smaller area than in the Euclidean lattice. The question arises of why the temporal behavior of the fluorescence intensity at room temperature depends on the observation wavelength. One possible explanation is that at room temperature the hydrogen bonds are apt to be broken.

3.13 Defects on cast PDA-4BCMU film

An important question is whether the carriers relax due to intrinsic defects or are trapped at extrinsic defects. Bloor *et al.* [17] concluded that the fluorescence in PDA-1OH is due to recombination at extrinsic sites, because their experimental data fitted well the one-dimensional trap mode. Wong *et al.* [45] suggested that the relaxation dynamics in PDA-1OH are strongly influenced by the structure of a few recombination sites including cross-linkages and dangling at chain ends, because three decay functions were needed to describe the observed fluorescence decay. Sixl and Warta [28] attributed the emission from PDA-FBS (R = $CH_2SO_3C_6H_4F$, *p*-fluorobenzene sulfonate), which has short conjugation lengths, to a polaron recombination. In addition, they observed weak exciton fluorescence.

Since the PDA-4BCMU cast film from CHCl$_3$ solution is not highly oriented and a chain of PDA-4BCMU forms a random coil in the CHCl$_3$ solution, the conjugation lengths of PDA-4BCMU in the cast film and solution are relatively short, as is known from their absorption spectra. In the case of the PDA-3BCMU cast film, the conjugation lengths are known to be relatively long from their absorption spectra. The lowest energy expected for optical absorption in an ordered polymer crystal is 2.0 eV (620 nm) from the theoretical calculation for the infinite conjugation length limit. The difference in the conjugation lengths between PDA-3BCMU and PDA-4BCMU is induced by stronger hydrogen-bond formation between side groups in the former than in the latter. The hydrogen-bond length in PDA-3BCMU is shorter than in PDA-4BCMU (Figure 7.45), since the N−H and C=O bonds are collinear to each other in the former and noncollinear in the latter. Sixl and Warta [28] proposed that in a shorter chain the butatrienic structure is more stable than the acetylene-like structure and in longer chains the situation is reversed. Four possible defects in PDA-4BCMU are shown in Figure 7.46. They are described as follows: (a) the conjugation of π-electrons on the backbone is broken by rotation around the single bond; (b) there are radicals at the chain ends; (c) there are radicals inside the chain; (d) there are oxygen defects. Among these (a), (b) and (c) are intrinsic defects and (d) is extrinsic, and the former are more probable than the latter. In the case of (d), it is difficult for oxygen atoms to come close to a carbon atom on the main chain because of the bulky side groups. Since the microscopic and macroscopic structures of PDA-4BCMU films are complicated unlike crystalline

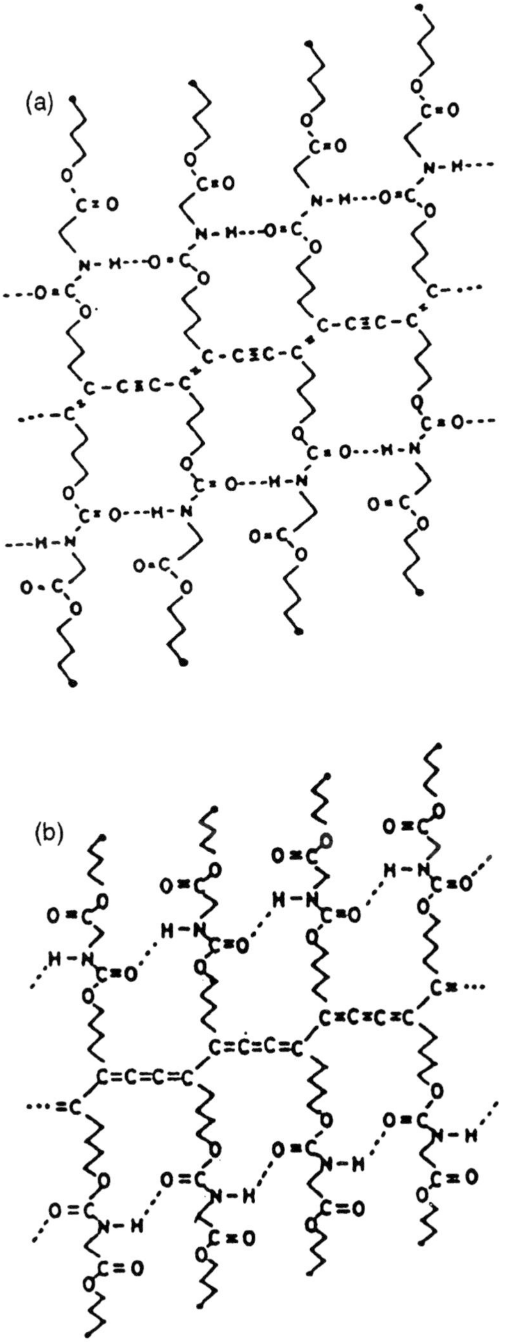

Figure 7.45. Hydrogen-bonded conformation of (a) PDA-3BCMU and (b) PDA-4BCMU as suggested by molecular models. The dashed lines between N−H and C=O indicate hydrogen bonds [51].

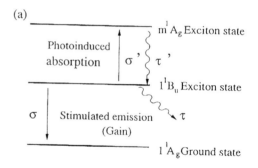

(a)

(b)

Figure 7.46. Schemes of the defects on the backbone chain of PDA: (a) rotation around single bonds; (b) radical in the chain end; (c) radicals inside the chain; (d) oxygen bonded [51].

PDAs, generation of the defect (a) seems to be more probable in the PDA-4BCMU films in crystalline PDAs.

3.14 Fluorescence spectra of a blue-phase PDA-C$_4$UC$_4$ obtained by probe saturation spectroscopy

In this section, the fluorescence of blue-phase PDA-C$_4$UC$_4$ is investigated by PSS (probe station spectroscopy). PDA films have two phases called the blue phase and the red phase according to their color as mentioned before. For red-phase PDAs, the stationary fluorescence spectra and the time dependence of fluorescence are observed [27,51,98]. For blue-phase PDAs, however, fluorescence spectrum has not been observed and the quantum efficiency is estimated to be extremely low ($<10^{-15}$) and cannot be detected by a conventional fluorometer. The fluorescence spectrum of blue-phase PDA has been determined by probe saturation spectroscopy (PSS) developed by the author's group for the first time.

The center wavelength of the amplified pulses is 630 nm (corresponding to 1.9 eV). The intensities of the pump and probe pulses are varied separately by neutral density filters. In the present study, the maximum probe photon density is 300-times larger than the minimum density and 2.7×10^{14} photons cm^{-2} integrated between 645 and 655 nm. The probe white continuum generated by self-phase modulation is detected with a polychromator/multichannel detector system and analyzed by a minicomputer. The thickness of the films of blue-phase PDA used in the experiment was typically about 60 nm.

The contributions of photoinduced absorption, bleaching of stationary absorption and stimulated emission generally coexist in transient difference transmission spectra near the optical gap region. They usually cannot be separated in the conventional pump–probe spectroscopy. However, if the dependence of their responses on probe light intensity is different, they can be separated by observing transient difference transmission under various probe light intensities.

A simplified model of energy levels and relaxation in PDA is shown in Figure 7.47a. Since this chapter considers the spectral region of no bleaching of ground-state absorption, the transient transmission spectrum consists of only stimulated emission and photoinduced absorption with cross-sections of σ and σ', respectively. The lowest ^1B_u (1^1B_u) exciton and the higher m^1A_g exciton have relaxation time constants τ and τ',

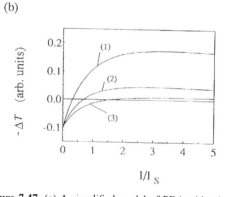

Figure 7.47. (a) A simplified model of PDA with stimulated emission and photoinduced absorption from the ^1B_u exciton level. Their cross-sections are σ and σ'. The lowest ^1B_u (1^1B_u) exciton and the higher m^1A_g exciton have relaxation time constants τ and τ', respectively. (b) The results of numerical calculation of difference transmission depending on the probe light intensity in the case of various τ' values. $\sigma/\sigma' = 0.9$; I is the probe light intensity; I_S is the saturation intensity given as $(\sigma\tau'_p)^{-1}$, where τ_p is the pulse duration (100 fs). The relaxation time constant of 1^1B_u excitons is 150 fs. (1), (2) and (3) are the cases $\tau_p = 10$, 50 and 100 fs, respectively.

respectively. Assume that all of the higher m^1A_g excitons relax to the lowest ^1B_u excitons. τ' is expected to be much shorter than τ. Figure 4.47b shows the results of the numerical calculation of difference transmission depending on the probe light intensity I in the cases $\tau' = 10$, 50 and 100 fs. The net difference transmission ΔT contains the contribution of stimulated emission and photoinduced absorption, ΔT_E and ΔT_A. From the data relaxation time τ of the nonthermal STE to quasi-thermal STE in several blue-phase polydiacetylene, τ is reasonably assumed to be 150 fs [98–100]. Since the signal $-\Delta T/T$ at a weak probe limit in the spectral region just below the stationary absorption edge is negative, σ'/σ is safely concluded to be smaller than unity. Hence σ'/σ is assumed to be 0.9, because of the efficient induced emission process, which is the reverse of the stationary absorption. Therefore, $\Delta T_E/\Delta T$ is calculated to be 0.1. I_S is the saturation intensity of the probe light given as $(\sigma\tau_p)^{-1}$, where τ_p is the probe pulse duration, and is 100 fs. In the case $\tau_p = 100$ fs, $|\Delta T|$ at $I = I_S$ is much smaller than $|\Delta T|$ at $I = 0$. The experimental results show that the sign of transient difference transmission changes in the region of $I < I_S$ near 1.8 eV. $-\delta T/T$ equals 0.02 at $I \sim I_S$ and $-\Delta T/T$ equals 0.04 at $I \ll I_S$. $-\Delta T/T$ equals -0.62 at $I \ll T_S$ at 1.9 eV; it is expected that the case realized in this experiment corresponds to case (1) or (2) in Figure 7.47b. So, τ' is estimated to be much shorter than 50 fs. When the transition rate σ' from the ^1B_u exciton state to the m^1A_g exciton state by the probe is smaller than τ^{-1}, τ'^{-1} is much larger than $I\sigma'$ and the population of the m^1A_g exciton state can be neglected all the time. When the relaxation time constant of the lowest ^1B_u excitons, τ, is of the same order of the pulse duration τ_p, or longer than τ_p, and the probe light intensity I is weaker than I_S, which is the saturation intensity given by $(\sigma\tau_p)^{-1}$, the contributions of stimulated emission and photoinduced absorption to the net difference transmission, ΔT_E and ΔT_A, are given in the approximation that the denominators in the following equations are taken to the first order of I

$$\Delta T_E(I) = \frac{\Delta T_{E0}}{1 + I/I_S} \tag{7.30}$$

$$\Delta T_A(I) = \frac{\Delta T_{A0} - \Delta T_{A\infty}}{1 + I/I_S} + \Delta T_{A\infty} \tag{7.31}$$

ΔT_{E0} and ΔT_{A0} are the contributions of stimulated emission and photoinduced absorption without any saturation in the limit of the weak probe light intensity, and are proportional to σ and σ', respectively. $\Delta T_{A\infty}$ is the contribution of photoinduced absorption in the

limit of strong probe light intensity, and the relation between T_{A0} and $T_{A\infty}$ is

$$\Delta T_{A\infty} = \Delta T_0/2 \tag{7.32}$$

Equations (7.30) and (7.31) are dependent on the probe light intensity when the homogeneous width is dominant over the inhomogeneous width. The contributions of stimulated emission and photoinduced absorption to the net difference transmission, ΔT_E and ΔT_A, are expected to show different probe light-intensity dependence. When there is no interference between the transitions, the net difference transmission $\Delta T(I)$ is given by the sum of $\Delta T_A(I) < 0$ and $\Delta T_E(I) > 0$.

Therefore, from the measurement of ΔT under the condition of different probe light intensities $I_1, \ldots, I_2$, I_S can be determined and ΔT_{E0} and ΔT_{A0} can be separated by using equations (7.30) and (7.31). ΔT at $I_1(I_1 \ll I_S)$ is

$$\Delta T(I_1) \simeq \Delta T_{E0} + \Delta T_{A0} \tag{7.33}$$

ΔT_{E0} and ΔT_{A0} are calculated using equations (7.30)–(7.33) for $I_2 \sim I_S$ as follows

$$\Delta T_{E0} = \frac{\Delta T(I_2) - [(1 + I_2/I_S)^{-1} + 1]\Delta T(I_1)/2}{[(1 + I_2/I_S)^{-1} - 1]/2} \tag{7.34}$$

$$\Delta T_{A0} = \frac{(1 + I_2/I_S)^{-1}\Delta T(I_1) - \Delta T(I_2)}{[(1 + I_2/I_S)^{-1} - 1]/2} \tag{7.35}$$

The spectra of stimulated emission and photoinduced absorption in the limit of weak probe light intensity can be separately determined by transient difference transmission spectra under the conditions of $I \ll I_S$ and $I \sim I_S$ using equations (7.34) and (7.35).

Figure 7.48a shows the stationary absorption and transient difference transmission spectra of the blue-phase oriented PDA-C_4UC_4 film at room temperature. The polarizations of the pump and probe pulses are parallel to the polymer chain. The peak photon energy and photon density of the pump pulses are 1.97 eV and 5.6×10^{14} photons cm^{-2}, respectively. The probe photon density is 2.7×10^{14} photons cm^{-2} integrated between 645 nm and 655 nm. It is expected that the stimulated emission appears in this region and the homogeneous width is about 10 nm.

There are several features in the transient difference transmission spectra shown in Figure 7.48a. The broad bleaching above 1.9 eV is attributed to the absorption saturation of the ^1B_u excitons. The growth of the photoinduced absorption peak near 1.8 eV with time from 0.0 to 0.5 ps is attributed to the relaxation from

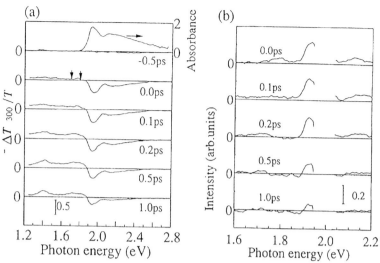

Figure 7.48. (a) The transient transmission spectra in a blue-phase oriented PDA-C$_4$UC$_4$ film at room temperature. The pump and probe pulses are polarized parallel to the polymer chain. The photon density of the 1.97 eV pump pulse is 5.6×10^{14} photons cm^{-2}. The probe photon density is about 2.7×10^{14} photons cm^{-2} integrated between 645 nm and 655 nm. (b) Stimulated emission spectra in a blue-phase oriented PDA-C$_4$UC$_4$ film at room temperature obtained from the transient transmission spectra shown in (a) and observed under probe light intensity 300 times as weak as in (a).

FEs to STEs. They are observed also in other PDAs such as PDA-nBCMU (n-butoxycarbonylmethyl-urethane, $n = 3,4$), PTs such as P3DT and P3MT and PTV [129].

Figure 7.48b shows the stimulated emission spectra determined from the transient difference transmission spectrum ΔT_{300} shown in Figure 7.48a and ΔT_1 observed under the condition of probe light intensity 300-times as weak as in Figure 7.48a by using equation (34). The saturation intensity of the simulated emission at 1.9 eV is determined as 3.1×10^{14} photons cm^{-2} by the measurement of ΔT under several different probe light intensities. The signal above 2.1 eV is considered to be due to the saturation of bleaching. In the case of weak probe light intensity, the scattering pump laser makes the data around the laser photon energy of 1.97 eV unavailable. The peaks at 1.8 and 1.9 eV decay rapidly until 0.5 ps after photoexcitation.

Figure 7.49 shows the time dependence of the stimulated emission intensity shown in Figure 7.48b at a photon energy of 1.91 eV. The time dependence of simulated emission at 1.91 eV can be fitted to an exponential function with the time constant of 140 ± 40 fs plus a constant component due to a long-lived species. The probe photon energy of 1.91 eV is near to the transition energy from the ground state to free exciton. This fast exponential-decay component is expected to be due to the stimulated emission from FEs

and/or nonthermal STEs, in which the binding energy of the STEs remains as the kinetic energy of the lattice vibration [98–100]. The peak at 1.8 eV is expected to be a phonon side band. The decay kinetics are due to the intrachain vibrational relaxation from FEs and nonthermal STEs to quasi-thermal STEs as observed in the transient difference transmission spectra with a lifetime of 100–150 fs [98–100]. The long-lived component is considered to be induced by the saturation of bleaching due to the population of triplet

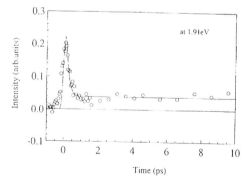

Figure 7.49. Time dependence of the stimulated emission in a blue-phase oriented PDA-C$_4$UC$_4$ film at room temperature at 1.91 eV. The fitted curve is the combination of an exponential function with a time constant of 140 fs and a long-lived component.

exciton states with much longer lifetimes than 100 ps (few tens of microseconds) [23].

The spectrum shown in Figure 7.48b is considered to be the combination of the stimulated emission spectrum and bleaching spectrum. The net stimulated emission spectrum at 0.0 ps is separated by subtracting the normalized bleaching spectrum. The normalized bleaching spectrum is obtained from the transient difference transmission spectrum at 0.5 ps since it is expected that the stimulated emission from free excitons and nonthermal STEs vanishes until 0.5 ps. The multiplying factor to obtain the normalized bleaching spectrum from the transient difference transmission spectrum at 0.5 ps is obtained by calculating and averaging the ratio of the transient difference transmission at 0.0 and 0.5 ps above 2.1 eV, because the stimulated emission does not appear above 2.1 eV and the bleaching spectra can be assumed to be independent of the decay time.

Figure 7.50 shows a fluorescence spectrum calculated from the stimulated emission spectra shown in Figure 7.50b using the relation between Einstein's A- and B-coefficients, A and B as follows

$$(\hbar\omega^3/\pi^2 c^3)B = A \qquad (7.36)$$

A peak at 1.9 eV and a small peak around 1.8 eV are found. The absorption spectrum obtained from the bleaching spectrum is also shown in Figure 7.50. The fluorescence spectrum is not exactly a mirror image of the stationary absorption spectrum shown in Figure 7.48a: this is explained as follows. The stationary

absorption corresponds to the transition from the ground state in thermal equilibrium to the $1\,^1B_u$ exciton state. However, the fluorescence shown in Figure 7.48b corresponds to the transition from the nonthermal STE to the ground state. Since the spectral diffusion time is $\lesssim 100$ fs [51] the spectral diffusion is not completed until 140 fs, the relaxation time constant of nonthermal STEs. So, the absorption spectrum obtained from the bleaching is compared with the fluorescence spectrum. The fluorescence spectrum has approximately a mirror-image relation with the absorption spectrum. By this new method we could obtain the fluorescence spectra of materials with extremely short lifetimes and very low quantum efficiencies for the first time [159].

3.15 Time-resolved fluorescence spectra of partially-polymerized PDA-4BCMU film

Figure 7.51 shows the time dependence of fluorescence intensity of partially polymerized PDA-4BCMU films. There is only a small difference in the decay curves between the 80% converted film, and the 48% converted film. On the other hand, the decay curve of the 100% converted film is different from partially polymerized samples. The decay curves of the fluorescence in the partially polymerized films have two components. The lifetime of the short-lived

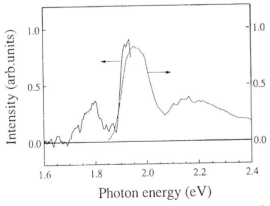

Figure 7.50. The absorption and fluorescence spectra of a blue-phase oriented PDA-C_4UC_4 film. The thick line (ie F) is the fluorescence spectrum at 0.0 ps in a blue-phase oriented PDA-C_4UC_4 film at room temperature calculated from the stimulated emission spectrum shown in Figure 7.47b. The thin line is the absorption spectrum obtained from the bleaching spectrum.

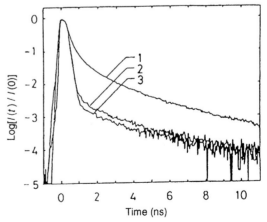

Figure 7.51. The time dependence of the fluorescence intensity from the fully polymerized films. Excitation photon energy and observed photon energy are 2.18 and 2.00 eV, respectively. 1: fully polymerized film; 2: 80%-converted film; 3: 48%-converted film [51].

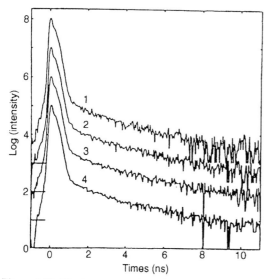

Figure 7.52. The temperature dependence of the fluorescence intensity of 48%-converted film. The excitation photon energy and the observed photon energy are 2.18 and 2.00 eV, respectively. 1: 300 K; 2: 200 K; 3: 100 K; 4: 4 K. The decay curves are arbitrarily offset [51].

is defined with calibration by the concentration as follows

$$\text{conversion rate} = \frac{S_p}{S_p + C_r S_0} \quad (7.37)$$

Here S_p and S_0 denote the areas under the curves in the GPC trace corresponding to polymer and oligomer, respectively; C_r is the concentration ratio between the oligomer and the polymer. The areas under the curve due to the polymer and the oligomer are separated from each other as shown in Figure 7.53. The distributions of the molecular masses of polymers at several conversion rates are similar to each other. The difference between Figure 7.53a and b is only the ratio of the square measure of the polymer to that of oligomer. According to Sixl and Warta [28], both the lifetime and quantum yield of the fluorescence from PDA decrease with increasing chain length of oligomer molecules. The decay curve in the partially polymerized PDA-4BCMU film is considered to be composed of fast and slow components, the former and the latter are due to oligomers and polymers, respectively.

component is too short for the decay function to be determined. In contrast, the slow component is parallel to that of the decay curve of the 100% converted film. The slow components were detectable owing to the very large dynamic range of the single-photon-counting system. There is no distinguishable change in the time dependence of fluorescence intensity among the data observed at temperatures between 300 and 4 K as shown in Figure 7.52. Oligomers were concluded to exist in partially polymerized films from the results of GPC using a $CHCl_3$ carrier and polystyrene elution standards, as shown in Figure 7.52. The conversion rate

3.16 Triplet exciton formation

3.16.1 Reflection spectra

The reflection spectra of the single crystal of PDA-MADF recorded with linearly polarized light oriented both parallel ($E \parallel b$) and perpendicular ($E \perp b$) to the polymer backbone at 302 ± 1 K are shown in Figure 7.54. For the parallel case, there is a peak in the reflection spectrum at 1.86 eV with a reflectance of 0.18. This value is smaller than the values measured for reflectance peaks at similar energies of single crystals of other polydiacetylenes in the visible region. For

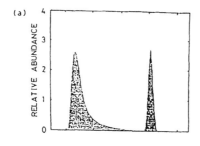

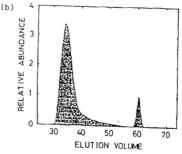

Figure 7.53. GPC chart of partially polymerized PDA-4BCMU; (a) 48% converted; (b) 80% converted [51].

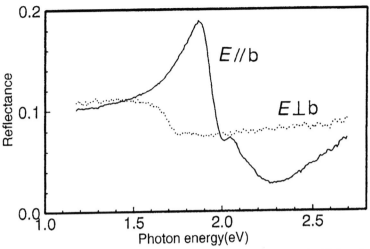

Figure 7.54. The reflection spectra of a single crystal of PDA-MADF at 302 K recorded with linearly polarized light oriented parallel ($E \parallel$ b) and perpendicular ($E \perp$ b) to the main chain [53].

instance, PDA-TS and PDA-DCHD (poly-1,6-di-(*N*-carbazolyl)-2,4-hexadiyne) were found to have the reflectance of 0.75 [60] and 0.46 [61], respectively. The reflection spectrum measured for the perpendicular case ($E \perp$ b) does not reveal any corresponding peaks. Hence, the transition dipole moment in the visible region should be highly aligned along the polymer backbone. For the parallel case, there are structures attributed to phonon sidebands which occur at photon energies higher than the peak at 1.86 eV. The energy of this reflection peak (1.86 eV) is closer to those of peaks found in the reflection spectra of other single crystals of blue-phase polydiacetylenes than red-phase ones. In particular the spectral features of the single crystal of PDA-TCDU-A (poly-5,7-decadiyne-1,12-diol-bis-phenylurethane) [67] are similar to that of PDA-MADF with respect to both the photon energy of the main peak and the appearance of resolved sideband structures at 2.1–2.3 eV.

Figure 7.55 displays a reflection spectrum measured at 30 ± 0.1 K. The spectrum is similar to that obtained at 302 K; however, the transition energies of the main peak and the phonon sidebands shift to the red by approximately 0.03 eV. With lower temperature all the structures in the spectrum become slightly better resolved. At 30 K an additional structure appears at a photon energy just below the main peak. Such a structure is also observed in the reflection spectrum of a single crystal of PDA-TCDU-A at 77 K [66], and it is considered to arise from lattice imperfection. Since the corresponding structure in PDA-MADF was dependent

on samples, the situation was probably the same as in the case of PDA-TCDU-A.

The absorption spectrum of PDA-MADF was computed using the Kramers–Kronig relation. The reflectance of PDA-MADF was assumed to be constant with respect to the photon frequency below 1.71 eV for the transformation, because in the energy region no prominent optical transition exists [130,131]. Ohsugi *et al.* [132] investigated the absorption spectrum of the microcrystalite of PDA-MADF. Although the absolute value of the absorption coefficient was not known, the spectral shape was able to serve as a criterion of validity of the transformation. The Kramers–Kronig transformation of the reflection spectrum was performed with several values of the power *p* in the approximation for

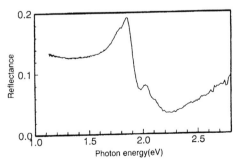

Figure 7.55. The reflection spectrum of PDA-MADF at 30 K for polarization of the light parallel ($E \parallel$ b) to the main chain [53].

the asymptotic value in the higher-energy region where there is no observation. The asymptonic value is

$$R(E_{ph}) = cE_{ph}^p \qquad (7.38)$$

where R is the reflectance, c is a fitting parameter, and E_{ph} is a photon energy. Each calculated spectrum was compared with the spectrum obtained by Ohsugi *et al.*. The power of 0.5 was found to give a reasonable absorption spectrum as shown in Figure 7.56. The peak positions and the calculated absorption coefficient at the energies near the main peak and phonon sidebands were not sensitive to the power of p in equation (7.38). For example, the uncertainty of the position of the peaks in the calculated absorption spectra was within a range of 0.007 eV and the absorption coefficient of the peaks varied by only about 10% to each other when the value of p is varied from 0 to 2. In contrast, the tail of the absorption spectrum is rather sensitive to the value of p. The calculated absorption coefficients at 1.5 eV of $p = 0$ and $p = 2$ were 1.8×10^4 cm^{-1} and 7.8×10^4 cm^{-1}, respectively.

The calculated spectrum at 302 K reveals an absorption band at 1.88 eV. The peak energies were found to shift to red by 0.03 eV when the temperature was decreased from 302 to 30 K. The peak energies of the phonon sidebands were estimated at 2.05 and 2.15 eV. In other polydiacetylenes, such as a single crystal of PDA-TS or a cast film of PDA-3BCMU, the two most intense sidebands at higher energies than the main peak were assigned to the C=C and C≡C

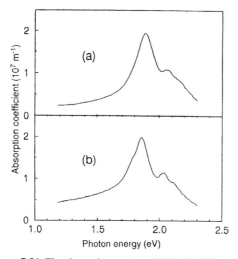

Figure 7.56. The absorption spectra of PDA-MADF at 302 K (a) and 30 K (b) for polarization of light parallel ($E \parallel$ b) to the main chain, as computed from reflectance data by Kramers–Kronig analysis [53].

stretching odes of the main chain. For PDA-3BCMU the energies of the C=C and C≡C vibrational modes were determined to be 1461 cm^{-1} (0.1811 eV) and 2079 cm^{-1} (0.2578 eV), respectively, by Raman spectroscopy [62] and for PDA-TS they are 1486 cm^{-1} (0.1842 eV) and 2192 cm^{-1} (0.2594 eV) also from the Raman spectrum [133]. By analogy with these polydiacetylenes, the two peaks of the phonon sidebands at about 0.2 and 0.3 eV higher than the main absorption peak were assigned to the vibrational bands of the C=C and C≡C stretching modes.

The residual absorption in the low-energy region below 1.6 eV can be seen in the calculated absorption spectra. However, a discussion or an assignment of the absorption in this energy region is difficult, since absorption coefficient in the low-energy region is rather sensitive to the absorption of the reflection spectra in the high-energy region, which is not obtained experimentally.

3.16.2   Transient reflection spectra

3.16.2.1   Photoinduced excitations

Figure 7.57 shows the transient photoinduced reflection spectra of a single crystal of PDA-MADF at 297 ± 1 K. The FWHM of the pump pulse was 0.04 eV. The pump energy of 1.98 eV is 0.01 eV higher than the energy of the excitonic absorption peak of PDA-MADF. The linear polarizations of both the pump and probe beams were parallel to the orientation of the main chain in the PDA-MADF crystal ($E_{ex} \parallel$ b and $E_{pr} \parallel$ b, where E_{ex} and E_{pr} are electric field vectors of pump and probe light, respectively). Since the spectral range of the polychromator was 120 nm, six different measurements were needed to obtain the whole spectrum shown in Figure 7.57. Because of the possible error in the determination of the temporal origin the six curves were not connected smoothly to each other where ultrafast time dependence was observed. When excited with perpendicularly polarized light, the threshold for optical damage was lower than the parallel case. With excitation light polarized perpendicular to the main chain ($E_{ex} \perp$ b) with a photon density of 2.4×10^{14} photons cm^{-2}, which was below the damage threshold for the perpendicular case, the induced reflectance change in the vicinity of 1.8 eV was less than 10% of the detected signal in the parallel case. Therefore, the transient reflection spectra recorded with the polarization of pump beam parallel to the main chain ($E_{ex} \parallel$ b) were employed for this study. The linear polarization

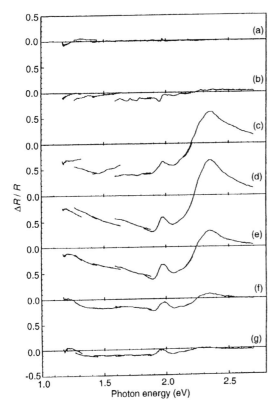

Figure 7.57. The transient photoinduced reflectance change $\Delta R/R$ of PDA-MADF vs. the probe photon energy at 297 K plotted at the following seven pump–probe delay times: (a) -0.4; (b) -0.2, (c) 0.0; (d) 0.2; (e) 0.4; (f) 5.0; and (g) 100 ps [53].

of the probe light was always maintained parallel to the polymer backbone ($E_{pr} \parallel b$).

At both delay times of -0.2 and 0.0 ps, two small structures with minima at 1.68 and 1.76 eV, respectively, were observed. These minima disappeared 0.2 ps after excitation. Although the interpretation of the reflectance change is not as straightforward as of the absorbance change, the localized structures in the transient reflection spectra tend to have corresponding structures at similar photon energies in the transient absorption spectra. The energy differences of pump photon energy from the peak energies are 1.98–1.76 = 0.22 eV and 1.98–1.68 = 0.30 eV. These differences are similar to the vibrational energies of 0.2 and 0.3 eV obtained in the calculated absorption spectra of PDA-MADF. Since Raman gain signals of C=C and C≡C stretching modes in the main chain of a polydiacetylene have been observed in the femtosecond time-resolved absorption spectrum of a thin film of

PDA-3BCMU [52], a similar Raman gain process due to these stretching modes in the main chain is probably responsible for these two structures in PDA-MADF. This occurs only when pump and probe pulses are overlapped with each other.

Both the positive signal in the reflection spectrum at the high-energy region in the vicinity of 2.4 eV and the negative signal at a low energy of about 1.2 eV show an initial rapid decay with a time constant of the order of a sub-picosecond. After the rapid decrease of the signal, the amplitude of the reflectance change decreases with a time constant of about a picosecond. At later delay times the signal remains relatively constant within the time span observed (200 ps).

The two peaks at 1.88 and 2.06 eV in the photoinduced spectra correspond to the exciton peak at 1.86 eV and phonon sideband at 2.03 eV, respectively, in the reflection spectrum.

The time-resolved reflectance change decay curves plotted over the first five picoseconds for six different probe energies are displayed in Figure 7.58. In the case of both low and high energies, an initial sub-picosecond relaxation occurs, as shown in curves (a) 2.6, (b) 2.2, (e) 1.4 and (f) 1.2 eV.

In contrast to higher and lower energy regions, in the vicinity of 2.0 eV, the photoinduced relaxation occurs on a picosecond time scale and are reflected in curves (c0 2.0 and (d) 1.8 eV. In all the curves except for the measurement at 2.6 eV, the long-lived component was observed. Figure 7.59 shows the reflectance change data at 1.8 eV up to 200 ps. It shows that the long-lived component relaxes with a time constant much longer than 200 ps.

In this paper it is assumed that the magnitude of reflectance change is proportional to the number density of the excited species related to the observed reflectance change in the sample.

A single-exponential-decay fitting form with a constant offset contribution for the long-lived signal was used for the photoinduced decay process. The decay curve was fitted by convoluting the experimentally determined cross-correlation function with an exponential decay as shown later in Figure 7.62. The exponential-decay time constant of the reflectance change at 1.8 eV was found to be $1.2 + 0.1$ ps. The observed transient curve in the picosecond time domain decays slightly faster than the best-fit curve. The fitted curves in this study were calculated using all the data within the observed time span (200 ps).

Figure 7.60 shows the time dependence of the reflectance change at 2.4 eV and 1.2 eV in which the ultrafast relaxation was exhibited. The decay curves

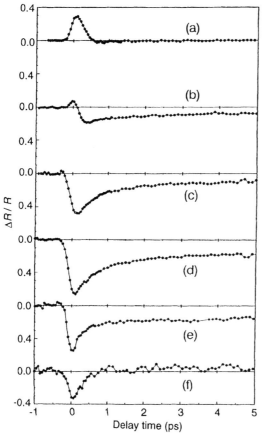

Figure 7.58. The time dependence of the reflectance change of PDA-MADF vs. the delay time at 297 K plotted at the following six photon energies: (a) 2.6; (b) 2.2; (c) 2.0; (d) 1.8; (e) 1.4; and (f) 1.2 eV [53].

were fitted by convoluting the cross-correlation function with the two exponential-decay components with the offset term, assuming a time constant of 1.2 ps

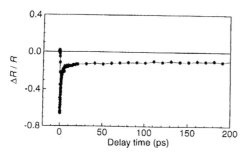

Figure 7.59. The time dependence of the photoinduced reflectance change of PDA-MADF at 297 K with a 200 ps delay time. The probe photon energy was 1.8 eV [53].

(observed in the decay process at 1.8 eV) as a given parameter for the slower process. From the best-fit curve with a fixed decay rate of 1.2 ps, the time constant of the sub-picosecond relaxation were estimated to be 220 ± 20 fs for 2.4 eV and 210 ± 30 fs for 1.2 eV. Since these two time constants are in close agreement with each other, the ultrafast decay of the reflectance change at 2.4 and 1.2 eV suggests that the origins of the sub-picosecond responses at different probe energies are the same. In short, the initial decay process with a time constant of approximately 200 fs occurred in PDA-MADF.

Figure 7.61 shows the calculated transient absorption spectra obtained by applying the Kramers–Kronig transformation to the data of transient reflectance change. The reflection spectrum after excitation (R^*) was obtained from the transient reflection spectrum ($\Delta R/R$) and the spectrum of the ordinary reflectance (R). The reflection spectrum after excitation (R^*) was then transformed into the absorption spectrum after excitation (A^*). The transient difference absorption spectra were given by calculating the difference between the absorption spectra with and without excitation ($\Delta A = A^* - A$). The detailed procedure of the Kramer–Kronig transformation was the same as that for the stationary absorption spectrum of PDA-MADF.

Above 1.6 eV the shape of the photoinduced absorbance change is similar to the absorption spectrum. The bleaching peak at 1.9 eV corresponds to the absorption peak at 1.88 eV, and there are also corresponding bleaching structures of phonon sidebands in the vicinity of 2.1 eV. Uniform bleaching signals of the absorption due to the lowest $1B_u$ exciton and its phonon sidebands in the femtosecond time domain have also been observed in PDA-3BCMU and PDA-4BCMU the [51,52,98], and in a single crystal of PDA-TS [134]. Greene et al. explained the exciton absorption saturation of PDA-TS in terms of the phase-space filling model, which had been known to be applicable to bulk and quantum confined GaAs [135,137]. In their model for polydiacetylenes, the existence of excitons suppresses the exciton absorption.

The author's group photoinduced absorption below 1.85 eV, and the absorption measured at 1.77 eV revealed a rise time of 150 fs in a cast film of PDA-3BCMU. The transient absorption below 1.85 eV was assigned to STEs [52,87]. This model, which explains the relaxation dynamics in PDA-3BCMU, is as follows.

As photoexcited free ^1B_u excitons (FEs) relax to STEs, absorption due to the transition from the STEs to the ^1A_g excitons and/or the electron–hole pairs at

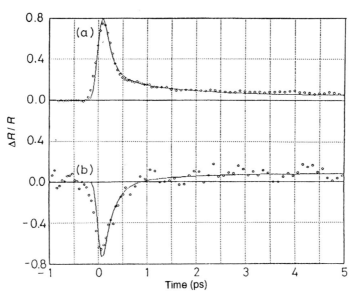

Figure 7.60. The time dependence of the reflectance change of PDA-MADF at 2.4 (a) and 1.2 eV (b) measured at 297 K. The best-fit curves shown with the solid lines were obtained by assuming two-exponential-decay components with a constant offset term and a fixed decay rate of 1.2 ps [53].

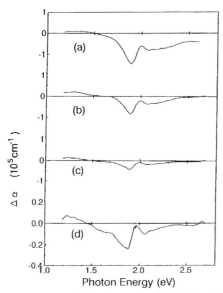

Figure 7.61. The calculated spectra of photoinduced absorption of PDA-MADF obtained by applying the Kramers–Kronig transformation to the data of transient reflectance at room temperature. The spectra are measured at the following four pump–probe delay times: (a) 0.2; (b) 1.0; (c) 5.0; and (d) 100 ps [53].

higher energies can be detected. While the free ^1B_u excitons and the STEs exist, the transition due to the lowest ^1B_u excitons is saturated, and the bleaching signal is observed. From the time dependence of the photoinduced absorption and the recovery curve of photoinduced bleaching, we estimated the formation and the decay times of the STEs as 150 ± 50 and 1.5 ± 0.2 ps, respectively [52]. The time constants were recorded at 290 K. Toyozawa reported that the formation time of the STE at 10 K was also 150 ± 50 fs, and the results are consistent with the theoretical work which suggests that in one-dimensional systems FEs relax to STEs without a barrier [136].

The interpretation of the transient absorption spectra of PDA-3BCMU in terms of STEs is also consistent with the work of Blanchard et al. [77]. They observed the phonon-mediated optical Stark effect in PDA-TS, which suggests strong exciton–phonon coupling.

The transient reflection of PDA-MADF also showed both ultrafast relaxation with time constant of 220 ± 20 fs and recovery with 1.2 ± 0.1 ps. These time constants are in good agreement with those observed in other polydiacetylenes, such as PDA-3BCMU [52], PDA-4BCMU [51,98], PDA-TS [29,137,138] and PDA-15,8 [138]. The STE model of relaxation dynamics in polydiacetylenes is reasonable introduced for the one-dimensional electron system [77,136]. It was applied for the photoinduced absorption in several polydiacetylenes [52]; time constants of the order of a hundred femtoseconds and of one or a few picoseconds are measured for several kinds of polydiacetylenes, the relaxation dynamics observed in this study are also explained essentially in terms of STEs.

The calculated absorption edge at the low-energy side (1.5 eV) in PDA-MADF is not steep compared to that of a PDA-3BCMU cast film, but has a long wavelength absorption. Therefore, in the calculated absorbance change of PDA-MADF the transient absorption of the STE is not clearly observed; it is probably hidden by the bleaching of the long-wavelength absorption. However, since the absorbance in this low-energy region is easily influenced by the assumption in the Kramers–Konig transformation, it is difficult to discuss the origin of the bleaching in this region. Below 1.4 eV there exists a 'positive' photo-induced absorption which lasts up to 100 ps. Within the delay time of several picoseconds, the STEs may contribute to the spectra. The spectrum at a delay time of 100 ps (Figure 7.61d) represents the transient absorption of the long-lived excitations, which shows photoinduced absorption around 1.3 eV.

As candidates of the long-lived excitations in polydiacetylenes, triplet excitons, polarons and bipolarons can be given. Kim et al. [139] reported the creation of long-lived, localized, charged and spinless excitations observed by photoinduced infrared absorption and photoinduced electron-spin resonance measurements on crystalline powders of PDA-9PA [R = R′ = $(CH_2)_9OCOCH_2C_6H_5$], PDA-4BCMU, and PDA-TS, with the pump photon energy of 2.7 eV. They assigned the excitations to bipolarons. However, the bipolarons cannot explain the long-lived component in PDA-MADF, since the lifetime of the bipolarons is extremely long (the order of minutes at 80 K) [139]. In addition, it is difficult to consider that the induced transient absorption of these long-lifetime excitations can be observed in a femtosecond measurement with a sampling frequency of 10 Hz where each decay point is a result of 100 sampled measurements. Furthermore, long-lived bipolarons are likely to have been created by polarons, which are generated separately in neighboring main chains by the excitation of light polarized perpendicular to the main chain ($E_{ex} \perp b$); however, the long-lived excitations observed in PDA-MADF were created by the light polarized parallel to the main chain ($E_{ex} \| b$).

Owing to the energy position of the photoinduced absorption of the long-lived excitations (around 1.3 eV) and the experimental configuration that the polarizations of the pump and the probe light were parallel to

the main chain, the long-lived excitations in PDA-MADF can be identified as triplet excitons. It is known that the generation efficiency of the triplet excitons strongly depends on the photon energy of excitations. In the femtosecond experiment the single crystal of PDA-MADF was excited by 1.98 eV light. This photon energy is less than the excitation energy used for other experiments [23,140–142]. Therefore as a result of a high-excitation photon density per unit time, it is considered that the process, which is not the direct result of a one-photon resonant one-quantum transition, has a significant role in the formation of triplet excitons in PDA-MADF. The pump intensity dependence of the long-lived component yields important information about the formation process of the triplet excitons.

3.16.2.2   Excitation intensity dependence of photoinduced reflectance change

The time dependence of the reflectance change probed at 1.8 eV for two different excitation intensities is shown in Figure 7.62. At higher excitation intensity the solid best-fit curve, calculated by assuming a single-exponential-decay fitting form with a constant offset contribution, exhibits a slight deviation from the observed decay curve. However, when the intensity was decreased to 9.4×10^{13} photons cm^{-2}, the decay curve was well fitted by the calculated solid curve shown in Figure 7.62b. At the lower pump intensity, the time constant was estimated at 1.8 ± 0.1 ps. Therefore, at the higher pump intensity, interactions between excitons most likely shortened the lifetime of the state and accelerated the recovery of the bleaching signal. Thus, a time constant of 1.8 ps reflects the more intrinsic lifetime compared with the value of 1.2 ps at the higher intensity.

The ratio between the amplitudes of the initial reflectance change and the long-lived component at a higher-excitation photon density was smaller than that observed at a lower photon density. Figure 7.63 shows the excitation intensity dependence of the photoinduced reflection change probed at 1.8 eV. The data were obtained from the decay curves which were measured at seven different excitation photon densities from 3.8×10^{13} to 2.4×10^{15} photon cm^{-2}. The time from the nonthermal STE to the quasi-thermal STE is estimated to be 100–200 fs, which is comparable to the cross-correlation time of the pump and probe pulses. Therefore, the reflectance change at 150 fs occurs as a result of the combined effects of nonthermal STE, quasi-thermal STE, and possibly (quasi-thermal) triplet

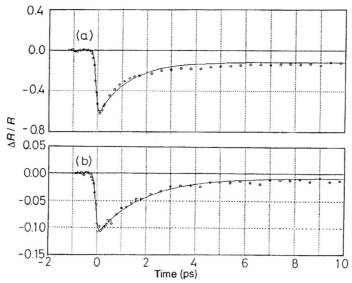

Figure 7.62. The time dependence of the photoinduced reflectance change of PDA-MADF at two different excitation intensities of 2.4×10^{15} photons cm^{-2} (a) and 9.4×10^{13} photons cm^{-2} (b) measured at 297 K with a probe photon energy of 1.8 eV. The solid best-fit curves were obtained assuming a single-exponential decay component with a constant offset term [53].

excitons. Since at the probe photon energy of 1.8 eV, the decay curves do not exhibit the ultrafast relaxation of 100–200 fs, the contribution of the quasi-thermal STEs to the reflectance change is considered to be the dominant effect. On the other hand, the reflectance change at a delay time longer than 50 ps is due to the (quasi-thermal) triplet exciton, since 50 ps is much longer than the decay time of the singlet ^1B_u STEs.

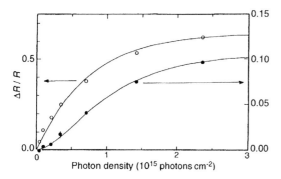

Figure 7.63. The excitation intensity dependence of the photoinduced reflectance change of PDA-MADF at 297 K with a probe photon energy of 1.8 eV. The open circles represent the reflectance change at 150 fs, and the closed circles represent an average of the amplitudes from 50 to 150 ps. The two solid curves fit to the open and closed circles are fitting curves using equations (7.43) and (7.44), respectively [53].

At photoexcitation densities below 3×10^{14} photons cm^{-2}, the amplitude of the fast-decay component was linearly proportional to the photon density, while the long-lived component varied quadratically with an increasing photon density. When the pump intensity was increased beyond the region which gave a linear signal dependence, the fast-decay component exhibited a saturation effect of the reflectance change. It is interesting to note that the reflectance change of the long-lived component was also saturated in the photon-density region. There is a combination effect of the saturation of the reflectance changes due to the fast component and to the long-lived component. This combination is an important feature of the intensity dependence of the relaxation dynamics. The combination suggests that the creation of the triplet excitons strongly depends on the population of the singlet excitons.

3.16.2.3 Temperature dependence of photoinduced reflectance change

Figure 7.64 shows the time dependence of the reflectance change at 1,8 eV for two different temperatures. The solid lines represent the calculated curves with a time constant of 1.0 ps for 297 K and 1.7 ps for 30 K. The interaction between excitons may contribute to the slight deviations from the calculated exponential

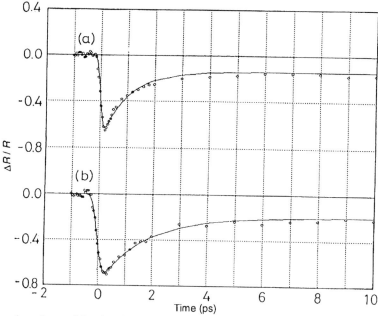

Figure 7.64. The time dependence of the photoinduced reflectance change of PDA-MADF at 1.8 eV measured at 297 K (a) and 30 K (b). The solid best-fit curves were obtained assuming a single-exponential-decay component with a constant offset term [53].

curves. When the temperature was decreased, the lifetime of the STE became longer, and the amplitude of the long-lived component due to the photogenerated triplet excitons increased.

The ultrafast relaxation of the reflectance change was also observed at 73 ± 1 K. With the same calculation used to obtain the ultrafast relaxation time at 297 K, the time constant of the sub-picosecond decay component at 2.2 eV was estimated as 210 ± 20 fs. This relaxation time is the same as that obtained for the free exciton at 297 K (220 ± 20 fs for 2.4 eV and 210 ± 30 fs for 1.2 eV). Therefore, the decay time from the free ^1B_u exciton to the STE was found to be independent of the temperature in the same way as PDA-3BCMU and PDA-4BCMU [52,98].

3.16.3   Discussion

3.16.3.1   Optical spectra

The main absorption peak of PDA-MADF located at 1.88 eV can be assigned to the ^1B_u exciton transition. The values of the calculated absorption coefficient at the exciton peak was 1.9×10^5 cm^{-1} and smaller than

those of PDA-TS ($a = 7.3 \times 10^5$ cm^{-1} [60] or PDA-DCHD ($a = 5.0 \times 10^5$ cm^{-1} [60]. The band width was approximately 0.2 eV and was wider than those of PDA-TS ($\Delta E = 0.11$ eV [60] or PDA-DCHD ($\Delta E = 0.14$ eV [61]). This broader band structure of PDA-MADF most likely means that the crystalline quality is not as good as those of PDA-TS or PDA-DCHD.

3.16.3.2   Relaxation dynamics of photoexcitations

When a single crystal of PDA-MADF is excited with a photon energy of 1.98 eV, free ^1B_u excitons are generated. The free excitons relax to STEs owing to electron–phonon coupling. The relaxation time is estimated as approximately 200 fs and it is so fast that fluorescence from the free ^1B_u is very weak.

The author's group [51,98] investigated the relaxation dynamics in cast films of PDA-3BCMU and PDA-4BCMU, and explained that the relaxation of the STEs to the ground state occurs mainly via a quantum mechanical tunneling process. We observed that the STEs in the blue-phase PDA-4BCMU decay exponentially with lifetimes of 1.6 ± 0.1 ps at 290 K and 2.1 ± 0.2 ps at 10 K. The decay curve in the red-phase PDA-4BCMU was not single exponential but was fitted to a biexponential function with time constants slightly shorter than 1 ps and about 5 ps [51,98].

A single crystal of PDA-MADF at low-excitation photon density (9.4×10^{13} photon cm^{-2}) fitted well the calculated exponential curve with the time constant of 1.7 ± 0.1 ps. Since PDA-MADF is in the blue phase, the relaxation dynamics of PDA-MADF is in agreement with the observation in a previous study [51]. The small temperature dependence of the lifetime of the STE in PDA-MADF (1.0 ± 0.1 ps at 297 K and 1.7 ± 0.1 ps at 30 K) is also consistent with the model, because the energy distribution of the excitons depends on the temperature, and excitons at different temperatures tunnel through the barrier of different thickness and height [51]. A similar temperature dependence of the lifetime can be observed in a cast film of PDA-3BCMU [52]. Therefore, the STEs in PDA-MADF are likely to relax by a tunneling process to the ground state.

3.16.3.3 Mechanism of saturation

Figures 7.62 and 7.63 demonstrate that the ratio of the initial signal amplitude (due to the singlet excitons) to the long-lived signal amplitude (due to the triplet) excitons) is dependent on the pump intensity of 1.98 eV excitation light. At a higher-excitation photon density, the ratio decreased. Similar intensity dependence was observed by Greene *et al.* [143] and the author's group [52,98]. On the other hand it is known that the generation efficiency of triplet excitons shows strong-excitation photon-energy dependence in PDA-TS [140,144]. Yoshizawa *et al.* [52] also observed that the long-lived component was more efficiently generated with a 3.94 eV pulse than a 1.97 eV pulse in PDA-3BCMU, and it was reported that the relative magnitude of the long-lived component in PDA-TS was not intensity dependent when excited with 3.94 eV light [145]. Austin *et al.* [145] investigated both excitation photon energy and intensity dependence of the triplet yield in PDA-4BCMU in the red and yellow phases and found that the yield was linearly proportional to the pump intensity when excited by 3.49 eV, while it was proportional to the square of the pump intensity when excited by 2.33 eV light in the red phase and 2.48 eV light in the yellow phase. These observations suggested

that it was necessary to access the higher-lying states such as singlet excitons or electron–hole pairs to efficiently form the triplet excitons.

For the observations stated above, the process to generate the triplet excitons with excitation of 1.9–2.5 eV light was generally believed to be a 'stepwise' two-photon absorption [52,143,145].

However, there may be an alternative mechanism other than the process of triplet exciton generation by 'stepwise' two-photon absorption when the lowest ^1B_u exciton is resonantly excited. Let us consider that the interaction between two singlet excitons generates a certain number of triplet exciton(s). If the process is not the dominant pathway of the relaxation, the intensity dependence of the triplet yield is also expected to be quadratic, but only at low intensities where the yield of the lowest singlet exciton is linearly proportional to the pump photon density. In this case, however, the population of the triplet exciton continues to be proportional to the square of the population of the singlet exciton even at higher pump intensities where the population of singlet excitons should exhibit saturation. Even if the 'stepwise' two-photon absorption does generate triplet excitons, the intensity dependence of the triplet excitons should be also influenced by the saturation of the population of the singlet excitons and may start to deviate from the quadratic curve at a higher intensity region. However, in this case it is not necessary for the triplet yield to be proportional to the square of the singlet yield at high pump intensities at all! This is an important point. Hence the relationship between the initial and the long-lived reflectance change, at the pump intensities where saturation occurs, is a crucial observation which provides insight for the formation process of the triplet excitons. It is not possible to distinguish the two triplet-state formation mechanisms of 'stepwise' two-photon absorption and an interaction between singlet excitons at the low excitation intensities. The previous intensity dependence measurements [145] have only been performed and interpreted at low excitation-intensity conditions.

In the present study, the dependence of the initial and the long-lived reflectance changes are investigated over a wide range of excitation intensities. If the population of the triplet exciton is proportional to the square of that of the singlet exciton, even at the high pump intensities where the initial reflectance change exhibits saturation, triplet generation by a 'stepwise' two-photon absorption can not be the primary mechanism. However, the interaction between the singlet excitons is totally consistent with the intensity dependence measurement

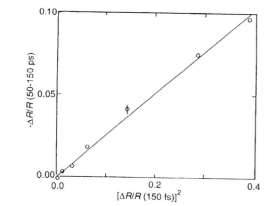

Figure 7.65. The plot of the photoinduced reflectance change of the long-lived component (the average of the amplitudes from 50 to 150 ps) vs. the square of the initial reflectance change (the amplitude at 150 fs) of PDA-MADF measured at 297 K with a probe photon energy of 1.8 eV. The straight line is best fitted by the least-square method [53].

and offers a reasonable explanation for the generation of the triplet exciton.

The plot of the reflectance change of the long-lived component vs. the square of the initial reflectance change is shown in Figure 7.65. The slope is well described by a straight line, including the data measured in the high pump intensity region. This plot clearly shows that the triplet excitons are created by a bimolecular interaction between optically excited excitons, and not by the 'stepwise' two-photon absorption as previously considered.

To demonstrate the saturation behavior of the initial and the long-lived reflectance change, a two-level model is used. At first let us consider the pump intensity dependence of a population in a two-level atomic system using a rate equation approximation. The equations describing the time dependence of the number densities of the excited state $N_e(t)$ and the ground state $N_g(t)$ are

$$\frac{d}{dt}N_e(t) = BI(t)N_g(t) - BI(t)N_e(t) - kN_e(t)$$

$$N_g + N_e = N \qquad (7.39)$$

where $I(t)$ is the energy density of a pump pulse, B is the rate constant of induced radiative transitions per light-energy density, k is a decay rate constant, and N is the total number density of the atoms. If a rectangular pulse profile

$$I(t) = a \qquad (-\tau_p/2 < t < \tau_p/2)$$
$$0 \qquad (t < -\tau_p/2, \tau_p/2 < t) \qquad (7.40)$$

is assumed, the population of excited state just after excitation

$$N_e(\tau_p/2) = \frac{NBa}{k + 2Ba}[1 - \exp(k - 2Ba)\tau_p] \qquad (7.41)$$

is given by substituting $t = \tau_p/2$ for the solution of equation (7.39). If the pulse width is much shorter than the lifetime of the excited state ($k\tau_p \ll 1$), and a pump rate is much higher than a decay rate ($k \ll Ba$), equation (7.41) gives

$$N_e(\tau_p/2) = (N/2)[1 - \exp(-2Ba\tau_p)] \qquad (7.42)$$

Equation (7.42) was applied to the intensity dependence of the initial photoinduced reflectance change $\Delta R/R(I)_i$ of PDA-MADF. For the application, $N/2$ and $1/(2 \cdot B)$ in equation (7.42) are replaced with the fitting parameters of $(\Delta R/R)_s$ and I_s, respectively. Then

$$\Delta R/R(I)_i = (\Delta R/R)_s[1 - \exp(-I/I_s)] \qquad (7.43)$$

where I is an excitation photon density, is obtained.

With this model the intensity dependence of the initial reflectance change is well described as shown in Figure 7.63. The top solid curve represents $\Delta R/R(I)_i$ with $(\Delta R/R)_s = 0.65$ and $I_s = 7.5 \times 10^{14}$ photons cm^{-2}. The square of the function is

$$\Delta R/R(I)_l = A - [\Delta R/R(I)_i]^2 \qquad (7.44)$$

which describes well the intensity dependence of the long-lived reflectance change. The lower solid curve in 7.63 represents this function with a fitting constant of $A = 0.25$.

From the curves in Figure 7.63, it is clearly found that the number densities of the triplet excitons are well described by the square of the number densities of the singlet excitons, even when the signals are saturated.

This is the first evidence that the interactions between the singlet excitons are the dominant mechanism in the formation of the triplet excitons when the lowest singlet exciton absorption edge is excited in a polydiacetylene.

3.16.3.4   Formation of triplet excitons

In the detailed investigation of the intensity dependence of the photoinduced reflectance change, it is proposed for the first time that interactions between the singlet excitons form the triplet excitons. The mechanism which represents this process is

$$S_0 + \hbar v \rightarrow S_1 \qquad (a)$$
$$S_1 + S_1 \rightarrow T_1 (+T_1) \qquad (b)$$

where S_0, S_1, T_1 and $\hbar v$ are the ground state, the lowest excited singlet state, the lowest excited triplet state and a photon, respectively. In previous works [52,143] the state kinetics to form the triplet excitons were thought to occur in the following manner

$$S_0 + \hbar v \rightarrow S_1, S_1 + \hbar v \rightarrow S_n \qquad (c)$$
$$S_1 + \hbar v \rightarrow e + h \qquad (c')$$
$$S_n \rightarrow T_1 (+T_1) \qquad (d)$$
$$e + h \rightarrow T_1 (+T_1) \qquad (d')$$

where S_n, e and h are a higher singlet excited state, an electron and a hole, respectively. The pairs of the processes, (c) and (d) and (c') and (d') are suggested in literatures [52,143], respectively. In short, it is found in our study that the second process is not an optical process [(c) or (c')] but a process via bimolecular interactions between excitons [process (b)].

Now process (b) will be discussed. Since triplet excitons are efficiently produced via higher-lying states, it is reasonable to consider that a higher-lying state is accessed by interactions between singlet excitons. One plausible way to access such a higher-lying state is the Auger process. In the interaction of two singlet excitons, the first exciton loses its energy nonradiatively. The second exciton obtains the energy released in the interaction and is excited into a higher-lying state. This is a type of the Auger process.

In the early stages of the study of triplet excitons an electron-hole pair was given as the likely candidate for the higher-lying state [140], since the photoconduction quantum yield showed a similar photon-energy dependence to the yield of the triplet excitons in a single crystal of PDA-TS [146]. However, Winter et al. [147] observed the triplet excitons in PDA-TS by pulsed ESR after photoexcitation at 3.68 eV and suggested that triplet excitons are generated from a triplet-pair state, which is directly excited via a one-photon absorption. Austin et al. [145] measured the magnetic field dependence of the triplet-exciton production and decay in PDA-4BCMU, and the result indicated that the triplet excitons are created by a fission process from higher-lying singlet excitons. Based on the observation of the intensity dependence of triplet yield, they proposed that the higher-lying singlet state is accessible by a two-photon process when the lowest singlet excitons are resonantly pumped.

The character of the state, which is accessible by the Auger effect from the lowest singlet excitons, may be different from that of the state allowed via a one-photon absorption from the ground state, since with the Auger effect the singlet exciton proceeds to a dipole-allowed higher-lying singlet state from the lowest excited singlet state. Therefore, higher-lying singlet excitons with the ^1A_g symmetry are likely candidates for the excitons to create triplet excitons rather than the one-photon allowed triplet-pair state. Two triplet excitons can be generated via a fission process of the higher-lying ^1A_g exciton. The fission process of a singlet exciton into two triplet excitons is known to occur in organic crystals such as tetracene crystals [148–150]. Hence a plausible mechanism of process (b) is

$$S_1 + S_1 \rightarrow S_0 + S_n \qquad (e)$$
$$S_n \rightarrow T_1 + T_1 \qquad (f)$$

The lifetime of the STEs at 30 K (1.7 ± 0.1 ps) was slightly longer than the lifetime at 297 K (1.0 ± 0.1 ps), and the amplitude of the long-lived reflectance change at 30 K was larger than at 297 K. On the other hand, the excitation intensity dependence of the lifetime of the STEs (the lifetime was shorter at higher excitation intensity) indicated interactions between them. These results suggest that interactions between the STE has a significant role in the formation of the triplet excitons.

Observations of the transient absorption of the singlet exciton S_n and decay kinetics of process (f) would confirm the validity of processes (e) and (f). If the rise time of the absorption due to S_n can be experimentally observed, the roles of the interactions between the free ^1B_u excitons and between the STEs can be clarified in the creation of the triplet excitons.

In summary, the proposed model of the mechanism of triplet exciton formation in polydiacetylenes under resonant excitation of the lowest ^1B_u exciton is as follows: The optically excited short-lived ^1B_u excitons interact with each other by the Auger effect. The higher-lying excitons, with probable ^1A_g symmetry, are generated with a number density proportional to the square of the number density of the singlet excitons. The triple excitons are then created by the fission process of the higher-lying singlet exciton into two triplet excitons. The proposed model is displayed in Figure 7.66.

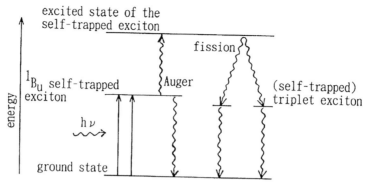

Figure 7.66. A proposed model for the mechanism of triplet exciton formation in polydiacetylenes [53].

3.17 Separation of spectral response due to various excitations in PDA-3BCMU epitaxially grown on a KCl single crystal

The photoinduced absorption spectra of PDA-3BCMU of 0.13 μm thickness at 290 K are shown in Figure 7.67 together with the stationary absorption and pump spectra. Both polarizations of the pump and probe light are parallel to the orientational direction of the polymer chain. The structure and kinetics of the transient absorbance change in the epitaxially grown PDA-3BCMU are very similar to those observed in other PDAs, i.e. cast films of PDA-3BCMU, vacuum-deposited films of PDA-4BCMU and PDA-(12,8) LB films [51,54]. A sharp bleaching peak at 1.97 eV is due to saturation of the excitonic absorption and coherent coupling between the pump polarization and probe field. Two negative peaks at 1.79 and 1.71 eV are observed at the delay time of 0.0 ps. They are due to the Raman gain process of the stretching vibrations of the C=C and C≡C bonds, respectively.

At 0.0 ps, the photoinduced absorbance change below 1.8 eV is larger at lower probe photon energies down to 1.2 eV. Then the absorption shifts to higher energy with time from 0.0 to 0.5 ps and has two peaks at 1.8 and 1.4 eV. This spectral change is explained by the geometrical relaxation of free excitons to STEs [51,54]. The photoexcited free excitons are coupled with the C−C stretching modes within the phonon periods of 10–20 fs and become nonthermal STEs. Then the nonthermal STEs thermalize and relax to the bottom of their potential surfaces. Therefore, the transient absorption is shifted to higher energy. The absorption peak at 1.8 eV is assigned to the transitions from the lowest ^1B_u exciton to a biexciton with ^1A_g symmetry, and the peak at 1.47 eV is assigned to the transition from a ^1B_u exciton to a higher ^1A_u exciton [51].

Figure 7.68 shows the time dependence of the absorbance changes. The bleaching at 1.95 eV appears just after photoexcitation, while the absorbance change at 1.84 eV increases slower than the resolution time. The transient response at 1.84 eV can be fitted to

$$\Delta A(t) = \Delta A_s \{\exp(-t/\tau_s) - \exp(-t/\tau_f)\} + \Delta A_c \quad (7.46)$$

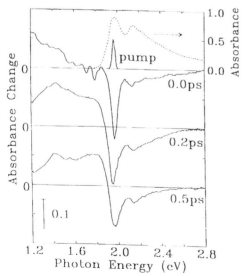

Figure 7.67. The transient absorption spectra of blue-phase PDA-3BCMU on the KCl substrate at 290 K. The stationary absorption (dotted curve) and pump spectra are shown [158].

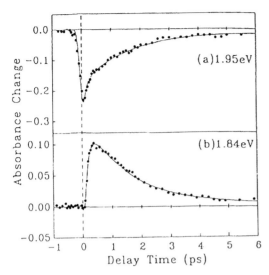

Figure 7.68. The time dependence of absorbance changes at (a) 1.95 and (b) 1.84 eV. The solid curves are the best fit of equations (7.46) and (7.47) to the data at 1.84 and 1.95 eV, respectively. The time constants τ_f and τ_s are (a) 150 fs and 1.4 ps and (b) 150 fs and 1.5 ps, respectively. The resolution times at 1.95 and 1.84 eV are obtained by a cross-correlation method as 200 and 150 fs, respectively [158].

where τ_f and τ_s are the rise and decay time constants, respectively, and ΔA_c is a long-lived component due to triplet excitons. The resolution time at 1.84 eV is obtained at 150 fs by the cross-correlation method and convoluted to equation (7.46) using a Gaussian function. The rise time τ_f corresponds to the time constant of the spectral change due to the thermalization of nonthermal STEs and is obtained as 150 ± 40 fs.

In red-phase PDA-4BCMU, another slower spectral change with the time constant of 1.1 ± 0.1 ps has been observed [98]. The fast spectral change is assigned to relaxation from nonthermal STEs to quasi-thermal STEs among the vibrational modes within a single chain of the polymer. The slow spectral change in red-phase PDA is assigned to relaxation from quasi-thermal STEs to thermal STEs that are associated with energy distribution induced by the coupling between the intrachain and interchain vibrational modes. These fast and slow processes were named in the previous papers as the relaxation from unrelaxed STEs to nonthermal STEs by phonon emission and the thermalization from nonthermal STEs to thermalized STEs, respectively [51,54]. Relaxation kinetics in molecular systems have similar two-step thermalization processes called intramolecular and intermolecular vibrational relaxations. However, the spectral change due to the relaxation from

the quasi-thermal STEs to the thermal STEs could not be observed in blue-phase PDA. The decay time constants τ_s corresponds to the relaxation time constant from the quasi-thermal STEs to the ground state and is obtained as $\tau_s = 1.5 \pm 0.1$ ps at 1.84 eV.

The signals due to nonthermal STEs and instantaneous nonlinear effects such as coherent coupling and Raman gain are expected to be contained in the absorbance change at 1.95 eV. Therefore, the time dependence at 1.95 eV is fitted to

$$\Delta A(t) = \Delta A_0 \delta(t) + \Delta A_f \exp(-t/\tau_f) \\ + \Delta A_s \{\exp(-t/\tau_s) - \exp(-t/\tau_f)\} + \Delta A_c \quad (7.47)$$

where the first term is the term which responds within the pulse duration and ΔA_f corresponds to the absorbance change due to nonthermal STEs. When the time constant τ_f is fixed to 150 fs, the time constant τ_s is obtained as 1.4 ± 0.1 ps. This is consistent with the decay time constant obtained at 1.84 eV. However, when the change in the transient absorption spectra due to thermalization takes place, the over-all decrease in the absorbance change due to depopulation of STEs also takes place. Therefore, the time constants determined at other photon energies are slightly different from those at 1.84 and 1.95 eV. The decay kinetics at 10 K are similar to those at 290 K and the time constants are estimated as $\tau_f = 130 \pm 40$ fs and $\tau_s = 2.0 \pm 0.2$ ps using the time dependence of the absorbance change at 1.82 eV.

The decay kinetics in the epitaxially grown PDA-3BCMU have the same time constants as the cast film of PDA-3BCMU [51]. All the blue-phase PDAs investigated by femtosecond spectroscopy show similar time-resolved spectral changes with similar decay kinetics, i.e. approximately the same time constants, $\tau_f = 150$ fs and $\tau_s = 1.5$ ps at 290 K, and $\tau_f = 2.0$ ps at 10 K [51,54]. Therefore, it can be concluded that the relaxation processes of the photoexcited excitons in blue-phase PDAs in the femtosecond to 10 picoseconds time region are very insensitive to the morphology and the side groups. This strongly supports the argument that the observed spectral changes in this time region are due to intrinsic processes, such as the free exciton self-trapping associated with geometrical relaxation, thermalization, and tunneling from the nonthermal STE to the ground state. These processes are not affected much by the structural defects or impurities even though their concentrations may be moderately high in disordered systems such as cast films.

3.18   Determination of nonlinear optical susceptibility from time-resolved pump–probe measurements

The third-order nonlinear optical susceptibilities have been measured by various methods, i.e. third-harmonic generation, DFWM, optical Kerr effect and absorption saturation. In this study, the nonlinear optical susceptibility is investigated by femtosecond time-resolved absorption spectroscopy. The optical susceptibility χ is defined in the c.g.s. unit system by

$$P = \chi^{(1)}E + \chi^{(2)}EE + \chi^{(3)}EEE + \cdots$$
$$+ \chi^{(n)} \overbrace{E \cdot E \cdot E \cdot E \cdots E}^{n} + \cdots, \quad (7.48)$$

where P and E are the polarization and electric field, respectively, and $\chi^{(n)}$ are the n-th-order nonlinear susceptibilities. The spectrum of the imaginary part of the third-order nonlinear susceptibility, $\text{Im}[-\chi^{(3)}(-\omega_2;\omega_2,-\omega_1,\omega_1)]$ can be obtained by pump–probe absorption spectroscopy, where ω_1 and ω_2 are the frequencies of the pump and probe lights, respectively. Higher-order nonlinear susceptibilities can also be obtained if the time-resolved difference spectrum is dependent on the pump intensity.

When the nonlinear optical susceptibility is to be determined by time-resolved spectroscopy, the value is usually evaluated using the peak intensity of the pulse. In this case, experiments of the same nonlinear optical system with different pump durations give different values of nonlinear susceptibilities. For example, the nonlinear susceptibilities observed by pulses with a duration much shorter than the response time are smaller than those obtained by longer pulses. This is because the signal is proportional not to the peak intensity but to the pulse energy, and shorter pulses give higher peak intensities for the same pulse energy. Therefore, the resolution time of the experiments should be considered in the calculation of the nonlinear susceptibilities.

The observed transient responses in PDA-3BCMU are resolved in time using equation (7.47). The δ-function term is due to instantaneous nonlinear effects which decay within the pulse duration of 100 fs. ΔA_f and ΔA_s correspond to the absorbance changes due to nonthermal STE and quasi-thermal STE, respectively. The time constants are $\tau_f = 150$ fs and $\tau_s = 1.5$ ps at 290 K. The long-lived components ΔA_c is due to the triplet excitons. The lifetime of the triplet excitons has been obtained as 44 μs in PDA-TS and is much longer than the life time constants of the STEs.

The nonlinear optical susceptibility of the transient response with lifetime τ is estimated as follows. When an excitation with lifetime τ is induced by an excitation pulse $I_{ex}(t)$ and probed by a probe pulse $I_{pr}(T)$, the observed absorbance change $\Delta A(t)$ is given by

$$\Delta A = \frac{\int_{-\infty}^{\infty} dt'' I_{pr}(t'' - t') \int_{-\infty}^{t''} dt' a I_{ex}(t') \exp[-(t'' - t')/\tau]}{\int_{-\infty}^{\infty} dt'' I_{pr}(t'')}$$

$$(7.49)$$

where a is a constant which is proportional to the imaginary part of the third-order nonlinear susceptibility.

The absorbance change observed by continuous excitation and probe light is obtained from equation (7.49) as

$$\Delta A_{cw} = a I_{ex} \tau. \quad (7.50)$$

In this case, the third-order nonlinear susceptibility measured by pump–probe spectroscopy is defined as

$$\text{Im}[\chi^{(3)}(-\omega_2; \omega_2, -\omega_1, \omega_1)] = \frac{2.303 \Delta A_{cw} n(\omega_1) n(\omega_2) c^2}{192\pi^2 \omega_2 I_{ex} L}$$

$$(7.51)$$

where $n(\omega_1)$ and $n(\omega_2)$ are the real refractive indices at the frequencies of ω_1 and ω_2, respectively, c is the velocity of light in vacuum, and L is the thickness of a sample.

When the lifetime τ is much shorter than the pulse duration t_p, the observed absorbance change at time zero $\Delta A(0)$ is equal to ΔA_{cw} and the nonlinear susceptibility can be calculated using equation (7.51). When the duration of the pump and probe pulses is nearly equal to or shorter than the lifetime ($t_p < \tau$), the absorbance change at time zero is different from ΔA_{cw}. Assuming that the excitation and probe pulse shapes are rectangular with duration t_p, the observed absorbance change at delay time zero is obtained as

$$\Delta A(0) = a I_{ex} \tau \{1 + [\exp(-t_p/\tau) - 1]\tau/t_p\} \quad (7.52)$$

Even if the pulse shape is not rectangular, the calculated result does not change much, since the pulse has a sech^2 shape. The absorbance change for a delay time longer than the pulse duration ($t > t_p$) is given by

$$\Delta A(t) = \Delta A_1 \exp(-t/\tau)$$
$$= \frac{4 a I_{ex} \tau^2}{t_p} \sinh^2(t_p/2\tau) \exp(-t/\tau) \quad (7.53)$$

Then the response for the continuous excitation and probe ΔA_{cw} can be obtained from equations (7.50) and (7.53) as

$$\Delta A_{cw} = \frac{\Delta A_i t_p}{4\tau \sinh^2\{t_p/2\tau\}} \qquad (7.54)$$

Equation (7.54) can be simplified for $t_p \ll \tau$ as

$$\Delta A_{cw} = \Delta A_i \tau / t_p \qquad (7.55)$$

The nonlinear susceptibility measured by the pulses with duration t_p can be estimated using equations (7.51) and (7.54). The third-order nonlinear optical properties of materials with response time τ are often evaluated by a figure of merit $F_m = \chi^{(3)}/\alpha\tau$, where α is the absorption coefficient at the laser wavelength. The figure of merit can be defined only when the response time and the absorption coefficient are finite. In this study, the nonlinear optical properties are investigated in the wide region from 1.2 eV to 2.8 eV by considering the pulse duration and response times.

The photoinduced absorbance change ΔA is usually proportional to the imaginary part of the nonlinear optical susceptibility as given by equation (7.51). However, the wavelength of the probe light is shifted by the change of the refractive index n. The shift is proportional to the time derivative as

$$\Delta\omega = \frac{\omega L}{c} \frac{dn}{dt} \qquad (7.56)$$

Then the derivative of the probe spectrum appears in the observed absorbance change as

$$\Delta A(\omega) = -\frac{\omega L}{2.303 c I_{pr}} \frac{dI_{pr}}{d\omega} \frac{dn}{dt} \qquad (7.57)$$

The transient transmission change due to the IPM observed in femtosecond spectroscopy of several materials such as red-phase PDA-4BCMU must be corrected. The transmission change due to IPM is observed even in a spectral region where there is no real absorbance change. The artificial absorbance change has been observed in neat liquid of CS_2 [151]. Care must be taken to obtain a real absorbance change spectrum by eliminating the effect of IPM and coherence artifacts, such as perturbed free induction decay [75].

Figure 7.69 shows the signal due to IPM that is expected to be obtained in this study. The white

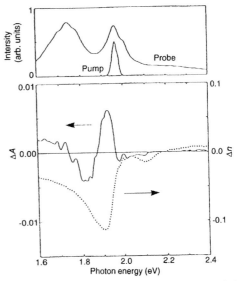

Figure 7.69. The transmittance change due to a photoinduced refractive index change Δn (dotted line) caused by the induced-phase modulation (solid curve). The pump and probe spectra are shown in the upper part [158].

continuum generated from the 1.97 eV pulse has a peak around 2 eV. After passing through a thin blue color filter the probe spectrum has two peaks at 1.97 and 1.75 eV. The refractive index change spectrum Δn shown in the figure is calculated from the absorbance change at 0.2 ps using the Kramers–Kronig (K–K) relations. The application of the K–K relation to obtain the refractive index is only possible when the transmittance change due to IPM is much smaller than the real absorbance change. This condition is found to be satisfied as discussed below. Then the expected signals due to IPM at 0.0 ps are calculated. Since the change of the refractive index due to photoexcited excitons is negative around 1.9 eV, the probe pulse shifts to higher frequency and the observed absorbance change has a positive peak at 1.92 eV. The oscillatory structure comes from the structure of the probe spectrum. The maximum signal due to IPM is expected to be 0.006 and is much smaller than the observed absorbance change in this study. Since the time derivative of the refractive index has a maximum value at 0.0 ps and becomes smaller at a longer delay time, the transmittance changes due to IPM can be neglected in the ΔA_f, ΔA_s, and ΔA_c spectra shown later.

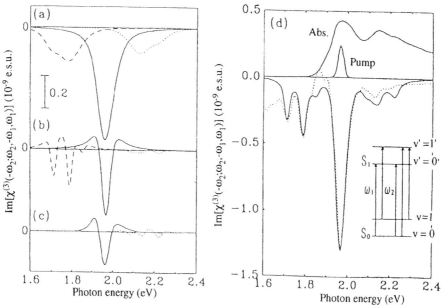

Figure 7.70. The third-order nonlinear susceptibilities calculated using the inset model of a four-level system. (a) Level population terms: equation (7.58a) (solid curve), equation (7.59a) (dashed curve) and equation (7.60a) (dotted curve). (b) Pump polarization coupling terms: equation (7.58b) (solid curve), equation (7.59b) (dashed curve) and equation (7.60b) (dotted curve). (c) Perturbed free-induction decay terms: equation (7.58c) (solid curve) and equations (7.60c) and (7.60d) (dotted curve). (d) The total calculated nonlinear susceptibilities (solid curve) and the observed nonlinear susceptibility of the 'instantaneous' processes (dotted curve). The stationary absorption and pump spectra are shown [158].

3.19 Nonlinear susceptibilities of various nonlinear optical processes with ultrafast response in PDA

The wavelength dependence of the imaginary part of the nonlinear susceptibility due to the instantaneous processes is calculated and shown in Figure 7.70d. Here, the transmittance change due to IPM is corrected in the ΔA_0 spectrum. The spectrum has two negative minima at 1.71 and 1.79 eV, and small dispersion-type structures at 2.15 and 2.23 eV. These structures can be explained by calculating $\chi^{(3)}$ using model shown in Figure 7.70d. Here, zero-phonon ($v=0$) and one-phonon ($v=1$) states of four major vibrational modes are taken into consideration in both exciton (S_1) and ground level (S_0). Since the spectral bandwidth of the pump pulse (0.03 eV) is much smaller than the phonon energies (0.11–0.26 eV) and the 1.97 eV pump pulse (ω_1) is resonant with the 0–0' and 1–1' transitions, the pump resonances with the 0–1' and 1–0' transitions are neglected. The nonlinear susceptibilities of three probe resonant transitions (0–0', 1–0' and 0–1') are calculated

using two-level and three-level density matrix models.

$\chi^{(3)}$ corresponding to the 0–0' exciton transitions is calculated using a two-level model and given by

$$\chi_{00'}^{(3)} = \frac{|\mu_{00'}|^4}{\hbar^3}\left\{ \frac{4\Gamma_x/\gamma_x}{[(\omega_1 - \omega_x)^2 + \Gamma_x^2](\omega_2 - \omega_x + i\Gamma_x)} \right.$$

$$(7.58a)$$

$$+ \frac{2}{(\omega_1 - \omega_x - i\Gamma_x)(\omega_2 - \omega_x + i\Gamma_x)(\omega_2 - \omega_1 + i\gamma_x)}$$

$$(7.58b)$$

$$\left. - \frac{2}{(\omega_2 - \omega_x + i\Gamma_x)^2(\omega_2 - \omega_1 + i\gamma_x)} \right\} \quad (7.58c)$$

where ω_1 and ω_2 are the frequencies of the pump and probe fields, respectively, and $\mu_{00'}$, ω_x, Γ_x and γ_x are the dipole moment, the transition frequency, the transverse relaxation rate and the longitudinal relaxation rate of the exciton, respectively.

$\chi^{(3)}$ corresponding to the 1–0′ transition is calculated using a three-level model as

$$\chi^{(3)}_{01'} = \frac{|\mu_{00'}|^2 |\mu_{01'}|^2}{\hbar^3}$$

$$+ \left\{ \frac{2\Gamma_x/\gamma_x}{[(\omega_1 - \omega_x)^2 + \Gamma_x^2](\omega_2 - \omega_x + \omega_i + i\Gamma_x)} \right. \tag{7.59a}$$

$$+ \frac{1}{(\omega_1 - \omega_x - i\Gamma_x)(\omega_2 - \omega_x + \omega_i + i\Gamma_x)} \left. \right\}$$
$$\times (\omega_2 - \omega_1 + \omega_i + i\Gamma_i) \tag{7.59b}$$

where $\mu_{10'}$ is the dipole moment of the 1–0′ transition and ω_i and Γ_i^{-1} are the frequency and the decay rate of each phonon modes, respectively.

Since the pump pulse is resonant with both the 0–0′ and 1–1′ transition, the nonlinear susceptibility for the 0–1′ transition is rather complicated and obtained as

$$\chi^{(3)}_{01'} = \frac{|\mu_{00'}|^2 |\mu_{01'}|^2}{\hbar^3}$$

$$\times \left\{ \frac{2\Gamma_x/\gamma_x}{[(\omega_1 - \omega_x)^2 + \Gamma_x^2](\omega_2 - \omega_x - \omega_i + i\Gamma_x)} \right. \tag{7.60a}$$

$$+ \frac{1}{(\omega_1 - \omega_x - i\Gamma_x)(\omega_2 - \omega_x - \omega_i + i\Gamma_x)}$$
$$es(\omega_2 - \omega_1 - \omega_i + i\Gamma_i) \tag{7.60b}$$

$$- \frac{1}{(\omega_2 - \omega_x + i\Gamma_x)^2(\omega_2 - \omega_1 - \omega_i + i\Gamma_x)} \left. \right\} \tag{7.60c}$$

$$- \frac{|\mu_{11'}|^2 |\mu_{01'}|^2}{\hbar^3}$$

$$\times \frac{1}{(\omega_2 - \omega_x - \omega_i + i\Gamma_x)^2(\omega_2 - \omega_1 - \omega_i + i\Gamma_i)} \tag{7.60d}$$

where $\mu_{11'}$ and $\mu_{01'}$ are the dipole moments of the 1–1′ and 0–1′ transitions, respectively. The common level of the transition in the terms (7.60a)–(7.60c) is the ground level (S_{00}), while the common level of the term (7.60d) is the excited level ($S_{11'}$).

The imaginary parts of each term and the total of the calculated nonlinear susceptibilities are shown in Figure 7.70 with the observed spectrum. Here, the

phonon energies of the C=C and C≡C stretching vibrational modes are determined from the observed Raman gain signal as 0.180 and 0.257 eV, respectively. The observed pump spectrum is also used for the calculation. The reported values of PDA-TS are used for the energies of the other two vibrational modes, the cross-sections of each transition, and $\Gamma_i = 0.003$ eV [152]. The relaxation rates Γ_x and γ_x and the width of the inhomogeneous broadening are used as variable parameters to fit both the third-order nonlinear susceptibility and stationary absorption spectra. Then the calculation with $\Gamma_x = 0.03$ eV and $\gamma_x = 0.03$ eV gives a good fit as shown in Figure 7.70d.

The terms (7.58a), (7.59a) and (7.60a) for which spectra are shown in Figure 7.70a correspond to the level population term of the transient transmittance change calculated using a two-level system. The three terms are hole burning, stimulated emission and hole burning of phonon sidebands, respectively. The terms (7.58b), (7.58b), (7.60b) correspond to the pump polarization coupling terms. The term (7.58b) is due to the coherent coupling between the pump polarization and probe field. The terms (7.59) and (7.60b) are the stimulated Raman signals in the Stokes (Raman gain) and anti-Stokes sides, respectively. The terms (7.58c) and (7.60c) correspond to the perturbed free-induction decay term. The last term (7.60d) is the inverse Raman scattering.

The peaks due to Raman gain at 1.71 and 1.79 eV are clearly reproduced in the calculated nonlinear susceptibility. The small peaks at 2.14 and 2.22 eV are due to both the phonon sideband hole and inverse Raman scattering. When the pump pulse is nearly resonant to the exciton transition, the inverse Raman signal has a dispersive structure. The inverse Raman signal in this study is diminished by inhomogeneous broadening of the exciton transition and the broad pump spectrum, because of the overlapping of the dispersive structure. The calculated spectrum differs slightly from the observed spectrum around 1.6 and 1.85 eV. It is mainly due to the deviation of the decay kinetics from the biexponential function given by equation (7.47). The relaxation from the nonthermal STE to the quasi-thermal STE has a time constant of 150 fs, but the transient absorbance change due to the spectral shift cannot be exactly fitted to the exponential function. Therefore, small deviations appear in the time-resolved spectrum.

The imaginary part of the third-order nonlinear susceptibility spectra $\mathrm{Im}[\chi^{(3)}_{1111}(-\omega_2; \omega_2, -\omega_1, \omega_1)]$ are estimated also from ΔA_f and ΔA_s spectra using equations (7.51) and (7.54). The nonlinear susceptibility $Im[\chi_f^{(3)}]$ shown in Figure 7.71a is due to the nonthermal STE with a response time of 150 fs. The spectrum has a negative peak at 1.93 eV and is positive below 1.8 eV. $Im[\chi_s^{(3)}]$ in Figure 7.71b is due to the quasi-thermal STE. The response time of $\chi_s^{(3)}$ is 1.5 ps. The spectrum has a positive peak at 1.82 eV and a negative peak at 1.96 eV. The peak value of $Im[\chi_s^{(3)}]$ is 1.0×10^{-8} esu. It is larger than $Im[\chi_f^{(3)}]$ because the response time of the quasi-thermal STE is ten times longer than that of the nonthermal STE. The nonlinear susceptibility of PDA-3BCMU epitaxially grown in KCl is larger than that of PDA-3BCMU cast film [51] because the main chains of the polymers align to the polarization of the pump and probe beams and the inhomogeneous broadening is narrower.

Figure 7.71c shows the nonlinear susceptibility spectrum $Im[\chi_{TE}^{(5)}]$ due to the triplet exciton. The $Im[\chi_{TE}^{(5)}]$ has a positive peak at 1.42 eV and a negative peak at 1.94 eV. Since the triplet excitons are generated by the fusion of two singlet excitons, ΔA_c increases proportionally with the square of the pump intensity. Therefore, the observed response corresponds to the fifth-order nonlinearity. Here, the lifetime of the triplet exciton is assumed to be 44 μs using the reported lifetime in PDA-TS. If pump pulses of 1 ps duration and 100 GHz repetition rate (10 ps interval) are used for excitation, the pulses with the

peak intensity of about 300 kW cm^{-2} give the same amount of signal due to triplet excitons as the ultrafast nonlinear responses due to free excitons and STEs. Since slow nonlinear optical effects prevent the ultrafast responses, attention must be paid to their properties for applications in devices at high repetition rate.

3.20 Femtosecond time-resolved resonance Raman gain spectroscopy of PDA-3BCMU

Time-resolved resonance Raman spectroscopy has been recognized as a powerful method for studying structures of transient species and electronic excited states. Terai et al. have calculated phonon modes of localized excited states in PA and predicted that solitons and polarons can be distinguished by Raman spectroscopy [153]. However, only a few time-resolved Raman experiments have been done in conjugated polymers [154–157] because of the difficulties due to the very short lifetime (about 1 ps) and near-infrared absorption of the excited states. Zheng et al. have investigated picosecond transient Raman scattering in PDA-4BCMU and observed the reduction of the Raman intensity due to the saturation of the Raman cross-section caused by photogenerated excitons [154]. Lanzani et al. have measured transient photoinduced resonance Raman scattering in PA using two-color picosecond dye lasers [155]. However, since the laser pulses are resonant with the absorption from the ground state, the observed signal $\Delta I/I$ is negative and has been explained using the phase space filling model. To the best of our knowledge new phonon modes of excited states in conjugated polymers have not been observed by the transient resonance Raman spectroscopy, because two-color ultrashort pulses, which are resonant with the absorption from the ground state and from the excited state, respectively, and a near-infrared detector with high sensitivity are needed for the experiment. In this section, the configuration of the main chain in a polydiacetylene PDA-3BCMU has been investigated by femtosecond time-resolved resonance Raman gain spectroscopy.

In the experiment the amplified pulse was split into three beams. The first beam (pump 1) was resonant with the excitonic absorption and generated excited states in PDA. The second one was focused in a 3 mm cell containing CCl$_4$ and a femtosecond white continuum was generated by self-phase modulation (SPM). A part of the white continuum was selected with a set of

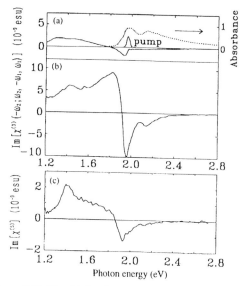

Figure 7.71. The third-order nonlinear susceptibilities due to (a) nonthermal self-trapped excitons, (b) quasi-thermal self-trapped excitons and (c) the fifth-order nonlinear susceptibility due to triplet excitons. The stationary absorption and pump spectra are shown [158].

prism pairs and a slit, and was amplified by a two-stage dye amplifier pumped by the second harmonic of the Q-switched Nd:YAG laser. An interference filter (COR-ION, S10-700-F) was inset between the amplification stage to suppress the amplified spontaneous emission and to stabilize the wavelength. The amplified pulse has a center phonon energy of 1.78 eV and a duration of 200 fs. The 1.78 eV pulse (pump 2) was resonant with the absorption from the STE and was used to the pump pulse of the Raman gain spectroscopy. The probe pulse was white continuum generated from the last beam by SPM and detected with a polychromator/CCD system. Using this technique the time dependence and spectra of photoinduced absorption, bleaching, stimulated emission and Raman gain were observed at the same time. The Raman gain signal was distinguished using the time dependence and sharp structure, because the photoinduced absorbance change and luminescence in PDA have a finite lifetime longer than 150 fs and broad spectra [98,100,114,158,159]. Polarizations of the three beams were parallel to the oriented polymer chains of PDA-3BCMU deposited on a KCl crystal [2]. The experiment was done at room temperature.

Figure 7.72 shows transient absorption spectra induced by the 1.97 eV pulse (pump 1), which is exactly resonant with an excitonic absorption peak. When the pump and probe pulses overlap in time at the sample, three sharp negative peaks are observed. The peak at 1.97 eV is due to the saturation of the excitonic

absorption, and coherent coupling between pump polarization and probe field [158]. The other two peaks at 1.79 and 1.71 eV are due to the Raman gain of the stretching vibration modes of C=C and C≡C bonds, respectively. The photoinduced absorbance change below 1.8 eV shifts to higher energy with time from 0.0 to 0.57 ps. This spectral change has been explained by the relaxation model shown in Figure 7.73 [98,100,158]. The FE (state 1) photoexcited by the pump 1 pulse is coupled with the stretching modes of carbon atoms within the phonon periods of 10–20 fs and becomes nonthermal STE (state 2). Then the nonthermal STE relaxes to the bottom of the potential surfaces (quasi-thermal STE, state 3) by the quasi-thermalization process. Therefore, the transient absorption is shifted to a higher energy and the transient fluorescence of the FE decays very rapidly as observed by newly developed probe saturation spectroscopy (PSS) [159]. The STE relaxes to the ground state (G) by tunneling in the configuration space before thermalization with the time constant of 1.5 ps. The 1.78 eV pulse (pump 2) is resonant with the transition observed at 1.8 eV. This absorption peak is assigned to the transition from the STE to a biexciton with A_g symmetry [100,158].

Transient absorption spectrum induced by the 1.78 eV pulse (pump 2) is shown in Figure 7.74. The

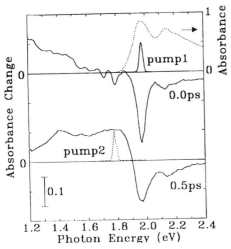

Figure 7.72. Transient absorption spectra induced by the 1.97 eV pulse in PDA-3BCMU. The stationary absorption spectrum and the spectra of the 1.97 eV (pump 1) and 1.78 eV (pump 2) pulses are shown. The excitation photon density is 1.7×10^{15} photons cm^{-2}.

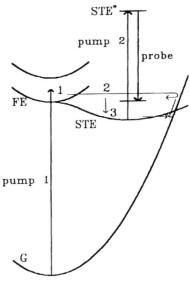

Figure 7.73. Relaxation model of polydiacetylene shown by the potential surfaces of the excitons. States 1, 2, 3 and G are free exciton (FE), nonthermal self-trapped exciton (STE), quasi-thermal STE and the ground state, respectively. STE* is a biexciton state or a higher excited state with A_g symmetry.

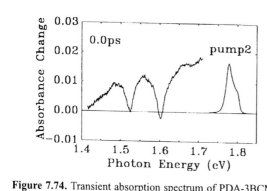

Figure 7.74. Transient absorption spectrum of PDA-3BCMU induced by the 1.78 eV pulse (pump 2). The excitation photon density is 9.0×10^{15} photons cm^{-2}.

pump 2 and probe pulses pass through the sample simultaneously. Two minima due to the Raman gain are clearly observed at 1.60 and 1.52 eV. The broad positive absorbance change is due to the photogenerated excitons, because the 1.78 eV pulse excites the edge of the excitonic absorption. The Raman gain spectra is obtained by subtracting the absorption spectra due to the excitons from the observed absorbance change. The absorption spectra of the excitons are estimated from the data in the regions at 1.42–1.46 eV and 1.67–1.71 eV using a quadratic equation, because the spectra have no sharp structure between 1.4 and 1.8 eV except the Raman gain signals as shown in Figure 7.72.

Figure 7.75a shows the normalized Raman gain spectra obtained using the 1.78 eV pulse at several delay times after the 1.97 eV photoexcitation. At −0.5 ps, two Raman gain peaks are observed at 1440 and 2060 cm^{-1}. They are assigned to the stretching vibrations of the C=C and C≡C bonds in the acetylene-like structure of the ground state. The spectrum at 0.0 ps has a broad signal below the 140 cm^{-1} Raman peak down to 1000 cm^{-1}. At a delay time longer than 0.2 ps the Raman signal has a clear peak at 1200 cm^{-1}. The spectral change of the Raman signal around 1200 cm^{-1} is reproducible and is observed also in PDA-C$_4$UC$_4$. The width of the 2060 cm^{-1} Raman signal becomes slightly broader after the photoexcitation, but no new Raman peak is observed around 2000 cm^{-1}.

Figure 7.75b shows the transient Raman gain change at 1200 and 1440 cm^{-1}. The negative change at 1440 cm^{-1} is explained by the depletion of the ground state due to the formation of the STE. The time dependence is consistent with the decay kinetics of the STE. The signal appears slightly slower than the 1.97 eV pump pulse and decays within several picoseconds. The solid curve is the best-fit curve using the formation time constant of 150 fs and 1.5 ps. The change at 2060 cm^{-1} is also negative and has similar time dependence with the 1440 cm^{-1} signal. The time dependence of the Raman signal at 1200 cm^{-1} has two components. The long-lived component decays within

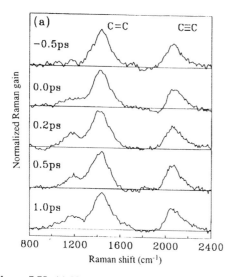

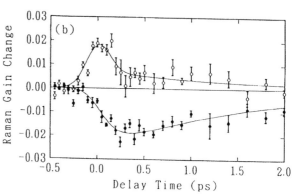

Figure 7.75. (a) Normalized resonance Raman gain spectra at several delay times and (b) transient Raman gain changes at 1200 cm^{-1} (open circles) and 1440 cm^{-1} (closed circles) of PDA-3BCMU after the 1.97 eV photoexcitation. The solid curves in (b) are the best-fitted curves with time constants of 150 fs and 1.5 ps. The resolution time is 300 fs. The excitation photon densities of the 1.97 eV and 1.78 eV pulses are 2.9×10^{15} and 9.0×10^{15} photons cm^{-2}, respectively.

several picoseconds and is assigned to the STE. The short-lived component has a time constant shorter than the present resolution time of 300 fs and is probably due to the nonthermal STE, because the 1.78 eV pulse can be resonant with the transition between the nonthermal STE and the ground state.

The Raman signal of the ground state in PDA has a weak peak assigned to the bending mode of the C=C bond around 1200 cm^{-1} [152,160]. Therefore, the observed 1200 cm^{-1} Raman signal may be explained by the increase in the Raman cross-section of the bending mode. However, since the 1200 cm^{-1} spectrum at 0.0 ps is broader than that at later times, this spectral change cannot be explained by the increase of the cross-section. It suggests that the geometrical relaxation of main-chain configuration due to the self-trapping process takes place after the 1.97 eV photo-excitation. Since the wave packet of the nonthermal STE oscillates in the configuration space as shown in Figure 7.73, the Raman signal due to the nonthermal STE is broad. After quasi-thermalization, the nonthermal STE relaxes to the bottom of the STE potential surface. Therefore, the oscillatory motion in the configuration space becomes smaller and the clear Raman peak is observed at 1200 cm^{-1}.

The theoretical calculation has predicted that the localized excitations in *trans*-PA have several Raman-active phonon modes [153]. The expected signal is the reduction of the stretching vibration modes and new Raman lines at lower frequencies than the stretching modes. The Raman signal observed in PDA is similar to this feature. However, the phonon modes of the STE in PDA have not been investigated. Here, the observed Raman frequency is compared with the stretching modes of center bonds in unsaturated hydrocarbons with four carbon atoms, i.e. repeat units of PDA [161]. The formation of the STE in PDA is expected to be the geometrical relaxation from the acetylene-like structure $(=CR-C\equiv C-CR=)_x$ to the butatriene-like structure $(-CR=C=C=CR-)_x$. The C=C bond in *trans*-butene-2 $(CH_3-CH=CH-CH_3)$ has a stretching mode at 1675 cm^{-1}, while the frequency of the C−C bond in *trans*-1,3-butadiene $(CH_2=CH-CH=CH_2)$ is 1202 cm^{-1}. Therefore, the 1200 cm^{-1} Raman peak can be assigned to the C−C bond in the butatriene-like structure. However, the Raman signal due to the C=C bond in the butatriene-like structure cannot be observed in this study. It may be explained by close frequencies of the stretching modes of the center C=C bond on butatriene $(CH_2=C=C=CH_2)$ and the C≡C bond in dimethylacetylene $(CH_3-C\equiv C-CH_3)$, which are 2079 and 2235 cm^{-1}, respectively. The expected new Raman

signal near the 2060 cm^{-1} peak is not resolved in this study because of the broad pump spectrum.

The intensity of the 1200 cm^{-1} peak is about one third of that of the 1440 cm^{-1} peak, as shown in Figure 7.75a and the change at 1200 cm^{-1} is also smaller than the change at 1440 cm^{-1}. The resonance Raman signal of the ground state pumped by the 1.78 eV pulse is estimated to be about 15% of that pumped by the 1.97 eV pulse using the homogeneous and inhomogeneous widths of the excitonic absorption of the PDA-3BCMU [98]. The absorbance of the sample at 1.97 eV is 0.8, while the maximum transient absorbance at 1.78 eV is 0.13 when the STE is generated by the 1.97 eV pulse with a density of 2.9×10^{15} photons cm^{-1}. Therefore, if the ground state and STE have the same Raman cross-section, the Raman signals due to the STE and the ground state are estimated to have almost the same intensity. The smaller 1200 cm^{-1} signal indicates that the Raman cross-section of the STE is smaller than that in the ground state. It is mainly due to the broader absorption spectrum of the transition from the STE to higher excited states than the excitonic absorption from the ground state.

3.21 Optical Stark shifts of Raman gain spectra in polydiacetylene

Optical Stark shifts of the exciton transition in the conjugated polymer polydiacetylene, PDA-pTS [77], have been shown to result from the interaction of a laser field with either the ^1B_u exciton $\leftarrow$ ^1A_g zero-phonon ground-state transition or the ^1B_u exciton $\leftarrow$ ^1A_g (v_1 and v_2) one-phonon transitions [162]. The present work extends the previous experimental studies of the optical Stark effects in 1D conjugated systems which employed nonresonant excitation conditions at low pump intensities [77]. In this investigation, the Stark shift behavior of the Stokes Raman signal in a PDA is studied using a near-resonant ultrashort laser excitation with high pump intensities.

Femtosecond pump–probe spectroscopy is performed with a 10 Hz amplified colliding-pulse mode-locked dye laser yielding about 100 fs pulses. White-light continuum generation of probe pulses is employed, where the linearly polarized pump and probe beams are oriented parallel to the conjugated chain direction in the polymer, and are overlapped on the free-standing crystal sample in air at 297 K. The relative intensity of the pump beam, with a beam diameter of 0.04 cm, is adjusted with a variable neutral density filter. Determination of the pump flux is hampered by experimental uncertainties in both the pulse-to-pulse intensity fluctuation and spot size, which are estimated to be 20–40% and 20%, respectively.

The experiments are performed with a pump photon energy, $\omega_p = 16\,010$ cm^{-1}(1.98 eV), which is lower in energy than the peak of the absorption band, at 18 270 cm^{-1}(2.27 eV), due to the ^1B_u exciton transition [67]. In the photoinduced absorption spectra of PDA-DFMP, at delay times from $t = -170$ to 200 fs, two peak minima can be observed and are attributed to Raman gain (RG) signals [163]. At the weakest pump intensity the gain signals that occur at probe photon energies ω_{pr} (I_{min}) are about 14 510 cm^{-1} (1.80 eV) and 13 845 cm^{-1} (1.72 eV). The measured RG Stokes shifts of about 2165 and 1500 cm^{-1} are in approximate agreement with the ground state stretching mode frequencies of about 2135 cm^{-1} (0.265 eV) for v_1 and 1500 cm^{-1} (0.186 eV) for v_2, previously obtained by resonant Raman scattering in PDA-DFMP [67]. The RG spectra for both v_1 and v_2 are measured for six

different pump intensities varying from $I_{min} = 19$ to $I_{max} = 58$ GW cm^{-2}, at a delay time of $t = 0$ ps. In Figure 7.76 the RG spectra are plotted as a function of the optical Stark shift of the peak probe energy, which is expressed as, $\omega_{pr}(I_{min}) - \omega_{pr}(I)$, where $\omega_{pr}(I)$ is the peak probe energy at a given pump intensity.

At the highest pump intensity, the peak probe photon energies for the RG signal $\omega_{pr}(I_{max})$ are about 13 900 cm^{-1} (1.72 eV) for v_1 and 14 550 cm^{-1} (1.80 eV) for v_2, corresponding to a Stark shift of the peak probe energy of approximately -55 and -40 cm^{-1}, respectively.

For the purposes of visual comparison of the RG spectra which vary in signal amplitude with the pump intensity, the spectral peaks of the RG signals in Figure 7.76 are displayed in a normalized fashion. Additional ultrafast photoinduced absorption signals are observed in PDA-DFMP at the probe energy regions near the RG signals. These photoinduced absorption signals are discussed elsewhere [51,129], and have been attributed

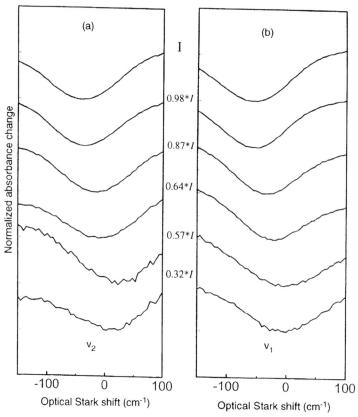

Figure 7.76. The experimental RG spectra plotted as the normalized absorbance change vs. the probe photon energy ω_{pr} at six different pump intensities (where $I = 58$ GW cm^{-2}) for (a) v_2 and (b) v_1.

to the formation of a STE state and decay process to the ground state, where the exponential formation and decay constants are about 100 fs and 1.3 ps, respectively. It is noted that the RG peaks can be both temporally and spectrally distinguished from other photoinduced absorption signals.

Before the presentation of the two schemes used to understand the origin of the observed optical Stark shifts, a brief examination of exciton optical coupling (EOC) and phonon-mediated optical coupling (PMOC), the two types of optical field-level interactions relevant to the RG process in polydiacetylene, is necessary. In this chapter, the energy levels of polydiacetylene are treated as a three-level system consisting of a zero-phonon ^1A_g ground state ($|a\rangle$), a thermally unpopulated ground state one-phonon level (either v_1 or v_2) ($|b\rangle$) and a ^1B_u exciton state ($|c\rangle$), where ω_{ac}, ω_{bc} and ω_{ab} are the transition energies.

In the EOC case, the pump laser field interacts with the $|a\rangle$ and $|c\rangle$ states, at a Rabi energy given by $R_2 = \mu_{ac}E$, where E is the applied field strength of the pump laser and μ_{ac} is the transition dipole moment. Over the range of experimental pump intensities, R_2 varies from 70 to 210 cm^{-1} (0.0087 to 0.0260 eV), where $|b\rangle$ is either v_1 or v_2 [164]. Under EOC, the $|a\rangle$ state can be Stark shifted to either lower or higher energies depending upon the sign of the excitonic detuning, while the $|b\rangle$ state is not coupled by the optical field. The effects of EOC in PDA have been previously observed by inverse Raman spectroscopy under small-signal conditions (i.e. I_p about 10^7 W cm^{-2}) [77]. Assuming a two-level system in the weak-field limit ($\Delta_{EOC} > R_2$), with the experimental parameters of a pump laser detuning of $\Delta_{EOC} = \omega_{ac} - \omega_p = 2260$ cm^{-1} and $R = 210$ cm^{-1}, a rough estimate of the maximum EOC shift is given by Δ about $(R_2^2/2\Delta_{EOC}) = 10$ cm^{-1}.

In the PMOC case, the Rabi energy R_3 optically couples the $|b\rangle$ and the $|c\rangle$ states. When the pump laser is both detuned from the exciton transition energy (i.e. $\omega_p < \omega_{ac}$) and the condition $\omega_p + \omega_{ab} \leqslant \omega_{ac}$ is satisfied, the $|b\rangle$ and $|c\rangle$ states are shifted to higher the lower energies, respectively, resulting in the *phonon-mediated optical Stark effect* [77,134]. Over the range of experimental pump intensities, R_3 varies from 93 to 280 cm^{-1} (0.0115 to 0.0347 eV) and 80 to 240 cm^{-1} (0.0099 to 0.0298 eV) for v_2 and v_1, respectively [77]. For v_2, where $R_3 = 280$ cm$^{-1} \geqslant \Delta_{ph}$ and where $\Delta_{ph} = \omega_{ac} - \omega_p - \omega_{ab} = 760$ cm^{-1}, Stark shifting of the $|b\rangle$ and $|c\rangle$ states is expected to be significant. We note that since neither the photoinduced absorption changes at the *exciton*-absorption

transition energy nor the inverse Raman signal can be experimentally detected, Stark-shift behavior due to PMOC cannot be directly monitored here.

To assess the relative contribution of EOC and PMOC to the data, two model calculations are performed. The first model includes the effect of EOC on the RG signal, while the second model includes the joint effects of both EOC and PMOC. The model calculations each yield a characteristic direction for the Stark shift of the RG signal, so the better model can be distinguished.

The effect of an intense pump field on the Stokes-detected Raman process has been the subject of numerous theoretical treatments [165–167]. In experiments where an ultrashort, high-intensity pump is nearly resonant with an electronic transition with a relatively large transition moment, the conventional approach of applying a specific order of the susceptibility in the perturbation expansion of the macroscopic polarization in powers of the optical field strength may not be particularly meaningful. A more unified, nonperturbative treatment for calculating an optically coupled few-level system with arbitrarily high pump intensities is required. The method developed by Dick and Hochstrasser [165,168] satisfies these requirements and is employed here.

In the model calculations presented below, the RG process in PDA in the presence of intense pump laser excitation is modeled by choosing the appropriate coupling of the near-resonant optical fields to a three-level system (i.e. EOC or EOC and PMOC). The technique employed here, to evaluate the optical-field interactions in a three-level system, closely follows the density matrix approach in refs. 168 and 165, which is briefly outlined as follows: under (near) resonant conditions, the Liouville equation of the density operator for a few-level system, including the potential in the dipole approximation, is transformed to a time-independent linear equation by removing all the rapidly varying terms. In the three-level system considered here, the time-independent (transformation) matrix representation of the Liouville superoperator $\mathbf{X}$ is a 9×9 matrix which includes the off-diagonal optical coupling Rabi energy terms R_1, R_2, and R_3. By assuming Markovian relaxation dynamics, damping parameters such as electronic dephasing rates (Γ_{ca} and Γ_{cb}), a vibrational relaxation rate, (Γ_{ba}) and the population decay rates (Γ_{cc} and Γ_{bb}) are introduced as off-diagonal elements in $\mathbf{X}$. The rotating frame for this three-level system assumes that the transition energies ω_{ca} and ω_{bc} (i.e. the Stokes emission) are considered to be nearly resonant with the pump energy

ω_p, and the probe energy ω_{pr}, respectively. For a given set of Rabi energy values, the solution of the steady-state equation of motion is found by numerically calculating the inverse of the full 9×9 Liouville superoperator, $\mathbf{X}'$ (given as equation (34) in ref. [168]) over a range of probe energies. The calculated spectrum $q(\omega_{pr})$ is obtained by evaluating the correlation functions (shown in ref. [165]) upon the inverse matrix elements of the Liouville superoperator $\mathbf{X}$ in the rotating frame. For a given set of Rabi field-strength values, the spectrum, $q(\omega_{pr})$, at a particular probe energy is evaluated as [165]

$$q(\omega_{pr}) = \mathrm{Re}\{\mathbf{X}'_{cbab}\rho^o_{ac} + \mathbf{X}'_{cbbb}\rho^o_{bc} - \mathbf{X}'_{cbcc}\rho^o_{bc}\} \quad (7.61)$$

where ρ^o_{ac} and ρ^o_{bc} are the steady-state density operators associated with the (Stokes) Raman process and the relevant coherence terms, respectively. Partitioning the spectrum into the (resonant) Raman and spontaneous emission (i.e. fluorescence) components, as identified by Shen [169], the above expression neglects contributions from the later components (i.e. population steady-state density operators ρ^o_{bb} and ρ^o_{cc}).

3.21.1 Model with only EOC

Previous investigations have examined the effect of high-pump excitation intensities on the Stokes Raman spectra of β-carotene [170] and hemoglobin [171]. The observed broadening of the Raman signal was found to be in good agreement with a model which included the optical coupling between $|a\rangle$ and $|c\rangle$ states in a three-level system using the analytical expression prescribed by ref. [165]. However, at smaller pump intensities of about 10^9 W cm^{-2}, Stark-induced shifts of the Raman signal were not reported.

In the first model, the $|c\rangle \leftarrow |a\rangle$ transition is optically coupled by the pump laser which is detuned from the exciton transition by $\Delta_{EOC} = 2260$ cm^{-1} (as shown in Figure 7.77. The 9×9 density matrix is evaluated using the parameters listed in Figure 7.77. Using equation (7.61) the calculated spectrum is plotted (in Figure 7.77) for the case where $|b\rangle$ is v_1. This model predicts that as the Rabi energy R_2 is varied from 70 to 210 cm^{-1}, a Stark shift of 16 cm^{-1} occurs for the RG signal of v_1, where the optical Stark shift for a given Rabi value R is defined by $\omega_p - \omega_{ac} - \omega_{pr}(\max)$, and $\omega_{pr}(\max)$ is the peak energy of the maximum (negative) signal. Using this definition, a positive value for the Stark shift corresponds to a (red)-shift of the RG signal to a lower probe energy, while a negative value corresponds to a blue shift. Similar

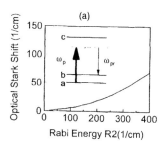

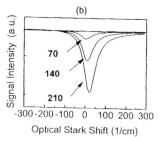

Figure 7.77. Model with EOC mechanism. (a) The calculated optical Stark shift as a function of the Rabi energy. (b) The calculated RG spectrum at various Rabi energies $R_2 = 70$, 140 and 210 cm^{-1} for the third-level model evaluated with the following parameters: $\omega_{ca} = 18\,270$ cm^{-1}; $\omega_{ab} = 2135$; $\omega_p = 16\,010$ cm^{-1}; $\Gamma_{ca} = \Gamma_{cb} = 500$ cm^{-1}; $\Gamma_{ba} = 30$ cm^{-1}; $\Gamma_{cc} = \Gamma_{bb} = 0$ cm^{-1}. In the inset, the pump laser field (thick arrow) is shown coupled to the excitonzero-phonon ground state and the probe transition (thin arrow) is also depicted.

shifts are found for v_2. Additionally, since R_2 exceeds the vibrational relaxation rate, Γ_{ba}, some spectral broadening of the RG signal is predicted (about 7 cm^{-1}), although no attempt has been made to account for either inhomogeneous broadening or the spectral width of the pump laser (about 150 cm^{-1}).

The spectrum for the RG signal, as evaluated by equation (7.61) has no contribution from either the second or the third terms when only the EOC mechanism is operable. It is noted that this calculation is equivalent to that of the analytical expression of the reduced 2×2 density matrix approach used by Dick and Hochstrasser [165].

The predicted red shift of the Stokes-detected Raman signal with increasing Rabi energy due to EOC only, is explained as follows. With a near-resonant pump, an optical Stark shift of the $|a\rangle$ and $|c\rangle$ states to lower and higher energies, respectively, results. The gain signal is monitored at the difference frequency between the pump (which is coincident with a virtual exciton $\leftarrow$ Stark shifted-$|a\rangle$ transition) and the (constant) phonon energy, ω_{ab}. As the $|a\rangle$ state undergoes a shift to a lower energy, the RG signals yield a red optical Stark shift. Since the direction of the calculated intensity-dependent Stark shift is inconsistent with the experimental results, *the observed Stark shift cannot be explained by only exciton optical coupling effects.*

3.21.2 Model with both EOC and PMOC

In this scheme, the same pump laser, which optically couples the $|c\rangle \leftarrow |a\rangle$ and the $|c\rangle \leftarrow |b\rangle$ transitions, is detuned by $\Delta_{EOC} = 2260$ cm^{-1} (for both v_1 and v_2) and $\Delta_{PMOC} = 125$ and 760 cm^{-1} for v_1 and v_2, respectively (as shown in the inset of Figures 7.78a and b). The 9×9 density matrix is separately evaluated with $|b\rangle$ as either v_1 and v_2 using the parameters listed in Figure 7.78 and the following range of values for the Rabi energies: R_2 is varied from 0 to 210 cm^{-1} (for either v_1 and v_2) and R_3 is varied from 0 to 240 cm^{-1} and 0, to 280 $^{-1}$ for v_1 and v_2, respectively. The calculated spectrum, $q(\omega_{pr})$, is evaluated using equation (61).

Over the range of the experimental pump intensities is found that by increasing the Rabi energies the peak of the RG signal is found to shift to higher probe photon energies by 53 and 58 cm^{-1}, for v_1 and v_2, respectively. Calculated spectra for v_1 and v_2 are shown in Figures 7.78c and d, respectively. The direction of the resulting blue shift is consistent with the experiment, and the magnitude of the calculated shift is in semi-quantitative agreement with the data. The agreement between the experimental and calculated values is quite reasonable given the experimental uncertainty in the pump intensity (i.e. Rabi energy). Any broadening effects due to assumed broadening parameters have a negligible effect on the spectral width of the RG signal, since the experimental spectral width is dominated by inhomogeneous broadening and the spectral width of the pump laser, which are not included in the calculation.

In the lower Rabi energy regime ($R_3 \leqslant 140$ cm^{-1}), the first term of equation 7.78, the resonant Raman term, dominates the calculated spectral signal, resulting in a Lorentzian-like peak. With increasing Rabi energies, the peak of this component slightly shifts to the blue region by about 8 cm^{-1}. As $R_3 \geqslant 140$ cm^{-1}, this component spectrally broadens producing a tail only on the red side of the spectral peak. Contributions from the coherent components, the second and third terms of equation 7.78, to the spectral profile only become significant in the moderate Rabi-energy regime

($R_3 \geqslant 140$ cm^{-1}). However, the most important contribution to the blue shift of the peak is from these coherent terms. As the Rabi energy increases, the peak of this component shifts to higher photon energies. The second term has a pronounced asymmetric shape, with positive (absorptive) and negative (emissive) going spectral components at higher and lower photon energies, respectively. When $R_3 \geqslant 200$ cm^{-1}, the large amplitude of the absorptive component of this term effectively produces a dip on the red side of the calculated spectrum.

Although the EOC mechanism is intrinsic to the RG process in the presence of a near-resonant high-intensity pump, due to the smaller pump detuning for PMOC relative to the EOC process (i.e. $\Delta_{PMOC} < \Delta_{EOC}$), the former process can dominate the Stark-shift behavior. The larger effect of PMOC relative to EOC is consistent with the inverse Raman spectroscopy of polydiacetylene-p-toluenesulfonate (PDA-TS) measured by Blanchard et al. [77] in the small-signal limit. At sufficiently high Rabi energy, R_3, $|b\rangle$ shifts to lower energy. Since the relative size of the shift of $|b\rangle$ should be larger than that of $|a\rangle$, this results in an increase in the peak probe energy of the RG spectra. The sign of the optical Stark shift predicted by this mechanism is consistent with the observed direction of the shift. A check of the calculation, where the pump energy is charged such that $\omega_p > \omega_{ac}$, yields a positive Stark shift.

Additionally, a slight difference in the calculated optical Stark-shift behavior for v_1 and v_2 is controlled by two factors; the size of the transition moment μ_{bc}, and the detuning energy Δ_{PMOC}. The larger transition moment μ_{bc} for v_2 (and hence a larger R_3) than for v_1 would suggest a greater shift for v_2, while a smaller pump detuning Δ_{PMOC} for v_1 than for v_2 would suggest a greater shift for v_1. The model calculation suggests that the former effect is the more important one.

In comparison with previous treatments of coherent nonlinear optical effects, the models discussed here differ in two respects. First, these three-level system models include a mechanism in which the combined effects of EOC and PMOC are present. The net result is that all the states in the three-level system are optically coupled and can each exhibit level shifts. Second, the calculational technique employed here can be evaluated with an arbitrarily high pump intensity for an optically coupled few-level system without the need to associate a specific order of the susceptibility in the expansion of the macroscopic polarization to the nonlinear optical effect. Previous treatments of optical Stark effects have been explained using three-level systems where only

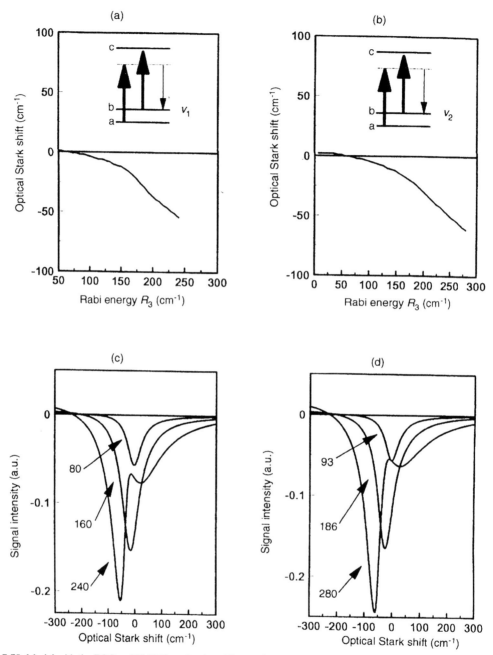

Figure 7.78. Model with the EOC and PMOC mechanism. The maximum optical Stark shift as a function of Rabi energy R_3 for v_1 (a) and v_2 (b) calculated with the following parameters: $\omega_{ca} = 18\,270$ cm^{-1}; $\omega_{ab} = 2135$ and 1500 cm^{-1} for v_1 and v_2, respectively; $\omega_p = 16\,010$ cm^{-1}; $\Gamma_{ca} = \Gamma_{cb} = 500$ cm^{-1}; $\Gamma_{ba} = 30$ cm^{-1}; $\Gamma_{cc} = \Gamma_{bb} = 0$ cm^{-1}. The calculated spectrum vs. the optical Stark shift for (c) v_1 at $R_3 = 80$, 160 and 240 cm^{-1} and (d) v_2 at $R_3 = 93$, 186 and 280 cm^{-1}.

two of the states are optically coupled with a pump laser. Analogous to the inverse Raman process, the effects of PMOC have been explained using a three-level $\chi^{(3)}$ density matrix calculation. For example, Saikan *et al.* [172] have shown that with moderate Rabi energies and when $\omega_p + \omega_{ab} \approx \omega_{ac}$ (i.e. conditions for PMOC), the splitting of the $|b\rangle$ and $|c\rangle$ states give rise to the 'hole' in the spectrum, similar to experimental observations by Greene *et al.* in PDA-TS [134]. Additionally, three-level $\chi^{(3)}$ density matrix models have been used to explain optical Stark effects in both Cu_2O [80] and multiple quantum well structures [80], when an intense field can optically couple with the $|c\rangle \leftarrow |b\rangle$ transition. The features predicted by the $\chi^{(3)}$ three-level system models have been found to be consistent with the nonlinear optical behavior calculated using the more rigorous theory of phonon-mediated changes in susceptibility [77,134].

3.22 Identification and separation of spectral change due to a temperature increase in nanosecond transient absorption spectra of PDA-3BCMU

In nanosecond time-resolved spectroscopy, it is important to identify and separate the effect of a temperature increase from the photoinduced signal of the sample. Both the transient reflection and absorption are often affected by the thermal effects. Usually the thermal effect can be distinguished by the methods described below. One is a comparison between the photoinduced absorption and thermoabsorption spectra. However, sometimes there is a possibility that the photoinduced absorption, which has no contributions from thermal effects, shows a similar spectrum to the thermoabsorption one. Therefore, this method is not always useful. Another method is a comparison of either the absolute value or the time dependence of the absorbance change between the observed and calculated values. This method is often not practical, since the physical constants of the sample such as the heat capacity and thermal conductivity, are usually not well known.

A newly designed method is proposed here. This method is applicable for a sample which can be doped uniformly in a polymer matrix such as polymethylmethacrylate (PMMA). This method was applied for photoinduced absorption of PDA-3BCMU.

In this section, unless indicated otherwise, a cast film means a film of PDA-3BCMU, which is not doped in PMMA.

Cross-sectional schematics of the samples are shown in Figure 7.79. Thin films of PDA-3BCMU doped in PMMA are formed on PMMA substrates. The doped films have the following properties:

(1) the density of PDA-3BCMU is small enough, and the thermal properties of the entire sample including the PMMA substrate are considered to be uniform;
(2) the absorbance of a film is small enough, and the number density of excitation is considered to be

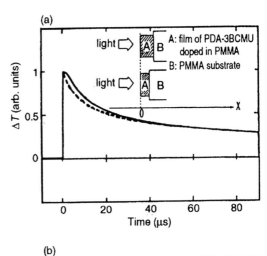

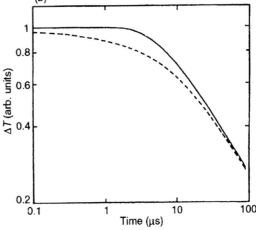

Figure 7.79. The calculated decay curves of the temperature increase ΔT. The solid and dotted curves represent the time dependence of ΔT at the surface of the sample and the averaged temperature increase in the excited volume, respectively. The cross-sectional schematics of the thin films of PDA-3BCMU doped in PMMA matrices on PMMA substrates are shown. Two samples with different thickness were prepared as schematically shown in the inset of (a).

constant along the x-coordinate defined in Figure 7.79.

If these conditions are satisfied, the temperature increase at a position x of the sample and time t after excitation is

$$U(x, t) = \frac{\theta}{2}\left[\phi\left(\frac{d-x}{2at^{1/2}}\right) + \phi\left(\frac{d+x}{2at^{1/2}}\right)\right] \quad (0 \leqslant x \leqslant d)$$
(7.62)

$$a^2 = K/\rho C_p$$
(7.63)

$$\phi(x) = \frac{2}{\pi^{1/2}}\int_0^x \exp(-\xi^2)\mathrm{d}\xi$$
(7.64)

where d is the film thickness, θ is the temperature increase of the film at $t=0$, and ρ, C_p, and K are the mass density, the specific heat capacity, and the thermal conductivity of PMMA, respectively. The temperature increase at the surface $x=0$ is obtained from equation (7.62)

$$U(0, t) = \theta\phi[(t/\tau)^{-1/2}]$$
(7.65)

where the characteristic time is given by

$$\tau = d^2/4a^2$$
(7.66)

If the time constant τ obtained from the observed decay curve depends on the sample thickness according to equation (7.66), then the component of the time constant τ is due to the thermal effect.

Therefore, the contribution of the thermal effect can be distinguished. The merit of this method is as follows:

(1) The time constant τ can be estimated using only the properties of PMMA which are well known.
(2) By preparing samples with two different thicknesses, the thermal effect can be clearly distinguished from the thickness dependence of the time constant.

In this analysis, attention must be paid to the following three points: (i) The temperature dependence of the values of K, C_p and ρ. (ii) The relation between the temperature increase at ΔT and the local absorbance change ΔA. (iii) The absorbance change at each position x, due to the thermal effect, is assumed to be proportional to ΔT; therefore, the observed time dependence of the absorbance change ΔA should be proportional to

$$U_{\mathrm{av}}(t) = \frac{1}{d}\int_0^d U(x, t)\mathrm{d}x$$
(7.67)

and the time constant τ in equations (7.65) must be obtained from the observation of the time dependence of the $U_{\mathrm{av}}(t)$.

In the experiment carried out here θ is estimated at approximately 10 K; therefore, the temperature dependence of the values of K, C_p and ρ are negligible, and ΔA is considered to be proportional to ΔT within the temperature change.

Figure 7.79a shows the calculated time dependence of the temperature increase of the sample. The solid and dotted curves represent the temperature change $U(0, t)$ at the front surface of the film and the averaged temperature increase $U_{\mathrm{av}}(t)$, respectively. The logarithmic plots of the time dependence are displayed in Figure 7.79b. The curves are calculated using the sample thickness $d = 1.63$ mm, $\rho = 1.18 \times 10^3$ kg m^{-1}, $C_p = 1.47 \times 10^3$ J kg^{-1}K^{-1} and $K = 0.21$ W m^{-1} K^{-1} [23]. The time constant of the curve was calculated to be 5.5 μs. The decay curves of $U(0, t)$ and $U_{\mathrm{av}}(t)$ have the same value initially and at long delay times. Therefore, if the curve $U(0, t)$ is varied by changing the time constant τ to fit the observed decay curve, which may be due to the thermal effect, in particular at initial and long decay times, the time constant τ can be estimated. In the experiment the difference of two curves was within statistical error.

Cast films of the PDA-3BCMU dispersed in PMMA matrices (3.8 and 0.67 wt%) were formed on PMMA substrates. Samples of 2.4 and 1.6 μm thickness were prepared for the 3.8% film.

Nanosecond time-resolved absorption spectroscopy was performed using the pump and probe method described in ref [52]. The excitation photon energy was 2.33 eV and the photon number density was approximately 2×10^{16} photons cm^{-2}. The photoinduced absorption was recorded with the sample in vacuo below 3.5×10^{-5} Torr at room temperature. The diameter of the excited area was approximately 3 mm. With these conditions thermal diffusion into the substrate along the x-coordinate was the dominant process for the heat transfer from and within the film.

Figure 7.80 shows the time dependence of the photoinduced absorption of PDA-3BCMU dispersed in PMMA measured for the 2.4 and 1.6 μm films. The absorbance change was observed at 1.97 eV. Fast and slow decays were found as shown in Figure 7.80a. The fast component was independent of the thickness of the films, while the long component exhibits clear thickness dependence as shown in Figure 7.80b. This dependence indicates that the longer component is due to the thermal effect.

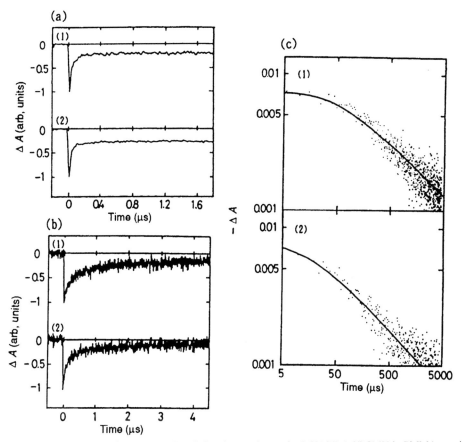

Figure 7.80. The time dependence of the photoinduced absorbance change for 3.8% PDA-3BCMU in PMMA matrix at different time scales: 2 µs (a) and 5 ms (b) and (c). The thicker [2.4 µm (1)] and thinner [1.6 µm (2)] film samples were measured at room temperature. Logarithmic plots for the longer time scale for the thicker (1) and thinner (2) films are shown in (c). The solid lines are the best-fit curves using equation (7.65).

The logarithmic plots of the time dependence of the long component are displayed in Figure 7.80c. From the fitting curves using equations (7.65), the time constants were 43 and 16 µs for the thicknesses of 2.4 and 1.6 µm, respectively, suggesting the thermal effect, because the ratio of the time constants, 2.7, is close to the square of the thickness ratio, 2.3.

The calculated time constants were 12 and 5.3 µs for the thick (2.4 µm) and thin (1.6 µm) samples, respectively. These values were calculated using the material constants of PMMA and the thicknesses of the samples. The deviation of the two calculated constants (12 and 5.3 µs) from the observed values (43 and 16 µs) was mainly due to the error of the thickness estimation. The estimated values were probably smaller than the real ones, because the measured

absorption coefficient of a cast film and the weight density of the PDA-3BCMU in the PMMA matrix were used for the estimation. In general in a cast film of PDA-3BCMU, the polymers are thought to be stacked parallel to the surface of the substrate; therefore, a cast film is expected to yield a larger absorption coefficient than a PMMA film in which the orientations of the PDA-3BCMU polymer main chains are distributed more randomly, even though the densities of PDA-3BCMU of both films per unit area are the same.

The important point here is that the calculated time constants vary quadratically with the thickness of the sample. From the thickness dependence of the time constant, it was clearly found that the slow component was due to the temperature increase of the sample.

The time dependence of the photoinduced absorption of the 0.67% PDA-3BCMU film is shown in Figure 7.81a. Figure 7.81b shows the logarithmic plot of the fast component of the decay curve. Since the plot yields a straight line, the time dependence is $\Delta A \propto t_b$.

The value of the power b was estimated to be -0.51 ± 0.02 from the best-fit curve represented by solid curve in the figure. In the film of a low-density PDA-3BCMU, the chains of PDA-3BCMU were considered to be well isolated from each other. Therefore, neither intermolecular relaxation processes nor thermal diffusion process from large aggregates of PDA-3BCMU polymers need to be considered. The excitations which relax via a geminate recombination process after one-dimensional random walk show a decay process that closely fits a $t^{-1/2}$ time dependence. Hence, the fast component which reveals a $t^{-1/2}$ time

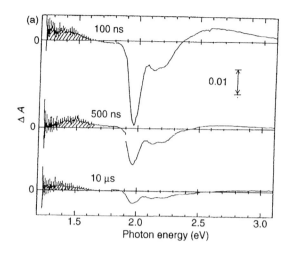

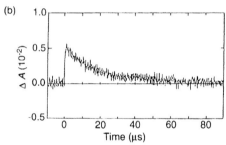

Figure 7.82. (a) The photoinduced absorbance change of the PDA-3BCMU cast film at 100 ns, 500 ns and 10 μs measured at room temperature. The photon-energy region smaller than 1.6 eV is assigned to the photoinduced absorption of triplet excitons. (b) The time dependence of the photoinduced absorbance change of a PDA-3BCMU cast film at 1.44 eV measured at room temperature.

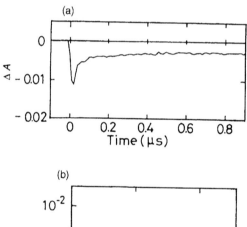

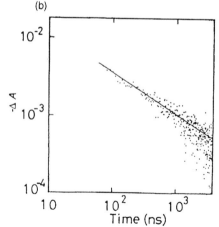

Figure 7.81. Linear (a) and logarithmic (b) plots of the time dependence of the photoinduced absorbance change of the PMMA film doped with a small weight per cent (0.67%) of PDA-3BCMU at room temperature. The long-lived component was subtracted for the logarithmic plot. The solid curve represents the best-fit linear approximation with the slope of 0.51.

dependence indicates the creation of excitations which relax by geminate recombination within a chain.

When a pure film of PDA-3BCMU was excited by light of 3.49 eV at a photon density of 1.1×10^{16} photons cm^{-2}, the transient absorption due to triplet excitons was observed at around 1.4 eV as shown in Figure 82a. It was measured with apparatus described in literature [51]. Figure 7.82b shows the time dependence of triplet exciton absorption. The decay kinetics were well described with an exponential curve with a time constant of about 20 μs, and no signal could be observed around 1.4 eV when the same was excited with a 2.33 eV light at a photon density of 1.8×10^{16} photons cm^{-2}. Therefore, the excitons in 0.67% PDA-3BCMU in PMMA pumped with 2.337 eV light are not triplet excitons.

3.23 Confinement effect of photogenerated soliton–antisoliton pair on the ultrafast relaxation in a substituted polyacetylene

Since Su, Schrieffer and Heeger (SSH) [173,174] predicted the existence of a soliton excitation, i.e. a domain wall separating the two dimerization phases corresponding to the doubly degenerate ground states, and Su and Schrieffer [175] demonstrated the evolution of an electron–hole pair photoinjected into the main chain of $trans$-(CH)$_x$, a great deal of attention has been paid to the excited-state dynamics of $trans$-(CH)$_x$. Due to its strong electron–phonon coupling and one dimensionality, the photogenerated electron and hole induce a local distortion of the lattice spontaneously and rapidly separate to relax. Thereby a spinless charged soliton–antisoliton pair is formed. The soliton has a localized electronic state deep in the midgap [7] and associated infrared-active vibrations (IRAV) [176], which are both spectroscopically accessible. Photo-induced absorption (PA) and photoinduced bleaching (PB) of $trans$-(CH)$_x$ under quasi-static conditions [177] have experimentally confirmed the existence of solitons, and their relaxation dynamics have been also investigated so far with picosecond [178–181] and femtosecond [5,182,183] time resolution.

Vardeny et $al.$ [178] observed the PB of interband transition in $trans$-(CH)$_x$ with picosecond time resolution and found that the bleaching decayed with time as $t^{-1/2}$ up to 50 ps. They also estimated the diffusion constant of charged excitations as 2×10^{-2} cm^2 s^{-1} from the decay of the polarization memory. Shank et $al.$ [5] investigated the femtosecond dynamics of the PA in $trans$-(CH)$_x$ and cis-(CH)$_x$. They observed that in $trans$-(CH)$_x$ the PA appeared at 1.45 eV within less than 150 fs after photoexcitation, and that its recovery dynamics could be modeled by geminate recombination controlled by 1D diffusion, though the identification of the PA peak was remained unclear. Direct measurement of the lower energy (LE) in-gap absorption in the sub-picosecond time regime was recently demonstrated by Rothberg et $al.$ [183], who observed the direct generation of charged solitons within 500 fs after photoexcitation. They found that over 95% recombined geminately in about 2 ps.

However, there are still difficulties in identifying the observed PA peak near 1.4 eV especially on the ultrashort time scale. Isolated charged solitons are somewhat doubtful because the feature appears at different energy from that expected by the SSH model. Although the PA associated with neutral solitons is known to appear above the midgap, by taking into account the Coulomb interaction between soliton sites [184], it does not seem to be the case here because the branching ratio of neutral to charged solitons is less than 10^{-2} in the photogeneration processes [185]. It is also unlikely, from the fast rise of the PA, that isolated neutral solitons are created by a reaction between photogenerated charged species.

In this section, to clarify the complicated feature and also to elucidate the ultrafast relaxation dynamics, the stress is placed on the importance of the $confinement$ of charge carriers photogenerated into the main chain of 1D conjugated polymers. The possibility for an elementary excitation to take the form of a $confined$ $soliton–antisoliton$ $pair$ is also discussed. Since conjugated polymers inevitably have some chemical impurities and structural defects which will limit the motion of solitons, the ultrafast relaxation dynamics in such a confined system is considered to be different from the picture predicted by the SSH model, which is based on an independent electron picture in the $infinite$ 1D chain.

For the purposes of investigating the confinement effects of a photogenerated soliton–antisoliton pair on its ultrafast relaxation dynamics, the suthor's team have performed femtosecond time-resolved absorption measurements with a pump–probe technique on poly-[o-(trimethylsilyl)phenylacetylene] (referred as PMSPA hereafter), which has the same backbone geometry as that of $trans$-(CH)$_x$, but an $ortho$-substituted phenyl ring is attached to every second carbon sites. This introduces a reduction of an electronic conjugation length because of a distortion of the main chain due to steric hindrance, resulting in the confinement of the soliton and antisoliton. The experimental results show that the spectrum and decay kinetics of the observed PA can be well accounted for by the presented picture of a confined soliton–antisoliton pair. Although the use of the term solitons for incomplete degenerate systems such as PMSPA seems to be somewhat misleading, it is thought that PMSPA has basically the same electronic excitations as $trans$-(CH)$_x$ even with the small perturbation introduced by the side groups. The confined soliton–antisoliton pair is referred to, in this paper, as an overall neutral pair of charged excitations with changes of dimerization pattern.

The classification is also made for various polymers in terms of a degree of confinement, and a picture of a 'confined excitation', which takes a form of an isolated soliton, a confined soliton–antisoliton pair, or an exciton according to the degree of confinement is proposed. Due to the similarity of the main-chain geometry, the results obtained here will give hints to elucidate the relaxation mechanism in *trans*-(CH)$_x$ by comparing the photoinduced features and the temporal behaviors of both materials.

An interchain photoexcitation process is also discussed. So far there are two proposed mechanism that explain how polarons generated by the interchain photoexcitations relax to form solitons. Firstly, as has been proposed by Orenstein *et al.* [140], the neutral defects, S^0, which exist inherently in *trans*-(CH)$_x$ and also in PMSPA, act as trapping sites, and the polarons encountering the S^0 are converted to charged soliton states according to the following reaction: $P^+ + S^0 \rightarrow S^+$. Rothberg *et al.* [180] ascribed the increase of the LE in-gap absorption, observed after 20 ps following the photoexcitation, to charged solitons that were converted from polarons by this mechanism. The mechanism is expected to be dominant in the sample with a larger number of neutral defects. This is supported by the fact that the LE in-gap absorption observed by Rotherg *et al.* [180] was greatly suppressed in *trans*-(CH)$_x$ where fewer neutral defects were observed in electron spin resonance spectroscopy (ESR) measurement. Secondly, when more than two polarons are generated on the same chain by interchain photoexcitation, two likely charged polarons on the same chain are expected to be converted to a likely charged soliton pair by the energetically favored reaction: $P^+ + P^+ \rightarrow S^+ + S^+$ [11]. This time, it has been found that the observed component due to long-lived species in PA and PB results from the interchain photoexcitation, and that the amplitude of the long-lived component has quadratic dependence on the pump intensity, implying that the latter mechanism is applicable to PMSPA.

3.23.1 Photoinduced features and their decay kinetics

Stationary absorption spectra of PMSPA thin films at two different temperatures (297 and 10 K are shown in Figure 7.83 with its chemical structure as an inset. They have a steep rise at $\pi \rightarrow \pi^*$ band-gap transition energy (E_g) and no absorption below it; these features are often seen for this group of 1D conjugated polymers. They have a well-defined peak at 2.3 eV with an onset around 1.9 eV. The absorption profile is shifted to higher energy by about 0.5 eV as compared with that of *trans*-(CH)$_x$. This probably results from a reduction of the conjugation length of PMSPA due to distortions

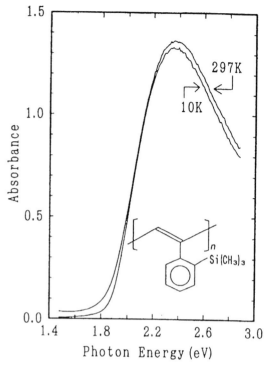

Figure 7.83. Absorption spectra of PMSPA measured at 297 and 10 K. Inset: chemical structure of PMSPA.

of the main chain. They are caused by a steric hindrance between bulky substituent groups attached to the same side of the backbone. As a result of fitting with the following absorption profile, which is characteristic of a one-dimension electronic system [186]

$$\alpha(\omega) = A_0 \frac{E_g/\omega}{\sqrt{(\omega^2 - E_g^2)}} \tag{7.68}$$

E_g was determined as 2.2 eV and an inhomogeneous broadening was estimated as 0.25 eV, mainly due to a statistical distribution of the conjugation length. No significant change was observed with changes of temperature, implying the rigid chemical structure. Unlike the other polymers such as P3MT [90] and PTV [187], no features appeared at the rising edge even at low temperature within the special resolution. The photocurrent action spectrum of PMSPA measured by Kang *et al.* [188] roughly follows the absorption spectrum and extends somewhat beyond the absorption edge, implying that the transition is not due to excitons.

Photoinduced differential absorbance spectra measured at 10 K are shown in Figure 7.84 up to 1 ps after photoexcitation at 1.98 eV. The pump duration and excitation density were 100 fs, and 2×10^{16} photons cm^{-2}, respectively. The following features can be seen from this figure. Firstly, PA rises around 1.2 eV simultaneously with photoexcitation and shows a gradual blue shift to a peak near 1.8 eV at a delay time of 0.15 ps. The signal amplitude is constant during the shift. Secondly, PB is observed above 2.0 eV. The region is expanded to higher photon energy with an increase in delay time. Since the PB peak at 2.3 eV occurs at the same energy as that of the stationary absorption spectra (Figure 7.83), this PB is due to the saturation of the interband transition. Thirdly, a peak minimum (gain signal) at 1.82 eV is observed as indicated by RG in Figure 7.84. Since it appears only around a delay time of 0.0 ps, i.e. when the pump and probe pulses temporally overlap, it is distinguishable from the noise. The signal was also observed for several other polymers by the author's group for the first time [52,99]. It is due to the Raman gain and appears at the Stokes side of the pump frequency at 1410 cm^{-1}. It corresponds to the C=C stretching mode of the ground-state configuration. Finally, a sharp dispersion-type feature can be seen at small positive delay time as indicated with IPM in the same figure. It is due to the induced phase modulation [101]. It was also observed for other polymers by the author's group for the first time [98]. At this delay time a probe pulse experiences the frequency blue shift due to a modulation of the refractive index of the sample caused by a trailing tail of a strong pump pulse. Since the probe spectrum has a peak at the fundamental frequency of the laser (2.0 eV), the shift of the probe spectrum results in the dispersion-type feature. It should be noted that such an effect (change of refractive index) will be always superimposed more or less on the population-induced absorbance change (change of extinction coefficient) in this time scale.

Although the femtosecond time regime is too short for the heat given by the pump flux to be transferred to the lattice, it is important to evaluate the magnitude of the thermal effect on the absorbance change. The estimation of the effect is done below, though it is only a rough estimation because detailed values for the weight density ρ, specific heat capacity C_p and film thickness d of the sample have not been obtained. From the pump photon density $N = 2 \times 10^{16}$ photons cm^{-2} and optical density at the pump frequency $A = 0.5$, the actual absorbed energy density per pulse is calculated as $Q = 4.4 \times 10^{-3}$ J cm^{-2}. Using the rela-

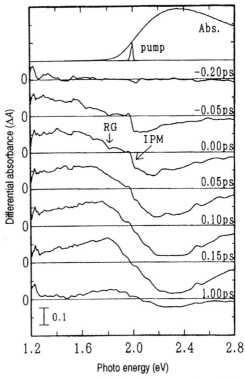

Figure 7.84. Differential absorbance spectra of PMSPA measured at 10 K up to 1 ps after photoexcitation at 1.98 eV. The pump duration and excitation density are 100 fs and 2×10^{16} photons cm^{-2}, respectively. The positive change shows photoinduced absorption, while the negative one shows photoinduced bleaching. The absorption spectrum (Abs.) of PMSPA and pump laser spectrum (pump) are also shown for comparison at the top. RG: peak minimum due to the Raman gain, IPM: sharp dispersion-type feature due to induced phase modulation.

tion $Q = C_b \rho d \Delta T$, the temperature increase ΔT of the sample is estimated as 24 K. Here, the parameters for PMMA, $\rho = 1.2$ g cm^{-3} and $C_p = 1.5$J (gK)$^{-1}$ are used and $d = 1$ μm is estimated. By combining ΔT and the temperature variation of the absorbance $dA/dT = 1.7 \times 10^{-4}$ K^{-1} (at 1.8 eV), the absorbance change due to the thermal effect is evaluated to be about 4×10^{-3} at the largest, which is still two orders of magnitude smaller than the obtained signal, and hence can be neglected.

Figure 7.85 shows the temporal decay profiles at three different photon energies. In the lower probe-photon-energy region near 1.43 eV, the PA almost disappears by 1 ps and the ground state is recovered to some extent. Assuming an exponential decay, the time constant of the decay is 135 fs. Since it seems to follow

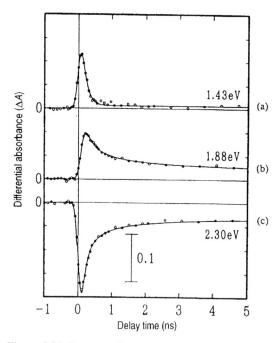

Figure 7.85. Decay profiles at three different probe photon energies. The open circles represent the experimental data, and the solid lines show the fitting curves with an exponential function for (a) and a power-law function for (b) and (c) (See text).

the pump pulse profile, it is possible that the actual decay rate at this energy is still faster. In contrast, the temporal behavior around the PA peak at 1.88 eV has some finite rising component, which corresponds to the initial blue shift of the PA peak mentioned above. The following decay rate is observed to become much slower than that at the lower energy region, which clearly can be seen by comparing (a) and (b) of Figure 7.85. Furthermore, the existence of a long-lived component is also confirmed because no significant reduction of the signal can be observed at a delay time of more than 20 ps. At the PB peak [2.3 eV, (c)] again, a long-lived component is observed after the initial fast decay.

Since it was not possible to fit the decay profile (b) in Figure 7.85 to the exponential function, and in turn, a linear behavior vs. delay time was obtained almost three orders of magnitude in the log–log plot, the PA signal can be considered to follow a power-law decay. So, the fitting of the decay profile (b) with the following function was made

$$\Delta A(t) = A\,\mathrm{erf}[(\sigma t)^{-n}] + C \qquad (7.69)$$

where erf( ) is an error function. The first term represents the component showing the power-law decay, while the second term, C, represents the long-lived component (baseline signal) mentioned above. The fitting was performed by convoluting equation (7.69) with the pump pulse profile which was assumed as a Gaussian type. The least-squares method was used to determine all the parameters at the same time, so that an ambiguity of subtracting the baseline was avoided.

As a result of this fitting, the power (exponent) and the value of the baseline signal for the PA peak at 1.88 eV were obtained as $n = 0.65 \pm 0.05$ and $C = (1.5 \pm 0.18) \times 10^{-2}$, respectively. The variation of n is less than 0.05 around the PA peak. The fitted curves are also shown with solid lines in Figure 7.85.

3.23.2 Confinement of the soliton–antisoliton pair

From a theoretical point of view, the power-law behavior corresponds to a geminate recombination process between two excitations after their random walks in a one dimensional chain [189]. Thus it is implied that the observed PA is associated with photogenerated soliton–antisoliton pairs and that its decay results from the geminate recombination of the soliton and antisoliton after encountering each other. Such a power-law behavior was also observed in the experiments on *trans*-(CH)$_x$ demonstrated by Shank *et al.* on the same time scale [5,182]. They found that the decay profile of PA can be fitted with $n = 0.5$ which is expected for an ideal one-dimensional system. However, the power, n, obtained in present work of the author's team is slightly larger than 0.5 for the PA peak. One of reasons why such a discrepancy in n is obtained even with the same backbone geometry between PMSPA and *trans*-(CH)$_x$ is considered to be the difference in the initial distance between the soliton and antisoliton photogenerated. In the case of *trans*-(CH)$_x$, a photogenerated electron–hole pair can induce a distortion of the surrounding lattice to strong electron–phonon coupling and relaxes to form a soliton–antisoliton pair within an order of 10^{-13} s according to the SSH mechanism. Since *trans*-(CH)$_x$ has doubly degenerate ground states, the total energy of a soliton–antisoliton pair is independent of the distance between them and therefore they are free to dissociate. So, after separation by the use of the excess energy of photoexcitation, the soliton and antisoliton can exist isolated from each other, and therefore their decay is considered to follow the power-law with the power n being near to 0.5.

In contrast, in the case of PMSPA, it is supposed that the photogenerated soliton and antisoliton pair are confined and cannot separate at a long enough distance compared to the width of the soliton wavefunction. The possible reasons for this are described below. Firstly, since the conjugation length of PMSPA is rather short as mentioned above (and also discussed later again) a soliton and its antisoliton cannot separate without a loss of electronic coherence, and therefore they are confined in a conjugation-length-limited segment of the main chain. Secondly, unlike the *trans*-(CH)$_x$, an *ortho*-substituted phenyl group attached to the main chain of PMSPA still has rotational freedom against the plane of the main chain, which leads to a small lifting of the ground-state degeneracy. Though this effect may become obscured as the size of the system under consideration increases, if the separation of the soliton and antisoliton is restricted within the above-mentioned segment, the ground state does not necessarily become degenerate. If it is assumed that nearly degenerate ground states consist of an A phase with energy E_0 and a B phase with energy $E_0 + \Delta E(\Delta E > 0)$, then the energetically lowest state of this system consists of purely A phase, and photogeneration of a soliton–antisoliton pair is accompanied by the appearance of the energetically disfavored B phase. The existence of the B phase will raise the total energy of this system by [186]

$$\Delta E = \left(\frac{\delta E}{a} \right) L \qquad (7.70)$$

where a is a unit site length, $\Delta E / a$ represents an energy difference per site, and L is the length of the B segment. An energy loss proportional to the soliton–antisoliton distance prevents further separation. Lastly, the existence of soliton–antisoliton pairs may obstruct the other pairs on the same chain.

3.23.3 Estimation of the distance between the soliton and antisoliton

From the reasons discussed in the previous section, the photogenerated soliton and antisoliton are considered to exist not as separated solitons in the distance but as confined soliton–antisoliton pairs in PMSPA. Estimation of the distance between them has been done below according to the results of

Fesser *et al.* [186]. In the continuum model proposed by Takayama *et al.* [190], the SSH Hamiltonian is transformed as

$$H = \sum_s \int \frac{dy}{a} \left[\frac{\omega_0^2}{2g^2} \Delta^2(y) + \Psi_s(y) \right. \\ \left. \times \left[-iv_F \sigma_3 \frac{d}{dy} + \Delta(y)\sigma_1 \right] \Psi_s(y) \right] \qquad (7.71)$$

Here, $\omega_Q^2 / 2g^2 = K/4\alpha^2$, $v_F = 2at_0$, where K, α, a are a spring constant of a σ-bond, an electron–phonon coupling constant and a lattice constant, respectively, and t_0 is a hopping matrix element of a π-electron. The lattice displacement pattern is described by $\Delta(y)$ and the electronic field is described by spinor $\Psi_s(y)$. In our nearly degenerate case, $\Delta(y)$ can be assumed to consist of the following two components [191]

$$\Delta(y) = \Delta_i(y) + \Delta_e \qquad (7.72)$$

Here, $\Delta_i(y)$ denotes the intrinsic component sensitive to an electron–phonon coupling, while Δ_e represents the constant displacement due to extrinsic origin. Now let the intrinsic gap parameter of a soliton–antisoliton pair with a distance $2d$ be given by

$$\Delta_i(y) = \Delta_0 - K_0 V_F \{ \tanh[K_0(y+d)] \\ - \tanh\{[K_0(y-d)]\} \qquad (7.73)$$

with

$$\tanh(2K_0 d) = K_0 \xi \qquad (7.74)$$

The parameter K_0 exhibiting the extension of a soliton is related to d by equation (7.74), and Δ_0 is the gap parameter corresponding to the ground-state configuration of *trans*-(CH)$_x$. When the Coulomb interaction between a soliton and its antisoliton is negligible, equation (7.71) yields two bounds states for this dimerization pattern as the lowest excited state with energy separation from the gap center at $+\omega_0$ and $-\omega_0$. This is shown in the inset of Figure 7.86, where

$$\omega_0 = (\Delta^2 - K_0^2 v_F^2)^{1/2} \qquad (7.75)$$

is a half confinement energy.

The ratio ω_0 / Δ as a function of d can be calculated with equations (7.74) and (7.75), the result being shown in Figure 7.86. Its behavior is qualitatively interpreted as follows. When the soliton and antisoliton are far apart, the confinement energy becomes asymptotically zero and the associated gap states appear in the middle of band gap with twofold degeneracy. On the other hand, when the soliton and antisoliton get closer, the confinement energy suddenly increases due to an

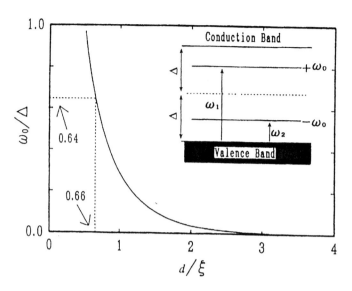

Figure 7.86. Plot of the half confinement energy (ω_0) as a function of the distance between the soliton and antisoliton (d). Inset: energy diagram associated with a soliton–antisoliton pair.

overlapping of wavefunctions of the two solitons, and an upper gap state comes near to the conduction band. The latter extreme case corresponds to an exciton picture, where a photogenerated electron and hole cannot separate from each other due to a strong confinement and Coulomb binding. This is followed by the formation of a STE [52] with a geometrical relaxation due to an electron–phonon coupling, just as occurs in the soliton–antisoliton pair formation in the system with the degenerate ground states.

The two gap states ($\pm\omega_0$) will give the PA peaks in the transparent region. One state will appear just below the conduction band and the other just above the valence band in the case of strong confinement. Since the observed PA peak near 1.8 eV appears only 0.4 eV below the optical gap and its decay profile shows the power-law behavior, the transition associated with the confined soliton–antisoliton pairs is the most relevant to this peak. Furthermore, its rather broad feature suggests that it involves a transition from (to) to the continuum states. Therefore, the observed PA peak near 1.8 eV can be assigned to the ω_1 transition from the valence band to the upper gap state as schematically drawn in the inset of Figure 7.86. From experimental results of the author's group, $E_g = 2\Delta = 2.2$ eV and $\omega_0 = \omega_1 - \Delta = 0.7$ eV have been obtained. This leads to the ratio $\omega_0/\Delta = 0.64$, corresponding to $d/\xi = 0.66$, as shown in Figure 7.86.

If the substituents have only a little effect on a and t_0, an extension of a soliton is nearly proportional to Δ^{-1}, and the soliton width in PMSPA turns out to be

$$\xi = \xi_0 \frac{\Delta_0}{\Delta} = 4.5a \qquad (7.76)$$

Here, $\xi_0 = 7_a$ and $\Delta_0 = 0.7$ eV are used, which are both appropriate for *trans*-$(CH)_x$. Thus the distance between the soliton and antisoliton can be estimated as

$$L = 2d = 5.9a \qquad (7.77)$$

This is a fairly small separation; it implies the strong confinement of the photogenerated soliton–antisoliton pair, as expected from the large steric hindrance. It is consistent with the fact that the power n obtained above is slightly larger than 0.5, because the probability for the soliton and antisoliton to encounter each other becomes larger as the initial distance gets smaller. Another estimation of the separation was also done according to the theory presented in ref. [192]. When the soliton wavelength is overlapped, the degenerate soliton states at the gap center will be lifted by

$$\Delta E = 4\Delta e^{-L/\xi} \qquad (7.78)$$

where L is the distance between the soliton and antisoliton. Substitution with $\Delta E = 2\omega_0 = 1.4$ eV and $\Delta = 1.1$ eV leads to $L = 5.1a$, which is consistent with the former estimation [equation (7.69)]. If the length of

C−C and C=C bonding is assumed to be 1.463 and 1.368 Å, respectively, and the zig-zag structure of the main chain is taken into account, $L = 5.9a$ corresponds to the size of 14.4 Å. This size should be compared with 11.6 Å which is the size of a certain regular structure indicated to exist in PMSPA by X-ray diffraction measurements [193]. It may be possible that the main chain of PMSPA is divided into segments with 14.4 Å by some structural distortions, and that the photogenerated soliton and antisoliton are locked in the segments.

Since the above experimentally determined separation is not larger than the size of the lattice deformation, one may suspect the diffusion picture in PMSPA. On this point, it should, at first, be noted that the estimation does not include the effect of Coulomb interaction. It might shift the energy spectra associated with soliton–antisoliton pairs and lead to a different separation. Therefore the obtained results are not quantitatively correct, but give a measure of the degree of confinement. Secondly, the spectral shift observed in ΔA spectra may have some relations with the diffusional motion of the solitons. As stated [182] in the formation process of the two in-gap states associated with the two interacting excitations, they are expected to move through the band gap, resulting in the spectral shift of the in-gap absorption as well. Since the formation process occurs in the same time scale as the experimental time resolution used by the author's group, it is believed that the observed shift is associated with the formation process accompanied by the diffusion of two interacting solitons and that the shift may indicate the translational freedom of solitons.

Another feature that should be noted is that the PB recovers faster than the PA as shown in Figure 7.85. Concerning this point, it is possible that the bleaching may recover faster than the absorption signal due to species like intrachain polaron pairs or polaron–excitons which can not escape from the initial fast recombination. They should also give absorption signals below the optical gap. Although the author's group has not yet succeeded in extracting their spectra, because of the rather broad ΔA spectra, it is conceivable that a fast decay observed at a lower-photon-energy region below 1.6 eV (Figure 7.85) may correspond to that of the polaron pairs or polaron–excitons. The initial absorption at lower photon energy region is mainly associated with species like polaron–excitons which can be regarded as a single quasi-particle. This is followed by a separation of those charged particles which can escape from the initial recombination to form soliton–antisoliton pairs with the observed

spectral shift. Therefore the fast exponential decay seen in Figure 7.85a may be included in the recovery of the bleaching signal and this may make the bleaching signal decay faster. The bleaching signal is probably composed of both the fast exponential decay seen in Figure 7.85a and the power-law decay with $n = 0.65$.

3.23.4 Comparison with other polymers

The confinement parameter γ [90,186] defined as

$$\gamma = \frac{1}{2\lambda}\frac{\Delta_e}{\Delta_0 + \Delta_e} = \frac{\frac{\omega_0}{\Delta}\sin^{-1}\left(\frac{\omega_0}{\Delta}\right)}{\sqrt{\left[1 - \left(\frac{\omega_0}{\Delta}\right)^2\right]}} \tag{7.79}$$

indicates an extrinsic contribution to the displacement, where λ is an electron–phonon coupling constant, and Δ and ω_0 are both experimentally accessible. For example, in ideal $trans$-(CH)$_x$ with no confinement, soliton states should appear in the center of the gap ($\omega_0 = 0$) and γ should be zero. On the other hand, in our material PMSPA, γ is calculated as 0.57, again showing that the strong confinement is the case.

In general, the ultrafast relaxation dynamics after photoexcitation can be considered to depend on the initial distance between photogenerated charge carriers. In systems where soliton formation can take place, these electrons and holes will relax to isolated solitons [in the case of $trans$-(CH)$_x$] or confined soliton–antisoliton pairs, and will subsequently separate more or less according to the degree of confinement. The following decay dynamics will be dominated by a geminate recombination process of these solitons, showing the power-law behavior. In the other extreme, for example in polydiacetylene, photoexcited electrons and holes cannot separate due to the strong confinement caused by a highly nondegenerate ground state and Coulomb binding. Then the relaxation dynamics are dominated by that of excitons which decay mono-molecularly with an exponential behavior, as has been already observed [51,52].

The confinement of charge carriers is caused partly by a distortion of the main-chain structure introduced by the steric hindrance between side groups, as in the case of PMSPA, and partly by the nondegeneracy of the ground-state configuration. [199] In other words, the energy difference between the ground state and the alternate dimerization phase, $\Delta E = E_B - E_A$, determines the degree of confinement. Thus the one-dimensional π-electron conjugated polymers with carbon backbone structure can be classified using the confinement parameter γ as an index, and the relaxation of the photoexcited state including its functional type and characteristic decay time can be mapped along the 'γ-axis' for every system from trans-(CH)$_x$ to polydiacetylene. The boundary between a power-law and exponential behavior is ill-defined and further investigations are required to discuss it in full detail.

3.23.5 Interchain photoexcitation

Although the discussion so far have been mainly on the results of intrachain photoexcitations, *interchain* photoexcitation can also occur due to the nonvanishing interchain transfer integral, which generates a positive polaron (P$^+$) on one chain and a negative polaron (P$^-$) on the neighboring chain. Unlike a soliton which cannot hop to another chain because of its topological feature, these polarons can move separately on each chain or hop onto another chain and therefore show three-dimensional diffusion [11]. They are considered to play an important role in transport phenomena such as photoconductivity [194] or quasi-static PIA [11,177, 195] as have been reported so far on trans-(CH)$_x$.

The experimental results and discussion presented here have been published elsewhere (refs. 196–199 and those cited already in the text).

Since the relaxation from the polaron states requires interchain hopping, the polarons are expected to have a long lifetime, probably an order of a microsecond or longer. Therefore, it is natural to associate the long-lived component, described by the second term of equation (7.69) with these polaron states. The amplitude of the long-lived baseline signal at both PA and PB peaks is plotted vs. the square pump intensity in Figure 7.87. Since the data plotted in the figure are separate from those in Figures 7.84 and 7.85, the amplitude of both data cannot be compared directly. A quadratic dependence of the baseline signal on the pump intensity is clearly seen from this figure, implying that it is caused by either a two-photon process or a reaction between two excitations, which themselves are generated by a one-photon process. Furthermore, for both PA and PB, these long-lived components have their peaks at the same energy as those of the fast components which are due to intrachain soliton–antisoliton pairs. One of the possible ways to account for these two experimental facts is to associate the long-lived component with confined pairs of likely charged

solitons. However here, these soliton pairs are considered to be generated by a different process, i.e. at first, the polarons are generated by interchain photoexcitations, and likely charged polarons on the same chain are converted to likely charged solitons pairs after encountering each other by the following reaction

$$P^\pm + P^\pm \rightarrow S^\pm + S^\pm \tag{7.80}$$

which is energetically favored. It should be noted that the confined pair of likely charged solitons is very similar to bipolarons from the viewpoint of lattice deformation, and that the term soliton is used here in contrast with the intrachain soliton–antisoliton pair. The quadratic rather than linear dependence for both the absorption and bleaching baseline, seen in Figure 7.87 may indicate the efficient conversion of polarons by the above mechanism or a relatively long lifetime of the soliton pairs compared to that of the polarons which is probably due to the higher mobility of the polarons than that of the soliton pairs.

However direct two-photon absorption as a mechanism for the baseline signal cannot be ruled out from the experimental data of the author's team alone. In this case, the creation of long-lived species like triplet excitons could occur.

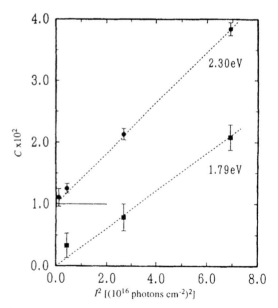

Figure 7.87. Pump intensity dependence of the amplitude of the long-lived components at both PA and PB peaks. Data at 2.30 eV are shifted by 0.01 [197].

4 CONCLUSIONS

The relaxation dynamics of the photoexcitations in films of PDA-3BCMU and P3MT have been studied using femtosecond time-resolved absorption spectroscopy. The absorbance change due to photogenerated excitons, hole burning, Raman gain and the optical Stark effect have been observed in both polymers.

The cast film of PDA-3BCMU showed a broad absorption ($\Delta E \approx 0.2$ eV) due to the ^1B_u exciton at 2.0 eV and the width of the absorption is closer to that of PDA-CPDO than to other polydiacetylenes such as PDA-TS. The electroabsorption spectrum of the cast film of PDA-3BCMU showed a Stark red shift, while the electroreflection spectrum of the single crystal of PDA-CPDO revealed a spectral broadening that is attributed to the Franz–Keldysh effect. This result supports the interpretation that the lowest optical absorption of PDA-CPDO is due to an interband transition.

In the femtosecond transient spectroscopy of PDA-3BCMU, bleaching due to the ^1B_u excitons appears at 1.92 eV instantaneously with the 1.97 eV pump pulse, while the photoinduced absorption below 1.85 eV appears rather slowly and finally has a peak at 1.80 eV. The kinetics of the absorbance changes are explained in terms of the self-trappings of the photoexcited ^1B_u excitons. The formation and decay times of the STEs generated by a 1.97 eV pump pulse at 10 K are 150 ± 50 fs and 2.0 ± 0.1 ps, respectively. The formation and decay time constants at room temperature are 150 ± 50 fs and 1.5 ± 0.1 ps, respectively. When the 3.94 eV pump pulse was used, the formation time of the STE could not be resolved because of the poorer resolution (about 300 fs). The decay time is no different from that of the 1.97 eV excitation. The most STE relax into the ground state nonradiatively. There is no distinct difference between the decay and formation time constants of the STE in PDA-3BCMU between 10 K (2.0 ps) and room temperature (290 K) (1.5 ps). This indicates that the contribution of the thermal activation process associated with a potential barrier is not dominant in both the formation and relaxation processes of the STE in PDA-3BCMU. The dominant relaxation channel is the tunneling through the potential barrier between the STE and the ground state.

The dependence on the excitation photon density is shown by the absorption saturation of the excitons at a pump photon energy of 1.97 eV. The triplet excitons are generated by 3.94 eV excitation and also by the two-photon absorption of the 1.97 eV pump pulse. The peak absorption of the triplet excitons is at 1.5 eV, and the lifetime is far longer than the time range studied by femtosecond spectroscopy. It was measured to be 18 μs by nanosecond spectroscopy.

Similar results were obtained for PDA-4BCMU when excited by a 3.94 eV femtosecond pulse. The decay time of the STE is 3.0 ± 0.3 ps at 10 K and 2.1 ± 0.2 ps at 290 K.

The singlet excitons are strongly coupled with the C–C stretching modes in the backbone chain, but the formation time is much longer than the phonon oscillation period. Study of the formation processes of the STE in PDAs with different side groups and in other conjugated polymers will elucidate the mechanisms of self-trapping of excitons.

Data from the femtosecond transient spectrum of P3MT indicate the dynamics after excitation with a pulse at 1.97 eV near the absorption edge of P3MT. The self-trapping of free excitons takes place with a time constant of 70 ± 50 fs at 10 K. The relaxation time constant of the STE is 800 ± 100 fs at 10 K. The formation and relaxation times of the STE at room temperature (290 K) are almost the same as those at 10 K. The shorter formation time of the STE in P3MT than in PDA-3BCMU is explained by the difference in the chemical structure of these polymers. P3MT has a rigid ring structure and has small conformation change; this is expected to be associated with electronic excitation. PDA-3BCMU has bulky side groups with hydrogen bonds, which maintain the planarity of the polymer chain. A change in the bond order in the main chain of the polymer is associated with a conformation change in the bulky side groups. Therefore PDA-3BCMU is expected to have a long formation time for the STE. The STE relaxes mainly nonradiatively, again into the ground state. The remaining long-lived absorbance change is due to triplet excitons, polarons, and/or bipolarons.

The time dependences of fluorescence intensity have been measured for the investigation of the relaxation dynamics in PDA-4BCMU film. The relaxation dynamics in soluble PDAs is more complex than other well-oriented PDAs, such as single crystals. The dynamics are well described by a random walk in the fractal dimension.

In this work it was revealed that PDA-4BCMU, in common with other organic polymers, can take a fractal structure. An explanation of the decay of the fluorescence intensity from the PDA-4BCMU films in terms of the random walk on the fractal structure model at lower temperatures was attempted. The partially polymerized film, which has highly polymerized parts and oligomers, has a component which decays faster than the resolution time of the detection system and also has a similar decay to that of the fully polymerized film at a longer delay time. The fluorescence decay kinetics consist of contributions from emissions from both oligomers and polymers.

Spectral changes due to several nonlinear optical processes, i.e. hole burning, Raman gain and the dynamic Stark effect, have also been observed. The third-order susceptibility has been determined for these nonlinear processes, such as absorption saturation at the exciton transition, Raman gain and the resonant Kerr effect from the observed spectral change. The third-order susceptibility corresponding to absorption saturation in PDA-3BCMU and P3MT was obtained as $\mathrm{Im}[\chi^{(3)}_{1111}(-\omega;\omega,-\omega,\omega)] = -2.6 \times 10^{-9}$ esu and -3.5×10^{-10} esu, respectively, for $\hbar\omega = 1.97$ eV. The third-order susceptibility corresponding to the Raman gain in PDA-3BCMU and in P3MT was determined as $\mathrm{IM}[\chi^{(3)}_{1111}(-\omega_2;\omega_2,-\omega_1, \omega_1)] = -5.8 \times 10^{-10}$ esu, respectively, for $\hbar\omega_1 = 1.97$ eV and $\hbar\omega_2 = 1.79$ eV. From the resonant Kerr experiment in the P3MT film, $|\Delta\chi^{(3)}| = -|\chi^{(3)}_{1111}(-\omega_2;\omega_2,-\omega_1,\omega_1) - \chi^{(3)}_{1122}(-\omega_2;\omega_2,-\omega_1,\omega_1)| = 3.9 \times 10^{-11}$ esu was determined for $\hbar\omega_1 = 1.97$ eV and $\hbar\omega_2 = 1.88$ eV.

The ultrafast nonlinear optical responses of epitaxially grown blue-phase PDA-3BCMU on KCl single crystal are time-resolved and separated into four components including an instantaneous nonlinear response term which decays within 100 fs in a nonthermal STE, quasi-thermal STE and triplet exciton. The spectra of the nonlinear optical susceptibilities are determined, taking into account the pulse duration and response times. The spectrum of the instantaneous nonlinear optical response is explained using a model of the two-level system with vibrational modes coupled to the excitonic transition. The instantaneous nonlinear response consists of several nonlinear effects, i.e. hole burning, coherent coupling between pump polarization and probe field, Raman gain, hole burning of phonon sidebands and inverse Raman scattering. The contribution of the induced-phase modulation to the observed absorbance change is also estimated.

The nonlinear optical effects due to the nonthermal and quasi-thermal STEs are the saturation of the excitonic absorption and the transition from the photoexcited lowest ^1B_u exciton to a biexciton and/or a higher exciton with ^1A_g symmetry. The response times of the nonlinear processes are 150 fs and 1.5 ps at 290 K, respectively. The imaginary part of $\chi^{(3)}(-\omega_2; \omega_2,-\omega_1,\omega_1)$ due to the quasi-thermal STE has a peak

minimum of $-1.0 \times^{-8}$ esu at $\hbar\omega_1 = 1.97$ eV and $\hbar\omega_2 = 1.95$ eV.

Since stimulated emission and photoinduced absorption have different dependences on the probe light intensity, their spectra can be separated by the measurement of transient transmission spectra under the conditions of different probe intensities. This method, named 'probe saturation spectroscopy' (PSP) has been applied to a vacuum-deposited film of blue-phase PDA-C_4UC_4 to separate the stimulated emission spectrum and calculate the fluorescence spectrum from the stimulated emission spectrum. The fluorescence spectrum of material such as PDAs with a lifetime as short as 140 fs and a quantum efficiency smaller than $10 - 5$ have been obtained. The calculated fluorescence spectrum has peaks at 1.9 and 1.8 eV and has approximately a mirror-image relation with the absorption spectrum. They correspond to the transition from free excitons and nonthermal STEs to quasi-thermal STEs and the phonon sideband.

The single crystal of recently synthesized polydiacetylene PDA-MADF was investigated with both conventional and time-resolved spectroscopies.

The single crystal of PDA-MADF showed a similar polarized reflection spectrum to that of the conventional blue-phase polydiacetylenes, such as PDA-TS, PDA-DCHD and PDA-TCDU-A. The calculated spectrum, using the Kramer–Kronig transformation, revealed an absorption peak due to the lowest ^1B_u exciton at 1.88 eV with phonon sidebands at 0.2 and 0.3 eV above the main peak.

In the time-resolved spectroscopy with an excitation photon energy of 1.98 eV, the time dependence of the photoinduced reflectance change in PDA-MADF exhibits three decay components, the lifetimes of which are approximately 200 fs, 1–2 ps and much longer than 200 ps. These components are assigned to the relaxation of free ^1B_u excitons, STEs and triplet excitons.

The analysis of the excitation intensity dependence of the photoinduced reflectance change of PDA-MADF clarified the mechanism of the triplet exciton formation in the case when the lowest singlet excitons are resonantly excited. Instead of the 'stepwise' two-photon absorption, interactions between singlet excitons have a significant role in the formation of triplet excitons. Based on the interactions between the lowest singlet excitons, a model for the formation of triplet excitons in polydiacetylene was proposed. The Auger process between the two lowest singlet excitons can excite one singlet exciton into the higher-lying singlet state, which in turn creates two triplet excitons by a fission process.

The new Raman peak due to photoexcited states in PDA has been observed at 1200 cm^{-1} for the first time by the femtosecond time-resolved Raman gain spectroscopy. The observed Raman signals indicate that the acetylene-like structure of the main chain relaxes to the butatriene-like structure due to the formation of the STE with geometrical relaxation. The formation and decay kinetics of the Raman signals are consistent with the relaxation processes of excitons observed by femtosecond absorption and fluorescence spectroscopies.

The optical Stark shift of the Raman gain spectra for v_1 and v_2 in a polydiacetylene under near-resonant pump excitation has been observed. The experimental shifts are not consistent with a mechanism which includes either EOC or PMOC alone. Both the direction and size of the observed shift are shown to be in semiquantitative agreement with the predictions of a three-level system model which includes both EOC and PMOC. The observed effect may be potentially useful in a nondegenerate four-wave mixing ultrafast optical switching scheme. Similar optical Stark behavior is expected for near-resonant high-intensity pump excitation in other molecular systems which have a relatively large transition moment between the intermediate and excited states.

Pump–probe experiments on PMSPA thin film with femtosecond time resolution have been performed. The photoinduced absorption peaked at 1.8 eV, the interband bleaching centered at 2.3 eV and the decay kinetics at the photoinduced absorption peak obeyed the power-law behavior with power $n = 0.65 \pm 0.05$. This implies that the geminate recombination process between the soliton and antisoliton generated by intrachain photoexcitation is most relevant. By taking account of a significant reduction in a conjugation length of the polymer, caused by a distortion of the main chain structure, it is concluded that these two features result from the intrachain soliton–antisoliton pairs, which are confined in conjugation-length-limited segments of the main chain. The distance between the confined soliton and antisoliton was estimated as about six repeat unit lengths. This is consistent with the size of a certain regular structure indicated to exist in PMSPA by X-ray diffraction measurements.

A long-lived component was also observed at the same energy as that of the intrachain feature. Since the amplitude of the signal had a quadratic dependence on the pump intensity, it is concluded to be due to confined pairs of likely charged solitons, which are generated by a reaction between likely charged polarons created by interchain photoexcitations.

Various polymers can be classified by their confinement parameter γ. Small γ indicates that the photogenerated charge carriers can separate. The decay dynamics tend to obey a power-law behavior with a power nearly equal to 0.5. On the other hand, large γ corresponds to strong confinement in which the decay dynamics can be described by either a power-law function with a power larger than 0.5 or an exponential function.

Femtosecond time-resolved absorption measurement was performed on a thin film of a PMSPA. The temporal behavior of the photoinduced absorption observed over 1.2–2.0 eV could be successfully reproduced by a sum of the exponentially decaying component with a time constant of 0.135 ps and the time-delayed power-law component. The former peaking at 1.2–1.4 eV corresponds to a nonthermal singlet exciton polaron directly generated by intrachain photoexcitation, while the latter peaking at 1.8 eV corresponds to an oppositely charged, overall neutral, siliton–antisoliton pair confined in a segmented conjugation chain. The spectrum of each component has been clearly separated. The obtained time delay between the two components could explain the observed blue shift of photoinduced absorption spectra. The formation time of the soliton–antisoliton pair was resolved and determined to be 98 ± 8 fs for the first time in the present work.

5   ACKNOWLEDGMENTS

The author thanks Drs M. Yoshizawa, H. Scott, K. Ichimura and M. Taiji; and S. Takeuchi, Mrs. Hasegawa and A. Yasuda for their collaboration in the femtosecond experiment; Dr H. Uchiki and Mr T. Gima for the early stage of the construction of the dye amplifier; Professor T. Kotaka and Dr K. Se of Osaka University for providing samples of PDA-3BCMU and PDA-4BCMU, and the GPC data of PDA-4BCMU; and Mr Y. Hattori for the sample preparation of PDA-4BCMU on KCl substrate. Valuable discussions with Professors E. Hanamura and Y. Toyozawa are also gratefully acknowledged. This work was partly supported by a Grant-in-Aid for Special Distinguished Researches (56222005 and 05102002) from the Ministry of Education, Science and Culture of Japan, and a grant from the Toray Science Foundation to T.K., and the Kurata Research Grant from the Kurata Foundation to M. Yoshizawa and T.K.

6 REFERENCES

1. T. Kobayashi, *Nonlinear Optics of Organics and Semiconductors*, Springer-Verlag, Berlin, 1989.

2. M. Sinclair, D. Moses, K. Akagi, and A.J. Heeger, *Phys. Rev.* B **38**, 10724–10733 (1988).

3. W.S. Fann, S. Benson, J.M.J. Madey, S. Etemad, G.L. Baker, and F. Kajzar, *Phys. Rev. Lett.* **62**, 1492–1495 (1989).

4. W.P. Su and J.R. Schrieffer, *Proc. Natl. Acad. Sci.* **77**, 5626–5629 (1980).

5. C.V. Shank, R. Yen, R.L. Fork, J. Orenstein, and G.L. Baker, *Phys. Rev. Lett.* **49**, 1660–1663 (1982).

6. L. Rothberg, T.M. Jedju, S. Etemad, and G.L. Baker, *IEEE J. Quantum Electron.* **QE-24**, 311–314 (1988).

7. A.J. Heeger, S. Kivelson, J.R. Schrieffer, and W.P. Su, *Rev. Mod. Phys.* **60**, 781–850 (1988).

8. M. Yoshizawa, T. Kobayashi, H. Fujimoto, and J. Tanaka, *J. Phys. Soc. Jpn* **56**, 768–780 (1987).

9. M. Yoshizawa, T. Kobayashi, H. Fujimoto, and J. Tanaka, *Phys. Rev.* B **37**, 8988 (1988).

10. M. Yoshizawa, T. Kobayashi, K. Akagi, and H. Shirakawa, *Phys. Rev.* B **37**, 10301–10307 (1988).

11. P.D. Townsend and R.L. Friend, *Phys. Rev.* B **40**, 3112–3120 (1989).

12. G. Wegner, *Makromol. Chem.* **145**, 85–94 (1971).

13. H. S. Nalwa, *Adv. Mater.* **5**, 341 (1993).

14. R.H. Baughman and K.C. Yee, *J. Polym. Sci., Macromol. Rev.* **13**, 219–239 (1978).

15. R.H. Baughman, *J. Polym. Sci., Polym. Ed.* **12**, 1511–1535 (1974).

16. D. Bloor, in *Polydiacetylenes* (eds. D. Bloor and R.R. Chance), Martinus Nijhoff, Dordrecht, 1985.

17. H. S. Nalwa and S. Miyata, (eds.) *Nonlinear Optics of Organic Molecules and Polymers*, CRC Press, Boca Raton, Florida (1997).

18. K.C. Lim, A. Kapitunik, R. Zacher, and A.J. Heeger, *J. Chem. Phys.* **82**, 516–521 (1985).

19. K.C. Lim and A.J. Heeger, *J. Chem. Phys.* **82**, 522–530 (1985).

20. R.R. Chance, G. Patel, and J.D. Witt, *J. Chem. Phys.* **71**, 206–211 (1979).

21. G.N. Patel, R.R. Chance, and J.D. Witt, *J. Chem. Phys.* **70**, 4387–4392 (1979).

22. T. Kanetake, Y. Tokura, T. Koda, T. Kotaka, and H. Ohnuma, *J. Phys. Soc. Jpn* **54**, 4014–4026 (1985).

23. T. Kobayashi, K. Ogasawara, S. Koshihara, K. Ichimura, and R. Hara, in *Primary Processes in Photobiology* (ed. T. Kobayashi), Springer-Verlag, Berlin, Heidelberg, 1987, pp. 125.

24. A. Kobayashi, H. Kobayashi, Y. Tokura, T. Kanetake, and T. Koda, *J. Chem. Phys.* **87**, 4962 (1987).

25. T. Hattori and T. Kobayashi, *Chem. Phys. Lett.* **133**, 230 (1987).

26. J. Swiatkiewicz, X. Mi, P. Chopra, and P.N. Prasad, *J. Chem. Phys.* **87**, 1882 (1987).

27. S. Koshihara, T. Kobayashi, H. Uchiki, T. Kotaka, and H. Ohnuma, *Chem. Phys. Lett.* **114**, 446 (1985).

28. H. Sixl and R. Warta, in *Electronic Properties of Polymers and Related Compounds* (eds. H. Kuzmany, M. Mehring and S. Roth), Springer-Verlag, Berlin, 1985, pp. 246–248.

29. G.M. Carter, J.V. Hryniewicz, M.K. Thakur, Y.J. Chen, and S.E. Meyloer, *Appl. Phys. Lett.*, **49**, 998–1000 (1986).

30. P.P. Ho, N.L. Yang, T. Jimbo, Q.Z. Wang, and R.R. Alfano, *J. Opt. Soc. Am* B **4**, 1025–1029 (1987).

31. B.I. Greene, J. Orenstein, R.R. Millard, and L.R. Williams, *Chem. Phys. Lett.* **139**, 381–385 (1987).

32. T. Kobayashi, M. Yoshizawa, K. Ichimura, and M. Taiji, in *Ultrafast Phenomena VI* (eds. T. Yajima *et al.*), Springer-Verlag, Berlin, 1988.

33. P.W. Smith and W.J. Tomlinson, *IEEE Spectrum* **18**, 26–33 (1981).

34. T. Kanetake, K. Ishikawa, T. Koda, Y. Tokura, and K. Takeda, *Appl. Phys. Lett.* **51**, 1957 (1987).

35. T. Kobayashi and M. Hasegawa, unpublished.

36. T.C. Chung, J.H. Kaufman, A.J. Heeger, and F. Wudl, *Phys. Rev.* B **30**, 702–710 (1984).

37. Z. Vardeny, E. Ehrenfreund, O. Brafman, M. Nowak, H. Schaffer, A.J. Heeger, and F. Wudl, *Phys. Rev. Lett.* **56**, 671–674 (1986).

38. K. Kaneto, S. Hayashi, and K. Yoshino, *J. Phys. Soc. Jpn* **57**, 1119–1126 (1988).

39. N. Colaneri, M. Nowak, D. Spiegel, S. Hotta, and A.J. Heeger, *Phys. Rev.* B **36**, 7964–7968 (1987).

40. Y.H. Kim, S. Hotta, and A.J. Heeger, *Phys. Rev.* B **36**, 7486–7490 (1987).

41. K. Kaneto, F. Uesugi, and K. Yoshino, *Solid State Commun.* **65**, 783–786 (1987).

42. F. Moses, H. Schaffer, M. Kobayashi, A.J. Heeger, and F. Wudl, *Phys. Rev.* B **30**, 2948–2950 (1984).

43. H.E. Schaffer and A.J. Heeger, *Solid State Commun.* **59**, 415–421 (1986).

44. K. Kaneto, F. Uesugi, and K. Yoshino, *Solid State Commun.* **64**, 1195–1198 (1987).

45. K.S. Wong, W. Hayes, T. Hattori, R.A. Taylor, J.F. Ryan, K. Kaneto, K. Yoshino, and D. Bloor, *J. Phys. C* **18**, L843–L847 (1985).

46. H. Murata, S. Tokito, T. Tsutsui, and S. Saito, *Synth. Met.* **36**, 95 (1990).

47. U. Rauscher, H. Bassler, D.D.C. Bradley, and H. Hennecke, *Phys. Rev.* B **42**, 9830 (1990).

48. A.J. Brassett, N.F. Colaneri, D.D.C. Bradley, R.A. Lawrence, R.H. Friend, H. Murata, S. Tokito, T. Tsutsui, and S. Saito, *Phys. Rev.* B **41**, 10586 (1990).

49. I.D.W. Samuel, K.E. Meyer, D.D.C. Bradley, R.H. Friend, H. Murata, T. Tsutsui, and S. Saito, *Synth. Met.* **41**, 1377 (1991).

50. T. Kobayashi, J. Iwai, and M. Yoshizawa, *Chem. Phys. Lett.* **112**, 360–364 (1984).

51. T. Kobayashi, M. Yoshizawa, U. Stamm, M. Taiji, and M. Hasegawa, *J. Opt. Soc. Am.* B **7**, 1558 (1990).

52. M. Yoshizawa, M. Taiji, and T. Kobayashi, *IEEE J. Quantum Electron.* **QE-25**, 2532–2539 (1989).

53. K. Ichimura, M. Yoshizawa, H. Matsuda, S. Okada, M. Osugi, S. Hourai, H. Nakanishi, and T. Kobayashi, *J. Chem. Phys.* **99**, 7404 (1993).

54. M. Yoshizawa and T. Kobayashi, *IEEE J. Quantum Electron.* **QE-20**, 797–803 (1984).

55. R.L. Fork, O.E. Martinez, and J.P. Gordon, *Opt. Lett.* **9**, 150–152 (1984).

56. T. Gima, H. Uchiki, and T. Kobayashi, *Japan. J. Appl. Phys.* **23**, 1017–1019 (1984).

57. H. Ohnuma, K. Inoue, K. Se, and T. Kotaka, *Macromolecules* **17**, 1285–1287 (1984).

58. S. Okada, M. Ohsugi, A. Masaki, H. Matsuda, S. Takaragi, and H. Nakanishi, *Mol. Cryst. Liq. Cryst.* **183**, 81 (1990).

59. K. Kaneto, S. Hayashi, S. Ura, and K. Yoshino, *J. Phys. Soc. Jpn* **54**, 1146–1153 (1985).

60. B. Reimer, H. Baessler, J. Hesse, and G. Beiser, *Phys. Stat. Sol.* B **73**, 709 (1976).

61. R.J. Hood, H. Muller, C.J. Eckhardt, R.R. Chance, and K.C. Yee, *Chem. Phys. Lett.* **54**, 295 (1978).

62. K. Se, H. Ohnuma, and T. Kotaka, *Macromol.* **16**, 1581–1587 (1983).

63. L. Sebastian and G. Weiser, *Chem. Phys.* **62**, 447 (1981).

64. L. Sebastian and G. Weiser, *Chem. Phys. Lett.* **64**, 396 (1979).

65. L. Sebastian and G. Weiser, *Phys. Rev. Lett.* **46**, 1156 (1981).

66. Y. Tokura, O. Yukihito, T. Koda, R. Baughman, and H. Ray, *Chem. Phys.* **88**, 437 (1988).

67. Y. Tokura, T. Koba, A. Itsubo, M. Miyabayashi, K. Okuhara, and A. Ueda, *J. Chem. Phys.* **85**, 99 (1986).

68. T. Hasegawa, K. Ishikawa, T. Kanetake, T. Koda, K. Takeda, H. Kobayashi, and K. Kubodera, *Chem. Phys. Lett.* **171**, 239 (1990).

69. G.M. Carter, M.K. Thakur, Y.J. Chen, and J.V. Hryniewicz, *Appl. Phys. Lett.* **47**, 457 (1985).

70. T. Kanetake, T. Tokura, and T. Koda, *Solid State Commun.* **56**, 803 (1985).

71. K.C. Yee and R.R. Chance, *J. Polym. Sci., Polym. Phys. Ed.* **16**, 431 (1978).

72. B.Y. Lao, J.D. Dow, and F.C. Wienstein, *Phys. Rev.* B **4**, 4424 (1971).

73. Y. Tokura, K. Ishikawa, T. Kanetake, and T. Koda, *Phys. Rev.* B **36**, 2913–2915 (1987).

74. J.P. Sokoloff, M. Joffre, B. Fluegel, D. Hulin, M. Lindberg, S.W. Koch, A. Migus, A. Antonetti, and N. Peyghambarian, *Phys. Rev.* B **38**, 7615–7621 (1988).

75. C.H. Brito-Cruz, J.P. Gordon, P.C. Becker, R.L. Fork, and C.V. Shank, *IEEE J. Quantum Electron.* **24**, 261 (1988).

76. D.N. Batchelder, in *Polydiacetylenes* (eds. D. Bloor and R.R. Chance), Martinus Nijhoff, Dordrecht, 1985.

77. G.J. Blanchard, J.P. Heritage, A.C.V. Lehman, M.K. Kelly, G.L. Baker, and S. Etemad, *Phys. Rev. Lett.* **63**, 887–890 (1989).

78. F. Charra and J.M. Nunzi, in *Nonlinear Optical Effects in Organic Polymers* (ed. J. Messier), Kluwer, Amsterdam, 1991, p. 359.

79. A.M. Bonch-Bruevich and V.A. Khodovoi, *Usp. Fiz. Nauk. [Sov. Phys. Usp.]* **93[10]**, 71–110 [637–657] (1967 [1968]).

80. D. Frohlich, A.T. Nothe, and K. Reimann, *Phys. Rev. Lett.* **55**, 1748 (1985).

81. A. Mysyrowicz, D. Hulin, A. Antonetti, A. Migus, W.R. Masselink, and H.M. Morkoc, *Phys. Rev. Lett.* **56**, 2748–2751 (1986).

82. B.I. Greene, J.F. Mueller, J. Orenstein, D.H. Rapkine, S. Schmitt-Rink, and M. Thakur, *Phys. Rev. Lett.* **61**, 325–328 (1988).

83. H. Tanaka, M. Inoue, and E. Hanamura, *Solid State Commun.* **63**, 103–107 (1987).

84. J. Jortner, in *Excitons* (eds. E.I. Rashba and M.D. Sturge), North-Holland, Amsterdam, 1987.

85. E.I. Rashba, in *Excitons* (eds. E.I. Rashba and M.D. Sturge), North-Holland, Amsterdam, 1987, pp. 273–332.

86. M.N. Wybourne, B.J. Kiff, and D.N. Batchelder, *Phys. Rev. Lett.* **53**, 580–583 (1984).

87. J. Jortner, in *Vacuum Ultraviolet Radiation Physics* (eds. E.E. Koch, R. Haensel, and C. Kunz), Pergamon, Braunschweig, 1974, p. 263.

88. W. Rehwald, A. Vonlanthen, and W. Meyer, *Physica Status Solidi* (a) **75**, 219–226 (1983).

89. K. Nasu, *Luminescence* **38**, 90–92 (1987).

90. U. Stamm, M. Taiji, M. Yoshizawa, T. Kobayashi, and K. Yoshino, *Mol. Cryst. Liq. Cryst.* **182A**, 147–156 (1990).

91. M. Joffre, D. Hulin, A. Migus, A. Antonetti, C. Benoit a la Guillaume, N. Peyghambarian, M. Lindberg, and S.W. Koch, *Opt. Lett.* **13**, 276–278 (1988).

92. N. Peyghambarian and S.W. Koch, *Rev. Phys. Appl.* **22**, 1711–1714 (1987).

93. B. Fluegel, N. Peyghambarian, G. Olbright, M. Lindberg, S.W. Koch, M. Joffre, D. Hulin, A. Migus, and A. Antonetti, *Phys. Rev. Lett.* **59**, 2588–2591 (1987).

94. M. Lindberg and S.W. Koch, *J. Opt. Soc. Am.* B **5**, 139–146 (1988).

95. E.F. Steigmeier, H. Auderset, and W. Kobel, *Synth. Met.* **18**, 219–224 (1987).

96. D. McBranch, A. Heys, M. Sinclair, D. Moses, and A.J. Heeger, *Phys. Rev.* B **42**, 3011 (1990).

97. Z. Vardeny, H.T. Grahn, A.J. Heeger, and F. Wudl, *Synth. Met.* **28**, C299 (1989).

98. M. Yoshizawa, A. Yasuda, and T. Kobayashi, *Appl. Phys.* B **53**, 296 (1991).

99. T. Kobayashi and M. Yoshizawa, *Synth. Met.* **41–43**, 3129 (1991).

100. T. Kobayashi, *Nonlinear Opt.* **2**, 101 (1992).

101. P.L. Baldeck, R.R. Alfano, and G.P. Agrawal, *Appl. Phys. Lett.* **52**, 1939 (1988).

102. P.L. Baldeck, P.P. Ho, and R.R. Alfano, in *The Supercontinuum Laser Source* (ed. R.R. Alfano), Springer, New York, 1989.

103. Y. Furukawa, A. Sakamoto, and M. Tasumi, *J. Phys. Chem.* **93**, 5354 (1989).

104. J.Y. Mevellec, J.P. Buisson, S. LeFrant, and H. Eckhardt, *Synth. Met.* **41**, 283 (1991).

105. O.M. Gelson, D.D.C. Bradley, H. Murata, T. Tsutsui, J. Ruhe, and G. Wegner, *Synth. Met.* **41**, 875 (1991).

106. C.X. Cui and M. Kertesz, *Phys. Rev.* **B40**, 9661 (1989).

107. G. Gustafson, O. Inganas, H. Osterholm, and J. Laakso, private communication.

108. H. Sumi, M. Georgier, and A. Sumi, *Rev. Solid State Sci.* **4**, 209 (1990).

109. Y. Toyozawa, *J. Phys. Soc. Jpn* **58**, 2626 (1989).

110. Y. Toyozawa, Electron induced lattice instabilities, unpublished results.

111. R.H. Friend, D.D.C. Bradley, and P.D. Townsend, *J. Phys.* **D20**, 1367 (1987).

112. J.H. Burroughes, D.D.C. Bradley, A.R. Brown, R.N. Marks, K. Mackay, R.H. Friend, P.L. Burns, and A.B. Holmes, *Nature* **347**, 538 (1990).

113. M. Movaghar, G.W. Sauger, D. Wurtz, and D.L. Huber, *Solid State Commun.* **39**, 1179–1182 (1981).

114. M. Yoshizawa, K. Nishiyama, M. Fujihira, and T. Kobayashi, *Chem. Phys. Lett.* **207**, 46i (1993).

115. D. Birnbaum, B.E. Kohler, and C.W. Spangler, *J. Chem. Phys.* **94**, 1684 (1991).

116. G.W. Hayden and E.J. Mele, *Phys. Rev.* **B 34**, 5484 (1986).

117. P.L. Danielsen and R.C. Ball, *J. Physique* **46**, 1611 (1985).

118. E. Gutsce, *Phys. Stat. Sol.* (B) **109**, 583 (1982).

119. K.S. Song and C.H. Leung, *Solid State Commun.* **32**, 565 (1979).

120. T.H. Keil, *Phys. Rev.* **140A**, 601 (1965).

121. W.P. Su, *Phys. Rev.* **B 36**, 6040 (1987).

122. I.G. Hunt, D. Bloor, and B. Movagher, *J. Phys.* C **16**, L623–L628 (1983).

123. A. Blumen, G. Zumofen, and J. Klafter, *J. Phys.* **46C7**, 3–8 (1985).

124. G. Zumofen and A. Blumen, *Chem. Phys. Lett.* **88**, 63–67 (1982).

125. J. Klafter, G. Zumofen, and A. Blumen, *J. Phys. Lett.* **45**, L49–L56 (1984).

126. J. Klafter, A. Blumen, and G. Zumofen, *J. Stat. Phys.* **36**, 561–577 (1984).

127. P. Grassberger and I. Procaccia, *J. Chem. Phys.* **77**, 6281–6284 (1982).

128. A. Blumen, K. Klafter, B.S. White, and G. Zumofen, *Phys. Rev. Lett.* **53**, 1301–1304 (1984).

129. S.D. Halle, M. Yoshizawa, H. Murata, T. Tsutsui, S. Saito, and T. Kobayashi, *Synth. Met.* **50**, 429 (1992).

130. W. Krug, E. Miao, and M. Derstine, *J. Opt. Soc. Am. B* **6**, 726 (1989).

131. F.L. Pratt, K.S. Wong, W. Hayes, and D. Bloor, *J. Phys.* C **20**, L141 (1987).

132. M. Ohsugi, S. Takaragi, H. Matsuda, S. Okada, A. Masaki, and H. Nakanishi, *Polymer Preprints, Jpn.* **38**, 2683 (1989).

133. D. Bloor and F.H. Preston, *Phys. Stat. Sol.* (A) **39**, 607 (1977).

134. B.I. Greene, J. Orenstein, and S. Schmitt-Rink, *Science* **247**, 679 (1990).

135. S. Schmitt-Rink, D.S. Chemla, and D.A.B. Miller, *Phys. Rev.* **B 32**, 6601 (1985).

136. Y. Toyozawa, in *Highlights of Condensed-Matter Theory* (eds. F. Bassani, F. Fumi, and M.P. Tosi), North-Holland, Amsterdam, 1985, p. 798.

137. B.I. Greene, J. Orenstein, R.R. Millard, and L.R. Williams, *Phys. Rev. Lett.* **58**, 2750 (1987).

138. J.M. Huxley, P. Mataloni, R.W. Schoenlein, J.G. Fujimoto, E.P. Ippen, and G.M. Carter, *Appl. Phys. Lett.* **56**, 1600 (1990).

139. Y.H. Kim, M. Nowak, Z.G. Soos, and A.J. Heeger, *J. Phys.* C **21**, L503 (1988).

140. J. Orenstein, S. Etemad, and G.L. Baker, *J. Phys.* C **17**, L297 (1984).

141. T. Hattori, W. Hayes, and D. Bloor, *J. Phys.* C **17**, L881 (1984).

142. L. Robins, J. Orenstein, and R. Superfine, *Phys. Rev. Lett.* **56**, 1850 (1986).

143. B.I. Greene, J. Orenstein, R.R. Millard, and L.R. Williams, *Chem. Phys. Lett.* **139**, 381 (1987).

144. T. Kobayashi and H. Ikeda, *Chem. Phys. Lett.* **133**, 54 (1987).

145. R.H. Austin, G.L. Baker, S. Etemad, and R. Thompson, *J. Chem. Phys.* **90**, 6642 (1989).

146. K. Lochner, H. Bassler, B. Tieke, and G. Wegner, *Phys. Stat. Sol.* (B) **88**, 653 (1978).

147. M. Winter, A. Grupp, M. Mehring, and H. Sixl, *Chem. Phys. Lett.* **133**, 482 (1987).

148. R.E. Merrifield, P. Avakian, and R.P. Groff, *Chem. Phys. Lett.* **3**, 155 (1969).

149. N. Geacintov, M. Pope, and F. Vogel, *Phys. Rev. Lett.* **22**, 593 (1969).

150. C.E. Swenberg and N.E. Geacintov, in *Organic Molecular Photophysics* (ed. J.B. Birks), John Wiley and Sons, London, 1973, pp. 489.

151. E. Tokunaga, A. Terasaki, K. Tsunetomo, Y. Osaka, and T. Kobayashi, *J. Opt. Soc. Am.* B, **10**, 2364 (1993).

152. D.N. Batchelder and D. Bloor, *J. Phys.* C **15**, 3005 (1982).

153. A. Terai, Y. Ono, and Y. Wada, *J. Phys. Soc. Jpn.* **58**, 3798 (1989).

154. L.X. Zheng, R.E. Benner, Z.V. Vardeny, and G.L. Baker, *Synth. Met.* **49**, 313 (1992).

155. G. Lanzani, L.X. Zheng, G. Figari, R.E. Benner, and Z.V. Vardeny, *Phys. Rev. Lett.* **68**, 3104 (1992).

156. G. Lanzani, L.X. Zheng, R.E. Benner, and Z.V. Vardeny, *Synth. Met.* **49**, 321 (1992).

157. D.L. Weidman and D.B. Fitchen, *Proc. 10th Int. Conf. Raman Spectroscopy*, University of Oregon, Eugene, 1986, p. 12.

158. M. Yoshizawa, Y. Hattori, and T. Kobayashi, *Phys. Rev. B* **47**, 3882 (1993).

159. A. Yasuda, M. Yoshizawa, and T. Kobayashi, *Chem. Phys. Lett.* **209**, 281 (1993).

160. W.F. Lewis and D.N. Batchelder, *Chem. Phys. Lett.* **60**, 232 (1979).

161. L.M. Sverdlov, M.A. Kovner, and E.P. Krainov, *Vibrational Spectra of Polyatomic Molecules*, John Wiley & Sons, New York, 1970, p. 282.

162. G.J. Blanchard and J.P. Heritage, *J. Chem. Phys.* **93**, 4377 (1990).

163. S.D. Halle, M. Yoshizawa, H. Matsuda, S. Okada, H. Nakanishi, and T. Kobayashi, in *Time-Resolved Vibrational Spectroscopy* (ed. H. Takahashi), Springer-Verlag, Berlin, 1992, p. 260.

164. D.N. Batchelder and D. Bloor, *J. Phys. C: Solid State Phys.* **15**, 3005 (1982).

165. B. Dick and R.M. Hochstrasser, *J. Chem. Phys.* **81**, 2897 (1984).

166. E. Courtens and A. Szoke, *Phys. Rev. A* **1588**, 1977 (15).

167. C. Cohen-Tannoudji and S. Reynaud, *J. Phys. B* **10**, 365 (1977).

168. B. Dick and R.M. Hochstrasser, *Chem. Phys.* **75**, 133 (1983).

169. Y.R. Shen, *Phys. Rev. B* **9**, 622 (1974).

170. P.J. Carroll and L.E. Brus, *J. Chem. Phys.* **86**, 6584 (1987).

171. R.G. Alden, M.R. Ondrias, S. Courtney, E.W. Findsen, and J.M. Friedman, *J. Opt. Soc. Am. B* **7**, 1579 (1990).

172. S. Saikan, N. Hashimoto, T. Kushida, and K. Namba, *J. Chem. Phys.* **82**, 5409 (1985).

173. W.P. Su, J.R. Schrieffer, and A.J. Heeger, *Phys. Rev. Lett.* **42**, 1698 (1979).

174. W.P. Su, J.R. Schrieffer, and A.J. Heeger, *Phys. Rev. B* **22**, 2099 (1980).

175. W.P. Su and J.R. Schrieffer, *Proc. Natl. Acad. Sci. USA* **77**, 5626 (1980).

176. Z. Vardeny, J. Orenstein, and G.L. Baker, *Phys. Rev. Lett.* **50**, 2032 (1983).

177. J. Orenstein and G.L. Baker, *Phys. Rev. Lett.* **49**, 1043 (1982).

178. Z. Vardeny, D.M.J. Strait, T.C. Chung, and A.J. Heeger, *Phys. Rev. Lett.* **49**, 1657 (1982).

179. Z. Vardeny, *Physica* **127**B, 338 (1984).

180. L. Rothberg, T.M. Jedju, S. Etemad, and G.L. Baker, *Phys. Rev. Lett.* **57**, 3229 (1985).

181. L. Rothberg, T.M. Jedju, S. Etemad, and G.L. Baker, *Phys. Rev. B* **36**, 7529 (1987).

182. C.V. Shank, R. Yen, J. Orenstein, and G.L. Baker, *Phys. Rev. B* **28**, 6095 (1983).

183. L. Rothberg, T.M. Jedju, P.D. Townsend, S. Etemad, and G.L. Baker, *Mol. Cryst. Liq. Cryst.* **194**, 1 (1991).

184. D. Baeriswyl, D.K. Campbell, and S. Mazumdar, *Phys. Rev. Lett.* **56**, 1509 (1986).

185. J.D. Flood and A.J. Heeger, *Phys. Rev. B* **28**, 2356 (1983).

186. K. Fesser, A.R. Bishop, and D.K. Campbell, *Phys. Rev. B* **27**, 4804 (1983).

187. C.M. Foster, Y.H. Kim, N. Uotani, and A.J. Heeger, *Synth. Met.* **29**, E135 (1989).

188. E.T. Kang, K.G. Neoh, T. Masuda, T. Higashimura, and M. Yamamoto, *Polymer* **1989**, 1328 (1989).

189. A.S. Sididiqui, *J. Phys. (Paris)* **83**, C3–495 (1983).

190. T. Takayama, Y.R. Lin-Liu, and K. Maki, *Phys. Rev. B* **21**, 2388 (1980).

191. S.A. Brazovski and N.N. Kirova, *JETP Lett.* **33**, 4 (1981).

192. Y.R. Lin-Lui and K. Maki, *Phys. Rev. B* **22**, 5754 (1980).

193. T. Masuda and T. Higashimura, *Advances in Chemistry* **224**, 641 (1990).

194. T. Tani, P.M. Grant, W.D. Gill, G.B. Street, and T.C. Clarke, *Solid State Commun.* **33**, 499 (1980).

195. P.D. Townsend and R.H. Friend, *J. Phys. C* **20**, 4221 (1987).

196. S.D. Hall, M. Yoshizawa, H. Matsuda, S. Okada, H. Nakanishi, and T. Kobayashi, *J. Opt. Soc. Am. B* **11** (1994).

197. S. Takeuchi, M. Yoshizawa, T. Masuda, T. Higashimura, and T. Kobayashi, *IEEE J. Quantum Electron.* **QE28**, 2054 (1992).

198. S. Takeuchi, T. Masuda, T. Higashimura, and T. Kobayashi, *Solid State Commun.* **87**, 655 (1993).

199. S. Takeuchi, T. Masuda, and T. Kobayashi *J. Chem. Phys.*, **105**, 2859 (1996).

CHAPTER 8

Electropolymerized Phthalocyanines and their Applications

Thomas F. Guarr

Gentex Corporation, Zeeland, Michigan, USA

1 INTRODUCTION

During the last 30 years, literally thousands of literature reports have detailed the rich chemistry of phthalocyanines (Pcs) in such diverse areas as wastewater treatment [1], fuel cell development [2], photodynamic cancer therapy [3], gas sensing [4], optical recording [5], acoustic wave sensing [6], information display [7], and even the removal of nitrogen oxides from cigarette smoke [8]. Pcs have also been used extensively for the catalytic oxidation of thiols in hydrocarbon fuels (Merox sweetening) [9], and in the sensitive electrochemical detection of biologically important analytes [10]. The incorporation of Pcs into polymeric systems has further extended their utility by providing the clever chemist with another means of controlling physical and chemical properties at the molecular level.

Numerous methods for preparing Pc/polymer systems have been described, including simple attachment of Pcs to a polymeric mix by electrostatic or covalent means, direct synthesis of ring-linked Pc polymers (i.e. from tetracyanobenzene), chemical polymerization (often via condensation reactions) of simple monomeric Pcs, and synthesis of ligand-bridged cofacially stacked polymeric metallo-Pcs [11–17]. Much of this work, from the laboratories of Wöhrle, [11,12], Kenney [13], Hanack [14,15], and Marks [16,17], among others, has resulted in new materials with useful and intriguing properties such as increased catalytic activity, improved durability or high electronic conductivity. These materials have been the subject of several excellent recent reviews [11,14] and are beyond the scope of this chapter. Also not included in the present treatment is the elegant work of Sajii and coworkers [18,19] on the

Handbook of Organic Conductive Molecules and Polymers: Vol. 4. Conductive Polymers: Transport, Photophysics and Applications.
Edited by H. S. Nalwa. © 1997 John Wiley & Sons Ltd

fabrication of monomeric Pc thin films by the electrochemical disruption of micellar aggregates composed of ferrocenyl-containing surfactants. Rather, this review focuses on the properties and applications of phthalocyanine polymers produced by electrochemical polymerization.

Electropolymerization offers several distinct advantages over other polymerization methods, particularly when applied to the fabrication of chemically modified electrodes (CMEs) [20]. First, the method can be used with a wide variety of surfaces, including methals, conducting metal oxides and numerous carbon-based electrode materials. Electropolymerization of appropriate substrates onto redox conductive or electronically conductive polymer supports has also been achieved, resulting in the formation of polymeric bilayer materials with interesting electrical and mechanical properties [21,22]. Second, electropolymerization methods generally lead to the direct deposition of a uniform surface coating, and film thickness can be easily and reproducibly manipulated through appropriate control of the polymerization conditions (i.e. scan rate, potential range, polymerization time) [23]. Moreover, electropolymerized films, even at thicknesses of 20 monolayers or less, tend to be largely devoid of pinholes, thus reducing concerns regarding the role of defect sites in determining the overall performance of the CME [23].

Interest in the electropolymerization of phthalocyanines stems in large part from the well-known utility of various metal Pcs as redox catalysts [9,24]. The high conductivity of many one-dimensional metal Pc polymers [11,14–17,25] suggests that such systems will be capable of supporting reasonably rapid charge transport across thin films and therefore useful as immobilized electrocatalysts. Indeed, single crystal $NiPc(I_3)_{0.33}$ is a

metallic conductor, showing an increase in conductivity from 300–700 $S\,cm^{-1}$ at room temperature to 2000–5000 $S\,cm^{-1}$ at 25 K [25]. Since the measured conductivity of one-dimensional polymers is often limited by interchain hopping, the prospect of attaining very high conductivities in two-dimensional sheet polymers via electrochemical polymerization of multifunctional metallo-Pcs is certainly enticing. Venkatachalem et al. have already shown that the thermal polymerization of nickel phthalocyanine tetracarboxylic acid produces a sheet-like material which displays a room temperature conductivity of 2.3 $S\,cm^{-1}$ upon further thermal treatment [26,27].

2 POLYMER FORMATION AND PROPERTIES

2.1 Tetraaminophthalocyanines

2.1.1 Monomer synthesis

Electropolymerized films of metallo-Pcs were first reported by Li and Guarr in 1986 [28]. These authors extended the work of Murray and coworkers [29,30] on oxidatively electropolymerized metalloporphyrins to the cobalt, nickel and zinc complexes of 2,9,16,23-tetraaminophthalocyanine (T4APc, I). T4APc derivatives were prepared via sodium sulfide reduction of the corresponding tetranitrophthalocyanines, as reported by Achar et al. [31,32]. Using this procedure, T4APc complexes of cobalt, nickel, zinc, palladium and copper are readily prepared in good yields (MT4APc where M stands for a metal). Synthesis of the iron derivative by this route is more problematic, but small quantities of it

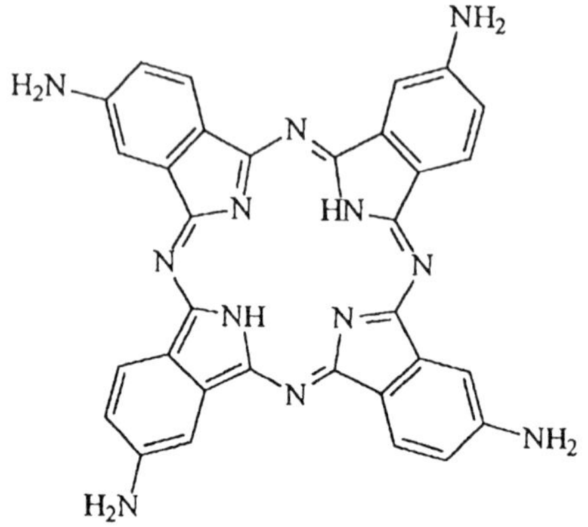

I 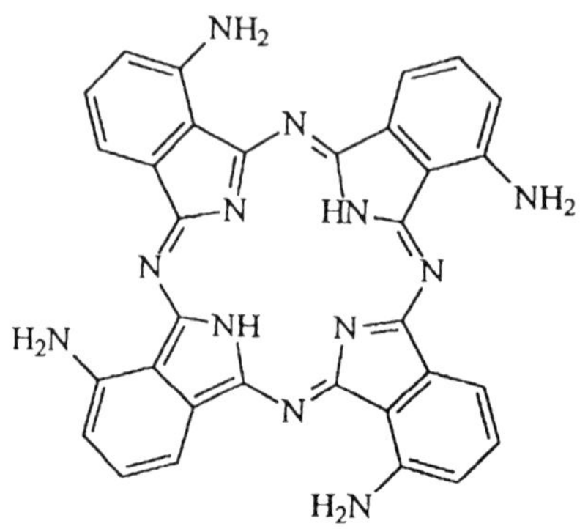   II

have been prepared [33]. An alternative preparation of CoT4APc (via 5-nitro-1,3-diiminoisoindoline) has been reported by Tse [34]. In this case sodium sulfide was also employed to reduce the intermediate cobalt tetranitrophthalocyanine.

It is worth noting that both synthetic routes give (presumably) a mixture of four conformational isomers [35,36] (only the structure of C_{4h} symmetry is shown in **I**). There have been no reported attempts to separate the components prior to electrochemical polymerization.

Because of steric constraints, the synthesis of **II** (1,8,15,22-tetraamino-phthalocyanine, T3APc) might be expected to preferentially yield the C_{4h} isomer, although such selectivity has not been conclusively demonstrated in this case. Metal complexes of **II** do undergo oxidative electropolymerization, albeit somewhat less efficiently than their MT4APc counterparts [35].

2.1.2 Electropolymerization mechanism

The electrochemistry of MT4APc complexes is dependent on the metal [28,35,37,38], but redox-inactive central metal typically shows two chemically and electrochemically reversible one-electron ring-centered reductions in dimethylsulfoxide (DMSO) or dimethylformamide (DMF) (see Figure 8.1 and Table 8.1). A ligand-centered oxidation occurs near +0.4 V vs SCE and is also generally reversible. Finally, at more positive potentials, a chemically reversible anodic wave is observed. This latter process is thought to involve oxidation of the amine functionalities and gives a

relatively clean four-electron wave for unmetalated T4APc [37], with somewhat less well-defined peaks often seen for other MT4APc derivatives [28,35].

Upon continuously cycling over a potential range including amine oxidation, Li and Guarr [28] observed a gradual increase in both anodic and cathodic peak current (Figure 8.2). Such behavior was attributed to the deposition of an electroactive film on the electrode surface. Note that, at least for M = Ni, extending the potential range leads not only to increased faradaic current at the peaks, but also to enhanced charging current in the 'valleys' (Figure 8.3), suggesting that poly(NiT4APc) films exhibit significant electronic conductivity [28]. Coverages up to 8×10^{-8} mol cm^{-2} were readily obtained, and film thicknesses, measured by Rutherford backscattering spectrometry (RBS), ranged from a few hundred to a few thousand Ångströms [28,38]. As Figure 8.4 illustrates, both the surface coverage (measured electrochemically) and film thickness (measured by RBS) vary linearly with the number of polymerization cycles over the range studied (5 to 60 cycles). RBS studies also give an average oxidation level of +0.5 and +0.3 (per monomer unit) for as-prepared films made in DMSO and DMF, respectively [35,39].

Li [35] has suggested that the chemically irreversible nature of ligand oxidation is caused by coupling of the ring-centered radical species generated at the electrode surface to produce an adherent polymer film. (Scheme 8.1).

Subsequent examination of the polymer electrochemistry and vibrational spectroscopy has generally supported this hypothesis [40,41]; however, on the basis of Raman data, Zheng et al. [42] have suggested that the electropolymerization of CuT4APc proceeds via formation of azo linkages. Tachikawa et al. also measured the Raman spectra of poly(MT4APc) polymers, but

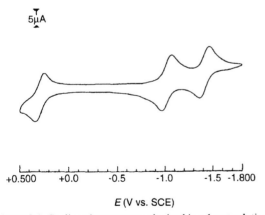

5 μA

+0.500   +0.0   -0.5   -1.0   -1.5 -1.800

E (V vs. SCE)

Figure 8.1. Cyclic voltammogram obtained in a 1 mM solution of CuT4APc in DMSO/0.1 M tetraethylammonium perchlorate. The working electrode is glassy carbon and the scan rate is 100 mV s^{-1}. (Reproduced by permission from ref. 41.)

Table 8.1. Redox potentials of tetraaminophthalocyanines*

Complex	$E_{red}(2)$	$E_{red}(1)$	$E_{ox}(1)$	$E_{ox}(2)$	Ref.
CoT4APc	−1.66	−0.52	+0.19	+0.44	28
NiT4APc	−1.45	−1.00	+0.23	+0.60	28
PdT4APc	−1.43	−0.95	+0.36	+0.60	28
CuT4APc	−1.45	−1.04	+0.29	+0.60	41
ZnT4APc	−1.60	−1.15	+0.27	+0.62	35
CoT3APc	−1.47	−0.50	+0.29	+0.52	35
NiT3APc	−1.29	−0.87	+0.31	+0.50	35

*All potentials are measured in DMSO/0.1 M tetraethylammonium perchlorate and referenced to SCE. All redox processes are localized on the Pc ring, except for $E_{red}(1)$ and $E_{ox}(1)$ of CoT4APc and CoT3APc. Note that the first oxidation of NiT4APc was incorrectly assigned as a metal-centered process in ref. 28.

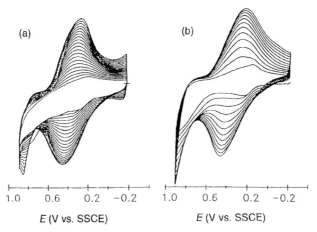

Figure 8.2. Repetitive cyclic voltammograms (−0.2 to +0.9 V vs. SCE) obtained at a glassy carbon electrode in DMSO/0.1 M tetraethylammonium perchlorate containing (a) 1 mM CoT4APc and (b) 1 mM NiT4APc. Scan rate = 200 mV s⁻¹; every fifth scan is recorded in (a); every scan is recorded in (b). (Reproduced by permission of the Royal Society of Chemistry from H. Li and T.F. Guarr, *J. Chem. Soc. Chem. Commun.* 1989, 832 (a) and ref. 35 (b).)

conclude only that most of the amine functionalities have reacted during polymer formation. [43].

The proposed polymerization mechanism (Scheme 8.1) is analogous to that invoked for the formation of polyaniline [44,45], with loss of 2H⁺ following the initial coupling step. However, *para*-coupling is observed almost exclusively in the formation of polyaniline, while *ortho*-coupling represents only a minor side reaction [46,47]. The *para*-position is unavailable in MT4APcs, and coupling is thought to involve only the *ortho*-carbon. Thus, the formation of extended polyaniline-like chains is not possible, so the

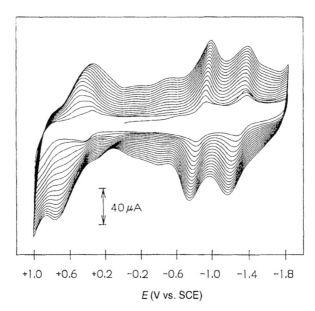

Figure 8.3. Repetitive cyclic voltammograms (+1.0 to −1.8 V vs. SCE) obtained at a glassy carbon electrode in DMSO/0.1 M tetraethylammonium perchlorate containing 1 mM NiT4APc. Scan rate = 200 mV s⁻¹. (Reproduced by permission from ref. 35.)

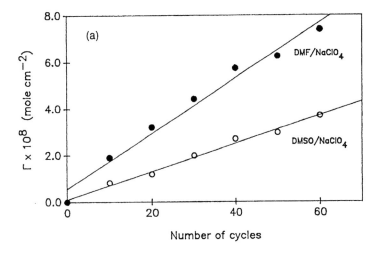

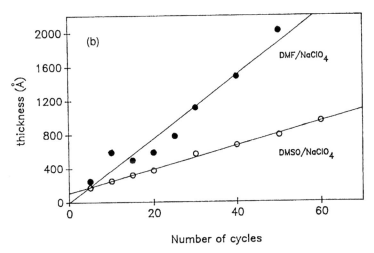

Figure 8.4. (A) Plots of film coverage (measured coulometrically) vs. number of potential cycles employed during electropolymerization. (B) Plots of film thickness (measured by RBS) vs. number of potential cycles employed during electropolymerization. Other conditions are as in Figure 8.2. (Reproduced by permission from ref. 35.)

presence of these linkages should not dominate the electrical properties of the MT4APc polymers. Indeed, unlike polyaniline, the poly(MT4APc) materials investigated to date behave as *n*-doped rather than *p*-doped conductors [28,37,39–41,48–50].

It is also worth noting that NiT4APc and PdT4APc can be cathodically electropolymerized in the presence of oxygen [35] (Figure 8.5). The mechanism of this process has not been adequately examined, but given

the well-known catalytic activity of metal Pcs, the participation of electrochemically generated peroxide (perhaps simply functioning as an oxidant) is one possibility. To date, there is no evidence of the formation of peroxo bridges, and the poly(NiT4APc) and poly(PdT4APc) films obtained from this route are electrochemically and spectroscopically indistinguishable from those produced anodically. Interestingly, CoT4APc and ZnT4APc do not undergo cathodic

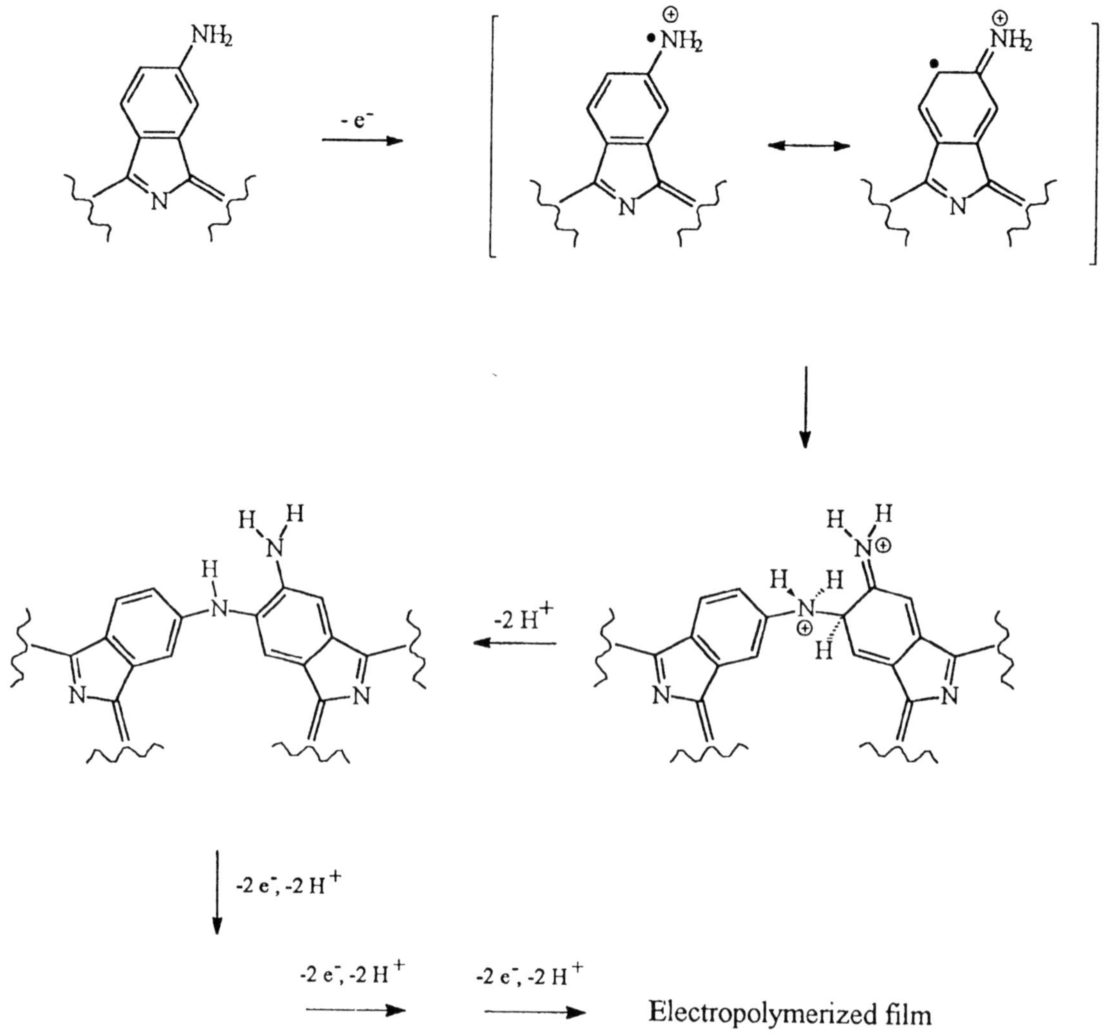

Scheme 8.1

electropolymerization from air- or O_2-saturated solution [35].

2.1.3 Polymerization rate

The polymerization efficiency is similar for all single-ring MT4APcs with redox-inactive central metals. The polymerization of CoT4APc is about ten times slower (see Figure 8.2), perhaps because ligand oxidation occurs only after oxidation of the central metal to Co (**III**); thus the diradical coupling step in Scheme 8.1 is

more strongly electrostatically inhibited. The polymerization rate of CoT4APc has also been reported to decrease with sample aging [40,51], possibly due to air oxidation of the solid.

At low electrolyte concentrations, the rate of MT4APc polymerization increases linearly with [electrolyte], then reaches a plateau at higher ionic strength. Interestingly, the polymerization rate was also found to increase with increasing size of the electrolyte cation, at least for tetraalkylammonium eletrolytes at concentrations greater than 0.1 M [40]. As with other conductive polymers (e.g. polypyrrole), the polymerization rate is

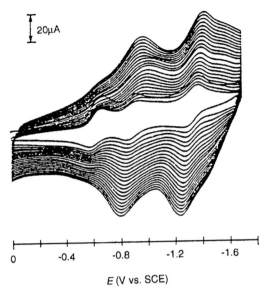

Figure 8.5. Repetitive cyclic voltammograms (0 to -1.8 V vs. SCE) obtained in air-saturated DMSO/0.1 M tetraethylammonium perchlorate containing 1 mM NiT4APc.The working electrode is glassy carbon; scan rate $= 200$ mV s^{-1}. (Reproduced by permission from ref. 35.)

appears to be the rate-controlling step, while mass transport becomes rate limiting with more positive limits. A decrease in apparent polymerization rate is observed with very positive potential limits ($\geqslant 1.0$ V vs. SCE); this behavior has been attributed to oxidative degradation [40]. Potentiostatically grown poly-(MT4APc) films show a similar dependence on applied potential (Figure 8.6).

Film formation in DMF is about 1.5 times faster than in DMSO, a fact that was attributed to the difference in solvent viscosity [40]. The intentional addition of water (up to 25%) produces only minor effects on film deposition; larger quantities cause a decrease in polymerization rate which is ascribed to decreased monomer solubility. In addition, decreasing the solution pH from 7.0 to 2.5 produces a gradual increase in polymerization rate, while further decreases in pH lead to lower rates, possibly owing to protonation of the monomer [40].

Although the *absolute* current efficiency depends, among other factors, on the identity of the electrolyte, electrolyte concentration, monomer concentration and the potential limits employed, it remains invariant throughout the course of the polymerization process. That is, for a given set of experimental conditions, film thickness increases linearly with the net charge passed (at least for $Q \geqslant 5$ mC).

Electropolymerization of the 'sandwich-type' Lu-(T4APc)$_2$ is reported to be very slow and extremely sensitive to experimental conditions [52], especially scan range. Reasons for this unusual behavior have not yet been established.

roughly proportional to the monomer concentration; however, note that MT4APc solubility constraints place a rather low upper limit on the accessible concentration range in such studies.

The relationship between the polymerization rate and the positive scan limit is somewhat more complicated. When a low limit is employed, monomer oxidation

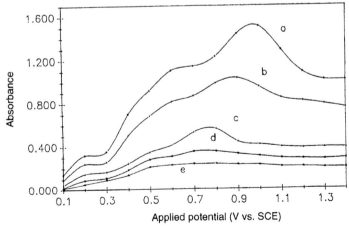

Figure 8.6. Effect of applied potential on potentiostatic polymerization rate. The reaction times were (a) 10, (b) 5, (c) 2, (d) 1 and (e) 0.5 min. [NiT4APc] $= 2.0$ mM in DMSO/0.1 M tetraethylammonium perchlorate. (Reproduced by permission from ref. 40.)

2.1.4 Film morphology

The effect of polymerization conditions on the morphology of the resultant polymer films has been investigated in detail [40,53]. Film roughness was found to increase with increasing electrolyte concentration as well as with increasing MT4APc concentration. Rough films are also favored by low scan rates and high positive potential limits. Likewise, film morphology depends on the solvent used for polymerization, with larger topographical features and more globular films being produced in DMF (as compared to DMSO under identical conditions) [40,53].

Although the nature of the electrolyte cation appears to have little effect on film morphology, a change in the electrolyte anion from perchlorate to dodecylsulfate results in a much smoother film surface, as observed for polypyrrole and polyaniline films. When the sodium salt of phthalocyanine tetracarboxylic acid was used as an electrolyte, only very thin (about 200 Å) light blue films could be prepared [40].

The morphology of the electrode coatings can be altered after the electropolymerization process [40,53]. While such behavior has been previously reported for polypyrrole [54,55] ('overoxidation', vide infra), the poly(MT4APc) films appear unique in that a more porous, open microstructure is obtained upon controlled reduction at -1.3 V vs. SCE. Unlike the oxidative process (which also serves to increase the porosity of MT4APc polymers), cathodic post-treatment preserves the electronic properties of these materials. Thus, by judicious choice of experimental parameters, it may be possible to 'fine-tune' the ionic conductivity and size-exclusion properties of poly(MT4APc) films while maintaining electronic conductivity. Such control might have important implications in many different applications, including electrocatalysis, electrochromism, fuel cell technology and the development of so-called 'supercapacitors'.

2.1.5 Conductivity

Following transferal of a poly(MT4APc)-coated electrode to fresh electrolyte, the poly(MT4APc)-modified electrodes exhibit a stable electrochemical response when cycled to the negative potentials [28]. A typical cyclic voltammogram consists of two reversible couples (with potentials near those of the related MT4APc monomer) superimposed on a broad background current (Figure 8.7). No change in the magnitude of either the peak currents or the background current is observed, even after several hundred cycles [35], provided that the upper limit of the scan range is $\leqslant 0$ V vs. SCE. However, for M = Ni, the peaks become less well defined when the scan range is extended to positive potentials, eventually yielding a broad featureless voltammogram (Figure 8.8) [28].

The shapes of the I–E curves for poly(NiT4APc) reflect the formation of an electronically conductive polymer and can be understood in terms of Feldberg's model for electrochemical switching between the oxidized (insulating) and reduced (conducting) states [48,56]. In this case, the shape of the voltammogram is dominated by the large capacitive current with relatively small contributions from faradaic processes within the film. However, it is noteworthy that the poly(NiT4APc) system differs from most previously reported conductive polymers in two important ways.

First, despite its formation via electrochemical oxidation, doping is achieved by reduction, making poly(NiT4APc) a rare example of a chemically stable n-doped conductive polymer. Second, poly(NiT4APc) displays electrical conductivity over an exceptionally broad potential window (spanning nearly 3 V!), suggesting that this material may find applications in elecroanalysis, electrocatalysis and other fields (vide infra) where the restricted potential range of previously available conductive polymers has limited their utility.

Poly(MT4APc) polymers containing other central metals show similar conductivity windows, but tend to exhibit smaller capacitive currents and lower conduc-

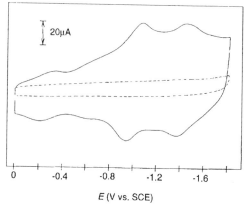

Figure 8.7. Cyclic voltammograms (0 to -1.8 V vs. SCE) obtained at a bare glassy carbon electrode (dashed line) and at a poly(NiT4APc)-modified glassy carbon electrode (solid line) following transferal to fresh DMSO/0.1 M tetraethylammonium perchlorate electrolyte. Scan rate = 200 mV s^{-1} in both cases. (Reproduced by permission from ref. 35.)

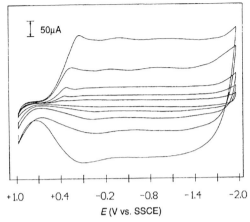

Figure 8.8. Cyclic voltammograms (+1.0 to −1.8 V vs. SCE) obtained at a poly(NiT4APc)-modified glassy carbon electrode following several cycles in fresh DMSO/0.1 M tetraethylammonium perchlorate electrolyte. Scan rates are 400, 200, 100, 50 and 20 mV s⁻¹ from high to low. (Reproduced by permission of the Royal Society of Chemistry from H. Li and T.F. Guarr, *J. Chem. Soc. Chem. Commun.* 1989, 832.)

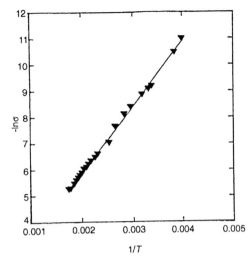

Figure 8.9. Temperature dependence of the conductivity of a poly(NiT4APc) thin film prepared from DMSO/0.1 M tetraethylammonium perchlorate. (Reproduced by permission from ref. 41.)

tivities. Still, the conductivities of these materials are generally sufficient for electrochemical applications as thin films, and the presence of a redox-active metal (especially cobalt) can sometimes offer additional advantages, such as redox conduction at potentials outside the electronic conduction window, enhanced catalytic activity or improved electrochemical properties (*vide infra*).

The room temperature conductivity of poly(NiT4APc) films is on the order of 10^{-4} S cm⁻¹ [40,41], a value which is similar to that reported for other polymeric phthalocyanines. Interestingly, measurements of poly(NiT4APc) conductivity as a function of temperature yield an apparent activation energy E_a of just 0.2 eV (Figure 8.9). This has been interpreted in terms of a fluctuation-induced carrier tunneling model [17,41,57], with electrical conduction resulting from tunneling through potential barriers between large conducting segments rather than via hopping between localized sites.

2.1.6 Overoxidation

Like other anodically polymerized conductive polymers, poly(MT4APc) films are susceptible to oxidative degradation processes which are commonly collectively termed 'overoxidation'. Overoxidation is reported to be a multielectron process for polypyrrole and polythio-

phene, but the relationship between the oxidation level of the polymer and the applied potential has not yet been clearly established. In general, oxidative degradation is a rather complex process, and leads to loss of conductivity and increased film porosity. Such changes have been exploited in efforts to improve the permeability and permselectivity of conductive polymer membranes in various applications [53,54,58].

Several early studies of poly(MT4APc) electropolymerization were hampered by partial overoxidation of the polymers during film formation. By restricting the positive potential limit to ≤ +0.7 V vs. SCE, Xu [40] was able to obtain poly(NiT4APc) films without significant levels of overoxidation (compare Figure 8.2 with Figure 8.10). Modified electrodes prepared in this fashion still show two well-defined reversible reductions, but these couples are shifted to significantly more positive potentials than those of the corresponding monomers. As reported earlier, additional cycling to +1.0 V causes flattening of these waves and the eventual development of a broad, featureless current envelope. During the initial stages of overoxidation, a large cathodic wave at *c.* −0.5 to −0.7 V reminiscent of charge-trapping behavior is observed; this wave quickly broadens into the background.

Overoxidation of poly(MT4APc) films has also been reported to cause changes in the electronic and vibrational spectra [40]. The IR spectrum of an overoxidized film shows a decrease in absorption in

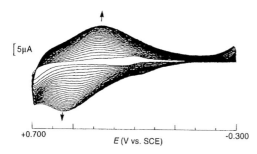

Figure 8.10. Cyclic voltammograms recorded during electropolymerization of NiT4APc from DMSO/0.1 M tetraethylammonium perchlorate. Scan rate = 100 mV s^{-1}; [NiT4APc] = 2.0 mM. (Reproduced by permission from ref. 40.)

the range of 1250–1300 cm^{-1} along with increased absorption above 1630 cm^{-1} and at 1511 cm^{-1}; these data have been taken as evidence for the formation of additional C=N—C and C=N—H bonds. A dramatic decrease in the strength of the 1027 cm^{-1} peak was attributed to perturbation of the Pc skeleton. Overoxidation causes increased broadening of the UV vis spectrum, coupled with loss of the 450 nm shoulder.

Chronocoulometry was used to establish the relationship between the oxidation level of the polymer and the applied potential [40]. These studies indicate that up to 10 e$^-$ per MT4APc monomer can be passed at very positive potentials, suggesting that overoxidation can be extensive and may be associated with the oxidation of 'dangling' amino groups and/or the formation of phenazine-like linkages between Pc rings (Scheme 8.2).

2.1.7 Poly(MT4APc) composites

Copolymerization of various MT4APc derivatives with pyrrole has been carried out with the intention of improving the mechanical properties of the poly-(MT4APc) films while preserving their electronic conductivity [41]. Despite the fact that pyrrole itself could not be polymerized under the conditions employed, CuT4APc/pyrrole copolymers were readily prepared by oxidative electropolymerization in DMSO. CuT4APc polymerization proceeds in typical MT4APc fashion; however, addition of 0.1 M pyrrole induces the growth of an additional redox couple near +0.2 V upon continuous cycling. This new electrochemical feature was ascribed to the formation of polypyrrole-type functionalities within the resultant film [41]. Similar behavior was found for NiT4APc/pyrrole solutions, but no copolymer was formed in the case of CoT4APc.

The rate of polypyrrole (PPy) incorporation into the copolymer was determined by measuring the cathodic current for the PPy redox couple and was found to vary linearly with CuT4APc concentration. The rate of PPy incorporation also increases with increasing [pyrrole] in the polymerization medium, but levels off above 0.1 M (Figure 8.11) [41].

Following transferal to fresh electrolyte (containing no CuT4APc or pyrrole), potential cycling of a poly(CuT4APc)/PPy-modified electrode results in a dramatic increase in the potential separation between the anodic and cathodic waves associated with PPy redox. Such behavior was attributed to the electrical isolation of the PPy regions in the copolymer from the electrode; PPy can then undergo oxidation/reduction only when the process is mediated by the CuT4APc backbone [41].

2.2 Octacyanophthalocyanines

Very recently, Vijayanathan et al. reported the anodic polymerization of unmetalated 2,3,9,10,16,17,23,24-octacyanophthalocyanine (OCPc, **III**) [59].

III

A gradual increase in the faradaic current was observed upon cycling between 0 and +1.6 V in solutions of **III** (even in the absence of supporting electrolyte!); this behavior was attributed to the formation of poly(OCPc) films. The authors found that dark green polymeric films of **III** could also be prepared from CH$_3$CN by constant potential electrolysis at +1.6 V vs. SCE, and proposed a mechanism involving coupling of terminal cyano groups as shown in Scheme 8.3.

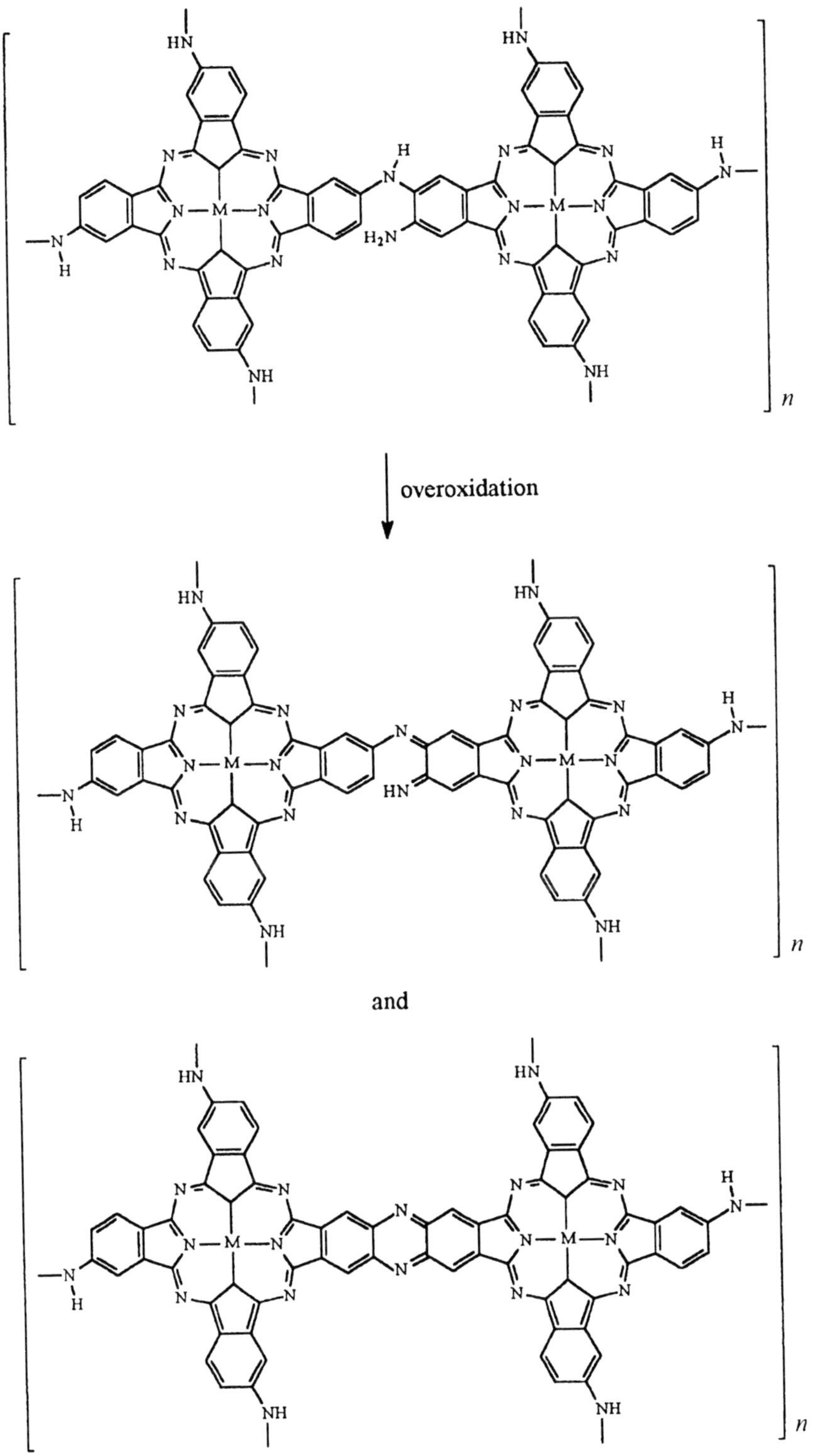

Scheme 8.2

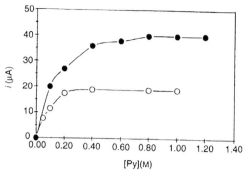

Figure 8.11. Plots of the cathodic current for PPy redox (measured on the fifth cycle) during the copolymerization of CuT4APc and pyrrole. ○ [CuT4APc] = 1 mM. ● [CuT4APc] = 2 mM. (Reproduced by permission from ref. 41.)

Evidence for this mechanism is the appearance of a broad IR absorption between 1600 and 1700 cm^{-1} which is characteristic of oligoisoindolenes formed from 1,2-dicyanobenzene [60]. The presence of triazine linkages, as well as unreacted terminal nitriles, was also noted [59].

Poly(OCPc) films prepared in this fashion exhibit a stable electrochemical response in aqueous electrolyte (a single couple is shown, with broad anodic and cathodic waves of approximately equivalent integrated intensity). However, the redox properties of the polymer are extremely sensitive to the nature of the electrolyte, with the largest peak separation (700 mV) observed for HCl and the smallest (100 mV) seen for tetrabutylam-

monium tetrafluoroborate [61]. In addition, the electrochemical behavior of poly(OCPc) films was found to depend on the presence of a supporting electrolyte during deposition, with those formed in the absence of electrolyte yielding the more stable electrochemical response. Films obtained 'in the presence of LiCl when cycled in 0.1 M KCl' [59] are reported to show peak currents which are proportional to the square root of the scan rate, but no further information regarding the range of film thickness or sweep rate over which this relationship holds is provided.

Surprisingly, poly(OCPc) films were shown to be permeable to rather large charge-compensating ions [59] such as Fe(phen)$_3^{2+}$ (undefined in ref. 59, but presumably the tris(1,10-phenanthrolino)iron(*II*) complex) upon reduction. When the poly(OCPc)-modifying layer is electrochemically reduced in aqueous sodium nitrate containing 1 mM Fe(phen)$_3^{2+}$, the iron complex is incorporated for charge balance, as evidenced by the appearance of the Fe(phen)$_3^{3+/2+}$ couple in cyclic voltammograms recorded following transferral to fresh electrolyte. No explanation for the apparently overwhelming preference for Fe(phen)$_3^{2+}$ over the more readily available Na$^+$ ions is provided. Interestingly, the peak current I_p for the Fe(phen)$_3^{3+/2+}$ couple within the polymer layer varies linearly with scan rate, suggesting thin film behavior, yet a diffusion-like difference in peak potentials $\Delta E_p = 60$ mV is reported. Ejection of the Fe(phen)$_3^{n+}$ ions occurs only upon film oxidation; no leaching of the ions was observed when the applied potential was held at 0 V vs. SCE [59].

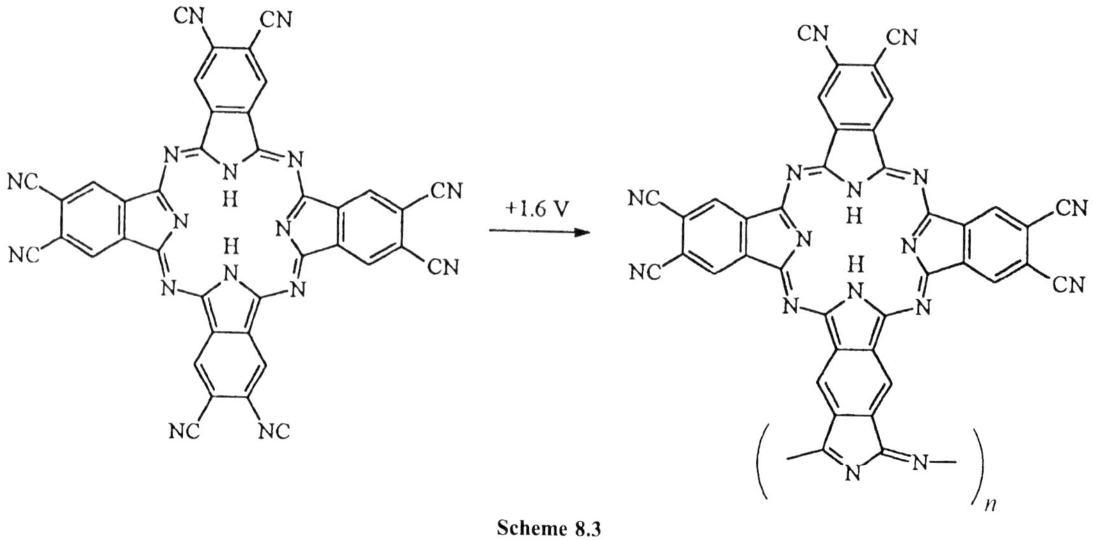

Scheme 8.3

3 APPLICATIONS

3.1 Photovoltaic devices

Although the use of various metallo-Pcs in photovoltaic applications is well established [62–64], comparable studies using electropolymerized Pc films remain virtually nonexistent. In the sole literature report to date, significantly greater photovoltages and photocurrents were recorded at poly(CuT4APc)-modified indium/tin oxide (ITO) surfaces as compared to ITO electrodes coated with monomeric CuT4APc [42].

3.2 Electrochromism

Although the electrochromic properties of rare earth diphthalocyanines have been recognized for more than 20 years [7,65], the simpler and more readily accessible Pc complexes of the first-row transition metals (e.g. CoPc, CuPc, FePc) have received comparatively little attention in this regard. One reason for the dearth of activity in this area is that the much lower bulk conductivities of such materials [typically 10^6 times smaller than that of $Lu(Pc)_2$] do not allow for the rapid electrochemical 'switching' between redox states required for the fabrication of a practical electrochromic device [7]. Green and Faulkner have demonstrated the reversible oxidation and re-reduction of thin films of single-ring metallo-Pcs, but reported a rapid decrease in the amount of charge passed on successive anodic sweeps [66]. Such behavior was attributed to the formation of electrically isolated domains within the films.

The spectroscopic properties of thin films (M = Fe, Co) were also examined by LeMoigne and Even [67], who reported significant changes in the optical spectra upon chemical reduction. These changes were completely reversed by the introduction of atmospheric oxygen into the cell. Interestingly, these authors also noted increases in conductivity of up to ten orders of magnitude following reduction, suggesting the development of n-doped semiconducting properties in these materials.

The reversible electrochromism of poly(MT4APc) films (M = Co, Ni, Pd) was demonstrated by Li and Guarr [68]. As expected, the nickel and palladium derivatives behave similarly, changing from green to blue over a rather broad potential range centered near −1.1 V vs. SCE. Excursion to more negative applied potentials causes a second transition to purple. By comparison with published MPc^{n-} spectra, the blue

and purple states were attributed to local transitions of the one-electron and two-electron MPc reduction products, respectively. Both of the electrochemical processes are thought to be localized on the Pc ring.

The poly(CoT4APc) films displayed equally interesting electrochromic properties [68]. Initial reduction (at c. −0.5 V vs. SCE) occurs at the central metal, leading to the development of a strong metal-to-ligand charge transfer at 440 nm and a corresponding change in hue from blue-green to yellow-brown (see Figure 8.12). A second, ring-centered reduction at very negative potentials (at c. −1.65 V vs. SCE) causes a further change to deep pink.

Electrochemical response times were also measured for the poly(CoT4APc) and poly(NiT4APc) films [68]. Surprisingly, the higher intrinsic conductivity of the nickel derivative did not manifest itself in these measurements; rather, the poly(CoT4APc) films exhibited the shortest switching times, as illustrated by the excellent correlation between the applied potential waveform and the optical absorbance at 470 nm (Figure 8.13). Both the magnitude and the speed of the absorbance change were found to be critically dependent on the nature of the electrolyte cation, suggesting that the switching times are controlled, at least in part, by the kinetics of counterion incorporation (Figure 8.14). Tetramethylammonium and tetraethylammonium perchlorate yield the largest and most rapid response, while the slowest coloration rates were found for tetrabutylammonium perchlorate. Response times were also surprisingly slow for $LiClO_4$ and $NaClO_4$ electrolytes, a fact which was attributed to the strong solvation of alkali metal cations (especially lithium) in DMSO [68].

The electrochromic properties of poly(CuT4APc) thin films have also been examined recently [69]. As with the analogous cobalt, nickel and palladium derivatives, the color changes which occur upon reduction can be correlated with spectroelectrochemical data on the corresponding monomers. Likewise, switching times show a comparable dependence on the size of the electrolyte cation. However, the copper system is unique in that it exhibits an unusual near-colorless (very light gray) state upon addition of 0.5 e$^-$ per monomer site. One possible explanation for such behavior is the formation of a low-bandgap semiconductor, but this hypothesis has not yet been fully tested.

The electrochromism of poly[Lu(T4APc)_2] is complicated by adsorption phenomena which seem to inhibit the polymerization and/or deposition processes [52]. In fact, $Lu(T4APc)_2$ appears completely electroinactive upon initial voltammetric scans at ITO, gold

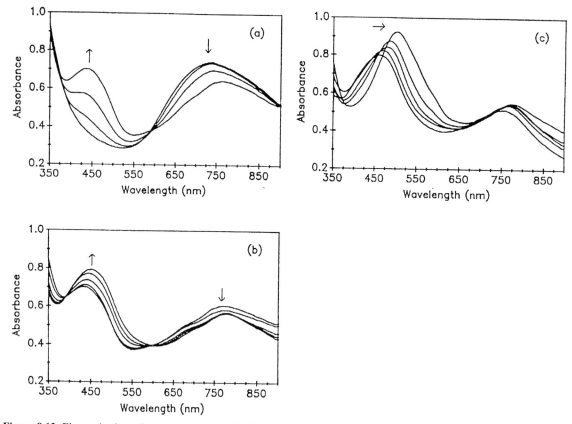

Figure 8.12. Electronic absorption spectra obtained during the reduction of a poly(CoT4APc) film coated on an ITO substrate. Arrows indicate the direction of change upon application of increasingly negative potentials as indicated: (a) -200, -400, -500 and $-600\,\text{mV}$ vs.SCE; (b) -700, -800, -1000, -1100 and $-1200\,\text{mV}$ vs.SCE; (c) -1300, -1500, -1600, -1700 and $-1800\,\text{mV}$ vs.SCE. (Reproduced by permission of Elsevier Science SA from H. Li and T.F. Guarr, *J. Electroanal. Chem.* **297**, 1991, 169).

and platinum electrodes. Nonetheless, through careful control of scan range, thin films of poly[Lu(T4APc)$_2$] can be electrodeposited on ITO after several hours of continuous cycling. The polymer-modified electrodes produced in such a manner show two broad reduction waves (both ascribed to ring-centered redox processes) superimposed on a large capacitive current which extends from $c.$ $-0.7\,\text{V}$ vs. SCE (Figure 8.15). As expected, color changes are observed upon scanning to negative potentials, with rapid transitions from green to gray to blue associated with the first and second reductions, respectively (Figure 8.16). Switching time plots for these films reflect multiple kinetic components, possibly because the broadness of both the optical absorption bands and voltammetric waves causes overlap in terms of both spectral properties and redox states [52]. As with the nickel and cobalt

systems described above, the use of smaller electrolyte cations favors rapid response times. Unfortunately, although aqueous electrolytes were utilized in most previous studies of lanthanide Pc electrochromism, poly[Lu(T4APc)$_2$] films are stable to repeated cycling within only a narrow potential range in aqueous solvents.

3.3 Metal deposition

The electrodeposition of metals onto the surface of various conductive polymers can prove advantageous in certain applications, including electrocatalytic processes [70,71] (i.e. methanol oxidation) and the development of practical electrochromic devices. The use of 'sandwich' electrodes (polymer-modified elec-

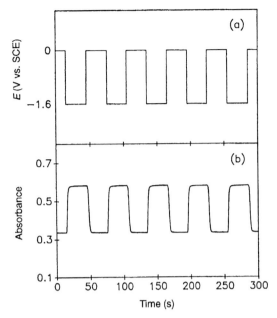

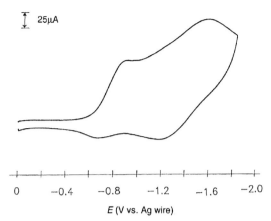

Figure 8.15. Cyclic voltammogram obtained upon transfer of poly[Lu(T4APc)$_2$]-modified ITO electrode to fresh electrolyte (DMSO/0.1 M tetraethylammonium perchlorate). The reference electrode is a silver wire; scan rate = 50 mV s^{-1}. (Reproduced by permission of Elsevier Science SA from D.J. Moore and T.F. Guarr, *J. Electroanal. Chem.* **314**, 1991, 313.)

Figure 8.13. (a) Applied potential waveform for $0 \leftrightarrow -1.6$ V vs. SCE switching time experiments. (b) Absorbance (at 470 nm) of a poly(CoT4APc)-modified ITO electrode in contact with DMSO/0.1 M tetraethylammonium perchlorate electrolyte; applied potential waveform as in (a). (Reproduced by permission of Elsevier Science SA from H. Li and T.F. Guarr, *J. Electroanal. Chem.* **297**, 1991, 169.)

trodes coated with a metal overlayer) in electronic devices has also been proposed [72]. Accordingly, the deposition of lead onto poly(NiT4APc)-modified electrodes has been examined in detail, along with somewhat less extensive investigations of the deposition processes of several other metals [40].

The electrodeposition of lead yields distinctly different results at the bare and polymer-modified

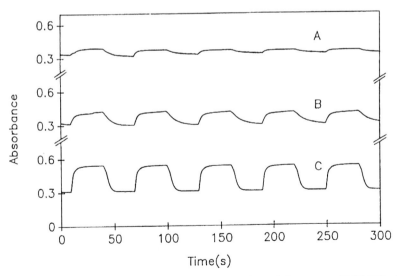

Figure 8.14. Absorbance (at 470 nm) of a poly(CoT4APc)-modified ITO electrode in contact with DMSO solution containing (A) 0.1 M Pr$_4$NClO$_4$, (B) 0.1 M NaClO$_4$, (C) 0.1 M Et$_4$NClO$_4$ (Pr = propyl, Et = ethyl). The applied potential is a $0 \leftrightarrow -1.6$ V vs. SCE square wave as shown in Figure 8.13. (Reproduced by permission of Elsevier Science SA from H. Li and T.F. Guarr, *J. Electroanal. Chem.* **297**, 1991, 169.)

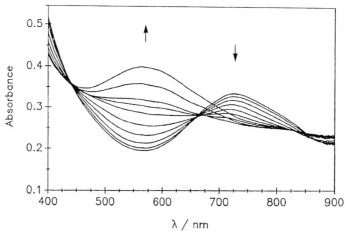

Figure 8.16. Electronic absorption spectra obtained during electrochemical reduction of a poly[Lu(T4APc)$_2$] film (coated on an ITO electrode). Arrows indicate the direction of change upon application of increasingly negative potentials (from -0.5 to -1.4 V vs. Ag in 100 mV steps). Spectra are recorded after a 30 s delay at each potential. The electrolyte is DMSO/0.1 M tetraethylammonium perchlorate. (Reproduced by permission of Elsevier Science SA from D.J. Moore and T.F. Guarr, *J. Electroanal. Chem.* **314**, 1991, 313.)

electrodes. The presence of the poly(NiT4APc) film results in a shift of the deposition peak to more negative potentials; this increase in the overpotential for deposition was ascribed to decreases in both the active site density and the average nucleation rate [40]. Also evident is a change in the shape of the voltammogram which is consistent with a two-stage deposition process. The two-stage nature of deposition on the polymer-modified electrode is clearly illustrated by chrono-amperometry data [40].

Two stripping peaks are also observed. The more negative of these decreases in size with increasing film thickness and is thus ascribed to the stripping of lead from the underlying solid electrode surface. The accumulation of lead deposits at the electrode/polymer interface occurs at the first cathodic wave and was attributed to a combination of film porosity and Pb^{2+} diffusion [40].

The second stripping peak (at *c.* -0.15 V vs. SCE) is significantly broader than the first and is not observed at bare carbon or platinum electrodes. Its magnitude increases with increasing film thickness at very low coverages, but reaches a limiting value for films $\geq c.$ 200 Å. This peak is associated with the second deposition stage and is insensitive to the nature of the underlying surface (platinum vs. glassy carbon). Such behavior suggests that this peak involved lead stripping within the film or at the polymer/solution interface. Scanning the electron microscopy (SEM) and energy-dispersive X-ray (EDX) analysis confirm

this hypothesis, and can be used to follow the deposition/stripping process. Interestingly, the SEM images show that the lead which deposits at the polymer/solution interface is in the shape of hexagonal flakes rather than in the globular form as found for other conductive polymers [70,73]. Further, the cyclic voltammetry, SEM and EDX data are all consistent with the intriguing notion that, through appropriate potential control, lead can be deposited at the electrode/polymer interface with minimal effect on the outer metal layer (at the polymer/solution interface) [40].

Other metals, including silver, copper, platinum, nickel and cobalt have also been deposited on poly(NiT4APc)-modified electrodes [40]. In all cases, the deposition potential increases with increasing film thickness, but the details of the deposition process are sensitive to film morphology and preparation conditions. Metal deposition was found to increase adhesion of the polymer to the underlying electrode surface, suggesting that it may prove useful for the mechanical protection of poly(MT4APc)-modified electrodes. The stripping behavior associated with other metals is similar to that seen with lead [40].

3.4 Redox Mediation

Since poly(MT4APc) films display electronic conductivity over such a broad potential range, their utility as a new type of solid electrode has been investigated. In some cases, surface effects seem to dominate the observed results, and the main *electrochemical* function of the polymer layer is to serve as an immobilized redox shuttle. The term 'redox mediation' is used here to describe those systems in which rate enhancements (and other phenomena) are linked to physical rather than chemical effects and is meant to distinguish them from electrocatalytic processes, where the chemical composition of the polymer layer plays an integral part in the observed electrochemistry.

Li and Guarr showed that the peak currents associated with a solution-phase redox couple which normally exhibits rapid heterogeneous electron transfer kinetics at a bare electrode (i.e. ferrocene/ferrocenium, methyl viologen) is virtually unaffected by the presence of the poly(MT4APc) coating [28,50]. A compelling example of such behavior is illustrated in Figure 8.17, where the voltammogram obtained at a poly(NiT4APc)-modified electrode immersed in a 'redox soup' (a mixture of compound yielding six well-separated reversible couples in CH_3CN solution) is nearly superimposable on the voltammogram obtained at a bare electrode under the same conditions [35]. In this case, the only discernible influence of the electrode coating is an increase in background current attributable to the underlying capacitive component. Poly-(NiT4APc) films have been reported to improve the reversibility of numerous redox processes, including the reduction of $Mn_2(CO)_{10}$ in CH_3CN as well as the oxidations of hydroquinone, catechol and ascorbate in aqueous electrolytes [50].

An extensive study of heterogeneous electron transfer at poly(NiT4APc)-modified electrodes has been reported by Mu and Schultz [48]. These authors found that the presence of a poly(NiT4APc) film caused a small decrease in the observed heterogeneous electron transfer rate (k_{het}). The attenuation of k_{het} was found to be linearly related to film thickness and was attributed to the internal resistance of the polymer (measured to be 4–16 MΩ cm under these conditions). Mu and Schultz interpreted this behavior in terms of a tightly packed porous conductor, where heterogeneous electron transfer is relatively rapid and solution phase reactants do not penetrate beyond the polymer/solution interface to any significant degree. This explanation is also consistent with the large measured capacitance of poly(NiT4APc). Slight decreases in film resistivity with increasing film thickness were explained in terms of increasing film porosity, as supported by SEM data [48].

Mu and Schultz demonstrated enhanced reversibility for the oxidation of $[W_2(\mu\text{-SPh})_2(CO)_8]^{2-}$ at poly-

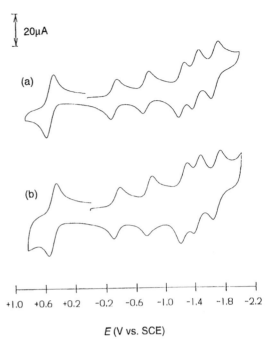

Figure 8.17. Cyclic voltammograms obtained in $CH_3CN/$ 0.1 M tetraethylammonium perchlorate electrolyte containing 1 mM ferrocene, 1 mM methyl viologen and 1 mM Ru(2,2'-bipyridine)$_3{}^{2+}$ at (a) bare glassy carbon and (b) a poly-(NiT4APc)-modified glassy carbon electrode. Scan rate = 100 mV s^{-1} for both. (Reproduced by permission from ref. 35.)

(NiT4APc)-modified electrodes, and suggested that the coating functions in such a manner by 'covering inactive regions on the surface, blocking adsorption of impurities or preventing adsorption of the reactant in an orientation unfavorable for electron transfer' [48]. They concluded that surface effects of this type were responsible for the enhanced heterogeneous electron transfer rates found at poly(MT4APc)-modified electrodes for numerous redox processes, including the oxidations of hydroquinone, catechol and ascorbate [48].

The oxidation of ascorbate at both poly(CoT4APc)- and poly(NiT4APc)-modified electrodes was investigated in terms of Savéant's model for catalytic reactions at modified electrodes [35,74–76]. The data from rotating disk electrochemistry (RDE) experiments with various substrate concentrations and poly(CoT4APc) film thicknesses indicate that this process falls in the $(S + E)$ kinetic regime, meaning that the overall rate is controlled by substrate permeation and the rate of internal charge transport within the polymer. By

contrast, the same reaction at a poly(NiT4APc)-modified electrode belongs to the (SR) class, suggesting that, for this polymer, the reaction rate is controlled by a combination of substrate permeation and the heterogeneous electron transfer rate. No kinetic effects of internal charge transport were found, providing further evidence of faster charge transport in the nickel-containing film. It should also be noted that Koutecky–Levich plots of the RDE data show conclusively that the degree of substrate permeation is very low for both cobalt- and nickel-containing films, with the reaction occurring nearly exclusively at the polymer/solution interface.

The poly(OCPc)-modified electrodes of Vijayanathan et al. [59] were reported to exhibit electrocatalytic activity for the oxidation of $Fe(CN)_6^{4-}$ and a rate retardation effect for the reduction of $IrCl_6^{2-}$, but the data shown do not appear to support either of these conclusions.

3.5 Electrocatalysis

3.5.1 Oxidations

For some electrochemical processes, poly(MT4APc) CMEs seem to play a more active role than simple prevention of undesirable adsorption. Those cases in which the poly(MPc) layer appears to exhibit a *specific catalytic* function are collected under this heading.

3.5.1.1 Oxalic acid

While the peak potential for the oxidation of oxalic acid is *c.* 300 mV less positive at poly(NiT4APc)- and poly(ZnT4APc)-modified glassy carbon electrodes as compared to the bare surface, an extraordinary shift of 780 mV is seen at poly(CoT4APc)-modified electrodes [77] (Figure 8.18). Thus, although the nickel- and zinc-containing polymers furnish some improvement in electrode kinetics (possibly through the surface effects noted above), the cobalt system provides an additional, catalytic pathway involving the central metal. A detailed kinetic analysis in terms of the Savéant model for heterogeneous electron transfer at modified electrodes [74–76] reveals that this process falls in the (SR) kinetic regime (at least for films of micrometer thickness) [77]. That is, the overall rate is controlled by a combination of substrate permeation and the rate of the catalytic step. As with the ascorbate oxidation study, the rate of substrate permeation was extremely slow, and only small deviations from linearity were

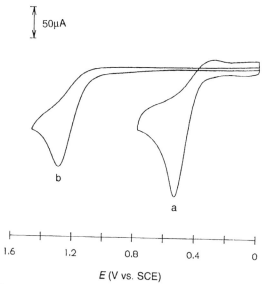

Figure 8.18. Cyclic voltammograms obtained in pH 2.13 phosphate buffer solution containing 5 mM oxalic acid at (a) a poly(CoT4APc)-modified glassy carbon electrode and (b) bare glassy carbon electrode. Scan rate = 100 mV s^{-1} for both. (Reproduced by permission of Elsevier Science SA from H. Li and T.F. Guarr, *J. Electroanal. Chem.* **317**, 1991, 189.)

observed in Levich plots of the RDE data, even at high rotation rates.

3.5.1.2 Hydrazine

The overpotential associated with the oxidation of hydrazine in aqueous solution depends on the nature of the electrode as well as on the electrode pretreatment procedure employed. The electrocatalytic oxidation of hydrazine by various metallo-Pcs has been studied by several groups, with cobalt and iron Pcs generally yielding the best results [78,79]. Likewise, poly-(CoT4APc) films display strong activity for N_2H_4 oxidation, yielding a well-defined chemically irreversible anodic peak at −0.30 V vs. SCE in pH 13 aqueous electrolyte [80]. The position of this wave, along with the inactivity of analogous poly(MT4APc) films, suggests involvement of the Co(II)/Co(I) couple, in accord with Zagal's previously proposed mechanism describing hydrazine oxidation at pyrolytic graphite electrodes modified with cobalt tetrasulfonatophthalocyanine (CoTsPc) [78]. The peak potential varies linearly with pH in very basic solution, and the observed slope of *c.* 250 mV per pH unit is consistent with the earlier Zagal mechanism [78] (overall 4 e$^-$

oxidation). Peak currents vary linearly with $v^{1/2}$ and with N_2H_4 concentration. Analysis of RDE data within the context of the Savéant model [74–76] demonstrates that the overall kinetics are controlled by the catalytic rate, with contribution from slow internal charge transport only for films of thickness > 2000 Å [80].

3.5.1.3   Hydroxylamines

Poly(CoT4APc)-modified electrodes have been utilized as effective catalysts for the electrochemical oxidation of numerous hydroxylamines [81]. Although such compounds are normally oxidized only at high positive potentials (> +1.0 V vs. Ag/AgCl) using bare glassy carbon electrodes, NH_2OH as well as various N-mono-, N,N-di-, and O-substituted derivatives were readily oxidized using modest potentials (+0.25 to +0.55 V) at poly(CoT4APc)-modified surfaces. Product analysis (by ^1H NMR) following bulk electrolysis suggests multiple reaction pathways leading to the formation of oximes, azoxy compounds, and dimeric species. Likewise, coulometric data are also suggestive of a mechanistically complex process, with passage of 1.2–1.6 electrons per molecule, depending on the conditions. The poly(CoT4APc)-modified electrodes (on reticulated vitreous carbon) were remarkably stable when employed for bulk electrolysis studies, and these authors estimated a turnover number in excess of 10^4 [81].

Poly(CoT4APc)-modified electrodes were also used as amperometric detectors for liquid chromatography, exhibiting excellent stability over several weeks after an initial decline in performance during the first few hours. Detection limits ranged from 0.4 pmol for hydroxylamine to 40 pmol for N,O-dimethylhydroxylamine. Interestingly, by operating the detector at +0.20 V vs. Ag/AgCl in this mode, the observed response was selective for NH_2OH and N-mono-substituted hydroxylamines [81].

3.5.1.4   Thiols

Poly(CoT4APc) also shows a strong electrocatalytic effect for the oxidation of various sulfur-bearing compounds, including L-cysteine, N-acetylcysteine, glutathione, and 6-mercaptopurine, among others [82]. This process represents the basis for the electrochemical detection of numerous thiols following liquid chromatography, with picomole detection limits achieved at modest potentials. Unlike previously reported carbon paste/CoPc systems, the poly-

(CoT4APc)-modified electrodes are compatible with both aqueous and nonaqueous eluents. In addition, the electrode response is extremely stable, with retention of greater than 80% of the initial activity after 40 hours of continuous use. [82].

3.5.1.5   Sulfide

In 1994, Lever *et al.* utilized the electrodeposition of N,N',N'',N'''-tetramethyltetra-3,4-pyridinoporphyrazinocobalt (II) (CoTmtppa) to develop a new type of ion-selective electrode (ISE) for the determination of S^{2-} [83]. However, one drawback of the CoTmtppa-based sensors is their tendency to undergo decomposition in basic solution (pH > 8). In an effort to avoid such difficulties, these same workers recently reported the fabrication of a similar ISE using a poly(CoT4APc)-modified electrode [84]. The poly(CoT4APc)-based devices proved useful for the quantitative determination of both sulfide and 2-mercaptoethanol, and displayed good stability up to at least pH 13. The effect of polymeric cation- and anion-exchange overlayers was also investigated.

Although no sulfide oxidation is observed at a bare glassy carbon electrode under the test conditions (pH 9 aqueous phosphate buffer), a well-defined anodic wave near −0.1 V vs. SCE is apparent at a poly(CoT4APc)-modified glassy carbon electrode. Upon reversing the direction of the potential sweep, a new cathodic wave near −0.6 V vs. SCE is seen. On the basis of their earlier work, Tse *et al.* [84] assign the anodic wave to the two-electron oxidation of S^{2-} to elemental sulfur; the cathodic wave is attributed to reduction of polysulfide ions formed by the reaction of the electrochemically produced sulfur with free sulfide ions in solution. The peak current for the oxidation is proportional to the square root of the scan rate, indicating that the anodic process is diffusion controlled.

The peak potential for sulfide oxidation varies linearly with sulfide concentration, although the slopes of the Nernst plots tend to increase with increasing film thickness. In addition, the slopes of the Nernst plots are sensitive to solution pH, varying from about −70 mV per decade at pH 7 to about −40 mV per decade at pH 13. Response times at very low Na_2S concentrations (i.e. < 8.5 μM) increased slightly when the poly(CoT4APc) was coated with a protective ion-exchange overlayer; however, rapid response times (< 20 s) were maintained for Na_2S concentrations above 10 μM. Likewise, response times were found to increase with increasing pH, but only at very low Na_2S concentrations. Unlike most ISEs, the addition of potentially interfering anions such as Cl^-, Br^-, I^- and SO_3^{2-} does not induce curvature of the Nernst plots: rather, linearity is maintained with only a slight change in slope [84].

3.5.1.6   Peroxide

The electrochemical oxidation of H_2O_2 normally requires large overpotentials, occurring only at potentials in excess of $+0.6\,V$ and $+0.9\,V$ vs. SCE at platinum and glassy carbon electrodes, respectively. However, modification of a glassy carbon electrode with a thin (c. 350 Å) layer of poly(CoT4APc) decreased the overpotential associated with H_2O_2 oxidation by 700 mV (anodic peak potential $E_{p,a} = +0.15\,V$ vs. SCE at pH 7) [85]. It should be noted that this effect is substantially larger than that observed for several previously reported CoPc-based electrode modification schemes. Surprisingly, only poly(CoT4APc) films showed such activity; polymers prepared from NiT4APc and CuT4APc were completely ineffective [86].

Peak currents for H_2O_2 at poly(CoT4APc)-modified electrodes vary linearly with the square root of the scan rate in the range of $v = 10$–$150\,mV\,s^{-1}$, at least for relatively low film coverages ($\Gamma \geqslant 2.1 \times 10^{-8}$ $mol\,cm^{-2}$). Peak potentials vary linearly with pH over the range 2–9, with near-Nernstian slopes of 54–56 mV per pH unit. Peak currents also vary linearly with H_2O_2 concentration over a wide range, with a voltammetric detection limit of $17\,\mu M$. Poly(CoT4APc)-coated electrodes maintain 74% of their activity for H_2O_2 oxidation after two weeks [85].

3.5.2   Reductions

3.5.2.1   Peroxide

The direct electrochemical reduction of H_2O_2, like its oxidation, is characterized by slow kinetics at conventional electrodes. Cross and Guarr found that both poly(NiT4APc) and poly(CuT4APc) modifying layers cause a 600 mV decrease in the overpotential for H_2O_2 reduction, while the cobalt-containing polymer showed almost no activity [85]. However, the RDE data of Tse [34] conflict with these results, showing a well-defined two-electron reduction of H_2O_2 at a poly(CoT4APc)-modified electrode near $-0.4\,V$ vs. SCE in pH 1.65 aqueous buffer. Detailed reasons for this discrepancy are not clear, but the sensitivity of poly(CoT4APc)

properties to the preparation conditions and to sample aging effects has already been noted [40,86].

Like the poly(CoT4APc)-catalyzed peroxide oxidation, peak currents for H_2O_2 reduction at poly(NiT4APc)- and poly(CuT4APc)-modified electrodes vary linearly with the scan rate at low film coverages ($\Gamma \geqslant 2.1 \times 10^{-8}\,mol\,cm^{-2}$). Linear calibration curves are also observed, with a detection limit of $80\,\mu M$ obtained at a poly(NiT4APc)-modified electrode [85]. The electrochemical stability of these systems is also quite good, with a retention of 94% of the initial peak current for H_2O_2 reduction at a poly(CuT4APc)-modified electrode after 100 cycles.

3.5.2.2   Oxygen

Poly(CoT4APc) has been utilized to catalyze the electrochemical reduction of O_2 by both a two-electron and a four-electron pathway [34]. In an oxygen-saturated buffer electrolyte, two irreversible cathodic waves are observed at poly(CoT4APc)-modified electrodes. The first of these shows a weak dependence on pH, shifting from $+0.10\,V$ vs. SCE at pH 1.65 to $-0.25\,V$ vs. SCE at pH 13, and is assigned to the two-electron conversion of O_2 to H_2O_2. The second wave is much broader and exhibits a somewhat stronger pH dependence, with peak potentials of $-0.36\,V$ and $-0.83\,V$ vs. SCE at pH 1.65 and pH 13, respectively. The second wave has been assigned to the net four-electron reduction of O_2 to H_2O_2 on the basis of rotating ring-disk electrochemical studies [34]. Peak currents for both waves vary linearly with the square root of the scan rate, as expected for diffusional processes.

Tse has shown that the two-electron reduction of O_2 at poly(CoT4APc)-modified electrodes proceeds through a one-electron rate-determining step in acidic media and implicates an intermediate oxygen adduct [34].

$$Co(II)T4APc + O_2 \Leftrightarrow (O_2^-)Co(III)T4APc$$

The n value (where n represents the number of electrons transferred per molecule of substrate) calculated from RDE data for the second reduction process ($n = 3.5$) is slightly less than the theoretical value of 4, suggesting that a slow chemical process may control the overall reaction kinetics. Tse suggests that the four-electron process proceeds through formation of H_2O_2 and that the reduction of this intermediate is catalyzed by the Co(II)/Co(I) couple [34].

3.6 Biological applications

3.6.1 NADH oxidation

While hundreds of enzymatic processes rely on the solution phase redox chemistry of NADH (reduced nicotinamide-adenine glycohydrolase), the electrochemical oxidation of NADH at unmodified solid electrodes is characterized by extremely slow kinetics and correspondingly large overpotentials. Thus, the electrocatalytic oxidation of NADH at poly(MT4APc)-modified electrodes has been investigated [51].

Three different metallo-Pcs (M = Co, Ni and Zn) were examined in this study, and all were effective electrocatalysts [51]. Peak potentials for NADH oxidation are several hundred millivolts less positive than those obtained at bare electrodes, and show only slight variation among the three different materials. Such behavior is consistent with Mu and Schultz's proposed model for improving heterogeneous electron transfer kinetics at polymer-modified electrodes through prevention of undesirable adsorption processes [48]. This scenario seems likely, since Blankespoor and Miller demonstrated that the adsorption of NAD^+ leads to increased overpotentials and broad voltammetric waves for the NADH oxidation process [87]. Direct tests of NAD^+ adsorption suggest that its effect on the electrochemical properties of polymer-modified electrodes is much less severe than on bare electrode surfaces [51]. In this case, 'ideal' film thicknesses are on the order of a few hundred ångströms and probably represent a compromise between the need for complete surface coverage and the fact that increasing ohmic resistance causes a gradual shift of the anodic wave to more positive potentials for thicker films. Poly-(MT4APc)-modified electrodes display a durable and reproducible electrochemical response for NADH oxidation, with retention of 70% of the initially obtained peak current after 200 cycles [51,86].

3.6.2 Direct electrochemistry of redox proteins

A unique demonstration of the surface effects on the heterogeneous electron transfer rates of biological molecules is provided by a comparison of the electrochemistry of cytochrome c at bare and poly-(CoT4APc)-modified glassy carbon electrodes [88] (Figure 8.19). This protein yields no detectable voltammetric response at an unmodified metal electrode (without pretreatment), yet displays a chemically and electrochemically reversible couple at the polymer-modified surface (ratio of anodic to cathodic peak current $i_{pa}/i_{pc} = 1$; $E_{pa} - E_{pc} = 71$ mV at a scan rate of 25 mV s^{-1}). Plots of the peak current vs. $\nu^{1/2}$ are linear up to a scan rate of 150 mV s^{-1}, indicating that the electron transfer process is controlled by mass transport under these conditions. The diffusion coefficient determined from these data ($D_0 = 1.3 \times 10^{-6}$ cm^2 s^{-1}) is in agreement with previously reported values [89,90]. The heterogeneous electron transfer rate constant for the reaction at the polymer-modified surface, $k_s = 4.5 \times 10^{-3}$ cm s^{-1}, is similar to values obtained for the reaction of cytochrome c at promoter-modified gold electrodes [91]. The poly(CoT4APc)-modified electrode showed only 10% loss of anodic peak current even after 350 cycles in a 25 μM cytochrome c solution. Covalent attachment of cytochrome c to the poly(CoT4APc)-modified surface resulted in an unfavorable orientation for electron transfer and consequently the complete loss of cytochrome c electroactivity [86].

Attempts to directly observe the redox activity of myoglobin at poly(CoT4APc)-modified electrodes were less successful, although an irreversible cathodic wave at $c.\ -0.4$ V vs. SCE was attributed to the reduction of the prosthetic heme [86] (Figure 8.20). The return wave displays the symmetric shape characteristic of a

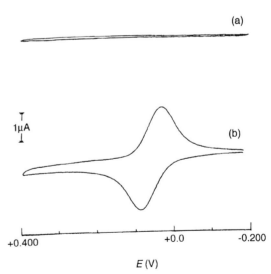

Figure 8.19. Cyclic voltammograms obtained in a $50\ \mu$M solution of cytochrome c (in pH 7 aqueous buffer) at (a) bare and (b) poly(CoT4APc)-modified glassy carbon electrodes. Scan rate = 50 mV s^{-1} for both; the background current obtained in the absence of cytochrome c has been subtracted in each case. (Reproduced by permission from ref. 86.)

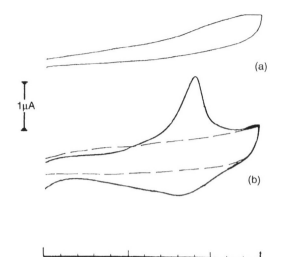

Figure 8.20. Cyclic voltammograms obtained in a 50 μM solution of myoglobin (in pH 7 aqueous buffer) at (a) bare and (b) poly(CoT4APc)-modified glassy carbon electrodes. Scan rate = 50 mV s^{-1} for both. In (b), both the voltammo-gram recorded in the presence (solid line) and absence (dashed line) of myoglobin are shown. (Reproduced by permission from ref. 86.)

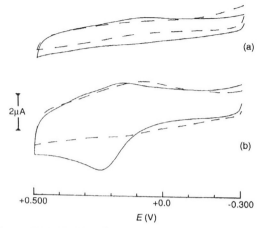

Figure 8.21. Cyclic voltammograms obtained in a 50 μM solution of Cu,ZnSOD (in pH 5.5 aqueous buffer) at (a) bare and (b) poly(CoT4APc)-modified gold disk electrodes. Voltammograms recorded in the presence (solid line) and absence (dashed line) of Cu,ZnSOD are shown. Scan rate = 50 mV s^{-1} in all cases. (Reproduced by permission from ref. 86.)

surface-confined species, perhaps suggestive of strong adsorption.

Cross also examined the electrochemistry of the important oxidoreductase copper, zinc superoxide dismustase at poly(CoT4APc)-modified electrodes [86]. The voltammetry is characterized by a largely irreversible anodic wave at +0.22 V vs. SCE (Figure 8.21); neither the apoenzyme nor the zinc-only reconstituted material displayed any detectable electro-activity under these conditions. Peak current varied linearly with the square root of the scan rate, indicating a diffusion-controlled process. Both peak current and peak potential are sensitive to solution pH, with an optimal response observed at pH 4–5 [86]. The effects of replacing zinc with magnesium and of arginine-141 modification were also investigated.

3.6.3 Amperometric biosensors

3.6.3.1 Glucose sensor

Sun and Tachikawa have reported the use of poly-(CuT4APc) and poly(CoT4APc) films in the fabrication of bilayer conducting polymer electrodes for the amperometric detection of glucose [92]. In their design,

a glassy carbon electrode was first derivatized with a thin layer of poly(MT4APc). An outer layer consisting of glucose oxidase (GOD) entrapped in PPy was then deposited via electrochemical polymerization of pyrrole from an aqueous solution of GOD containing no other added electrolyte. Electrodes prepared in this fashion exhibited a strong catalytic response when immersed in an aqueous solution containing glucose. Moreover, the bilayer sensors could be operated at much lower applied potentials than those required to elicit a comparable response at PPy-GOD single-layer electrodes. Max-imum catalytic currents were obtained at +0.4–0.5 V vs. SCE and +0.6 V vs. SCE for poly(CoT4APc)- and poly(CuT4APc)-containing systems, respectively. No response was observed in deoxygenated solution, suggesting that these electrodes function by catalytic oxidation of H$_2$O$_2$ rather than by direct electrochemical communication between the polymeric Pc layer and the encapsulated GOD enzyme [92].

Response currents varied linearly with glucose concentration at low concentrations (<5 mM), but deviated slightly from linearity at higher levels. Detection limits of 0.25 and 0.5 mM were obtained for the cobalt- and copper-containing electrodes, respec-tively, and both types of sensor showed excellent response times of only a few seconds. Despite the absence of a protective membrane overlayer, the electrodes exhibited a relatively stable current response, with retention of 80% of the initial activity after one week [92].

The low operating voltage of these sensors serves to suppress the effects of many common interferants (i.e. chlorpromazine). However, the tendency of poly-(MT4APc) layers to facilitate the heterogeneous electron transfer reactions of numerous organic substrates can complicate matters, as illustrated by the enhancement of uric acid interference at poly-(CuT4APc)/PPy-GOD electrodes.

3.6.3.2   Ethanol sensor

Cross and Guarr have exploited the ability of poly-(MT4APc)-modified electrodes to facilitate the electro-chemical oxidation of NADH in the development of amperometric biosensors for ethanol based on alcohol dehydrogenase (ADH) [93]. Various enzyme immobilization schemes were investigated with respect to their stability and the elimination of common interferants.

Simple adsorption of ADH onto the electrode surface produced a biosensor which was characterized by a rapid response time (<5 s) and rather low sensitivity. Further, the stability of the electrochemical response was poor, with a retention of less than 10% of the initally observed current density after standing in buffer for 6 h [93]. This loss of response is common for biosensors relying on adsorptive immobilization of the active component, and is undoubtedly caused by desorption of the enzyme from the electrode surface.

Glutaraldehyde cross-linking of ADH to albumin produced a sensor with a linear current response to ethanol over the concentration range from 0.35 to 4.0 mM. Sensitivity was also significantly improved, at the expense of a modest increase in response time (c. 8 s). Moreover, such sensors exhibit a stable and reproducible response to 1 mM ethanol over 36 hours [93].

Chemical immobilization of ADH was also achieved via carbodiimide coupling of glutamate and/or aspartate residues to the residual amine functionalities of the poly(MT4APc) surface. Unfortunately, the low sensitivity of electrodes fabricated in such a manner precludes their use in practical amperometric devices [86].

The physical entrapment of ADH in several thin polymeric overlayers was also investigated. Three different entrapment matrices were chosen for study: polypyrrole, Nafion™ and Kodak's AQ29s polymer. Plots of current density vs. ethanol concentration for all three systems are illustrated in Figure 8.22. Measured

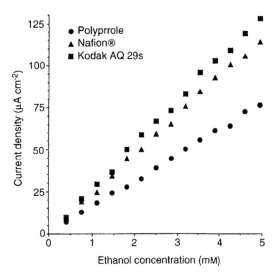

Figure 8.22. Correlation between current response and ethanol concentration for various methods of ADH immobilization at a poly(NiT4APc)-modified glassy carbon electrode (see text). The electrolyte is pH 7.4 phosphate buffer containing 100 mM NaCl as added supporting electrolyte. Applied potential = +0.050 V vs. SCE; electrode rotation rate = 500 rpm. (Reproduced by permission from ref. 86.)

response times were approximately 48, 15 and 12 s for electrodes incorporating PPy, Nafion™ and AQ29s overlayers, respectively [93].

Immobilization using either Nafion™ or AQ29s resulted in nearly identical sensitivity, and both of these materials exhibited similar rejection characteristics toward common anionic interferants such as acetaminophen, L-tyrosine, L-cysteine, uric acid, lactic acid and ascorbate. For example, the use of Nafion™ or AQ29s ionomer overlayers resulted in a nearly 50-fold decrease in the faradaic current associated with ascorbate oxidation at poly(NiT4APc)-modified electrodes [93]. Still, the total elimination of the amperometric signal from these interferants was not achieved.

The stability of poly(NiT4APc)/AQ29s sensors was also tested (see Figure 8.23). After an initial decrease in response during the first 24 h, only a small but steady decline in performance was noted over 14 days. At this point, measured current levels were about 75% of the initial value, and response times remained constant at about 12 s. Continued loss of current was accompanied by gradually increasing response times after 14 days. Poly(NiT4APc)/Nafion™ sensors exhibited similar behavior at a slightly accelerated rate (over a 7 day period) [93].

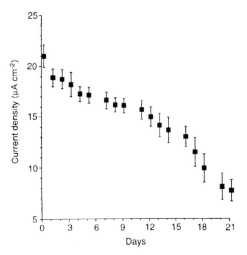

Figure 8.23. Stability and reproducibility ($n = 4$) of the current response from a poly(NiT4APc)/AQ29s/ADH-modified glassy carbon electrode in a 50 mM pH 7.4 phosphate buffer containing 100 mM NaCl and 0.7 mM ethanol. The electrode was stored in phosphate buffer at 4°C between analyses. (Reproduced by permission from ref. 86.)

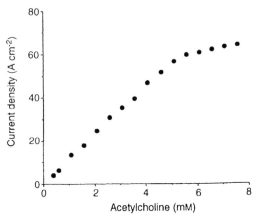

Figure 8.24. Observed change in the current response of poly(CoT4APc)-based acetylcholine biosensor (see text) upon successive additions of acetylcholine. Enzyme immobilization (30 units each of choline oxidase and acetyl cholinesterase) was achieved by physical entrapment of the enzymes in AQ29s. The electrolyte is pH 7 phosphate buffer containing 100 mM NaCl. Applied potential = $+0.200$ V vs. SCE; electrode rotation rate = 500 rpm. (Reproduced by permission from ref. 86.)

A poly(MT4APc)-based ethanol sensor has also been reported by Tachikawa *et al.* [43]. These authors immobilized the ADH onto poly(MT4APc)-modified electrodes (M = Co, Ni, Cu) by simply immersing the CME into a buffered (pH 7.4) solution of the enzyme at low temperature. Under these conditions, immobilization took place within a few seconds, as monitored by electrochemical quartz crystal microbalance measurements. The current response of such sensors was found to depend on MT4APc polymerization conditions, the ADH concentration employed during enzyme immobilization, pH, applied potential and the concentration of NAD^+ in the analyte solution. Nonlinear calibration curves were obtained, and a detection limit of 0.25 mM was quoted.

3.6.3.3 *Acetylcholine sensor*

The ability of poly(MT4APc) layers to facilitate the heterogeneous electron transfer reactions of H_2O_2 was utlized in the fabrication of an amperometric biosensor for the important neurotransmitter acetylcholine [94]. In this design, enzymatic hydrolysis of the substrate to choline and acetate occurred in a polymeric ionomer overlayer (Kodak AQ29s) containing immobilized acetylcholinesterase. Peroxide was produced by the action of choline oxidase (immobilized in the same polymeric overlayer) under aerobic conditions and the

resultant peroxide was detected via amperometry ($E_{applied} = +200$ mV vs. SCE) at the underlying poly(MT4APc)-modified electrode. This arrangement yields a rapid response and the measured current values vary linearly over a wide range of acetylcholine concentrations (*c.* 0.2 to 5 mM, see Figure 8.24) [94]. The permselectivity of the AQ29s layer affords some protection from the exposure of the electrochemically active surface to common anionic interferants, while the poly(MT4APc) coating serves to improve the kinetics for peroxide electrooxidation, thereby allowing operation at less positive potentials, where the contribution of the interfering species to the overall current is less pronounced. The poly(MT4APc) layer also retards electrode fouling, a primary failure mode in related systems.

4   FUTURE DIRECTIONS

It is clear that electropolymerized phthalocyanines hold tremendous promise in a wide variety of applications. Research efforts will continue into electroanalytic and electroanalytical uses, with particular emphasis on biologically relevant studies, including the development of practical biosensors. The electrodeposition of metals onto electrodes modified with electropolymerized Pcs poses some intriguing possibilities, but has not yet been fully explored. Likewise, the electrochromic properties of electropolymerized lanthanide Pcs are certainly worthy of further investigation. Fundamental studies of factors governing the electronic and ionic conductivity in these systems are also needed, as well as a detailed examination of their thermal, chemical and electrochemical stability.

However, this author believes that the work described above represents only the tip of the iceberg and that electropolymerized Pcs will find a growing number of potential uses. Among the areas which have received little attention to date are alternative electropolymerization chemistries, photovoltaic applications, waste treatment, gas sensing and electrosynthesis. Given the unique physical, electronic and chemical properties of such materials, a great deal of exciting chemistry undoubtedly lies just around the corner.

5 ACKNOWLEDGMENT

The author is deeply indebted to a series of outstanding graduate and undergraduate students who have toiled in his laboratory over the last few years. Financial support from the University of Kentucky, the National Science Foundation and the Department of Energy is also gratefully acknowledged. Finally, the author wishes to thank Professor A.B.P. Lever for supplying a preprint and for helpful discussions.

6 REFERENCES

1. L.J. Boucher, E.J. Dadey and D.R. Spears, *Proc. Int. Conf. Coal. Sci.* 763 (1983).
2. O. Hirabaru, T. Nakase, K. Hanabusa, H. Shirai, K. Takemoto and N. Hojo, *J. Chem. Soc. Dalton Trans.* 1485 (1984).
3. N. Brasseur, H. Ali, R. Langlois, J.R. Wagner, J. Rousseau and J.E. Van Lier, *Photochem. Photobiol.* **45**, 581 (1987).
4. W.R. Barger, H. Wohltjen, A.W. Snow, J. Lint and N.L. Jarvis, *ACS Symp. Ser.* No. 309, 1986, p. 155.
5. M.F. Dautartas, S.Y. Suh, S.R. Forrest, M.L. Kaplan, A.J. Lovinger and P.H. Schmidt, *Appl. Phys.* A**36**, 71 (1985).
6. M.S. Nieuwenhuizen, A.J. Nederlof and A.W. Barendz, *Anal. Chem.* **60**, 230 (1988).
7. M.M. Nicholson, in *Phthalocyanines: Properties and Applications*, Vol. 3 (Eds.) C.C. Leznoff and A.B.P. Lever, VCH Publishers, New York, 1993.
8. I. Ohizumi and H. Okayasu, Eur. Pat. Appl. EP 121,436 (1984).
9. F.H. Moser and A.L. Thomas, in *The Phthalocyanines*, CRC Press, Boca Raton 1983.
10. L.M. Santos and R.P. Baldwin, *Anal. Chem.* **59**, 1755 (1987).
11. D. Wöhrle, in *Phthalocyanines: Properties and Applications*, 3 (Eds.) C.C. Leznoff and A.B.P. Lever, VCH Publishers, New York, 1989.
12. D. Wöhrle, *Adv. Polym. Sci.* **50**, 46 (1983).
13. R.S. Nohr, P.M. Kuznesof, K.J. Wynne, M.E. Kenney and P.G. Siebenmann, *J. Am. Chem. Soc.* **103**, 4371 (1981).
14. O. Schneider and M. Hanack, *Angew. Chem. Int. Ed. Engl.* **19**, 392 (1980).
15. O. Schneider, J. Metz and M. Hanack, *Mol. Cryst. Liq. Cryst.* **81**, 273 (1982)
16. T.J. Marks, *Angew. Chem. Int. Ed. Engl.* **29**, 857 (1990).
17. B.N. Diel, T. Inabe, J.W. Lyding, K.F. Schoch, Jr, C.R. Kannewurf and T.J. Marks, *J. Am. Chem. Soc.* **105**, 1551 (1983).
18. T. Saji, in *Phthalocyanines: Properties and Applications*, Vol. 2 (Eds.) C.C. Leznoff and A.B.P. Lever, VCH Publishers, New York, 1993.
19. T. Saji, K. Hoshino, Y. Ishii and M. Goto, *J. Am. Chem. Soc.* **113**, 450 (1991).
20. A. Merz, *Topics Curr. Chem.* **152**, 49 (1990).
21. P. Denisevich, K.W. Willman and R.W. Murray, *J. Am. Chem. Soc.* **103**, 4727 (1981).
22. P.G. Pickup and R.W. Murray, *J. Electrochem. Soc.* **131**, 833 (1984).
23. R.W. Murray, *Ann. Rev. Mater. Sci.* **14**, 145 (1984).
24. K. Hanabusa and H. Shirai, in *Phthalocyanines: Properties and Applications*, Vol. 2 (Eds.) C.C. Leznoff and A.B.P. Lever, VCH Publishers, New York, 1993.
25. J. Martinsen, R.L. Greene, S.M. Palmer and B.M. Hoffman, *J. Am. Chem. Soc.* **105**, 677 (1983).
26. S.Venkatachalam, K.V.C. Rao and P.T. Manoharan, *Synth. Met.* **26**, 237 (1988).
27. B.N. Achar, G.M. Fohlen and J.A. Parker, *J. Polym. Sci. Polym. Chem. Ed.* **21**, 589 (1983).
28. H. Li and T.F. Guarr, *J. Chem. Soc., Chem. Commun.* 832 (1989)
29. A. Bettelheim, B.A. White, S.A. Raybuck and R.W. Murray, *Inorg. Chem.* **26**, 1009 (1987).
30. B.A. White and R.W. Murray, *J. Electroanal. Chem.* **189**, 345 (1985).
31. B.N. Achar, G.M. Fohlen and J.A. Parker, *J. Polym. Sci.* **20**, 2773 (1982).
32. B.N. Achar, G.M. Fohlen, J.A. Parker and J. Keshavayya, *J. Polym. Sci.* A **25**, 443 (1987).
33. H. Li and T.F. Guarr, unpublished results.
34. Y.-H. Tse, PhD Dissertation, York University, 1994.
35. H. Li, PhD Dissertation, University of Kentucky, 1991.
36. C.C. Leznoff, *Phthalocyanines: Properties and Applications*, (Eds.) C.C. Leznoff and A.B.P. Lever, VCH Publishers, New York, 1993.
37. Z. Jin, PhD Dissertation, York University, 1994.

38. A.B.P. Lever, E.R. Milaeva and G. Speier, in *Phthalocyanines: Properties and Applications*, Vol. 3 (Eds.) C.C. Leznoff and A.B.P. Lever, VCH Publishers, New York, 1993.

39. E.M. Baum, H. Li, T.F. Guarr and J.D. Robertson, *Nucl. Inst. Meth.* **B56/57**, 761 (1991).

40. F.Xu, PhD Dissertation, University of Kentucky, 1994.

41. Q. Peng, PhD Dissertation, University of Kentucky, 1994.

42. B. Zheng, M. Zhan and Z. Chen, *Xiamen Daxue Xuebao, Ziran, Kexueban* **31**, 392 (1992).

43. H. Tachikawa, Z. Dai and Z. Sun, *J. Polym. Mater. Sci. Eng.* **71**, 378 (1994).

44. T. Kobayashi, H. Yoneyama and H. Tamura, *J. Electroanal. Chem.* **161**, 419 (1984).

45. A.F. Diaz and J.A. Logan, *J. Electroanal. Chem.* **111**, 111 (1980).

46. E.M. Genies, A. Boyle, M. Lapkowski and C. Tsintavis, *Synth. Met.* **36**, 139 (1990).

47. F. Wudl, O.R. Angus, Jr, F.L. Fu, P.M. Allemand, D.J. Vachon, M. Nowak, Z.X. Liu and A.J. Heeger, *J. Am. Chem. Soc.* **109**, 3677 (1987).

48. X.H. Mu and F.A. Schultz, *J. Electroanal. Chem.* **361**, 49 (1993).

49. C. Hable, M.S. Wrighton, H. Li and T.F. Guarr, unpublished results.

50. H. Li and T.F. Guarr, *Synth. Met.* **38**, 243 (1990).

51. F. Xu, H. Li, S.J. Cross and T.F. Guarr, *J. Electroanal. Chem.* **368**, 221 (1994).

52. D.J. Moore and T.F. Guarr, *J. Electroanal. Chem.* **314**, 213 (1991).

53. F. Xu, H. Li, Q. Peng and T.F. Guarr, *Synth. Met.* **55–57**, 1668 (1993).

54. H. Naarmann, *Synth. Met.* **41–43**, 1 (1991).

55. J. Wang, S. Chen and M. Lin, *J. Electroanal. Chem.* **273**, 231 (1989).

56. S.W. Feldberg, *J. Am. Chem. Soc.* **106**, 4671 (1984).

57. P. Sheng, *Phys. Rev.* B **21**, 2180 (1980).

58. A. Witkowski, M.S. Freund and A. Brajter-Toth, *Anal. Chem.* **63**, 622 (1991).

59. V. Vijayanathan, S. Venkatachalam and V.N. Krishnamurthy, *Synth. Met.* **73**, 87 (1995).

60. R. Liepins, D. Campbell and C. Walker, *J. Polym. Sci. Part A* **6**, 3059 (1968).

61. Since these experiments were reportedly performed in aqueous media, the effect of Bu_4NBF_4 solubility on the observed results requires closer examination.

62. D. Wöhrle, R. Bannehr, B. Schumann, G. Meyer and N. Jaeger, *J. Mol. Cat.* **21**, 255 (1983).

63. P. Leempoel, F.-R.F. Fan and A.J. Bard, *J.Phys.Chem.* **87**, 2948 (1983).

64. F.-R. Fan and L.R. Faulkner, *J. Chem. Phys.* **69**, 3334 (1978).

65. P.N. Moskalev and I.S. Kirin, *Russ. J. Phys. Chem. (Engl. Trans.)* **46**, 1019 (1972).

66. J.M. Green and L.R. Faulkner, *J. Am. Chem. Soc.* **105**, 2950 (1983).

67. J. LeMoigne and R. Even, *J. Chem. Phys.* **83**, 6472 (1985).

68. H. Li and T.F. Guarr, *J. Electroanal. Chem.* **297**, 169 (1991).

69. Q. Peng, S.M. Crump and T.F. Guarr, submitted for publication.

70. G. Tourillon and F. Garnier, *J. Phys. Chem.* **88**, 5281 (1984).

71. K. Doblhofer and W. Durr, *J. Electrochem. Soc.* **127**, 1041 (1980).

72. P.G. Pickup, K.N. Kuo and R.W. Murray, *J. Electrochem. Soc.* **130**, 2205 (1983).

73. W. Kao and T. Kuwana, *J. Am. Chem. Soc.* **106**, 473 (1984).

74. C.P. Andrieux, J.M. Dumas-Bouchiat and J.-M. Savéant, *J. Electroanal. Chem.* **131**, 1 (1982).

75. C.P. Andrieux and J.-M. Savéant, *J. Electroanal. Chem.* **134**, 163 (1982).

76. C.P. Andrieux, J.M. Dumas-Bouchiat and J.-M. Savéant, *J. Electroanal. Chem.* **114**, 159 (1980).

77. H. Li and T.F. Guarr, *J. Electroanal. Chem.* **317**, 189 (1991).

78. J.H. Zagal, E. Munoz and S. Ureta-Zanarta, *Electrochim. Acta.* **27**, 1373 (1982).

79. K.M. Korfhage, K. Ravichandran and R.P. Baldwin, *Anal. Chem.* **56**, 1514 (1984).

80. Q. Peng and T.F. Guarr, *Electrochim. Acta.* **39**, 2629 (1994).

81. X. Qi and R.P. Baldwin, *Electroanalysis* **6**, 353 (1994).

82. X. Qi, R.P. Baldwin, H. Li and T.F. Guarr, *Electroanalysis* **3**, 119 (1991).

83. Y.-H. Tse, P. Janda and A.B.P. Lever, *Anal. Chem.* **66**, 384 (1994).

84. Y.-H. Tse, P. Janda and A.B.P. Lever, *Anal. Chem.* **67**, 981 (1995).

85. S.J. Cross and T.F. Guarr, manuscript in preparation.

86. S.J. Cross, PhD Dissertation, University of Kentucky, 1994.

87. R.L. Blankespoor and L.L. Miller, *J. Electroanal. Chem.* **171**, 231 (1984).

88. S.J. Cross and T.F. Guarr, manuscript in preparation.

89. G. Chottard and D. Lexa, *J. Electroanal. Chem.* **278**, 387 (1990).

90. K. DiGleria, H.A.O. Hill, V.J. Lowe and D.J. Page, *J. Electroanal. Chem.* **213**, 333 (1986).

91. H.A.O. Hill and D. Whitford, *J. Electroanal. Chem.* **235**, 153 (1987).

92. Z. Sun and H. Tachikawa, *Anal. Chem.* **64**, 1112 (1992).

93. S.J. Cross and T.F. Guarr, manuscript in preparation.

94. S.J. Cross and T.F. Guarr, manuscript in preparation.

CHAPTER 9

Characterization and Applications of Poly(*p*-phenylene) and Poly(*p*-phenylenevinylene)

Carita Kvarnström and Ari Ivaska
Åbo Akademi University, Åbo-Turku, Finland

INTRODUCTION

The first syntheses of poly(*p*-phenylene), PPP, were made using different chemical routes [1–4]. The material obtained was in powder form, which was difficult to process due to its insolubility in all common solvents.

Early studies on the preparation of PPP by electrochemical polymerization of benzene consisted of observations of unidentified layers on the electrode [5] or of insoluble polyphenylenes which precipitated into the solution [6]. The electropolymerization of benzene was described more thoroughly in the 1980s [7–10].

Despite the facts that PPP shows a high stability towards air oxidation (also at high temperatures), can be converted quite easily to both *n*- and *p*-type material, and thus reaches rather high conductivities, the interest in PPP has been quite low in comparison to other conductive polymers. One reason might be the relatively extreme electrochemical conditions needed even when dimer or substituted phenylenes were used

Handbook of Organic Conductive Molecules and Polymers: Vol. 4. Conductive Polymers: Transport, Photophysics and Applications.
Edited by H. S. Nalwa. © 1997 John Wiley & Sons Ltd

in the preparation. However, to date electrochemical synthesis procedures have been reported that yield adherent good-quality films [11–14], and PPP films with a large area has been synthesized by the precursor polymer route [15,16].

The first attempts to synthesize poly(p-phenylenevinylene), PPV, were through direct chemical polymerization reactions [17–19]. The material obtained in this way was in powder form, as in the case for PPP. PPV started to gain more interest when synthesis via a precursor polymer was reported [20–22]. Simultaneous stretching and thermal elimination of the sulfonium salt precursor polymer gave films with high crystallinity and good tensile strength in the drawn direction [23,24]. It was then possible to make transparent PPV films of large areas by the solution-processing technique. This process offered some potential to the flat-panel display technology.

PPP and PPV share some common features: they both can be n- and p-doped, they have a non-degenerate ground state, their electrochemical oxidation and reduction potentials are quite similar and they both show electroluminescence. Additionally, PPV shows non-linear optical activity [25–27].

In this chapter we give a brief overview of the properties and possible applications of PPP and PPV.

2 CHARACTERIZATION OF PPP

The properties of PPP are highly dependent on the way the polymer is synthesized. A detailed survey of the different chemical and electrochemical routes for the synthesis of PPP is beyond the scope of this chapter and only a few examples of different ways to obtain PPP is mentioned. We instead focus on the characterization of the material, and try to summarize the properties of PPP and on its potential for different applications.

The two most common chemical synthesis routes are the Kovacic method [1], where the starting material is benzene, and the Yamamoto method [2], where 1,4-dibromobenzene is used. Flexible films can be obtained when the synthesis is started from a precursor polymer and a phosphoric acid-catalyzed pyrolysis process is used [28]. A more extreme way to obtain PPP oligomeric material has been reported in ref. 29 where p-benzoquinone reacts on a reduced $TiO_2(001)$ surface to form volatile phenylene oligomers.

A large variety of electrochemical syntheses has been reported over the years, for example PPP films have been obtained by polymerization of benzene at a rather low potential in different acidic media such as

CF_3SO_3H [30], SbF_5 [31] and in $AlCl_3$ [32]. There are also results on PPP films obtained by polymerizing the monomer in an aqueous solution of an acidic microemulsion of benzene/sulfuric acid and sodium dodecyl sulphate [33]. Electroreductive polymerization of 1,4-dibromobenzene has also been reported to give PPP [34,35]. Good-quality PPP films have been obtained by electrochemical polymerization of biphenyl in different organic media [11–13].

2.1 Electrochemical properties

The electroanalytical methods most frequently used to study conductive polymers are potential or current pulse methods, cyclic voltammetry and impedance measurements. Cyclic voltammetry (cv) has been used very frequently both to study the electrochemical film growth and to examine the properties of the polymer in the potential range where it is electrically conductive.

PPP films prepared in different ways usually show variations in their redox response which can be examined easily by cv. Differences in the cv responses of the films are usually due to variations in the structure, conductivity, chain length and morphology of the films obtained during different synthesis routes. The influence of different parameters such as electrolyte, solvent and presence of residual water on the final film has been studied both in organic media [11] and in acidic aqueous solution [36]. Also the influence of temperature on both the film formation [37] and on the redox response [38] has been determined. The anodic oxidation (p-doping) of a PPP film usually takes place in the potential range 0.7–1.0 V (vs. Ag/AgCl reference electrode) when a standard three-electrode electrochemical cell setup is used. Reduction of the oxidized film during the reverse scan takes place in the range 0.7–0.4 V. Due to the relatively high oxidation potential of PPP compared with some other polymers, the degree of doping, i.e. the density of the charges on the polymer chain, is lower than in other conductive polymers [39]. Typical cyclic voltammograms of an electrochemically made PPP film on a platinum electrode cycled in acetonitrile are shown in Figure 9.1. The film in this particular case has been made by polymerizing 0.05 M biphenyl in 0.1 M TBABF$_4$–acetonitrile solution (TBA = tetrabutyl ammonium). For experimental details, see the caption. Figure 9.1a shows three consecutive scans at three different scan rates. The film relaxes quite slowly to its neutral state after each cycle. This can be seen especially at scan rates of 100 mV s^{-1} where the peak current decreases with the number of scans indicating that more and more charge is left in the polymer after each cycle. At slower scan rates or with a definite waiting period before starting a new scan, a stable and reproducible redox response of the film can be observed in the potential range studied; this is shown in Figure 9.1a. Goldenberg and Lacaze [36] investigated the influence of the waiting period on the redox response of a PPP film in a highly acidic medium. They used 80% sulfuric acid as the electrolyte medium, and found that a waiting time longer than 1000 s at 0 V broadened the oxidation and reduction peaks of the polymer indicating that a chemical modification of the polymer has taken place. When the scan is extended

into the positive direction beyond 1.2 V, overoxidation of the film can be seen in Figure 9.1b as a large irreversible peak at around 1.8–2.0 V, and upon further cycling the redox response of the film is very much damped or is not present at all. The stability of PPP during oxidation has been shown to depend on the nucleophilicity of the medium used [40]. PPP was found not to be electroactive in solvents with a donor number (representing nucleophilicity of the solvent) higher than 14. Overoxidation of PPP is assumed to take place through the formation of a polymer radical cation salt that upon further oxidation forms a dication with a high reactivity toward residual water in the

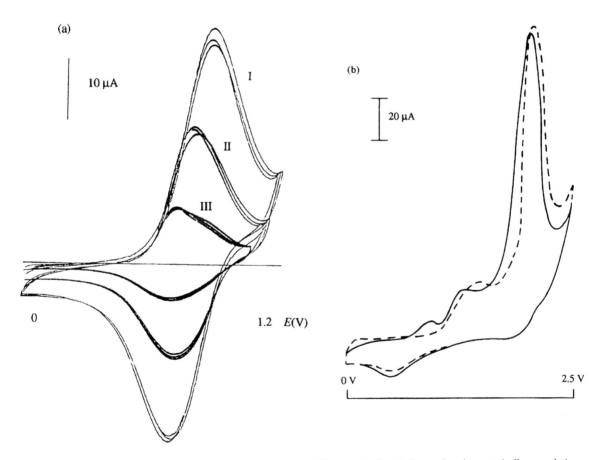

Figure 9.1. (a) Cyclic voltammograms of a PPP film in 0.1 M TBABF$_4$–acetonitrile solution made galvanostatically on a platinum electrode at a current density of 1.4 mA cm^{-2} in a 0.05 M biphenyl–0.1 M TBABF$_4$–acetonitrile solution. The numbers I, II and III indicate the cycles obtained at 20, 50 and 100 mV s^{-1}. (b) Two consecutive cyclic voltammograms of a freshly prepared thin PPP film on a platinum electrode in a 0.1 M TBABF$_4$–acetonitrile solution containing 0.8 M biphenyl. The solid line represents the first scan to 2.5 V and the dashed line the second scan. The scan rate is 100 mV s^{-1}. In the first scan the oxidation of the film can be seen at 0.8 V, and at around 1.6 V the oxidation of biphenyl in the solution can be seen. At around 2 V the overoxidation of the film takes place leading to deactivation of the film. The second scan does not show any electrochemical response from the film any more. (Reproduced by permission of Elsevier Science SA, from ref. 43, *Synth. Met.* **62**, 125, 1994)

solvent. As a result, a phenolic compound is formed first leading to a *p*-benzoquinone-type compound upon further oxidation. Total break down of the chains takes place by C–C bond cleavage followed by evolution of CO_2 and the formation of aromatic carboxylic acid [41]. The results obtained on overoxidation of conducting polymers including PPP have been summarized by Pud [42].

Sometimes the film gives rise to quite broad oxidation and reduction peaks upon charging and discharging [44]. Meerholz and Heinze [45] studied oligomers of the *p*-phenylene series in the solid state and found that during reduction of the oligomers the number of redox steps increased with increasing chain

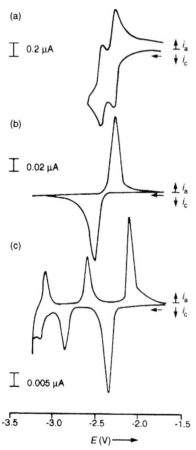

length, see Figure 9.2. This was explained by assuming that it is more difficult to reach a good conjugational conformation of the phenyl rings in the oligomers when polymerization is done in the solid rather than in the liquid state. This assumption may also partly explain the broad peak observed in some PPP material obtained by electrochemical polymerization indicating the presence of several different redox steps [44].

The very negative potential needed for *n*-doping of the PPP film has made it difficult to study the electrochemical reduction of PPP. Tabata *et al.* [46] studied the *n*-doping by measuring simultaneously the cyclic voltammogram and the absorption spectra, see Figure 9.3. The films studied were made from benzene in nitrobenzene containing (0.1 M) $CuCl_2$ and $LiAsF_6$. The reduction peak that can be seen in Figure 9.3 at -1.5 V has not been interpreted and the oxidation peak at -1.2 V is, according to the authors, connected with a reduction process that takes place at more negative potentials than -2.0 V. Similar voltammograms were also obtained with films prepared from benzene in $TBABF_4$–SO_2 medium by cyclic voltammetry in 0.1 M $TBABF_4$–tetrahydrofuran (THF) electrolyte [47]. Hérald *et al.* [48] studied the *n*-doping of PPP with Li^+, Na^+ and K^+ in solid state at temperatures of 60–80°C. When a slow scan rate was used with these materials a reversible cv response was obtained.

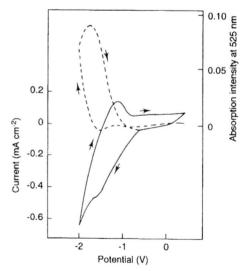

Figure 9.2. Cyclic voltammogram in Me_2NH–0.1 M TBABr of (a) *p*-quaterphenyl (saturated solution), $T = -10°C$, $v = 100$ mV s^{-1}; (b) *p*-quaterphenyl (thin layer on a platinum electrode), $T = -75°C$, $v = 10$ mV s^{-1}; (c) *p*-sexiphenyl (thin layer on platinum), $T = -75°C$, $v = 10$ mV s^{-1}. (Reproduced by permission of VCH-Verlag, Weinheim from ref. 45.)

Figure 9.3. The cyclic voltammogram (solid line) and the absorption intensity change (dotted line) at 525 nm of a PPP film in 0.1 M $TBAClO_4$–acetonitrile solution. Potential scan rate, 200 mV s^{-1}; potentials are given vs. Ag/Ag$^+$ reference electrode. (Reproduced by permission of the Institute of Physics from ref. 46.)

PPP films have also been studied by the electrochemical impedance technique [14,49,50]. A modified Randles-type equivalent circuit was used to describe the electrochemical behavior of the film [50]. Influence of the electrolyte and the film thickness on the impedance data were studied and the data obtained in different electrolytes are given in Table 9.1. The lowest value of the charge transfer resistance (R_{ct}) was obtained when BF_4^- and the highest value when ClO_4^- were used as the electrolyte anions. These results were explained by the fact that PPP films with perchlorate as the counter anion had a more compact structure [11].

2.2 Conductivity

Conductivity measurements with PPP has been used mainly to check the quality of the material produced in different synthesis procedures. Doping of the polymer takes place simultaneously with electrochemical polymerization of the film. However, when the polymer is made chemically, a linear or slightly crosslinked structure is observed and the polymer is undoped, i.e. non-conductive and has to be doped afterwards, either chemically or electrochemically. However, chemically prepared poly(*m*-phenylene) forms networks of polymer chains and therefore already shows a rather high conductivity, such as 10^{-4} S cm^{-1} at 300 K, in its undoped form [51]. Some values of the conductivities obtained for PPP synthesized by different chemical and electrochemical routes starting either from benzene or biphenyl are listed in ref. 52. Depending on the route of polymerization and on the dopant used the conductivity of the material was found to vary between 10^{-7} and 100 S cm^{-1}.

The conductivities of sexiphenyl and polyphenylene were compared by measuring simultaneously the cyclic voltammetric response and the conductivity *in situ* [53]. Both sexiphenylene and polyphenylene were cycled in a 0.1 M TBAPF$_6$–CH$_2$Cl$_2$ solution. The conductivity of the polymer obtained at a low charging level was almost the same as the conductivity of the sexiphenylene oligomers. Upon further charging of the polymer the conductivity increased rapidly. This was due to two different conductivity mechanisms: (1) hopping conductivity in the case of sexiphenylene; and (2) an improved conductivity along the chains, additional to the hopping conductivity, is obtained when the material is further charged.

Doping by ion implantation and the influence of different implantation parameters on the conductivity of chemically prepared PPP material have also been studied [54–57]. Using optimal conditions of implantation the conductivity of PPP was found to remain stable in air for over one year. The importance of the mass of the implantation ion on the resulting conductivity has been studied by using ions of different rare gases such as argon and xenon. Implantation with alkali and halogen ions has also been used in the conductivity studies. At low energies of implantation (below 150 keV) the conductivity increased with increasing ion mass. The experimental results showed two types of energy domains. With low-energy domain implantation (less than 100 keV) electronic conductivity through *n*-type doping with alkali ions and *p*-type doping with halogen ions was obtained. In the high-energy domain (higher than 150 keV) the conductivity was found to increase independently of the ion used and was therefore assumed to be the result of damage to the polymer backbone caused by the high-energy ions. They form in the polymer and act as traps for electrons.

Table 9.1. The effect of the electrolyte on the different parameters obtained with impedance measurement on PPP films. The same electrolyte was used for both film deposition and for impedance study. E_{app} is the applied potential vs. the Ag/AgCl double liquid junction electrode; R_s is the sum of the uncompensated ohmic drop from the film and the electrolyte solution; R_{ct} is the charge transfer resistance; j_0 is the exchange current density; C_{dl} is the double layer capacitance; C_L is the limiting capacitance; τ_D is the diffusion time constant; R_D is the diffusion resistance; and D is the diffusion coefficient. (Reprinted from *Electrochim. Acta*, **39**, ref. 50, Electrochemical impedance spectroscopic study of electropolymerized poly(paraphenylene) film on platinum electrode surface, 1419, Copyright 1994, with kind permission from Elsevier Science Ltd, The Boulevard, Langford Lane, Kidlington OX5 1GB, UK.)

Electrolyte 0.1 M	Thickness (μm)	E_{app} (V)	R_s (Ω)	R_{ct} (Ω)	j_0 (mA cm^{-2})	C_{dl} (μF)	C_L (μF)	τ_D (s)	R_D (Ω)	$10^7 D$ (cm^2 s^{-1})
LiBF$_4$	1.2	1.2	100	200	1.84	0.5	70	0.02	143	7.2
TBABF$_4$	1.2	1.0	100	300	1.22	0.7	110	0.05	450	2.0
TBAPF$_6$	1.0	1.1	110	380	0.97	0.6	200	0.10	500	0.74
LiClO$_4$	1.0	1.2	120	10000	0.036	0.5	600	2.0	3300	0.05
TBAClO$_4$	0.9	1.2	110	15000	0.024	0.4	700	10	14300	0.0081

Defects in the polymer can also decrease the conductivity. This was found when n-doping of PPP was done by implanting with Cs^+ ions. When a low current density for implanting was used, the conductivity of the PPP material was found to be higher than when high current density was used [55]. In ref. 56 it was shown that a reversible doping was obtained by implanting I^+-doped samples (p-doped) with cesium resulting in n-doped material. The conductivity obtained for these materials was above 10^{-3} S cm^{-1}.

Öpik et al. studied the doping of PPP with alkali metals [58] and halogens [59] at high temperatures. The conclusion from their experiments was that highly conducting complexes between PPP and a halogen were formed at temperatures of 400–450°C [59]. The interaction between the polymer and the dopant is much weaker at lower temperatures. The effect of the temperature was explained by assuming that the crystallinity in the polymer structure changes due to thermal treatment, and that the PPP material becomes more porous, thus increasing the number of places that can be doped. Conductivities as high as 150 S cm^{-1} were obtained.

2.3   FTIR and Raman spectroscopy studies

IR spectra of the precursors to PPP: benzene [60], biphenyl [61,62] and higher oligomers [63,64] have been thoroughly studied and reported. This simplifies the interpretation of the IR spectra of the polymer. Theoretical calculations of the frequencies and intensities of IR absorption bands of PPP have also been made [65–67]. The characteristic IR vibrations for polyphenylenes are listed in Table 9.2. An interesting experiment on PPP, prepared by the Kovacic method [1], was done by Hanfland et al. [68]. They pressurized the polymer and simultaneously recorded IR spectra. Interchain coupling was observed in the material when the spectra were interpreted. This coupling was caused by the high pressure applied on the PPP material which resulted in strong intermolecular interactions of the phenyl rings with the neighboring chains.

Polymer chain length and degree of crosslinking are two parameters that vary strongly with the way of synthesis. Different methods to estimate the chain length from the IR spectra have been reported. Kovacic and Jones [69] showed that the CH out-of-plane vibrations of 1,4-substituted rings shifted towards lower wavenumbers with increasing chain length. In ref. 1 the chain length was estimated from the ratio of the out-of-plane vibrations at around 800 cm^{-1} to the sum of the intensities of the two peaks from the out-of-plane vibrations of terminal phenyl rings around 760 and 690 cm^{-1}. Other modified equations based on the ratio of the intensities of the same vibrations have also been used [70]. Zerbi [71] found, when studying PPP and its oligomers, that from the CH stretching modes near 3100 cm^{-1}, the ratio of the peaks at 3060 cm^{-1} and at 3034 cm^{-1} decreased with the chain length. This indicated that absorption at the higher frequency is from the CH stretching at the ends of the chains.

The chemical synthesis of PPP through the reaction of biphenyl with $CuCl_2$ as oxidant and $AlCl_3$ as catalyst is known to lead to ortho-, meta- and para-polyphenylenes [72]. Simitzis et al. [73] studied the influence of the different structures of PPP on its conductivity. The different polymer structures were obtained by

Table 9.2. Typical IR vibrations of polyphenylenes

Wavenumber (cm^{-1})	Characteristic bonds from polyphenylenes	Ref.
3060	C−H stretching mode, aromatic	67
3034	-"-, for number of rings < 7	63, 67
3029	C−H stretching mode, aromatic	63, 67
1480		
1402		
1385	C−C vibrations	62, 63
1255		
1120		
999		
870–845	1,2,4-coupled phenyl rings	62, 76
817–805	C−H out-of-plane vibr. para disubst. phenyl rings	78
760	C−H out-of-plane vibr. mono subst. phenyl ring	84
750	1,2-coupled phenyl rings	69
692	C−H out-of-plane vibr. mono subst. phenyl ring	66

varying the concentrations of the compounds in the synthesis route in ref. 72. It was found that the material with the highest semiconductivity also had a high ratio *para*/*meta* couplings and that it also contained crystalline regions. Crosslinking, or *meta*- and *ortho*-coupling, was found to take place simultaneously with *para*-coupling of phenyl rings in the electrochemical synthesis of PPP from biphenyl in 0.1 M TBABF$_4$–CH$_2$Cl$_2$ [74], to some extent in 0.1 M TBABF$_4$–liquid SO$_2$ solution [75] and in 0.1 M TBABF$_4$–CH$_3$CN [43,76] medium as well. The influence of the current density on crosslinking was studied in ref. 43. The results showed that the extent of crosslinking increased with increasing current density used for polymerization. The external reflection spectra taken *in situ* during polymerization at two different current densities are shown in Figure 9.4. In the spectra from the experiments with high current, the peaks from 1,2,4-substituted compounds [77] are clearly visible in the range 1175–1125 cm^{-1} but not in the spectra of PPP obtained with low current density. This is a clear indication that high current density produces material with extensive crosslinking. The presence of peaks at 750 and 870 cm^{-1}, which are also from the crosslinked structure, are visible in the film made with the high current density. The spectra in Figure 9.4 are recorded *in situ* during polymerization and therefore bands originating from vibrations of the shorter oligomers in the 900–600 cm^{-1} region strongly overlap the PPP bands at 750 and 870 cm^{-1}.

The spectra obtained with different reflection techniques and taken *in situ* during electrochemical polymerization are usually rather complex in comparison to the spectra obtained from chemically made polymers. This is due to doping of the polymer that takes place simultaneously with polymerization. Intercalation of ions and solvent molecules causes structural changes in the polymer skeleton and also gives rise to additional vibration bands. Yli-Lahti *et al.* [79] studied, both by IR technique and by the method of Pron *et al.* [80], the doping level of PPP (obtained by the Kovacic method [1]) doped chemically by FeCl$_3$. They obtained spectra of PPP at different levels of doping and found new vibrations at 990, 1180, 1275 and 1530 cm^{-1}. These bands are interpreted to be such dopant-induced bands which are independent of the dopant and are related to changes taking place in the benzenoid structure towards the quinoid structure during doping. The doping-induced IR bands in PPP are listed in Table 9.3.

Oxidation and reduction of electrochemically made PPP from biphenyl in 0.1 M TBABF$_4$–CH$_3$CN medium has been investigated by the *in situ* external reflection FTIR technique [84]. The spectra were taken as the difference between the spectra taken at a reference potential, where the polymer is in its neutral state, and at increasing potentials in the range where the polymer reaches the conducting state. The spectra are shown in Figure 9.5.

At 0.9 V the dopant-induced bands are developed at wavenumbers 1530, 1275, 1180 and 980 cm^{-1}. The broad band in the range of 1550 to 1225 cm^{-1} is assigned to be overlapped by the vibrations of transitions from the benzoid form into the quinoid form. At 1.0 V a clear change in the IR baseline is observed in the wavenumber region 4000–3000 cm^{-1}. This is due to the increase in the concentration of free carriers in the polymer. The movement of ions in and

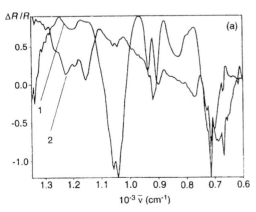
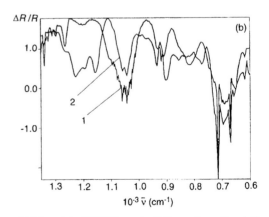

Figure 9.4. *In situ* FTIR spectra taken during the polymerization of biphenyl in TBABF$_4$–acetonitrile medium. Resolution, 8 cm^{-1}; number of scans, 250. (a) 0.5 and (b) 1.0 µm films made with current densities of 1. 78 and 2. 260 µA cm^{-2}. (Reproduced by permission of Elsevier Science SA from ref. 43, *Synth. Met.* **62**, 125, 1994.)

Table 9.3. Doping-induced IR bands in PPP

Wavenumber (cm^{-1})	Ref.
1225/1243/ 1275	79,81,83
1280	
1150/1180	79,81
1530	79,76,81
1343/1360	79,82,83
980–990	76,79,81

out of the film can also be followed by studying the bands between 1160 and 1000 cm^{-1} for the anion and the vibrations at 2960 and 2940 cm^{-1} for the cation. Intensities of these vibrations do not increase upon further scanning, indicating that a constant amount of anions and cations are exchanged during each scan.

Interpretation of Raman-active vibrations in oligophenyls and poly(p-phenylene) was reported by Krichene et al. [85,86]. They used a vibrational model based on force constants defined in terms of internal coordinates. The model they use gives a fit between the calculated and experimentally measured frequencies of better than 1%. The experimental Raman data were obtained with samples of PPP synthesized by the methods of Kovacic and Yamamoto [1,2]. Dependent on the excitation wavelength of the laser, the spectra of the undoped sample showed several lines in the wavenumber region 1200–1400 cm^{-1}. By using an excitation wavelength closer to the maximum of the π–π transition (i.e. $\lambda_L = 363.8$ nm) three main lines at 1595, 1280 and 1220 cm^{-1} can be seen in the Raman spectra [85,87]. The ratio between the lines at 1200 and 1280 cm^{-1} increases with the number of phenyl rings in the chain. Two new vibrations at 1244 and 1355 cm^{-1} were created upon p-doping of the PPP material. These new bands originate from structural modifications in the polymer backbone [86]. Furukawa et al. characterized the Raman spectra of sodium-doped (n-doped) PPP and compared them with the spectra of radical anions and dianions of p-oligophenylenes [88]. Negative polarons and bipolarons could be identified in n-doped PPP and the negative bipolaron was concluded to extend over four to six phenylene rings. The band at 1280 cm^{-1} is from the inter-ring C–C stretch [89] and this band was found to shift towards higher wavenumbers (1344–1353 cm^{-1}) upon doping. This doping-induced band is interpreted to rise from structural changes in the polymer, i.e. the change from benzenoid to quinoid form [88].

3 APPLICATIONS OF PPP

3.1 Batteries

The n- and p-doping property of PPP has gained a lot of interest in applications where the polymer has been used as electrode material in non-aqueous batteries [90–94]. PPP to be used as electrode material was synthesized as a loose brown powder with a high surface area (approx. 50 m^2 g^{-1}), pressed into pellets and predoped chemically by exposure to AsF_6. The predoped PPP was combined with a lithium counter electrode in a cell with $LiClO_4$–THF as the electrolyte. This cell showed a rather low open circuit voltage of 0.9–0.4 V. A higher voltage (4.4 V vs. Li) was obtained when PPP was used as the cathode in combination with a lithium anode in a lithium salt propylene carbonate (PC) medium, where the anions used were AsF_6^-, BF_4^- or PF_6^- [90]. PPP mixed with carbon black and doped with alkali metals (Li^+, Na^+ and K^+) has also been studied as an electrode in a battery application [92,93].

Thick, self-standing PPP films have been prepared electrochemically by the method of Yoshino [94] in nitrobenzene–$LiAsF_6$–$CuCl_2$ medium. The charging and discharging process of such an electrode was measured in PC with different electrolytes [95]. Reversibility of the process was found to depend on the anion used and it increased in the following order: $CF_3SO_3^- < ClO_4^- < PF_6^- < BF_4^- < AsF_6^-$. This order reflects the influence of the doping salt, $LiAsF_6$, used during polymerization of the film. The discharge capacity of the film was found to be 27 A h kg^{-1} at initial cycling and after 100 cycles it was still 23 A h kg^{-1}. PPP material was also used as an additive in an alkali-metal-alloy electrode. This improved both the reversibility and the rate of the charging–discharging process of the electrode. The reason for the improvement was that the introduction of PPP into the

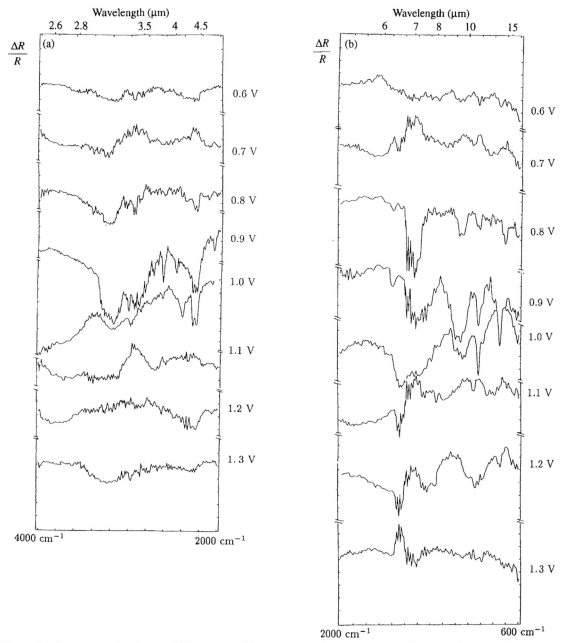

Figure 9.5. *In situ* external reflectance FTIR spectra of the oxidation (*p*-doping) of an approximately 0.1 μm PPP film: (a) 4000–2000 cm^{-1}; (b) 2000–600 cm^{-1} on the forward anodic scan from 0.6 to 1.3 V. (Reproduced by permission of Elsevier Science SA from ref. 43, *Synth. Met.* **62**, 133, 1994.)

electrode material increased the active surface area and the porosity of the alloy electrode. Also, the polymer functioned as a matrix for the alloy powder, and the conductivity of the polymer helped to maintain electronic and ionic conduction between the components of the electrode (electrode, alloy and electrolyte) [49]. PPP and other conductive polymers have been used as additives to the positive electrode in lead–acid batteries. Cyclability of such a composite electrode was improved and the maximum concentration of the polymer that could be used in the electrode was approx. 1 wt% [96].

3.2   Fibers and fibrils

Oriented fibers from PPP powder can be fabricated by solid state extrusion [97]. The die was filled with the polymer powder, heated up and pressurized. This extrusion process gave PPP fibers that were easy to handle and had a length of several meters. Polymer chain orientation in the fibers were confirmed by X-ray-scattering measurements. The fibers were afterwards doped with AsF_5 in the gas phase at different pressures. The doping process was followed optically by using a conventional light source and two lenses. When the picture of the fibers was magnified on a screen, swelling of the fibers could be seen. It was also found that doping of the fibers was rather inhomogeneous. The electrical conductivity was found to be highest in the direction of the chains. PPP fibrils have also been produced electrochemically by using benzene as the precursor [98]. The fibers started to grow parallel to each other but upon prolonged polymerization they started to grow together and formed a film with a high tensile strength along the fibers, i.e. 78 kg cm^{-2} in the temperature range 25–250°C.

3.3   Catalytic behavior, sensors and bilayers

PPP films made by electropolymerizing benzene in a microemulsion show electrocatalytic behavior towards the oxidation of ascorbic acid [33]. Cyclic voltammograms of ascorbic acid on platinum and Pt/PPP electrodes are shown in Figure 9.6. The catalytic effect can be seen in the separation of the peak potentials of the oxidation and reduction processes, ΔE_p, which decreases from 250 mV on the platinum electrode to 60 mV on the PPP film electrode. The overpotential decreased by 135 mV when the working electrode was changed from platinum to PPP film. Oxidation of

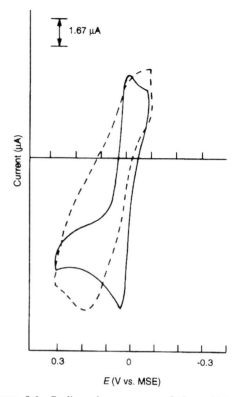

Figure 9.6. Cyclic voltammograms of the oxidation of ascorbic acid in 15 mol dm^{-3} sulfuric acid solution at the platinum electrode (dashed line) and at Pt/PPP (solid line). The scan rate was 50 mV s^{-1}. (Reproduced by permission of the Royal Society of Chemistry from ref. 33.)

ascorbic acid at the PPP electrode was interpreted to take place by a mediated electron transfer at the electrode surface.

Ashley *et al.* [99] reported that reduction of O_2 on a PPP-modified electrode took place at a potential approximately 300 mV less negative in comparison to a bare platinum electrode. Decrease in the peak separation was also found. It changed from 800 mV to less than 300 mV when PPP film was used as the working electrode.

The reduction of bromine has been studied on platinum and on both active and overoxidized PPP film electrodes [100]. It was found, that the reduction of bromine took place with a higher rate at overoxidized PPP than at active PPP or platinum. This behavior was explained by assuming that the hydroquinone groups formed during overoxidation act as electron transfer agents and therefore facilitate the overall electrode process.

Chemically reactive species have been trapped in PPP material by polymerizing biphenyl in the presence of those species. Ferrocene was trapped in this way in a PPP film and response from the ferrocene/ferrocenium couple could be seen in the cyclic voltammogram of the polymer even after several hundreds of scans [14]. The authors also incorporated anthracene and 9,10-diphenylanthracene in a similar way in a PPP matrix. Iron and manganese tetraphenylporphyrins have also been incorporated in PPP films. These groups were found to act as catalysts for oxygen insertion reactions. The polymer matrix was found to prevent porphyrins from undergoing side reactions to form μ-oxo-dimers. Such reactions, however, take place easily in free-standing porphyrins [14].

Yoshino and Gu have studied PPP as a sensor for ammonia [101]. The conductivity of the PPP film was found to increase by several orders of magnitude when exposed to NH_3. The change in electrical conductivity of PPP during exposure to NH_3 can be seen in Figure 9.7a, and the dependence of the electrical conductivity of PPP on the partial pressure of NH_3 is shown in Figure 9.7b. The change in conductivity was found to

take place relatively fast and to be reversible. When the partial pressure of NH_3 was reduced by evacuating the sample chamber the electrical conductivity of the PPP film decreased in a similar but reversed way to that shown in Figure 9.7a. From these results it could be concluded that NH_3 acts as an effective doping agent in n-doping of PPP. Similar results were obtained when PPP films were exposed to diethylamine and triethylamine [101].

Ye *et al.* polymerized acrylamine in 2 M H_2SO_4 on the top of a PPP layer [102]. Polyacrylamine, PAA, as such is not conducting, but when polymerized in the network structure of PPP the stability of PAA is increased. The structure of the PPP–PAA film was found to be porous and therefore the ion transport in the material was facilitated. The amino groups in the PAA part of the film showed high selectivity towards silver ions. A clear redox response for the silver–amino complex could be seen in the cyclic voltammogram of the PPP–PAA film electrode in a solution of silver ions. The formed metallic silver was deposited in the polymer network.

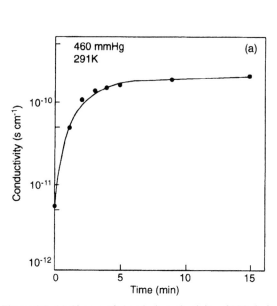
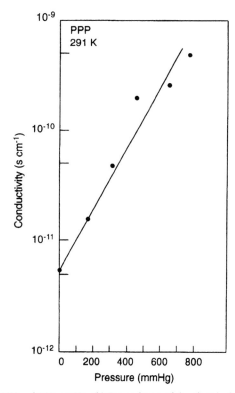

Figure 9.7. (a) Change of electrical conductivity of PPP during exposure of NH_3 of 460 mmHg. (b) Dependence of the electrical conductivity of PPP on the pressure of NH_3. (Reproduced by permission of the Japanese Journal of Applied Physics from ref. 101.)

Goldenberg *et al.* showed that it is also possible to make PPP films by electrochemical polymerization of benzene in fuming sulfuric acid in the presence of PVC which is a non-conducting material [103].

The electrochemical properties of PPP films in bilayers with either polypyrrole (PPy) or poly-(3-octylthiophene) (POT) has also been studied [104]. The bilayers were formed by electrochemical polymerization of the individual layers separately on platinum and on top of each other. In the combination of Pt/PPP/PPy some degree of charge trapping in the outer layer was observed due to the lower oxidation potential of the PPy layer. In the combination of Pt/PPP/POT the total charge of the bilayer was found to be higher than the sum of the charges of the individual layers of the same thicknesses, if charged separately.

3.4 PPP-based light-emitting diodes and Schottky-type devices

Nguyen *et al.* studied thin films of the Al/PPP/Al structure [105]. PPP was synthesized chemically via a precursor polymer route by the method of Ballard *et al.* [15]. Compact films with high processability were obtained by this way due to the solubility of the prepolymer. The polymer films were cast on an aluminium layer on a glass substrate. The aluminium layer on top of the PPP film was thereafter deposited at ambient temperature. The thickness of the metal layers was approx. 100 nm. It was found from conductance and capacitance measurements that the electrode behavior resembles that of a device where a contact resistance is in series with a bulk sample. The conduction mechanism was found to be thermally activated. By using an equivalent circuit analysis the authors could estimate the contributions of the bulk and the contact parts to the dielectric response of the device. By studying the aluminium/polymer interface with X-ray photoelectron spectroscopy, it was found that a metal–oxygen–polymer complex was formed at the interface [105].

The precursor polymer route of Ballard *et al.* [15] was also used in preparing electroluminescence devices of PPP [106–108]. Grem *et al.* constructed a device consisting of a PPP film between an indium tin oxide (ITO) substrate and a by-evaporation-deposited aluminium layer, ITO/PPP/Al [106,107]. They studied the blue-light emission from this device. The quantum efficiency of the device was approximately 1×10^{-4} photons per electron and the electroluminescence maximum for the device was found to be at 2.6 eV.

Absorption, electroluminescence and photoluminescence of PPP are shown in Figure 9.8.

The function of the ITO/PPP/Al device was explained by assuming a tunneling process in combination with an electrical-field-induced process [106]. The authors also give equations for the tunneling probability of electrons through an Al–oxide–polymer complex barrier in such a device. The characteristics of a sandwich-type ITO/PPP/Al device has also been studied by Miyashita and Kaneko [109].

Conjugation length in the PPP material, synthesized by the method of Ballard *et al.* [15], extends over approximately 10–12 phenyl units [110]. Leising *et al.* discuss the possibility to change the conjugation length of the polymer by introducing side groups in the precursor polymer [108a]. According to their experimental results this procedure changed the bandgap of the polymer and thus also the emission color [108b]. Phenylene oligomers and PPP-type ladder polymers have also been used in LED devices [111,112].

Schottky-type devices have been made of structures such as In/PPP/Pt [113] and Pb/PPP/Pt [114]. In both cases the constructed diodes showed asymmetric strong non-ohmic characteristics.

4 CHARACTERIZATION OF PPV

In the first studies on PPV as a conductive polymer, the material was chemically synthesized by direct synthetic

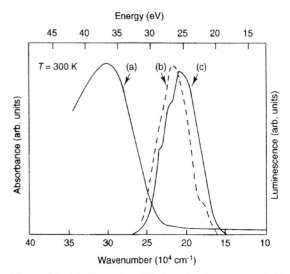

Figure 9.8. (a) Absorption; (b) electroluminescence and (c) photoluminescence of PPP at 300 K (Reproduced by permission of Elsevier Science SA from ref. 106, *Synth. Met.* **51**, 383, 1992.)

routes [17–19]. The material obtained in this way was of oligomeric character and in powder form. The highest conductivity obtained in this material was approximately 3 S cm^{-1}. In order to obtain higher chain length and higher conductivity in PPV a two-step synthesis method was employed using a precursor polymer [20–22,115–117]. The starting material was a sulfonium polyelectrolyte prepared in an aqueous solution. Polymerization of the monomer was initiated by adding a strong base. The resulting precursor polymer was cast as a film on a substrate or as a freestanding film and thereafter it was thermally converted to PPV. The length of conjugation that can be achieved in the PPV material by this process is highly dependent on the thermal treatment of the precursor polymer [118]. Some side reactions take place usually during thermal conversion, and C–Cl and C–S bonds are formed in the vinyl group [119]. Defect structures in PPV material obtained from the precursor polymer were studied by the XPS technique [120]. Results of such an analysis showed that the material obtained by thermal conversion contained traces of both chlorine and sulfur. PPV obtained by the treatment of the precursor polymer with H_2SO_4 or H_2SO_4 and SO_3 contained sulphur, and only the sodium-implanted PPV film was such that it contained neither traces of chlorine nor sulfur [120]. Besides thermal processing the precursor polymer can also be treated by irradiation with Ar$^+$ laser [121]. To increase the absorbance of the precursor polymer at the wavelength of the laser source, the polymer was impregnated with sulfonic acid dyes that had strong absorption at this wavelength. This laser-induced elimination reaction was reported to give fully conjugated PPV- and PPV-type films.

A high degree of orientation of the polymer chains could be obtained by stretching the precursor polymer [122]. A highly stretched film of PPV was reported to have a conductivity as high as 1.12×10^4 S cm^{-1} after treatment with concentrated sulfuric acid [123]. Patil *et al.* introduced the way of mixing the solution of polyelectrolyte (precursor polymer) with an excess of sodium salts, NaX (X = BF$_4$, AsF$_6$, I and PF$_6$) [124]. This process resulted in, what the authors call, an incipient doping of the material. By changing the counter ion by ion exchange i.e. Cl$^-$ to BF$_4^-$, a precursor polymer soluble in organic solvents was obtained [125].

4.1 Electrochemical response

Stenger-Smith *et al.* [126] compared the electrochemical behavior of PPV prepared from two different precursor polymers synthesized by the method of Lenz *et al.* [117]. The polymers were synthesized via precursor polymers starting from either acyclic or cyclic sulfonium salt monomers. The cyclic sulfides are known to give higher yields and higher molecular weight of the precursor polymer, and the elimination reaction during conversion is more efficient than that obtained with the acyclic monomer [117]. The precursor polymer made from the cyclic sulfonium salt monomer can be converted thermally to PPV at a temperature that is approx. 100°C lower than the temperature required for conversion of the precursor polymer made from the acyclic monomer. Due to the lower temperature during conversion of the precursor polymer the extent of disturbing side reactions is decreased.

Cyclic voltammograms of PPV films have been studied in dichloromethane–TBAClO$_4$ solutions [126]. The films were made by casting the prepolymers on the electrode substrate and then heating under vacuum. Cyclic voltammetric responses obtained with the PPV films made from both cyclic and acyclic sulfonium monomers were rather similar. The film showed a redox couple with the anodic and cathodic peak potentials at 1.05 V and 0.85 V (vs. Ag/AgCl), respectively. A reduction peak was observed at −1.8 V (vs. Ag/AgCl). All the potentials were measured from voltammograms taken at a scan rate of 100 mV s^{-1}. The potential values of the redox couple are in good agreement with the theoretical values reported [127]. When cycling the film to higher positive potentials, an irreversible oxidation peak was observed at 1.6 V [126,128]. The cyclic voltammograms of a PPV film made from the cyclic sulfonium salt precursor polymer are shown in Figure 9.9.

Onoda *et al.* [129] obtained almost an irreversible redox response from PPV film when it was cycled in LiBF$_4$–propylenecarbonate medium. They explained the behavior as an interaction between the solvent and the film electrode which takes place at higher potentials. By changing the solvent to nitrobenzene they obtained cv responses that are in good agreement with the potentials reported by other researchers [126].

The *p*- and *n*-doping of PPV films in TBABF$_4$–acetonitrile medium has also been investigated [130]. The films were made electrochemically by reduction of $\alpha,\alpha,\alpha',\alpha'$-tetrabromo-*p*-xylene in aprotic solvents. The electrochemically prepared PPV material is known not to be as pure as the one obtained by the precursor polymer route but the advantage with the electrochemical synthesis is that no thermal treatment is needed. The cyclic voltammograms of both chemically and electrochemically synthesized PPV material are rather similar [130].

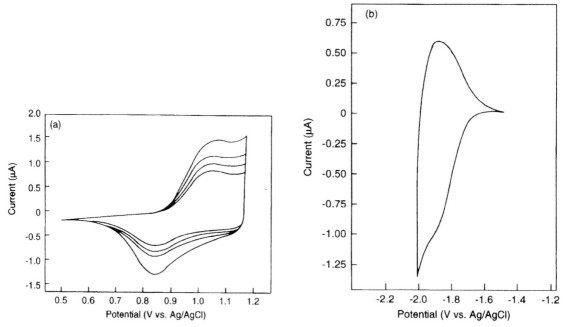

Figure 9.9. (a) Cyclic voltammograms from the electrochemical oxidation of a PPV film made chemically from the cyclic sulfonium salt precursor polymer. The electrochemical *p*-doping takes place in TBAClO$_4$–dichloromethane solution at a scan rate of 100 mV s^{-1}. The peak potential for the oxidation is at 1.05 V and for the reduction at 0.85 V. (b) A cyclic voltammogram of the *n*-doping of the PPV film made as in (a). A reduction wave can be seen at −1.8 V [vs. Ag(Ag/Cl)]. (Reprinted from *Polymer*, **30**, J.D. Stenger-Smith *et al.*, Spectroscopic and cyclic voltammetric studies on PPV, 1048, Copyright 1989, with kind permission from Butterworth-Heinemann journals, Elsevier Science Ltd, The Boulevard, Langford Lane, Kidlington OX5 1GB, UK.)

ac impedance measurements have also been used in studying the electrochemical properties of electrochemically made PPV [130]. The electrochemical redox response was analyzed on the bases of the charge transfer resistance, R_{ct}. As could be expected, a decrease in R_{ct} was observed in the potential ranges where the cation and anion doping process take place.

Heinze *et al.* [131] observed, with cv experiments during reduction (*n*-doping), that there exists different redox states in PPV oligomers, as in PPP. From the results, they draw the conclusion that the energetically lowest bipolaron state in a PPV oligomer involves at least eight monomeric subunits.

4.2 Electrical conductivity and photoconductivity

As mentioned earlier, stretching of a PPV film gives highly ordered material with high electrical conductivity [123]. The electrical conductivity showed high anisotropy in stretched PPV films [132]. Some results

have been reported on the temperature dependence of the dc electrical conductivity of *p*-doped PPV films [133,134]. PPV films, produced by the method of Lentz *et al.* [117], were stretched during thermal treatment of the precursor polymer and thereafter doped by immersing the films in a FeCl$_3$–nitromethane solution. At room temperature the conductivity of both stretched and unstretched doped PPV films increased with increasing doping time. The maximum value obtained for the unstretched film was 35 S cm^{-1} and for the stretched film 230 S cm^{-1}. Dependence of temperature on conductivity according to the Arrhenius law is shown in Figure 9.10 [134]. Due to the poor fit of the experimental results to the Arrhenius plot the authors of ref. 134 started to use Mott's formula that describes the conductivity in unstretched films by a three-dimensional variable-range hopping mechanism. A good fit was obtained independently of the draw ratio or the doping level by plotting log $\sigma T^{1/2}$ vs. $T^{-1/4}$. The conductivity in stretched films was explained by the fluctuation-induced tunneling conduction model as introduced by Sheng [135].

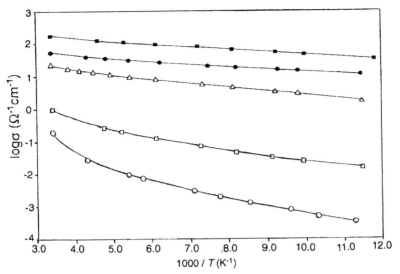

Figure 9.10. Arrhenius plot of the conductivity of unstretched PPV films doped for 10 s (○), 90 s (□), 240 s (△) and stretched films doped for 60 s (●) and 120 s (■) in 0.1 M FeCl$_3$ nitromethane solution (Reproduced by permission of Elsevier Science SA from *Synth. Met.* **55–57**, 1993, 3538.)

Lucas *et al.* studied the dependence of temperature on dc and ac conductivities in ion-implanted PPV films [136]. Samples were implanted with Cs$^+$ ions at different energies and fluency. At implantation with low energy the existence of different conducting mechanisms could be found by studying the changes in slope of the conductivity vs. temperature plots. At low temperatures the characteristics of the Arrhenius plot (log σ vs. T^{-1}) indicated a change from a variable-range hopping conduction [137] to a conduction in the polaronic [138] or in the band tail localized states. At high implantation fluence the Arrhenius plot was found to be almost linear over the temperature range of 140–380 K. This would indicate either conductivity in the polaronic localized states or in the band tails. The ac conductivity was also studied and it could be stated that at high temperatures the mechanism of this conductivity was the same as for the dc conductivity [136].

The photoconductivity was studied by comparing the photocurrent spectra with the optical absorption spectra of PPV film electrodes. It was shown that at room temperature the threshold of the photocurrent coincided with the absorption edge at 2.45 eV [139]. The formation of photocarriers was considered to comprise an initial creation of singlet excitons. The excitons migrate to the surface of the sample and dissociate at the surface defects to yield charge carriers. The photoresponse was observed to be a linear function of the electrical field applied and was also dependent on the temperature [139].

The response in transient photocurrent experiments can be divided into a fast decay followed by a slowly decreasing decay. Photoconductivity of the long-lived photocurrent components has been quite extensively studied [139–143]. Transient currents were observed under a positive potential on the illuminated electrode, which indicated a hole transport. Drift mobility of the holes was found to increase with increasing temperature. The mobility of the charge carriers in the slow interval of the decay process was reported to be in the range 10^{-3}–10^{-4} cm^2 V^{-1} s^{-1} [139]. Rentsch *et al.* estimated the charge carrier mobility in phenylated PPV films by integrating the charge density during the slow and the fast decay processes [144]. Mobility of the charge carriers was reported to be 4×10^{-3} cm^2 V^{-1} s^{-1} in the slow interval and twice that value in the fast interval [144]. The short-lived transient photoconductivity responses were studied by Bleier *et al.* [145] in highly ordered PPV films. Variations in the short-lived peak photocurrent with the applied dc-bias field followed the same behavior as observed for the long-lived response. The peak photocurrent was found to increase with the intensity of the incident light in a sublinear way [145].

4.3   Structural characterization

4.3.1   IR and Raman spectroscopy studies

IR and Raman spectroscopy have been proved to be useful techniques to study the molecular and electronic structure of PPV [146–160]. Frequencies of the IR-active modes in PPV are listed in Table 9.4.

IR spectra of highly oriented PPV films produced by stretching the precursor polymer during thermal treatment have been studied by Bradley *et al.* [146]. The degree of orientation was calculated from the dichroic ratios of the main IR modes.

PPV and low-molecular-weight model compounds have been studied both by IR and Raman techniques [148–156]. The model compounds consisted of *para*-substituted phenylene–vinylene units that were *para*-substituted with methyl groups at both ends of the molecule. The PPV unit has two local centers of symmetry, one in the center of the phenylene ring and the other one in the vinylene group. The in-phase CH out-of-plane IR bands at 972 cm^{-1} are strong in the solid state of the model compounds but shift to 964 cm^{-1} when measured in solution. This would indicate distortion of the vinylene groups in the liquid state. The authors also draw the conclusion that the vinylene groups in PPV are slightly distorted in the solid state because the vibration mode at 964 cm^{-1} is

visible in the spectra. When the polymer is in a planar *trans*-form the vinylene C=C stretch at 1629 cm^{-1} is inactive in the IR mode but becomes active in the distorted form. Similarly the band at 964 cm^{-1} is inactive in the Raman mode in a planar structure of PPV but becomes Raman active in a distorted form of PPV. The presence of the vinylene C=C stretch in the reported IR spectra and of the in-phase CH out-of-plane wag in the Raman spectra points towards the distorted form of the PPV. This would lead to a loosely helical polymer chain in the solid state. The dependence of the C=C stretching mode and the ring mode frequencies on the chain length indicate that there is a π-electron conjugation between the phenylene and the vinylene groups [148,156].

Theoretical studies of the IR vibrations in PPV are based on a single-periodic-chain model and harmonic potential with parameters transferred from the *p*-divinylbenzene molecule. Calculated and experimental IR vibration values of the *cis*- and *trans*-forms of PPV have been compared and found to be rather similar [66,158,161].

Buisson *et al.* [147,151,154] analyzed the IR and Raman vibrations by using a dynamical model based on valency-force-field calculations for PPV. The force constants were defined in terms of internal coordinates [21]. Planar geometry was assumed for both the polymer and the model compounds. In the work of

Table 9.4. IR vibrations of PPV

Wavenumber (cm^{-1})	Characteristic bond in PPV	Ref.
3105		
3077	Aromatic C–H stretch	147
3045		
2950		
2920	Aliphatic C–H stretch	147
2852		
1650–1630	C=C stretch	147,149,150
1594		
1561		
1519	C–C ring stretch	147, 148, 150, 155, 157
1426		
1339		
1267		
1211		
1176	C–H in-plane bend	147, 148, 150,155
1108		
1013		
972–965	*trans*-vinylene C–H out-of-plane bend	147, 149, 155, 157
837	C–H out-of-plane bend phenylene ring	147, 149, 157
784	out-of-plane vibr.	147,148
768	phenylene ring	

Buisson *et al.* the calculations were done on the in-plane modes and the force constants for the C—C and C=C associated bonds were determined.

When studying the Raman spectra of PPV it could be concluded that the observed Raman peaks were not dependent on the excitation wavelength of the laser [148,149]. This was explained by assuming that the π-electrons are strongly localized on the phenyl rings. It was also observed that the Raman spectrum was quite insensitive to the length of conjugation in PPV [148]. The main Raman vibrations observed in pristine PPV are at 966, 1174, 1304, 1330, 1550, 1586 and 1628 cm^{-1}. Upon doping new vibrations are observed at 1120, 1174, 1300, 1530 and 1567 cm^{-1} [154].

Zerbi *et al.* [153,162] applied the 'effective conjugation coordinate theory' [163,164] to interpret the vibrational spectra of PPV, and especially to interpret the doping induced IR bands. An effective conjugation coordinate, $\mathcal{R}$, was defined to describe the movement of the conjugated electron system along the polymer chain and its change from the benzenoid to the quinoid structure. By the way that the lattice dynamics were expressed using $\mathcal{R}$, the force constant $f_\mathcal{R}$ became a measure of effective conjugation. The smaller the force constant, the larger is the effective conjugation length. Most of the vibrations containing $\mathcal{R}$ shift to lower wavenumbers when the conjugation length was increased. The fit of the calculated values to the experimental vibrations obtained from Raman spectra, the doped PPV IR spectra, as well as from the photoexcited IR spectra is shown in Figure 9.11 [153].

Raman spectroscopy has been used to characterize polarons and bipolarons in PPV [155,157,165,166]. Such charged oligomers that show structural deformation similar to the localized excitations are used as model compounds. Self-localized excitations and charged oligomers are schematically presented in Figure 9.12. The Raman spectra of sodium- and potassium-doped PPV have also been analysed. Spectra of the doped PPV film show high dependence on the excitation wavelength in contrast to neutral PPV. This was explained by the resonant Raman effect [155], i.e. when the excitation wavelength is in resonance with the absorption wavelength of the existing polarons and bipolarons, the intensities of the Raman bands of the species associated with that absorption are selectively enhanced [157].

From the experiments in ref. 157 it could be concluded that the lightly sodium-doped PPV contained a bipolaron localized close to PV3 (see the caption of Figure 9.12 for explanation of the abbreviations) and two polarons of a length close to PV2 and PV3. When

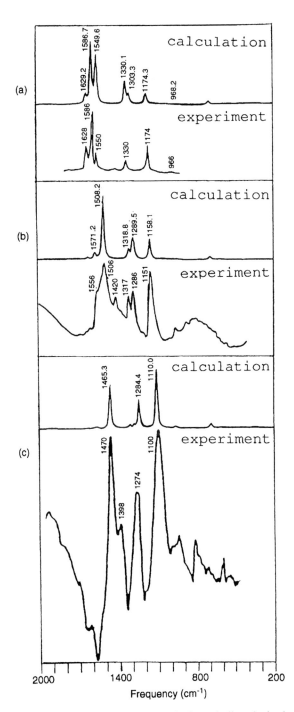

Figure 9.11. Experimentally and theoretically obtained spectra of (a) pristine PPV, (b) doped PPV and (c) photoexcited PPV (Reproduced by permission of Elsevier Science SA from *Synth. Met.* **41–43**, 1991, 255.)

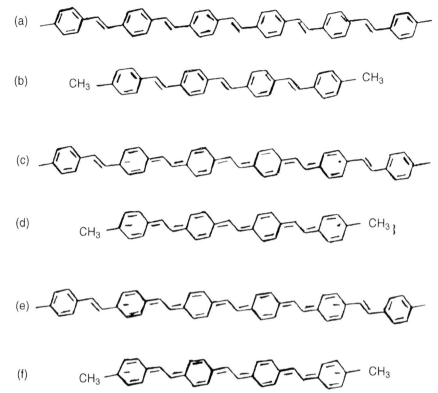

Figure 9.12. Schematic diagrams of self-localized excitations and charged oligomers (a) poly(1,4-phenylenevinylene) (PPV), (b) oligomer (PV3), (c) negative polaron, (d) radical anion, (e) negative bipolaron and (f) divalent anion (Reproduced by permission of Elsevier Science SA from *Synth. Met.* **49–50**, 1992, 335.)

comparing sodium and potassium doping some difference in the spectral region at $1271-1246$ cm^{-1} was found. This was explained by assuming the formation of channel structures in the PPV material upon doping. Alkali ions fill one-dimensional columns that are formed by three surrounding polymer chains forming a channel. The distance between the polymer chain and the metal was different in sodium- and potassium-doped material [167].

4.3.2 UV VIS spectroscopy studies

The UV vis spectrum of PPV films have been studied by Obrzut *et al.* [168]. The interpretation of the experimental result obtained was performed with semiempirical calculations both at the INDO [169] and Pariser–Pople–Parr level by using the valence effective Hamiltonian (VEH) techniques [170]. Spectra of oligo(phenylenevinylene)s are shown in Figure 9.13. During the elimination process of the precursor polymer the number of sulfonium salt units, *s*,

decreases and the number of phenylene vinylene units, *n*, increases. As the elimination process progresses the ratio *n/s* increases and the absorptions approach the wavelengths assigned for the longer conjugated segments: curves b and c. The absorption band at 200.2 nm shows a hypochromic shift and the band at 230.8 nm a bathochromic shift due to the exchange of the arylsulfonium salt for the phenylene unit. Changes in the shape of a broad band starting at 270 nm indicates that a conjugated π-electron system is formed. This band is, according to the authors, strongly perturbed by intramolecular electronic vibrational coupling [168].

Schenk *et al.* studied the charged oligo(phenylenevinylene)s by using UV/vis/NIR techniques [171]. Monoanions and dianions of soluble oligo(phenylenevinylene)s were generated and investigated. The monoanions were found to give rise to two absorption bands and the dianion to one band only. All absorption maxima showed a bathochromic shift with increasing chain length.

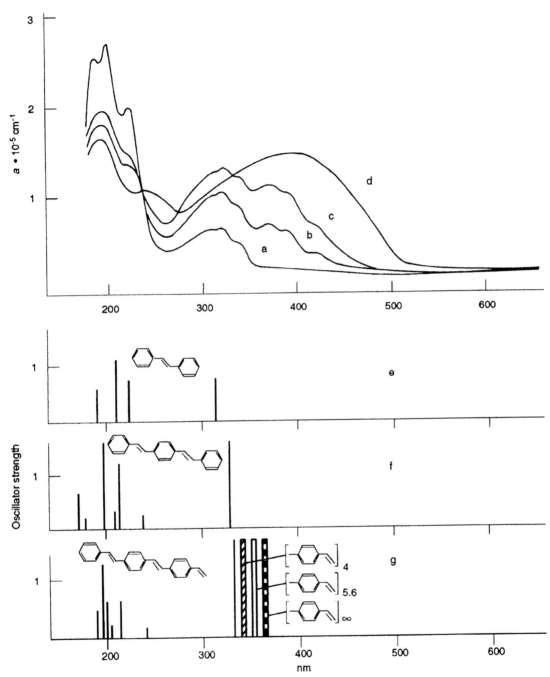

Figure 9.13. Spectra of PPV with different numbers of phenylene vinylene units, *n*, in relation to the sulfonium salt units, *s*. (a) $n:s = 1:2$, (b) $n:s = 2:1$, (c) $11:1 < n:s < 25:1$, (d) PPV, (e) calculated electronic transitions below 6 eV for stilbene, (f) for 1,4-bis(2-phenyl ethenyl) benzene, (g) for 1(2-phenyl ethenyl)-4(4-ethenyl-2-phenyl ethenyl) benzene, and the lowest $\pi \rightarrow \pi^*$ transitions calculated for *p*-phenylene vinylene oligomers (Reproduced by permission of the American Institute of Physics from ref. 168.)

Deussen and Bössler [172,173] studied the absorption spectra of chemically doped, p- and n-doped, oligo(phenylenevinylene)s and PPV. They did not find any significant difference in the spectra between the anion- and cation-doped oligomers or polymer. The authors report that there are three transitions originating from the monovalent species. The high-energy transition overlaps with the neutral 'educt' absorption. The configurational changes taking place in the polymer during charging are not detectable in the ion absorption spectra because the absorption spectra sense the electronic energy band scheme of the chromophore first after bond length adjustments have taken place. The conclusion was that the spectroscopy of charged conjugated polymers and oligomers should be analyzed by using highest occupied molecular orbital–lowest unoccupied molecular orbital (HOMO–LUMO) setting rather than using the polaron picture of semiconducting physics [172,173].

4.3.3 NMR spectroscopy studies

Solid state ^{13}C NMR spectroscopy has been used to characterize neutral PPV [174,175] and AsF_6-doped PPV films [174]. Results indicating a 180° rotational jump of the phenylene rings of PPV about the 1,4-phenylene ring axis (ring flips) has been reported both from ^2H and ^{13}C NMR spectra [176–178]. Results obtained with the ^{13}C measurements showed that the ring flip takes place in the crystalline part of PPV. Simpson et al. [179] showed that the dependence of the median jump rate on temperature could be described by the Arrhenius equation. The median activation energy was found to be 15 kcal mol^{-1}.

Sodium-doped, ring-deuterated PPV, PPV-d_4, was studied by ^2H quadrupole NMR spectrometry [180]. The polymer chains were found to be aligned parallel in a crystalline unit cell of PPV. When viewed from the end, the polymer chains are packed in a herring-bone pattern [181,182]. A structural change in the unit cell has been found to take place upon sodium doping; it becomes a structure where three PPV chains surround a column of sodium ions [183]. Three phases of different doping levels of sodium were observed. The color changes between these phases were from yellow-orange (neutral polymer), to blue (slightly doped polymer), to metallic gold and upon long doping times to grey-black.

Chen et al. found that the NMR spectra of the blue-phase sodium-doped PPV and the neutral PPV were identical [183]. This indicates that only a very low doping level was achieved, i.e. the concentration of sodium ions was not high enough to give any changes in the structure of PPV. The gold-phase material was found to be very brittle and to have a high reflective surface. Films of a thickness of 30 μm were completely opaque. It was concluded that the grey phase was due to the partly destroyed PPV structure. This was caused by the high level of sodium ions resulting in a distortion of the planar configuration of the polymer chains. The lineshape of a PPV-d_4 film doped to the gold-phase level was found to consist of a broad component (splitting of 131 kHz) originating from static deuterons still present in the phenyl rings. Doping-induced changes in the electronic environment lead to a reduction of the magnitude of the electrical field gradient which leads to a drop in the quadrupolar splitting from 133 to 131 kHz. A second component in the lineshape (64 kHz splitting) was studied in the temperature range −125 to 175°C and was found to be insensitive to temperature changes in contrast to undoped PPV-d_4 [179] or with H_2SO_4-doped PPV-d_4 [184]. This was explained by the fact that the doping process raises the activation of the phenylene ring flips so much that the rings cannot flip any longer on the NMR time scale, not even at 175°C. A central component of the lineshape was assigned to the formation of sodium deuteride molecules [180].

5 APPLICATIONS OF PPV

The interest in using conducting polymers as electronic materials has greatly increased. A lot of research interest has been focused on PPV ever since the observations of electroluminescence from light-emitting diodes, LED, based on this polymer were reported [185–187].

5.1 Polymer–substrate interface

The interface between a metal and a polymer layer is the basic component of LED devices, and the charge injection and the charge transfer taking place at the interface determines the characteristics of these devices. The nature of these interfaces has been the subject of several studies [188–192]. Nguyen et al. [189] studied the interfaces between PPV thin films and aluminium or chromium in a similar way as they did with PPP [105]. PPV films synthesized by the method of Wessling and Zimmerman [193] were studied and the precursor polymer was cast on a glass substrate with a predeposited chromium layer. After conversion of the precursor to a PPV film the top metallic (aluminium or chromium) layer was deposited on the electrode by evaporation. Characterization of the interfacial structures was made by X-ray photoelectron spectroscopy, XPS. The results indicated, that during deposition of chromium and aluminium on PPV films at room temperature the top electrode reacts with carbon in the presence of oxygen. The resulting complex was a metallic oxide carbide. The presence of oxygen was due to contamination, i.e. leaks in the vacuum chamber. No metallic oxide–PPV interaction could be detected between the bottom layer of chromium and the polymer. The compact PPV layer prevented oxygen from diffusing through the polymer to the underlying interface. Plots of the dark current density vs. the applied electric field of the Cr/PPV/Cr and Cr/PPV/Al devices showed asymmetrical characteristics and a rectifying behavior. Electron transfer directed from the top electrodes to the *p*-type polymer was proposed. The asymmetric *I–V* characteristics observed were due to the complicated structure of the polymer–metal interface. The rectifying effect was found to be greater in the Cr–PPV contact than in the Al–PPV contact and was related to the formation of metal carbides at the Cr–PPV interface.

The interface between aluminium and PPV has also been studied by Dannetun *et al.* [190,191]. The experimental results were obtained by XPS measurements and the theoretical approach was taken through the density of valence states generated from *ab initio* Hartree–Fock quantum chemical calculations [194]. The conclusions were that aluminium reacts selectively with the vinylene segments in PPV forming covalent bonds and that sp³-carbon sites are formed; this leads to a twist in the molecule at these sites and to a break in the π-conjugation. Mixing of the s- and p-atomic orbitals of aluminium with the HOMO and LUMO of the pure PPV system (before deposition of aluminium) led to the formation of a new localized energy state near the Fermi level.

Temperature dependence of electroluminescence was used to study the charge injection across the electrode/polymer interface in LEDs of the type ITO/PPV and then with either aluminium, magnesium or calcium as the negative electrode [192]. Two different processes could be distinguished: a thermally activated process at low electric field and a field emission process at high electric field. Based on the data obtained, field emission is suggested to originate from holes tunneling across the ITO/PPV interface. The low field process is proposed to be a space charge limited current, SCL. The presence of traps in the polymer material can be detected by using the SCL model for a trap-free material. This was the case for all the top electrode combinations with the three metals. Values for the charge mobility were estimated and were found to be 10^{-9} cm^2 V^{-1} s^{-1} for magnesium and calcium, and 10^{-10} cm^2 V^{-1} s^{-1} for aluminium at high temperatures. These values are small and the authors state that the SCL current is filamentous and restricted only to a small area of the sample.

5.2 Characterization of PPV-based light-emitting diode

Nguyen *et al.* [195,196] studied the dependence of temperature on the frequency response of a Cr/PPV/Cr structure. The dielectric loss of the structure was presented by a Cole–Cole plot $\varepsilon'' = f(\varepsilon')$. It could be concluded from the analysis that there were three relaxation mechanisms involved in the Cr/PPV/Cr structure. The relaxation mechanism at high frequencies was concluded to originate from the bulk mechanism. Relaxation in the middle frequency range was related to the interface between the bottom contact and the polymer. These two mechanisms were found to be slightly dependent on temperature and were treated as the bulk mechanism at high and middle frequencies. The third relaxation at low frequencies (could not be described by a Cole–Cole graph) was found to resemble a system with interfacial layers and was strongly dependent on temperature. Earlier studies by the same authors [197] showed that the adherence of the metallic layer to the polymer and therefore also the contact between the polymer and the metal were influenced by temperature. This led to the conclusion that the low frequency loss originates from the interface between the polymer and the top metallic layer. The complex formed at the metal–polymer interface was not found to be uniform and therefore no simple representation of the potential barrier could be applied to these structures.

When Cr/PPV/Al structures were plasma treated with O_2 and N_2, the current in the *I–V* plot was found to increase drastically [198]. Because plasma treatment is known not to change the nature of the bulk as ion implantation does [199], this method was proposed to be used on PPV films whenever a charge carrier confinement layer was desired to be included in the structure. This layer would improve the photoluminescent efficiency of the LED devices due to an easier charge injection into the polymer.

Karg *et al.* [200,201] showed that the *I–V* characteristics in the dark of an Al/PPV/ITO device is asymmetric, and rectification ratios of 10^4–10^6 were obtained. White light illumination of the device induced a photovoltaic effect. Current vs. applied voltage characteristics at room temperature both in dark and under illumination for an Al/PPV/ITO device are presented in Figure 9.14. Open-circuit voltages of around 1 V were measured and a short-circuit current density of 80 μA cm^{-2} was obtained under white light illumination with the effect of 100 mW cm^{-2}.

Impedance measurements were made on the Al/PPV/ITO structure and analyzed by the equivalent circuit model presented in the insert in Figure 9.15 [200]. Schottky barrier diodes are generally described by this circuit; the elements C_j and R_j describe the junction and the elements C_b and R_b describe the undepleted bulk. The real and imaginary parts of the complex impedance spectra for different applied bias voltages are shown in Figure 9.15. The resistive layer of the undepleted bulk (the left semicircle in Figure 9.15) remains almost unchanged, and was found to be present even when a strong forward bias was applied. Under conditions where the contribution from the bulk could be neglected (high temperatures and at low frequencies) the junction capacitance was measured under reverse bias. When the Schottky model was used the ionized acceptor dopant concentration, N_a, was determined to be 10^{17} cm^{-3} and the built-in potential, V_d, was found

to be 1 V. The width of the depletion layer was determined to be 0.1 μm. It could then be concluded from these experiments that at room temperatures a Schottky barrier exists in devices of the structure Al/PPV/ITO. At low temperatures (77 K) the *I–V* characteristics of the same device were found to be symmetrical, which indicates that the device shows electroluminescence (caused by the injection of carriers) in both current directions at about 35 V and the Schottky junction does not exist any longer. In this case the carriers are injected into the PPV layer from both contact electrodes [200].

Krag *et al.* [201] found that a two-month-old device stored in air at room temperature showed, after 16 h of evacuation, similar characteristics under 6 V forward bias as a new device under 4 V. From the results it could be concluded that exposure to air did not cause irreversible changes in the device but affects strongly the electronic properties requiring a higher bias voltage than a new device.

Braun and Heeger [186] reported that the light generation efficiency of PPV-based LEDs could be increased by approximately 1% photon/electron when calcium was used as the negative electrode. Measurements of lifetime and characterization of ITO/PPV/Ca devices were reported by Holmes *et al.* [202–204]. The response was studied under constant voltage, constant current and ac voltage. Continuous operation with square-wave biasing (switched between 5.5 V and 6.9 V; the initial voltage was set to ±5.5 V and was raised to higher values at later times; the frequency used was 1070 Hz) and a current density in the range 300–800 mA cm^{-2} was achieved over 1200 h. Electroluminescence emission spectra were recorded from the device after various periods of operation time and only very slight changes in the strength of the bands was observed. The shift in energy of the bands was less than 5 nm. The interface structure between alkaline earths and PPV has not to date been completely characterized. It is considered that the divalent calcium ion is less mobile than the monovalent alkali metal ions. The results reported in ref. 202 showed that long-term degradation of the PPV-based LEDs were not due to structural changes. This is also an indication that the diffusion rate of calcium is very low. Low diffusion rates are also expected for aluminium, which forms covalent bonds between the metal and PPV [190,191]. Stable long-term operating diodes can not be expected when alkali metals are used as the electrodes. This prediction can be based on the results reported on sodium where charge transfer from metal to PPV could be detected. Formation of a charge transfer complex

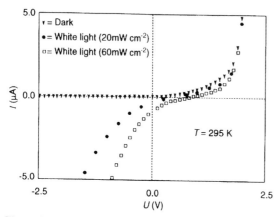

Figure 9.14. Current vs. applied voltage characteristics for the Al/PPV/ITO diode (thickness 0.5 μm). Results are obtained at room temperature in the dark and under white light illumination (Reproduced by permission of Elsevier Science SA from *Synth. Met.* **54**, 427, 1993.)

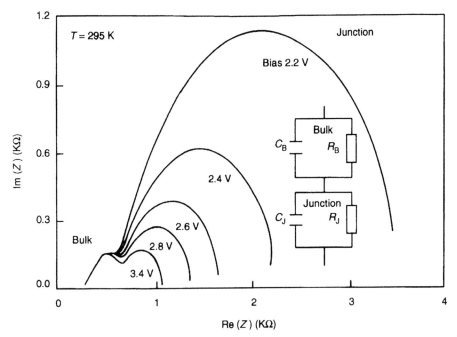

Figure 9.15. Impedance plots under forward bias for an Al/PPV/ITO Schottky diode with an active area of 4 mm². The Schottky barrier is described by the equivalent circuit depicted in the figure (Reproduced by permission of Elsevier Science SA from *Synth. Met.* **54**, 427, 1993.)

leads to the diffusion of metal ions into the PPV surface layer structure therefore shortening the lifetime of the device [205].

By shortening the effective conjugation length of PPV, a shift in the emission spectra from green to blue is expected due to the larger bandgap of PPV. Woo *et al.* [206] studied the PPV oligomer 1,4-distyrylbenzene sandwiched between ITO and indium. To increase the electron injection a layer of strontium was additionally inserted into the device giving the structure In/Sr/ oligomer/ITO. The addition of the strontium layer lowered the rectifying voltage to 5 V and the current at 5 V was 100 times higher than for the device without the strontium layer. A blue-violet emission was observed from this device at room temperature and at a bias of 5 V.

To increase the hole transport in a PPV-LED device, a polyimide layer was placed between the anode and the light-emitting PPV layer [207]. The LED structure studied was ITO/polyimide/PPV/Mg–Ag-alloy. The light-emitting PPV layer was a Langmuir–Blodgett (LB) film which was shown to have a higher order of orientation and less defects in the π-conjugated system than other PPV layers [208]. A uniform and stable

emission light at room temperature was observed. The polyimide layer acted as a hole transport layer and reduced the drive field (threshold voltage/total polymer layer thickness) in the device, which indicated that the charge injection was improved.

In order to increase the light generation efficiency of the PPV-LED devices attempts to improve the emissive properties of the polymer by forming copolymers has been undertaken [209–211]. Based on the knowledge that conjugated molecules with low molar mass and with photoluminescence maxima in the blue region often show parallel electroluminescence around the same wavelength [212] Yang *et al.* [213] synthesized a block copolymer that consisted of alternating rigid and flexible blocks. The rigid part was made by PPV-type segments forming the conjugated part and the flexible part was polyethylene contributing to solubility and film-forming properties of the polymer. The resulting polymer, poly(1,8-octanedioxy-2,6-dimethoxy-1,4-phenylene-1,2-etheylene-1,4-phenylene-1,2-etenylene-3,5-dimethoxy-1,4-phenylene), was found to emit blue light with the maximum at 465 nm in the electroluminescence spectra. Monoalkoxy-substituted PPV as the emitting layer (yellow light) has been reported to have its peak wavelength of electroluminescence (550 nm) very close to that of photoluminescence [214].

A comparison between the LED structures of ITO/PPV–cop/Al (cop indicates a partially conjugated copolymer from the PPV precursor with a methoxy leaving group) and ITO/PPV/Ca were studied by the X-band photoluminescence, electroluminescence, conductivity and photoconductivity methods [215]. Photoluminescence and electroluminescence efficiency of alkoxy, alkyl and cyano derivates of PPV have also been studied [216]. Dialkyl-PPV was used as a hole transport layer in between cyano-PPV films and the ITO substrate when indium was used as the top contact. In the alkyl- and alkoxy-PPV film LED devices no hole transport layer was used and the top contact was calcium. Photoluminescence efficiency was found to increase with the fraction of non-conjugated units while electroluminescence efficiency was dependent on the radiative recombination efficiency of the light-emitting material as well as on carrier transport into and inside the material.

5.3   Optical waveguides

Optical waveguides have been made of a copolymer consisting of PPV and 2,5-dimethoxy-PPV [217]. The precursor polymer is composed of two types of segments, one with a sulfonium leaving group and the other with a methoxy leaving group. The two segments require different conversion conditions; the sulfonium group needs only heat and the methoxy group needs heat and the presence of an acid. Hydrochloric acid is liberated as a by-product of the thermal conversion of

the sulfonium segment. The authors used the liberated acid in a selectively catalyzed conversion of the methoxy segment. A special polymer pattern was obtained by placing a 'capping' layer of gold or aluminium, which is impermeable to HCl, over the precursor polymer. Under thermal conversion the volatile acid by-product from the sulfonium-substituted (tetrahydrothiophene) segments cannot escape and they act therefore as catalysts for the conversion of the methoxy segments. The conversion of the uncapped regions does not lead to a fully conjugated polymer and therefore the conjugation length and also the bandgap can be varied by the composition of the copolymer and the capping patterning. The structure of a PPV/poly(2,5-dimethoxy-p-phenylenevinylene) optical waveguide is shown in Figure 9.16. The substrate onto which the precursor polymer was spin coated was single crystal silicon. The metal layer was removed after conversion and a transparent cladding layer was used to reduce the optical propagation losses. The smallest size of the waveguide structure was 1 μm where the thickness of the copolymer film was 0.15 μm. This type of waveguide was demonstrated to represent the short propagation distance devices.

Material with special waveguiding behavior can be made by combining materials with different properties. Embs et al. combined the optical waveguiding behavior of silica with high third-order non-linear susceptibility, χ^3, of PPV [218]. PPV–silica composite thin films were prepared via the sol-gel process [219]. The water-soluble PPV precursor was incorporated in the silica-sol phase, the solvent was evaporated and the spin coated or quiescently cast films were converted to

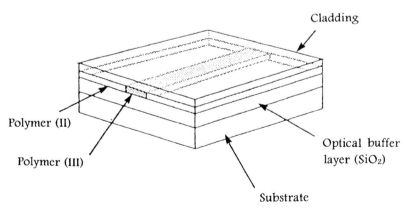

Figure 9.16. Diffused channel optical waveguide in poly(2,5-dimethoxy-p-phenylenevinylene)/PPV copolymer. Polymer (II) indicates the partially converted copolymer, i.e. only the sulfonium-leaving segments are converted; polymer (III) is the fully conjugated copolymer (Reproduced by permission of Elsevier Science SA from *Synth. Met.* **55–57**, 3683, 1993.)

PPV–silica composites. The resulting structure was found to be a nanophase-separated morphology composed of amorphous silica and paracrystalline PPV. The waveguiding behavior was highly improved compared to pure PPV films and the χ^3 value observed was only slightly lower than that for pure PPV. Sol–gel processed PPV/V_2O_5 composites also exhibit a large χ^3 value and the refractive index was higher than that for the PPV/silica composite [220].

5.4 Microdisk array and sensors

Tabei *et al.* [221] reported the fabrication of a microdisk array electrode from pyrolyzed PPV film. PPV precursor polymer was cast onto a silicon wafer covered with SiO_2. Conversion to PPV was done thermally. The heat treatment was then continued at a higher temperature to form carbonized PPV. A photo-resist film with a carbon layer was then placed on top of the carbonized PPV. The resist film was then exposed to UV light through a photomask and developed in the uncovered areas. Finally the electrode was heated. The conductivity of the carbonized PPV film was 200 S cm^{-1}. A high-density microdisk array with good microelectrode features and reproducibility of electrode shapes could be obtained with the PPV-based carbon film.

Yoshino and Gu [101] studied the effect of ammonium gas on the electrical properties of PPV in a similar manner as for PPP. The conductivity of PPV was found to increase upon exposure to ammonium gas which was explained to be due to *n*-doping of PPV. The authors proposed application of such a device as a gas sensor.

6 ACKNOWLEDGMENTS

Technical support by Mrs Maarit Hammert is kindly acknowledged.

7 REFERENCES

1. P. Kovacic and J. Oziomek, *J. Org. Chem.* **29**, 100 (1964).
2. T. Yamamoto, Y. Hayashi and A. Yamamoto, *Bull. Chem. Soc. Jpn.* **51**, 2091 (1978).
3. J.F. Fauvarque, M.A. Petit, F. Pfluger, A. Jutland, C. Chevrot and M. Troupel, *Makromol. Chem. Rapid Commun.* **4**, 455 (1983).
4. T.W. Campbell and R.N. McDonald, *J. Org. Chem.* **24**, 730 (1959).
5. A.F. Shepard and B.F. Dannels, *J. Polym. Sci., Polym. Chem. Ed.* **4**, 511 (1966).
6. T. Osa, A. Yildiz and T. Kuwana, *J. Am. Chem. Soc.* **91**, 3994 (1969).
7. F. Beck and A. Pruss, *Electrochim. Acta* **28**, 1847 (1983).
8. I. Rubinstein, *J. Electrochem. Soc.* **130**, 1506 (1983).
9. K. Kaeriyama, M. Soto, K. Someno and S. Tanaka, *J. Chem. Soc., Chem. Commun.* 1199 (1984).
10. M. Satoh, M. Tabata, K. Kaneto and K. Yoshino, *J. Electroanal. Chem.* **195**, 203 (1985).
11. S. Aeiyach and P.C. Lacaze, *J. Polym. Sci. Part A* **27**, 515 (1989).
12. P. Soubiran, S. Aeiyach and P.C. Lacaze, *J. Electroanal. Chem.* **303**, 125 (1991).
13. C. Kvarnström and A. Ivaska, *Synth. Met.* **41–43**, 2917 (1991).
14. J.F. McAleer, K. Ashley, J.J. Smith, S. Bandyopadhyay, J. Ghoroghchian, E.M. Eyring, S. Pons, H.B. Mark, Jr., and G. Dunmore, *J. Molec. Electro.* **2**, 183 (1986).
15. D.G.H. Ballard, A. Courtis, I.M. Shirley and S.C. Taylor, *J. Chem. Soc., Chem. Commun.* 954 (1983).
16. D.G.H. Ballard, A. Courtis, I.M. Shirley and S.C. Taylor, *Macromolecules* **21**, 294 (1988).
17. G.E. Wnek, J.C.W. Chien, F.E. Karasz and C.P. Lillya, *Polymer Commun.* **20**, 1441 (1979).
18. J.R. Reynolds, F.E. Karasz, J.C.W. Chien, K.D. Gourley and C.P. Lillya, *J. Phys.* **44**, C3-693 (1983).
19. K.D. Gourley, C.P. Lillya, J.R. Reynolds and J.C.W. Chien, *Macromolecules* **17**, 1025 (1984).
20. J.D. Capistran, D.R. Gagnon, S. Antoun, R.W. Lenz and F.E. Karasz, *ACS Polym. Prep.* **25**, 282 (1984).
21. I. Murase, T. Ohnishi, T. Noguchi and M. Hirooka, *Polym. Commun.* **25**, 327 (1984).
22. F.E. Karasz, J.D. Capistran, D.R. Gagnon and R.W. Lenz, *Mol. Cryst. Liq. Cryst.* **118**, 327 (1985).
23. J.M. Machado, F.E. Karasz, R.F. Kovar, J.M. Burnett and M.A. Druy, *New Polym. Mat.* **1**, 189 (1989).
24. J.M. Machado, M.S. Masse and F.E. Karasz, *Polymer* **30**, 1992 (1989).
25. D. McBranch, M. Sinclair, A.J. Heeger, A.O. Patil, S. Shi, S. Askari and F. Wudl, *Synth. Met.* **29**, E85 (1989).
26. J. Swiatkiewicz, P.N. Prasad, F.E. Karasz, M.A. Druy and P.J. Glatkowski, *J. Appl. Phys. Lett.* **56**, 892 (1990).
27. B.P. Singh, P.N. Prasad and F.E. Karasz, *Polymer* **29**, 1940 (1988).
28. D.L. Gin, J.K. Avlyanov and A.G. MacDiarmid, *Synth. Met.* **66**, 169 (1994).
29. H. Idriss and M.A. Barteau, *Langmuir* **10**, 3693 (1994).
30. S. Hara, S. Aeiyach and P.C. Lacaze, *J. Electroanal. Chem.* **364**, 223 (1994).
31. S. Aeiyach, P. Soubiran, P.C. Lacaze, G. Froyer and Y. Pelous, *Synth. Met.* **32**, 103 (1989).

32. L.M. Goldenberg, A.E. Palekh, V.I. Krinichnyi, O.S. Roshchupkina, A.F. Zueva, R.N. Lyubovskaya and O.N. Efimov, *Synth. Met.* **36**, 217 (1990).

33. K.L.N. Phani, S. Pitchumani, S. Ravichandra, S. Tamil Selvan and S. Bharathey, *J. Chem. Soc., Chem. Commun.* 179 (1993).

34. J.F. Fauvarque, M.A. Petit, A. Digua and G. Froyer, *Makromol. Chem.* **188**, 1833 (1987).

35. G. Froyer, G. Ollovier, C. Chevrot and A. Siove, *J. Electroanal. Chem.* **327**, 159 (1992).

36. L.M. Goldenberg and P.C. Lacaze, *Synth. Met.* **55–57**, 1365 (1993).

37. C. Kvarnström, R. Bilger, J. Heinze and A. Ivaska, *Electrochim. Acta*, submitted.

38. S. Hara, S. Aeiyach and P.C. Lacaze, *J. Electroanal. Chem.* **364**, 223 (1994).

39. J. Heinze, *Topics in Current Chemistry*, Vol. 152 Springer-Verlag, Berlin, 1990.

40. E. Yu Pisarevskaya and M.D. Levi, *Electrokhimiya* **27**, 496 (1991).

41. F. Beck and A. Pruss, *J. Electroanal. Chem.* **216**, 157 (1987).

42. A.A. Pud, *Synth. Met.* **66**, 1 (1994).

43. C. Kvarnström and A. Ivaska, *Synth. Met.* **62**, 125 (1994).

44. L.M. Goldenberg, S. Aeiyach and P.C. Lacaze, *J. Electroanal. Chem.* **335**, 151 (1992).

45. K. Meerholz and J. Heinze, *Angew. Chem.* **102**, 695 (1990).

46. M. Tabata, M. Satoh, K. Kaneto and K. Yoshino, *J. Phys. Part C* **19**, L101 (1986).

47. S. Aeiyach, P. Soubiran and P.C. Lacaze, *Synth. Met.* **32**, 103 (1989).

48. C. Hérold, D. Billaud and R. Yazami, *Solid State Ionics* **40–41**, 985 (1990).

49. T.R. Jow, L.W. Shacklette, M. Maxfield and O. Vernick, *J. Electrochem. Soc.* **134**, 1730 (1987).

50. Z. Gao, C. Kvarnström and A. Ivaska, *Electrochimica Acta* **39**, 1419 (1994).

51. P. Nagels, H. Krikor and M. Rotti, *Synth. Met.* **29**, E29 (1989).

52. L.M. Goldenberg and P.C. Lacaze, *Synth. Met.* **58**, 271 (1993).

53. K. Meerholz and J. Heinze, *Solid-State Sci.* **107**, 130 (1992).

54. J.L. Duroux, A. Moliton and G. Froyer, *Nucl. Inst. Meth. Phys. Res.* B **34**, 450 (1988).

55. J.L. Duroux, A. Moliton and G. Froyer, *Makromol. Chem., Macromol. Symp.* **24**, 162 (1989).

56. C. Le hüe, A. Moliton, B. Lucas and G. Froyer, *Adv. Mater. Opt. Electr.* **1**, 173 (1992).

57. A. Nejim and G. Carter, *Nucl. Inst. Meth. Phys. Res.* B **61**, 502 (1991).

58. A. Öpik, I. Golovtsov, A. Lobanov and K. Kerm, *Synth. Met.* **55**, 4924 (1993).

59. A. Öpik and T. Ahven, *Solid State Commun.* **73**, 661 (1990).

60. D.H. Son, C.H. Choi and K. Kim, *Vib. Spectrosc.* **4**, 349 (1993).

61. G. Zerbi and S. Sandroni, *Spectrochim. Acta* **24A**, 483 (1968).

62. G. Zerbi and S. Sandroni, *Spectrochim. Acta* **24A**, 511 (1968).

63. J. Dale, *Acta Chem. Scand.* **11**, 640 (1957).

64. C. Castiglioni, M. Gussoni, J.T.L. Navarrete and G. Zerbi, *Mikrochim. Acta* **I**, 247 (1988).

65. D. Rakovic, I. Bozovic, S.A. Stepanyan and L.A. Gribov, *Solid State Commun.* **43**, 127 (1982).

66. D. Rakovic, R. Kostic, L.A. Gribov, S.A. Stepanyan and I.E. Davidova, *Synth. Met.* **41–43**, 275 (1991).

67. C. Castiglioni, M. Gussoni and G. Zerbi, *Synth. Met.* **29**, E1 (1989).

68. M. Hanfland, A. Brillante, K. Syassen, M. Stamm and J. Fink, *Synth. Met.* **29**, E13 (1989).

69. P. Kovacic and M.B. Jones, *Chem. Rev.* **87**, 375 (1987).

70. S. Aeiyach and P.C. Lacaze, *J. Polym. Sci., Polym. Chem. Ed.* **27**, 515 (1989).

71. G. Zerbi, in *Advances in Applied Fourier Transform Infrared Spectroscopy* (ed. M.W. Mackenzie), John Wiley & Sons, Chichester, 1988.

72. J. Simitzis and C. Dimopoulou, *Makromol. Chem.* **185**, 2553 (1984).

73. J. Simitzis, D. Tzevelekis, A. Stamboulis and G. Hinrichsen, *Acta Polymer.* **44**, 294 (1993).

74. M.-C. Pham, S. Aeiyach, J. Moslih, P. Soubrian and P.-C. Lacaze, *J. Electroanal. Chem.* **277**, 327 (1990).

75. P. Soubiran, S. Aeiyach, J.J. Aaron, M. Delamar and P.C. Lacaze, *J. Electroanal. Chem.* **251**, 89 (1988).

76. C. Kvarnström, A.-S. Nybäck and A. Ivaska, *Synth. Met.* **55–57**, 503 (1993).

77. L.J. Bellamy, *The Infra-red Spectra of Complex Molecules*, John Wiley Sons, New York 1956.

78. G. Froyer, F. Maurice, J.Y. Goblot, J.F. Fauvarque, M.A. Petit and A. Digua, *Mol. Cryst. Liq. Cryst.* **118**, 267 (1985).

79. P. Yli-Lahti, H. Stubb, H. Isotalo, P. Kuivalainen and L. Kalervo, *Mol. Cryst. Liq. Cryst.* **118**, 305 (1985).

80. A. Pron, D. Billaud, I. Kulzewicz, C. Budrowski, J. Przyluski and J. Suwalski, *Mat. Res. Bull.* **16**, 1229 (1981).

81. G. Zannoni and G. Zerbi, *J. Chem. Phys.* **82**, 31 (1985).

82. L.W. Schacklette, H. Eckardt, R.R. Chance, G.G. Miller, D.M. Ivory and R.H. Baughman, *J. Chem. Phys.* **73**, 4098 (1980).

83. H. Krikor, R. Mertens, P. Nagels, R. Callaerts and G. Remaut, *Lower-Dimensional Systems and Molecular Electronics* (eds. R.M. Metzger, P. Day and G.C. Papavassiliou), Plenum Press, New York, 1991.

84. C. Kvarnström and A. Ivaska, *Synth. Met.* **62**, 133 (1994).

85. S. Krichene, J.P. Buisson, S. Lefrant, G. Froyer, F. Maurice, J.Y. Goblot, Y. Pelous and C. Fabre, *Mol. Cryst. Liq. Cryst.* **118**, 301 (1985).

86. S. Krichene, J.P. Buisson and S. Lefrant, *Synth. Met.* **17**, 589 (1987).

87. J.P. Lère-Porte, M. Radi, C. Chorro, J. Petrissans, J.L. Sauvajol, D. Gonbeau, G. Pfister-Guillouzo, G. Louarn and S. Lefrant, *Synth. Met.* **59**, 141 (1993).

88. Y. Furukawa, H. Ohtsuka and M. Tasumi, *Synth. Met.* **55**, 516 (1993).

89. S. Lefrant, J.P. Buisson and H. Eckhardt, *Synth. Met.* **37**, 91 (1990).

90. L.W. Shacklette, R.L. Esenbaumer, R.R. Chance, J.M. Sowa, D.M. Ivory, G.G. Miller and R.H. Baughman, *J. Chem. Soc., Chem. Commun.* 361 (1982).

91. M. Dietrich, J. Mortensen and J. Heinze, *J. Chem. Soc., Chem. Commun.* 1131 (1986).

92. L.W. Shacklette, N.S. Murthy and R.H. Baughman, *Mol. Cryst. Liq. Cryst.* **121**, 201 (1985).

93. L.W. Shacklette, J.E. Toth, N.S. Murthy and R.H. Baughman, *J. Electrochem. Soc.* **132**, 1529 (1985).

94. M. Satoh, K. Kaneto and K. Yoshino, *J. Chem. Soc., Chem. Commun.* 1629 (1985).

95. M. Morita, K. Komaguchi, H. Tsutsumi and Y. Matsuda, *Electrochim. Acta* **37**, 1093 (1992).

96. B.Z. Lubentsov, G.I. Zvereva, V.E. Dmitrienko and M.L. Khidekel, *Mater. Sci. Forum* **62–64**, 487 (1990).

97. M. Stamm, *Mol. Cryst. Liq. Cryst.* **105**, 259 (1984).

98. G. Shi, G. Xue, C. Li, S. Jin and B. Yu, *Macromolecules* **27**, 3678 (1994).

99. K. Ashley, D.B. Parry, J.M. Harris, S. Pons, D.N. Bennion, R. LaFollette, J. Jones and E.J. King, *Electrochim. Acta* **34**, 599 (1989).

100. M.D. Levi and E. Yu. Pisarevskaya, *Electrokhimiya* **27**, 1267 (1991).

101. K. Yoshino and H.B. Gu, *Japan. J. Appl. Phys.* **25**, 1064 (1986).

102. J.H. Ye, Y.Z. Chen and Z.W. Tian, *J. Electroanal. Chem.* **229**, 215 (1987).

103. L.M. Goldenberg, S. Aeiyach and P.C. Lacaze, *Synth. Met.* **51**, 339 (1992).

104. C. Kvarnström and A. Ivaska, *J. Electroanal. Chem.* accepted.

105. T.P. Nguyen, H. Ettaik, S. Lefrant, G. Leising and F. Stelzer, *Synth. Met.* **38**, 69 (1990).

106. G. Grem, G. Leditzky, B. Ullrich and G. Leising, *Synth. Met.* **51**, 383 (1992).

107. G. Grem and G. Leising, *Synth. Met.* **57**, 4105 (1993).

108. (a) G. Leising, G. Grem, G. Leditzky and U. Scherf, *SPIE Int. Soc. Opt. Eng.* 1910 (1993).
    (b) G. Leising, G. Grem, G. Leditzky, J. Stampfl and U. Scherf, *Frontiers of Polymers and Advanced Materials* (ed. P.N. Prasad), Plenum, New York, 1993.

109. K. Miyashita and M. Kaneko, *Synth. Met.* **68**, 161 (1995).

110. G. Leising, T. Verdon, G. Louarn and S. Lefrant, *Synth. Met.* **41**, 279 (1991).

111. J. Huber, K. Müllen, J. Salbeck, H. Schenk, U. Scherf, T. Stehlin and R. Stern, *Acta Polymer* **45**, 244 (1994).

112. G. Grem, V. Martin, F. Meghdadi, C. Paar, J. Stampfl, J. Sturm, S. Tasch and G. Leising, *Synth. Met.* **71**, 2193 (1995).

113. L.M. Goldenberg, V.I. Krinichnyi and I.B. Nazarova, *Synth. Met.* **44**, 199 (1991).

114. I.B. Nazarova, V.I. Krinichnyi and L.M. Goldenberg, *Synth. Met.* **53**, 399 (1993).

115. D.R. Gagnon, J.D. Capistran, F.E. Karasz, R.W. Lenz and S. Antoun, *Polymer* **28**, 567 (1987).

116. R.A. Wessling, *J. Polym. Sci., Polym. Symp.* **72**, 55 (1986).

117. R.W. Lenz, C.C. Han, J.D. Stenger-Smith and F.E. Karasz, *J. Polym. Sci., Polym. Chem. Ed.* **26**, 3241 (1988).

118. D. Bradley, G.P. Evans and R.H. Friend, *Synth. Met.* **17**, 651 (1987).

119. J. Obrzut, M.J. Obrzut and F.E. Karasz, *Synth. Met.* **29**, E109 (1989).

120. V.H. Tran, T.P. Nguyen, V. Massardier, J. Davenas and G. Boiteux, *Synth. Met.* **69**, 435 (1995).

121. A. Torres-Filho and R.W. Lenz, *J. Polym. Sci B* **31**, 959 (1993).

122. G. Leising, *Polym. Bull.* **11**, 401 (1984).

123. M. Hirooka, I. Murase, T. Ohnisi and T. Noguchi, *Frontiers of Macromolecular Science* (ed. T. Saegusa), Blackwell Scientific Publications, Oxford, 1989, p. 425.

124. A.O. Patil, S.D.D.V. Rughooputh and F. Wudl, *Synth. Met.* **29**, E115 (1989).

125. M. Machado, F.R. Denton, J.B. Schlenoff, F.E. Karasz and P.M. Lahti, *J. Polym. Sci., Polym. Phys. Ed.* **27**, 199 (1989).

126. J.D. Stenger-Smith, R.W. Lenz and G. Wegner, *Polymer* **30**, 1048 (1989).

127. H.H. Hörhold and M. Helbig, *Makromol. Chem. Makromol. Symp.* **12**, 229 (1987).

128. J. Obrzut and F.E. Karasz, *J. Chem. Phys.* **87**, 6178 (1987).

129. M. Onoda, H. Nakayama, T. Tanaka, K. Amakawa and K. Yoshino, *Kenkyu Hokoku-Himeji Kogyo Daigaku Kogakubu* **43**, 23 (1990).

130. H. Nishihara, M. Akasaka, M. Tateishi and K. Aramaki, *Chem. Lett.* 2061 (1992).

131. J. Heinze, M. Dietrich and J. Mortensen, *Makromol. Chem., Macromol. Symp.* **8**, 73 (1987).

132. D.R. Gagnon, F.E. Karasz, E.L. Thomas and R.W. Lenz, *Synth. Met.* **20**, 85 (1987).

133. J.M. Madsen, B.R. Johnson, X.L. Hua, R.B. Hallock, M.A. Masse and F.E. Karasz, *Phys. Rev. B* **40**, 11751 (1989).

134. R. Mertens, P. Nagels, R. Callaerts, J. Briers and H.J. Geise, *Synth. Met.* **55–57**, 3538 (1993).

135. P. Sheng, *Phys. Rev. B* **21**, 2180 (1980).

136. B. Lucas, B. Ratier, A. Moliton, C. Moreau and R.H. Friend, *Synth. Met.* **55–57**, 4912 (1993).

137. N.F. Mott and E.A. Davis, in *Electronic Processes in Non-crystalline Materials*, Clarendon, Oxford, 1979.

138. D. Emin and K.L. Ngai, *J. Phys.* **44**, C3–471 (1983).

139. J. Obrzut, M.J. Obrzut and F.E. Karasz, *Synth. Met.* **29**, E103 (1989).

140. H.H. Hörhold and J. Opfermann, *Makromol. Chem.* **131**, 105 (1970).

141. S. Tokito, T. Tsutsui, R. Tanaka and S. Saito, *Japan J. Appl. Phys.* **25**, L680 (1986).

142. T. Takiguchi, D.H. Park, H. Ueno, K. Yoshino and R. Sugimoto, *Synth. Met.* **17**, 657 (1987).

143. A.V. Vannikov, A. Yu. Kryukov and H.H. Hörhold, *Synth. Met.* **41–43**, 331 (1991).

144. S. Rentsch, J.P. Tang, H.L. Li, M. Lenzer and H. Bergner, *Synth. Met.* **41–43**, 1369 (1991).

145. H. Bleier, Y.Q. Shen, D.D.C. Bradley, H. Lindenberger and S. Roth, *Synth. Met.* **29**, E73 (1989).

146. D.D.C. Bradley, R.H. Friend, H. Lindenberger and S. Roth, *Polymer* **27**, 1709 (1986).

147. S. Lefrant, E. Perrin, J.P. Buisson, H. Eckhardt and C.C. Han, *Synth. Met.* **29**, E91 (1989).

148. Y. Furukawa, A. Sakamoto and M. Tasumi, *J. Phys. Chem.* **93**, 5354 (1989).

149. S. Lefrant, J.P. Buisson and H. Eckhardt, *Synth. Met.* **37**, 91 (1990).

150. S.C. Graham, D.D.C. Bradley and R.H. Friend, *Synth. Met.* **41–43** 1277 (1991).

151. J.P. Buisson, J.Y. Mevellec, S. Zeraoui and S. Lefrant, *Synth. Met.* **41–43**, 287 (1991).

152. Y. Furukawa, H. Ohta, A. Sakamoto and M. Tasumi, *Spectrochim. Acta* **47A**, 1367 (1991).

153. B. Tian and G. Zerbi, *Synth. Met.* **41–43**, 255 (1991).

154. J.P. Buisson, S. Lefrant, G. Louarn, J.Y. Mévellec, I. Orion and H. Eckhardt, *Synth. Met.* **49–50**, 305 (1992).

155. Y. Furukawa, A. Sakamoto, H. Ohta and M. Tasumi, *Synth. Met.* **49–50**, 335 (1992).

156. A. Sakamoto, Y. Furukawa and M. Tasumi, *J. Phys. Chem.* **96**, 1490 (1992).

157. A.A Sakamoto, Y. Furukawa and M. Tasumi, *Synth. Met.* **55–57**, 593 (1993).

158. D. Rakovic, R. Kostic, I.E. Davidova and L.A. Gribov, *Synth. Met.* **55–57**, 541 (1993).

159. C. Moreau, R.H. Friend, G.J. Sarnecki, B. Lucas, A. Moliton, B. Ratier and C. Belorgeot, *Synth. Met.* **55–57**, 224 (1993).

160. G.J. Lee, S.K. Yu, D. Kim, J.-I. Lee and H.-K. Shim, *Synth. Met.* **69**, 431 (1995).

161. D. Rakovic, R. Kostic, L.A. Gribov and I.E. Davidova, *Phys. Rev. B* **41**, 10744 (1990).

162. G. Zerbi, M. Gussoni and C. Castiglioni, in *Conjugated Polymers* (eds. J.L. Brédas and R. Silbey), Kluwer Academic Publishers, Dordrecht, 1991, p. 487.

163. C. Castiglioni, M. Gussoni, J.T. Lopez-Navarrete and G. Zerbi, *Solid State Commun.* **65**, D359 (1989).

164. G. Zerbi, C. Castiglioni, J.T. Lopez-Navarrete, B. Tian and M. Sussoni, *Synth. Met.* **28**, D359 (1989).

165. A. Sakamoto, Y. Furikawa and M. Tasumi, *J. Phys. Chem.* **96**, 3870 (1992).

166. Y. Furukawa, *Solid-State Sci.* **107**, 137 (1992).

167. M.J. Winokur, D. Chen and F.E. Karasz, *Synth. Met.* **41**, 341 (1991).

168. J. Obrzut and F.E. Karasz, *J. Chem. Phys.* **87**, 2349 (1987).

169. J. Lipinski, A. Nowek and H. Chojnacki, *Acta Phys. Polon.* A **53**, 229 (1978).

170. G. Nicolas and Ph. Durand, *J. Chem. Phys.* **72**, 453 (1980).

171. R. Schenk, H. Gregorius and K. Müllen, *Adv. Mater.* **3**, 492 (1991).

172. M. Deussen and H. Bässler, *Chem. Phys.* **164**, 247 (1992).

173. M. Deussen and H. Bässler, *Synth. Met.* **54**, 49 (1993).

174. J. Grobelny, J. Obrzut and F.E. Karasz, *Synth. Met.* **29**, E97 (1989).

175. A. Pron, F. Genoud, M. Nechtschein and A. Rousseau, *Synth. Met.* **31**, 147 (1989).

176. J.H. Simpson, N. Egger, M.A. Masse, D.M. Rice and F.E. Karasz, *J. Polym. Sci., Polym. Phys. Ed.* **28**, 1859 (1990).

177. J.H. Simpson, D.M. Rice and F.E. Karasz, *Macromolecules* **25**, 2099 (1992).

178. J.H. Simpson, W. Liang, D.M. Rice and F.E. Karasz, *Macromolecules* **25**, 3068 (1992).

179. J.H. Simpson, D.M. Rice and F.E. Karasz, *J. Polym. Sci., Polym. Phys. Ed.* **30**, 11 (1992).

180. J.H. Simpson, D.M. Rice, F.E. Karasz, F.C. Rossitto and P. Lahti, *Polymer* **34**, 4595 (1993).

181. T. Granier, E.L. Thomas, D.R. Gagnon, F.E. Karasz and R.W. Lenz, *J. Polym. Sci., Polym. Phys. Ed.* **24**, 2793 (1986).

182. T. Granier, E.L. Thomas, D.R. Gagnon and F.E. Karasz, *J. Polym. Sci., Polym. Phys. Ed.* **27**, 469 (1989).

183. D. Chen, M.J. Winokur, M.A. Masse and F.E. Karasz, *Phys. Rev. B* **41**, 6759 (1990).

184. J.H. Simpson, D.M. Rice and F.E. Karasz, *Polymer* **32**, 2340 (1992).

185. J.H. Burroughes, D.D.C. Bradley, A.R. Brown, R.N. Marks, K. Mackay, R.H. Friend, P.L. Burn and A.B. Holmes, *Nature (London)* **347**, 539 (1990).

186. D. Braun and A.J. Heeger, *Appl. Phys. Lett.* **58**, 1982 (1991).

187. G. Grem, G. Leditzky, B. Ullrich and G. Leising, *Adv. Mater.* **4**, 36 (1992).

188. V.H. Tran, V. Massardier, A. Guyot and T.P. Nguyen, *Polymer* **34**, 3179 (1993).

189. T.P. Nguyen, V.H. Tran and V. Massardier, *J. Phys. Condensed Mater.* **5**, 1 (1993).

190. P. Dannetun, M. Lögdlund, M. Fahlman, M. Boman, S. Stafström, W.R. Salaneck, R. Lazzaroni, C. Fredriksson, J.L. Brédas, S. Graham, R.H. Friend, A.B. Holmes, R. Zamboni and C. Taliani, *Synth. Met.* **55–57**, 212 (1993).

191. P. Dannetun, M. Lögdlund, W.R. Salaneck, C. Fredriksson, S. Stafström, A.B. Holmes, A. Brown, S. Graham, R.H. Friend and O. Lhost, *Mol. Cryst. Liq. Cryst.* **228**, 43 (1993).

192. R.N. Marks, D.D.C. Bradley, R.W. Jackson, P.L. Burn and A.B. Holmes, *Synth. Met.* **55–57**, 4128 (1993).

193. R.A. Wessling and R.G. Zimmerman, US Patent 3401 152 (1968).

194. C. Fredriksson, R. Lazzaroni, J.L. Brédas, P. Dannetun, M. Lögdlund and W.R. Salaneck, *Synth. Met.* **55–57**, 4632 (1993).

195. T.P. Nguyen and V.H. Tran, *Mat. Sci. Eng.* B **31**, 255 (1995).

196. T.P. Nguyen, V.H. Tran and S. Lefrant, *Synth. Met.* **69**, 443 (1995).

197. T.P. Nguyen, K. Amgaad, M. Cailler and V.H. Tran, *J. Adhes. Sci. Technol.* **8**, 821 (1994).

198. T.P. Nguyen, P. Le Rendu, K. Amgaad, M. Cailler and V.H. Tran, *Synth. Met.* **72**, 35 (1995).

199. T. Venkatesan, *Nucl. Instrum. Methods*, Sect. B **1**, 599 (1984).

200. S. Karg, W. Riess, V. Dyakonov and M. Schwoerer, *Synth. Met.* **54**, 427 (1993).

201. S. Karg, W. Reiss, M. Meier and M. Schwoerer, *Synth. Met.* **55–57**, 4186 (1993).

202. F. Cacialli, R.H. Friend, S.C. Moratti and A.B. Holmes, *Synth. Met.* **67**, 157 (1994).

203. A.R. Brown, J.H. Burroughes, N. Greenham, R.R. Friend, D.D.C. Bradley, P.L. Burn, A. Kraft and A.B. Holmes, *Appl. Phys. Lett.* **61**, 2793 (1992).

204. A.R. Brown, K. Pichler, N.C. Greenham, D.D.C. Bradley, R.H. Friend, P.L. Burn and A.B. Holmes, *Synth. Met.* **55–57**, 4117 (1993).

205. M. Fahlman, D. Beljonne, M. Lögdlund, A.B. Holmes, R.H. Friend, J.L. Brédas and W.R. Salaneck, *Chem. Phys. Lett.* **214**, 327 (1993).

206. H.S. Woo, J.G. Lee, H.K. Min, E.J. Oh, S.J. Park, K.W. Lee, J.H. Lee, S.H. Cho, T.W. Kim and C.H. Park, *Synth. Met.* **71**, 2173 (1995).

207. A. Wu, M. Jikei, M.-A. Kakimoto, Y. Imai, S. Ukishima and Y. Takahashi, *Chem. Lett.* 2913 (1994).

208. Y. Nishikata, S. Fukui, M. Kakimoto and Y. Imai, *Thin Solid Film* **210**, 296 (1992).

209. P.L. Burn, A.B. Holmes, A. Kraft, D.D.C. Bradley, A.R. Brown and R.H. Friend, *J. Chem. Soc., Chem. Commun.* 32 (1992).

210. D.D.C. Bradley, P.L. Burn, R.H. Friend, A.B. Holmes and A. Kraft, in *Electronic Properties of Conjugated Polymers* (ed. H. Kuzman), Springer Series on Solid State Sciences, Springer-Verlag, New York, 1991.

211. P.L. Burn, A.B. Holmes, A. Kraft, D.D.C. Bradley, A.R. Brown, R.H. Friend and R.W. Gymer, *Nature* **365**, 47 (1992).

212. C. Adachi, T. Tsutsui and S. Sato, *Appl. Phys. Lett.* **56**, 799 (1990).

213. Z. Yang, I. Sokolik and F.E. Karasz, *Macromolecules* **26**, 1188 (1993).

214. T. Zyung, J.-J. Kim, W.-Y. Hwang, D.H. Hwang and H.K. Shim, *Synth. Met.* **71**, 2167 (1995).

215. L.S. Swanson, J. Shinar, A.R. Brown, D.D.C. Bradley, R.H. Friend, P.L. Burn, A. Kraft and A.B. Holmes, *Synth. Met.* **55–57**, 241 (1993).

216. E.G.J. Staring, R.C.J.E. Demandt, D. Braun, G.L.J. Rikken, Y.A.R.R. Kessener, A.H.J. Venhuizen, M.M.F. van Knippenberg and M. Bouwmans, *Synth. Met.* **71**, 2179 (1995).

217. R.W. Gymer, R.H. Friend, H. Ahmed, P.L. Burn, A.M. Kraft and A.B. Holmes, *Synth. Met.* **55–57**, 3683 (1993).

218. F.W. Embs, E.L. Thomas, C.J. Wung and P.N. Prasad, *Polymer* **34**, 4607 (1993).

219. P.N. Prasad, F.E. Karasz, Y. Pang and C.J. Wung, U.S. Patent 5,130,362 (1992).

220. C.J. Wung, W.M.K.P. Wijekoon and P.N. Prasad, *Polymer* **34**, 1174 (1993).

221. H. Tabei, O. Niwa, T. Horiuchi and M. Morita, *Denki Kagaku Oyobi Kogyo Butsui Kagaku* **61**, 820 (1993).

CHAPTER 10

Artificial Muscles, Electrodissolution and Redox Processes in Conducting Polymers

Toribio Fernandez Otero

University of the Basque Country, San Sebastián, Spain

Handbook of Organic Conductive Molecules and Polymers: Vol. 4. Conductive Polymers: Transport, Photophysics and Applications.
Edited by H. S. Nalwa. © 1997 John Wiley & Sons Ltd

1   INTRODUCTION

Human development has been founded on the control of different energetic sources and the transformation between different types of energy. A long road has been travelled from the control and use of fire until the control and use of atomic energy. Human efforts are joined in the search for devices able to produce transformations between two or more kinds of energy. Nowadays macroscopic machines are being reduced to microscopic dimensions, and our intellectual effort is focused on natural machines developed by evolution in living organisms. Those natural machines are based on energetic transformations at the molecular level. Following this idea natural and artificial molecules that are able to produce transformations between mechanical, chemical, electrical, electromagnetic, thermal or pressure energies are being investigated.

The aim of this chapter is to review those devices able to transform electric energy into mechanical energy through electrochemical reactions occurring at the molecular level in solid conducting polymers. Those electrochemomechanical devices are named 'artificial muscles'. In order to check if this definition is only a literary metaphor, so usual in science nowadays, or if it corresponds closely to the performance of natural muscles, an overview will be attempted. This will look at other 'artificial muscles' in the literature based on synthetic polymer gels and

proteins, the morphology of natural muscles, the molecular structure of skeletal muscles, the triggering of muscle contraction and the conformational changes model (sliding filament), which is able to explain the transformation of chemical energy to mechanical work.

The characteristics of natural muscles will be studied in conducting polymers: ionic fluxes during work, conformational changes, changes of volume and electrochemomechanical properties. We will present how those microscopic properties, based on chemical transformations at the molecular level, were translated in our laboratory to macroscopic angular movements (more than 360 degrees in a few seconds and able to trail a hanging mass of 1000 times the polymer weight) through a bilayer or multilayer systems. Macroscopic and microscopic models were developed in order to explain electrochemomechanical properties, the influence of chemical and electrical parameters on this property, lifetimes, degradation processes and perfect control of the movement. New elegant devices developed by MacDiarmid, Inganäs and coworkers are presented. Similarities and differences between artificial and natural muscles are stated, together with the emerging applications in microrobotics, micromachinery, medical instrumentation at the end of probes, actuators and sensors.

Most of those devices were constructed from electrogenerated films removed from the electrode. Only soluble anilines or oligothiophenes can be used to prepare films from solution. A new method, electro-dissolution, has been developed in our laboratory, in collaboration with a group from the University of Barcelona. Poly-2,5-di-(2-thienyl)-pyrrole can be electrochemically polymerized following a faradaic process and electrochemically solved following a different faradaic process. Electrodissolution is not the reverse of electropolymerization, but, both processes mimic electrodeposition and electroerosion of metals, opening the way for technological applications.

Artificial muscles, electrodissolution processes, electrochromic changes, smart porosity in membranes, charge storage, etc. are properties based on redox processes in conducting polymers. These redox processes have been envisaged until now as pure electrochemical reactions, and have been treated mathematically as processes occurring on a metal–electrolyte interface. It is generally accepted in the field of electrochemistry that electrochemical responses such as chronoamperograms or chronopotentiograms do not contain any structural information on the electrode. This belief is due to the lack of scientific surface models in electrochemistry and all the other sciences based on surfaces. Our systems, nevertheless are based on electrochemical reactions in solid state or, more precisely, on a three-dimensional electrode: the conducting polymer layer. The flux of a current requires, or induces a flow of ions and solvent across the polymer–solution interface, conformational changes on the polymeric chains, and generation (or destruction) of free volume. Under specific conditions those conformational changes can be the controlling step in the electrochemical processes. My group has developed a conformational relaxation electrochemical model, taking as a reference the thermocurrents of depolarization relaxation theory, and has been able to simulate and predict the electrochemical responses and experimental behavior of artificial muscles. The aim is to devise a base from which to develop a model able to integrate electrochemistry and polymer science, which will explain and predict any behavior or device based on the redox properties of conducting polymers.

2   MUSCLES

2.1   Muscles: basic points

Muscles are elegant devices that have developed over millions of years of evolution to obtain mechanical energy and heat from chemical energy, at a constant temperature. Mechanical work is related to a change of volume in muscular fibers. In good agreement with thermodynamic laws a fraction of the chemical energy is transformed to heat. This heat allows the maintenance of a constant temperature in different animal species. The excess of heat has to be eliminated in order to avoid any lethal increase on temperature.

These natural devices are under the control of electric pulses arriving from the brain through the nervous system. Muscle work is triggered by an electric pulse. Their mechanical tension is kept under the control of electric pulses arriving from the brain. They are relaxed by nervous pulses as well. Once relaxed they recover the initial position only by the work of a complementary muscle. An initial approach to these natural actuators evidences the interaction between electric currents, chemical reactions, mechanical work and heat production (Figure 10.1). They can be considered as an electrochemomechanical system working under constant temperature and able to relax to the initial state.

The evidence of a similar nature between nervous pulses and electric currents was shown by Galvani's experiment two-hundred years ago: the nervous pulse

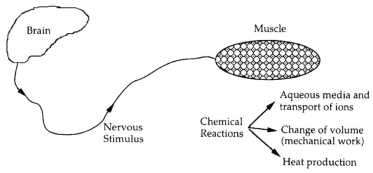

Figure 10.1. Muscle: a nervous stimulus (ionic pulse) arriving from the brain through the nervous system triggers ions and water transport, chemical reactions and changes of volume in a muscle with the production of mechanical energy and heat.

can be simulated by an electric current. An electric pulse sent through the nerves of a dissectioned frog's claw was able to promote the contraction of the claw. This experiment does not work at all if the dissected frog is dehydrated: nervous pulses and chemical reactions in muscular fibers are related to ionic fluxes in aqueous media and through cellular membranes. So a second approach to natural muscles allows a closer definition of those actuators as electrochemomechanical devices, working in aqueous media under constant temperature; the work involves ionic interchanges.

Researchers have been trying to reproduce such a device for a long time. An initial approach to mimic natural muscles guides our attention to chemical reactions linked to changes of volume and taking place over a short time. A large number of chemical reactions are based on the fast change of volume occurring with the reaction. Internal combustion engines (ICE) are based on those reactions. Nevertheless these engines can not be considered similar to artificial muscles. The main reason for that is the important change of temperature observed during the work. The conversion from chemical energy to mechanical work is defined by the Carnot's cycle applied to devices working in a gradient of temperature, but the search is for a device that works at a constant temperature. This is the main difference between a muscle and an ICE.

If we accept that a muscle involves electric pulses linked to chemical reactions, which promote a change of volume at constant temperature, these preliminary considerations point to electrochemistry as the most promising field in which to look for reactions able to mimic muscles.

2.2 Electrochemical systems

An electrochemical system is formed by electronic equipment and metal wires, in contact with ionic conductors (Figure 10.2). The electronic equipment allows the production of electric pulses which are sent,

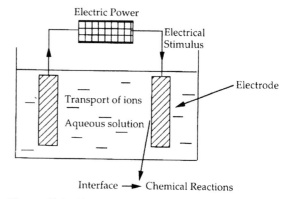

Figure 10.2. Electrochemical system: an electric flow is generated by the electric power equipment through wires and electrodes to give electrochemical reactions, movements of ions and water molecules at the metal–solution interface and heat, by the Joule effect in the solution.

through wires, to the metal electrode–solution interfaces. This part mimics brain and nerves (Figure 10.3).

At the electrode surface the electron flow is interrupted due to the presence of an ionic conductor on the other side of the interface: ions, from a salt, in a solvent. The energy level of electrons in the metal can be controlled from the electronic equipment in such a way that the electrons can be transferred from the electrode to the chemical species (molecules, ions) in the solution (electrochemical reduction) or from the chemical species present in the solution to the electrode (electrochemical oxidation). New reactive ions or radical ions are formed. A concomitant movement of counterions occurs in the solution in order to keep the electroneutrality The reactivity of the new species initiates chemical transformations in the solution.

A process of electrolysis requires at least two electrodes (anode and cathode) (Figure 10.2). The electrical energy is applied to the cell where it is transformed to chemical energy at a constant temperature. Taking into account that the internal resistance of

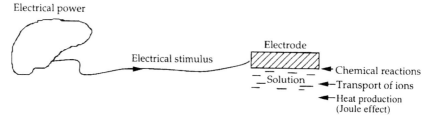

Figure 10.3. Similarities between an electrochemical system and a nervous system (Figure 10.1). The only difference is a change of volume related to the electrochemical reactions.

the system is not null, a fraction of the electrical energy is transformed to heat, which has to be eliminated in order to maintain a constant temperature. So, an initial approach to an electrochemical system allows the detection of connections between electric currents, chemical reactions and ionic fluxes at a constant temperature.

In order to mimic muscles, an electrochemical system is needed that gives a change of volume during the electrochemical reaction able to produce mechanical work. A second important aspect is the relaxation of the system to the initial position allowing repetitive mechanical work.

2.2.1 Change of volume during electrochemical reactions

Historically there are two kinds of electrochemical reactions able to promote changes of volume: the *electrogeneration of gases* from liquid solutions, and the *electrochemical oxidation* of metals to metal oxides (or the oxidation of solid metallic compounds to more oxidized states).

Not much effort has been spent on the electrochemical production of gases able to move pistons during expansion. The reason could be that these systems do not seem to be competitive with ICE because gases are, as well, produced from the chemicals. At the moment it is more efficient to consume electricity to produce pressure through pumps, rather than by electrochemical reactions (Figure 10.4a).

Related to metals, oxides or other solid metallic compounds are rigid and fragile. It is possible to construct actuators on the basis of a low change of volume but, at the moment, they do not seem adequate as materials from which to construct artificial muscles.

2.2.2 Relaxation to the initial state

Different electrochemical processes, as well as the properties related to them, can be reversed by changing the direction of the current flow through the electrochemical cell. Consequently some metal oxides can move toward more oxidized states by means of an anodic current (on the anode). During oxidation hydrated ions have to penetrate from the solution into the oxide in order to maintain the electroneutrality: an increase of volume occurs in parallel to this process. During reduction, the flow of a cathodic current, produces a reverse change of volume and ionic

movements occur. The initial volume and the initial oxidation state are recovered.

Related to other electrochemical systems the initial states can only be recovered by means of a reverse device. During electrolysis of water, gaseous hydrogen and oxygen are formed at constant temperatures giving a concomitant increase of volume (Figure 10.4A). The relaxation of the system to the original position requires a different device able to transform chemical energy to electrical energy: a fuel cell (Figure 10.4B). When gaseous hydrogen and oxygen are supplied, respectively, to each of the electrodes of a fuel cell the gases are transformed to water and electrical energy. The initial state is recovered with some losses due to the internal resistance of the cells.

Unfortunately, slow kinetics, catalytic problems and lack of knowledge or interest have blocked the development of any of the above reversible ways to direct the interchange of electrical energy and mechanical energy.

2.3 Artificial muscles in literature

In spite of a good similarity between the above electrochemical processes and muscles, an important difference can be observed: they have different chemical natures. The main component of muscles are macromolecules (proteins), whereas the most common compounds involved in electrochemical reactions are metal, metal oxides and gases. Proteins are molecular engines designed by chemical evolution for the conversion of different energy sources into useful mechanical motion.

In literature, systems other than electrochemical have been treated to transform different kinds of molecular energy to mechanical work. Most of the works are focused on systems whose main components are macromolecules.

Over the last decades, different organic systems, able to transform molecular energy to mechanical work, have been investigated. As the main components of muscles are proteins, efforts are being devoted to the study of the transduction of temperature, pressure, chemical potential, electrochemical potential and electromagnetic radiation to mechanical forces using proteins or proteins-based polymers as molecular machines [1].

Other polymers like polyelectrolyte gels, fibers and membranes suffer expansion and contraction in different electrolytes or solvents [2]. Ion exchange membranes change their dimensions upon exchanging a

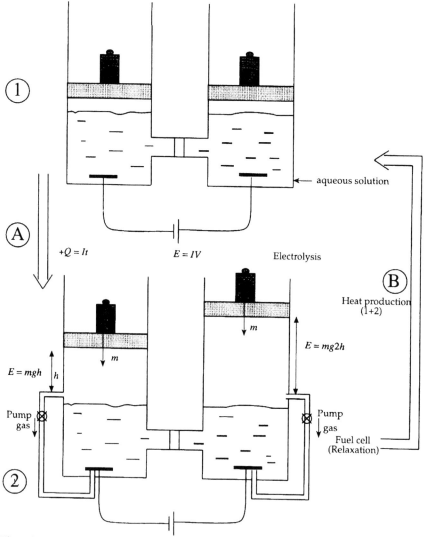

Figure 10.4. Electrolysis of water (A) with the production of mechanical work (*mgh* by the anode compartment, *mg2h* by the cathode) (B) Relaxation to the initial state generates a current in the fuel cell.

monovalent counterion for a divalent one [3]. They can also be used as materials able to give changes of volume like partially crystalline polymers which undergo melting of the crystallites with concurrent contractions [4]. Some fibrilar proteins, such as collagen or keratin, perform chemomechanical contractions upon interaction with strong salt solutions [5]. Polymeric membranes can change their volume by cooperative association with 'complementary' polymers [6].

2.3.1   *Polymer gels*

Most polymers that are able to transduce molecular energies to mechanical energies are gels. A gel is a three-dimensional network of a crosslinked polymer containing an important volume of solvent. Polymer gels can react to diverse stimuli, such as chemical, thermal, electrical or photonic excitation, by deformation, change of volume, hardening/softening or change in permeability. In this way they can be envisaged as potential chemomechanical, thermomechanical, photomechanical or electrochemomechanical transducers.

2.3.2 'Chemomechanical' actuators based on polymeric gels

The basic idea of using polymer gels as a muscle-like actuator originates from Kuhn et al. [2]. Their work involved the use of collagen fibers whose dimensions changed reversibly on transition from cyclic helices to random coils when they were immersed cyclically in salt solution and water. This change of conformation promotes a change of volume which can be applied to produce a mechanical work. Different materials like poly(methacrylic acid) and poly(ethylene glycol) [6], poly(vinyl alcohol) [7–9], poly(acrylonitrile) [8], poly-(vinyl methyl ether) [9], poly(acrylamide) [10], poly(N-isopropyl-acrylamide) [11] and others [12,13] can behave in a similar way.

All those polymers work by swelling and shrinking processes. Mechanical energy is generated from different changes occurring at the molecular level (structural, conformational, polymer–polymer interactions, or crosslinked structure) in the polymer gel. The physicochemical properties are associated with interactions between the macromolecular network and the liquid. Rigorously writing they are not chemical transformations with a reversible formation and destruction of bonds in chemical compounds. During swelling and shrinking they use conformational energies, van der Waals interactions or coulombic interactions to produce mechanical work. From this point of view, only those gels with reversible changes in the degree of crosslinking can be considered as chemomechanical transducers.

The swelling behavior of charged gels was described by Kuhn et al. [2], Flory and Rehner [14] and Flory [15] as resulting from a balance between the elastic energy of the network and the osmotic pressure of ions. In the presence of a salt, the osmotic pressure is associated with the establishment of Donnan equilibrium. Kuhn et al. also considered the effect of electrostatic interactions on the fixed charges on the polymer chains.

2.3.3 Thermomechanical systems

It is possible to convert thermal into mechanical energy using polymers that change volume when exposed to varying temperature. Thermosensitive gels of poly-(vinylmethyl ether) (PVME) are the most representative examples. An aqueous solution of this polymer has a lower critical solution temperature (LCST) of 37°C. Therefore, PVME is soluble in water below this temperature, but insoluble above it. So, when a gel of PVME is put into water, it swells below 37°C and shrinks above this temperature. The process is reversible; it gives reversible changes of volume and transforms thermal energy into mechanical work [16]. Similar effects are shown by poly(N-alkylacrylamide) [17].

2.3.4 Photomechanical systems

Those systems transform photons (radiation energy) into mechanical energy by photoestimulation of conformational levels in gels, giving reversible macroscopic changes of volume. The idea to use polymers for direct conversion of photons into mechanical work was proposed by Lovrien in 1967 [18]. On the basis of Lovrien's idea, Van der Veen and Prins reported the first photomechanical transducer, consisting of water-swollen gels of poly(2-hydroxyethyl methacrylate) [19]. The photoestimulated contraction of the gel was 1.2% of the initial volume. Since then, many materials have been reported to be suitable for the conversion of photon energy into mechanical motion [20]. Thus, polyacrylamide gels with triphenylmethane, show several cycles of dilatation and contraction by photoirradiation, with large deformation [21].

2.3.5 'Electrochemomechanical' systems

Different kind of systems are discussed in the literature under this denomination. These systems, usually a gel of polyelectrolytes, can bend backwards and forwards by application of an alternating electrical field. Here, water and ions migrate towards the electrode, bearing a charge opposite in sign to the net charge in the gel. This coupling of electroosmosis and electrophoresis is thought to be responsible for the observed chemomechanical behavior. They are models of electrically activated artificial muscles, working in an aerobic and aqueous system, because the gel contracts and expands reversibly by means of an electric stimulus under isothermal conditions. The electrical control makes use of crosslinked polyelectrolyte gels. On the other hand, the system is quite simple; for instance, a water-swollen polymer gel is inserted between a pair of electrodes connected to a dc source; the rate of volume change is proportional to the amount of current passed.

Other molecular bases were also used as electrochemomechanical devices. Tanaka *et al.* [22] showed that a partially hydrolyzed polyacrylamide gel undergoes phase transition with the application of an electric field, and collapses if the gel is placed in some solvents, e.g. a 50% acetone–water mixture. Yannes and Grodzinsky [23] showed that collagen fibers immersed in an electrolyte bath can deform in the presence of an external electric field, and thereby perform mechanical work. If the fiber is held at a constant length, the electric field generates a stress on the fibers. De Rossi *et al.* [24], who used poly(vinylalcohol)–poly(acrylic acid) composite membranes, found that a change in the shape of the membrane occurred when dc current was applied. Different materials, devices and treatments can be found in the literature [25–34].

As a final conclusion, all the named 'electrochemomechanical devices use electric currents, allowing electroosmosis, electrophoresis, phase transitions or physical deformations, at constant temperature with the flow of ions. The chemical reactions are not produced inside the device, but at the electrode–electrolyte interface.

2.3.6 *Remarks*

Devices named chemomechanical or electrochemomechanical actuators are based on the transformation of a molecular energy (generally a conformational energy) into mechanical energy. In most of these devices there is no chemical transformation with the formation of new covalent or ionic bonds.

With respect to the practical applications, in spite of the intense work performed from the fifties, researchers face difficulties to improve the properties and response rates. At present, the responses obtained from those systems are rather slow, generally in the order of minutes, and only small movements can be observed. Lack of mechanical toughness and long-term durability are other problems to be solved. These facts minimize the possibilities of applications of such systems as actuators in robotic systems. Essential improvements in the efficiency of energy conversion should also be made for the practical applications of the named chemomechanical and electrochemomechanical systems based on polymer gels.

All these aspects are important in the search for new polymers able to link reversible chemical reactions to changes of volume and electric currents, to give electrochemomechanical actuators.

2.4 Muscles

Movement is an intrinsic property associated with all living creatures. It occurs from different structural levels, including vectorial processes like ions transfer through membranes, separation of replicated chromosomes, beating of cilia and flagella or, the most common, contraction of muscles. Muscle contractions enable the organism to carry out organized and sophisticated movements, such as walking, running, flying or swimming.

Muscles were defined above as elegant devices able to transform chemical energy into mechanical energy. The direct source of energy for muscles movement is adenosine triphosphate (ATP) probably through the reaction

$$ATP^{4-} + H_2O \rightarrow ADP^{3+} + P_i^{2-} + 4H^+$$

$$\Delta G^0 = -7.3 \text{ kcal mol}^{-1}$$

where ADP is adenosine diphosphate. ATP is recovered by taking the energy from glucose (Figure 10.5).

They are three main eukaryotic motility systems that are driven by ATP [35–39]. In higher eukaryotes *muscle contraction* is mediated by the sliding of interdigitized *myosin* and *actin filaments*. Interactions between myosin and actin are responsible for active movements. The *beating of cilia* depends on the interplay of a different pair of proteins *dynein* and *tubulin. Movement of vesicles* and microtubules are mediated as well, by *kinesin.*

The binding of ATP to myosin, dynein and kinesin induces *conformational transitions* in these motor proteins. The binding of ATP and its subsequent hydrolysis control conformational changes that result in the sliding of one molecule relative to another.

2.4.1 *The morphology of muscle*

Four different types of muscles (Figure 10.6) are found in animals (eukaryotic): *skeletal muscles, cardiac* (heart) *muscle, smooth muscle* and *myoepithelial cells.* The cells of skeletal muscles are long and multinucleated and are referred as *muscle fibers.* At the microscopic level, skeletal and cardiac muscle display alternating light and dark bands, and for this reason are often referred to as *striated muscles.*

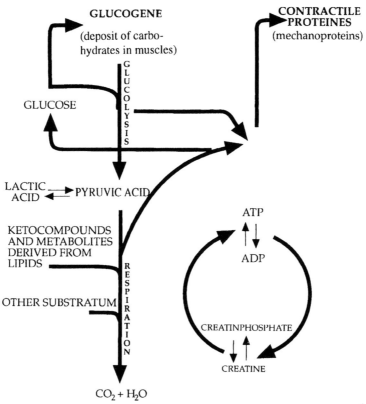

Figure 10.5. Schema for the transformation of chemical energy from glucose ($\Delta G° = -686$ kcal mol^{-1}) to mechanical energy through ATP restoration from ADP, at constant temperature.

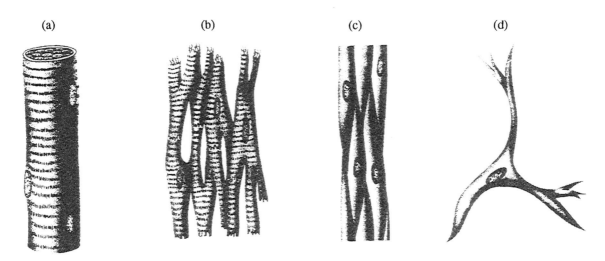

Figure 10.6. The four classes of muscle in mammals. Skeletal muscle and cardiac muscle are striated. Cardiac muscle, smooth muscle, and myoepithelial cells are mononucleated, whereas skeletal muscle is multinucleated.

2.4.2 Skeletal muscles: the molecular structure

Skeletal muscles in higher animals consist of 100 μm diameter fiber bundles, some as long as the muscle itself. Each of these muscle fibers contain hundreds of myofibrils. Examination of myofibrils at the microscopic level reveals a banded or striated structure (Figure 10.7a). The bands are traditionally identified by letters (Figure 10.7b). Regions of high density, denoted *A bands* (A comes from *anisotrope* under polarized light), alternate with regions of low electron density, the *I bands* (I, from *isotrope*). Small, dark *Z lines* lie in the middle of the I band, marking the end of the sarcomere. The central region of the A band,

termed the *H zone*, is less dense than the rest of the band. The H zone contains a central *M line*.

Two kinds of interacting *protein filaments* are revealed from electron micrographs of the cross-section. The *thick filaments* (Figure 10.8) have diameters of about 15 nm, whereas the *thin filaments* have diameters of about 9 nm. The thick filaments are formed by myosin (Figure 10.9). The thin filaments contain actin, tropomyosin and the troponim complex.

Each thin filament has three neighboring thick filaments and each thick filament is encircled by six thin filaments. Both, thick and thin filaments interact by cross-bridges, which are domains of myosin molecules. The muscle contraction is accomplished by the sliding of the cross-bridges along the thin filaments, a

(a)

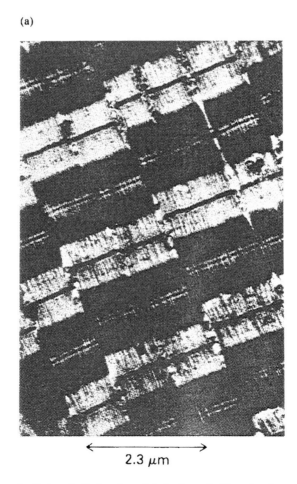

2.3 μm

Figure10.7(a) Electron micrograph of a longitudinal section of skeletal muscle fiber, showing several myofibrils extending from upper to lower right. (Reproduced by permission of Editorial Reverté from ref. 37.)

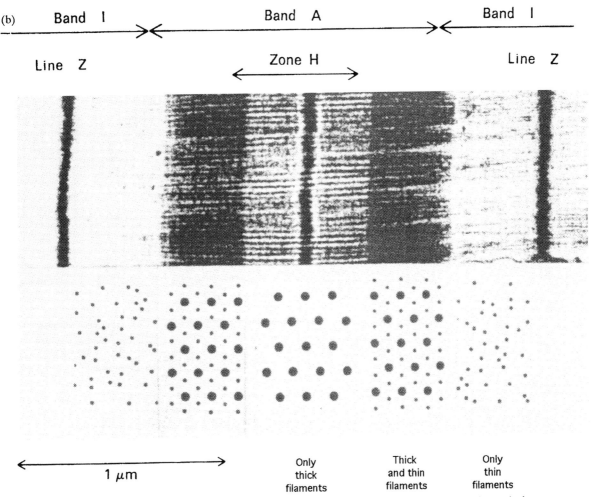

Figure10.7 (b) Electron micrograph (top) of a longitudinal section of a skeletal muscle myofibril, showing a single sarcomere. Schematic diagrams of cross-sections are shown below the corresponding regions in the micrograph. (Reproduced by permission of Editorial Reverté from ref. 37.)

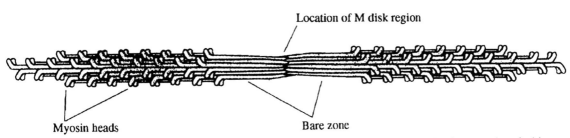

Figure 10.8. The packing of myosin molecules in a thick filament. Adjoining molecules are offset by approximately 14 μm, a distance corresponding to 98 residues of the coil.

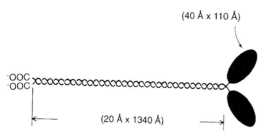

(40 Å x 110 Å)

⁻OOC
⁻OOC

(20 Å x 1340 Å)

Figure 10.9. Schematic diagram of a myosin molecule. (Reproduced by permission of Editorial Reverté from ref. 37.)

mechanical movement driven by the free energy of ATP hydrolysis.

2.4.3 Trigger of muscle contraction

The muscle contraction is triggered by a nervous pulse. It is generally accepted that the nervous pulse promotes an increase in Ca^{2+} concentration in the vicinity of the muscle fibers. This increase in Ca^{2+} is due to their flow through calcium channels. The channels display voltage-dependent gating and are selective for divalent cations over monovalent cations. They presumably undergo voltage-dependent conformation changes. The Ca^{2+} concentration increases inside the myofibril from 10^{-7} M to 10^{-3} M, which is enough to promote conformational changes in the troponim–tropomyosin allowing the muscle contraction.

When the nervous excitation is switched off, Ca^{2+} is pumped out the myofibril and the tropomyosin recovers the rest: conformation and relaxation of the muscle occurs.

2.4.4 The sliding filament model

Under contraction thick *myosin filaments slide* along the thin actin filaments (Figure 10.10). Sarcomeres

decrease in length during contraction (Figure 10.11). This decrease is due to the decrease in the width of both the I and A band. The length of both thin and thick filaments is constant during contraction.

The shorting of the myofibrils length involves steadying motions in opposing directions at the two ends of the myosin thick filament. The *free energy of ATP hydrolysis* is *translated* into *conformation changes* in the myosin heads. This dissociates myosin and actin allowing a rebinding of myosin and actin to occur with stepwise movement of the myosin S1 head along the actin filament (Figure 10.12). The conformation change in the myosin head is driven by the hydrolysis of ATP.

The release of phosphate from the hydrolysis of ATP is followed by the *crucial conformational change* by the S1 myosin heads—the so called power stroke—and ADP dissociation. In the power stroke the myosin heads tilt by approximately $45°$ and the conformational energy of the myosin heads is lowered by about 29 KJ mol^{-1}. This moves the thick filament approximately 10 nm along the thin filament. Subsequent binding and hydrolysis of ATP cause dissociation from the thin filament. As a consequence the myosin heads shift back to their high energy conformation. The heads may then begin another cycle. This cycle is repeated at rates of up to five per second in two typical skeletal muscle contractions. The conformational changes occurring in this cycle are the secret of the energy coupling.

2.4.5 Remarks

The sliding model was proposed in 1954 by two different groups and from then the majority of scientists in the field have followed this model. Water and other cations and anions play an important role.

Several points can be stressed.

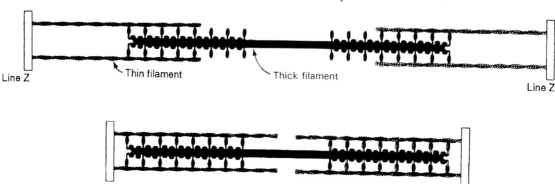

Line Z   Thin filament   Thick filament

Line Z

Figure 10.10. Schematic diagram showing the interaction of thick and thin filaments in skeletal muscle contraction. (Reproduced by permission of Editorial Reverté from ref. 37.)

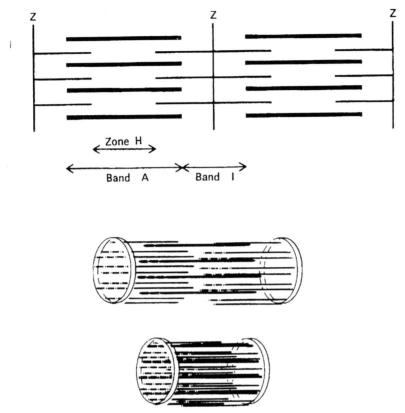

Figure 10.11. The sliding filament model of skeletal muscle contraction. The decrease in sarcomere length is due to decreases in the width of the I band and H zone, with no change in the width of the A band. These observations mean that the lengths of both the thick and thin filaments do not change during contraction. Rather, the thick and thin filaments slide along one another. (Reproduced by permission of Editorial Reverté from ref. 37.)

(1) Nervous pulses promote *ionic interchanges* between the myofibril and the surrounding in a few microseconds.

(2) The *free energy* of ATP hydrolysis *drives* a *conformational change* in the myosin heads resulting in the net movement of myosin along the actin filament.

(3) The movement occurs through the *formation* of *myosin–actin complexes* through ATP and subsequent *dissociation* of those complexes.

(4) The high *energetic content of ATP is restored* from ADP by transfering energy from *glycolysis.*

(5) The *contraction is relaxed by* Ca^{2+} release from the myofibrils.

(6) *Interchanges of water* and *ions other* than Ca^{2+} play *an important role* in muscles contraction.

(7) The final result is the *generation of mechanical energy from chemical energy through conformational changes* promoted by the formation and dissociation of ionic complexes bridged by ATP^{4+} ions.

2.5 Intrinsically conducting polymers

In muscles nervous pulses (*tens of mV*) trigger muscle contraction and enable myofibrils to react within microseconds. In this way the contraction mechanism is fast and simultaneous.

Polymer gel and proteins used as artificial muscles are not electronic conductors. Inside an electric field they are ionic conductors: electroosmotic and diffusion processes have to be used as intermediates to promote conformational changes in the polymeric structures. The concomitant movements are slow. Strong electric fields are applied (*tens of volts*) giving electrochemical reactions, at the electrode/water interface, with gaseous evolution.

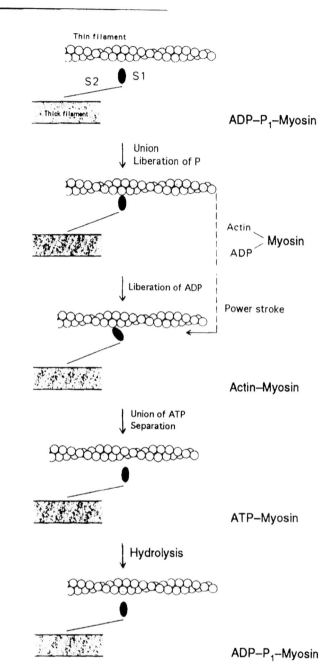

Figure 10.12. Proposed mechanism for the generation of force by interaction of an S1 unit of a myosin filament with an actin filament. In the power stroke, the thin filament moves relative to the thick filament when S1 undergoes conformational changes accompanying the release of ADP. (Reproduced by permission of Editorial Reverté from ref. 37.)

In order to mimic muscles *new materials are required*, which are able to conduct both electrons and ions. *Ions*, as in muscles, *have to interact with the polymeric molecules* promoting ionic links. The *electric pulse*, like the nervous pulse, has to *arrive at any point* in the material *simultaneously*. The *range of the electric pulse* has to be from hundreds to more than one thousand of mV. In order to keep a similar molecular structure these materials should have macromolecular components. Intrinsically conducting polymers, or composites of those conducting polymers, seem to be the closest to fulfilling these requirements; they are electron and ion conductors [40,41]. When used as electrodes in ionic media they can be oxidized and reduced in a reversible way (Figure 10.13). Linked to this electrochemical process, a reversible change of volume takes place thereby opening new possibilities to develop molecular machines. In this way a new system is available that is able to mimic most of the elemental components of a natural muscle. The brain and the nervous system, produce and drive a nervous pulse to the nerve/myofibril interface (Figure 10.1); this system is simulated by electrochemical equipment and metallic wires which generate and drive an electric pulse to the metal/polymer interface (Figure 10.3).

2.5.1 Trigger of muscle work: transduction of an electric pulse to an ionic flux

The nervous pulse triggers an important ion transfer (mainly Ca^{2+}) inside the myofibril promoting important conformational changes in the proteins of the myofibrils (mainly in the myosin heads). Those conformational changes are mediated by ATP and glycolysis, and give

an overall transformation from the chemical energy in glucose to mechanical work.

When the electric pulse arrives at the metal/polymer interface (Figure 10.14), it passes through the conducting polymer and arrives at the polymer/electrolyte interface. If this is an anodic pulse electrons are extracted from the polymeric chains and positive charges (radical cations–polarons or dications–bipolarons) are stored. At the same time ionic transfer occurs between the polymer and the solution in order to maintain the electroneutrality in the solid. The nature of this movement depends on the polymer i.e. initial loss of protons from polyaniline, loss of cations (or anions) from polyelectrolytes (polyanions or polycations, respectively), or from polyelectrolyte-conducting polymer composites; or entrance of anions in non-substituted polypyrrole, polythiophene, etc. Even in the last cases an initial loss of cations is followed by an entrance of anions. The nature and direction of the movement change, even for the same polymer, when the solvent is changed. This agrees with a strong influence on the solvent movement, in and out of the polymer, together with the ions. The nature and direction of the movement also change when different electrolytes are used.

In order to simplify this chapter and clarify movements and concomitant properties it will be accepted initially that when a polypyrrole film is oxidized negative charges (small solvated anions) come from the solution inside the solid polymer. Working in aqueous solutions the high dipolar moment of water molecules interacts with positive charges in the polymer promoting a flux of water inside the film.

If we take into account the fact that in the neutral state strong attractive polymer–polymer Van der Waals

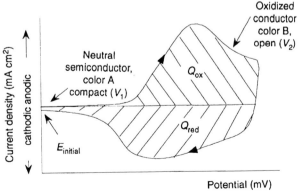

Figure 10.13. Voltammogram obtained from a polypyrrole film in a 0.1 M $LiClO_4$ aqueous solution at 20 mV s^{-1}, using a platinum sheet as counterelectrode. This figure shows the reversibility of both anodic and cathodic processes.

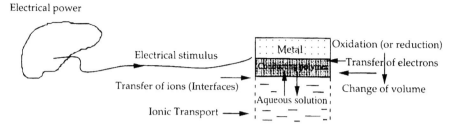

Figure 10.14. Electric stimulus generated by the electric power system, and sent through wires connections and transduced by electrochemical reactions at the metal–polymer interface to an ion pulse through the polymer–solution interface.

forces are present, the solid presents a compact structure. The question now is how anions and water molecules penetrate this compact structure during oxidation.

2.5.2   Conformational changes

The only way to maintain electroneutrality in the solid following our simplified model is to allow the penetration of anions: the polymeric structure has to be opened by the formation of channels which allow the penetration of solvated counterions [42,43]. Positive charges along the chains at the polymer–solution interface promote coulombic repulsions between those chains and thereby provide enough energy to allow conformational changes (Figure 10.15) [44]. Those conformational movements open enough space between the chains to allow the penetration of counterions from the solution side of the double layer (Figure 10.16) [45]. The process expands in a uniform way from the polymer–solution interface to the inside of the polymer. Counterions and water molecules diffuse from the solution through the oxidized region until the limit between oxidized and neutral zones.

The reduction of the polymer at different cathodic potentials promotes kinetic control of the subsequent oxidation by conformational relaxation of the polymer (Figure 10.17). This allows the redox processes to be modelled by our group through an electrochemical conformational relaxation theory. I will return to this later.

The opening of channels in this polymer allows the penetration of solvated counterions and water molecules. During oxidation an increase of volume occurs.

2.5.3   Change of volume

Dry polypyrrole films, electrogenerated on platinum electrodes, were weighed, using an ultramicrobalance with a precision of 10^{-7} g, under both oxidized (the counterion being hydrated ClO_4^-) and neutral states [45]. The weight of the oxidized films was always 35% greater than that of the reduced one. Different dry films were removed from the metal and were used to determine densities by flotation in $CHCl_3/CCl_4$ mixtures. Similar results were obtained using neutral and oxidized polypyrrole: 1.51 g cm^{-3}. Those results confirm compaction during the reduction giving, at least, a 35% reduction in volume. Using the polymer weight, the polymer density and the area of the pristine-coated metal, an average thickness of 10 μm was obtained for the oxidized film.

Baughman et al. [46] state, from the study by X-ray diffraction of polyacetylene doped with alkali metals, a large volume expansion of 12.5 cm^3 faraday^{-1}, which is about 27% of the molar volume of potasium. This corresponds to a volume change for the polymer of 6.6% [47–49], or a 1.06% change in volume per per cent change in dopant concentration. This value is close to the volume change measured by bulk dimensional changes for sodium [50] by Francois et al. (1.5) or for potassium by Plichta [51] using a doping solution of K^+–naphtalide complex in 2-methyltetrahydrofuran (1%).

When dopant ions are solvated enormous dimensional changes can result. Okabayashi et al. [52] found that about three propylene carbonate molecules are reversibly inserted with each ClO_4^- ion during the oxidation of polyaniline. The associated volume change of the polyaniline is 297 cm^3 faraday^{-1}, corresponding to a volume increase of the polyaniline by a factor of 2.2 over the total observed doping range.

In situ bulk measurements by Slama and Tanguy [53] give a reversible volume change on the oxidation of polypyrrole in propylene carbonate/LiClO$_4$ of 272 cm^{3}faraday^{-1} similar to that observed by Okabayashi for polyaniline.

Polypyrrole samples measured by De Rossi et al. show fractional length variations up to 2% with most

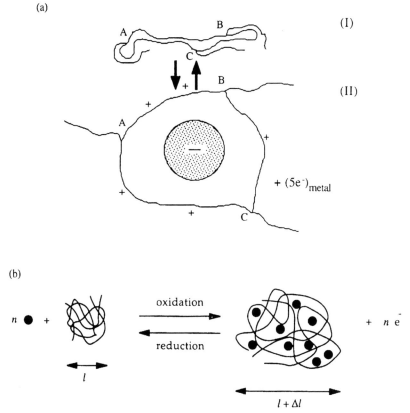

Figure 10.15. (a) (I) Entangled structure after reduction at high cathodic potentials (two dimensions). (II) High anodic potentials are needed to inject positive charges, open the structure and allow counterions penetration. (Reprinted with permission of Kluwer Academic Publishers from ref. 45.). (b) A compact element of volume (with a length l) increases its length $l + \Delta l$ during oxidation: electrons are lost from the polymer, channels are opened and hydrated counterions from the solution penetrate to maintain electroneutrality.

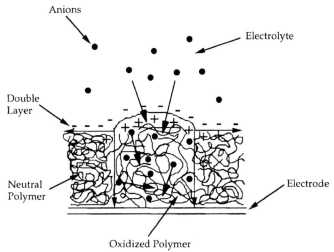

Figure 10.16. Conformational movements of the polymeric chains in the neutral polymer open the compact entanglement (by repulsion between positive charges) generating free volume and allowing the penetration of counterions during oxidation.

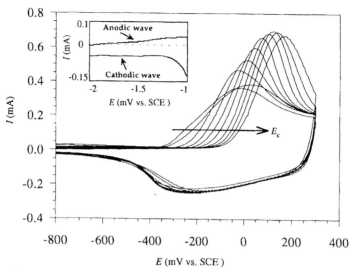

Figure 10.17. Voltammetric behavior of the polypyrrole in 0.1 M LiClO$_4$/propylene carbonate solution. The potential sweep was carried out at 30 mV s^{-1} between different cathodic potentials and 300 mV vs. SCE. The cathodic potentials were from left to right: -800, -1200, -1400, -1600, -1800, -2000, -2200, -2400, -2600 and -2800 mV vs. SCE. The inset shows the sweep from -1 to -2 V: cathodic current is present at all the potential values during the cathodic sweep.

drastic variations of the Young's modulus, up to 80% [54]. The use of the bending beam method allowed Pie and Inganäs to determine a volume change in the range 0.65–3.4% for cation-insertion charge compensation in tosylate-doped polypyrrole [55,56].

Polypyrrole–polyacrylic acid (PAA)–polyvinyl alcohol (PVA) composites were swollen and shrunk by acidic or basic additives and showed length variations in the range 5–10%; meanwhile PAA–PVA composites shift between 20 and 50% [57]. Polypyrrole was chemically synthesized in the PAA–PVA composite from gas catalyzed by FeCl$_3$.

2.5.4   Relaxation to the initial volume

If the oxidized polymer is submitted now to a cathodic potential or a cathodic constant current, reverse processes occur.

- At the metal/polymer interface electrons, generated by the electrochemical equipment and sent through the connection wires, are injected into the polymer chains: a flow of electrons from the metal to the polymer occurs.
- At the level of the polymer chains positive charges are cancelled: polymer–water and polymer–counterions forces become null, meanwhile polymer–polymer interactions become strong.

- The polymer shrinks, and counterions and water molecules cross the polymer/solution interface toward the solution.

As a consequence the initial volume is recovered. So, a conducting polymer can be envisaged as a transducer of electrical charge to changes of volume, linked to a flow of ions and water molecules through an electrochemical reaction.

All those processes: conformational changes, concentration of positive charges in the solid polymer, counterion concentration in the solid polymer and water concentration in the solid are related to the electric pulse created by the electrochemical equipment. This points to a control of the change in volume, or any other property related to the redox state, through electrical variables.

2.5.5   Electrochemomechanical properties

A search, as discussed in previous sections, to find suitable materials was performed. Conducting polymers are able to transduce any electric pulse to a change of volume, working at constant temperature, through a chemical reaction of oxidation or reduction linked to both ionic and solvent flow. This transduction can be useful from the point of view of links between a current flow and a flow of ions giving sensors, dosing elements or artificial nerves. Changes in the channel diameters and channel charges are used for smart membranes. Changes in the concentration of positive charges in chains give electrochromic properties.

Now changes of volume are discussed. From this point of view an *electric* pulse gives a *chemical* reaction of oxidation or reduction promoting, through polymeric conformational changes and flow of ions and solvent, a *mechanical* transformation of the solid. This can be named a true *electrochemomechanical* property. The problem now was how to construct artificial muscles from adequate conducting polymers.

2.5.6 The synthesis

Conducting polymers are available as chemically synthesized powders, electrochemically generated films or soluble oligomers. In our laboratory the most usual were electrogenerated films.

The main problem related to the films was that electrogeneration is a complex, though fast, mechanism to obtain a mixed material [58]. Another chapter of this book discusses the literature on this problem showing how the ideal radical coupling mechanism, giving ideal molecules, coexists with either proton-initiated chemical polymerization (giving non-conducting polymer encapsulated in the film), degradation, crosslinking and overoxidation processes. Moreover the mechanism can change with thickening of the film as shown by the variation of the ability to store electrical charge per milligram of polymer [59,60]. But complex, once studied, means flexible. In our laboratory most of the time was dedicated to the study of the kinetics of electrochemically initiated polymerization processes. Empirical kinetics were followed by *ex situ* ultramicrogravimetry of the electrogenerated polymer. In parallel, the influence of different variables (monomer or electrolyte concentration, temperature, electric potential, current density, water content, etc.) and different polymerization times on the ability to store electrical charge (defined as millicoulomb stored per milligram of electrogenerated polymer) was checked for each film. The polymer weight was obtained by ultramicrogravimetric difference between the uncoated and coated electrodes. The stored charge was obtained by voltammetric control [58].

A thick film (> 3 μm) with the highest ability to store charge is needed; this means the highest number of counterions and water molecules are interchanged during redox processes giving the highest relative variation of volume. This fact requires the avoidance

of any parallel proton-initiated chemical polymerization, and the attainment of the lowest degradation and overoxidation of the polymer during polymerization.

Low degradation and overoxidation can be attained at low potentials in organic solvents. Nevertheless the polymerization through a polycondensation of radical cations gives two protons by the incorporation of the monomeric molecule in the polymer. This process promotes a fast pyrrole protonation and chemical polymerization around the electrode: a dense cloud is observed when acetonitrile was used as solvent. This is avoided using a 2% water content due to the higher interaction between water and protons than between protons and pyrrole [60]. This is a solution of compromise, between the lower degradation processes in acetonitrile, but higher proton-polymerization and a greater degradation in the presence of water but proton elimination.

2.5.7 The problem

Once synthesized, the polymer films present a new problem. Conformational changes and molecular movements are able to produce small changes of volume. The question now was how to translate these molecular movements to macroscopic movements. This question was solved for small movements by means of tensile films or fibers as unimorph or bimorph cells [46]. Bimorph cells use two different films, fibers or conducting polymers.

The original idea in our laboratory was based on the bimetallic thermometer: two sheet of metals with different expansion coefficients are welded side to side. An increase of temperature gives a faster separation of atoms in one network than on the other one. As a consequence a greater increase in length occurs on the first metal sheet than on the second one and a stress gradient appears at the interface. The macroscopic result is that the bilayer bends, the sheet with the greater expansion coefficient makes up the convex side. If one of the bilayer's ends is fixed, the bottom of the bilayer describes an angular movement proportional to the temperature gradient. A decrease in temperature promotes a reverse movement.

Thus, the change of volume of a film of conducting polymer in an electrolyte is controlled through an electric current at constant temperature. Any adherent, non-conducting and flexible polymeric material can be used as the second film [61–65]. Almost any of the tapes in the market fulfill these requirements.

2.6   Bilayer muscles

In this way bilayers of 3 cm by 1 cm of polypyrrole were electrochemically synthesized on stainless steel [45, 61–65]. Different thicknesses, between 3 μm and 100 μm were obtained. Both sides of the stainless steel electrode were uniformly covered. The uniformity on the electric field was maintained using two stainless steel counterelectrodes.

The electrogeneration was performed at different anodic potentials. The process was stopped and the polymer was polarized at 200 mV. Subsequently the polypyrrole films were partially oxidized. Once extracted from the solution the film is rinsed with acetonitrile and dried (Figure 10.18a).

Using one side of the coated steel, a tape was stuck to the dry polypyrrole (b) and the formed bilayer was removed from the electrode (c). A bilayer device of 3 cm × 1 cm is formed (d).

A metallic clamp fixes to one end of the bilayer and keeps electric contact with the film of conducting polymer. A fraction of the bilayer is put into an electrolyte. The bilayer is used with a platinum counterelectrode and a reference electrode (Figure 10.19).

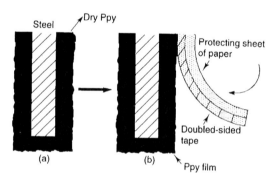

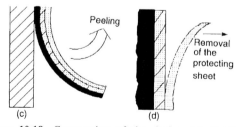

Figure 10.18. Construction of the device. (a) Oxidized polypyrrole film electrogenerated on a steel electrode. (b) A tape was adhered to the dry polypyrrole (Ppy). (c) The bilayer device with a protecting film is removed from the electrode. (d) The protecting sheet is peeled and the bilayer is ready to use.

2.6.1   The working muscle: macroscopic model

When a cathodic current flows through the conducting film, electrons are injected into the polymer, positive charges along chains are compensated and counterions are expelled in order to maintain the electroneutrality. Polymer–polymer Van der Waals interactions increase quickly, thereby closing channels and diminishing the volume. At the conducting polymer/non-conducting polymer interface the closing of channels in the conducting films attempts to trail fibers of the non-conducting and elastic film thereby promoting a stress gradient of contraction across every point of the interface.

The macroscopic consequence is the bending of the bilayer: the conducting polymer is inside (concave side) (Figure 10.19) and the non-conducting film is outside (convex side). The top of the bilayer was fixed by the clamp and the bottom describes an angular movement around the fixed top. This angular movement can be greater than 360° (Figure 10.20).

The flow of an anodic current promotes reverse processes with controlled swelling of the conducting film. The bottom of the bilayer moves in the opposite direction: the opening of channels allows the polymer/polymer interface to return to the starting position: the free end of the bilayer just hangs straight down from the fixed end.

As the bilayer was constructed under partial oxidation of the conducting film, which was, as well, completely dry, the vertical position was recovered under partial oxidation of the conducting film. If the anodic flow of the current continues most counterions and water molecules penetrate the solid. The expansion of the conducting polymer continues and expansion stresses are formed at every point of the polymer/polymer interface. The macroscopic consequence is the bend of the bilayer, the conducting film being outside and the non-conducting film being inside the arc (Figure 10.21). The free end of the bilayer describes an angular movement in the opposite sense to that observed during the flow of cathodic currents. The overall movement can be observed in Figure 10.22.

2.6.2   The bilayer as an electrochemopositioning device

The angular movement described by the free end of the bilayer is the result of the oxidation state of the conducting film. The oxidation process is an electro-

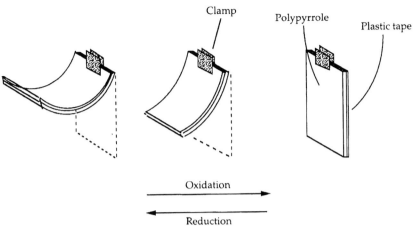

Figure 10.19. The bilayer movement when a reduction (right) or oxidation (left) potential was applied. (Reproduced by permission of Elsevier Science SA from *Synth. Met.* **55–57**, 3716 1993.)

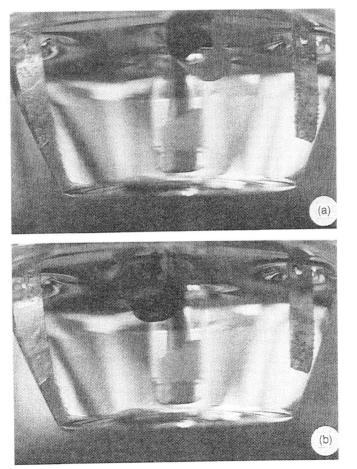

Figure 10.20. Movement of a polypyrrole/non-conducting polymer bilayer in 1 M LiClO$_4$ aqueous solution: (a) under anodic flow through the polypyrrole film; (b) under cathodic flow.

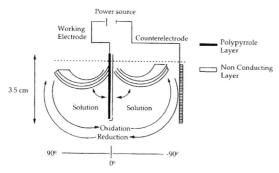

Figure 10.21. The movement of an artificial muscle formed by a bilayer polypyrrole film/non-conducting plastic tape during current flow, starting from the vertical position. The flow of a cathodic current through the conducting film promotes its reduction with the expulsion of counterions and water. Polypyrrole shrinks and the bilayer bends to the left. The reverse process occurs during oxidation.

chemical reaction in solid state that, starting from a neutral polymer, can reach and be stopped at any intermediate composition. In this case most of the electrogenerated films have an ability to store electrical charge around 380 mC mg^{-1}, which means

$$380 \text{ mC mg}^{-1} \times \frac{1 \text{ C}}{1000 \text{ mC}} \times \frac{1 \text{ F}}{96\,500 \text{ C}}$$
$$\times \frac{6.023 \times 10^{23} e^-}{1 \text{ F}} = 23.7 \times 10^{17} e^- \text{ mg}^{-1}$$

Taking into account that every lost electron during polymer oxidation generates a positive charge on the chains and every positive charge is compensated by a counterion, around 24×10^{17} anions penetrate 1 mg of neutral polymer to produce an overall oxidation.

Starting from the neutral polymer any intermediate composition between zero anions per milligram of polymer and 24×10^{17} anions per milligram can be attained. This hypothesis is confirmed by experimental results. Oxidation and reduction processes performed by voltammetry support the previous ideas (Figure 10.13). The oxidation processes begin at -650 mV and from there a continuous oxidation occurs until 200 mV. The potential sweep can be stopped at any point and the oxidation processes will stop, as well. If a voltammogram is performed at very slow sweep rates (i.e. 0.2 mV s^{-1}) the oxidation state at every potential can be considered as the subsequent steady state. Limits for the oxidation process of the bilayer and consumed charges change when different solvents are used [66].

This means that the number of counterions per milligram of neutral polymer corresponds to the steady state at every potential and increases continuously with this potential. As the increase of the conducting film volume depends on the number of counterions, a continuous increase in volume is present. Subsequently for every potential there is a different position of the free end of the bilayer [45]. Taking this position as the angle described related to the vertical, the results observed in Table 10.1 were obtained.

Experimental voltammograms show that the process is reversed and ions are expelled during reduction. If the original potential is applied the free end of the bilayer recovers after a few seconds, depending on the values of some physical and chemical variables, as will be studied later. This is an electrochemopositioning device. Through an electrochemical reaction controlled by a potential, or a charge, a position in an electrolyte can be defined.

Those devices can find applications in multiple macroscopic or microscopic systems.

2.6.3   Action and reproducibility

In order to test the reproducibility of these devices four different bilayers were constructed and checked in 0.1 M LiClO$_4$ aqueous solution [67]. Under oxidation to 1000 mV (vs. SCE), the bottom of the bilayer was bent as the polypyrrole film pushed the plastic tape. When the bottom described an angle of 90° relative to the vertical position, being closer to the solution surface, the potential was stopped at -1000 mV. Electrochemical reactions and the subsequent direction of the movement were now reversed. The bottom of the device described a movement of 180°.

The average time (for the four bilayers checked independently) required to describe the 90° and 180° movements during the oxidation processes were 4.5 ± 0.31 s and 11.2 ± 0.54 s, respectively.

Under reduction (by polarization at -1000 mV) the movement of those bilayers was faster: 2.56 ± 0.005 s were required to reach 90° and 6.09 ± 0.46 s, to describe 180°. The reproducibility is good. The highest deviation of the average rate for the movement was achieved under reduction through 180°.

2.6.4   Movement and reproducibility when trailing a hanging mass

The weight of polypyrrole in the bilayer was 6 mg (2.5 cm × 1.5 cm). The large mass of the adhesive rubber coated with a cellulose film was 51 mg. To

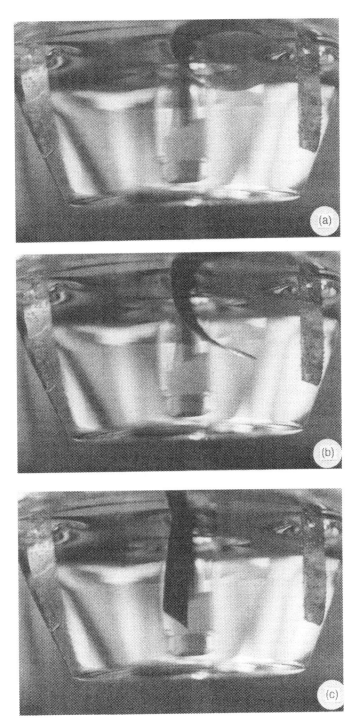

Figure 10.22. Movement of the free end of a bilayer during an anodic flow of current through the conducting polymer (a) to (e). The flow of a cathodic current promotes a reverse movement. The movement stops immediately if the current flow is stopped.

Figure 10.22. (*continued*)

observe the ability to perform mechanical work, a glass rod weighing 0.1 g (~17 times the polypyrrole mass) was stuck to the bottom of the bilayer. In order to observe the reproducibility of the mechanical work the above experiment was repeated [67] for every charged bilayer (Figure 10.23).

The time required to describe the 90° and 180° movements under oxidation was 5.38 ± 0.95 s and 13.9 ± 2.7 s, respectively. Under reduction a time of 2.27 ± 0.41 s was taken to reach 90° and 12.2 ± 1.75 s

was taken to reach 180°. The reproducibility is lower when carrying a weight. Deviations up to 20% can be attained. Bilayers were able to trail 1000 times their own weight [45].

2.6.5 The bilayer device as an electrochemical reaction

Movement and mechanical work in this bilayer muscle has been assumed to be linked to electrochemical

Table 10.1. Angular movement described by the free end of the bilayer from the initial vertical position when submitted to a potential sweep from 400 mV to −170 mV at 1 mV s⁻¹ in 0.1 M LiClO₄ aqueous solution at ambient temperature (Reprinted by permission of Kluwer Academic Publishers from ref. 45.).

E (mV vs. SCE)	400	340	285	210	165	90	20	30	−90	−175
Angle vs. vertical (°)	0	10	20	30	40	50	60	70	80	90

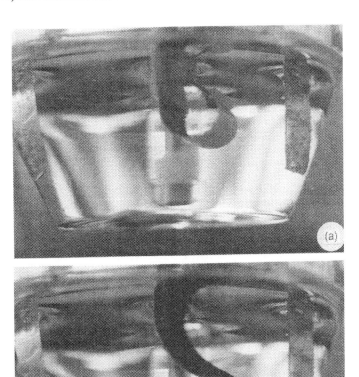

Figure 10.23. A bilayer device (6 mg of polypyrrole) was charged with a steel pin (0.1 g). The sequence from (a) to (f) shows the movement described by the flow of an anodic current through the device. The opposite movement is described during the flow of cathodic current. Reversing the current flow at any point makes the movement reverse.

processes occurring in the solid film. However, electrochemical processes and molecular conformation changes have to be taken into account. The conformational changes may be the controlling process.

Attention is focused on the electrochemical reaction. In the ideal case the following reaction is assumed:

$$(\text{Polymer})_s + n(\text{ClO}_4^{-})_{aq}$$
$$\rightleftharpoons [\text{Polymer}^{n+}(\text{ClO}_4^{-})_n]_s + ne_m^{-} \qquad (10.1)$$

where s, aq and m denote solid film, aqueous solution and metal, respectively.

Any physical or chemical variable acting on the kinetics of the electrochemical reaction will affect the concomitant mechanical movement. The physical variables acting on the oxidation or reduction rates of this solid film will be anodic overvoltage, anodic current density, cathodic overvoltage and cathodic current density. Two are due to the chemical components of the reaction at constant temperature: the concentration of counterions in the solution and the concentration of positive charges, under equilibrium, in the polymer. This second variable is also a function of the anodic overvoltage.

2.6.5.1 Influence of the anodic overvoltage

A bilayer was constructed such that, under polarization at -200 mV in 0.1 M LiClO$_4$ aqueous solution, the free end stops at $-90°$ to the vertical [45]. When the

Figure 10.23. (*continuea*)

potential was increased to 600 mV the free end describes a movement of 180°, from −90° through zero to +90°, where upon the movement stops. The time required to describe this movement was 115 s and 467 mC were consumed during this time.

Returning to −200 mV, the original position of −90° was recovered. When the potential was stepped to increasing anodic potentials lower times were required to cross over 180° (Table 10.2).

Anodic steps up to 10 V were performed and the time needed to cross over 180° drops to a few seconds. The reversibility of the process was high, whatever was the anodic potential, if the polarization is switched off when the free end arrived at 90°: the reversibility is maintained if the oxidation only attains a certain level, whatever the potential of polarization.

If the polarization at a high potential is maintained, the movement follows on and describes an angular

Table 10.2. Time needed and electrochemical charge consumed to describe an angular movement of 180° (moving up and down) by the free end of the bilayer (3 × 1 cm) when submitted to step potentials from −200 mV to different anodic potentials in 0.1 M LiClO₄ aqueous solution (Reprinted by permission of Kluwer Academic Publishers from ref. 45.).

E (mV vs. SCE)	600	700	800	900	1000	1200	1500	1800	2000
Time over 180° (s)	115	100	70	53	45	37	30	25	22
Charge cons. (mC)	467	485	480	477	481	499	545	588	614

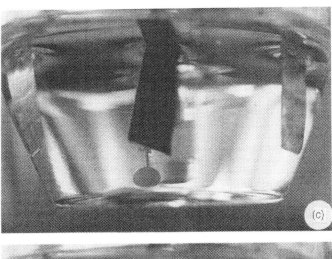

Figure 10.23. (*continued*)

movement greater than 360° (the lateral side of the bilayer describes more than 360°). A degradation process occurs at such deep oxidation and for such long periods of anodic polarization, and the stress gradient between the layers relaxes. The free end recovers an angle lower than 90°. The muscle dies without response to any ulterior potentiostatic stimulation.

Experimental results support the conclusion that a constant oxidation level is required to describe the same angular movement. The charge consumed to cross over 180° was the same (around 480 mC) for lower potentials than 1500 mV. The charge and potential are lower for thinner films [67]. An increase of the consumed charge is observed at higher potentials and at the same time bubbles are present on the polypyrrole layer. The production of oxygen on the polypyrrole

does not affect the reversibility of the process if the polarization is interrupted at the same oxidation level in the polymer (the same angular movement of the free end of the bilayer).

In conclusion, an angular movement is linked to an oxidation level. Greater current densities flow through the system at increasing overvoltages (Figure 10.24) and shorter times are needed to consume the same charge, describing the same angle.

2.6.5.2 Muscle working under constant current density

The rate of an electrochemical reaction can be controlled by the current density flowing through the system. That is true if counterions diffusion control, or

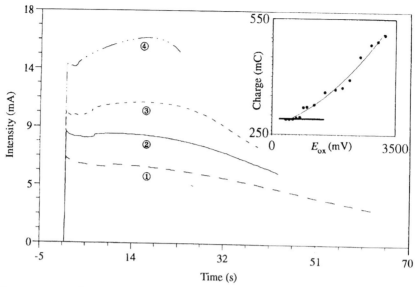

Figure 10.24. Chronoamperometric responses obtained when a bilayer was submitted to potential steps from -200 mV to different anodic potentials: (1) 600 mV; (2) 800 mV; (3) 10000 mV; (4) 1200 mV, in a 0.1 M LiClO$_4$/water solution.

conformational changes control, of the process are not present.

To check this variable the free end of the bilayer is put on the vertical related to the electric contact and a different anodic current is sent to the bilayer every time [68]. The rate of the movement increases when the current density rises. The time required to cross over 180° decreases when the current density increases (Figure 10.31). Experiments were performed this time in 1 M LiClO$_4$ aqueous solution.

2.6.5.3   Reverse movement

The movement of the bilayer is linked to an electrochemical reaction. This reaction can be reversed by changing the direction of the polarization or the direction of the current flow. The influence of the cathodic overvoltage or the cathodic current density on the reverse movement will be explained below.

2.6.5.4   Influence of the cathodic overvoltage or cathodic current density

A bilayer is polarized at 600 mV. The polypyrrole layer is partially oxidized and the free end is positioned at 90° to the vertical. A potential step to different cathodic potentials from this position showed that the times required to cross over 180° (following the direction from +90° across the vertical to -90°) decreases with

increasing cathodic overvoltages (Table 10.3). This fact is linked to the flow of increasing current densities when the device was submitted to increasing cathodic potentials (Figure 10.26).

In a similar way, the times required to cross over 90° (t_{red}) in 1 M LiClO$_4$, decreases with increasing current densities (I):

$$t_{red} = 194\, I^{-0.94} \quad (\text{regression} = 0.998) \quad (10.2)$$

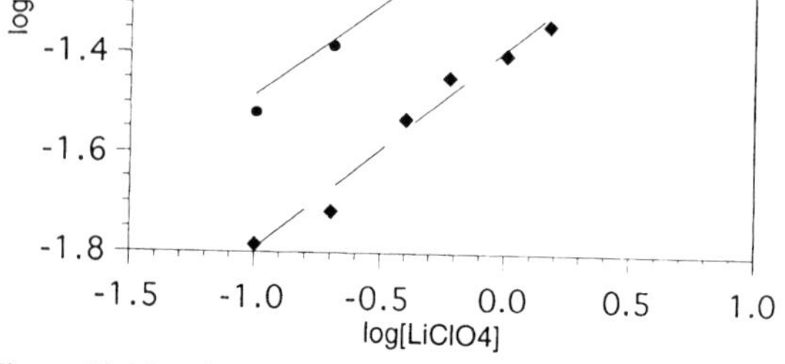

—— $y = -1{,}1204 + 0{,}36285x$ R = 0,98702
-- $y = -1.3911 + 0{,}40179x$ R = 0,99032

Figure 10.25. Influence of the electrolyte concentration on the time required (t_{red}) to cross over 180° by the free end of the bilayer when submitted to potential steps between -200 and 400 mV (●) or reverse (■).

Table 10.3. Time needed and electrical charges consumed to describe an angular movement of 180° by the free end of a bilayer when submitted to step potentials from 800 mV to different cathodic potentials in 0.1 M LiClO$_4$ solution (Reprinted by permission of Kluwer Academic Publishers from ref. 45.).

E (mV vs. SCE)	−400	−500	−700	−1000	−1200	−1500	−1800	−2000
Time over 180° (s)	65	59	45	33	29	24	21	19
Charge cons. (mC)	520	529	536	545	558	596	633	661

2.6.5.5   Influence of the counterions concentration

The empirical kinetics for the electrochemical reaction (10.1) is

$$r = ke^{-\frac{\alpha RT}{nF}} \eta [\text{ClO}_4{}^-]_{aq}^{\alpha} [\text{polymer*}]_s^{\beta} \qquad (10.3)$$

where [polymer*] represent the concentration of positive charges (moles) stored per liter of neutral polypyrrole at constant overvoltage (η) under equilibrium, and [ClO$_4{}^-$]$_{aq}$ represents the concentration of perchlorate anions (counterions in general) present in aqueous solution. r is the oxidation rate.

If the movement of the bilayer device is due to this electrochemical reaction and the salt concentration influences the kinetics, the influence of the salt on the time required to cross 180° (under constant overvoltages) can be studied empirically.

A bilayer was studied in different concentrations of LiClO$_4$. It was submitted to potential steps from −200 mV to 400 mV. The times required to complete an angular movement of 180° are shown in Figure 10.25 (T. F. Otero, H. Grande and J. Rodriguez, unpublished results).

Those times are shorter at higher concentrations of salt. The movements related to anodic processes are slower than the correlative movements during reduction. This fact appears to be linked to the extra energy required to open the molecular entanglement and to allow the penetration of counterions during the oxidation process. Nevertheless those relative times seem to be linked, as well, to the oxidation level of the conducting polymer when the bilayer was prepared. A high oxidation level promotes slow movements when a small cathodic gradient of potential (i.e. $\Delta E = 600$ mV as above) was applied because the film suffers only a partial reduction at this potential of −200 mV. A reverse anodic gradient applied subsequently leads to a

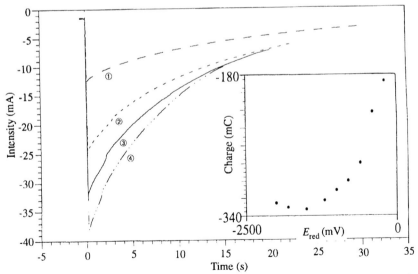

Figure 10.26. Chronoamperograms obtained when a polypyrrole/non-conducting film bilayer was submitted to potential steps (in a 0.1 M LiClO$_4$ aqueous solution) from 400 mV, to different cathodic potentials: (1) −200 mV; (2) −600 mV; (3) −1000 mV; (4) −1500 mV.

partially oxidized conducting polymer with an open structure. No extra energy has been consumed to open the structure. A subsequent fast rate of counterions penetration occurs giving a faster movement than during reduction. This fact was experimentally proved in our laboratory.

As both oxidation and reduction processes power the muscle, similar to that of any of the electrochemical properties linked to redox processes (storage of energy, electrochromism, smart membranes, etc.), it should be possible to obtain kinetic parameters such as order dependencies or activation energy, as well as the influence of the polymer structure, polymer–polymer, polymer–solvent, and polymer–ion interactions, etc. on the rate of the electrochemical processes. Some of these subjects will be considered from a theoretical point of view later.

2.6.5.6   Consumed energy and generated power

The energy required to move the muscle is an electric energy:

$$P = i \int_0^t E \, dt \quad (mJ = mA \, V \, s) \text{ working at constant } i$$

$$(10.4)$$

where P is the energy, i the current flowing by the bilayer, E the electric potential related to the equilibrium potential and t the time of current flow.

The response of the muscle to a flow of constant current density is shown in chronoamperograms. By integration of these curves from $E = E_{eq}$ (the potential attained after one day in the $LiClO_4$ aqueous solution under nitrogen), he consumed energy is obtained.

The consumed electric power is

$$W(t) = I \, E(t) \, (mW = mA \, V) \qquad (10.5)$$

The mechanical energy produced to trail a mass (m) along a height (h) is $m'gh$; m' is the difference between the mass in air and the Arquimedes effect. The percentage of consumed electric energy transformed to mechanical work can be calculated and is always lower than 1%. A lot of work has to be done in order to optimize the materials, thicknesses and conditions of control.

2.6.5.7   Macroscopic model

A macroscopic approach can be performed from the first idea to produce a macroscopic device able to visualize swelling and shrinking processes in conduct-

ing polymers. Following the ideal model (polypyrrole electrogenerated from $LiClO_4$ aqueous solutions and checked in the background solution responds quite well to this model), the conducting films swell during oxidation and shrink under the reduction process.

From a macroscopic point of view the continuous expansion of the conducting polymer under anodic currents promotes an overall increase of length of the film. The non-conducting film is not affected by the current flow; at the polymer/polymer interface a stress gradient of expansion appears. If the adherent layer is elastic; the stress gradient spreads out on both sides of the interface, as shown in Figure 10.27, promoting the bilayer bending.

During the reduction process the length of the conducting film decreases and becomes shorter than the non-conducting film (if the conducting polymer was in an intermediate oxidation state when the bilayer was created). A stress gradient of compression appears across the polymer/polymer interface giving a bilayer bending in the opposite direction to that above.

When a uniform and homogeneous adherent film is used, compression and expansion stresses have a circular symmetry giving a concave or convex spheric

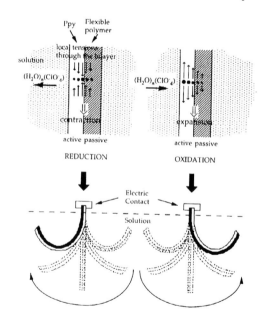

Figure 10.27. Artificial muscle under work. The weight of the polypyrrole film was 6 mg and the weight of the non-conducting plastic tape 10 mg. This bilayer trails a steel sheet of weight 200 mg through 180°, under the flow of a cathodic current of 5 mA. The time required was 20 s, when the electrolyte ($LiClO_4$) concentration was 1 M. (Reprinted by permission of Kluwer Academic Publishers from ref. 45.)

cap. Because most of the commercial tapes where stretched, the symmetry of the properties was lost in most of the cases with preferential directions, like most of the used to obtain bilayers bending uniformly in a longitudinal (described until now) or transverse direction (Figure 10.28.

2.6.5.8 Theoretical approach

Volume and length variation
An attempt will be made to develop a theoretical model able to predict experimental results (T. F. Otero, H. Grande and J. Rodriguez, unpublished results). A double layer will be considered: an electroactive, homogeneous and electroelastic film of a polyconjugated material stuck to a non-electroactive, and flexible film. The bilayer is placed in aqueous solution of an electrolyte. Anodic or cathodic currents pass through the polyconjugated film to allow electrochemical oxidation or reduction. The redox state of the film is a stationary state related to the consumed charge.

The oxidation of a microscopic element of volume is shown by Figure 10.15. The polymer presents a compact structure in the neutral state and is opened during oxidation due to penetration of counterions. Under equilibrium conditions the relative increase in the length of this element (Δl) is only a function of the charge consumed during the oxidation process: $\Delta l = f(Q)$. The overall volume change of the polymer can be split into two components:

(1) During oxidation solvated counterions penetrate the polymer and electrons are lost from polymeric chains. So the relative change in the length of a polymeric segment will be proportional to the number of electrons flowing through the external circuit. This number is equal to the number of solvated counterions penetrating the polymer:

$$\frac{\Delta l_1}{l} = h_1 \frac{Q_{segment}}{l^3} \quad (10.6)$$

where h_1 is a constant related to the volume of the hydrated anions and $Q_{segment}$ is the anodic charge required to oxidize a unitary volume of polyconjugated polymer.

(2) The second component is related to the charged polymer–charged polymer, charged polymer–anion, anion–anion, polymer–solvent and anion–solvent interactions. Based on the theory of swelling of crosslinked gels

$$\frac{\Delta l_2}{l} = h_2 \frac{Q_{segment}}{l^3} \quad (10.7)$$

where h_2 is a constant including interaction parameters, the dielectric constant of the solvent, screen effects, etc. As a result of these interactions, a shrinking process occurs.

By addition of both expressions

$$\Delta l = (h_1 + h_2)\frac{Q_{segment}}{l^2} = h\frac{Q_{segment}}{l^2} \quad (10.8)$$

This equation relates the variation of length (Δl) with the charge consumed per segment.

Asymmetric effects on the bilayer
According to equation (10.8) the relative increase (swelling) or decrease (shrinking) of length is proportional to the electrical charge consumed during the concomitant oxidation or reduction processes. When the conducting polymer film is adhered to a flexible and non-electroactive film a stress gradient appears across the polymer during the oxidation process (Figure 10.29): at the interface the electron loss induced by anodic potential produces a swelling process, but the fibers are adhered to the non-conducting polymer and the swelling is hindered by an opposite mechanical force. Due to the crosslinking of the polyconjugated material the stress gradient reaches a thickness d. For longer distances than d the polyconjugated film expands as follows:

$$\Delta l = h\frac{Q_{seg}}{l^2} = h\frac{Q}{l^2}\cdot\frac{V_{seg}}{V_{pol}} = h\frac{Q}{l^2}\cdot\frac{l^3}{Ae} = h\frac{Ql}{Ae} \quad (10.9)$$

where Q is the overall charge consumed to oxidize the film, V_{seg} is the volume of the considered segment, V_{pol} is the overall volume of the studied polyconjugated material, A is the area of the polymer–polymer interface and e is the thickness of the conducting layer.

Assuming a linear strain gradient from the interface to d, the elongation of an element of volume present at a distance x from the interface will be

$$\Delta l\frac{x}{\delta} = h\frac{Ql}{Ae}\cdot\frac{x}{\delta} \quad (10.10)$$

The concomitant length variation in our bilayer (ΔL) will be

ΔL = length increment per segment

$$\times \text{ number of segments along } L = \Delta l\,\frac{L}{l} \quad (10.11)$$

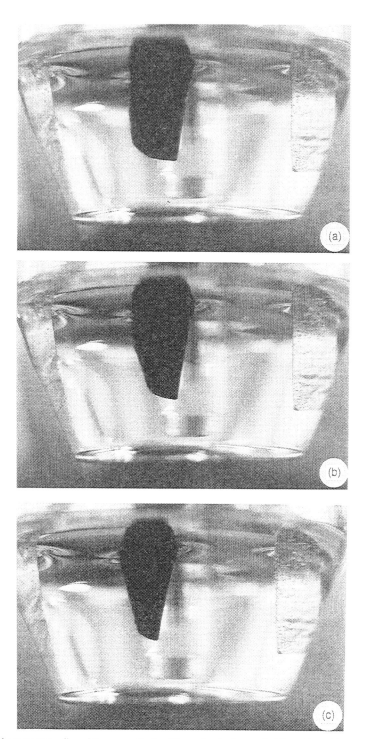

Figure 10.28. When the tape was adhered transversally the bilayer bends forwards (a) to (c) and backwards.

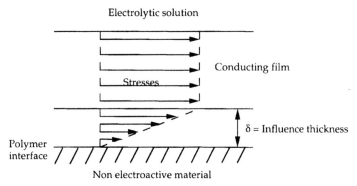

Figure 10.29. The stress gradient appearing at the polymer–polymer interface as a consequence of the volume change in the conducting film during oxidation.

So

$$\Delta L = h\frac{QL}{Ae} \quad \text{if } x > \delta$$

$$h\frac{QL}{Ae}\cdot\frac{x}{\delta} \quad \text{if } x < \delta \tag{10.12}$$

Taking into account the change of length as a function of the charge and the distance to the interface, changes in the angle described by the free end of the bilayer as a function of the consumed charge can be obtained (Figure 10.30). The arc of circumference (L) and the radius (r) are related through the bending angle α

$$\left.\begin{array}{l} L = \alpha r \\ L + \Delta L = \alpha(r + e + d) \end{array}\right\} \rightarrow \alpha = \frac{\Delta L}{e + d} \tag{10.13}$$

Including ΔL in equation (10.8)

$$\alpha = h\frac{QL}{Ae(e + d)} \quad \text{if } e > \delta$$

$$h\frac{QL}{A\delta(e + d)} \quad \text{if } e < \delta \tag{10.14}$$

These expressions can be written as a function of the doping level (positive charges per volume unit): $Y = Q/$

V_{pol}. So the expressions for α become

$$\alpha = h\frac{L}{e + d}Y \quad \text{if } e > \delta$$

$$h\frac{Le}{\delta(e + d)}Y \quad \text{if } e < \delta \tag{10.15}$$

So for a constant doping level (Y), α is a function of the thickness of the polyconjugated film (Figure 10.32). From an electrochemical point of view d is the optimum thickness of the conducting film to form a bilayer. A greater thickness gives electroactive material that consumes charge and does not participate in the mechanical stress. A lower thickness gives a low ΔL value and a concomitant low value of α.

Dynamic behavior at constant current
Equation (10.14) can be summarized as

$$\alpha = kQ \text{ being} \quad k = \frac{hL}{Ae(e + d)} \quad \text{if } e > \delta$$

$$\text{and} \quad k = \frac{hL}{A\delta(e + d)} \quad \text{if } e < \delta \tag{10.16}$$

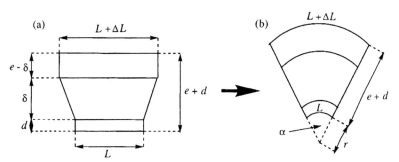

Figure 10.30. Change of length (a) and concomitant bending (b) in a bilayer muscle. Here d represents the thickness of the non-electroactive film; δ is the thickness influenced by the stress gradient, e is the overall thickness of the polypyrrole layer and α is the bending angle.

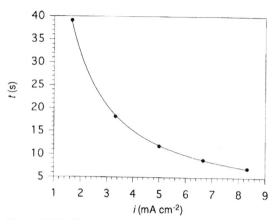

Figure 10.31. Time required to cross over 180° by the free end of a bilayer during oxidation at different current densities. A 1 M LiClO$_4$ aqueous solution was used as the electrolyte and a platinum sheet as the counterelectrode.

Consequently the variation of α as a function of the duration of current flow gives

$$\frac{d\alpha}{dt} = kI \qquad (10.17)$$

and by integration under constant current

$$\alpha = kIt \qquad (10.18)$$

The time (t) required to describe a constant angle is

$$t = \frac{\alpha}{k}\frac{1}{I} \qquad (10.19)$$

As can be seen, this time decreases when I increases as was proved experimentally (Figure 10.31).

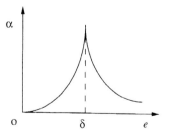

Figure 10.32. Variation of the bending angle of a muscle, for a constant consumed charge, as a function of the polypyrrole thickness, following equation (10.10). The optimum thickness (δ) is shown on the figure.

2.6.5.9 Control of the movement

The rate of the movement depends on every variable influencing the kinetics of the electrochemical reaction taking place in the solid film. Working at constant temperature and constant concentration of electrolyte, the rate of the movement is controlled through the current density. If very high current densities are used controlling steps like ionic diffusion in the solid or slow conformational relaxations induce a resistance, and the potential of the polymer increases to a new value where a new reaction (oxygen or hydrogen evolution) consumes the current flow. The upper limit depends both on the conditions of synthesis of the conducting polymer and on the condition for bilayer work. In our laboratory films able to accept current densities up to 100 mA cm^{-2} were constructed. The manipulation has to be careful to avoid overoxidation processes or combustion by Joule effects.

Working at constant current densities lower than 30 mA cm^{-2} the rate of the movement is proportional to the current density as was proved above. The direction of the movement changes when the current flow is reversed and the movement is stopped when the current flow is stopped. Thus by means of the sense and intensity of the current we control sense and rate of the movements.

2.6.5.10 Lifetime

This parameter has not yet been optimized. Partial studies were performed in our laboratory by means of continuous square waves of potential or of current that promoted consecutive cycles of movement (a movement from −90°, related to the vertical, to +90° and return to −90° is considered a cycle). Sufficiently high current densities, or potentials, to produce a rate of 10 s per cycle, give between 200 and 1000 cycles of life [53].

The main problem appeared at the metal/polymer borders, around the electric contact. After several hundred cycles a fissure appeared around the metallic contact. In previous cycles an important Joule effect is present, as can be observed by the water evaporation. At high current densities even incandescent points are observed [65].

2.6.5.11 Degradation

Artificial muscles degrade during cycling. Nevertheless, most of the electrogenerated films of conducting polymers are, already partially degraded. During polymerization the electrogeneration of lineal poly-meric molecules coexist with degradation processes, crosslinking processes and chemical generation (most of them initiated by protonation) of non-conducting chains in the reaction layer [58]. Most of the chemically generated polymer is adsorbed on the growing film before being finally included in it. Degradation, or electrochemical overoxidation is usually initiated at lower potentials than those used for electropolymeriza-tion [59,60]. The key idea in this work is that electropolymerization is a fast, though complex, mechanism to obtain a mixed material: a network of crosslinked conducting chains, containing non-con-ducting molecules of chemically generated polymer and lakes of degraded material [58]. The composition and properties of the electrogenerated film have to be adequate for every application and can be controlled by means of the conditions of synthesis. In the case of artificial muscles a conducting film without chemically generated non-conducting molecules, with an adequate degree of crosslinking to optimize stress gradients and without degraded regions is needed.

Once the bilayer is formed, the degradation of the conducting film has to be avoided. Degradation processes seem to occur through two main ways: Joule effects and overoxidation. The Joule effect is due to the semiconducting nature of the neutral polymer. This makes it difficult to conduct high anodic or cathodic currents. The flow of those current through a resistance generates heat, water bubbling and oxygen reaction.

Overoxidation seems to be related to electrochemical reactions of solvent oxidation. Radicals are formed that are able to react with polarons promoting a conjugation loss.

Overoxidation processes can be followed by voltam-metry. A bilayer is submitted to a potential sweep between -100 mV (vs. SCE) and 3000 mV in 0.1 M LiClO$_4$ aqueous solutions (Figure 10.33). In order to achieve an equilibrium state at every potential a low sweep rate of 3 mV s^{-1} was employed. During the oxidation process from -100 mV to 650–700 mV a movement of 180° is observed. If the potential sweep is reversed from any potential previous to 750 mV the movement is reversed. At potentials higher than 750 mV an overoxidation–degradation process takes place and a slow relaxation of the bilayer is observed to $+45°$, when the potential arrives at 2 V. This act points to a decrease of stress gradients at the polymer–polymer interface, probably due to a generation of oxygen. All the electrochemical and mechanical processes become irreversible. Current densities de-crease with higher overpotentials, pointing to a decrease in conductivity.

2.6.5.12   Microscopic model

The insolubility of both oxidized and neutral states of most of the electrogenerated conducting polymers supports the existence of a crosslinked structure. In

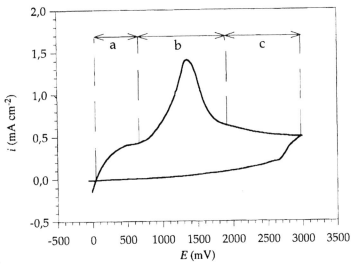

Figure 10.33. Voltammogram obtained from a bilayer between -100 mV (vs. SCE) and 3 V at 3 mV s^{-1}, in a 1 M LiClO$_4$ aqueous solution. Region a is related to the reversible oxidation process; region b is linked to an overoxidation–degradation process and region c is a remaining oxidation process at high overvoltages.

the neutral state (a theoretical state where only non-charged polymeric fibers are present), strong polymer–polymer interactions exist giving a compact structure. This compact structure is confirmed by the movement of the muscles, a result of the increase in compactness during reduction, as well as by the determination of the density by flotation: in spite of the loss of counterions and solvent during reduction the density of polypyrrole, obtained by flotation, is the same in the neutral and oxidized states [53].

An anodic polarization promotes the formation of positive charges along the chains at the polymer–solution interface due to coulombic repulsions with the positive metal and between polarons. The simultaneous presence of coulombic repulsions between charged chains, coulombic attractions between charged chains and counterions and ion–dipole interactions between charged chains and water molecules promotes the opening of the polymeric entanglement at the polymer–solution interface (Figure 10.34).

The initial movement perturbs and induces conformation movements on the surrounding chains, which favors the expansion of the oxidized domain. As a consequence of the compact structure of the neutral film and the open structure of the oxidized domains a stress appears at the interface of the neutral and oxidized domains favoring the oxidation process.

The expansion of the oxidized domains progresses along the electric field favored by ionic migration in this field and transversely favored by the interfacial stress between oxidized and neutral domains, assisted by counterions and solvent diffusion.

During reduction by a cathodic current, reverse processes occur: electrons are injected into the polymer, positive charges along chains are compensated, counterions and solvent molecules are expelled by ionic migration, the electroosmotic effect and stress compres-sion of the polymeric fibres takes place. This compression is due to increasing polymer–polymer interactions.

A mathematical model is in progress based on the electrochemical conformational relaxation model presented later.

2.7 Other configurations. A triple layer

The bilayer structure is a useful device. Nevertheless, the electrical current flowing through the system needs a counterelectrode to allow this flow. At the interface, solution–counterelectrode, an electrochemical reaction is triggered by this flow i.e. oxygen or hydrogen production. The evolution of a gas is dangerous, specially if our artificial muscle is used to move a scalpel at the bottom of a medical probe. A second technical aspect is related to the loss of the electrical energy consumed to produce the counterelectrode reactions.

The configuration that allows the use of either anodic and cathodic processes is a triple layer: conducting polymer/double-sided tape/conducting polymer (Figure 10.35) [63,65]. The experimental procedure is the same as that used to obtain a bilayer. A steel sheet is coated, by electropolymerization from pyrrole in acetonitrile containing LiClO$_4$ and 1% water. Once rinsed and dried a double-sided tape, protected on one side, is adhered to the partially oxidized polypyrrole film coating on one side of the stainless steel. The polypyrrole/double-sided tape/protecting plastic layer is removed from the stainless steel. Then the protecting plastic is removed and the bilayer polypyrrole/double-sided plastic tape is stuck to the remaining coated face of the stainless steel. Then, the final triple layer polypyrrole/double-sided plastic tape/polypyrrole is removed and is ready to be used.

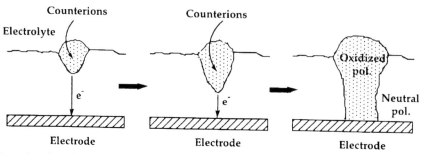

Figure 10.34. Lateral section of the polymeric film during the nucleation and growth of the conducting zones after a potential step. (Reproduced by permission of Elsevier Science SA from *J. Electroanal. Chem.* **394**, 211, 1995.)

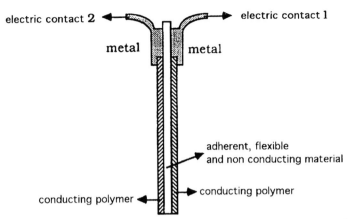

Figure 10.35. A triple-layer device formed by a conducting polymer/a double-sided tape/and a second conducting film [63].

The double-sided plastic tape acts both, as an electronic insulator between the polypyrrole films and as a support for stress gradients during muscle work. Those stress gradients and electrical connections are shown in Figure 10.36. One of the conducting films is connected to the anode, thereby swelling and pushing the free bottom of the triple layer. The second conducting film is connected to the cathode thereby shrinking, by reduction processes, and trailing the triple layer. No electric contact exists between the films due to the presence of the non-conducting layer in the middle. The electric flow between the conducting films takes place by ionic conduction through the solution. Figure 10.37 shows the movement of a triple layer in an aqueous solution of $LiClO_4$.

The advantage of the triple layer is that it is a compact system, where the electrode and counter-electrode are packed in the same multilayer. The second advantage is that current is used twice, thus allowing a greater efficiency and avoiding any parallel reaction.

This structure opens new possibilities to the use of multilayers with a polyelectrolyte between two triple layers. Thicker structure presents, nevertheless, an important rigidity.

2.7.1 Compact and closed multilayer systems

A second and most useful possibility developed in the laboratory is a multilayer system containing the triple layer muscles, surrounded by a gel film containing an electrolyte, or a solid and flexible polyelectrolyte allowing ionic contact between the conducting films. The multilayer system is sealed by a polymeric film

allowing electric contact (Figure 10.38). In this way a compact multilayer, able to work in air, gases or pure liquid media (provided the sealing film is stable in the media), has been produced.

2.7.2 Artificial muscles and actuators based on conducting polymers in literature

Changes of volume in conducting polymers have been envisaged as a base to produce smart systems, actuators and artificial muscles. In 1991, Baughman *et al.* [46] underlined the great interest in developing material technologies suitable for the construction of electrochemical actuators with very small dimensions. The goal proposed is to make the same transition in scale for mechanical devices that has already been made for electronic devices.

A design of both unimorph and bimorph electrochemical cells, analogous to bimorph structures for piezoelectric polymers, is proposed. A simple electrochemical bimorph cell will consist of a polymer electrode strip and a polymer counterelectrode strip cemented together by a polymeric electrolyte, which electronically separates the electrode elements.

Different microactuators, such as devices for controlling fluid flow (Figure 10.39), tweezers (Figure 10.40), conducting polymers with four or five electrode actuators for the two-dimensional movement of optical cable (Figure 10.41) or a Bourdon tube electromechanical actuator (Figure 10.42), were proposed.

Following this idea, Pei and Inganäs developed complex multilayer systems based on non-conducting polymer films like polyethylene (PE) or polyimide (PI). Those films (45 μm thick) were coated with chromium

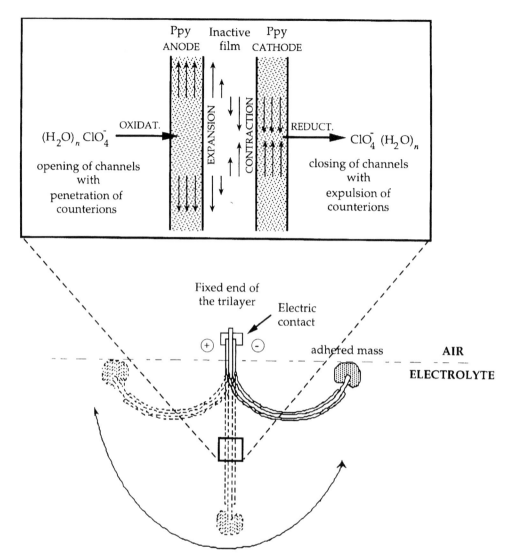

Figure 10.36. Artifical muscle formed by a triple layer polypyrrole/non-conducting tape/polypyrrole. The consumed charge works twice in this device: when polypyrrole I is oxidized (anodic process) pushing the free end of the layer, polypyrrole II is reduced (cathodic process) thereby trailing the layer. Stresses at the polymer–polymer interfaces are summarized in the square box.

and later with a 2 μm thick gold layer by vacuum evaporation. The films were used like anodes to electrogenerate polypyrrole (40 μm thick) films from aqueous solutions at 0.7 V, or polythiophene films from acetonitrile solutions at 1.7 V. Films from alkylthiophene derivates were prepared, using soluble oligomers, by solution casting [69–71].

Multilayer strips of 4 cm length and 0.3 cm width, were used in the sensor test for ammonia gas or iodine vapors. The degree of bending was calculated from the movement of the free ends of the strips [72].

Polypyrrole can reversibly absorb/desorb ammonia gas, other than chemical compensation of the doped PPy by ammonia. Iodine vapor doping of polythiophenes induces general volume dilation in the polymer, the volume change depending on the long alkyl substituents. The same triple layer, PE/Au/PPy, was used as artificial muscles. Electrochemically controlled movements through very small angles were detected. The reason for those small movements seems to be the condition of synthesis of the polypyrrole (from aqueous solution with a strong participation of parallel degrada-

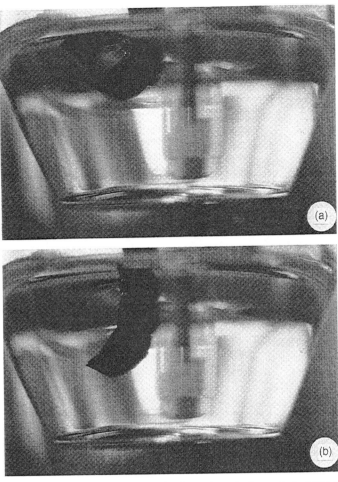

Figure 10.37. Movement of a triple layer polypyrrole/non-conducting polymer/polypyrrole in 0.5 M LiClO$_4$ aqueous solution. During the sequence (a) to (d) an anodic current flows through the polypyrrole film on the left side. Reverse movement takes place if the sense of the current is reversed.

tion processes) [58]. A video tape was presented by Inganäs at the Göteborg's meeting of the International Conference of Science and Technology of Synthetic Metals (ICSM '92) showing changes of curvature from zero to 0.16°. A video tape showing three generations of our group's artificial muscle moving 180° with overall control of the movement by means of the direction of the current flow, shot one year earlier and presented in a different meeting [65] was also presented.

Experimental procedures were improved by differential adhesion; the bilayer polypyrrole/Au–Cr pulls part of the completed device by the tensile stress generated during doping of the polymer [73]. As a non-conducting polymer base was used here, silicon gives a low silicon–chromium adherence.

Using this bilayer actuator, an elegant center-mounted box enclosing two small grains of sand was constructed. The structure is claimed to be able to change shape in 0.5 to 10 s, depending on the thickness of PPy, when the voltage was changed abruptly between −1 V and 0.35 V in aqueous Na$^+$DBS$^-$ electrolyte solutions.

Following a similar procedure and using a gold-coated polyimide film as the electrode, MacDiarmid et al. electrogenerated a polyaniline film [74,75] that can be used as an actuator. The multilayer bending was followed by a laser displacement meter.

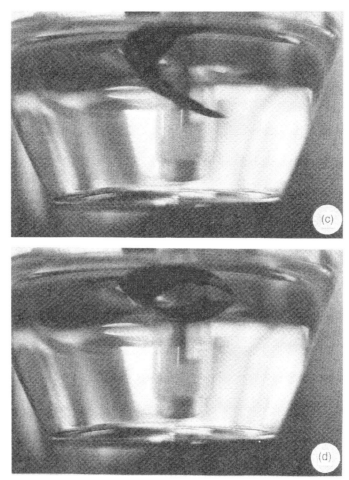

Figure 10.37. (*continued*)

Two triple layers, a 'backbone-type' actuator (similar to my group's triple layer), and a fascinating polyaniline/adhesive solid polymer electrolyte/polyaniline shell-type (artificial muscle) were constructed using casting films from soluble emeraldine. The films (50–100 μm thick) were peeled and stretched to about four times the original length. The films were treated with 1 M HCl to obtain emeraldine salt films, which were used to construct both devices.

The 'shell-type' artificial muscle (similar to Figure 10.35 but using a polyelectrolyte instead of the two-sided plastic tape) was operated in air using a conventional 1.5 V battery. This 'all solid' muscle took several seconds to reach the maximum bending and easily lifted two paper clips of *c.* 0.5 g each.

2.8   Muscles and artificial muscles

In muscles a nervous pulse (ionic) arrives from the brain through the nervous system and triggers chemical reactions, mediated by ATP, in every muscular fiber. A change of volume is promoted by these reactions giving a concomitant mechanical movement. Both, chemical reactions and changes of volume are linked to both the conformational changes on the myosin heads, and the flow of ions, and water, through the fiber membrane. All these processes occur at a constant temperature, but entropic variations during the chemical transformations generate heat, which has to be eliminated in order to keep a constant temperature. A new nervous pulse promotes the relaxation of the muscles, which returns

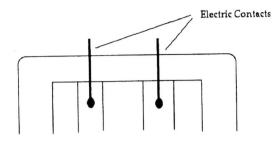

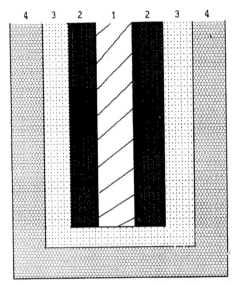

Figure 10.38. Compact and solid multilayer device. (a) Double-sided tape; (2) films of conducting polymer; (3) polyelectrolyte or gel; (4) sealing film [63].

to the orignal position assisted by the contraction of another muscle.

In artificial muscles based on conducting polymers an electric pulse, generated by electrochemical equipment, arrives via metal wires to the conducting polymer films. There, an electrochemical reaction is driven by the current flow. The electrochemical reaction promotes a change in the volume of the film giving a concomitant macroscopic movement of the bilayer and mechanical energy. The electrochemical reaction is linked to conformational changes of the polymer chains together with a flow of ions and water through the polymer–solution interface. All the electrochemical processes occur at constant temperature. The internal resistance of an electrochemical system is always different to zero; some heat is produced by the Joule effect. The overall

variation of temperature is very slow in these artificial muscles. It can be assumed that work was developed under constant temperature. A reverse electric pulse promotes reverse reaction and the muscle returns to the original position.

2.8.1 Similarities

The similarities between natural and artificial muscles can be summarized as follows.

- An electric pulse linked to chemical reactions is involved.
- A change of volume is observed during work.
- A controlled mechanical movement is observed.
- The generated mechanical energy comes, directly, from the transformed chemical energy.
- The mediation of any mechanical piece between both energies does not exist.
- Conformational changes in polymeric chains are linked to the work of the muscles.
- Flow of ions and water molecules through interfaces or membranes are required.
- The systems have to be immersed in aqueous solutions of ionized salts.
- The systems work under constant temperature (they are not Carnot engines).
- A fraction of the involved energy is transformed into heat during work which has to be eliminated.

2.8.2 Differences

In spite of important improvements to artificial muscles based on polymer gels, in relation to natural muscles, some differences still persist.

- The driving power in muscles is chemical energy produced by combustion, at constant temperature, of glucose; the nervous pulse being a trigger. The driving power in artificial muscles is the consumed electric charge; the polymer oxidation and reduction reactions are mediators.
- Muscles only work under contractions due to the irreversibility of the chemical reactions. The work of a second muscle is required to allow the studied muscle to return to the original position. Artificial muscles based on conducting polymers work under contraction and expansion because the electrochemical reaction is reversed when the direction of the current flow is changed.

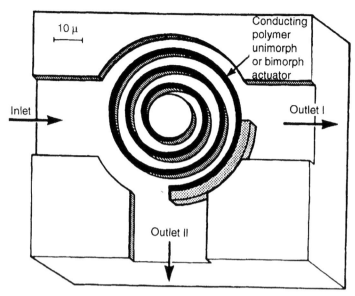

Figure 10.39. Micromechanical actuator device for controlling fluid flow, which utilizes unimorph or bimorph conduction polymers strips. (Reprinted by permission of Kluwer Academic Publishers from ref. 46.)

2.9   Emerging applications

The present level of development allows the construction of reproducible devices for microrobotics and medical instrumentation at the end of probes for the exploration and repair of the human body. Short-range goals include devices such as microtweezers, microvalves, micropositioners for optical instrumentation or fine mechanics, sensors for the detection of chemical compounds able to oxidize, or interact with, the conducting film.

Two different methods have been developed to date: macroscopic using large electrodes (up to 10 cm) and microscopic using silicon-deposition techniques. The field is open, taking into account that conformational changes (at molecular level) controlled through electrochemical or chemical redox processes are the base for these devices. Only technical difficulties limit the miniaturization of molecular muscles, sensors and actuators. The addition of molecular components give a macroscopic system: the ability to build volumetric devices that will mimic and repair natural muscle is only limited by difficulties in synthesis and our imagination. From the tens of well-known conducting polymers, only polypyrrole, polyaniline and polythiophene have been explored at a preliminary level. New developments and surprising applications will appear in the next few years.

3.   ELECTRODISSOLUTION

3.1   Introduction

Conducting polymers are considered in literature as one of the most important components of organic metals. Nevertheless they are considered by many scientists as hybrids of opposite nature. Many of the studied conducting polymers seem to lack some of the most specific polymeric properties: solubility and fusibility. Related to interactions between electrochemistry and metals, conducting polymers can be electrochemically generated like metals, but do not seem to have one of the most important properties: the electrodissolution of metals.

The deficits cause important restrictions in the application of these materials in different fields. Insolubility and infusibility means, from a technological point of view, lack of processability. This fact, together with an extended idea about easy degradation, causes a loss of confidence in most of the conducting polymers, in spite of their fascinating properties, amongst engineers, physicists and condensed matter scientists.

During the last ten years an important effort has been made in order to produce soluble oligomers for important technological applications [76–79]. Nevertheless, not many soluble polymers with more than 24 alternating double bonds have been synthesized.

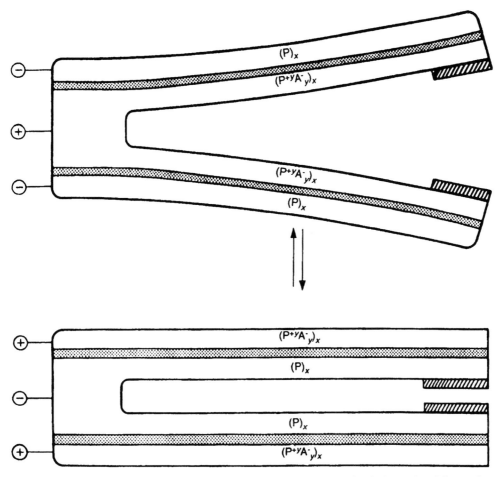

Figure 10.40. Paired bimorph actuators used as micromechanical tweezers. Electrochemical transfer of dopant from the outer layer to the inner layer of each bimorph causes the opening of the tweezers. (Reprinted by permission of Kluwer Academic Publishers from ref. 46.)

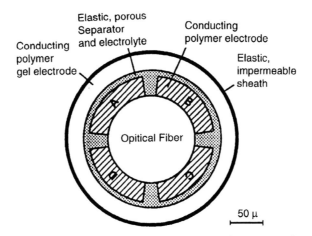

Figure 10.41. Conducting polymer five-electrode actuator for the two-dimensional movement of an optical cable. (Reprinted by permission of Kluwer Academic Publishers from ref. 46.)

Cross-section view of expansion tube

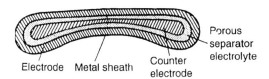

Electrode   Metal sheath   Counter   Porous
                           electrode   separator
                                        electrolyte

Figure 10.42. A Bourdon tube electromechanical actuator, which operates by the net volume change of electrode, counter electrode, and electrolyte upon electrochemical reduction and oxidation. (Reprinted by permission of Kluwer Academic Publishers from ref. 46.)

For several years our group's attention was attracted by the lack of electrodissolution of conducting polymers. We were interested in the degradation and crosslinking processes occurring together with electropolymerization [80,81]. Another important fact was related to the solubility, or insolubility, or to the relative value of the crossed interactions between neutral polymer chains, charged polymer chains, ions and solvent molecules. The aim was to get a non-crosslinked and non-degraded polymer film with strong interactions between the molecular and ionic units present in the film (polymer, charged polymer and ions) and the solvent molecules. Under those conditions soluble polymers will be obtained.

In order to produce polymers able to be electrochemically solved, the aim was to find conducting polymers with oxidized and reduced states which show different solubilities. This depends on the properties of the polymer and the electrolyte. The electrolyte is necessary to allow the current flow because a solubilization during (and proportional to) a current flow is desired.

The most direct way to attain this target seemed to be through the study of some of the known soluble polymers obtained by electrogeneration. Between them, trimers formed by different combinations of pyrrole (N), thiophene (S), indole (O), etc. are well known in literature [82–89]. The second aspect to be solved is the synthesis of monomers with high enough mass to allow the kinetics of both electrogeneration and electrodissolution processes to be followed. This is important in order to obtain solubility in different solvents.

The monomers SNS, SOS, SSS all give soluble polymers, but only SNS has been found to have important differences in solubility between the neutral and oxidized forms in some solvents. The work has focused on this monomer in order to describe the synthesis of both monomer and polymer, the electro-

dissolution processes and characterization of the polymer.

3.2   Synthesis of monomer

The availability of important quantities of monomer was possible by improving the method described by Wynberg and Matselaar [84]. The intermediate 1,4-di-(2-thienyl)-1,4-butanedione was kept under reflux with ammonium acetate, glacial acetic acid and acetic anhydric overnight under a nitrogen atmosphere [90]. The reaction mixture was then poured into 250 ml of distilled water and the resulting dark-green solid was analyzed by chromatography using a silica gel column, and a mixture of dichloromethane and hexane (3 : 2) for elution. A 75% yield of SNS, as pale-yellow crystals, was obtained.

Several grams of the synthesized monomer can be obtained every time; the solid was stored under nitrogen.

3.3.   Electropolymerization: soluble or insoluble films

The electropolymerization of polySNS was performed on 1 cm^2 platinum electrodes either by cyclic voltammetry, constant potential polarization or flow of a constant current density.

Voltammograms obtained in 5 mM SNS and 0.1 M LiClO$_4$ acetonitrile solutions showed a monomeric oxidation initiated at 300 mV (Figure 10.43). This oxidation promotes the coating of the electrode with a violet-dark polymeric film. Films generated up to an anodic potential just lower than 0.75 V were solved along the cathodic potential sweep giving a yellow-green cloud around the electrode. Films generated until more anodic potentials limits remain unsolved during the cathodic potential excursion, as well as during consecutive potential cycles [90,91].

Films generated by polarization of platinum electrodes at constant potential showed a powdered morphology, with low adherence. This fact makes manipulation and quantitative determination of polymeric masses difficult.

The flow of constant current densities gives uniform and adherent films. The flow of low current densities, such as 0.5 mA cm^2, gives, under these experimental conditions, a constant potential, during the polymerization time, of 0.6 V (Figure 10.44). Higher current densities give two or more (up to four) potential steps, as can be observed on the same figure. The different

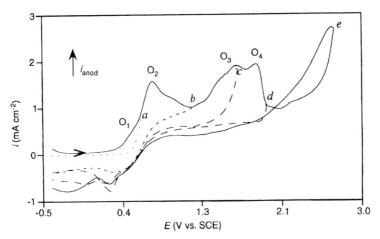

Figure 10.43. Cyclic voltammograms obtained in a 5 mM SNS and 0.1 M LiClO$_4$/acetonitrile solutions using a 1 cm^2 platinum electrode. Scan rate: 50 mV s^{-1}; temperature: 20°C; initial and final potential: -0.4 V; reversal potentials: (a) 0.60, (b) 1.15, (c) 1.65, (d) 1.96 and (e) 2.70 V. (Reproduced by permission of Elsevier Science SA from *J. Electroanal. Chem.* **370**, 231, 1994.)

steps of the potential correlate quite well with maxima on the voltammograms of Figure 10.43. A similar correlation is also seen with the solubilities: films obtained during the first potential step by the flow of a constant current, are soluble in different organic solvents (methanol, acetone, tetrahydrofuran, propylene carbonate or acetonitrile) and were insoluble in water or in LiClO$_4$ acetonitrile solutions. Films obtained when the potential stopped on the second step ($E > 0.9$ V) were partially soluble. When the potential reached the third or fourth step the films were insoluble.

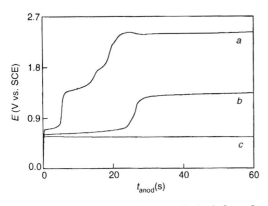

Figure 10.44. Chronopotentiograms obtained from 5 mM SNS in 0.1 M LiClO$_4$ aqueous acetonitrile solution with a 1% (v/v) water content on platinum at a constant anodic current density of (a) 2.0, (b) 1.0, (c) 0.5 mA cm^{-2}. Temperature: 20°C. (Reproduced by permission of Elsevier Science SA from *J. Electroanal. Chem.* **370**, 231, 1994.)

3.4   Kinetics of electropolymerization: a faradaic process

Polymeric films were generated at different constant current densities during different polymerization times. The polymer mass was determined every time by ultramicrogravimetric analysis of the rinsed and dried film. Masses were determined with a precision of 10^{-7} g. All the experimental points fall on a straight line when the masses are plotted vs. the consumed charges (Figure 10.45). This relation is independent of the current density used, or the polymerization time. The slope of this line is the productivity of the current (2.1×10^{-3} mg mC^{-1}). The same result was obtained with solutions containing different concentrations of monomer or electrolyte. The conclusion is that the electropolymerization of SNS is a faradaic process.

From the experimental masses (W) and consumed charges (Q_{pol}), the number of electrons consumed to incorporate a monomeric unit into the polymer can be obtained

$$n = \frac{M}{F} \frac{Q_{pol}}{W} \tag{10.20}$$

where M is the molecular mass of SNS and F is the Faraday constant. A value of n close to 1 (Table 10.4) was obtained from each experiment, which does not suggest a radical–cation polycondensation mechanism for the polymerization ($n = 2.25$).

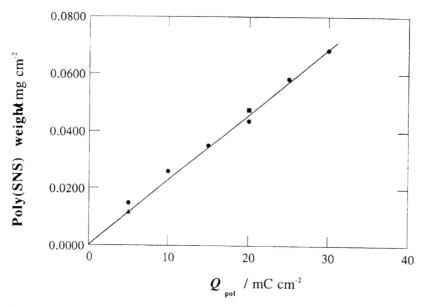

Figure 10.45. Linear dependence of the weight of the electrogenerated poly[2,5-di-(2-thienyl)pyrrole] film on the consumed anodic charge (Q_{pol}) during polymerization in a 5 mM SNS and 0.1 M LiClO$_4$/acetonitrile solution, at different current densities: ▲ 0.25; ● 0.5; ■ 1.0 mA cm^{-2}. Temperature: 20°C. (Reproduced by permission of Elsevier Science SA from *J. Electroanal. Chem.* **370**, 231, 1994.)

3.5   Solubility of the oxidized polymer

Polymeric films once obtained are in an oxidized state. After rinsing and drying they were solved in organic solvents until saturation: an unsolved film remains overnight in the solution, but solubilizes immediately when translated into pure solvent. From those saturated solutions 1 ml was taken and the solvent was eliminated by evaporation. The mass of the dry polymer was obtained.

The solubility of the oxidized polymer (g l^{-1}) decreases (Table 10.5) from dimethylsulfoxide (DMSO) to acetone; it is very low in acetonitrile and

the polymer is insoluble in water or organo-aqueous solutions (< 50% water).

3.6   Electrochemical reduction of the polymer: electrodissolution

The low solubility of polySNS in acetonitrile drops to zero when different salts are solved in its solvent. This is the case of 0.1 M LiClO$_4$ acetonitrile solution being the presence of the salt the origin of this insolubility. Lower solubilities of the oxidized polymer are obtained in 0.1 M LiClO$_4$ and 1% water in acetonitrile; mean-

Table 10.4. Anodic electrogeneration and cathodic stripping of poly(SNS) films at 20°C (Reproduced by permission of Elsevier Science SA from *J. Electroanal. Chem.* **370**, 1994, 231–239.).

t_{anod} (s)	Weight of poly(SNS) (mg cm^{-2})	Q_{pol} (mC cm^{-2})	Q_{strip} (mC cm^{-2})	$\lvert Q_{strip} \rvert Q_{pol}$	n
10	0.0149	5	−1.38	0.28	0.80
20	0.0259	10	−2.90	0.29	0.93
30	0.0350	15	−4.32	0.29	1.03
40	0.0436	20	−5.86	0.29	1.13
50	0.0582	25	−7.22	0.29	1.03
60	0.0682	30	−8.60	0.29	1.05

Table 10.5. Density, conductivity and solubility in different solvents for the oxidized and reduced forms of poly(SNS). Temperature: 20°C (Reproduced by permission of Elsevier Science SA from *J. Electroanal. Chem.* **392**, 392, 1995.).

Poly(SNS)	ρ (g cm^{-3})	σ (S cm^{-1})	Solubility (mg (100 ml)$^{-1}$)		
			DMSO	Acetone	Acetonitrile
Oxidized form	0.986	5×10^{-5}	28	18	4
Reduced form	0.998	5×10^{-9}	420	13	55

while electropolymerization in the same background solution gives very thick conducting film [92].

When a platinum electrode coated with polySNS film was submitted to polarization at potentials more negative than 0.2 V (vs. SCE) a yellow cloud of soluble polymer is observed around the electrode: electrodissolution (Figure 10.46). The dissolution is very fast (~ 2 s) at potentials lower than -0.2 V. When the coated electrode is subjected to polarization at anodic potentials higher than 0.4 V no electrodissolution was observed: the oxidized film is insoluble.

A potential sweep from 0.4 V to 2 V shows three new oxidation processes of the oxidized polymer (Figure 10.47). Processes O_3 and O_4 give an insoluble and passive (redox processes become irreversible) polymeric film. Nevertheless if the potential sweep is

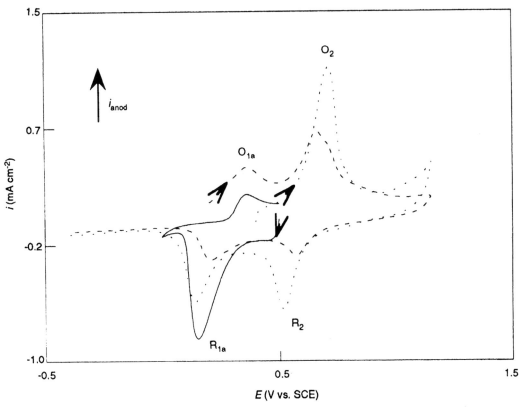

Figure 10.46. Control voltammograms of poly(SNS) films in 0.1 M LiClO$_4$/acetonitrile solution at 50 mV s^{-1} and 20°C: $----$ initial and final potential: 0.5 V; reversal potential: 0 V; $-----$ initial potential: 0.2 V; reversal potential: 1.15 V; final potential: -0.4 V; .... initial potential: 0.4 V; reversal potential: 1.15 V; final potential: -0.4 V. The films were generated from a 5 mM SNS and 0.1 M LiClO$_4$/acetonitrile solution at $j_{anod} = 0.5$ mA cm^{-2} for 20 s. (Reproduced by permission of Elsevier Science SA from *J. Electroanal. Chem.* **370**, 231, 1994.)

reversed before the start of the O_3 process, two reduction maxima are observed, related to bipolaronic and polaronic processes: R_2 and R_1. During R_1 the electrodissolution process occurs (Figure 10.46).

If the oxidized film is submitted to a constant cathodic current density, the concomitant chronopotentiogram shows (Figure 10.48) a very slow decrease in potential while the electrodissolution process is observed. When all the polymer was solved and the platinum electrode recovers its original shine a sharp potential jump is observed on the chronopotentiogram.

The insolubility of the polymer in water (or aqueous solutions) allows the study of the redox processes in 0.1 M $LiClO_4$ aqueous solutions [93]. Both oxidation processes, O_1 and O_2, seem to overlap when studied by either cyclic voltammetry (Figure 10.49) or by a constant cathodic current (Figure 10.50); and the polymer becomes reduced at -0.3 V.

Ratios between the charge consumed during polymerization and those consumed during voltammetric reduction (Q_{pol}/Q_{red}) remains constant for different masses of polymer and equals 0.22 ± 0.01. The masses of the dried reduced films were $21.5 \pm 1\%$ lower than those of the correlative oxidized films.

3.7   Solubility of the reduced polymer

The reduced polymer is available by reduction of the just electrogenerated films from aqueous acetonitrile solutions. Once dried these films are solved in different organic solvents in order to determine solubilities, using the above method. Solubilities in DMSO, acetone and acetonitrile are shown in Table 10.5. The solubility of the reduced polymer in DMSO is 15 times that of the oxidized polymer. In acetone both states have similar solubilities. In acetonitrile the neutral polymer is 11 times more soluble than the oxidized polymer.

Under these conditions electrodissolution can be understood. The low solubility of the oxidized polymer (an ionic compound) in acetonitrile becomes zero by the effect of a common ion in the solution. Molecules of the neutral polymer interact with acetonitrile molecules and become soluble at the same time as they are formed by reduction. This is an important structural conclusion related to the electrochemical reduction–dissolution and it takes place at the polymer–solvent interface. The unsolved film remains compact and adhered to the back metal. This fact can not be generalized to crosslinked or insoluble film.

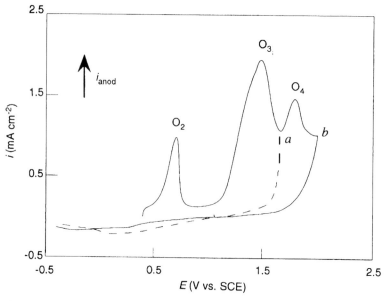

Figure 10.47. Control voltammograms of poly(SNS) films in 0.1 M $LiClO_4$/acetonitrile solution at 50 mV s^{-1} and 20°C. Initial potential: 0.4 V; reversal potential: (a) 1.6 and (b) 2 V, and final potential, -0.4 V. The films were generated from a 5 mM SNS and 0.1 M $LiClO_4$/acetonitrile solution at $j_{anod} = 0.5$ mA cm^{-2} for 20 s. (Reproduced by permission of Elsevier Science SA from *J. Electroanal. Chem.* **370**, 231, 1994.)

3.8 Electrodissolution kinetics: a faradaic process

To follow the kinetics of the electrodissolution process, reproducible polymer films (one for every experimental point) weighing 0.150 ± 0.001 mg were electrogenerated by flow of 0.5 mA cm^{-2} [92] for 160 s on a 1 cm^2 platinum electrode. To solve the polymer, cathodic constant current densities of 0.5, 0.2, 0.1 and 0.05 mA cm^{-2} were passed through the 0.1 M LiClO$_4$ acetonitrile solution. A lineal variation of the solved mass (obtained by the difference between the masses of the dry coated electrodes before electrodissolution and after every electrodissolution time) with increasing times of current flow is observed (Figure 10.51). The slopes are proportional to the employed cathodic currents giving a faradaic process: 7.2×10^{-3} mg of polymer are solved per millicoulomb of consumed cathodic charge. Similar results were obtained from different thicknesses, current densities or background solutions [90,91].

As electrogeneration and electrodissolution are both faradaic processes, a constant (Q_{disol}/Q_{pol}) ratio of 0.29 ± 0.01 is obtained. Thirty per cent of the charge consumed during generation is required to solve the polymer; this means one electron every 3.3 monomeric units (10 rings).

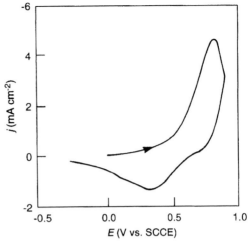

Figure 10.49. Control voltammogram of a reduced poly(SNS) film in 0.1 M LiClO$_4$ aqueous solution at 50 mV s^{-1} and 20°C. Initial potential: 0 V; reversal potential: 0.90 V and final potential: -30 V. The films were electrogenerated on a 1 cm^2 platinum electrode from a 5 mM SNS and 0.1 M LiClO$_4$/acetonitrile solution at $j_{anod} = 0.5$ mA cm^{-2} for 60 s and further reduced in 0.1 M LiClO$_4$ aqueous solution at $j_{cat} = -0.2$ mA cm^{-2} until $j_{red} = 31$ s. (Reproduced by permission of Elsevier Science SA from *J. Electroanal. Chem.* **392**, 392, 1995.)

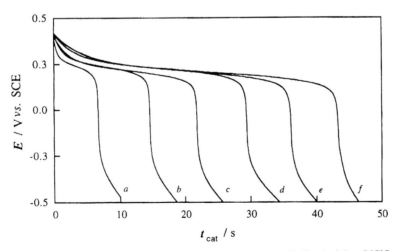

Figure 10.48. Chronopotentiograms obtained for electrodissolution of a poly(SNS) film in 0.1 M LiClO$_4$ aqueous acetonitrile solution containing 1% (v/v) water on platinum at a constant cathodic current density of -0.1 mA cm^{-2}. Each film was electrogenerated on a 1 cm^2 platinum anode from a 5 mM SNS at $i_{anod} = 0.5$ mA cm^{-2} for t_{anod}: (a) 20, (b) 60, (c) 100, (d) 120, (e) 160 s. Temperature: 20°C. (Reproduced by permission of Elsevier Science SA from *J. Electroanal. Chem.* **370**, 231, 1994.)

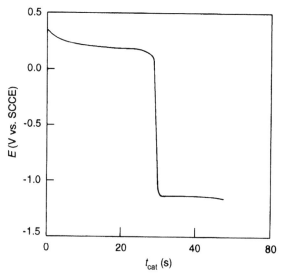

Figure 10.50. Chronopotentiogram recorded for the electro-reduction of a poly(SNS) film in 0.1 M LiClO$_4$ aqueous solution at $j_{cat} = -0.2$ mA cm^{-2} and 20°C. The films was previously electrogenerated on a 1 cm^2 platinum electrode from a 5 mM SNS and 0.1 M LiClO$_4$/acetonitrile solution at $j_{anod} = 0.5$ mA cm^{-2} for 60 s. (Reproduced by permission of Elsevier Science SA from *J. Electroanal. Chem.* **392**, 392, 1995.)

3.9 Densities and conductivities

Several dried films obtained from different concentrations of monomer, some in an oxidized state, others after electrochemical reduction, were peeled from the platinum electrodes. The densities were determined by flotation from methanol/water mixtures in which polymers were insoluble. All the samples have the same density of 0.986 ± 0.002 g cm^{-3}, whatever the conditions of synthesis.

To determine the electrical conductivity, a thick film, 1.5 cm × 0.70 cm × 1.5 μm, was prepared from a 0.1 M LiClO$_4$ and 10 mM monomer in acetonitrile. A non-conducting plastic tape was adhered to the dried polymeric film and the bilayer was removed from the platinum electrode. A conductivity of 5×10^{-5} S cm^{-1} was obtained from oxidized films and 5×10^{-7} S cm^{-2} was obtained from reduced films using the four-points method.

3.10 Analysis and characterization

The elemental analysis of the oxidized polymer [93] indicates a salt structure that contains 0.65 perchlorate ions per monomeric unit. The formula of this compound is

$$[(SNS^{0.65+})_n(ClO_4^-)_{0.65n}]_s$$

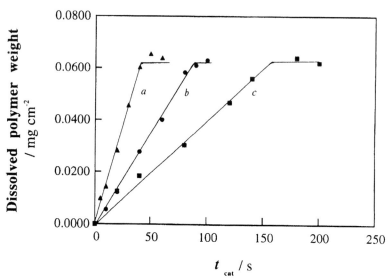

Figure 10.51. Evolution of the weight of disolved poly(SNS) with time when a cathodic current density of: (a) −0.2, (b) −0.1, (c) −0.05 mA cm^{-2} was applied to a film in 0.1 M LiClO$_4$ + acetonitrile solution at 20°C. Each film was generated on a 1 cm^2 platinum electrode from a 5 mM SNS and 0.1 M LiClO$_4$/acetonitrile solution at $j_{anod} = 0.5$ mA cm^{-2} for 60 s, the film weight being 0.0685 ± 0.0005 mg cm^{-2}. (Reproduced by permission of Elsevier Science SA from *J. Electroanal. Chem.* **370**, 231, 1994.)

This means that a positive charge is stored every 1.54 monomeric units.

Comparison of the FTIR reflectance spectra of the polymeric film and FTIR spectra of the monomer shows the disappearance of the monomeric band present at 691 cm^{-1}, and related to α-hydrogen atoms. This fact points to the formation of a lineal polymer.

UV/vis spectra (Figure 10.52) show two bipolaronic states and a self-reduction process of the polymer in solution: the absorptions decrease with time on the bipolaronic bands and increase on both the polaronic band and on the M $\rightarrow$ M* transition band.

The determination of the molar masses by fast atom bombardment mass spectroscopy (FAB-MS) allowed

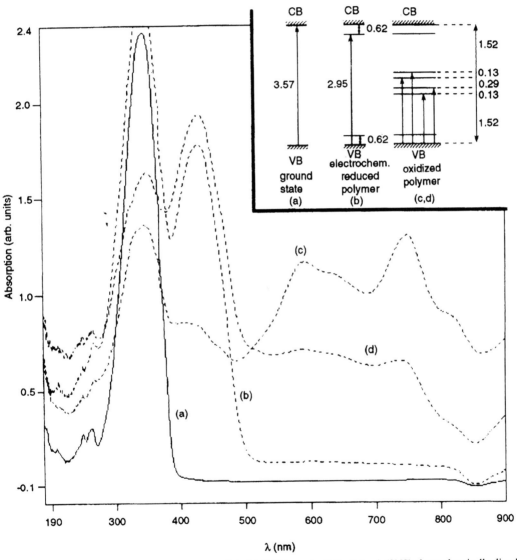

Figure 10.52. UV/vis spectra (absorption in arbitrary units): (a) SNS in acetonitrile; (b) poly(SNS) electrochemically dissolved in 0.1 M LiClO$_4$/acetonitrile solution; (c) poly(SNS) dissolved in acetonitrile; (d) as (c) but 20 min later. Band structures (related to the higher density of states for each band) for each spectrum are given in the inset. In (c) and (d) only the new transitions are depicted. Transitions are depicted in eV. (Reproduced by permission of Elsevier Science SA from *J. Electroanal. Chem.* **270**, 231, 1994.)

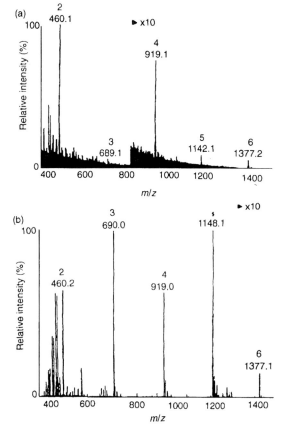

Figure 10.53. FAB-MS positive ionization mode for poly(SNS) in a thioglycerol matrix; (a) oxidized form (previously saturated in DMSO); (b) reduced form. Mass peaks given are the maximum peaks associated with dimer 2, trimer 3, tetramer 4, pentamer 5 and hexamer 6 proceding from electropolymerization of monomer 1. (Reproduced by permission of Elsevier Science SA from *J. Electroanal. Chem.* **392**, 392, 1995.)

the indentification of oligomers in the soluble polymer up to six monomeric (SNS) units (1377 a.u.). Peaks were also present that corresponded to the heptamer

and octamer, but the intensities were so small that it was difficult to analyze. The same soluble oligomers as for the oxidized form (Figure 10.53) were also detected for the reduced polySNS. Experimental results up to hexamers are collected on Table 10.6. Higher molecular masses can not be detected using this technique.

3.11 Conclusions

Electrodissolution is a new way to process conducting polymers and it is suitable for technological applications. PolySNS can be electrochemically generated, following a faradaic process, and electrochemically reduced, following a different faradaic process. These processes mimic electrodeposition and electrodissolution of inorganic metals, although polymeric processes (electropolymerization and electrodissolution) are not symmetric processes. This fact allows similar applications: electroreprography, electrophotography, electrolithography, electromachining, electroerosion and electropolishing.

Polymeric electrodissolution is supported on soluble polymers with different solubilities between the oxidized and reduced states. Effort has to be devoted to obtaining new soluble conducting polymers by electrogeneration and to the study of the solubilities of their oxidized and reduced states in different solvents, solvent mixtures and electrolytic solutions.

A second technological point of interest is the availability of solutions from conducting polymers than are able to give conducting films by casting.

4 REDOX PROCESSES

4.1 Introduction

Pure conducting polymers like polypyrrole, polythiophene, polyaniline, etc. can be considered as *neutral*

Table 10.6. Molecular mass of n times the monomer mass A, theoretical masses of oligomers $(M)_n$ accepting a polycondensation mechanism, $nA - (2n - 2)$, and experimental maximum mass peaks for the oligomers of the oxidized and reduced poly(SNS) obtained from FAB-MS (n is the number of monomeric units present in the oligomer) (Reproduced by permission of Elsevier Science SA fro *J .Electroanal. Chem.* **392**, 392, 1995.)

Molecular mass (a.u.)	Monomer	Dimer	Trimer	Tetramer	Pentamer	Hexamer
nA	231	462	693	924	1155	1386
$nA - (2n - 2)$		460	689	918	1147	1376
Oxidized poly(SNS)		460	689	919	1148	1377
Reduced poly(SNS)		460	690	919	1148	1377

organic materials. They have a low conductivity and can be oxidized both by chemical or electrochemical ways to give *non-stoichiometric ionic composites*: cationic polymeric chain/anions. Those anions can be either inorganic or organic. The weight percentage of anions in the composite can be changed in a continuous way from zero up to fifty per cent in some cases.

The electrochemical manipulation of this process allows perfect control of the oxidation and any intermediate concentration of the non-stequiometric composite can be attained. The second and well-known important fact is that the electrochemical process can be reversed and the original neutral conducting polymer can be recovered by reduction.

Oxidation/reduction processes involving a non-stequiometric compound is the base for all the electrochemical applications of conducting polymers linked to electrochemical properties. Reversible redox processes and properties based on neutral polypyrrole can be summarized as follows.

$$[\text{Pyrrole}]_s + [n(\text{ClO}_4^{-})(\text{H}_2\text{O})_a]_{aq} \underset{\text{Red.}}{\overset{\text{Oxid.}}{\rightleftarrows}}$$

$$[(\text{Polypyrrole})^{n+}\{(\text{ClO}_4^{-})(\text{H}_2\text{O})_a\}_n]_s + (ne^-)_m$$

Neutral $\longleftrightarrow$ *Solid state reaction* $\longleftrightarrow$ Oxidized
Semiconducting $\longleftrightarrow$ *Conductivity* $\longleftrightarrow$ Conducting
Color a $\longleftrightarrow$ *Electrochromic* $\longleftrightarrow$ Color b
Volume 1 $\longleftrightarrow$ *Electrochemomechanical* $\longleftrightarrow$ Volume 2
Compact $\longleftrightarrow$ *Electropermeability* $\longleftrightarrow$ Expanded
Soluble $\longleftrightarrow$ *Electrodissolution* $\longleftrightarrow$ Insoluble

This is a solid state reaction because the change of composition takes place in a solid polymer. The composition of the non-stequiometric compound is assumed to be uniform whatever the percentage of the oxidized state. The exception to this behavior is the oxidation of the polymer under conformational relaxation control of the polymer fibres. A nucleus of oxidized polymer is formed which expands by consumption of the reduced structure [45]. These processes will be covered later. Uniform composition and continuous processes will be discussed now, i.e. changes of the uniform composition during an oxidation process, which are linked to continuous changes of the concomitant properties.

The conductivity of the oxidized film changes continuously as the color does, the permeability and the volume. The electrodissolution is related to the change of solubility between a reduced and an oxidized polymer (allowing a faradaic control of the solubility).

4.2   Theoretical models for redox processes in literature

The most simple and oldest model is focused on the role of penetrating counterions in the electrochemical switching of conducting polymers. In this sense, experimental data can be analyzed using a diffusion model (Cottrell equation) or, more adequately, by means of a migration model, e.g. the single-pore model of a porous electrode proposed by Posey and Morozumi [94]. This last model is valid for porous conducting polymers with open structures, so no dependence of the switching rate on the polymer morphology is included. In order to simulate chronoamperometric curves, some electrical magnitudes of the film are needed: the electronic resistance of the solid phase (which is assumed to be negligible), the ionic resistance of the electrolytic phase and the double-layer capacitance. To simulate experimental voltammograms, the variation of these magnitudes with potential is required. Simulated chronoamperometric curves predict a monotonous decay of intensity during oxidation, depending on both the magnitude of the potential step and the electric features of the film. This mode is not able to predict the shape of chronoamperograms performed from high cathodic potentials, or the hysteresis effect on voltammetric curves.

Aoki and coworkers [95,96] simulated chronoamperometric and voltammetric oxidation curves relying on the supposition that the conductive domain propagates from the metal–polymer interface to the polymer–solution interface. The conductive domain behaves as a metal electrode which reacts electrochemically with the non-conductive part. The rate at which the polymer becomes conductive is assumed to be controlled by a charge-transfer step at the interface between the conductive and non-conductive polymer, following a Butler–Volmer equation. Diffusion of counterions on the polymer matrix does not affect the oxidation rate. The appearance of a peak followed by an exponential decay is predicted for the *i*–*t* response after a potential step, the peak time being proportional to the thickness of the film. A semilogarithmic relation between the peak time and the applied anodic potential is also predicted. As the model does not include structural magnitudes of he polymer, it cannot predict the shifting of the oxidation peak to longer times when the polymer is submitted to more cathodic potentials of prepolarization.

Otherwise, electrochemical transition from oxidized to reduced states was simulated by Aoki and coworkers [97,98] by a two-dimensional square lattice model in the context of electrical percolation. In this model, the conducting to insulator conversion is assumed to occur at any position in the bulk of the film. The number and size of the conducting zones decrease rapidly during the time of cathodic polarization, but they do not completely disappear in the first moments of reduction. The conducting to insulator process stops at a percolation threshold, when the electric contact between conducting clusters and the electrode is ceased. Conducting zones left in the film are now slowly transported to the electrode by diffusion during the time of cathodic polarization. This rearrangement of the conducting zones allows reduction to continue. In order to simulate this process, the conducting polymer is assumed to be an isotropic medium, without crosslinks or entanglements between chains, without crystal structure and so porous that the counterions are sufficiently supplied to the redox sites. From geometrical (not structural) considerations, a semilogarithmic dependence between the number of clusters remaining conductive and the wait-time at a given cathodic potential is obtained. As a semilogarithmic relation between the peak potential or intensity and the wait-time can be observed in voltammetric experiments [99], it is deduced that this slow relaxation process is responsible for these anomalous effects, but no direct evidence is provided.

Feldberg [100] introduces capacitive currents to explain the form of the voltammetric curves, making the assumption that the experimental value for the faradaic charge spent during complete oxidation of conducting polymers can not be directly obtained from simple electrochemical experiments (e.g. cyclic voltammetry or chronoamperometry). The differential capacitance of the conducting film is assumed to be proportional to the amount of oxidized polypyrrole and independent of potential. In order to obtain an expression for the faradaic charge involved in the redox process, it is considered as a simple reversible one-electron transfer. In a similar way to the other models, this treatment pays no attention to the structural features of the polymer. In this way, only simple behaviors can be simulated.

Another type of model is based on the conservation of mass and charge along the different regions in the electrode. In this sense, White and coworkers [101,102] developed a mathematical model able to simulate cyclic voltammograms of polypyrrole on rotating disk electrodes. The polypyrrole film is treated as a porous electrode with a high surface-to-volume ratio and a large double-layer capacitance, which is proportional to the amount of oxidized film, in a similar way to the

model of Feldberg. The mathematical model includes migration of charged species in an electric field, diffusion of charged and uncharged species and the electrochemical reaction occurring within the porous film. A comparison of the simulated and experimental cyclic voltammograms shows quantitative agreement. Spatial and time dependences of concentration, potential and stored charge within the film are predicted by the model. These profiles show that the switching process is mainly governed by the rate of ionic charge transport in the film, which depends on the network structure of the film. Nevertheless, no structural characteristics of the polymer are included, so the model is unable to explain the anomalous effects.

Finally, Otero and coworkers [42,44] considered the switching transition as an electrochemical reaction in solid state. Redox sites (the concentration depends on the electric potential) and counterions are assumed to be the reactants. The activation energy for the rate constant is obtained, assuming an Arrhenius-type law. Reaction orders for both electrolyte and redox sites concentration can be also reduced from this semiempirical model. The polymeric structure affects the counterions mobility within the film and hence its concentration near the oxidation centers. Consequently, the reaction rate is controlled by structural parameters, but this dependence is not quantified as the model is not predictive.

4.3. Conducting polymers as three-dimensional electrodes

A conducting polymer can be considered like a crosslinked network; the polymeric structure makes it insoluble. During oxidation processes, electrons are lost from the polymeric network; positive charges are stored along chains and a flux of ions across the polymer–solution interface takes place in order to maintain the electroneutrality inside the polymer. Positive charges and counterions distribute uniformly inside the film: this is a three-dimensional electrode.

4.4 Generation and destruction of free volume

During the oxidation processes other changes occur inside the film. Neutral chains have strong polymer–polymer interactions and low polymer–water or polymer–ion interactions: the polymeric structure of neutral polypyrrole is compact, as was proved by experimental determination of the density or *in situ* variations of volume. Counterions, required to compensate positive charges generated during oxidation, need the formation of a free volume inside the polymer able to allow their penetration and storage. The kind of ions flowing inside or outside the polymer can be controlled by the generation of free volume at each stage of the oxidation.

The question now is how this free volume can be generated. During oxidation new positive charges are stored along the chains. So polymer–polymer Van der Waals attraction forces between neutral chains are transformed to coulombic repulsion forces; polymer–water interactions become (ion–dipole) attractive forces and polymer–ion interactions become ion–ion coulombic attractions. Under these forces, conformational movement of the polymeric chains forming the network open channels of free volume starting at the polymer–solution interface and advancing through all the film. The free volume is immediately occupied by counterions. If conformational movements are slow and ion-couples remain trapped in the polymer small cations can be initially expelled along the small free volume in the network. During oxidation an increase of volume is present.

At high oxidation levels new chemical forces can be present that allow crosslinking or degradation processes. The forces also influence the evolution of free volume. Those aspects will not be considered here.

Our single approach is based on a pyrrole-type polymeric network. The use of a substituted pyrrole-type with ionic groups, or a polypyrrole–polyelectrolyte type composite promotes new kinds of interactions and subsequently different fluxes of ions and solvents. Polyaniline-type polymers also have a different behavior; oxidation gives deprotonation steps and polaronic and bipolaronic steps. In order to simplify this treatment pure polypyrrole type polymers are discussed.

Following the ideal polypyrrole-type film during reduction in $LiClO_4$ aqueous solution a continuous decrease of volume is observed using artificial muscles, pointing to a continuous destruction of free volume and the expulsion of counterions and water due to interactions opposite to those described above.

4.5 Oxidation controlled by polymeric conformational changes

If the presence of strong polymer–polymer interchain and intrachain forces in a neutral polymer is accepted, the most compact structure will be attained under the total elimination of positive charges along polymeric chains in the 'ideal' crosslinked polypyrrole type. Experimental results from artificial muscles agree with this assumption. Experimental results performed by cyclic voltammetry using coated electrodes show that a cathodic limit of potential does not exist when the cathodic current becomes zero (Figure 10.17). Using an electrolyte with a large enough potential window (e.g. propylene carbonate/$LiClO_4$) the polymer can be reduced up to -3.5 V vs. SCE. Voltammograms performed from different cathodic potentials show increasing oxidation potentials, increasing currents on the oxidation maxima and the same consumed charges (Figure 10.17). Reduction processes show overlapping branches but the cathodic currents are different to zero, even at -3 V.

. These results point to the fact that, according to our hypothesis about the free volume, ionic movements through a compact structure are difficult when the polymer is closed in the neutral state [103–105]. High electric fields are needed to move ions (negative ions out of, or positive ions into) across the compact film. Starting from more cathodic potentials extra energy is required to open the close structure of chains during the anodic sweep. By working under constant temperature, the only way to give energy to the polymeric chains is through the electric potential: increased anodic potentials are required to open more tightly packed structures, while more cathodic potentials compact the structures further. This extra energy required to open the polymeric structure during anodic sweeps is consumed by the conformational changes required to generate free volume [105].

4.6 Electrochemical responses include structural information

These experimental facts were described in literature several years ago and are known as relaxation processes, wait-time processes or the effect of the first voltammogram. Theoretical treatments are based on electrochemistry. But electrochemical magnitudes are macroscopic magnitudes: current densities, overpotentials, molar concentrations and they do not include any structural information. In this way it seems a difficult task to look for a theoretical treatment that includes structural information. Thus, attempts are criticized by many electrochemists.

The reason for this rejection seems to be founded on the lack of theoretical models that include surface structure, in the field of theoretical electrochemistry. Nevertheless it is generally accepted that electrochemical responses include structural information about the electric double layer inside the solution, and this structure can be deduced from electrochemical measurements able to affect this structure.

In a similar way structural information about the polymer can be included in this theoretical model. All the information about the polymer is deduced from the field of polymer science and the electrochemical response has to follow laws from the field of electrochemistry. The task thus is to attempt an integration of polymer science and electrochemistry.

4.7   Electrochemically stimulated conformational relaxation model (ESCR)

The increase of oxidation potential observed when the same film was polarized at the same time to different cathodic potentials [44] (Figure 10.17) or was polarized to the same cathodic potential for different wait-times [99,106–109] was related to the more compact structure of the polymer and the greater energy needed for conformational relaxation changes. Thus structural information on the electrochemical response can be included in a relaxation model.

Reference can be made to the thermocurrents of the depolarization relaxation theory [110–113]. Here a fused polymer is subjected to an electric field. Dipoles in the polymeric molecules are orientated in the field. The temperature is then decreased and the sample is frozen under the electric field. Now the electric field is eliminated and substituted by a capacitor which is charged by the orientated dipoles. An increasing temperature sweep allows the detection of conformational changes in the polymer when they are allowed. This is the glass transition temperature (T_g). Here dipoles lose the uniform orientation to attain a Boltzman distribution. Charges induced by oriented dipoles on the capacitor plates decrease as conformational movements liberate this orientation. A thermocurrent maximum is observed on the external circuit during the temperature sweep.

Our group's system is conceptually similar. There the model was a frozen oriented structure; here the model is a compact neutral polymer network. There free volume was created by giving thermal energy to the polymer, thus allowing conformational changes and dipoles movement; here free volume was created by giving electrical energy (at constant temperature), which changes the charges distribution along the chain backbones. There free volume and conformational relaxation with free movement of dipoles induce a

thermocurrent; here the new free volume and conformational relaxations with the change of the electronic distributions allows electrochemical reactions and electrochemical responses. There the thermocurrent includes structural information about transitions from a frozen solid to a glass; here the electrochemical response includes informations about transitions from a neutral compact structure to an oxidized and open structure.

4.7.1   Relaxation time

To develop an electrochemical theory of conformational relaxation a relaxation time (τ) is defined as the time required to change the conformation of a polymeric segment subjected to a cathodic potential E_{cat} when it is oxidized to an anodic potential E_{an}.

4.7.2   Polymeric segment

A polymeric segment is the minimum chain length for which conformational changes allow ionic interchanges between the polymer and the solution.

4.7.3   Required energy

The relaxation time is related to the energy ΔH required to complete conformational changes by an Arhenius law:

$$\tau = \tau_0' \exp(\Delta H/RT) \qquad (10.21)$$

Up to this point an overlap with other relaxation models is present. In our electrochemically stimulated conformational relaxation model, the energy required for each individual relaxation process can be expressed as the sum of three terms: a thermal component (ΔH^*) present in the absence of any external electric field; a component including the increment of conformational energy (stored energy) due to the closure of the polymeric matrix during cathodic polarization (ΔH_c); and a third term that contains the decrease of stored conformational energy during electrochemical relaxation by anodic polarization ($-\Delta H_e$):

$$\Delta H = \Delta H^* + \Delta H_c + \Delta H_e \qquad (10.22)$$

Any of those three components is a molar conformational energy. The closure of the polymeric matrix is proportional to a cathodic overpotential η_c related to a potential E_s at which the polymer structure closes. Under the simplified model for polypyrrole oxidation, this means that any increment in the oxidation level of the polymer starting from a potential more anodic than E_s is not controlled by energetic consumption to open the polymeric network: it is open and only counterion diffusion control exists. Starting from a more cathodic potential than E_s extra energy is required to open the polymeric structure (ΔH_c) before the diffusion of counterions is allowed. This concept can be easily understood in biological systems. Cell membranes in muscles have an organic structure with ionic channels. In order to be opened Ca^{2+}, K^+ or Na^+ channels require a minimum potential gradient across the membrane. Under those conditions channels are opened by conformational relaxations of the polymeric components of the cell membranes and then ions fluxes are allowed.

Recovering our proportionality:

$$\Delta H_c = z_c \eta_c \quad (CV = joules) \qquad (10.23)$$

where z_c, the coefficient of cathodic polarization, is the *charge required to compact* one mole of polymeric segments (number of electrons injected into the polymeric segment). ΔH_c is the minimum energy required to allow counterions penetration after polarization at E_c:

$$\Delta H_c = z_c \eta_c = z_c(E_s - E_c) \quad \text{where } \eta_c = E_s - E_c \qquad (10.24)$$

In a similar way, once the penetration of counterions is allowed, conformational relaxations in the conducting polymer, occur with oxidation processes. As the depth of this oxidation depends on the anodic overvoltage (η) the increase in volume (generation of free volume to allow counterions arrangement) requires new conformational changes and the consumption of electrochemical conformational energy (ΔH_e) proportional to η:

$$\Delta H_e = -z_r \eta = -z_r(E - E_0) \quad (CV = joules) \quad (10.25)$$

where z_r, the coefficient of electrochemical *relaxation*, is the charge (coulombs) needed to relax one mole of polymeric segments and E_0 is the usual potential of polymer oxidation (neutral polymer and not compacted).

Thus

$$\tau = \tau_0 \exp[(\Delta H^* + z_c \eta_c - z_r \eta)/RT] \qquad (10.26)$$

is the desired equation. It includes magnitudes related to the polymeric structure, as relaxation time, and the conformational energy needed to relax one mole of polymeric segments in the absence of any electric field (ΔH^*), together with charges linked to compaction and relaxation (z_c, z_r) and, most importantly, overpotentials

(η_0, η). The hope is to integrate, through τ, τ_0 and ΔH^*, the polymer science, and, by means of z_c, z_r, η_c and η_a, the electrochemistry. In this way electrochemical responses and structural parameters of the polymer will be linked and translated to membrane channels, nervous pulses, enzymatic (redox) processes and other biological processes.

The relaxation time can be written:

$$\tau = \tau_0' \exp[(\Delta H^* + z_c(E_s - E_c) - z_r(E - E_0))/RT] \qquad (10.27)$$

In order to simplify the mathematical treatment an electrochemical relaxation function is defined

$$\psi(E, E_c) = (-z_c E_c - z_r E)/RT \qquad (10.28)$$

By including this in equation (10.27) all the other terms are energy ratios, which together with τ_0' are included in τ_0:

$$\tau = \tau_0 \exp[\psi(E, E_c)] \qquad (10.29)$$

4.7.4 Relaxation through nucleated regions

The closed entanglement of a polymer that has been polarized at very cathodic potentials is not uniformly open when submitted to an anodic overvoltage: the oxidation is initiated on singular points and expands like pits similar to the pitting corrosion processes of metals. This model was proposed by my laboratory several years ago based on experimental observations shown in a video [65]. Taking into account the fact that polymer oxidation models are a controversial subject more detailed experimental evidence is given below.

A polypyrrole film was electrogenerated on a flat platinum sheet by cyclic voltammetry between −100 mV and 750 mV in 0.1 M $LiClO_4$ and 0.1 M pyrrole acetonitrile solution containing 1% of water. A uniform and electrochromic film is obtained. Consecutive cyclic voltammetry allows the permanent control of electrochromism during growth. The reduced film shows a uniform yellow color (Figure 10.54a). The oxidized film is a uniform dark-blue color (Figure 10.54d).

The reduced coated electrode was rinsed in propylene carbonate and put inside a 0.1 M $LiClO_4$ propylene carbonate solution. There the oxidation process was studied by potential steps from different cathodic potentials ranging form −3.5 V, where the electrode was kept for 120 s every time, until the same anodic

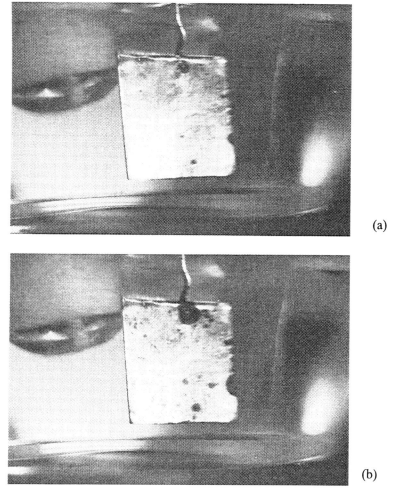

Figure 10.54. Photographs of a mirror-polished platinum electrode coated with a polypyrrole film in 0.1 M LiClO$_4$/propylene carbonate solution (a) reduced at -3 V; (b) and (c) at increasing times after a potential step to 200 mV; (d) oxidized at 200 mV for a long polarization time.

potential 300 mV. Figure 10.55 shows the experimental chronoamperograms. From an electrochemical point of view the presence of a minimum on the chronoamperograms performed from more cathodic potentials than -1000 mV/ECS (E_s) is linked to a nucleation process or a nucleation-like process. Chronoamperograms initiated at higher potentials than -1000 mV show a continuous decrease of current which is linked to diffusion-controlled processes (see the chronoamperograms performed from -800 mV on the figure).

When a film that had been prepolarized at -3.5 V was observed at different anodic polarization times, the photographs shown in Figure 10.54b–d were

obtained. A nucleation process of the oxidized polymer on the reduced film occurs. If the experiment is repeated the same sequences of photographs are obtained: nucleation always occur at the same points. This means that irregularities exist on the polymer structure generated during polymerization: the nucleation detects these irregularities like the pitting corrosion of metals detects inclusions, grains borders, or concentration depletion points on borders. Due to polymeric crosslinking, the structure is recovered every time that it returns to the same cathodic potential.

The structure is controlled by the metal base. A perfect mirror-polished platinum sheet gives a uniform polymer film during electropolymerization and only a

(c)

(d)

Figure 10.54. (*continued*)

few nuclei (3–4 per cm^2) are generated by oxidation. Irregular and old platinum sheets do not allow the differentiation of nucleations. Chronoamperograms, nevertheless are similar in both cases, the difference being the number of nuclei per square centimeter.

4.8 Relaxation–nucleation

When the cathodically polarized polymer is subjected to an anodic overpotential, electrons are extracted from the semiconducting neutral polymer film coating the platinum electrode. As the polymer structure is compact counterions can not penetrate inside the polymer and those positive charges concentrate on the polymer surface where they can be compensated by counterions

in the solution side of the EDL electrical double layer. At specific points on the surface where irregularities exist (irregularity here means easy free conformational movements of the polymeric network), the penetration of counterions into the polymer and the network expansion through conformational movements are initiated (Figure 10.56). The polymeric oxidation induces a clear border between the expanded and oxidized polymer and the neutral and compact film. As mechanical stress appears on this border, which favors conformational relaxation, the nucleus of the oxidized polymer has a tendency to expand like a sphere. Nevertheless the back neutral film thickness decreases giving increasing potential gradients between the oxidized nucleus and the back metal: the oxidation

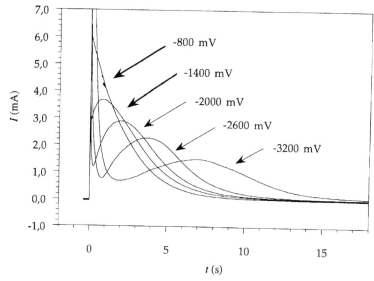

Figure 10.55. The potential steps carried out on a polypyrrole electrode in a 0.1 M LiClO$_4$/propylene carbonate solution from different cathodic potentials, indicated on the figure, to 300 mV vs. SCE. (Reproduced by permission of Elsevier Science SA from *J. Electroanal. Chem.* **394**, 211, 1995.)

progresses very quickly to the metal–polymer interface (Figure 10.54) giving a cylinder of oxidized polymer.

4.8.1 Oxidation rate under relaxation–nucleation control

Once formed, the cylinders of oxidized polymer expand (Figure 10.54), as can be observed on photographs.

Counterions penetrate the polymer by the oxidized polymer–electrolyte interphase and diffuse, through the oxidized cylinder, to the oxidized polymer–neutral polymer interphase.

The radius of the cylinder increases by λ in a time τ

$$\frac{\mathrm{d}r}{\mathrm{d}t} = \frac{\lambda}{\tau} = \frac{\lambda}{\tau_0}\exp[-\psi(E, E_\mathrm{c})] \tag{10.30}$$

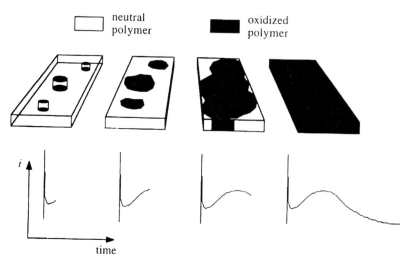

Figure 10.56. Evolution of the surface aspect of the polypyrrole film during the oxidation reaction initiated from high cathodic potentials ($E < -800$ mV vs. SCE). The chronoamperometric response is shown at the bottom. (Reproduced by permission of Elsevier Science SA from *Solid State Ionica* **63–65**, 800, 1993.)

where λ is the length of a monolayer of polymeric segments. By integration

$$r = \frac{\lambda}{\tau_0}\exp[-\psi(E, E_c)]t \qquad (10.31)$$

This shows a constant expansion rate of each cylinder during the polarization time. The expansion rate decreases with increasing negative starting potentials (E_c) or for decreasing positive anodic potentials (E) in good agreement with experimental results (Figure 10.55 and Figure 10.57).

All the previous experimental results and radius variation rates include the following hypothesis:

(1) A nucleated relaxation is present, and the relaxed nuclei expand as concentric cylinders until they coalesce.
(2) During this time an oxidized phase (inside cylinders) and a neutral polymer (outside cylinders) are present with clear separation surfaces (the lateral area of cylinders).
(3) Oxidized regions have a uniform composition at every polarization time and subsequently a constant charge density. Regions of neutral polymer also have a uniform composition.
(4) The relaxation of a mole of segments in the oxidized/neutral polymer borders involves the loss of z_r electrons, the subsequent storage of z_r positive charges, and the flow of z_r monovalent (solvated) anions from the solution.

(5) The relaxation of an elemental segment consumes a time τ. Thus, the relaxation process approaches a step function (Figure 10.58).
(6) N_0 nuclei are initiated per square centimeter during an oxidation process at constant temperature, and at constant concentration of a defined electrolyte in a specific solvent.
(7) The overall oxidation charge at E has two components: the charge required to relax the compact structure and to allow counterion penetration (Q_r) (under relaxation control); and the charge consumed to complete the oxidation–relaxation under diffusional control (Q_d)

$$Q_t = Q_r + Q_d \qquad (10.32)$$

(8) If the oxidation process is interrupted when the nuclei are growing the relaxation process continues pushed by the stress gradient between oxidized and reduced regions. A rearrangement of positive charges in the chains and counterions in the solid takes place and finishes when all the film has the same composition: the average composition of the stop time.

4.9   Response to potential steps

The polymer film will be oxidized at constant anodic potential (E) by potential steps from a cathodic

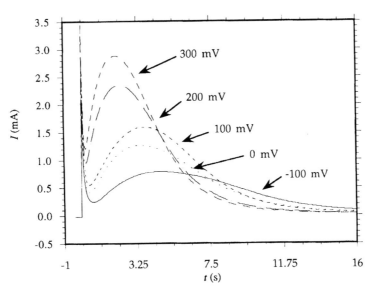

Figure 10.57. Chronoamperometric responses after application of potential steps from -2000 mV until different anodic potentials to a polypyrrole-coated electrode in a 0.1 M LiClO$_4$/polypyrrole carbonate solution.

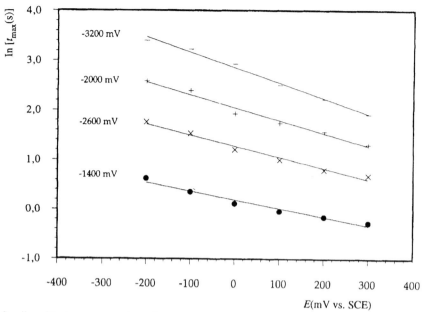

Figure 10.58. Semilogarithmic representation of t_{max} vs. anodic potential from series of potential steps. Each series was performed between a cathodic potential and different anodic potentials in a 0.1 M LiClO$_4$/propylene carbonate solution. (Reproduced by permission of Elsevier Science SA from *J. Electroanal. Chem.* **394**, 211, 1995.)

potential more cathodic than E_s. This means that a conformational relaxation control of the oxidation process will be present.

Under E a stationary density of charges δ_t (C cm^{-3}) will be stored in the polymer. The formation of these charges was controlled by both the conformational relaxation, δ_r, and by diffusion-controlled processes, δ_d:

$$\delta_t = \delta_r + \delta_d \quad \text{C cm}^3)\qquad(10.33)$$

The concentration of polymeric elemental segments that were relaxed by conformational movements is

$$C_{vol} = \frac{\delta_r}{z_r} = \frac{\text{C cm}^{-3}}{\text{C mol}^{-1}} = \text{mol cm}^{-3}\qquad(10.34)$$

z_r is the charge consumed to relax a mol of elemental segments (C mol^{-1}) and C_{vol} is the mol of polymeric segment per cubic centimeter.

If a cylinder is considered that is formed by a nucleus of radius r and a height h (the thickness of the film), in the next τ seconds the nucleus will expand to a cylinder of radius $r+\lambda$ (Figure 10.56), λ being the length of the segment.

To relax the polymer inside the cylinder, an electrical charge $Q_r(t)$ was consumed

$$Q_r = \pi r^2 h C_{vol} Z_r F$$
$$= \pi r^2 h \delta_r \quad (\text{cm}^3 \text{ mol cm}^{-3} \text{ C mol}^{-1} = \text{C})\quad(10.35)$$

The current flowing through the system was:

$$I_r = \frac{dQ}{dt} = 2\pi r h \delta_r \frac{dr}{dt}\qquad(10.36)$$

and taking $dr/dt = \lambda/\tau$

$$I_r = 2\pi r h \delta_r \frac{\lambda}{\tau}\qquad(10.37)$$

Thus the current flowing through the system to relax the growing cylinder as a function of the relaxation time (τ) is obtained from equation (10.30). By integration:

$$r = \frac{\lambda}{\tau_0} \exp[\psi(E, E_c)]t + C\qquad(10.38)$$

when $t = 0$ no nuclei exist so $C = 0$.

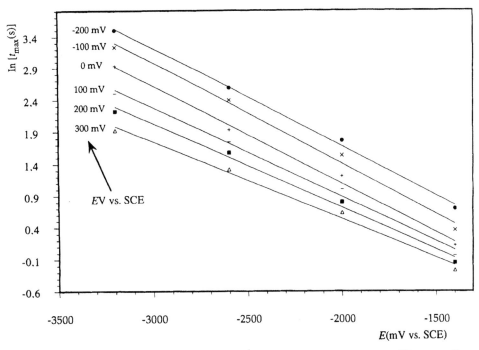

Figure 10.59. Semilogarithmic representation of t_{max} vs. cathodic potential for potential steps to different anodic potentials. The coefficient of cathodic polarization (z_c) was calculated from the slopes. (Reproduced by permission of Elsevier Science SA from *J. Electroanal. Chem.* **394**, 211, 1995.)

If N_0 nuclei are growing simultaneously

$$I_r(t) = \frac{2\pi h \delta_r \lambda^2 N_0}{\tau_0^2} \exp[-2\psi(E, E_c)]t$$
$$= \frac{2\pi N_0 \lambda^2}{\tau_0^2} Q_r \exp[-2\psi(E, E_c)]t \qquad (10.39)$$

A is the area of the electrode and

$$Q_r = Ah\delta_r \ (\mathrm{cm}^2 \ \mathrm{cm} \ \mathrm{C} \ \mathrm{cm}^{-3} = \mathrm{C})$$

This equation works until adjacent nuclei coalesce. Taking Avrami's formulation and considering the symmetry of the processes the three-dimensional system is reduced to a problem of two dimensions and

$$dS = (1 - S)dS_{ext} \rightarrow \begin{cases} S = 1 - e^{-S_{ext}} \\ I_r = I_{ext}e^{-S_{ext}} \end{cases} \qquad (10.40)$$

where I_r is the current flowing after coalescence; I_{ext} is I_r without coalescence ($I)_r$; S is the area of the oxidized regions after coalescence; and S_{ext} is the area if coalescence not occur.

$$S_{ext} = \frac{1}{A} N_0 \pi [r(E)]^2 = \frac{\pi N_0 l^2}{A \tau_0^2} e^{-\psi(E, E_c)} t^2 \qquad (10.41)$$

by substitution of equations (10.41) and (10.39) into equation (10.40):

$$I_r(t) = \frac{2\pi h \delta_r \lambda^2 N_0}{\tau_0^2} \exp[-2\psi(E, E_c)]t$$
$$= \frac{2\pi N_0 \lambda^2}{A\tau_0^2} Q_r \exp[-2\psi(E, E_c)]t \qquad (10.42)$$

and by integration along t:

$$Q_r(t) = Q_r \left[1 - \exp\left[-\frac{\pi N_0 \lambda^2}{\tau_0^2 A} \exp[-2\psi(E, E_c)]t^2 \right] \right] \qquad (10.43)$$

where ψ (E, E_c) comes from equation (10.28).

Equation (10.42) gives the theoretical chronoamperogram obtained when a polymer film is subjected to a potential step from E_c to E, (E_c is more cathodic than E_s) which promotes conformational relaxation control of the oxidation processes, through nucleation of oxidized zones on a reduced film.

A new constant a can be defined:

$$a = \pi N_0 \lambda^2 / \tau_0^2 A \exp[-2\psi(E_{ox}, E_{cat})] \qquad (10.44)$$

The constant a includes structural magnitudes such as the length of an elemental nucleus (λ); the number of nuclei for a defined film (N_0); the relaxation time (τ_0); and the area of the polymer film (A). It also includes electrochemical magnitudes as cathodic (η_c) or anodic (η) overpotentials. All those magnitudes can be determined experimentally, only λ/τ_0 has to be calculated.

As has been pointed out above, when the relaxation of the polymer has been initiated and the oxidized zones are growing, a second process is present: the transport of ions from the solution through the oxidized polymer by diffusion in the solid polymer matrix.

The inclusion of the diffusion equation in the model results in a good simulation of the chronoamperograms, as will be developed below, but introduces mathematical complications for the extraction of the magnitudes related to the maximum. Assuming that diffusion, once the structure was open, is in equilibrium with the remaining conformational changes required to complete the oxidation of the volume, the shape of the chronoamperograms and their main points will be controlled by the initial conformational relaxation process. The equilibrium will be present for high electrolyte concentrations (c. 0.1 M) thin films ($< 5~\mu m$) and slow relaxation processes (high cathodic overvoltages and low anodic overvoltages), as in most of the experimental conditions.

Therefore the time t_{max} at which the current density reaches its maximum can be determined by differentiating equation (10.42)

$$t_{max} = \frac{1}{\sqrt{(2a)}} = \sqrt{\left(\frac{A}{2\pi N_0}\frac{\tau_0}{\lambda}\right)} \exp[\psi(E, E_c)] \quad (10.45)$$

The model predicts, from equations (10.45) and (10.28), a semilogarithmic dependence of t_{max} on the upper and lower potentials:

$$t_{max} = C + \psi(E, E_c) = C - (z_c/RT)E_c - (z_r/RT)E \quad (10.46)$$

4.10   Correlation between experiments and theory

Polypyrrole films were electrogenerated from 0.1 M LiClO$_4$ and 0.1 M pyrrole/acetonitrile solutions with a 2% water content. A 1 cm^{-2} platinum electrode polished until it had a mirror finish, was submitted to a potential step from -300 (vs. SCE) to 800 mV. The anodic potential was maintained for 100 s. During this time 120 mC flowed through the system and a polypyrrole film average thickness 2.3 μm was generated. This is an electrochromic film that showed a pale yellow color in the neutral state and a dark-blue color in the oxidized state.

4.10.1   Influence of the cathodic potential

Once electrogenerated, the coated electrode was rinsed in the reduced state with propylene carbonate. Then was transferred into 0.1 M LiClO$_4$ in propylene carbonate. There it was subjected to four series of cathodic potential steps: -800, -1400, -2000, -2600 and -3200 mV, and four different anodic potentials: -100, 0, 100, 200 and 300 mV.

In order to illustrate the influence of the cathodic potential the series of chronoamperograms obtained in response to potential steps from different cathodic potentials to 300 mV was depicted in Figure 10.55. Polarization at -800 mV gives an open polymeric structure if the fact that the potential required to close the studied polymer ($E_s = -1000$ mV) is taken into account. The subsequent potential step gives a continuous decrease of the current density after a sharp initial step related to the charge of the electrical double layers. This is a diffusion-controlled process without conformational relaxation control.

Chronoamperograms obtained in steps starting from higher cathodic potential show the initial sharp peak of current density linked to conformational relaxation control. A higher cathodic potential of prepolarization lead to a more compact initial state of the polymer. Also longer times were required to produce conformational movements sufficient to allow the polymeric oxidation–nucleation: the maximum controlled by conformational relaxation appears at longer times and lower current densities. This maximum is due to the coalescence between different expanding nuclei of the oxidized polymer (Figures 10.54 and 10.56).

Polarization times required to attain these maxima are semilogarithmically dependent on the cathodic potential of prepolarization (Figure 10.59). Shorter times are needed to attain the maximum related to the conformational control of oxidation when a series of experiments were performed until increasing anodic potentials from the same cathodic potentials.

More conformational energy is transferred to the polymer chains per unit of time from increasing anodic potentials. These results show good agreement between experimental results and predictions from the model [equation (10.46)].

Slopes from the lines on Figure 10.59 are related to z_c, the charge consumed to promote conformational changes and close one mole of open polymeric segments. This figure shows that the slopes from different experimental series are the same: the coefficient of the cathodic polarization (z_c) is independent of the anodic potential of polarization, as predicted by the model. A value of 4626 C mol^{-1} was obtained from the slopes.

The constant C in equation (10.46) includes the potential E_s for the closure of the polymer structure. This potential is the limit between mainly diffusion-controlled and relaxation-controlled processes in the subsequent anodic step. A value of -1000 mV/SCE was obtained for E_s from experimental data (Figure 10.59).

4.10.2 Influence of the anodic potential

Keeping the cathodic potential of prepolarization constant at -2000 mV, which was maintained for 160 s in order to allow the structure to become closed, potential steps were performed up to different anodic potentials. Experimental chronoamperograms obtained by steps up to -100, 0, 100, 200 and 300 mV for four different cathodic potentials can be observed in Figure 10.57. Nucleation–relaxation processes occur more

quickly as more energy is given to the chains through anodic polarization (η). The polarization times required to attain the maximum are shorter at increasing anodic potentials for the oxidation process.

These times follow a semilogarithmic dependence on the anodic potential of polarization (Figure 10.58), as predicted by equation (10.46), and longer times are required when the steps were performed from more cathodic potentials. The coefficient of electrochemical relaxation is obtained directly from the slopes. This coefficient shows a linear dependence on the starting cathodic potential, as shown in Figure 10.60: z_r increases as the initial potential becomes more negative. More negative initial potentials promote a more compact polymeric entanglement and more charge is consumed to produce conformational changes and polymer oxidation during any anodic polarization. The relaxation coefficients obtained range between 4342 C mol^{-1}, when the initial potential was -1400 mV, and 7726 C mol^{-1}, when it was -3200 mV. The equation of the straight line is

$$z_r = 1825 - 1820\,E_c \qquad (10.47)$$

Here E_c takes negative values and it is expressed in volts. This means that a minimum value of z_r is related to the potential of closure ($E_c = -1000$ mV) of 3650 C mol^{-1}.

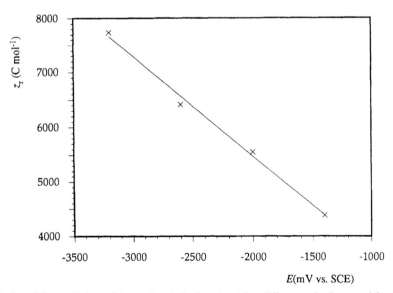

Figure 10.60. Evolution of the coefficient of electrochemical relaxation (z_r) at different cathodic potentials of departure. Values of z_r were obtained from the slopes in Figure 10.57. Reproduced by permission of Elsevier Science SA from *J. Electroanal. Chem.* **394**, 211, 1995.)

This result does not seem to agree with equation (10.46), which has to be rewritten as

$$\ln t_{max} = C - \frac{z_c}{RT}E_c - \left(\frac{1825 - 1820 E_c}{RT}\right)E$$
$$= C - \left(\frac{z_c - 1825 E}{RT}\right)E_c - \frac{z_r}{RT}E \quad (10.48)$$

Taking this fact into account and calculating z_c again from experimental results, a value of 5225 C mol^{-1} was obtained.

As z_r is a function of E_c, this dependence has to be included in equations (10.26) and (10.27). This inclusion does not modify either the model or the conclusions. τ remains a function of z_c, z_r, E, E_c, ΔH^* and T. The electrochemical conformational relaxation function, equation (10.28), also remains a function of both E and E_c. Equations giving current densities related to conformational relaxation control process, $I_r(t)$ (10.42), consumed charges involved, $Q_r(t)$ (10.43) and polarization times at which conformational relaxation–nucleation maxima appears, t_{max} (10.45) remain functions of the same variables.

4.10.3   Influence of the electrolyte concentration

Counterions are needed to compensate positive charges generated by anodic polarizations along polymeric chains: this is an electrochemical reaction in solid state, equation (10.1). One of the components of this reaction depends exponentially on the concentration of electrolyte, equation (10.3). Under constant temperature, and constant anodic and cathodic potentials of polarization, equation (10.3) becomes

$$\text{oxidation rate} = k'[\text{anion}]^\alpha \quad (10.49)$$

because the electrochemical kinetics are influenced by the relaxation time (reciprocal dependence): low conformational relaxation times give faster kinetics and long relaxation times give slow kinetics. So, equation (10.49) can be rewritten as

$$1/\tau = k'[\text{anion}]^\alpha; \quad \log \tau = -\log k' - \alpha \log [\text{anion}] \quad (10.50)$$

From equations (10.29) and (10.45)

$$t_{max} = \sqrt{\left(\frac{A}{2\pi N_0}\right)}\frac{\tau}{\lambda} \quad (10.51)$$

A double logarithmic dependence of t_{max} on the electrolyte concentration is obtained:

$$\log t_{max} = C - \alpha \log [\text{anion}] \quad (10.52)$$

Experimental chronoamperograms were performed by potential steps for -2000 mV to -100, 100 and 300 mV (using a film obtained as described above) in propylene carbonate solutions with different concen-

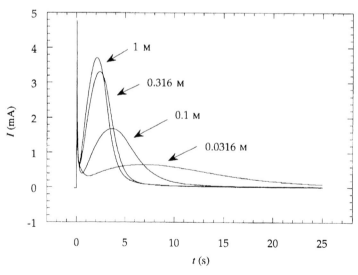

Figure 10.61. Chronoamperometric responses after application of potential steps for -2000 mV to 100 mV vs. SCE to a polypyrrole-coated platinum electrode in $LiClO_4$/propylene carbonate solutions (concentrations indicated on the figure).

trations of LiClO$_4$ (0.03, 0.1, 0.3 and 1 M). Chronoamperograms obtained by potentials steps up to 100 mV can be observed in Figure 10.61.

As predicted by this model, nucleation–relaxation processes are faster (i.e. the maximum appears at shorter polarization times) at higher concentrations of electrolyte. In accordance with equation (10.52), a double logarithmic plot of t_{max} vs. the electrolyte concentration give three parallel straight lines as shown in Figure 10.62. From the slopes of the lines, a value of 0.3 for the reaction order α was obtained, which agrees with the empirical values reported in literature for the oxidation rate dQ/dt [44].

4.10.4 Influence of temperature

The oxidation of a film of conducting polymer under conformational relaxation control is affected by temperature as shown by equation (10.27). Temperature influences the opening of the polymeric entanglement through the coefficient z_c/RT: higher temperatures give more conformational energy to the chains, favoring molecular movement, penetration of counterions and shortening of the relaxation time. In a similar way, the closing of the polymeric entanglement during reduction (through the coefficient z_r/RT) is favored: the relaxation times decrease as the temperature is raised. The affected

conformational movements in the absence of any electrochemical reaction through $\Delta H^*/RT$ are analogous. This fact was proved when polarizations giving relaxation–nucleation were stopped by switching of the potential before the nuclei coalesce. Oxidized (blue) nuclei expand on the reduced film (yellow) faster if the temperature was higher. Charges and ions in the oxidized nuclei spread across the film and the process does not consume any charge.

In order to quantify the influence of the temperature on the oxidation–relaxation process independently of the reduction–compaction process, an experimental procedure following different steps was designed. The coated platinum electrode, which showed electrochromic responses, was reduced every time in 0.1 M LiClO$_4$ propylene carbonate solutions, at ambient temperature, by polarization at -2000 mV for 120 s; the steady cathodic current was around 40 μA after this time for every experiment. Under an inert atmosphere the temperature is moved to the experimental value and kept there for 5 min. Then a potential step between -2000 mV (1 s) and 100 mV is performed. The chronoamperograms depicted in Figure 10.63 were obtained. As predicted by our model, t_{max} is semilogarithmically dependent on T^{-1} (Figure 10.64). From the slope an average activation energy for the oxidation processes under conformational relaxation control of 28.2 kJ mol^{-1} was obtained. A

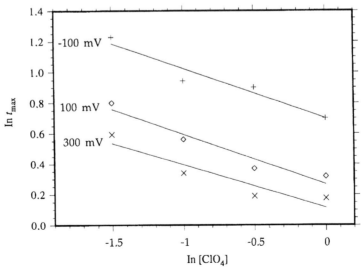

Figure 10.62. Calculation of the reaction order for the electrolyte concentration, at different upper potentials (-100, 100 and 300 mV vs. SCE), starting from the same lower potential of -2600 mV. (Reproduced by permission of Elsevier Science SA from *J. Electroanal. Chem.* **394**, 211, 1995.)

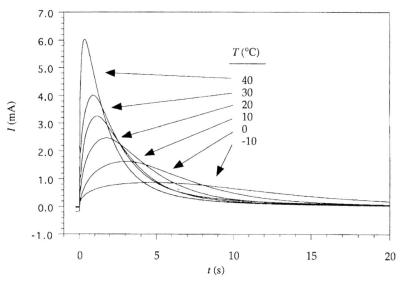

Figure 10.63. Chronoamperometric responses to the application of potential steps from -2000 mV to 300 mV in a 0.1 M LiClO$_4$/propylene carbonate solution, at the different temperatures indicated on the figure. Cathodic prepolarization temperature was always 25°C (room temperature).

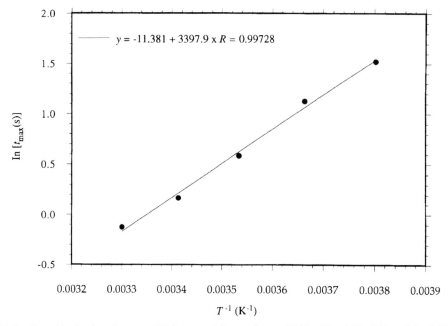

Figure 10.64. Semilogarithmic plot of t_{max} vs. $1/T$ for potential steps from -2000 mV to 300 mV in a 0.1 M LiClO$_4$/propylene carbonate solution. The activation energy for the conformational relaxation process was calculated from the slope.

ΔH° value of 28.35 kJ mol^{-1} was obtained $(E_a = \Delta H^{\circ} - z_r\eta + z_c\eta_c)$.

4.10.5 Conclusions about the model

This conformational relaxation electrochemical model [equations (10.42) and (10.43)] has been checked with the experimental results obtained when conformational relaxation processes control the oxidation process. The following conclusion can be stated:

(1) The model predicts a semilogarithmic dependence of t_{max}, the time of the chronoamperometric maximum performed under conformational relaxation control, on the cathodic or anodic potentials. This is in good agreement with experimental results (Figures 10.57 and 10.59).
(2) These variations allow z_c and z_r, the charges required to compact, or relax, one mole of polymeric segments, to be obtained. The relaxation coefficient z_r is a function of the cathodic potential ($z_r = a - bE_c$): more charge is required to open one mole of polymer than was closed by more cathodic (negative) potentials.
(3) The potential of closure of the polymeric network is also obtained from the straight lines, giving a theoretical value that fits experimental results. ($E_s = -1000$ mV at ambient temperature).
(4) Higher average activation energies are required to open polymeric structures that were compacted at higher cathodic overvoltages; this fits the experimental results obtained from oxidation rates.
(5) The theoretical reaction order for the solid state oxidation process, obtained from the predicted variation of t_{max} as a function of the electrolyte concentration, fits experimental values obtained from oxidation rates.
(6) Equations (10.42) and (10.43) predict t_{max} but are not adequate to simulate chronoamperograms and chronocoulograms due to the lack of diffusion and double layers components. Diffusion processes will be discussed in the next section to check the ability of the model to predict experimental chronoamperograms as a function of the different variables.

4.11 Diffusion-controlled oxidation completion

The equilibrium between conformational relaxation and diffusion processes, the polymeric oxidation, (when the structure is opened above t_{max}) is controlled by ionic diffusion in the polymeric network. Anodic potential steps initiated from more anodic potentials than E_s also give diffusion-controlled processes. We will attempt to complete the simplified model by including diffusion processes.

The transport of ions has two components: transport through the polymer–solution double layer and diffusion through the oxidized polymer towards the oxidation centers. The flow of ions into the polymer $N(t)$ can be assumed to be proportional to the difference between the charge density in the polymer when the film attains the oxidation steady state at large polarization times C_d^{ss} and the charge density consumed after a given oxidation time $C_d(t)$:

$$N(t) = D[C_d^{ss} - C_d(t)]/\xi \quad (\text{C cm}^{-2}) \qquad (10.53)$$

where D is the diffusion coefficient of ions in the film and ξ is the equivalent length of diffusion across the film. The infinitesimal increment of stored charge in the polymer during diffusion control will be

$$dQ_d(t) = N(t)A(t)dt \qquad (10.54)$$

If the oxidized film volume is $V(t) = A(t)h$ at a given oxidation time:

$$dC_d(t) = \frac{dQ_d(t)}{V} = \frac{N(t)}{h}dt \quad (\text{C cm}^{-3}\text{ s}) \qquad (10.55)$$

By substitution $N(t)$ from equation (10.53):

$$dC_d(t) = \frac{D}{\xi h}[C_d^{ss} - C_d(t)]dt \qquad (10.56)$$

or

$$\frac{dC_d(t)}{dt} = \frac{D}{\xi h}[C_d^{ss} - C_d(t)] \qquad (10.57)$$

By integration:

$$C_d(t) = C_d^{ss}(1 - e^{-Dt/\xi h}) \qquad (10.58)$$

C_d^{ss} and $C_d(t)$ can be rewritten as

$$C_d^{ss} = \frac{Q_d}{Ah}; \qquad (10.59)$$

$$C_d(t) = \frac{Q_d(t)}{A(t)h} \qquad (10.60)$$

So from equations (10.60), (10.58) and (10.59):

$$Q_d(t) = Q_d(1 - e^{-Dt/\xi h})\frac{A(t)}{A} \qquad (10.61)$$

Using $A(t)$ obtained from equations (10.40) and (10.41) and taking into account the constant a (10.44):

$$A(t) = A(1 - e^{-at^2}) \qquad (10.62)$$

and defining a new constant $b = D/\xi h$ for a film of constant thickness, h, from equations (10.62) and (10.61), the evolution of the consumed charge involved in diffusion processes, as a function of the polarization time can be obtained.

$$Q_d(t) = Q_d(1 - e^{-bt})(1 - e^{-at^2}) \qquad (10.63)$$

The current density due to diffusion control is

$$I_d(t) = dQ_d(t)/dt$$
$$= bQ_d e^{-bt}(1 - e^{-at^2}) + 2aQ_d t(1 - e^{-bt})e^{-at^2}$$
$$(10.64)$$

The current density flowing through the system, when the polymeric structure was open under conformational relaxation, also depends on the conformational changes needed to complete oxidation processes through the constant a. Nevertheless those are not yet the controlling terms of the kinetics: thus, diffusion-controlled oxidation is obtained.

If the charge consumed during the initial polarization times to charge the electrical double layer is considered: metal–polymer and polymer–electrolyte (Q_c), the overall $Q(t)$ and $I(t)$ will be represented by the addition of three components: electrical double layers (Q_c, I_c), relaxation (Q_r, I_r) and transport processes (Q_d, I_d):

$$Q(t) = Q_c + [Q_r + Q_d(1 - e^{-bt})](1 - e^{-at^2}) \quad (10.65)$$

$$I(t) = Q_c d(t) + 2atQ_r e^{-at^2} + bQ_d e^{-bt}(1 - e^{-at})$$
$$+ 2a\, Q_d t(1 - e^{-bt})e^{-at^2} \qquad (10.66)$$

Equations (10.65) and (10.66) show the consumed charge and flow of current as a function of the anodic polarization time during the oxidation of a conducting polymer film, under relaxation and/or diffusion control, by a potential step from E_c to E, where

$$b = \frac{D}{\xi h} \quad \text{and} \quad a = \frac{\pi N_0 \lambda^2}{\tau_0^2 A} \exp[-2\psi(E, E_c)]$$
$$(10.67)$$

Equations (10.65) and (10.66) define the electrochemical oxidation process of a three-volume conducting polymer film under the control of conformational movements in the polymeric network. When the initial potential for the potential step is more anodic than E_s, the cathodic overpotential becomes zero and a contains

only anodic overpotential, i.e. the process becomes diffusion controlled.

The simplified model for polymer oxidation allows both charge and current to be obtained, as a function of the polarization time. It depends on structural magnitudes such as segment length (l), relaxation time (τ_0), z_c and z_r charges, a diffusion constant b, which is a function of both the diffusion coefficient and oxidation deepness (δ) and the number of nuclei where oxidation is initiated (N_0); and macroscopic magnitudes, such as the area of the film (A), thickness (h) of the film and electrochemical variables, such as cathodic (η_c) and anodic overpotential (η) So polymer oxidation is described as a function of electrochemical variables and structural variables. This simplified development, has integrated both polymer science and electrochemistry through the electrochemistry of conducting polymers.

Taking into account the fact that equation (10.66) represents the theoretical chronoamperograms, predicted by a conformational relaxation–nucleation model, the ability of the model to predict experimental chronoamperograms obtained under the influence of different experimental variables can now be checked (Figures 10.54, 10.58, 10.61 and 10.63).

4.12 Predicted chronoamperograms: influence of the studied variables

To obtain theoretical chronoamperograms it is necessary to include the current densities implicated to change the electrical double layers and to relax the polymeric structure by conformational movements, and the charge needed to complete oxidation processes, controlled by diffusion, once the structure is opened. So equation (10.66) is required. Taking into account the fact that the initial sharp maximum is related to the charge of the double layers, metal–polymer and polymer–solvent, the kinetics for the separation of positive and negative charges in the solid compact polymer are responsible for this maximum. This requires a different physical treatment and is outside the scope of this chapter. An attempt will be made to simulate the second maximum linked to conformational–relaxation and diffusion processes, as well as the influence of anodic or cathodic potentials of prepolarization, anion concentration and temperature.

Under these conditions theoretical chronoamperograms were obtained by simulation of the current flowing through a film of polypyrrole consuming a charge to be oxidized that is a function of the anodic potential:

$$Q(E) = 9.76 + 0.015\, E \quad \text{(mC)} \qquad (10.68)$$

Q_r, the charge consumed to allow conformational relaxation is $0.6\, Q$; Q_d the charge consumed to complete the oxidation process under diffusion control is $0.4\, Q$; the thickness of the layer h, is 2.7 μm; $\xi = {}^1\!/_2 h$; the number of nuclei where the oxidation is initiated, N_0 is 7; the area of the electrode, A, is 1 cm^2; $E_0 = -400$ mV; $E_s = -1000$ mV; $z_c = 4600$ C mol^{-1}; $z_r = 1825 - 1.827\, E_c$; and $D = 1.29 \times 10^{-11}$ cm^2 s^{-1}. λ/τ_0 was calculated to be 0.075 cm s^{-1}, using these experimental parameters.

The influence of the cathodic potential on the anodic potential of 300 mV can be observed in Figure 10.65. The chronoamperograms obtained [equation (10.66)] from -2000 mV to different anodic potentials are depicted in Figure 10.66. The influence of the electrolyte concentration the chronoamperogram simulated by the conformational relaxation model is shown in Figure 10.67; and the chronoamperograms simulated by potential steps from -200 mV to 800 mV at different temperatures when the polymer was closed at -2000 mV and 25°C, can be observed in Figure 10.68.

4.12.1 Correlation between experiments and theory

The model of the oxidation of conducting polymers through nucleation–conformational relaxation pro-

cesses, shows good correlation between theoretical and experimental influence of the different variables on the time of the maximum. This agreement allows different constant such as E, E_s, z_r and z_c to be obtained.

The inclusion of the diffusion charge and diffusion current, when the structure was open, allows the simulation of the majority of the chronoamperograms related to the nucleation–conformational relaxation and diffusion charges. A good agreement exists between experimental and theoretical chronoamperograms performed (a) from different cathodic potentials (Figures 10.55 and 10.65, respectively); (b) up to different anodic potentials (Figures 10.58 and 10.66, respectively); (c) in different concentrations of electrolyte (Figures 10.61 and 10.67, respectively); or (d) at different temperatures (Figures 10.63 and 10.68, respectively). A similar qualitative agreement exists when experimental chronocoulometric curves (Q vs. t) are compared with simulated ones, equation (10.65).

In order to translate this qualitative agreement of the chronoamperograms and the shift (as a function of the different variables) to quantitative relationships, experimental current densities were plotted vs. theoretical values. Results related to the influence of the cathodic potential of prepolarization are depicted in Figure 10.69; Figure 10.70 shows the correlation obtained from steps to different anodic potentials, and the correlations obtained as a consequence of the concentration of electrolyte or the temperature are depicted in Figures 10.71 and 10.72, respectively.

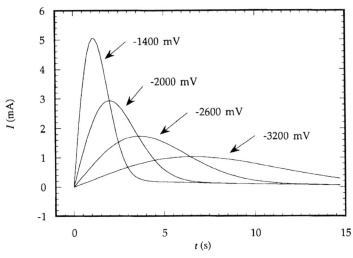

Figure 10.65. Theoretical chronoamperograms calculated for polypyrrole films, from different cathodic potentials of prepolarization (indicated on the figure) to 300 mV in a 0.1 M LiClO$_4$/propylene carbonate solution at room temperature.

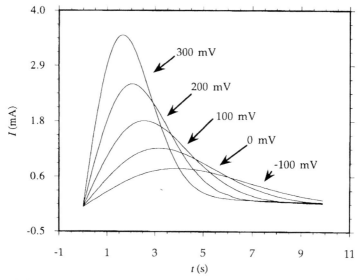

Figure 10.66. Theoretical chronoamperograms obtained for polypyrrole films, from -2000 mV to different anodic potentials ranging between -100 to 300 mV in a 0.1 M LiClO$_4$/propylene carbonate solution at room temperature.

Those results show good correlation for different anodic potentials (Figure 10.70). A positive shift of theoretical values appears at high anodic potentials $E > 200$ mV. The correlations are not so good for the other variables.

This model also allows the simulation of voltammograms and voltamperograms shifts as a function of different experimental variables; starting cathodic potential for the potential sweeps, different polarization times at the same initial potential, different temperatures and different concentration of electrolytes.

4.13   Improving the model

We have presented my group's simplified model for the oxidation of conducting polymers (transition from a neutral state to an oxidized state) under conformational

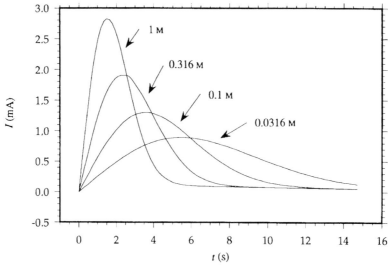

Figure 10.67. Theoretical chronoamperograms obtained for polypyrrole films, from -2000 mV to 100 mV at different LiClO$_4$ concentrations (indicated on the figure) in propylene carbonate solution at room temperature.

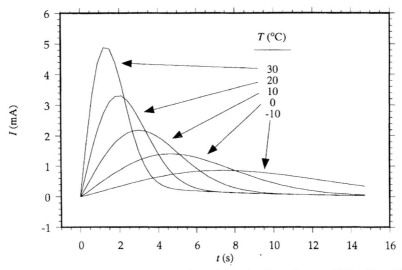

Figure 10.68. Theoretical chronoamperograms obtained for polypyrrole films, from -2000 mV to 100 mV at different temperature of anodic polarization (ranging between -10 and $30°C$). Temperature of cathodic prepolarization was always $25°C$ (room temperature).

relaxation–nucleation control. This is an open model where the physical chemistry of every specific polymer–solvent–electrolyte system is implicit and need to be explicited in subsequent developments. Molecular interactions like polymer–polymer, polymer–solvent and polymer-ion interactions, in both neutral or oxidized polymer states; parameters linked to the polymer, such as the degree of crosslinking or electroactivity per unit of volume; or parameters linked to swelling and shrinking processes, such as the

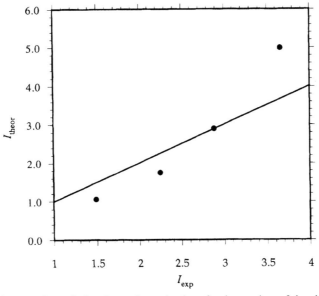

Figure 10.69. Correlation between theoretical and experimental values for the maxima of the chronoamperometric curves, at different cathodic potentials of prepolarization. The points were obtained from Figures 10.65 and 10.54, respectively.

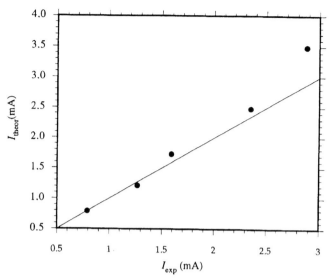

Figure 10.70. Correlation between theoretical and experimental values for the maxima of the chronoamperometric curves, at different anodic potentials of prepolarization. The points were obtained from Figures 10.66 and 10.58, respectively.

generation or destruction of free volume, were also included in the determination of the relaxation time, and hence in ΔH^* and $\Delta l/\tau_0$.

In spite of this oversimplification both experimental chronoamperograms and the influence of different variables on chronoamperogram shifts can be simulated

through the model. Nevertheless these assumptions promote a poor correlation between experimental and theoretical current densities on the maxima linked to conformational relaxation–nucleation.

Taking all of this into account the model can be improved by the following: instantaneous nucleation

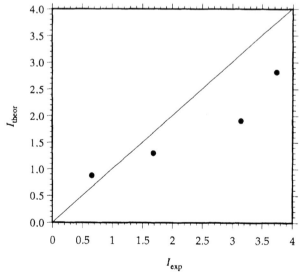

Figure 10.71. Correlation between theoretical and experimental values for the maxima of the chronoamperometric curves, at different concentrations of the LiClO$_4$/propylene carbonate solutions. The points were obtained from Figures 10.67 and 10.61, respectively.

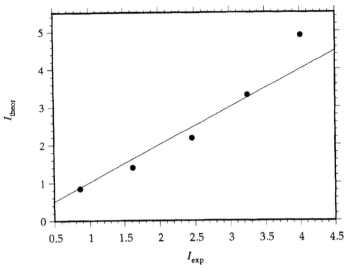

Figure 10.72. Correlation between theoretical and experimental values for the maxima of the chronoamperometric curves, at different temperatures. The points were obtained from Figures 10.68 and 10.63, respectively.

processes have to be checked and probably changed by a progressive nucleation; the relaxation time has to be written as a function of the polymer–polymer, polymer–solvent and polymer–ion interactions: the degree of crosslinking and the electroactivity of each film will also influence τ.

4.14 Technological consequences

The existence of a closure potential, E_s, has important technological consequences when transitions are performed from more cathodic potentials than E_s:

- current densities are low within a short time (batteries);
- the transition process requires a long time (electrochromic windows, filters or screens; sensors, actuators, smart membranes; or artificial muscles)

In order to improve the response times to charge and discharge batteries, to switch electrochromic devices, and to allow sensors, actuators, smart membranes or artificial muscles to respond quickly, it is necessary to work outside the range of potentials that give conformational relaxation control for the redox process, i.e. at more anodic potentials that E_s. The switching of an electrochromic film by polarization up to cathodic potentials higher than E_c will give long switching times

during the subsequent anodic polarization. Response times for the other electrochemical applications of conducting polymers are influenced in a similar way.

Compact structures also have a great technological interest. A compact material needs high energies to allow oxidation processes, through conformational movements and penetration of counterions or gaseous oxygen molecules, to occur. As a consequence, films compacted by cathodic polarization can be stored in air for long periods of time (days or months) without oxidation–degradation of the polymer bulk: only a monolayer of oxidized polymer is present at the polymer–air interface. Films, generated, compacted and dried in the laboratory were used, weeks or months later, for ion implantation experiments by means of low-energy ion beams. Before implantation the reduced state of every film was verified by conductivity measurements.

Close structures require high overvoltages to be opened (around 300 mV, Figure 10.52) or any equivalent chemical energy. The combination of oxygen and ambient temperature are not energetic enough to promote conformational changes in a compacted film, but they are able to oxidize and degrade a film that was reduced at anodic potentials higher than E_s. This application can be generalized to any conducting polymer used in any technological application: any device that has to be stored for long periods of time before use is protected against degradation by electrochemical compaction of the conducting films used. Before use every stored device needs an activation process.

A second important application of these compact films was discovered on electronic applications: the attainment of air-stable p–n junctions by ion implantation. The close structure allows the penetration of low-energy accelerated ions [114,115]. The structure recovers its compact state once the ions dope the polymer underneath. In this way p–n junctions were obtained by ion implantation. The compact polymer present between the doped polymer and the air hinders any subsequent spontaneous oxidation in air, as occurs with polymer films reduced at potentials around E_s.

4.15 Conformational relaxation model and muscles

The work of a muscle is due to a stress gradient between the conducting and non-conducting films. This stress gradient is the conformational energy developed during polymer oxidation or reduction. Thus the bilayer or triple layer devices work if a δ thickness of the conducting film (Figure 10.29) is under conformational relaxation control whatever the depth of oxidation. Under these conditions the current flowing when the bilayer is subjected to a potential step has to follow equation (10.43); the angle α is a function of the overall charge through equation (10.14). By combination

$$\alpha = h\frac{L}{Ae(e+d)}$$
$$\times Q_r\left[1 - \exp\left[-\frac{\pi N_0 \lambda^2}{\tau_0^2 A}\exp[-2\psi(E, E_c)]t^2\right]\right]$$

$$(10.69)$$

The equation gives the angle described (α) by a bilayer device during time (t) after a potential step from E_c to E.

In this equation h includes the volume of hydrated counterions, interaction parameters (polymer–polymer, polymer–solvent, polymer–ions) and the dielectric constant of the polymer; τ_0, λ and N_0 include structural and relaxation magnitudes; A, e and d includes geometrical parameters; and E, E_c and Q are electrochemical variables.

5 REFERENCES

1. D.W. Hurry, in *Biomolecular Materials by Design* (eds. M. Alper, H. Bayley, D. Kaplan and M. Navia), vol. 330, pp. 321–332, Material Research Society. Symposium, 1994.

2. W. Kuhn, B. Hargitay, A. Katchalsky and H. Eisemberg, *Nature* **165**, 514 (1950).

3. A. Katchalschi and M. Zwick, *J. Polym. Sci.* **16**, 221 (1955).

4. I.Z. Steinberg, A. Oplatka and A. Katchalsky, *Nature* **210**, 568 (1966).

5. H. Yuki, S. Sakakibara, T. Tani and H. Tani, *Bull. Chem. Sci. Jpn* **29**, 664 (1956).

6. Y. Osada and M. Sato, *Polymer* **21**, 1057 (1980).

7. F. Horkay and M. Zriyi, *Macromol. Chem. Macromol. Symp.* **30**, 133 (1989).

8. S. Umemoto, N. Okui and T. Sakai, in *Polymer Gels* (eds. D. de Rossi *et al.*) Plenum Press, New York, 1991, p. 257.

9. M. Suzuki and O. Hirasa, *Adv. Polym. Sci.* **110**, 241 (1993).

10. T. Tanaka, *J. Chem. Phys.* **40**, 820 (1978).

11. S. Fujishige, K. Kubota and I. Andu, *J. Phys. Chem.* **93**, 3311 (1988).

12. Y. Osaka, in *Advances in Polymer Science* (eds. S. Olive and G. Henrici-Olive), Springer, Berlin, 1987.

13. D. de Rossi, M. Susuki, Y. Osada and P. Morassi, *J. Intell. Mat. Syst. Struct.* **3**, 75 (1992).

14. P.J. Flory and J. Retner, *J. Chem. Phys.* **11**, 512 (1943).

15. P.J. Flory, *Principles of Polymer Chemistry* Cornell University Press, Ithaca, 1953.

16. O. Hirasa, S. Ito, A. Yamuchi, S. Fujishige and H. Ighijo, in *Polymer Gels* (eds. D. de Rossi *et al.*) Plenum Press, New York, 1991 pp. 247–256.

17. M.R. Jin, C.F. Wu, P.V. Lin and W. Hou, *J. Appl. Polym. Sci.* **56**, 285 (1995).

18. R. Lovrien, *Proc. Nat. Acad. Sci. USA* **57**, 236 (1967).

19. G. Van der Veen and W. Prins, *Phys. Sci.* **230**, 70 (1971).

20. H.S. Blair and H.I. Pogue, *Polymer* **23**, 779 (1982).

21. M. Irie and D. Kungwatchakun, *Makromol. Chem. Rapid. Commun.* **5**, 829 (1985).

22. T. Tanaka, I. Nishio, S.T. Sun and S.V. Nishio, *Science* **218**, 467 (1982).

23. I.V. Yannas and A.J. Grodzinsky, *J. Mechanochem. Cell. Mobility* **2**, 113 (1973).

24. D. de Rossi, P. Chiarell, G. Buzzigoli and C. Domenici, *Trans. Am. Soc. Artific. Inter. Organs* **32**, 157 (1986).

25. K. Okabayashi, F. Goto, K. Abe and T. Yushida, *Synth. Met.* **18**, 3718 (1993).

26. T. Shiga and T. Kurauchi, *J. Appl. Polym. Sci.* **39**, 2305 (1990).

27. R. Kishi, N. Hasebe, M. Hara and Y. Osada, *Polym. Adv. Tech.* **1**, 19 (1990).

28. Y. Osada *Adv. Mat.* **3**, 107 (1991).

29. T. Karauchi, T. Shiga, Y. Hirose and A. Okada, in *Polymer Gels* (eds. D. de Rossi *et al.*, Plenum Press, New York, 1991, pp. 237–246.

30. Y. Osada, J.P. Gong and K. Sawahata, *J. Macromol. Sci: Chem.* **A28**, 1189 (1991).

31. Y. Osada, H. Okuzaki and H. Hori, *Nature* **355**, 242 (1992).
32. H. Okuzaki and Y. Osada, *J. Biomater. Sci. Polymer. Ed.* **5**, 485 (1994).
33. Z. Hu, X. Zang and Y. Li, *Nature* **269**, 525 (1995).
34. K. Asaka, K. Oguro, Y. Nishimura, M. Mizuhata and H. Takenaka, *Polym. J.* **27**, 436 (1995).
35. R.H. Garret and Ch.M. Grisham, *Molecular Aspects of Cell Biology* Saunders College Publishing and Harcourt Brace College Publishers.
36. D. Boet and J Voet, *Ediciones* Omega S A, Barcelona.
37. L. Stryyer, *Biquímica*, Ed. Reverté, Barcelona (1988).
38. A.J. Fisher, C.A. Smith, J. Thoden, R. Smith, K. Sutoh, H.M. Holden and I. Rayment, *Biophys. J.* **68**, 19s (1995).
39. D.D. Thomas, S. Ramachandran, O. Rooprarine, D.W. Hayden and E.M. Ostap, *Biophys. J.* **68**, 135s (1995).
40. A. Oplatka, *Chemtracks: Biochem. Mol. Biol.* **5**, 120 (1994).
41. P. Burgmayer, R.W. Murray, in *Handbook of Conducting Polymers* (eds. T.A. Skotheim), Marcel Dekker, New York, 1986, pp. 507–523.
42. T.F. Otero, C. Santamaría, E. Angulo and J. Rodríguez, *Synth. Met.* **41–43**, 2947 (1991).
43. T.F. Otero and C. Santamaría, *Solid State Ionics* **63–65**, 810 (1993).
44. T.F. Otero and E. Angulo, *Solid State Ionics* **63–65**, 803 (1993).
45. T.F. Otero and J. Rodríguez, in *Intrinsically Conducting Polymers: an Emerging Technology* (ed. M. Aldissi), Kluwer, Dordrecht, 1993, pp. 179–190.
46. R.H. Baughman, L.W. Shacklette, R.L. Elsenbaumer, E. Plichta and C. Becht, in *Conjugated Polymer Materials: Opportunities in Electronics, Optoelectronics and Molecular Electronics* (eds. J.L. Bredas and R.R. Chance), Kluwer, Dordrecht, 1990, pp. 559–582.
47. R.H. Baughman, L.W. Shacklette, N.S. Murthy, G.G. Miller and R.L. Elsenbaumer, *Mol. Cryst. Liq. Cryst.* **118**, 253 (1985).
48. L.W. Shacklette and J.E. Toth, *Phys. Rev.* **B32**, 5892 (1985).
49. N.S. Murthy, L.W. Shacklette and R.H. Baughman, *J. Chem. Phys.* **87**, 2346 (1987).
50. B. Françoise, N. Mermilliod and L. Zoppiroli, *Synth. Met.* **4**. 131 (1981).
51. E.J. Plichta, Mater Thesis, Rutgers University, New Brunswick, New Jersey, 1989.
52. K. Okabayashi, F. Goto, K. Abe and T. Yoshida, *Synth. Met.* **18**, 365 (1987).
53. M. Slama and J .Tanguy, *Synth. Met.* **28**, C171 (1989).
54. P. Chiarelli, D. de Rossi, A. Della Santa and A. Mazzoldi, *Polymer Gels and Networks* **2**, 289 (1994).
55. Q. Pei and O. Inganäs, *J. Phys. Chem.* **96**, 10507 (1992).
56. Q. Pei and O. Inganäs. *J. Phys. Chem.* **97**, 6034 (1993).
57. P. Chiarelli, K. Umezawa and D. de Rossi, in *Polymer Gels* (eds. D. de Rossi *et al.*, Plenum Press, New York, 1991.
58. T.F. Otero and J. Rodríguez, *Electrochim. Acta* **39**, 245 (1994).
59. T.F. Otero, C. Santamaría, E. Angulo and J. Rodríguez, *Synth. Met.* **55**, 1574 (1993).
60. T.F. Otero, C. Santamaría and J. Rodríguez, *Mat. Res. Soc. Symp.* **328**, 805 (1994).
61. T.F. Otero and J. Rodríguez, *Synth. Met.* **55**, 1418 (1993).
62. T.F. Otero, E. Angulo, J. Rodríguez and C. Santamaría, Patent EP9200095, 1992.
63. T.F. Otero, J. Rodríguez and C. Santamaría, Patent EP9202628, 1992.
64. T.F. Otero, E. Angulo, J. Rodríguez and C. Santamaría, *J. Electroanal. Chem.* **341**, 369 (1992).
65. T.F. Otero, Artificial muscles, electrochemomechanical devices using conducting polymers. Presentation through videos, p. 138, 4th BRITE-EURAM Conference, Seville, Spain. Commission of the European Communities, 1992.
66. T.F. Otero, J. Rodríguez, E. Angulo and C. Santamaría, *Synth. Met.* **55**, 1418 (1993).
67. T.F. Otero and J.M Sansiñena, *J. Bioelectrochem. Bioenerg.* in press.
68. T.F. Otero, J. Rodríguez and C. Santamaría. *Synth. Met.* **55–57**, 3713 (1993).
69. Q. Pei and O. Inganäs, *Synth. Met.* **55–57**, 3718 (1993).
70. Q. Pei and O. Inganäs, *Synth. Met.* **55–57**, 3724 (1993).
71. Q. Pei and O. Inganäs, *Adv. Mat. Commun.* **5**, 630 (1993).
72. Q. Pei and O. Inganäs. *Synth. Met.* **55–57**, 3730 (1993).
73. E. Smela, O. Inganäs and I. Lundström, *Science* **268**, 1735 (1995).
74. T. Takashima, M. Kaneto, K. Kaneto and A.G. McDiarmid, *Synth. Met.* **71**, 2265 (1995).
75. K. Kaneto, M. Kaneto, Y. Min and A.G. McDiarmid, *Synth. Met.* **71**, 2211 (1995).
76. J.R. Reynolds and M. Pomerantz, in *Electroresponsive Molecular and Polymeric Systems* (ed. T.A. Skotheim), Marcel Dekker, New York, 1991, pp. 187–256.
77. A.J. Heeger and P. Smith, in *Conjugated Polymers* (eds. J.L. Bredas and R. Silbey), Kluwer, Dordrecht, pp. 141–210; J.P. Aime, in ibid. pp. 224–314; G. Guftanson, O. Inganäs, W.R. Salanck, J. Laaso, M. Loponen, T. Taka, J.E. Österholm, H. Stabb and T.H. Jertberg, in ibid. pp. 315–362.
78. V.G. Kulkarni, in *Intrinsically Conducting Polymers: an Emerging Technology* (ed. M. Aldissi), Kluwer, Dordrecht, 1993, pp. 45–50; E.J. Epstein, J. Ju, G.Y. Wu, A. Benatar, C.F. Fairst, Jr, J. Zegarski and A.G. MacDiarmid in ibid. pp. 165–178.
79. D. Fichou, G. Horowitz, B. Xu and F. Garnier, *Synth. Met.* **41**, 463 (1991).

80. H. Naarman, in *Intrinsically Conducting Polymers: an Emerging Technology* (ed. M. Aldissi), Kluwer, Dordrecht, 1993, pp. 1–12.

81. T.F. Otero *et al*, *Synth. Met.* **41–43**, 2947 (1991); **54**, 217 (1993).

82. G.G. McLeod, M.G.B. Mahboubian-Jones, R.A. Pethrick, S.D. Watson, N.D. Truong, J.C. Galin and J. François, *Polymer* **27**, 455 (1986).

83. J.P. Ferraris and T.R. Hanlon, *Polymer* **30**, 1319 (1989).

84. H. Wynberg and J. Matselaar, *Synth. Commum* **14**, 1 (1984).

85. M. Montilla, Licenciature Thesis, University of Barcelona. Faculty of Chemistry, Barcelona, Spain, 1995.

86. J. Roncali, F. Garnier, M. Lemaire and M. Garreau, *Synth. Met.* **15**, 323 (1987).

87. J.P. Ferraris and G.D. Skiles, *Polymer* **28**, 179 (1987).

88. G. Kofsmehl, *Makromol. Chem. Makromol. Symp.* **4**, 45 (1986).

89. Y. Yumoto, K. Morishita and S. Yoshimura, *Synth. Met.* **18**, 203 (1987).

90. J. Carrasco, A. Figueras, T.F. Otero and E. Brillas, *Synth. Met.* **61**, 253 (1993).

91. T.F. Otero, J. Carrasco, A. Figueras and E. Brillas, *J. Electroanal. Chem.* **370**, 231 (1993).

92. T.F. Otero, E. Brillas, J. Carrasco and A. Figueras, *Mat. Res. Soc. Symp. Proc.* **328**, 799 (1994).

93. E. Brillas, J. Carrasco, A. Figueras, F. Urpi and T.F. Otero, *J. Electroanal. Chem.* **392**, 55 (1995).

94. F.A. Posey and T. Morozumi, *J. Electrochem. Soc.* **113**, 176 (1966).

95. K. Aoki and Y. Tezuka, *J. Electroanal. Chem.* **267**, 55 (1989).

96. K. Aoki, Y. Tezuka, K. Shinozaki and H. Sato, *Electrochem. Soc. Jpn.* **57**, 397 (1989).

97. K. Aoki, J. Cao and Y. Hoshino, *Electrochimica Acta* **39**, 2291 (1994).

98. K. Aoki, *J. Electroanal. Chem.* **373**, 67 (1994).

99. C. Odin and M. Nechtschein, *Synth. Met.* **55–57**, 1281 (1993).

100. S.W. Feldberg, *J. Am. Chem. Soc.* **106**, 4671 (1984).

101. T. Yeu, K.M. Yin, J. Carbajal and R.E. White, *J. Electrochem. Soc.* **138**, 2869 (1991).

102. T. Yeu, T.V. Nguyen and E. White, *J. Electrochem. Soc.* **135**, 1971 (1988).

103. T.F. Otero, H. Grande and J. Rodriguez, *J. Electroanal. Chem.* **394**, 211 (1995).

104. T.F. Otero, H. Grande and J. Rodriguez, *Synth. Met.* in press.

105. H. Grande, Thesis de Licenciatura, Faculty of Chemistry, University of the Basque Country, San Sebastian, Spain, 1994.

106. C. Odin and M. Nechtschein, *Synth. Met.* **43**, 2943 (1991).

107. C. Odin and M. Nechtschein, *Synth. Met.* **55–57**, 1287 (1993).

108. Q. Pei and O. Inganäs, *J. Phys. Chem.* **97**, 6034 (1993).

109. B. Francois, N. Mermilliod and L. Zuppiroli, *Synth. Met.* **4**, 131 (1981).

110. C. Bucci and R. Fieschi, *Phys. Rev. Lett*, **12**, 16 (1964).

111. J. Van Turnhout, *Thermally Stimulated Discharge of Polymers Electrodes*, Elsevier, Amsterdam, 1975.

112. Thermally stimulated relaxation in solids, in *Topics in Applied Physics*, vol. 37. (ed. P. Braunlich), Springer-Verlag, New York, 1979.

113. P.K.C. Pillai, Polymeric Electrets, in *Ferroelectric Polymers: Chemistry, Physics and Applications*, (ed. H.S. Nalwa), Marcel Dekker, New York, 1995, Chapter 1. pp. 1–61.

114. B. Lucas, B. Ratier, A. Moliton, J.P. Moliton, T.F. Otero, C. Santamaria, E. Angulo and J. Rodriguez, *Synth. Met.* **57**, 1459 (1993).

115. B. Lucas, J. Rodriguez, T.F. Otero, B. Guille and A. Moliton, *Advanced Materials for Optics and Electronics*, in press.

CHAPTER 11

Conducting Polymers for Batteries, Supercapacitors and Optical Devices

Catia Arbizzani*, Marina Mastragostino[†] and Bruno Scrosati[‡]
*University of Bologna, Italy; †University of Palermo, Italy; ‡University La Sapienza of Rome, Italy

1 INTRODUCTION

The discovery in the late seventies that certain types of polymers, though intrinsically poor conductors acquire, following chemical or electrochemical treatment, electric conductivity approaching that of metals, opened up an exciting new area of materials research [1,2]. The essential structural characteristic needed by polymers to attain this significant conductivity change is a conjugated π-system extending over a large number of monomer units, a trait common to polyacetylenes, to polyheterocycles such as polypyrroles and polythiophenes, and to polyanilines. However, since polyacetylene suffers from high environmental instability, a major obstacle to practical applications, most studies are today focused on heterocyclic polymers and polyanilines.

The processes that switch conjugated polymers from the insulating to the conducting state are redox reactions, whether chemically or electrochemically driven. They are called doping processes, *p*-doping or *n*-doping in relation to the positive or negative sign of the injected charge in the polymer chain. For instance,

by anodically polarizing a conjugated polymer film placed in an electrochemical cell with a suitable electrolyte (e.g. a solution of lithium or alkylammonium salt in an organic solvent such as propylene carbonate, PC) and a suitable counter-electrode (e.g. platinum foil), the polymer becomes oxidized (*p*-doped), whereas by cathodically polarizing the polymer, it becomes reduced (*n*-doped), as shown in Scheme 11.1 with polythiophene as the conjugated polymer. The most important feature of these doping processes is their reversibility, i.e. conducting polymers can be switched repeatedly between their conducting (doped) and insulating (undoped) states by electrochemical oxidations and reductions.

The electrolyte ions involved in these processes are called the dopant anion and the dopant cation, respectively, while *y*, which represents the ratio between the dopant ion and the polymer repeat unit, is commonly referred to as the doping level of the polymer electrode. Both the oxidation (*p*-doping) and reduction (*n*-doping) processes of conjugated polymers lead to the formation of charged complexes which include the polycation or polyanion and the dopant ion

Handbook of Organic Conductive Molecules and Polymers: Vol. 4. Conductive Polymers: Transport, Photophysics and Applications.
Edited by H. S. Nalwa. © 1997 John Wiley & Sons Ltd

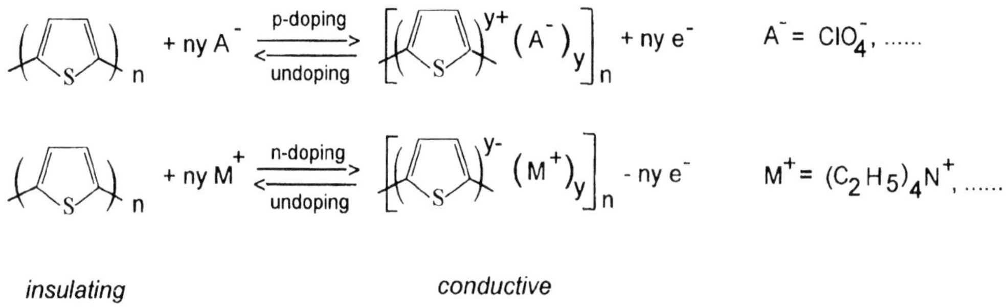

insulating   *conductive*

Scheme 11.1

in a combined structure where the electronic transport is accompanied by the diffusion of the dopant ion into the polymer. However, as cation motion can also be involved in the *p*-doping process, the mechanism is sometimes more complicated than simply the dopant anion entering the polymer during anodic polarization and leaving it during the undoping process [3].

Conjugated polymers have an electronic band structure, and the energy gap between the highest occupied π-electron band (valence band) and the lowest unoccupied one (conduction band) determines the intrinsic electric and optical properties of the polymer. The doping processes modify the polymer's electronic band structure by generating new electronic states in the energy gap which cause the colour changes. Doping shifts the optical absorption towards lower energies; the colour contrast between the doped and undoped forms is related to the energy gap [4].

The fact that conjugated polymers may repeatedly undergo electrochemical doping–undoping processes involving a relatively large amount of electronic charge makes conjugated polymers attractive electroactive materials, and much effort has been devoted to their use in advanced electrochemical devices, where the replacement of the conventional electrode materials with these polymers can lead to substantial improvements in design, versatility, reliability and cost efficiency. Conjugated polymers are an interesting class of ion-insertion electrochemical materials that act during the charge–discharge process as mixed electronic and ionic conductors and perform like conventional electrodes based on inorganic intercalation compounds [5]. They thus offer the extra advantage of being highly conductive in addition to those advantages typical of their polymeric nature, i.e. plasticity, chemical stability, ease of manufacturing and low cost. A great deal of attention is currently being focused on exploiting this unique combination for new-concept, plastic-like batteries and a part of this chapter will be devoted to

the results reported in this area by academic and industrial research laboratories.

A second trait typical of conducting polymers is linked to their morphology. Depending upon the preparation conditions, thick polymer films with an open structure can be routinely prepared, an important feature for electrode materials in which the kinetics of charge–discharge are generally controlled by the diffusion of the dopant ions into and out of the polymer matrix, particularly for high-power performance devices like supercapacitors. These devices are already of significant importance in the consumer electronics market and are highly regarded for their potential in sustaining electric vehicle (EV) operation by providing the high-power pulses required for acceleration [6]. Accordingly, the second part of this chapter will deal with polymer-based supercapacitors.

Another very important feature of conducting polymers is related to the colour changes induced and controlled by electrochemical doping–undoping. It enables conducting polymers to be used in the manufacture of advanced optical devices such as multichromic displays or 'smart' electrochromic windows. This again places conducting polymers in an important technological area since these optical devices today play a well-established role in the automobile industry (as self-regulating rear-view mirrors) and are considered very promising for energy control in buildings [7]. These optically active polymer devices will be treated in the final part of this chapter.

2 SYNTHESIS PROCEDURE

The properties and the response of the conducting polymers are greatly influenced by their synthesis conditions. Therefore, before discussing applications, certain details of the synthesis procedures will be mentioned and evaluated. Conducting polymers may be prepared either chemically or electrochemically. The chemical methods imply the use of reagents of various nature and, in some cases, even of a catalyst, so that the morphological, chemical and physical properties of the final material greatly depend upon the synthesis conditions, e.g. type and concentration of the reagents, catalyst, solvent, temperature and so forth.

The preferred chemical process can vary from case to case. For instance, polypyrrole can be obtained by chemical oxidation of the pyrrole using different types of oxidizing agents, the most common being ferric salts such as $FeCl_3$ and $Fe(ClO_4)_3$ [8]. When polypyrrole is used as battery electrode material, specific oxidizing agents are recommended so as to increase the compatibility of the polymer with the type of battery's electrolyte. For instance, in lithium batteries where $LiBF_4$ is routinely used as the electrolyte salt, fluorinated oxidizing agents such as $Cu(BF_4)_2$ appear to be a convenient choice [9].

Polythiophenes can also be obtained by polymerization with such oxidizing agents as $Cu(ClO_4)_2$ [10], $FeCl_3$ [11] and, in polythiophene preparation for lithium battery applications, $NOBF_4$ [12]. Polyanilines can be prepared by a polymerization route promoted by specific oxidizing agents like $(NH_4)_2S_2O_8$ [13]. For the synthesis of lithium battery grade polyanilines, $Fe(ClO_4)_3$ and $Cu(BF_4)_2$ are both suitable oxidizing agents since $LiClO_4$, like $LiBF_4$, is also a common lithium battery electrolyte salt.

As an alternative to the chemical route, conducting polymers can by synthesized by the electrochemical method. The latter, starting from the monomer, promotes the formation of the doped conductive film on a suitable substrate. By controlling the charge in the cell, one can easily monitor the polymer's thickness, which can range from a few ångströms to many micrometres or even to millimetres. Furthermore, by changing the nature of the dopant ion (i.e. by changing the supporting electrolyte in solution) or the value of the electrodeposition current density or voltage, the physicochemical properties and the morphology of the final polymer can be modified. Electrochemical polymerization has thus rapidly expanded since 1979 [14], when a highly conductive free-standing film of polypyrrole was first prepared, and electropolymerization is still the most popular method of preparing conducting polymers.

The electrochemical synthesis of a given conducting polymer can be easily achieved by galvanostatic method in an electrochemical cell consisting of a working electrode (i.e. the substrate electrode for polymer film deposition) and a counter-electrode, both immersed in a solution, generally non-aqueous, containing the monomer and a supporting electrolyte, the latter operating both as the conductive medium and as the source of the dopant anions. An important feature of the oxidative electropolymerization reaction is that it proceeds with electrochemical stoichiometry, with values generally in the 2.1–2.6 F mol^{-1} range depending on the type of polymer and electrosynthesis conditions. Polymer formation requires two electron/monomer and the excess charge corresponds to the p-doping of the polymer. The electrochemical process can basically be described as the sequence of several steps, as shown for polypyrrole in Scheme 11.2. The first step involves the oxidation of the monomer to the radical cation, the second consists of the α–α' coupling of two radical cations, which leads to a dimer after re-aromatization and the loss of two protons. Electropolymerization proceeds with chain propagation and the formation of a p-doped polymer.

This simple electrochemical method can be easily scaled-up from laboratory samples to the synthesis of large-area polymer sheets for industrial uses. One successful example is the production of continuous polypyrrole sheeting using an electrochemical reactor and a rotating substrate electrode, a method first developed in the late eighties by Naarmann [15], and recently brought back to the fore [16].

Like polypyrrole, polythiophenes can also be electrochemically prepared. Yet, owing to the drastic electrosynthesis conditions needed to oxidize the thiophene, the starting molecule is usually the dimer 2,2'-bithiophene. The electrochemical method has been the most widely used to synthesize all the thiophene-based polymers tested for different applications, especially when very thin and homogeneous films are required, as in electrochromic electrodes.

Electropolymerization is a very convenient approach for good-quality, highly electroactive films, even in the case of polyaniline. Since the electroactivity of the latter is promoted by protons, all preparation methods require acid solution media. The most common procedures involve the deposition of polyaniline films from highly acidic (e.g. H_2SO_4 or $HClO_4$) aqueous solutions [17,18]. Polyaniline films can also be obtained in cells with aqueous 1.0 M $NaHSO_4$ (pH = 1) solutions containing aniline at an applied voltage of 0.9 V vs. SCE [19]. However, in view of the lithium-battery application, the polymerization in aqueous solutions is something of a drawback since it requires the additional step of complete film drying prior to use. Direct synthesis from the same non-aqueous solutions commonly used as lithium-battery

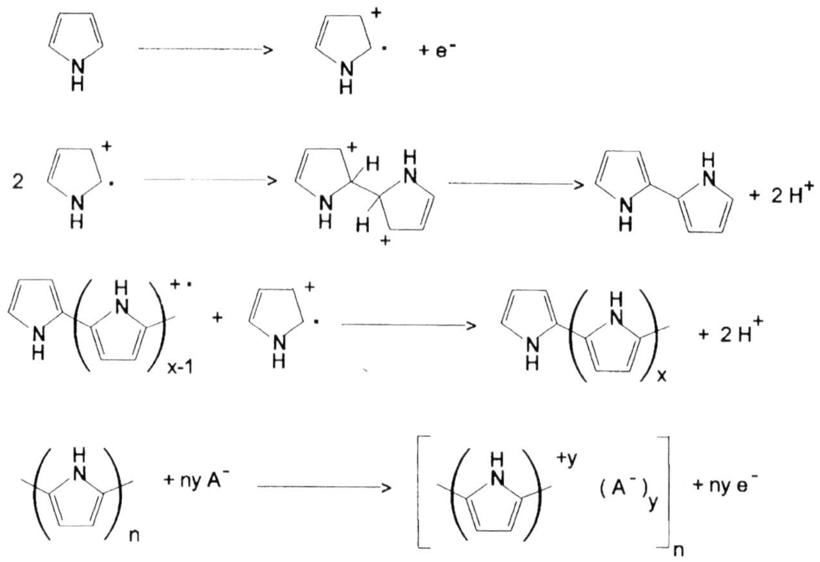

Scheme 11.2

electrolytes would certainly be more convenient. One such successful approach involves polymerization from a $LiClO_4$–PC solution, acidified by CF_3COOH, containing aniline [20].

3   LITHIUM–POLYMER BATTERIES

3.1   General principles

The use of conducting polymers as battery electrodes relies on their redox (doping) processes being driven electrochemically. A given polymer can be repeatedly cycled between different oxidation states, thereby acting as a reversible electrode for a rechargeable battery. Although in principle conducting polymers can be used both as anodes (i.e. by exploiting their reduction or n-doping processes) or as cathodes (i.e. by exploiting their oxidation or p-doping processes), most battery applications are confined to the latter case. This chapter will focus on batteries using polymer cathodes mainly in combination with lithium, the metal with the highest specific capacity (3.86 Ah g^{-1}), so as to attain high-energy batteries.

The p-doping process for polypyrrole (pPy) has been extensively exploited in cathode material for batteries with a lithium anode and a suitable electrolyte, e.g. a $LiClO_4$–PC solution. The electrochemical charge–discharge process of this battery can be written as

$$(Py)_n + nyLiClO_4 \underset{\text{discharge}}{\overset{\text{charge}}{\rightleftharpoons}} [(Py)^{y+}(ClO_4^-)_y]_n + nyLi$$

(11.1)

The charge process essentially involves the p-doping of the polymer, i.e. its oxidation and the formation of a polycation whose positive charge is counterbalanced by the electrolyte dopant anion (ClO_4^- in the above example) which diffuses into the polymer matrix. The pPy's oxidation at the positive electrode is accompanied at the negative electrode by the reduction of lithium ions, which deposit as lithium metal on a given substrate. In the discharge process the electroactive polymer cathode releases the anions and lithium ions are stripped from the metal anode to restore the initial electrolyte concentration.

The electrochemical process in this, as in most of the lithium–polymer batteries, involves the electrolyte salt to an extent which is defined by the doping level y, i.e. the extent of the oxidation state reached by the polymer electrode. Thus, the doping level, which represents the percentage of moles of the dopant anion over the moles of monomer units (e.g. moles of Py), is proportional to the amount of charge involved and directly related to battery capacity, as measured in ampere hours. The maximum achievable doping level depends upon the given polymer, varying from 25–35% for polypyrrole, from 7–25% for polythiophene and from 50–60% for polyaniline (pAni). Very simply put, the energy in a lithium–polymer battery is supplied during charge to activate electrochemically the polymer electrode by promoting its doping process up to a level y and the energy is released during discharge by the undoping process; this cycle can be repeated several times.

Figure 11.1 illustrates a typical charge–discharge cycle of a lithium–pPy battery in $LiClO_4$-PC [21]. The electrochemical process driving this battery involves the injection (charge)–removal (discharge) of ClO_4^- ionic species. Accordingly, in situ electrogravimetric analysis of the pPy electrode shows a linear relationship between the amount of charge supplied and the electrode's weight (Figure 11.2) [22]. Lithium–polymer batteries act as concentration cells, the value of voltage depending upon the polymer's oxidation (doping level) state, as shown in Figure 11.1. This voltage-controlling mechanism in turn influences the energy density of the battery, the latter (measured in watt hours per kilogram, $Wh\ kg^{-1}$) is determined by the capacity (i.e. doping level y) and by the average voltage acquired during the cyclic process. In these lithium–polymer batteries, the concentration of dopants in the electrolyte solution decreases during battery charge [see equation (11.1)]. Yet the energy density depends also, and particularly, on the electrolyte concentration: high values can only be achieved using electrolyte solutions with both a high dopant concentration and a high conductivity throughout the charge–discharge cycle.

Table 11.1 reports the values of the average voltage and of the theoretical energy density (i.e. the maximum energy achievable per unit weight) of lithium–polymer batteries in comparison to those of other lithium (inorganic intercalation cathodes) and standard (e.g. nickel–cadmium, Ni–Cd, nickel–metal hydride, Ni–MH, and lead–acid, $Pb–H_2SO_4$) batteries. The theore-

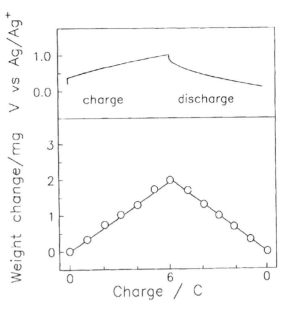

Figure 11.2. Galvanostatic charge–discharge cycle at 0.5 mA cm^{-2} of a pPy electrode (electrosynthesized with 20 C cm^{-2}) (top) and change of the polymer electrode's mass during charge and discharge (bottom). (Reproduced by permission of the Electrochemical Society from ref. 22.)

tical energy density of the former compares well to that of conventional systems, although it is usually lower than that of lithium–inorganic intercalation batteries. The former are further penalized by an excess of electrolyte (to assure both the electrochemical balance and the electric transport across the cell) required for their operation. Thus, the practical energy density (i.e. that associated with the battery in its complete assemblage, including electrolyte, current collectors and hardware) ultimately decreases to lower values. For instance, practical energy density values around 80–100 $Wh\ kg^{-1}$ are reported for lithium–intercalation electrode batteries [23], whereas the values for lithium–pPy prototypes do not exceed 30 $Wh\ kg^{-1}$ (Table 11.2) [24].

Another particular feature of lithium–polymer batteries is that the electrode kinetics are generally controlled by the diffusion of the dopant anions throughout the polymer structure. As expected, the charge–discharge rate of polymer electrodes greatly depends upon the nature (size and charge density) of the dopant anion and upon electrode morphology, which in turn depends on the electrosynthesis conditions [25–28].

An interesting approach to enhance the electrode rates involves the formation of polymer structures

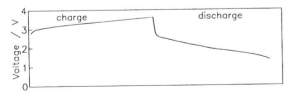

Figure 11.1. Galvanostatic charge–discharge cycle (0.1 mA cm^{-2}) of a $Li/LiClO_4$–PC/pPy cell at room temperature. $Q_{charge} = 42\ mC\ cm^{-2}$. (Reprinted from *Electrochim. Acta* S. Panero *et al.*, **32**, 1007, Copyright 1987, with kind permission from Elsevier Science Ltd, The Boulevard, Langford Lane, Kidlington OX5 1GB, UK.)

Table 11.1. Average voltage and theoretical energy density values of some selected batteries

Battery	Average Voltage (V)	Energy density (Wh kg^{-1})		Doping level y(%) or x
		Theoretical	Practical	
Li–pPy(ClO$_4$)$_y$	3.0	240	20–30	30
Li–pT(ClO$_4$)$_y$	3.9	200		20
Li–pAni(ClO$_4$)$_y$	3.0	380		60
Li–Li$_x$TiS$_2$	2.1	480	80–100	1
Li–Li$_x$MnO$_2$	2.8	415	120	1
Ni–Cd	1.2	215	30–40	
Ni–MH	1.2	270	45–70	
Pb–H$_2$SO$_4$	2	170	30–35	

pPy = polypyrrole; pT = polythiophene; pAni = polyaniline.

Table 11.2. Characteristics of lithium–polypyrrole batteries developed at BASF–VARTA Companies. (Reprinted from *Synth. Met.* H. Münstedt *et al.*, **18**, 259, 1987, with kind permission from Elsevier Science SA, Lausanne, Switzerland.)

	Cylindrical	Sandwich I	Sandwich II
Total mass (g)	15	44	49
Energy density (Wh kg^{-1})	15	20	30

incorporating large anions. This is achieved by employing, in the electrosynthesis, supporting electrolytes such as sodium dodecysulfate NaDS [29] or sodium poly-(4-styrenesulfonate) NaPSS [27,30]. The large anions, once inserted into the polymer structure, are difficult to remove; the electrochemical process is then likely to involve the incorporation of electrolyte cations rather than the release of the large anions. Since the former generally have a smaller size, their diffuion is faster than that of the anions. The promotion of electrochemical processes preferably involving cation transport can ultimately enhance the battery charge–discharge rate.

The rate of self-discharge is also an important concern in lithium–polymer batteries. Most polymer electrodes show poor charge retention in common organic liquid electrolytes. For instance, a self-discharge rate of about 1% per day over 30 days for the lithium-pPy battery is reported [24] and a significantly higher self-discharge rate was observed with charged poly(2,2′-bithiophene) electrodes [31]. A slower decay has been found for lithium-pAni batteries, which showed self-discharge rates of about 5–7% per month [32,33], a fact accounting in part for the industrial interest in this system (see below).

Various studies indicate that self-discharge involves a spontaneous undoping process, without irreversible degradation of the polymer electrode, induced by an as-yet unexplained mechanism [28]. A plausible assumption is that the self-undoping (reduction) of the polymer is accompanied by a concurrent (oxidation) process of the solvent and/or impurities. Improved shelf-life values are reported with very carefully purified media [31]. A suitable choice of electrolyte medium, carefully purified, is crucial to ensure stability of the polymer electrodes and long shelf-life of the lithium–polymer batteries.

Despite the problem noted above, the lithium–polymer electrode configuration retains such specific advantages as flexibility in geometry and design, compatibility with the environment and projected low cost, factors which make them competitive for small-sized, low-rate button prototypes for the microelectronics consumer market.

3.2 Battery prototypes

Considerable interest has been focused since the mid-1980s on the marketing of lithium–polymer batteries, especially of prototypes using pAni as the cathode material. Here, the electrochemical process involves the reaction of the pAni emeraldine form with lithium (e.g. in a cell using the usual LiClO$_4$–PC electrolyte solution) and the formation of the emeraldine hydroperchlorate salt, as shown in Scheme 11.3.

Figure 11.3 shows a typical charge–discharge cycle of a lithium–pAni cell using a polymer film originally

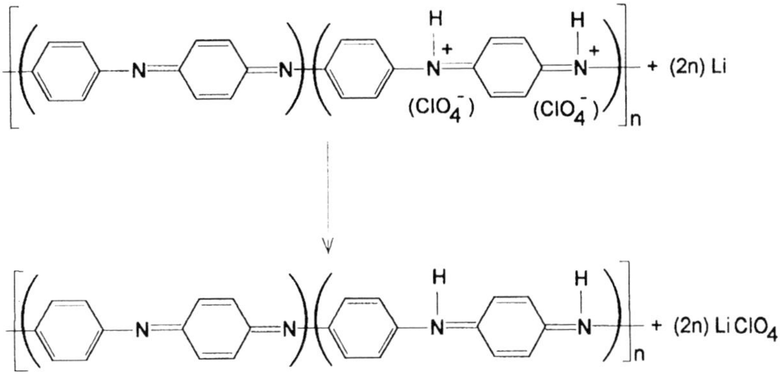

Scheme 11.3

electrosynthesized in the same $LiClO_4$–PC solution [20]. One can clearly see that the charge and discharge curves are almost symmetrical, thus confirming the reversibility of the film's doping–undoping reaction. A theoretical energy density of 380 Wh kg^{-1} was found for this $Li/LiClO_4$–PC/pAni cell [34].

The lithium–pAni system was exploited by Japan's Bridgestone Company in the large-scale production of a high-cycle-life, button-type three-volt battery. The battery features an electrochemically prepared pAni cathode and a lithium–aluminium alloy anode [35]. A cross-section of the battery and the charge–discharge cycle of a typical prototype (PR 2016) is shown in Figure 11.4 [2,36].

Other Japanese industrial laboratories have investigated button-type lithium–polymer batteries: Figure 11.5 illustrates typical charge–discharge cycles of various types using different polymer cathodes, as reported by Sanyo Laboratories [12]. Even the German Varta and BASF industries jointly attempted the development of lithium–pPy, cylindrical and sandwich

types, three-volt batteries [24]. Their characteristics are listed in Table 11.2 and the structure of the cylindrical type is schematically shown in Figure 11.6. Note that lithium–pPy cells are generally characterized by a high charge–discharge efficiency (Figure 11.7) [21].

The initial enthusiasm for lithium–polymer batteries, which in the eighties was accompanied by more or less concrete marketing campaigns, has dampened in more recent years to very limited manufacturing levels or even to remove from production. The main reason behind this change of strategy in the consumer electronics market is that lithium–polymer batteries are not competitive with such established power sources as the nickel–cadmium battery. The main drawback of the former is their comparatively low practical energy density (see Table 11.1), limited power density and high self-discharge rate.

Yet, if one imagines a more specialized market niche in which the specific and unique characteristics of polymer materials can be adequately exploited, e.g. one wherein size and flexibility in design are more important requisites than energy density and power rate, then lithium–polymer batteries may acquire a renewed importance. However, to be really successful consumer products, reliability must also be assured. Thus, the replacement of the liquid with a solid electrolyte for a fully solid-state configuration is here one of the key challenges. The absence of liquids or gases would rule out leaking or pressure build-up, especially important when safety becomes an essential requirement. Thin-film, solid-state configurations also provide plasticity, a highly welcome feature when solid polymer electrolytes are used.

This important concept was first exploited with a poly(ethylene) (PEO)–based ionically conducting polymer as the solid, plastic separator of a thin-film,

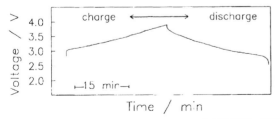

Figure 11.3. Charge–discharge curve of $Li/LiClO_4$–PC/ pAni cell at 2 mA cm^{-2}. The cell was charged to a 60% doping level and the pAni electrode was prepared with 20 C cm^{-2}. (Reproduced by permission of the Electrochemical Society from ref. 20.)

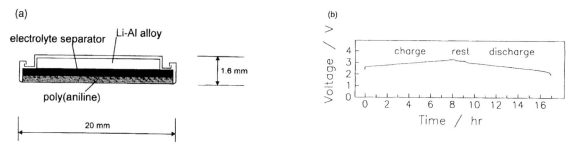

Figure 11.4. Cross-sectional view (a) and typical discharge curve (b) of the Bridgestone (PR 2016) button-type battery Li-Al/ LiBF$_4$–PC–DME/pAni (DME = dimethoxyethane) [2,36].

lithium–polymer cell [26,37–41]. This cell type can deliver a large number of cycles at reasonable rates with a high charge–discharge efficiency. Yet these positive characteristics are offset by the high operating temperature (about 70°C) required of the polymer electrolyte to meet suitable conductivity values, the major drawback of these otherwise promising solid-state power sources. The replacement of the traditional PEO-based electrolyte with an improved, highly conducting polymer membrane should greatly improve their versatility and competitiveness.

This approach, pursued by various academic and industrial laboratories has achieved interesting results [42,43] by combining a PEO–SEO–LiClO$_4$ polymer electrolyte, an interpenetrating network conceived for room-temperature applications [44], with new 'tailor made' poly(N-oxyalkylpyrrole) electrodes, designed to form a mixed ionic–electronic conducting matrix [45,46]. The addition of SEO (a styrenic macromonomer of PEO) enhances the amorphous domain, and hence the ionic conductivity, of the polymer electrolyte

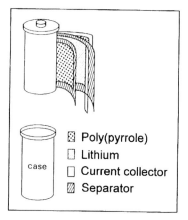

Figure 11.6. Structure of the cylindrical-type lithium–pPy battery developed at BASF–VARTA Companies. (Reprinted from *Synth. Met.* **18**, H. Münstedt *et al.*, 259, 1987, with kind permission from Elsevier Science SA, Lausanne, Switzerland.)

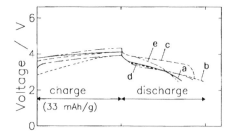

Figure 11.5. Charge–discharge curves of lithium–polymer cells using chemically synthesized polymers: (a) polyacetylene, (b) polypyrrole, (c) polythiophene, (d) polyiminodibenzyl, (e) polycarbazole. (Reprinted from *J. Power Sources* K. Nishio *et al.*, **34**, 153, 1991, with kind permission from Elsevier Science SA, Lausanne, Switzerland.)

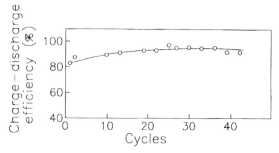

Figure 11.7. Charge–discharge efficiency of a Li/LiClO$_4$– PC/pPy cell. Current density: 0.1 mA cm^{-2}. (Reprinted from *Electrochim. Acta* S. Panero *et al.*, **32**, 1007, Copyright 1987, with kind permission from Elsevier Science Ltd, The Boulevard, Langford Lane, Kidlington OX5 1GB, UK.)

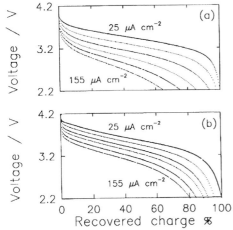

Figure 11.8. Discharge curves at 25, 50, 75, 100, 130 and 155 μA cm^{-2} of (a) Li/(PEO–SEO)$_{20}$LiClO$_4$/pNDPy (1.2 × 10^{-6} mol NDPy cm^{-2}, charged with 15 mC cm^{-2}) at 25°C and (b) Li/(PEO–SEO)$_{20}$LiClO$_4$/pNTPy (1.2 × 10^{-6} mol NTPy cm^{-2}, charged with 18 mC cm^{-2}) at 36°C. pNTPy = poly(N-3,6-dioxaheptylpyrrole), pNTPy = poly(N-3,6,9-trioxadecylpyrrole). (Reprinted from *Electrochim. Acta* C. Arbizzani *et al.*, **37**, 1631, Copyright 1992, with kind permission from Elsevier Science Ltd, The Boulevard, Langford Lane, Kidlington OX5 1GB, UK.)

and ether groups tailored to the pyrrole ring improve the electrode's kinetics. Batteries of Li/(PEO–SEO)LiClO$_4$/poly(N-oxyalkylpyrrole) can thus discharge quickly at room temperature while retaining good charge recovery (see Figure 11.8).

Significant results are also reported using highly conducting electrolyte membranes prepared by gelification of liquid solutions (e.g. LiClO$_4$ in PC) in a polymer matrix [e.g. poly(acrylonitrile), PAN, or poly(methylmethacrylate), PMMA] [47,48]. Cells assembled by laminating a lithium metal anode, a PMMA-based electrolyte membrane and a pPy cathode have already been tested [49,50]. Two types of electrosynthesized pPy films were used as cathode materials. One, electropolymerized at 4.2 V vs. lithium, had a rough, porous morphology and the other, electropolymerized at 3.9 V vs. lithium, had a flat, dense morphology. The former yielded the best results as to response, thereby confirming the role of polymer-electrode morphology in the kinetics of the electrochemical process. The Li/PMMA-based membrane/pPy button-type battery showed very promising features, including a very high charge–discharge coulombic efficiency (Figure 11.9a) and long cyclability (Figure 11.9b).

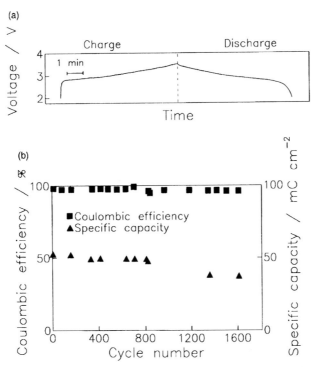

Figure 11.9. Charge–discharge cycle (a) and coulombic efficiency and specific capacity (b) of a Li/PMMA–EC–PC–LiClO$_4$/pPy button-type battery. Charge–discharge current density: 0.1 mA cm^{-2} [50].

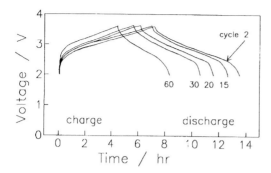

Figure 11.10. Charge–discharge cycles of a carbon/LiClO₄–EC–DMC/pPy rocking-chair battery. Charge–discharge current density: 0.25 mA cm^{-2}. (Reproduced by permission of the Electrochemical Society from ref. 52.)

Yet even these new thin-layer lithium–polymer batteries suffer from the general drawback of all lithium cells, i.e. instability of the metal electrode–electrolyte interface. This problems has been obviated for lithium batteries in general by the so-called rocking chair cells, in which the lithium metal anode is replaced by a carbon-based lithium intercalated electrode [51]. This concept has also been exploited in the specific case of lithium–polymer batteries by assembling a cell which in the discharged state has a graphite anode, a lithium ion-conducting electrolyte (a solution of LiClO₄ in an ethylene carbonate–dimethylcarbonate, EC–DMC, mixture) and a pPy film cathode [52]. By charging this cell, Li$^+$ cations enter the graphite structure and ClO₄- anions simultaneously inject into the pPy structure:

$$nC_6 + (Py)_n + nyLiClO_4 \underset{\text{discharge}}{\overset{\text{charge}}{\rightleftarrows}}$$

$$[(Py)^{y+}(ClO_4^-)_y]_n + n + Li_yC_6 \quad (11.2)$$

The good cyclability of this configuration is confirmed in Figure 11.10, which shows the trends of a typical charge–discharge of the carbon/pPy rocking-chair cell. A total energy density of 300 Wh kg^{-1} was found for this battery.

These results show that proper choice of battery components enables the intrinsic potentialities of the polymer electrode materials to be exploited for the development of revolutionary electrochemical devices. Let us recall here that in addition to lithium batteries, these potentialities are today being investigated for high-performance supercapacitors designed to meet the peak power needs of electric vehicles and for light-modulating optical windows designed for energy

control in buildings. A number of laboratories are currently seeking to enhance the electrochemical and optical properties of conducting polymers by designing suitable materials, the final goal being to optimize their response in the given device. The achievements in supercapacitors and optical devices will be discussed in the following sections.

4 POLYMER SUPERCAPACITORS

4.1 General Principles

Another recently suggested application for conducting polymers is as electrode-active materials for supercapacitors, i.e. high-power, energy-storage devices that are now being accorded increased attention [53]. Supercapacitors provide a higher power density than batteries and, despite their lower operating voltage, a higher energy density than conventional dielectric capacitors because of the high capacitance of their electrode materials. Although industrial research began in the United States in the 1960s, by the early 1970s the technology had been leased to Japanese firms. They developed high-performance devices for two types of applications: military systems, which typically require short (milliseconds) but intense bursts of power, and computer memory backup, which requires less power but needs to be supplied for longer periods of time.

The recent interest in supercapacitors is stimulated by their potential application as power-storage devices operating in parallel with the battery in an electric vehicle. The supercapacitors are expected to provide pulses of peak power during acceleration and on uphill gradients, and be recharged during deceleration and travel at constant speed by regenerative braking or by battery. This application demands less power than the military one but with out-power maintained for a longer time (up to half a minute), so that devices for EV application fall between those for memory backup and very high-power military devices. Many academic and industrial laboratories are currently investigating two types of supercapacitors: the double layer and the redox. The charge-storage mode is different in each: the capacitance in the former is electrostatic in origin, i.e. it arises from the separation of electron and ionic charges at the interface between high specific-area carbon electrodes and an aqueous or an organic electrolyte [54]; in the latter, fast Faradaic charge transfers take place at the electrode materials, as in a battery, and produced what is called pseudocapacitance [55].

The objectives are the same, regardless of the technology: to develop electrode materials with high specific capacitance (F g^{-1} and F cm^{-3}), for maximizing energy storage, and a lower electrical resistance (Ω cm^2), for maximizing power. In addition, market success demands that the electrode materials be stable to assure long life of the device, and have a high performance-to-cost ratio. The electrolyte in supercapacitors should ideally have a high breakdown voltage for greater energy storage and low electrical resistance for greater power. Low-resistance aqueous electrolytes lead to devices with good power but whose energy storage is limited by voltage decomposition of about 1 V. Since organic electrolytes break down at about 3 V, energy increases up to about one order of magnitude, given its dependence on the square of the voltage, but they have higher resistance, which limits power.

Attention is presently focused on two classes of electrode materials for redox supercapacitors: the noble metal oxides or hydrous oxides, which work in aqueous electrolytes, and the conducting polymers which work both in aqueous and non-aqueous electrolytes. Ruthenium oxide is the best example of this inorganic class of pseudocapacitive materials, and a hydrous form recently prepared by a sol-gel process at low temperatures shows capacitance values over 720 F g^{-1} [56].

Conducting polymers are the other promising class of pseudocapacitive materials. This because the kinetics of the charge–discharge processes, i.e. the doping–undoping processes, of polymers with suitable morphology are fast; the charge is stored throughout the volume of the material; and they can generally be produced at significantly lower cost than noble metal oxides. Simple conducting polymers like polypyrrole, polyaniline and polythiophene can be manufacturing at costs comparable to those of carbon capacitor electrodes.

As Rudge et al. have noted [57], supercapacitors based on conducting polymers might equally well be described as batteries. Yet there are criteria for preferring the terminology of supercapacitor: (i) the fundamental nature of the charge–discharge process which involves only translation of electronic and ionic charges; (ii) the typical combinations of energy and power density levels which these devices provide; and (iii) the nature of the low-frequency impedance.

The versatility of conducting polymers makes possible at least three different types of polymer supercapacitor, with an increasing storage–charge capacity and operating potential range from type to type: (I) a symmetric supercapacitor in which a p-dopable polymer is the active material on both electrodes; (II) an unsymmetric supercapacitor based on two p-dopable polymers selected by virtue of the difference in potential ranges over which they become p-doped; (III) a symmetric supercapacitor based on a p- and n-dopable polymer [58].

Type I involves, in the completely charged state, one polymer electrode in the fully p-doped form and the other in the undoped form; the cell voltage is typically 1 V. Since in the completely discharged supercapacitor both polymer electrodes are at a half p-doped state, only half of the polymer's total p-doping charge can be delivered by the operating supercapacitor. Type II, in the completely charged state, has the polymer of the positive electrode in the fully p-doped state and that of the negative electrode in the undoped state and, when it is discharged, the polymers of both electrodes are partially p-doped. The operating potential and the charge involved in this type of supercapacitor are greater than those of type I, being related to the difference between the p-doping potential domains of the two polymers. Type III, in the fully discharged state, has both electrodes in the undoped states, whereas in the charged state one electrode is p-doped and the other n-doped. The advantage of this configuration is that all the doping charge can be delivered by the supercapacitor at high potential because of the separation between p- and n-doping potential domains, which is related to the polymer energy gap. Type III is thus the most promising in terms of energy density; it has a further advantage over types I and II in that the charged supercapacitor has both electrodes in the doped state, i.e. in the conducting state, and instantaneous power density is greater. In addition, types III and I, unlike type II, have a symmetrical configuration, a further advantage in device manufacture and use. Yet even the advantages of type III are partially offset by the difficulty of achieving polymers than can be efficiently n-doped.

Types I and II can be assembled using such conventional heterocyclic polymers as polypyrrole, polyaniline and polythiophene, which are efficient p-dopable polymers and can easily be chemically or electrochemically synthesized from inexpensive commercially available monomers. By contrast, the very negative potentials required for n-doping of conventional conjugated polymers makes the n-doped forms of these polymers very reactive, a fact determining high self-discharge and poor electrode cycle life. Indeed, polypyrrole cannot be n-doped because the injection of the negative charge would require potentials so negative as to be incompatible with polymer and electrolyte stability. The smaller energy gap of polythiophene enables *n*-doping of this polymer with suitable tetra-lkylammonium salt–solvent combinations having a wide electrochemical window, although the *n*-doped form is very reactive [59].

The development of type III thus requires new materials featuring a smaller energy gap than that of polythiophene and a comparable oxidation potential so as to attain negative charge injection at less negative potentials. Different strategies have been pursued for tailoring polymers that can be *n*-doped in an efficient way.

Improved *n*-doping at less negative potentials have been achieved with polymers electrosynthesized from a number of 3-(*p*-X-phenyl)thiophenes, where $X = -H$, $-F, -Cl, -Br, -CF_3, -CH_3, -OCH_3, -C(CH_3), -SO_2C-H_3$ [60,61]. Of these polymers, Rudge *et al.* [61] demonstrated that the poly(3-[4-fluorophenyl]thiophene) (pFPT) performs best as an electrode material for type III supercapacitors. Another successful approach, which has led to a polymer with a narrow energy gap wherein the *n*-doping process occurs at less negative potentials than in polythiophene and performs well in type III supercapacitors, involves the use of a starting monomer with specially designed fused thiophene rings such as poly(dithieno[3,4-*b*:3′,4′-*d*]thiophene) (pDTT) [62].

High-performance devices, particularly those involving the *n*-doping process, require, in addition to polymer electrode materials of optimized morphology, careful purification of the electrolyte as well as careful

optimization of the salt, solvent and depth of charge. As alkali metal salts in standard organic solvents impede the ability to *n*-dope polythiophene and its derivatives, alkylammonium salts have to be used. The right choice of alkylammonium salt significantly improves the *n*-doping process of pFPT, as illustrated in Figure 11.11. This polymer, in $(CH_3)_4NCF_3SO_3$–acetonitrile (ACN) solution, instead of $(C_4H_9)_4NPF_6$–ACN, can be *n*- and *p*-doped at the same doping levels. This equivalence is a benefit not only in terms of energy and power density, but also in device manufacture and use since an equal amount of polymer can be used for each electrode in a type III supercapacitor, leading to easier bipolar plate manufacture and the ability to reverse the device's polarity without affecting performance.

Cyclic voltammetry and impedance spectroscopy, in the three- and two-electrode modes, have been widely used to evaluate the capacitance values of the electrode materials and of supercapacitors. The cyclic voltammetries that are scaled by dividing by the scan rate enable the direct evaluation of device capacitance, as shown by the typical example of Figure 11.12 related to a pPy-based type I supercapacitor [63]. The electric response of this device resembles that of a circuit with a capacitor in series with a resistor and a time constant approaching the experimental time scale of the highest sweep-rate cyclic voltammetries.

Impedance spectroscopy is a very useful technique for characterizing supercapacitors because, in addition to capacitance values, it provides at high frequency the

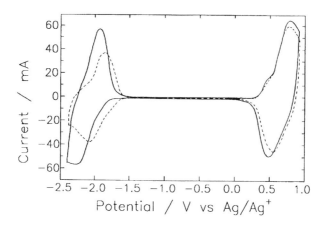

Figure 11.11. Cyclic voltammograms at 25 mV s^{-1} for a film of pFPT in 1 M $(CH_3)_4NCF_3SO_3$–ACN (solid line) and in 1 M $(C_4H_9)_4NPF_6$–ACN (dotted line). The pFPT film was galvanostatically grown on carbon paper (cross-sectional area: 1 cm^2) with 8 C cm^{-2}. (Reprinted from *Electrochim. Acta A.* Rudge *et al.*, **39**, 273, Copyright 1994, with kind permission from Elsevier Science Ltd, The Boulevard, Langford Lane, Kidlington OX5 1GB, UK.)

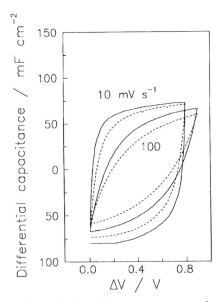

Figure 11.12. Cyclic voltammograms, expressed as differential capacitance, of a pPy/pPy supercapacitor (two-electrode mode) in 1 M LiClO$_4$–PC at 10, 25, 63 and 100 mV s^{-1}. The pPy films were galvanostatically grown on carbon paper with 2.5 C cm^{-2}. (Reproduced by permission of the Materials Research Society from ref. 63.)

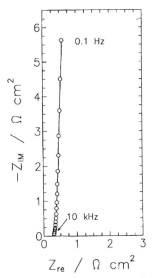

Figure 11.13. Impedance spectrum of a pAni-based supercapacitor in 1 M MHClO$_4$.＋3 M NaClO$_4$. The pAni films were galvanostatically grown on platinized platinum with 5 C cm^{-2} [64].

impedance generally associated with the electrolyte layer between the electrodes and at zero frequency the impedance that must be overcome to extract all the device's capacitance. Knowing the equivalent series resistance is a useful indicator of the energy efficiency that the device can achieve. When polymer electrodes have an optimized morphology, the impedance spectra of polymer supercapacitors very closely approach those of an ideal capacitor with a high capacitance in series with a small resistance, as in the case of the type I system based on pAni [64] shown in Figure 11.13. However, the spectra can substantially deviate from those of an ideal capacitor, as shown in Figure 11.14 [65], and the equivalent circuit is determined by the complex distribution (which can differ from case to case) of capacitances and resistances, the latter in turn being related to ion-transport and charge-transfer phenomena. Charge transfer dominates at high frequency, leading to a semicircle in the complex impedance diagram; ion diffusion dominates at intermediate frequencies, leading to Warburg-type behaviour; and highly capacitive behaviour is observed at low frequencies owing to the finite thickness of the polymer film.

It has been frequently observed that the low-frequency capacitance values measured with these polymer materials by impedance spectroscopy are smaller than those yielded by cyclic voltammetry, even

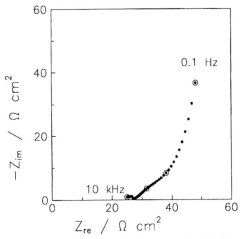

Figure 11.14. Impedance spectrum of a pPy/p3MT supercapacitor (two-electrode mode) at $\Delta V = 0.8$ V in 1 M LiClO$_4$–PC. Both pPy and p3MT were galvanostatically grown on carbon paper with 2.5 C cm^{-2}. p3MT = poly(3-methylthiophene). (Reprinted from *Electrochim. Acta* C. Arbizzani *et al.*, **41**, 21, Copyright 1996, with kind permission of Elsevier Science Ltd, The Boulevard, Langford Lane, Kidlington OX5 1GB, UK.)

when the time scale of the experiments is the same. This discrepancy between data evaluated by techniques using different potential perturbation amplitudes is real, and the conformational changes of the polymer from a twisted one in the undoped to a planar one in the doped state are responsible for this potential amplitude dependence of the capacitance values [66].

Performance tests of polymer supercapacitors have generally been run on prototypes with electrochemically synthesized polymer electrodes with a growth charge of several Coulombs per square centimetre. When these thick films are efficiently doped, a fact strongly dependent on polymer morphology which in turn depends on electrosynthesis conditions, capacitance values of the order of 1 F cm^{-2} are to be expected.

Given that the charge in polymer supercapacitors is stored throughout the volume of the electrode materials and the charge–discharge kinetics can be controlled by the diffusion of the dopant ions into and out of the polymer matrix, comparative performance tests of different systems should be run among prototypes with comparable values of polymer electrode thickness. In addition, to determine whether these devices meet the power requisites for EV load-levelling applications, the test time scale has to be compatible with the discharge time of the EV supercapacitors, which is of the order of half a minute. The performance of polymer supercapacitors has generally been evaluated on the basis of the most significant parameters for this type of device, i.e. nominal voltage, nominal capacity, nominal energy density, experimental energy densities, equivalent series resistance, instantaneous power density and experimental power densities.

4.2 Supercapacitors prototypes

The field of polymer supercapacitors is younger than that of lithium–polymer batteries and heretofore prototypes of a small geometric area have been developed only in academic laboratories. A comparative performance study of the three types of polymer supercapacitors has been carried out by Rudge et al. [58] using pPy for type I, pPy and poly(2,2'-bithiophene) (pBT) for type II, and pFPT for type III. The electrolyte was 1 M (CH$_3$)$_4$NCF$_3$SO$_3$–ACN in all three systems. Polymer electrodes were electrochemically prepared on a carbon paper support with a growth charge of 20 C cm^{-2} for pPy and pFPT films, and 15 C cm^{-2} for pBT film to assure approximately the same doping charge as that of the pPy electrode. The voltage–time curves of

the three devices recorded during discharge at 1 mA cm^{-2} are shown in Figure 11.15 and the values of energy density per unit mass of capacitor active material calculated ($E = i\int V\mathrm{d}t$) from these curves are reported in Table 11.3. The performance of the three supercapacitor types are in the expected order: type III is better than type II, which in turn is better than type I. While an energy value of 11 Wh kg^{-1} for the pPy-based type I system is satisfactory, the energy density per kilogram of a complete device, e.g. the practical energy density, would not represent a major improvement over the values of high specific-area carbon double-layer supercapacitors, which currently stand at 2 Wh kg^{-1}. The improvement in energy density of type II supercapacitor over type I is significant, and system II, like system I, has the advantage of employing inexpensive conducting polymers. The 39 Wh kg^{-1} achieved with type III, in which an optimized pFPT is the active material, is a very promising result in view of the US Department of Energy's (DOE) goal for devices in EV application (5 Wh kg^{-1} of device). In addition, it has been estimated that for the pFPT-based supercapacitor the equivalent series resistance of the total cell will be no more than 4 Ω cm^2, a promising value in that high power density has been estimated as 35 kW kg^{-1} of the active material on both electrodes, which is more than an order of magnitude higher than the power density of the DOE's goal (> 500 W kg^{-1} of device).

The DOE requires a life of 100,000 cycles for devices in EV application, and long cycle-life tests on polymer supercapacitors have been run to date on thick pPy-based type I prototypes working in 1 M (CH$_3$)$_4$NCF$_3$SO$_3$–ACN and on pAni prototypes working in liquid aqueous electrolyte (1 M HClO$_4$ + 3 M NaClO$_4$) [64]. Discharge curves at different current densities of these prototypes charged at constant voltage of 1.0 V for the pPy prototype and at 0.6 V for the pAni prototype are shown in Figure 11.16 and 11.17; the capacity and the discharge energy values during constant current multicycle tests for pPy and pAni prototypes are shown in Figure 11.18 and 11.19. The results clearly indicate that these supercapacitors meet the cycle-life requisites set by the DOE.

Comparative performance data of three supercapacitor types are also reported in PC with lithium and alkylammonium salts [65]. Unlike ACN, PC is a nontoxic solvent of limited environmental impact and is preferable for industrial applications. Figure 11.20 shows the discharge curves at different current densities of the pPy-based (type I), pPy/poly(3-methylthiophene) (p3MT) (type II) and pDTT-based (type III) prototypes.

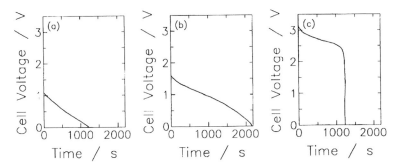

Figure 11.15. Galvanostatic discharge curves at 1 mA cm^{-2} in 1 M (CH$_3$)$_4$NCF$_3$SO$_3$–ACN of (a) pPy/pPy (type I); (b) pPy/pBT (type II); (c) pFPT/pFPT (type III) supercapacitors. (Reproduced by kind permission of the Electrochemical Society from ref. 58.)

Table 11.3. Energy density for the three schemes of supercapacitor. (Reproduced by permission of the Electrochemical Society from ref. 58.)

Scheme	Voltage (V)	Energy density		
		J cm^{-2}*	J g^{-1}†	Wh kg^{-1}†
I	1.0	0.56	41	11
II	1.5	1.9	100	27
III	3.1	3.5	140	39

*Energy densities calculated per geometric square centimetre of the carbon paper electrodes.
†Energy densities calculated per gram of active material on both electrodes in the capacitor configuration.

All the polymers were electrochemically grown on carbon paper electrodes with a growth change ranging from 2.5 to 10 C cm^{-2}. Discharge data clearly evince that type III performs best in terms of energy and power values.

Figure 11.21 shows two Ragone plots in which power and energy densities are reported in terms of mW cm^{-2} and mJ cm^{-2} (top) and in terms of W l^{-1} and Wh l^{-1} (bottom). The performance of prototypes type I and

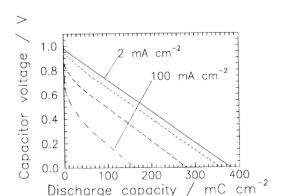

Figure 11.16. Galvanostatic discharge curves of a pPy/pPy (type I) supercapacitor in 1 M (CH$_3$)$_4$NCF$_3$SO$_3$–ACN at 2, 10, 40 and 100 mA cm^{-2}. The pPy electrodes were galvanostatically grown with 10 C cm^{-2} on carbon paper [64].

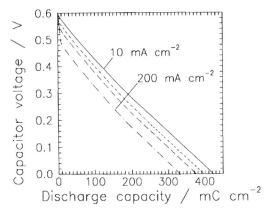

Figure 11.17. Galvanostatic discharge curves of a pAni/pAni (type I) supercapacitor in 1 M HClO$_4$ + 3 M NaClO$_4$ (aq) at 10, 50, 100 and 200 mA cm^{-2}. The pAni electrodes were galvanostatically grown with 10 C cm^{-2} on carbon paper [64].

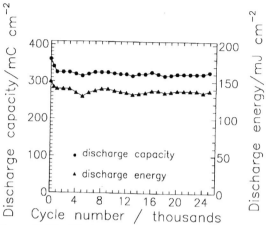

Figure 11.18. The long-term performance during constant current multicycle testing of a pPy/pPy (type I) supercapacitor at 10 mA cm^{-2}, between 0.0 V and 1.0 V bias in 1 M (CH$_3$)$_4$NCF$_3$SO$_3$–ACN. The pPy electrodes were galvanostatically grown with 10 C cm^{-2} on carbon paper [64].

type II with thick electrodes was evaluated in 1 M (C$_2$H$_5$)$_4$NBF$_4$–PC, LiClO$_4$ being replaced to meet demands in widespread industrial application. The parameters of a double-layer supercapacitors with active-carbon electrons of 2500 m^2 g^{-1} in the same electrolytic medium, are also shown for comparison. Increasing the polymer thickness performs more positively in II than in I. Comparing the two Ragone plots indicates that the double-layer supercapacitor is

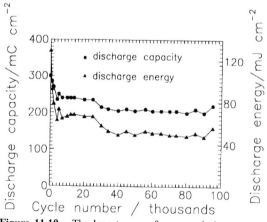

Figure 11.19. The long-term performance during constant current multicycle testing of a pAni/pAni (type I) supercapacitor at 50 mA cm^{-2}, between 0.0 V and 0.75 V bias, in 1 M MClO$_4$ + 3 M NaClO$_4$ (aq). The pAni electrodes were galvanostatically grown with 5 C cm^{-2} on platinized platinum foil [64].

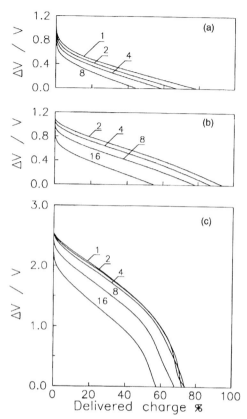

Figure 11.20. Galvanostatic discharge curves at different current densities (mA cm^{-2}) of supercapacitors. (a) pPy/pPy (type I) in 1 M LiClO$_4$–PC, charged at $\Delta V = 1.1$ V with 0.11 C cm^{-2}; (b) pPy/pMeT (type II) in 1 M LiClO$_4$–PC, charged at $\Delta V = 1.15$ V with 0.12 C cm^{-2}; and (c) pDTT/pDTT (type III) in 0.2 M (C$_2$H$_5$)$_4$NBF$_4$–PC, charged at $\Delta V = 3.0$ V with 0.23 C cm^{-2}. The polymer electrodes were galvanostatically grown with 2.5 C cm^{-2}. (Reprinted from *Electrochim. Acta* C. Arbizzani *et al.*, **41**, 21, Copyright 1996, with kind permission from Elsevier Science Ltd, The Boulevard, Langford Lane, Kidlington OX5 1GB, UK.)

adversely affected by the pronounced thickness of the carbon electrodes. While the performance of I and II is comparable to that of the carbon supercapacitor, the performance of type III is significantly better, suggesting that future research should be focused on developing new polymer materials that can be *n*-doped in a truly efficient way.

Performance data of a type I button-prototype of the supercapacitor based on pPy that was chemically prepared by a patented procedure [68] have been very recently reported [69]. The electrodes are a composite material of pPy, carbon black and Teflon, and the

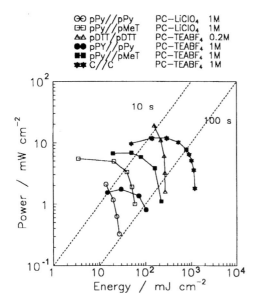

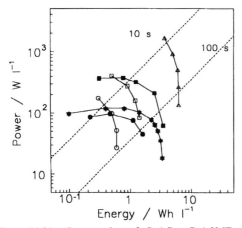

Figure 11.21. Ragone plots of pPy/pPy, pPy/p3MT, pDTT/pDTT redox supercapacitors, with polymer electrodes of different electrode thicknesses (O, □, △, growth charge 2.5 C cm^{-1}; ●, ■. 10 C cm^{-2}) and C/C (★) double-layer supercapacitor in different electrolytes (Reproduced by permission of VCH Verlagsgesellschaft from ref. 67)

separator is a polyamide felt wetted by 1 M LiClO$_4$–PC electrolyte. A capacity of 50 F cm^{-3} and an energy density of 4 Wh kg^{-1} are reported along with a good cyclability and a low self-discharge, although the time constant is relatively high (120 s).

As for lithium–polymer batteries, a solid-state configuration has been proposed also for polymer-based supercapacitors. A type III pDTT-based supercapacitor with a hybrid PEO–PC–(C$_2$H$_5$)$_4$NBF$_4$ electrolyte [70] and a type I pPy-based one [71] with a

PMMA–EC–PC–LiClO$_4$ gel electrolyte have been tested.

Although the self-discharge phenomenon can also greatly affect the supercapacitor's life, this issue has not yet been treated in a very exhaustive way. The main problem related to self-discharge is the charge imbalance of the electrodes (especially types I and II) during shelf life; less important is the self-discharge of type III, as it can be kept with both electrodes in the undoped state. When the supercapacitors are operating, their states of charge (to a given voltage) can be maintained by providing a 'float current' equal to the rate of self-discharge. Some self-discharge data are available only for the pPy-based typed I supercapacitor, for which a 'leakage current' of 1 μA/F.V is reported [69].

5 ELECTROCHROMIC DEVICES

5.1 General principles

An electrochromic material is one whose colour changes in a persistent yet reversible manner through an electrochemical reaction. Accordingly, conducting polymers that can be repeatedly switched electrochemically from the doped to the undoped states with high contrast in colour have emerged over the last ten years as an extremely versatile class of materials for electrochromic device technology [72,73].

An electrochromic device is essentially an electrochemical cell in which an electrochromic electrode is separated by a suitable electrolyte (liquid or solid) from a charge-balancing counter-electrode, and colour changes occur by charging and discharging the electrochemical cell by a few volt square-waved pulses. The electrochromic electrode, which can operate in transmissive or in reflective mode, is a conductive, transparent glass, generally indium tin oxide (ITO) coated, on which a thin film of the electrochromic material is deposited. The counter-electrode can be of any material that provides a reversible electrochemical reaction in a device operating in the reflective mode, such as electrochromic displays. However, in variable light transmission devices, such as electrochromic windows, where the entire system is in the optical path, it has to be transparent, optically passive or electrochromic in a complementary mode to the primary electrochromic material. Transparent electrolytes are required in transmissive devices.

Electrochromic technology has very widespread applications. The most exciting is probably in energy-management 'smart' windows for buildings, cars, boats and aircraft to control the flow of light and heat. Other popular applications are in switchable rear-view mirrors and visors for cars, in large-area displays for stock boards and airline terminals, and in eyeglasses. Numerous private industry and academic research groups are currently working on component materials to develop the best electrochromic devices for specific applications.

Many of the electrochromic materials discovered and developed in the past two decades can be divided into two main classes: inorganic materials, generally the transition metal oxides, and conducting polymers. The most thoroughly investigated as well as the most widely applied electrochromic material to date is tungsten trioxide, WO_3 [74]. A great variety of colour contrast can be achieved with conducting polymers—the most significant characteristic of these ion-insertion polymer electrochromic materials with respect to their inorganic counterparts.

The colour changes elicited by the doping are due to modification of the polymer's band electronic structure. The polymer's unidimensional character promotes the localization on its chain of the charge created by doping and the relaxation of the lattice around this charge. The charge confinement then creates the defects that produce, in the energy gap, new electronic states which cause the colour changes. In polymers that have an aromatic-like structure, such as heterocyclic ones, lattice relaxation is towards the quinoid structure and the defects, polarons (single-charged, $^1/_2$ spin) and bipolarons (double-charged, spinless), are related to the doping level. As the doping level increases, the bipolaron states overlap in bipolaron bands and the bipolaron bands can merge with the valence and conduction bands, leading to a metallic-like state [75].

As an example of the spectra evolution of this class of polymers, Figure 11.22 shows visible and near-infrared spectra of p3MT at increased p-doping level values [76]. The undoped polymer's strong absorption band, with maximum at 2.34 eV is typical of $\pi-\pi^*$ interband transition (a). This band decreases upon p-doping and two new bands appear at low energy, as is consistent with the presence of the two bipolaron bands (b–e). The characteristic absorption pattern of the metallic-like state finally appears when the bipolaron bands merge with the valence and conduction bands (f).

Undoped p3MT, with onset of the optical absorption at 1.89 eV and maximum absorption at 2.34 eV, is highly absorbent in the visible region and purple in colour, whereas after doping the free-carrier-state absorption is weak in the visible region and the polymer is transparent pale blue.

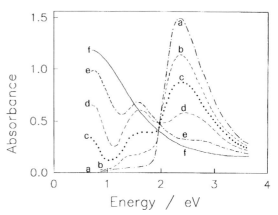

Figure 11.22. Visible and near-infrared spectra of p3MT recorded *in situ* during electrochemical *p*-doping: (a) undoped, (b–f) doped at increased doping levels. (Reprinted from *Synth. Met.* M. Mastragostino *et al.*, **28**, C501, 1989, with kind permission from Elsevier Science SA, Lausanne, Switzerland.)

Doping shifts the absorption towards low energy and the colour contrast is related to the magnitude of the energy gap, which in turn depends on the polymer's effective conjugation length. Polymers with energy gap, evaluated from the maximum absorption greater than 3 eV are transparent or slightly coloured in the undoped form and generally highly absorbent in the visible region in the doped form. By contrast, polymers with small energy gap are highly absorbent in the undoped form, the colour being related to the wavelength of the maximum absorption, but after doping they can become almost transparent in the visible region. Table 11.4 lists the energy gaps, as well as the colour of the undoped and doped forms, of some conventional polymers and their switching potential values.

Polymers with a tailored energy gap, and hence tailored electrochromic properties, can be synthesized using suitable starting molecules. Various strategies have been pursued for tuning the colour contrast of thiophene-based polymers. Poly(3-alkylthiophenes) with a very fine gradation of colour contrast, as shown in Table 11.5, have been electrochemically synthesized starting from isomeric oligomers of different length, whether symmetric (equivalent α–α' positions) or unsymmetric (non-equivalent α–α' positions) molecules, with alkyl groups of different length [77–79]. The conformational differences deriving from the steric hindrance of the alkyl groups, which are arranged so as to originate various patterns in the resulting chains, play the main role in tailoring the effective conjugation length of the polymers, to which the energy gap is related. Differences in the optical properties of 3-

Table 11.4. Energy-gap values (E_g) from the maximum absorption of the $\pi-\pi^*$ interband transition, colours of the undoped and doped forms of some conventional polymers and their switching potential values

Polymer	E_g (eV)	Colour		Switch potentials	
		Undoped	Doped	V vs. Li	
polypyrrole	3.2	yellow	blue-violet	3.0	3.7
poly(2,2'-bithiophene)	2.6	orange-red	blue	2.5	4.2
poly(3-methylthiophene)	2.3	purple	pale blue	2.5	4.0

methylthiophene-based polymers make it possible to design a variable light transmission electrochromic device with two regiochemically and conformationally different polymers, electrosynthesized respectively from 3-methylthiophene and 3,3'-methyl-2,2'-bithiophene, operating in the complementary mode [80]. The modulation of the conjugation length via copolymerization of 3-alkyl and 3,4-dialkylthiophenes has also been pursued [81].

It is well known that alkoxy- and dialkoxythiophene-based polymers have a lower energy gap than polyalkylthiophenes because of the two electron-donating oxygen atoms adjacent to the thiophene unit [82–84] and that poly(cycloalkylenedioxythiophene) [85] is much more stable in the doped state than ordinary poly(3-alkylthiophene)s. Recently, a polymer with promising electrochromic properties has been electrochemically prepared from 3,4-ethylenedioxythio-

phene. In the undoped state it has the onset of optical absorption at about 1.5 eV and maximum absorption at 2 eV and is dark purple/blue, and in the doped state it is highly transparent with a sky-blue colour [86,87].

In addition to their n-doping capacity, conjugated polymers with narrow energy gap are currently coming under close scrutiny because of their electrochromic properties. As rule, the energy gap decreases as the quinoid character of the polymer backbone increases at the expense of its aromatic character. The poly(isothionaphthene) (pITN) is the forerunner of the thiophene-derivative polymers with a narrow energy gap [88]. In this polymer, which has an oxidation potential comparable to that of conventional polythiophenes but significantly lower energy gap, the injection of change of both signs is possible and the colour changes from deep blue to colourless. A symmetric electrochromic device with both electrodes of pITN, which become p-

Table 11.5. Wavelengths of the maximum absorption ($\lambda_{\pi-\pi^*}$) and colours of the undoped and doped forms of poly(3-alkylthiophenes) from different starting molecules. (Reproduced by permission of VCH Verlagsgesellschaft from ref. 79.)

Starting molecules	R = methyl			R = hexyl		
	$\lambda_{\pi-\pi^*}$ (nm)	Colour		$\lambda_{\pi-\pi^*}$ (nm)	Colour	
		Undoped	Doped		Undoped	Doped
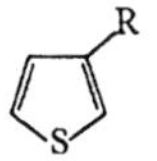 *(thiophene, R)*	530	purple	pale blue	520	red	pale blue
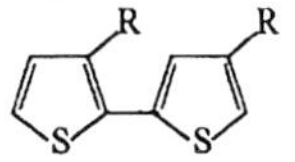 *(bithiophene, R R)*	505	red	blue	455	orange-red	pale blue
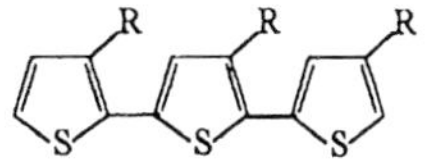 *(terthiophene, R R R)*	500	red	blue	470	red	pale blue
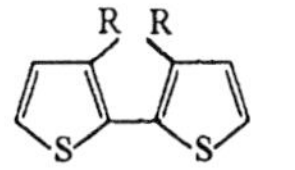 *(fused bithiophene, R R)*	415	yellow	blue-violet	390	pale green	pale blue
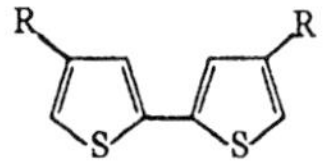 *(bithiophene isomer, R R)*	445	orange-red	blue	390	pale green	pale blue

and *n*-doped as in the type III supercapacitor, has been recently suggested [89].

The use of fused thiophene rings like dithieno[3,4-*b*:3′,4′-*d*]thiophene and dithieno[3,4-*b*:3′,2′-*d*]thiophene has proved a successful strategy in developing polymers with the onset of the maximum absorption at 1.08 eV and 1.13 eV, and maximum 2.05 eV (606 nm) and 1.63 eV (759 nm). They are blue-violet and blue-green in the undoped, and pale grey and pale grey-green in the doped states, respectively. These polymers, which have oxidation potentials less positive than those of conventional polythiophenes but significantly lower energy gap values, can be efficiently *n*- and *p*-doped, as reported above. Symmetric electrochromic devices that have the same type of poly(dithienothiophene) on both electrodes, as in the type III supercapacitor, are under study [90].

A very recent strategy for band-gap reduction involves the synthesis of bridged bithiophene polymers: Ferraris *et al.* are investigating a family of polymers starting from designed molecules which have electron-withdrawing groups at sp^2 carbon bridging the 3,3′ positions of a 2,2′-bithiophene. One of these polymers, the poly(cyclopenta[2,1-*b*;4,3-*b*′]dithiophen-4-(cyano, nonafluorobutylsulfonyl)methylidene), which is efficiently *p*- and *n*-dopable, has also been suggested as an electrochromic material for symmetric devices [91].

Although many spectroelectrochemical studies are to be found in the literature, relatively few reports thus far have been devoted to ascertaining the performance of conducting polymers for use in electrochromic device technology. High electrochromic efficiency (defined as the variation at specific wavelength in optical density to injected charge as a function of unit area, expressed in C^{-1} cm^2), short switching time (defined as the time from fully coloured state 1 to fully coloured state 2), long switching life (defined as the stability of the material to repeated switching) and optical memory (defined as the persistence of the coloured state even after the driving voltage is removed) are the most important requisites for electrochromic materials in device technology.

The electric power needed to operate an electrochromic device is related to the electrochromic efficiency, an important parameter in discriminating among electrochromic systems for large area windows like those for energy saving in buildings. (The electrochromic efficiency at the wavelength of maximum absorption of the undoped polymers is generally high, e.g. p3MT has a value of about 0.25 mC^{-1} cm^2 at 530 nm.) Although the switching time of conducting polymers, which are ion-insertion electrochromic materials, is faster than that of their inorganic counter-

parts, it cannot be expected to be better than that of liquid crystals. Long switching life is assured only when the coulombic efficiency of the cell's reaction is 100%. Note, however, that different technological applications of electrochromic devices can require different electrochromic switching times and switching lives. A time range from several to a few hundred milliseconds and a life of up to 10^6–10^7 switches are demanded for display applications, while a few seconds and 10^5 switches can be satisfactory for variable light transmission devices such as electrochromic windows.

5.2 Prototypes of electrochromic devices

Industries in North America, Japan and Europe are currently involved in the research and development of electrochromic devices for different applications. Their attention is mainly focused on devices with inorganic electrochromic materials; eyeglasses and rear-view mirrors are the most visible market examples [92,93]. Prototypes exploiting the electrochromism of conducting polymers have so far been developed only in academic laboratories, the exception being one based on pAni and WO$_3$ developed for car-window applications by the Toyota Central R&D Laboratories [94].

The prototypes with conducting polymers as working electrodes operate with counter-electrodes of either polymer or inorganic materials; the electrolytes are both liquid and solid, the latter generally being ionic conducting polymers. The performance data of two variable light transmission devices with p3MT electrochemically synthesized by monomer oxidation as electrochromic material and an optically passive counter-electrode have been reported: one uses an ITO counter-electrode [95] and the other an ITO glass covered by lithium-intercalated nickel oxide (Li$_x$NiO$_z$/ITO), which acts as a colourless optically passive counter-electrode [96]. The cell reactions can be rendered as

$$(3MT)_n + ITO + nyLiClO_4 \; \underset{\longleftarrow}{\overset{\longrightarrow}{}}$$
purple

$$Li_{ny}ITO + [(3MT)^{+y}(ClO_4^-)_y]_n \quad (11.3)$$
pale blue

$$(3MT)_n + nLi_xNiO_z + nyLiClO_4 \; \underset{\longleftarrow}{\overset{\longrightarrow}{}}$$
purple

$$nLi_{(x+y)}NiO_z + [(3MT)^{+y}(ClO_4^-)_y]_n \quad (11.4)$$
pale blue

and the colour change is from purple to transparent pale blue for both devices. Data for repeated square-wave potential switches show that both devices have almost the same response time (20 s for full bleaching and significantly less for darkening), and a small decrease of the maximum difference in the transmitted light after 100 cycles, which is due not to an irreversible degradation of the polymer material but to a coulombic efficiency of the electrochemical reaction that is less than 100%.

Electrochromic window prototypes, in which a conducting polymer like pPy or pAni works in complementary mode with an inorganic electrochromic material like WO_3 or Prussian Blue (PB), have also been developed. For windows based on pAni and PB, electrochromic compatibility was achieved by combining the oxidized coloured state of pAni with the dark state of PB and the bleached reduced state of the polymer with the oxidized form (pale green) of the coordination compound, known as Berlin Green. The performance data of prototypes operating with aqueous electrolyte [97] as well with polymer electrolyte [98] have been reported.

For windows based on pAni and WO_3, which have been developed both with a liquid electrolyte like $LiClO_4$–PC [94] at Toyota Laboratories and with a solid polymer electrolyte such as poly(2-acrylamido-2-methylpropanesulfonic acid) (pAMPS) by Jelle et al. [99,100], the schematic cell reaction, including the colour changes, can be written as

$$nWO_3 + (Ani)_n + nyMA \rightleftharpoons$$
$$\textit{transparent}$$

$$nM_yWO_3 + [(Ani)^{+y}(A^-)_y]_n \quad (11.5)$$
$$\textit{blue}$$

The test results at the Toyota Laboratories indicate transmittance changes from 80 to 30% in 1 s, maintained for 10^6 cycles, and from 80 to 40% in 1 min, indicating that pAni–WO_3 is a very promising

system for device technology. Figure 11.23 compares transmission spectra at different applied potentials, in the 290–3300 nm range, of the solid-state pAni–WO_3 prototype to those of an advanced version [101,102] in which PB is electrodeposited on a pAni electrode. The addition of PB blocks out considerably more light, while retaining approximately the same transparency during bleaching of the window: 49% of the total solar energy can be regulated by this device.

Device prototypes based on pPy doped with dodecysulfate (pPy–DS) and WO_3 have been developed both with 1 M $LiClO_4$–PC [103] and a solid polymer electrolyte based on poly(ethyleneoxide-co-epiclorydrine [104]. Cell reaction switches the colour from pale yellow to blue. This system, which maintains its optical properties over 10^4 cycles, significantly regulates the transmitted light in the visible–near-infrared region.

Performance data of prototypes in liquid and solid polymer electrolytes, which have both electrodes of polymer material and combine the colour of different configurational versions of 3-methylthiophene-based polymers, have been reported [80,105]. The prototypes were assembled by facing off an electrode of p3MT in the doped state and an electrode of undoped poly(3,3'-dimethyl-2,2'-bithiophene) (p33'DMBT). The electrochemical process that results from switching the applied voltage between ± 0.9 V is

$$(33'DMBT)_n + [(3MT)^{+y}(ClO_4^-)_y]_n \rightleftharpoons$$
$$\textit{pale green}$$

$$[(33'DMBT)^{+y}(ClO_4^-)_y]_n + (3MT)_n \quad (11.6)$$
$$\textit{dark violet}$$

and the colour changes from pale green to dark violet. The p33'DMBT, despite a very tilted structure due to steric hindrance of head-to-head linked 3-methylthio-

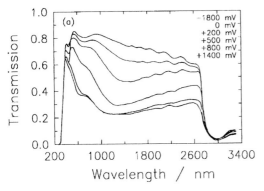
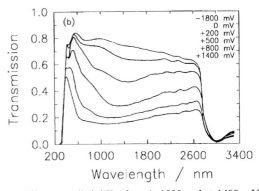

Figure 11.23. Transmission spectra to two electrochromic windows at different applied ΔV values (-1800 and $+1400$ mV correspond to the upper and the lower transmission curves, respectively): (a) pAni/WO_3 and (b) pAni–PB/WO_3. (Reproduced by permission of the Electrochemical Society from ref. 102.)

phene units, meets the most important requisites for use in electrochromic devices.

Smart windows for energy saving in buildings is the most market-attractive application of electrochromic devices, and Figure 11.24 shows how some of the above prototypes regulate solar energy flux [106]. The solar irradiance vs. wavelength [107], together with the calculated product of the solar irradiance and the transmission in the window at the switching potentials, are given in the figure. These results clearly show the feasibility of the prototypes based on pPy–DS/WO$_3$ and pAni–PB/WO$_3$ for applications in building windows, where the energy management scheme preferably

requires the simultaneous modulation of visible and infrared transmission. The all-polymer prototype (p3MT/p33'DMBT) can also be of practical value if the management scheme requires cut-off of the infrared radiation and modulation of the visible light, as in tropical climates. The prototype p3MT/ITO, which modulates simultaneously the visible and infrared radiation in opposite directions, does not meet the requisites for building energy-saving technology and its use should be directed to other applications.

Note that the optical memory of polymer electrochromic devices is related to self-discharge phenomena, which more adversely affect the performance of optical

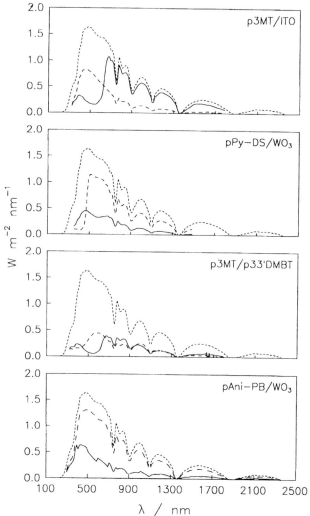

Figure 11.24.   Solar irradiance (dotted line) and transmitted solar energy flux by each device in the bleached (dashed line) and coloured (solid line) states.

devices than that of energy-storage devices because even a small loss of the injected charge causes significant optical changes in conducting polymers. This is a very crucial point and requires every effort to optimize all the device components, especially for building windows as they can only expect a successful market impact with a ten-year working life.

6 REFERENCES

1. T.A. Skotheim (ed.) *Handbook of Conducting Polymers*. Vol. 1 and 2, Marcel Dekker, New York, 1986.
2. M.G. Kanatzidis, *Chem. Eng. News*, **68**, 36 (1990).
3. F. Chao, J.L. Baudoin, M. Costa and P. Lang, *Makromol. Chem. Macromol. Symp.* **8**, 173 (1987).
4. J.M. André, J Delhalle and J.L. Bredas, *Quantum Chemistry Aided Design of Organic Polymers*, World Scientific, Singapore, 1991.
5. C. Julien and G.A. Nazri (eds.) *Solid State Batteries: Materials Design and Optimization*, Kluwer Academic Publisher, Boston, 1994.
6. B.E. Conway, in *Proc. Fall Meeting Electrochem Soc.*, Chicago, 8–13 October, 1995, Abstr. #43.
7. G.C. Granqvist (ed.) *Materials Sciences for Solar Energy Conversion Systems*, Pergamon Press, Oxford, 1991.
8. J.P. Travers, P. Audebert and G. Bidan, *Mol. Cryst. Liq. Cryst.*, **118**, 149 (1985).
9. N. Furukawa and K. Nishio, in *Application of Electroactive Polymers* (Ed. B. Scrosati), Chapman & Hall, London, 1993, pp. 150–181.
10. M.B. Inoue, E.F. Velazquez and M. Inoue, *Synth. Met.*, **24**, 223 (1988).
11. S. Hotta, M. Soga and N. Sonoda, *Synth. Met.*, **26**, 267 (1988).
12. K. Nishio, M. Fujimoto, N. Yoshinaga, O. Ando, H. Ono and T. Suzuki, *J. Power Sources*, **34**, 153 (1991).
13. E.M. Genies, P. Hany and C. Santier, *J. Appl. Electrochem,*, **18**, 751 (1988).
14. A.F. Diaz, K.K. Kanazawa and G.P. Gardini, *J. Chem. Soc., Chem. Commun.*, 635, (1979).
15. H. Naarmann, *Makromol. Chem. Macromol. Symp.*, **8**, 1 (1987).
16. P. Kathirgamanathan and M.M. Qayyum, *J. Electrochem. Soc.*, **141**, 147 (1994).
17. C.K. Chang and A.G. MacDiarmid, *Synth. Met.*, **13**, 193 (1986).
18. A. Kitani, M. Kaya and K. Sasaki, *J. Electrochem. Soc.*, **133**, 1069 (1986).
19. G. Zotti and G. Schiavon, *Synth. Met.*, **30**, 151 (1989).
20. T. Osaka, S. Ogano, K. Naoi and N. Oyama, *J. Electrochem. Soc.*, **136**, 306 (1989).
21. S. Panero, P. Prosperi, F. Bonino, B. Scrosati, A. Corradini and M. Mastragostino, *Electrochim. Acta*, **32**, 1007 (1987).
22. K. Okabayashi, F. Goto, K. Abe and T. Yoshida, *J. Electrochem. Soc.*, **136**, 1986 (1989).
23. S. Hossain, in *Handbook of Batteries*, 2nd edn, (ed. D. Linden), McGraw-Hill, New York, (1985) p. 36.
24. H. Müstedt, G. Köhler, H. Möhwald, D. Naegele, R. Bitthin, G. Ely and H. Meissner, *Synth. Met.*, **18**, 259 (1987).
25. T. Osaka, K. Naoi and S. Ogano, *J. Electrochem. Soc.*, **135**, 1071 (1988).
26. C. Arbizzani and M. Mastragostino, in *Proc. Second International Symposium on Polymer Electrolytes* (Ed. B. Scrosati) Elsevier Applied Science, London, 1990, pp. 373–381.
27. W.H. Smyrl and M. Lien, in *Application of Electroactive Polymers* (Ed. B. Scrosati), Chapman & Hall, London, 1993, pp. 29–74.
28. B. Scrosati, in *Solid State Electrochemistry* (Ed. P.G. Bruce) Cambridge University Press, Cambridge, pp. 229–263.
29. M.A. De Paoli, M. Panero, P. Prosperi and B. Scrosati, *Electrochim. Acta*, **35**, 1145 (1990).
30. K. Naoi, M. Lien and W.H. Smyrl, *J. Electrochem. Soc.*, **138**, 440 (1991).
31. M. Mastragostino, A.M. Marinangeli, A. Corradini and C. Arbizzani, *Electrochim. Acta*, **32**, 1589 (1987).
32. F. Goto, K. Abe, K. Okabayashi, T. Yoshida and H. Morimoto, *J. Power Sources*, **20**, 243 (1987).
33. K. Nishio, M. Fujimoto, M. Yoshinaga, N. Furukawa, O. Ando, H. Ono and T. Suzuki, *Proc. 40th ISE Meeting*, Kyoto, 17–22 September, 1989, Ext. Abstr. Vol. 1, p. 553.
34. T. Osaka, T. Nakajima, T. Naoi and B.B. Owens, *J. Electrochem. Soc.*, **137**, 2139 (1990).
35. T. Nakajima and T. Kawagoe, *Synth. Met.*, **28**, C269 (1989).
36. Technical Research Laboratory Bridgestone Company, Tokyo 187, Japan.
37. S. Panero, P. Prosperi and B. Scrosati, *Electrochim. Acta*, **32**, 1461 (1987).
38. P. Novak, O. Inganäs and R. Bjorklund, *J. Power Sources*, **21**, 17 (1987).
39. P. Novak and O. Inganäs, *J. Electrochem. Soc.*, **135**, 2485 (1988).
40. C. Arbizzani, M. Mastragostino, S. Panero, P. Prosperi and B. Scrosati, *Synth. Met.*, **28**, C663 (1989).
41. C. Arbizzani and M. Mastragostino, *Electrochim. Acta*, **35**, 251 (1990).
42. C. Arbizzani, M. Mastragostino, L. Meneghello, T. Hamaide and A. Guyot, *Electrochim. Acta*, **37**, 1631 (1992).
43. C. Arbizzani, A.M. Marinangeli, M. Mastragostino, L. Meneghello, T. Hamaide and A. Guyot, *J. Power Sources*, **43–44**, 453 (1993).
44. T. Hamaide, C. Carré and A. Guyot, *Solid State Ionics*, **39**, 173 (1990).
45. M.G. Minnet and J.R. Owen, *Solid State Ionics*, **28–30**, 1192 (1988).

46. T. Hamaide, *Synth. Commun.*, **20**, 2913 (1990).

47. K.M. Abraham, in *Application of Electroactive Polymers* (Ed. B. Scrosati), Chapman & Hall, London, 1993, pp. 75–112.

48. F. Croce, F. Gerace, G. Dautzenberg, S. Passerini, G.B. Appetecchi and B. Scrosati, *Electrochim. Acta*, **39**, 2187 (1994).

49. S. Kakuda, T. Momma, T. Osaka, G.B. Appetecchi and B. Scrosati, *J. Electrochem. Soc.*, **142**, L1 (1995).

50. H. Ito, Y. Veda, T. Momma, T. Osaka and B. Scrosati, in *Proc. 35th Battery Symposium of Japan*, Kyoto, 1995, p. 313.

51. S. Megahed and B. Scrosati, *J. Power Sources*, **51**, 79 (1994).

52. S. Panero, E. Spila and B. Scrosati, *J. Electrochem. Soc.*, **143**, L24 (1996).

53. *Proc. Third International Seminar on Double-Layer Capacitors and Similar Energy Storage Devices*, Deerfield Beach, Florida, 6–8 December, 1993.

54. A. Nishino, in *Proc. Electrochem. Soc.* (Ed. B.M. Barnett, E. Dowgiallo, G. Halpert, Y. Matsuda and Z-I Takeara), Vol. 93-23, The Electrochemical Society, Pennington, 1993, pp. 1–14.

55. B.E. Conway, in *Proc. Electrochem. Soc.* (Ed. B.M. Barnett, E. Dowgiallo, G. Halpert, Y. Matsuda and Z-I Takehara), Vol. 93-23, The Electrochemical Society, Pennington, 1993, pp. 15–37.

56. J.P. Zheng, P.G. Cygan and T.R. Low, *J. Electrochem. Soc.*, **142**, 2699 (1995).

57. A. Rudge, J. Davey, I. Raistrick, S. Gottesfeld and J.P. Ferraris, *J. Power Sources*, **47**, 89 (1994).

58. A. Rudge, J. Davey, S. Gottesfeld and J.P. Ferraris, in *Proc. Electrochem. Soc.* (Eds. B.M. Barnett, E. Dowgiallo, G. Halpert, Y. Matsuda and Z-I Takehara, Vol. 93–23, The Electrochemical Society, Pennington, 1993, pp. 74–85.

59. M. Mastragostino and L. Soddu, *Electrochim. Acta*, **35**, 463 (1990).

60. M. Sato, S. Tanaka and K. Kaeriyama, *Makromol. Chem.*, **190**, 1233 (1989).

61. A. Rudge, I. Raistrick, S. Gottesfeld and J.P. Ferraris, *Electrochim. Acta*, **39**, 273 (1994).

62. C. Arbizzani, M. Catellani, M. Mastragostino and G. Mingazzini, *Electrochem. Acta*, **40**, 1871 (1995).

63. C. Arbizzani, M. Mastragostino and L. Meneghello, in *Materials Research Society Symposium Proc.* (Eds. G.A. Nazri, J.M. Tarascon and M. Schreiber), Vol. 369, Materials Research Society, Pittsburgh, 1994, pp. 605–612.

64. A. Rudge, J. Davey, F. Uribe, J. Landeros, Jr, and S. Gottesfeld, in *Proc. Third International Seminar on Double-Layer Capacitors and Similar Energy Storage Devices*, Deerfield Beach, Florida, 6–8 December, 1993.

65. C. Arbizzani, M. Mastragostino and L. Meneghello, *Electrochem. Acta*, **41**, 21 (1996).

66. X. Ren and P.G. Pickup, *J. Electroanal. Chem.*, **372**, 289 (1994).

67. C. Arbizzani, M. Mastragostino, L. Meneghello and R. Paraventi, *Adv. Mat.*, **8**, 331 (1996).

68. F. Hannecart, E. Destryker, J.F. Fauvarque, A. De Guibert and X. Andrieu, European Patent 90.202 099.9 (20 August 1990); French Patent 89.109 52 (14 August 1989).

69. X. Andrieu, L. Josset and J.F. Fauvarque, *J. Chim. Phys.*, **92**, 879 (1995).

70. C. Arbizzani, M. Mastragostino and L. Meneghello, *Electrochim. Acta*, **40**, 2223 (1995).

71. A. Clemente, S. Panero, B. Scrosati and E. Spila, in *Proc. Giornate dell'Elettrochimica Italiana*, Riccione, 25–27 September, 1995, Abstr. O8.

72. M. Mastragostino, in *Application of Electroactive Polymers* (Ed. B. Scrosati), Chapman & Hall, London, 1993, pp. 223–249.

73. B. Scrosati, in *Application of Electroactive Polymers* (Ed. B. Scrosati), Chapman & Hall, London, 1993, pp. 250–282.

74. C.G. Granqvist, in *Material Science for Solar Energy Conversion Systems* (Ed. C.G. Granqvist), Pergamon Press, Oxford, 1990, pp. 106–167.

75. J.L. Bredas, B. Thémans, J.G. Fripiat, J.M. André, R.R. Chance, *Phys. Rev.*, **B 29**, 6761 (1984).

76. M. Mastragostino, A.M. Marinangeli, A. Corradini and S. Giacobbe, *Synth. Met.*, **28**, C501 (1989).

77. C. Arbizzani, G. Barbarella, A. Bongini, M. Mastragostino and M. Zambianchi, *Synth. Met*, **52**, 329 (1992).

78. M. Mastragostino, C. Arbizzani, A. Bongini, G. Barbarella and M. Zambianchi, *Electrochim. Acta*, **38**, 135 (1993).

79. C. Arbizzani, A. Bongini, M. Mastragostino, A. Zanelli, G. Barbarella and M. Zambianchi, *Adv. Mat.*, **7**, 571 (1995).

80. C. Arbizzani, M. Mastragostino, L. Meneghello, X. Andrieu and T. Vicedo, in *Materials Research Society Symposium Proc.* (Eds. G.A. Nazri, J.M. Tarascon and M. Armand), Vol. 293, Materials Research Society, Pittsburgh, 1993, pp. 169–178.

81. M. Catellani, C. Arbizzani, M. Mastragostino and A. Zanelli, *Synth. Met.* , **69**, 373 (1995).

82. T. Hagiwara, M. Yamaura, K. Sato, M. Hirasaka and K. Iwata, *Synth. Met.*, **32**, 367 (1989).

83. M. Leclerc and G. Daoust, *J. Chem. Soc. Chem. Commun.*, 273 (1990).

84. M. Dietrich and J. Heinze, *Synth. Met.*, **41–43**, 503 (1991).

85. G. Heywang and F. Jonas, *Adv. Mat.*, **4**, 116 (1992).

86. J.C. Gustafsson, B. Liedberg and O. Inganäs, *Solid State Ionics*, **69**, 145 (1994).

87. Q. Pei, G. Zuccarello, M. Ahlskog and O. Inganäs, *Polymer*, **35**, 1347 (1994).

88. H. Yashima, M. Kobayashi, K.B. Lee, D. Chung, A.J. Heeger and F. Wudl, *J. Electrochem. Soc.*, **134**, 46 (1987).

89. M. Onoda, H. Nakayama, S. Morita and K. Yoshino, *J. Electrochem. Soc.*, **141**, 338 (1994).

90. C. Arbizzani, M.G. Cerroni and M. Mastragostino, *Solar En. Mat. Solar Cells*, to be published.

91. J.P. Ferraris, C. Henderson, D. Torres and D. Meeker, *Synth. Met.*, **72**, 147 (1995).

92. K. Bange and T. Gambke, *Adv. Mat.*, **2**, 10 (1990).

93. H.J. Byker, Gentex Corporation, US Patent 4902108 (1990).

94. T. Asaoka, T. Okabayashi, T. Abe and T. Yoshida, 40th ISE Meeting, Kyoto, 17–22 September 1989, Ext. Abstr. Vol.1, p. 245.

95. A. Corradini, A.M. Marinangeli and M. Mastragostino, *Electrochim. Acta*, **35**, 1757 (1990).

96. C. Arbizzani, M. Mastragostino, S. Passerini, R. Pileggi and B. Scrosati, *Electrochim. Acta*, **36**, 837 (1991).

97. E.A.R. Duek, M.A. De Paoli and M. Mastragostino, *Adv. Mat.*, **4**, 287 (1992).

98. E.A.R. Duek, M.A. De Paoli and M. Mastragostino, *Adv. Mat.*, **5**, 650 (1993).

99. B.P. Jelle, G. Hagen and R. Odegård, *Electrochim. Acta*, **37**, 1377 (1992).

100. B.P. Jelle, G. Hagen, S.M. Hesjevik and R. Odegård, *Mat. Sci. Eng.*, B **13**, 239 (1992).

101. B.P. Jelle, G. Hagen and S. Nodland, *Electrochim. Acta*, **38**, 1497 (1993).

102. B.P. Jelle and G. Hagen, *J. Electrochem. Soc.*, **140**, 3560 (1993).

103. A.M. Rocco, M.A. De Paoli, A. Zanelli and M. Mastragostino, *Electrochim. Acta*, **41**, 2805 (1996).

104. A.M. Rocco, M.A. De Paoli and M. Mastragostino, *Electrochim. Acta*, submitted.

105. C. Arbizzani, M. Mastragostino, L. Meneghello, M. Morselli and A. Zanelli, *J. Appl. Electrochem.* **26**, 121 (1996).

106. C. Arbizzani, M. Mastragostino and A. Zanelli, *Solar En. Mat. Solar Cells* **39**, 213 (1995).

107. R.C. Weast (ed.) *Handbook of Chemistry and Physics*, 56th Edition, CRC Press, Cleveland, 1975, p. F-196.

CHAPTER 12

Molecular Recognition for Chemical Sensing: General Survey and the Specific Role of π-Conjugated Systems

Wolfgang Göpel and Klaus-Dieter Schierbaum
University of Tübingen, Germany

1 INTRODUCTION

Thin layers of molecular, supramolecular and polymeric compounds gain increasing interest as sensitive, selective and stable ('3s') coatings of transducers for chemical sensors for *small inorganic compounds (such as O_2, Cl_2, NO_2 or CO_2) and volatile organic compounds* ('VOCs') in the gas phase [1]. The main reason is their huge flexibility for tailoring recognition structures by both, controlled chemical synthesis and controlled thin film preparation techniques [1–8]. The molecular recognition occurs in thin film devices by utilizing one of the various possible mechanisms to convert chemical information about the concentrations of molecules into electronic information by the use of suitable transducers. This is illustrated schematically in Figure 12.1. The sensor-active materials utilize either *surface reactions* or *bulk absorption* of free molecules.

Commonly used transducers are shown in Figure 12.2.

In this chapter, molecular recognition with supramolecular and polymeric compounds is reviewed. Particular emphasis is put on π-conjugated systems as chemically active materials since they are particularly suitable to monitor sensitively and selectively oxidizing molecules like NO_2 and Cl_2 [9–11]. This results from the fact that extremely low concentrations of these molecules in the gas phase may be monitored by conductivity changes. The physical origin of this sensitivity of organic semiconductors is similar to doping effects in inorganic semiconductors, which leads to a drastic influence of electro-active dopants on the conductivity of semiconductors. Typical examples of the latter are phosphorus or boron dopants *in the bulk* of silicon and oxygen or NO_2 dopants *at the surface* of SnO_2.

Handbook of Organic Conductive Molecules and Polymers: Vol. 4. Conductive Polymers: Transport, Photophysics and Applications.
Edited by H. S. Nalwa. © 1997 John Wiley & Sons Ltd

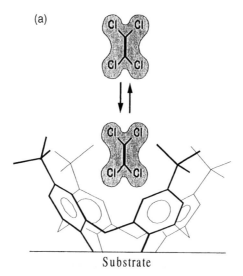

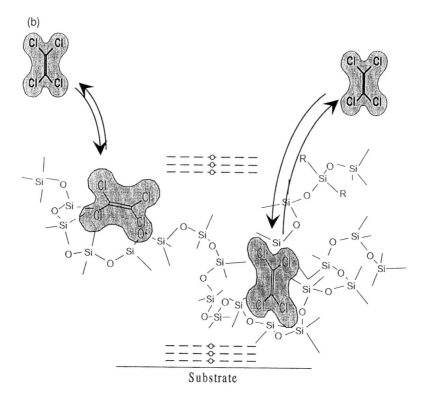

Figure 12.1. (a) Interaction of a typical VOC (here tetrachloroethylene, C_2Cl_4) with a simple supramolecular structure (here tert-butyl-calix[4]arene) with its *surface interaction site* facing towards the gas phase. Because of the rigid recognition structure, the simple 'key/lock' principle describes the interaction mechanism in this situation. (b) Dissolution of the VOCs into the *bulk* of polymeric thin films (here C_2Cl_4 in polydimethylsiloxane PDMS). Because of the large fluctuations in the time-dependent structure, the term 'induced fit' describes the interaction mechanism in this situation.

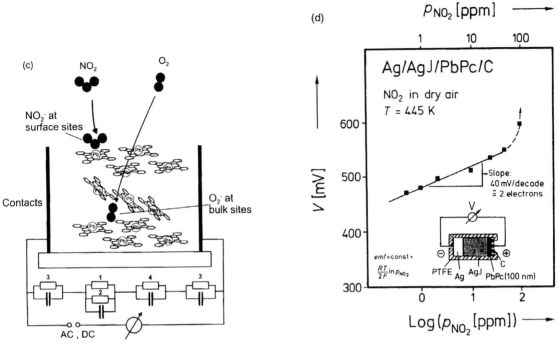

Figure 12.1. (c) Charge transfer reactions during the interaction of O_2 with the bulk, and of NO_2 with the surface of lead phthalocyanine (PbPc) thin film structures. For further details, see Figure 12.4. (d) Sensor signals from controlled interface reactions involving electron and ion conductors: sensor response V of a potentiometric Ag/AgI/PbPc/C solid electrolyte sensor as a function of the NO_2 partial pressure in air. For details, see Figure 12.18.

Sensor-active materials, in general, also include *inorganic oxide-based compounds* with different metal dopants which show charge transfer by electronic, mixed or ionic conduction (such as SnO_2, TiO_2, $SrTiO_3$, ZrO_2 etc.) and *inorganic cage compounds* which show occupation of cages that is influenced by the gas phase (such as zeolites) [12]. However, for the selective detection of organic molecules the inorganic materials offer far less flexibility in their atomic structure and less independent preparation parameters (when preparing thin layers with the aim of tailoring their recognition properties) as compared to the organic compounds treated here. The other extreme in terms of adjustable structures and parameters to tailor the recognition properties are *components of biomolecular systems* (such as receptors, transport proteins, etc.). Their large sizes and molecular weights exhibit far too much flexibility in their preparation. They are characterized by complex molecular structures [13,14] which evidently (at least today) can not be controlled unequivocally on the molecular scale. The organic materials discussed represent a compromise between the two extremes of too low and too high flexibility.

They are therefore particularly suitable for the design of chemical sensors for monitoring molecules with molecular weights in the order of 100 a.m.u.

To illustrate the general trends in research and development in this field, and to give an overview of the present state of the art, typical 'prototype materials' are selected which allow the detection of small inorganic molecules and VOCs in the gas phase [15]. These materials have been developed in recent years by two complementary approaches. Suitable polymeric compounds have usually been derived from *empirical 'trial and error' studies* whereas suitable molecular and supramolecular compounds have also been derived from *theoretical concepts and systematic studies of molecular recognition*.

The chapter is organized as follows.

- Commonly used *transducers* that may be applied to convert the information about the chemical composition in the gas phase into an electronic signal are listed in Section 2.
- A survey of *typical sensor materials and thin film preparation techniques* is given in Section 3.

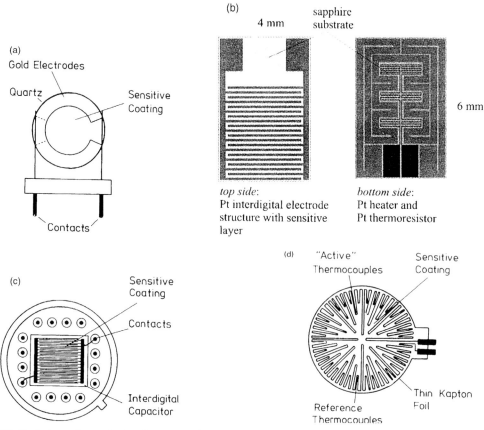

Figure 12.2. Typical transducers for chemical sensors: (a) quartz microbalance oscillator, (b) interdigital comb electrode structure, (c) capacitance device and (d) thermopile are shown that monitor changes of frequencies (Δf), conductivites ($\Delta \sigma$), capacitances (ΔC) and temperatures (ΔT).

- Typical examples of our *theoretical understanding* of the specific molecular recognition processes and sensing mechanisms are discussed briefly in Section 4.
- The fine-tuning with respect to sensitive, selective and stable ('3s') molecular recognition requires a detailed *characterization of the sensor structures* adjusted by different thin film preparation or test conditions. This characterization requires complementary microscopic and spectroscopic studies. Particular emphasis must be put on the structural analysis of interfaces in chemically sensitive films. Concepts and typical examples are discussed in Section 5.
- These electronic sensor signals are monitored under conditions of thermodynamic equilibria or of kinetically controlled non-equilibria. Typical *examples of chemical sensors* are discussed in Section 6.
- *Recent trends* focus on the synthesis of new molecules, at sensor arrays with pattern recognition

and at miniaturization of sensors down to the nanometer scale. This will be outlined briefly in Section 7.

2   TYPICAL SENSOR TRANSDUCERS

The different types of chemical sensors may be classified according to the different properties used for particle detection. The different transducers coated with sensor-active materials monitor the interaction of molecules from the gas and liquid phase by changes in the

- conductance (dc) of electrons or ions under direct current conditions,
- capacitance,
- conductance (ac) at different frequencies (complex impedance),

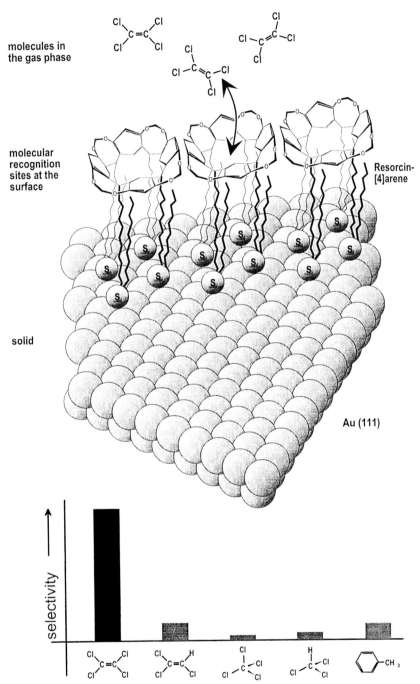

Figure 12.3. Schematic illustration of the covalent coupling of resorcinarenes via sulfide groups to Au(111) as a perfect monolayer coating of a bulk acoustic wave (BAW) device. Selectivities (frequency changes of the devices at identical gas phase concentrations in arbitrary units) for different VOCs monitored at room temperature are compared in the lower part [17].

- work function,
- temperature,
- optical thickness,
- optical absorption, or
- mass.

Typical examples of transducers are given in Figure 12.2. For a comparison of the different transducer principles and details, see, e.g. ref. [16]).

3 TYPICAL SENSOR MATERIALS AND THIN FILM PREPARATION TECHNIQUES

Two approaches may, in principle, lead to perfect sensors: the *empirical optimization* of new sensor-active materials including their controlled synthesis and thin film preparation on suitable transducers (to be discussed in this section); and the *theoretical understanding* of the sensing mechanisms (to be discussed in the next section). Ideally, both approaches should be chosen in parallel to reduce the time and cost of successful sensor research and development with limited resources.

A variety of *molecular, supramolecular and polymeric compounds* has been tested so far as coatings on suitable transducer substrates for gas sensing. Their different structures may tentatively be classified as

- semiconducting compounds,
- molecular crystals,
- molecular liquid crystals and molecular liquids,
- extramolecular cages (clathrates),
- intramolecular cages (coronands and cryptands (such as crownethers), cyclophanes (such as cyclodextrins), calixarenes, cryptophanes, cavitands, carcerands, etc.),
- polymers, and
- polymers modified by intramolecular cages.

Owing to the existing huge market of certain *standard monomers (such as phthalocyanines) and polymers* and an extensive knowledge concerning their synthesis, film preparation, structure and chemical reactivity, their use as chemically active sensor coatings has been and will be of major practical importance. However, next-generation sensors will require better sensitivities, selectivities and stabilities. This will require 'fine-tuning' of the molecular structure itself and the thin film formation. As an example, Figure 12.3

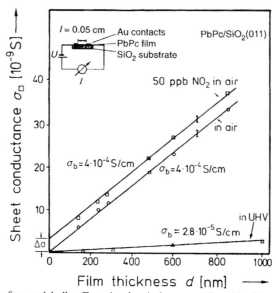

Figure 12.4. Separation of surface and bulk effects in chemical sensors. For thin films of PbPc on SiO_2 substrates, the conductance per unit area (i.e. the sheet conductance) $\sigma_\square = \sigma_b\, d + \Delta\sigma$ is plotted as a function of film thickness, d. The intrinsic bulk conductivity, σ_b, is obtained from the slope under ultra-high vacuum (UHV) conditions. Exposure to air leads to a bulk interaction with oxygen, a change in the slope and hence in the bulk conductivity, σ_b. Additional exposure to NO_2 leads to a surface reaction (chemisorption), a change in the axis intercept and hence in the surface excess conductivity, $\Delta\sigma$ [11].

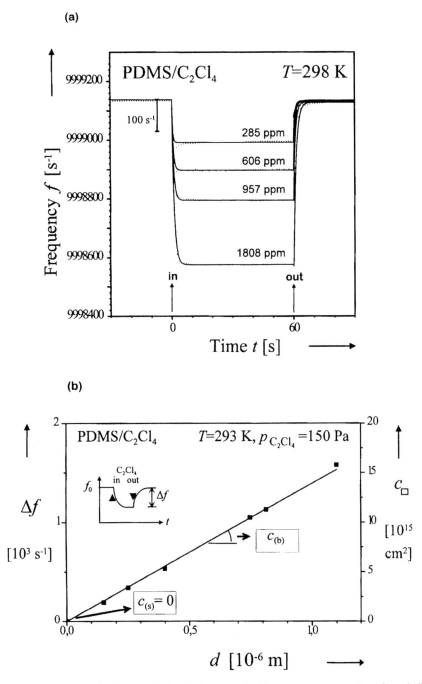

Figure 12.5. Separation of surface and bulk effects in chemical sensors. (a) Frequency change vs. time for polydimethylsiloxane (PDMS) interacting with C_2Cl_4 as monitored with a BAW device at room temperature in air. (b) For PDMS interacting with C_2Cl_4, the experimentally determined concentration per unit area $c_\square = c_{(b)}d + c_{(s)}$ is plotted as a function of thickness. Here, $c_{(b)}$ denotes the bulk concentration per unit volume and $c_{(s)}$ the surface excess concentration per unit area. The data are derived from equilibrium values of frequency variations $-\Delta f \sim m \sim c_\square$ monitored by means of a BAW quartz crystal oscillator as a function of time. The real response time of these polymers is very fast (<1 s) and is, in this particular experiment only, determined by the flow system. The results clearly indicate bulk dissolution with negligible surface effects [19].

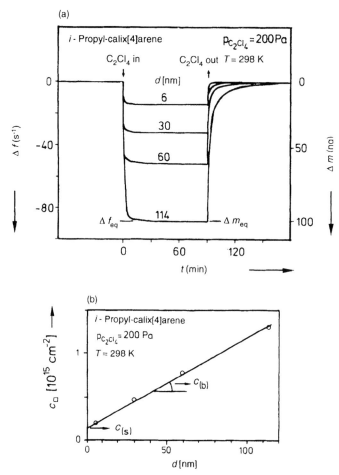

Figure 12.6. Separation of surface and bulk effects in chemical sensors. (a) Frequency variations, Δf, monitored with a BAW transducer during the interaction of iso-propylcalix(4)arene with C_2Cl_4. Increased thicknesses, d, lead to increased recovery times after pumping off the gas, because of slower bulk diffusion. (b) Sheet concentration, $c_\square$, as determined from the equilibrium values Δf as a function of film thickness, d, obtained from the results in Figure 12.6a. In addition to the bulk concentration, $c_{(b)}$, per unit volume, an excess surface concentration per unit area, $c_{(s)}$, is observed [18].

shows supramolecular structures of resorcinarenes bound to Au(111) surfaces by self-organization from solution, and their controlled assembly in a monolayer with their recognition sites perfectly ordered and oriented towards the gas phase [17]. The resorcinarene macrocycles provide molecular recognition sites for C_2Cl_4 molecules with a high selectivity. The covalent coupling to the gold surface (in this case, the electrode of a quartz microbalance transducer) results from strong S–Au bonds which terminate the aliphatic chains. The latter acts as a spacer between the gold surface and recognition sites. This spacer is stiff, is optimized to the length of both alkyl groups adjusted to the sulfur atoms, and has the same diameter as the cage itself.

A clear trend is to be seen in the research and development (R + D) of chemical sensing with molecular materials in general. This trend is similar in other potential fields of future application of the same molecules including, for instance, molecular electronics or integrated optics. It concerns the use of *ultra-pure monomers and oligomers* instead of polymers as the starting materials and the *growth of well-ordered monolayer or multilayer films* with well-defined molecular or supramolecular units (compare Figure 12.3). The systematic improvement of sensor performances, failure analysis, and unequivocal understanding of chemical sensing in this next R + D step requires complete control of the chemical compositions, geome-

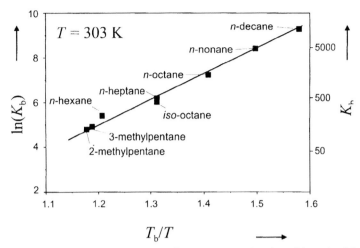

Figure 12.7. Experimentally (BAW) determined partition coefficients, K_b, as a function of the ratio of the boiling temperature and the absolute temperature, T_b/T, of different VOCs [20] and polysiloxanes. The experimental data shown here are taken at $T = 303$ K to allow the comparison of different VOCs. The linear slope provides a good approximation for PDMS in order to explain the temperature dependence of K_b for each individual VOC. Characteristic deviations from the straight line occur for modified polysiloxanes and will be discussed in Section 6.

trical arrangement of the atoms, and electronic and dynamic structures with particular emphasis on the structure of the interfaces to the gas phase and transducer substrate.

The *sensor preparation* must therefore include

- the elaboration of reproducible experimental conditions which lead to atomically flat and chemically homogeneous transducer substrates like oxides (e.g. SiO_2), semiconductors (e.g. silicon) or metals (e.g. gold);

$$E_{tot} = \sum E_{vdw} + \sum E_{elec} + \sum E_{tors} + \sum E_{oop} + \sum E_{str} + \sum E_{bend}$$

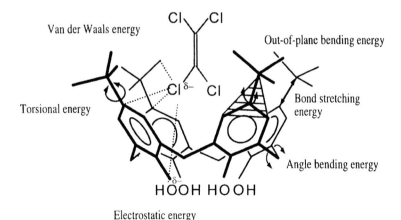

Figure 12.8. Schematic illustration of the different energy contributions, E_i, to the total energy E_{tot}, in force field calculations which are based on different molecular potentials to estimate the key lock [C_2Cl_4/calix[4]arene] interaction energy in the 'solid state limit' [22]. The indices in the equation denote the van der Waals energy (vdw), the electrostatic energy (elec), the torsional energy (tors), the out-of-plane energy (oop), the bond stretching energy (str) and the angle bending energy (bend).

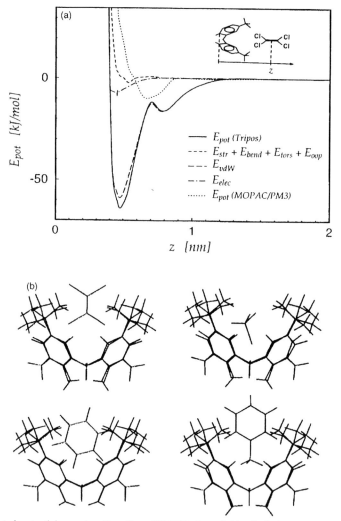

Figure 12.9. (a) Calculated potential energies, E_{pot}, (from TRIPOS force-field calculations) and the different contributions E_{vdW}, E_{elec}, and $(E_{str} + E_{bend} + E_{oop} + E_{tors})$ as a function of the distance, z, between a tert-butyl-calix[4]arene unit and one perchlorethylene (C_2Cl_4) molecule. For the definition of energy contributions, see Figure 12.8. For comparison, the E_{pot} values (as determined quantum-chemically from MOPAC/PM3) are also shown [22]. (b) Geometric presentation of different supramolecular tert-butyl-calix[4]arene/VOC complexes as determined theoretically with (1) tetrachloroethylene, (2) chloroform, (3) benzene and (4) toluene.

- the extraction, preparation or synthesis of well-characterized molecules for chemical sensing and their subsequent purification by carrier gas preparation, ultra-high vacuum (UHV) sublimation in a temperature gradient, zone refining, chromatographic cleaning techniques etc.;
- the subsequent preparation of layer structures with molecular or supramolecular units for which many different technologies are available, including Langmuir–Blodgett (LB) film formation, Knudsen evaporation from the gas phase, self-assembly from solution, electrochemical deposition, spray or spin-coating deposition, and a variety of deposition techniques with an additional subsequent crosslinking by thermal, photon- or plasma-assisted, or other treatments, and;
- the application of a suitable periphery such as contacts for electrical transducers.

The latter aspect requires the identification of experimental conditions for thermodynamically controlled or kinetically hindered interactions of thin organic films with metal atoms, with electronically conducting metal oxides or with conducting organic molecules.

In the next step of sensor development, the sensor responses are monitored as a function of time ('kinetics') but also if at all possible under equilibrium conditions ('thermodynamics') during the exposure to one, two and more molecules in the gas phase at different concentrations and temperatures. The final experimental condition is to monitor their response in the real environment of their future practical application. These studies aim to determine the conditions for reproducible (dynamic or static) sensor responses. Finally, the systematic and non-systematic drifts must

be estimated in order to determine the required time intervals of recalibration and the expected lifetime of the sensor. In many cases, these design goals can only be achieved by identifying the underlying sensing mechanisms down to the atomic scale, as will be discussed briefly in the following section.

4   SENSING MECHANISMS

The theoretical approach to understanding and optimizing sensing mechanisms can be performed with a variety of different models, all of which, however, have to be adapted to the specific sensor material and transducer parameter.

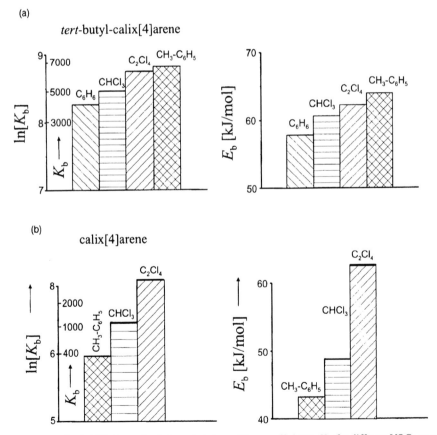

Figure 12.10. (a) Experimentally (BAW) determined characteristic partition coefficients, K_b, for different VOCs as compared with theoretically determined binding energies, E_b, from force field calculations of tert-butyl-calix[4]arenes [19,22]. (b) The modification of calixarenes leads to modified selectivities with a typical example shown here. Modified selectivities make it possible to build sensor arrays and electronic noses for the selective detection of molecules in mixtures by subsequent pattern recognition techniques.

There is a principal difference between surface and bulk chemical sensors. Thick films are preferred for sensors based on bulk absorption, and thin monolayer films are preferred for sensors based on surface reactions (see Figure 12.1). The important distinction between bulk and surface reactions in sensor-active materials may be deduced from sensor response signals monitored as a function of film thickness for identical gas compositions and temperature. Non-zero axis intercepts (deduced by extrapolating the sensor response signal to zero thickness) indicate surface reactions, and non-zero slopes indicate bulk absorption effects [18,19].

Three examples illustrate this approach.

- An exceptional case in which both surface and bulk sensing effects have been observed for the same material, lead phthalocyanine (PbPc) but for two different gases is shown in Figure 12.4 [11].
- The great advantage of a variety of polymers is shown in the typical results in Figure 12.5. For selected compounds and operation conditions their sensor response is a pure bulk effect [19,20].
- In most experimental situations, however, a more complex sensor response is observed. As a typical example, Figure 12.6 shows the results of calixarene films. The thickness dependence indicates an influence from both surface and bulk effects, the absolute values of which depend strongly on the preparation

conditions of the thin films. In this particular example, the latter includes the choice of substrate temperature, vacuum conditions and evaporation rate during the gas-phase deposition of the thin films under Knudsen conditions [18].

The example in Figure 12.3 illustrates the obvious importance of a perfect control of the geometric arrangement of sensor-active molecular or supramolecular structures in ordered thin films to obtain a controlled sensor response.

A few theoretical approaches to understand sensor parameters are outlined briefly in the following. The first example concerns the *conductivity mechanisms in semiconducting molecular crystals and their changes upon gas exposure*, which are usually not understood on a theoretical basis at a sufficient level of sophistication [11,21]. As one specific aspect, we choose the sheet conductance, $\sigma_\square$, as a typical sensor property. It is defined as conductivity per unit surface area. Specific data are shown in Figure 12.4 for lead phthalocyanine sandwich structures PbPc/SiO$_2$. Under UHV conditions, negligible sheet conductance is observed at the axis intercept ($d = 0$). We therefore conclude that negligible excess charge carriers occur at the surface and at the PbPc/SiO$_2$ interface. Subsequent O$_2$ exposure leads to a change in the slope and hence indicates bulk interactions. The simultaneous formation of defect electrons (holes) can be deduced from the

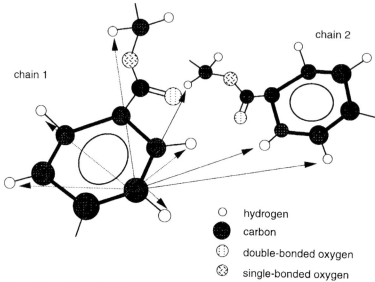

○   hydrogen

●   carbon

◉   double-bonded oxygen

⊗   single-bonded oxygen

Figure 12.11. Typical interchain and intrachain interactions in a polymer (here polyhydroxyterephthalate) which are treated in molecular dynamics calculations to estimate time- and spatial-dependent structures before and after incorporation of VOCs [34].

change in the bulk position of the valence band maximum, E_V, with respect to the Fermi energy, E_F (i.e. $\Delta(E_V - E_F)_b$) in ultraviolet photoemission spectroscopy (UPS, compare Section 5). Additional exposure to small amounts of NO_2 leads to changes in the axis intercept and hence indicates the occurrence of surface contributions to the overall sheet conductance. In suitable temperature ranges, both the effects of bulk (O_2^-) doping and surface (NO_2^-) doping are reversible. The absolute value of conductivity changes upon 'doping'; however, it cannot be described theoretically in terms of absolute values of dopants (here O_2^- ions in the PbPc lattice) even if the incorporation of the dopant can be calculated by quantum-chemical approaches to determine the structures, energies and electronic orbitals of the corresponding bulk complex. The key problem in describing theoretically the changes in conductivities as the experimentally monitored sensor parameters is our weak theoretical understanding of electronic mobilities in molecular crystals and in polycrystalline materials. Similar difficulties exist to interpret theoretically the changes in capacitances or, more generally, in the real and imaginary part of the complex impedance, or to interpret the changes in work functions during chemical sensing.

The second example concerns a sensor parameter which is more suitable for general applications. In contrast to conductivity changes, *mass changes* upon gas exposure to polymeric (Figure 12.5) and supramolecular compounds (Figure 12.6) always occur and may be interpreted in a thermodynamically straightforward approach. From the mass change of sensor-active coating with different thicknesses, d, upon exposure to well-known molecules at given pressure and temperature, the concentrations

$$c_\square = c_{(b)}d + c_{(s)} \qquad (12.1)$$

and hence the partition coefficients of these molecules in the bulk, K_b, or at the surface, K_s, are determined directly. Here, $K_{b,s}$ is defined as the ratio of the concentrations of molecules in the bulk (index b) or surface (index s) and the gas phase (index g), i.e.

$$K_b = \frac{c_{(b)}}{c_g} \text{ and } K_s = \frac{c_{(s)}}{c_g} \qquad (12.2)$$

From the temperature and pressure dependence of the partition coefficients, the corresponding heats of *ad*sorption or *ab*sorption, ΔH°, are obtained [6,20]. These heats may be monitored independently by the

temperature changes of thermopile transducers or determined indirectly from temperature maxima in thermodesorption experiments (see Section 5).

The partition coefficients are key parameters that describe the sensitivity of a certain polymer to detect a certain molecule 'i' which in many cases is found to be independent of the absolute value of the concentration. The values $K_{b(i)}$ hence determine the changes in the standard Gibbs enthalpy $\Delta G_{(i)}^\circ$ according to

$$\Delta G_{(i)}^\circ = \Delta H_{(i)}^\circ - T\Delta S_{(i)}^\circ = -RT \ln K_{b(i)} \qquad (12.3)$$

This equation describes the bulk absorption of solvent molecules with index 'i' within the polymeric matrix. The K_b values for each gas 'i' can be easily estimated for simple unsubstituted polymers such as polydimethylsiloxane (PDMS) by assuming that interaction enthalpies $\Delta H_{(i)}^\circ$ between the different solvent mole-

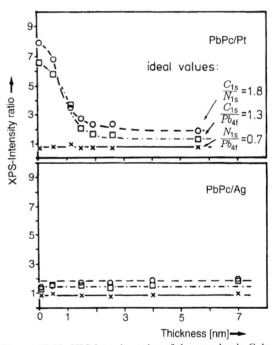

Figure 12.12. XPS-intensity ratios of the core levels C 1s, N 1s, and Pb 4f for PbPc/Pt and PbPc/Ag thin film sandwich structures as a function of PbPc thicknesses. Thickness-dependent ratios indicate deviations from the bulk stoichiometry and hence deviations from a constant ratio given by the relative elemental concentrations and XPS cross-sections. Chemical reactions at the PbPc/Pt interface lead to a drastic increase of carbon-containing compounds. In highly oriented molecular materials with large unit cells, corrections must be made for the different attenuation depths of different elements within the unit cell. These effects are, however, an order of magnitude smaller than those in the upper part of this figure [27].

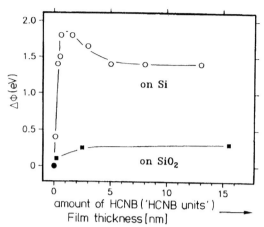

Figure 12.13. Work function changes $\Delta\phi$ for hexacyanobenzene (HCNB) on Si(100) and SiO_2 as a function of the mean film thickness expressed in numbers of HCNB monolayers, 'amount of HCNB' [28].

cules and the polymeric macromolecule are to a first approximation proportional to the interaction energies between the solvent molecules in their liquid state, and that changes in the entropy are identical for different solvents. In this case a linear dependence is expected of ln K_b on the boiling temperature divided by the absolute temperature T_b/T (Figure 12.7). Selectivities of these sensors may be improved by monitoring and comparing the response signals at different sensor temperatures.

The thermodynamic data contain energy and entropy contributions which may then be interpreted theoretically by means of force field or molecular dynamics approaches. For this first-approximation interpretation two completely different models are chosen.

- The host/guest interaction in small supramolecular structures is treated in the *rigid lattice* ('*solid state*') approach to describe the 'key/lock' interaction by means of *force field calculations* [22]. In a first approach, the energetics of individual supramolecules may be considered by neglecting interactions between the different 'locks'. Examples are illustrated in Figures 12.8 and 12.9. The different contributions to the total interaction energy, E_b, are compared in Figure 12.10 with experimental data on partition coefficients, K_b, for different molecules. The latter were measured under identical conditions of partial pressures and temperatures. These results are also in line with independent experimental data on interaction energies from activation energies of desorption in thermodesorption spectroscopy (TDS) measurements (see Section 5) [20].

- The host/guest interaction in polymeric structures is treated as an embedding effect in a *flexible lattice (induced fit in the 'liquid state')* (see the example in Figure 12.11) by *molecular dynamics calculations*. This approach to characterize interaction energies is far more time consuming and less precise. The energy as well as the entropy of the total system has to be calculated as a function of time for various atomic configurations before and after incorporation of organic molecules into the flexible matrix. Gibbs energy minima have to be calculated by considering certain boundary conditions, which are determined by the coordination of atoms, the different inter- and intramolecular forces, the finite temperature, and the chemical potential of the molecules in the gas phase. So far, this approach could only be performed by optimizing structures over short time scales and calculating atomic motions in relatively small systems with a small number of atoms in a closed volume with adapted boundary conditions at opposite edges of this volume.

A final remark concerns the gap between theory and experiments: There is a pronounced contrast between our weak theoretical understanding of the phenomenological sensor parameters like conductivity, capacitance, complex impedance, or work function in particular of semiconducting π-conjugated systems like PbPc, and the accuracy in the experimental determination of sensor parameters. These parameters may often be monitored with high precision and may therefore be formally described satisfactorily by the various model parameters as characterized in the last section.

The sensor responses may subsequently be evaluated by

- certain mathematical algorithms,
- the determination of the components of electric equivalent circuits ('the electroengineer's modelling of sensor response behavior') or
- kinetic and thermodynamic theories.

These three different approaches of data evaluation for sensors and sensor arrays avoid any atomistic understanding of the sensing mechanisms. The atomistic understanding and fine-tuning, however, of the model parameters, (e.g. the components of the electrical equivalent circuits) do require the additional input from the microscopies and spectroscopies to be discussed in the next section.

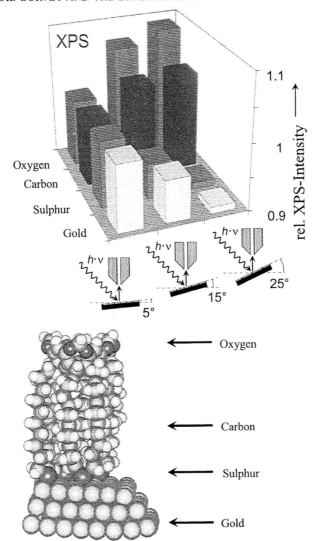

Figure 12.14. Characterization of the elemental composition and orientation of resorcinarene monolayers on Au(111) from the angle-dependent XPS photoemission of those core level electrons which result from oxygen, carbon, sulfur and gold atoms in surface and subsurface layers. Since the information depth of this technique is of the order of the layer thickness, angle-dependent XPS data indicate the locations of the different atoms in the top layer and in subsurface regions (see also Figure 12.3) [17].

5 MICROSCOPIC AND SPECTROSCOPIC ANALYSIS OF THIN FILM STRUCTURES

The *sample characterization* and also the *failure analysis* during different preparation steps, before or after the practical use of the complete chemical sensor requires the application of a variety of different tools of surface and interface analysis [23–25]. The structures

have to be investigated microscopically, spectroscopically and with respect to those phenomenological properties which are monitored by appropriate transducers (see Section 2).

The most important spectroscopic techniques used in the laboratory are based on the use of electrons, photons, ions and atoms as probes to monitor the geometric, electronic and dynamic structure as well as the chemical composition. Others require the applica-

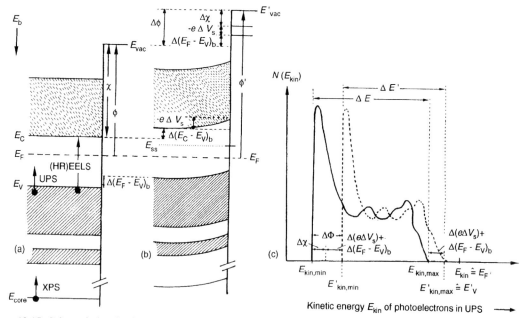

Figure 12.15. Schematic band scheme with typical binding energies, E_b, of electrons in a semiconductor, and the schematic principles of the most common surface spectroscopies XPS, UPS, and EELS to identify (a) ideal surface electronic structures; (b) their changes upon formation of electronic surface states, E_{SS}, e.g. by surface reactions such as chemisorption or the formation of intrinsic defects; and (c) the corresponding photoemission UPS spectrum with the number of the photoemitted electrons, $N(E_{kin})$, as a function of the kinetic energy, E_{kin}, of the photoelectrons. Also indicated is the evaluation scheme to separate the different contributions to the work function changes, $\Delta\phi$, which occur between situations (a) and (b). The separation of bulk effects $(E_F - E_V)_b$ requires additional measurements (e.g. UPS measurements for different bulk dopings or for different information depths of the emitted electrons). To a first approximation, the band bending, $-e\Delta V_S$, is assumed to be the same for all electronic states and the bandgap is assumed to be independent of the doping i.e. $\Delta(E_F - E_V)_b = \Delta(E_F - E_V)_s$.

tion of synchrotron radiation (e.g. near-edge X-ray absorption fine structure (NEXAFS), X-ray diffraction at surfaces under grazing incidence conditions) or neutron sources (e.g. small-angle neutron scattering). Of increasing interest for characterizing and also structuring thin films are scanning tunneling micro-

scopy ('STM') techniques and a variety of different modifications thereof ('SXM'). These techniques monitor various spatial variations, e.g. electron densities attributed to different energetic states, attractive or repulsive forces perpendicular or parallel to the surface, ion currents, heats, optical or magnetic properties. For

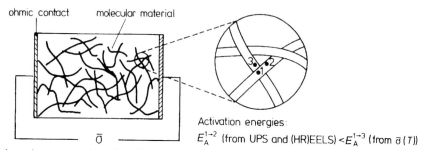

Activation energies:
$E_A^{1 \to 2}$ (from UPS and (HR)EELS) $< E_A^{1 \to 3}$ (from $\bar{\sigma}(T)$)

Figure 12.16. Schematic representation of conductivity measurements in polymeric thin films with apparent activation energies, $E_A^{1 \to 3}$, of mean conductivities, $\bar{\sigma}$, and activation energies, $E_A^{1 \to 2}$, of conductivities along the chains. $E_A^{1 \to 2}$ may be estimated indirectly by experimentally determining $E_F - E_V$ from UPS and $E_g = (E_C - E_V)$ from HREELS measurements. $E_A^{1 \to 3}$ may be estimated from temperature-dependent conductivity measurements. A comparison of both requires assumptions about the mobilities [29].

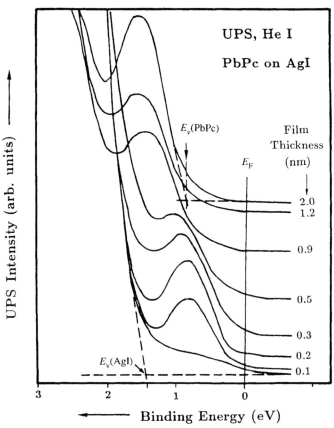

Figure 12.17. UPS intensities as a function of the binding energy referred to at the Fermi level $[E_b(E_F) = 0 - E_{kin}$ in Figure 12.5c] near the valence band edge, E_v, for different film thicknesses of PbPc on silver ionide AgI substrates. The PbPc bulk position of E_V (PbPc) is obtained for 2 nm thick films [30].

further details on SXM approaches, see, e.g. ref 26. Unfortunately, some commonly used and powerful techniques for the detailed structural analysis of macromolecules (including in particular two-dimensional Fourier transform NMR or X-ray crystallography) are not (yet) sensitive enough to analyze thin film structures at the interface with the required resolution down to the molecular scale.

The critical film preparation steps should be analyzed with the techniques mentioned above to control the correlations between structures and functions of the thin film samples in the different stages of their synthesis. In addition, the same systematic studies are also required to then investigate the influence of all those molecules or ions from the gas or liquid phase which are present under the various experimental conditions of the sample preparation and of the subsequent device application. These studies require

again the combined use of the spectroscopic or microscopic techniques of interface analysis and those techniques monitoring phenomenological layer properties with appropriate transducers [1].

Of particular importance in this context are comparative investigations performed under clean room or even ultra-high-vacuum conditions on the one hand (characterization of 'model systems') and under controlled atmospheric pressure or electrochemical conditions on the other hand (characterization of 'real systems'). If at all possible, *in situ* techniques should be applied which monitor structures under both ideal conditions as well as under the conditions in which the final device will be used. These include, in particular, various optical techniques such as absorption and reflection spectroscopy between the IR and UV range, measurements of dichroic ratios, Raman spectroscopy, surface plasmon resonance spectroscopy, spectroelectrochemistry, or SXM techniques.

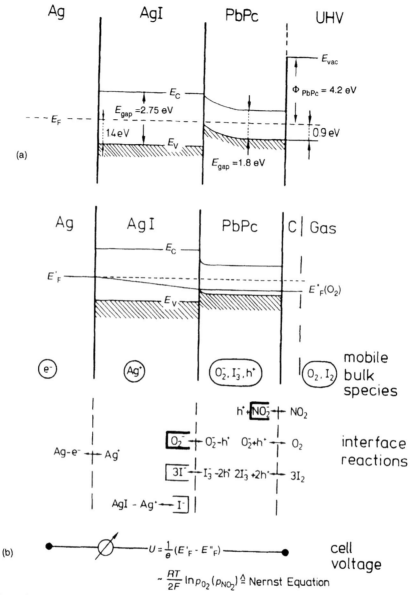

Figure 12.18. Schematic energy band diagram of a potentiometric cell to monitor O_2 and NO_2 in the gas phase on the right hand side. The sandwich structure makes use of electron conduction (e^-) across the silver and carbon contacts on both sides of the sensor. The carbon contacts are structured so that PbPc/C/gas three-phase boundaries become effective. Ionic conduction of Ag^+ occurs in AgI, and mixed conduction of O_2^-, I_3^-, h^+ in PbPc. The concentrations of interface species O_2^{2-} and of surface species NO_2^- are determined by the corresponding pressures p_{O_2} and p_{NO_2} in the gas phase, both of which determine the Nernst response of the cell voltage, $U = f(p_{O_2}, p_{NO_2})$. For further details, see the text and ref. 30.

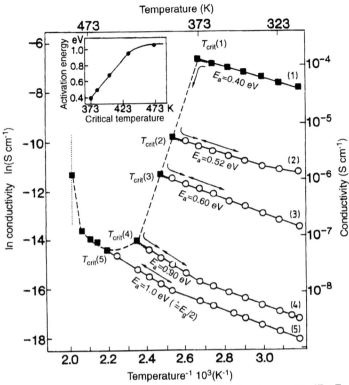

Figure 12.19. Conductivities as a function of temperature and corresponding low-temperature ($T < T_{crit}$) activation energies E_a for O_2-doped PbPc with increasing doping curves (1 to 5). E_a denotes activation energies and E_g denotes the bandgap determining the intrinsic conduction. Below T_{crit}, the conductivity values become independent of p_{O_2}. A special p_{O_2} determines a specific value of T_{crit}, activation energy, E_a, and conductivity or temperature behavior for $T < T_{crit}$. Around $T = T_{crit}$, long-term drifts are observed if p_{O_2} is not the equilibrium value. Above $T = T_{crit}$, PbPc can be used as an oxygen sensor.

5.1 Thickness-dependent film properties

The preparation of all the film structures to be discussed starts with the preparation and characterization of those *substrate* materials that show the desired metallic, semiconducting, or insulating electrical properties which can only be achieved by controlled surface atom geometries.

Subsequent systematic measurements of thickness-dependent film properties during the *formation of organic overlayers* then make it possible to obtain an overview of the different bonding types of individual organic molecules and molecular aggregates assembled at the surface.

• In the submonolayer range, individual molecules and their bonds to the substrate may contain van der Waals, hydrogen bridge, covalent and/or ionic

contributions with specific intermolecular changes if compared with free molecules. These changes may show up as occupied or empty electronic states [as deduced from UV photoelectron spectroscopy (UPS) or high-resolution electron energy loss spectroscopy (HREELS)], in vibrational modes [as deduced from HREELS or Fourier transform infrared spectroscopy (FTIR)], or in chemical shifts [as deduced from X-ray photoelectron spectroscopy (XPS)]. Cluster calculations simulating molecule/surface complexes before and after bond formation to the substrate can be used to identify these changes.

• In the monolayer range, two-dimensional molecular systems can be studied with their specific substrate as well as intermolecular bonding. Preparation conditions may be optimized carefully to obtain ideal two-dimensional order. Often, the molecule/substrate interaction is stronger than the molecule/molecule

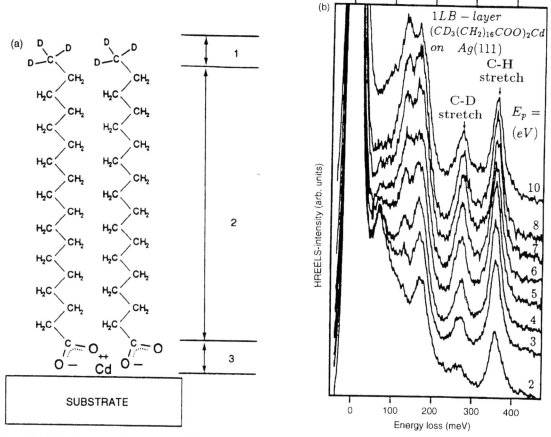

Figure 12.20. Electron energy losses obtained from HREELS of one Langmuir–Blodgett layer of cadmium arachidate on Ag(111) with a characteristic attenuation of the C–D stretch if compared with the C–H stretch vibration at lower primary energies, $E_p < 10$ eV, of the impinging electrons [31].

interaction within the layer. In this situation, mono-layer two-dimensional systems may be prepared easily under thermodynamic equilibrium conditions by adjusting partial pressures or concentrations in the gas phase or in the solution at a constant temperature. If, in addition, the surface mobility of the molecules is reduced strongly at monolayer saturation coverage, it is possible to investigate their molecular order by low energy electron diffraction (LEED) or scanning tunneling microscopy (STM).

• At larger thicknesses than the monolayer, intrinsic properties of the molecular material itself begin to dominate increasingly the film properties. The growth mode may be determined by the formation of homo-geneous epitaxial or amorphous films, by cluster growth, etc. This leads to typical influences of the substrate on the overall electrical and spectroscopic

properties which decrease in a characteristic manner as a function of the mean film thickness. This means that film thickness is determined unequivocally from the total amount of deposited organic molecules.

5.2   Typical spectroscopic results (case studies)

The follow specific examples illustrate the usefulness of spectroscopic techniques for designing chemical sensors based upon key lock interactions of molecular compounds.

Surface-sensitive UPS, XPS and HREELS are particularly useful as 'workhorse' techniques to study elemental compositions, and chemical as well as electronic structures.

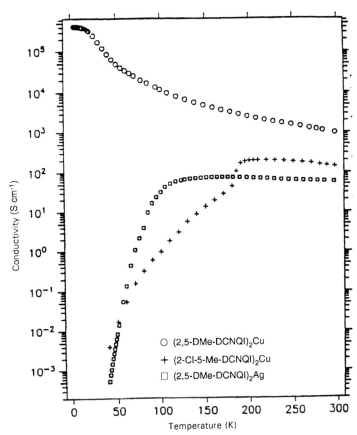

Figure 12.21. Conductivity as a function of temperature for different DCNQI salts with metallic, semiconductor and metal/semiconductor transition of 2,5-dimethyl (DMe), 2-chloro-5-methyl (2-Cl-5-Me), and 2,5-dimethyl (2,5-DMe) DCNQI$_2$Ag and copper salts [32].

(1) The control of interfacial bonding and stability of molecules in the monolayer range is of particular importance. Uncontrolled chemical reactions at interfaces that lead to the decomposition of organic molecules must be avoided. As these reactions are restricted to the first monolayer only, they can be identified in thickness-dependent signals from any thin film sensitive technique. The example in Figure 12.12 illustrates chemically stable PbPc/Ag and unstable PbPc/Pt interfaces [27]. The example in Figure 12.13 shows how the specific interface reactions of hexacyanobenzene (HCNB) causes drastic work function changes $\Delta\phi$ upon formation of the first monolayer [28]. These reactions involve the reactive dangling bond (sp^3-derived) electronic states in the bandgap of Si(100) single crystal surfaces. The figure also indicates the non-reactive interactions that occur with

SiO$_2$ surfaces because these surfaces exhibit oxygen-saturated dangling bonds of silicon and a large bandgap without electronic surface states. Bond saturation of silicon with oxygen (as in this example), and also with hydrogen or nitrogen, is a common preparation step to avoid chemical reactions between organic molecules and single-crystal silicon surfaces for organic/inorganic sandwich structures ('passivation of reactive dangling bonds').

(2) The next example concerns the characterization of the elemental composition of resorcinarene monolayers (compare Figure 12.3) with particular emphasis on the preparation of conditions for a perfect orientation of the oxygen-containing head groups facing the gas phase as compared to the sulfur-containing tail groups facing the transducer (from XPS measurements, see Figure 12.14 [17]).

(3) The study of the electronic structures of occupied and empty states in the bandgap and hence the highest occupied molecular orbital–lowest unoccupied molecular orbital HOMO–LUMO range may be done with UPS and HREELS. Typical information obtained from these spectroscopies is best pictured in a band scheme; Figure 12.15 shows typical inorganic or organic semiconducting molecular materials. As well as providing information on the characteristic densities of states in the valence-band (from UPS) or core-level range (from XPS), the kinetic energy, E_{kin}, of the photoelectrons yields information about (i) the Fermi level relative to the valence-band edge, $(E_F - E_V)_s$, at the surface, (ii) the work function, ϕ, (iii) changes resulting from band bending, $-e\Delta V_s$, bulk doping, $(E_F - E_V)_b$ and surface dipoles, including the electron affinity changes, $\Delta\chi$. One great advantage of UPS and HREELS is illustrated in Figure 12.16 for an extremely disordered polymeric sample. Both examples can be used to estimate intrinsic conductivities of the molecular unit itself, even under conditions where the overall conductivity may differ by orders of

magnitudes because of the strong influence from electronic hopping effects across grain boundaries, fibrilles or different molecular chains (as shown here with activation energies, $E_A^{1\to3} \gg E_A^{1\to2}$) [29]. For these systems, the UPS and HREELS data make it possible to at least estimate roughly the intrinsic charge-carrier concentrations within the molecular units. This is possible, even if perfect contacting of these units and hence direct measurements of their electrical conductivities are not possible or are not yet solved technologically, e.g. by applying microcontacts. As well as the possibility to determine bandgap energies, E_g, the technique HREELS has the additional advantage that other loss mechanisms resulting from

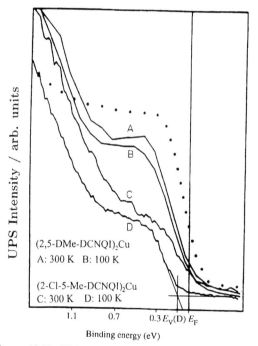

Figure 12.22. UPS spectra to determine the valence band edges, E_V, of different DCNQI salts near the Fermi energy, E_F. The dotted line corresponds to the reference (metallic silver Fermi edge) spectra, curves A–C characterize metallic behavior with $E_V = E_F$ and curve D shows semiconducting behavior, $E_{CV}(D) < E_F$ [32].

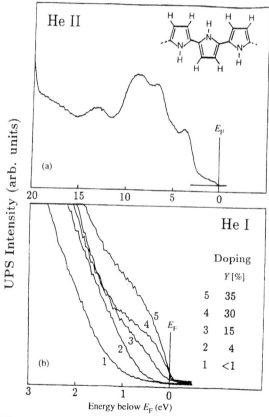

Figure 12.23. Helium II-excited UPS spectrum of polypyrrole for different percentages of doping Y od tosylate. The magnified lower part clearly indicates the transition from semiconducting to metallic (polaron) conduction at doping concentrations above 5%. The spectroscopic determination of polaronic gap states and bands is possible by taking additional HREELS results into account [29].

(a) MO's

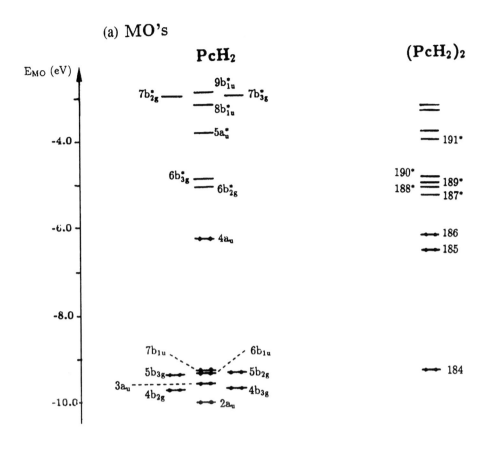

(b) On-top view

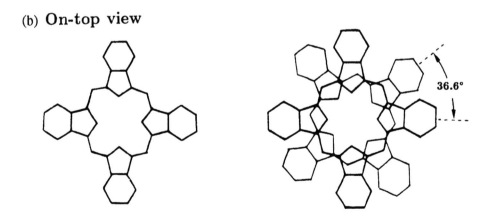

Figure 12.24. Binding energies of molecular orbitals (MOs) (a) in the valence band range of PcH_2, $(PcH_2)_2$ and Pc_2Lu that indicate the closed shell system of ordinary Pc dimers and the open shell system of Pc_2Lu. The latter determines the low bandgap behavior and an asymmetric bending of the ring shown in (c). The on-top view (b) was deduced from the bulk X-ray data of the Pc monomers and dimers, respectively.

(c) **side view**

Pc₂Lu

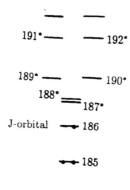

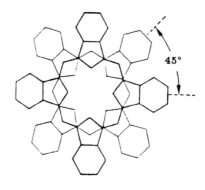

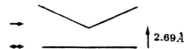

Figure 12.24. (*continued*)

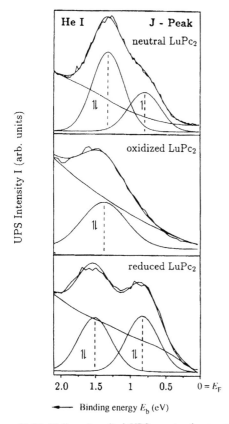

Figure 12.25. Helium I-excited UPS spectra (upper curves) and Gaussian fits (lower symmetric curves with zero base lines) of the two highest MOs including the HOMO orbital (J peak) in neutral $LuPc_2$, oxidized $LuPc_2^{(+)}$, and reduced $LuPc_2^{(-)}$ [33]. The number of electrons in the orbital is proportional to its peak area in reference to the area of the next peak at higher binding energy.

the excitation of vibrational modes, plasmons, excitons, or polarons can be determined with the same instrument.

(4) Surface or interface charge transfer of free carriers during the formation of ionic species strongly influences the conductivity of semiconducting molecular thin films because of band-bending effects (see Figure 12.15) [23]. The conductivity parallel to and across the molecular semiconductor/inorganic interface is determined by energetic discontinuities of the valence and conduction band at the interface, as they can be deduced from thickness-dependent UPS and HREELS studies. Typical results are given in Figure 12.17. The overlayer formation of PbPc on AgI (as silver-ion conducting substrate) was monitored with UPS at film

thicknesses between 0 and 2 nm. The changes $\Delta(E_V - E_F)$ and $\Delta\phi$ (from $\Delta E_{kin,min}$, see Figure 12.15c) are identical. This indicates negligible electron affinity changes, $\Delta\chi$. The corresponding band scheme of the PbPc/AgJ interface is the central part of a sandwich structure (Figure 12.18) [30].

(5) A formal separation of bulk and surface or interface properties may be performed by following the general thermodynamic definition of 'Gibbs excess interface properties'. This is done experimentally by monitoring film properties as a function of film thickness. Typical examples have been shown in Figures 12.4–12.6. The specific data obtained for PbPc/SiO$_2$ sandwich structures exhibit reversible effects of bulk (O_2^-) doping and surface (NO_2^-) doping. Reversibility of bulk doping does, however, require elevated temperatures to be adjusted (Figure 12.19). In many earlier studies, such dopants had been incorporated unintentionally during the thin film formation under insufficient vacuum. The same results can be obtained after the first exposure of UHV-prepared films to reactive gases in the air; in experiments of this kind, an identification of electroactive dopants is possible. For many systems, these dopants cannot be changed reversibly by just changing the thermodynamic conditions in the gas phase under atmospheric pressure.

(6) Calibrations are usually required in the application of surface-sensitive techniques if thin film structures are to be characterized quantitatively. The characteristic information depth which, for example, is probed by the emitted electrons in UPS, XPS, or HREELS experiments depends on their energy and on the material itself. In XPS, as a specific example, this information depth may be calculated from the measurements of the characteristic attenuation of the substrate emission with systematically varied overlayer thicknesses of LB films. The information depths of electrons in the 1 keV range are typically a factor of two larger than for inorganic materials [30]. As another specific example, the information depth in HREELS may be deduced from the relative C—H and C—D vibrational mode intensities of one selectively deuterated LB layer in which the top carbon/hydrogen unit only (region 1 in Figure 12.20) was deuterated. The primary energy (E_P) dependence of the data shown in Figure 12.20 makes it possible to calculate the corresponding information depth when (as in XPS) it is larger than the corresponding depth obtained for inorganic materials.

(7) The extrapolation of thickness-dependent measurements to conditions at infinite layer thicknesses or measurements on single crystals lead to the bulk

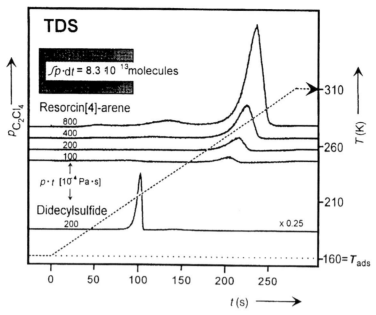

Figure 12.26. Typical thermodesorption spectrum obtained from C_2Cl_4 desorbing from resorcinarenes (upper part) and didecylsulfide (lower part) during the linear increase of the temperature after exposure of both substrates to C_2Cl_4 at 160 K under UHV conditions. From the thermodesorption maximum, activation energies of desorption are estimated. The area under the desorption peak correlates with the amount of adsorbed species. The calibration is obtained from $\int p\,dt$. Evidently, the interaction of C_2Cl_4 with didecylsulfide is much weaker because desorption occurs at significantly lower temperatures [17].

properties of molecular materials. A first approach to characterize electrical bulk properties, well-known from inorganic semiconductor studies, is to monitor temperature-dependent conductivities and mobilities. A correct choice of contacts makes it possible to monitor electronic and ionic contributions separately. This does,

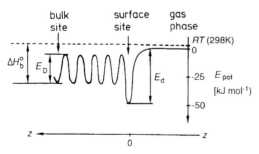

Figure 12.27. Changes of the potential energy, E_{pot}, as a function of distance, z, in thin layers as determined from the time dependence and temperature dependence of mass changes during gas exposure (monitored with BAW) and from the corresponding results of thermodesorption spectroscopy of the C_2Cl_4/calix[4]arene system. Also indicated is the heat of bulk dissolution, ΔH_b^o, and the activation energy of bulk diffusion, E_d [19].

however, often require the use of elevated temperatures and very specific contact materials (see, e.g. the example in Figure 12.4). Electronic conductivity data along certain crystal orientations, corresponding mobility data and comparative spectroscopic results on electronic states may be obtained for individual molecules and/or bulk material of a well-known geometric order. Typical examples for the latter are shown in Figure 12.21 for different salts of N, N'-dicyano-p-quinonediimines (DCNQI) [32]. Corresponding theoretical models then explain the intrinsic or, after controlled doping, the extrinsic conduction mechanism in the bulk. In this context, usually five problems occur which are discussed in the following examples.

(8) Certain classes of molecular materials with high conductivities such as phthalocyanines or DCNQI salts contain metal atoms which may determine their electronic conduction. The first problem is that the assignment of the molecular orbitals near the Fermi edge (E_F), in general, and the energetic position of the metal atom-derived states, in particular, are often not known. Hence, the conductivity mechanism, in general, and the role of metals in the lower-dimensional

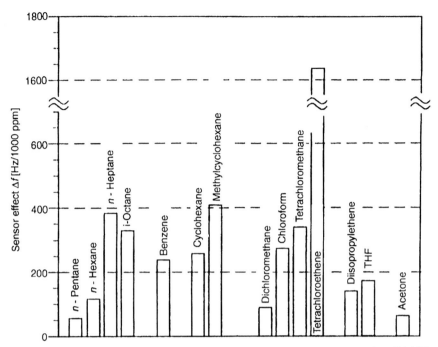

Figure 12.28. Frequency response of a modified surface acoustic wave (SAW) resonator operated at 420 MHz to 1000 ppm of solvent vapor interacting with a thin layer of trimethyl-Si groups covalently anchored at a quartz surface [7]. The sensitivities correlate roughly with the data obtained for bulk PDMS (Figure 12.7). Achievable stabilities, sensitivities and reproducibilities, however, are rather limited in these monolayer films. They become increasingly better with increasing thickness of such films because the sensor responses are based upon bulk effects (see also Figure 12.5).

electronic conduction of molecular materials, in particular, cannot be deduced from experimental data which might confirm theoretical arguments based upon bonding. In the past the understanding of conduction mechanisms could often only be deduced from 'chemical intuition'. This problem can be solved satisfactorily by monitoring photon energy and angle-dependent valence-band densities of states, e.g. around the resonances of the metal atom transitions and by comparison with theoretical MO- or band-structure calculations. The onset of the valence-band emission in particular may be used to clarify the type of intrinsic (electron or hole) conduction if independent data are available on the bandgap, E_g, (from conductivity measurements, HREELS, or optical spectroscopies). Typical examples are shown in Figure 12.22 [32]. The same approach may also be used to characterize the additional influence of doping. Typical examples are shown in Figure 12.23. Here, the metallic contacts to monitor 'intrinsic' conductivities would cause problems as the material is polymeric (compare the Figure 12.16).

(9) One problem concerns investigations on insulating or low-conductivity materials with small concentra-tions of free charges. In photoemission results, addi-tional relaxation effects must be corrected for, e.g. the determination of $E_V - E_F$. In this context, the meaning of the experimentally determined Fermi level has to be reconsidered too. In addition, the unequivocal separa-tion of charge-carrier concentrations and mobilities in the overall conductivity and their interpretations in terms of bond or band pictures for different molecular structures becomes a delicate problem.

(10) Intrinsic conductivities can often not be mea-sured because of influences from uncontrolled con-tamination. As a typical example, the conductivity of PbPc thin films increases by three orders of magnitude even at 300 K, after film preparation in UHV, during the first addition of oxygen with pressures of only 10^{-6} mbar. Conductivity values depend on the partial pressure of O_2. This effect is utilized in PbPc gas sensors to monitor O_2 (see Figure 12.18 and 12.19). Oxygen forms mobile O_2^- ions and holes in PbPc and thereby creates mixed (O_2^-) ionic and electronic conduction with holes a predominant charge carriers (compare also Figure 12.4). Similar effects occur for the many other well-known Pc bulk dopants such as

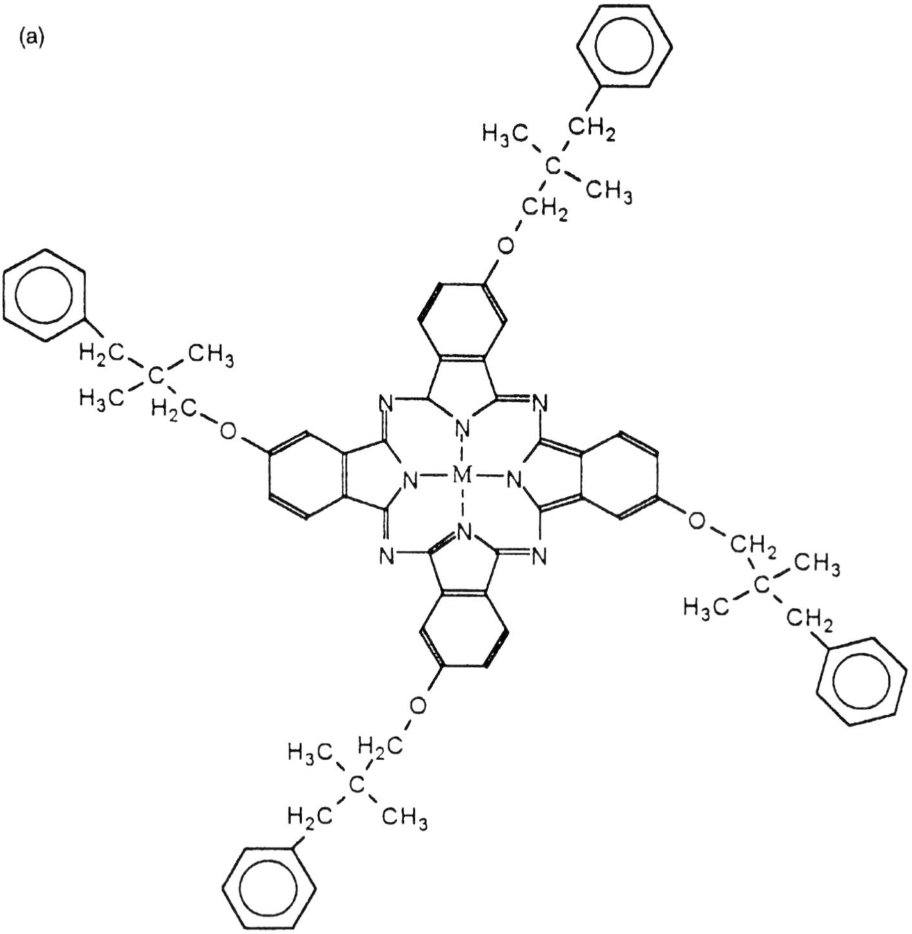

(a)

Figure 12.29. (a) Typical monomeric (H$_2$ and transition metal) phthalocyanines [tetra-(2,2-dimethyl-3-phenyl-propoxy)phthalocyanines] used as BAW coatings to monitor VOCs [35]. (b) Extramolecular cages are examples of a clathrate-type supramolecular bulk structure: covalently bound fullerene (C$_{60}$) derivatives are self-assembled in a multilayer thin film of a SAW.

halogens and AsF$_5$. These may be added intentionally by gas or electrochemical treatments. The diffusion rate (as characterized by the activation energy) and the degree of doping (as characterized by the free enthalpy in the bulk if compared with the gas phase) varies significantly for different molecular material/dopant systems. As activation energies of bulk diffusion are often larger than thermal energies and always decrease with increasing doping, drifts occur in the conductivity of highly doped materials whenever the concentration of the doping component is reduced to below its equilibrium value in the surrounding atmosphere of the electrolyte.

(11) The great advantage of narrow-bandgap materials, such as Lu(Pc)$_2$, is their high intrinsic conduction and hence low sensitivity to influences from contaminations [33]. Molecular orbital schemes are given in Figure 12.24 [34]. In Lu(Pc)$_2$ the intrinsic conduction occurs via the formation of a Lu(Pc)$_2^+$/Lu(Pc)$_2^-$ intermediate donor/acceptor complex. These donor and acceptor states of Lu(Pc)$_2$ can be prepared in homogeneous bulk material by gas or electrochemical treatment under reducing or oxidizing conditions. This removes or adds one electron from or to the HOMO-(J) state in Figure 12.24. This J peak appears in the typical UPS results of Figure 12.25 [33]. Incorporation of other

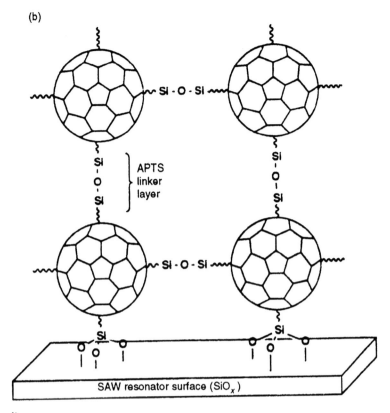

Figure 12.29. (*continued*)

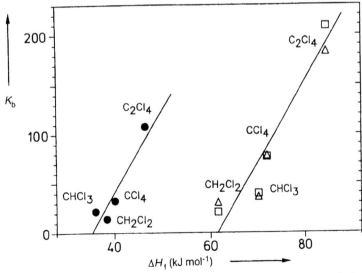

Figure 12.30. Correlation between the sensor effect (characterized by the partition coefficient, K_b; for its definition, see Section 4) of a BAW transducer coated with different molecular cages (examples 1 (left) and 2 (right) in Figure 12.29c) and the theoretically calculated heat of complex formation, ΔH_f [7].

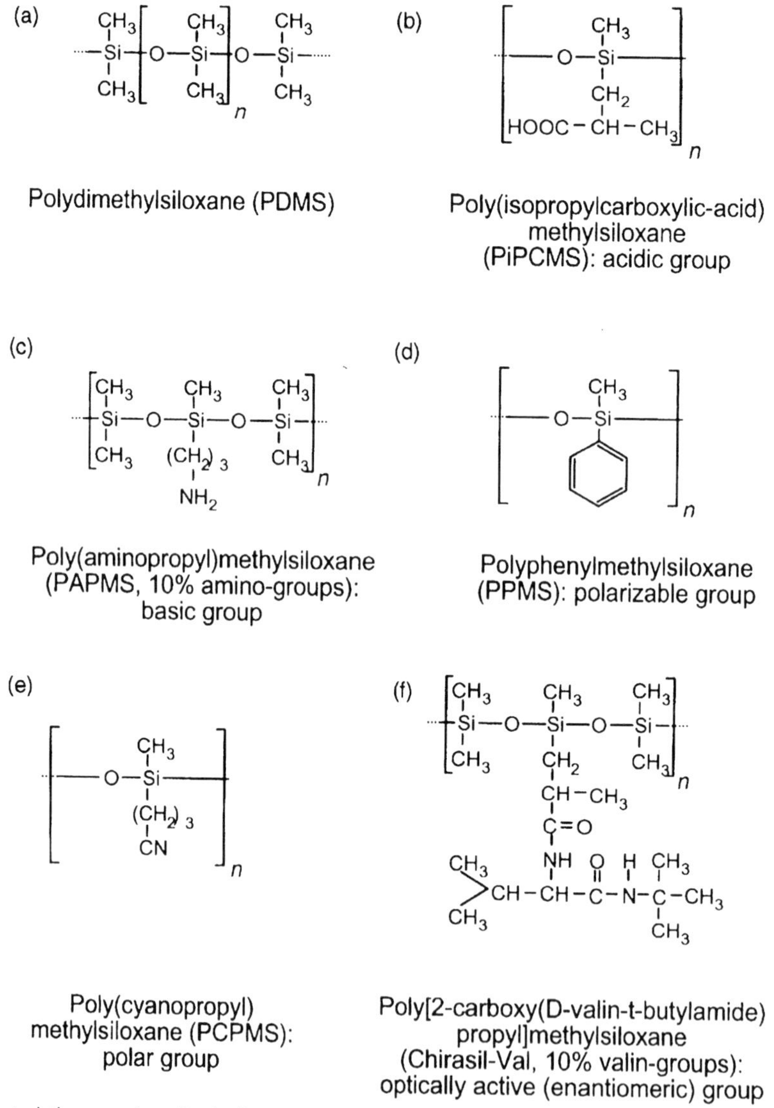

(a)

Polydimethylsiloxane (PDMS)

(b)

Poly(isopropylcarboxylic-acid)
methylsiloxane
(PiPCMS): acidic group

(c)

Poly(aminopropyl)methylsiloxane
(PAPMS, 10% amino-groups):
basic group

(d)

Polyphenylmethylsiloxane
(PPMS): polarizable group

(e)

Poly(cyanopropyl)
methylsiloxane (PCPMS):
polar group

(f)

Poly[2-carboxy(D-valin-t-butylamide)
propyl]methylsiloxane
(Chirasil-Val, 10% valin-groups):
optically active (enantiomeric) group

Figure 12.31. Characteristic examples of polysiloxanes and typical modifications used as chemically sensitive coatings: (a)–(f) simple modifications; (g) and (h) modification with sensor-active transition metals (here nickel and europium); and (i) modification by covalently bound intramolecular cages (here β-cyclodextrin to form Chirasil-DEX).

counter ions in Lu(Pc)$_2$ by electrochemical treatment makes it possible to adjust effective partial charges between -3 and $+3$, which correspond to different effective bandgaps between 0.2 and 1.2 eV and redox potentials between -1.92 and $+1.43$ eV. This is associated with characteristic electrochromic effects in the visible range and can be of use in electrochromic thin film displays based upon the partial ion conduction in Lu(Pc)$_2$, e.g. it makes it possible to dope reversibly this predominantly electron-conducting material.

(12) The identification of bulk crystallographic structures particularly of thin films is essential in crystalline materials. These structures must be known to identify the conductivity and mobility tensors in homogeneous molecular materials. As well as the usual X-ray or electron diffraction, corresponding small-angle diffraction or LEED and a variety of other techniques may be used. As one example, HREELS and FTIR results and a subsequent internal vibrational mode analysis, corrected for the influence from external

(g) Ni-Polymer:

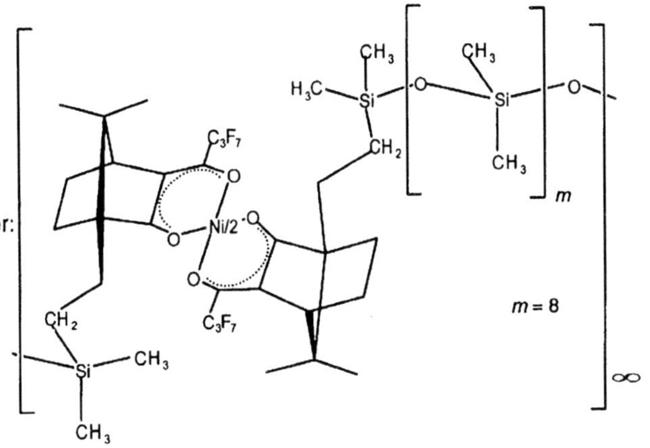

(h) Eu-Polymer:

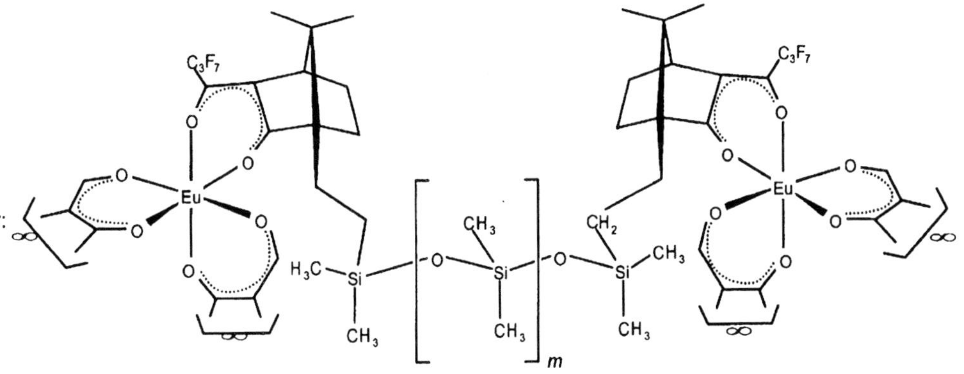

(i)

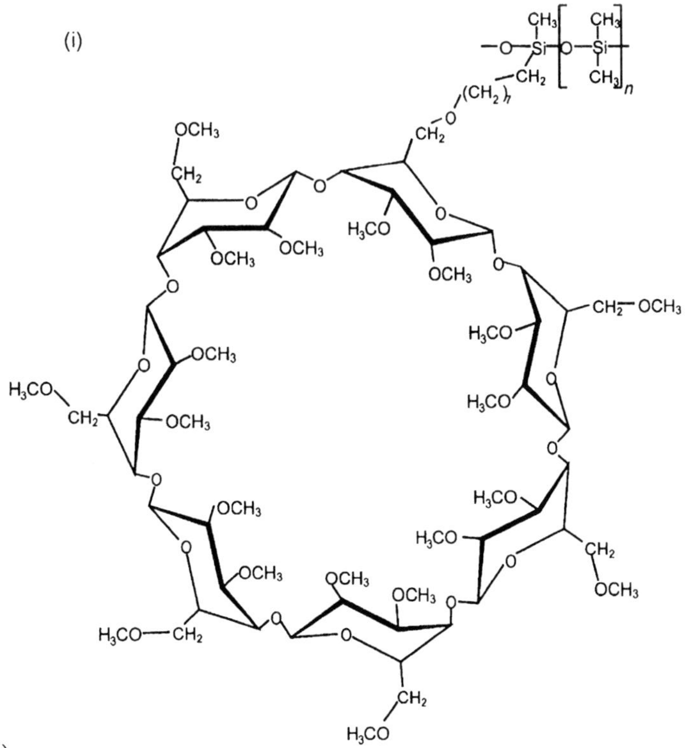

Figure 12.31. (*continued*)

modes, may be used to obtain detailed structural information.

(13) The next problem concerns the influence from real structures of molecular materials, often polycrystalline, which show a variety of defects and grain boundaries, and which hence may not be easily contacted for electrical measurements. All of these irregularities determine the overall transport properties in general, and the electronic properties in particular. The intentional avoidance or control of these irregularities is an extremely difficult experimental task. As a result, the conductivity and mobility tensors of most molecular materials are not known yet. The recent developments of microcontacts, the preparation of ideally structured materials, and the application of local probe spectroscopies (see, e.g. Figure 12.16) now aim to solve these problems more systematically.

(14) The next example, illustrated in Figure 12.26, concerns the characterization of binding energies of organic molecules within the molecular cage as compared to their binding energies in suitably chosen 'non-sensing' spacer groups (from *mass spectrometric thermal desorption (TDS)* experiments) [17].

(15) The last example concerns the optimization of Knudsen-evaporated thin films of calixarenes towards reproducible interaction energies at the surface and in the bulk with minimum activation energies of bulk diffusion (from TDS experiments). Results of the latter are in line with mass-sensitive quartz microbalance (QMB) results on the uptake of C_2Cl_4 at room temperature (Figure 12.27).

6 EXAMPLES OF CHEMICAL SENSORS

6.1 Sensing by bulk absorption

The following 'prototype materials' exhibit different bulk absorption mechanisms.

- Molecular crystals of PbPc monitor oxygen by *conductivity changes* due to electronic charge transfer from the highest occupied molecular orbital of the conjugated π-system of PbPc to form O_2^-. This leads to *p*-type conduction (Figure 12.4) in the partly filled π-band [11]. Sidechain-modified phthalocyanines exhibit negligible electrical conductivity, but better solubility in organic solvents (Figure 12.29a). The latter can be monitored in the gas phase with bulk acoustic wave (BAW) devices by

studying *mass changes* in thin films of these materials [35].

- Liquid crystals of side chain-modified polymers of siloxanes monitor benzene by the host/guest interaction via *changes in the mean capacitance* due to the disturbed nematic order [36].

- Molecular liquids with dissolved dye molecules show *color changes* upon association with gas molecules because of electronic charge transfer interactions [7,37–40].

- Extramolecular or intramolecular cages of supramolecular structures [7,17,41–47] show *mass changes* upon host/guest interactions (Figure 12.29a and b). The latter are based on intermolecular, non-covalent forces. Small host molecules that form supramolecular structures may be treated as rigid cages ('solid state limit'). Their total energy is modified during the incorporation of guests. This leads to a certain concentration ('sensor effect') at a given pressure and temperature (see, e.g. Figure 12.30). The heat of complex formation may be calculated in the force field approximation (see Section 4). Other sensor parameters monitored to date with these materials include *changes in optical absorption* [48] and *dielectric constants* [7].

- In contrast to these small host molecules the larger supramolecular structures (and also polymers, see Figure 12.31) offer time-fluctuating and spatially fluctuating interaction sites for gas molecules ('induced fits') [6,20,49–52].

- For polymers, this 'liquid state limit' to describe chemical sensing by 'induced fits' may be treated in the molecular dynamics approach (see Section 4). As typical phenomena, fast bulk diffusion (Figure 12.5a) and less selective absorption are observed in the polymers if compared to the situation in highly ordered films of rigid supramolecular structures. The sensor effects may be monitored by *changes in the mass* [6,20,35,41,42,49–56], *conductance* [11,37], *capacitance* [6,57] or *optical properties* [57–59]. Selectivities to detect certain VOCs may be tailored by adding specific groups which lead to specific local interactions (Figure 12.31a and b).

- Polymers modified with intramolecular cages combine two advantages, i.e. the fast bulk diffusion coefficients of gas molecules in the open network of the polymers and the usually higher selectivity of rigid supramolecular cages (see the example in Figure 12.31c). An additional advantage for practical applications is that contaminations with high-molecular weight in the gas phase do not contaminate the sensor-active sites in the bulk of the polymeric structure.

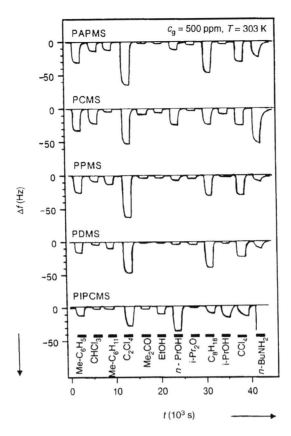

Figure 12.32. Typical time-dependent frequency variations monitored with a BAW device for different VOCs and organic polymers (compare Figure 12.31 a–e) with poly(aminopropyl)methylsiloxane (PAPMS), polycyanomethylsiloxane (PCMS), polyphenylmethylsiloxane (PPMS), polydimethylsiloxane (PDMS), poly(iso-propylcarboxylic acid)methylsiloxane (PIPCMS) [19,34].

The most reliable results for practical gas sensing particularly of VOCs have been obtained to date with polymer sensors. Empirical studies have been started from well-established polymeric materials which, from other applications, are known to be stable in the long-term. As a typical example, the results of modified polysiloxanes (Figure 12.31) interacting with organic solvent molecules are shown in the Figures 12.32 and 12.33. Characteristic deviations from the simple logarithmic relationship ($\ln[K_b] \sim T_b/T$, compare Section 4) occur whenever characteristic functional groups are introduced into the polymer. The latter may induce new and specific molecular interactions, such as induced dipole–induced dipole, dipole–induced dipole and dipole–dipole, or interactions due to different basicity and acicity. Some data have been presented already in Figure 12.7. In order to characterize

unequivocally the influence of different functional groups on the sensitivity of different polymers, the partition coefficients of different individual sensors in a sensor array are normalized. The typical results in Figure 12.34 demonstrate that modified polysiloxanes with their different functional groups may be used to advantage in sensor arrays which monitor mass changes with quartz microbalance oscillators. This makes it possible to identify different individual gases or even gas mixtures with their different molecules. To improve the performance of such sensor systems for specific applications, other modifications of polymers may be envisaged, e.g. by introducing new functional groups [20,43,44,60] (see, e.g. Figure 12.31g and h), optically active enantiomeric sites (see e.g. Figure 12.31f) and labels for optical transducers [58], or supramolecular cages (see, e.g. Figure 12.31i).

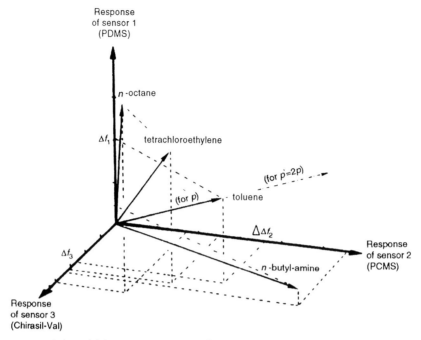

Figure 12.33. Frequency variations, Δf_i, i.e. sensor responses of three differently coated BAW sensors upon exposure to four different VOCs. The vector orientation (given by the ratio of the different axis intercepts) is characteristic for the VOC. Over a large variation of gas concentrations the length of the vector is, to a good approximation, proportional to its concentration [19,34].

6.2   Sensing by surface reactions

The π-conjugated conducting phthalocyanines like PbPc, CuPc, or $(AlPcF))_n$ are particularly suitable for gas sensing applications. Extremely low detection limits for NO_2 ($p_{NO_2} \approx 1$ ppb) and for Cl_2 ($p_{Cl_2} \geqslant 200$ ppb) by PbPc-based conductance sensors have been reported [9,10]. Small cross-sensitivites are reported for H_2S, NH_3, SO_2, H_2, CH_4 and CO.

Typical results of the time-dependent conductances G (proportional to $\sigma_\square$ as determined in Section 4) during exposure to NO_2 in air are shown in Figure 12.35 [11]. They indicate an acceptor-type NO_2 chemisorption at the surface which increases the defect electron concentration in the bulk PbPc. The existence of a surface effect was deduced from the thickness-dependent chemisorption measurements with thin film PbPc sensors (compare Figure 12.4). The detection mechanisms of NO_2 and O_2 by surface and bulk reactions, respectively, make it possible to optimize the sensor's sensitivity by choosing thick or thin layers. By choosing two sensors with different

thicknesses, cross-sensitivites between these two molecules can be eliminated mathematically.

PbPc has even been applied in room-temperature NO_2 sensors. Their preparation requires a well-defined annealing procedure of thin PbPc films [61]. This leads to drastic variations of the surface composition (inorganic oxides are formed) and hence of the microstructure with a formation of small and stable crystallites. Achievable response and decay times are of the order of 10 s and 2 min, respectively.

Fast and selective chemical sensing is expected for monolayer thin film sensors with well-oriented recognition centers at the surface. Typical examples for monolayer coatings of suitable transducers include

- monomeric and oligomeric phthalocyanines prepared by Knudsen evaporation or the Langmuir–Blodgett technique [62];
- organosilanes (e.g. *N,N*-dimethylaminotrimethylsilane) covalently coupled from solution on quartz substrates to monitor solvent vapors with surface

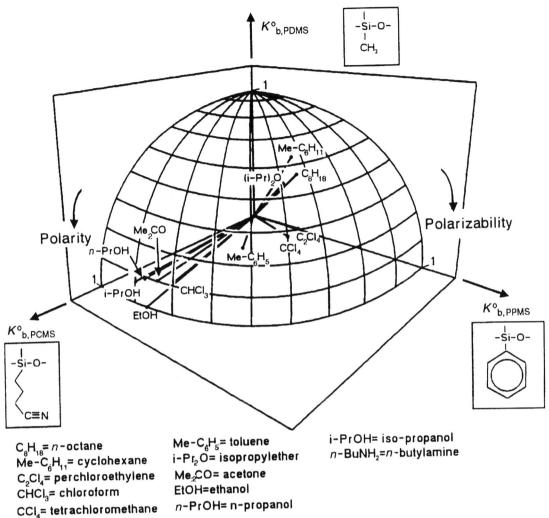

C$_8$H$_{18}$= n-octane
Me-C$_6$H$_{11}$= cyclohexane
C$_2$Cl$_4$= perchloroethylene
CHCl$_3$= chloroform
CCl$_4$= tetrachloromethane

Me-C$_6$H$_5$= toluene
i-Pr$_2$O= isopropylether
Me$_2$CO= acetone
EtOH=ethanol
n-PrOH= n-propanol

i-PrOH= iso-propanol
n-BuNH$_2$=n-butylamine

Figure 12.34. Selectivities of three modified polysiloxanes to monitor the different organic VOCs listed in the lower part. The three different partition coefficients, K_b, obtained for one gas with the three different sensors were normalized to K_b^0 as indicated in the lower part in order to make direct comparisons of the orientations possible in one figure. Polarizabilities and polarities of modifying side groups in these polymers change the selectivities in a characteristic manner (compare Figure 12.31). This is a prerequisite for the development of different sensors to be used in sensor arrays for an analysis of gas mixtures [19,34].

acoustic wave (SAW) devices (see results in Figure 12.28) [63];

- substituted tetraazacyclophanes ('Stetter-type') with a covalently attached lipid layer to monitor perchloroethylene with a SAW device [63];
- cyclodextrin films prepared by self-organization from solution to monitor chloroform with a SAW device [63];
- various cyclodextrins and calixarenes non-covalently coupled, prepared by Knudsen evaporation onto

quartz substrates to monitor a variety of VOCs with BAW devices; and

- resorcinarenes attached by self-organization through covalent links of sulfur groups on Au(111) substrates to monitor VOCs (see Figure 12.3) [17].

Published data of the first three studies illustrated qualitatively the suitability of monolayer sensing, but has not yet led to reproducible results which fulfill the 'sss' criteria of practical devices.

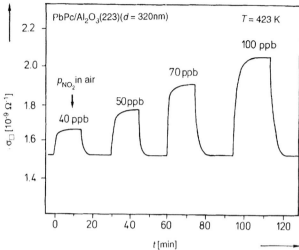

Figure 12.35. Typical time-dependent measurements of the surface conductivity of PbPc films during exposure to different partial pressures p_{NO_2} in air [12].

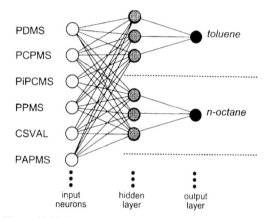

Figure 12.36. An example of pattern recognition with sensor arrays by means of neural networks: schematical drawing of the network architecture for the correlation of sensor signals using the polymers listed in Figure 12.31a–f with different VOC analyte concentrations. For details see ref. 56.

6.3   Sensing by controlled interface reactions

The only extensively studied prototype system makes use of controlled dc transport properties across interfaces to detect O_2 and NO_2 with a thin film sandwich-structure (PbPc/AgI) (Nernst-type) potentiometric sensor. As shown schematically in Figure 12.18, this device makes use of changes in electronic potential drops across inner interfaces for O_2 detection, and

measurements lead to details about chemical compositions, geometric structures, layer thicknesses, identification of mobile bulk species, i.e. free charged particles (indicated by circles) interface reactions with trapped ionic species (given in brackets: O_2^-, I^-, I_3^- NO_2^-), positions of occupied and empty electronic states including the valence band edge, E_V, conductance band edge, E_C, the bandgap, E_g, work function, ϕ, and Fermi level, E_F. For different operation modes, these details have been determined under UHV (a) and different atmospheric gas conditions (b) by means of XPS, UPS and HREELS [in (a)] as well as electrical measurements [in (a) and (b)]. The valence band discontinuity between PbPc and AgI in particular was determined from UPS (see Figure 12.17). For simplification, a few mobile bulk species are not shown in this drawing (for instance J_x^- with $x = 3$ in PbPc). This sandwich structure operates as an electrochemical potentiometric cell. The cell voltage, U, is determined by the partial pressures p_{O_2} and p_{NO_2} in the Nernst equation, as described in the lower part of Figure 12.18.

7   TRENDS

- *UHV, glove box, and related clean-room technologies* make it possible to systematically tailor the interface properties of ultraclean molecular materials in general and of π-conjugated materials. An atomistic understanding of controlled electron and ion transport parallel and perpendicular to the

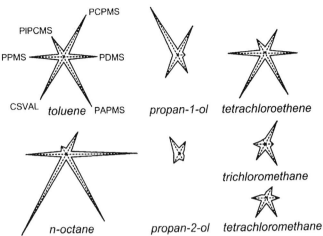

Figure 12.37. Star symbol plot for the partial sensitivities of the sensor array in Figure 12.36 to different organic solvent vapors. The polymer coatings PDMS, PCPMS, PIPCMS, PPMS, SCVAL (Chirasil-Val), PAPMS (for abbreviations compare Figure 12.31a–f) are separated by angles of 60° (dashed lines). The corners of the stars mark the partial sensitivities of these materials to 800 ppm of toluene: *n*-octane, propan-1-ol, propan-2-ol, tetrachloroethene, trichloromethane, and tetrachloromethane. The star symbol for tetrachloroethene is scaled to $^1/_5$ of its real size. For PPMS, the partial sensitivity to propan-1-ol and propan-2-ol was not measured during calibration [56].

surface normal of extended sandwich structures is the key to many practical applications of chemical sensors now and in the future.

- The *optimum film thickness* of the sensors is determined by the compound and its specific sensing mechanisms. Thick films are preferred for sensors based on bulk absorption and thin monolayer films for sensors based on surface reactions. For bulk absorption sensors, the diffusion of molecules must be sufficiently fast.

- The principal limits of miniaturization of sensors are currently checked by modifying the tips of either STM or related SXM setups by different chemical manipulations. Modified tips may also be used to manipulate molecules or atoms. Modifications of this experimental approach might lead to 'electronic noses' monitoring individual molecules. An important task in this context is to determine the minimum number of molecules required to produce an electronic signal with sufficient signal-to-noise ratio [64]. This is a prerequisite to identify the ultimate limits in the miniaturization of any chemical or physical sensor.

- *Sensor arrays* are of increasing importance. Different stable sensor elements are used which show high selectivities and sensitivities towards the detection of different molecules. An 'electronic-nose approach' with data analysis from sensor arrays and subsequent pattern recognition will make it possible to char-

acterize gas mixtures quantitatively or odors qualitatively [56,65–70] (see Figures 12.36 and 12.37).

- The miniaturization of transducers requires a more detailed understanding of sensing mechanisms. The results of these efforts will then lead to even cheaper and more complex sensor arrays. Thus it will be possible to collect a huge amount of data in parallel in real time by making use of many different host/guest interactions.

- The characterization of 'patterns' generated in such sensor arrays with their differently coated transducers will lead to the function-determined characterization of existing and the function-determined synthesis of new molecular structures.

8 REFERENCES

1. W. Göpel, *Sensors and Actuators* B **18**, 1 (1994).
2. J.M. Lehn, *Angew. Chem.* **100**, 91 (1988).
3. D.J. Cram, *Angew. Chem.* **100**, 1041 (1988).
4. F. Diederich, *Angew. Chem.* **100**, 372 (1988).
5. W. Göpel, State and Perspectives of Research on Surfaces and Interfaces, Report for DG XII, Commission of the European Community, Luxemburg, Report EUR 13108 EN, 1990.
6. K.D. Schierbaum and W. Göpel, *Synth. Met.* **61**, 37 (1993).
7. F.L. Dickert, U.P.A. Bäumler and G.K. Zwissler, *Synth. Met.* **61**, 47 (1993).

8. F. Vögtle, *Supramolekulare Chemie*, B.G. Teubner, Stuttgart 1992.

9. C. Maleysson, M. Passard, V. Battut, J.P. Germain, A. Pauly, V. Demarne, A. Grisel, C. Tiret and R. Planade, *Sensors and Actuators B* **26–27**, 144 (1995).

10. J. Souto, J.A. de Saja, M.I. Gobernado-Mitre, M.L. Rodriguez and R. Aroca, *Sensors and Actuators B* **15–16, 306 (1993).**

11. H. Mockert, D. Schmeißer and W. Göpel, *Sensors and Actuators* **19**, 159 (1989).

12. W. Göpel and K.D. Schierbaum, Specific molecular interactions and detection principles, in *Sensors—A Comprehensive Survey, Vol. 2, Chemical and Biochemical Sensors* (Part I) (eds. W. Göpel, J. Hesse, J.N. Zemel), VCH-Verlag, Weinheim, 1991.

13. W. Göpel and P. Heiduschka, *Biosensors and Bioelectronics* **9**, iii (1994).

14. W. Göpel, *Biosensors and Bioelectronics* **10**, 35 (1994).

15. W. Göpel, *Sensors and Actuators B* **4**, 7 (1991).

16. W. Göpel and K.D. Schierbaum, Definitions and typical examples, in *SensorsA Comprehensive Survey, Vol. 2, Chemical and Biochemical Sensors* (Part I) (eds. W. Göpel, J. Hesse, J.N. Zemel), VCH-Verlag, Weinheim, 1991.

17. K.D. Schierbaum, T. Weiss, E.U. Thoden van Velzen, J.F.J. Engbersen, D.N. Reinhoudt and W. Göpel, *Science* **265**, 1413 (1994).

18. K.D. Schierbaum, A. Gerlach, W. Göpel, W.M. Müller, F. Vögtle, A. Dominik and H.J. Roth, *Fresenius' Z. Anal. Chem.* **349**, 372 (1994).

19. K.D. Schierbaum and W. Göpel, Selective chemical sensing: molecular recognition with cage compounds and polymeric permselective layers, in *Polymer Films and Sensor Technologies* (ed. G. Harsányi), Technomic, Lancester, 1994.

20. J.W. Grate and M.H. Abraham, *Sensors and Actuators B* **3**, 85 (1991).

21. V. Gutmann, *The Donor–Acceptor Approach to Molecular Interactions*, Plenum Press, New York (1978).

22. A. Dominik, H.J. Roth, K.D. Schierbaum and W. Göpel, *Supramolecular Science* **1**, 11 (1994).

23. W. Göpel, Interface engineering of molecular materials, in *Nanostructures based on Molecular Materials* (eds W. Göpel and Ch. Ziegler), VCH, Weinheim, 1992, p. 349.

24. W. Göpel, *Sensors and Actuators* **16**, 167 (1989).

25. M. Henzler and W. Göpel, *Oberflächenphysik*, Teubner, Stuttgart, 1991.

26. D. Bonnell, *Scanning Tunneling Microscopy*, VCH, Weinheim, 1993.

27. A. Rager, B. Gompf, L. Dürselen, H. Mockert, D. Schmeißer and W. Göpel, *J. Molecular Electronics* **5**, 227 (1989).

28. D. Schmeißer and W. Göpel, *Ber. Bunsenges. Phys. Chem.* **97**, 372 (1993).

29. P. Bätz, D. Schmeißer and W. Göpel, *Phys. Rev. B* **43**, 9178 (1991).

30. H.D. Wiemhöfer, D. Schmeißer and W. Göpel, *Solid State Ionics* **40/41**, 1009 (1990).

31. M. Schreck, M. Abraham, W. Göpel and H. Schier, *Surf. Sci. Lett.* **237**, 405 (1990).

32. D. Schmeißer, U. Langohr, J.U.V. Schütz, H.C. Wolf and W. Göpel, *Synth. Met.* **41–43**, 1805 (1991).

33. J. Bufler, Ch. Ziegler and W. Göpel, *Synth. Met.* **61**, 127 (1993).

34. W. Göpel, *Sensors and Actuators B* **24–25**, 17 (1995).

35. K.D. Schierbaum, R. Zhou, S. Knecht, R. Dieing, M. Hanack and W. Göpel, *Sensors and Actuators B* **24–25**, 69 (1995).

36. F.L. Dickert, G.K. Zwissler and E. Obermeier, *Ber. Bunsenges. Phys. Chem.* **97**, 184 (1993).

37. C. Reichardt, P. Milart and G. Schäfer, *Liebigs Ann. Chem.* 441 (1990).

38. V.R. Belosludov, M.Y. Lavrentiev and Y.A. Dyadin, *J. Incl. Phenom.* **10**, 399 (1991).

39. A. Ehlen, C. Immer, E. Weber and J. Bargon, *Angew. Chem., Int. Ed. Engl.* **32**, 110 (1993).

40. S.A. Bourne, L. Johnson, C. Marais, L.R. Nassimbeni, E. Weber, K. Skobridis and F. Toda, *J. Chem. Soc., Perkin Trans. 2* 1707 (1991).

41. F.L. Dickert and O. Schuster, *Adv. Mater.* **5**, 826 (1993).

42. F.L. Dickert, A. Haunschild and V. Maune, *Sensors and Actuators B* **12**, 169 (1993).

43. M.A.F. Elmosalamy, G.J. Moody, J.D.R. Rhomas, F.A. Kohnke and J.F. Stoddart, *Analytical Proceedings* **26**, 12 (1989).

44. F.H. Kohnke, A.M.Z. Slawin, J.F. Stoddart and D.J. Williams, *Angew. Chem., Int. Ed. Engl.* **26**, 892 (1987).

45. P.D. Beer, Molecular and Ionic Recognition by Chemical Methods, in *Chemical Sensors* (ed. T.E. Edmonds), Blackie and Son, Glasgow, 1988.

46. D.Q. Li and B.I. Swanson, *Langmuir* **9**, 3341 (1993).

47. P.R. Ashton, N.S. Isaacs, F.H. Kohnke, A.M.Z. Slawin, C.M. Spencer, J.F. Stoddart and D.H. Williams, *Helv. Chim. Acta*, in preparation.

48. F.H. Kohnke, J.P. Mathias and J.F. Stoddart, *Adv. Mater.* **1**, 275 (1989).

49. E. Dalcanale and J. Hartmann, *Sensors and Actuators B* **24**, 39 (1995).

50. C. Déjous, D. Rebière, J. Pistré, C. Tiret, R. Planade and M. Hommady, *Sensors and Actuators B* **24**, 58 (1995).

51. F. Benmakroha, R. Boudjerda, R. Boufenar, A. Fouwuki and J.J. McCallum, *Conf. Proc. 5th Int. Meeting on Chemical Sensors*, Rome, 1994.

52. A.J. Ricco, C. Xu, R.E. Allred and R.M. Crooks, *Conf. Proc. 5th Int. Meeting on Chemical Sensors*, Rome (1994).

53. Y. Tomita, M.H. Ho and G.G. Guilbault, *Anal. Chem.* **51**, 1475 (1979).

54. T.E. Edmonds and T.S. West, *Anal. Chim. Acta* **117**, 147 (1980).

55. C.S.I. Lai, G.J. Moody, J.D.R. Thomas, D.C. Mulligan, J.F. Stoddart and R. Zarzycki, *J. Chem. Soc., Perkin Trans. 2* 319 (1989).

56. A. Hierlemann, U. Weimar, G. Kraus, G. Gauglitz and W. Göpel, *Sensors and Materials* **7**, 13 (1994).

57. K.D. Schierbaum, M. Haug, W. Nahm, G. Gauglitz and W. Göpel, *Sensors and Actuators B* **11**, 383 (1993).

58. G. Gauglitz and G. Kraus, *Fresenius' J. Anal. Chem.* **346**, 572 (1993).

59. Y. Sadaoka, M. Matsuguchi, Y. Sakai and Y.U. Murata, *Chem. Lett.* 53 (1992).

60. R. Zhou, M. Haug, K.E. Geckeler and W. Göpel, *Sensors and Actuators B* **15–16**, 312 (1993).

61. Y. Sadaoka, T.A. Jones and W. Göpel, *J. Material Science Lett.* **8**, 1095 (1989).

62. J. Souto, J.A. de Saja, M.I. Gobernado-Mitre, M.L. Rodriguez and R. Aroca, *Sensors and Actuators B* **15–16**, 306 (1993).

63. F.L. Dickert and A. Haunschild, *Adv. Mater.* **5**, 887 (1993).

64. W. Göpel, *Nanostructured Sensors for Molecular Recognition, Proc. of NATO Advanced Research Workshop*, Kluwer, Cambridge, 1994.

65. M. Otto, T. George, C. Schierle and W. Wegschneider, *Pure Appl. Chem.* **64**, 497 (1992).

66. U. Weimar, S. Vaihinger, K.D. Schierbaum and W. Göpel, Multicomponent analysis in chemical sensing, in *Chem. Sensor Technol., Vol. III* (ed. N. Yamazoe), Kodansha, Tokyo, 1991, p. 51.

67. T.C. Pearce, J.W. Gardner, S. Friel, P.N. Bartlett and N. Blair, *Analyst*, 118 (1993).

68. P.N. Bartlett and J.W. Gardner, Odour sensors for an electronic nose, in *Sensors and Sensory Systems for an Electronic Nose*, Kluwer, Dordrecht, 1992, pp. 31–51.

69. F. Davide, C. Di Natale and A. D'Amico, *Sensors and Actuators B* **18–19**, 244 (1994).

70. F. Davide, C. Di Natale, A. D'Amico, W. Göpel and U. Weimar, *Sensors and Actuators B* **18–19**, 654 (1994).

CHAPTER 13

Photoelectric Conversion by Polymeric and Organic Materials

Masao Kaneko

Ibaraki University, Mito, Japan

1   INTRODUCTION

Conductive polymers and organic molecules can be applied to photoelectric conversion devices which convert photon energy into electricity. Commercial photoelectric conversion devices, important in our social life and industry, are mostly made of inorganic semiconductors; they are used as solar batteries to produce electricity from solar irradiation and as various sensors to convert light information to an electrical signal. These inorganic semiconductor-based devices are limited by the kind of inorganic compounds; consequently the development of new devices with respect to new functions and high cost–performance is also restricted. Many polymeric and organic materials are suitable for use in photoelectric conversion devices and they are easy to fabricate into devices. Thus they are promising for future photoelectric conversion. A

considerable amount of research has been carried out on these materials in the past two decades. These studies are aimed at creating new devices for photonics in the 21st century [1], and also low-cost solar batteries as a renewable energy source. The latter is particularly significant in respect to the protection of the environment [2]. Although the application of these polymeric and organic materials causes problems in relation to their efficiency, reliability and long-term stability, they are potentially suitable for practical use in the future.

In this chapter photoelectric conversion by polymeric and organic materials is reviewed. Non-conducting as well as conducting materials are important for fabricating such devices, so both are described. Since the scientific principle for photoelectric conversion is important for designing the devices, it will be mentioned briefly in the next section. There are two major principles for photoelectric conversion: one

Handbook of Organic Conductive Molecules and Polymers: Vol. 4. Conductive Polymers: Transport, Photophysics and Applications.
Edited by H. S. Nalwa. © 1997 John Wiley & Sons Ltd

utilizes junctions of semiconducting materials such as *pn* or Schottky, and the other is based on the photochemical process which produces electrical output from a photochemical reaction. The following sections are divided according to these principles. Section 3 describes photoelectric conversion by solid-state junctions of semiconducting polymeric and organic materials, Section 4 details liquid-junction-type devices, Section 5 covers inorganic semiconductors in combination with polymeric and organic films and Section 6 describes photochemical processes for photoelectric conversion.

2 PRINCIPLES OF DESIGNING PHOTOELECTRIC CONVERSION DEVICES AND FUNDAMENTAL STUDIES

There are two principles of photoelectric conversion (light-to-electricity): one utilizes photophysical pro-

cesses at semiconductor (SC) junctions such as *pn* and Schottky, and the other is based on photochemical processes between sensitizer (photoexcitation center) and acceptor and/or donor (Table 13.1). For both the most important point is the separation of the photoproduced charges (electrons and holes).

In the photophysical devices, junctions serve this purpose. A *pn* junction is formed when *p*-type and *n*-type SCs are contacted. When they are contacted, electrons flow from the *n*-SC to *p*-SC making the Fermi level come to the same height, which induces bending of the valence band (VB) and the conduction band (CB). Irradiation of photons onto the interface between the *p*- and *n*-SCs induces electron transition from VB to CB (thereby a hole is produced in the VB). Because of the band bending at the *p/n* interface (depletion layer), the electrons in CB and the holes in VB formed by the photon irradiation diffuse in different directions thus accomplishing charge separation. The photoseparated charges flow to an outer circuit through conducting

Table 13.1. Principles for photoelectric conversion

Principal processes	Photophysical processes		Photochemical processes
	Band mechanism	Exciton mechanism	
Excitation by irradiation (interaction of the photons and the electric field of a substance, and the subsequent change of the electron energy from the ground to the excited states)	Electron transition from the valence band to the conduction band	Exciton generation	Electron transition from the ground to the excited states
Charge separation		Ion pair formation by ionization of the exciton, applied potential or reaction with impurities	Electron transfer reaction of the excited state with an acceptor or donor
Transport of the separated charges	Transport of the separated charges through the gradient of the electric field		Successive electron transfer reaction with the second donor or acceptor
Output (1) Electricity	Electron transport between the electrodes and the separated charges		
(2) Chemical energy	Catalytic reduction and oxidation reactions by the separated charges		
Representative schemes			

E_f, Fermi level; P, photoreaction center; D, donor; A, acceptor.

electrodes attached on the *p*- and *n*-SC surfaces to produce electricity.

A Schottky junction is formed between *p*- (or *n*-SC) and metal with a proper Fermi level similar to the *pn*-junction (see also Table 13.1). An electrolyte solution can also be used instead of a metal, i.e. *p*- or *n*-SC is dipped in an electrolyte solution to form a kind of Schottky junction; it is called a liquid junction in this case. Such a liquid junction is especially useful for achieving photoinduced chemical reaction on the SC surface (mentioned in Sections 4 and 5). In order to achieve high conversion efficiency, it is important to suppress charge recombination.

There can be a big difference between inorganic and organic SCs in the mechanism of carrier generation. In inorganic SC, the lattice force is strong and an electron/ hole pair is formed directly by light absorption. In organic SC, the force between the molecules is weak (*c.* 0.1 eV at the most) and perturbation of the energies by defects or impurities is small, so that absorbed light induces only the formation of trapped excitons of a Frenkel type and direct carrier generation does not take place. The mechanism of carrier generation is always a matter for argument in organic semiconductors. When an exciton is formed in organic SCs by irradiation, this exciton is ionized into an electron and hole by internal electric field or by applied potentials.

Practical photoelectrical conversion devices all utilize photophysical processes at semiconductor junctions. However, photochemical processes can also bring about photoelectrical conversion if one or both of the photochemical reaction product(s) exchange(s) electrons and holes at separate electrodes. In a typical example, electron donor (D), sensitizer (photoexcitation center; P), and electron acceptor (A) produce photochemically oxidized D and reduced A via photoexcitation of P as shown in equations (13.1) and (13.2).

$$D + P + A + h_\nu \rightarrow D + P^* + A$$
$$\rightarrow D + P^+ + A^-$$
$$\text{(oxidative quenching)}$$
$$\rightarrow D^+ + P + A^- \qquad (13.1)$$

$$D + P + A + h_\nu \rightarrow D + P^* + A$$
$$\rightarrow D^+ + P^- + A$$
$$\text{(reductive quenching)}$$
$$\rightarrow D^+ + P + A^- \qquad (13.2)$$

If the three components are arranged unidirectionally by utilizing multilayered polymer membranes, Langmuir–Blodgett membranes, etc. it is possible to obtain electricity at the outer circuit when the products, D^+

and A^-, exchange the charges at the applied electrodes. In the photochemical process it is also possible to have two-component systems for the conversion, i.e. D and P, or P and A, instead of the D–P–A system. It should be noted here that it is especially important for the photochemical process to suppress the charge recombination which is usually predominant in photochemical events.

In the photochemical process, the redox potentials of the components (for D and A at ground state, and for P at ground as well as at excited states), and the kinetics for the forward and backward reactions are important factors in the design.

Photoelectric conversion described in this chapter is classified in Table 13.2 according to the type and material of the devices. Details can be found in the corresponding sections.

For cell fabrication, transparent conductive glass such as indium tin oxide-coated glass (ITO) and tin oxide-coated glass (NESA) are often used. Polymers are coated onto the substrate by casting its solution, simple dipping, spin-coating, or by electropolymerization to form films. Dyes are often layered by vapor deposition or coated as a film by mixing with polymers. In solid-state cells, metal electrodes are formed by vapor deposition as a thin layer so that it is semitransparent in the incident light, which leads to a sandwich-type cell typically of the configuration: transparent electrode/SC layer(s)/semi-transparent metal electrode.

The characteristics of photoelectric conversion devices are typically represented by Figure 13.1. To argue the efficiency of the device, open circuit photovoltage (V_{oc}), short circuit photocurrent (I_{sc}), fill factor (FF) and conversion efficiency are of importance. The fill factor is defined by equation (13.3) (see also Figure 13.1).

$$FF = (\text{real maximum output } V_{max} \times I_{max})/(V_{oc} \times I_{sc})$$
$$(13.3)$$

A typical inorganic SC solar battery gives a FF of around 0.75. There are two definitions of the conversion efficiency:

Engineering conversion efficiency ($Y\%$)
$$= (\text{output energy/total incident energy}) \times 100 \quad (13.4)$$

Quantum efficiency (Q)
$$= \text{photon numbers effectively used/photon}$$
$$\text{numbers absorbed}$$
$$= \text{output energy/absorbed energy.} \qquad (13.5)$$

In a practical device $Y\%$ is usually used, and Q is used often in fundamental study. To determine Q, mono-

Table 13.2. Photoelectric conversion devices composed of polymeric and organic materials

Type of devices	Typical examples
Photophysical devices	
Polymeric and organic SC	
Solid-state { *pn*-junction	$p\text{-}(CH)_n\text{-}/n\text{-}(CH)_n(PA)$
	$p\text{-}(CH)_n\text{-}/n\text{-}CdS$
	Perylene/M–phthalocyanine
Schottky junction	Phthalocyanine/Al
Liquid junction	Polypyrrole (PP)/liquid
	Polyaniline (PAn)/liquid
	Polythiophene (PT)/liquid
	M–phthalocyanine/liquid
Inorganic SC with polymeric and organic films	
Stabilization of small bandgap SC	$n\text{-}Si/PP/liquid$
	$n\text{-}GaAs/polymer\text{–}Ru\ (bpy)_3{}^{2+}$
Sensitization of large bandgap SC	$n\text{-}TiO_2/Ru$ complex
Organic films with SC particles	$(CdS + Pt)/Nafion$
Photochemical devices	
Photoredox systems	$Thionine/Fe^{2+}$
Non-photoredox systems	$Ru(bpy)_3{}^{2+}/MV^{2+}$
	$Ru(bpy)_3{}^{2+}/Prussian\ Blue$
	M–phthalocyanine/O_2
Other photochemical and photobiological systems	*cis*/*trans* isomerization
	Bacteriorhodopsin

chromatic light is usually used. Q is larger than $Y/100$ in a general case since it is only based on the absorbed light.

Conventional tungsten, tungsten halogen or a xenon lamp is used for visible light, and a mercury lamp for ultraviolet light. The spectrum of the xenon lamp is similar to solar irradiation so it is often used for studying solar battery in laboratories. AM 1 is a standard light intensity (100 mW cm^{-2}); it represents the average solar intensity on the sea surface when the sun irradiates at right angles. AM 2 is a light intensity of 75 mW cm^{-2}. It should be noted that in polymeric/organic devices the photoelectric conversion efficiency is, in many cases, strongly dependent on light intensity. Weak light usually gives a high conversion efficiency.

3 SOLID-STATE DEVICES OF POLYMERIC AND ORGANIC SEMICONDUCTORS

Most of the practical photoelectric conversion devices, especially solar batteries, are made of inorganic SCs. The solar battery is the most likely candidate to replace fossil fuels as an energy source in the next century. Thus, the Earth's environment would be protected against the green house effect, acid rain, etc. A problem for this objective is the high cost of SC production. After the so-called oil crisis in 1973, polymeric and organic SCs received much attention in the search to produce low-cost photoelectric conversion devices. This is because devices based on these materials can be produced at low costs. However, no good polymeric and organic material can achieve a high conversion

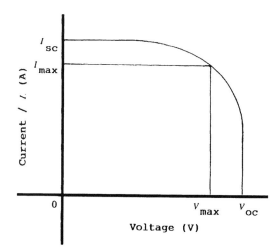

Figure 13.1. Current–voltage characteristics of a photoelectrical conversion cell. The maximum output of the cell equals $V_{max}I_{max}\ (= P_{max})$, and $FF = P_{max}/V_{oc}I_{sc}$.

efficiency; one of the reasons was the low conductivity of polymeric and organic SCs in former times.

The creation of new materials is also important for the future development of photoelectric conversion devices. However, the types of inorganic SCs are very much limited. Polymeric/organic compound-based SC is also attracting attention in this respect.

In the next section photoelectric conversion by solid-state devices made from polymeric and organic SCs is reviewed.

3.1 Semiconducting polymers

When it was reported that acetylene can be polymerized as a film [3] and that the formed polyacetylene (PA; **1**) can show high conductivity by doping with donor or acceptor [4], PA attracted much attention as a candidate to produce a high cost–performance solar battery.

$$\{CH=CH\}_n \qquad (PA; \ 1 \ )$$

p-Type PA is formed by doping with an acceptor, such as halogens (Cl_2, Br_2, I_2 etc.), Lewis acid, AsF_5, etc. and a n-type is formed by doping with a donor, such as alkali metals. A pn-junction was obtained by the p- and n-type PA films. Since the dopants migrate with each other at the p/n interface in a homo pn-junction of PA, the stability is a problem.

Hetero pn-junction devices have been studied which combine inorganic SC with PA. An n-Si/p-PA device exhibited $V_{oc} = 0.53$ V, $I_{sc} = 18.18$ mA cm^{-2} and $FF = 0.32$ under 72 mW cm^{-2} light intensity with engineering conversion efficiency (Y) = 4.3% [5]. The Schottky-junction device, Al/p-PA, gave $V_{oc} = 0.3$ V, $I_{sc} = 6$ µA cm^{-2} under 3.5 mW cm^{-2} irradiation with $Y = 0.3$% [6]. The output of such PA-based devices is, at the most, of the order of V_{oc} $c.$ 0.8 V, $I_{sc} = 20$ mA cm^{-2}, $FF = 0.5$ and $Y = 7$%, which is inferior to inorganic SC-based devices.

Since PA is not soluble in solvents, it can not be processed as a film after synthesis. Solution-processible poly-$(CH_3)_3$Si-cyclooctatetraene has been prepared [7]. This polymer film was formed on a n-doped silicone by casting its solution, and was then doped with iodine to fabricate a pn-heterojunction solar cell. Under 9 mW cm^{-2} irradiation, it gave $V_{oc} = 0.5$ V, $I_{sc} = 0.4$ mA cm^{-2} with $Y = 1.5$%.

Migration of dopants at the homojunction pn-PA interface is one of the problems of obtaining stable

characteristics. By sodium ion implantation a stable pn-junction PA was fabricated in high density PA [8]. Considerable numbers of papers have appeared on the formation, structure, basic characteristics, electric conductivity and pn-junctions of PA. Another major problem about PA is its stability against O_2 oxidation. It degrades rapidly under air; thus the characteristics are changed drastically. Because of its simple −CH− based structure, PA is still a material for fundamental studies, but its practical application does not seem to be easy. Electronic properties of polypyrrole (PP; **2**)/n-Si heterojunctions [9a], polyaniline (PAn; **3**)/Al [9b] and poly(3-octylthiophene)/Al [9c] Schottky junctions have been reported, but their photoelectrical behavior was not given.

$(PP; \ 2 \ )$

$(PAn; \ 3 \ )$

ITO/polythiophene (PT; **4**)/Al Schottky-junction cell gave $V_{oc} = 1.07$ V and $I_{sc} = 1.35$ µA cm^{-2} [10]. The poly(p-phenylene) (PPP; **5**)/n-Si pn-heterojunction cell showed a conversion efficiency of 3.2% at a light intensity of 7 mW cm^{-2} and $c.$ 1% under AM 1 illumination [11a]. The PP/n-Si cell gave $V_{oc} = 0.29$ V, $I_{sc} = 9$ mA cm^{-2}, $FF = 0.46$ and $Y = 1.24$% under AM 1 [11b].

$(PT; \ 4 \ )$

$(PPP; \ 5 \ )$

A single poly(3-methylthiophene) (PMT) film was doped at first with cation and then with anion to give a pn-homojunction diode film [11c]. The cell, Al/pn-PMT/Au, exhibited $V_{oc} = 0.23$ V, $I_{sc} = 160$ nA cm^{-2} and $FF = 0.30$.

Some other conductive polymers have been used for obtaining photovoltaic response. Poly(N-vinylcarba-

(PVCz; **6**), a photoconducting material, was transformed into an electrically conducting polymer by electrochemical doping of ClO_4^-.

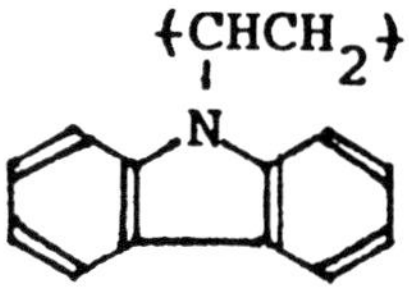

(PVCz; 6)

The doped PVCz (c. 10 μm thick) was sandwiched between semi-transparent aluminum and gold metals by vacuum deposition of the metals onto the front and back surface of the film. When it was illuminated from the aluminium side with 366 nm monochromatic light (transmittance 4.5%), it generated photocurrent: $V_{oc} = 1.0$ V, $I_{sc} = 182$ nA cm^{-2}, $FF = 0.237$ and $Y = 0.028\%$ with 1.08 mW cm^{-2} incident light [12].

The conducting complex of poly(ethyleneoxide) (PEO; **7**) with sodium polyiodide was sandwiched between ITO and platinum electrodes and then illuminated through ITO to give photovoltaic effect [13].

$\{CH_2CH_2O\}_n$ (PEO; 7)

A sandwich cell was fabricated with NESA glass and iodine complex of poly(2-vinylpyridine) (P2VP; **8**); NESA/P2VP–I_2/Pt. Values of V_{oc} around 92 mV and $I_{sc} = 100$ μA cm^{-2} were obtained with $Y = 0.05\%$ at 2 mW cm^{-2} illumination [14].

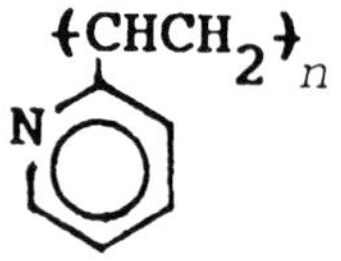

(P2VP; 8)

A layer of chlorophyll (Chl) was contacted with wet poly(vinyl alcohol) (PVA; **9**) and they were sandwiched between two gold electrodes to give the device, Au/Chl/PVA/Au [15]. Values of $V_{oc} = 130$ mV, $I_{sc} = 27$ nA, $FF = 0.39$ and $Y = 0.01\%$ were reported under 743 nm light. A kind of Schottky junction was suggested between p-type Chl and wet PVA.

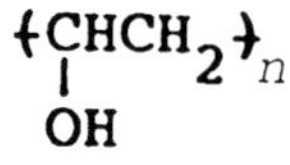

(PVA; 9)

3.2 Organic dyes

Since many organic dyes are semiconducting, they can form Schottky-type junctions with metals, especially metals with low work functions. The typical dyes used for this purpose were metal-free phthalocyanine (H_2Pc; **10**), metal phthalocyanine (MPc; **11**), porphyrins (P; **12**), metal–tetraphenylporphyrins (MTPP; **13**), merocyanines (MC; **14**), coumarins (CM), etc. These dyes were also contacted with inorganic n-SCs to form pn heterojunctions.

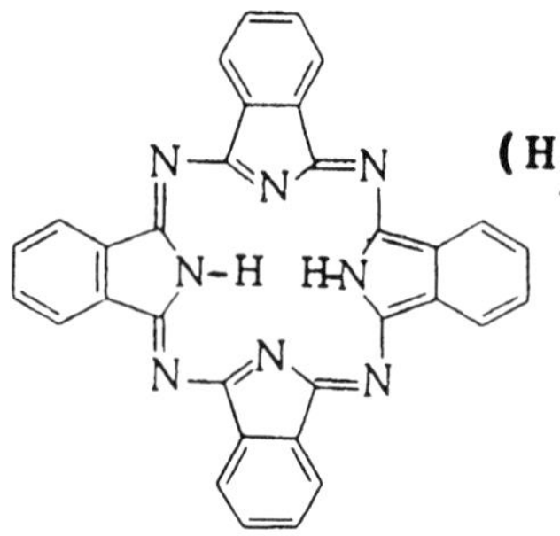

(H$_2$Pc; 10)

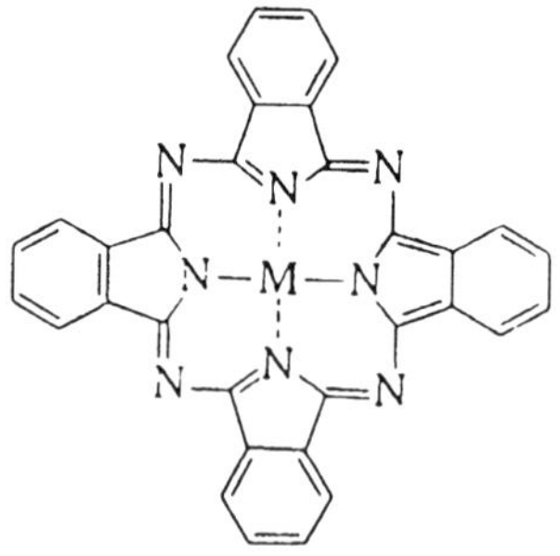

(MPc; 11)

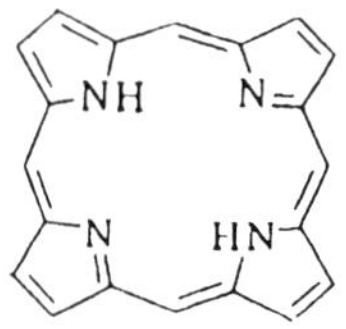

(P; 12)

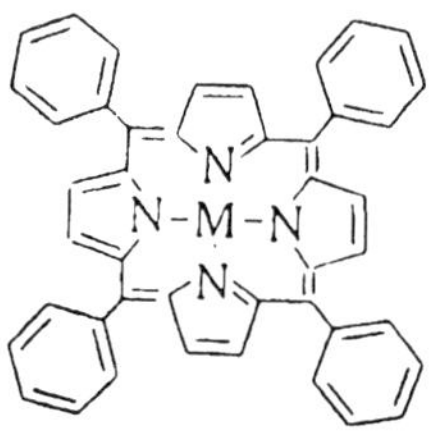

(MTPP; 13)

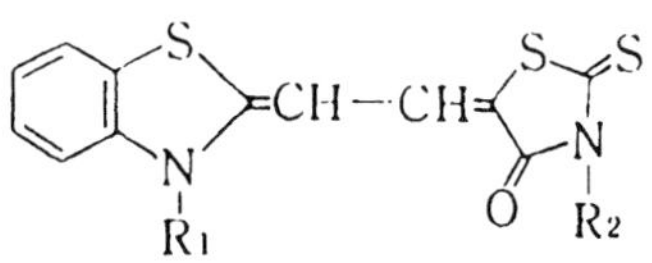

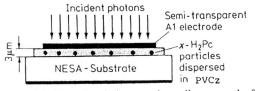

(MC; 14)

When such dyes are dispersed in a polymer film, the conversion characteristics can be enhanced. x-Crystal form H_2Pc was dispersed in a PVCz **6** film and sandwiched between NESA and aluminium electrodes; it worked as a Schottky-type photovoltaic cell [16]. See Figure 13.2 for the cell configuration. The dark and photo I–V curves are shown in Figure 13.3 for the cell, NESA/x-H_2Pc(PVA)/Al. This cell gave $V_{oc} = 0.86$ V, $I_{sc} = 1.4$ μA cm^{-2}, $FF = 0.33$ and $Y = 6.5\%$ (based on the transmitted light intensity through the semi-transparent aluminium electrode) under 6 μW cm^{-2} light at 670 nm. But when the data were extrapolated to the 100 mW cm^{-2} irradiation intensity of 670 nm light, the calculated value of Y decreased to 0.02%. This decrease of conversion efficiency at higher input light intensity was attributed to a limitation in I_{sc}. The decrease of conversion efficiency is a problem in this kind of cell. Another problem is the absorption of light by the semi-transparent metal electrode. Since the Schottky junction is formed at the interface of the dye and the metal, the light must be illuminated from the metal side, which causes transmission of the incident light to reach the interface only partly.

The polymer-dispersed dye gave much higher conversion efficiency than the neat phthalocyanine, e.g. the efficiency was optimal at 60 wt % x-H_2Pc in a PVA film. This could be ascribed to facilitate charge separation from the excitons formed by irradiation in the microenvironment provided by the polar polymer matrix.

The action spectrum for the device, In/H_2Pc, shown in Figure 13.4, shows evidently that excitation of the phthalocyanine induces the photoeffect [17]. The p-type phthalocyanine forms a blocking contact at aluminium with a relatively low work function as

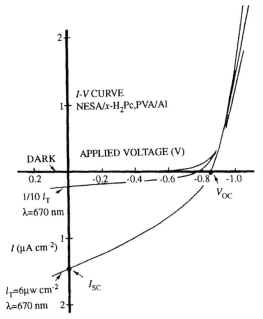

Figure 13.3. Dark and photo current–voltage curves of the NESA/H_2Pc(PVA)/Al cell. I_T is the incident light intensity of 670 nm. (Reproduced by permission of the American Institute of Physics from ref. 16.)

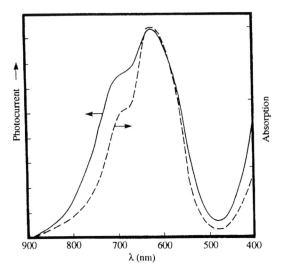

Figure 13.2. Photoelectrical conversion cell composed of a polymer film containing dispersed phthalocyanine. (Reproduced by permission of the American Institute of Physics from ref. 16.)

Figure 13.4. Action and absorption spectrum of the phthalocyanine cell. (Reproduced by permission of the National Research Council of Canada from ref. 17.)

shown in Figure 13.5. After the contact, electrons flow from aluminium into the SC until the Fermi levels of the two materials are equal. This results in a depletion of holes and a distribution of negative charge inside H_2Pc. The charge distribution extends a distance d from the contact and defines the barrier width. The ohmic contact is provided by conductive materials such as ITO or NESA, and by other metals with a relatively high work function such as gold or platinum at the another surface.

The cell, Sn/AlPcF/Au, exhibited $V_{oc} = 320$ V, $I_{sc} = 32$ μA cm^{-2}, $FF = 0.3$ and $Y = 0.4\%$ under 0.76 mW cm^{-2} irradiation of the light of wavelength greater than 480 nm [18]. The effects of binder polymers have been investigated on the performance of a In_2O_3/polymer (1 to 2 μm thickness), H_2Pc/Al sandwich cell [19]. Polymers with polar groups such as poly(vinylidenefluoride) and poly(vinylidenechloride) gave higher efficiency. This was ascribed to the enhancement of exciton dissociation due to a large electric field formed by the polar groups. For poly(vinylidenechloride) $V_{oc} = 0.46$ V, $I_{sc} = 490$ nA cm^{-2}, $FF = 0.34$ and $Y = 0.77\%$ were reported based on the transmitted 617 nm light of 10 μW cm^{-2}.

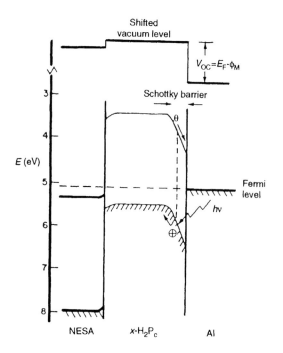

Figure 13.5. Energy diagram of the valence, conduction and Fermi level (E_F) of the components after contact. ϕ_M is the work function of the metal (Al). (Reproduced by permission of American Institute of Physics from ref. 16.)

intensity. Surfactant AlPc was electrodeposited to produce the cell, Al/AlPc/Ag, which gave 0.7% efficiency at 638 nm [20]. This method is useful for preparing a large area device in contrast to the vapor deposition of neat pigment compounds.

Thinner dye layers favor the conversion efficiency; thicker layers tend to cause charge recombination. The addition of acceptors leads to a higher conversion efficiency by increasing the conductivity. Higher temperature also increases the efficiency by 4 to 5% K^{-1} due to higher concentration of carriers.

Other dyes such as MTPP **13** [21], MC **14** derivatives [22], and CM [23] have also been used to fabricate Schottky-type devices with metals. Correlations between the cell performance and the molecular structure of the dyes have been studied, and the desirable properties of the dyes discussed [23]. A flexible photovoltaic cell was fabricated by vapor deposition of MC dye on a transparent polyester film coated with ITO [24]. As for the carrier generation in this kind of Schottky devices, the excitons that migrate to the Schottky barrier region are dissociated into an electron and a hole. Therefore, the barrier thickness and the exciton migration distance are important factors. In an Al/porphyrin/Au device, the exciton migration distance was estimated as 15 nm for H_2TPP and 3 nm for ZnTPP; the barrier thickness wad 39 nm for H_2TPP and 14 nm for ZnTPP [25]. A different mechanism for carrier generation, which was dependent on the dye, was also reported. In the cell, Al/TPP or ZnTPP/Au, the electron and hole are generated through an exciplex formed by singlet exciton and an acceptor for TPP, but for ZnTPP direct carrier generation following light absorption was suggested [26]. The efficiency under AM 1 sunlight was 0.1%.

Photovoltaic behaviors were reported for the cells, Ag/triphenodioxazine(TPDO)/Al and Al/6,13-dichloro-triphenodioxazine (DCITPDO)/ITO [27]. The carriers are produced at the interface, Ag/TPDO or DCITPDO/ITO, and both pigments are n-type SCs. The values are $V_{oc} = 0.52$ V, $I_{sc} = 91$ nA cm^{-2}, $FF = 0.16$ and $Y = 3.7 \times 10^{-6}\%$ under 200 mW cm^{-2} white light irradiation.

The effect of molecular oxygen (O_2) on these cells is one of the discussion points. The effect of O_2 on the surface photovoltage of NiPc and CuPc has been studied [28]. The observed effects were accounted for with a model which involves the transfer of charge from the Pc to the adsorbed O_2. When O_2 is adsorbed on NiPc, there are two forms: one is irreversible and the other is reversible. The reversibly adsorbed O_2 can be removed by evacuating the ambient O_2, while the irreversibly adsorbed O_2 can only be removed by heating to 433 K [29]. The irreversibly adsorbed O_2 caused a one order of magnitude increase in the photovoltage, which was ascribed to an increased quantum efficiency of minority carrier injection. In the cell, Al/MgTTP/Ag, the presence of O_2 and water was necessary, thus suggesting that a metal/insulator/semiconductor junction (MIS type) of Al/AlO$_x$/MgTPP is effective for the device [21]. It is suggested that the cation radical of porphyrin formed photochemically can oxidize aluminium with water (a kind of corrosion) producing photovoltage. The effect of O_2 should still be studied in these devices.

Some examples, other than the Schottky-type cell, have been reported using organic dyes. The symmetrical cell, ITO/liquid crystal porphyrin (LCP, zinc octakis(beta-octyloxyethyl)porphyrin)/ITO, induced photoeffect [30]. The illuminated electrode acts as a photoanode, and stable photocurrents of c. 0.4 mA cm^{-2} have been obtained under 150 mW cm^{-2} white light. This photovoltaic effect was interpreted as resulting from exciton dissociation at the illuminated electrode leading to the preferential photoinjection of electrons into the ITO and holes into the porphyrin. The predominance of the photoinjection process in these liquid crystal films was attributed to the single-crystal-like character of the LCP films. The heterojunction cell, ITO/n-CdS/Chl a/Ag, produced $V_{oc} = 0.3$–0.4 V, $I_{sc} = 150$–200 nA cm^{-2}, $FF = 0.26$ and $Y = 0.17\%$ under 20 µW cm^{-2} intensity of 740 nm light [31]. The use of CdS instead of aluminium has led to an improvement in the performance, which was attributed to the elimination of the insulating layer of Al_2O_3 existing at the Al/Chl a interface.

Since many organic dyes are p-conducting, this has been a limitation for the fabrication of pn-junction cells by the use of only organic dyes. n-Type perylene tetracarboxylic derivative (PV; 15) was first used to form pn-junctions with copper phthalocyanine (CuPc; 16). The two-layer organic photovoltaic cell, ITO/CuPc/PV/Ag, exhibited a high conversion efficiency of $V_{oc} = 450$ mV, $I_{sc} = 2.3$ mA cm^{-2}, $FF = 0.65$ and $Y = 0.95\%$ under AM 2 illumination [32]. The I–V characteristics of the cell is shown in Figure 13.6. This conversion efficiency was remarkably high in comparison with the known organic photovoltaic cells under white light hitherto reported. Consequently, it evoked much attention and expectation in relation to organic solar cells. The fill factor, as well as the efficiency, is much higher than the hitherto known organic devices. This was attributed to the charge generation efficiency,

which is relatively independent of the bias voltage. The action spectrum showed that the absorption by both the compounds is effective for the photophysical event. Both the CuPc and PV layers will generate excitons, which dissociate into electrons and holes at the interface. The electrons are preferentially transported in the PV layer, while the holes are preferentially transported in the CuPc layer. The effectiveness of the exciton dissociation was assumed to be associated with a high built-in field of unknown origin, perhaps a dipole field or a field due to trapped charges at the interface. If the strength of this built-in field is sufficiently high (around 10^6 V cm^{-1}), the carrier generation efficiency would be virtually saturated and is determined by the diffusion of excitons into the interface region. Superposition of a small external voltage would then moderately affect the carrier collection, thus leading to a weak bias dependence of the collection efficiency.

(PV; 15)

(CuPc; 16)

A three-layered organic cell with an interlayer of codeposited pigments of n-type perylene tetracarboxylic derivative (Me-PTC; 17) and p-H_2Pc in between the respective pigments gave twice as much photocurrent when compared to the double-layered cell without the interlayer [33]. Typical data for the cell, ITO/Me-PTC/Me-PTC-H_2Pc/H_2Pc/Au, are $V_{oc} = 0.51$ V, $I_{sc} = 2.14$ mA cm^{-2}, $FF = 0.48$ and $Y = 0.7\%$ under AM 1 illumination.

It was suggested that the charge is photogenerated via the exciplex of H_2Pc and Me-PTC formed in the

(Me-PTC; 17

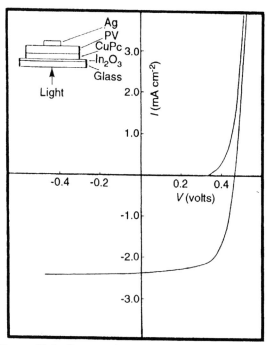

Figure 13.6. Current–voltage characteristics of an ITO/CuPc(25 nm)/PV(45 nm)/Ag cell under AM 2 illumination. (Reproduced by permission of the American Institute of Physics from ref. 32.)

3.3 Solid polymer electrolytes

As described in Section 4, semiconductors form a liquid junction (a kind of Schottky junction) when dipped in an electrolyte solution. This type of cell is easy to fabricate, but the stability of inorganic SCs, especially with small bandgap to passivate or dissolve in solution, is a problem. In order to overcome this problem, polymer solid electrolyte can be used for photoelectric conversion in combination with inorganic SCs to form a kind of Schottky-type junction. The advantages are the ease of fabrication, the stability of the inorganic SC material, and the easy fabrication of multispectral thin film cells. Alkali ion conducting PEO (7) was used to fabricate the cell, p-InP/PEO-NaSCN, Na_2S, S/n-CdS [36]. Photopotentials up to 540 mV and 0.3 μA currents are obtained under AM 1 illumination. Recombination of photogenerated carriers is a problem. This type of cell was applied to the storage of photogenerated electrochemical energy. The cell, p-InP/Nafion-porphine/Nafion-Ru(bpy)$_3^{2+}$/n-CdS (where bpy stands for bipyridine), showed more than 1 V photopotential with a charge current of 30 μA cm^{-2} under AM 1 conditions [37,38], where Nafion (Nf; **18**) is a sulfonated perfluoroalkyl polymer with cation exchange ability.

$$ \left(CF_2 - CF \left(CF_2 - CF_2 \right)_1 \right)_n $$

$$ \begin{array}{c} O \\ | \\ CF_2 \\ | \\ CFCF_3 \\ | \\ O_m \\ | \\ CF_2CF_2SO_3H \end{array} $$

(Nf; 18)

The n-Si/(KI + I$_2$)PEO/ITO cell was fabricated [39]. In order to improve the rates of charge transfer between the SC and the redox couples in the polymer film, the surface of the SC was modified. The cell, n-Si/Pt/PP (**2**)/(KI + I$_2$)PEO/Pt/ITO, gave $V_{oc} = 0.75$ V and $I_{sc} = 1.5$ mA cm^{-2} under AM 1 illumination [40]. Such surface modification of the SC overcame large activation energy barriers to allow efficient charge transfer between the SC and the redox ions in the solid polymer electrolyte. The efficiency of the cell is limited by a high surface recombination associated with the

interlayer, and that built-in potential is effectively provided in the interlayer forming a p-i-n like junction [34].

The photovoltaic property of an organic SC modulated by an exciton-dissociating film was studied [35a]. In the cell, ITO/N,N'–diphenyl-N,N'-ditolyl benzidine (TPD) dispersed in poly(vinylpyridine) film/perylene-bis(phenethylimide) (PPEI)/In, the exciton fluorescence is quenched by the diamine film. The presence of the film inverts the photovoltaic effect relative to that in its absence, and causes the front electrode to become more photoactive than the back (Ag); this was interpreted as resulting from interfacial exciton dissociation.

Photodiodes containing microscopic pn-junctions of organic dyes were fabricated using the Langmuir–Blodgett (LB) technique. The LB film, composed of surfactant MC (**14**) dye and surfactant viologen, showed photoinduced transient voltage response to impulsive laser light irradiation [35b]. The signal was assignable to the photoinduced electron transfer from n-dye (viologen) to p-dye (MC).

surface states of the SC. A tandem photovoltaic cell, n-CdS/(Na$_2$S$_4$)PEO/p-CdTe, gave $V_{oc} = 625$ mV, $I_{sc} = 35$ μA cm^{-2} and $FF = 0.25$ under AM 1 [41].

The cells, CdS/poly(vinylferrocene) (PVF; **19**)/Au and CdS/PP/Au, have been fabricated [42]. It was important for the performance that the redox couple is in a partially oxidized state.

$$\{CHCH_2\}_n$$

Fe   (PVF; 19)

A typical result of the latter cell when PP is partially oxidized is $V_{oc} = 0.34$ V, I_{sc} around 0.6 mA cm^{-2} and $FF = 0.56$ under 200 mW cm^{-2} illumination. The performance was controlled by the extent of the oxidation state of the polymer.

4 LIQUID-JUNCTION DEVICES OF POLYMERIC AND ORGANIC SEMICONDUCTORS

When SC is dipped in an electrolyte solution, a kind of Schottky junction called a liquid junction is formed at the SC/liquid interface, which brings about photo-electric conversion. The conversion efficiency is usually not high in these devices, but a variety of photo-responsive devices are fabricated easily based on this principle, which has evoked much interest. Films of semiconducting polymers, metal complexes and organic dyes are used to fabricate this type of devices by immersing a film-coated electrode into a solution, typically aqueous containing electrolyte. In most cases only fundamental photoresponsive properties have been studied. The characteristics of these devices as solar cells have not been studied much.

4.1 Semiconducting polymers

Semiconducting polymers, such as PA (**1**), PP (**2**), PAn (**3**), PT (**4**) and PPP (**5**) can be photoresponsive when they are simply dipped in an electrolyte solution. In most cases aqueous solution is used.

When PA film was dipped into an aqueous sodium polysulfide solution, it gave $V_{oc} = 0.3$ V and $I_{sc} = 40$ μA cm^{-2} under AM 1 [43]. A similar photoelectrochemical cell was fabricated with *trans*-PA by dipping it into an aqueous methylviologen (MV^{2+}; **20**) solution [44].

$$_3HC-N\overset{+}{\bigcirc}-\bigcirc\overset{+}{N}-CH_3 \quad (MV^{2+}; 20)$$
$$Cl^- \qquad\qquad Cl^-$$

Since electrochemical doping takes place in SC polymer films dipped in an electrolyte solution, the liquid-junction device composed of polymer SC films can be doped photoelectrochemically under irradiation.

p-Polymer $+ e^- + M^+ \rightarrow p$-polymer-M$^+$,

$$(M^+ = cation) \quad (13.6)$$

n-Polymer $+ h^+ + X^- \rightarrow n$-polymer-X$^-$,

$$(X^- = anion) \quad (13.7)$$

Photodoping of a PA film by dipping it in a CH$_3$CN solution of Et$_4$NClO$_4$ has been reported [45], where Et stands for the ethyl group. The PA was doped with Et$_4$N$^+$ in the potential range above -1.2 V upon irradiation with visible light. Photochemical doping of PP was carried out in methylenedichloride containing diphenyliodonium, which was hexafluoroarsenated upon irradiation with UV light [46].

Liquid-junction PP photoelectrochemical cells have been fabricated by dipping PP films into aqueous solutions of redox reagents such as MV^{2+}, hydro-quinone, oxygen, Cu^{2+}, I$^-$ and Br$^-$ [47]. Cathodic photocurrents were obtained at potentials more negative than 0 V vs. Ag with redox reactions on the PP films. Photocurrent generation by a liquid-junction PP film coated on ITO glass was studied [48]. A device, ITO/PP/Al, showed that the PP is p-type. The cell, ITO/PP film/aqueous electrolyte, gave a higher cathodic photo-current (45 μA cm^{-2}) than the undoped one, and a higher concentration of the electrolyte (LiClO$_4$) produced a larger photocurrent. Slow current changes induced by on and off of the incident light indicate that some slow process, probably doping/undoping, is involved in the photoelectrochemical event.

PT (**4**) also generated photocurrent by the device, ITO/PT film/aqueous electrolyte [49]. In the region more negative than the onset potential of 0.5 V vs. Ag/AgCl, it gave cathodic photocurrent up to 55 μA cm^{-2}. The photoelectrochemical response was rapid in this case. Semiconducting polymers have been used in combination with inorganic semiconductors for fabri-cating solar cells. A poly(phenylenesulfide) film was used for liquid-junction photoelectrochemical cells by dipping it into a CH_3CN electrolyte solution [50]. The presence of MV^{2+} in the solution induced a more stable photocurrent.

4.2 Metal complexes and organic dyes

Layers of metal complexes and organic dyes with semiconducting properties are used also for liquid-junction devices when the film is dipped in an electrolyte solution. The compounds used are typically phthalocyanines (**10**, **11**), porphyrins (**12**, **13**), etc. Redox couple often exists in the liquid phase. It is sometimes difficult to distinguish if a Schottky-type junction is effective for the photoelectric process or if excited state reactions of the metal complex (or dye) induce photoelectrochemical response.

Thin films of chlorogallium Pc, vanadyl Pc and other Pc s prepared by vapor deposition gave a photoelectrical response when dipped into an aqueous electrolyte solution [51–56]. The cell consists of GaPc-Cl thin film backed by a conductive metal substrate; in contact with an electrolyte solution it develops a photopotential which is determined by the difference in the Fermi potential of the metal and the electrode in the redox couple. Typical cyclic voltammograms in the dark and under illumination of the cell, Au/GaPc-Cl/hydro-quinone aqueous solution, are shown in Figure 13.7 [53]. The cell, Au/GaPc-Cl/Fe(CN)$_6^{3-/4-}$ aqueous solution/Ga-Pc-Cl/Pt, gave $V_{oc} = 0.55$ V, $I_{sc} = 0.1$ mA cm^{-2}, $FF = 0.55$ and $Y = 0.03–0.05\%$ under AM 1 illumination [54].

The photoelectrochemical properties of NESA/H$_2$Pc in contact with various redox couple aqueous solutions

were studied [57]. The performance of the cell is given in Table 13.3 in relation to the redox couple. For the Fe(CN)$_6^{4-/3-}$ couple, the cell gave $V_{oc} = 0.33$ V, $I_{sc} = 0.63$ mA cm^{-2}, $FF = 0.25$ and $Y = 0.07\%$ under AM 2 illumination. The low efficiency of the carrier collection was attributed to the high series resistance, surface recombination of the carriers and the slow charge exchange at the counter electrode.

In the Pt/MPc/quinone–hydroquinone aqueous solution cell, doping of the phthalocyanine with an electron acceptor such as *o*-chloranil improved the behavior to $V_{oc} = 49.5$ mV, $I_{sc} = 28$ μA, $FF = 0.33$ and $Q = 0.46$ [58]. A thin film of CuPc (**16**) coated on gold and dipped in an aqueous electrolyte responded to near-infrared light of 1100 nm [59]. Electrodeposited surfactant AlPc formed a solid-junction cell in contact with an aqueous solution containing Fe(CN)$_6^{3-/4-}$, benzoquinone/hydroquinone, I$_3^-$/I$^-$, Fe$^{3+/2+}$, Sn$^{4+/2+}$ or Fe(EDTA)$^{1-/2-}$ [60], where EDTA stands for ethylenediaminetetraacetic acid. V_{oc} was 0.13 V, I_{sc} increases as the sheet resistivity of the ohmic electrode decreases and FF was 0.25. The maximum Q is 0.04 at 630 nm and 0.26 mW cm^{-2}, and Y is *c.* 0.01% under AM 1 irradiation. Improvement of the photoelectrochemical reactivity of chloro-AlPc films by anion uptake and structural modifications were reported [61]. In I$_3^-$/I$^-$, V_{oc} of 110 mV and I_{sc} of 1.0 mA cm^{-2} were obtained under 35 mW cm^{-2} irradiation.

A cosensitization effect was observed based on *p–p* isotype Schottky junctions composed of ITO/ZnPc/ ZnTPP [62]. A *pn-* junction was formed between *p*-ZnPc and *n*-type tetra(4-pyridyl)porphyrin (TPyP), and the cell, ITO/ZnPc/TPyP/electrolyte solution, exhibited much higher photocurrent at the Soret band than the simple ITO/TPyP cell [63].

Various metal porphyrin compounds have also been used to fabricate liquid-junction devices [64–67]. Mn- and Cd-TPP (**13**) in contact with a Fe(CN)$_6^{3-/4-}$ aqueous solution gave $V_{oc} = 1$ V and $Q = 0.2$ under monochromatic light [68]. The photovoltage decreased in the order, Pd > H$_2$ > Zn > Cu > Mg > Pb > Ni > Co. The photoresponse of a VTPP-coated platinum electrode depends on the pH of the solution with which the metal complex is in contact. A cathodic photocurrent was obtained at lower pH, whereas anodic photocurrent is obtained at higher pH [69]. This was explained as the combined effect of band-bending at the Pt/VTPP interface and the adsorption of both hydrogen and hydroxy ions at the VTPP/solution interface.

When MgTPP is coated on ITO for a photoelectrochemical cell, it acts only as a photocathode. The

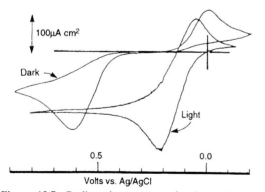

Figure 13.7. Cyclic voltammograms for the oxidation of 1 mM H$_2$Q in a pH 4 buffer in the dark and under illumination on GaPc–Cl/Au at a scan rate of 20 mV s^{-1}. (Reprinted with permission from ref. 53. Copyright 1984 American Chemical Society.)

Table 13.3. Effect of the redox couple formal potential on the performance of NESA/H$_2$Pc cathodes

Redox system	E° (V vs. SCE)	V_{ph} (V)	I_{sc} (μA cm^{-2})
Co(CN)$_6$$^{4-/3-}$	-0.83	0.44	80
Cr$^{3+/2+}$	-0.65	0.25	1.0
PQ^{2+}/PQ$^+$	-0.66	0.21	37
BQ/QH$_2$	-0.05	0.21	120
Fe(CN)$_6$$^{4-/3-}$	0.20	0.165	200
I$_2$/I$_3$$^-$	0.26	0.15	120
Fe$^{3+/2+}$	$+0.52$	0.15	30
Co(CH$_3$COCHCOCH$_3$)$_3$$^{3+/2+}$	1.84	0.24	8

ITO coated with ZnTPP, CuTPP or H$_2$TPP, which have oxidation potentials more positive than that of MgTPP, works as both a photocathode and a photoanode, depending on the electrode potential and the direction of the irradiation [70]. The cathodic photocurrent is attributed to the downward band-bending in the space charge layer at the TPP/electrolyte interface, and the anodic one to that at the ITO/TPP contact.

The complex film can be prepared by various methods, as well as by conventional vapor deposition or solution casting. Porphyrin thin films were prepared using the electrolytic micelle disruption method [71]. An aqueous mixture of non-ionic surfactant (11-ferrocenyl)-undecyltridecaethyleneglycol ether, metal porphyrin and LiBr was sonicated, stirred and centrifuged. The supernatant solution was electrolyzed at 0.5 V vs. SCE to form a film on an ITO electrode dipped in the solution. The resultant ITO/ZnTPP dipped in an aqueous solution of I$_3$$^-$/I$^-$ induced photocurrent in the quantum yield of 0.126 at 425 nm Soret peak wavelength; this was 20 times higher than that generated by a conventional vacuum sublimed film.

The orientation of porphyrins and phthalocyanines can be controlled by epitaxial vacuum evaporation onto a KCl crystal. After separating an oriented TPP film (150 nm thick) from KCl substrate in water, it was transferred onto an ITO electrode. When the ITO/TPP was dipped in an aqueous I$_3$$^-$/I$^-$ solution, the oriented film gave a photocurrent quantum yield three times as high as the polycrystalline film (see the action spectra in Figure 13.8) [72]. The enhancement was ascribed to the decrease of electrical resistivity due to the extending overlap between the oriented TPP molecules as well as to the low probability of electron–hole recombination at the grain boundaries and structural defects.

The photoelectrochemical behavior of Chl a dispersed in a poly(vinyl acetate) film was studied. When the film cast from dimethylsulfoxide (DMSO) solution

was immersed in water, it showed a transformation from a Chl a aggregate to the monomer or from a Chl a–DMSO aggregate to dimeric Chl a. The former induced photocathodic current, while the latter induced photoanodic current [73]. The cell, composed of a NESA/TPP photocathode and a NESA/Victorial Blue B photoanode gave $V_{oc} = 1$ V, $I_{sc} = 10$ μA and $Q = 0.01$ [74].

MC (**14**) derivatives coated on ITO formed aggregate, and gave cathodic photocurrent of $Q = 0.108$ under 550 nm light irradiation [75]. Photoelectrochemistry of a one-dimensional metal–dithiolene complex (**21**) thin film coated on a platinum electrode was studied [76,77].

$$\begin{bmatrix} NC & S & S & CN \\ & >\!\!=\!\!< M >\!\!=\!\!< & \\ NC & S & S & CN \end{bmatrix}^{2-} \qquad (21)$$

Electrochemically grown dithiooxamido Cu(II) (copper rubeanate) films on a copper electrode gave a cathodic photocurrent [78].

Bilayer lipid membranes (BLM) containing dyes can also induce photopotential when one surface is contacted with an electron donor and another surface with an electron acceptor. Photodriven (uphill) electron transfer took place across the lipid bilayers containing magnesium octaethylporphyrin (MgOEP) which separates aqueous ascorbate (donor) and ferricyanide (or MV^{2+}) solutions [79]; i.e. aqueous ascorbate solution/lipid bilayer(MgOEP)/aqueous MV^{2+} solution. The MeOEP cation formed under excitation at the acceptor side does not contribute to electron transport across the membrane, and the charge carrier is likely to be the neutral protonated MgOEP. The membrane transport appears to be rate limiting when the reaction is

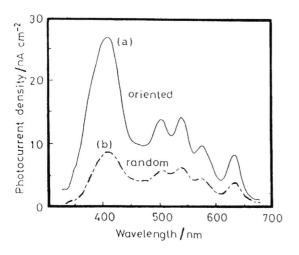

Figure 13.8. Action spectra of ITO/TPP/I$_3^-$, I$^-$/Pt cells using (a) epitaxial and (b) polycrystalline TPP films deposited on KCl and muscovite, respectively. (Reproduced by permission of the Chemical Society of Japan from ref. 72.)

photoinitiated and the interfacial reaction appears to be limiting when the reaction is photodriven. The magnesium porphyrin is both sensitizer and trans-membrane redox mediator. The induced photovoltage was of the order of 1 mV

5 LIQUID-JUNCTION DEVICES OF INORGANIC SEMICONDUCTORS IN COMBINATION WITH COATED POLYMERIC OR ORGANIC FILMS

UV light irradiation on n-titanium dioxide (TiO$_2$) SC photoanode dipped in an aqueous electrolyte solution generated photocurrent, and at the same time induced water cleavage, dioxygen evolving at the TiO$_2$ and dihydrogen at the counter platinum electrode [80]. This system is represented in Figure 13.9, where the n-TiO$_2$ forms a liquid junction at the solid/liquid interface with a depletion layer near the TiO$_2$ surface. The Fermi level at which the band is flat is called flat band potential (E_{fb}).

This report evoked much interest in utilizing such liquid-junction SCs to produce energy from sunshine and water [81]. However, because of its large bandgap of 3.0 eV, it can absorb only UV light of less than

400 nm wavelength, and the main part of the solar spectrum (visible region from 400 to 760 nms) can not be utilized by TiO$_2$. The bandgap (E_g) and the energy levels of the valence and conduction bands (VB and CB) are shown in Figure 13.10 with the redox potentials of proton reduction (H$^+$/H$_2$) and water oxidation (O$_2$/H$_2$O) at pH 7 vs. NHE (normal hydrogen electrode). In order to cleave water by visible light for H$_2$ and O$_2$ production, the bandgap should be below 3.0 eV, the lower edge of CB more negative than the H$^+$/H$_2$ level (-0.41 V), and the upper edge of VB more positive than the O$_2$/H$_2$O level (0.82 V). The candidate SCs able to satisfy these conditions are limited to n-type SCs, such as n-CdS, n-CdSe, etc. but these small bandgap n-SCs degrade readily under irradiation; they are either corroded and dissolved in water, or inactivated by the formation of oxide films on the surface. To overcome this problem, two approaches have been carried out as follows.

(1) Stabilization of a small bandgap SC against corrosion in water by coating with metal layers, polymer films or organic compounds.
(2) Sensitization of a UV region SC by coating with sensitizers (dyes).

In this section they are reviewed with additional systems composed of SC particles incorporated in membranes of polymer, LB and lipid bilayers.

The liquid-junction device has the merit of driving some chemical reactions on the SC surface and the counter electrode. If electricity generation is the aim of this device, an appropriate redox couple should be put in the liquid phase as shown in Figure 13.11. By reacting on both the SC and counter electrodes oxidatively and reductively, the redox couple transports charges across the liquid to result in photoelectrical conversion without the production of net photochemical products. This type of cell is called a photoregenerative cell since the redox couple is always regenerated at the SC and counter electrodes.

As a parameter to evaluate the junction characteristics, the junction ideality factor, n, is obtained from

$$V_{oc} \simeq (nkT/q) \ln (I_{sc}/I_o) \qquad (13.8)$$

where k is Boltzman's constant, q is the electron charge and I_o is the reverse saturation current density. When I_{sc} is proportional to the incident light intensity and $I_{sc} \gg I_o$, plots of V_{oc} vs. ln(light intensity) should yield a straight line with slope n.

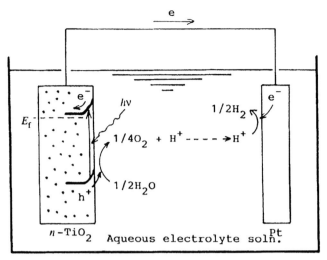

Figure 13.9. Ultraviolet light cleavage of water by the cell, n-TiO$_2$/aqueous electrolyte solution/Pt to produce H$_2$ and O$_2$.

5.1   Stabilization of small bandgap inorganic semiconductors of the liquid-junction type

Since photocorrosion of small bandgap inorganic n-SC in water takes place in the presence of water, the prevention of water contact is a promising approach to stabilize against corrosion. Coating of the SC with thin metals, polymeric films, etc. has been reported for this purpose. It is important that the coating on SC prevents water contact and that the coated material is able to transport charges from the SC surface to the redox couple in the liquid phase.

Polypyrrole (PP; **2**), an electrochemically polymerized film, was effective for stabilizing small bandgap n-SC photoanodes. Such film is often coated on an n-SC by oxidative polymerization of the monomer under irradiation of the SC. A liquid-junction n-GaAs electrode was stabilized against photocorrosion by coating with a thin film of PP which was oxidatively polymerized on the SC under irradiation [82]. The flatband potential (V_{fb}) values were equivalent for both the bare and PP-coated GaAs. The cell, PP-coated GaAs/water[Fe(CN)$_6^{3-/4-}$, NaCN, NaOH]/Pt gave $V_{oc} = 1.37$ V, $I_{sc} = 19$ mA cm^{-2}, $FF = 0.70$ and $Y = 10.5\%$ under 170 mW cm^{-2} irradiation.

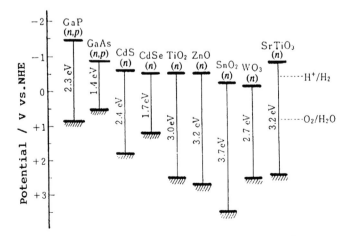

Figure 13.10. Energy levels of the valence and conduction band of semiconductors. H$^+$/H$_2$ and O$_2$/H$_2$O represent the redox potentials of the proton reduction and water oxidation reactions, respectively.

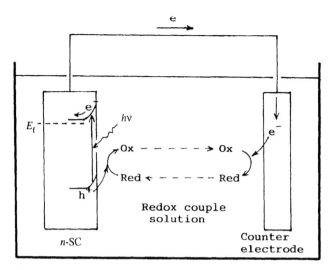

Figure 13.11. Photoregenerative cell with the configuration, SC/redox couple solution/counter electrode. Red and Ox are the reduced and oxidized forms of the redox couple.

n-Si photoanode was also stabilized by an electro-chemically grown PP film coat [83,84]. The cell, PP-coated n-Si/water (I_2/I_3^-)/Pt gave $V_{oc} = c.$ 0.34 V, $I_{sc} = 8$ mA cm^{-2}, $FF = 0.58$ and $Y = 2.8\%$ under 60 mW cm^{-2} irradiation [83]. Stabilization of liquid-junction n-Si was substantially improved when a metal underlayer such as platinum was at first coated on the silicon followed by coating with PP giving a Y value of 5.5% under 55 mW cm^{-2} intensity [85]. The main factor limiting the conversion efficiency was ascribed to low I_{sc} due to light absorption in the concentrated I_3^-/I_2 solution. The ideality factor (n) of 1.1 has been achieved showing an almost ideal junction behavior. Coating of n-Si with a thin layer of gold followed by PP film was also effective for the stabilization [86].

Durability of PP-coated n-Si was improved by the application of N-(3-(trimethoxysilyl)propyl)pyrrole (**22**) polymer coating [87]. Longer-term stability of the modified n-Si (Figure 13.12) was interpreted by covalent attachment of the PP film on the silicon photoanode, which must protect the SC against water/electrolyte undermining of the polymer.

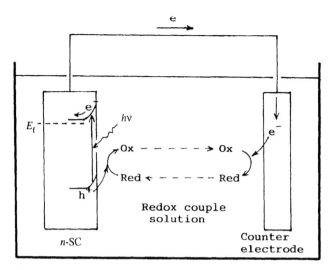

The electron transfer at the PP/Si interface has been studied [88]. Facile electron transfer between the silicon surface states and the polymer film was shown. The polymer film equilibrates rapidly with the ferrocene redox couple in solution. The PP thus provides electronic communication between the silicon electrode and the solution species. It has been shown that counter ions have a profound influence on the E_{fb} of PP films coated on p-Si, with E_{fb} changing from more positive to less positive values in the order: $ClO_4 > BF_4$, $PF_6 >$ tetra-n-ethyl-p-toluenesulfonate [89]. This observation was in concurrence with the predictions from conductivity values.

Such a stabilized n-SC photoanode could induce visible light water cleavage if a catalyst coexists on the coated surface and no redox electrolyte is present in the aqueous phase, under the conditions when an appropriate SC for water cleavage is used. It has been reported that the incorporation of ruthenium dioxide (RuO_2) in PP films coated on a n-GaP photoanode brings about visible light water cleavage in a $HClO_4$ electrolyte aqueous solution [90]. O_2 bubbles were observed on the PP-coated surface, but the photocurrent decreased after each scan, and the film bursted by the O_2. Visible light water cleavage has also been reported by PP-coated n-CdS photoanode on which RuO_2 was attached [91]. For this system corrosion of the SC also took place in parallel with the water oxidation. It has been claimed that the photochemical diode, RuO_2(polystyrene film)/PP/CdS/Pt black (polystyrene film), dipped in aqueous electrolyte solution, induced water cleavage under visible light [92]. Visible light water cleavage by a modified SCL photoanode is problematic

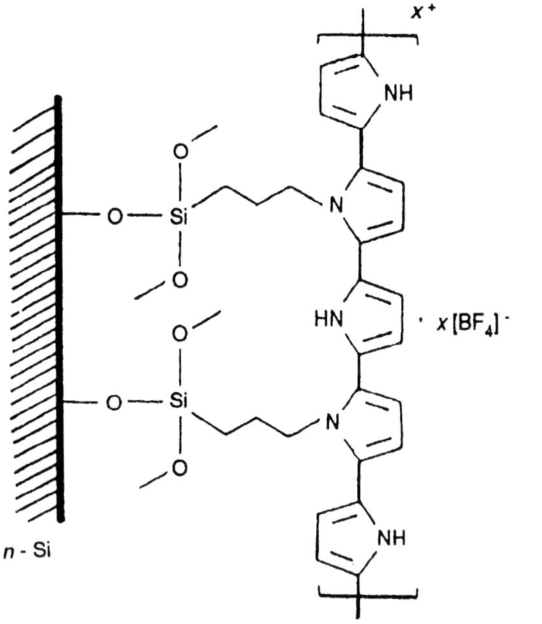

Figure 13.12. Stabilization of the *n*-Si photoanode by modifying it with **22** and polymerizing pyrrole electrochemically. (Reprinted with permission from ref. 87. Copyright 1982 American Chemical Society.)

junction characteristics were not altered appreciably as long as the polymer is doped to the metallically conductive state. The redox species were found to shift the potential of the polymer and have only a minor effect on the barrier height of the CdS/PMT junction. Other oxidatively polymerized films of polyaniline and poly(*o*-phenylenediamine) were used for stabilizing *n*-CdSe [95] and *n*-WSe₂ [96], respectively.

Various polymers can be used for stabilizing liquid-junction-type small bandgap SCs when redox centers exist in the coated polymer film. Polymer-pendant Ru(bpy)₃²⁺ (Poly-Ru; **23**) was used to coat *n*-GaAs, and the modified GaAs showed stable characteristics in an aqueous solution containing redox electrolyte, as shown in Table 13.4 [97].

The photovoltage is shown as a function of the redox potentials of various redox electrolytes in Figure 13.13. The observed correspondence implies the absence of Fermi level pinning at the interface. The ideality of the modified GaAs/electrolyte interface was examined from the relation between I_{sc} and the light intensity (see Figure 13.14) using equation (13.8). The ideality factor obtained from the V_{oc} vs. ln(light intensity) relation was 1.3 close to unity, typical for the SC/electrolyte interface wherein carrier recombination losses are negligible. The same polymer complex (**23**) that incorporated RuO₂ catalyst in the film stabilized *n*-CdS in an aqueous solution containing halogen ions [98a]. When the CdS is at first coated with RuO₂ powders followed by coating with the polymer ruthenium complex, the CdS forms ohmic contact at the CdS/RuO₂ interface, and the Fe(CN)₆⁴⁻/³⁻ redox electrolyte gives ohmic response with CdS through partial ohmic contact while giving also photoelectrochemical response through CdS/electrolyte junction. The cyclic voltammograms in the dark and under illumination (Figure 13.15) are explained only on the basis of the Schottky barrier plus partial ohmic contact [98b]. The anionic redox couple is presumably bound electrostatically in the cationic polymer layer and would react there.

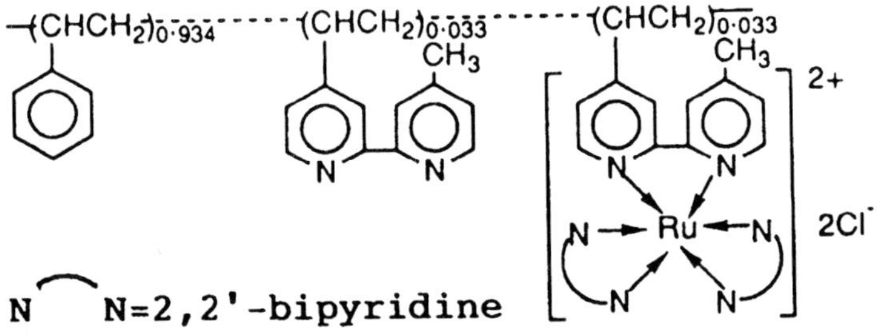

by virtue of its stability and the quantitative formation of O₂ and H₂.

Long-term stabilization of a PT(**4**)-protected *n*-GaAs photoanode dipped in Fe(CN)₆⁴⁻/³⁻ aqueous solution was achieved to give $V_{oc} = 0.655$ V, $I_{sc} = 3.52$ mA cm⁻², $FF = 0.655$ and $Y10.9\%$ under 14 mW cm⁻² irradiation [93]. The solid-state properties of the junction between highly doped metallic-like poly(3-methylthiophene) (PMT) and *n*-CdS are reported [94]. The characteristics of the CdS/PMT/Au solid-state cell compared favorably with those of the CdS/metal Schottky barrier device. When the cell, CdS/PMT, contacted various redox electrolyte solutions, the

Table 13.4. Typical performance parameters of regenerative PEC* systems based on polymer-coated n-GaAs photoanodes

Redox couple	$V_{oc}^{\dagger}$ (V)	I_{sc} (mA cm^{-2})	P_{max} (mW cm^{-2})	FF	Y (%)
$Fe^{2+/3+}$	1.09 (1.10)‡	10.89 (15.11)	8.41 (11.54)	0.71 (0.69)	12.0 (11.5)
$Fe^{2+/3+}$–EDTA	0.47	8.33	1.07	0.27	1.5
I^-/I_3^-	0.72	11.67	2.33	0.28	3.3

V_{oc}, open-circuit voltage; I_{sc}, short-circuit current density; P_{max}, maximum power output; FF, fill factor; Y, optical-to-electrical conversion efficiency.
* Light intensity: 70 mW cm^{-2}.
$\dagger$ Taken as the difference between the potential onset of the photocurrent and the equilibrium redox potential.
$\ddagger$ Parameters in parentheses correspond to AM1 illumination ($\sim$ 100 mW cm^{-2}).

The similar modified CdS coated with **23**, which contains dispersed RuO$_2$ catalyst, brought about water cleavage under visible light irradiation when no redox couple is present in the aqueous phase [99]. The catalytic ability of RuO$_2$ for water oxidation is not so high, which causes insufficient water cleavage. An active water oxidation catalyst of oxo-bridged trinuclear ammine ruthenium complex called Ru-red (**24**) was attached on the CdS photoanode by adsorbing it electrostatically into a Nafion (**18**) film coated on the CdS; visible light water cleavage was carried out with KNO$_3$/HNO$_3$ electrolytes [100].

The photoelectrochemical cell, CdS/Ru-red(Nafion)/KNO$_3$, HNO$_3$/Pt, produced O$_2$ and H$_2$ under illumination, and their yields were 16% based on the irradiated monochromatic light at 496 nm. Although the stability and reproducibility of this kind of photoelectrochemical cells, for obtaining photoreaction products, should still

be improved, such a molecule-based design of the catalysis represented in Figure 13.16, in combination with a SC device, would lead to new photoelectrochemical systems for photoenergy conversion and storage.

Poly(4-vinylpyridine) (PVP; **25**) complex of Ru(bpy)$_2$Cl , coated on n-MoSe$_2$ and n-WS$_2$, catalyzed the hole transfer from the SC surface to a redox ion in the solution [101]. A Nafion film containing electrostatically incorporated tetrathiafulvalenium (TTF$^+$) stabilized n-Si photoanode in Fe(II) solution [102].

Coating of n-Si by cationic poly(o-xylylviologen) (**27**) substantially improved the photoanode against corrosion in the presence of electroactive anions, but there was no improvement for positively charged ions [103]. This was ascribed to electrostatic binding of the anions, which provided efficient scavenging of the photogenerated holes at n-Si.

Adsorption of other compounds on the liquid-junction SC can also stabilize it against degradation. Exposure of the n-GaAs electrode in an aqueous Co(NH$_3$)$_5$(H$_2$O)$^{3+}$ solution brought about attachment of the complex onto the GaAs surface and dramatically improved the I–V characteristics of the n-GaAs/KOH-Se$^{-/2-}$ junction giving $V_{oc} = 740$ mV, $I_{sc} = 23$ mA cm^{-2}, $FF = 0.70$ and $Y = 12.5\%$ under 88 mW cm^{-2} illumination [104]. The stability of n-CdS in a Fe(CN)$_6^{4-/3-}$ electrolyte solution was ascribed to the formation of a CdFe(CN)$_6^{2-/1-}$ overlayer, and mediated hole transfer through the overlayer to the solution, Fe(CN)$_6^{4-}$, was proposed [105]. The energetics and kinetics are very sensitive to the supporting electrolyte cation. A supporting electrolyte containing both K$^+$ and Cs$^+$ is found to maximize the cell performance. The correct cation induces good overlap between electrochemically active valence band states and filled redox states in the overlap layer, thus promoting mediated hole transfer. A thin film of MPc (**11**) or H$_2$Pc (**10**) was coated on n-Si, resulting in

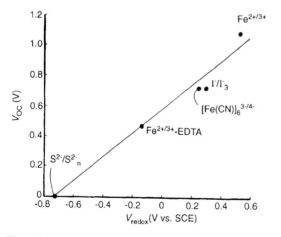

Figure 13.13. Correlation between V_{oc} and the redox potential (V_{redox}) of the redox couple in the aqueous solution for the polymer(**23**)-coated n-GaAs. (Reproduced by permission of the Electrochemical Society Inc. from ref. 97.)

$[(NH_3)_5Ru-O-Ru(NH_3)_4-O-Ru(NH_3)_5]^{6+}6X^-$ (Ru-red; 24)

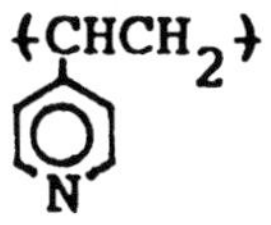

$\{CHCH_2\}$ (PVP; 25)

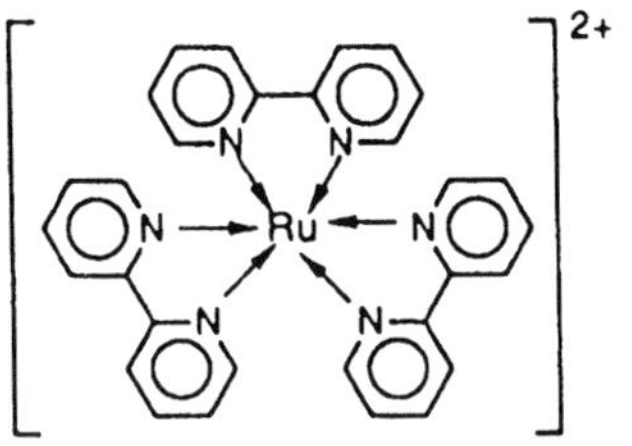

$[Ru(bpy)_3^{2+}; 26)$

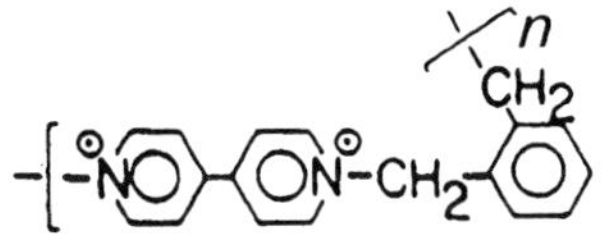

(Poly-OXV^{2+}; 27)

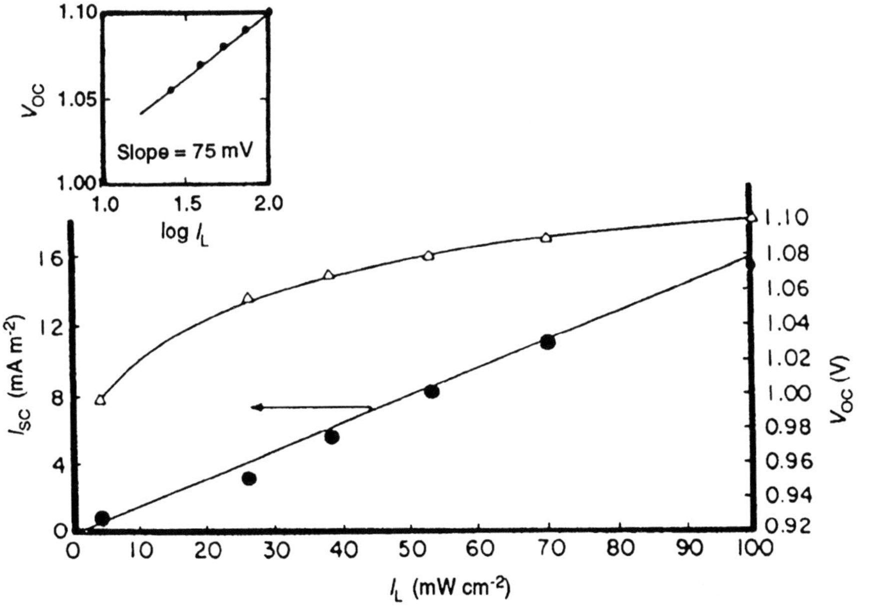

Figure 13.14. Dependence of I_{sc} and V_{oc} on the light intensity (I_L) for the polymer(**23**)-coated n-GaAs in contact with the Fe$^{2+/}$ $^{3+}$ redox electrolyte aqueous solution. (Reproduced by permission of the Electrochemical Society Inc. from ref. 97.)

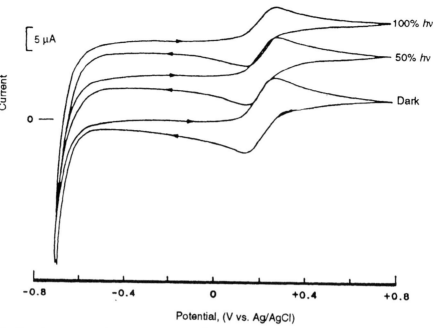

Figure 13.15. Cyclic voltammograms (scan rate: 100 mV s^{-1}) for an *n*-CdS photoanode modified at first with RuO$_2$, and then covered with **23** polymer film and dipped in Fe(CN)$_6^{3-/4-}$ aqueous solution. (Reprinted with permission from ref. 98b. Copyright 1985 American Chemical Society.)

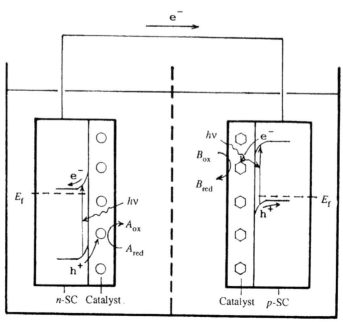

Figure 13.16. Photoelectric devices for photoenergy conversion and storage with semiconductors in combination with molecule-based catalysts applied by means of polymer coatings.

stabilization of the photoanode in aqueous redox solutions [106]. The photoelectrochemical cell, n-Si/CuPc/Fe$^{3+/2+}$ aqueous solution, gave better I–V behavior than a solid photocell, n-Si/CuPc/Pd.

5.2 Sensitization of large bandgap inorganic semiconductors of the liquid-junction type

One of the promising approaches to visible light photoenergy conversion by liquid-junction inorganic SCs is to sensitize large bandgap SCs by coating them with sensitizer, such as polymeric and organic colored compounds. The mechanism of the sensitization of large bandgap n-SC is based on electron injection from the excited dye, attached to the SC, to the conduction band, followed by electron injection to the oxidized dye molecule from the electron donor present in the solution. Ru(bpy)$_3$$^{2+}$ (26) was attached to n-SiO$_2$ via condensation of surface hydroxyl groups with Ru(4-trichlosilylethyl)-4'-methyl-2,2'-bipyridine)bis(2,2'-bipyridine)bis(hexafluorophosphate) to achieve sensitization [107].

A polymer film containing Ru(bpy)$_3$$^{2+}$ (26) pendant groups was formed on a SnO$_2$ electrode by polymerizing the mixture of Ru(bpy)$_3$$^+$ cinnamate and PVA (9) cinnamate on the electrode to obtain a crosslinked polymer coating [108]. The coated SnO$_2$ gave a sensitized photocurrent when dipped in an aqueous solution containing EDTA. An n-SiO$_2$ coated with Nafion (18), which electrostatically bound the Ru complex (26) generated, also sensitized the photocurrent [109]. Monolayer assemblies composed of a Ru(bpy)$_3$$^{2+}$-containing surfactant molecule (28) and arachidic acid sensitized the anodic photocurrent on n-SnO$_2$ [110].

The efficiencies of these systems were not high, mainly because of the small quantum yield for the charge injection from the excited state of the sensitizer to the conduction band of the large bandgap SC

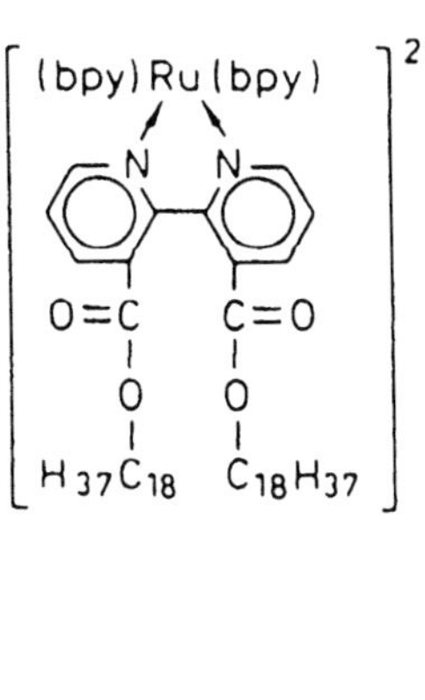

photoanode. Efficient sensitized photoelectrochemical cells have been fabricated based on colloidal TiO$_2$ films [111]. Dispersion of TiO$_2$ fine particles (average size 15 nm) was spread on a conductive glass to give a TiO$_2$ membrane with a large surface area of 10 µm thickness. It was covered with a monolayer of the ruthenium complex derivative (29) as a sensitizer. The photoregenerative cell (see Figure 13.17), conducting glass/TiO$_2$/dye (29)/(I$^-$–I$_3$$^-$)ethylenecarbonate–acetonitrile solution/conducting glass, gave $V_{oc} = 0.68$ V, $I_{sc} = 11.2$ mA cm^{-2}, $FF = 0.68$ and $Y = 7.1\%$ under AM 2 irradiation, and $Y = 12\%$ under room light.

(29)

Under 2 months' operation, the dye turned over five million times and the photocurrent changed by only 10%. The action spectrum for the photocurrent agreed with the absorption of the dye. Detailed studies have been reported on this highly efficient cell [111].

The light-harvesting sensitization by the dye (29) has been studied [112a]. Emission and excitation spectra show that in this complex the light energy absorbed by the terminal Ru(bpy)$_2$(CN)$_2$ (antenna) groups is efficiently funneled to the central Ru(bpy(COO)$_2$)$_2$$^-$ (sensitizer) fragment. Diffuse reflectance studies have been undertaken to investigate the photochemical behavior of a ruthenium complex, bis(2,2'-bipyridine)(4,4'-dicarboxy-2,2'-bipyridine)ruthenium(II)-[Ru(bpy)$_2$(dcbpy)$^{2+}$], on the surface of TiO$_2$ particles [112b]. Decreased emission yield and lifetime of the excited complex on the TiO$_2$ surface indicated dominance of the charge injection process from the excited complex to TiO$_2$. The rate constant for the heterogeneous electron transfer measured from the luminescence decay was in the range of 1.0–5.5 × 10^8 s^{-1}. Upon irradiation with visible light, the degassed sample of the complex-coated TiO$_2$ particles

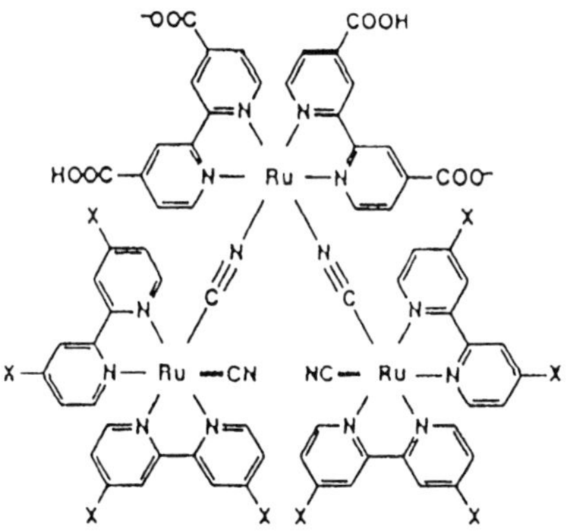

(28)

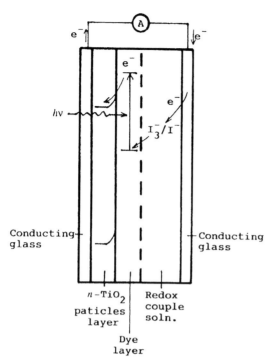

Figure 13.17. Dye (**29**)-sensitized TiO₂ photoregenerative cell with the I_3^-/I^- redox couple in organic medium.

turned blue as photogenerated electrons got trapped at the TiO₂ surface. The blue color disappeared instantly upon exposure to air. The energetic distribution of the interfacial states on a single crystalline TiO₂ electrode modified with Ru(dcbpy)₂ has been studied [112c]. The use of other sensitizers such as Chl derivatives and related natural porphyrins were reported [113]. An important point relating to this kind of cell is the durability of not only the dye but also other components such as the redox electrolyte and the solvent which should not be consumed during the photoelectrochemical event. The details should still be studied.

Chlorophyll has been used to fabricate photovoltaic cells [114–117]. Zinc or Mn–tetrakis(4-sulfophenyl)-porphyrin was incorporated into a film of PVA (**9**) cinnamate coated on an In₂O₃ electrode [118]. The

presence of ascorbic acid in the solution enhanced the anodic photocurrent. Films of various MTPPs (**12**) coated on NESA glass by dipping or vacuum sublimation were effective for the sensitization of the anodic photocurrent [119]. Cobalt, copper and NiTPP were ineffective for the sensitization. The study of film thickness showed that multilayer activity takes place. A film-type photovoltaic cell was fabricated with Zn-tetramethylpyridiniumporphyrin dispersed in a conventional polymer film such as polyacrylonitrile, celluloseacetate, PVP (**21**), etc. [120a]. Thus, the cell, ITO-coated polyester/polymer film containing sensitizer and electrolyte/polyester coated with gold, was fabricated. Electrons excited within the thin layer of *p*-type SC, poly(2,5-furylenvinylene), coated on ZnO were injected into the conduction band of ZnO [120b].

Films of CuPc (**16**) vacuum-deposited on NESA glass generated sensitized anodic photocurrent [121,122]. They also gave cathodic photocurrents which are attributed to excitons generated and dissociated in the bulk CuPc. MC (**14**) and Rhodamine B (Rh B; **30**) were incorporated into polyethyleneimine (PEI; **31**) and PVA (**9**) by using diethylphosphoryl cyanide or 1-methyl-2-chloropyridinium iodide as the coupling agent [123].

The SnO₂ coated with these films exhibited remarkably sensitized photocurrents. MC-coated NESA showed an enhanced photocurrent due to the MC aggregation [124]. The quantum efficiencies of the sensitized photocurrent obtained at ZnO were an order of magnitude higher for the surface-formed *J* aggregates of the cyanine dyes (**32**) than for those adsorbed from solution [125].

Sensitized photocurrent spectra of ZnO electrodes show distinct peaks at the *J*-band of cyanide dyes adsorbed on the ZnO [126]. Spectral sensitization of a SnO₂ thin film electrode by a cyanine dye (33) coated by a Langmuir–Blodgett technique was examined [126b].

Arachidic acid monolayers containing the dye were used for the coating. A distinct *J*-band was observed in the action spectra of the sensitized anodic and cathodic photocurrents. The following mechanism involving a molecular exciton of the *J*-aggregates is proposed for the sensitized photocurrent. The anodic photocurrent is

(C₂H₅)₂N — O — N⁺(C₂H₅)₂ Cl⁻ **(Rh B; 30)**

C
COOH

$\{CH_2CH_2\underset{H}{N}\}_n$ (PEI; 31)

(32)

(33)

caused by hole trapping by some reducing agent and concomitant injection of the electron from molecular exciton to the conduction band of SnO_2. Extraction of a conduction band electron of SnO_2 by a molecular exciton and its supply to an oxidizing agent, such as dissolved oxygen, are responsible for the cathodic photocurrent.

The quantum yields of the sensitized photocurrent were studied on a NESA glass coated with Rh B (30) or Rose Bengal [127]. The quantum yield (Q) was dependent on the donor intensity (N_d) of the semiconductor and showed an optimum point. The strong dependence of Q on N_d may be interpreted by considering the decay process of the electrons injected from the excited dye through possible surface states. The increase in N_d causes an increase in the electric field of the space charge region of the semiconductor and hence suppresses these decays up to the optimum N_d. The decrease in Q for N_d may be attributed to a leakage of the conduction electrons by tunneling through the thin space charge layer.

Arene (e.g. naphthalene, anthracene, pyrene, etc.)-derivatized electrodes gave sensitized photocurrents in the presence of Rh B and hydroquinone in solution [128]. The attached molecules were considered to function as energy or electron relays.

A simple model for the electron transfer from the excited dye to the conduction band of SC was proposed [129]. The equation for other electron transfer does not contain the Franck–Condon term, but the state density of the SC plays an important role. The energy gap dependence of the electron transfer rates, which were quite different from the molecular systems, was explained by this model. The electron exchange energy

was determined for $Ru(bpy)_3^{2+}$ (26), $Ru(bpz)_3^{2+}$ and Rh B (30). Temperature dependence of the electron exchange energy term, caused by thermal motion of the adsorbed dyes, was suggested as a new origin of the activation energy of the electron transfer on the surface. The required characteristics of an ideal photoelectric converter were summarized [130]. The four unavoidable loss mechanisms inherent in single-junction photoconverters-lack of absorption of sub-bandgap photons, thermalization of ultra-bandgap photons, the difference between the available energy and internal energy of the thermalized excited states, and the small loss of excited states by radiative decay were qualified. A discussion was included on the poorer performance of photoconverters based on molecular chromophores when compared with those based on SCs; this arises because of the difference between the two broadened Einstein relations, and between the absorption spectra and transport properties of each system.

5.3 Organic membranes in combination with inorganic semiconductor particles

Membranes containing SC particles have been used for photocatalytic reactions [131–133]. Bilayer lipid membranes (BLM) consisting of a phosphatidylcholine and phosphatidylethanolamine mixture incorporated both MC (14) and CdSe particles. The cell, $CuSO_4$ aqueous solution/BLM(MC + CdSe)/Na_2S, induced a stable and higher photocurrent than the cell without CdSe [132].

6 PHOTOCHEMICAL DEVICES

Most photoelectric conversion devices hitherto used and reported are based on the photovoltaic processes of inorganic and organic semiconducting materials. However, photoelectric conversion can also be achieved by utilizing photochemical reactions if the photochemically separated charges (electrons and holes) are taken out and exchanged at the anode and cathode, respectively. This kind of cells in which a photochemical reaction is involved are called 'photogalvanic cells' in comparison to the 'photovoltaic cells' based on semiconductors.

The photochemical reactions used for the photogalvanic cells are divided into three major categories.

(1) The so-called photoredox reaction for which the equilibrium of the system shifts under illumination toward photochemical products because of the

relatively slow back reaction. It returns to the original state when the light is turned off.

(2) The non-photoredox reaction in which illumination does not shift the equilibrium towards the products because of the rapid back reaction.

(3) Other photochemical reactions such as photoisomerization.

In this section photogalvanic cells, based on these three types of reactions, are described. Although the conversion efficiency based on the photochemical reaction is not high at the present time, many types of reaction couple have allowed the design of various cells. In many cases the cell has the configuration of a wet type; sensitizer-coated electrode/electron acceptor or donor in a liquid/counter electrode.

6.1   Photoredox systems

A typical example of the photoredox reaction is the thionine (TH$^+$; **34**) and Fe^{2+} for which the photochemical reaction is [134,135]

$$TH^+ + Fe^{2+} \underset{dark}{\overset{h\nu}{\rightleftarrows}} TH^{+\cdot} + Fe^{3+} \qquad (13.9)$$
$$\text{(purple)} \qquad \text{(colorless)}$$

When the solution is in a photogalvanic cell composed of light and dark chambers, a photovoltage of about 150 mV is induced between the two electrodes dipped in the solution [135,136].

(TH$^+$; 34)

The TH$^+$/Fe^{2+} photoredox system was incorporated into polymeric gel films to fabricate the cell [137], SnO$_2$/TH$^+$–Fe^{2+}(polymer gel)/Pt. Polymer gels such PVA (**9**) and PEI (**31**) were used as carriers for the redox couple. The TH$^+$ polymer gel (**35**) was prepared

from poly(epichlorohydrin), TH$^+$ and triethylamine, and was used as a film in combination with Fe^{2+} ions.

If both the photochemical reaction products react at an electrode equally, the cell system can not give a photoresponse because of the short-circuit reaction occurring on the electrode. It is desirable therefore that the electrode discriminates between the two photochemical products in order to generate photoeffect at the electrode. It has been reported that TH$^+$-coated SnO$_2$ and platinum electrodes are suitable for a photogalvanic cell in this respect [138,139]. TH$^+$ was coated on an electrode by applying the potential of 1.1–1.5 V for several minutes in a solution containing TH$^+$. The TH$^+$-coated electrode showed selectivity for the reaction of TH$^+$.

Polymer-pendant TH$^+$ was used for electrode coating. One example is poly(acrylamidomethylthionine-co-methyloylacrylamide), which was coated on a platinum electrode by casting as a thick film of about 10 μm [140]. The cell composed of this polymer-coated platinum and counter electrode dipped in ferrous sulfate aqueous solution gave $V_{oc} = 37$ mV and $I_{sc} = 2.8$ μA cm^{-2}. Another study showed, however, that coating the electrode with TH$^+$ reduces the power conversion efficiency of the cell by a factor of 19 in comparison with the efficiency of 4×10^{-4}% for the uncoated gold electrode [141].

A PP(**2**)-coated electrode was selective for the Fe$^{2+/3+}$ couple, giving improved output in the TH$^+$/Fe^{2+} photogalvanic cell [142]. The cell gave $V_{oc} = 138$ mV and $I_{sc} = 26.3$ μA cm^{-2}. TH$^+$ attached to poly(N-methylolacrylamide-co-acrylic acid) was coated on a platinum electrode. Its cyclic voltammogram showed the formation of a complex between the coated dye and ferric/ferrous cyanide present in the solution. When the electrode is illuminated in a Fe^{2+} aqueous solution, cathodic polarization at the coated electrode is observed, in contrast to the bare electrode dipped in a mixture of TH$^+$ and Fe^{2+}, which gave an anodic response at the electrode [143]. It was proposed that, for the polymer-coated electrode, the excited states of TH$^+$ and Fe^{2+} form a complex, which is stabilized by the polymer network and accepts

(35)

electrons from the electrode. A flash photolysis study showed the formation of such a complex [144].

Protonated PVP (**25**) was coated on a NESA glass-bound Rose Bengal electrode electrostatically. This modified electrode, when dipped in an aqueous solution containing a redox reagent such as Fe^{2+}, induced an anodic photogalvanic effect [145]. A photovoltage of 135 mV and a photocurrent of 1.24 μA cm^{-2} were reported. Enhanced photogalvanic performance was attributed to (i) the presence of a high dye concentration at the electrode surface, (ii) a longer-lived excited state of the dye in the polymer film, (iii) the electrostatic characteristics of the polymer in controlling the access of a charged redox couple to the illuminated electrode.

A photogalvanic cell was fabricated from ion-exchange resins. Cation- and anion-exchange membranes adsorb TH$^+$ and ascorbic acid, respectively. These membranes are confined and irradiated to produce a photopotential of about 30 mV [146]. The couple of thiazine and/or phenothiazine dye/aliphatic amine, such as EDTA or triethanolamine, gave high photopotential, as exemplified by the photopotential of 844 mV by the phenosafranine/EDTA couple [147]. Chl a was deposited on a SnO$_2$ electrode, and dipped in an aqueous solution containing Safranine T and triethanolamine [148]. The safranine T had a great effect on the V_{oc} (853 mV) and I_{sc} (7.4 μA cm^{-2}) of the photoelectrochemical cell. Although such amines seem to undergo decomposition during the photoreaction, they are regarded as model compounds to study photogalvanic effects.

Thionine was incorporated into a clay-modified electrode and photoelectrochemical investigations have been reported [145,149]. The modified electrode gave photopotential in the presence of EDTA in the solution. The irreversible photochemical reaction of the excited state of Ru[(PVP)$_2$(bpy)$_2$]$^{2+}$ coated on a platinum electrode with Co(OX$_3$)$^{3-}$ (where OX is oxalate, C$_2$O$_4$$^{2-}$) present in solution induced a cathodic photocurrent of the order of 45 μA cm^{-2} at the applied potential of 0.3 V vs. SSCE [150]. A blue-green alga,

Anabena cylindrica, immobilized on SnO$_2$ coated with MV^{2+}/poly(styrene sulfonate)/PVA showed an anodic photocurrent [151].

A redox bilayer system composed of poly(cobaltocene) and poly(ferrocene) was used to fabricate a molecule-based diode, ITO/poly(cobaltocene)/poly(ferrocene)/ITO, and the irradiation induced photocurrents [152].

6.2 Non-photoredox systems

In most photochemical electron transfer reactions between two compounds, back electron transfer is so rapid that the net products can not be obtained under steady state as long as the reaction is reversible. In these photochemical reactions it is not simple to obtain photoresponse at electrodes applied to the photoreaction systems. However, when one or both of the photochemical reaction components are attached to the electrode, there is a chance that the photochemical products can exchange charges at the electrode before the back reaction takes place, which then gives photoresponse at the electrode. Since such charge exchange at the electrode is competitive with the rapid back electron transfer, the photoresponse at the electrode has not been found to be very large to date.

A typical photoinduced electron transfer from the excited state Ru(bpy)$_3$$^{2+}$ (**26**) to MV^{2+} (**20**) attracted much attention as one of the promising approaches toward visible light water cleavage [1,2]. As shown in equation (13.10), the back electron transfer is so rapid that the photochemical reaction of the components in a homogeneous solution can not induce photoresponse at the electrodes, which are simply dipped in the solution.

$$Ru(bpy)_3^{2+} + MV^{2+} \xrightarrow{h\nu} Ru(bpy)_3^{2+*} + MV^{2+}$$
$$\overset{\longleftarrow}{} Ru(bpy)_3^{3+} + MV^+ \longleftarrow \qquad (13.10)$$

When the ruthenium complex was coated on an electrode, such as a basal plane pyrolytic graphite (BPG) and ITO as a form of polymer-pendant species

$$-\!\!\!\left(CHCH_2\right)_{\!0.59}\!\cdots\!\cdots\!\left(CHCH_2\right)_{\!0.07}\!\cdots\!\cdots\!\left(CHCH_2\right)_{\!0.34}$$

(**3 6**)

as **23**, a photoresponse was obtained in the presence of MV^{2+} in the aqueous solution when the coated electrode is dipped [153,154]. The action spectrum for the photocurrent showed that the photoelectrical response is induced by the excitation of the ruthenium complex which has an absorption maximum at 450 nm. MV^{2+} polymer (**36**) was also coated on the top of the ruthenium complex layer to form a bilayer-coated device [155].

The typical cell configurations are BPG/polymer–$Ru(bpy)_3^{2+}/MV^{2+}$(water)/Pt, and BPG/polymer–$Ru(bpy)_3^{2+}$/polymer–MV^{2+}/aqueous electrolyte/Pt. The presence of O_2 greatly enhances the photocurrent by oxidizing the formed MV^+.

As for the coating, polysiloxane-pendant complex (**37**) [156], monomeric $Ru(bpy)_3^{2+}$ adsorbed in poly-(styrene sulfonate) or Nafion (**18**) were also used.

The monolayer-coated system containing only the ruthenium complex in the coating and MV^{2+} in the solution, induced both cathodic and anodic photocurrents depending on the applied potential [154]. This was ascribed to the electrode reaction of both the photochemically produced $Ru(bpy)_3^{3+}$ and MV^+, the latter can be present also near the electrode surface due to adsorption in the polymer layer. However, the above bilayer-coated system induced mainly cathodic photo-

currents due to the separated structure of electrode/$Ru(bpy)_3^{2+}/MV^{2+}$. The platinum-coated film of chlorosulfonated polystyrene-pendant $Ru(bpy)_2(phen)^{2+}$, where phen is 1,10-phenanthroline, induced anodic photocurrent in the presence of MV^{2+} in the solution due to the incorporation of the acceptor in the film [157]. The coexistence of triethanolamine in the solution greatly enhanced the photocurrent, which is due to the sacrificial reduction of the formed $Ru(bpy)_3^{3+}$ by the amine.

A solid-state cell was fabricated by using bilayers of the polymer-pendant ruthenium complex (**37**) and the polymer-pendant MV^{2+} (**38**), as shown in Figure 13.18 [156]. Although the cell efficiency ($V_{oc} = 0.25$ V, $I_{sc} = 0.48$ μA cm^{-2}, $FF = 0.25$ under 34 mW cm^{-2} irradiation) is not high, this result might lead to the production of a new photoelectric conversion system that utilizes photochemical reactions in a solid state. A similar, semidry solid-state cell, ITO/$Ru(bpy)_3^{2+}$(Nafion)/MV^{2+}(Nafion)/ITO, was fabricated in which the Nafion films contain aqueous electrolyte solution [158]. The presence of O_2 greatly enhances the cathodic photocurrent by reacting with the formed MV^+. In this kind of photochemistry-based device, the excited state reaction is very rapid, and the charge transport in the solid phase is limiting, which causes rapid back electron transfer (charge recombination).

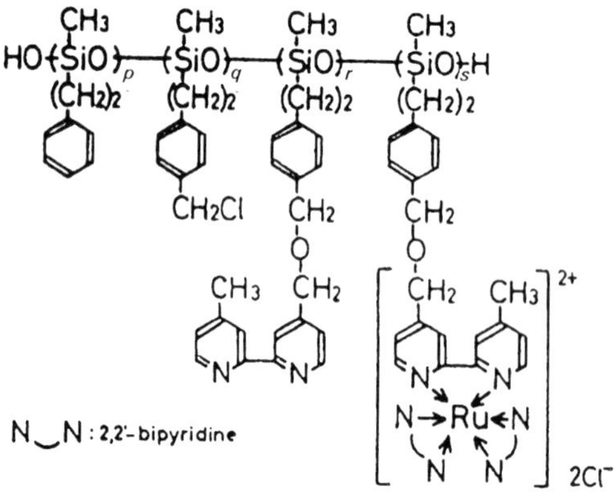

(37)

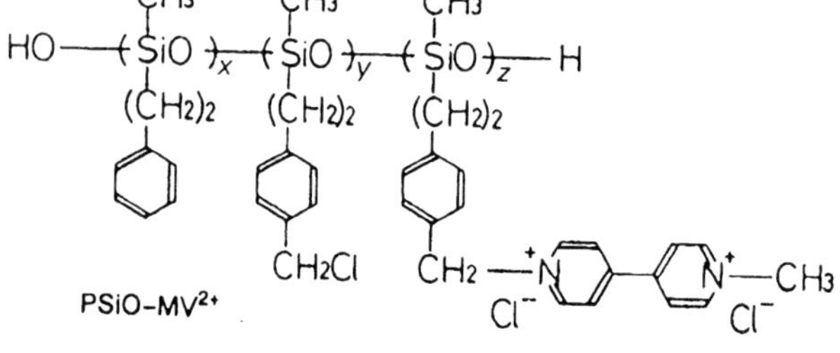

(38)

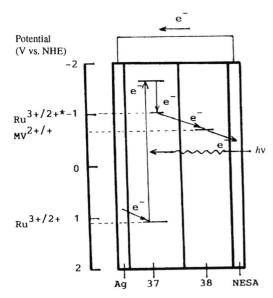

$$CH_3 \quad (CH_2)_2-N \qquad \qquad (40)$$

Figure 13.18. Solid-state cell based on photochemical reaction between poly(siloxane)-pendant $Ru(bpy)_3^{2+}$ (**37**) and poly(siloxane)-pendant MV^{2+}. (Adapted from ref. 156.)

Other compounds can also work as an electron donor or acceptor for the photoexcited $Ru(bpy)_3^{2+}$. Prussian Blue, a mixed valent Fe^{2+}/Fe^{3+} polynuclear complex with a high molecular weight, whose unit structure is represented by **39**, works both as an acceptor and a donor for the photoexcited $Ru(bpy)_3^{2+}$. The bilayer-coated BPG electrodes, $BPG/polymer-Ru(bpy)_3^{2+}$ (**23**)/PB and $BPG/PB/polymer-Ru(bpy)_3^{2+}$ (**23**), induced photocurrents depending on the applied potential [159,160]. The PB film alone gave photocurrent when coated on a BPG electrode [161].

Molecular oxygen works also as an electron acceptor for the excited $Ru(bpy)_3^{2+}$, thus allowing the generation of cathodic photocurrent [161]. The coating of viologen polymer as a second layer on the top of the ruthenium complex layer enhanced the photocurrent, the viologen working as an electron mediator between the excited ruthenium complex and O_2. A film of electropolymerized $Ru(bpy)_3^{2+}$ polymer containing pyrrole groups (the monomer is **40**) gave a cathodic photocurrent when irradiated in the presence of 4-methylbenzenediazonium tetrafluoroborate in acetonitrile [162].

The increase of the film thickness favored the photocurrent intensity, but this effect was canceled by the electron diffusion limitation by increasing the thickness more.

Anodic polarity was observed under irradiation at a platinum electrode coated with a Nafion film that contained adsorbed $Ru(bpy)_3^{2+}$ in the presence of Fe^{3+} in the aqueous solution. Cathodic polarity was exhibited at a platinum electrode coated with montmorillonite clay that contained adsorbed $Ru(bpy)_3^{2+}$ under irradiation [163]. A new photogalvanic cell, $Pt/Ru(bpy)_3^{2+}(Nafion)/Fe^{3+}$ aqueous solution/$Ru(bpy)_3^{2+}(montmorillonite)/Pt$, was fabricated by these two coated electrodes. The different behaviors of these two electrodes were ascribed to adsorption of Fe^{3+} in the Nafion film where the photochemically formed Fe^{2+} donates electron to the electrode, while at the clay-coated layer the reduction of Fe^{3+} to Fe^{2+} takes place at the clay/water interface and the formed $Ru(bpy)_3^{3+}$ in the coated clay accepts an electron from the electrode.

Other metal complexes such as metal–phthalocyanine (MPc; **11**) and metaltetraphenylporphyrin (MTPP; **12**) can induce photocurrent based on photochemical reaction with an acceptor or donor molecule. Viologen is also a good acceptor for the excited state of such macrocyclic compounds that generate photocurrent. When the complex is dispersed in a polymer film such as PVCz (**6**) and poly(vinylidenefluoride) and is coated on the electrode, better photochemical behavior is obtained. A bilayer-coated electrode composed of PVCz film containing ZnPc and polymer-pendant benzylviologen (**38**) film generated cathodic photocurrent [164]. The cell, $ITO/ZnPc(PVCz)/viologen(38)/O_2$ (KCl aqueous solution)/Pt, showed the efficiency, $V_{oc} = 0.58$ V, $I_{sc} = 5.5$ μA cm^{-2} and $FF = 0.40$ under 22 mW cm^{-2} irradiation.

A linked relay system composed of ZnPc and anthraquinone (AQ) (ZnPc–AQ) was prepared and incorporated into a phosphatidylcholine bilayer lipid membrane (BLM). The cell, ascorbic acid aqueous solution/ZnPc–AQ(BLM)/Ce^{4+} aqueous solution, in-

$$Fe^{3+}{}_4[Fe^{II}(CN)_6]^{4-}{}_3 \qquad (PB; \ 39)$$

duced a photopotential which is much higher than the system composed of separate ZnPc and AQ molecules [165a]. This was attributed to the facilitated electron transfer from the excited ZnPc to the linked AQ. Bilayer phosphatidylserine membrane containing ZnTPP (12) and separating two solutions of ascorbate and Fe^{3+} generated steady-state photocurrents [165b]. Photoelectrochemical properties of thin films of zinc porphyrin derivatives dipped in quinhydrone aqueous solution have been studied, and the effect of molecular arrangement of the porphyrin was discussed [166]. LB films containing arranged donor(ferrocene)/sensitizer(pyrene)/acceptor(viologen) molecule (41) coated on ITO induced photocurrents [167].

Molecular oxygen works also as an electron acceptor for the excited macrocyclic metal complexes in order to yield photoresponse at the complex-coated electrode. Photoelectroreduction of O_2 by H_2Pc thin films has been studied in alkaline solutions [168]. Thin films of ZnPc (11) coated on an ITO induced cathodic photocurrent in the presence of O_2 in the aqueous phase due to electron transfer from the excited ZnPc to O_2 [169]. Dispersion of the complex in a polymer matrix such as PVCz (6) brings about higher efficiency and better stability [170]. Various metal Pcs, porphyrins (12, 13) and polymer matrices have been studied [170]. It is still unsolved if the complex film works as a molecule-based sensitizer or as a p-type SC. The cell, ITO/Cu/Cu_2–p-diethynylbenzene complex film/2,6-dianthraquinone disulfonate, LiOH (water)/Pt, gave $V_{oc} = 0.275$ V, $I_{sc} = 60$ μA cm^{-2}, $FF = 0.4$ and $Y = 0.02\%$ based on 35 mW cm^{-2} incident light [171].

Other organic compounds can be sensitizers for generating photochemistry-based currents. A polymer-bilayer-coated electrode, which consisted of an electron mediator film composed of PEI-pendant anthraquinone as an inner layer and a sensitizer film composed of PEI-pendant merocyanine (MC) as an outer layer, produced anodic photocurrents in the presence of a sacrificial electron donor (triethanolamine) [172]. A new photoelectrochemical analysis cell for high performance liquid chromatograph was developed that utilized photoreduction of polystyrene-pendant anthraquinone film (coated on ITO) by various amino acids [173]. Organic radicals can also be sensitizers in the presence of an acceptor. A new photogalvanic effect of the MV^{2+} cation radical was observed; it is formed electrochemically on the surface of an ITO electrode dipped in the aqueous solution of MV^{2+} [174]. This effect was ascribed to the oxidation of the photoexcited cation radical (absorption maximum at 610 nm) with O_2. The photoelectrochemical response of MV^{2+} cation radicals that were incorporated in an ITO-coated polysiloxane film was also observed [175].

In these photogalvanic cells, sacrificial redox reagents such as O_2, amines and other organic compounds are sometimes used. In these cases the photocurrents flow only when the unreacted agents are present, but this photoelectrochemical system can be applied to devices such as sensors.

6.3 Other photochemical and photobiological systems

Other photochemical and photobiological reactions induce photoelectric conversion, such as a change of membrane potentials by illumination. Photochromic change of organic compounds or photoisomerization of compounds, such as visual pigment rhodopsin, are

(41)

often utilized to obtain membrane potential changes. Biological materials are also used as functional centers by extracting them from natural substances and incorporating them into synthetic membranes. They are classified into charge carrier transport membranes and polarized membranes [176]. Chloroplast BLM is an example of the former; irradiation induces flows of charge carriers that result in photocurrent. An example of the latter is an asymmetric BLM adsorbing cyanine dye on one side, which induces only transient potential change by irradiation.

Photoinduced potential changes across plasticized polyvinyl chloride (PVC; **42**)–crown ether membranes were reversibly induced by utilizing an incorporated photoresponsive crown ether that contained a *trans*-azobenzene group (**43**) [177a].

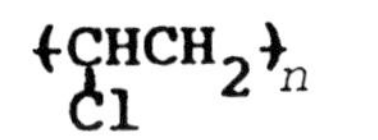

$$+CHCH_2+_n \qquad (PVC; \ 42)$$
$$|$$
$$Cl$$

The azobenzene is isomerized into a *cis*-form by UV irradiation; under visible light it returns to the *trans*-

form. A negative shift of the membrane potential was observed under UV irradiation, and the initial potential was recovered rapidly by irradiating visible light (Figure 13.19). The presence of KCl was important for the effect, and NaCl was ineffective. It is suggested that the potential decrease under UV irradiation is caused by the enhanced K^+ flux in the membrane due to the increase in K^+ uptake by the *cis*-form crown ether. Higher and more rapid photoinduced potential change was observed across a PVP/spirobenzopyran (**44**) membrane [177b].

A remarkable negative shift of the potential (*c.* 100 mV) was obtained under UV irradiation, and visible light irradiation recovered the potential to the original value. This shift followed the absorption spectral change of **44**. This was ascribed to the polarity change of the membrane which incorporates compound **44**; UV light induces opening of the spirobenzopyran ring which results in the formation of an ionic structure, and visible light causes it to return to the non-polar structure. The potential change by this PVP/**44** system was investigated [178] and the behavior was ascribed to the hydrogen ion selectivity of the membrane; the

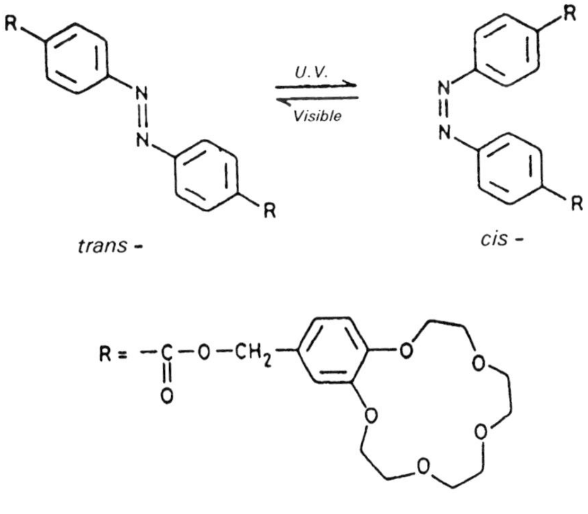

(43)

trans -   cis -

R = —C—O—CH₂— ...

(44)

Closed -   Opened -

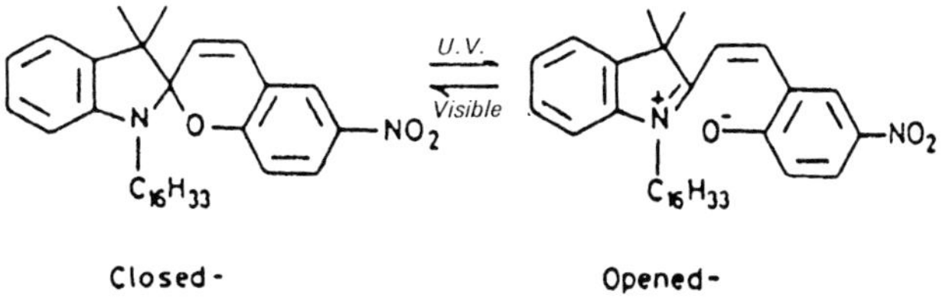

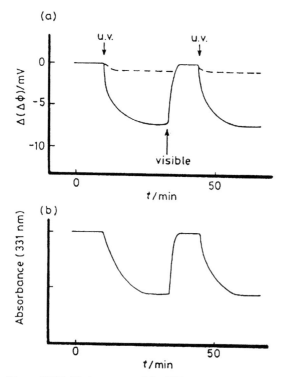

Figure 13.19. Photoresponse of the PVP membrane incorporating crown ether with azobenzene groups under alternative UV and visible light irradiation ($c_1 = 1$ mM, $c_2 = 500$ mM NaCl ——— or KCl ———). (a) Potential change and (b) absorbance change. (Reproduced by permission of the Royal Society of Chemistry from ref. 117a.)

closed form is a very weak base and the open form is a much stronger base. An LB membrane incorporating **44** also induced the similar potential changes [179]. A poly(methacrylic acid) membrane containing spirobenzopyran also photoinduced a potential change [180].

Photoelectric conversion was obtained by photoisomerization of spiropyran (SP) monolayers immobilized onto gold electrodes by the self-assembly technique [181]. Reversible photoisomerization of the monolayer between the SP and protonated merocyanine state (MRH^+) that was induced photochemically produced capacity current changes. A mixed monolayer composed of SP and pyrroloquinolinoquinone (PQQ) on a gold electrode photostimulated oxidation of nicotinamide-adenine dinucleotide phosphate (NADPH) cofactors in the presence of Ca^{2+} ions (Figure 13.20). While the PQQ + SP monolayer electrode exhibits effective electrocatalytic properties towards the oxidation of NADPH, the PQQ + MRH^+

monolayer electrode reveals poor electrocatalytic properties.

A thin film comprised of fragments of bacteriorhodopsin(bR)-containing purple membrane was formed on a NESA electrode by the LB method [182]. The film was put into contact with a thin, aqueous electrolyte gel to construct an electrochemical sandwich-type photocell (Figure 13.21), NESA/bR LB membrane/electrolyte aqueous solution (gel)/Au.

Under visible light irradiation, this cell generated transient rectified photocurrent (Figure 13.22) that proved to have differential responsivity to light intensity, which is characteristic of *in vivo* biological photoreceptors. It works as an image sensor, and senses both light intensity and movement rate. This cell simulates the mechanism of visual functions *in vivo* and also leads to the construction of intelligent image sensors.

A photoelectric device was prepared using LB films of photosynthetic reaction centers (RCs). Monolayers of RCs from *Rhodopseudomonas viridis* were deposited on an ITO electrode; the orientation of the RCs on the electrode was controlled using various substrates with different surface wettabilities. The degree of alignment was evaluated by measuring the polarities of the light-induced electric responses of the cell, ITO/RC(LB film)/Au [183]. This cell gave steady-state photocurrents, anodic or cathodic, based on the direction of the alignment. Colloidal platinum was precipitated onto the surface of thylakoid membranes, and these platinized chloroplasts were immobilized on a filter paper and sandwiched between metal gauze electrodes [184,185]. The direction of the photocurrent was consistent with the vectorial model of photosynthesis, which predicts that the electric potential of the external surface of the photosynthetic membrane swings negative with respect to the internal surface.

Photoresponse of Chl/ascorbic acid and Chl/buffer acetate junctions has been studied, and a model was proposed for the charge transfer at Chl/electrolyte junctions where photoinduced isoenergetic charge transfer is supposed to take place between the triplet excited state of Chl and the oxidized energy level of the electrolyte [186]. Photoelectron transfer was obtained at a lipid bilayer that contained a positively charged Chl *b* cholyl hydrazone separating two salt solutions, one of which contains ferricyanide; this resulted in the generation of a photopotential of about 20 mV [187].

Biophotoelectrochemical fuel cells that contain photosynthetic microorganisms, *Anabaena variabilis M-2*, and 2-hydroxy-1,4-naphthoquinone were fabricated by using the suspension solution of the micro-

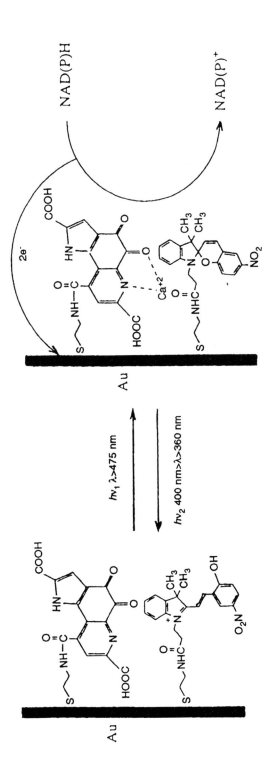

Figure 13.20. A mixed PQQ + SP monolayer prepared on a gold electrode by photostimulated oxidation of NADPH. (Reproduced by permission of Elsevier Science SA from *J. Electroanal. Chem.* **382**, 1995, 25–31.)

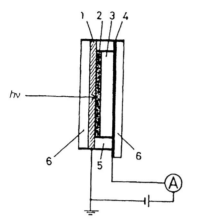

Figure 13.21. Cross-section of the bR-immobilized thin cell: 1, SnO$_2$ layer (450 nm thick); 2, purple membrane LB film (six to ten layers); 3, aqueous electrolyte gel layer comprising 6% carboxymethylchitin and 1 M KCl (200 μm thick); 4, gold layer (< 100 nm thick); 5, Teflon ring spacer; 6, glass substrate. (Reprinted with permission from T. Miyasaka *et al.*, *Science* **255**, 342–344, 1992. Copyright 1992 American Association for the Advancement of Science.)

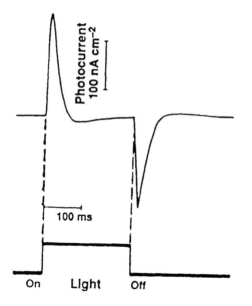

Figure 13.22. Typical photoresponse of bR-induced photo-current for the cell given in Figure 13.21 illuminated with 540 nm light (half-width 130 nm) at 10 mW cm^{-2}. The cathodic photocurrent was obtained at 0.6 V bias against the counter electrode. (Reprinted with permission from T. Miyasaka *et al.*, *Science* **255**, 342–344, 1992. Copyright 1992 American Association for the Advancement of Science.)

organisms [188]. *Synechococcus sp.* was also used for this type of biofuel cell [189].

7 CONCLUSIONS AND FUTURE SCOPE

Photoelectric conversion by polymeric and organic materials in the past two decades (1975–1995) has been reviewed. There are a lot of variations in the design of such devices, which are made from a limited range of inorganic materials. Although the photoelectric conversion efficiency of these polymeric and organic devices is not high, they can be used as sensors and other photoelectric devices. Recent advances in the efficiency of solar cells, based on organic dyes, encourage the development of devices with high cost–performance. When using polymeric and organic material-based devices, durability is another problem. However, when considering the organic nature of the natural biological compounds and also many commercial materials, such as liquid crystals, pigments, etc., the problem with durability could be overcome in the future by both fundamental and application-directed research. The variety of polymeric and organic materials should promise to open a new era of photoelectric conversion devices.

8 REFERENCES

1. (a) R.-J. Lin and M. Kaneko, *Molecular Electronics and Molecular Electronic Devices* (ed. K. Sienicki), CRC Press, Boca Raton, 1993, pp. 207–241.
   (b) M. Kaneko and A. Yamada, *Adv. Polym. Sci.* **55**, 1 (1984).
   (c) M. Kaneko and D. Woehrle, *Adv. Polym. Sci.* **84**, 141 (1988).
2. (a) M. Graetzel (ed.) *Energy Resources through Photochemistry and Catalysis*, Academic Press, New York, 1983.
   (b) K. Kalyanasundaram and M. Graetzel (eds.) *Photosensitization and Photocatalysis Using Inorganic and Organometallic Compounds*, Kluwer, Dordrecht, 1993.
3. H. Shirakawa and S. Ikeda, *Polym. J.* **2**, 231 (1971).
4. H. Shirakawa, E.J. Louis, A.G. MacDiarmid, C.K. Chiang and A.J. Heeger, *J. Chem. Soc., Chem. Commun.* **1977**, 578 (1977).
5. H. Shirakawa and S. Ikeda, *Kobunshi* **28**, 369 (1979).
6. B.R. Weinberger, S.C. Gau and Z. Kiss, *Appl. Phys. Lett.* **38**, 555 (1981).
7. M.J. Sailor, E.J. Ginsburg, C.B. Gorman, A. Kumar, R.H. Grubbs and N.S. Lewis, *Science* **249**, 1146 (1990).

8. T. Wada, A. Takeno, M. Iwaki, H. Sasabe and Y. Kobayashi, *J. Chem. Soc., Chem. Commun.* **1985**, 1194 (1985).

9. (a) T. Watanabe, S. Murakami, K. Mori and Y. Kashiwaba, *Macromolecules* **22**, 4231 (1989).
   (b) S.C.K. Misra, M.K. Ram, S.S. Pandey, B.D. Malhotra and S. Chandra, *Appl. Phys. Lett.* **61**, 1219 (1992).
   (c) A. Assadi, C. Svensson, M. Willander and O. Inganaes, *J. Appl. Phys.* **72**, 2900 (1992).

10. M. Kaneko, A. Yamada, E. Tsuchida and T. Kenmochi, *J. Polym. Sci., Polym. Lett. Ed.* **23**, 629 (1985).

11. (a) O. Niwa, M. Hikita and T. Tamamura, *Polym. Prepr. Jpn* **33**, 2515 (1984).
    (b) T. Skotheim, O. Inganaes, J. Prejza and I. Lundstroem, *Mol. Cryst. Liq. Cryst.* **83**, 329 (1982).
    (c) Y. Kunugi, Y. Harima and K. Yamashita, *J. Chem. Soc., Chem. Commun.* **1995**, 787 (1995).

12. Y. Shirota, T. Kakuta, H. Kanega and H. Mikawa, *J. Chem. Soc., Chem. Commun.* **1985**, 1201 (1985).

13. L.C. Hardy and D.F. Shriver, *J. Am. Chem. Soc.* **108**, 2887 (1986).

14. E.V. Donckt, B. Noirhomme, J. Kanicki, R. Eeltour and G. Gusman, *J. Appl. Polym. Sci.* **27**, 1 (1982).

15. K. Uehara, T. Yoshikawa, M. Katoh, M. Tanaka and N. Isomatsu, *Chem. Lett.* **1984**, 1499 (1984).

16. R.O. Loutfy and J.H. Sharp, *J. Chem. Phys.* **71**, 1211 (1979).

17. R.O. Loutfy, C.K. Hsiao and R. Ho, *Can. J. Phys.* **61**, 1416 (1983).

18. M. Shimura and A. Toyoda, *Japan. J. Appl. Phys.* **23**, 1462 (1984).

19. N. Minami, K. Sasaki and K. Tsuda, *J. Appl. Phys.* **54**, 6764 (1983).

20. J.P. Dodelet, H.P. Pommier and M. Ringuet, *J. Appl. Phys.* **53**, 4270 (1982).

21. B.J. Stanbery, M. Gouterman and R.M. Burges, *J. Phys. Chem.* **89**, 4950 (1985).

22. A.P. Piechowski, G.R. Bird, D.L. Morel and E.L. Stogryn, *J. Phys. Chem.* **88**, 934 (1984).

23. (a) S. Iijima, F. Mizutani and Y. Tanaka, *Bull. Chem. Soc. Jpn* **58**, 1585 (1985).
    (b) D.L. Morel, E.L. Stogryn, A.K. Ghosh, T. Feng, P.E. Purwin, R.F. Shaw, G.R. Bird and A.P. Piechowski, *J. Phys. Chem.* **88**, 923 (1984).

24. T. Moriizumi and K. Kudo, *Appl. Phys. Lett.* **38**, 85 (1981).

25. K. Yamashita, Y. Harima and H. Iwashima, *J. Phys. Chem.* **91**, 3055 (1987).

26. Y. Harima, K. Yamamoto, K. Takeda and K. Yamashita, *Bull. Chem. Soc. Jpn* **62**, 1458 (1989).

27. S. Yoshida, K. Kozawa and T. Uchida, *Bull. Chem. Soc. Jpn* **68** 738 (1995).

28. S.C. Dahlberg and M.E. Musser, *J. Chem. Phys.* **72**, 6706 (1980).

29. M.E. Musser and S.C. Dahlberg, *Surface Sci.* **100**, 605 (1980).

30. B.A. Gregg, M.A. Fox and A.J. Bard, *J. Phys. Chem.* **94**, 1586 (1990).

31. J. Segui, S. Hotchandani, D. Baddou and R.M. Leblanc, *J. Phys. Chem.* **95**, 8807 (1991).

32. C.W. Tang, *Appl. Phys. Lett.* **48**, 183 (1986).

33. M. Hiramoto, H. Fujiwara and M. Yokoyama, *Appl. Phys. Lett.* **58**, 1062 (1991).

34. M. Hiramoto, H. Fujiwara and M. Yokoyama, *J. Appl. Phys.* **72**, 3781 (1992).

35. (a) B.A. Gregg, *Appl. Phys. Lett.* **67**, 1271 (1995).
    (b) K. Saito and M. Sugi, *Appl. Phys. Lett.* **61**, 116 (1992).

36. A.F. Sammells and P.G.P. Ang, *J. Electrochem. Soc.* **131**, 617 (1984).

37. A.F. Sammels and S.K. Schmidt, *J. Electrochem. Soc.* **132**, 520 (1985).

38. R.L. Cook and A.F. Sammells, *J. Electrochem. Soc.* **132** 2429 (1985).

39. T.A. Skotheim and I. Lundstroem, *J. Electrochem. Soc.* **129**, 894 (1982).

40. T.A. Skotheim and O. Inganaes, *J. Electrochem. Soc.* **132**, 2116 (1985).

41. T.A. Skotheim, *Appl. Phys. Lett.* **38**, 712 (1981).

42. M.P. Hagemeister and H.S. White, *J. Phys. Chem.* **91**, 150 (1987).

43. S.N. Chen, A.J. Heeger, Z. Kiss, A.G. MacDiarmid, S.C. Gau andD.L. Peebles, *Appl. Phys. Lett.* **36**, 96 (1980).

44. T. Yamase, H. Harada, T. Ikawa, S. Ikeda and H. Shirakawa, *Bull. Chem. Soc. Jpn* **54**, 2817 (1981).

45. M. Aizawa, S. Watanabe, H. Shinohara and H. Shirakawa, *J. Chem. Soc., Chem. Commun.* **1985**, 62 (1985).

46. S. Pitchumani and F. Willig, *J. Chem. Soc., Chem. Commun.* **1983**, 809 (1983).

47. G. Horowitz and F. Garnier, *J. Electrochem. Soc.* **132**, 634 (1985).

48. M. Kaneko, K. Okuzumi and A. Yamada, *J. Electroanal. Chem.* **183**, 407 (1985).

49. T. Kenmochi, E. Tsuchida, M. Kaneko and A. Yamada, *Electrochim. Acta* **30**, 1405 (1985).

50. P.V. Kamat and R.A. Basheer, *Chem. Phys. Lett.* **103**, 503 (1984).

51. F.-R.F. Fan and L.R. Faulkner, *J. Am. Chem. Soc.* **101**, 4779 (1979).

52. P.C. Rieke and N.R. Armstrong, *J. Am. Chem. Soc.* **106**, 47 (1984).

53. P.C. Rieke, C.A. Linkous and N.R. Armstrong, *J. Phys. Chem.* **88**, 1351 (1984).

54. W.J. Buttner, P.C. Rieke and N.R. Armstrong, *J. Am. Chem. Soc.* **107**, 3738 (1985).

55. T.J. Klofta, C.A. Linkous and N.R. Armstrong, *J. Electroanal. Chem.* **185**, 73 (1985).

56. T.J. Klofta, C.A. Buttner, A. Nanthakumar, T.D. Mewborn and N.R. Armstrong, *J. Electrochem. Soc.* **132**, 2134 (1985).

57. R.O. Loutfy and F. McIntyre, *Can. J. Chem.* **61**, 72 (1983).

58. P. Leempoel, F.-R.F. Fan and A.J. Bard, *J. Phys. Chem.* **87**, 2948 (1983).

59. N. Minami, *J. Chem. Phys.* **72**, 6317 (1980).

60. D. Belanger, J.P. Dodelet, L.H. Dao and B.A. Lombos, *J. Phys. Chem.* **88**, 4288 (1984).

61. L. Gastonguay, G. Veilleux, R. Cote, R.G.S. Jacques and J.P. Dodelet, *J. Electrochem. Soc.* **139**, 337 (1992).

62. Y. Harima and K. Yamashita, *J. Phys. Chem.* **89**, 5325 (1985).

63. Y. Harima and K. Yamashita, *J. Electroanal. Chem.* **186**, 313 (1985).

64. T. Kawai, K. Tanimura and T. Sakata, *Chem. Phys. Lett.* **56**, 541 (1978).

65. F.J. Kampas, K. Yamashita and J. Fajer, *Nature* **284**, 40 (1980).

66. H. Jimbo, H. Yoneyama and H. Tamura, *Photochem. Photobiol.* **32**, 319 (1980).

67. C.H. Langford, B.R. Hollebone and D. Nadezhdin, *Can. J. Chem.* **59**, 652 (1981).

68. T. Katsu, K. Tamagake and Y. Fujita, *Chem. Lett.* **1980**, 289 (1980).

69. J. Basu, A. Bhattacharya, K. Das, A.B. Chatterjee, K.K. Kundu and K.K.R. Mukherjee, *Ind. J. Chem.* **22A**, 695 (1983).

70. K. Yamashita, Y. Harima and Y. Matsumura, *Bull. Chem. Soc. Jpn* **58**, 1761 (1985).

71. K. Takeda, Y. Harima, S. Yokoyama and K. Yamashita, *Japan. J. Appl. Phys.* **28**, L141 (1989).

72. H. Yanagi, M. Ashida, Y. Harima and K. Yamashita, *Chem. Lett.* **1990**, 385 (1990).

73. K. Uehara, K. Shibata and M. Tanaka, *Chem. Lett.* **1985**, 897 (1985).

74. H.T. Tien and J. Higgins, *Phys. Lett.* **93**, 276 (1982).

75. F. Mizutani, S. Iijima, K. Sasaki and Y. Shimura, *Ber. Bunsenges. Phys. Chem.* **86**, 907 (1982).

76. Y. Umezawa, T. Yamamura and A. Kobayashi, *J. Electrochem. Soc.* **129**, 2378 (1982).

77. L. Persaud and C.H. Langford, *Inorg. Chem.* **24**, 3562 (1985).

78. F. Decker, M.F. Decker, G. Zotti and G. Mengoli, *Electrochim. Acta* **30**, 1147 (1985).

79. A. Ilani, M. Woodle and D. Mauzerall, *Photochem. Photobiol.* **49**, 673 (1989).

80. A. Fujishima and K. Honda, *Nature*, **238**, 37 (1972).

81. E. Pelizzetti and M. Schiavello (eds.) *Photochemical Conversion and Storage of Solar Energy*, Kluwer, Dordrecht, 1991.

82. R.N. Noufi, D. Tench and L.F. Warren, *J. Electrochem. Soc.* **127**, 2310 (1980).

83. T. Skotheim, I. Lundstroem and J. Prejza, *J. Electrochem. Soc.* **128**, 1625 (1981).

84. R.N. Noufi, A.J. Frank and A.J. Nozik, *J. Am. Chem. Soc.* **103**, 1849 (1981).

85. (a) I. Lundstroem, *J. Electrochem. Soc.* **129**, 1737 (1982).

(b) T. Skotheim, O. Inganaes, J. Prejza and I. Lundstroem, *Mol. Cryst. Liq. Cryst.* **83**, 329 (1982).

86. F.-R.F. Fan, B.L. Wheeler, A.J. Bard and R.N. Noufi, *J. Electrochem. Soc.* **128**, 2042 (1981).

87. R.A. Simon, A.J. Ricco and M.S. Wrighton, *J. Am. Chem. Soc.* **104**, 2031 (1982).

88. S. Holdcroft and B.L. Funt, *J. Electrochem. Soc.* **135**, 3106 (1988).

89. G. Nagasubramanian, S. DiStefano and J. Moacanin, *J. Electrochem. Soc.* **133**, 305 (1986).

90. R.N. Noufi, *J. Electrochem. Soc.* **130**, 2126 (1983).

91. A.J. Frank and K. Honda, *J. Phys. Chem.* **86**, 1933 (1982).

92. K. Honda and A.J. Frank, *J. Phys. Chem.* **88**, 5577 (1984).

93. G. Horowitz and F. Garnier, *J. Electrochem. Soc.* **132**, 634 (1985).

94. A.J. Frank, S. Glenis and A.J. Nelson, *J. Phys. Chem.* **93**, 3818 (1989).

95. R.N. Noufi, A.J. Nozik, J. White and L.F. Warren, *J. Electrochem. Soc.* **129**, 2261 (1982).

96. H.S. White, H.D. Abruna and A.J. Bard, *J. Electrochem. Soc.* **129**, 265 (1982).

97. K. Rajeshwar, M. Kaneko and A. Yamada, *J. Electrochem. Soc.* **130**, 38 (1983).

98. (a) K. Rajeshwar, M. Kaneko, A. Yamada and R.N. Noufi, *J. Phys. Chem.* **89**, 806 (1985).

(b) K. Rajeshwar and M. Kaneko, *J. Phys. Chem.* **89**, 3587 (1985).

99. M. Kaneko, T. Okada, S. Teratani and K. Taya, *Electrochim. Acta* **32**, 1405 (1987).

100. M. Kaneko, G.-J. Yao and A. Kira, *J. Chem. Soc., Chem. Commun.* **1989**, 1099 (1989).

101. O. Haas, N. Mueller and H. Gerischer, *Electrochim. Acta* **27**, 991 (1982).

102. T.P. Henning, H.S. White and A.J. Bard, *J. Am. Chem. Soc.* **104**, 5862 (1982).

103. M.D. Rosenblum and N.S. Lewis, *J. Phys. Chem.* **88**, 3103 (1984).

104. I.L. Abrahams, B.J. Tufts and N.S. Lewis, *J. Am. Chem. Soc.* **109**, 3472 (1987).

105. H.D. Rubin, D.J. Arent, B.D. Humphrey and A.B. Bocarsly, *J. Electrochem. Soc.* **134**, 93 (1987).

106. Y. Nakato, M. Shioji and H. Tsubomura, *J. Phys. Chem.* **85**, 1670 (1981).

107. P.K. Ghosh and T.G. Spiro, *J. Am. Chem. Soc.* **102**, 5543 (1980).

108. W. Kawai and S. Yamamura, *Nippon Kagaku Kaishi* **1981**, 1217 (1981).

109. M. Krishnan, X. Zhang and A.J. Bard, *J. Am. Chem. Soc.* **106**, 7371 (1984).

110. R. Memming and F. Schroeppel, *Chem. Phys. Lett.* **62**, 207 (1979).

111. (a) B. O'Regan and M. Graetzel, *Nature* **353**, 737 (1991).

(b) M. Graetzel, *Platinum Metals Rev.* **38**, 151 (1994).

112. (a) R. Amadelli, R. Argazzi, C.A. Bignozzi and F. Scandola, *J. Am. Chem. Soc.* **112**, *7099 (1990).*
    (b) K. Vinodgopal, X. Hua, R.L. Dahlgren, A.G. Lappin, L.K. Patterson and P.V. Kamat, *J. Phys. Chem.* **99**, 10883 (1995).
    (c) K.H. Liao and D.H. Waldeck, *J. Phys. Chem.* **49**, 4569 (1995).

113. A. Kay and M. Graetzel, *J. Phys. Chem.* **97**, 6272 (1993).

114. H.T. Tien, *Photochem. Photobiol.* **24**, 97 (1976).

115. T. Miyasaka, T. Watanabe, A. Fujishima and K. Honda, *Nature* **227**, 638 (1978).

116. T. Tow and J.B. Wagner, *J. Electrochem. Soc.* **125**, 613 (1978).

117. K. Tennekone and W.M.R. Divigalpitiya, *Japan. J. Appl. Phys.* **20**, 299 (1981).

118. S. Yamamura and W. Kawai, *Nippon Kagaku Kaishi* **1982**, 1287 (1982).

119. P.A. Breddels and G. Blasse, *J. Chem. Soc., Faraday Trans. 2*, **80**, 1055 (1984).

120. (a) T. Shimidzu, T. Iyoda and Y. Koida, *Polym. J.* **16**, 919 (1984).
    (b) X.-P. Li and H. Tributsch, *Photochem. Photobiol.* **50**, 531 (1989).

121. N. Minami, T. Watanabe, A. Fujishima and K. Honda, *Ber. Bunsenges. Phys. Chem.* **83**, 476 (1979).

122. P. Leempoel, F.-R.F. Fan and A.J. Bard, *J. Phys. Chem.* **87**, 2948 (1983).

123. Y. Morishima, M. Isono, Y. Itoh and S. Nozakura, *Chem. Lett.* **1981**, 1149 (1981).

124. K. Iriyama, F. Mizutani and M. Yoshiura, *Chem. Lett.* **1980**, 1399 (1980).

125. L.M. Natoli, M.A. Ryan and M.Y. Spitler, *J. Phys. Chem.* **89**, 1488 (1985).

126. (a) H. Hiroshi, Y. Yonezawa and H. Inabe, *Chem. Lett.* **1980**, 467 (1980).
    (b) A. Haraguchi, Y. Yonezawa and R. Hanawa, *Photochem. Photobiol.* **52**, 307 (1990).

127. M. Nakao, Y. Itoh and K. Honda, *J. Phys. Chem.* **88**, 4906 (1984).

128. M.A. Fox, F.J. Nobs and T.A. Voynick, *J. Am. Chem. Soc.* **102**, 4036 (1980).

129. T. Sakata, K. Hashimoto and M. Hiramoto, *J. Phys. Chem.* **94**, 3040 (1990).

130. M.D. Archer and J.R. Bolton, *J. Phys. Chem.* **94**, 8028 (1990).

131. (a) D. Meissner, R. Memming and B. Kastening, *Chem. Phys. Lett.* **96**, 34 (1983).
    (b) J.P. Kuczynski, B.H. Milosavljevic and J.K. Thomas, *J. Phys. Chem.* **88**, 980 (1984).
    (c) N. Kakuta, J.M. White, A. Campion, A.J. Bard, M.A. Fox and S.E. Webber, *J. Phys. Chem.* **89**, 48 (1985).
    (d) N. Kakuta, K.H. Park, M.F. Finlayson, A.J. Bard, A. Campion, M.A. Fox, S.E. Webber and J.M. White, *J. Phys. Chem.* **89**, 5028 (1985).

132. J. Kutnik and H.T. Tien, *Photochem. Photobiol.* **46**, 413 (1987).

133. R. Roland, D. Ricci and O. Brandt, *J. Phys. Chem.* **96**, 6783 (1992).

134. E. Rabinowitch, *J. Chem. Phys.* **8**, 551, 560 (1940).

135. M. Einsenberg and H.P. Silverman, *Electrochim. Acta* **5**, 1 (1961).

136. M. Kaneko and A. Yamada, *Rep. Inst. Phys. Chem. Res.* **52**, 210 (1976).

137. K. Shigehara, M. Nishimura and E. Tsuchida, *Bull. Chem. Soc. Jpn* **50**, 3397 (1977).

138. W.J. Albrey, A.W. Foulds, K.J. Hall, R.G. Egdell and A.F. Orchard, *Nature* **282**, 793 (1979).

139. W.J. Albrey, A.W. Foulds, K.J. Hall and A.R. Hillman, *J. Electrochem. Soc.* **127**, 654 (1980).

140. R. Tamailarasan and P. Natarajan, *Nature* **292**, 224 (1981).

141. T.I. Quickenden and I.R. Harrison, *J. Electrochem. Soc.* **132**, 81 (1985).

142. A.S.N. Murthy and K.S. Reddy, *Electrochim. Acta* **28**, 473 (1983).

143. R. Tamilarasan, R. Ramaraj, R. Subramanian and P. Natrajan, *J. Chem. Soc., Faraday Trans. 1* **80**, 2405 (1984).

144. R. Ramaraj, R. Tamilarasan and P. Natarajan, *J. Chem. Soc., Faraday Trans. 1* **81**, 2763 (1985).

145. P.V. Kamat and M.A. Fox, *J. Electrochem. Soc.* **131**, 1032 (1984).

146. M. Yoshida and I. Oshida, *Japan. J. Appl. Phys.* **33**, 34 (1964).

147. M. Kaneko and A. Yamada, *J. Phys. Chem.* **81**, 1213 (1977).

148. R.-L. Zhou, Y.-G. Yang and Y.-Y. Han, *J. Photochem. Photobiol. A, Chem.* **81**, 59 (1994).

149. P.V. Kamat, *J. Electroanal. Chem.* **163**, 389 (1984).

150. T.D. Westmoreland, J.M. Calvert, R.W. Murray and T.J. Meyer, *J. Chem. Soc., Chem. Commun.* **1983**, 65 (1983).

151. K. Kobayashi, T. Sagara, M. Okada and K. Niki, *Chem. Lett.* **1983**, 373 (1983).

152. H. Nishihara, K. Sakamoto, H. Motohashi and K. Aramaki, *Denki Kagaku* **61**, 883 (1993).

153. M. Kaneko, M. Ochiai and A. Yamada, *Makromol. Chem. Rapid Commun.* **3**, 299 (1982).

154. M. Kaneko, A. Yamada, N. Oyama and S. Yamaguchi, *Makromol. Chem. Rapid Commun.* **3**, 769 (1982).

155. M. Kaneko, S. Moriya, A. Yamada, H. Yamamoto and N. Oyama, *Electrochim. Acta* **29**, 115 (1984).

156. K. Yamada, N. Kobayashi, K. Ikeda, R. Hirohashi and M. Kaneko, *Japan. J. Appl. Phys.* **33**, L544 (1994).

157. J.T. Hupp, J.P. Otruba, S.J. Parus and T.J. Meyer, *J. Electroanal. Chem.* **190**, 287 (1985).

158. G.-J. Yao, T. Onikubo and M. Kaneko, *Electrochim. Acta* **38**, 1093 (1993).

159. M. Kaneko, *J. Macromol. Sci. Chem.* **A24**, 357 (1987).

160. M. Kaneko and A. Yamada, *Electrochim. Acta* **31**, 273 (1986).

161. M. Kaneko, S. Hara and A. Yamada, *J. Electroanal. Chem.* **194**, 165 (1985).

162. S. Cosnier, A. Deronizer and J.C. Moutet, *J. Phys. Chem.* **89**, 4895 (1985).

163. K. Gobi and R. Ramaraj, *J. Electroanal. Chem.* **368**, 77 (1994).

164. K. Yamada, Y. Ueno, K. Ikeda, N. Takamiya and M. Kaneko, *Denki Kagaku* **57**, 1129 (1989).

165. (a) H. Xu, T. Shen, Q. Zhou, S. Shen, J. Liu and L. Li, *J. Photochem. Photobiol.* **65**, 267 (1992).
     (b) E. Bienvenue, P. Seta, A. Hofmanova and C. Gavach, *J. Electroanal. Chem.* **162**, 275 (1984).

166. K. Takahashi, T. Komura and H. Imanaga, *Bull. Chem. Soc. Jpn* **62**, 386 (1989).

167. (a) M. Fujihira, *Organic Thin Films and Surfaces*, Vol. 1, Academic Press, Boston, 1994.
     (b) M. Fujihira, M. Sakomura and T. Kamei, *Thin Solid Films, 179, 471 (1989)*.

168. O. Contamin, E. Levart, G. Magner, M. Savy and G. Scarber, *J. Electroanal. Chem.* **237**, 39 (1987).

169. M. Kaneko, D. Woehrle, D. Schlettwein and V. Schmidt, *Makromol. Chem.* **189**, 2419 (1988).

170. D. Schlettwein, M. Kaneko, A. Yamada, D. Woehrle and N.I. Jaeger, *J. Phys. Chem.* **95**, 1748 (1991).

171. S. Cattarin, G. Zotti, M.M. Musiani and G. Mengoli, *J. Electroanal. Chem.* **207**, 247 (1986).

172. Y. Morishima, Y. Fukushima and S. Nozakura, *J. Chem. Soc., Chem. Commun.* **1985**, 912 (1985).

173. N. Egashira, T. Fujisawa and K. Ohga, *Denki Kagaku* **57**, 65 (1989).

174. M. Kaneko and D. Woehrle, *J. Electroanal. Chem.* **307**, 209 (1991).

175. K. Yamada, C.B. Lin, N. Kobayashi, K. Ikeda, R. Hirohashi and M. Kaneko, *J. Electroanal. Chem.* **370**, 59 (1994).

176. D.S. Berns, *Photochem. Photobiol.* **24**, 117 (1976).

177. (a) J. Anzai, H. Sasaki, A. Ueno and T. Osa, *J. Chem. Soc., Chem. Commun.* **1983**, 1045 (1983).
     (b) J. Anzai, A. Ueno and T. Osa, *J. Chem. Soc., Chem. Commun.* **1984**, 688 (1984).

178. O. Ryba and J. Petranek, *Makromol. Chem. Rapid Commun.* **9**, 125 (1988).

179. J. Anzai, K. Sakamura and T. Osa, *J. Chem. Soc., Chem. Commun.* **1992**, 888 (1992).

180. M. Irie, A. Menju and K. Hayashi, *Nippon Kagaku Kaishi* **1984**, 227 (1984).

181. E. Katz, M.-L. Dagan and I. Willner, *J. Electroanal. Chem.* **382**, 25 (1995).

182. T. Miyasaka, K. Koyama and I. Itoh, *Science* **255**, 342 (1992).

183. Y. Yasuda, H. Sugino, H. Toyotama, Y. Hirata, M. Hara and J. Miyake, *Bioelectrochem. Bioenerg.* **34**, 135 (1994).

184. E. Greenbaum, *Bioelectrochem. Bioenerg.* **21**, 171 (1989).

185. E. Greenbaum, *J. Phys. Chem.* **96**, 514 (1992).

186. S. Chandra, B.B. Srivastava and N. Khare, *Solid State Commun.* **56**, 975 (1985).

187. A. Losev and D. Mauzerall, *Photochem. Photobiol.* **38**, 355 (1983).

188. K. Tanaka, N. Kashiwagi and T. Ogawa, *J. Chem. Tech. Biotechnol.* **42**, 235 (1988).

189. T. Yagishita, T. Horigome and K. Tanaka, *J. Jpn. Solar Energy Soc.* **20**, 59 (1994).

Index

Index compiled by Geoffrey C. Jones

Volume 1 Contents

Volume 2 Contents

Volume 2 Contents

Volume 3 Contents

Volume 3 Contents

Volume 4 Contents